D0026023

Manufacturing Processes and Equipment

Manufacturing Processes and Equipment

Jiri Tlusty
University of Florida

Prentice Hall
Upper Saddle River, New Jersey 07458

Library of Congress Cataloging-in-Publication Data

Tlusty, Jiri.
Manufacturing processes and equipment / Jiri Tlusty.
p. cm.
Includes bibliographical references and index.
ISBN 0-201-49865-0
1. Manufacturing processes. 2. Machine-tools. I. Title.
TS183.T57 1999 99-32425
670.42—dc21 CIP

Acquisitions editor: Michael Slaugter
Editor in Chief: Marcia Horton
Production editor: Caroline Jumper
Executive managing editor: Vince O'Brien
Assistant managing editor: Eileen Clark
Art director: Jayne Conte
Cover design: Joe Sengotta
Manufacturing manager: Trudy Pisciotti
Assistant vice president of production and manufacturing: David W. Riccardi

Prentice-Hall, Inc.
Upper Saddle River, NJ 07458

Printed in the United States of America

10 9 8 7 6 5 4 3 2

ISBN 0-201-49865-0

Prentice-Hall International (UK) Limited, *London*
Prentice-Hall of Australia Pty. Limited, *Sydney*
Prentice-Hall Canada, Inc., *Toronto*
Prentice-Hall Hispanoamericana, S.A., *Mexico*
Prentice-Hall of India Private Limited, *New Delhi*
Prentice-Hall of Japan, Inc., *Tokyo*
Prentice-Hall Pte Ltd., *Singapore*
Editora Prentice-Hall do Brazil, Ltda., *Rio de Janeiro*

About the Author

Jiri (George) Tlusty was born in Prague, Czechoslovakia. He obtained the Dipl.Ing, Ph.D. and Dr.Sc. degrees from the Technical University in Prague. He first worked in the engineering design office of a machine tool company and then was hired by a large group of machine tool companies to run a small laboratory to test prototypes of new machine tools in the group. Under his guidance the lab grew into a research institute for machine tools and machining technology including nontraditional processes. The institute served its member firms by conducting research and development of accuracy, dynamics of structures, and control of drives of a wide variety of machine tool types. Since the 1950s, the work included development of NC machine tools. During his 20 years as Director of Research, the institute VUOSO became established internationally, serving not only the machine tool industry but also the manufacturing industries, contributing to the breadth of Tlusty's manufacturing expertise.

In 1971, after a 2-year stay at the University in Manchester England as a research fellow, Tlusty became a professor at McMaster University in Hamilton, Ontario. Apart from teaching, he established a laboratory for machine tools and robots with extensive research and industrial contract work. In 1984, he moved to the University of Florida where he holds the rank of Graduate Research Professor. He built a laboratory that evolved as a machine tool research center, with activities including the development of techniques and machines for high speed milling of aluminum aircraft parts in conjunction with McDonnell Douglas (now Boeing), further high speed milling of hardened steel parts, and also improvement of technology of machining titanium helicopter parts. He is president of MLI Inc., a small company specializing in instrumentation and controls and design and manufacture of special machines.

His book *Self-Excited Vibrations in Machining,* published in Prague in 1954, was, with a series of revisions, published in Moscow, Berlin, and with a further revision in Oxford in 1970 and in Tokyo in 1972. He contributed a large chapter to a book on high speed machining and has published close to a hundred research papers. He was the principal advisor to successful graduation of 27 Ph.D. students and 41 M.S. students. He has been a member of the International Institution for Production Engineering Research CIRP since 1957, and was its president in 1968. He also is a founding member and past president of NAMRI, Fellow of ASME, Fellow of SME, and a registered professional engineer of Ontario.

He received the Czechoslovak State Prize of 1954, the 1979 SME Gold Medal, 1980 ASME Centenial Medaillon, 1990 ASME Blackall Award, and the 1997 ASME W. T. Ennor Manufacturing Technology Award. He speaks Czech, German, Russian, French, and English.

Contents

Preface

I spent the first half of my professional life as designer and researcher in the machine tool industry in Europe. In 1971 I emigrated to Canada and joined the mechanical engineering faculty at McMaster University in the heavily industrial town of Hamilton, Ontario. While my primary theoretical background was in structural dynamics and controls, I also taught elasticity and plasticity, design, analytical methods, and heat transfer. However, my main responsibility was for the required course in Manufacturing Engineering (MfgE).

At that time there were a number of very good textbooks in MfgE available. I used Doyle et al., *Manufacturing Processes and Materials for Engineers.* It was an excellent comprehensive collection of descriptions of techniques, materials, processes, and machines including well-selected illustrations, accompanied by qualitative judgments. Several other books were similar. However, there was very little analytical development of the subjects and no connection or relationship was established between existing books and the traditional mechanical engineering textbooks on vibrations, controls, heat transfer, etc. On the other hand, those other "machine science" courses contained very little physical interpretation and application material that might help to better understand MfgE. Why should the students learn all of the theory if they were not shown that it was useful or needed in the vast and significant field of manufacturing engineering?

MfgE was at that time not scientifically esteemed or well developed, and many of the teachers concentrated their lectures on little more than "shop practice." Since then the situation has changed: There is now a large academic community engaged in extensive research in MfgE. The profile of the teachers has changed substantially and I am happy to have been heavily involved in that transition. Back then, however, I began to write analytical notes as handouts to complement the textbook in my teaching of the course. At the same time I started to learn about computational approaches to solve many of the problems and included them in my teaching. I continued the practice after moving to the University of Florida in 1984, where I met colleagues who joined me in using these handouts in teaching the MfgE course. I continued developing the material that finally was condensed to become the book presented here.

It is important to teach students how to formulate problems in MfgE and to show them how to use all the knowledge gained in the specialty background courses to solve these problems.This belief stems from my experience as an industrial researcher and is also based on the 28 years of teaching the topic, giving homework assignments and exams based on analytical exercises, and finding favorable acceptance by the students. It is not enough to simply include large numbers of ready-made equations with little or no analytical development in otherwise descriptive texts on Mfg processes, as has been done by several authors. If the students do not understand how a formula has been obtained they will not be able to use it. It is necessary to present derivations in some detail and to involve the students in developing the corresponding computer programs. This enables them to creatively use other new theoretical and computational approaches to problems that will challenge them in their future work as production engineers.

What is the essence of this book, and in what ways is it novel and different from the others on the market? One-half of its mission is to describe and explain existing production processes and machinery whereas the other half is to show how to use the powerful analytical tools of machine science and apply them to the solution of manufacturing problems to further development of the state of the art. In comparison with other books, there is more emphasis on analytical development and application of engineering theory to MfgE problems. This book is unique in its inclusion of computer solutions to these problems. Methods of heat transfer are used in analysing thermal fields in chip and tool in machining and in the zone around the arc in welding. Plasticity is applied to metal forming, fluid flow to plastics processing. Metrology and vibrations theory are applied to machine tool design and performance analysis. Control theory is used to understand computer control of machine tools as well as the behavior of the arc welding processes. Computer programs to be derived by the students are predominantly used rather than packaged software. So, for instance, students are shown how to write their own 1D transient and 2D steady-state finite difference programs for the thermal field solutions. Matlab® general software is used throughout, but simulations of servo systems and of vibrations are created by the students from scratch instead of using Simulink Toolbox. The latter would, of course, be more comfortable and efficient but does not offer proper insight and could rather be used in future work once the student fully understands the problems.

This book is intended as an undergraduate textbook for mechanical and industrial engineers. In the typical undergraduate curriculum, the MfgE course as well as that in Engineering Design are the two applied engineering courses closely related to the work the student will perform after graduation. It is useful to present them in the junior/senior years for recapitulating and summarizing material that the student has learned earlier. I and my colleagues have successfully used the precursor notes in a senior level MfgE course. The book assumes some prerequisite theoretical knowledge and does not derive common formulations such as Mohr's circles or the Laplace transform or the Nyquist theorem. On the other hand, most of the prerequisite material is briefly reviewed, such as basics of vibrations and first-order control systems. The mathematical level is restricted to such general skills as differential equations with constant coefficients, systems of linear equations (Gauss-Seidel), Euler-type integration, etc. For field problems, finite difference and not finite elements are used, since the latter is not yet universal in the ME curricula.

More than any other textbook available, this book includes chapters on machine tools and other production equipment, discussing the aspects of performance and design of drives, structures, and controls. For example, while a positional servo system should have been fully explained in a previous course on controls, the inclusion of a structural spring mass system in the loop, which is specifically important for machine tools, is treated here showing the use of both feedback and feedforward compensations. It is essential for today's students to learn about production machinery. Looking back at the development of productivity in the 20th century, more than half of its increase was due to improvements and automation of equipment rather than to modifications of the techniques and conditions of the processes and advances in tool materials and geometries. The fresh graduate will very likely be up to the task when asked to specify, select, install, and efficiently utilize new investments in equipment.

Although this book concentrates on the traditional processes of forming, cutting, assembly, and welding and the corresponding equipment, the related topics of manufacturing management as well as the prerequisite topics of materials, primary metallurgical processes, and the more specialized nontraditional processes are included, although in a more descriptive, nonanalytical way. The book does not seek to be encyclopedic or a comprehensive handbook. Instead, such aspects of the various processes have been selected and analyzed in more detailed ways that are most significant in each of them. So, for instance, instead of dwelling on the chip formation mechanism to try to derive the cutting force, an empirical approach is chosen based on extensive experimental data, and more attention is given to the understanding of the temperatures on the tool face—one of the decisive factors in the core problem of tool wear in the process of metal cutting. In another instance, for structures of machine tools, both forced and self-excited vibrations are discussed. For welding, the arc welding process is chosen and the control of the arc as well as the temperature field around the pool and the resulting residual stresses and distortions are dealt with. Encouraged by the reviewers of the manuscript, I look forward to a sympathetic reception of the concept of the book.

Teaching This Course

Some potential users of this book might doubt whether it is possible to cover all this material in a one-semester course of 13 weeks, 3 hours per week, a typical volume for a general MfgE course in many MechEng curricula. My colleagues S. Smith, J. Schuller, J. Ziegert, and I have done so for the past 16 years using the notes that have become this book.

We assign 12 homework sets (each with two or three problems, many based on computing) selected from the Problems sections at the end of Chapters 5–6 and 8–11, plus a few items from the Questions sections that appear in every chapter. We give one midterm and one final examination, each including four or five problems that do not require the use of computers (although that may change in a few years when we have computer terminals in our examination rooms). The Teacher's Manual contains examples of the homework sets and the test problems.

We have found that students are quite capable of handling this course, with the usual grade distributions resulting. Over the years students have become more adept at computing, and they enjoy it. The computer exercises in the book are written in Matlab, but of course it is possible to use the same algorithms in any other language.

Finally, it is always up to the teacher to select which topics to cover from the wide selection presented.

Acknowledgments

There are many friends and former and current associates whom I want to mention with gratitude. First of all, I gladly remember those with whom I worked in the Research Institute VUOSO in Prague: They contributed to my professional development by their innovative ideas, advice, discussions, and research accomplishments. For all of them, let me just name a few: Alois Kanka, Milos Polacek, Karel Stepanek, Jaromir Zeleny.

The next category of people who helped me to grow further and influenced my writing this book are the graduate students who worked with me on academic research projects, both at McMaster and at the University of Florida. I have been blessed by being able to spend my life working with all these talented, motivated, and dedicated young engineers and it makes me happy to continue doing it. We have all had fun in solving problems and discovering new methods and designs. I cannot mention them by name because the list is rather long but if they read this they will know that I hold them dear to my heart and memory. I should also mention the undergraduate students in my classes of manufacturing engineering at both the universities. Many of them gave me pleasure by their active approach to homework and by their questions and comments. And quite a number of the best in class went on to join me as graduate students. I enjoy meeting those working now in industry as well as those in academia. Some of the latter might even start teaching out of this book and spread the knowledge contained in it.

I thank my colleagues in the M.E. department at the University of Florida who used the precursor notes to this book in teaching the MfgE course and gave me their suggestions for improvement: Scott Smith, John Schueller, John Ziegert.

I appreciate very much the permission from all the publishers to reproduce many illustrations from a number of authors as well as the courtesy of many industrial companies that provided photographs and drawings of their products to be included in this book. The credits for all this help are given in a coded form in the captions under the illustrations concerned and complete details about the individual sources are printed at the end of each chapter, following the lists of references.

Jointly with the publisher, I wish to thank the following reviewers for their comments, advice, and suggestions during the development of the manuscript: M. Khairul Alam, Ohio University; M. Elbestawi, McMaster University; Ratna Babu Chinnam, North Dakota State; L. R. Cornwell, Texas A&M University; Paul Dawson, Cornell University; David Dornfeld, University of California at Berkeley; Elijah Kannatey-Assibu, University of Michigan; Ranga Komanduri, Oklahoma State University; Patrick Kwon, Michigan State University; Steven Liang, Georgia Institute of Technology; John E. Mayer, Jr., Texas A&M University; Anthony Okafor, University of Missouri–Rolla; Jay M. Samuel, University of Wisconsin; Steven Schmid, University of Notre Dame; Rajiv Shivpuri, Ohio State University; J. W. Sutherland, Michigan Technological University.

I appreciate the work and help of the staff at Addison Wesley Longman. I thank editor-in-chief Chuck Iossi and senior acquisitions editor Michael Slaughter for their trust and advice, development editor Liz Fisher for her suggestions for improvements to the text and figure captions, associate editors Susan Slater and Colleen Kelly for capably managing my various submissions, Harold Moorehead for his efforts in preparing the electronic transmissions of many pieces of art, Pattie Myers and Kamila Storr for their help in producing the book. I am especially grateful to my production editor Caroline Jumper. I enjoyed her responsiveness in working through the reviews of copy edits and of page proofs, her advice, and her reliability. Obviously, as in manufacturing a piece of machinery, the product is as much the creation of the designers as of the manufacturing engineers—it takes a lot of special knowledge and skills and effort to convert the author's manuscript into a book. I am happy to give credit to those who contributed to this transformation.

Two of the graduate students who helped me with the graphics deserve special mention for the quality of their work: Rob Ford and Scott Leeb.

My thanks also go to those at Prentice Hall who took over the final stages and who will have to cope with the marketing and sales of the fruit of my efforts.

Last but not least I acknowledge with love the patient support of my wife Hana. She saved my life when I was very sick a few years ago. She looks after me in many ways and has not only given me moral support and encouragement, but also helped with some of the many chores associated with preparing a manuscript.

Jiri Tlusty

PART 1

Background Matters

1 Manufacturing Management

This is the first of three chapters that form Part 1. It has been included to emphasize the point that the main text, after Part 1, deals with only one aspect of manufacturing engineering, that of processes and equipment, and it uses the analytical techniques traditionally taught in the Mechanical Engineering (ME) Departments. This aspect is called production engineering in the context of this chapter. Another aspect of manufacturing engineering that deals with management activities, such as production planning and control, uses the analytical techniques traditionally taught in Industrial Engineering Departments. Both these aspects are combined and overlap in the real manufacturing world. In most engineering colleges the two departments are separate, and most textbooks deal with one aspect or the other. With the trend toward interdisciplinary education, some colleges include both mechanical and industrial engineering in one department. Even so, specialized courses are still offered; in ME, for example, there are courses in mechanics, dynamics, heat transfer, and so on, each with its own textbooks. These are synthesized in the two subjects of design and manufacturing, and it is understood that both are parts of concurrent engineering. While I understand these relationships, I believe that a course focusing on the production engineering aspect is well justified. The industrial engineering aspect is thus only briefly reviewed in this introductory chapter.

1.1 MECHANICAL PRODUCTION

The wording of the title of this section is not commonly used; it is borrowed from the French "production mecanique" to indicate essentially manufacturing plants, or factories that produce machinery of all kinds, from airplanes to cars, ships, bicycles, and so on. It also indicates factories that use the same kinds of processes of sheet-metal forming, machining, welding, and so on, as in building machines, to produce office equipment, electrical machines, or computers. We are not trying to formulate a definition of what does and does not belong in the text; what we have in mind will become obvious as we proceed. What is *not* included is manufacturing of food or clothing, or the processing of chemicals. Primary manufacturing processes that produce materials and those that shape the materials into the various primary items, such as steel plates and sheet, forgings and castings, are only briefly discussed because of the different technologies based essentially on metallurgy. From the point of view of the traditional specialization of engineering colleges, this book deals primarily with the domain of mechanical engineers and to some extent, of industrial engineers.

Another way of indicating the economic sector that is the subject of this text is to show its share in the overall economy of the country. The table in Fig. 1.1 was compiled from data contained in the 1994 *Statistical Abstract of the United States* published by the Bureau of the Census.

The overall gross national product reached $7.576 trillion in 1996. The federal budget was 18.6% of it, and defense costs were 4.6%. The breakdown of the individual industries is interesting. The largest components are finances (banking, insurance, and real estate), which amount to $1 trillion, and services (e.g., hotels, personal services, auto repair and service, motion pictures, amusement and recreation, health services [representing $408 billion], legal services, and education), which amount again to more than $1 trillion. The focus of this book is on the production of durable goods, for which a selective breakdown is given on the right side of the table. Nondurable goods include food, tobacco, textiles, chemicals, and petroleum.

The volume of mechanical manufactures is presented both in value added and shipments. It is more realistic to consider value added, because otherwise the individual categories would overlap. For example, the category of primary metal production, which includes steel mills among others, is then distributed in almost all of the other categories that use steel in their products and include its cost in the value of shipments.

National Economy of the U.S.A., 1996 (billions)

GDP (gross domestic product)	$7,576	
Federal budget	1,406	18.6% of GDP
Defense	346	4.56% of GDP

Industry, 1994 (GDP $6,604)	
Agriculture	$115
Mining	97
Construction	253
Durable goods	658
Nondurable goods	510
Transp. and utilities	585
Communications	182
Wholesale trade	450
Retail trade	595
Finance, Ins., Real Est.	1,192
Services	1,249
Total	$3,876

Mechanical Manufactures, 1995 (selected branches)	Value Added	Shipments
Primary metal	$70	$180
Engines and turbines	9	23
Farm machinery	9	21
Construction machinery	17	38
Metalworking machinery	22	35
Computer and office eq.	36	90
Electrical and Electronic	174	299
Automobiles and equipment	105	326
Aircraft and parts	42	84
Ship and boat bldg.	8	15
Missiles and space vehicles	11	19
	$503	$1,130

Figure 1.1
National Economy. Data extracted from the *Statistical Abstract of the United States, 1997*. US Department of Commerce, Bureau of the Census. Manufacture of machinery, electrical equipment, and computers amounts to only about 15% of the GDP, but its products are decisive for the performance of all other industries.

The largest category is electrical ($174 billion), which includes electrical distribution equipment, electric motors, household appliances, audio/video equipment, and so on. The other substantial categories are motor vehicles, (i.e., cars, trucks, and equipment), and aircraft, which contains both civilian and military airplanes. Metalworking machinery, including machine tools, presses, and cutting tools, is a medium-size industry, yet significant, because this machinery is used in the production of all the other categories, and it is decisive for their productivity and quality.

To illustrate the aims of mechanical manufacturing, it is useful to briefly describe some of the products of the mechanical industries. The first case is a jet engine, one of the most fascinating examples of modern sophisticated machinery. Within the rather mature field of mechanical engineering, where most of the classes of machines have existed for about a century (e.g., combustion engines of the reciprocating type, automobiles, water turbines, steam engines, diesel engines, steam turbines), jet-engine propulsion of airplanes is relatively recent. High-performance jet engines have been used both for military and commercial aircraft for only about forty years. Figure 1.2 shows a drawing of the F117-PW-100 turbofan engine, used in the C-17 military transport. Most of the parts are made of exotic materials, medium- and high-temperature-resistant-alloys, and they are of high precision with complex shapes. The temperature in the combustor reaches about 1,250° C, and the vanes and blades of both turbines are exposed to hot gases. The big fan at the front is driven by the turbine at the very end of the engine via the hollow shaft that stretches all along the length of the engine. The rotational speed of these parts is 4,000 rpm, and the peripheral speed on the ends of the

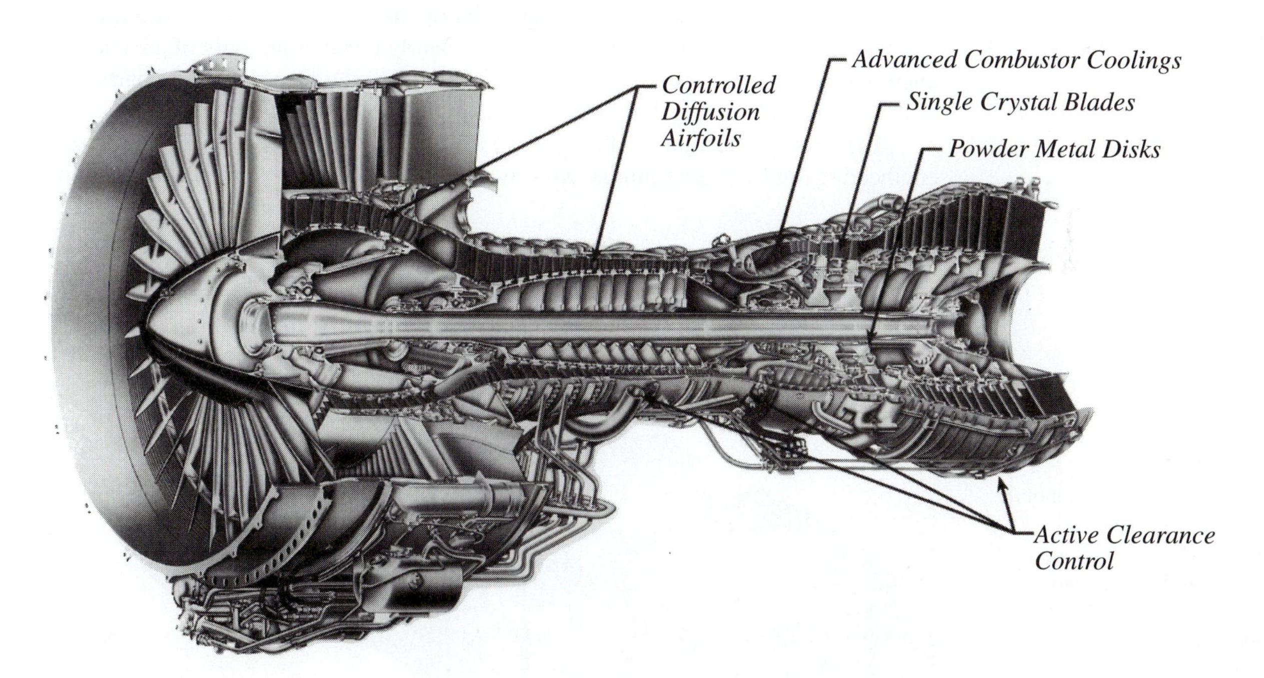

Figure 1.2
A Pratt & Whitney turbofan jet engine. (Courtesy of United Technologies, P&W) Jet engines represent the high-tech class of machinery with their extreme demands on strength of parts under high-temperature conditions. The various parts shown are the main shaft, inlet case, outer case, fan, low- and high-pressure compressor, high- and low-pressure turbine, all with a variety of blades.

fan blades is 440 m/sec. This results in a centrifugal acceleration on the tips of the blades of about 200,000 m/sec^2 (20,000 g). Every gram of mass at the blade tip seems to weigh 20 kg. These blades are forged from a titanium alloy that has a strength comparable to alloy steel but is about two times lighter. The fan provides more than half of the thrust force, and the rest is provided by the gas jet.

Four stages of compressor blades and vanes (controlled diffusion airfoils) are attached to and rotate with the fan. A second set of airfoils in the middle of the engine is mounted on discs welded together by an electron beam to form a drum. This is the high-pressure compressor, which delivers air into the combustor. This set rotates at about 11,000 rpm together with the front part of the turbine, labeled as single crystal blades. The blades in the turbine must endure high centrifugal stresses at high temperature. They are cast as single crystals from a high-temperature-resistant nickel alloy, are hollow and drilled for cooling holes, and coated with a ceramic. The whole combustion chamber is ceramic-coated. The blades of the fan and of the low-pressure compressor, as well as the discs of the drum, are made of forged titanium. All the vanes and blades have complex aerodynamic shapes, and they are machined on five-axis, numerically controlled milling machines, except for the short blades of the rear stage of the compressor, which are made as extrusions. The stationary housing and casings are mostly made of titanium castings or forgings, but some of these parts are welded of sheet metal. This brief description reveals that the manufacture of the engine requires a high level of expertise and first-class manufacturing equipment and processes. The titanium alloys and the nickel alloys are very difficult to forge and to machine because of their low thermal conductivity, which causes high local temperatures during processing. While the engine as a whole is made in small quantities, some of its parts (e.g., the compressor blades) are made using methods that permit economic manufacture of greater quantities. Producing such engines is clearly a matter not only of advanced production processes but also of complex organization in the manufacture of parts as well as in the assembly and testing processes.

The second illustration (Fig. 1.3) is a cutaway picture of an automobile. Many engineering students are familiar with its parts. Let us discuss some of them from the

Figure 1.3
A picture of an automobile (Buick Park Avenue) showing familiar parts such as the engine, radiator, transmission, and suspension, as well as the sheet metal body parts. The car is certainly the most popular piece of machinery. (COURTESY MADISON FORD PHOTOGRAPHIC)

point of view of manufacture. The engine block is cast from cast iron or from an aluminum alloy and then undergoes a large number of machining operations on a transfer line. Its top is face-milled or broached flat, cylinders and other openings are bored, many other attachment surfaces are milled, and a multitude of threaded holes are drilled and tapped. The cylinder head is processed in a very similar way. The pistons are die cast of aluminum and diamond turned. Piston pins are centerless-ground. Connecting rods start either as castings or as forgings and are then machined, as are crankshafts and camshafts, which are both finish-ground on machines with multiple grinding wheels. Then there are valves, valve lifters, pumps, seals, and so on. Many parts are made of plastics, such as dashboards and other parts of the interior, and bumpers. Some car bodies are made of fiber-reinforced plastics. Mostly, however, the body panels (e.g., hood, roof, side panels, doors, and trunk lid), are stamped from steel sheet and then spot-welded together. The transmission contains gears that are forged into the basic shape, gear cut, and shaved or ground. The picture shows other parts such as suspensions consisting of springs, shock absorbers, arms, knuckles and linkages, and disc brakes and drum brakes. Automobiles are made in large quantities, hundreds of thousands of each model per year, and the methods used are those of mass production. Not only is it necessary to master the production processes and to design and build the various types of production machinery, but it is also necessary to organize the flow of parts through the multitude of processes that lead to the assembly line, as well as the procurement of parts from subcontractors ("out-sourcing") in proper quantities at proper times and of satisfactory quality. It is also necessary to provide the financial and human resources. Production techniques, on one hand, and planning, control, and management, on the other, are outlined in this chapter as two aspects of manufacturing engineering.

Most machinery products are not made in large quantities, so they do not fall into the category of mass production. One, ten, or several thousand may be made of a product per year. Their manufacture is divided into small (1–20) or medium lots (20–200), and each lot may be produced several times per year. Typical examples of such products are bulldozers, graders, excavators, turbines, ships, and machine tools. One type of a machine tool, a horizontal-spindle machining center, is shown in Fig. 1.4, rendered in a semi-cutaway fashion. It is a three-axis, numerically controlled (NC) machine tool intended for milling, drilling, tapping, and boring of box-type workpieces. Its frame is made of iron castings or else of steel plates flame-cut to shape and welded. It consists of the bed (B) and of a column (C) that rides on the x-coordinate guideways (GX) located on the top of the bed. This motion is driven by the x-servomotor (SMX). On the front of the column is another set of guideways (GY). The headstock (H) moves vertically in the y-coordinate on these guideways. It carries the spindle (SP), its drive transmission (DT), and motor (M). The motor speed is steplessly variable. Typically, the speed range of the spindle is 50–5,000 rpm, and over most of the range a power of 25 kW may be available. At the side of the machine is a chain-type tool magazine (TM). Face mills, end mills, drills, reamers, taps, and boring bars of different sizes are stored in this magazine, and a mechanism is available (not visible) to carry out automatic change of the tools into and out of the spindle, where they are automatically clamped and unclamped. The bed represents the bar of a letter **T**, its stem formed by the front bed (FB) with guideways (GZ). A saddle rides in parallel with the spindle axis and carries a rotary indexing table (T). On its top sits the pallet (P) that normally has a workpiece clamped on it (not shown). The pallet can be shuttled and exchanged with another one (P2), waiting in the front, which sits on a rotary table and can be deposited in station S2 and replaced by the one in station S1. The whole work zone is enclosed by

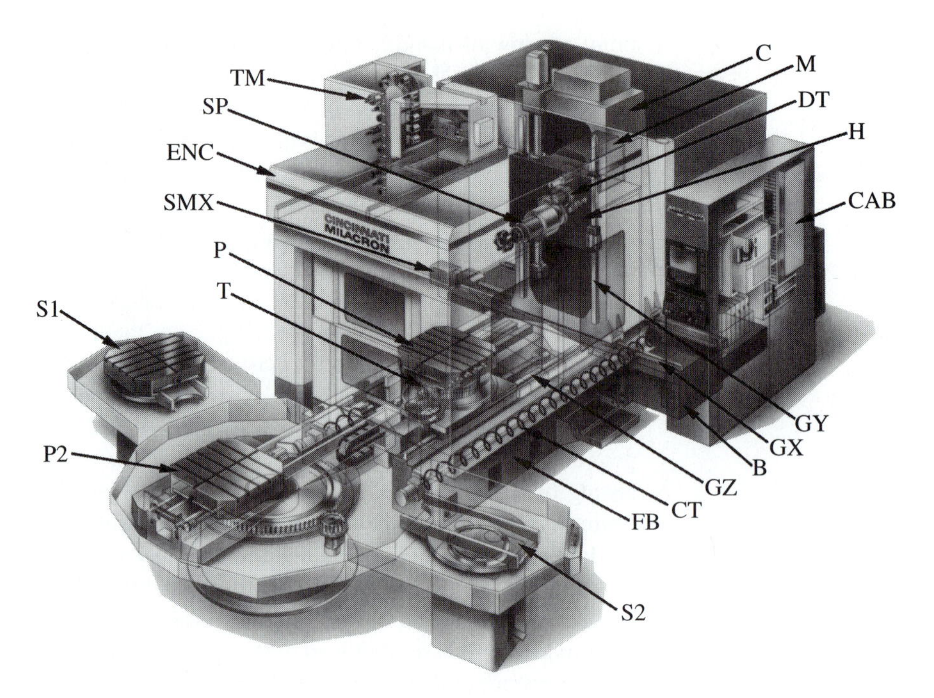

Figure 1.4
Horizontal-spindle machining center. This is the most universal of computerized machine tools, very significant for machinery production. Among the individual parts described in the text, the prominent ones are bed B, front bed FB, column C, spindle SP mounted in headstock H and driven by electromotor M, table T with pallet P, guideways GX, GY, GZ, tool magazine TM. [CINCINNATI MACHINE]

enclosure ENC, and chips are moved out by the screw-type transporter CT. All the motions including 3D-synchronized coordinate motions X,Y, and Z are controlled from a computer located in the cabinet (CAB) with an operator interface panel on its front. The system of guideways is the basis for the precision of the 3D work. They must be very straight and square to each other. The lengths of the coordinate motions are derived from precision scales or from rotary encoders attached to the servomotors and providing positional feedback signals for the servomechanisms. In some high-precision machines laser interferometers are built in instead of the scales. Although these machines are assembled in medium lots, many of the elements (e.g., servomotors, recirculating, ball-type leadscrew/nut transmissions, encoders, controller, coolant supply system) are subcontracted by the machine-tool builder to specialized suppliers, where these elements are manufactured in larger quantities and with high quality. Again, while a great deal of mechanical engineering and production engineering work is involved, there is also a large effort devoted to planning, organizing, scheduling, and overall controlling of all the production activities, to the procurement of subcontracted work, and to the management of personnel of various skills and various specialties, all of which are based on the techniques of industrial engineering.

The fourth and last example is a military aircraft, the strike fighter plane F/A-18E/F Hornet. Figure 1.5 shows it getting ready to take off by flying straight up into the sky while accelerating with a 7-*g* (seven times gravity) acceleration. From the point of view of design and of materials used, a fighter plane is rather different from commercial passenger airplanes. The need for high acceleration and maneuverability places even more emphasis on the strength-to-weight ratio of all the parts. While the structure of a commercial airplane is made almost entirely of high-strength aluminum alloys, their use in the fighter airplane is more limited. The F/A-18C/D airplane uses aluminum for about 50% of its structural weight, and the F/A-18E/F only 29%. The other

Figure 1.5
F/A-18E/F Hornet fighter airplane. (COURTESY BOEING CO.) Aircraft manufacture equals that of cars as one of the largest classes of mechanical production. A fighter plane is an amazing high-performance structure propelled by the jet engine and controlled by a highly skilled person aided by a high-performance computer.

major constituents of the latter model are alloy steel, 13%; titanium alloys, 15%; and carbon fiber–epoxy composite, 21%; with a variety of other materials representing the remaining 22%.

In discussing both the jet engine and the machine tool, not much space was devoted to the assembly of all the individual parts, although that final operation is a demanding task. As regards the automobile, the public has been aware of the importance of the assembly process since Henry Ford introduced the moving assembly line in the 1920s. In the case of the fighter plane, it is helpful to start from the schematic representation in Figure 1.6 of the use of fixtures in the low-rate expandable tooling (LRET) concept. Only a part of the whole system is reproduced here. This system is a flexible way of facilitating the drilling, reaming, and joining operations. At the start of production of a new airplane model, when the rate of production is very low, say less than one per month, only the minimum necessary part of the system of fixtures and jigs is prepared and used. As the production rate increases to something like four per month, the system is expanded to cope with the need for higher productivity. Separate modules are built for subassemblies that are added to the main assembly.

Apart from expressing the significant role of the assembly operations, Fig. 1.6 also points out the main parts of the airplane structure. They are, left to right, the forward fuselage, the center fuselage, the inner wing, the outer wing, the aft center keel, and the vertical tails. The horizontal stabilizers that are also attached to the aft center keel are not shown.

Let us finally have a closer look at one of the main parts, the center fuselage, shown in Fig. 1.7. The structure is viewed from the back (bottom of the picture), that would be attached to the aft keel, toward the front, to be later attached to the forward fuselage (top of picture). It consists primarily of a succession of ten bulkheads, located parallel to each other, crosswise to the axis of the airplane. The bulkheads are panels with rectangular pockets milled into them so as to create ribs that are about 60 to 90 mm high and rise from the thin floors (webs) of the pockets. The thickness of the ribs and of the webs is only about 2 mm. As a result the bulkheads are strong and at the same time light, monolithic (integral) structures. Counting from the back, the first five bulkheads are made of aluminum alloy, and the last five, the so-called carrythrough frames, are made of titanium. The inner wings will be attached to their side faces where the triangular assembly tooling angled structures are seen. The bulkheads are topped by flat and curved bands called the longerons, and they are longitudinally connected by the long central box that has a number of openings at the top and by the side panels, as

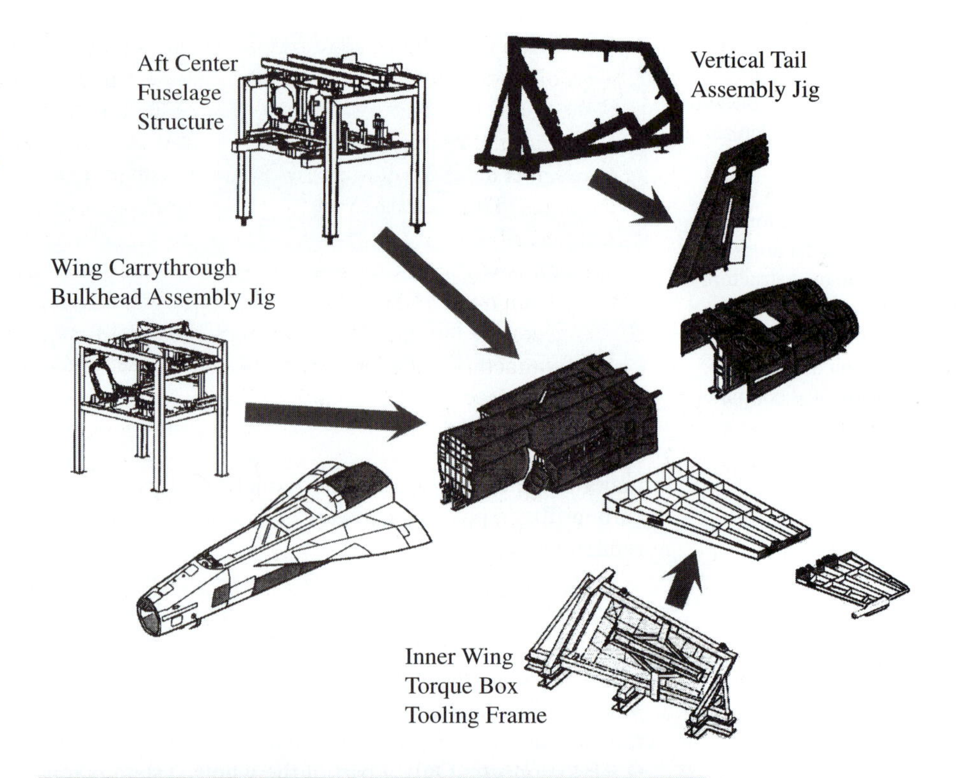

Figure 1.6
Part of a tooling concept for assembling the Hornet airplane. (Courtesy Boeing Co.) Fixtures for assembling parts of fuselage and of wings are shown.

Figure 1.7
The aft center fuselage of the Hornet airplane. (Courtesy Boeing Co.)

well as by several longitudinal strips called stringers. From the point of view of production, the bulkheads and the longerons are made of forgings that are further heavily processed by end milling, which removes a major part of the material of the forgings to produce the thin-ribbed pockets. In Chapter 9 a section is devoted to the design of the milling machines used in these operations. The longerons and the stringers are attached to the bulkheads by means of pin-and-collar fasteners made of titanium and also of stainless steel. For this purpose a large amount of drilling and of reaming the holes for the insertion of the fasteners is carried out as part of the assembly process. The whole structure of the fuselage and of the wings is covered by sheets called skin, and these are attached by means of fasteners inserted into still more drilled holes. Altogether, more than a million holes are drilled in the production of one plane. Some of the skin is made of aluminum alloys, but a major part is made of carbon fiber–epoxy composites. The wings are designed as box-type structures made of spars (along the wing) and ribs (across the wing) connected together by fasteners. Spars and ribs are again beams with pockets milled in them. The cost of such an airplane derives almost equally from making the structure, making the jet engines, and acquiring the computer that is necessary for the pilot to fly such a powerful piece of machinery.

Manufacturing the parts that make up a product is only one aspect of the life cycle of a product (see Fig. 1.8). A strong link exists between design and manufacturing, and the actual shape, dimensions, and tolerances of a part may change if the designer consults with a production engineer to decide how the part will be made, and which process and equipment should be used. When the computer is used in both these activities, they are known as computer-aided design (CAD) and computer-aided manufacturing (CAM) and share the database in CAD/CAM. This strong link is then best extended to all the other types of activities, from the identification of need to training

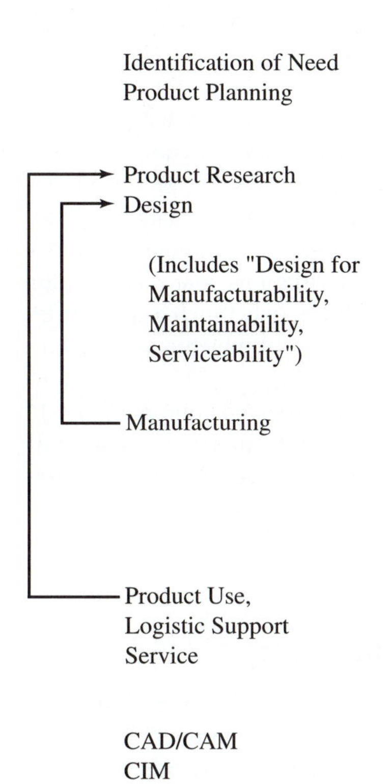

Figure 1.8
The product life cycle and its main components. Methods used are computer-aided design/computer-aided manufacturing (CAD/CAM), computer-integrated manufacturing, and concurrent engineering.

in the use of the product. We are then dealing with computer-integrated manufacturing (CIM) and with concurrent engineering (CE).

The various products—the electric switches, coffee makers, lawn mowers, bicycles, appliances, machine tools, cars, and airplanes, each category produced in a variety of models—differ in complexity, being composed of perhaps ten or more than ten thousand parts. They differ in length from less than an inch to hundreds of feet. They differ in numbers produced per year. They differ in materials of which their parts are made, whether they be metals, plastics, ceramics, composites. Obviously, the factories and the manufacturing equipment, the manufacturing processes, material-handling equipment, and the assembly methods vary tremendously. We may distinguish various types of manufacturing systems. With regard to *quantity*, it is customary to distinguish job shops, medium-lot and large-lot production, and, finally, mass production.

In job shops various parts are made in small quantities, down to only one, and they may change all the time. The equipment often consists of metal-cutting machine tools, because they are versatile and there is minimal need for special tooling, dies, molds, and fixtures. Manual welding is often used. The machine tools are mostly manually operated, although more and more of these shops use numerically controlled (NC) machine tools.

The great majority of factories produce parts in small, medium, and large batches. Production equipment is universal, and each machine can perform a variety of operations. Flexible automation, mostly NC, is commonly used. The yearly quantity of parts is manufactured in batches of tens and hundreds, even thousands, and each batch is repeatedly produced several times per year, so as to distribute the inventory of completed products (dead capital), which would be large if the yearly quota was made in one month and stored and sold throughout the year. Production usually requires special tooling such as dies or molds, fixtures, and NC programs. A more detailed discussion of the costs and benefits of automation and of the significance of setup actions and costs is presented at the beginning of Chapter 10.

Then there is mass productions—of cars, cameras, appliances, ball bearings, and so on. Typically, a car engine may be produced in quantities of 5,000 per day. Production equipment is specialized: a special machine carries out a particular operation repeatedly, with no change. An automobile cylinder block may be machined on a transfer line consisting of fifty stations, many with multiple spindles. Hundreds of tools may work on a block simultaneously. Such a transfer line is a complex piece of "rigid automation" equipment, the cost of which is only justified by the large quantity of the identical parts being made.

The manufacturing processes and equipment used for a particular part depend to a great extent on the *material* of the part. Clearly, aircraft parts or automobile parts will be differently manufactured, as some of them are made of aluminum, of titanium, of composites, of sheet metal, or of plastics.

1.2 INDUSTRIAL AND PRODUCTION ENGINEERING

There is a lot of engineering activity behind all successful manufacturing. Some of it is what we will call production engineering (PE). This term does not enjoy a standard usage in the United States, and sometimes manufacturing engineering is used instead. However, we will use it in the same sense as it has been used for many years in

England, where they actually have a Society of Production Engineers. PEs deal with the activities related to the specification, design, selection, testing, and usage of production equipment, such as machine tools, presses and hammers, materials-handling equipment, robots, and so on, as well as with the specification and design of tooling and fixtures. They are involved in process planning to determine the manufacturing processes/equipment and tooling for all the parts to be produced. This includes the preparation of NC programs, selection and design of such manufacturing processes as cutting and forming, and of their parameters, and also research to improve existing processes and to develop new ones.

These activities are all related to the engineering design function the results of which are the drawings of all the parts and the bill of materials. For the past decade the symbiosis and cooperation of designers and production engineers, recognized as concurrent engineering, has been evolving into a continuously closer relationship that is supported and made possible by the development of computer-aided design systems as well as of computer-aided manufacturing systems, and finally the combined CAD/CAM systems. The definition of the design of a product and of all its parts is stored in the CAD system where it is also available for analytical work such as finite element computations. The same database is used by the CAM software to prepare the NC programs for all the parts to be machined, flame-cut, or electro-discharge-wire cut, and so on. More recently an overall concept of computer-integrated manufacturing (CIM) combines all the software activities of an enterprise into one large system.

It is obvious that in order to make one part or even one piece of machinery, it is sufficient to have the machine tools and other tools, and to possess the qualifications of a production engineer. This background is based heavily on the knowledge obtained in the various fundamental subjects of the mechanical engineering curriculum such as strength of materials, elasticity and plasticity, thermodynamics, heat transfer, kinematics and dynamics, and automatic controls.

When it comes to producing a variety of products with all their different parts, we still need the production engineer to develop the automation systems, the material-handling equipment, and the assembly equipment. However, now we have a great number of operations going on simultaneously, competing for the availability of pieces of production equipment, to be completed at particular dates, and all of this must be organized efficiently. This is then the responsibility of staff trained and knowledgeable in what the Germans call *Betriebsorganisation* (organization of production), and what is in general known as manufacturing management or production planning and control. The methods used in these activities are mostly developed in a general fashion so as to apply not only to mechanical production but also to insurance, health organization, and so on. Thus, for instance, when it comes to optimizing inventory and queuing, it does not matter whether we deal with the manufacture of clothing or of machines. These methods and activities are typically addressed by industrial engineering (IE). Manufacturing engineering thus comprises two basic engineering disciplines: production engineering and industrial engineering (see Fig. 1.9). The profession of IE uses some different techniques and is based on a different kind of education than that of mechanical engineering (ME). It uses statistics, operations research, artificial intelligence and, most important, it also includes human behavior, psychology, and sociology.

Manufacturing engineering is the most commonly used term in the United States for both PE and IE. However, the remaining part of this book deals essentially with PE only. Those who want to learn more about IE may study texts devoted to this subject, several of which are listed in the references at the end of Chapter 1.

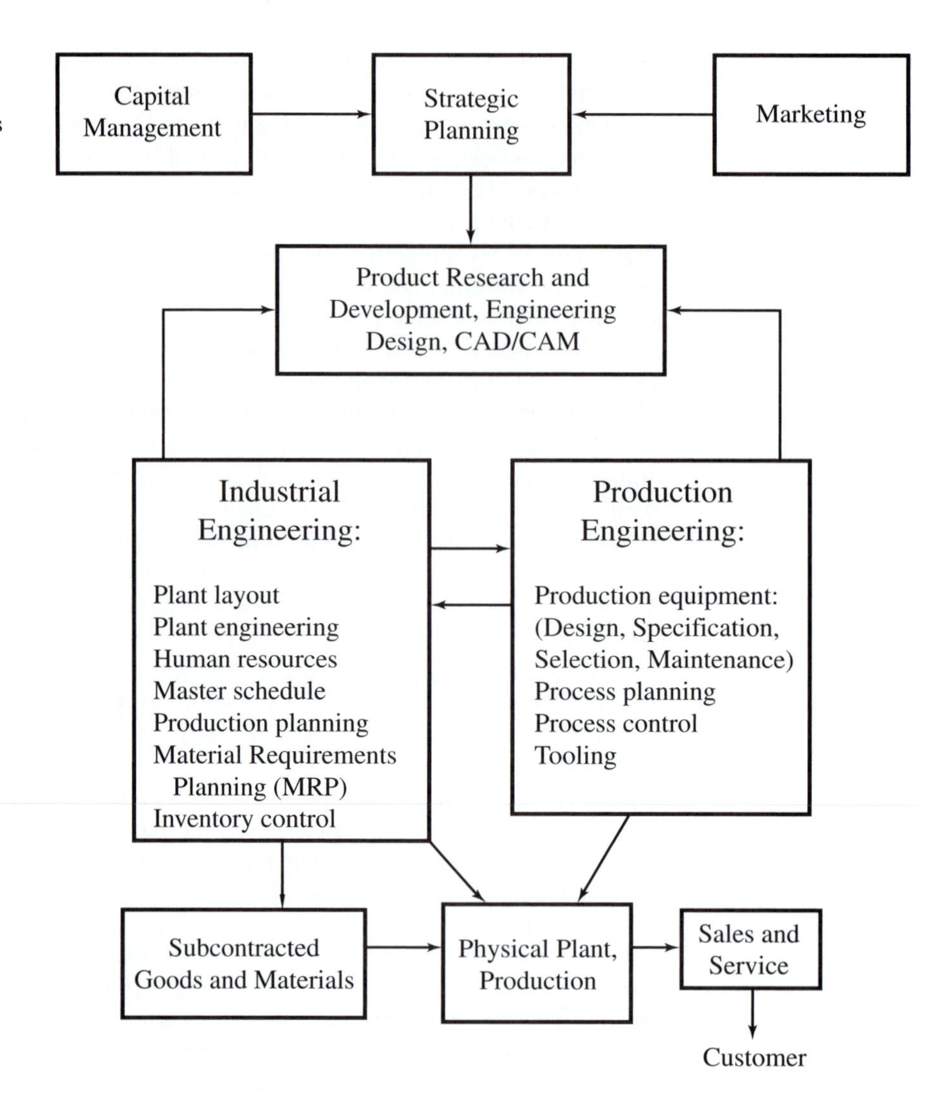

Figure 1.9
Industrial and production engineering, and their roles in the manufacturing enterprise.

1.3 INDUSTRIAL ENGINEERING: PRODUCTION PLANNING AND CONTROL

Industrial engineering emerged as a profession from the work on methods engineering and scientific management at the beginning of this century. F. W. Taylor, Frank B. Gilbreth, and Henry L. Gatt are well known for their contributions to analyzing the work in factories and designing methods for improving its efficiency. During the 1920s and 1930s work was done on economic aspects of managerial decisions, inventory control, incentive plans, factory layout, and principles of organization. During World War II the theory of operations research was established and then widely applied to problems of industrial engineering. In the recent past the profession turned to the use of computers and started to develop the system of computer-integrated manufacturing (CIM). Extensive work has been devoted to the computerization of the individual building blocks of the system, such as computer-aided process planning (CAPP) and material requirements planning (MRP). Methods of artificial intelligence have been successfully applied to some of the tasks.

We will discuss the basic building blocks of the manufacturing system, emphasizing the computer-aided versions of these activities. The diagram in Fig. 1.10 indicates the composition of a manufacturing system that would apply to a typical mid- to large-size manufacturing company. The diagram concentrates on management and engineering activities, but it does not define the actual organization of the various layers of management.

At the top is the strategic planning activity, the purpose of which is to determine a long-term (several years) production program by types and models of the individual products and their quantities throughout the planning horizon. The input to this activity comes primarily from marketing, which may use forecasting algorithms to process past sales statistics, as well as perform an analysis of the market and of the performance of competitors. Additional input is obtained from the financial department and the per-

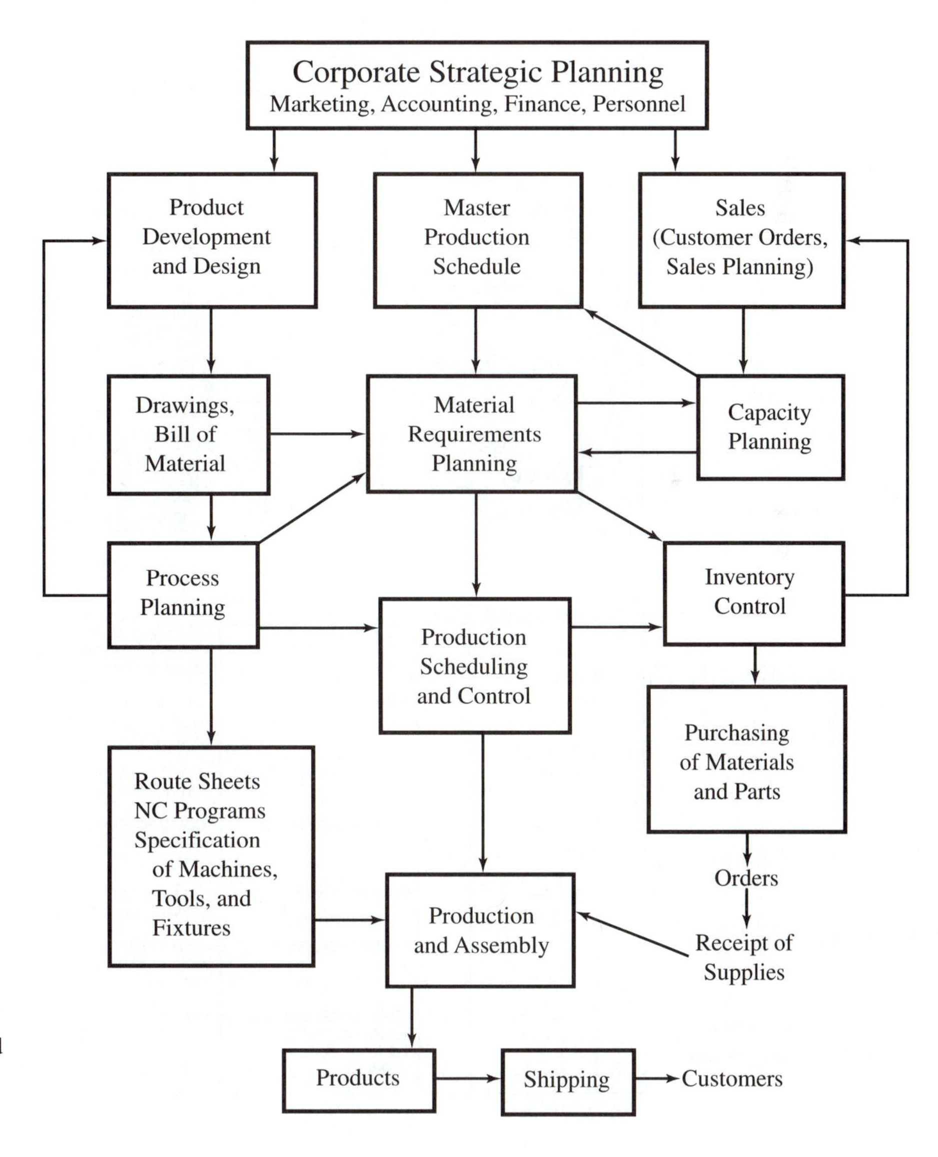

Figure 1.10
Manufacturing planning and control in a mid- to large-size company.

sonnel department. The result of this activity is the master production schedule. It is not very detailed, and it lists just the individual products and their quantities per month. The system provides a check of the schedule from the capacity planning module. If there is a discrepancy, then either the production capacity must be changed or the schedule modified.

The engineering department provides data about the *product design*, usually from an existing design, but each product is usually further developed, and new documentation is prepared. This documentation consists primarily of assembly drawings and part drawings, commonly produced by a CAD system with a plotter. However, some companies do not produce the drawings; process planners just access the CAD contents in the computer. Another important part of the design documentation is the bill of materials. Figure 1.11 shows an example of the form. The figure shows a drawing of a leadscrew thrust bearing subassembly, which is a part of a machine tool. Some of the components are labeled in another view of the subassembly. Below the drawing is a small part of the bill of materials. The complete bill of materials for the whole machine is rather long. This small section of it refers to those item numbers associated with the drawing. It contains the number of the part, the quantity of the part in the product, the type and size of material of which the part is made (e.g., bar stock, casting, forging,

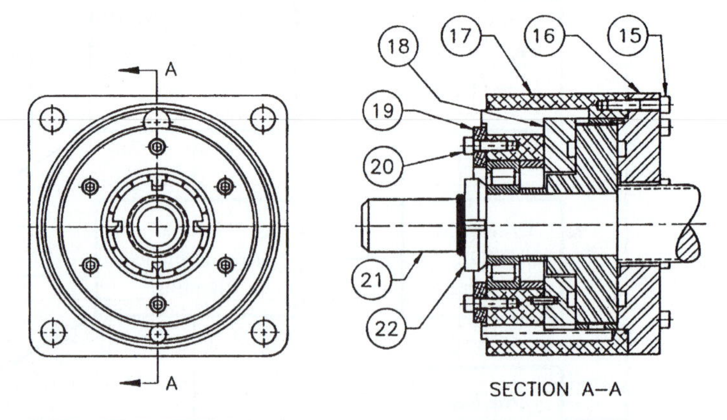

ITEM	PART NUMBER	QTY	DESCRIPTION	MATERIAL
15		9	CAP SCREW	18-8 STAINLESS STEEL #10-UNF SOCKET HEAD CAP SCREW, 1" LONG
16	1121-4597	1	THRUST BEARING COVER	TOOL STEEL O6 OIL HARDENABLE BAR STOCK 5.5" OD, 2" LONG
17	1121-4593	1	THRUST BEARING HOUSING	CAST IRON CASTING PART NUMBER 2-378
18	1121-4523	1	INTERNAL THRUST PAD	TOOL STEEL O6 OIL HARDENABLE BAR STOCK 5" OD, 2" LONG
19	1121-4590	1	RETAINING RING	4140 ANNEALED BAR STOCK 5"OD,1" LONG
20		6	#10-UNF SHCS, .625" L	ALLOY STEEL #10-UNF SOCKET HEAD CAP SCREW, .625" LONG
21		1	1.500X.200 GROUND BALL SCREW ASSEMBLY,	THOMSON, TYPE SSP, 1.500X.200, ORDER #7820375
22		1	LOCKNUT	STEEL LOCKNUT, SKF, SIZE 6, SNW 06

Figure 1.11
A section of a bill of materials. Its individual items are identified in the assembly drawing of the bearing group of a leadscrew.

piece of sheet metal). In the case of standard components that are purchased ready-made (screws, bolts, nuts, electric motors, ball bearings, seals) the designation of the part is entered in the material column, and no part number is issued.

The process planning (PP) function must select and record, for each part listed in the bill of materials, the manufacturing processes, machine tools, fixtures, and other tools to use in its production, as well as the conditions defining the individual processes. For a machining process like turning, for example, the type of lathe, the method of part clamping, and the particular cutting tools must be determined, as well as spindle speeds, depths of cut, and feeds per revolution. A sequence of several processes is usually obtained, such as a cold-forming operation followed by several machining operations. For NC machining, the NC program must be prepared. From all these data, the time for each operation and, as appropriate, the handling time and setup time are calculated. This determines the lead time needed for production. An example of a simple route sheet is shown in Fig. 1.12. Another result of the PP activity is the list of machine tools, tools, and fixtures needed. If none of the existing means can be used, the request to design and make new ones is issued, which affects the overall lead time for the product involved. The diagram in Fig. 1.10 shows feedback from PP to design. This represents the input of production engineers into the design process to facilitate the manufacture of all the parts in the product.

The role of the material requirements planning (MRP) module is to expand the master production schedule into a plan determining the needs for the individual parts, subassemblies, and assemblies expressed in quantities and dates. The primary inputs for this task are the bill of materials and the route sheets obtained in process planning.

Additional input is obtained from inventory control, where the status of the various parts and materials in storage and parts in-process is obtained. These inputs result in two lists: the list of materials and standard parts to purchase, how many of each, and

OPERATION SHEET (Process Plan)

Part No. S576-67   Material Cast Iron 9lb/pc

Part Name Eccentric Strap Cap Half

Orig. ______   Changes ______

Checked ______   Approved ______

No.	Operation Description	Machine	Setup Description	Operate Min./Unit
5	Rough and finish mill 2 mating surfaces	Cinc. Mill (Kender #136)	Gang 6 castings in fixture	75
10	Spotface and drill two holes 33/84 in. D; drill 27/84 in. D pipe hole; tap 1/4 in. pipe thread	Multispindle drill press	Piece on table Piece in 35° drill jig	120
15	Rough and finish bore 6 1/4 in. D; bore 6 1/2 in. D x 3/8 in. wide groove	Bullard Vert. Boring Mill (Kender #335)	Clamp to Eccentric Connector Half (S563-5), then mount both parts in 4-jaw chuck	170

Figure 1.12
Part of a process plan (route sheet), showing a sequence of machining operations for a particular component. [Chang Wysk Wang]

when to purchase them is derived, taking into account the lead times for the individual supplies, and a list of the various parts to produce, how many of each, and when, taking into account the production lead times. This results further in the calculation of needed capacities in labor hours and in production equipment. This is called capacity planning, and any discrepancy between this need and available capacity must be resolved either by overtime work, or by hiring additional workers, or subcontracting or purchasing additional production equipment. Otherwise the master schedule must be modified.

Next in Fig. 1.10 is a block for production scheduling and control. The production orders released from MRP are now allocated to the individual machines. This is called machine loading. There is a tremendous competition between the hundreds of orders and the limited number of facilities. Each machine has a queue of orders waiting to be processed. Making the decisions about priorities and the order in which jobs will be taken on is called job sequencing. A number of rules to use have been developed such as "first come, first serve," "shortest processing time," "earliest due date," and so on. The progress of work in the shop is monitored, and the information is fed back to the other activities. It happens commonly that there are disruptions in the schedule due to poor process planning or machine or tool failure. Jobs have to be rescheduled, process plans changed, and so on. This is called production control.

The brief review of the manufacturing system related to Fig. 1.10 is expressed in an alternative form in Fig. 1.13. The left half of the diagram is the domain of industrial engineering, and the right half gathers the activities of what we have decided to call production engineering. Both of them are also involved in quality control.

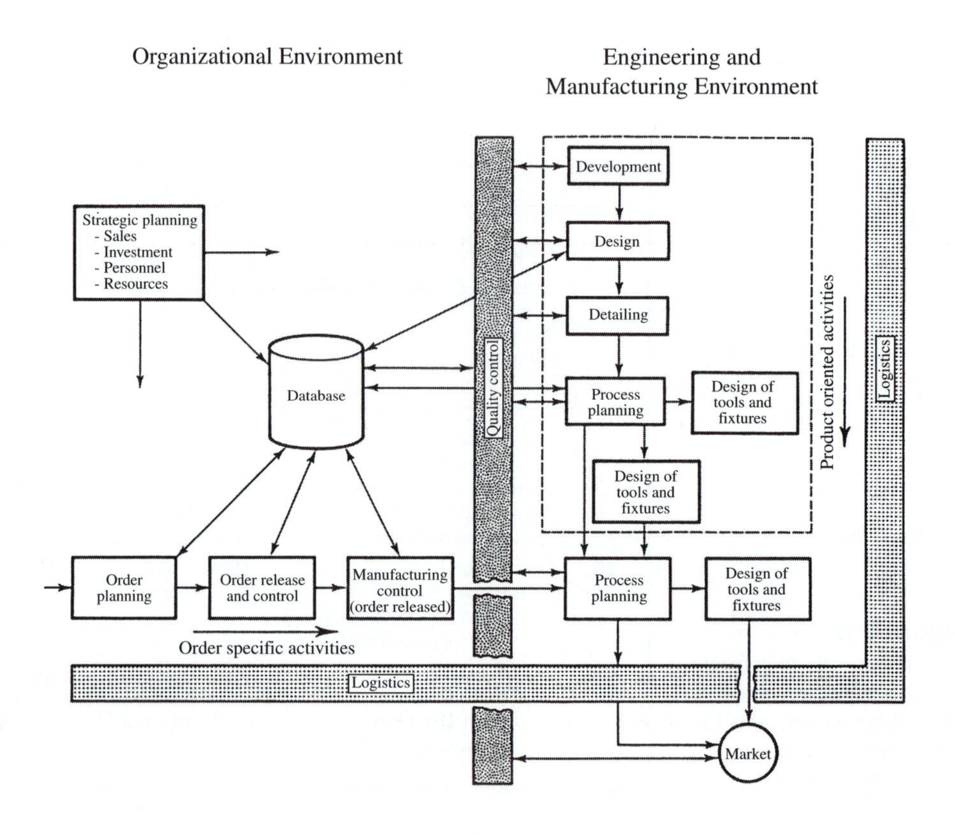

Figure 1.13
Organizational, engineering, and manufacturing environments. [REMBOLD]

So far, we have not specifically distinguished between manual and computerized activities, although the usefulness of the computer must have been obvious for most of them. In building the full, computerized CIM system various models and corresponding software have been developed by a number of companies. Early on IBM got involved. Their effort has been briefly described in [1]. Theirs was the first broadly conceived effort to create a general type of hardware and software for CIM. The system presented in the early 1970s was called communication oriented production information and control system (COPICS; see Fig. 1.14). The diagram shows how all-encompassing the system was; it included the administrative functions of purchasing and sales, plant maintenance, engineering data, and cost control, as well as the various components of production planning and control from the master schedule to resource planning, inventory control, production planning, ordering, and scheduling. Strong emphasis was placed on communication, database management, and presentation. IBM has been continuously engaged in modifying its CIM products, and it currently offers a variety of data processing systems for manufacturing companies. It emphasizes the significance of the computerized database. Among other developments, activities carried out at Digital Equipment Corporation (DEC) are presented in Fig. 1.15. This system has a number of functions similar to those of the IBM system; however, DEC emphasizes the all-important aspect of quality that interacts with all the main activities

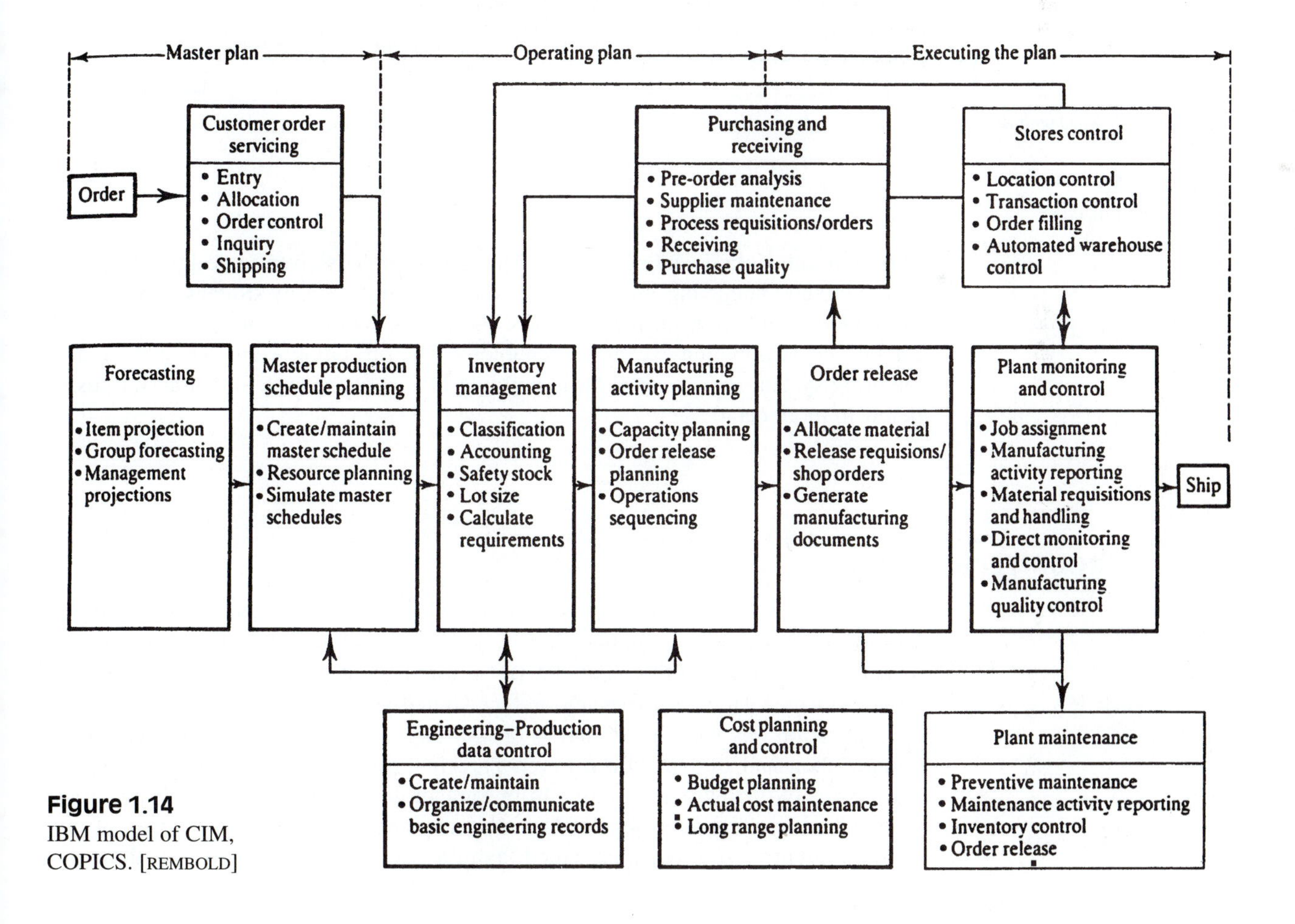

Figure 1.14
IBM model of CIM, COPICS. [REMBOLD]

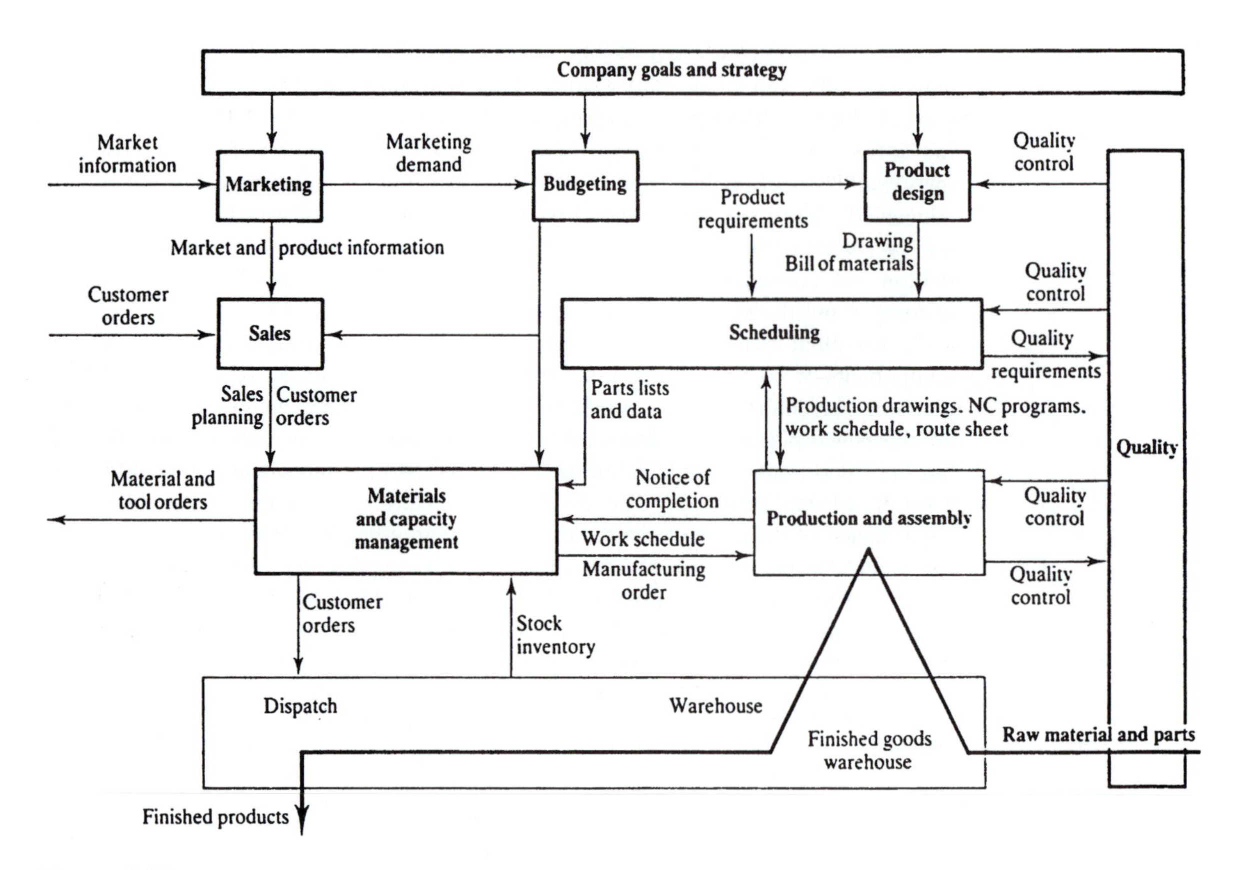

Figure 1.15
DEC model of CIM.
[REMBOLD]

of design, scheduling, and production. The primary role is given to design, along with marketing and budgeting. Inputs are derived from the company strategy, from market information, and from customer orders. All the activities are supported by a well-established information-exchange technology. The preceding examples of the manufacturing system show several basic building blocks of the system that are discussed below in more detail.

1.4 GROUP TECHNOLOGY

The concept of group technology (GT) in manufacturing originated in the late 1950s. One of the first significant publications was a 1958 book by a Russian engineer, S. P. Mitrofanov. In the 1960s and 1970s a number of GT systems appeared in Czechoslovakia, Holland, Germany, and the United States. The basic idea derives from the realization that while there may be a very large number, say five thousand to one hundred thousand, of nominally different parts in the manufacturing program of a company, many of them differ very little from some of the other parts. Great advantage can be obtained by creating groups of parts that are similar primarily in the way of processing. Often these groups are called "families" of parts. They will pass through a similar sequence of manufacturing operations and will be processed on the same machine tools. While each part may be made in rather small batches, the technological similarity is such that the setup of an automated machine only requires small changes between the individual batches, and the sum of the individual setup times will be drastically reduced. The emphasis here is on technological similarity. Thus, while the cylindrical parts may look almost the same, the difference in the materials, tolerances, and surface finishes may cause them to be made on different machine tools. One such part made of carbon steel may be completed in one operation on an automatic lathe (e.g., the "screw automatic"), another one made of stainless steel and requiring tighter tolerances will first also be turned, but at a much lower speed because of the low machinability of stainless steel. Then it will be heat treated and finally centerless ground. On the contrary, the parts in Fig. 1.16a, while they differ in their geometry, will all be made on the same NC lathe, and the parts in Fig. 1.16b will all be made on the same machining center. They are of almost the same size and require milling and drilling on three sides. One may create a family of parts around a composite component (see Fig. 1.17), where the individual family members may have less than all the eight features of the basic component.

In the GT system parts are classified, coded, and stored in a database that is available to all the activities in the manufacturing system. Classification benefits a number of aspects of the system. Primarily, it is useful in *design*, where it plays a role in the retrieval system. The designer may design a part of a machine—for instance, a gear. This is the preliminary step, and the part is encoded. He then retrieves existing parts from the same group and may find that one of them will fully satisfy his need. Or he may design a new part as a modification of an existing one. This kind of activity has enabled various companies to achieve substantial reductions in the number of different parts (e.g., of different gears). Next, GT is used in *process planning*, where it reduces effort substantially, because the new part may use an existing route sheet with only slight modifications. Existing tools and fixtures may be used. Finally, there is significant benefit in the production itself. Rather large families of parts may be created, so that it is economically justifiable to set up specific groups of machine tools in *manufacturing cells*. All parts in the corresponding family can then be produced in the same cell, with almost identical routing. This minimizes the problem of handling of parts between the machines.

Various classification schemes and GT coding systems have been developed, and over one hundred such systems are currently in use. One such system was created in

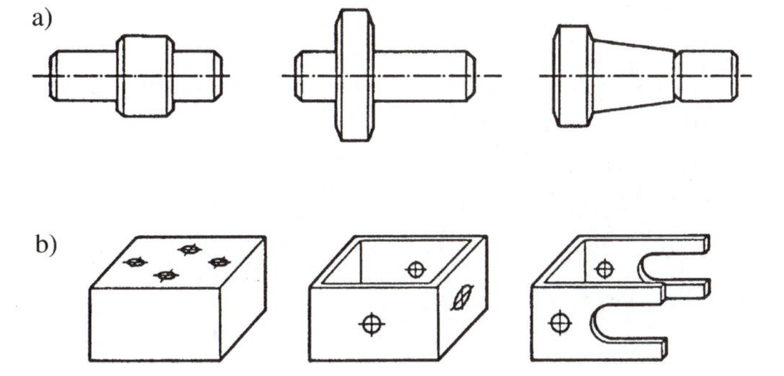

Figure 1.16
Parts of different geometry but similar technology. [REMBOLD]

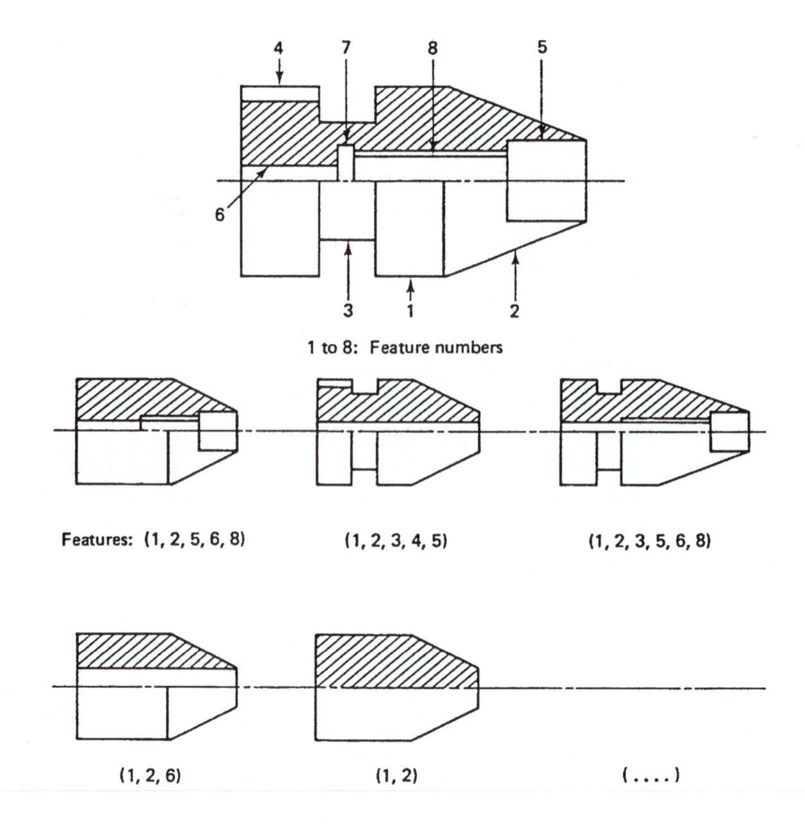

Figure 1.17
A composite component.
[CHANG WYSK WANG]

Germany at the Technische Hochschule Aachen in the institute of Professor Opitz. This system is in the public domain. The code consists of nine digits and four letters:

Form code: 1 2 3 4 5

Supplementary code: 6 7 8 9

Secondary code: A B C D

The basic structure is shown in Fig. 1.18. The first five columns define the part geometry, main dimensions, and main geometric features. For rotational parts, *L* means length, and *D* means diameter. For nonrotational, box-type parts, *A*, *B*, and *C* are the dimensions of the box enclosing the part. So, category 6 generally includes plates, category 7 includes long parts, and category 8 includes boxes. The next four columns deal with material, tolerances, raw material form, and surface finish. A more detailed specification of the first five columns for rotational parts appears in Fig. 1.19. For instance, a code number 12303 designates a rotational part with *L/D* between 0.5 and 3, stepped to one end, with a thread on one outer diameter, a smooth hole with a functional groove, no surface plane, and with a radial hole. The composite part of Fig. 1.17 would be encoded as 17330. Of the many other systems, let us mention the VUOSO Praha code, the Japanese K3 system, the MICLASS system of the institute TNO in Holland, and the DCLASS system of D. Allen of Brigham Young University [10]. A comprehensive bibliography of group technology, including 480 items covering the period 1955–1975, appears in [11].

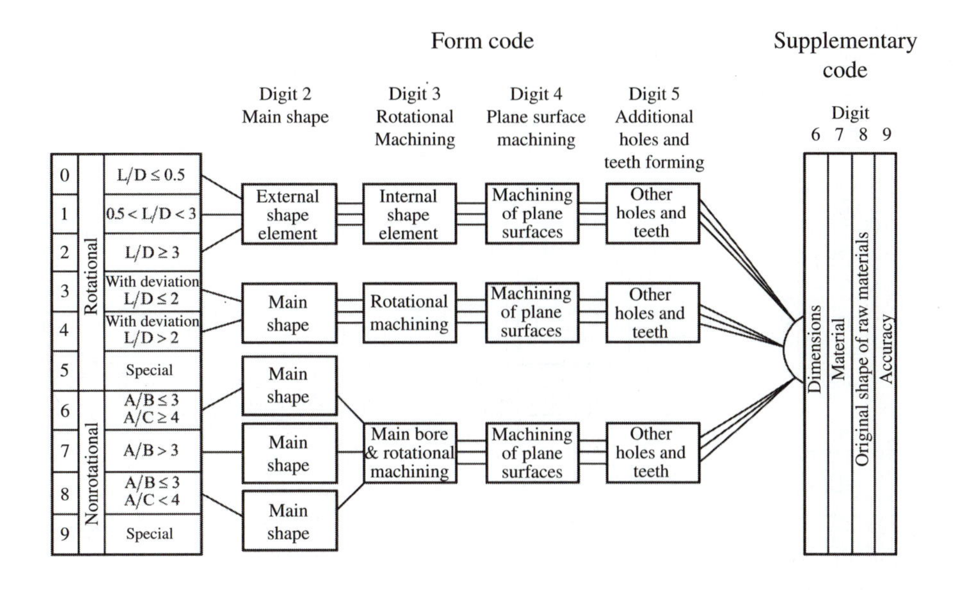

Figure 1.18
Group technology classification developed by Opitz.

	Digit 1		Digit 2		Digit 3		Digit 4	Digit 5	
	Part class		External shape, external shape elements		Internal shape, internal shape elements		Plane surface machining	Auxiliary holes and gear teeth	
0	Rotational parts	$L/D \leqslant 0.5$		Smooth, no shape elements		No hole, no breakthrough	No surface machining	No gear teeth	No auxiliary hole
1		$0.5 < L/D < 3$	Stepped to one end or smooth	No shape elements	Smooth or stepped to one end	No shape elements	Surface plane and/or curved in one direction, external		Axial, not on pitch circle diameter
2		$L/D \geqslant 3$		Thread		Thread	External plane surface related by graduation around a circle		Axial on pitch circle diameter
3				Functional groove		Functional groove	External groove and/or slot		Radial, not on pitch circle diameter
4			Stepped to both ends	No shape elements	Stepped to both ends	No shape elements	External spline (polygon)		Axial and/or radial and/or other direction
5				Thread		Thread	External plane surface and/or slot, external spline		Axial and/or radial on pitch circle diameter (p.c.d.) and/or other directions
6	Nonrotational parts			Functional groove		Functional groove	Internal plane surface and/or slot	With gear teeth	Spur gear teeth
7				Functional cone		Functional cone	Internal spline (polygon)		Bevel gear teeth
8				Operating thread		Operating thread	Internal and external polygon, groove and/or slot		Other gear teeth
9				All others		All others	All others		All others

Figure 1.19
Detailed classification as part of the system of Fig.1.18.

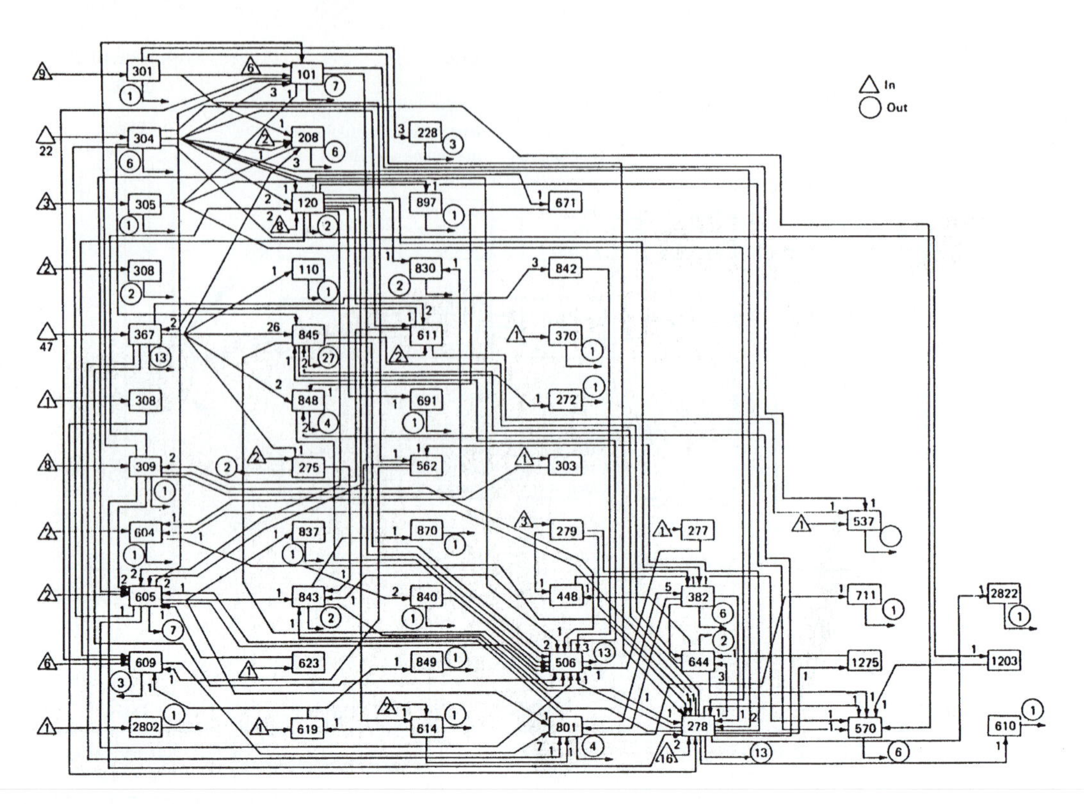

Figure 1.20
Old routing of 150 very similar parts. [TEICHHOLZ ORR]

Let us look at an interesting example introduced in [3]. In this concept a code leads automatically to a configuration of a particular manufacturing cell. A case of a manufacturing facility is mentioned where 150 very similar parts were routed over 51 machine tools. There were 87 different process plans for these similar parts. By analyzing these process plans, it was possible to develop a group technology cell comprising only 8 machine tools, in which all 150 parts could be made using only 31 routings. These conditions and the improvements achieved are easily appreciated by comparing the "before" routings in Fig. 1.20 and the "after" routings in Fig. 1.21. Looking at this example, one wonders, "Why didn't I think of this a long time ago?" Of course, the answer is that computers and methods of database management first had to be developed to a stage that permitted these techniques.

1.5 PROCESS PLANNING: COMPUTER-AIDED PROCESS PLANNING

The role of process planing (PP) is to determine, for each part, the manufacturing processes to use (forming, turning, milling, grinding, etc.) and their sequence. For each process, we must select the conditions (sequence of passes, speed, feed, depth of cut), the machine tool to use, the fixtures, tools, and mechanical handling devices. If an NC machine tool is selected, the NC program must be written. Technically, the NC programmers form a group separate from process planners, although they interrelate closely. Obviously, the process planners must have thorough knowledge of the various processes, machine tools, and tools. The remaining chapters of our book are about this kind of knowledge.

The process plan is recorded on an operation (route) sheet. Figure 1.22 shows a simple example of such a sheet. Process planning as a speciality has existed for over fifty years, and route sheets have been the backbone of production planning and control. Parts move through the shop in boxes with the route sheets attached, from machine to intermediate storage and then to the next machine, all handled by human dispatchers. A process planner's role is very important, and a good or bad plan may make a several-fold difference in production cost and lead time. A process planner must be

Figure 1.21
New routing of parts of Fig. 1.20, after application of group technology. [TEICHHOLZ ORR]

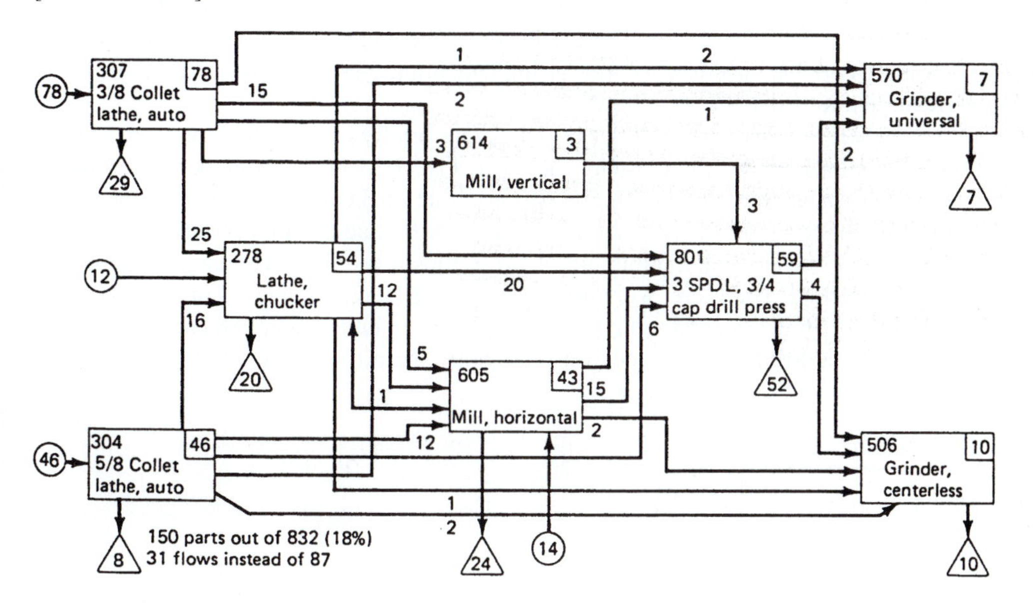

Figure 1.22
A manual operation sheet. [REMBOLD]

Operation sheet No.:				Data:		
Part No.:		Part name:		Drawing No.:		
Orig:		Checked:	Changes:		Approved:	
Pieces:		Matl:		Weight:		
Op No.	Operation	Machine Tool	Tools	Fixtures	Set-up time (hrs)	Operation time (hrs)
5	Rough turning	Lathe 4	T5	Chuck	0.2	0.2
10	Fine turning	Lathe 2	T3	Chuck	0.1	0.2
15	Drilling	D Press 2	D2	Drill jig	0.15	0.1
20	Chamfer	D Press 2	Ch3	Drill jig	0.1	0.07
25	Counterboring	D Press 2	D1	Drill jig	0.1	0.09
30	Heat-treat	Furnace			0.15	0.09
35	Grinding	Grind S			0.15	0.06
⋮	⋮	⋮			⋮	⋮

familiar with the various processes and the machine tools in the shop. Often he has been recruited from among the best machinists. There is now a growing shortage of people with good shop experience. They will have to be replaced by engineers capable of finding, interpreting, and using accumulated expertise, and the computer is a great help in this endeavor. Several researchers have developed artificial intelligence (AI) methods, creating expert systems for the task of computer-aided process planning (CAPP). However, the rapid developments in the field of AI have also had negative effects on the progress of process planning development. Van Houten [8] states, "Prototype systems have been built by knowledge engineers without engineering knowledge. Most of these prototypes incorporate a small set of mostly trivial but sometimes even incorrect manufacturing rules. These systems usually require more effort to generate the input than it takes to produce the plan by hand. It has been shown by subsequent generations of procedural systems that the operations planning task can be automated without the use of AI. Knowledge gained from years of scientific research in manufacturing processes has been compacted into mathematical models with more expressive power than any set of loosely connected and ambiguous rules." However, if AI is accepted as just another way of implementing these models, it can improve the flexibility and adaptability of CAPP systems.

Before we discuss CAPP, let us once more briefly explain the procedure of manual PP. Consider the sample part shown in Fig. 1.23 and a brief derivation of the process plan as presented in [4]. The part is a rotational component, and a lathe should most likely be used. The part datum surfaces are S3 and S7. Precut bar stock is used. Since bar stock is used, a minimum of two setups is required. The threaded area cannot be chucked. Therefore, we must cut surfaces S1, S2, S3, and S4 in the first setup, and S5 and S6 in a second setup. S7 can be drilled in either setup. S8 cannot be drilled on a lathe; a drill press is needed for these holes, requiring yet another setup.

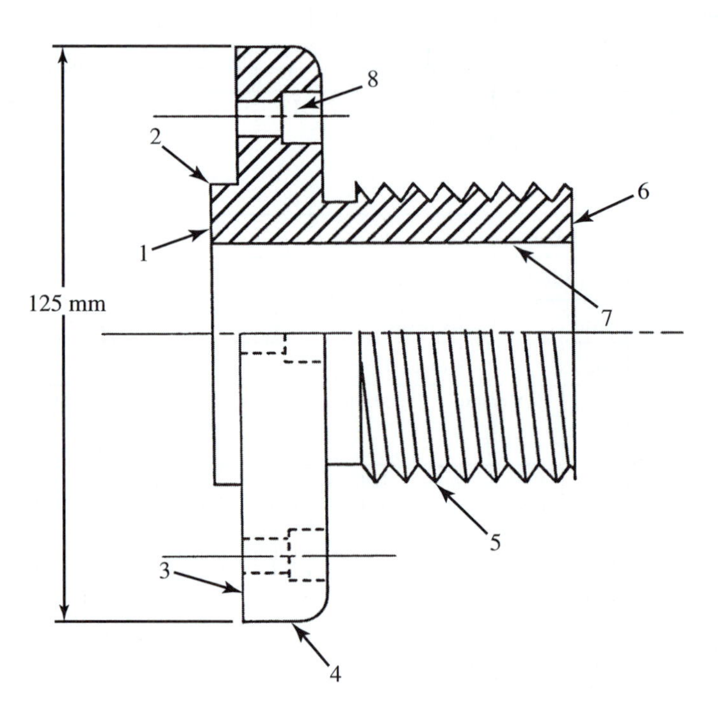

Figure 1.23
Sample part for a process plan, showing the individual surfaces to be machined. [CHANG WYSK WANG]

S1, S3, and S6 require facing operations. S2 and S4 require turning. S5 requires turning followed by threading and an undercut. Drilling operations are needed for S7 and S8. The following is the sequence of operations determined to produce the part.

Setup 1 Chuck the workpiece.
Turn S4 to a 5.25-in. diameter.
Turn S2 to a 2.751-in. diameter.
Face S1 and then S3.
Core drill and drill S7.

Setup 2 Chuck the workpiece on S4.
Turn S5 to 2.75-in. diameter.
Thread S5.
Undercut the neck.
Face S6.
Remove the part and move it to a drill press.

Setup 3 Locate the workpiece using S3 and S7.
Mark and center-drill four holes, S8.
Drill and counterbore four holes, S8.

We stop our planning here and let the reader work out the tool and parameter selection for each operation.

In general, the steps in preparing the process plan are as follows:

Inspect the drawing. Identify the manufacturing features.
Determine the initial material shape and dimensions.
Identify reference surfaces. Determine individual setups and machine tools to use.
Determine the sequence of operations.
Select tools. Determine cutting conditions for each operation.
Select or design fixtures for each setup.
Produce the route sheet.
Describe these steps as they apply to the sample part.

The initial development of CAPP used the computer to assist the process planner in these tasks by providing access to various databases (e.g., one listing the capabilities of the machine tools in the plant; one for available types, sizes, and dimensions of tools; one with recommended cutting conditions, etc.) and helping in the formal task of printing the route sheets. This by itself was found very valuable, and it is the way most process planning activities are currently carried out; however, many developments have gone beyond this stage.

One of the best-known automated systems is the CAPP system prepared by CAM-I [10]. The method used in this system is called *variant planning*. It is related to the group technology classification, where for each GT family there exists a master process plan. By encoding the part and fitting it to a GT family, one retrieves the master plan and modifies it to satisfy the requirements for the new part. The new plan is stored for further use. The procedure of variant planning is shown in Fig. 1.24. The procedure is

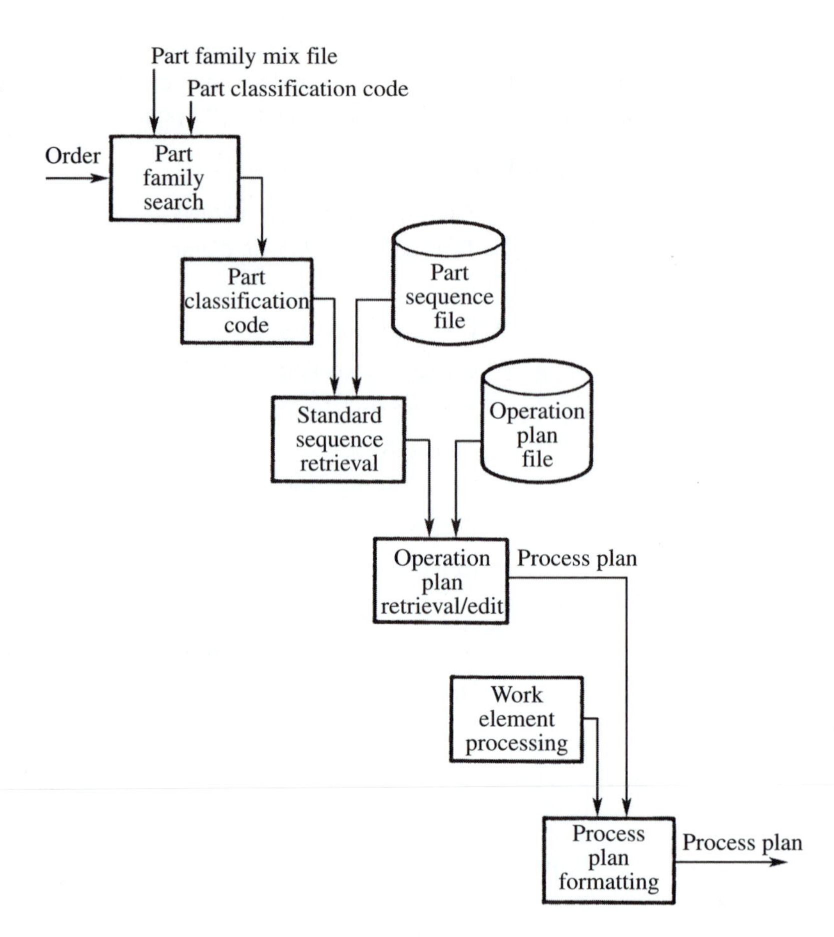

Figure 1.24 Procedure of variant process planning. [REMBOLD]

explained in detail in [4]. Apart from the system developed by CAM-I, other variant systems exist, such as DCLASSL, MIPLAN, and SOPS [8].

Many other researchers rejected this approach and worked on developing systems in which the process plan is created automatically without referring to existing plans. This method is called *generative process planning* (see Fig. 1.25). There is currently no truly automatic generative process planner in use. Existing systems still require human interaction and only work for noncomplex geometric shapes. Most of the advances have been achieved by using AI expert systems; see [9]. We will not discuss these techniques because they demand knowledge of the AI background. In general, the most difficult problems are those related to automatic design interpretation and to automatic decision making.

Automatic design interpretation means that the computer should automatically process the CAD part definition to produce information needed for process planning. One way to do this is to define and recognize geometric features of the part from its drawing. Systems are being created that can automatically extract features from geometry and topology and then use codes to associate each feature with the appropriate manufacturing process. This procedure is shown in Fig. 1.26. The design data from a regular CAD system are passed through a neutral data-conversion system, such as Initial Graphics Exchange Specification (IGES), and a data restructuring system that converts them to the form of a computable set of vertices, edges, and faces. In the fea-

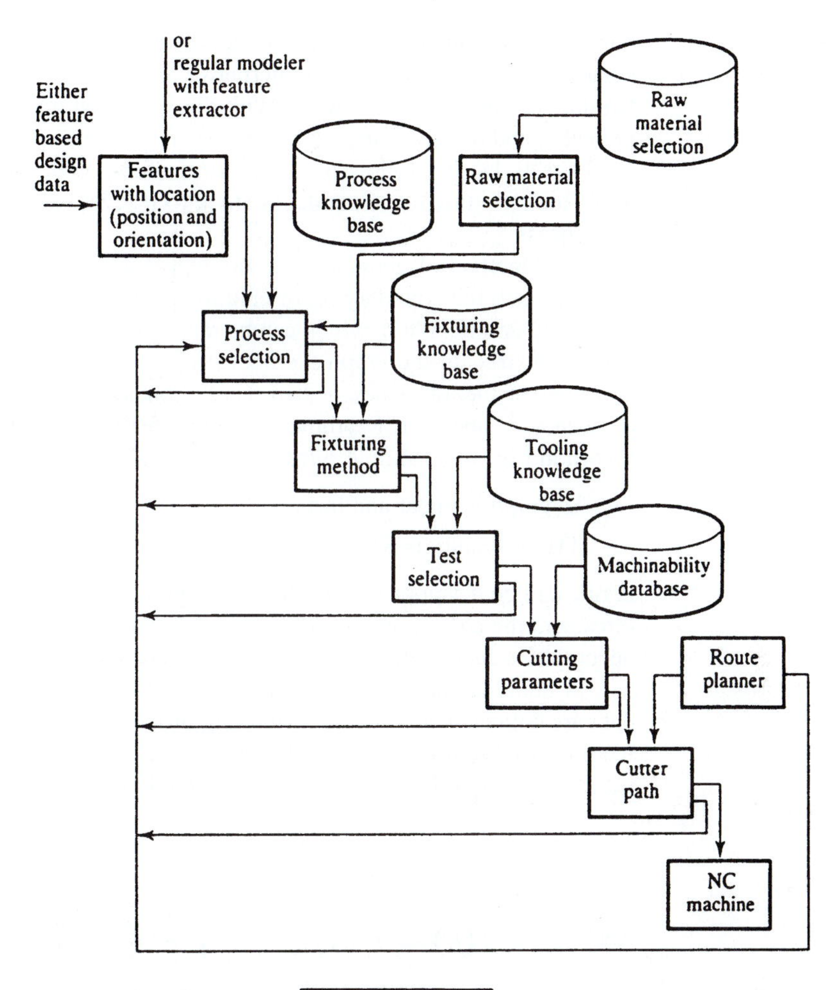

Figure 1.25
Procedure of generative process planning. [REMBOLD]

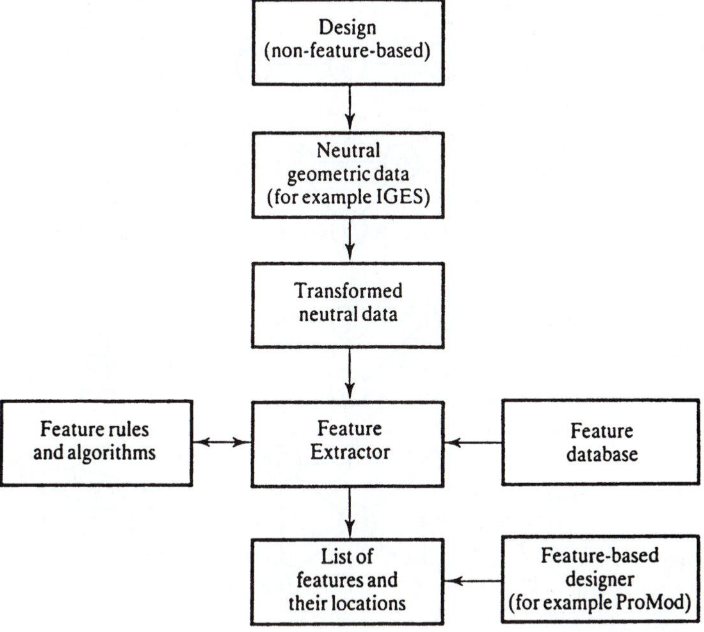

Figure 1.26
Feature extraction methods. [REMBOLD]

ture extraction module this set of data is matched against a database of features using some rules and algorithms. An alternative approach is based on using a design with a feature system such as Pro Mod. With this approach the designer labels features symbolically, and the modeling system places them at the appropriate locations of the part.

The tasks that follow associate the features with particular processes, machine tools, tools, fixtures, and blanks or stock to use, and also determine the conditions of the process.

One method of geometric feature extraction has been proposed by Joshi [5]. In this approach the boundary representation of the part is transformed into an attributed adjacency graph (AAG), which can be represented in the computer in the form of a triangular matrix. In the graph the nodes represent faces, and the arcs are assigned the attributes of the adjacency of two faces. Concave adjacency is labeled 0, and convex adjacency is labeled 1. Examples of simple features and the corresponding graphs are shown in Fig. 1.27. Simple rules can be formulated such as the following:

IF graph is linear and has exactly two nodes with an arc marked 0

THEN feature is STEP.

A rule can be developed for each feature type. An actual part may have numerous features, and the corresponding AAG may be rather complex. A procedure is applied that decomposes such a graph into subgraphs representing the individual features.

A detailed classification of the feature requires additional steps in which all the data are collected, so that the machining process for each feature is fully defined. The feature extraction process is only one step in an automated process planning system such as the Quick Turnaround Cell project at Purdue University [5]. The operation flow

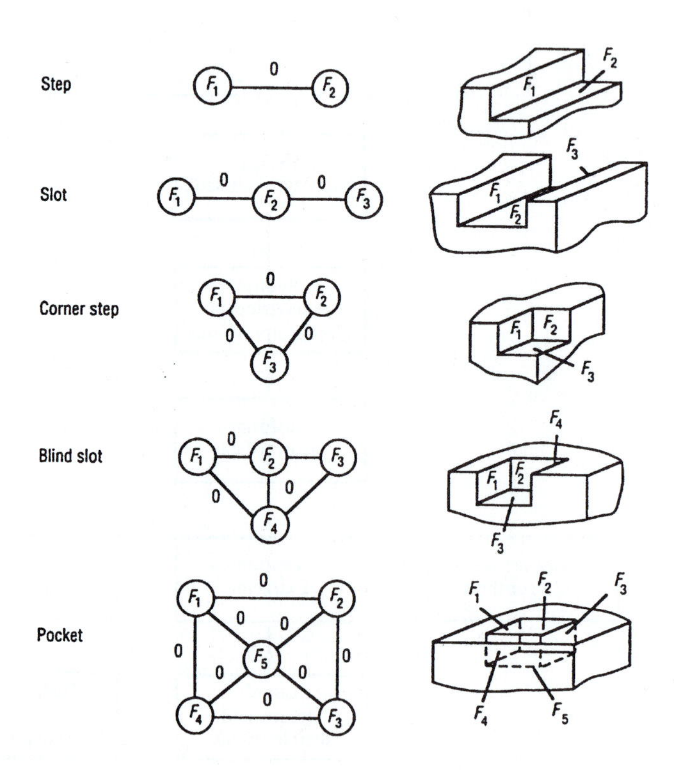

Figure 1.27
Features represented by attributed adjacency graph (AAG). [CHANG]

diagram of this system is shown in Fig. 1.28. The system is intended for one-of-a-kind prismatic parts to be machined on a machining center, and it is in the research stage. The authors admit that it is still very complex, and many problems must be treated as special cases for which special algorithms must be developed.

The generative method for CAPP has been researched in a number of institutions. Kusiak [9] has worked on an AI system based on decomposition of the whole task of machining a part into simple machinable volumes for which the process plan is formulated. Eversheim at the Technical University Aachen in Germany has developed the AUTAP system. It is intended for rotational parts that are produced on turning centers. It performs material selection, process sequencing, machine tool selection, tool selection, chuck selection, and NC program generation. Parts are represented by a subset of constructive solid geometry (CSG) using Boolean operations. The diagram of the

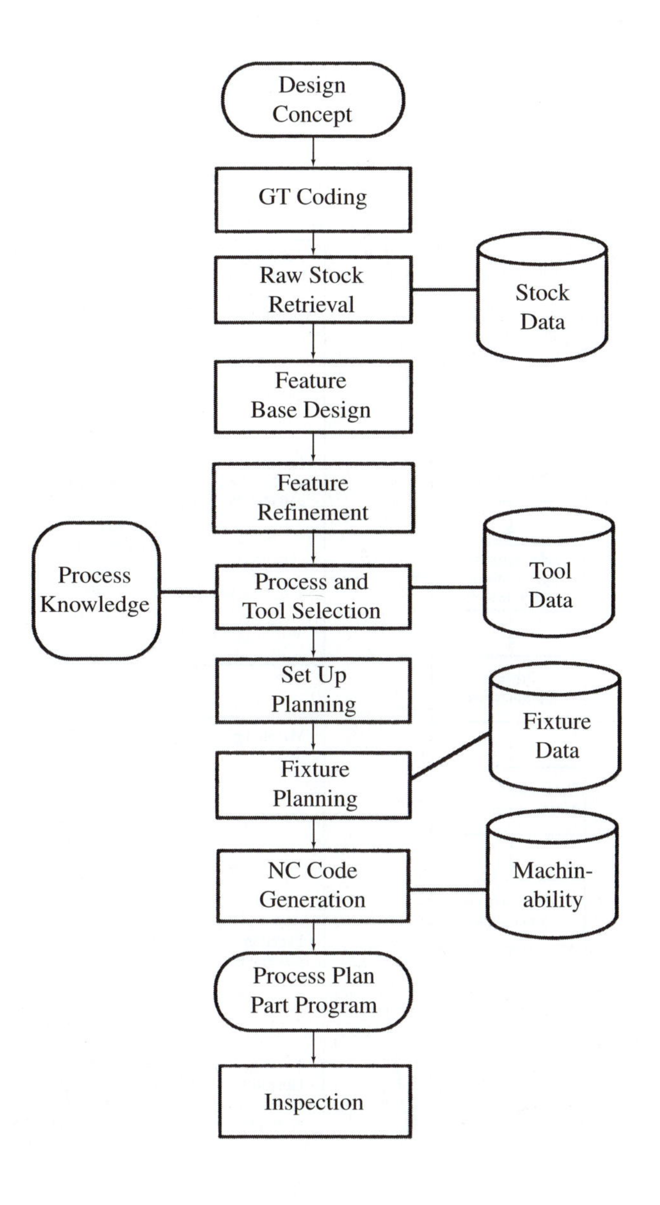

Figure 1.28
Operation flow diagram for Purdue University's Quick Turnaround Cell. [CHANG]

system appears in Fig. 1.29. In this process it is possible to include realistic and significant characteristics such as geometric tolerances, lot sizes, and so on. For the selection of processes and machine tools, the system uses decision tables. It is reported that this system has found wide acceptance in the German industry.

Decision tables specify various actions (decisions) corresponding to various combinations of conditions (inputs). An example of a table used for the selection of hole-making processes appears in Fig. 1.30. The letter *T* stands for true, and *F* for false. So, for instance, a hole of diameter 0.25 with a diameter tolerance of 0.001 (see the encircled choices) will have to be drilled, semifinish-bored, and finish-bored, whereas a hole of the same diameter but with a true position error of 0.02 (choices in diamonds) will be drilled and reamed. Another form expressing the same kind of relationship is the decision tree (Fig. 1.31). There is a single root from which emanate branches. The branches represent the conditions, and they carry values or expressions. Each branch stands for an IF statement, and branches in series represent a logical AND. The actions A_i are listed at the ends of the terminal branches. The decision tree can be directly translated into a computer program flowchart. This is shown for the example of a rota-

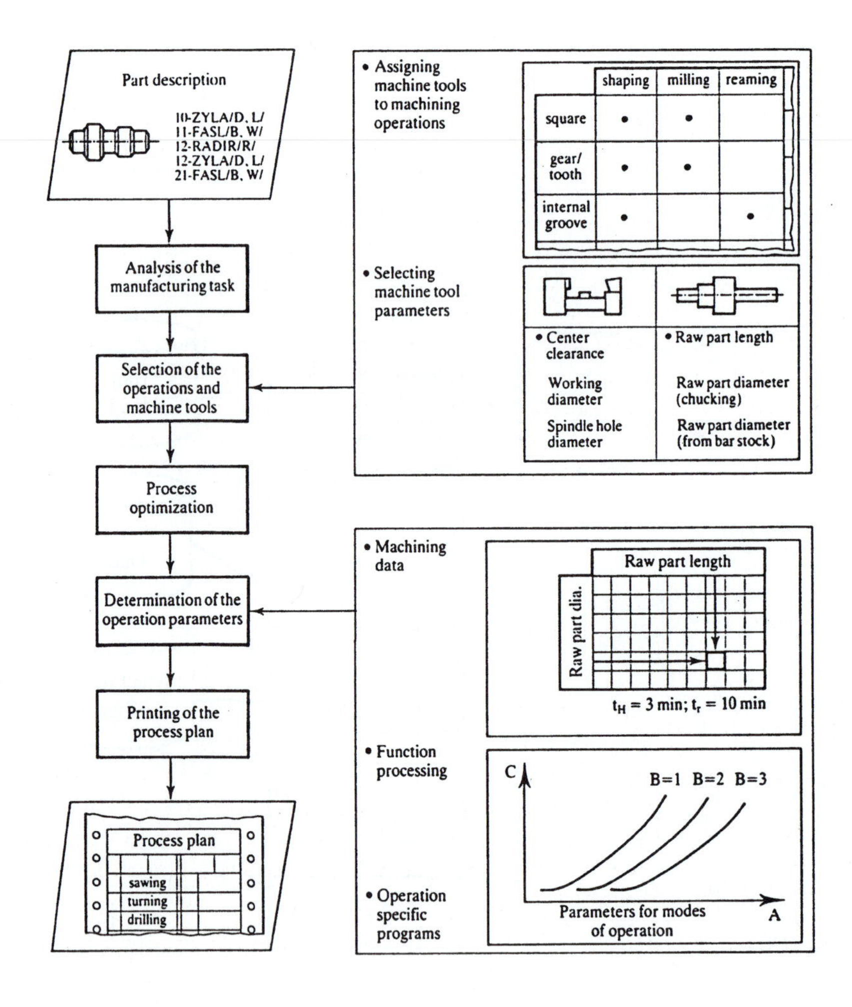

Figure 1.29
The AUTAP generative system of Technical University Aachen.

Figure 1.30
Decision table for hole making. [CHANG WYSK WANG]

Dia. ≤ 0.5	T	⟨T⟩	T	(T)	F	F	F	F	F	F	F	F
0.5 < Dia. ≤ 1.0	F	F	F	F	T	T	T	T	T	F	F	F
1.0 < Dia. ≤ 2.0	F	F	F	F	F	F	F	F	F			
10 < Dia.	F	F	F	F	F	F	F	F	F			
T.P. ≤ 0.002	F	F			F	F	F	F	T	F	F	
0.002 < T.P. ≤ 0.01	F	F			F	F	T	T	F	F		
0.01 < T.P.	T	⟨T⟩	F	F	T	T	F	F	F	T		
Tol ≤ 0.002	F		F	(T)	F		F	T	T	F	F	T
0.002 < Tol ≤ 0.01	F			F	F		T	F	F	F	T	F
0.01 < Tol	T	F		F	T	F	F	F	F	T	F	F
Drill	1	⟨1⟩	1	(1)	1	1	1	1	1	1	1	1
Ream		⟨2⟩										
Semifinish bore				(2)				2	2			2
Finish bore			2	(3)		2	2	3	3		2	3
Rapid travel			3						4		3	4

T.P., true position.

tional part with a hole of tolerance > 0.01. The root is the start node, and each branch carries a decision statement, true or false. The true branches lead finally to various individual actions. Here we have illustrated the scope of the two alternatives—the variant and the generative CAPP—and briefly presented some of the techniques used.

Process planning translates the design definition into complete manufacturing instructions. The quality of this activity has a tremendous influence on the productivity, economy, and quality of production. The use of the computer should help to assure this and eliminate the dependence of this process on the individual experience of the process planning personnel. The variant method of CAPP is widely accepted in the industry and has caused a fundamental change in the process, especially in the environment of numerical control. The various developments of the generative method will undoubtedly lead to useful, practical systems in the near future.

1.6 MANUFACTURING RESOURCES PLANNING

Since the 1970s systems have been evolving that use the growing capabilities of the computer to organize and control the acquisition of purchased parts and raw materials. These systems have been called materials requirement planning (MRP). Eventually, especially with the inclusion of capacities planning, the name was upgraded to manufacturing resources planning, also commonly designated as MRPII. All these systems are based on linking the individual parts to their parent structure, the particular product, as expressed by the bill of materials, and meeting their individual needs by the corresponding manufacturing or ordering lead times to provide a definition of material flow on the time base.

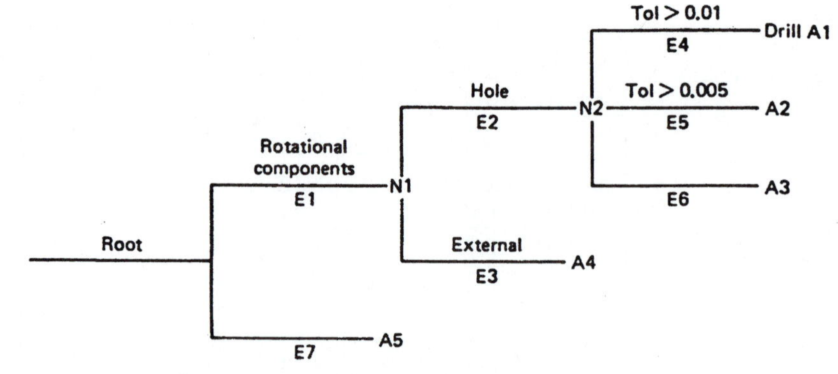

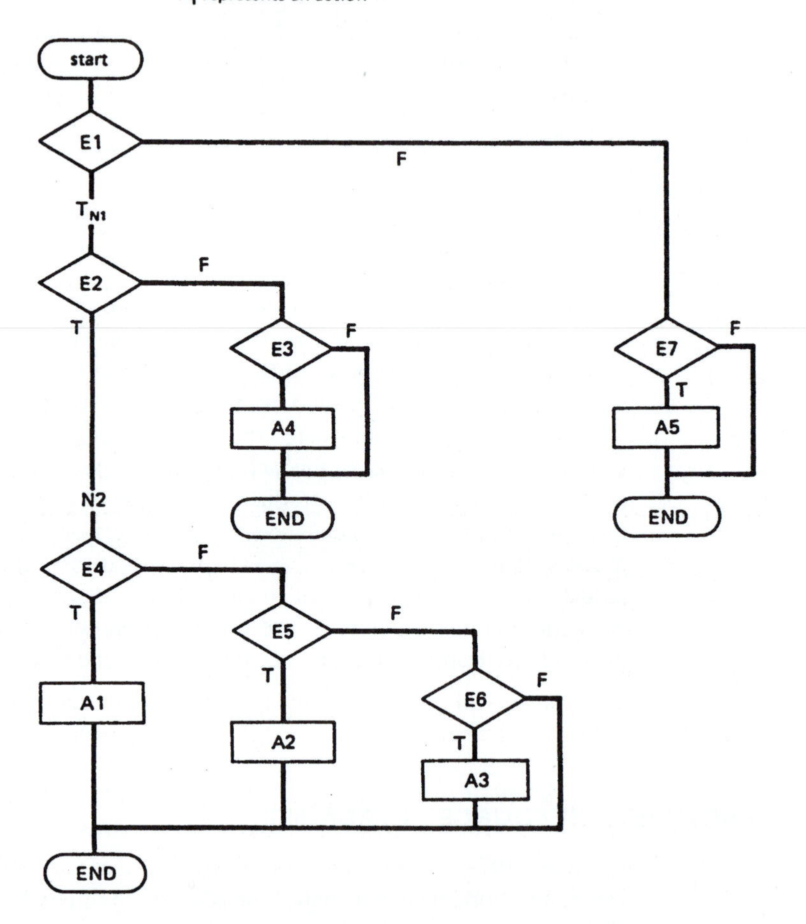

Figure 1.31
Decision tree and its implementation. [CHANG WYSK WANG]

Manufacturing resource planning encompasses the development of the master production schedule, the capacity planning activity, the determination of material requirements planning (MRP), and production-order release and scheduling. At the input is the bill of materials database and the process planning database, and at the output are the shop orders and purchasing orders. Figure 1.32 shows an example of the

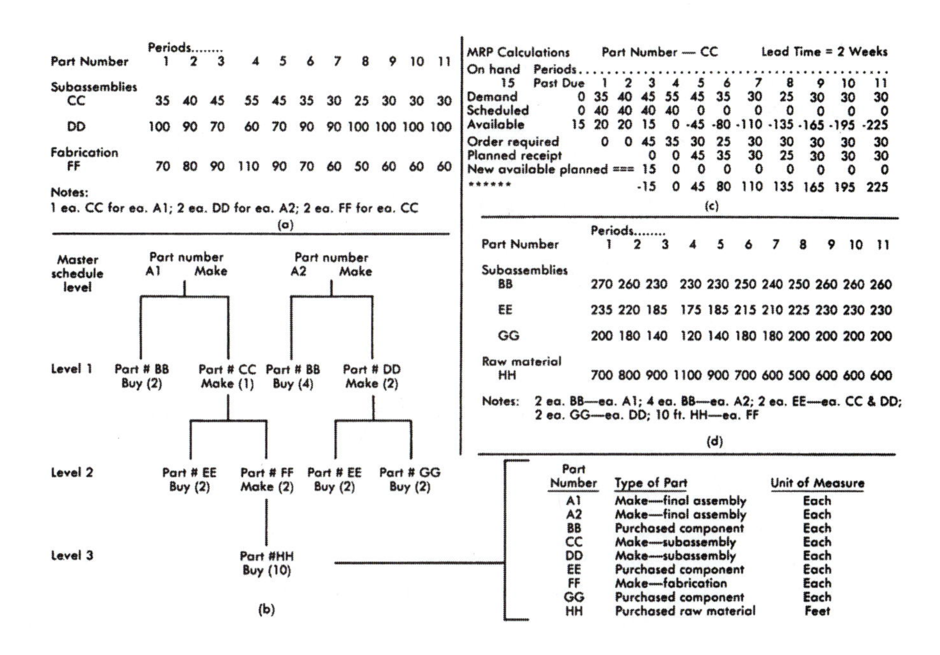

Part Number	1	2	3	4	5	6	7	8	9	10	11
Subassemblies											
CC	35	40	45	55	45	35	30	25	30	30	30
DD	100	90	70	60	70	90	90	100	100	100	100
Fabrication											
FF	70	80	90	110	90	70	60	50	60	60	60

(Periods........)

Notes:
1 ea. CC for ea. A1; 2 ea. DD for ea. A2; 2 ea. FF for ea. CC

(a)

MRP Calculations   Part Number — CC   Lead Time = 2 Weeks

On hand 15	Past Due	1	2	3	4	5	6	7	8	9	10	11
Demand	0	35	40	45	55	45	35	30	25	30	30	30
Scheduled	0	40	40	40	40	0	0	0	0	0	0	0
Available	15	20	20	15	0	-45	-80	-110	-135	-165	-195	-225
Order required		0	0	45	35	30	25	30	30	30	30	30
Planned receipt				0	0	45	35	30	25	30	30	30
New available planned ===				15	0	0	0	0	0	0	0	0
******				-15	0	45	80	110	135	165	195	225

(c)

Part Number	1	2	3	4	5	6	7	8	9	10	11
Subassemblies											
BB	270	260	230	230	230	250	240	250	260	260	260
EE	235	220	185	175	185	215	210	225	230	230	230
GG	200	180	140	120	140	180	180	200	200	200	200
Raw material											
HH	700	800	900	1100	900	700	600	500	600	600	600

(Periods........)

Notes: 2 ea. BB—ea. A1; 4 ea. BB—ea. A2; 2 ea. EE—ea. CC & DD; 2 ea. GG—ea. DD; 10 ft. HH—ea. FF

(d)

Part Number	Type of Part	Unit of Measure
A1	Make—final assembly	Each
A2	Make—final assembly	Each
BB	Purchased component	Each
CC	Make—subassembly	Each
DD	Make—subassembly	Each
EE	Purchased component	Each
FF	Make—fabrication	Each
GG	Purchased component	Each
HH	Purchased raw material	Feet

Figure 1.32
Structure of MRPII (manufacturing resource planning). [SME5 20-12]

MRPII structure. In part (a) demands for three selected manufactured parts—subassemblies CC and DD and a fabricated part FF—are listed as derived from the master schedule. The hierarchy of the bill of materials is presented in part (b), showing the need of one subassembly CC for the product A1 and two subassemblies DD for product A2; there are two fabricated parts FF for each subassembly CC. The orders for CC are derived in part (c), where the manufacturing lead time of two weeks is calculated and shown between the orders and receipts. The demand numbers in line 1 are obtained from part (a). There are still some parts available in the inventory and some parts scheduled to be received from previous orders. The difference between the old inventory plus the new receipts minus the demand yields the new available inventory to which the receipts are added starting in period 5. Figure 1.32d shows the demands for purchased materials. Parts BB and EE are shared between products A1 and A2, while four parts GG are needed for each product A2. Thus the demand for GG is double that of DD in part (a). This example, on a small scale, illustrates how sensitive the whole MRP system is to variations of the individual input parameters, such as the actual lead time for part CC.

An important component of MRP is the control of inventory. On one hand, inventory is necessary as a safety stock; on the other hand it is an evil because it ties up funds that could be used for current needs. Modern methods of production control such as the just-in-time (JIT) system described later in this chapter strive not only to minimize inventory but to eliminate it completely.

In any case, to compensate for the inability to determine precisely all the inevitable variations, there is a need for inventory to ensure that items will be available when needed. One of the tasks of inventory control is to determine the desired level of inventory and, as it is being consumed, to determine the point at which new supplies should be ordered. Various methods have been developed to ensure that materials are available when needed while minimizing the cost of stocking items.

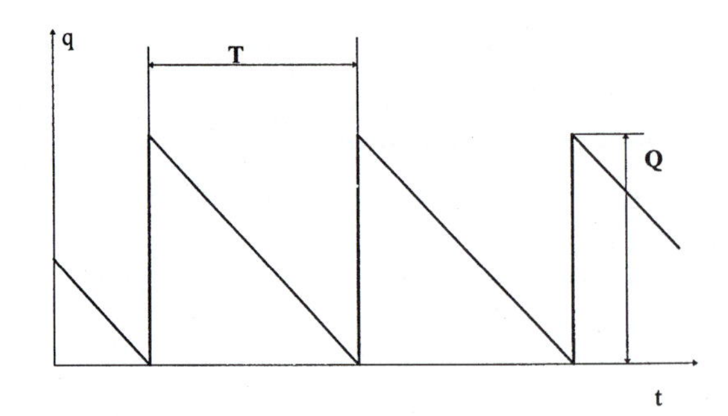

Figure 1.33
Inventory contents versus time.

One of these methods determines the economic lot size Q that gives the lowest inventory cost. The calculation of Q in a simplified form assumes a constant inventory reorder period and a uniform consumption; see Fig. 1.33, in which the inventory level q is plotted versus time. Parts are manufactured repeatedly in lots of Q. At the beginning of the production of the lot, the machine must be set up for the particular operation (see also Fig. 10.2). Such a setup consists mainly of loading and changing the necessary tools, clamping devices, or fixtures. At the beginning of period T, all the Q parts are stocked in inventory, after which they are uniformly used. The average number of parts in the inventory is $Q/2$. This scheme is valid under the assumption that the time for the production of the lot is much shorter than the consumption time T. Meanwhile, the particular machine would be producing many other parts.

The total cost C of these parts in the inventory consists of

C_{su}, the setup costs, \$/lot

C_c, the inventory carrying cost, \$/part/year

C_m, the cost of making the part, \$/part

In a particular period, say one year, y parts are needed. Then

$$C = \frac{y}{Q} C_{su} + \frac{Q}{2} C_c + yC_m \tag{1.1}$$

Eq. (1.1) is illustrated in Fig. 1.34, where the individual terms are denoted 1, 2, 3. There is a minimum for a particular Q_{opt}.

To obtain the number Q_{opt} for a minimum C, we set

$$\frac{\partial C}{\partial Q} = -\frac{yC_{su}}{Q^2} + \frac{C_c}{2} = 0 \tag{1.2}$$

and

$$Q_{opt} = \sqrt{\frac{2yC_{su}}{C_c}} \tag{1.3}$$

The minimum cost is

$$C_{min} = \sqrt{2yC_{su}C_c} + yC_m \tag{1.4}$$

The economically optimum reorder quantity increases with the total number of components y and with the setup cost C_{su}, and it decreases with the magnitude of the inventory carrying costs C_c. The formula indicates that by decreasing the setup cost, smaller

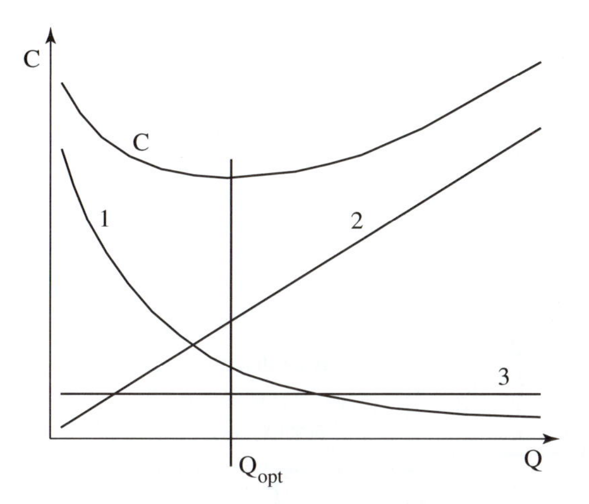

Figure 1.34
Optimum inventory. Plot of overall inventory cost for different values of reorder quantity.

inventories are achieved. This method of determining the reorder period is not very practical in the actual production environment where the demand is not constant over time. However, it shows the significance of the individual parameters involved. More complex techniques exist to address this limitation.

1.7 PRODUCTION SCHEDULING, MONITORING, AND CONTROL

Let us now check the stage we have reached in the discussion of the manufacturing system, and recall the diagram in Fig. 1.10, and review the diagram in Fig. 1.35. From the master production schedule and, in parallel, from the bill of materials and process planning, we have dealt with material requirements planning. This will have been checked against capacity planning and inventory control. At this stage the individual jobs are ready for manufacture. Hundreds or thousands of job orders may be waiting for execution, and it is necessary to decide in which sequence they will be loaded onto the available production machines and utilize the necessary resources. This is called *scheduling*. Various computer algorithms have been worked out to help optimize the flow of orders through the production facility. These deal with problems like full utilization of the production equipment, optimum lot size for a minimum cost by minimizing setup times, and so on. For instance, it may be found that a group of different parts representing an assembly may best be scheduled in lots of only one of each to be processed on a machining center. This eliminates the need for multiple fixtures and delivers all the parts directly to assembly with a minimum of buffer storage, while there is almost no need for tool magazine changes; thus the time for the whole setup shrinks to pallet shuttle time. There is substantial saving in that only one pallet fixture is needed for each part. (This example is illustrated in Chapter 10, Section 10.4, in reference to Fig. 10.29, Case A and Fig. 10.30.) If parts were processed in lots of ten, for example, then ten identical pallet fixtures would be needed.

In reality, even with the best scheduling procedures, disturbances occur in the form of failures of production equipment, breakage of tools, or inadequacy of the

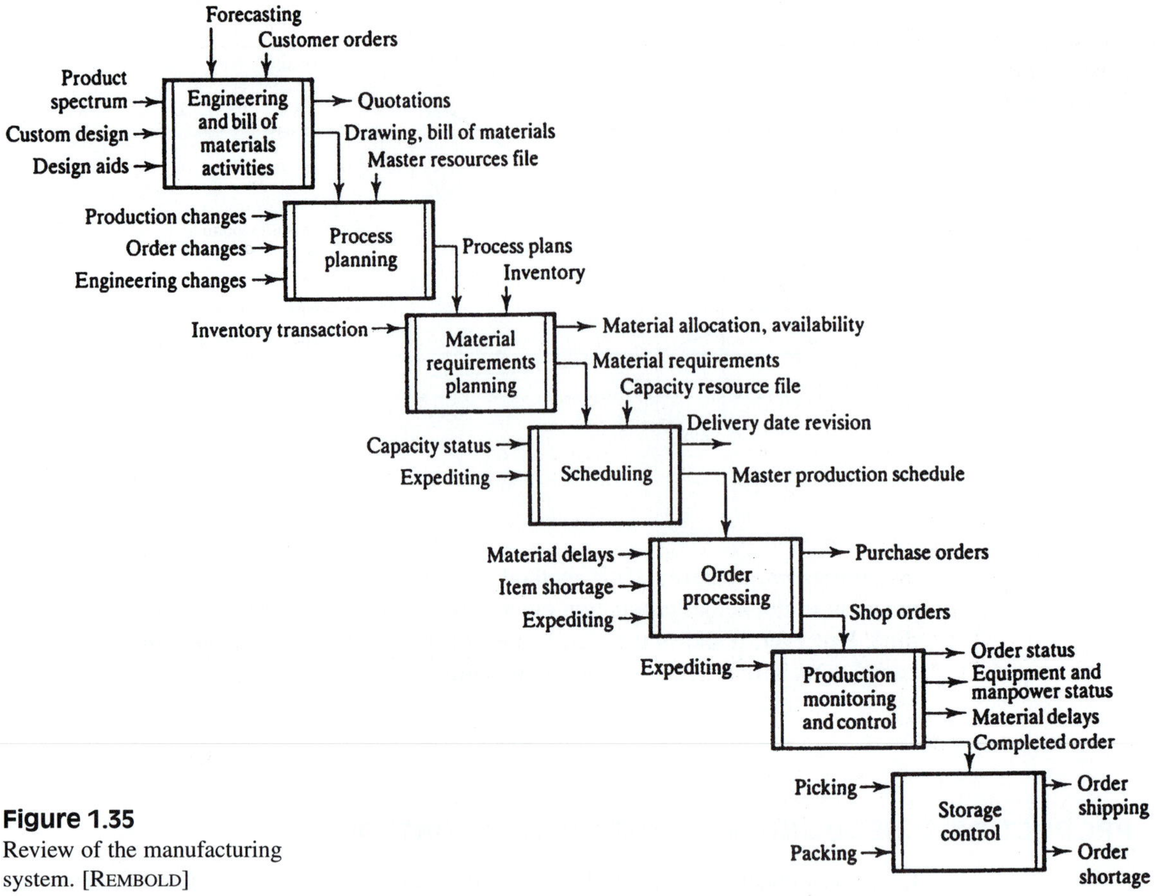

Figure 1.35
Review of the manufacturing system. [REMBOLD]

process plan. It is the role of the *production control* system to react in a flexible way and change the schedule so as to minimize losses. An important part of the system is computerized plant floor monitoring. Typically, computer consoles are distributed throughout the facility and connected through a local area network (LAN). The diagram in Fig. 1.36 outlines the inputs and outputs of a computerized monitoring and control system. The outputs can be compiled into various kinds of reports to be used by production management. The control function makes sure that material flow and operations are executed according to schedule and that materials as well as NC programs are supplied when needed.

1.8 JUST IN TIME

The just-in-time (JIT) manufacturing philosophy discussed in this section differs substantially from all that has been discussed so far in this chapter. Instead of trying to build a comprehensive, complex CIM system, it tends to simplify and break the whole system into small, autonomous units. The author of *Toyota Production Systems* [6]

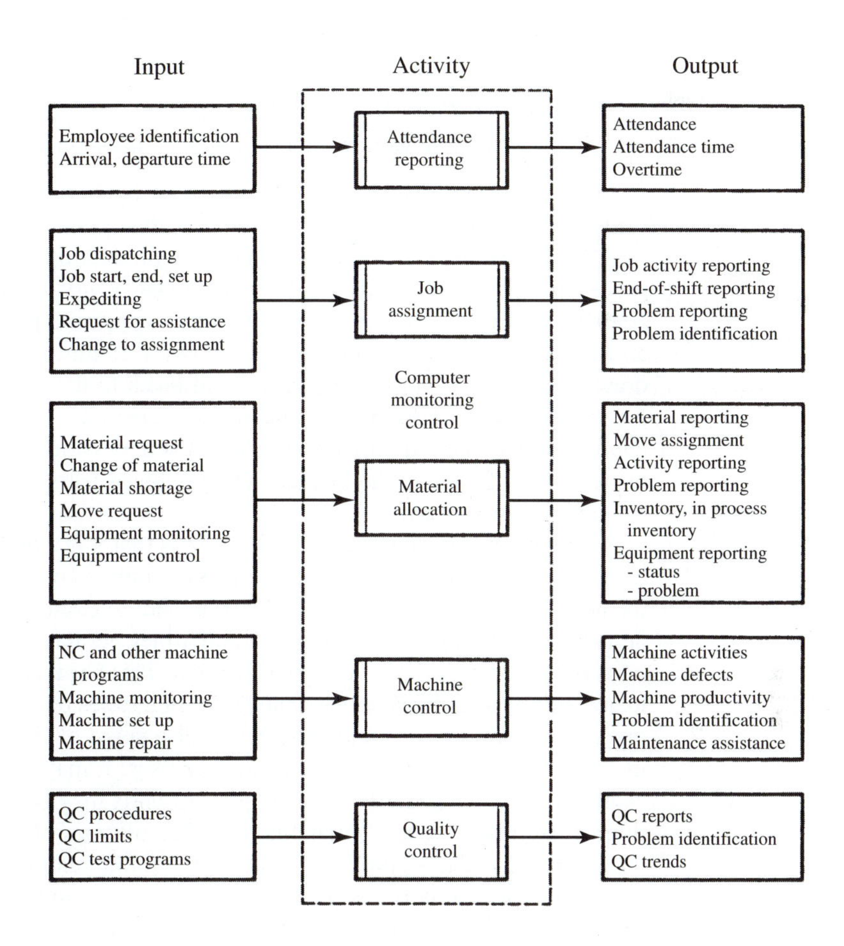

Figure 1.36
Computerized plant floor monitoring and control system. [REMBOLD]

describes it as follows: "The JIT manufacturing control philosophy has evolved over the last thirty years. Manufacturers in the U.S. have only begun to understand the dramatic productivity and quality results available through the use of JIT techniques. In essence the Japanese have rejected the Western obsession with complex management programs and control, computers and information processing and with mathematical modeling. The Japanese way is to simplify the problem." Actually, the JIT philosophy originated at the Toyota automobile manufacturing company, which is surprising because, contrary to what we know about other car companies, where mass production still prevails, Toyota emphasizes small-lot production. Thus, we are considering a special small-lot production system with rather small variations in the production flow.

The main features of the JIT system are (1) reduction of inventory down to zero and (2) nonacceptance of incorrect parts at any stage of their processing. In a more general way, the goal is the elimination of waste in all its forms; for example, producing more parts than needed at any time period, employing more people on a particular job than absolutely needed, and producing faulty parts. Obviously, the intention is to increase productivity and quality. The system consists of several elements (subsystems):

1. The *kanban* system to maintain JIT production
2. Production smoothing

3. Shortening of setup times so as to reduce lead times of components
4. Standardization of operations
5. The "flexible worker" concept and a corresponding machine layout

The whole philosophy is well presented by Monden [6]. We will discuss only some of the elements, beginning with the *kanban* system.

A *kanban* is a card used to control production quantities in the individual processes. Two main kinds are the withdrawal *kanban* and the production-ordering *kanban* (see Fig. 1.37). The figure shows two sequential processes: a machine line and an assembly line. They could as well be two different machining stations. The withdrawal *kanban* is collected at post (1) and taken to the store of parts machined in the preceding process. This indicates the need of a certain number of parts. The production *kanban* goes to the ordering post (5), and it is understood as an order to produce a certain number of parts at the line (7). The produced parts go to storage, where they are available for withdrawal.

The main characteristic of this process is that it is a *pull*-type process. The assembly station orders from the preceding process only as many parts as will be needed in the next time period. This period may be a tenth or twentieth part of the day. The cycle depicted in the diagram repeats that many times per day. Consequently the number of parts produced over the basic period is small. This permits flexibility, and it is possible to accommodate a demand that changes several times per day. If demand increases, the preceding station must produce faster or else some of the increased number of parts must be made in overtime. If demand decreases, it may be necessary to stop the preceding process temporarily. Several types of parts may be involved between these stations if, for instance, the demand is for different types of the product to be assembled during the day. The significant feature is the minimization of in-process inventory. In spite of the flexibility of the process, the JIT system aims to "smooth production." The whole system is laid out so as to adapt to production variation with minimum disruption. Another feature is the absolute quality assurance. Defective parts must not be passed to the next process, and the worker must make sure that the machine is immediately adjusted to eliminate the defect. Another basic feature is the design of fixtures

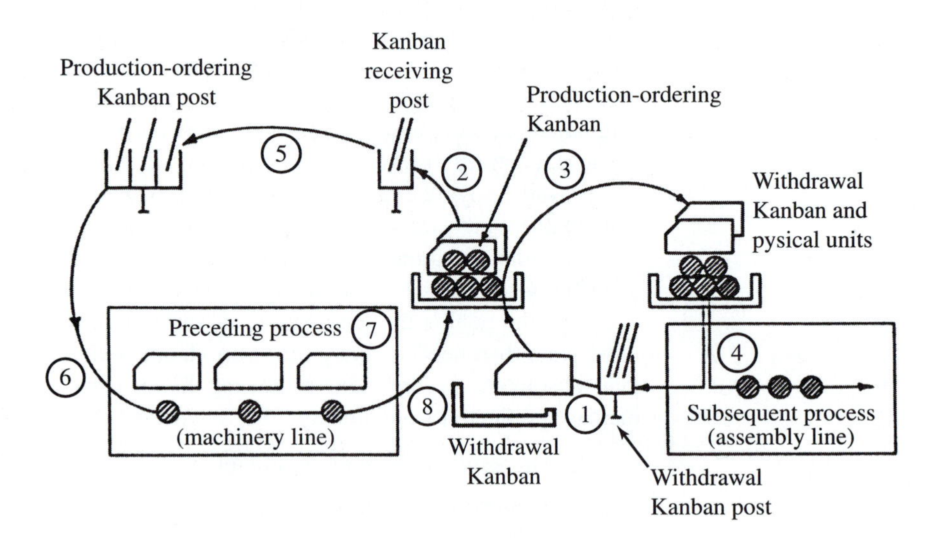

Figure 1.37
Steps involved in using *kanban*. [MONDEN]

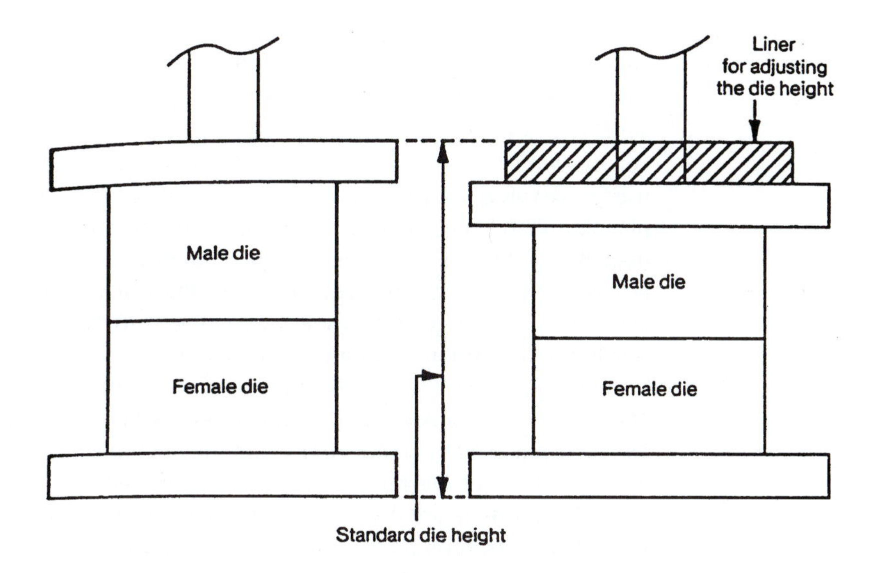

Figure 1.38
Press die adjustment for quick setup. [MONDEN]

and machine modification to achieve the shortest setup time possible. This may be achieved by using NC machining stations with small numbers of tools and by designing the product to require a minimum number of different cutting tools. For instance, the variety of thread sizes must be minimized to require a small number of drill, reamer, and tap sizes. A well-publicized example is the decrease of die-changing time for presswork. This time used to be 2–3 hours in the period 1945 to 1954. It was reduced to 15 minutes between 1955 and 1964, and after 1970 it dropped to 3 minutes. An example of a measure making this possible appears in Fig. 1.38: a "liner" (spacer) is used to adjust the height of a die to a standard one. This eliminates the need to adjust the stroke of the press. The JIT system of Toyota includes subcontracting to suppliers of various components, and there is a special "supplier *kanban*" used in the process.

The pull character of JIT is fundamentally different from the push character of the MRP system, where the flow of parts goes in the same direction as the flow of the control information. In the JIT system the main control element, the withdrawal *kanban*, moves in the direction opposite the flow of parts. In the overall picture one can imagine this flow from customer demands back to the initial materials of production. The MRP system is derived from the master production schedule, which is put together from estimates of the customer demand. However, even at Toyota plans are needed for the creation of the layout of the production units. Thus, in both the United States and Japan the best organization combines the two concepts.

1.9 CONCLUSION

We have briefly discussed the various activities and techniques used in production management. These techniques are taught in Departments of Industrial Engineering, and graduates of these departments are generally hired into manufacturing management. Unfortunately, it is quite common to encounter the attitude that the activities on the shop floor are well understood and that the professional content of production engineering is trivial and best left to the "man at the machine who knows best." This foolish statement was made recently by a high-level manufacturing manager in a large aerospace company in a lecture given as part of a graduate series of seminars for university students by experts from industry. Since F. W. Taylor, who in the early 1900s tried to develop a science of metal cutting so as to generate a better basis for process planning than standing with a stopwatch behind an operator, research in the field of manufacturing systems, processes, and equipment has been carried out in industry and academia. Knowledge has been developed that must be used in manufacturing planning and control if the most efficient production is to be achieved. Process planning and scheduling require a thorough understanding of the effect of individual process parameters on productivity and the quality of each particular operation. This is especially evident for work done on NC machine tools, where the core of the process plan is the NC program. Finally, productivity and the quality of production depend primarily on the capabilities of machine tools, fixtures, and tools used.

There is no doubt that the organization of production, from the master schedule to production scheduling and control, has a tremendous influence on the economy, productivity, and quality of the whole manufacturing enterprise. The computer has brought these activities to a completely new level, and practicing the art of manufacturing management is now mainly a matter of developing and employing CIM (computer-integrated manufacturing). However, it is equally important to use highly productive, efficient, and accurate production equipment—especially equipment with the appropriate type and degree of automation. It is important to run the various production processes under conditions leading to maximum output and best quality, and to keep developing new and better processes. In all of this, numerical control and, more generally, CAM (computer-aided manufacturing) play a decisive role. Both high-quality industrial engineering and mechanical engineering graduates are needed for successful and competitive manufacturing.

REFERENCES

1 Rembold, U.; B. O. Nnaji; and A. Storr. *Computer-Integrated Manufacturing and Engineering*. Addison-Wesley, 1993.

2 Singh, N. *Systems Approach to Computer-Integrated Design and Manufacturing*. John Wiley & Sons, 1996.

3 Teicholz, E., and J. N. Orr. *Computer-Integrated Manufacturing Handbook*. McGraw-Hill, 1987.

4 Chang, T. C.; R. A. Wysk; and H. P. Wang. *Computer-Aided Manufacturing*. Prentice Hall, 1992.

5 Chang, T. C. *Expert Process Planning for Manufacturing*. Addison-Wesley, 1990.

6 Monden, Y. *Toyota Production System*. Industrial Engineering and Management Press, 1983.

7 Tool and Manufacturing Engineers Handbook. Vol. 5, *Manufacturing Management*. 4th ed., Society of Manufacturing Engineers, 1988.

8. Van Houten, F. J. A. M. *PART: A Computer-Aided Process Planning System*. Ph.D. Dissertation, University of Twente, Netherlands, 1991.

9 Kusiak, A. *Intelligent Manufacturing Systems*. Prentice Hall, 1990.

10 Ham, I.; K. Hitomi; and T. Yoshida. *Group Technology: Applications to Production Management*. Kluwer, Nijhokk Publishing, 1985.

11 De Vries, M. F.; S. M. Harvey; and V. A. Tipnis. *Group Technology: An Overview and Bibliography*. Publication No. MDC 76–601, Machinability Data Center, Defense Supply Agency, MetCut Assoc., 1976.

Illustration Sources

[CHANG] Reprinted from: Tien-Chien Chang, *Expert Process Planning for Manufacturing*, Addison-Wesley Publishing Co, 1990, with permission of Addison Wesley Longman.

[CHANG WYSK WANG] Reprinted from: Tien-Chien Chang, Wysk, Richard A., Hsu-Pin Wang, *Computer-Aided Manufacturing*, Prentice Hall, A Division of Simon & Schuster, Englewood Cliffs, NJ, 1991 with permission of Prentice Hall, Inc.

[MONDEN] Reprinted from: Yasuhiro Monden, *Toyota Production System*, Industrial Engineering and Management Press, Norcross, GA, Institute of Industrial Engineers, 1983, with permission by Yasuhiro Monden.

[REMBOLD] Reprinted from: Rembold, U., Nnaji, B. O., Storr, A., *Computer Integrated Manufacturing and Engineering*, Addison-Wesley Publishing Co, 1993, with permission of Addison Wesley Longman.

[SME5] Reprinted from: *Tool and Manufacturing Engineers Handbook*, Vol. 5-*Manufacturing Management,* 4th Ed., R. F. Veilleux, Louis W. Pietro Editors, Soc. of Manufacturing Engineers, Dearborn, MI, 1988 with permission of SME.

[TEICHHOLZ ORR] Reprinted from: Teichholz, E., Orr, J. N., *Computer Integrated Manufacturing Handbook*, McGraw-Hill Book Co, New York, 1987 with permission of McGraw-Hill.

Industrial Contributor

[CINCINNATI MACHINE] Courtesy of Cincinnati Machine, a Unova Company, 4701 Marburg Avenue, Cincinnati, OH 45209-1025

QUESTIONS

Q1.1 Name nine different products of mechanical production, three from each of the following: **(a)** mass production, **(b)** medium-lot production, **(c)** small-lot production.

Q1.2 What manufacturing equipment—**(a)** a transfer line, **(b)** an NC machine tool, **(c)** a manually operated machine tool—would you assign to production of lot sizes of (1) 20/yr; (2) 1,000/yr; (3) 1 million/yr?

Q1.3 **(a)** Which three essential characteristics are entered for each part on the bill of materials (BM)?

(b) Does the BM contain all the parts of a product?

(c) Does this include also the purchased parts?

(d) Are the BMs of the various products in a company combined?

(e) If so, why? As an example, assume the production of several automobile models in an assembly plant.

(f) Which typical subassembly may be included in BMs of many models?

Q1.4 **(a)** What is group technology?

(b) Which characteristics are common to all parts in a family?

(c) Explain the impact of GT on design, on process planning, and on the layout of the shop.

Q1.5 **(a)** Name an example of a GT classification system.

(b) Give examples of ten characteristics used in a classification system.

(c) Which activities of those in Fig. 1.10 depend on access to the GT classified database?

Q1.6 **(a)** Define the task of process planning.

(b) Which types of instructions and which numerical values does a route sheet contain?

(c) Which production control activities use the route sheet as input?

(d) Explain the variant method of CAPP.

(e) Explain the generative method. Which design characteristics must be determined as inputs?

Q1.7 **(a)** Which kind of CAM software is commonly used in a CAD/CAM system?

(b) In which aspects is the NC program much more detailed than the route sheet for manual operations?

(c) Beyond CAD/CAM, which other industrial engineering activities are included in a CIM system?

(d) Which other functions beyond CAD/CAM constitute concurrent engineering? As a whole, how are all these functions related to the product? Which cycle do they represent?

Q1.8 **(a)** Explain the difference between the push and the pull systems of production control.

(b) Which software and/or hardware systems are the basis for each system?

(c) What is one of the main goals of the JIT system, and what other goals are based on it?

(d) Name the main prerequisites for a successful JIT system.

Q1.9 **(a)** Explain the difference in the missions of production engineers and industrial engineers.

(b) Which typical theoretical disciplines are the basis of the work of each of these two specialities?

(c) Is there a lot of interaction between these two groups of engineers within a manufacturing company? Should there be more? If so, under which system can this happen?

(d) In which activities of the manufacturing management system, such as those shown in Fig. 1.10, is the real-time input of production engineering most needed?

Q1.10 **(a)** Express the difference between process planning and production scheduling.

(b) Name some of the rules commonly used in machine loading (allocation of jobs).

(c) Which events or conditions may disrupt the production schedule?

(d) How is feedback provided to make production control possible?

(e) Which kind of hardware is used for feedback?

(f) Name typical contents of this feedback.

2 Engineering Materials and Their Properties

2.1 INTRODUCTION

This chapter is not intended to replace a proper course on Material Science devoted to engineering materials. Indeed, it is considered a significant advantage for manufacturing engineers to have a deeper understanding of materials and their properties. For students who have had such a separate course, this chapter is a brief summary of the knowledge useful for our course. For those who have not had such a separate course, this review provides what is necessary to understand the role of materials in manufacturing processes.

The emphasis here is on those material properties that are significant for the processing of workpieces and for the functioning of tools. For illustration, let us mention the significance of the yield strength of workpiece material for the cutting force applied in machining and in forming operations; the significance of specific heat and thermal conductivity of the workpiece material for the temperature in the cut and, consequently, for the wear rate of the cutting tool; the importance of hardness and toughness at high temperature for cutting-tool materials; and the effect of hardenability of steel on welding. Many of these material properties are also important for the service and function of the workpiece as part of a machine and they must be considered by the designer of the part. Some other properties important for the designer are not so significant for the manufacturing engineer, like fatigue strength or creep resistance. However, because the designer and the manufacturing engineer must work together to develop machines and their parts that are functionally correct and easy to manufacture, a general understanding of materials and their properties is needed.

There are four basic groups of solid materials to be considered for workpieces and for tools: metals, plastics, ceramics, and composites. In addition, fluids and gases may play important roles in the various operations: as coolants and lubricants in metal cutting and metal forming, as protective media in welding, and as electrolytes in electrical machining. Only solid materials are discussed in this chapter; the operational fluids and gases are dealt with in conjunction with the respective processes.

Metals represent the largest group of interest. Most manufactured parts are made of various types of steels, cast irons, aluminum alloys, titanium alloys, bronzes, brasses, and other alloys in which many different metals are combined. They offer high strength, stiffness, and wear resistance at normal and often at elevated temperatures, or good thermal and electrical conductivity.

The use of plastics has been steadily increasing. They may combine a number of useful characteristics: they are light and reasonably strong; transparent, translucent, or opaque; obtainable in many colors; good electrical insulators, and easy to mold and form.

Ceramics are mostly hard and heat-resistant, but brittle. Their main use in manufacturing has been for lining furnaces. However, they are also important as cutting-tool materials, and they are being considered for use in parts of combustion engines and of gas turbines.

Composite materials include various structural combinations of metals, plastics, and ceramics, and they range from sintered, silver-graphite brushes for electric motors to fiberglass for boat skins. Lately, the use of glass-fiber-reinforced plastics and carbon-fiber-reinforced plastics has been rapidly increasing for sports equipment such as golf clubs, tennis rackets, and skis, and significantly in the aerospace industries for certain structural parts of airframes.

2.2 MECHANICAL PROPERTIES

The various mechanical properties are defined and specified by corresponding standard test procedures.

2.2.1 The Tensile Test

The most widely used test is the tensile test, in which a simple form of a specimen is subjected to uniaxial tension. The form of specimen used for testing metals is shown in Fig. 2.1. The standard specimen has a diameter $d_0 = 0.5$ in $= 12.7$ mm and a gage length of $L_0 = 2.0$ in $= 50.8$ mm. It is clamped in the jaws of a testing machine and subjected to a slowly increasing tensile load. An extensometer may be attached to it for the measurement of changes in the gage length. Both the force and the extension are continuously recorded. A diagram of the "engineering stress" S versus "engineering strain" e is plotted; an example is shown in Fig. 2.2a.

It represents

$$S = \frac{F}{A_0} = \frac{F}{\pi d_0^2/4} \tag{2.1}$$

First, the specimen elongates uniformly along its entire length, and the diameter correspondingly decreases to d_1 while the gage length extends to L_1. This strain is initially elastic and it is proportional to the stress, where the slope is the modulus

$$e = \frac{L_1 - L_0}{L_0} = \frac{\Delta L}{L_0} \tag{2.2}$$

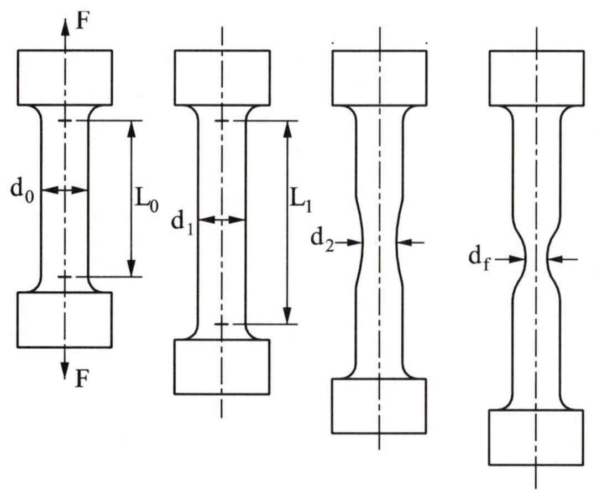

Figure 2.1
The tensile test specimen.

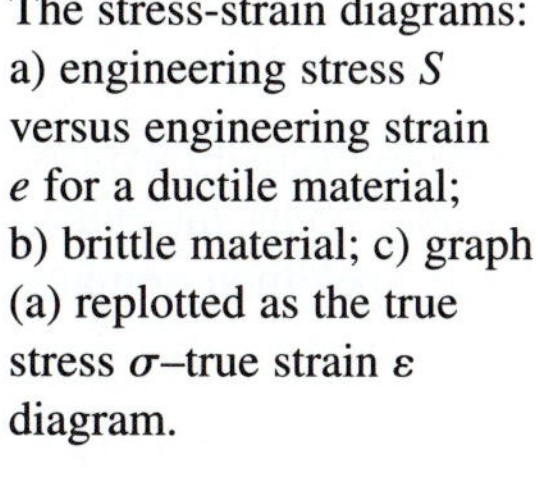
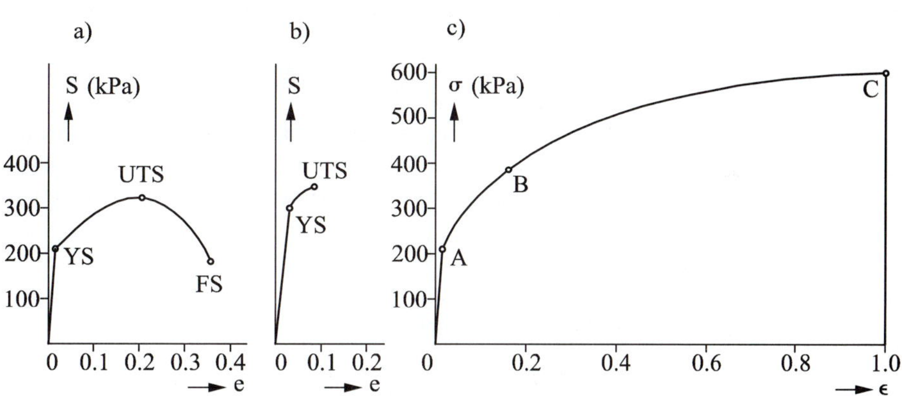

Figure 2.2
The stress-strain diagrams: a) engineering stress S versus engineering strain e for a ductile material; b) brittle material; c) graph (a) replotted as the true stress σ–true strain ε diagram.

of elasticity $E = \Delta S/\Delta L$. When the stress reaches the value indicated in Fig. 2.2a as *YS, yield strength,* plastic extension is started. Elastic strain is recoverable. Within the elastic range, if the load is removed the specimen returns to its original length. However, the plastic strain is permanent. If a specimen is loaded beyond the elastic limit and the load is removed, only the elastic part of the strain is recovered. The point *YS* is seen with some metals, like mild carbon steels, as a distinct discontinuity in the diagram. With other metals the transition between the elastic and plastic parts of the stress-strain diagram is gradual. Usually, the yield point is defined to be point at which a permanent, plastic strain reaches the value $e = 0.002$.

The specimen elongates further uniformly while the stress S is increasing until it reaches a maximum at the point denoted *UTS, ultimate tensile strength.* From this point on, the specimen elongates further, but the engineering stress decreases along the portion *UTS* to *FS,* the *fracture strength.* In this way, the engineering stress-strain diagram is completed. The strain at fracture e_f is also called simply elongation, and it is a measure of the ductility of the material. Another material, which behaves in the manner shown in Fig. 2.2b, has very little ductility; it is brittle.

The decreasing portion of the diagram in Fig. 2.2a has no real meaning in terms of decreasing stress. What is actually happening is that the specimen starts to plastically deform locally. It forms a "neck" (see Fig. 2.1), which is narrowing down fast to diameter d_2 and which fractures as its diameter reaches the value d_f. Correspondingly, it is more proper to evaluate, for the entire test, the "true" stress σ and the "true" strain ε, and to relate these values to the region of the neck only, once it starts to form.

The true stress is obtained as follows:

$$\sigma = \frac{F}{A} \tag{2.3}$$

where A is the actual area at any instant, and the true strain is obtained by formulating its increment,

$$d\varepsilon = \frac{dL}{L} \tag{2.4}$$

where L is the gage length at any instant, and dL is its infinitesimal increment. The total strain is obtained by integrating:

$$\varepsilon = \int_{L_0}^{L} \frac{dL}{L} = \ln\left(\frac{L}{L_0}\right) \tag{2.5}$$

Formula (2.5) is applicable only during the uniform deformation. In order to express the strain in the local area of the neck, we refer to cross-sectional areas instead of lengths. In plastic deformations the volume of any part of the specimen remains constant,

$$V = L_1A_1 = L_2A_2 \tag{2.6}$$

and so on. Therefore, it may be written

$$\varepsilon = \ln\left(\frac{A_0}{A}\right) \tag{2.7}$$

and this expression may be used also to express the true strain at fracture. Using typical values for mild steel of $d_0 = 12.7$ mm, $d_f = 7.46$ mm, it is

$$\varepsilon_f = \ln\left(\frac{d_0}{d_f}\right)^2 = 1.064$$

Then the engineering stress-strain diagram of Fig. 2.2a is replotted as the true stress–true strain diagram of Fig. 2.2c. The points A, B, and C correspond to the yield point, the *UTS* point, and the fracture point, respectively.

Although the true stress–true strain diagram is more realistic, the majority of tensile test data is published in the form of Fig. 2.2a, and the three most important parameters characterizing a metal are *YS, UTS*, and the elongation e_f, which is usually denoted simply e. If only one value is given, it is *UTS*. Although it is a parameter with little fundamental meaning, it is a good practical comparative characteristic. Also the elongation e is commonly used as a measure of *ductility*, although it consists of both the uniform and local strains. It is useful to remember typical values of *UTS* and e for various metals. Here are some values for steels:

Mild steel: *UTS* = 60,000 psi = 410 MPa; e = 35%

Medium-carbon steel: *UTS* = 85,000 psi = 590 MPa; e = 26%

High-strength alloy steel: *UTS* = 180,000 psi = 1,240 MPa; e = 6%

The significance of the area under the *A, B, C* curve in Fig. 2.2c is obtained by expressing the increment of work related to volume:

$$\mathrm{d}W = \frac{FdL}{AL} = \left(\frac{F}{A}\right)\left(\frac{dL}{L}\right) = \sigma\,\mathrm{d}\varepsilon \tag{2.8}$$

$$W_s = \int_0^{\varepsilon} \mathrm{d}W \tag{2.9}$$

which is the specific (per volume) work to fracture and a measure of the *toughness* of the material.

Toughness and ductility are closely related, and they are often used interchangeably. If they are distinguished, ductility is understood as the ability of a material to undergo large plastic deformations before fracture. It is very useful in machinery where local overloads lead to local yielding and subsequently to more uniform distribution of stresses without fractures. It is also very useful in metal forming. Toughness is often understood as the ability of a material to absorb impacts and to dissipate the corresponding kinetic energy in plastic deformation without failure.

The ductility or toughness parameters as obtained from the tensile test are not directly applicable to the various functional situations of a part or to the various forming operations because they are obtained under the particular stress situation in the neck of the tensile specimen, and the material will sustain different ultimate strains under different stress conditions. Therefore, other tests leading to the stress-strain diagram have been developed, like the various compression and torsion tests discussed in Chapter 5.

Another phenomenon that can be interpreted from the tensile stress-strain diagram is *strain hardening.* If, as shown in Fig. 2.3, the test is interrupted at point *D* and the load is removed, the material recovers the elastic part of the strain and retains the plastic part. In the diagram the test returns to point 2. We may now consider the test specimen as a new piece of material and start loading again from the new origin 2. The material will now yield at a stress $YS_2 > YS_1$. The effect of the preceding loading cycle and of the associated cold plastic strain is such that the material is stronger and harder. Later on in this chapter we will discuss this again. Strain hardening may occur as a result of various kinds of cold forming operations with stress conditions different than in the tensile test. This is discussed further in Chapter 5.

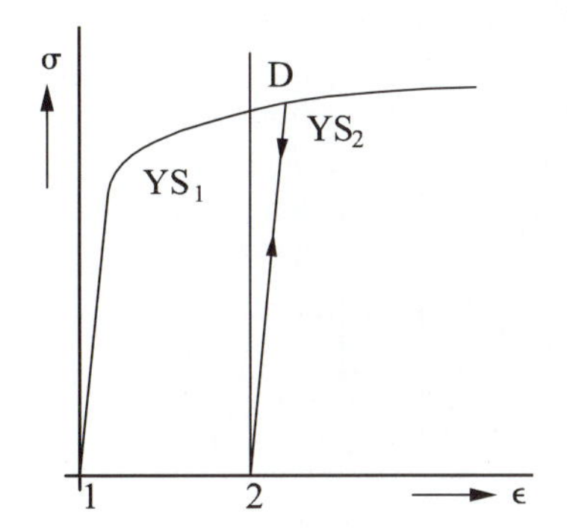

Figure 2.3
Strain-hardening in an interrupted and restarted tensile test.

2.2.2 Hardness Testing

Several standard hardness tests are in practical use. They are rather simple to execute and are therefore the most common tests carried out in quality control of both incoming material and finished parts. All hardness tests are based on pressing an indenter into the surface of the material tested. Correspondingly, hardness expresses the resistance of the material surface to indentation. This resistance is closely related to the yield strength of the material, and it is further affected by its susceptibility to strain hardening. There is a good relationship between the strength parameters of the tensile test and the results of hardness tests for most materials. The theory of indentation is discussed in detail in Chapter 5.

The *Brinell hardness* tester forces a hardened steel ball, or a carbide ball, of diameter 10 mm, into the material (see Fig. 2.4a). A standard load of 3,000 kg force (29,430 N) is used for stronger metals and of 500 kg force (4,905 N) for softer metals. The diameter of the indentation is measured by means of a microscope and the Brinell Hardness Number is expressed as

$$BHN = \frac{F}{A}\,(\text{kgf/mm}^2) \tag{2.10}$$

where F is the loading force, and A is the surface of the spherical imprint as derived from the measured diameter. The obsolete units of kgf (kilogram force) are being used because of the familiar past usage. Although hardness is related essentially to yield strength, there is a commonly used relationship between *UTS* of steels and their hardness. It is based on an average ratio of *UTS/YS* of about 1.2 for a strain-hardened steel. This relationship is

$$UTS\,(\text{lb/in}^2) = 500\,BHN\,(\text{kgf/mm}^2) \tag{2.11}$$

If it is transformed to equal units, it is

$$BHN\,(\text{N/mm}^2) = 2.84\,UTS\,(\text{N/mm}^2)$$

This ratio is theoretically explained in Sec. 5.2 of Chapter 5.

The relationship between *BHN* and *UTS* is utilized to quickly check and distinguish among various kinds of steel stock. For this purpose a portable Brinell hardness tester can be used. Typical values of *BHN* (kgf/mm^2) are

Mild steel: 120

Medium-carbon steel: 165

Cast iron: 150–200

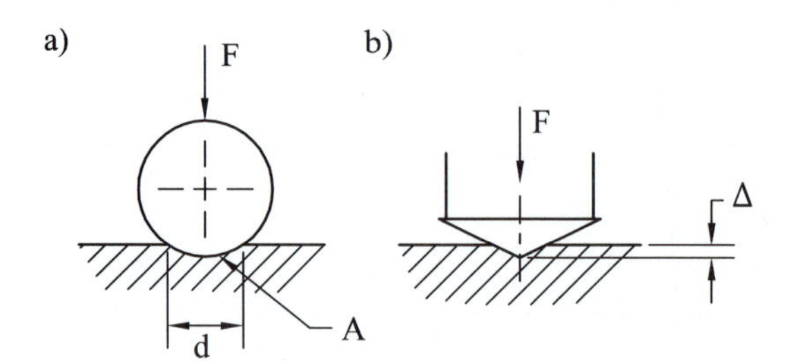

Figure 2.4
Hardness testing: a) Brinell indenter; b) Rockwell C indenter.

TABLE 2.1 Comparison of Hardness Numbers

BHN (10 mm ball, 3000 kg force)	Rockwell *C* (diamond cone, 150 kg force)	Rockwell *B* (1/16 in. ball, 100 kg force)	*UTS* of steel (ksi)	*UTS* of steel (MPa)
	68			
	65			
653	60			
563	55			
483	50		300	2070
422	45		215	1484
371	40		182	1256
336	35		157	1083
285	30		138	952
258	25		125	863
226	20	98	108	745
200		91	93	642
175		88	85	587
150		81	74	511
130		74	65	449
110		65	55	380

The limitations of the Brinell test are several. The first one is due to the use of a ball for the indenter; for very hard materials, the shallow imprint is inaccurate. The second limitation, for materials of lower hardness, is due to the rather large size of the indenter and of the force and, therefore, of the imprint. Correspondingly, the value obtained is the average of all the material grains involved. If a part with a thin surface hardened layer is to be tested, this layer will be deformed and the underlying softer material will be involved, which affects the result of the measurement.

Harder materials are tested by means of several specifications of the *Rockwell hardness* test. The material of the indenter is diamond, and in the most popular Rockwell *C* scale it has the shape of a cone (see Fig. 2.4b). The tester is so designed that the vertical penetration Δ between an initial small load and full load is the measure of hardness. For the *C* scale, the full load is 150 kg force. The *B* scale uses a ball of 1/16 in. diameter and 100 kg force, and it is intended for softer materials.

There are several microhardness testers that use pyramidal diamond indenters and small or very small loads. The size of the indentation is measured under a microscope. It is thus possible to measure the hardness of very small areas of the material surface and distinguish the hardness of individual constituents.

A comparison of the various hardness parameters is given in Table 2.1.

2.2.3 Notched Bar Impact Tests

A typical test of this kind is the *Charpy* test (see Fig. 2.5). A pendulum-type hammer is used to strike a specimen in which a standard shape of a notch is provided. The difference in height of fall and of rise after the strike determines the energy spent on the fracturing of the specimen. This energy is not directly comparable to the specific

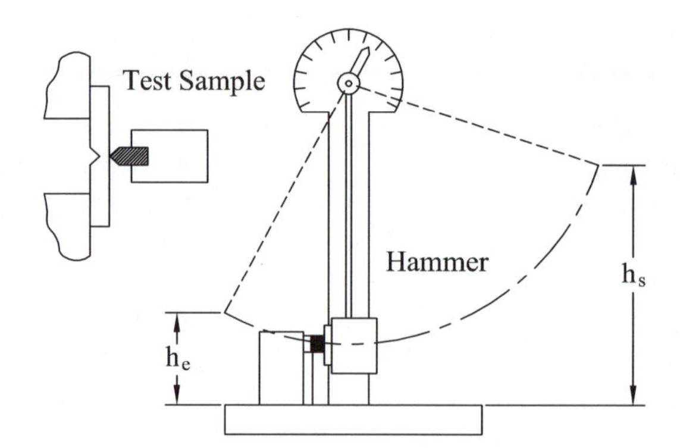

Figure 2.5
Charpy impact testing of toughness using a notched specimen. The difference in the initial and final potential energies defines the fracture energy.

energy to fracture in the tensile test because different materials are differently sensitive to the stress concentration around the notch. The test is mostly used for evaluating the effect of various heat treatments of steels on their toughness. The fractured specimens can be analyzed by fractographic criteria in which the appearance of the fractured surface indicates the mechanism by which it was created.

2.2.4 High-Temperature Tests

Elevated temperature affects the properties of metals. These effects may be determined by carrying out tests like the tensile test at various temperatures while keeping the specimen inside of a furnace. In other instances upset hot forging of test workpieces is carried out. These kinds of tests deal with short-time plastic behavior of materials.

If long time periods are considered, the phenomenon of *creep* becomes significant. Materials yield at elevated temperatures very slowly under stresses considerably lower than those needed for a fast plastic deformation. This is described as viscous behavior; for some materials (e.g., plastics) it is observed also at room temperature. The time scales involved are usually measured in months and years. The tests are, however, arranged in an accelerated manner and still permit extrapolation for long service times. These tests will not be discussed here in any detail because they have practically no significance for manufacturing processes.

2.2.5 Fatigue Testing

Fatigue failure occurs as a result of cyclic loads at stress levels that, when applied statically, would not produce any yielding or failure. The level of stress at which fatigue failure happens decreases with the number of duty cycles. However, for ferrous alloys, an endurance limit is observed as a stress below which failure does not occur for any number of cycles. This limit is reached approximately after 10^6 to 10^7 cycles. Fatigue develops in ductile materials by propagation of cracks and by spreading of defects from highly stressed areas. Eventually the final fracture is of the brittle type. Again, fatigue strength has little significance for manufacturing processes; however, manufacturing processes may affect the fatigue strength of parts produced, and fatigue testing is a way of determining surface damage caused by the various manufacturing processes.

2.3 STRUCTURES AND TRANSFORMATIONS IN METALS AND ALLOYS

Metals and their alloys are the most common engineering materials. Most are solid at room temperature and exhibit elastic behavior up to the yield stress. Their *YS*, *UTS*, and hardness are rather high, and they also mostly possess significant ductility. Their strength is retained often to elevated temperatures, and the melting points of most metals are in the 1000°C to 2000°C range. They are good electrical and thermal conductors.

Their atoms are held strongly together by the metallic bond which involves loosely held valence electrons that are "free" to move through the structure as an "electron cloud" shared by adjacent atoms. The mobility of electrons is the reason for the good electrical and thermal conductivity. The atoms in solid metals are arranged in long-range order regular patterns of crystals. At the early stage of manufacture every metal is in the molten stage, and as it is cooled it starts to solidify simultaneously at many points. From these initial sites the individual crystals start to grow, each with a different orientation, and eventually they meet, forming the individual grains of the material. Each of these grains is one crystal (see Fig. 2.6). At the grain boundaries the regular lattices of the adjacent crystals are mismatched in thin intergranular layers. The grains may, in different materials, differ in size, having an average diameter from several micrometers up to several millimeters. The fine-grained materials are preferable because they have a better combination of strength and ductility.

2.3.1 Crystal Structures

The crystal structure consists of unit cells which repeat in a regular pattern to form the crystal lattice. Several types of such unit cells exist. Two of these types involve the so-called close-packed cells. They have a layer of a plane filled with the individual atoms in the closest possible contact with each other (see Fig. 2.7a) [3]. In this model each atom is represented by a ball, each one surrounded by six balls touching each other. The next layer, above the one shown in solid-line circles, is shown in dashed-line circles. It is located over one half of the valleys of the first layer. The third layer can be arranged in two different ways: either exactly above the first one, as shown by the open dot, resulting in a sequence of arrangements ABAB . . ., or over the valleys in between, as shown by the black dot, giving a sequence ABCABC. . . . The former is known as a hexagonal close-packed (*HCP*) structure (see Fig. 2.7b) and the latter as face-centered cubic (*FCC*; Fig. 2.7c) [1]. In both these structures 74% of space is filled with the balls of atoms. Another very common structure is the body-centered cubic cell (*BCC*) shown in Fig. 2.7f. The "packing factor" of this structure is 68%. In the upper part of the

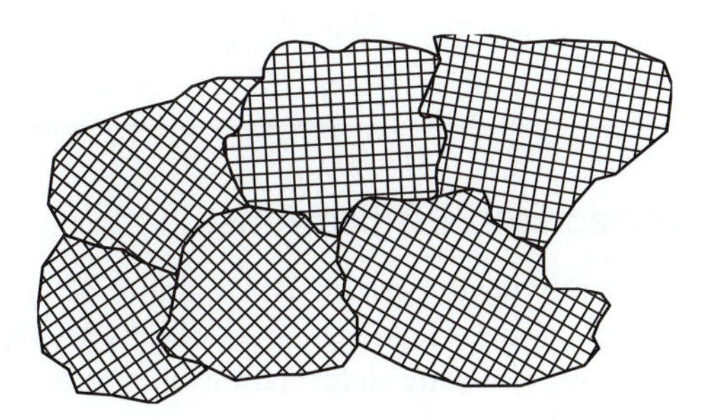

Figure 2.6
Crystal grains of a metal.

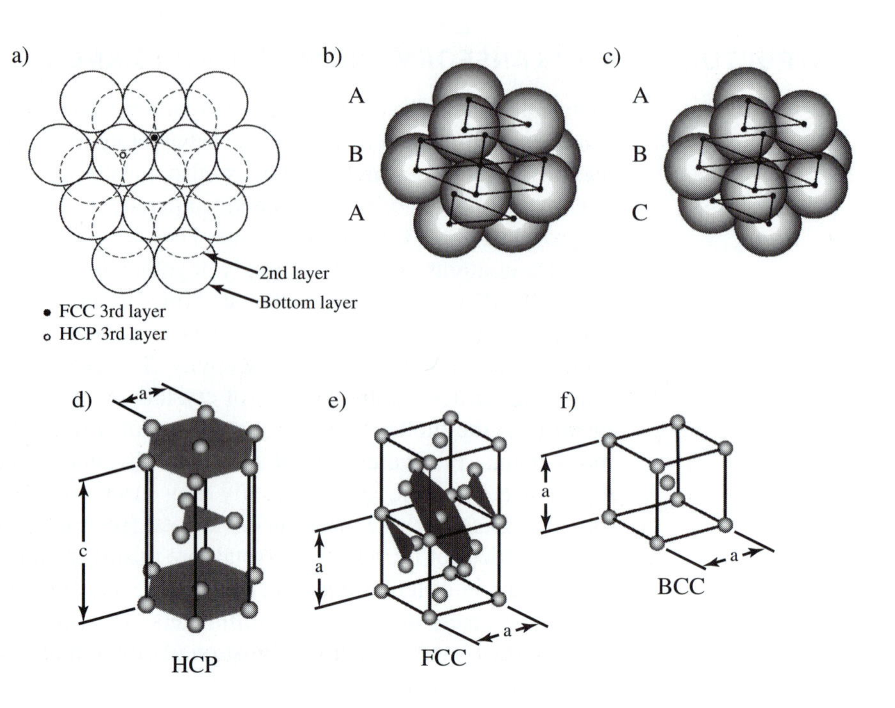

Figure 2.7
The HCP, FCC, and BCC unit cells. [Brick Gordon Phillips], [Van Vlack]

figure the cells are represented by "hard ball" models of atoms. In the bottom row, for better visibility of locations, only the centers of atoms are indicated, by shaded dots. The atoms are held in every metal at well-defined distances from each other. These interatomic distances are given in Table 2.2 for several selected elements in angstroms ($1Å = 1 \times 10^{-10}$ m).

The following metals have the HCP structure: Be, Mg, Co, Zn, Y, Zr, Ru, Hf, Re, Os. Because of the limited number of slip planes of this structure, these metals have limited ductility. The metals that crystallize in the FCC structure are, on the contrary, most ductile: Al, Ca, Sc, Ni, Cu, Sr, Rh, Pd, Ag, Ir, Pt, Au, PB, Th. Metals of the BCC structure have medium ductility: Li, Na, K, V, Cr, Rb, Nb, Mo, Cs, Ba, Ta, W.

Two metals have a property called allotropy, which means that they can exist in two different structures, depending mainly on temperature. Thus, Fe is BCC at room temperature and FCC above 723°C, and Ti has an HCP structure at room temperature and transforms to BCC at 880°C.

TABLE 2.2 Interatomic Distances of Selected Crystals (Å)

Al	Cu	Fe	Zn	Cr	Ti	C
2.862	2.556	2.4824	2.665	2.498	2.9503	1.545

2.3.2 Crystal Imperfections: Dislocations

Crystals are practically never perfect. The material behavior actually depends mainly on the various kinds of imperfections, structural disorders, and impurities in the structures. Most importantly, plastic deformation of metals is entirely dependent on line imperfections called dislocations.

First, let us mention the point defects in crystals. These exist as vacancies or interstitial atoms or else as interstitial and substitutional impurities (Fig. 2.8). These defects change the distribution of energy in the crystal. The strain energy around their sites distorts the lattice, and these distortions may influence movements of atoms. The linear defects are the dislocations, which are discussed in a little more detail in several parts of the following text. The area defects are mainly at the grain boundaries with their mismatched crystal orientations.

Sliding of whole atomic planes in a shear type of plastic deformation would require much higher stresses than those encountered in practice. Such very high strengths can only be achieved in crystals without imperfections. Such crystals can be made in laboratory setups in the special form of very fine (and mostly short) fibers called whiskers. Actually, in all regularly manufactured metals, each crystal contains a large number of dislocations, and plastic deformations result from the motion of dislocations. A tremendous amount of research has been devoted in the past thirty years to the study of dislocations and of their effects on the behavior of materials, and a good number of texts discuss them in great detail. We will mention only the fundamentals of this topic. Dislocations exist in two basic types: the edge dislocation and the screw dislocation; and other types are combinations of these. The edge dislocation is shown in Fig. 2.9a. It has the form of an extra plane of atoms above the plane AA on which a shear stress is applied. The dislocation is the carrier of the corresponding plastic shear deformation. This does not happen by a simultaneous slip of all the atomic planes above AA. As shown in Fig. 2.9b there is only one exchange: the extra plane attaches itself to the plane below AA and the dislocation passes onto the vertical plane. Fast successive actions of this type result in the overall slip along AA as shown in Fig. 2.9c.

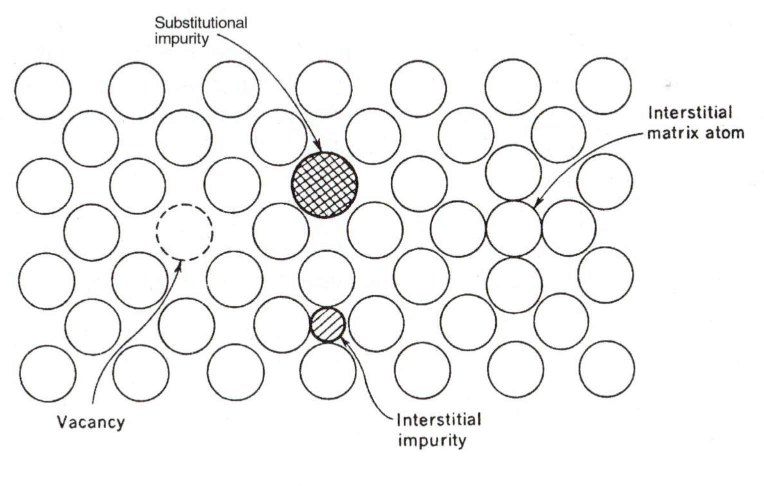

Figure 2.8
Point defects in crystals. [SHEWMON]

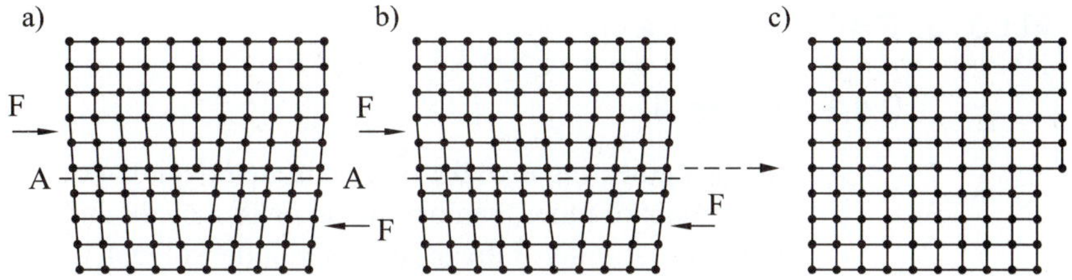

Figure 2.9
The edge dislocation and its movement.

There are a great number of dislocations in a material, and any stress will encounter enough of them to carry out the plastic deformations. Dislocations can climb, cross-slip, expand, and multiply. During plastic deformation their number grows; they interact with each other and with other barriers, especially with grain boundaries and with precipitate particles and foreign atoms, where they pile up and interlock. These are the mechanisms of strain hardening. While dislocations are needed in sufficient numbers to enable plastic flow, their movement is impeded if there are too many of them. The dislocation density (length per volume) of an annealed crystal is 10^5 to 10^6 cm^{-2}, and it increases to 10^{10} to 10^{12} cm^{-2} in a cold-worked metal.

2.3.3 Grain Boundaries and Deformation

Our discussion of plastic deformation so far has considered slip in a crystal. These slips occur most easily on densely packed planes that have definite orientations in each type of the crystal structure. Correspondingly the yield stress is different in different directions of deformations in a crystal. However, as mentioned before, every engineering material is polycrystalline and consists of a great number of grains with different structural orientations. Therefore, the anisotropic stress-strain behavior of a single crystal is lost and averaged out over the whole piece of material. Strain must continue across grain boundaries, and although there is a great amount of variation in local strain within the grains and at and across the boundaries, it all combines to produce the deformation of the whole volume concerned.

The grain boundaries are regions of disturbed lattice only a few atomic diameters wide. Generally, there is an abrupt change of crystal orientation at a boundary. There are many instances of low-angle boundaries between parts of crystal where the orientation differs by less than 1°, and these may be considered as a succession of edge dislocations. A grain boundary has rather high surface energy and is thus the locality for preferential precipitation of foreign atoms. This is one more obstacle to dislocation movement at grain boundaries. The yield stress of a material increases with the decrease of grain size mainly because for finer grains, the grain boundary surface area per volume increases. The Hall-Petch relationship states that the tensile yield stress is related to grain size as follows:

$$Y = \sigma_i + k'D^{0.5} \tag{2.12}$$

where σ_i is friction stress opposing the motion of dislocations, k' is the "unpinning constant" expressing the extent to which dislocations are piled up at barriers, and D is grain diameter.

2.4 ALLOYS: PHASE DIAGRAMS

2.4.1 General

Most engineering metallic materials are not pure metals of one kind but alloys that contain atoms of more than one metal or of a metal and a nonmetal. In an alloy, the combination of atoms of the basic metal A, the solvent, and of the added metal B, the solute, may be a solid solution or it may be a compound, and in a metal the grains representing different solid solutions or compounds may be mixed together.

Solid solutions are of two kinds:

In a *substitutional* solution, the atoms of the solute replace some atoms of the solvent in its crystal cells. Such a solution is formed under the following conditions:

1. The atoms of the two metals do not differ in diameter by more than 15%.
2. Their space lattices are similar.
3. They are close to another in the electromotoric series.

An example is given in Fig. 2.10a of BCC brass, called β-brass, which contains equal numbers of Cu and Zn atoms. The diameter of the latter (see Table 2.2) is only 4% larger than the former, and the two metals are adjacent in the periodic table.

An *interstitial solid* solution is formed if the atoms of the solute occupy positions in between the atoms of the solvent in its cell. Obviously, this is only possible if the solute atoms are much smaller than those of the solvent. Even then the solubility is limited. The most common example is that of steel, where the C atoms are located in the Fe cells. At room temperature Fe has the BCC structure (see Fig. 2.10b), which offers little interstitial space. The edge a of the cube is $(4/3)^{-1/2}R$, where the Fe atom radius R is 1.23 Å, $a = 2.86$ Å, and the radius of an interstitial hole is 0.36 Å. The radius of the C atom is 0.772 Å and, correspondingly, it distorts the lattice if it occupies the hole. Two of the four possible sites on one face are shown. Solubility is limited to 0.025% C. At temperatures above the recrystallization temperature of Fe, its cells are FCC, and they provide more space for interstitial C atoms (see Fig. 2.10c). The edge of the FCC cube is 3.56 Å, and the radius of the interstitial hole is 0.52 Å, which is much larger than that in the BCC structure. Correspondingly, solubility of C in this form of Fe is higher; depending on temperature it ranges from 0.8 to 2.06% C. The interstitial atoms of the solute are highly mobile in the structure of the solvent, and the C atoms may easily diffuse between parts of the Fe body as well as in and out of it. Solid solubility increases with temperature as it does with liquid solutions. Correspondingly, solute atoms precipitate on cooling (by diffusion) out of the atoms of the solvent.

Substitutional solid solutions in which there is a long-range order of the two metallic atoms and the composition is well defined in simple ratios of atom fractions are considered as *intermetallic compounds.* They have a sharp melting point and other characteristics of distinct chemical species. Very often, such compounds have higher melting points than either of the elements, and they are hard and brittle.

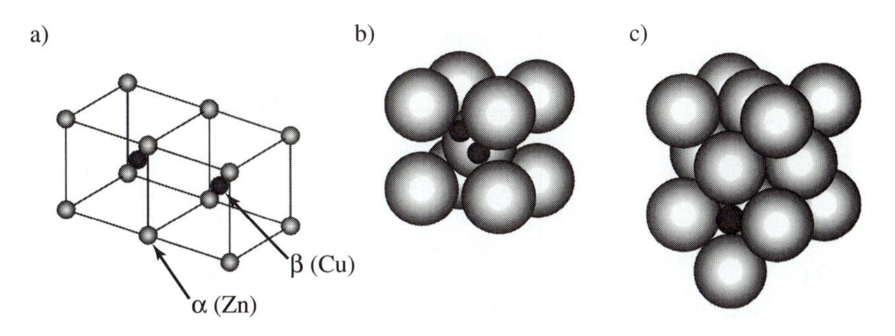

Figure 2.10
a) Substitutional, (β brass). [VAN VLACK]; and b), c) interstitial (BCC and FCC lattices) solid solutions. [BRICK GORDON PHILLIPS]

The transition between the solid and liquid states of an alloy is different from that of pure elements. Depending on the composition, the melting and freezing temperatures vary and, for a given composition, there is a temperature range over which it is partly solid and partly liquid.

A solid alloy usually contains several *phases* which differ in structure and composition. The most useful form of presenting information about the structures and combination of phases is the *phase diagram*, which is plotted in composition-temperature coordinates. It is also called an equilibrium diagram because it expresses states obtained by slow temperature changes—so slow that all the diffusion processes necessary for the rearrangement of phases can be completed and balanced states are obtained. Later on we will learn about the differences obtained by rapid cooling. Phase diagrams may be drawn for compositions of two elements (binary diagrams) or of three elements (ternary diagrams) or more. We will only deal with binary diagrams. There are several types of phase diagrams according to the solid solubility of the constituents.

Let us first discuss a diagram of an alloy of Cu and Ni, which are mutually completely solid-soluble. This diagram is presented in Fig. 2.11. It contains two lines, the liquidus and the solidus. Above the former all compositions represent a single liquid phase and below the latter they represent a single solid phase, which is a substitutional solid solution denoted α. All grains in this solid contain the same type of crystals. Between the two lines the liquid and solid phases coexist. In this two-phase field the composition of each of the phases is obtained by the application of the *inverse lever rule*. Let us illustrate it using the example of a 52% Ni alloy. Starting from a liquid of this composition and cooling down, at the temperature of point *a*, some parts of the liquid start to solidify. The solid phase at this temperature can only exist as a composition *b*, of the 30/70 Cu-Ni ratio. The crystals that are initially formed have this composition, which is rich in the higher melting-point Ni component. As the temperature is further decreased, more of the alloy solidifies. At 1300°C it is a mixture of a liquid of composition *c* (47% Ni) and of a solid with composition *e* (63% Ni). If the cooling rate was very slow the original Ni-richer crystals of composition *b* would lose some of the Ni by diffusion. At realistic cooling rates, however, there will be gradients of composition in the solid. The lever rule can be used to determine the weight proportions of the two components at 1300°C as follows:

$$\%\text{ liquid (47 Ni)} = \frac{e - d}{e - c} \times 100 = \frac{62 - 52}{62 - 47} \times 100 = 66.67\%$$

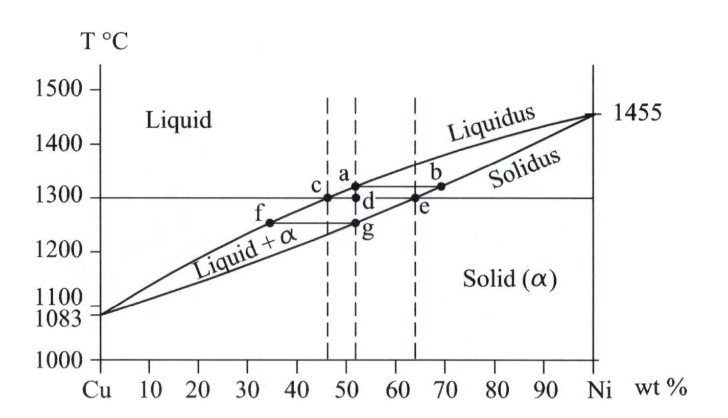

Figure 2.11
The Cu-Ni phase diagram.

and

$$\% \text{ solid (63 Ni)} = \frac{d - c}{e - c} \times 100 = 33.33\%$$

This rule is called the lever rule because it is similar to calculating forces on a lever supported at d with forces acting at c and e, and it is called inverse because the contents of an element are obtained by setting the inverse part of the lever ($e - d$ for liquid) in the numerator of the expression. It can be used in all the phase diagrams, in their two-phase fields, on a horizontal (constant-temperature) line with endpoints on the line separating the two-phase field from the adjacent single-phase fields.

To obtain the 52% Ni composition, on further cooling we arrive at point g, where the last parts of the Ni-poor liquid with composition f of 33% Ni solidify into a single-phase solid with composition g, of 52% Ni. Correspondingly, during this process the amount of the liquid phase was decreasing while it was becoming poorer in Ni from its initial 52% Ni composition, and the amount of the solid phase was increasing, and its composition was changing from the initial 70% Ni (at point b) down to 52% Ni.

In another basic type of a phase diagram, the two components A and B are fully mutually soluble in the liquid phase but completely unsoluble in the solid state. This idealized diagram is shown in Fig. 2.12. The melting temperatures of the alloys decrease from both ends, and there is a particular composition called the *eutectic*, which has the lowest melting temperature. Also, this composition transforms fully from the liquid to the solid state, and in the reverse direction, at one single temperature T_1. Any other composition transforms over a range of temperatures. Let us assume that the eutectic has 62% B and let us follow the transformation of a composition of 25% B, starting in the liquid single-phase state. At point a the first crystals of phase α, which is pure metal A, start to form. The liquid is becoming poorer in A, richer in B. At the temperature T_2 there is a mixture of solid crystals α (point b) and of liquid with composition d (44% B). The contents are distributed according to the lever rule:

$$\%\alpha = \frac{d - c}{d - b} \times 100 = \frac{44 - 25}{44} = 43\%$$

$$\% \text{ liquid (44\%B)} = 57\%$$

On further cooling, more and more of α the phase solidifies, and the liquid follows the line a-f, becoming poorer in A and richer in B. At point e all of the alloy is solid with the 75% α/25% β composition. The eutectic may exist as a well-defined

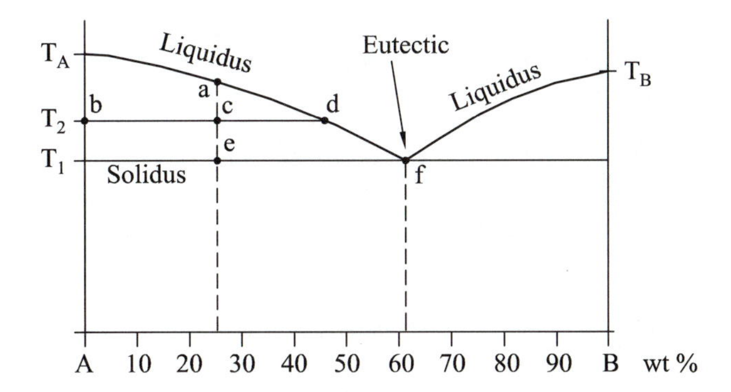

Figure 2.12
Phase diagram of components with no solid solubility.

phase of grains with a characteristic structure containing 38% α and 62% β. The contents at e may then be expressed as

$$\% \ \alpha = \frac{62 - 65}{62} \times 100 = 59.7\% \ \alpha$$

$$\% \text{ eutectic} = 40.3\%$$

2.4.2 The Fe-C Phase Diagram

The phase diagrams of actual alloys differ from the idealized forms because there is always some partial solid solubility available. The most significant phase diagram is the one of Fe-C alloys, which is presented in Fig. 2.13.

This is actually the Fe-Fe_3C diagram because no more than the amount of carbon in the iron carbide can be associated with Fe. The compound Fe_3C contains 6.67 wt% C. Iron carbide is hard, brittle, and white. As a constituent of the Fe-C alloys, it is called *cementite*.

Pure iron solidifies at 1539°C. Over a short range of high temperatures it has the BCC structure called the δ iron. When cooled to 1400°C the structure changes to FCC γ iron. Below 910°C the structure changes again to a BCC structure called the α iron. This shows that iron is allotropic, and this property is the basis for heat treatment of steels, by which their strength and hardness can be substantially increased. Heat treatment will be discussed later. Iron also changes from a nonmagnetic material at high temperatures to a magnetic one below 771°C. The α phase can dissolve a maximum of 0.025% C at 723°C. The solid solution of C in BCC is called *ferrite*.

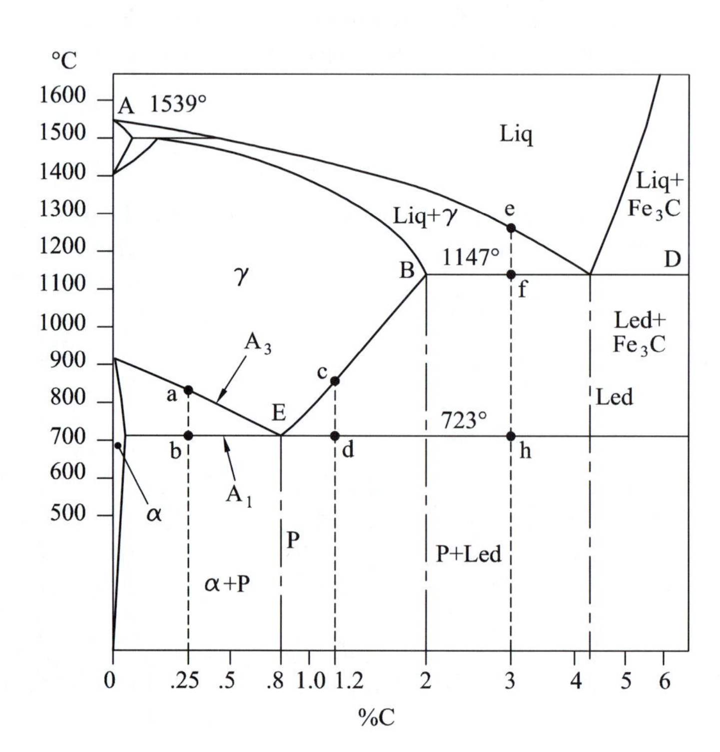

Figure 2.13
The Fe-Fe_3C phase diagram.

We will now neglect the lines and fields at the upper left corner of the diagram and follow the liquidus lines. The 4.2% composition is the *eutectic*, and it has the lowest melting point of all Fe-C compositions at 1147°C. This is approximately the composition of pig iron as obtained from the blast furnace because it is difficult to generate higher temperatures in the stack of the furnace. The structure of the solidified eutectic is called *ledeburite.* The solidus line slopes down from the point *A* of the low-carbon compositions to the point *B* at 2.06% C and 1130°C, and then it remains horizontal to point *D*. There is another horizontal line in the diagram which is commonly denoted A_1. It is the line below which all γ iron is transformed to α iron. The lines A_3 and A_{cm} are analogous to the liquidus lines and the line A_1 to the solidus line of the eutectic, except that they indicate the γ-α transformation instead of the liquid-solid transformation. Because of the analogy, the point *E* is called the *eutectoid* point, at 0.8% C and 723°C. The composition of 0.8% C, which consists of ferrite and cementite in a very characteristic lamellar configuration, is called *pearlite (p).* The solid solutions of C in γ (FCC) iron, which are bounded by the lines of the solidus (*A-B*) and the line A_3 and A_{cm}, are called *austenite*.

The microstructure of pearlite is shown in 650× magnification in the lighter fields of Fig. 2.14a2. The whiter background field is ferrite, and the darker bands are cross

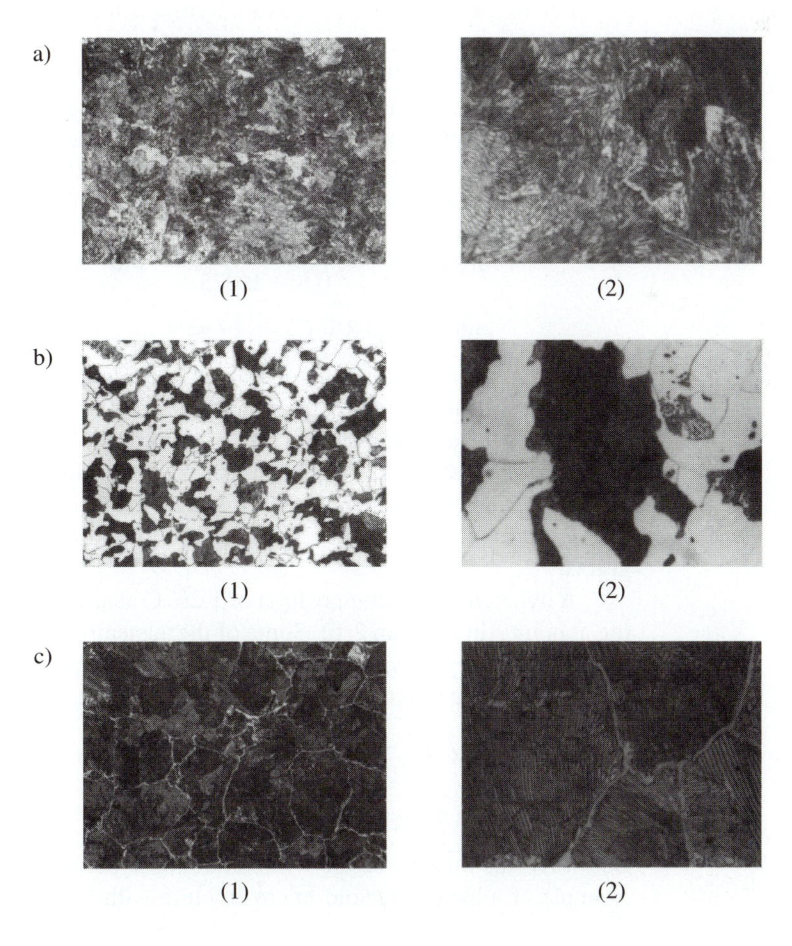

Figure 2.14
Microstructure of a) pearlite, shown in a1) at 130× and in a2) at 650× magnifications. The pearlitic structure is well recognizable in the lighter fields of a2); b) 0.35% C hypoeutectoid carbon steel, shown in b1) at 130× and in b2) at 650× magnifications. The dark areas are pearlite, the white areas are ferrite, and the dark lines in them are grain boundaries; c) 1.3% C hypereutectoid carbon steel: c1) is 130× and c2) 650×. The dark areas are pearlite and the white ones are carbide. (Courtesy Dr. D. A. R. Kay, McMaster University)

sections of the lamellae of cementite. The darker parts of the picture are where the lamellae enter the surface at a small angle. Depending on the cooling rate and the presence of some alloying elements, the spacing of the pearlite lamellae may range from several micrometers (coarse pearlite) down to medium and fine pearlite. The interlayers of the two components reinforce each other. Ferrite has a strength of about 300 MPa and elongation of 40%, and cementite has a strength of about 2000 MPa and negligible elongation. Thus ferrite imparts ductility and cementite strength, and the combination results in strength between 800 to 1000 MPa and elongation of 10–15%.

Austenite can exist at elevated temperatures (see Fig. 2.13) in a range of compositions up to 2% C. It is soft and ductile and is therefore very suitable for hot forming processes like forging and hot rolling. The eutectic of 0.8% C in the lamellar form of pearlite is only obtained by cooling from the 0.8% C austenite by its simultaneous transformation to ferrite and cementite. Pearlite is treated in practice as if it was a particular phase, and one speaks of grains of pearlite, although they contain a mixture of two different crystals. Ferrite is BCC, and cementite has an orthorhombic structure.

Compositions with less than 2% C are generally called *steels*, and those above this limit are called *cast irons.* The latter cannot be forged or hot rolled because they are brittle at all temperatures below the solidus line, and they are also brittle at room temperature. Steel compositions with less than 0.8% C are called *hypoeutectoid*, and those with more than 0.8% C are call *hypereutectoid.*

Let us consider a hypoeutectoid composition of 0.25% C. Heated up to, say, 1000°C, it is austenitic. On cooling, at point a some of the crystals start changing from γ to α (from FCC to BCC), and at the same time they lose the interstitial carbon, which diffuses out from them to the remaining austenitic grains, which become richer in C. This process continues between points a and b, where there will be grains of 0.025% C ferrite and grains of austenite that have changed their composition from 0.25% C to 0.8% C. The proportion will be, according to the lever rule,

$$\alpha(0.025\%) = \frac{0.8 - 0.25}{0.8 - 0.025} = 71 \text{ wt\%}$$

$$\text{eutectoid } (0.8\% \text{ C}) = 29 \text{ wt\%}$$

The remaining austenite changes, on further cooling, to pearlite. At temperatures below A_1 the microstructure will consist of 71% grains of ferrite and 29% grains of pearlite. An example of a hypoeutectoid microstructure with 0.35% C is shown in 130× and 650× magnifications in Fig. 2.14b. The white grains are ferrite, and the dark grains pearlite. Its lamellar structure is not distinguishable at the lower magnification, but it is recognizable in part b2. The dark lines are grain boundaries that delineate the ferrite grains.

A hypereutectoid composition of 1.2% C austenite will start transforming on cooling at point c in diagram 2.13. Some of the austenitic grains will lose some of their carbon to adjacent grains, which will start transforming to Fe_3C cementite. This process continues from c to d, where we find

$$Fe_3C(6.67\% \text{ C}) = \frac{0.8}{6.67} = 12\%$$

$$\gamma(0.8\% \text{ C}) = 88\%$$

Below A_1, the microstructure will contain 88% pearlite grains and 12% cementite. An example of a hypereutectoid microstructure with 1.3% C is shown in Fig. 2.14c. The magnifications are again 130× and 650×. The pearlite grains appear dark. The carbide is white, and it is seen to have formed mainly at the boundaries of the original austenitic grains.

The amount of carbon in steel affects its strength and its ductility. Low-carbon steels with less than 0.1% C are very ductile and easy to cold form; they are used for cans, automobile body panels, and many other sheet-metal parts for appliances and other products. Their strength is about 350 MPa, and their elongation is 35%. Steel with about 0.2% C is used for structural beams with strength of 420 MPa and $e = 30\%$. Medium carbon (0.3 to 0.45% C) steels are used for machine parts like shafts and gears. They may be hardened to some extent. Steels with 0.7% C are used for railroad rails. Their strength is of the order of 800 MPa, and elongation is about 10%.

Grain size is important; fine-grained steel is stronger and tougher than steel with coarse grains. The grain size is determined in the austenitic range and depends on the conditions of manufacture (rolling, forging) and on heat treatment. This is discussed in more detail in Chapter 5.

We may now discuss solidification of cast irons in a similar way as for steels, referring again to Fig. 2.13. For instance, if we consider a 3.0% C composition in the liquid, some crystals of austenite with composition g start to solidify, on cooling, at point e. During cooling from e to f the austenite crystals become richer in C, and so does the liquid. At point f, on solidification the alloy consists of grains of 2.0% C austenite and eutectic 4.2% C ledeburite. On further cooling, more carbide is rejected from the austenite. At point h at 723°C the material consists of 0.8% C austenite, eutectic 4.2% C ledeburite, and 6.67% cementite.

Cementite in cast irons is not very stable, and if cooling is slow, graphite is formed instead of carbide. The most common form of cast iron, gray cast iron, contains ferrite, graphite, and some pearlite. Later on we will discuss the various forms of cast irons.

2.5 HEAT TREATMENT OF METALS

2.5.1 Allotropic Metals: Steels

The transformations discussed in conjunction with the Fe-C phase diagram occur as described only if the cooling rates are very slow, because most of the changes are accomplished by diffusion of carbon between the grains of the material, which takes time. It is not within the scope of our text to deal in any detail with the diffusion processes. It should suffice to say that the rate of diffusion depends very strongly on temperature, and at lower temperatures the time required to reach the equilibrium state is very long.

However, the γ to α transformation itself, based on the change of crystal structure from FCC to BCC, is practically instantaneous because it is accomplished by a shear type of shift of atom planes over a small distance. If the cooling rate of a steel from an austenitic region, say of the eutectoid composition of 0.8% C, is fast across the A_1 line, then the crystal structure changes, but the carbon atoms cannot diffuse out of the γ iron to produce ferrite and cementite; thus they are trapped in the BCC structure, which is supersaturated with C and therefore highly distorted. This structure is called *martensite.* It is very hard and brittle. The microstructure of martensite, which is acicular (needlelike), is shown in Fig. 2.15. It is shown at 130× magnification for steel with 0.35% C.

The allotropy of iron combined with the large difference in C solubility in γ and α irons is thus the basis of hardening of steel. The process is accomplished by first heating the steel into an austenitic range, holding it there long enough to obtain austenite grains with uniform composition and then quenching the part for fast cooling. If the cooling rate is not fast enough, the martensite will not form fully or not at all. This is best illustrated by means of a diagram constructed from observations of isothermal transformations. Such a diagram is called a *time-temperature transformation* (*TTT*) diagram. This diagram, for an eutectoid carbon steel, is shown in Fig. 2.16. It is obtained by carrying out experiments like the one indicated by the line *A-E.* The austenitized material is first quenched into a salt bath of temperature T_1, line *A-B.* Then it is held at this temperature for a certain time, say, until point *C,* and then quenched to room temperature. Next time it is held at T_1 until point *D,* next time until *E,* and so on. The microstructures so obtained are studied. The next series of experiments follows the path *A-F-G,* and so on, for isothermal changes at temperature T_2, and so on. Two curves, 1 and 2, are obtained, the first one for the start of a transformation and the second one for its end. Transformations in which the initial quench is interrupted above the temperature T_p lead to a pearlitic structure. Those which end between the temperatures M_s and T_p lead to a structure called *bainite.* Bainite is an intermediate structure between pearlite and martensite as far as hardness is concerned. It has a feathery appearance.

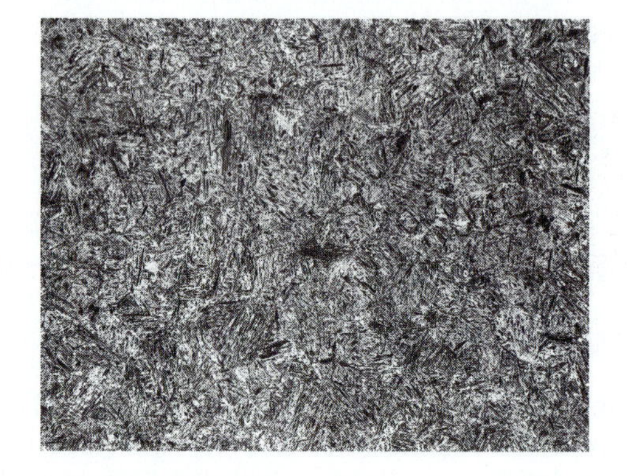

Figure 2.15
Microstructure of martensite in steel with 0.35% C. Magnification 130×. (COURTESY DR. D. A. R. KAY, MCMASTER UNIVERSITY)

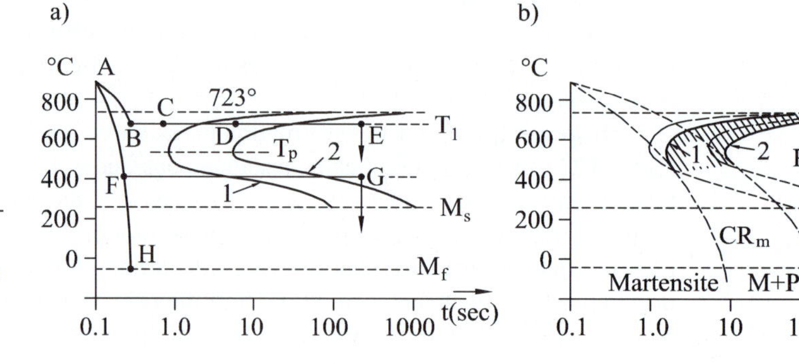

Figure 2.16
a) The isothermal TTT (time-temperature transformation) diagram for eutectoid carbon steel. b) The continuous cooling graph.

If the fast cooling is not interrupted, curve *A-H,* martensite starts to form at the temperature M_s and all of the austenite is transformed to martensite at the temperature M_f. The TTT diagram, which applies to isothermal transformations, may be modified into a graph of continuous cooling transformation, as shown in Fig. 2.16b. Two curves are important. Any cooling process carried out to the left of the curve CR_m (minimum cooling rate) can lead to the formation of martensite, and any process to the right of CR_p (maximum cooling rate for pearlite) will produce a full pearlitic structure. In a eutectoid carbon steel the rates are CR_m = 200°C/sec and CR_p = 50°C/sec. In steels with other additional alloying elements these rates may be considerably slower, and martensite may be obtained by much slower cooling. This leads to less distortion of the hardened steel and to the possibility of hardening thicker sections.

In quenching various parts the fastest cooling rate is obtained on the surface of the part; the cooling rate decreases into the interior of the part. Therefore, depending on the thickness or on the diameter of the section of the part, that hardening may be obtained only to a certain depth. If "through" hardening is required, it is necessary to use suitable alloy steels.

The ability of a steel to attain martensitic transformation also in thicker sections, or at slower cooling rates, is called hardenability. A standard test, which was specified by W. Jominy, makes it possible to compare hardenability (see Fig. 2.17). A test bar of standard dimensions is heated to a uniform austenitic state in a furnace and then placed in the fixture. A cold water jet is directed against its end face. This produces different cooling rates at different distances from the end. After the bar has cooled down, a flat

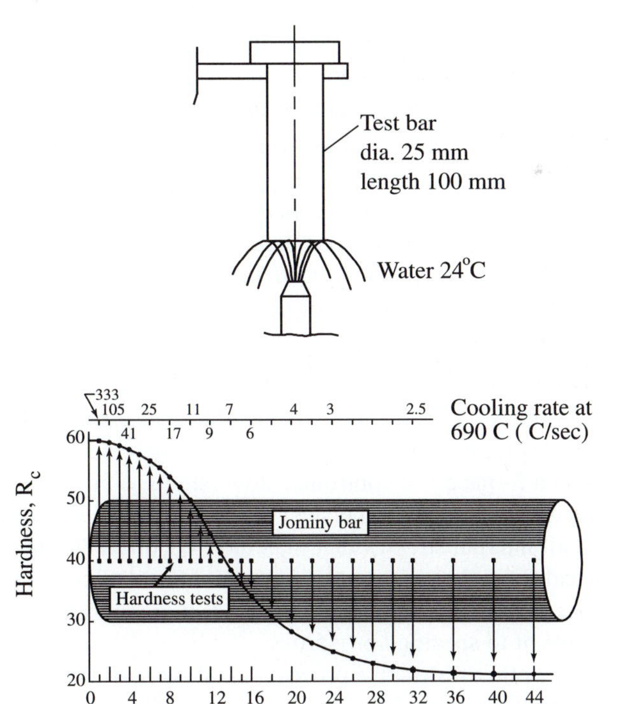

Figure 2.17
TOP: The Jominy test of hardenability. BOTTOM: Typical hardness distribution in Jominy bars. (A. G. Guy, "Elements of Physical Metallurgy," 2nd Ed., Addison-Wesley Publishing Co., Inc., Reading, MA, 1959, Fig. 14-14, p. 484)

is ground on its side and R_c hardness measured every 1/16 in. (0.159 mm). A typical plot of a result of this test is shown in the bottom panel of the figure, where the different cooling rates obtained in the test are also noted.

The hardness obtainable in quenched steels depends on the amount of carbon. Steels with less than 0.3% cannot be successfully hardened. For steels with 0.3% to 0.6% carbon, the maximum obtainable hardness increases from about 48 R_c to about 66 R_c.

The steel quenched through the martensite transformation is very hard and brittle and, because of the brittleness, it cannot in most instances be used in this state. It is, therefore, subjected to another heat-treatment process called *tempering*. This is done by heating the part to some temperature below 723°C, holding it at this temperature for a certain period of time, and quenching it again. During tempering the martensitic tetragonal, highly distorted structure reverts to regular BCC and allows the carbon atoms to migrate and form some very fine iron carbide crystals. Hardness decreases but ductility improves.

The changes in hardness and ductility depend on the temperature and time of tempering. The temperature has a much larger effect than time. Therefore, in practical terms the temperature is given assuming that the time will be of the order of one half to several hours. There are several typical temperature ranges of tempering:

- 100°–200°C. Upon reaching 200°C the martensite is fully decomposed. Structural changes are obtained on a very fine scale and cannot be detected under a microscope. Hardness almost does not change at all; some ductility is obtained. This is called fine-tempered martensite.
- 200°–360°C. Structure is still tempered martensite, but hardness decreases to 55–58 R_c. However, in this range, due to precipitation of oxides and nitrides a decrease of toughness may occur. This is called blue brittleness because it occurs at temperatures that leave a blue oxide of film on the steel.
- 360°–723°. The higher the temperature in this range, the coarser the cementite particles. At or above 650°C these particles can easily be resolved under an optical microscope. Because of their spherical shape, the structure is called *spheroidite*. The decrease of hardness and increase of ductility are almost linearly proportional to tempering temperature. The range of hardness extends from about 60 R_c for martensite down to about 20 R_c for spheroidite.

All the structural changes obtained by hardening and tempering may be eliminated by *annealing*. The so-called full annealing is done by heating to about 50°C above the recrystallization temperature (i.e., above A_3 for hypoeutectoid steel and above A_1 for hypereutectoid steels), holding it there for uniform heating, and then cooling in a furnace at a controlled slow rate to room temperature. An equilibrium type of structure is obtained. The steel returns to minimum hardness and maximum ductility, and all internal stresses that may have occurred during hardening are eliminated. Full annealing is also done to eliminate the effects of hot and cold forming and to eliminate grain distortions and restore fine grain. It prepares the steel for cold forming or for heat treatment to specified properties.

Another restorative process is call *normalizing*. It consists of heating to the same temperatures as in full annealing but cooling in still air instead of in a furnace. Its effects are not as complete and predictable as in full annealing because some parts of the structure may cool too fast to reach the full pearlite state, but it is less expensive.

Still another process is called *commercial annealing* or *stress relieving*. It consists of heating to slightly below the recrystallization temperature and cooling slowly. The main result is the elimination of internal stresses. It is mostly done on weldments right after welding to prevent structural distortions that would otherwise develop due to internal stresses at the welds, which have gone through the cooling process from the high temperatures with different temperature gradients in the various zones.

Some structural parts should preferably be hard on the surface but soft and ductile inside: shafts, gears, guideways of machine tools. They may be treated by various *surface hardening* processes. These consist of using a medium carbon steel or cast iron part, heating it on the surface only, and quenching it. Surface heating is done by using gas torches in *flame hardening* or by using high-frequency electric current and coils adapted to the shape of the surface to be heated in *induction hardening*. The same effect of hardening only a surface layer of the part is obtained by *case hardening*, in which a low-carbon steel, not hardenable by itself, is used for the part. The parts are inserted in boxes (cases) filled with carbonaceous material, heated in a furnace to austenitizing temperature, and held for a time period during which the surface is enriched in carbon to a desired depth. Carburizing may also be achieved by using CO gas instead of a solid carburizing material. The carburized part is then quenched for surface hardening.

2.5.2 Phase Diagram for Al Alloys: Precipitation Hardening

Pure aluminum is soft and ductile. By using some alloying elements and corresponding heat treatments, either alone or in combination with cold working, substantial increase of hardness (strength) can be obtained. As an example of an alloy that can be strengthened by *precipitation hardening,* also called *age hardening,* we will use an Al-Cu alloy. The Al-Cu phase diagram is shown in Fig. 2.18. Of particular interest is its Al-rich end, showing the α phase, which is a solid solution of Cu in Al. The Cu-rich end is given by the intermetallic compound $CuAl_2$ designated as θ phase. A maximum of 5.6% Cu can exist in the α phase at about 550°C. The solid solubility of Cu decreases to about 0.5% at room temperature. An alloy denoted 2014 contains 4.5% Cu and about 0.5% Mg.

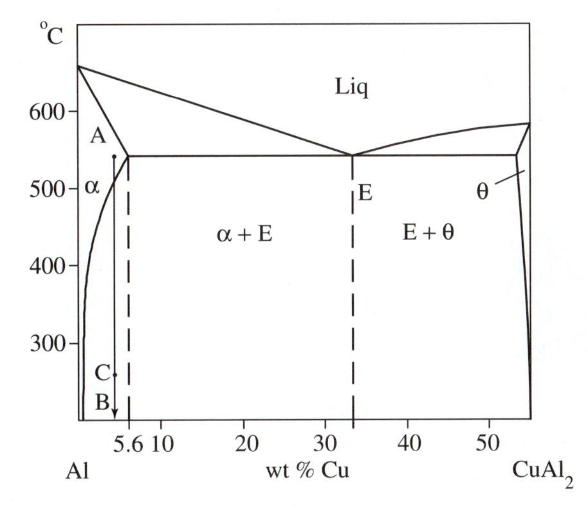

Figure 2.18
The Al-Cu phase diagram.

Precipitation hardening is achieved in three steps:

1. The alloy is heated to about 550°C and held until a homogeneous solid solution forms; see point *A* in Fig. 2.18.
2. It is quenched in water to room temperature, point *B*.
3. It is aged, that is, held at a temperature, point *C*, chosen so that the desired precipitation of the θ phase is reached during a convenient time period.

The processes occurring during this treatment are illustrated in Fig. 2.19. At the start the microstructure consists of grains of the α phase (solid solution of 0.5% Cu in Al) and coarse particles of the $CuAl_2$ phase. By heating to temperature T_1 and holding, a uniform solid solution is obtained. By quenching from T_1 to room temperature, the Cu atoms are trapped in the Al matrix, and the result is a supersaturated solid solution of 4.5% Cu in Al. This state of the alloy is soft. In the next stage it is heated to temperature T_2 and held for "aging." At this elevated temperature the diffusion of the Cu atoms out of the supersaturated solid solution is much faster than at room temperature. Fine precipitates of Cu particles "coherent" with the Al matrix are obtained. The matrix is highly distorted around these particles, which creates obstacles to the motion of dislocations. The state of the alloy is much stronger than after the quench. When the desired strength (hardness) is reached, the alloy is cooled to room temperature. Aging may be overdone by holding the alloy at the aging temperature too long. The coherent precipitates would then collapse, and large, noncoherent Cu particles would form. This would lead to loss of strength and hardness.

The effects of the aging process are illustrated in Fig. 2.20, which gives the yield strength of the 2014 aluminum alloy versus aging time at different temperatures. After the quench the alloy has YS = 40,000 psi = 276 MPa. At T_1 = 260°C maximum strength of YS = 48,000 psi (330 MPa) is obtained after 5 minutes. If it is held at this

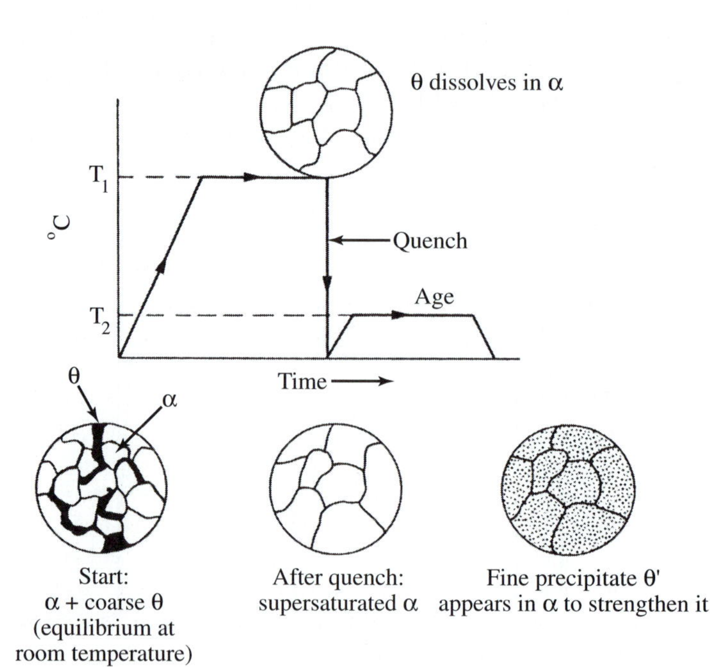

Figure 2.19
Precipitation hardening. Heat-treatment chart and microstructures. [FLINN TROJAN]

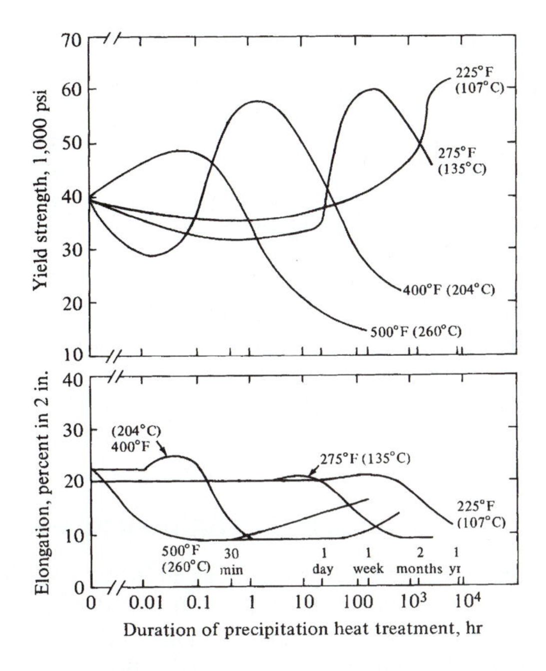

Figure 2.20
Effects of age hardening on mechanical properties of 4.5% copper-aluminum alloy (2014). [FLINN TROJAN]

temperature longer than 10 minutes, strength will start to decrease. For $T_1 = 204°C$ maximum strength of $YS = 58{,}000$ psi (400 MPa) is reached after about 2 hours, and it would start to decrease after about 4 hours. For $T_1 = 135°C$ maximum hardness of $YS = 62{,}000$ psi (425 MPa) is reached after about 150 hours and starts to decline after about 500 hours at this temperature. When aging at 107°C about 2,000 hours would be needed, and so on. The corresponding changes of ductility as expressed by elongation are given in the diagram at the bottom of Fig. 2.20.

Other aluminum alloys like Al-Si (12.5% Si) and Al-Zn (5.6% Zn) can also be treated by precipitation hardening because they also show the same feature of forming a homogeneous solid solution at a certain temperature below which precipitation of a coherent second phase occurs. Alloys of other metals can also be treated in this way.

2.5.3 Solid Solution Treatment

In some metals the ability to form a solid solution with an alloying element at room temperature may by itself have a significant effect on strength. Two examples are Al-Mn (1.3% Mn) and Al-Mg (2.5% Mg). These alloys in the annealed condition are much stronger than pure aluminum. They are also stronger in the cold-worked condition. Their microstructure shows simple grains of a single phase; however, the substitutional atoms generate enough internal distortion of the lattice to hinder dislocation motions and produce an increase of both the yield and ultimate strengths.

The effect of various alloying elements and their solute concentrations is shown in Fig. 2.21 for alloys of Fe (part a), and Cu (part b). The strengthening effect increases with the amount of the solute element, expressed here in atomic percent. Different solute atoms act differently depending on the mismatch in atomic size and on their deformability. Clearly, certain solid solutions increase the hardness and strength of the basic metal considerably.

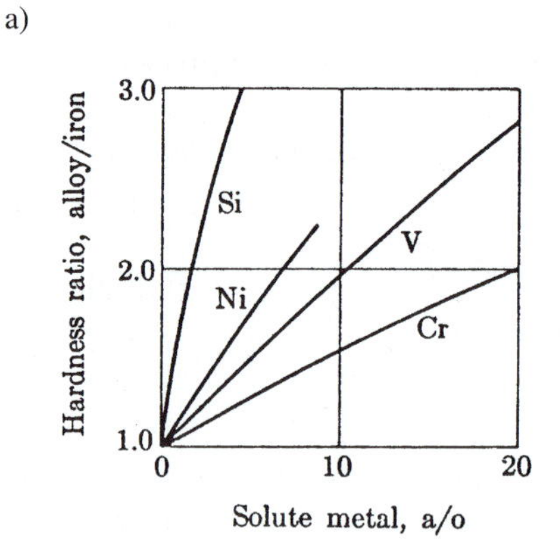

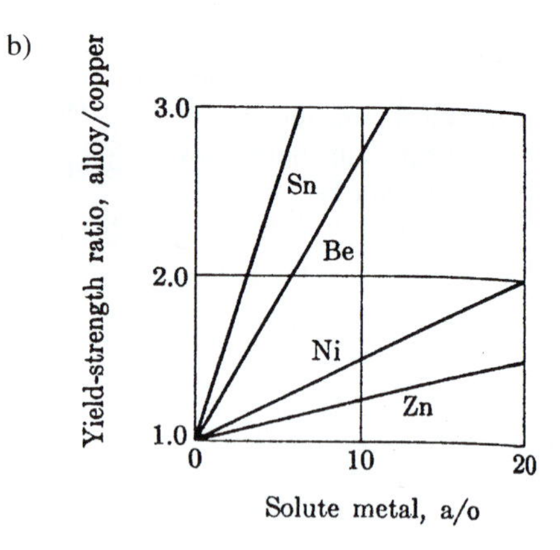

Figure 2.21
Effects of solid solution treatment on strength of alloys: a) Iron; b) Copper. The hardness and strength ratios for the alloy versus the unalloyed metal depend on the size mismatch of solute versus solvent atoms and the stress field that results. [VAN VLACK]

2.5.4 Summarizing Methods of Strengthening Metals

Strengthening is meant here primarily as an increase of the yield strength. It can also be expressed as an increase of hardness. It is understood that, generally, a simultaneous decrease of ductility is obtained, although not with the same intensity for the different methods recapitulated here. We are not discussing here the resistance to fracture, whether of the brittle or of the ductile type, because there is no direct relationship between yield strength and fracture strength. The latter depends on the state of triaxial stress and on the presence of pores and impurities, which would hardly affect yield strength.

Increasing yield strength means increasing the load-carrying capacity of elements of machines and of their structures or, from the reverse point of view, of allowing the use of elements with smaller cross section, which are consequently lighter, for the same loading.

The methods for strengthening metals as discussed in the preceding sections are all based on the various mechanisms of interfering with dislocation movements. A summary of the various methods follows.

- *Obtaining fine-grain material.* This is the most general method applicable to any metal. The finer the grain, the better the combination of strength and ductility, or toughness, that is obtained. It is the method in which increasing strength does not necessarily mean substantial loss of ductility. Fine grain is obtained by a proper combination of hot working and recrystallization and, mainly, by annealing after cold work. It is also obtained by the use of some alloying elements; for instance, aluminum is added to steel during its production. A dispersion of Al_20_3 particles arises, which prevents grain growth during solidification and during reheating. These procedures are discussed in more detail in Chapter 5.
- *Cold work (strain hardening).* This is also a rather universal method applicable to almost all metals with the exception of those that are so brittle that they cannot be cold-worked, like cast iron. The rate of strain hardening is different for different metals. Also the total amount of strain, which is limited by a total loss of ductility, is different for different metals. This is discussed in more detail in Chapter 5.

- *Solid-solution treatment.* This method is limited to alloys in which a sufficiently solute-rich solid-solution phase exists at room temperature (at the temperature at which the alloy is used). This treatment may be combined with cold work, with the two effects being additive.
- *Precipitation hardening.* This treatment is limited to alloys in which substantial solid solubility exists at an elevated temperature and decreases with temperature decrease in such a way that heating to a homogeneous solid solution followed by quenching yields a supersaturated solution that, on aging, releases fine, coherent precipitate. This treatment can also be combined with cold work.
- *Allotropic hardening.* This is limited to steels only. However, because steels are the most common materials in mechanical manufacturing, it is an important method. Heating up to homogeneous austenite and quenching leads to a martensitic structure which is very hard and brittle. Hardening is, therefore, usually followed by tempering, during which some of the hardness is lost and some ductility gained.

2.6 ENGINEERING METALS

This section describes the most common metallic alloys, their properties, and their usage. The alloys considered are steels, cast irons, alloys of aluminum, copper, titanium, and Ni- and Co-based superalloys.

The production volumes in tons per year in the United States of the individual classes of metals are given below:

Steels and cast irons	100 million
Aluminum alloys	36 million
Copper alloys	1 million
Ni-based alloys	< 100,000

The significance of the individual classes of metals is not entirely expressed by these tonnages. For example, the Ni-based alloys are extremely important for gas turbines and jet engines; without them modern aircraft would not exist, and there would not be such a great need for aluminum alloys.

2.6.1 Steels

Steels are available in a great number of grades for numerous applications. This great variety is obtained by using various alloying elements and various methods of processing. We will briefly discuss some of the most important classes of steels.

Carbon Steels

This class is represented by steels that do not contain any appreciable amount of intentionally added alloying elements other than C. They do contain small amounts of other elements that are residual from the ores in iron making and from scrap used in steel making. Some of these are considered impurities because they have an adverse effect on the properties of steels, mainly on its ductility and toughness. These are mainly Si, S, and P. Some other elements are beneficial. The most important of these is Mn, which is contained in practically all steels in amounts of 0.3 to 1.0%. Its effect is such that it improves hardenability. Many carbon steels are never heat-treated to attain a martensitic structure because of the cost of this treatment or because of the size of the parts. An important distinction is made between hot-rolled (*HR*) and cold-rolled (*CR*) steels: the latter are generally stronger.

Carbon steels may be considered in two large subclasses: low-carbon (nonhardenable) steels, also called mild steels, and medium- and high-carbon hardenable steels.

Low-Carbon Steels

Low-carbon steels have less than 0.2% C, very often less than 0.1% C. Optimum properties are achieved by the control of grain size in their manufacture and by the combinations of cooling rates, cold work, and annealing cycles in their production. A large proportion of these steels are made and used as thin *sheet steel* for car bodies, for appliances, and for sidings of houses. Some of this production is finished with enamel coating. A large amount is produced as tin plate for the production of cans. The thickness of sheet metal for all these applications ranges from 125 μm (0.005 in.) to 2 mm (0.080 in.).

Typical mechanical properties of these steels are YS = 240 to 700 MPa (35 to 100 ksi), UTS = 300 to 700 MPa (44 to 100 ksi), and e = 35 to 1%, depending on the amount of cold work for a typical composition of 0.12% C, 0.2 to 0.6% Mn, 0.01% Si max, 0.05% S max, 0.02% P max, 0.2% Cu max.

Another large part of mild steel production is in the form of heavy steel plates for ships and tanks, and for the structures and frames of heavy machinery. The thickness ranges from 6 mm (0.25 in) to 100 mm (4 in.). These steels must not embrittle at low environmental temperatures—a characteristic achieved by grain-size control. They are mostly used in fabrications involving welding. Therefore, they must have low carbon content in order not to produce martensitic structures in the melting and cooling cycle of the welding process.

Hardenable Carbon Steels

Hardenable carbon steels contain more than 0.3% C. They are heat-treated by quenching in water and tempering. Their hardenability is poor, and only thin sections, less than 10 mm thick, can be hardened through. Because of the rapid quenching, large temperature gradients, and differences in microstructure obtained at different points due to different cooling rates, these parts may distort during heat treatment. Often only certain portions of a machine part are quenched to the martensitic structure, with other portions becoming pearlitic.

Hot-rolled carbon steels are supplied as thicker plates, structural sections (**I** and **U** beams, rails), and round bars. They are available in low, medium, and high carbon compositions in a range of strengths with *UTS* values from 275 to 830 MPa (40 to 120 ksi) and elongations between 30% and 10%. These steels are also available as round bars in diameters from 152 mm and more. *Cold-worked carbon steels* are supplied, apart from the low-carbon, thin-sheet stock, also as cold-rolled thin plates and as cold-rolled and/or cold-drawn bars and wires of a variety of low and medium carbon compositions. They are used either as drawn and stress-relieved or as quenched, tempered, and then drawn. Depending on the carbon content and amount of cold work, *UTS* values from 410 to 1000 MPa (60 ksi to 150 ksi) are obtained.

Low-Alloy Steels

These steels contain relatively small amounts of alloying elements. They are almost always used in a heat-treated (quenched and tempered) state. The main purpose of the alloying elements is to enhance hardenability. Most of these are quenched in oil instead of in water, as the carbon steels are. Cooling rates are lower, as are the temperature gradients and differential expansions that lead to distortions and cracks in carbon steels. It is possible to obtain tempered martensite also in thicker sections with the resulting higher yield strength combined with good toughness. Some of these alloy steels contain carbides of Cr, Mo, or V, which soften at higher temperatures than the iron carbide. This leads to more efficient stress-relieving while obtaining a specific tempered hardness at higher temperatures than in carbon steels.

The Society of Automotive Engineers (SAE) and the American Iron and Steel Institute (AISI) have developed specifications and a designation code for the low-alloy steels. The code consists of four to five digits. The last two to three digits indicate the carbon content in hundredths of a percent. The first two digits indicate the type of alloying, that is, the alloying elements used and their combinations. The code also includes carbon steels, for which the first two digits are 10 and 11. The 11 grade contains sulfides of molybdenum and/or lead to enhance machinability (free machining steels). Correspondingly, 1045 is a carbon steel with 0.45% carbon. Selected grades and compositions of the SAE-AISI standard steels are given in Table 2.3.

The various degrees of hardenability for four of these alloy steels are illustrated in Fig. 2.22. The hardness at the martensitic end is the same for all four steels shown, but it differs strongly away from the end. The microstructure at the distance of 2 in. (50 mm) from the end is ferrite and pearlite in the 1040 and 2340 steels, ferrite, pearlite and bainite in the 4140 steel, and martensite plus bainite in the 4340 steel.

Correspondingly, depending on the size of the sections of the part, different combinations of strength and toughness can be obtained using these different steels. In order to illustrate the mechanical properties obtainable, the values for 4340 steel are given in the Table 2.4. They apply to bars 25 mm in diameter, normalized at 870°C, quenched from 830°C, and tested as bars with 12.83 mm diameter.

TABLE 2.3 Specifications of Selected SAE-AISI Low-Alloy Steels

Type	Grade	C	Mn	Ni	Cr	Mo	Other
C	1020	0.20	0.4				
C (f,m)	1112	0.12	0.8				0.25
Mn	1330	0.30	1.8				
3% Ni	2340	0.40	0.8	3.5			
5% Ni	2515	0.15	0.5	5.0			
NiCr	3140	0.40	0.8	1.3	0.65		
Mo	4032	0.32	0.8			0.25	
CrMo	4140	0.40	0.8		1.0	0.20	
NiCrMo	4340	0.40	0.7	1.8	0.8	0.25	
NiMo	4640	0.40	0.7	1.8		0.25	
Cr	52100	1.00	0.4		1.4		
CrV	6150	0.50	0.8		1.0		0.15 min V
Low NiCrMo	8640	0.40	1.0	6.0	0.55	0.20	

In all these steels P and S are 0.04 max, and Si is between 0.20% and 0.30%.

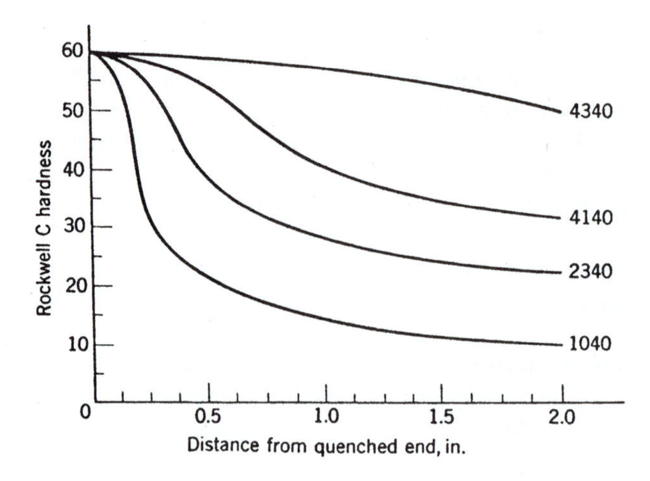

Figure 2.22
Hardenability of four steels. [BRICK GORDON PHILLIPS]

TABLE 2.4 Properties of 4340 Steel

Tempering Temperature (°C)	*YS* (ksi)	*UTS* (ksi)	*YS* (MPa)	*UTS* (MPa)	Elongation (%)
200	245	270	1690	1860	10
350	215	230	1485	1590	10
550	155	175	1070	1210	13
700	110	135	760	930	23

High-Alloy Steels

High-alloy steels contain well over 5% of alloying elements. Their effect consists of strongly changing the austenite field in the phase diagrams. In some of these steels the austenite field extends to room temperature; in other types the austenite field is strongly contracted. For example, Ni widens the field, and Cr narrows it. In combination, very different properties are obtained. We will discuss two classes: stainless steels and tool steels.

Stainless Steels

Stainless steels are of three kinds: ferritic, martensitic, and austenitic. All of them have resistance to oxidizing corrosion, which is based mainly on the contents of Cr and increases with increasing percentages of Cr. Alloying with Ni improves mechanical properties and increases resistance to neutral chloride solutions and to acids of low oxidizing capacity. Alloying with Mo improves resistance to corrosion in hot sulphuric and sulphurous acids and to neutral chlorides, including sea water.

The austenitic types are vulnerable to intergranular corrosive attack. This can be avoided by maintaining carbon contents below 0.03% or by alloying with Cb or Ti and Ta.

Selected grades and their typical compositions and properties are presented in Table 2.5. The *ferritic* grades contain more than 13% Cr, which has the effect of completely suppressing the austenitic field. They have good resistance to corrosion and are moderately strong when cold-worked. They are used for automotive trim.

TABLE 2.5 Stainless Steels

Type	Grade	C	Mn	Cr	Ni	Other	*YS* (MPa)	*UTS* (MPa)	Elongation (%)
Ferritic	430	0.15m*	1.0	16			570	640	20
Martensitic	410	0.15m	1.0	12			712	1000	20
Martensitic	440C	1.0	1.0	17		0.75Mo	1960	2030	2
Austenitic	301	0.15m	2.0	17	7		1000	1280	9
Austenitic	304	0.08m	2.0	19	10		534	783	12
Austenitic	347	0.08m	2.0	18	11	Cb,Ta	250	640	45

*m stands for max.

The *martensitic* grades have either less than 13% Cr at low carbon content, like the 410, or more than 13% Cr with high carbon content, like the 440C. In both instances they can be heated up to the austenite field and cooled to form martensite. They harden so well that in some cases it is sufficient to cool in air to obtain martensite. The strength and elongation values in Table 2.5 are given as obtained in the quenched and tempered conditions. The low-carbon grade is used for springs; the high-carbon grade, which attains high hardness, is used for instruments, cutlery, valves, and so on.

The *austenitic* grades are soft and very ductile (elongation 40–60%) in the annealed state. The high content of Ni extends the austenite field down to room temperature. However, they achieve high strength by cold work. This is especially true for the 301 grade, which has less Ni and contains some ferrite. The 347 grade is used for welding because of its low carbon content and additions of Cb and Ta, which combine with the carbon and prevent formation of chromium carbide, which would be susceptible to intergranular corrosion.

Tool Steels

Tool steels exist in many types which are useful in a variety of applications depending on whether they work at room temperature, as in punching and cold-forming dies, or at high temperatures, as in forging dies and cutting tools, and depending also on the magnitude of local loads, which are high at cutting edges and in bulk forming, and on the life expectancy and economy as related to the numbers of parts produced in a given die. In all instances high hardness is required, in some instances high "hot hardness." Depending on the type of loading, some degree of toughness is also required. All the tool steels have rather high carbon content, at least 0.6%, to achieve the high hardness required. The role of the alloying elements is to improve hardenability and, for hot-working tools, to achieve high hot hardness. Specifically, Cr is a strong carbide former and improves hardenability. Alloying with W or Mo leads to hot hardness. Both these elements form hard carbides that resist softening at low red temperatures. The hardest carbide is V_4C_3. It does not dissolve in austenite and remains unchanged through the heat-treatment cycles. It produces high wear resistance. However, it is expensive, and the tool steels alloyed with V are difficult to grind. Specifications of selected tool steels are given in Table 2.6.

All these steels, especially the high-speed steels (HSS), need unusually high austenitizing temperatures T_a. The reason for this is that the austenite field is restricted,

TABLE 2.6 Tool Steels

Grade	C(%)	Cr(%)	Mo(%)	W(%)	V(%)	T_a(°C)	T_t(°C)	Comment
W1	0.6–14					760–840	100–340	Water quenching Cold-heading dies Woodcutting tools
01	0.9	0.5		0.5		815	170–260	Oil quenching Dies for cold work
02	1.5	12.0	1.0			980	200–500	High C, high Cr Gages, blanking dies
H11	0.35	5.0	1.5		0.4	1010	540–650	Forging dies, die-casting dies
M1	0.8	4.0	8.5	1.5	1.0	1200	540	High-speed steel Cutting tools
T1	0.7	4.0		18	1.0	1290	540	High-speed steel Cutting tools
T15	1.5	4.0		12	5.0	1300	540	High-speed steel Cutting tools

and the eutectic temperature at which maximum solid solubility of C in austenite is reached is much higher than for carbon or low-alloy steels. The tempering of HSS is also different from the usual practice of low-alloy steels. The structure after quenching contains some retained austenite, which can be converted to martensite by tempering at $T_t = 540°C$. In order to achieve the full effect, repeated tempering is required.

2.6.2 Cast Irons

Cast iron is used in four basic structural types: gray, white, ductile (nodular), and malleable. They contain between 2.25% and 4.4% C and from 1.15% to 3% Si, which is the most common additional alloying element. They are used as important structural materials. Structures and frames of machine tools, presses, and rolling mills, and the housings of water turbines and of large diesel engines are made with iron castings. As an illustration of the usefulness of cast irons one may consider parts of an automobile engine: the cylinder block and head are gray iron castings, the crankshaft is a ductile iron casting.

White cast iron structure consists of grains of cementite (iron carbide) and of pearlite or ferrite, and it is obtained at fast cooling rates. At slow cooling rates, especially of cast iron with added silicon, iron carbides decompose into ferrite and graphite, resulting in *gray cast iron.* Graphite in gray iron has the form of flakes. Figure 2.23 shows an example of the microstructure of gray iron.

White iron is very hard and brittle (500 *BHN*); gray cast iron is much softer (150 to 200 *BHN*) and brittle. The lack of ductility of gray cast iron is due to the notch effects at graphite flake edges and the softness of graphite. Fracture proceeds from flake to flake. The fractured surface appears gray because most of the fracture occurred in the graphite, which represents almost all of the broken surface. The tensile strength of gray cast iron varies between 140 MPa and 415 MPa, depending on the refinement of the flake size and on the achievement of a fine pearlite matrix. This is a matter of close control of chemical composition and pouring temperatures. Gray cast iron with

140 MPa tensile strength has 580 MPa compressive strength. In compression the matrix carries the load without the detrimental effects of the notches provided by the graphite flakes.

White and gray cast irons may be alloyed with Cr, Ni, Mo, and V with similar effects as these alloys had in steels. They can also be heat-treated to obtain martensite in the matrix. Surface hardening of guideways on cast iron beds of machine tools is a common operation.

The detrimental effect of graphite flakes on the ductility and tensile strength of gray cast iron is eliminated in *ductile* or *nodular* cast iron. This is obtained by adding alloying elements such as Mg, Ce, Ca, or Na during casting of an iron with low S (0.03% max). The graphite inclusions then form as tiny balls, or spherulites, instead of flakes. The microstructure of this type of cast iron is shown in Fig. 2.24. The most common nodularizing agent is Mg. The resulting nodular cast iron typically has a composition of 0.03 to 0.08% Mg, 1.0 to 1.5% Ni, and less than 0.01% S. The tensile strength of nodular cast iron is in the range of 550 MPa, and it has 5–12% elongation.

Another type of cast iron which has considerable ductility is called *malleable iron.* Typically, it has tensile strength of 380 MPa and elongation of 12–18%. It is obtained by heat treatment of white cast iron. Heating is done below the eutectic temperature, at 950°C, usually for 12 hours. The structure of austenite plus carbide changes to austenite plus compact particles of graphite called tempered carbon. Subsequently the alloy is cooled to 760°C and then very slowly to 649°C. The slow cooling in this phase is necessary to obtain also tempered carbon as rejected from austenite, instead of iron carbide.

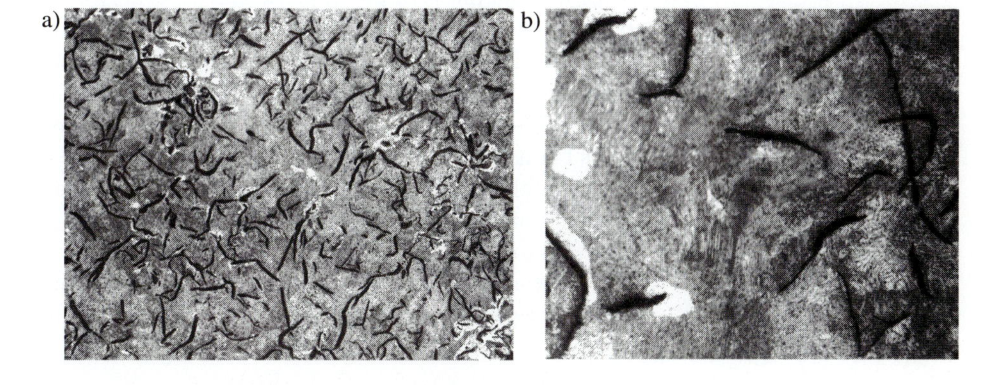

Figure 2.23
Microstructure of gray cast iron, magnification a) 30×; b) 650×. (COURTESY DR. D. A. R. KAY, MCMASTER UNIVERSITY)

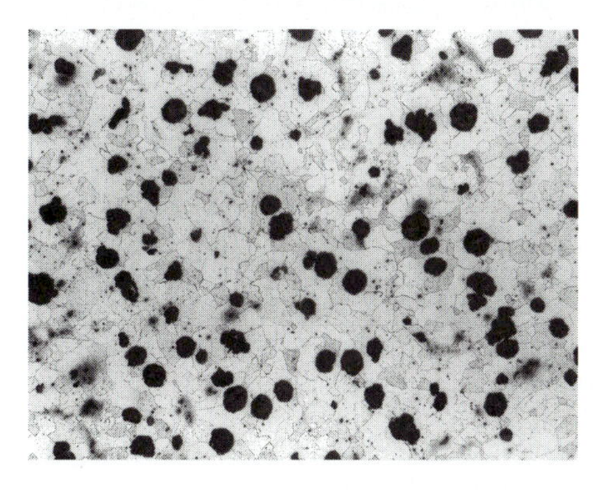

Figure 2.24
Microstructure of nodular cast iron, magnification 65×. (COURTESY DR. D. A. R. KAY, MCMASTER UNIVERSITY)

2.6.3 Aluminum Alloys and Magnesium Alloys

Pure aluminum is light (specific mass 2.7 g/cm^3, about three times lighter than iron), soft, and easy to form (YS = 28 MPa, e = 42%). It is an excellent electric conductor (electrical resistivity 1.7 times that of silver, 1.6 times that of copper). The unit cell is FCC, which leads to excellent ductility. Its modulus of elasticity is about three times lower than that of steel. It has good resistance to corrosion. Its melting temperature is $T_m = 660°C$.

Unalloyed aluminum can be considerably strengthened by cold work, and it is used for many products, from electric wire to extruded structural shapes for housing construction. However, it is usually used as an alloy. The most common alloying elements are Si, Cu, Mn, and Mg. Some of these alloys can be precipitation-hardened, and others can be solution-treated. Through combination of heat treatment and cold work, some of the alloys may be made rather strong and achieve YS = 570 MPa. Their strength/weight ratio is better than that of most steels. This, together with much easier forging and machining of these alloys than of steels, makes aluminum alloys the preferred material of aircraft structures.

Many aluminum products are made as castings: for example, the engine blocks of some automobiles and motorcycles. Many of these castings are cast in sand molds; others are die-cast. Alloying is important in these instances because, apart from improving mechanical properties, it improves the fluidity of the melt. The best castings with least porosity are obtained from high Si alloys. The Al-Si phase diagram shows an eutectic at about 12% Si, at 560°C. The eutectic alloy is used for automobile engine castings because of its superior castability; however, it is difficult to machine because of the high hardness of the Si phase and especially because some of the inclusions are in the form of the very hard SiO_2. Sintered diamond cutting tools must be used. Alloys of Al-Cu with higher Cu contents are also easily castable. A summary of selected aluminum casting alloys is presented in Table 2.7. The table shows that the eutectic Al-Si alloy A132-T65, which has the best castability, also has the highest strength (and hardness), but it is very brittle. It also has a low coefficient of thermal expansion. The 195-T6 is the classical Al-Cu alloy suitable for age hardening. The Al-Mg alloy 220-T4 combines high tensile strength with high ductility and is therefore suitable for castings subjected to high loading, even impact loading. It can be used for some aircraft parts.

Most aluminum alloy products are made from rolled bars and plates, from extruded sections, and from forgings. These wrought alloys are identified by numerical

TABLE 2.7 Aluminum Casting Alloys (properties given for sand castings)

Grade	Cu (%)	Si (%)	Mg (%)	Other (%)	Condition	*YS* (MPa)	*UTS* (MPa)	Elongation (%)
112-F	7	—	—	1.7 Zn	C	106	170	1.5
195-T6	4.5	0.8	—	—	HT	170	256	5
310-F	3.5	6.3	—	—	C	128	192	2
219-T6	3.5	6.3	—	—	HT	170	256	2
356-T7	—	—	—	—	HT	214	242	2
220-T4	—	—	10	—	HT	178	328	14
A132-T65	0.8	12	1.2	0.8 Fe 2.5 Ni	HT	306	334	0.5

Notes: C: as cast; HT: heat-treated (aged)

designations followed by a letter and number combination which specifies the type of heat treatment and/or cold work. The heat treatment code is as follows:

T3	Solution heat-treated and cold-worked. Degree of cold work is indicated by a second digit.
T4	Solution heat treatment followed by natural aging at room temperature.
T5	Artificial aging after an elevated-temperature, rapid-cooling process like casting or hot extrusion.
T6	Solution treatment followed by artificial aging.
T7	Solution treatment plus stabilization.
T8	Solution treatment, cold work, artificial aging.
T9	Solution treatment, artificial aging, cold work.
0	Annealed.
F	As fabricated (as rolled, as cast, etc.).

The cold-working code designations are the following:

H1X	Cold-worked. The higher the number, between 11 and 19, the greater the degree of cold work.
H2X	Cold-worked and annealed.
H3X	Cold-worked and stabilized. Stabilizing means heating to 30–35°C above the maximum service temperature so that softening does not occur in service.

Selected wrought aluminum alloys and their properties are presented in Table 2.8.

The difference in strength of annealed pure aluminum, condition 0, and the one with a large amount of cold work, condition H18, illustrates the significance of strengthening obtainable by cold work. The very high strength of 2014-T6 and 7075-T6 illustrates the effect of age hardening. The strength of the latter in ratio to specific mass can be compared to the strength of the 4340 steel tempered at 350°C, which has comparable ductility (*UTS* = 1900 MPa, three times heavier). On this basis we find that the 7075-T6 attains 93% of the strength of 4340.

The different alloys shown in Table 2.8 as well as other alloys omitted here differ in the ease with which they are cold-formed or hot-formed (forged), and in their corrosion resistance. Details about these properties and their relationship to microstructure as well as about precautions to be taken in processing the various alloys may be found in more specialized texts.

Magnesium is very light; its density is only two-thirds that of aluminum. It is very reactive, and in powder form it burns in air. Its melting temperature is 650°C. It corrodes in many media, such as seawater. Its unit cell is HCP and, correspondingly, it is difficult to cold form, but it can be hot extruded quite well. Also deep drawing of magnesium sheet metal is possible at temperatures of about 200°C. Forging is carried out between 300 and 400°C. There are abundant quantities of Mg in seawater; therefore, it appears to have a promising future when compared with other metals, which are mostly made from ores whose supplies are diminishing.

Magnesium is alloyed with Al, Zn, Mn, Ce, Zr, and Ag. Many of these alloys can be precipitation-hardened. As for aluminum, magnesium alloys are defined in two groups, casting and wrought alloys. Their main use is in the aerospace industries. The mechanical properties of Mg alloys are, on average, in a similar range as those of alu-

TABLE 2.8 Wrought Aluminum Alloys

Grade	Cu (%)	Si (%)	Mg (%)	Mn (%)	Other (%)	*YS* (MPa)	*UTS* (MPa)	Elongation (%)	Comment
1100-0	—	—	—	—	—	28	85	40	Pure Al
1100-H18	—	—	—	—	—	140	157	7	
3003-H18	—	—	—	1.2	—	185	206	6	
2014-0	4.4	0.8	0.4	0.8	—	100	192	18	
2014-T6	4.4	0.8	0.4	0.8	—	427	500	13	
4032-T6	0.9	12.5	1.0		0.9 Ni	327	390	9	
7075-0	1.6		2.5	0.2	5.6 Zn 0.3 Cr	107	235	17	
7075-T6	1.6		2.5	0.2	5.6 Zn 0.3 Cr	527	560	11	Strongest alloy

minum alloys, although none of them attains *UTS* over 350 MPa, while the strongest Al alloys reach *UTS* of 500 or 570 MPa.

2.6.4 Copper, Nickel, Zinc, and Their Alloys

Copper has very good thermal and electrical conductivity (second only to silver), high corrosion resistance, high ductility, and high formability. Its melting temperature is 1084°C. It crystallizes into FCC cells. It forms important alloys with a number of elements. Over thousands of years the bronzes, Cu-Sn, and brasses, Cu-Zn, have been known and used. Like the Al alloys, all copper alloys can be divided into solid-solution and multiphase alloys.

Solid solutions of Cu-Ni are obtained over the whole range of compositions. The more important alloys are those richer in Ni; they are discussed later. An important solid-solution effect is obtained with only about 0.01% Ag, which raises the softening temperature of cold-worked Cu over 100°C. This makes it possible to soft-solder it without losing the strength gained by previous cold work. The most common solid solutions are the Cu-Zn brasses, with between 30% and 35% Zn.

Of the multiphase alloys, a variety of bronzes are produced as cast alloys. Lead bronze, with 5% Zn and 5% Pb, is used for plain bearings. Lead is soluble in liquid Cu but not in the solid state, where the particles of lead in the structure act as solid lubricants. Tin forms with copper a hard, intermetallic compound denoted as the δ phase. This phase is surrounded by ductile α Cu and, in this way, a structure is obtained which mates very well with hardened steel in a typical combination of steel worm–bronze worm wheel transmission. Similarly, aluminum bronze forms a hard γ phase in the α Cu matrix.

A 2% beryllium alloy can be age-hardened to high yield strength and hardness comparable to steel alloys. It is used for springs where corrosion resistance is needed or where high electrical conductivity is required, as in electrical relays. Selected Cu alloys are presented in Table 2.9.

Nickel is similar to iron with almost the same density and melting temperature. Its unit cell is FCC. Its alloys have excellent resistance to corrosion and strength at elevated temperatures. Some of these will be discussed later in the Section on superalloys.

TABLE 2.9 Selected Copper and Nickel Alloys

Alloy Designation	Composition (%)	Condition	*YS* (MPa)	*UTS* (MPa)	Elongation (%)
Cu wrought alloys					
110, ETP Cu	99.9 Cu	A	70	228	45
		CW	285	355	6
268 Yellow brass	65 Cu, 35 Zn	A	100	330	65
		CW	430	527	8
614, Al bronze	91 Cu, 7 Al, 2 Fe	CW	285	584	35
715, Cupnickel	70 Cu, 30 Ni	CW	485	534	12
172, Beryllium bronze	98 Cu, 2 Be	PH	1000	1250	7
Cu cast alloys					
937, for bearings	80 Cu, 10 Sn, 10 Pb	AC	130	250	20
824, for tools	98 Cu, 2 Be	PH	1000	1070	1
953, for gears	89 Cu, 10 Al, 1 Fe	AC	190	530	25
		HT	300	605	15
Ni wrought alloys					
Nickel 200	99.5 Ni	A	157	460	47
		CW	655	855	8
Inconel 600	78 Ni, 15 Cr, 7 Fe	CW	890	1070	15
Dura nickel 301	94 Ni, 4.5 Al, 0.5 Ti	CW,AH	1280	1400	8
Monel K500	65 Ni, 2.8 Al	CW,AH	1100	1320	7

Notes: A: annealed; CW: cold worked; PH: precipitation-hardened; AC: as cast; HT: heat-treated; AH: age-hardened.

Here we will mention only the corrosion resistant alloys. The primary alloying elements are Cu, Cr for the solid solution alloys and Al, Ti, Si for the age hardening alloys. Selected grades are presented in Table 2.9. The ETP Cu is "electrolytic tough pitch," an almost pure copper.

Obviously, a great variety of Cu and Ni alloys exists, with a large range of properties. Some of them achieve high strength combined with good ductility: for example, 172 beryllium bronze, Dura nickel 301, and even cast Inconel 610. They combine good mechanical properties with special anticorrosive characteristics.

Zinc has corrosion resistance to atmospheric conditions, and it is above iron in the galvanic series. Therefore, one of its most important uses is as a coating applied to steel wire and sheet. If exposed to high humidity, it develops superficial corrosion, which may be prevented by an acid chromate treatment. Zinc is alloyed with Mg, Al, and Cu, which improve its corrosion resistance and strength. Zinc alloys have a rather low melting temperature of about 410°C and excellent castability; thus, they are the most common die-casting materials. Typical alloys and their properties follow in Table 2.10.

TABLE 2.10 Zinc Die-Casting Alloys

Designation	Composition (%)	*E* (GPa)	*UTS* (MPa)	Elongation (%)
BS1004, A	4.0 Al, 0.05 Mg	103	255–300	15–25
BS1004, B	4.0 Al, 0.05 Mg, 1.0 Cu	103	300–340	9–14

Zinc die castings can be coated with a variety of decorative and/or protective finishes. The most frequently used finishing is electroplating with layers of Cu, Ni, and Cr, in that sequence. The copper prevents the zinc alloy from contaminating the Ni solution, the nickel provides most of the corrosion resistance, and the chromium contributes a bright finish and further protection.

2.6.5 Titanium Alloys

Titanium has a very high strength-to-weight ratio, a high melting point, and very good corrosion resistance. It is important as a structural material for supersonic aircraft and for spacecraft. Its density is only 57% the density of iron, while the strength of Ti alloys is comparable to that of alloy steels. The melting temperature is $T_m = 1812°C$. The Ti alloys retain useful strength up to the 400–500°C range.

Titantium is allotropic. Pure Ti has the HCP cell (α Ti) up to 882°C when it changes to BCC (β Ti). Various alloying elements enlarge either the α field or the β field; thus, the basic classification is into α, β or $\alpha + \beta$ alloys. The α alloys have good resistance to oxidation at higher temperatures but poor formability, while the β alloys have good formability but oxidize easily. The properties of the $\alpha + \beta$ alloys are a compromise of both.

The most common of the single-phase α type are Ti-Al alloys, which acquire high strength by solid-solution hardening. Tin is used as an additional element in these alloys. The $\alpha + \beta$ alloys are obtained using Mn as the β stabilizer, and the highest strength is obtained in the Ti-Al-Mn alloys, which can be heat-treated by first using a solution treatment in which the $\alpha + \beta$ structure is obtained, followed by age-hardening. The effect of temperature on the strength of these various alloys is illustrated in Fig. 2.25, where they are also compared with an aluminum alloy and with a stainless steel in terms of the *YS*/density ratio.

Selected Ti alloys and their properties are presented in Table 2.11.

The heat treatment of the $\alpha + \beta$ alloys consists of a "solution" treatment still in the $\alpha + \beta$ field followed by an aging heating cycle. The mechanical properties are given in Table 2.11 at room temperature, and the yield strength is also given for 315°C. On the basis of the strength/weight ratio, these properties are equal to or better than those of 4340 or those of stainless steels.

Titanium has poor thermal conductivity, which contributes to making it difficult to machine.

TABLE 2.11 Titanium Alloys

Type	Composition	Condition	Room Temperature		Elongation	At 315°C
	(%)		*UTS* (MPa)	*YS* (MPa)	(%)	*YS* (MPa)
All α	5 Al, 2.5 Sn	A	900	864	18	470
$\alpha + \beta$	8 Mn	A	993	900	15	598
$\alpha + \beta$	4 Al, 4 Mn	A	1065	958	16	650
		HT	1166	1000	9	720
$\alpha + \beta$	6 Al, 4 V	A	900	864	11	684
		HT	1224	1070	7	756
All β	3 Al, 13 V, 11 C	HT	1296	1224	6	1044

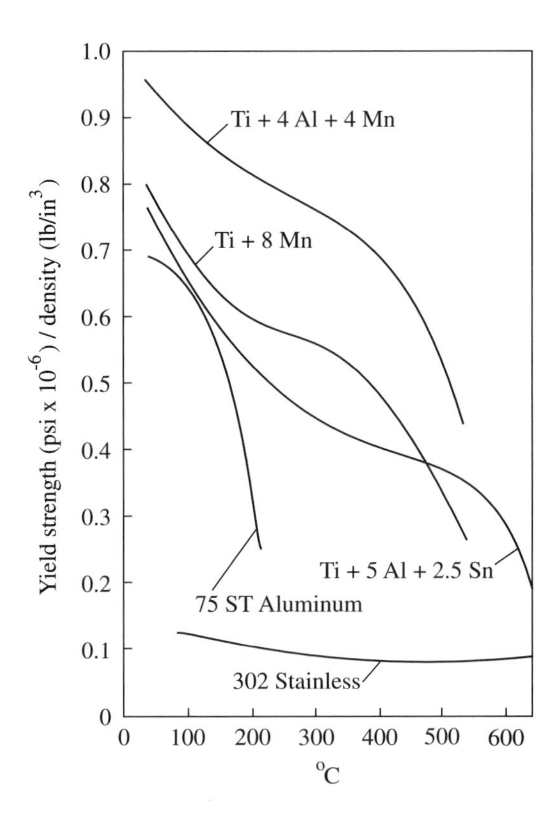

Figure 2.25
Specific yield strength (per density) of Ti alloys compared with a strong Al alloy and with type 302 austenitic stainless steel. [BRICK GORDON PHILLIPS]

2.6.6 Superalloys

The development of gas turbines for jet engines went hand in hand with the development of metals that would work as parts of the turbine (e.g., blades, discs, and shrouds) at high temperatures. The efficiency of the turbine and, correspondingly, the flight speeds which can be achieved increase with the temperature of the hot gases. During the past thirty years these temperatures have increased from 750°C to about 1000°C. This was only made possible by the development of new alloys called the superalloys. The requirements imposed on these alloys follow:

- Adequate strength at the service temperature
- Resistance to oxidation at the service temperature
- Adequate ductility both at room and at service temperature
- Ability to withstand the "thermal shocks" associated with the start-ups and shutdowns of engines
- Ability to be manufactured and processed into desired shapes

The following comments are added to explain these requirements. *Strength* cannot be characterized by the short-term tensile test values either at room temperature or at service temperature. Instead the long-term creep test and stress-rupture values apply. This involves both the amount of plastic deformation and the eventual failure.

TABLE 2.12 Superalloys

Designation	Composition							100-hr Rupture Strength (MPa)	
								At 870°C	At 980°C
Ni-Based Alloys	C (%)	Cr (%)	Al (%)	Ti (%)	Mo (%)	Co (%)	B (%)		
Inconel X	0.6	16	0.6	2.5	—	—	—	90	—
Inconel 700	0.10	17	3.3	2.2	3.0	3.0	0.008	200	—
Astroloy	0.10	15	4.2	4.0	5.0	15.5	0.03	260	—
Rene 41	0.09	19	1.5	3.1	10.0	11.0	0.005	200	—
Co-Based Alloys	C (%)	Cr (%)	Ni (%)	Mo (%)	W (%)	Fe (%)	B (%)		
HS 21	0.25	27	2.5	5.5	—	—	0.007B	85	67
HS 31	0.5	25.5	10.5	—	7.5	2	—	135	80
H 151	0.05	20	1.0	—	12.8	2	0.15 Ti	200	90
SM 302	0.85	22	—	—	10	2	9 Ta	230	103

Resistance to oxidation is also meant to be viewed as a rate of oxidation at service temperature; it is important that whatever oxide layers form on the surface do not permit penetration of oxygen to the inner metal. Even rather thin oxidated layers may significantly affect the inner metal. *Ductility* at both the service and room temperature is emphasized because many high-melting-temperature metals may be already in the brittle range at room temperature. The thermal shock aspect is involved because various alloys may undergo phase changes between the room and service temperatures, and this may strongly disturb their strength. The *manufacturability* involves the difficulties of casting related to the high melting temperatures of some metals, difficulties of rolling and forging due to the high yield strength of the superalloys at the forging temperatures and the corresponding problems of die life. From some of these alloys, parts can only be made by casting; the "lost wax" methods of investment casting are used for the production of turbine blades, for example.

The structures and heat treatment of the superalloys are beyond the scope of our text. Instead, only examples of compositions and strength data will be given. The two basic classes of superalloys are the Ni-based and the Co-based alloys. The compositions of selected alloys are given in Table 2.12, together with the rupture strengths in 100 hrs at 870°C and at 980°C.

2.6.7 Refractory Metals

Refractory metals have very high melting temperatures. Densities and melting temperatures of refractory metals are given in Table 2.13. These metals may have future use in alloys intended for very high temperature service. For example, one alloy of molybdenum with 25% W and 0.1% Zr has a strength of 533 MPa at 1315°C . However, because of difficulties with oxidation at high temperatures and difficulties with manufacturing, most of the refractory metals are not yet used for machinery parts. Their

TABLE 2.13 Refractory Metals

Metal	Ti	Cb	Mo	Ta	W	Graphite
Density (g/cm^3)	4.5	8.57	10.22	16.6	19.3	2.25
T_m (°C)	1812	2468	2610	2996	3410	3727 (sublimes)

main use is as alloying elements for steels and for superalloys, for filaments in incandescent lamps, in electronic parts, and for chemical equipment. Tungsten is used on a larger scale as electrode material in nonconsumable-electrode arc welding and as sintered carbide WC, for inserts for cutting tools.

2.7 PLASTICS

With a few exceptions (e.g., natural rubber), plastics are synthetic, man-made polymers with very large organic molecules. Their composition and structure may be purposely designed so as to obtain a great variety of materials with a range and combination of properties like strength, stiffness, plasticity, color, transparency, and so on. In general, they are light, good electric insulators, and easily formed into intricate shapes. The very large molecules of these engineering materials have some similarity with natural materials like wool and wood, and with biological polymers. Their molecular composition always includes atoms of C in combination with atoms of H, F, Cl, Br, I, O, S, N, and Si.

Plastics are a creation of this century. The first plastic was celluloid, invented in 1868 for the purpose of replacing ivory for billiard balls. It was later used also for other products like denture plates, toothbrush handles, combs, and movie film. Forty years later another plastic was invented: the phenolformaldehyde resin with the trade name Bakelite. The next was cellulose acetate, and rapid growth of the plastics industry followed with the development of hundreds of compounds and thousands of products available now. They can be found everywhere, from nylon stockings and nylon ropes to the trim of aircraft cabins, housing, and parts of household appliances, automobile body and interior parts, electrical boxes, containers, cans, plastic bags, and so on.

In discussing the mechanical properties of plastics, it is first necessary to state that they are viscoelastic. This means that their deformation in both its elastic and plastic parts under a given stress depends very much on the strain rate, in a range of what ordinarily would be considered very low strain rates (on a time scale of hours, days, months). The plastic part of this behavior may be understood as creep. In short-term tests, one may determine properties like the modulus of elasticity and tensile strength. Their values are much lower than in metals, ranging at room temperature for *UTS* between 7 and 70 MPa and for E between 140 and 4200 MPa. Both their short-term and long-term mechanical properties depend strongly on temperature. In the upper temperature ranges they may be very plastic; in the middle range, rubbery (capable of large elastic strains); and at low temperatures, quite brittle. All these properties depend on the chemical composition, length of molecules, and structure.

Many other mechanical properties of plastics are inferior to those of metals. Their thermal expansion is large, and some of them swell as they absorb moisture. Most of them degrade by exposure to ultraviolet radiation; thus, for example, plastic pipes can be used for underground plumbing but not for gutters. Their operating temperature range is limited mostly to about 200–300°C, except for recent developments of grades with service temperatures up to 400°C.

On the positive side, apart from the overwhelming advantage of easy plastic forming, it is necessary to consider the strength parameters in the ratio to weight. In this respect, the strength of some plastics comes closer to that of metals. It is, however, mainly their use in composites (see Sec. 2.9) which leads to the creation of materials with very high strength/weight ratios, to be used in structures of modern aircraft and spacecraft.

All plastics can be classified in two large groups. The first group is that of *thermoplastics,* which can be repeatedly softened and made plastic by heating and brought back into a usefully stiff state by returning to room temperature (or to a low service temperature). The second group is that of *thermosetting plastics,* which, once formed by heating, generate so many bonds internally that they cannot anymore be made plastic.

2.7.1 Polymerization Methods, Bonding, and Structures

Polymerization is the process in which rather simple small molecules called *mers* (or monomers) are bonded into large molecules, the *polymers.* One such process is *addition polymerization.* In this process a double carbon bond of an unsaturated hydrocarbon monomer opens up and becomes available for covalent bonds with another monomer. This reaction is very fast and chainlike, and polymer chains containing thousands of mers can be generated in seconds or minutes. Many chainlike long molecules are created simultaneously in this process, providing the entangled, "spaghetti" structure of an amorphous thermoplastic. An example of the basic reaction is shown in Fig. 2.26. The original material, the hydrocarbon gas ethane C_2H_6 with all C bonds saturated (Fig. 2.26a), transforms into the unsaturated ethylene C_2H_4 (Fig. 2.26b), which shares two electrons between the C atoms in a double bond. Under heat and pressure and with the use of a catalyst, called the initiator, the double bond can be opened (Fig. 2.26c), and this form of monomer is suitable to generate the chainlike structure of the polymer (Fig. 2.26d).

In this case the ethylene mer is bifunctional because the C=C bond can react with two adjacent monomers. Similarly the N−H bond in an amine, the O−H in an alcohol, and the C−OH bond of an acid are *bifunctional.* In other materials the bonds may be *trifunctional* or multifunctional, and in polymerization they generate a *network of mers* bonded in all directions, creating a large volume of a single molecule that can no longer be released. This is the nature of the thermosetting plastics.

```
a)              b)             c)              d)

   H   H          H   H          H   H            H   H   H   H   H   H
   |   |          |   |          |   |          / |   |   |   |   |   | \
H - C - C - H      C = C        - C - C -       ( - C - C - C - C - C - C - )
   |   |          |   |          |   |          \ |   |   |   |   |   | /
   H   H          H   H          H   H            H   H   H   H   H   H    n
```

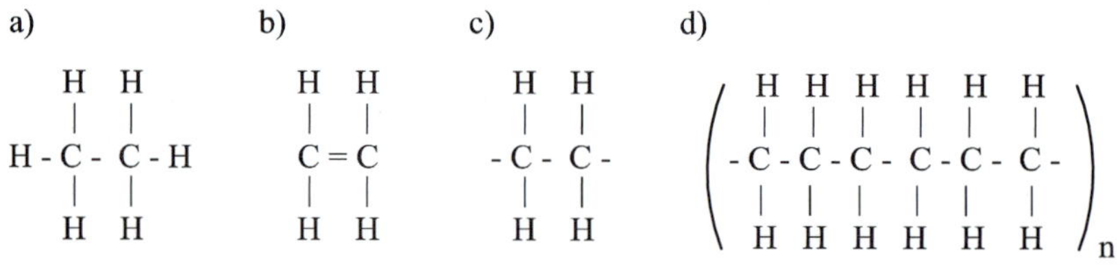

Figure 2.26
Addition polymerization of polyethylene.

The basic structure of an *amorphous* thermoplastic polymer is the above-mentioned "spaghetti" entanglement of long molecules, where every molecule is bonded internally by strong covalent bonds, but only rather weak forces of the Van der Waals type, or of the hydrogen bond type, or of ionic bonds act between the individual molecules. In between this type of structure and the tight network of the thermosets there are thermoplastic materials with *branches* from the main chains of the molecules or, in synthetic rubbers, with *crosslinks* that are provided by the addition of sulphur (vulcanization).

The type of the mers and the type of structure of the plastic determine its properties. Many thermoplastic materials have an amorphous structure with a haphazard arrangement of molecules. These may be very long and more or less aligned in strong and tough plastics. Materials with coiled or branched molecules are more flexible and ductile. Heat increases the atomic movements, separates the molecules, and softens the material. With increasing temperature the viscosity of the material decreases. There is no distinct transition between the solid and the liquid state. Some of the thermoplastics may develop structures that are in some parts *crystalline.* Although these are not equal to the highly organized atomic lattices of metallic crystals, the molecule chains are well aligned and densely held together. Fig. 2.27 shows diagrammatically the various types of organization of molecular chains: *(a)* the most random "spaghetti-type" arrangement of simple chains; *(b)* a random arrangement of chains with branches; *(c)* and *(d)* partially crystalline structures, and *(e)* the network of a thermoset. The crystalline thermoplastics have more clearly defined melting points; with an increasing degree of crystallinity, they are stronger and tougher.

Polymers exhibit two distinctly different types of mechanical behavior. When cooled they pass through the so-called *glass-transition temperature,* T_g. This temperature may be above room temperature for some plastics like the polystyrene with T_g = 52°C, but for most materials it is below room temperature. Below T_g they are stiff and brittle. It is believed that below T_g molecular motions cease, and only atomic vibratory movements are possible.

The thermosets are affected by temperature quite differently than either the amorphous or crystalline thermoplastics. Their polymerization in practical manufacturing is

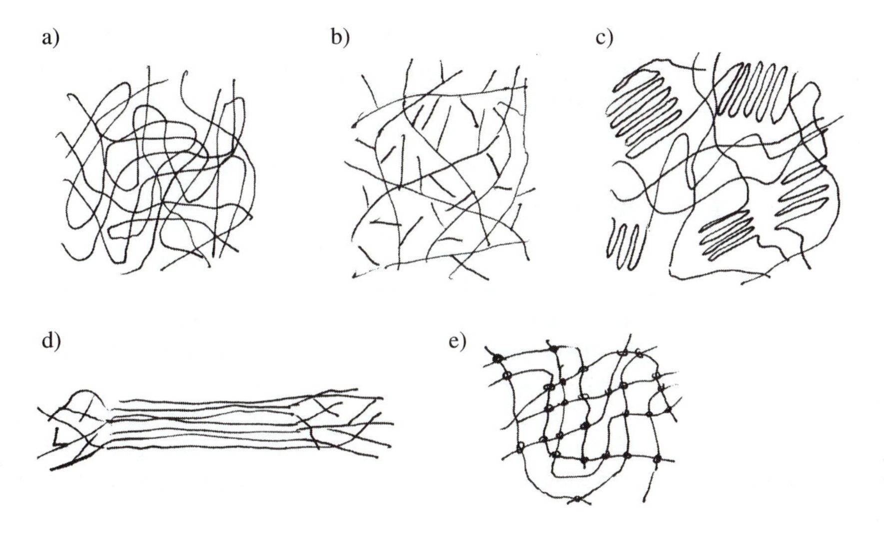

Figure 2.27
Various types of molecular chains.

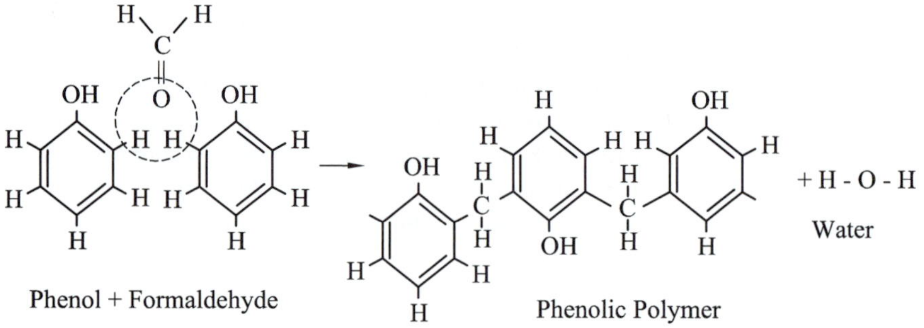

Figure 2.28
Condensation polymerization of phenol formaldehyde.

accomplished in two stages. The first stage is carried out in a primary processor in which only partial polymerization is achieved and a degree of plasticity is retained. The second stage and the completion of polymerization are carried out in the mold that gives the part its final shape and "sets" it. The whole part is then essentially a single, giant molecule with strong, primary, covalent links throughout the structure. This material will no longer soften with increased temperature. If it is heated too much or too long, it will decompose and char. Thermosets generally have better dimensional stability, heat resistance, chemical resistance, and electrical insulation properties than thermoplastics.

The second basic plastics manufacturing process is *condensation polymerization.* It proceeds in steps in which two or more molecules are joined into a larger molecule, and it is slower than addition polymerization. In these structure-forming steps, a by-product, mostly water, is condensed out. The product may be a chainlike molecule of thermoplastics or a network structure of thermosets. An example of condensation polymerization is given in Fig. 2.28. This example illustrates the formation of the thermosetting phenolformaldehyde.

So far, we have discussed polymers consisting of chains or networks of one type of mer. It is also possible to synthesize polymers consisting of different mers. An example is the ethylene-vinylchloride *copolymer*:

$$-(CH_2)_2 - C_2H_3Cl - (CH_2)_2 - C_2H_3Cl -$$

This copolymer has an alternating arrangement of the two constituting mers. Other possible arrangements are random; block type, with alternating short chains of the two polymers; and graft type, in which branches of one polymer are attached to the main chain of the other polymer.

2.7.2 Additives

Most plastics contain, apart from the large polymeric molecules, additional molecules or particles for the purpose of obtaining some particular characteristics. These additives are mixed with the basic plastic during the early manufacturing stages. They may serve as fillers, pigments or dyes, plasticizers, stabilizers, hardeners, lubricants, flame retardants, or antistatic agents.

Fillers affect the mechanical properties of the material. Some are added just to provide bulk and many others to increase strength. Fillers are used for most thermosets and for some thermoplastics. Common fillers are inorganic materials like calcium carbonate and asbestos, or fabric cuttings, cord, and wood chips. Glass or carbon fibers are added for exceptional strength; they will be discussed separately later under composites. *Pigments* are inorganic materials that are dispersed as small particles throughout the polymer. They add color and decrease the transparency of the plastic. *Dyes* are organic materials that dissolve in the material, color it, and have little effect on its transparency. *Plasticizers* are added to improve plasticity and flexibility of the material by weakening the secondary bonds between chains of large molecules. *Lubricants* are added to reduce the friction between the polymer and the dies in extruding or molding plastic parts. *Hardeners* are chemical substances which promote cross-linking between linear polymer molecules. All plastics burn, and particular fillers or other chemical agents act as *flame retardants. Antistatic agents* are added to reduce the build-up of electrostatic charge on the surface of plastics.

2.7.3 Thermoplastics

Examples of selected thermoplastics are presented in Fig. 2.29. The *polyethylenes* exist in a range of densities depending on the amount of branching in their structure. Those with a high degree of branching have a low degree of crystallinity, about 50%, and low density, less than 0.925 g/cm^3. Other grades may have as much as 90% crystallinity and densities above 0.94 g/cm^3. The former grades soften at 75°C; the latter ones remain hard and rigid up to 110°C, which allows them to be used in boiling water. The poly-

```
  H  H                       H  H
  |  |                       |  |
( -C - C- )   Polyethylene  ( -C - C- )    Polypropylene
  |  |                       |  |
  H  H   n                   H  CH3  n

  H  H                       Fl Fl
  |  |                       |  |
( -C - C- )   Polyvinyl     ( -C - C- )    Polytetrafluorethylene
  |  |        chloride       |  |              (PTFE)
  H  Cl  n    (PVC)          Fl Fl   n

  H  H                       H  CH3
  |  |                       |  |
( -C - C- )   Polystyrene   ( -C - C- )
  |  |                       |  |
  H  C6H5  n                 H  C=O        Acrylics
                                |
                                O
                                |
                               CH3

     H  H
     |  |
HO - C - C - O (H     HO) C - [benzene ring] - C - OH     Polyester
     |  |                 ||                   ||
     H  H                 O                    O
```

Figure 2.29
Compositions of selected thermoplastics.

ethylenes are the most common group of plastics, representing about 25% of the total market. They are used as clear sheet, in packaging as bags and bottles, for housewares, piping, and ducts.

The next most common type of plastics are the *polyvinyl chlorides.* Their share of the market is about 17%. They are classified as either rigid or flexible. The former grades are hard and tough with excellent electrical properties and good weather resistance and can be variously colored. They are used for floor and wall coverings, tubing, and as safety glass. The flexible grades are used for packaging and in fabrics for upholstery.

Another large group, with about 16% of the market, are the *polystyrenes.* They are low-cost materials, clear, rigid and brittle. They are used in containers and dinnerware, and for foams in packaging and heat insulating.

All the other types of thermoplastics are used in lesser quantities than the above three types, and their market share individually is less than 5%. The *polyesters* are rather strong and tough and have good dimensional stability, electrical properties, and chemical resistance. They exist also as thermosets. They are used as textile fibers, for magnetic tapes, and for films. The strongest plastics are in the *nylon* group. They are used as textile fibers, ropes, gears, and other machine parts. The *polypropylenes* are tough and ductile, lightweight, low-cost materials. They can be electroplated. They are very hard and glossy, and they are available in bright, transparent colors. They are used for windows, lenses, decoration, and bottles. The *fluoroplastics*, of which the most common type is *PTFE,* have excellent electrical chemical and heat resistance, low friction, and although they are rather soft, they have high resistance to friction wear but not to scratching. They are used for chemical wares, seals, gaskets, bearings, electrical insulation, and as nonstick coatings on pots and pans. Other popular types of thermoplastics are the *ABS* group (acrylonitrile-butadiene-styrene copolymer): *acetals, cellulosics, polycarbonates,* and *polysulphones.*

2.7.4 Thermosets

Examples of selected thermosetting plastics are given in Fig. 2.30. The *phenolic resins*, like phenol formaldehyde, are the most popular and common plastics. They have been used for many decades (Bakelite) for electrical equipment, and for housings, cabinets, and various containers. Their popularity is due to their low cost and excellent heat and electrical resistance. They are hard and rigid. They are always made with considerable filler content: wood floor, glass fibers, cotton flock, or asbestos. They are also made laminated with sheets of paper, cloth, or wood veneer, and molded as sheet, plate, channels, bars, and tubes. The *urea formaldehyde* is also a hard and rigid material with good heat and electrical resistance, as well as resistance to organic commercial solvents, weak acids, and weak alkalies. It can be colored to pastel shades. It is used for containers, cabinets, toilet seats, and as a low-cost surfacing for metal or wood. *Melamine resins* have properties similar to those of the phenolic and urea resins but over a greater range of temperatures. Their resistance to acids and alkalies is also stronger. They are used as tableware and in electric switchgear. The *polyester resins* (alkyds) have low moisture absorption and good resistance to weather, chemicals, and flame. A typical application of this resin, laminated with glass fibers, is in the construction of boat hulls and car bodies. The *epoxy* resins have good physical and mechanical properties and adhere well to metals and glass. They are resistant to temperatures up to 175°C. They are used widely as laminates in stamping dies for car body parts and for jigs and fix-

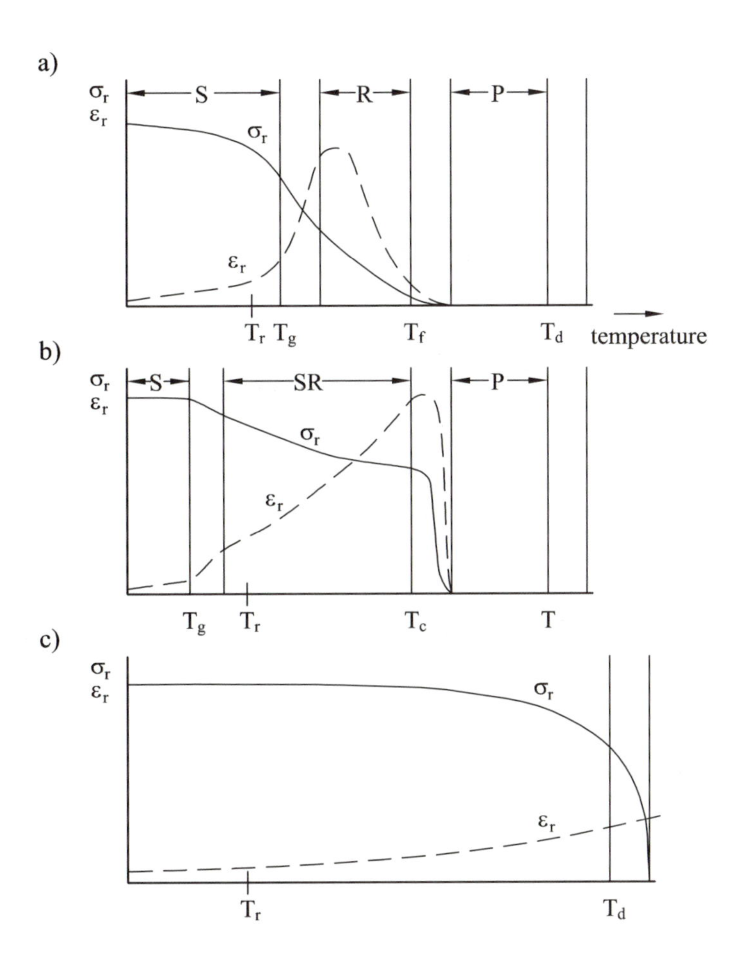

Figure 2.31
Variation of mechanical properties of plastics with temperature: a) This graph corresponds to amorphous thermoplasts (e.g., PVC).
b) This graph represents thermoplastic material with considerable crystallinity (e.g., dense PP).
c) This graph expresses characteristics of thermosets.

character of this latter effect by means of the graphs in Fig. 2.31. The graph in Fig. 2.31a corresponds to amorphous thermoplasts (e.g., PVC). The horizontal axis of the graph marks temperature, and the vertical axis represents tensile strength at rupture σ_r and strain at rupture ε_r (ductility). In this case the glass-transition temperature T_g is above room temperature T_r. Below T_g the material behaves like a rather stiff solid *S*. Some thermoplasts are very brittle below T_g, like polystyrene. Others, like PVC, retain some ductility also below room temperature because the motion of branches or of short chains in the structure is already released. The strength σ_r decreases slowly, and ductility ε_r increases with temperature in the *S* range. Over a certain range of temperatures just above T_g, σ_r drops strongly, and ductility ε_r increases considerably and reaches a maximum. Between T_g and the flow temperature T_f the material behaves in a viscoelastic manner. Its modulus of elasticity is low but it can sustain large strains. With increasing temperature the material softens and over a range at T_f becomes viscous plastic *P*. With further temperature rise it reaches the range of decomposition T_d. The graph in Fig. 2.31b represents a thermoplastic material that has considerable crystallinity (e.g., a dense polypropylene). Below T_g the material is rather brittle. Above T_g the amorphous parts of the material soften, but the crystalline parts remain rather strong. The material is also tough because of the softer parts, and it is indicated as solid rubbery, *SR*. Above the crystallization temperature T_c it becomes viscous plastic. Figure 2.31c expresses the characteristics of thermosets. These materials behave as rigid, rather brittle solids up to the decomposition temperature range T_d.

Another way of illustrating the effect of temperature on the properties of various plastics is used in the graph in Fig. 2.32. It gives the modulus of elasticity E as a function of temperature and shows how strongly it is affected. This contrasts with the behavior of metals, which retain their values of E over a wide range of temperatures above room temperature. We can see that there are large differences in softening of the various types of thermoplastics depending on their packing, branching, crystallinity.

The strain-versus-temperature behavior of a plastic is expressed in the graph of Fig. 2.33a. At the moment of application of the load an instantaneous elastic strain occurs, which is followed by a combination of viscous plastic flow and retarded elastic strain. On unloading there is an instantaneous elastic recovery, a retarded elastic recovery, and a permanent plastic set. This type of behavior can be modeled as shown in Fig. 2.33b with the spring k_1 representing the instant elasticity, the set of spring k_2 and dashpot c_2 in parallel representing the elastic viscous part, and the dashpot c_3 representing the plastic part. An example of a strain-versus-time graph obtained for polycarbonate at 23°C is given in Fig. 2.34, which shows that the viscous part of the strain develops over rather long time periods. It is therefore quite illustrative to consider strains obtained in short-time tests and use the standard parameters of modulus of elasticity E, ultimate strength *UTS*, and elongation e to characterize the individual materials. These parameters are given in Table 2.14 for selected plastics. The specific mass $\rho(\mathrm{g/cm^3})$ is also given. The hardness R is given as obtained on a special scale for plastics. It has a comparative value.

The table shows that the thermosets are as strong as the strongest thermoplasts, with the exception of nylon. The strength of nylon can be compared with that of cold-

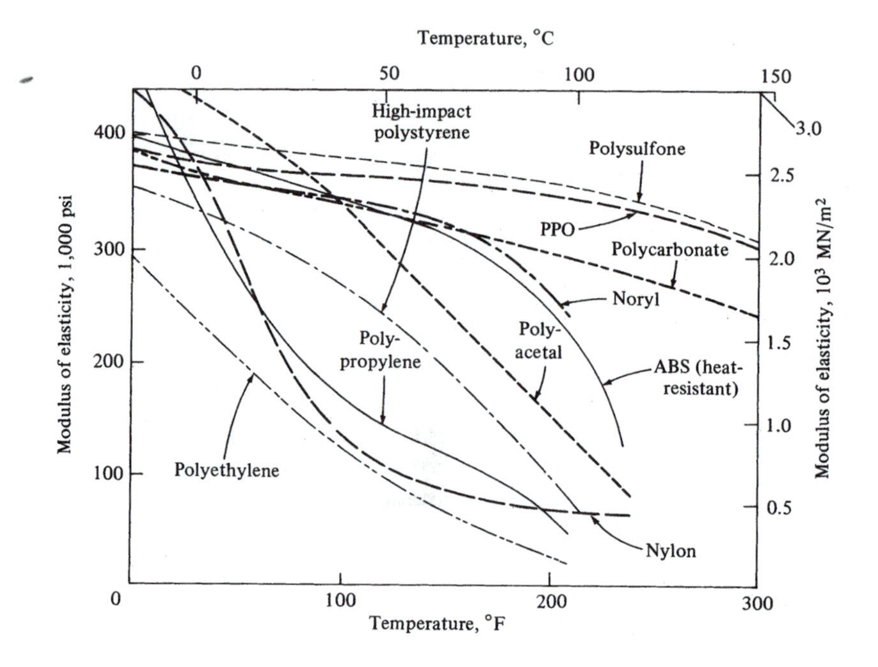

Figure 2.32
Modulus of elasticity of various resins as a function of temperature. [FLINN TROJAN]

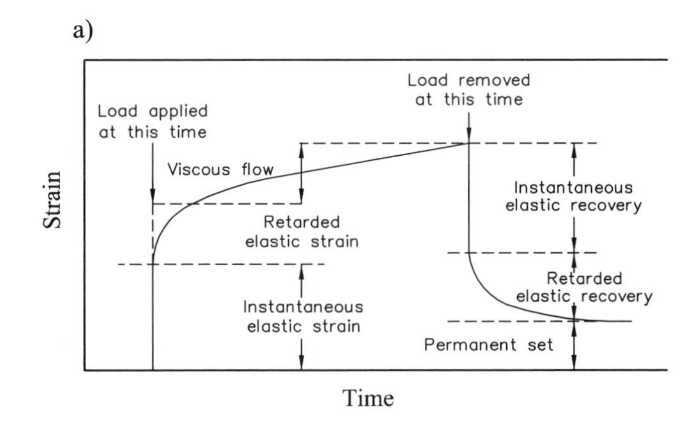

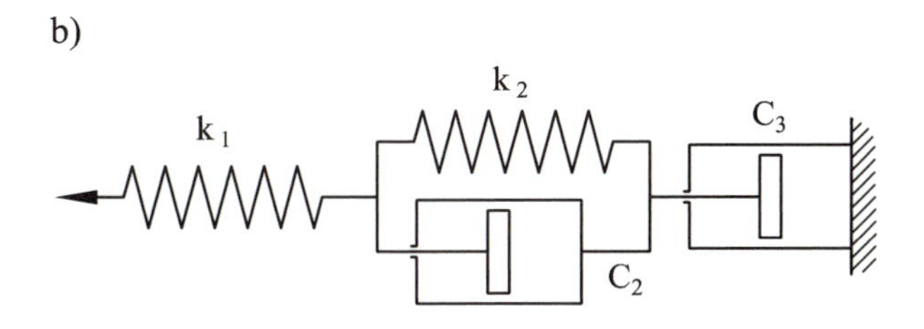

Figure 2.33
Viscoelastic and plastic nature of strains.

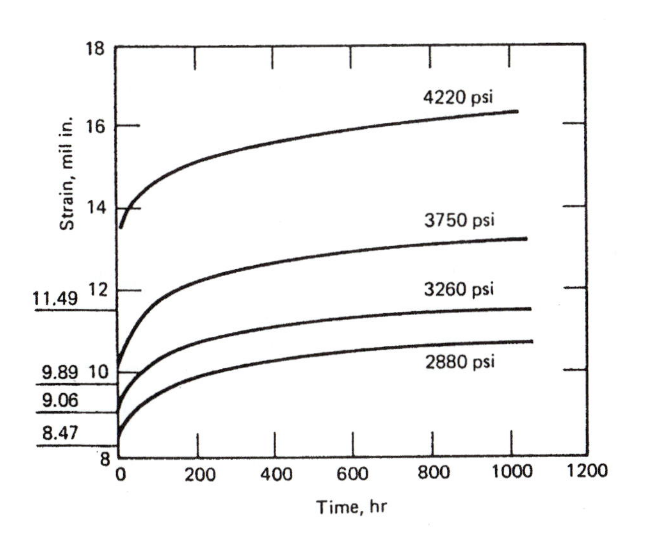

Figure 2.34
Strain versus time for a polycarbonate at 23°C. [BROWN]

drawn mild steel (UTS = 355 MPa psi) on the basis of the strength-to-weight ratio, with ρ = 1.1 g/cm^3 and ρ = 7.8 g/cm^3 for steel. On this basis, nylon is actually 1.66 times stronger. The ductility of low-density polyethylene is as high as that of natural rubber, while it reaches 3000% for butadiene styrene. Thermosets have practically no ductility at all. Table 2.14 expresses in a brief summary the wide range of mechanical properties that can be obtained from the various types of plastics, and it illustrates the great choice of materials available to the designer of parts made of plastics.

TABLE 2.14 Room-Temperature, Short-Time Parameters of Selected Plastics

Type of Plastic	E (MPa)	R	UTS (MPa)	Elongation (%)	ρ (g/cm^3)
Thermoplastics					
Polyethylene, low-density	180	10	7–14	90–800	0.92
Polyethylene, high-density	855	40	30–45	15–100	0.96
Polystyrene	3200	75	35–70	1–2	1.05
PVC	2850	110	35–60	2–30	1.40
PTFE	425	70	14–35	100–350	2.13
Polyamides (nylon)	3000	120	70–85	60	1.10
Polyesters	2400	120	40–70	300	1.30
Thermosets					
Phenolics	1,000	115	500–55	0	1.40
Epoxies	1,000	90	35–85	0	1.10
Elastomers		Durometer (A)			
Natural rubber, Isoprene		20–100	18–25	1,000	
Butadiene styrene		30–95	1.5–2.3	3,000	

2.8 CERAMICS

Ceramics were the first man-made materials in the form of fired clay used for cooking pots. As engineering materials they include concrete, glasses, brick, porcelanic insulators, and also materials for cutting tools. Chemically, they are *oxides* and *silicates*, and nonmetallic *carbides, nitrides*, and *graphite.* They have strict crystalline structures similar to those of metals; however, their atoms are held together not by metallic bonds but by ionic or covalent bonds. As examples of materials with ionic bonds, we have two of the most important ceramic materials: silica SiO_2 and alumina Al_2O_3. For an example of covalent bonding, consider the hardest material known, the diamond, which is also important as a cutting material. Ceramics are generally *hard* and completely *brittle.* They do not exhibit any plasticity and fail by brittle fracture. Not only pure compounds but also *solid solutions* are formed in ceramic materials, for example between MgO and FeO, and transformations of structures also exist, depending on temperatures.

In manufacturing engineering ceramics are significant as *linings of furnaces* in the primary manufacture of metals, as *sands in casting*, and as *cutting materials.* The former applications are discussed in Chapter 3 and the latter in Chapter 7 and in Chapter 8. Ceramics also have to be considered as important workpiece materials in the form of *glass* and *carbon* fibers in *composite materials*, discussed in the following section.

The most important characteristic of ceramics, their hardness, is expressed in Table 2.15 for a few selected materials. It is measured in Vickers hardness numbers (*VHN*), which are obtained by impressing a diamond pyramid into the material and measuring the ratio of the force to the surface of the impression in kgf/mm^2. In the lower range, between 250 and 600, these numbers coincide well with the Brinell hardness numbers.

Glass is actually noncrystalline. It is obtained by fusing SiO_2 sand in an electric arc and cooling rapidly. The basic cell, the tetrahedron SiO_4, is conserved in this

TABLE 2.15 Hardness of Some Ceramics

Material	Microhardness (*VHN*)
Hardened tool steel	600
SiO_2	1250
Al_2O_3	2800
SiC	3500
B_4C	3700
Diamond	~8000

process, but its regular crystalline arrangement as found in quartz sand is lost and replaced by a random network. Glass does not have a distinct melting point; on heating, it becomes viscous with gradually decreasing viscosity. In this condition it is easily formed and blown into shapes. The characteristic temperatures for glass are defined by viscosity measured in poise. The highest such point is the temperature common for forming the glass. It is called the working point, and it is such that viscosity is 10^4 poise. On cooling, the viscosity increases to the softening point at $10^{7.6}$ poise, the annealing point at $10^{13.4}$ poise, and the strain point at $10^{14.6}$ poise. At this point glass becomes rigid. Glass exists in many grades obtained by additions of other oxides to the basic material of SiO_2. Pure silica has a very high softening point of 1667°C and is therefore difficult to process. Soda-lime glass with 26% of additives like Na_2O, K_2O, and Al_2O_3, has a softening point of only 696°C and is much easier to form. Lead-alkali glass with 35% PbO, which also has a low softening point, has a very high refraction index and is therefore used as "crystal glass" for optics, as well as for dinnerware and decorative glass. Pure silica is thermal-shock-resistant, as is the easier-to-form borosilicate (Pyrex), which contains 12.9% B_2O_3, has a softening point of 820°C, and is used for cookware.

Carbon exists in several forms. *Polygranular carbon* is most commonly encountered. It is used for electrodes for electric furnaces, for linings of furnaces, in chemical equipment, nuclear reactors, and so on. It is produced by combining carbon powder with tar, which acts as a binder. In this condition the material can be molded into shapes, and it is then converted to all carbon by prolonged heating (coking) up to 1000°C in an inert atmosphere. Impregnating and heating is repeated several times to reduce porosity. Polygranular carbon is either used in the fired state, or it is additionally graphitized at temperatures between 2500 and 3000°C. *Vitreous carbon* is obtained from thermosetting polymers like polycondensates of phenol with formaldehyde, which are heated to the decomposition point in an inert atmosphere. It is used in melting crucibles, protective pipes, thermocouples, and medical implants. *Pyrocarbon* is made by decomposition of hydrocarbon gases passed over a hot surface, where a layer of carbon is deposited. This process is used to coat nuclear fuel particles. The jet pipes of rocket motors are also lined with pyrocarbon. *Carbon fibers* are made either by coking fibers of polymers like polyacrylonitrile or from coal tar pitch by melt-spinning and drawing of fibers. Carbon fibers have very high strength and stiffness in their longitudinal direction. We will discuss them in more detail in the next section.

The various types of carbon differ in their structures, mainly in the degree in which large graphite crystals are formed. This depends on the original material from which carbon was produced. Polygranular carbon does not contain large graphite crystals; vitreous carbon contains entangled narrow and thin graphite layers; and carbon fiber contains larger crystals axially oriented (see Fig. 2.35). The graphite layers are all oriented axially, and they have a high degree of continuity, especially in the outer layers outside of the core. This highly organized structure is responsible for the high strength and stiffness of carbon fibers. The tensile strength of polygranular carbon is about 22–72 MPa; for pyrocarbon and vitreous carbon it is about 100 MPa; and for carbon fiber it is 2100–4300 MPa.

The crystal cell of graphite is shown in Fig. 2.36. This layered structure has strong covalent bonds within the sheets but weaker, semimetallic bonds between the sheets. Because the sheets slide easily, it is an excellent lubricant and is used for seals and bearings, commutator brushes, and so on. Carbon and graphite materials are inert to chemical action, have high strength at high temperatures, excellent resistance to thermal shock, a high sublimating point (above 3000°C), low coefficient of friction, and good heat and electrical conductivity. The super-hard form of carbon is diamond. Its structure is shown in Fig. 2.37. There is perfect covalent bonding throughout the structure, which creates the high hardness. Diamond is a natural material, but it can be synthesized under high pressure and high temperature. Relatively small grains are obtained that can be used in grinding wheels or sintered into a polycrystalline form for cutting tools.

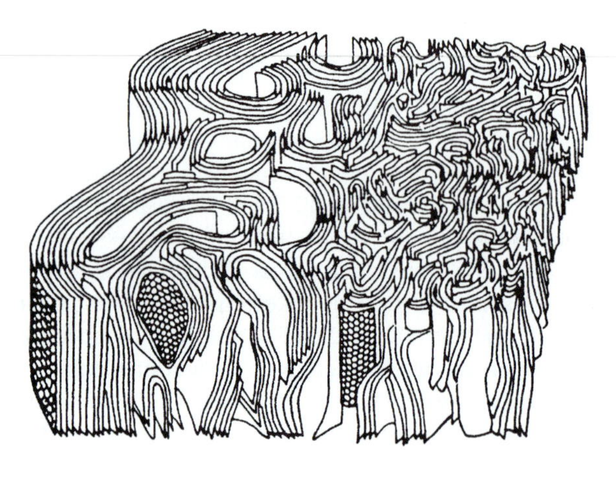

Figure 2.35
Structural model of a polyacrylonitrile (PAN) carbon fiber. [VDI]

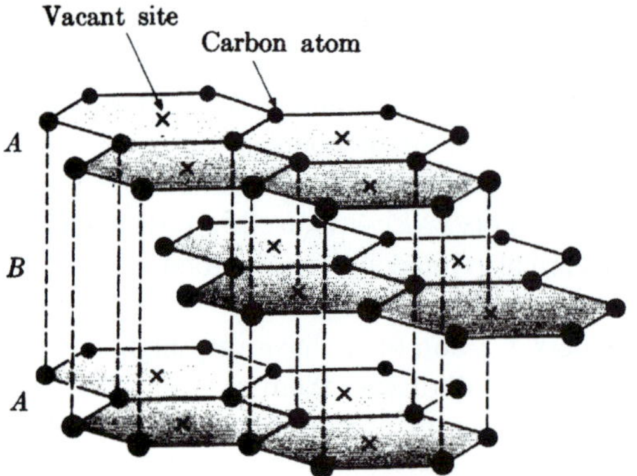

Figure 2.36
Crystal structure of graphite. [VAN VLACK]

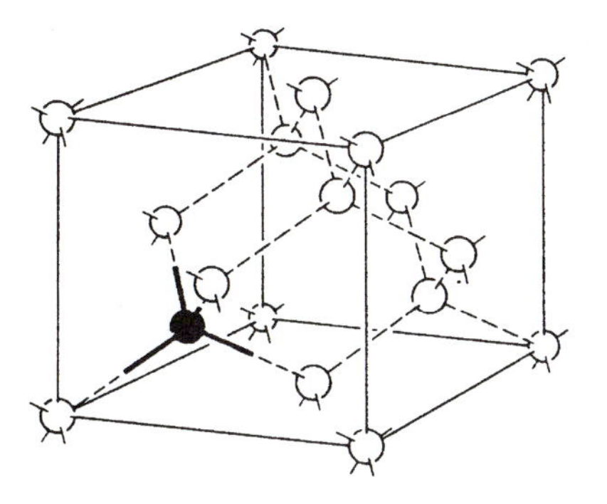

Figure 2.37
Crystal structure of diamond. [VAN VLACK]

2.9 COMPOSITE MATERIALS

Many materials in nature and in engineering may be classified as composites because they consist of a combination of particles, fibers or laminations of different materials. Wood is one of the natural composites, with its combination of fibers, tubular cells, and radial vascular rays. Among engineering materials, concrete is a mixture of cement and sand, and reinforced concrete is a purposeful design of steel-rod structure within the concrete. Other examples involve glass-coated sheet steel, steel-wire-reinforced rubber tires, and many other materials in which the properties of the components combine into the properties of the composite. We have already discussed how the lamellar structure of ferrite, which consists of hard, brittle carbide and soft, ductile ferrite, results in a good combination of strength and ductility. We have seen how the form of graphite in cast iron, whether flakelike or globular, strongly affects the properties of the different kinds of cast irons.

In this section we concentrate mainly on the modern composites that are being developed, which surpass previous materials in stiffness-to-weight and strength-to-weight ratios and, in some instances, in ease of processing. They are intended mainly for use in supersonic aircraft and in spacecraft, although they are also used in high-performance sport equipment, as parts of automobile transmissions, and for automobile and boat bodies. In most of the materials the component that gives stiffness and strength is in the form of thin fibers held together by a softer and ductile matrix of aluminum or, mainly, of plastics. The most common fibers are glass and graphite, but other materials, such as boron deposited on tungsten wire, silicon carbide on tungsten wire, and alumina, have been used.

The most widely used composites for which the production technology has been well developed are discussed below in three groups: glass-reinforced plastics (GRP), carbon-reinforced plastics (CRP), and carbon-reinforced carbon (CRC). In all these materials the resulting properties depend on the strength and stiffness of the fibers, on the strength-versus-temperature behavior of the matrix material, and on the cohesiveness between matrix and fiber. The fibers are much stiffer than the matrix and thus carry almost all the load. This does not cause problems in materials with filamentous (endless) fibers. Some materials are filled with chopped fibers only about 6–25 mm long, or shorter. In these materials the matrix has to transfer stresses between the fibers. The GRP and CRP materials are designed for use at moderate temperatures in parts of or over the whole range between −70°C and about 200°C. The CRC materials can be used for long-term time exposures to temperatures up to about 400°C, above which the carbon oxidizes. Impregnation with zinc phosphate extends this to about 550°C, and still higher temperatures may be reached by impregnation with silicon carbide. Short-term exposures up to 2000°C are possible.

The fibers of glass and of carbon are supplied in bundles consisting of 2000–4000 fibers with diameters from 5 to 20 μm. Filaments of these fibers are then bonded by the polymer matrix into tapes with uniaxial orientation of fibers, and sheets with two-dimensional orientations. These may be combined in multilayered fabrics. Composites filled with endless-filament fibers achieve very high strength and stiffness either in one direction or two directions, depending on the design of the fabric.

Other types of composites contain short fibers randomly oriented. This type of reinforcement results in a moderate increase of the strength, about 25–40% compared with the strength of the plastic used for the matrix, and a substantial increase of the modulus of elasticity (stiffness), by about 200–300%. These materials are less expensive than those with filaments. They are produced in the form of mats of various thicknesses that can be compression-molded into various shapes.

Figure 2.38 shows how tapes, sheets, and mats can be combined in multiple layers. Combining layers with different filament orientations provides different stiffness and strength values in different directions.

Techniques also exist for orienting the short, chopped fibers in composites, which then have improved strength and stiffness. Two-dimensional and three-dimensional structures are obtainable with various mixes of glass and graphite fibers. As the percentage of graphite increases, the modulus of elasticity increases, but the impact strength decreases. Chopped-fiber materials fall far short of achieving the strength and stiffness values of endless-fiber materials. They do however, offer some considerable advantages for certain applications: they can be easily molded into complex shapes, they can easily be made as hybrids (glass/carbon) for use in components where high impact strength is desired, and they can be made from waste of endless-fiber materials. Typical applications are covers, flaps, and compact, thick-walled parts.

The best parameters are achieved in materials with endless fibers. Some of these are presented in Table 2.16, where strength and stiffness are expressed both in absolute values (in ksi and in MPa) and in ratios to their densities. For comparison, aluminum and titanium alloys and high-strength steel are also included. The composites considered in this table have a volume fraction of fibers $V_f = 60\%$. The polymers used are epoxy resins. Note that, generally, even in absolute values, the strength of *GRP* and *CRP* long-fiber materials is comparable to that of alloy steels and Ti alloys. In the strength-to-weight ratio, GRP is slightly better than Ti alloys and high-strength steel, and *CRP* materials are two to four times better than the 4340 steel. The stiffness-to-weight ratio of the glass-reinforced plastics is only about 60% of that for Al, Ti, and 4340, but for the carbon-reinforced plastics it is up to three times better than for 4340.

The outstanding strength/weight and stiffness/weight parameters of the CRP endless-fiber materials make them very suitable for the structures of aircraft and spacecraft. They are also used for skis, golf clubs, fishing rods, and masts. Experimentally they have been used for drive shafts of cars, where both the high strength/weight and stiffness/weight ratios result in greater length without intermediate support, for a given critical speed (e.g., of 7200 rpm). In addition to the high parameters, one can choose a variety of fiber arrangements. For car bodies and boat hulls, the chopped fiber GRP or GRP/CRP hybrids are used.

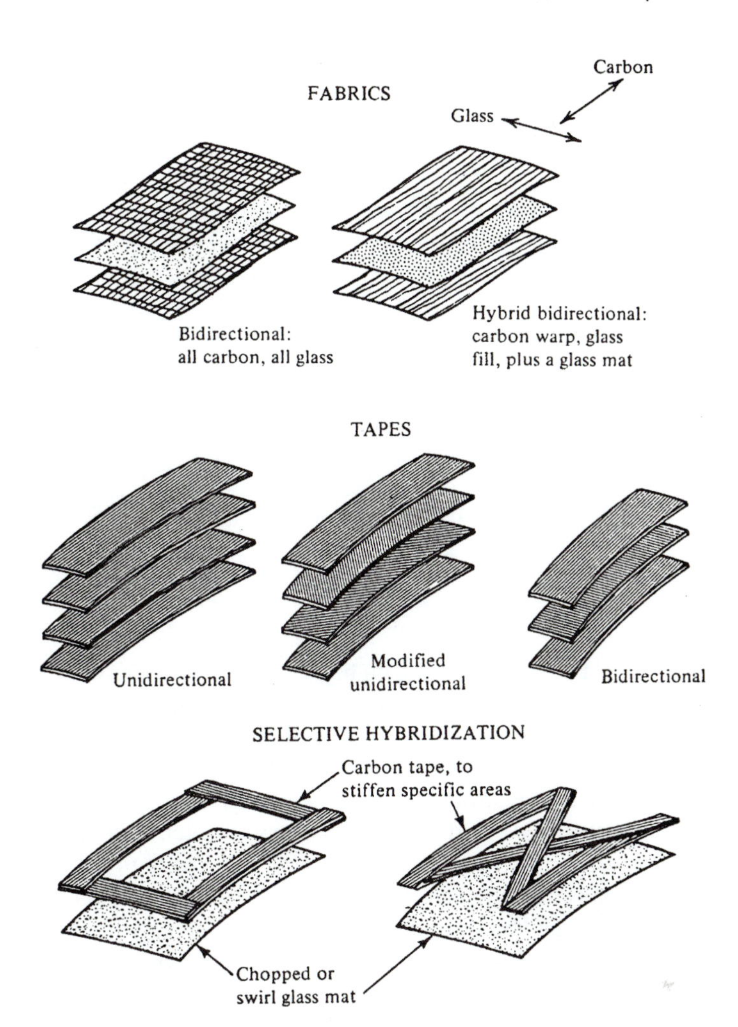

Figure 2.38
Designs of composite fabrics.

TABLE 2.16 Mechanical Properties of GRP and CRP in Comparison with Metals

Material	*UTS* (ksi)	*E* (ksi)	*UTS* (MPa)	*E* (MPa)	Density ρ(g/cm^3)	*UTS/ρ* (Nm/g)	E/ρ (Nm/g)
Aluminum 2014 T6	55	10×10^3	380	70×10^3	2.8	135	25,000
Titanium (Al6,V4)	140	15×10^3	970	105×10^3	4.5	215	23,300
4340 steel	230	30×10^3	1,580	210×10^3	7.8	202	27,000
GRP*	104	4.3×10^3	720	30×10^3	2.1	343	14,285
CRP, HT*	130	13×10^3	900	88×10^3	1.5	600	58,670
CRP, HM*	104	17×10^3	720	120×10^3	1.6	450	75,000
CRP, UD	200	19×10^3	1,800	130×10^3	1.55	1,160	83,870

*Fiber orientation, 0° 45° (50/50); $V_f = 60\%$. HT = high tensile strength; HM = high modulus of elasticity; UD = unidirectional tape.

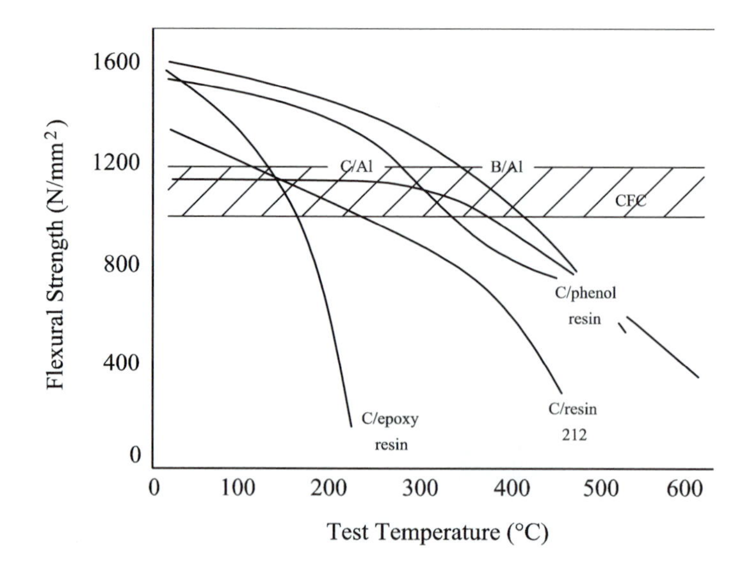

Figure 2.39
Strength versus temperature for different composites. [VDI]

For applications with higher service temperatures, the carbon-fiber-carbon (CFC) or carbon-reinforced carbon (CRC) composites are available. These are made from graphite endless fibers in a semigraphitized carbon matrix. There are no reactivity problems between the fibers and the matrix. The main problem in the manufacture is to obtain a matrix with little porosity, because during coking, carbon shrinks. The most successful method is to impregnate with pitch under pressure and at about 700°C. After cooling, coking is repeated at 1000°C. The high-temperature applicability of various composite materials is shown in Fig. 2.39. The graph includes carbon-reinforced aluminum and boron-reinforced aluminum, along with three different CRP materials. It is evident that the CRC material is superior to all the others above 450°C. It is used in the ablation shields that protect spacecraft or missiles during re-entry into the atmosphere, rocket noses, and wing edges of supersonic fighter aircraft and of the NASA space shuttle at points where high temperatures are generated by friction with the atmosphere. It is also used in the jet pipes and entry cones of solid fuel motors of rockets, brake pads and shoes of supersonic aircraft, and artificial hip-joint shafts.

Composite materials are still in early stages of development, and it is certain that new materials and combinations of them will be developed with improved properties. It is a challenge to manufacturing engineering to provide processes and machinery that can efficiently consolidate these materials into machinery parts and other products. Some of the manufacturing methods for processing composites are discussed in Chapter 6.

REFERENCES

1 Van Vlack, L. H. *Materials Science for Engineers.* Addison-Wesley, 1970.

2 Shewmon, P. G. *Transformations in Metals.* McGraw-Hill, 1969.

3 Brick, R. M.; R. B. Gordon; and A. Phillips. *Structure and Properties of Alloys.* McGraw-Hill, 1965.

4 Flinn, R. A.; and P. K. Trojan. *Engineering Materials and Their Applications.* Houghton Mifflin, 1975.

5 Brown, R. L. E. *Design and Manufacture of Plastic Parts.* John Wiley and Sons, 1980.

6 *Processing and Uses of Carbon-Fibre-Reinforced Plastics.* VDI Verlag, 1981.

7 Doyle, L. E. *Manufacturing Processes and Materials for Engineers.* Prentice Hall, 1969.

8 Lindberg, R. A. *Processes and Materials of Manufacture.* Allyn and Bacon, 1983.

9 American Society for Metals. *Metals Handbook.* 8th ed. American Society for Metals, 1961.

Illustration Sources

[BRICK GORDON PHILLIPS] Reprinted from: Brick, R. M., R. B. Gordon, A. Phillips, *Structure and Properties of Alloys*, McGraw-Hill, 1965, with permission by McGraw-Hill.

[BROWN] Reprinted from: Brown, R. L. E., *Design and Manufacture of Plastic Parts*, John Wiley and Sons, 1980.

[FLINN TROJAN] Reprinted with permission from: Flinn, R. A., P. K. Trojan, *Engineering Materials and their Applications*, John Wiley & Sons Inc., New York, 1975.

[SHEWMON] Reprinted from: Shewmon, P. G., *Transformations in Metals*, McGraw-Hill, 1969.

[VDI] Reprinted from: *Processing and Uses of Carbon Fibre Reinforced Plastics*, VDI Verlag, 1981.

[VAN VLACK] Reprinted from: Van Vlack, L. H., *Materials Science for Engineers*, Addison-Wesley, 1970 with permission of Addison Wesley Longman.

QUESTIONS

Q2.1 Name the individual standard tests for mechanical properties of materials and the corresponding parameters obtained from them. State the relationship between the *UTS* and *BHN* numbers for steels.

Q2.2 How do the three concepts of ductility, toughness, and impact toughness compare, and from which tests do they result?

Q2.3 Name the unit cell structures of iron, aluminum, and titanium, and their interatomic distances.

Q2.4 Name and explain imperfections in crystals.

Q2.5 Express the relationship between tensile yield strength and grain size.

Q2.6 Explain the difference between substitutional and interstitial solid solutions in metals and the conditions under which they form.

Q2.7 Draw the fundamental shape of two basic binary phase diagrams: (a) one where the two components are mutually completely soluble (give an example); (b) one with no mutual solubility of the two components.

Q2.8 Consider the iron-carbon phase diagram. State the following: maximum C in austenite, at which temperature; maximum C in ferrite, at which temperature; eutectic: composition and melting temperature; eutectoid: composition and γ to α transformation temperature. What are the phase components in pearlite, and what is their structural form?

Q2.9 Take a hypereutectoid composition with 1.0% C. What are the phase components and their percentages?

Q2.10 Discuss the TTT diagram. Explain the significance of cooling rates of CR_m and CR_p and the structure obtained for a cooling rate between them.

Q2.11 Explain the concept of hardenability. How is the hardenability of a steel raised? Which alloy steel is through-hardenable upon cooling in air?

Q2.12 Explain the concept of tempering. What is blue brittleness, and at which tempering temperature does it occur? What is the usual tempering time? Which structure is obtained in carbon steel on tempering at 700°C?

Q2.13 Explain annealing and normalizing as well as stress relieving.

Q2.14 Describe the process of precipitation hardening.

Q2.15 Name five methods for increasing the yield strength of metals. Which of them are universal, and which apply to certain metals only? Give an example for each of the latter kind.

Q2.16 Give typical properties of low-carbon steel and their fields of application.

Q2.17 Give typical tensile strength values for medium carbon hot-rolled and cold-rolled steels.

Q2.18 Select one from each of the following categories of alloy steels and give their compositions and *YS* and *UTS* values where available: (a) a well-hardenable SAE-AISI low-alloy steel; (b) austenitic stainless steel (Is it magnetic?); (c) tool steel for forging dies; (d) "high-speed" steel used for cutting tools.

Q2.19 Explain the differences in structure between gray, nodular, and malleable cast irons and the methods of obtaining each.

Q2.20 Select an aluminum alloy for an automobile engine and give its composition and mechanical properties. Why did you choose it? What are the problems with machining it? What kind of cutting tools are used?

Q2.21 Give the composition, mechanical properties, and type of treatment of a popular aluminum alloy used for aircraft structures, the 7075-T6.

Q2.22 Discuss the 6Al 4V titanium alloy and its yield strength at room temperature and at 315°C. Compare its *YS* at 150°C with that of the 75ST aluminum and 302 stainless steel.

Q2.23 Discuss the requirements imposed on superalloys and how they are satisfied with Inconel 700 and with the Co-based H151 alloy. Which of the refractory metals has the highest melting temperature, and what is its value?

Q2.24 Give one example in each of the three basic categories of plastics: thermoplastic, thermosetting, and elastomers, and their compositions.

Q2.25 Discuss the differences between amorphous and crystalline polymers and the significance of the glass-transition and melting temperatures.

Q2.26 Discuss the difference between addition and condensation polymerization.

Q2.27 Which are the three most common types of thermoplastic materials and their uses? Name three of the most common thermosets and their uses. Name three of the most common elastomers and their uses.

Q2.28 Discuss the effect of temperature on strength and strain at rupture of thermoplastic and of thermosetting polymers.

Q2.29 Explain the strain versus time behavior after load application and load removal as resulting from the viscoelastic and plastic model of polymers.

Q2.30 Name five types of ceramics and their uses.

Q2.31 What is the meaning of the abbreviations GRP, CRP, and CRC for composite materials? In which different forms are fibers applied to composites? Compare the strength/weight ratios of bag-fiber GRP and CRP with that of heat-treated 4340 steel.

3 Primary Metalworking

3.1 INTRODUCTION: IRON AND STEEL INDUSTRIES

In this chapter we discuss some of the processes that precede the secondary processes that are the main topics of the following chapters. The products of the primary processes are used as inputs to the secondary processes, in the form of rods, plates, castings, forgings, and so on, generally denoted by the term *blanks*, to be worked on by cold forming, machining, welding, or one of the nontraditional processes. This chapter is included for several reasons. First, it is useful to have a basic understanding of the variety of the blank forms available and the different characteristics that result as they are produced by the different primary processes. These characteristics encompass precision of form, strength, and toughness, as well as cost. Second, the primary processes and the equipment used in them are impressive and fascinating examples of manufacturing technology and of applications of mechanical design. Lessons learned from this technology may offer inspiration in the design of secondary technologies. Obviously, a mechanical engineering student should possess basic knowledge about the making of iron and steel, about rolling mills, and about castings. Thirdly, mechanical engineering graduates may find employment in primary metalworking plants, where they will need at least the introductory knowledge presented here.

In this chapter we also discuss the making of materials, but we limit the discussion to metals, and almost exclusively to iron and steel. We leave out plastics and composites and only deal briefly in Chapter 6 with the preparation of these materials. Furthermore, we concentrate on the hot primary processes of rolling, casting, and powder metallurgy. The other types of hot forming, such as forging and extrusion, are discussed together with their cold alternatives in Chapters 4 and 5.

Let us first present the big picture of the iron- and steel-making industry in a very simple manner. The bulk of the final production consists of steel of various kinds and in various forms. Only a small proportion of production goes just part of the way and ends as cast iron to be used in foundries. Even so, the development of this industry over the past one hundred years has resulted in processes in which almost all of the original source of iron, the various kinds of iron ore being oxides, is first converted into cast iron with 3.5–4.4% carbon and then, in a second step, processed further into steel with up to 2% but mostly less than 1% carbon. Simply expressed, the iron oxides are reduced by means of carbon obtained from coal/coke. Burning of coke supplies both the smelting energy and the carbon monoxide gas (CO) used to remove the oxygen from the ore. This process is accomplished in the marvelous device of the continuously working blast furnace, where the reducing reaction is overdone and yields an alloy of iron with a relatively high percentage of carbon, mostly approximately of the eutectic composition (see the Fe-C phase diagram in Fig. 2.13). There are several reasons for this intermediate step. First, it is most economical to run the blast furnace without attempting exact control of the reducing process, which would be extremely difficult on the large scale of the process. Secondly, the eutectic melts at a substantially lower temperature than the alloys with less carbon, which requires less energy and is thus easier to achieve and extends the life of the hearth lining. Thirdly, the excess carbon becomes a source of energy in the steel-making furnaces, where the goal is just to burn the carbon off and heat after heat, varying steel formulations can be more easily achieved.

Let us now review the processes included in the iron and steel business by looking at two companies, both located in Hamilton, Ontario, on the shores of Lake Ontario and Lake Erie. Their main supplies of iron ore are delivered from the Lake Superior region through the waterways of the Great Lakes and from Labrador and Quebec through the Saint Lawrence river. The history of one hundred years of these companies is similar to those of the steel-making industries in Pittsburgh and Bethlehem in Pennsylvania. All of them played key roles in the development of the railway systems, of the automobile industries, of pipes for oil drilling, and of pipelines for oil and gas transportation. Even with the increased use of plastics and composites, steel is still the most important engineering material.

The diagram in Fig. 3.1, prepared by the Stelco company, is an excellent condensation of the main processes in the steel-making industry. The blast furnace (3) is the initial device. The three components of the charge (1) that is delivered to its top are the iron ore in its natural form of small pieces of rock or else processed into suitable size chunks of pellets or sinter, the limestone, which is the source of flux, and coal, which is first converted to coke in coke ovens (2). A variety of by-products from the coke oven are ammonium sulphate, used in agricultural fertilizers; phenol, used for the manufacture of Bakelite and other chemicals; pitch for roofing and for road tars; benzene, toluen, and other chemicals. The fourth component of the inputs into the blast furnace is not shown in the diagram; it is the blast air necessary for the combustion of coke. The output from the blast furnace is pig iron, some part of which may be used for iron castings. Most of it is transported to steel-making furnaces, together with steel scrap, to the basic oxygen furnace (4) or the open-hearth furnace (5). Molten steel is poured (teemed) into the ingot molds (6) and then the ingots are stored in the soaking pit (7), or the steel may be continuously cast into slabs. The reheated ingots go into the blooming or slabbing mill (8). The blooms and billets (9) are hot rolled into bars, rods, or tube rounds which are hot rolled into seamless pipe. The bars and rods are further processed by cold drawing into bars and wire. The slabs are hot rolled into plates and strip to be further cold rolled, or else into skelp and plates welded into large-diameter pipes.

Figure 3.1
Iron- and steel-making operations with hot and cold rolling. (COURTESY STELCO, HAMILTON, ONT., CANADA)

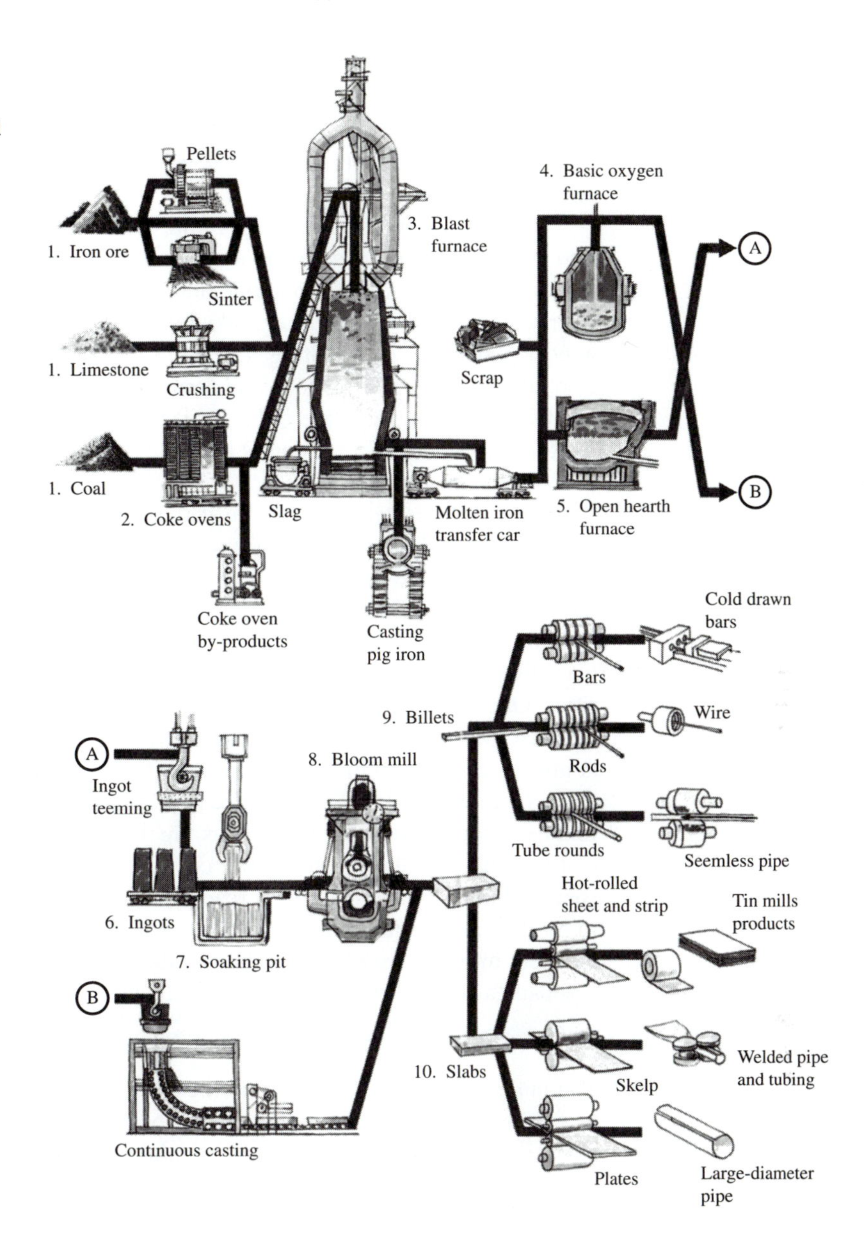

A slightly different presentation of a steel-making enterprise appears in a brochure by Dofasco. First, the operations described in the preceding paragraph, from charging the blast furnace to teeming ingots, are discussed:

> Three of the raw materials required to make steel—coal, limestone, and iron ore—are stockpiled at the stock yard. The limestone and ore are ready for the blast furnace when they arrive, but coal must first be converted to a more pure form called coke, by baking in coke ovens. Here gases and impurities are driven off, and many are later converted into synthetic products and fertilizers.

> The coke is mixed with the two other raw materials in a weight ratio of about 7 parts iron ore, 2-1/2 parts coke, and 1 part limestone. It takes 1-1/2 tons of iron ore, 1/2 ton of coke, and 1/5 ton of limestone, and 200 kg of steam to make 1 ton of iron. The ingredients are poured into the top of giant blast furnaces, which operate continuously, by a skip car. This is called charging the furnace.
>
> Inside the furnace, the temperature is about 700°C. Preheated air is forced into the furnace (hence the name "blast furnace") causing the coke to burn fiercely, forming carbon monoxide and raising the temperature even higher. The intense heat and carbon monoxide remove the oxygen from the ore, leaving almost pure iron. The melting limestone floats to the top of the heavier molten iron, attracting impurities to it and forming slag. When molten iron has accumulated in sufficient quantities, the furnace is "tapped," and the iron poured into molten-iron transfer cars. Slag is separated and treated to make a variety of useful building products.
>
> Iron requires refining to give it the toughness and durability of steel. This refining process essentially removes the carbon that was added to the iron by coke in the blast furnaces. At Dofasco, iron is refined in giant, pear-shaped oxygen furnaces that can make up to 150 tons of steel in 35 to 40 minutes. The molten iron is poured directly from ladle cars into transfer ladles, and then into the furnace. A quantity of steel scrap is added at this stage, along with some lime to attract impurities. Once charged, the furnace is tilted upright, and a hollow tube called a lance is thrust into the opening of the furnace. Pure oxygen is blown through the lance at great pressure into the molten iron mixture. The result is dramatic. Carbon and silicon in the mixture are oxidized, producing tremendous amounts of heat. Finally, after about 18 to 20 minutes, the lance is removed and tests are made to confirm that the steel matches the customer's needs, according to a number of specifications. When the steel is ready for pouring, the furnace tips first one way, to remove the top layer of slag while taking final tests, then another, discharging the steel into a massive ladle. After all the steel has been poured off, the furnace tips back a second time to empty it of all slag.

From here on, the narrative refers to Fig. 3.2:

> The ladles of steel have small pouring holes in the bottom, with a cover that can be opened (like pulling the plug in the bathtub). This vessel, filled with molten steel, is carried by a crane over a number of hollow ingot molds. "Pulling the plug" in the ladle allows the steel to flow into the ingot molds. In some cases, the steel may be poured into specially constructed brickwork which carries it down and under the molds, filling them from below. The manner in which the ingots are poured—from the top or from the bottom—has an important effect on the kind of steel produced. When the molten steel has cooled sufficiently inside the molds, the molds are removed, or "stripped," leaving red-hot ingots weighing as much as 15 tons each.
>
> Before being rolled flat, the ingots must be heated to an even temperature in the soaking pits. This process takes about 8 to 10 hours. When completed, the ingots are conveyed to the roughing mill, where they are squeezed until reduced in thickness and ready for hot rolling. Now the steel slab—still red-hot and weighing up to 15 tons—is rolled under tremendous pressure until it emerges about 2.5 mm thick, over 300 m long—and traveling approximately 8m/sec. It is coiled, then cooled, and can be shipped as hot-rolled plain or cleaned of scale by an acid

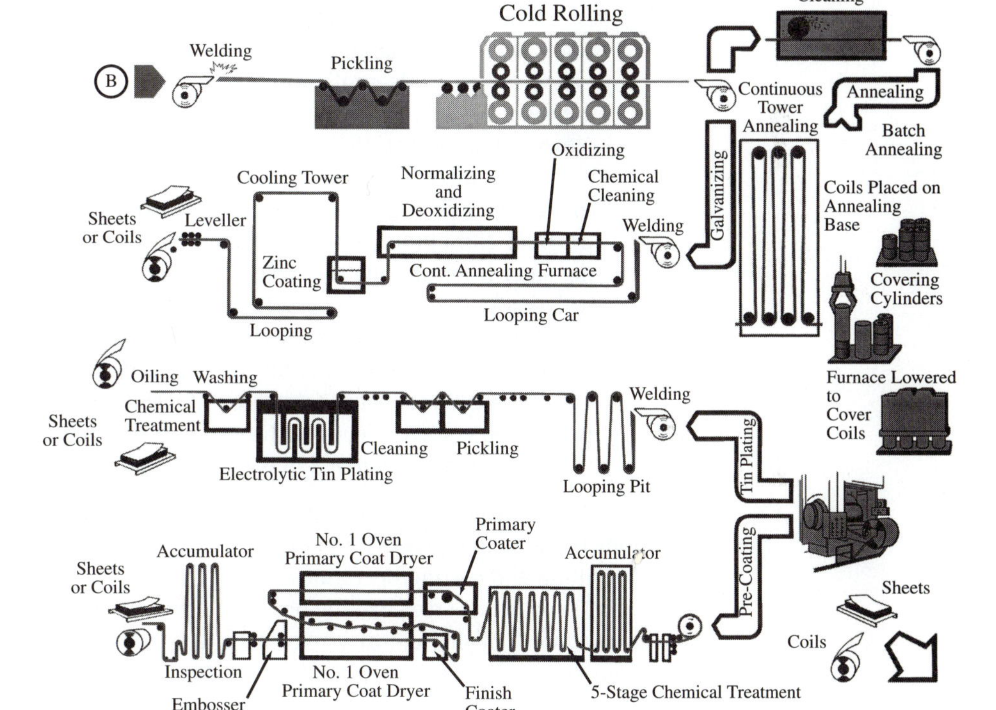

Figure 3.2
Cold rolling of steel and subsequent surface-finish treatments. (COURTESY DOFASCO, HAMILTON, ONT., CANADA)

treatment called pickling and shipped as hot-rolled pickled. Thirdly, it can receive further processing in the cold-rolling mill.

The mill consists of five four-high stands, and the sheet is stepwise reduced in thickness in them. Consequently, its speed increases stepwise and may reach 40 m/sec at the exit, where it is coiled on a reel. Interstand tension is produced by corresponding control of the rotational speed of the rolls in each stand, to relieve the pressure between the rolls and the sheet and reduce the roll-separating force as well as the driving torque. The sheet is reduced in thickness down to

1.2–0.6 mm for such uses as car bodies and appliance cabinets or further down to as little as 0.25 mm for tin plate used for canning food.

Cold rolling reduces the thickness of the steel through a combination of roll pressure, tension, and lubrication. But the effect on the steel is more than just a change in thickness. It becomes very brittle on the inside, and often acquires a sheen, or shiny surface, on the outside. Steel in this form (called "full hard") is not suitable for most uses that require forming or shaping. So the steel is first cleaned of lubricants picked up from cold rolling, then annealed to soften it to the quality required. Annealing consists of heating the steel in an oxygen-free atmosphere for a specific period of time.

To restore the hardness required, the steel is now tempered, which also provides the necessary finish and flatness. Tempering is achieved by passing the sheet through another four-high, one-stand rolling mill where, however, only a very small thickness reduction, about 1% for soft tempers and 2% for harder tempers, is produced. The steel sheet may now be shipped as cold-rolled steel for use in automobiles, appliances, and other products.

Cold-rolled steel may be specially coated, however, to make it more suitable for a number of applications. Each of the three basic coatings—zinc, tin, and baked-on paint—is applied in a continuous procedure, with the cold-rolled steel entering one end of the line and emerging, without stopping, at the other end, as either galvanized steel (zinc coating), tin plate, or pre-painted steel.

Galvanized steel is produced by dipping cold-rolled steel into a bath of molten zinc. The zinc, which lends a distinctive "spangled" surface to the steel, provides extra protection against corrosion. Tin plate, as used in tin cans, prevents the steel from altering the taste of foods and beverages. The tin is applied electrolytically to cold-rolled steel to a thickness of barely 0.4 μm.

Finally, the steel—either galvanized, tin plate, or cold-rolled—may receive a special baked-on paint coat that adds beauty and corrosion resistance. After special chemical treatment to prepare the surface, the steel receives a "prime" coat of paint, then is baked. The finish coat, in practically any color of the rainbow (plus some, like wood grain and brass, that are not!) is applied and baked on.

After this general introduction, we will now consider details of the individual sections of the whole process: the blast furnace and the steel-making operations and then the rolling and casting operations.

The iron and steel-making processes are discussed here in great detail because the development of this industry played such an important role in industrial history. The industry's ability to respond to the large demand for steel rail was decisive for opening the western United States, and its response to the demand for steel sheet made possible the rapid growth of the automobile industry. This development required the resolution of complex problems of metallurgy and of the design and manufacture of furnaces and the whole system of auxiliary units for the iron- and steel-making plants, as well as of the logistics of transporting iron ore and coal.

The study of these developments offers valuable lessons in methods of approaching complex problems. This was one of the significant cornerstones of industrial technology, as were the subsequent tasks of the development of jet engines and jet aircraft, of Numerically Controlled machine tools, and, most recently, the development of computer hardware and software.

3.2 BLAST FURNACE OPERATIONS

3.2.1 Design of the Furnace: Inputs and Outputs

The modern furnace is the result of about 500 years of development, and its modern design in the United States dates from about 1929 (see Fig.3.3). It has the shape of two opposing cones, with the largest diameter (approx. 10 m) at about a third of the total height of 35 m. The foundation is a 3-m-thick slab of reinforced concrete mounted on H beam piles driven to bedrock. It has to support a weight of approximately 10,000 tons. The bases of the furnace columns rest on the top surface of the foundation. This surface is covered by 60 mm of heat-resistant concrete, which provides a level working support for the bottom course of hearth bricks.

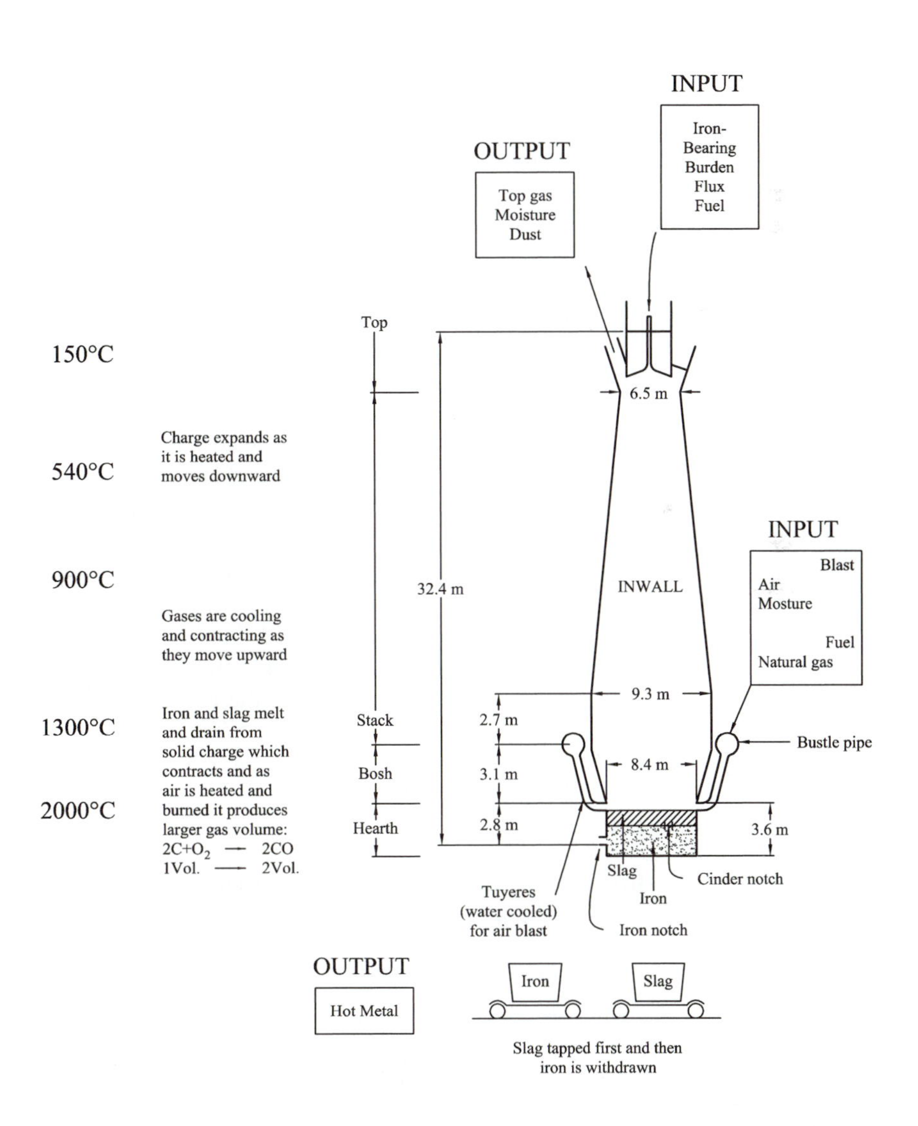

Figure 3.3
A typical blast furnace, showing inputs, outputs, dimensions, and temperatures developed in the furnace. [USS10]

There are 8–10 columns supporting the mantle, which is a heavy, horizontal steel ring and the shell of the furnace sits on the mantle. The shell has the form of a truncated cone, and it is made of welded steel plates. At its upper end is the top cone. The part from the top of the mantle to the top of the furnace is called the shaft. The lower opposing cone is the bosh, and the bottom is the hearth. At the very top is the double bell construction that permits the charging of the furnace during its operation and carries the gas uptake piping. The inside walls of the shaft, bosh, and the hearth are lined with refractory brick. The parts of the hearth that are most exposed to the hot metal and slag are made of carbon brick, and they are cooled on the outside by means of cast iron segments with internal cooling water jackets.

The iron produced in the smelting process is withdrawn at periodic intervals through the iron notch made in the upper hearth wall at a height of about 1 m above the top of the bottom brick blocks. The slag generated in the smelting process is lighter than iron, and it floats on the molten pool of iron. It is withdrawn through the slag notch located about 120 to 140 cm above the centerline of the iron notch. It consists of three water-cooled concentric cast copper cones.

The heated blast of air is delivered into the upper part of the hearth at a height of about 1 m above the slag notch (about 30 cm below the top of the hearth) through water-cooled copper nozzles known as tuyeres. There are twenty of these nozzle openings spaced uniformly in one horizontal plane. The blast is supplied to the tuyeres from the bustle pipe that encircles the outside of the hearth wall.

The proportions of the individual parts of the furnace have been optimized in the development process based on accumulated experience. The materials charged into the furnace are called the burden. They are delivered to the furnace from the top through a double bell housing. The double bell system makes it possible to charge the materials without opening the top of the furnace to the atmosphere. Both the upper, smaller, and lower, larger, bells have the form of inverted cones. First, with both bells closed the skip bucket charges the material in the hopper above the upper bell. Next, the upper bell opens and admits the charge to the hopper above the lower bell. Then the small bell closes and the large bell admits the charge into the furnace. Finally with both bells closed the cycle can start again. The rotating chute distributes the charge into the four quadrants of the hopper via seal valves that are effective against the pressure of the top gas. The charge is hoisted from the ground-level stock house in skip cars riding on tracks in an inclined, fabricated steel structure called the skip bridge. The charge consists of iron ore, coke, and flux material. *Iron ore* is the source of iron, *coke* provides the heat energy and the reducing gas, and limestone supplies the *flux* that converts to slag.

Iron Ore

Iron-bearing ores exist as oxides, carbonates, sulphides, and iron silicates. Oxide minerals are the most important sources of iron. Four different compositions exist:

1. Magnetite	Fe_3O_4	ferrous-ferric oxide
2. Hematite	Fe_2O_3	ferric oxide
3. Ilmenite	$Fe\ Ti\ O_3$	iron-titanium oxide
4. Limonite	$FeO\ (OH)$	hydrous iron oxides

These chemical compositions exist in pure minerals. The iron content of commercial ores or concentrates is lower because of impurities.

The magnetite formula corresponds to 72% iron and 28% oxygen. It is strongly magnetic. This permits exploration by magnetic methods and makes possible magnetic separation from gangue materials to produce high-quality concentrate. The hematite formula represents content of 70% iron and 30% oxygen. Hematite is the most abundant iron mineral. The hydrous oxides are goethite, with 63% iron, and lepidocrite. Ilmenite contains 37% iron, 31% titanium, and 32% oxygen. Iron ore reserves exist in many parts of the world, most importantly in the United States, Canada, Brazil, France, Sweden, the United Kingdom, South Africa, Russia, India, and Australia. Sufficient supplies of iron ore exist in the world for the foreseeable requirements for at least the next 100 years. Moreover, new sites will be discovered and made accessible in the future. Currently the need in the United States is for 235 million tons annually of iron ore. In the United States 80% of the iron ore reserve is in the Lake Superior district, and these supplies are easily shipped to the Lower Lakes areas in Chicago, Detroit, and Cleveland, and to Ontario in Canada. These are shipped partly as natural ore and partly after beneficiation, done by gravity separation of silica, which may originally compose up to 10% of the ore. Steel-making plants on the eastern seaboard may import ore from Labrador, West Africa, or Australia. Almost all the ore used in Alabama and Texas is imported, since transoceanic transportation of high-quality foreign ores on special large ships is rather economical.

Ores are mined mostly in open pits but also underground in some locations. Ores are graded by producers and may be mixed to supply the required chemistry and purity. Apart from iron content, the contents of ore are graded in terms of silica, phosphorus, manganese, and alumina, as well as the most deleterious constituent, sulphur.

The iron ore is processed to improve its chemical and physical properties using various methods such as crushing, screening, blending, grinding, concentrating, and agglomerating. The mining program at individual mines is set up to prepare a uniform product. Apart from crushing and screening this may be achieved by stacking on piles and reclaiming to achieve blending of individual layers of the heap. Many processes are used for concentrating the iron content. This is done first by washing, which removes a large proportion of the sand, clay, and rock. Then other methods are used, such as heavy-media separation (sink-and-float), in which a mixture of two minerals is loaded in a liquid with specific gravity between that of the minerals (e.g., quartz, specific gravity 2.65; hematite, 5.00); flotation, in which certain substances are added to suspensions of finely ground minerals that exhibit an affinity to air and cling to air bubbles; and magnetic and electrostatic concentration.

In the blast furnace the solid charge materials descend, and the hot reducing gases blow upward. The best contact between the solids and gas is obtained if the solids exist as small lumps, of rather uniform size between 6 and 12 mm. These lumps must be strong to withstand stockpiling, transportation, and the high temperature in the furnace. Strength is obtained in the processes of agglomeration: sintering, palletizing, briquetting, and nodulizing. In sintering fine ores, flue dust and other iron-bearing materials of very small particle sizes are converted into a granular form. The materials are thoroughly mixed with finely divided fuel such as coke and then moved on a traveling grate bed on which the fuel is gradually burned at 1300–1500°C and the fine ore particles are sintered into porous coherent lumps. Pelletizing differs from sintering in that a "green," unbaked pellet is first formed and then hardened by heating. The use of sinter and pellets increases the efficiency of the blast furnace.

Coke

Coke is obtained by destructive distillation of selected coking coals at temperatures in the range of 900–1100°C. Coal consists of the remains of vegetable matter partially decomposed in the presence of water and absence of air and subject to variations of pressure and temperature by geologic action. It is a complex mixture of organic compounds. The principal elements are carbon and hydrogen with smaller amounts of oxygen, nitrogen, and sulphur. Coal, when heated to high temperature in the absence of air, breaks down to gases, liquid and solid organic compounds of lower molecular weight, and a nonvolatile carbonaceous residue, the coke. There are three main kinds of coke: low, medium, and high-temperature, classified according to their manufacture. Coke for metallurgical use must be carbonized at higher temperatures. It must be strong and hard. It must be of uniform size, between 20 and 70 mm. Smaller pieces, commonly called *coke breeze,* are usually used for secondary purposes as boiler fuel or sinter fuel. It is produced in coke ovens where the heat for the distillation is obtained by burning blast furnace gas or coke oven gas.

Apart from coke a variety of volatile products are generated. The process consists of three steps: (1) Primary breakdown of coal at temperatures below 700°C, resulting in water, carbon oxides, hydrogen sulphide, hydroaromatic compounds, paraffins, olefin, and nitrogen compounds. Secondary reactions among these products as they pass through hot coke or over hot oven walls result in synthesis and degradation. Hydrogen, methane, and aromatic hydrocarbons occur above 700°C. Decomposition of the nitrogen compounds produces ammonia, hydrogen cyanide, pyridine bases, and nitrogen. Finally, removal of hydrogen from the residue produces hard coke. Typically, one ton of coal yields the following products:

Blast furnace coke	550–640 kg
Coke breeze	100 kg
Coke oven gas	3700 m^3
Tar	4–6 kg
Ammonium sulphate	10–14 kg
Ammonia liquor	4–6 l
Light oil	1 l

The coke oven by-products find uses beyond the iron and steel plants. The coke oven gas is used as fuel in the coke ovens as well as in the open-hearth furnace and the reheating furnaces in steel making. Ammonium sulphate is used in agricultural fertilizers. Coal tar and ammonia liquor are utilized to produce phenol (carbolic acid), which is used in the manufacture of thermosetting plastics. Ortho cresol and meta-para cresol from ammonia liquor are used in the production of synthetic resins, insecticides, and weed killers. Creosote from coal tar is used for impregnation of wood pilings, poles, and railroad ties. Benzene recovered from refining light oil is used in many chemical manufactures.

Fluxes and Slags

The function of the fluxes is (1) to form a fluid slag with the coke ash, ore gangue (waste material), and other charged impurities, and (2) to form a slag of such chemical composition that it will control the sulphur content of the iron. Many of the impurities associated with the iron ore are difficult to melt. The flux makes the melting easier. Other compounds combined with iron can be removed because they will preferentially combine with the flux. To remove basic impurities, acid flux must be used, and to remove acid components, a basic flux is needed.

The fluxes used in the blast furnace are basic: they are limestone, composed mainly of calcium carbonate ($CaCO_3$), and dolomite, which is essentially calcium-magnesium carbonate ([Ca, Mg] CO_3). When large amounts of sulphur must be removed, limestone is preferred. Limestone is widely distributed in the areas drained by the Mississippi and Ohio rivers. Dolomite is found in West Virginia. The materials require drying and sizing. The preferred size for the blast furnace is 50 to 100 mm.

The fluxes react with the alumina, silica, and phosphorous impurities to form the molten slag that can be removed from the blast furnace separately from the iron because it is lighter and floats on the molten iron pool. There it serves as a blanket to protect the metal from the action of the hot gases. The slag obtained in this process finds many uses in various industries. Some of the most significant uses are in making concrete, for road and airport runway construction, and as railroad ballast.

3.2.2 Chemistry of the Blast Furnace Reactions

The materials charged in the blast furnace at its top (the burden) are heated by the hot gases rising from the hearth. The preheated coke burns in contact with the hot air blast with great intensity at about 1700°C. The temperature at which iron and slag melt in the bosh is about 1200–1300°C, and as the gases ascend they are cooling gradually down to about 160°C at the top of the furnace. In the high temperature of the hearth any CO_2 that forms reacts immediately with C to form carbon monoxide, CO.

$$2C + O_2 \rightarrow 2CO$$

This reaction is the main source of heat in the furnace, and carbon monoxide is the main iron-oxide-reducing gas. There is always some moisture in the blast air that reacts with the coke to produce hydrogen

$$C + H_2O \rightarrow CO + H_2$$

This reaction does not produce heat but consumes it; however, it produces more reducing gas for every unit of carbon than is produced by carbon burning in air. In the upper part of the furnace where the temperature is below 950°C the gases start to reduce the iron oxide of the ore:

$$Fe_2O_3 + 3CO \rightarrow 2Fe + 3CO_2$$

$$Fe_3O_4 + 4CO \rightarrow 3Fe + 4CO_2$$

$$FeO + CO \rightarrow Fe + CO_2$$

$$CO_2 + C \rightarrow 2CO$$

$$Fe_2O_3 + 3H_2 \rightarrow 2Fe + 3H_2O$$

$$Fe_3O_4 + 4H_2 \rightarrow 3Fe + 4H_2O$$

$$FeO + H_2 = Fe + H_2O$$

The iron oxide that is not reduced in the upper part of the furnace is reduced in its lower part, where both CO_2 and H_2O are unstable so that the actual reduction is obtained as

$$FeO + C = Fe + CO$$

The heat for the process is only partially supplied by the burning of coke. The temperature of the blast may be as high as 900°C. Natural gas, fuel oil, or pulverized coal is injected into the blast and burns to CO and H_2.

It is necessary to at least briefly discuss reactions involving the impurities of manganese, phosphorus, silicon, and sulphur. Manganese oxides MnO_2 and Mn_3O_4 are reduced by CO to a lower oxide MnO in the upper part of the furnace; and in the lower part, at 1500°C, they are reduced further:

$$MnO + C = Mn + CO$$

The manganese that is reduced dissolves in the hot iron, while the unreduced portion becomes part of the slag.

Silicon oxide is reduced at very high temperatures:

$$SiO_2 + 2C = Si + 2CO$$

The unreduced portion of the oxide reacts with limestone in the flux and becomes a part of the slag.

Phosphorus oxide is also reduced at a very high temperature:

$$P_2O_5 + 5C = 2P + 5CO$$

Almost all of the phosphorus gets reduced and dissolves in the hot iron.

Sulphur enters the blast furnace mainly in the coke and is released into the gas stream as H_2S. Most of this combines with the lime in the flux, and some combines with the iron. It can then be removed at the high temperatures in the hearth by reduction of iron sulphide in the presence of basic flux such as lime (CaO):

$$FeS + CaO + C \rightarrow CaS + Fe + CO$$

The degree of sulphur removal depends on the temperature in the hearth and on the ratio of the basic oxides lime (CaO) and magnesia (MgO) to the acid oxides silica (SiO_2) and alumina (Al_2O_3) in the slag.

The result of all these reactions is *pig iron*; its typical compositions are shown in Table 3.1. The amounts of all these elements dissolved in iron vary. It is the role of the subsequent metallurgical processes such as steel-making and the melting of iron for

TABLE 3.1. Pig Iron Compositions

Type of Iron	Si	S	P	Mn	Fe
Basic pig iron	1.5 max	0.05 max	0.4–0.9	0.4–2.0	3.5–4.4
Basic pig for oxygen steel	0.2–2.0	0.05 max	0.4 max	0.4–2.5	3.5–4.4
Acid pig for open hearth	0.7–1.5	0.045 max	0.05 max	0.5–2.5	4.15–4.4

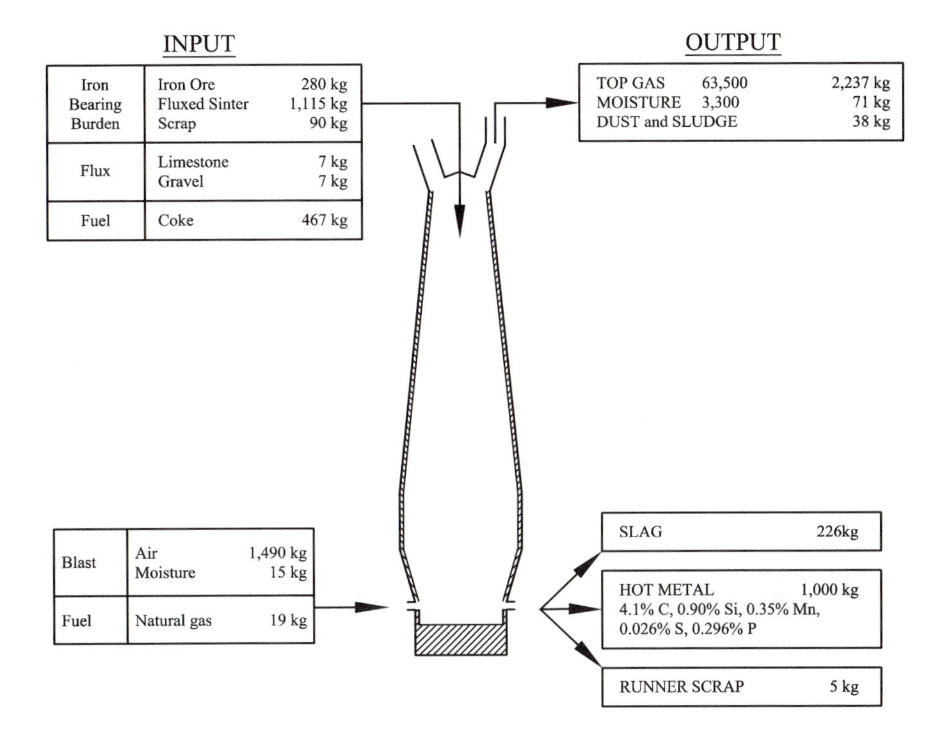

Figure 3.4
Blast furnace material balance per ton of hot metal. [USS10]

casting to regulate them further as well as to add and control the contents of other elements, alloying the iron according to the desired final compositions.

The typical balance of materials entering into and exiting from the blast furnace per ton of pig iron (hot metal) is presented in Fig. 3.4.

It is interesting to see that the components with the largest mass are the gases, both in the input as the blast air and at the output as the top gas. In the iron-bearing burden the largest component is the ore sinter, then natural ore, and finally scrap. The weight of coke is about a third of these, and the weight of flux is rather small. The two products, slag and pig iron, are in a weight proportion of about 1:4. Apart from the blast furnace, the iron-making plant consists of a number of other auxiliary units.

The iron ores and limestone are stored on a yard as they are brought in by the boats or by railway. The ore bridge over the yard carries the cranes that transfer these materials into the bins to be loaded into the skip buckets that will carry them to the top of the furnace. Coal is also delivered by boats or rail and stored on another yard, from which it is transported and charged in the coke ovens. These are usually also located on the grounds of the plant. Coke is then transported from the ovens to the raw material bins on a belt conveyor. The iron ores have usually been processed at the mine by crushing, screening, and blending and separated from dirt (gangue) by flotation or by magnetic or electrostatic means, and they are delivered in the form of particles of fairly uniform size (average diameter: 20 to 25 mm). Ore fines in the form of powder or dust undergo agglomeration by sintering or pelletizing. This is either done at the mine before delivery to the iron and steel plant, or some of it is a part of the blast furnace operation. In that case the ore fines, lime, and coke are transported also into the sintering plant.

Other inputs to the sintering plant are provided by the flue dust collected from the gas escaping from the top of the blast furnace and separated in the vessel of the dust catcher, and by the sludge separated in the Venturi washer. The cleaned gas passes on through the cooling tower and is then further distributed and put to several uses.

One such use is to heat two of the three stoves that have the form of tall round towers with dome-shaped tops. They have a diameter of about 9 m and height of 40 m. The function of the stoves is to preheat the blast before it enters the furnace. One part of the stove is the combustion chamber in which the gas rises from the burner located at the bottom to the dome. The second part is the checkerwork, which consists of many small-diameter, vertical holes through which the hot gases pass to the bottom and exit to the stack. The surface area of the checkerwork is about 30,000 m^2 of heating surface. The brick is first-grade firebrick with high thermal capacity and conductivity, and the brick lining lasts 15–20 years except for the upper part, which may have to be replaced every 5–7 years. While two stoves are "on gas," being heated, the third stove is "on blast," with the blast air passing through it in the direction opposite to the direction of the gas in the other phase of the cycle.

The blast is generated in the turbo blowers driven by the steam turbines powered by steam produced in the boiler house. The blast passes from the stove to the "bustle pipe" and through the tuyeres (nozzles) into the furnace. The second use of the blast furnace gas is to provide the fuel to the coke ovens. The third use is as fuel in the foundry core oven and in the annealing furnace.

The molten iron that exits from the tap hole in the hearth of the furnace goes into the hot metal ladle and from there via the hot metal mixer to the steel-making furnaces or else to the pig casting machine and pig iron storage. From there it will be transported to the foundries.

To illustrate the magnitude of the operation of such plants, Table 3.2 contains data about the individual utilities for a plant with two blast furnaces producing a total of 8,400 tons (metric) of pig iron per day. An approximate overall evaluation of this production, if it all were further processed into car-body sheet steel at the 1996 average wholesale price of $0.66/kg, would be $5.54 million per day.

TABLE 3.2 Utilities Requirements of a Self-Contained Blast Furnace Plant Producing 8,400 Tons of Iron per Day

Utility	Quantity per day
AC electric power, purchased	302,400 kWh
DC electric power, self-produced	84,000 kWh
Recirculating water	121,216 m^3
Make-up water	1,962 m^3
Other service water	25,180 m^3
Water to power generation	226,736 m^3
Potable water	590 m^3

The data in Table 3.2 can be summarized in two figures: the average electric power used is 16,100 kW, and the total water flow is 260,890 l/min. The overall energy balance including the chemical heat energy is 32,720 GJ/day, which converts to 379 MW.

3.3 STEEL-MAKING FURNACE OPERATIONS

The operations of a steel-making furnace may be distinguished as two kinds: those where the primary input is molten pig iron and the primary source of energy is the burning of the excess carbon, and those where the primary energy is electric. In the former class there are two types: (1) the open-hearth (OH) furnace, in which the main charge is pig iron and iron ore, and additional energy is derived from burning gas, and (2) the basic oxygen furnace (BOF), in which the charge is molten pig iron and steel scrap. In the BOF the burning of excess carbon and of silicon and other impurities in the iron is the only source of heat. The latter class consists of the electric arc furnace and of the electric induction furnace.

Apart from the concerns of manipulating the carbon content these processes have to deal with other constituents, either beneficial or detrimental. Some of these may originate from the ores, such as manganese, silicon, phosphorus, and sulphur from coke, and some arise in the steel-making process, such as nitrogen from air and too much oxygen. Other alloying elements must also be added to produce the desired steel compositions.

We will see that the refractory linings of the furnaces play an important role in dealing with these tasks. During the development of the processes, it was necessary to distinguish between the two kinds of refractories and, correspondingly, of slags: the acid and the basic formulations.

Historically, modern steel-making originated in the 1850s with the acid Bessemer process. It was discovered that oxidation of Si, Mn, and C was chemically preferential to that of Fe and that blowing air through molten pig iron produced enough energy from these oxidations to make molten steel without any additional heat source. Bessemer used iron smelted from Swedish ores low in P, high enough in Mn, and with enough Si to meet the thermal needs. Later, the addition of spiegeleisen (an alloy of iron with about 6.5% C and 16–30% Mn) was found to combine Mn with sulphur and become a part of the slag, and whatever portion of it remained dissolved in iron assumed a form that eliminated the detrimental effect of S on the brittleness of steel. Furthermore, Si and Mn helped in limiting the excess oxygen in steel. However, the acid lining of the furnace was not able to deal with higher contents of phosphorus. Therefore, with the depletion of the reserves of high-quality ore, low in S and P, the acid Bessemer process was gradually and, finally, completely abandoned around 1950.

At about the same time as the Bessemer process was invented in the 1850s, the open-hearth process was developed by Sir William Siemens in England. Later on, in the 1950s, the basic oxygen process was introduced. Both processes are used in current steel-making practice, with the basic oxygen furnace (BOF) gaining more and more usage.

3.3.1 The Open-Hearth (OH) Process

Originally known as the Siemens regenerative process and later modified by the Martin brothers in France, the open-hearth process is characterized by the regeneration of heat in the "checker" chambers and by the use of pig iron and ore (Siemens) or pig, ore, and scrap (Martin). The iron ore supplies the oxygen needed to lower the content of carbon in iron. The Martins diluted the pig iron with steel scrap and thus less oxidation was necessary. The smelting energy is supplied by pumping air and fuel for combustion into the hearth. In this way, the temperature is independent of purifying reactions, and the composition of the bath is under better control. Greater variety of scrap can be used than in the BOF. Increased yield of steel from the same quantity of pig iron is obtained because of the additional iron recovered from the iron ore. Phosphorus is eliminated before carbon due to different temperature conditions than in the BOF, where P is burned in the afterblow. By 1890 there were already sixteen OHFs in the United States.

Open-hearth furnaces are arranged lengthwise in a row, up to twelve of them, in a large steel building. Each furnace, such as the one shown in Fig. 3.5, is about 30 m long. In the front is the charging floor (left out in the picture; it would be just above the regeneration chambers), about 7 m above the level of the yard. At the back of the furnace is the pouring floor where the tapping spout is shown (but the floor itself is left out). There is a rail track on the charging floor on which diesel locomotives move the buggies that carry the charging boxes loaded with the solid materials to be charged into

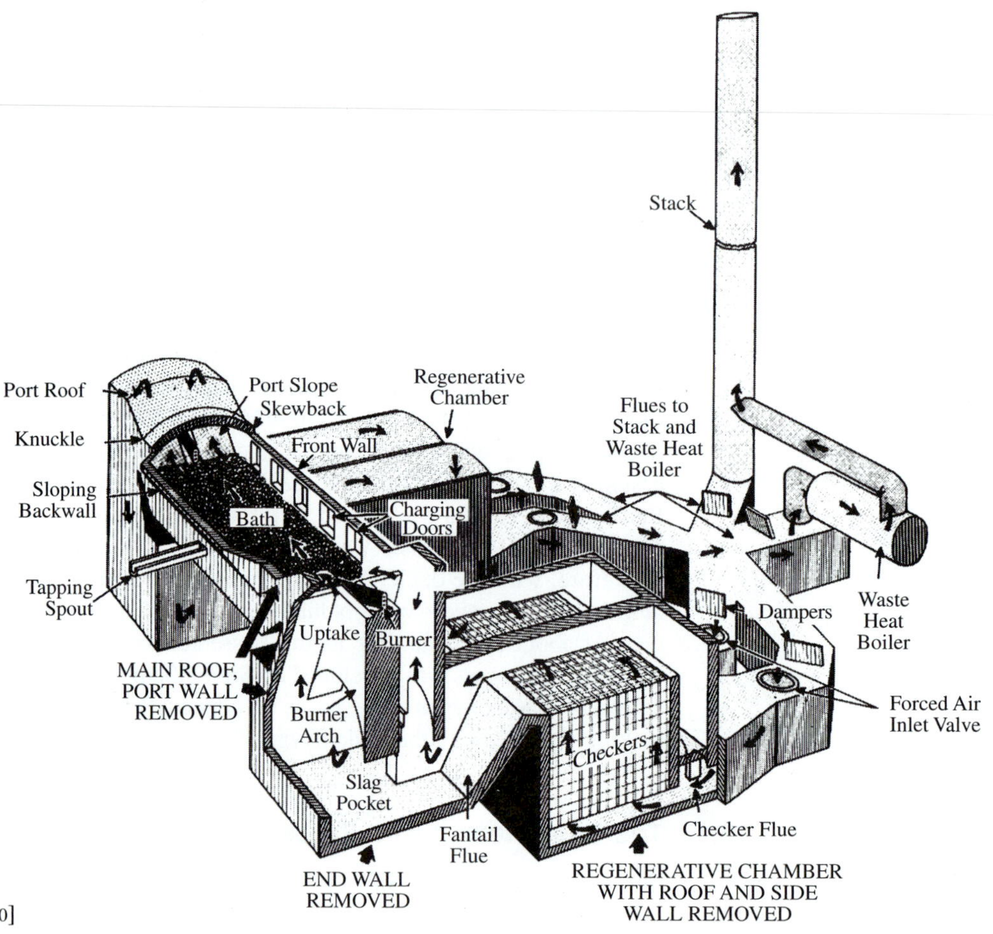

Figure 3.5
Passage of blown air and of burned gas through the regenerative system of the open-hearth furnace. [USS10]

the furnace. Another track, parallel with the first, is used by the charging machines that pick up the charging boxes one at a time, thrust them through an open door into the furnace, and turn them to dump the material on the hearth. Overhead cranes carry ladles of molten iron from the mixers and pour it into the furnace. Hot metal mixers with capacities of about 800 tons each are located in a separate building at the end of the charging floor. They serve as reservoirs of molten pig iron and also for mixing several lots of iron from the blast furnace to equalize the composition. The cranes can carry 100 to 150 tons of molten metal.

The pouring floor, or pit side, is about 25 m wide and serves for the transfer of the molten steel from the furnace to the ingot molds standing on mold cars along the wall opposite to the tapping side of the furnaces. Ingots are castings and the molds have been cast from pig iron. Ingots are made in various sizes from 300 tons weight for large forgings down to 10–40 tons for slab rolling. The ingots are allowed to solidify while the molds are held at the pouring platforms, and then they are moved to the strippers. The strippers are carried by stripper cranes. The ingots are then transferred to the adjacent building that contains the soaking pits. These are furnaces in which the ingots are kept at a temperature suitable for subsequent hot rolling, for instance at 1,250°C.

Both acid and basic OH furnaces are in use. Their construction is practically the same, but they differ in the kind of brick used for the bottom and the roof of the hearth. In the United States the majority of the OHFs are of the basic type. The OHF is both reverberatory and regenerative as regards the use of heat. The reverberatory effect is due to the radiation of the heat from the rather low roof of the furnace onto the charge in the hearth. The regenerative action is the one originally introduced by Siemens, which allows the heat capacity of the fuel to be used much more efficiently to achieve sufficiently high temperatures. The hot gases out of the furnace are passed through regenerative chambers to heat brick known as checkers. There are two sets of these chambers. Once one of the chambers is hot enough, the hot gases from the furnace are directed into the other chamber, and the first is used to heat up air to be blown into the furnace for combustion of the fuel. Thus the heat from the products of combustion is stored in the checkers and recovered in the air in the second half of the cycle. Using hot air causes a substantial increase in the temperature of the flame. The furnace is fired by liquid fuel entering through the burners shown at the front of Fig. 3.5 into the furnace, and the hot waste gases exit through the rear set of checkers into the stack. There are five doors at the front of the furnace, shown at the back in Fig. 3.5, while the tapping spout is shown in the foreground. The checker chambers are located under the charging floor. In the basic type of OHF the bottom is made of burned-in grain magnesite or of rammed magnesite. The hearth lasts for about five years, through about 3000–6000 heats. In the acid OHF the bottom of the hearth is made of silica brick.

Several sources of oxygen are available for the process: rust on scrap, CO_2 resulting from calcination of limestone, oxides in slags at different stages (FeO, MnO), oxygen from combustion air, roll scale collected from the hot rolling mills, and iron ore. Solid parts of the charge are placed in the furnace and heated for a sufficient time before pouring in molten iron. Limestone is charged first, iron ore is spread over it, then steel scrap and solid pieces of pig iron, discarded pieces of steel ingots, and broken pieces of cast iron ingot molds. The iron ore is charged as fine particles or sintered, modulized, and briquetted. After molten iron is added important chemical reactions occur, determined by the proportions of the various components of the charge. These cause the removal of Si and Mn, then C, and finally P. The oxides of these elements become parts of the slag. Sulphur is also transferred to the slag as CaS. With the progress of removal of C from all the liquid bath the temperature of the steel rises, and this has to be supported by increasing the supply of the fuel. At this stage when low-carbon steel is being produced, oxygen is blown on the bath through roof lances. During the whole process samples are taken to determine the chemical composition of the bath, which can then be corrected by additions of lime, fluorspar, and roll scale. The total heat time of a furnace with a 200-ton capacity fired with a liquid fuel such as tar or oil is about 10 hours.

Even during tapping and teeming of the steel into the ladle and from the ladle into the ingot molds, additions are made to remove oxygen dissolved in the metal, to adjust final composition, and give special properties to the steel, such as resistance to corrosion or good machinability. Elements that are not oxidizable, such as copper and nickel, may be added to the furnace before tapping, and chromium or manganese, if required, may also be added to the furnace just before tapping. During the filling of the molds deoxidization is achieved by adding aluminum or silicon.

3.3.2 Basic Oxygen Furnace (BOF)

In the basic oxygen furnace pure oxygen is blown on the surface of molten blast furnace iron bath. The oxygen gas is the only active agent, and burning of the impurities in the iron is the only source of heat added to the heat content of the bath. No external fuel is used, and these internal heat sources are more than enough to melt an added charge of steel, up to 30%. The elements involved in the chemical reactions are carbon, silicon, manganese, phosphorus, sulphur, nitrogen, and oxygen.

The furnace is barrel-shaped, about 10 m high, and mounted on pivots so that it can be tilted for charging scrap and pouring in the molten iron (Fig. 3.6). After charging it is brought back into the upright position, and the oxygen lance is lowered into the furnace and starts blowing oxygen. Then fluxes are added through a chute in the hood. Fumes are drawn off into the hood by a fan. The time between the charging of scrap and the start of oxygen blow is less than three minutes. When the blow is complete, the furnace is tilted horizontal to the charging side, and a sample of steel is taken for spectrographic analysis. Thereafter, it is returned to the vertical position and, as needed, scrap or limestone is added or an additional short oxygen blow executed. Then the furnace is tilted in the opposite direction for tapping the steel into a ladle; during this process alloying additions are made. Then the furnace is tilted back to the charging aisle and fully inverted to dump slag into the slag pot. Finally, the empty furnace is turned back to the charging position for the start of the next heat. The whole cycle is completed in about 20–40 minutes depending on the size of the furnace. Most of the furnaces in existence have capacities of 175–200 tons per heat; the largest furnaces built have capacities of 350 tons/heat.

The chemical reactions in the BOF are similar to those in the OHF, but they occur much more rapidly. Carbon is removed as carbon monoxide and carbon dioxide. Silicon is oxidized to silica and transferred to slag, and it is almost completely eliminated. Manganese is considered beneficial, and its residual amount depends on its level in the molten iron used in the charge. Phosphorus is efficiently removed in the slag. Sulphur is also efficiently removed, even better than in the OHF, where some enters in the fuel. Because oxygen is used and not air, the steel contains very little nitrogen. The BOF can operate on blast furnace iron with 0.2–2.0% Si, 0.4–2.5% Mn, and up to 0.3% or 0.4% P. With double slagging, iron with up to 2.0% P can be used.

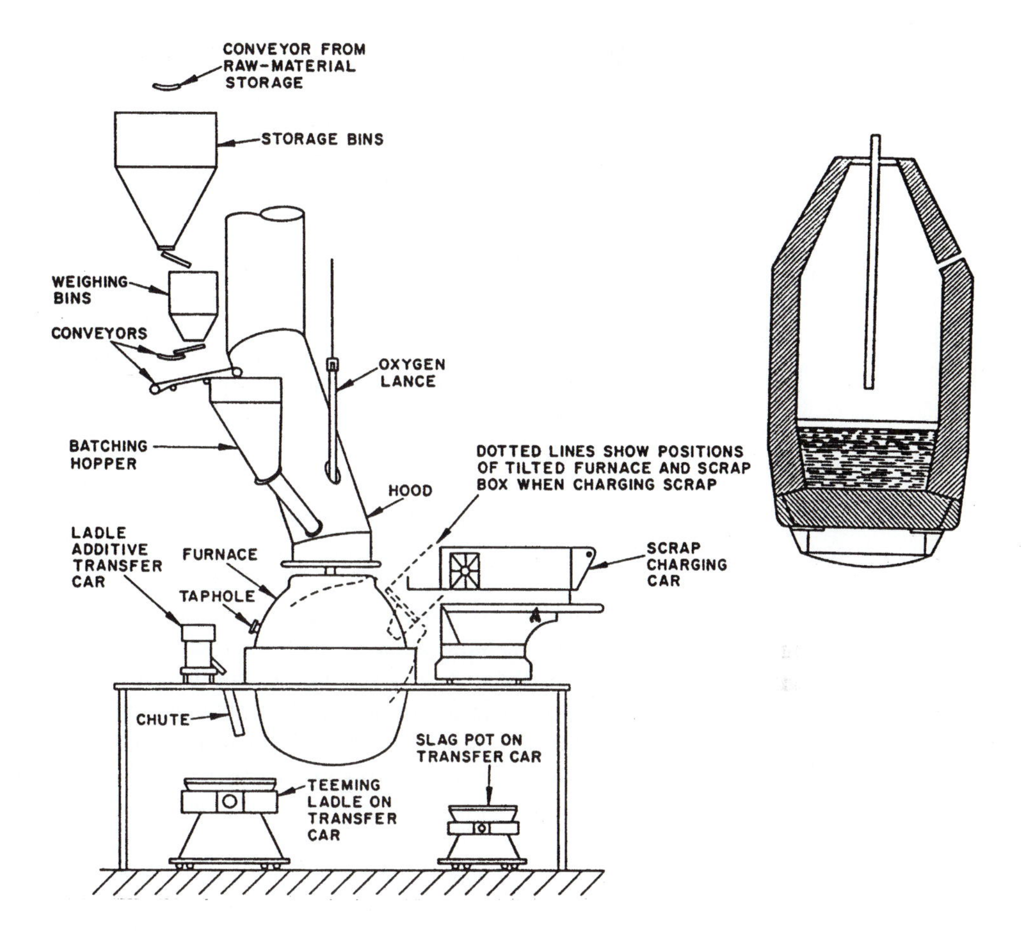

Figure 3.6
The basic oxygen furnace and its tilting capability. [USS10]

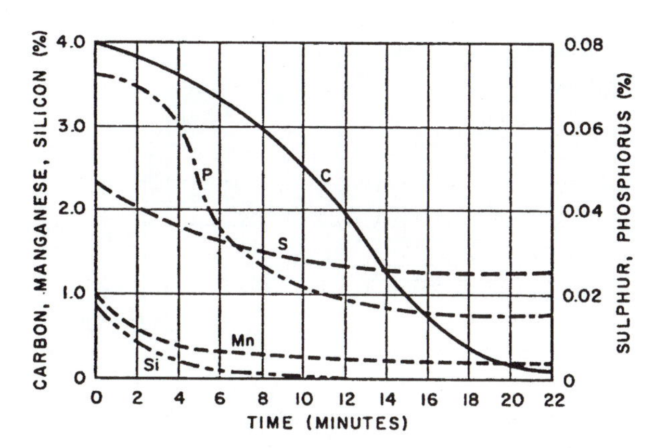

Figure 3.7
Schematic representation of the progress of refining the charge in a basic oxygen furnace. [USS10]

The progress of refining the charge is represented in the graph of Fig. 3.7, which shows that oxidation of Si, Mn, and P precedes that of C. Notice that scales for S and P are on the right side of the graph. Steels for a variety of applications are made in the BOF, for hot- and cold-rolled plates and sheets, bars, wires, and common alloy steels.

3.3.3 Electric Furnaces

Both basic and acid electric furnaces have been built. The latter were used in the past for melting steel in foundries. However, basic furnaces are currently used not only for producing steel for rolling but also primarily for casting applications. We will limit discussion to this latter type. Heat is generated either in the electric arc or by electric resistance in the induction-type furnaces. Vacuum melting furnaces for specialty steels also exist in both the electric arc variety, using consumable electrodes, and the induction heating design.

The basic electric arc furnace is a cylindrical vessel with a concave bottom and concave roof (see Fig. 3.8). The bottom is lined with burned magnesite brick and a layer of high magnesia on the top, which is the working surface. The walls are lined with magnesite-chrome brick, and the ceiling is covered by high alumina brick. The bottom slopes up to the tapping spout and on the opposite side is the slagging door, covered by fireclay brick. The furnace rests on curved, toothed rockers that permit up to 45° of tilt forward for tapping and 15° of tilt backward for deslagging. The tilting motion is moved by a rack-and-pinion mechanism driven by an electric motor. The roof can be lifted and swung aside to permit charging material into the furnace from the top. The charge consists mostly of scrap of all kinds, from machining chips, to medium-size steel parts and large pieces such as ingot butts and broken roll sections, all of them easily passing through the large open top of the surface.

The electric arc furnace works essentially from steel scrap. Three large electrodes pass vertically through the roof. They can be raised or lowered to bore into the scrap charge and maintain the proper length of the arc. These electrodes are made from strong,

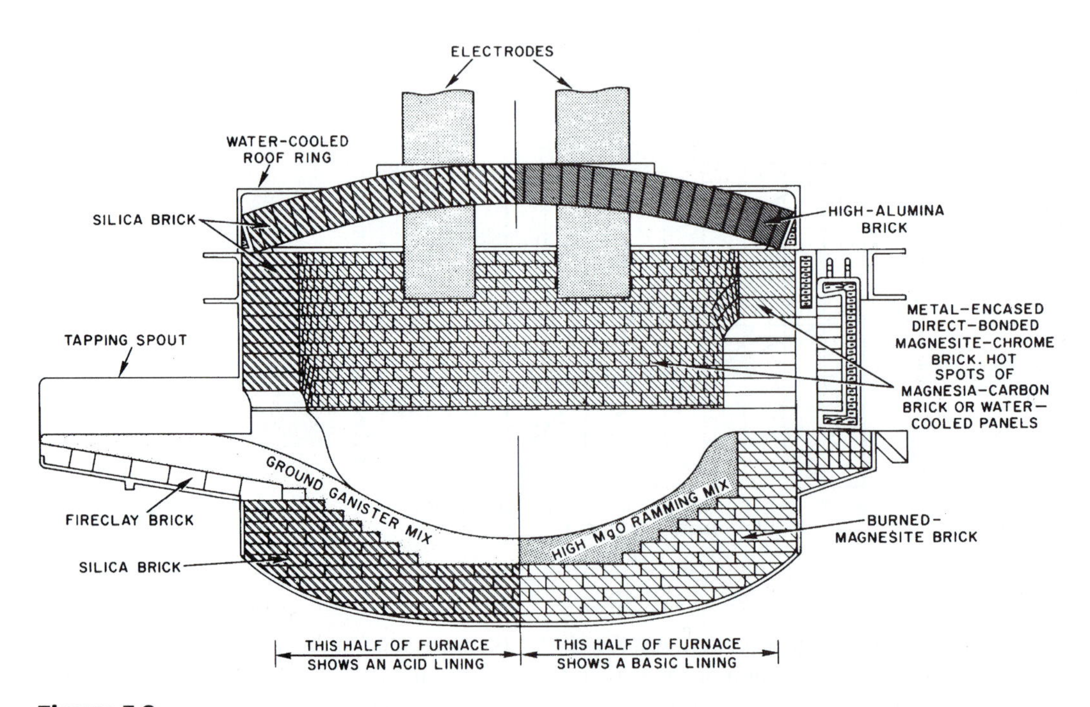

Figure 3.8
Cross section of the electric arc furnace. (Only two of the three electrodes are shown in this section). [USS10]

high-quality graphite capable of carrying large currents. Depending on the size of the furnace, electrode diameters range from 50 mm to 600 mm. A 500 mm-diameter electrode can carry up to 60,000 A.

The furnaces exist in a great range of sizes, with shell diameters from 2 m to 10 m and capacities from 32 tons to 400 tons. The corresponding transformer capacities range from 2,000 to 150,000 kVA of three-phase, 60 Hz power. The secondary winding is provided with taps to permit selection of different melting voltages, with the highest voltages ranging from 250 V to 700 V. The current in them is regulated by changing the arc gap and ranges from 30,000 to 100,000 A.

Practically all grades of steel can be produced, including plain carbon, low-alloy, high-manganese, high-silicon, all kinds of stainless, super alloy, high-speed, and other tool steels. The composition of the steel depends on the alloying elements in the charge and can be controlled by the choice of slags. The sources of oxygen for oxidizing of phosphorus, manganese, silicon, and carbon are calcination of limestone, oxides of alloying elements added to the charge, ore, cinder, and scale, and mainly oxygen gas injected into the bath when needed.

Induction-type electric furnaces are used to melt charges in a crucible. The charge is carefully selected to produce the desired composition with minimal further additions. Special alloys can be produced by melting in vacuum or in an inert gas atmosphere (see Fig. 3.9). A coreless, high-frequency induction unit is enclosed in a tank that can be either evacuated or filled with an atmosphere of any desired composition and pressure. Most vacuum furnaces are of one-quarter- or one-half-ton capacity, but larger units up to 65 tons have been used.

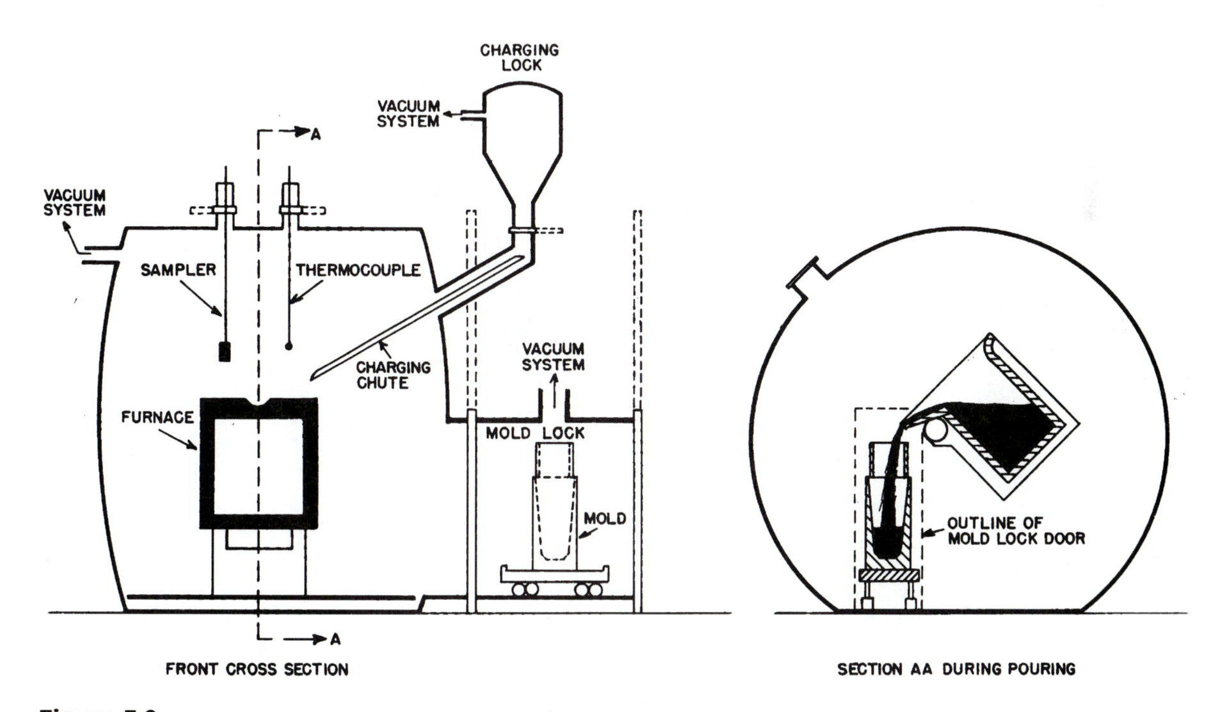

Figure 3.9
Schematic arrangement of electric induction furnace in vacuum chamber equipped with charging and mold locks. [USS10]

Another type of vacuum melting furnace is the consumable electrode type. The electrode chamber and a copper mold are surrounded by a water-cooled tank. Direct current is used to produce the arc with electrode-negative polarity. The material to be remelted is the electrode. During the process hydrogen, oxygen, and nitrogen contents in the molten metal are reduced, and center porosity and segregation in the ingot are eliminated. Mechanical properties of the steel are improved.

3.3.4 Summary of Steel Production

In 1910 the total steel production in the United States was 28 million tons, and it reached a peak of 115 million tons in 1956. The production decreased in the 1950s but recovered in the 1960s, and in 1969 it amounted to 128 million tons. Through the 1970s it stayed at the level of 120–130 million tons. The graph in Fig. 3.10 expresses the total production of raw steel in the United States from 1900 through 1980. The production for 1997 of 96 million tons is also given. The same graph plots the production of pig iron. The difference between the two graphs is due mainly to the use of scrap in steel production.

As regards the use of the three main processes, for many years the open-hearth process was dominant, but the basic oxygen process has become the leading one since 1969. The share of the electric processes has been steadily increasing. The largest basic OHF can produce heats up to 600 tons in 10 hours, whereas a large BOF produces heats up to 300 tons in three-quarters of an hour. The electric furnace produces heats up to 200 tons in 5 hours. The OHF and BOF produce carbon and alloy steels of the same general grades. Stainless steel can be made in the BOF but not in the OHF. Electric fur-

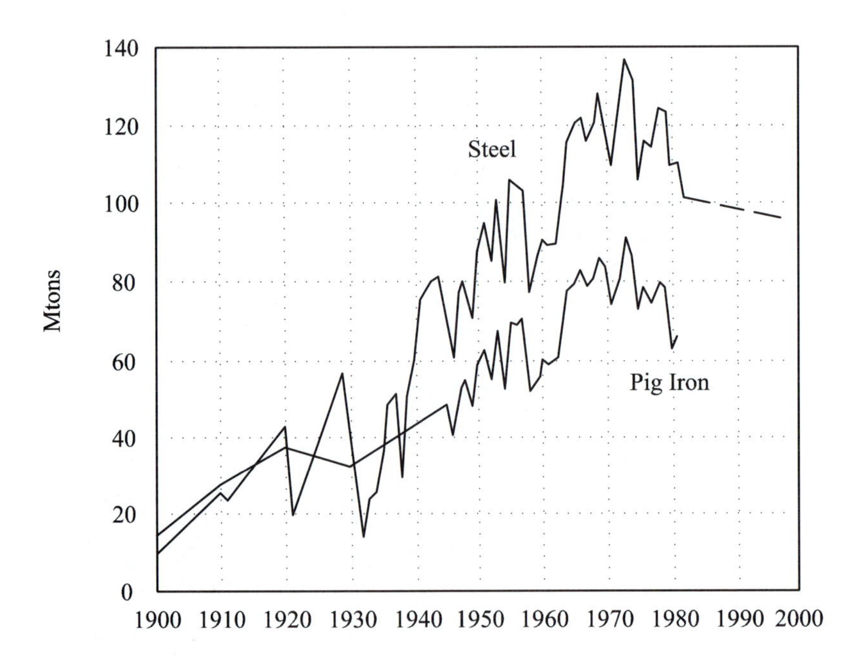

Figure 3.10
Production of iron and steel in the United States. [USS10]

naces can produce all kinds of alloy steels and they produce also carbon steels where there is no source of molten iron from the blast furnace. The BOF does not offer flexibility in the use of scrap as do the OHF and electric furnaces. The maximum percentage of scrap in the charge is 30% for the BOF, 55% for the OHF, and 100% for the electric furnace. The production of heats of steel at regular, short intervals is an advantage of the BOF when it is used to supply steel for the continuous casting process.

3.4 INGOTS: CONTINUOUS CASTING OF SLABS

Molten metal from the steel-making furnaces is in most instances cast into ingots. Alternatively, the casting of ingots and their initial hot rolling into slabs may be bypassed and slabs produced by continuous casting. Let us first discuss the various techniques of ingot casting and the various kinds of steel with respect to the subsequent rolling practice.

Ingots are made in a range of sizes, mostly 10–40 tons for rolling but up to 300 tons for open-die forging. They have shapes that are approximately square in cross section and slightly tapered along their height with the big end down or up. They are tapped from the transfer ladle into cast iron molds. After solidification they are stored in "soaking pits," where gas burners provide the heat necessary to equalize their temperature at the level suitable for the subsequent hot forming operations. The solidification of a 10-ton ingot takes about 2 hours.

There are gases dissolved in molten steel, primarily oxygen but also hydrogen and nitrogen. As the molten steel cools, the solubility of the gases decreases, and they are expelled from the metal. The chemical equilibrium between oxygen and carbon changes with decreasing temperature, and they react to produce carbon monoxide that is not soluble. Steel solidifies over a range of temperatures (see Fig. 2.13), and the gases evolved from the still-liquid portions may be trapped in the solidifying steel to produce blowholes in the ingot. Depending on the measures taken to deoxidize the steel in the heat before pouring into the ladle, or in the ladle, or while teeming the ingot, different kinds of steel are produced; these are called killed, semikilled, capped, or rimmed steel. The amount of oxygen dissolved in molten steel increases with the decreasing carbon content. To kill the oxygen in the low-carbon steels would require large amounts of deoxidizing elements, such as aluminum, magnesium, or silicon, and produce an excessive number of nonmetallic, hard inclusions in the steel. Therefore, these steels are mostly made as rimmed or capped, while steels with carbon content over 0.3% are produced as killed or semikilled.

The various forms of the trapped gas are illustrated in Fig. 3.11. The pictures are all of big-end-down ingots, although the killed steels are usually made with big end up; nevertheless, the drawings show the various features rather well. They range from a "dead" killed (no. 1) to a violently rimming case (no. 8). The dotted line indicates the height to which the ingot was poured. The fully killed ingot evolved no gas; its top is slightly concave and contains a shrinkage cavity called a pipe. Usually it is poured in big-top molds with refractory-lined hot tops, and the top portion with the pipe that has a shape as shown in (1) is sawn off before rolling. The semikilled ingots, (2) and (3), have blowholes at the upper portion, and (3) also has blowholes close to the surface on the side of the ingot. Ingot 4 produced so much gas that numerous honeycomb blow-

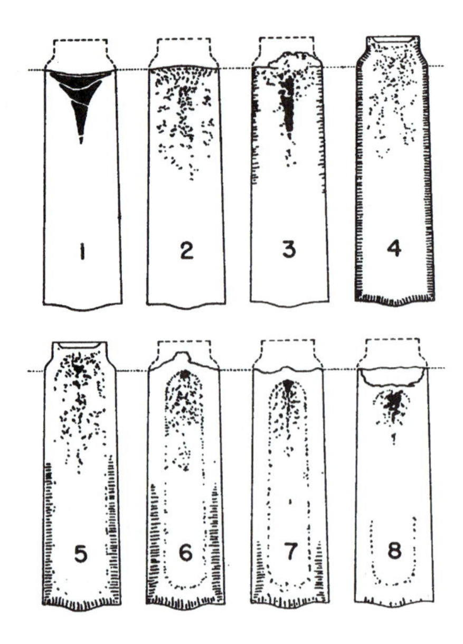

Figure 3.11
Various forms of trapped gas in a series of ingot structures. [USS10]

holes developed all over the surface of the ingot. The rising gas produced a boiling action at the tip called the rimming action that was stopped by a metal cap attached to the top of the mold. Ingot 5 is a similar case; however, there was so much gas that the strong currents along the sides swept away the side-surface blowholes in the upper part of the capped ingot. Ingots 6, 7, and 8 are all rimmed, not capped, with increasing amounts of evolving gas. The result is such that the blowholes at the sides are limited, and they do not reach the surface. Their internal surfaces will not be oxidized and not covered by scale. They will be successfully completely welded shut during rolling, unlike the defects that reach the surface and oxidize during the stay of the ingot in the soaking pit; these will not weld and will cause seams in the rolled plates and sheet.

Other phenomena, such as segregation and excessive columnar grain structure, must also be controlled by suitable pouring practices. Segregation means different compositions of steel in different parts of the ingot. Purer metal solidifies first. Correspondingly the center of the ingot would have more of the alloying elements. Turbulence due to gas evolution in the liquid steel increases the tendency of elements to segregate. Therefore, killed steels are less segregated than semikilled, and these are less segregated then capped or rimmed steels. The molten metal in contact with the cold walls of the mold freezes quickly, which results in a zone of small and randomly oriented crystals. From there on towards the center extends first a layer of long dendritic crystals; in the center, the dendrites are randomly oriented. However, all these crystals get crushed and reformed during the subsequent forging or rolling operations.

In summary, steels with 0.12–0.15% and even the 0.06–0.10% and those with less than 0.06% carbon are produced as rimmed steel, although the level of oxidations is differently adjusted in the furnace before pouring for each of these three classes. Capped steel is a variation of the rimming procedure. The capped ingot has a thin rim zone relatively free from blowholes and a core zone with less segregation than in the rimmed ingot. The procedure is used for steel with more than 0.15% C. Between 0.15 and 0.30% C semikilled steel is produced, and then the fully killed steel is made with higher C contents.

Gases absorbed by the liquid steel from the atmosphere can cause flaking, embrittlement, voids, inclusions, and other undesirable features in the solidified steel. This applies to hydrogen in particular and also to nitrogen. These problems are eliminated by exposing the surface of the liquid steel to low-pressure environment in various vacuum degassing procedures. The vacuum used is rather soft, just about 0.1 to 0.2 mm Hg. Several different systems are in use. One of them is shown in Fig. 3.12. In this procedure a stream of steel from a pony ladle to the mold is exposed to the vacuum in the surrounding tank. The droplets in the stream present a rather large surface area for the given volume of steel and provide for efficient removal of oxygen and of hydrogen. Carbon contents can be lowered further in steels with about 0.04% C, down to 0.01% by reaction with oxygen to form CO that is pumped out. In other, similar procedures the vacuum is applied to teeming from ladle to ladle or between the BOF and the ladle.

Processes exist that eliminate the need for ingots, soaking pits, and the primary stages of hot rolling of ingots to produce blooms and slabs. These primary products are usually allowed to cool down and are subsequently further hot rolled into plates, bars, and structural shapes. Blooms are of square cross sections between 150 mm $\times$ 150 mm and 300 mm $\times$ 300 mm, and slabs have rectangular cross sections 50–225 mm thick and 600–1500 mm wide.

The processes that bypass ingots and primary hot rolling are those of continuous casting. An example of a slab continuous-casting machine is shown in Fig. 3.13. It consists of a vertical section, a transitional curved section, and a horizontal section. The vessel at the top, called the tundish, is a container filled from ladles. In the machine of this example, it is 30 m above ground. The ladles are transported from the BOF to the tower of the casting machine. The steel may be, if required, vacuum degassed into other ladles, and the ladles are then hoisted in the tower above the tundish. This machine produces slabs up to 250 mm thick and 1900 mm wide.

To maintain continuous operation, two sets of ladle cars and ladle hoists are used on alternate heats. The tundish has multiple slide gates in its bottom and nozzles

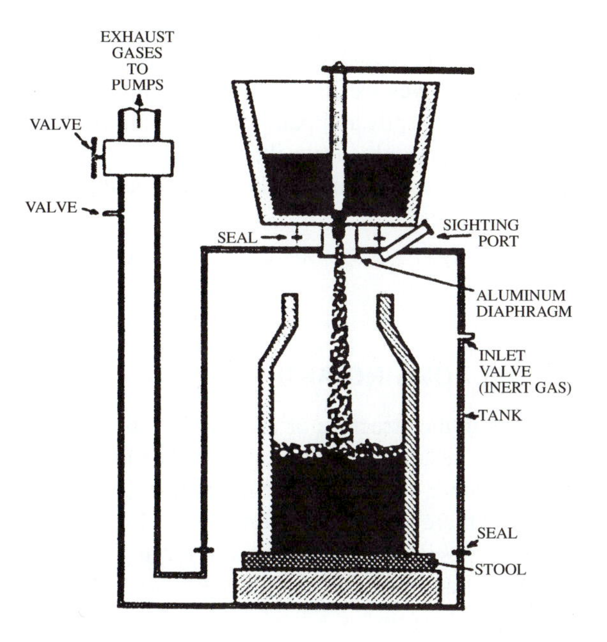

Figure 3.12
Schematic arrangement of vacuum-casting installation employing the stream-degassing technique. [USS10]

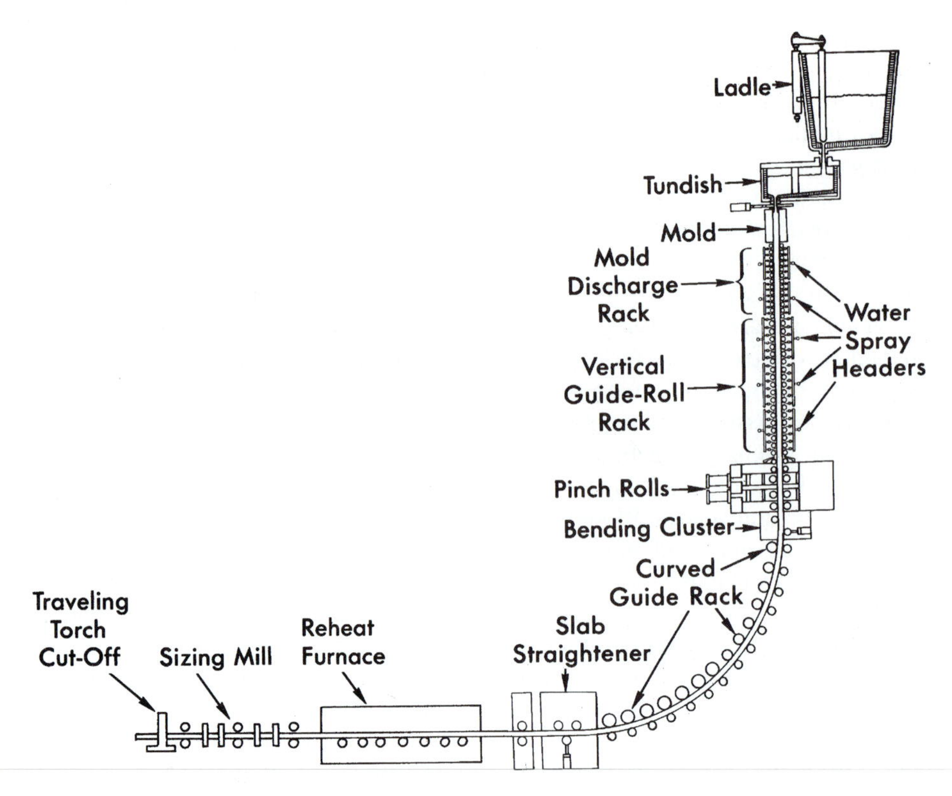

Figure 3.13
Continuous slab-casting machine. [USS10]

through which the molten steel enters into a water-cooled copper mold. In order to prevent the metal from adhering to the walls of the mold a thin skin solidifies on the metal, which is then further cooled by water sprays in the discharge rack and in the vertical guide-roll rack. Below this rack are the pinch rolls that provide the driving and determine the velocity of the movement of the slab, which is then bent. It moves on through the curved guide rack and slab straightener into a reheat furnace. On exiting the furnace the slab receives its final dimensions and shape in a sizing rolling mill with sets of pairs of vertical rolls and pairs of horizontal rolls. As they exit the rolling mill, the slabs are cut to length by a traveling torch. Continuous casting machines are used also for casting blooms. Carbon and alloy steel, including stainless, have been successfully cast in this manner.

3.5 HOT FORMING: OPEN-DIE FORGING AND ROLLING

The changes in the structure of the material as it is hot formed and the corresponding changes in the resulting properties as well as problems with overheating and scale are discussed in more detail at the beginning of Chapter 4. The various forging processes are described there as well, including those of open-die forging (ODF), followed by the many other types, including impression forging and closed-die forging. Thus, although ODF is an operation carried out on ingots taken out of the soaking pit, especially very large ones and therefore should be discussed in this chapter, we will postpone this subject to Chapter 4 for the sake of completeness in the discussion of forging. On the other hand, because rolling mills are traditionally located in the steel-making factories and practically nowhere else, both hot and cold rolling are discussed here except for the analytical treatment of cold rolling, which is discussed in Chapter 5.

3.5.1 Primary Hot Rolling

The primary hot rolling operations are used to produce slabs, blooms, and billets in sizes shown in Fig. 3.14. These primary products will be further hot rolled. Usually, however, they are first left to cool down and subjected to conditioning to remove various defects arising from casting and rolling, so as to prevent them from affecting the surface quality of finished products. As such, they are called semifinished steel. The defects to be removed are ingot cracks, from interdendritic zones, folds due to surging of the molten metal that produce transverse cracks, scabs that form as oxidized patches solidified in contact with the ingot mold wall, deep seams and other defects produced

Typical Cross-Section
and
Dimensional Characteristics*

SLAB

Always Oblong
Mostly 50-230 mm (2-9 in) thick
Mostly 610-1520 mm (24-60 in) wide

BLOOM

Square or Slightly Oblong
Mostly in the range 150 mm x 150 mm (6 in x 6 in)
to 300 mm x 300 mm (12 in x 12 in)

BILLET

Mostly Square
Mostly in the range 50 mm x 50 mm (2 in x 2 in)
to 125 mm x 125 mm (5 in x 5 in)

*Dimensions usually given to nearest round number.
All corners are rounded, as shown.

Figure 3.14
Comparison of the relative shapes and sizes of rolled steel governing nomenclature of products of primary and billet mills. [USS10]

in primary rolling such as cinder patches, burned steel arising from the flame in the soaking pit and ruptured in the mill, and laps derived from fins turned down in rolling. The conditioning procedures start with inspection, then pickling in chemicals to remove scale, grid blasting, hand chipping of defects using hammers and pneumatically driven chisels, peripheral milling, scarfing that consists of driving the surface defects away by using oxygen torches, and very efficient grinding with high-power, coarse-grain grinding wheels mounted on spindles attached to long arms that are free to move in the direction normal to the ground surface, under a constant load; the grinding wheel floats over the surface while removing a layer of the defective surface. After conditioning, the semifinished steel is heated again for further rolling into plates, bars, rods, and various profiles such as **I**, **U**, **H**, and railroad rail.

The primary roughing mill is mostly a two-high reversing mill (see Fig. 3.15). The ingot is rolled flat and simultaneously widens and lengthens. The amount of reduction of thickness $\Delta = h_1 - h_2$ is called *draft,* where h_1 is the thickness before and h_2 after one pass. The magnitude of the draft depends on the force, torque, and power of the mill and also on friction that is relied on to pull the slab into the roll gap (the "bite"; see Fig. 3.16).

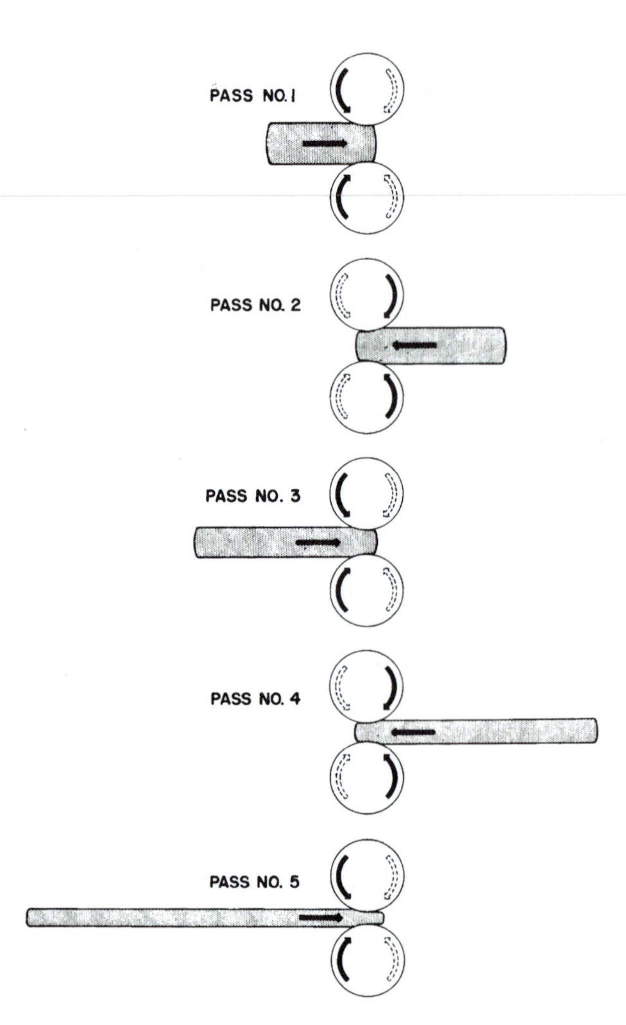

Figure 3.15
Diagrammatic representation of the sequence of rolling operations involved in reducing an ingot to a slab on a two-high reversing mill. [USS10]

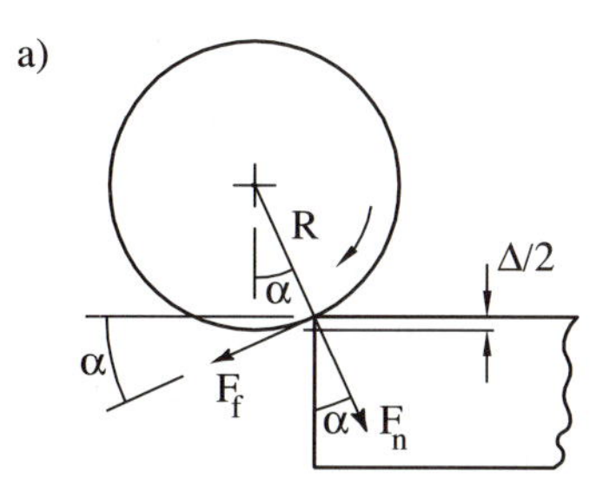

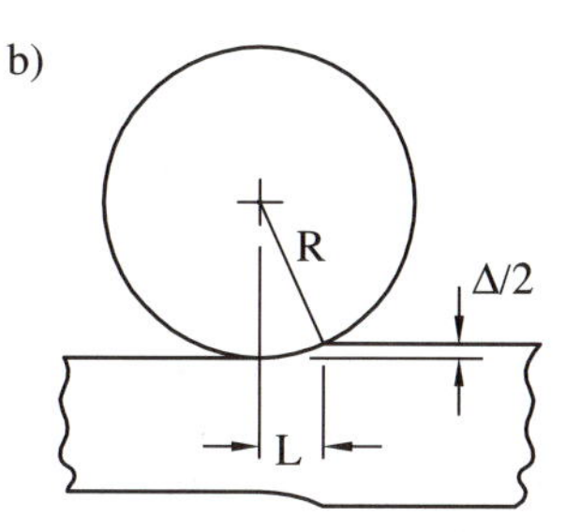

Figure 3.16
Forces and geometry of a) the "bite", and b) fully engaged rolling.

As shown in Fig. 3.16a, the block to roll will be drawn in between the rolls if the horizontal component of the friction force F_f is larger or, at least equal to the opposing horizontal component of the normal force F_n:

$$F_f \cos \alpha \geq F_n \sin \alpha \tag{3.1}$$

and it is

$$F_f = \mu F_n \tag{3.2}$$

where μ is the coefficient of friction,

$$\tan \alpha = \mu \tag{3.3}$$

It is further

$$\cos \alpha = \frac{R - \Delta/2}{R} = 1 - \frac{\Delta}{2R}$$

where $\Delta << R$, and $\sin \alpha = \sqrt{1 - \cos^2\alpha}$.

$$\sin \alpha = \sqrt{1 - 1 + \frac{\Delta}{R} - \left(\frac{\Delta}{2R}\right)^2}; \text{ neglect } \left(\frac{\Delta}{2R}\right)^2,$$

$$\sin \alpha \cong \sqrt{\frac{\Delta}{R}} \tag{3.4}$$

and

$$\tan \alpha = \sqrt{\frac{\frac{\Delta}{R}}{1 - \frac{\Delta}{R} + \left(\frac{\Delta}{2R}\right)^2}} \cong \sqrt{\frac{\Delta}{R - \Delta}} \approx \sqrt{\frac{\Delta}{R}} \tag{3.5}$$

So, approximately

$$(\tan \alpha)^2 = \mu^2 = \frac{\Delta}{R}, \text{ and } \Delta_{\max} = \mu^2 R \tag{3.6}$$

Equation (3.6) shows that for large Δ, the roll radius R should be large. For instance, if $\mu = 0.2$, $R = 250\Delta_{\max}$ and, for "sticking friction," well applicable to hot rolling, $\mu = 0.5$ and $R = 3.85\Delta_{\max}$.

On the other hand (see Fig. 3.16b), the rolling force F, once the block is already between the rolls, will be approximately

$$F = p\,L\,w \tag{3.7}$$

where L is the length of contact between the roll and the work, w is the width of the workpiece, and p is the forming pressure,

$$p = Q_f\,Y \tag{3.8}$$

where Y is yield strength and Q_f is a coefficient accounting for the effect of friction on the rolling pressure (see further discussion in Chap. 5.); let us choose $Q_f = 1.2$. The length L depends on draft Δ and roll radius R:

$$L^2 = 2R\frac{\Delta}{2} = R\Delta \tag{3.9}$$

The rolling force is then

$$F = 1.2\ Yw\ \sqrt{R\Delta} \tag{3.10}$$

and, assuming that the resulting force acts on an arm $L/2$, the torque T on one roll is

$$T = F\frac{L}{2} = 1.2\ Yw\ R/2 \tag{3.11}$$

The torque is proportional to the roll radius R. To minimize the force F that determines the cross sections of the uprights of the rolling mill frame and also the torque T that determines the size of the driving motor, the roll radius should be small. So, the demands on the magnitude of the roll radius are contradictory between the requirement of large first bite and that of small force and torque in steady-state rolling.

Indeed, for hot rolling, especially for the roughing work, producing slabs and blooms, the criterion of pulling the workpiece into the roll gap prevails, and large roll radii are used, typically about 500 mm. The yield strength of the hot steel is not very high and, correspondingly, the force and torques are acceptable. The pressure in the contact between roll and work is of the order of 100 MPa for carbon steels in the initial passes, at work temperature of 1100°C, and 285 MPa in the later passes, at 800°C. In the finishing operations, especially in cold rolling of sheet, the yield strength is high, two to three times higher than of the hot metal, and this is also due to the strain-hardening effect. The rolling pressures are of the order of 500 to 1000 MPa. The drafts are small, typically 1 mm; this would, from the condition (3.6), with $\mu = 0.15$, call for $R > 44$ mm. In the tandem mills with four-high stands the work rolls usually have $R = 250$ mm. Cluster mill rolls have small radii, down to 50 mm. The finishing rolls in hot plate rolling typically have $R = 350$ mm. The rolling power, from torques on both rolls, depends mainly on the speed v of the workpiece; it is given by

$$P = 2\,T\,\omega = 2T\,v\,/\,R = 1.2\,Y\,w\,v \tag{3.12}$$

and it does not depend on roll radius.

3.5.2 Rolling Mill Configurations

The various roll configurations are presented in Fig. 3.17. In the two-high pull-over mill (Fig. 3.17a), the workpiece passes between the rolls and is then returned over the top roll back to the start and is passed again through the rolls, which have meanwhile been adjusted for a smaller gap. In a two-high continuous mill (Fig. 3.17b), the workpiece passes through three stands, being stepwise reduced in thickness. A very common roughing mill arrangement is the reversing two-high mill shown in Fig. 3.17c. The

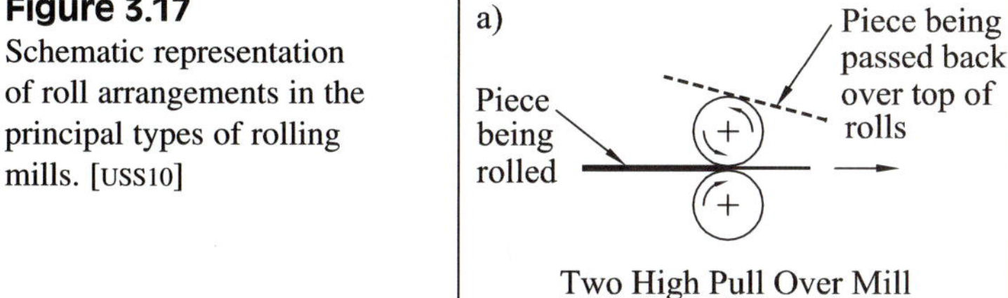

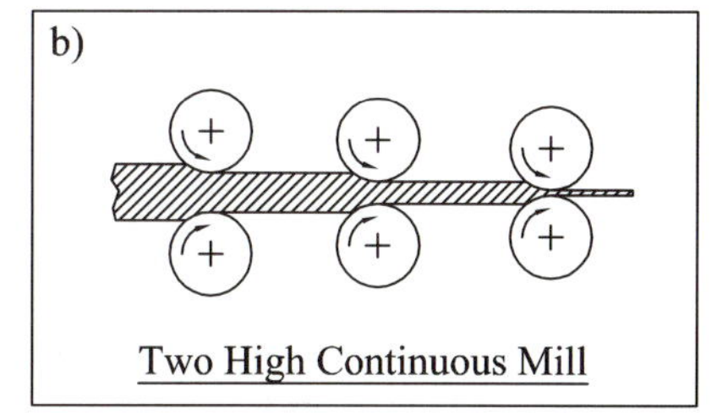

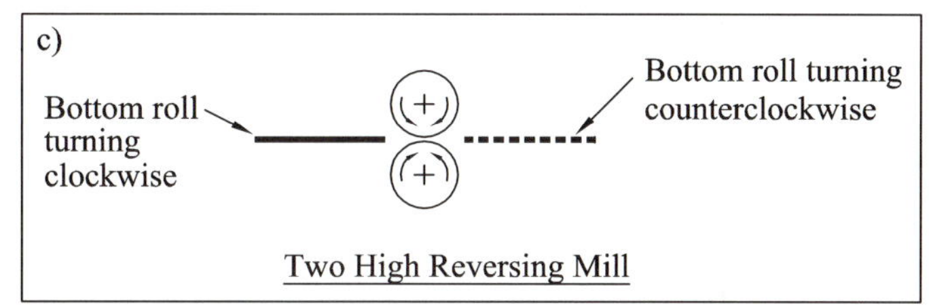

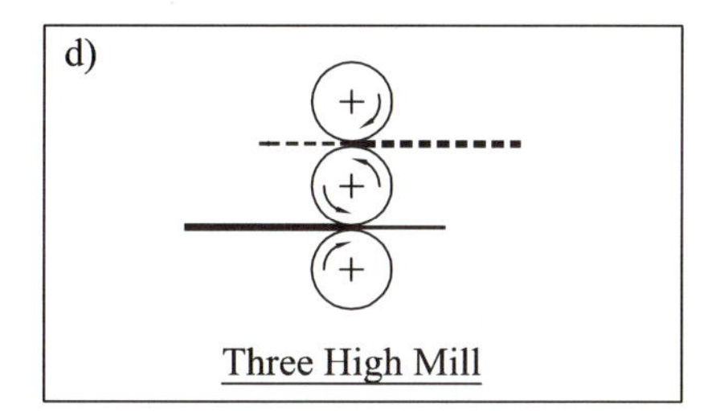

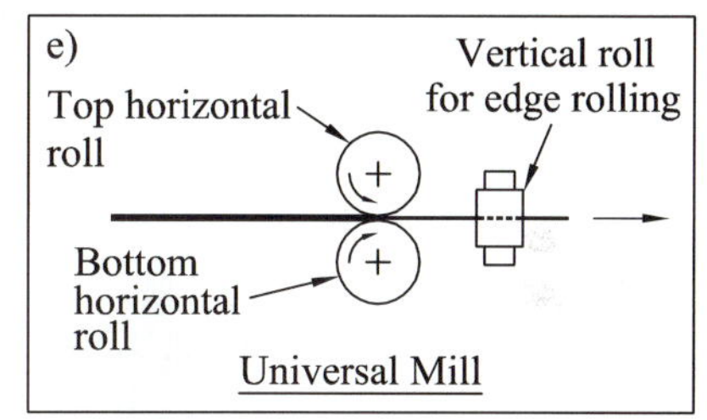

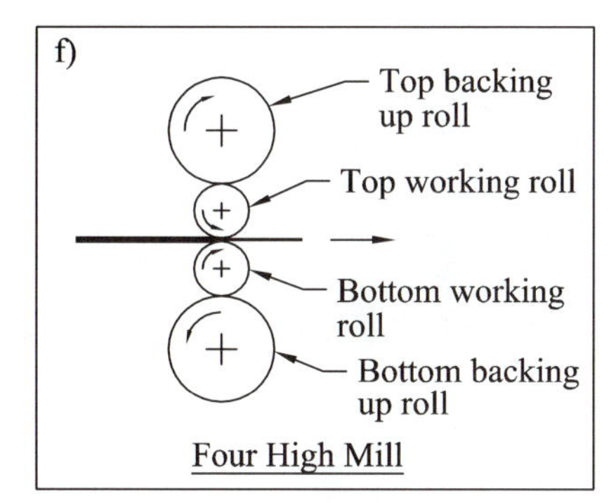

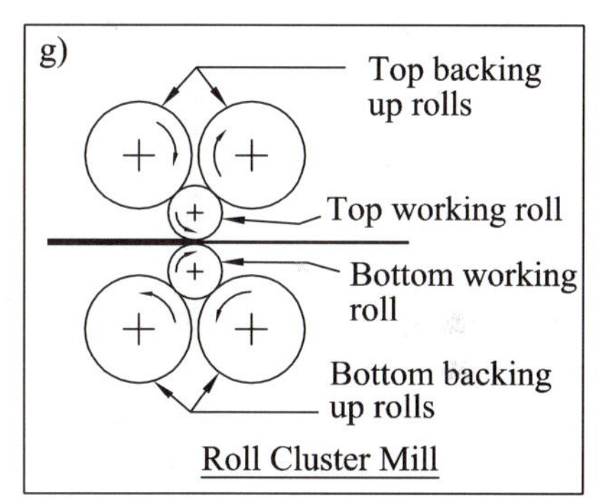

Figure 3.17
Schematic representation of roll arrangements in the principal types of rolling mills. [USS10]

workpiece passes back and forth through the gap, which keeps diminishing pass after pass. An alternative is the three-high mill shown in Fig. 3.17d. The workpiece passes between the middle and bottom rolls to the right onto a table which is then lifted to pass the workpiece back between the top and the middle roll.

In the production of slabs the two-high reversing mill is built in a "high-lift" design where the gap between the rolls can be set rather large to roll the edges of a slab rotated 90° to a vertical position. The same task can be performed without rotating the slabs on the "universal mill" shown in Fig. 3.17e. This mill has a pair of rolls with horizontal axes and another pair of rolls with vertical axes. The latter pair does the edge rolling. The four-high mill depicted in Fig. 3.17f uses work rolls with small diameters that are supported against bending by large-diameter back-up rolls. This configuration is used both for hot rolling of plates and cold rolling of sheet metal. They usually exist as tandem mills in which the workpiece passes continuously through four to six stands, being gradually reduced in thickness and moving faster and faster. An even smaller-diameter pair of rolls is used in the cluster mill (Fig. 3.17g), and each of these is supported by two back-up rolls. This arrangement is used for one-pass rolling of thin sheet. The principle of this arrangement is further developed in the Sendzimir cold-rolling mill, where each work roll is supported by altogether nine rolls in three tiers. The gap is adjusted not by a screw-down of the bearings of the top rolls, as in all the other setups, but by rotating eccentric rings at the saddles of the bearings of the back-up rolls. It is used for one-pass rolling of alloy steel sheets. A different design permitting the use of small-diameter rolls in hot rolling of thin plates is the planetary mill in.

On those mills where the workpiece passes in both directions, such as a two-high, high-lift mill or three-high mill, the slabs move out of the roll gap onto front and back tables, where they move on live (driven) rollers that decelerate the workpiece and accelerate it back for the next pass. The workpiece may be shifted sideways to move to the various openings between the rolls or turned by 90° or 180° on these tables by means of devices called manipulators. The passes applied to billets on a three-high mill are illustrated in Fig. 3.18. Since the pieces being rolled must be entirely out of the rolls

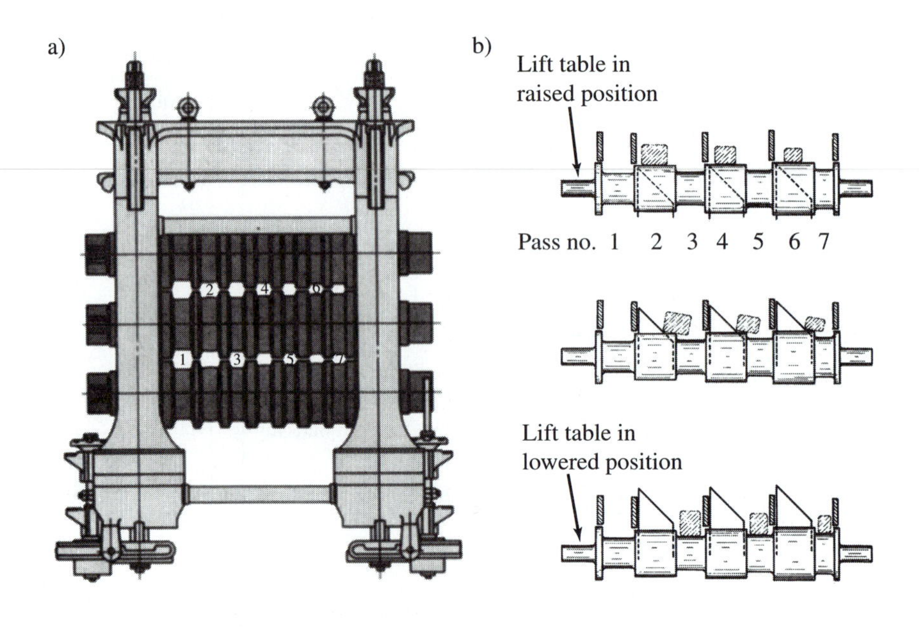

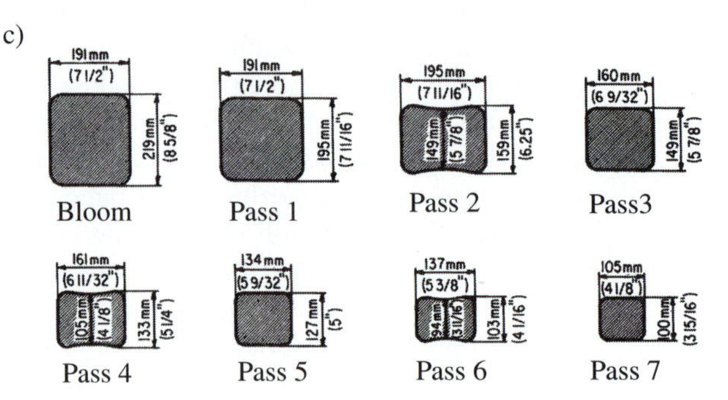

Figure 3.18
a) Passes in rolling billets in a three-high mill. b) Action of the fingers in turning the piece from pass to pass.
c) Cross sections arising in the individual passes. [USS9]

for the lift table to transfer them between the top and bottom passes, their length must be limited to that of the table. The blooms exiting from the blooming mill must be cut to such length that when they are elongated they will still fit on the tables of the billet mill. Another example of an application of a three-high rolling mill is the production of rails. Rails have been rolled for over a hundred years, and their manufacture has developed into a rather complex system. For only a small glimpse of this technology, an example of roll-pass contours for rails is given in Fig. 3.19.

The forces, torques, and powers in the hot rolling operations are expressed in the graph of Fig. 3.20 where the rolling pressures are given for different steels as functions of the work temperature. The temperatures range from 1300°C down to 700°C, covering from the roughing to the finishing operations. Over this range, the pressures

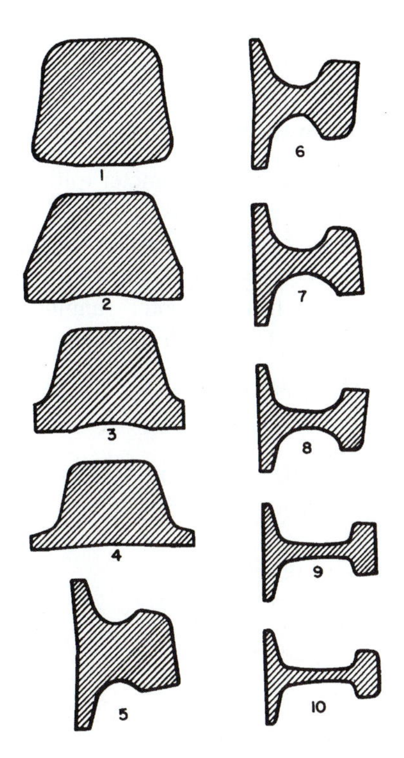

Figure 3.19
Roll-pass contours for producing a 132-pound RE rail. [USS9]

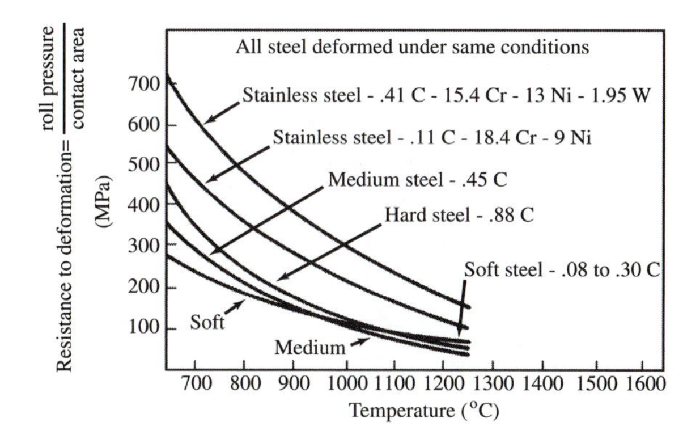

Figure 3.20
Specific rolling pressure as a function of temperature, for various steels. [USS10]

increase four to five times. Correspondingly, for a given torque and power available in the various mills, the draft will have to be reduced. Numerically, the range spans from 70 MPa to 350 MPa for carbon and low-alloy steels, and from 140 MPa to 700 MPa for stainless steels. This may be compared with the room-temperature yield strength of 300–600 MPa for most steels in the normalized condition.

Specific powers are listed in [1] as between 1.2 $W/cm^3/min$ for low-carbon steel at high temperatures and 3 $W/cm^3/min$ for high-alloy steels and lower temperatures. An interesting comparison can be drawn with Table 8.1 where specific forces and simultaneously specific powers are listed for cutting (machining) various metals. For a medium-carbon steel the specific force of 2200 N/mm^2 (MPa) is given, which translates into specific power of 37 $W/cm^3/min$. This illustrates that hot rolling forms the metal more than ten times more efficiently than machining. To illustrate, let us work out an example.

EXAMPLE 3.1 Specific Power in Rough Hot Rolling ▼

Let us assume a roll with $d = 1$m diameter rolling a slab $w = 1$m wide and $h =$ 250 mm thick, reducing the thickness down to 210 mm. This results in a draft of $\Delta = 40$ mm (see Fig. 3.21). Assume further the rolling pressure $p = 150\ N/mm^2$ (MPa) and a rolling speed of $v = 1$ m/sec. This corresponds to $n = 19.1$ rpm roll speed, which is derived from a motor running at 150 rpm through a gear-transmission ratio of 7.85. The length of contact is

$$L = \sqrt{R\Delta} = 141\ mm$$

and the rolling force F is

$$F = pLw = 150 \times 141 \times 1000 = 2.12e7\ \text{N}.$$

The torque driving each roll is

$$T = FL/2 = 1.496e6\ \text{Nm}$$

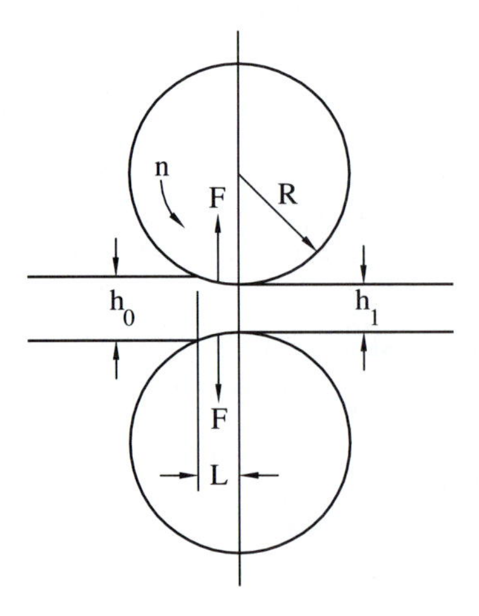

Figure 3.21
Drawing for Example 3.1, to use in determining specific power in rolling.

and the power driving each roll is

$$P = T\omega = T\,2\,\pi\,n/60 = 2{,}991 \text{ kW}$$

The volume of metal rolled per minute MRR is

$$MRR = \Delta\, w\, v = 4 \times 100 \times 100 \times 60 = 2.4e6 \text{ cm}^3/\text{min}$$

and specific power P_{sp} is

$$P_{sp} = P/MRR = 1.25 \text{ W/cm}^3/\text{min}.$$

The drives of hot rolling mills are derived mostly from DC motors with ratings between 2,000 and 10,000 kW. ▲

3.5.3 Hot Forming of Tubes and Pipes

Apart from the production of slabs, blooms, billets, plates, rails, structural profiled beams, and bars there is the significant and interesting process of hot forming of tubes and pipes for a great variety of applications: gas, oil, and water transportation; construction of railings, scaffolds, columns, and bridges; drilling for oil and natural gas; and for boilers and heat exchangers. Two different kinds of processes are used: in one of them butt-welding is applied to hot rolled skelp, and the other one produces seamless pipes.

Butt-welding of pipe is carried out in two different ways. In the first method pipe sizes between 12.5 mm diameter and 3 mm wall thickness and 100 mm diameter and up to 8.3 mm wall thickness are produced in a process in which welding is accomplished by a combination of pressure and temperature, that is, by hot compression welding. In the second alternative tubing with diameters up to 500 mm with wall thicknesses as thin as 2.4 mm and as thick as 12.5 mm is made by the application of electric resistance welding in a continuous fashion, or else even larger diameters are formed and welded piecewise.

The hot compression, continuous butt-welding process starts from slabs that are heated and rolled into strips between 3 mm and 8.3 mm thick and 22 mm up to 440 mm wide, called skelp. In this part of the process the skelp ends in coils. In the second part skelp is uncoiled, and the ends of many coils are welded together to produce a long, continuous strip that passes through a long furnace. Gas burners heat the edges of the skelp, and subsequently it passes through forming rolls, welding rolls, and reducing rolls. The resulting pipe is cut to lengths and cooled down.

In a particular case, the slab-heating furnace had a capacity of 120 tons/hour. The skelp mill consisted of one 400-mm scale-breaking edging stand, ten 475-mm-wide, two-high horizontal stands, and four 300-mm-wide edging stands. The motors driving the individual rolls ranged between 150 and 800 HP for a total of 7,450 HP driving power. The runout table of the mill was 25 m long, with a delivery speed from the mill of between 150 and 470 m/min. The finish, rolled skelp was turned on edge into a serpentine pattern on a conveyor that carried it to reel.

The individual skelp coils pass through an uncoiler, one by one, and the ends of the skelp of each coil pass through a flash-type welder, resulting in a large loop on the floor before entering the furnace. On exit from the furnace in which the edges have been heated to a welding temperature, the skelp is formed in a set of rollers and welded in a subsequent pair of rolls. It then passes through a set of reducing rolls. Their speeds are so regulated as to create enough tension to produce drawing of the pipe and the combination of the tension and roll profiles determines the diameter and wall thickness of the pipe. Finally, the pipe is cut to length and delivered to cooling racks where it is stacked.

An alternative process in which electric resistance welding is employed uses a set of rolls to cold form the pipe and feed it into the pressure roll where the electric current is applied to produce the heat across the contact of the edges of the circular cross section of the rolled pipe. For more detail about this process, see Chapter 11.

Large-diameter welded pipe is produced piecewise in plates of the desired lengths, widths, and thicknesses. The plate is cold-formed in a press to a **V** shape and then an **O** shape. Welding is carried out by the submerged arc process (see Chapter 11).

Seamless tubes, up to 650 mm diameter, are made using the Mannessman process (see Fig. 3.22), by hot helical rolling. The solid-bar workpiece is heated to 1200–1280°C and passed between two rolls that are double-conical in shape, and their axes are inclined by 6° and 12° out of parallel with the work axis, in mutually opposite directions. This rolling action rotates the workpiece and also pushes it axially over a projectile-shaped mandrel that also rotates but is held axially immovable. The compressive action of the rolls due to their diameter increasing from the entrance diameter towards the location of the piercing point of the mandrel would by itself produce a cavity in the center of the bar, but it would be small and irregular. This phenomenon can be explained by stress analysis, and it is rather simple to understand. If you press a round disk

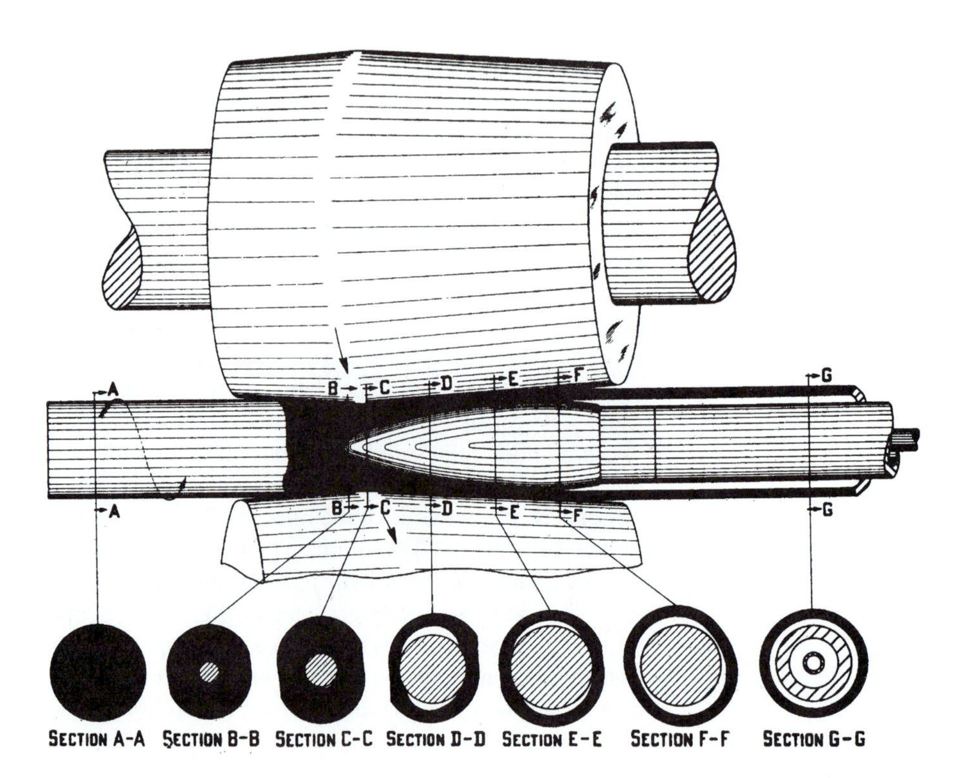

Figure 3.22
Action of a rotary piercing mill for production of seamless tubes. [USS10]

between two parallel plates across a diameter, this diameter is compressed, but the diameter at 90° to it wants to expand, and the disc becomes elliptical. A rather high tensile stress is generated in the direction perpendicular to the compressed diameter, causing a crack. This action makes the piercing work easier. In practice, depending on the size of the work, rolls between 500 and 700 mm long and 800 to 1100 mm in diameter rotate with a peripheral speed of 240 to 350 m/min and may be driven by motors with 750 to 5000 HP. For large-diameter pipes the action is executed sequentially twice in the double-piercing method.

Further processing of seamless tubes is carried out by cold drawing. Shapes other than round can also be made by hot rolling, hot extrusion, and cold drawing (see Fig. 3.23). All profiles are shown in the same scale that is indicated in the top center profile for the top of the figure and in the first pattern of the third row from the bottom, for the lower part of the figure.

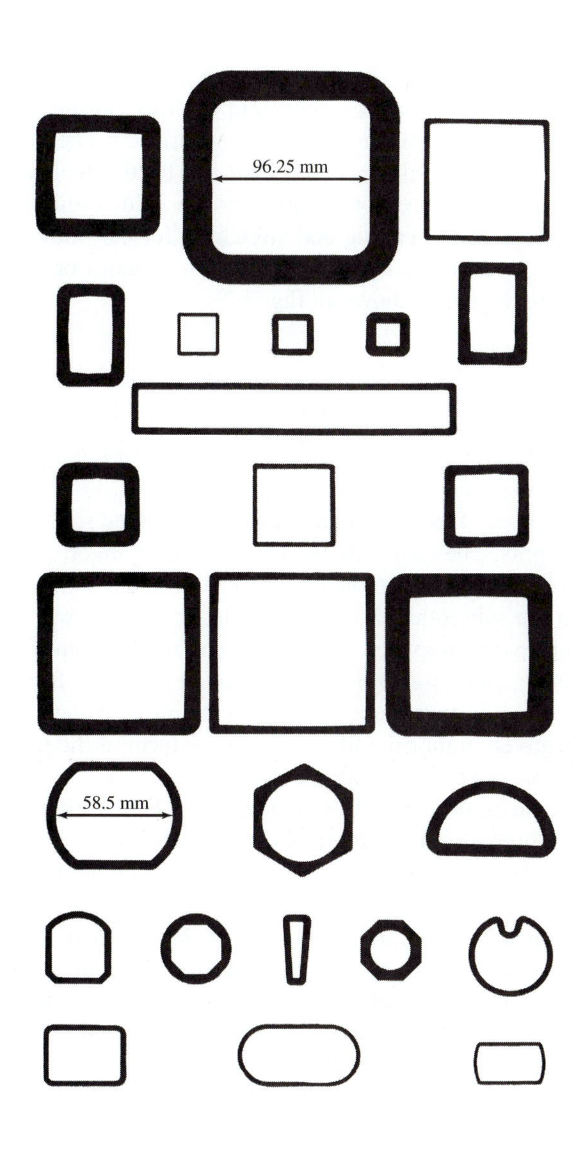

Figure 3.23
Assortment of unround seamless tubes. [USS10]

3.6 COLD ROLLING OF SHEET METAL

Sheet metal may be made from carbon, alloy, and stainless steels. Cold-rolled, carbon-steel sheets are produced in large quantities for many uses, mostly for automobile bodies. As pointed out in Fig. 3.2, it is often processed further after cold rolling by heat treatment and by various surface treatments and coatings. Coils of hot-rolled, pickled, cleaned, and oiled sheet about 2.5 mm thick are used as the input to four or five of four-high stands in a tandem cold-rolling mill. It is reduced to 0.6–1.6-mm thicknesses, in 750–1800-mm-wide sheet for car bodies, architectural use, and household appliances. Tin plate for cans is rolled down to 0.1–0.6-mm thickness in widths of 60–1000 mm.

All the mills have uncoiling reels and cradles, or boxes, from which the sheet is fed into the roll train. At the discharge end the sheet metal is re-coiled on a mandrel. Work rolls with diameters from 250 to 500 mm and back-up rolls with diameters from 1250 to 1400 mm are used. This means that work rolls in the last stand rotate at 5000 rpm for an exit speed of 40 m/sec. At the start of the operation the rolled metal is accelerated with about 0.25g. The cold-rolling process is further analyzed in Section 5.3.5.

After cold rolling, during which an oil-type lubricant is used, the sheet metal is cleaned by application of a detergent such as caustic soda, sodium orthosilicate, or other alkaline solutions and subsequently rinsed. The cold-forming, strain-hardening rolling operation produces rather hard and brittle steel with very little residual ductility, and it is therefore not suitable for the cold-forming operations that will be performed on it (e.g., bending and stretching for car bodies, or deep drawing for cans). Therefore, most of it is annealed, either in batches or in a continuous operation, and further processed, as shown in Fig. 3.2.

3.7 CASTING

The processes of rolling and hot forging discussed so far actually started with the casting of ingots. The rather coarse structure of an ingot, with its large grains, voids, and inclusions, has been completely crushed. Voids are eliminated and inclusions streamlined in the subsequent hot-forming plastic work in which the original ingot is drastically reformed into plates, bars, pipes, and forged shapes (for the latter, see Chapter 5). The casting processes are used to produce the final form of some primary products. They may be further processed by machining or by welding, but these processes produce relatively minor modifications of the form of the part. Some castings may not be further processed at all; they are used as cast.

All the casting processes include the steps of melting the metal in the desired composition, preparing the pattern and using it to produce a mold, pouring the metal into the mold, letting it cool down and solidify, and conditioning the casting mechanically by removing the sprue, runners, gates, and other auxiliary parts, and otherwise improving the surface. If required, the casting is heat treated. There are several different casting processes that are classified according to the material and structure of the mold, which is mainly either expendable or permanent. The choice of the process depends on the size and material of the part, the accuracy and surface finish required, and very strongly on the quantity of parts to be made.

Casting is an ancient art. We are all familiar with the large and complex statues of Buddha made more than a thousand years ago, the complex statues made by Michelangelo and other medieval sculptors, and the castings of bells for churches. All of these castings were made of bronze. The Great Buddha of Kamahura was cast in the thirteenth century and contains 120 tons of bronze [4]. Subsequently, the art was applied to making gun barrels as well as bells, and eventually came the industrial applications. We will concentrate on the industrial casting processes in the production of machinery. A great variety of metals are cast: gray iron, malleable and ductile iron, steel, aluminum alloys, zinc alloys, brasses, bronzes, and many others.

As already mentioned, hot forming improves the material properties of the metal, mainly its toughness. This is discussed further in Chapters 4 and 5, where it is also explained that cold forming improves the strength of metals. Material properties such as strength and ductility, or toughness, obtained by casting are generally inferior to those obtained by hot or cold rolling or by forging. Nevertheless, it is important to understand how the best possible properties are obtained by controlling grain size and orientation through the use of alloying elements, especially those that multiply the nucleation sites, by controlling cooling rates, and by using techniques that affect the form of inclusions such as segregated carbon in iron or silicon in aluminum.

The formation of grains during their solidification out of the melt differs essentially among pure metals, eutectic compositions, and other solid-solution alloys. A eutectic alloy solidifies at one particular temperature (see Fig. 2.12). On cooling, all of the liquid of the eutectic composition will take some time to solidify, depending on the cooling rate. During this time the temperature does not change until all of the liquid has changed to solid, and the composition of the solid parts remains the same during this process. A noneutectic solid solution (see Fig. 2.11), for example, a CuNi composition with 52% Ni, solidifies over a range of temperatures. At the start of solidification, point a in the graph, the first solid grains will be rich in the higher-melting-point component and contain 70% Ni. As the temperature decreases, more of the solid becomes less rich in Ni. At point g, when the last parts of the liquid turn to solid, it will have the final composition of 52% Ni, while the last drops of the liquid would contain only 33% Ni. If the cooling process is very slow, the excess of Ni in the initial solid crystals will diffuse out of them into the leaner liquid. Similarly the final solid crystals would be affected by the very lean liquid. Usually, there is not enough time for this equalizing diffusion, and the solid will contain grains of different composition. The composition of crystals within a grain may also vary; this is called microsegregation. Macrosegregation occurs when different parts of the casting solidify first and they are richer in the higher-melting-point element than other parts.

Just after solidification, below the freezing point, grains start and grow and disintegrate in a state of high agitation. It is only at much lower temperature that stable grains form and grow, and their size depends on the number of nucleation sites and on the cooling rate. Nucleating agents in a finely dispersed form may be added to the molten metal just before pouring. These are mostly intermetallic compounds; if they have a lattice structure similar to that of the cast metal, they easily start the growth of individual crystals.

In Section 2.5 it was explained that achieving fine grain in metallic structures is one of the most effective ways of strengthening metals while at the same time preserving high ductility. Eliminating hard, nonwetting inclusions and voids (porosity) increases the toughness of the materials. There are a few exceptions to the rule of small grain size. For high-temperature, creep-resistant applications, coarse grain is preferable. Jet engine turbine blades have better strength with large grains oriented longitudinally in the direction of the centrifugal stress, and the best results are obtained if the whole blade consists of a single crystal. Single-crystal ingots are also used for the manufacture of silicon wafers for electronic circuits.

In pure metals and in eutectics a plane solidification front propagates away from the walls of the mold, which is where the heat is extracted (see Fig. 3.24a). Close to the wall, where the cooling is fast, a layer of fine, equiaxial grains is produced. Further away from the wall long, columnar grains grow into the melt. Very little porosity is obtained and as a result of shrinkage of the material as it changes from liquid to solid and as the solid cools down, a rather deep depression called a pipe is generated at the top of material that has been poured into an open mold. In closed molds it is necessary to ensure a steady supply of molten metal from cavities additional to the functional mold, called risers, to ensure complete filling of the mold. Furthermore, the mold must be designed so as to ensure that freezing starts at walls and in cavities farthest from the inlet of the molten material, called the gate, and propagates towards the gate.

In noneutectic solid solutions with their wide range of solidification temperatures and the variations of the composition in the crystals throughout the solidification process there is also a fine-grain layer at the wall of the mold (see Fig. 3.24b), but farther away, the grains grow in the form of dendrites that resemble trees with branches

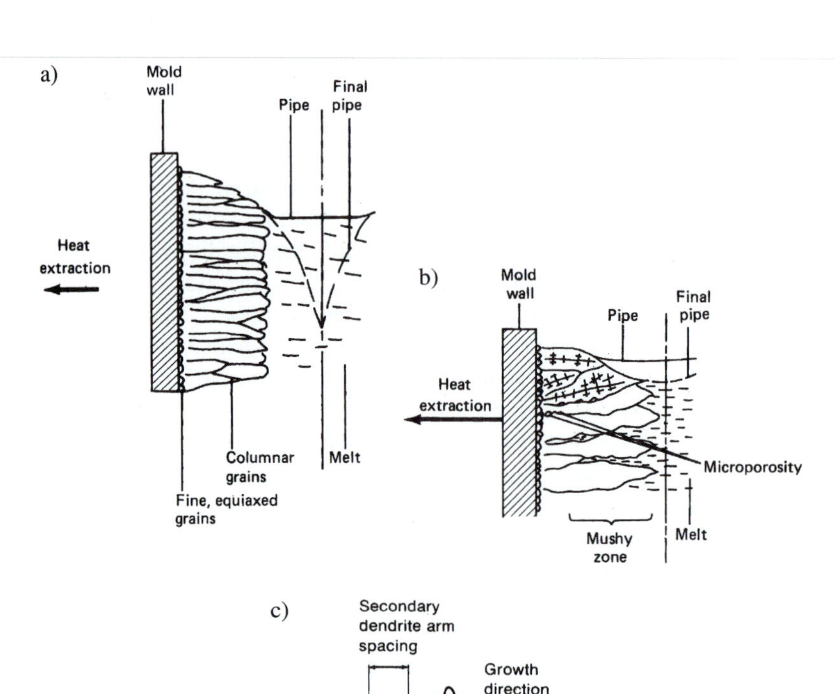

Figure 3.24
Solidification structures in a) columnar grains in eutectics, b) dendrites in non-eutectic solid solutions. Diagram c) shows the dendritic structure growing in crystallographically favorable directions. (REPRINTED WITH PERMISSION FROM J. SCHEY, "INTRODUCTION TO MANUFACTURING PROCESSES," 2ND ED., MCGRAW-HILL, 1987.)

(Fig. 3.24c). These forms are initially weak, and they can be broken by agitation to result in fine grains. As the dendrites interweave, they lock in place and the spaces between the arms may be starved of fluid due to local shrinkage. This results in microporosity, which is harmful to the toughness of the casting. The pipe in this case is much smaller due to the distributed shrinkage.

The properties of the cast metal may also be improved by treatments applied after casting. One of them is called high-temperature isostatic pressing (HIP), in which a neutral gas such as argon is used to pressurize the casting. Pressures up to 200 MPa and temperatures up to 2000°C are possible. The process is applied to superalloy and titanium castings. It eliminates porosity and improves toughness and fatigue strength. Other methods involve various kinds of heat treatment. Steel and iron castings may be quenched and tempered, and aluminum and titanium castings may be subjected to solid-solution or precipitation-hardening treatments. Annealing is often applied to various metals with the purpose of homogenizing the micro- and macrosegregation. Stress-relief heat treatment is also rather common.

In designing a casting process it is necessary to consider the fluidity of the metal, pressure and velocity distribution in the casting system, as well as heat extraction and the propagation of the solidification front. Advanced computer programs are now available to analyze all these aspects of the process so as to design the whole system of the mold and of the pouring and casting passageways for the production of successful and sound castings with the desired strength, ductility, and toughness. We will discuss the three aspects of fluidity, flow distribution, and solidification time in rather simplified ways.

Fluidity is understood as the ability to fill the various details of the mold cavity. It is the inverse of viscosity and is affected by the modes of the solidification front, by surface tension and oxide films, and by the thermal permeability of the mold material. Obviously it improves with the temperature of the molten metal and with the temperature of the mold, but at the cost of slower cooling and coarser grains. The mode of solidification plays a role in the ease of flow in channels. The dendritic mode has a tendency to clog the channel which must accordingly be made larger and with high flow velocity.

The velocity of the flow can be approximated by an assumption of lamellar, Newtonian flow as follows:

$$v = \sqrt{2hg}$$

where g is gravitational acceleration, and h is the height difference down from the level of the molten metal in the pouring basin. Correspondingly, the first channel that leads vertically down from the pouring basin, the sprue, is made conical with a decreasing diameter to accommodate the continuity of flow to the increasing flow velocity.

The solidification time T_s depends on the ratio of volume over surface area of the part being cast. It has been established that the relationship is approximately quadratic:

$$T_s = C(V/A)^2 \tag{3.13}$$

where V is volume, A is surface area, and the constant C depends (1) upon the thermal properties (e.g., specific heat, solidification heat, the heat conductivity of the metal cast) and (2) upon the material and surface condition of the mold that determines the coefficients of heat convection and conduction. This constant is established experimentally. The formula (3.13) leads to various rules for the design of the mold. So, for instance, the metal in the riser with its role of replenishing the metal lost in shrinkage must obviously remain liquid longer than the body of the casting. Correspondingly the riser must have a higher value of *V/A*.

With this general introduction to casting, the following sections describe the various casting methods and the corresponding features of the patterns and molds. Brief sections follow about the special characteristics of the various metals cast and about melting furnaces. There are many different casting processes in existence. We will describe only a limited selection of the most interesting ones.

3.7.1 Expendable-Mold Processes

Sand Casting

Sand casting is the most universal process. It is used economically for all kinds of metals, for small, medium, large, and very large castings made individually or in small and medium batches, or in large quantities. Specifically, the mold made of "green" sand can also withstand the temperature and pressure of molten steel. The word *green* is used to distinguish this process from those such as shell molding, in which the sand mold is fired in a furnace. There are various kinds of sand, and activities connected with preparing the molding sand are an important and necessary part of the casting technology. The basic ingredient of the various compositions is silica (SiO_2) sand. Natural sand contains only the binder clay as mined, with water added. *Semisynthetic* and *synthetic* sands are prepared as well-specified mixtures of various ingredients. Bentonite is used as binder material, and other additives may modify the moldability, hardness, and strength. Basic characteristics are specified as follows:

- Ability to withstand high temperatures
- Ability to retain shape under the action of metal flow
- Permeability, that is, the ability to permit passage of gases and vapors that developed during pouring,
- Collapsibility, permitting the mold to be broken up and separated from the casting.

For different metals, different sizes of the casting, different parts of the mold, and different shape detail, different sand formulations are used by selecting grain size and shape, binder type and content, and the amount of organic additives that burn out during casting and contribute to the permeability of the mold.

The basic components of the mold are indicated in Fig. 3.25. The mold is made in a box called the *flask*, and it consists of the bottom part, called the *drag*, and the top part, called the *cope*. They meet at the parting surface, which may be flat or a little more complex. The mold can be opened and the two parts separated to remove the *pattern* around which the mold has been compacted. The pattern is made of wood, plastic, or metal and has essentially the external form of the final casting. The internal form of the cavities in hollow castings or semi-enclosed cavities in hollow castings or semi-enclosed cavities on the outer surface such as cooling fins on a cylinder block of a motorcycle engine are determined by the corresponding shape of *cores* made of sand and inserted in the mold in cavities called core prints. The cores are prepared in wooden forms called core boxes that play an analogous role for the cores as the patterns do for the molds, except that patterns determine the inner surfaces of the molds and core

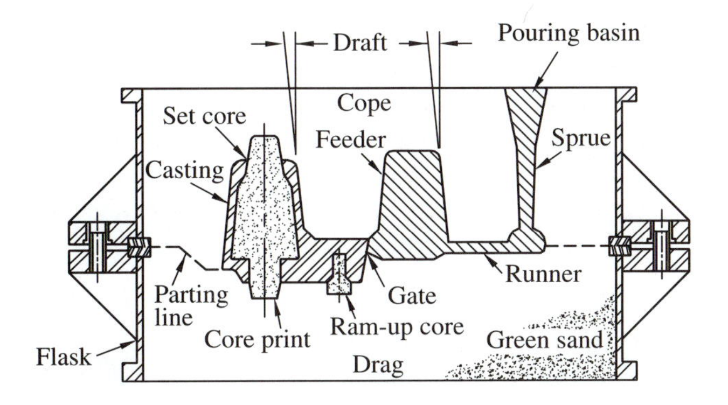

Figure 3.25
Typical parts of a green sand mold made in a flask. [ICS]

boxes define the outer surfaces of the cores. The sand for the cores must be stronger than for the mold because the core, often a long cylindrical beam located horizontally, unlike the vertical one in the picture, is subjected to forces of buoyancy that result from the much lower mass density of the cores compared to that of the metal. The actual dimensions and shapes of the patterns and of the core boxes differ from those of the casting by shrinkage allowances and by draft angles needed to facilitate the removal of the patterns from the mold. Furthermore, the mold contains cavities additional to those defining the final casting. They serve the purpose of feeding the molten metal from the pouring basin to the casting: the sprue, the runner, the gates, and the riser (feeder). The role of the latter is to supply the molten metal to compensate for the shrinkage of the casting. To make sure that the metal in the riser does not solidify before the main body of the casting its (V/A) value must be distinctly larger. Directional solidification in which the freezing front moves from the far ends towards the runner can be helped by strategic use of chills. These are also useful to eliminate local porosity by increasing the local cooling rate. External and internal chills in the form of metal plates are used. The former are included in the mold while the latter are inserted in the cavity created by the removal of the pattern. These must be made of the same metal as the casting and they will become its parts.

The molding process is illustrated in Fig. 3.26. The two halves of the mold are made in the two halves of the flask. The mold may be made by hand, but more often it is made on machines such as the jolt type. The pattern is mounted on a pattern plate that is clamped to the table of the machine, which is attached to the top of the operating pressure air piston. The flask is also clamped to the table and filled with sand. The jolt valve admits the air that lifts the cylinder; a port is then opened to let the air escape, and the table falls until the piston hits the bottom of the cylinder, producing a sharp jolt. This is repeated at a rate of several jots per second until the sand is compacted. The jolt-squeeze rollover machine adds the squeezing action. The flask is rolled over by air-operated arms, and the two halves are squeezed together by steady action of the of air piston. The cope is then raised, the pattern removed, the core inserted, and the mold is closed again and moved to the pouring station. Molds for very large castings are prepared in pits in the floor of the foundry instead of in flasks.

The most common sand castings are made of cast iron. Typical examples are machine tool structures; most of the machine tools shown in the photographs in chapters 7 and 10 are assembled of beds, columns, headstocks, base plates, angle plates, and

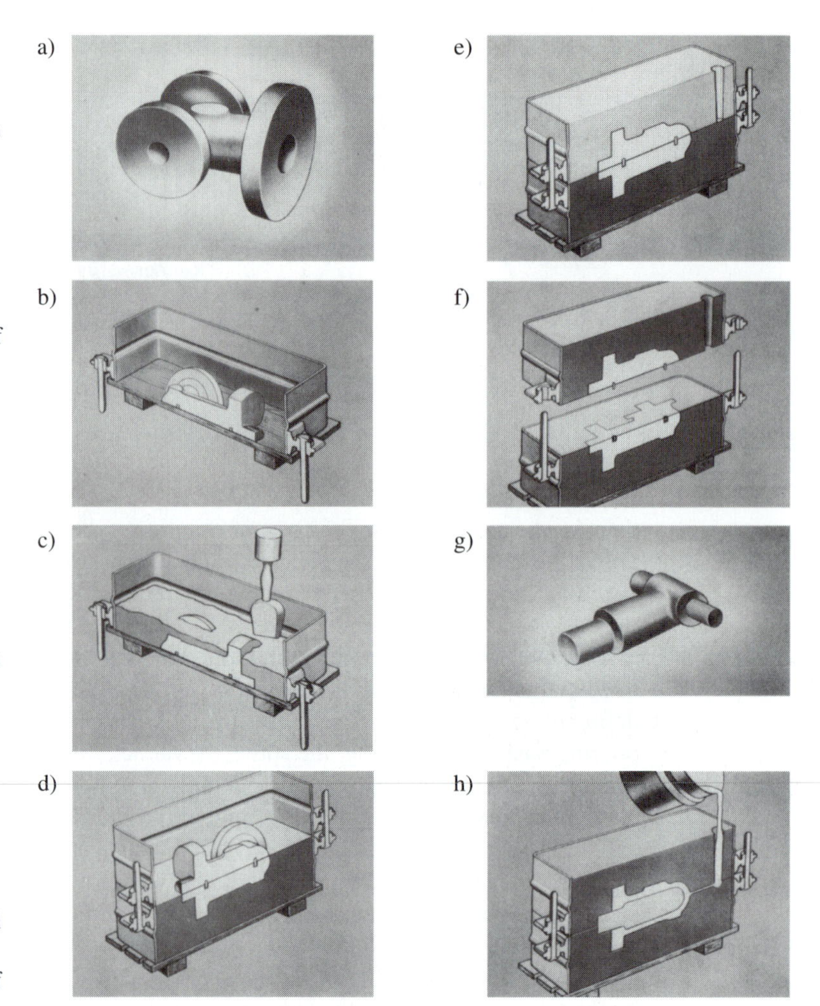

Figure 3.26
The main steps in the molding process. a) Iron casting to be produced in the subsequent illustrations of molding; b) cross section of the first step in making a green sand mold. Bottom half of the pattern is on the mold board and surrounded by the bottom or drag half of the flask; c) molding sand is rammed around the pattern in multiple steps to provide uniform density; d) after the bottom of the mold is filled, it is rolled upright and the top half of the pattern and flask are put in place to complete the mold; e) section through the completed mold with pattern still in place and the sprue hole formed for entrance of molten metal; f) cope and drag halves of mold are separated in order that the pattern may be removed. The gate channel is then cut from the sprue to the mold cavity; g) the core of bonded sand is made separately to form the internal passages of the casting; h) after placing core in the mold, it is closed and clamped to resist the pressure exerted by the molten metal when it is poured in the mold. [ICS]

so on, that are iron castings. The flexibility of the casting process makes it possible to design these structures with complex ribbing, resulting in high rigidity. Machines of small sizes and medium sizes with working motions 1 m long as well as very large machines with beds 10 m and 20 m long are made in this way. Another typical and popular example is that of automobile engines, whether cast of iron or aluminum. The photograph in Fig. 3.27, from an automobile company foundry, shows the manipulation of castings of cylinder blocks of an eight-cylinder engine.

Steel castings are used in a great variety of applications: in the transportation industries for parts of railway cars, highway trucks, and ships; in the mining industry for large excavators, shovels, ball mills, and crushers; in the construction industry as parts of road-building machinery; in agricultural machines; in power generation for the housings of turbines and for valve bodies; in steel plants for charging machines, rolls for rolling mills, and frames of rolling-mill stands. Some of the steel castings are very large. Several examples are shown here in impressive photographs. A railroad freight

Figure 3.27
Manipulation of castings of a V8 cylinder block. (COURTESY GMC)

Figure 3.28
Steel casting of a six-wheel motor truck frame for a diesel locomotive. [SCH]

car contains steel castings as the coupler, the stricker, truck side frame, and bolster. These parts benefit from the resistance to shock and fatigue, high strength, and toughness of cast steel. The coupler that connects the cars is made of an assembly of steel castings and is exposed to forces arising from starting and stopping a train of over a hundred cars. A six-wheel motor truck frame for a diesel locomotive with most of the associated assembly, including the motors, is shown in Fig. 3.28. Many components of a ship are made as steel castings: the rudder frame, hause pipes, anchor, and others. A stainless steel propeller 1.5 m in diameter is shown in Fig. 3.29 in the machined state. The stripping shovels and drag lines in strip mining contain many steel castings. The track shoe assembly in Fig. 3.30 is constructed for a 107 m^3 shovel; four track assemblies are required for each shovel. There are 38 shoes in each assembly, and each weighs over 2 tons. They are made of a low-alloy cast steel with 827 MPa strength. The total weight of the shoes for each shovel is 333 tons.

Figure 3.29
Stainless steel propeller casting with 1.5 m diameter. [SCH]

Figure 3.30
Assembly of track shoes cast in steel, for a 107 m^3 shovel. [SCH]

Shell-Mold Casting

In shell-mold casting a metal pattern is used to prepare a thin shell of a mold made of fine sand mixed with a thermosetting resin binder. The resin cures in contact with the preheated pattern. After the shell has been cured and stripped from the pattern any required cores are inserted, and the cope and drag halves of the mold are bonded together, placed in a flask, and backed up by green sand. The mold is then ready for the casting process. Smoother surfaces and better accuracy are achieved than in the regular sand mold. Of course the metal pattern is much more expensive to make than the wooden one commonly used in sand casting. Therefore, the process is economical only if used for large quantities of the casting. The open compressor wheel weighing 363 kg shown in Fig. 3.31 is an example of a precision casting made by the ceramic molding process.

Figure 3.31
Steel open compressor wheel cast in ceramic mold, weighs 363 kg. [SCH]

Polystyrene Expendable Pattern

The polystyrene expendable pattern is frequently used for simple forms not requiring very high accuracy when only one or a few pieces of a casting are required. The pattern may be assembled by cutting pieces of polystyrene foam plates and bonding them together so as to produce the shape of the desired casting. The mold is made of molding sand around the pattern, and when the molten metal is poured in, the pattern vaporizes. No draft is necessary as for wooden patterns that have to be taken out of the mold, and no parting planes are needed. For larger quantities of castings, the pattern is itself molded in a form to produce the desired shapes, in a way similar to the making of Styrofoam cups. Because of the elimination of the need to open the mold for taking the pattern out the process is simple and inexpensive.

Lost Wax Method

Another expendable-pattern process is the lost wax method, also called investment casting or precision casting. It uses a mold that is built up from ceramic slurry around a pattern made of wax. On pouring the metal in, the wax is burned out. The steps in making casting by the investment molding process are presented in Fig. 3.32. The pattern itself is repeatedly cast in a metal die, but both the mold and pattern are destroyed in the production of the casting. Mass production of complex shapes with very fine detail, good dimensional accuracy, and smooth surface is possible. Almost any metal can be cast because the mold can be made of refractory ceramic. Some of the metals have to be cast in vacuum or in an inert atmosphere. Castings of small sizes, up to 5–10 kg, are made, and tooling cost is high. This is an ancient process used in Egypt 3500 years ago. Bronze products were cast in ceramic molds, and the dies for the wax pattern were made of fired clay. The jewelers in medieval Italy used it in impressive ways, as in Cellini's golden rings with intricate miniature sculptures on them.

In current industrial applications, the mold can be made in a flask when the wax pattern is first dipped in a fine-grain slurry for a precoat and then put into the flask where it is backed up by coarser granulated refractory. Often the pattern is dragged

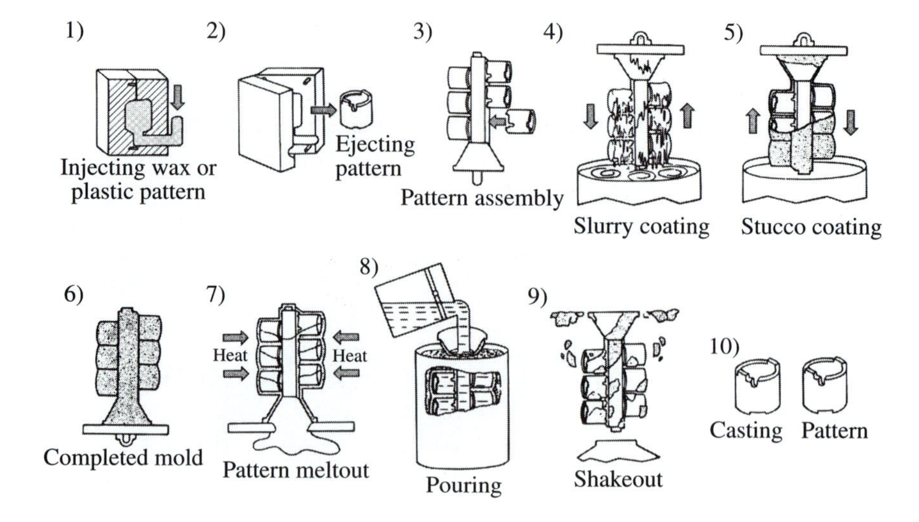

Figure 3.32
Steps in investment casting. [SCH]

through several coarser and thicker baths and the molds are built up in a tree branch fashion around a central sprue for casting multiple pieces.

The dies for making the wax patterns may be made by machining from solid or by casting. Their materials may be unhardened steel, aluminum, magnesium, or brass, when liquid or semisolid wax is used. For solid waxes or for patterns made of plastics, the pattern die is made of die steels such as those used in injection molding of plastics. Vegetable waxes or mineral and synthetic waxes with melting temperatures of 65–93°C are used, mixed with soft plasticizers, hard resins, and antioxidants.

3.7.2 Permanent-Mold Casting

The Basic Process

A permanent mold is used for production of castings in large quantities. A mold consisting of two or more parts is used repeatedly. Small parts with medium complexity are typical; otherwise the cost of the mold affects the economy of production too much. Either metal or ceramic cores are used. In the latter case the process is called semi-permanent-mold casting. The metals that can be cast include aluminum, magnesium, zinc, and copper alloys. Cast iron is seldom used because it erodes the mold, and steels, with their even higher melting temperature, are unsuitable.

Advantages of the process are good surface finish and dimensional control. Drawbacks are the limitations on the complexity of the casting because it must be possible to take it out without destroying the mold and the high cost of the mold. Typical successful parts are automotive pistons.

The process has been further developed into the rather important processes of die casting and of centrifugal casting.

Die Casting

In die casting the molten metal is forced into the metal molds called dies under rather high pressure and it is held under the pressure until it solidifies. This results in several substantial advantages of the process. Thinner walls and complex shapes can be obtained with good surface finish and dimensional accuracy. The process is fast, and high production rates are achieved especially when multiple-cavity dies are used. There is very little need for subsequent machining except for facing surfaces used for joining the casting to another body or drilling and tapping of holes. Thanks to minimal porosity, higher strength is obtained. Limitations of the process are due to the high cost of the machine and of the dies. Furthermore, satisfactory life of the dies is only obtained for casting metals with melting temperatures no higher than those of copper-based alloys. Two subsets of the process exist. One is called "hot-chamber," and the other is "cold-chamber," depending on whether the pressure cylinder is submerged in a bath of the molten metal or is filled periodically with molten metal just before injection. The former is used with lower-melting-temperature metals such as zinc alloys, and the latter is applied to aluminum and copper alloys. Finally, the process is limited to smaller castings, up to about 25 kg mass, and with best efficiency usually to about 5 kg.

The pressures applied in die casting range between 10 and 30 MPa for the hot-chamber process and between 15 and 160 MPa for the cold-chamber process in which the higher-melting-temperature metals are cast. The dies are made of alloy steels designated H for hot work; typically the H13 grade is used with 5% chromium, heat-treated to about 36 HRC for the lower-melting metals and to about 55 HRC for the higher-melting metals. The die cavities are often produced by the electro discharge machining process (see Chapter 12), which is applied to the hardened steel block. Otherwise, the block may be first annealed and the cavities produced by NC milling, then heat-treated and manually finished by honing and polishing. Recent methods of high-speed milling permit rough milling with fine-grain carbide tools and finish milling using CBN tools, both capable of machining the steel in the hardened state. In this case no hand finishing is necessary.

The overall design of a die-casting machine, specifically of the hot-chamber type, is shown in Fig. 3.33. The fixed half of the die is attached to the platen on the injection side. The other half is attached to a movable platen actuated by a toggle mechanism to close the die under a force sufficiently higher than the casting pressure times the largest

Figure 3.33
Hot-chamber die-casting machine. [ASM15]

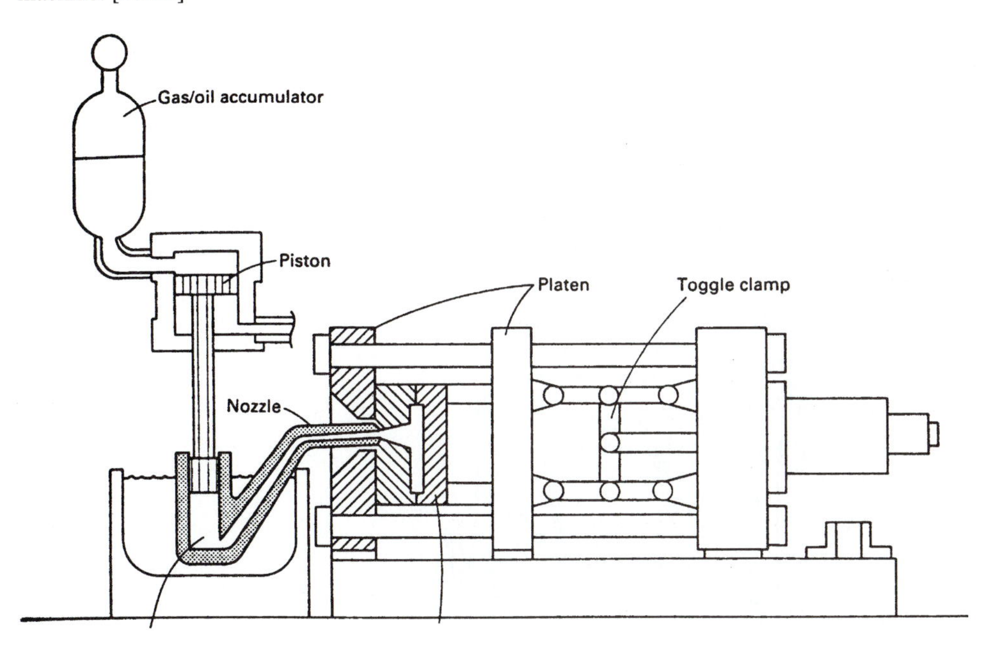

casting area, including the runners, for which the machine is normally rated. The toggle mechanism is driven by a hydraulic cylinder, and the whole frame is held together by strong tie rods that act also as guides for the movable platen.

The basic design of the hot-chamber casting mechanism is illustrated in Fig. 3.34. The pressure cylinder is submerged in a pot filled with molten metal and heated by burning gas in a steel shell lined with firebrick. The cylinder body and its extension leading to the nozzle and to the input port of the die has a gooseneck shape. There is an intake port at the top of the cylinder, and it is open when the plunger has been retracted. To inject the metal, the plunger is driven down, and pressure is held for a predetermined time needed for the solidification of the casting, upon which the plunger retracts. The whole cycle is rather short; shot rates range between 50, 500, up to 5000 shots per hour (one per minute up to more than one per second). A zipper-casting machine is claimed to achieve 5 shots per second.

The principle of the cold-chamber process is shown in Fig. 3.35. Molten metal is ladled into the shot chamber, and the plunger advances to cover the pouring hole. Then it moves rapidly to inject the metal in the die. The plunger is made of a nitrided alloy steel, and it may be hollow and water-cooled. The shot chamber is made of the H13 steel. The cold-chamber process is slower, and the molten metal is exposed to air and agitation and may suffer from oxidation. However, many successful operations exist with high production rates. The process has to be carefully designed, the ladling and injection temperatures as well as die temperatures closely controlled, and the runner, gating, and overflow system calculated for fast and complete filling.

Metal temperatures at the gates are typically 650°C for aluminum and 425°C for a zinc alloy. Die temperatures are maintained at 260°C and 205°C, respectively. Cavity fill

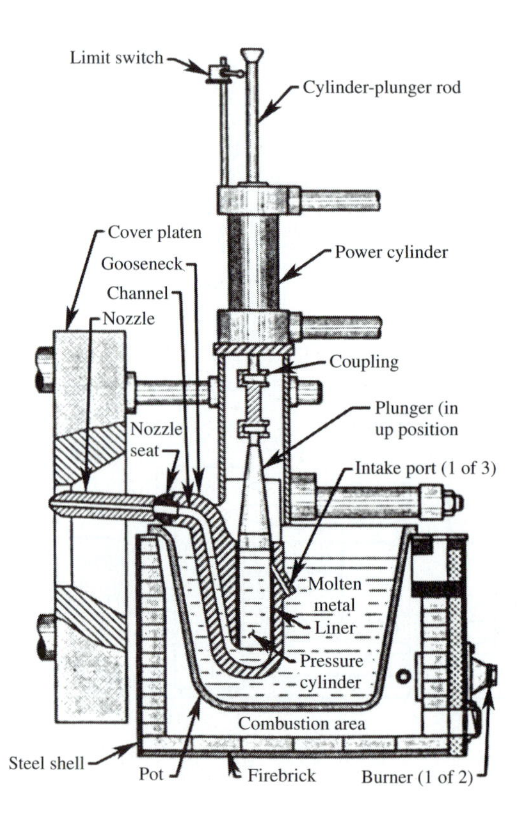

Figure 3.34
The hot-chamber casting mechanism. [ASM15]

times depend mainly on the wall thickness of the casting. They are indicated as 0.005 sec for Al and 0.002 sec for Zn with wall thickness of 1 mm and up to 0.25 sec for Al and 0.15 sec for Zn with wall thickness of 6 mm. The velocity of flow at the gate may be typically 30–50 m/sec. Die casting processes may be highly mechanized or completely automated. Interestingly, the first application of an industrial robot, the Unimate, was in 1969 for removing die castings out of the open die, an operation that was considered hazardous for a manual action. Various devices such as a mechanized ladle, air-pressure system, a vacuum system, or a pumping system have been designed for supplying the molten metal into the shot chamber of the cold-chamber, die-casting machines.

A particular application of a vacuum system to automate the filling of the shot chamber in the cold-chamber method is the vertical-vacuum, high-pressure-casting Gibbs process. It claims filling times of less than two seconds and improved casting microstructure and reduced porosity due to the evacuation of all the air from the die cavities and feed channels before injection of the metal. Thus both increased productivity and quality are obtained. Examples of castings are assembled in Fig. 3.36.

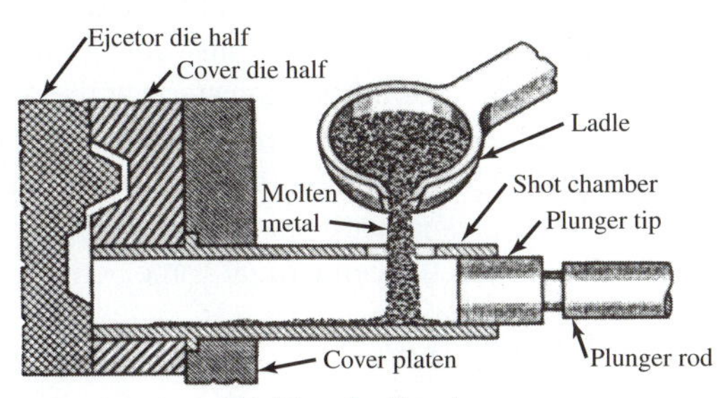

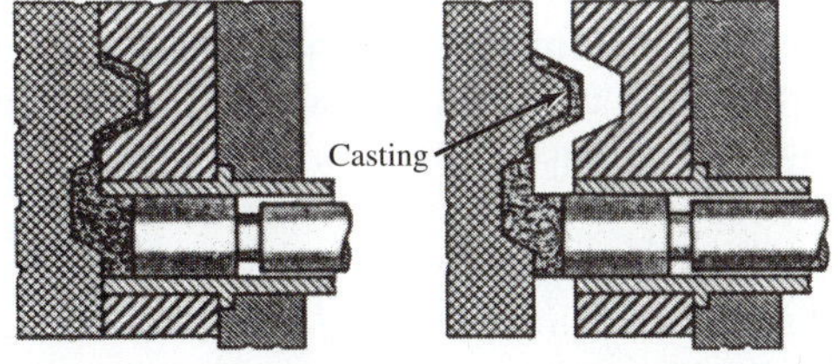

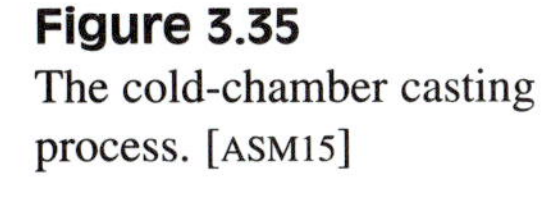

Figure 3.35
The cold-chamber casting process. [ASM15]

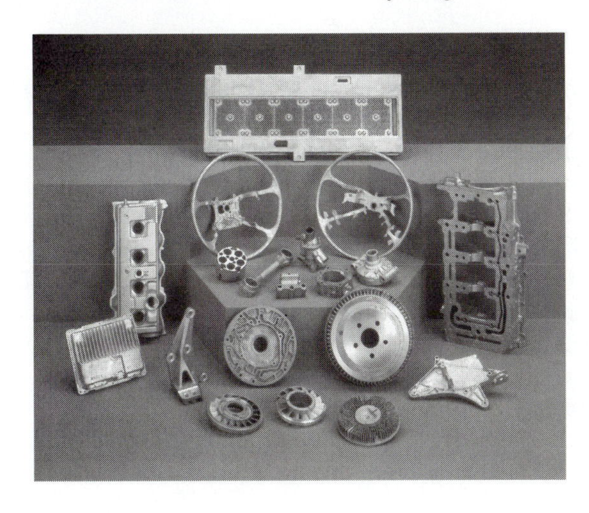

Figure 3.36
Examples of aluminum castings produced by the vertical-vacuum Gibbs die-casting process. (COURTESY OF THE GIBBS DIE CASTING ALUMINUM CORP., HENDERSON, KY)

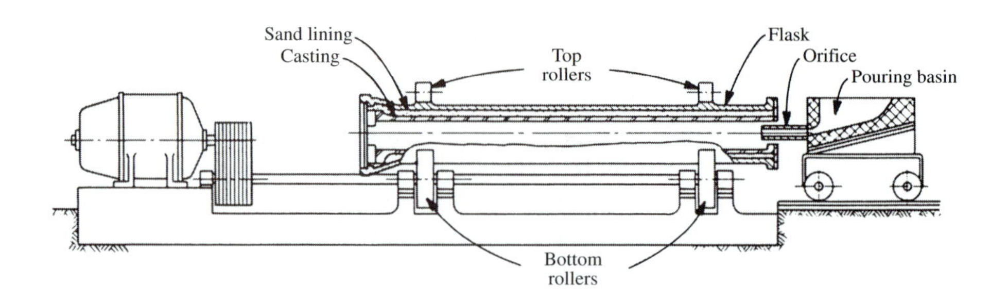

Figure 3.37
Horizontal centrifugal casting machine. [SME2]

Centrifugal Casting

Centrifugal casting is a rather simple method for producing tubular parts such as cast iron pipes (see Fig. 3.37). A tubular flask supported on rollers is rotated around a horizontal or vertical axis. The flask is lined to produce a mold made of sand or graphite. Metal molds can be used as well. The mold determines the outer shape of the product while the inner shape obtained without any core is simply cylindrical. No sprue, runner, or gates are necessary. Solidification starts at the outer surface and the thickness of the wall depends on the amount of molten poured in. The casting is strong and dense, and lighter impurities concentrate on the inner wall, from which they can subsequently be machined away. This is actually casting under pressure. The pressure developed by the rotation is obtained as the centrifugal force acting on a unit area:

$$p = h\,\rho\,r\,(2\pi n/60)^2$$

where h is the thickness of the wall of the casting, ρ is the density of the cast metal, r is the radius of the cylindrical casting, and n (rpm) is the rotational speed. For instance, if $h = 2$ cm, then ρ for cast iron is 7.8 g/cm^3, $r = 0.25$ m, $n = 2000$ rpm, and $p = 173$ kPa.

Apart from this, "true" centrifugal casting is commonly used for the manufacture of pipes, brake drums, bushings and rings from various materials. Rotation of the mold around a vertical axis is often used to improve the density of parts that are not hollow or of clusters of parts located on the periphery of a cylindrical mold while the sprue is vertical and runners are located radially. An example of such a part may be a spoked wheel or a pulley.

3.7.3 Casting Materials

The most common casting materials are the various kinds of iron: gray, ductile, and malleable, as described in Section 2.6.3. The fluidity of iron depends on composition and temperature. The melting and freezing points of eutectic and hypoeutectic compositions vary inversely with carbon and silicon contents from 1088°C for 4.4% C and 0.6% Si to 1250°C with 3.56% C and 2.4% Si. Shrinkage varies from 0.25% to 1.0%. It is the result of a combination of contraction and expansion during the cooling down from the superheated liquid to room-temperature solid. The contraction occurs during the transition to solidification, the actual solidification, and the cooling of austenite and ledeburite. This is followed by a large expansion due to graphitization of carbon between 1120°C and 1065°C, then contraction during cooling down to 720°C, and another expansion due to recrystallization from γ to α iron, and final contraction due to cooling down to room temperature. Another growth occurs during annealing, amounting to about 0.2%. Formation of graphite is enhanced by slow cooling. In gray iron graphite has the form of flakes. With the addition of Mg and Ce in minute amounts during the casting of iron with low S (0.03% max) ductile iron is obtained with graphite in spheroidal form. White iron results from rapid cooling. Carbon is contained in it in the form of cementite. White iron is very hard (400 BHN). It may be converted into malleable iron by heat treatment below the eutectic temperature, at 950°C, for 12 hours. The graphite assumes the form of compact particles called tempered carbon. Gray and white irons are brittle; ductile and malleable irons have enhanced tensile strength and ductility.

The iron for casting is most commonly melted in a furnace called the cupola. It resembles a scaled-down blast furnace in that the charge consists of layers of coke and lime and the source of iron which, however, is not iron ore but pig iron and steel scrap. The cupola has the form of a tall, cylindrical vessel charged through doors at the top level, and it has a tapping spout at the bottom to let out the molten iron. On a higher level there is a slag spout. As in the blast furnace, air is blasted at a low level to provide the source of oxygen to burn the coke and produce the heat needed for heating up and finally melting the pig iron and steel scrap charges as they descend through the furnace. The amount of coke is held at a minimum necessary, 8–10% of the weight of the metallic charge to keep down the absorption of sulphur by the metal. Carbon content may be increased or reduced depending on the initial amount present in the combined metal charge. In some foundries electric furnaces of either the arc or induction type are used instead of the cupola. Otherwise a combination of the cupola that provides the molten metal and the electric furnace to modify and adjust the composition of the metal is used for higher-quality castings.

Cast steels in both carbon and alloy grades are available. Generally, the compositions of 0.13–0.35 C, 0.5–1.0% Mn, 0.6–0.8% Si, 0.03–0.06% P, and 0.03–0.05% S exist for the various common use classifications, with 0.4–1.2% Mo and 0.5–2.0% Cr and, in the C classes, up to 10.0% Cr for the alloy steels. Further, higher-alloy steels contain more than 12% of Cr, Ni, Co, Mo, and W. Steels are melted in both electric arc and induction-type furnaces for the higher-alloy, corrosion-resistant, or heat-resistant steels. For large castings, molten steel from the open-hearth furnace may be used. Patterns are made to provide for 1.5–2% shrinkage. Many castings are subjected to various heat treatments such as annealing, normalizing, quenching and tempering, or flame hardening of some surfaces.

3.8 ALUMINUM: MANUFACTURE, USE, AND PROCESSING

3.8.1 Manufacture and Use

The usage of aluminum is second only to that of iron and steel. In 1994 the production of aluminum amounted to 19.114 million tons worldwide and 3.3 million tons in the United States. This compares to about 760 million and 96 million respectively for steel. Considering that aluminum is three times lighter the aluminum production by volume is about 10% of that of steel. Aluminum is the third most abundant element in the earth's crust, after oxygen and silicon, and the first among the metallic elements. It is contained in all common rocks, in clay, shale, slate, granite, and especially in an ore called bauxite after the town of Le Beaux in southern France near which it was first found and analyzed. Chemically it is hydrated alumina (Al_2O_3), and it contains 40 to 50% of Al.

The history of the manufacture of aluminum is very different from that of iron and steel. While the latter has been around for more than 2000 years, it was not until 1854 that a Frenchman, Henri Etienne Saint Claire Deville, after hundreds of years of other attempts, made small amounts of the metal. His work was continued by others, so that in 1855 bars of aluminum were displayed at the Paris Exposition by the Javel Company. However, the process they used was not economical. One kilogram of the metal cost the equivalent of current $598. Today, aluminum is sold in ingot form for about 50 cents per kilogram. A radical improvement of the process was developed in the 1880s by H. Y. Castner, but it was soon overshadowed by the electrolytic process. This process was discovered simultaneously by two researchers in 1886–89, C. M. Hall in the United States and P. L. T. Herault in France. The success of the process was helped by the fact that electric energy was becoming available at the same time. The first manufacture was begun by the Pittsburgh Reduction Company in 1888, producing about 20–50 pounds a day, first for $5 and later for $2 per pound with production of about 500 lb/day and, by 1894, of 2000 lb/day. In 1893 the producer made a contract with the Niagara Falls Power Co., and production was moved to Niagara Falls. In 1907 the Pittsburgh Reduction Co. changed its name to the Aluminum Company of America (Alcoa), which is still today one of the largest producers and processors of aluminum in the United States.

The Hall-Herault process uses alumina produced from bauxite processed by crushing, washing, and drying and refined by a process patented in Germany in 1888 by Karl Josef Bayer, in which bauxite is digested under pressure with hot sodium hydroxide solution, forming dissolved sodium aluminate. After a number of other steps it yields alumina. Alcoa has improved the process by employing a lime-soda-sinter cycle to enable ores with higher silica content to be processed economically. The basic discovery of Hall was to dissolve alumina in molten cryolite (sodium-aluminum-fluoride) instead of trying to use molten alumina, which has a rather high melting point of 2050°C. Cryolite offered a reasonably low melting point, a low operating voltage of about 6 volts, and low specific gravity, which causes the reduced aluminum to sink to the bottom of the pot from where it can be siphoned off.

The aluminum-smelting process is continuous. The smelting pots are deep, rectangular, steel shells (see Fig.3.38), lined with carbon through which the electric current flows. The layer of metallic aluminum deposited at the bottom of the pot serves as cathode. Current is introduced through carbon bars acting as anodes. Long rows of smelting pots known as "potlines" fill the factory. Some plants contain more than a hundred pots in one building. Approximately 0.5 kg of carbon is consumed for every kg of aluminum produced and the manufacture of carbon anodes is a significant auxiliary activity in the plant. Direct electric current is used, between 50,000 and 300,000 A. Developments are under way to reach a 1 mega-amp cell operating at more than 97% efficiency. The energy requirement has been reduced in the past 20 years from 17 down to 14.5 kWh/kg for aluminum.

The U.S. aluminum industry employs more than 130,000 people and produces 3.3 million tons of ingot and fabricated mill products, 1.15 million tons (35%) of which is secondary recycled metal. The significance of aluminum is due to its outstanding strength-to-weight ratio, outstanding stiffness-to-weight ratio (refer to Chap. 2), and

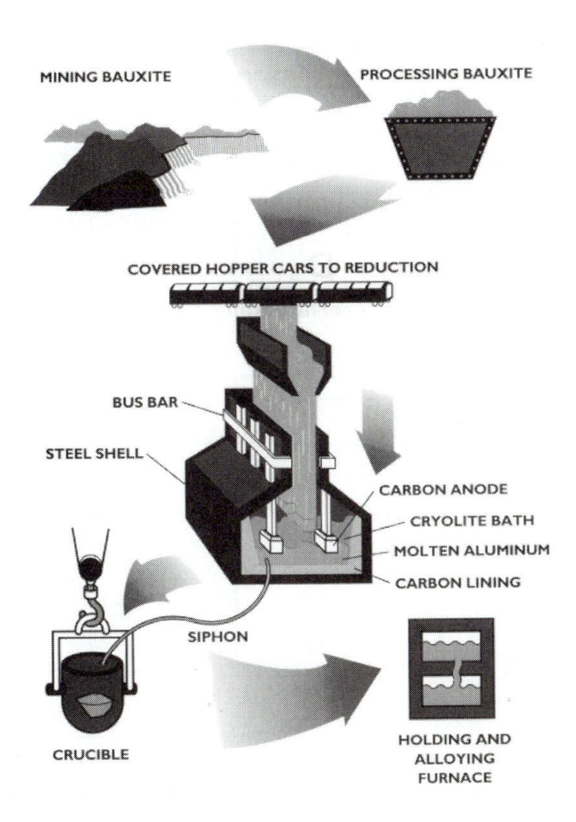

Figure 3.38 Processes involved in the making of aluminum. [AL21]

due to its corrosion resistance, ease of processing by forming and cutting, and ease of recycling. It is used in many industrial sectors. Primarily this is for packaging in the food industry, as 400,000 tons per year of aluminum wrapping foil, various forms of containers, and 100 billion beverage cans produced each year (60% of these are recycled). Soda-can recycling is by itself a 1 billion-dollars-per-year business.

The next sector is transportation, mainly cars and trucks. There is currently an average of about 100 kg of aluminum parts in an automobile. This involves engine blocks, cylinder heads, pistons, intake manifolds, heat exchangers, transmission housings, wheels, and a variety of smaller parts. Some recent automobiles use hoods made of aluminum, in an effort to reduce weight and improve recyclability. The structure (fuselage) of an airplane is almost completely assembled of aluminum stringers, frames, spars, and skin. The excellent electrical conductivity of aluminum has led to an overwhelming use of the metal in the electric power distribution industry. The high currents used in aluminum smelting are conducted in flat aluminum bars. This kind of bus was used in 1895 for a current of 20,000 A by the smelter in Niagara Falls, and it was in service until 1949 when the plant was retired. Nearly all producers of electrolytic sodium hydroxide and clarion use EC-grade aluminum bus for the direct current. Corrosion is negligible, and protective coatings are not necessary. More than 50 million kg per year of aluminum is used for electrical wire. There are many uses of aluminum in housing construction, and in highway and railroad bridges.

World aluminum production is indicated in Fig. 3.39. While the total production has increased more than four times between 1960 and 1994, the production in the United States has increased 1.8 times, and its share has dropped from 40.7% to 17.2%.

Figure 3.39
World aluminum production in 1960 and 1994. [AL/PF]

1960		
Country	Production Quantity	Percent of Total
Total World	**4,490**	**100.0**
United States	1,827	40.7
Canada	691	15.4
U.S.S.R.	640	14.3
France	239	5.3
West Germany	169	3.8
Norway	165	3.7
China, P.R.	80	1.8
Other	679	15.1

(Production Quantity x 1,000 metric tons)

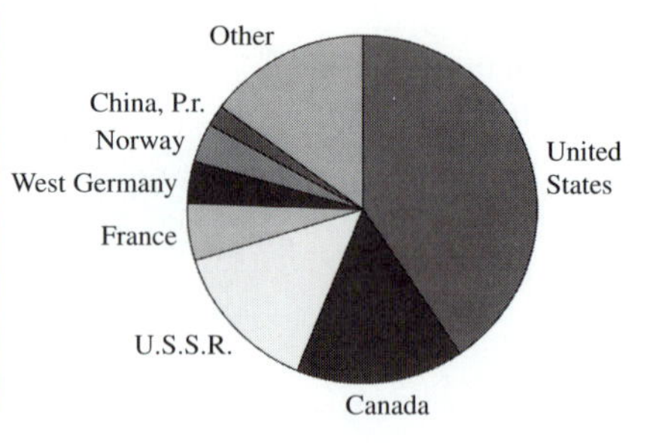

1994		
Country	Production Quantity	Percent of Total
Total World	**19,144**	**100.0**
United States	3,299	17.2
Russia	2,670	13.9
Canada	2,250	11.8
China, P.R.	1,450	7.6
Australia	1,320	6.9
Brazil	1,200	6.3
Norway	857	4.5
Venezuela	580	3.0
Other	5,518	28.8

(Production Quantity x 1,000 metric tons)

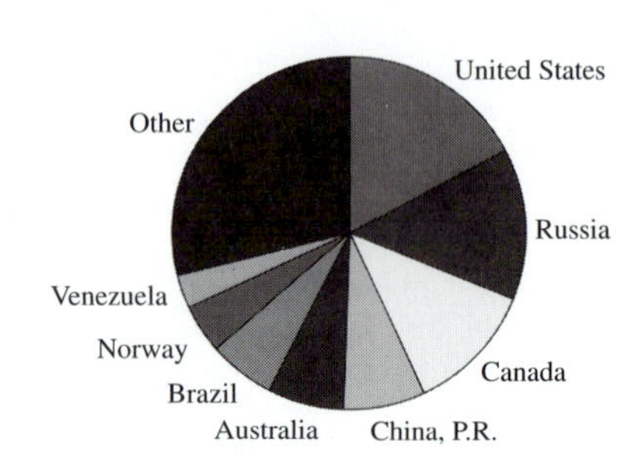

Energy is a major component of the cost of producing aluminum. The cost of electric energy in the United States has increased in the period 1972–1982 from 1.2 to 5 cents per kWh. Countries with low-cost hydroelectric power and those with natural gas reserves such as Canada, Russia, Venezuela, and Australia enjoyed lower power rates in 1992 than the United States. Moreover, Australia has now the largest production of bauxite ore. Current U.S. imports of metallurgical-grade bauxite come from Australia, Guinea, and Jamaica. This raw material is processed in refineries in Louisiana, Texas, and the U.S. Virgin Islands, but the supply of alumina is not sufficient to meet the demand of primary aluminum smelters. As a result, the United States supplements the domestic production of both alumina and aluminum by imports.

The aluminum industry consists of three sectors. The *raw materials sector* produces alumina and primary and secondary (recycled) molten metal and ingots. The *semifabricated product sector* produces plate, sheet, foil, forgings castings, wire, rod, bar, and extrusions. The third sector, *finished products*, manufactures and supplies parts for automobiles, airplanes, and complete products such as windows, doors, beverage cans, and so on.

3.8.2 Processing

All of the manufacturing processes discussed in the preceding sections for iron and steel are also used in the manufacture of products made of aluminum.

Hot and cold rolling is used to produce plates, sheet, and foil. Plates are made in thicknesses > 6.35 mm up to 150 mm and widths up to 1500 mm. Sheet is classified with thickness between 0.152 mm and 6.3 mm, and it can be made as wide as 2743 mm. Aluminum foil is rolled as thin as 4 microns and up to 150 microns. Many special grades of sheet and plate are supplied for specific applications. Among these are anodizing sheet; prepainted sheet with a baked, synthetic resin coating in a variety of colors; reflector sheet with a high-quality surface finish; litho sheet with a high degree of flatness for use in offset printing; vinyl-coated sheet; porcelain enameling sheet; rural roofing sheet in thicknesses from 0.444 mm to 0.61 mm; industrial roofing sheet; armor plate; stainless clad aluminum; and many other specialties. The foil is made with two sides bright or one bright and one satin-finished. The latter, common for thicknesses less than 25 microns, is produced by pack rolling in which two sheets on top of each other are passed through the rolls at the same time. The outer surfaces come out bright. Most foil applications are produced in the 1145 alloy and the O and H19 tempers. The O temper produces flexible foil, and the H19 results in a rigid foil. The foil can be coated or combined by laminating with many flexible and rigid materials. The hard foil is also used in adhesive-bonded honeycomb plates used for aircraft parts.

Rolling is also used to produce wire, rod, and bar, often in combination with subsequent drawing operations. Wire can be round, square, rectangular, or hexagonal in cross section. It is made in diameters between less than 25 microns and 9.5 mm. Larger diameters qualify as rod and bar.

Casting

Sand, permanent-mold, plaster, investment, and die casting are all used. Metal for foundry use, including ingot and scrap, is melted in reverberatory, crucible, or electric induction furnaces. In foundries that cast a variety of alloys, the crucible furnaces are the choice. For continuous production of one alloy, the electric induction furnace is preferred. Refractories can be selected with regard to the alloy cast. Automatic induction pumping and mechanized ladles are incorporated. The equipment is suitable for high production levels, such as in die casting.

Sand casting of aluminum uses the same type of molding equipment as described in Section 3.7 for iron casting, except that lower temperatures are involved. The process is used for small quantities of identical castings from single or multiple patterns, requiring intricate coring. Small- and medium-size castings are made.

Larger quantities of more precise castings may be obtained by *permanent-mold casting* using a mold made of die steel. Most castings weigh less than 10 kg, but larger ones weighing 25–100 kg are not uncommon.

For high production quantities and rates the best process is *die casting.* Cold-chamber machines are used as discussed in Section 3.7, and an excellent example of complexity, precision, and high production rate is obtained. Larger, cold-chamber die-casting machines with die-closing force of 2500 tons or more can produce castings weighing 25–60 kg. Multiple-cavity dies used for small parts can produce 400–1000 pieces per hour. For large castings, rates of 20–50 pieces per hour are achieved. Dies are made of hot-work die steels H11 and H13 hardened to 44 to 48 HRc. If designed for easy replacement of small cores, ejector pins, and inserts, they may last for production of from 200,000 up to 1,000,000 pieces. Small- and medium-size castings for automobiles are mostly die cast.

Extrusion: Impact Extrusion

Extrusion is currently suited to aluminum processing because of the lower processing temperatures and pressures on the die than in extruding steel. This results in long life of dies made from die steel. Most of the standard extrusion operations involve hot forming, while impact extrusion, which is actually a combination of extrusion and forging, is mostly performed cold.

Hydraulic extrusion presses are used, either horizontal or vertical. Presses with capacity of 1600–2500 tons are most common, but capacities exist of up to 14,000 tons. They are distinguished as rod or piercer presses; the latter are intended for the production of hollow extrusions. In Fig. 3.40a the two processes of direct and indirect extrusion are presented. Fig. 3.40b and c show that a seamless pipe can be extruded with the mandrel either moving with the ram or stationary. The ingot is charged into the breech end of the container. The stroke of the ram must equal at least the maximum ingot length plus the length of the container. Press containers operate at 300 to 500°C, and they are stressed to 1000 MPa or more. Their shell is made of two or more cylinders press-fitted together to induce compressive stresses in the liner so as to assure its integrity under the high pressures. Auxiliary equipment includes furnaces for reheating of ingots and of dies. Gas- or oil-fired or induction heating furnaces are used. The recommended press capacities range from 650 tons for a 110-mm-diameter container to 2500 tons for a 280-mm-diameter, for low- and medium-strength alloys, and from 1000 to 4200 tons for the same diameter ranges and high-strength alloys. The relative extrudability, expressing a combination of factors such as temperature, pressure, and other specific requirements, is referred to the 6063 alloy as 100% and ranges from 8% for the 7178 and 9% for 7075 to 20% for 2024, 135% for 1060, and 160% for the EC alloy. The extrusion temperature is limited on the high side by the development of hot short-

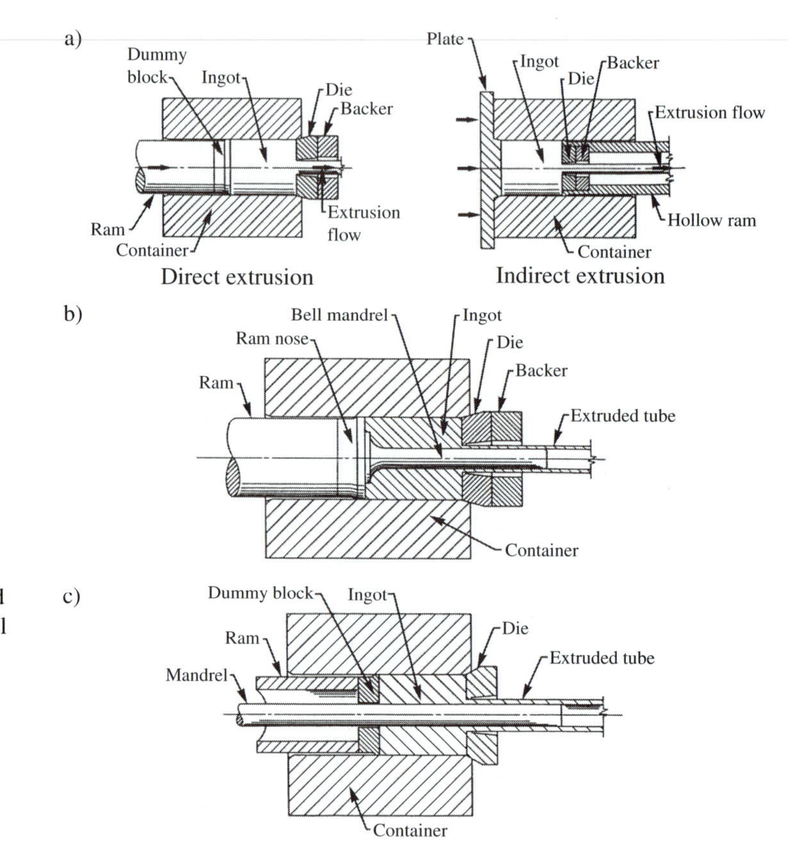

Figure 3.40
a) Tooling and metal flow for direct and indirect extrusion processes; b) tooling and metal flow for extrusion of seamless tube with mandrel attached to press piercer; c) mandrel operating through the hollow ram. (REPRINTED BY PERMISSION BY ASM INTERNATIONAL, FROM ALUMINUM, VOL. III, FABRICATION AND FINISHING, 1967)

ness and by tear phenomena on the surface due to longitudinal tensile stresses induced by friction between work and the die. Commercial-purity aluminum and the low-strength alloys are extruded at the highest temperatures, up to 510°C, while the high-strength alloys such as the 7075 and 7178 are processed at the low end of the temperature range, at 320°C.

Examples of how extruded shapes replace shapes made by other methods are shown in Fig. 3.41. In *(a)* better strength is obtained than for a rolled shape; in *(b)* the extruded shape offers lower cost than a welded one; in *(c)* the extruded shape replaces an assembly of rolled ones; in *(d)* the extrusion saves machining needed for the rolled shape; and in *(e)* castings or forgings are replaced by sawing off parts from an extrusion.

Impact extrusion was originated as a process for making collapsible tubes out of soft metals and it was further developed as a very efficient method for the manufacture of parts that combine a kind of hollow tube with a solid bottom or flange. It is carried out either cold or hot. Fig. 3.42 summarizes the four basic types of the process: *(a)* reverse impact, *(b)* forward impact, *(c)* lateral impact, and *(d)* combination of forward and reverse impact. The slugs used as input to the operation are blanked out or sawed of rolled plate, although individually cast slugs are also used. Their dimensions should be kept rather accurate to avoid additional trimming or machining. Mechanical properties attainable are such as can be achieved by alloying, cold work, or heat treatment and their combination. The as-extruded F temper serves adequately where work-hardening alloy is used, but no alloy can combine well the required strength and good extrusion characteristics. Therefore the added cost of subsequent heat treatment and artificial aging is needed for most applications.

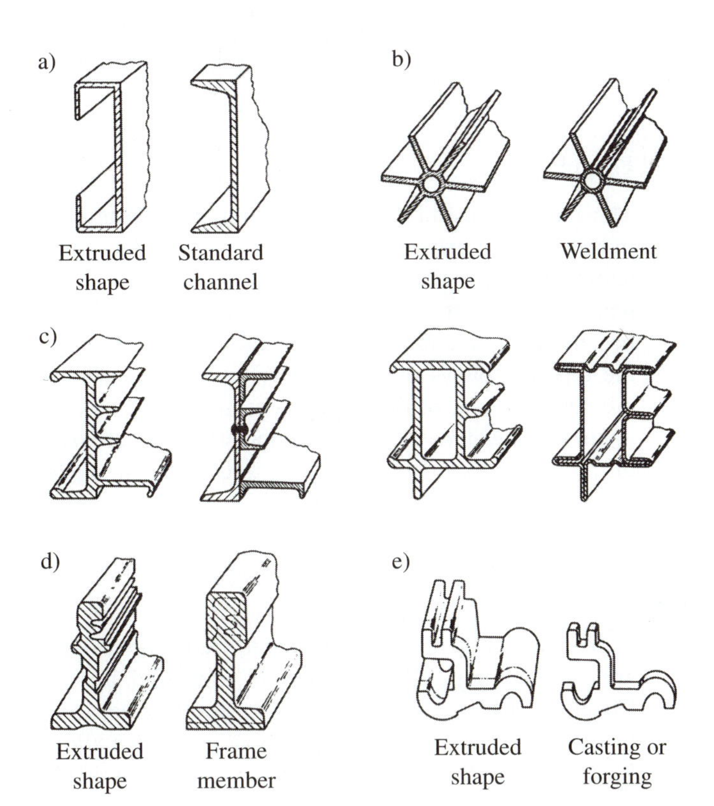

Figure 3.41
Examples where extruded shapes are economical for various reasons; see text discussion. (Reprinted with permission by ASM International, from Aluminum, Vol. II, Design and Application, 1967)

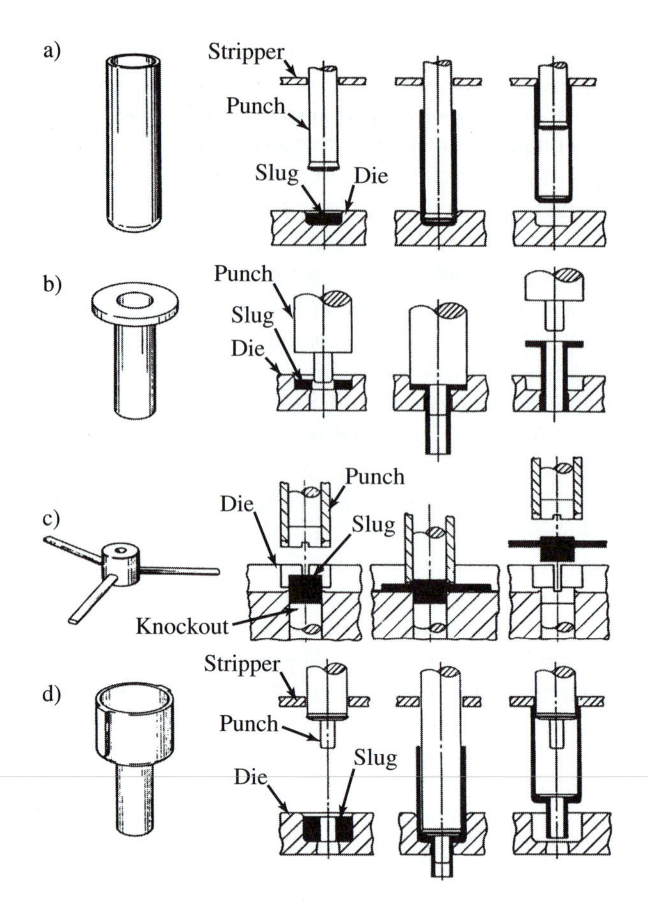

Figure 3.42
Tooling and metal flow for four basic types of impact extrusions: a) reverse impact; b) forward impact; c) lateral impact; d) combination forward and reverse impact. (REPRINTED WITH PERMISSION BY ASM INTERNATIONAL, FROM ALUMINUM, VOL. III, FABRICATION AND FINISHING, 1967)

Forging

The manufacture of forged aluminum parts was initially introduced in the aircraft industry where many small and large forgings have been applied. It has spread to other applications that require the advantages of higher strength and toughness due to the dense structure of the material. Hydraulic presses are the preferred type of equipment because many aluminum alloys forge better under the continuous squeeze than under the repeated blows of a hammer. However, mechanical presses, upsetting (heading) presses, and ring rolling equipment are used as well. All kinds of forging methods, open-die as well as closed and impression die, are used. As for iron and steel, further discussion of forging is postponed to Chapters 4 and 5.

3.9 OTHER METALS

Primary attention has been devoted to the making, shaping, and treatment of iron, steel, and of aluminum. Other metals of industrial importance have been mentioned in Chapter 2. Copper is used in high purity for electric wire and other forms of conductors, and in copper alloys such as brasses (Cu-Zn alloys), bronzes (Cu-Sn; Cu-Al; Cu-Si-Zn, Cu-Be alloys). It finds a wide variety of uses. Magnesium alloys and zinc alloys are used for parts of intricate forms where the light weight and not much strength are the main requirements. Ease of manufacture by permanent mold and die casting as well as by machining are advantages. Titanium alloys provide rather high strength even at elevated temperatures (up to about 450°C). Nickel-based alloys offer high strength and corrosion resistance at high temperatures.

It is not the purpose of this textbook to discuss the extraction metallurgy of all these metals. Many of the techniques described in the former part of this chapter are applied to their production and processing. They originate mostly in ore mines as oxides or sulfides that have to be crushed, ground, or dissolved, and further processed by application of heat, heat and pressure, or electric current. Brief discussion is included here of copper, magnesium, and titanium because the processing of each of these includes some special features.

The principal copper ores are sulfides and ferrosulfates with a rather low content of about 1% only. The rest is gangue (impurities) in the form of Si, Al, and Ca oxides. Therefore, the first step consists of concentration. The ore is crushed and ground in ball mills and coated with oil. The oil wets the sulfide particle but not the gangue. The mixture is fed into flotation tanks where it is agitated. The oil picks up small air bubbles and the copper sulfides, as well as iron sulfide particles float on the surface, while the gangue settles at the bottom. The foam is skimmed and filtered resulting in particles with about 40% Cu. This material is passed into roasting furnaces where many impurities are oxidized and blown away as gases. The result is a mixture of Cu_2S and FeS, SiO_2 as the main impurities and traces of gold and silver as by-products.

The roasted ore is mixed with limestone flux and smelted in furnaces of the open-hearth type to produce sulfides or iron and copper as a compound called matte, which is then poured into a Bessemer type converter where air is blown over the bath. The sulphur is oxidized and vented out, and the iron oxide reacts with the flux and is removed as slag. The copper sulfide is converted to copper oxide and copper sulfate, and then into 98–99% pure copper and sulphur dioxide. The "tough-pitch" copper is further refined electrolytically to higher purity, remelted, and cast into bars.

Magnesium is recovered from seawater that is treated in a tank with milk of lime produced from oyster shells in a kiln at about 1320°C. Magnesium hydrate slurry settles at the bottom and is further converted to $Mg\ Cl_2$ by the addition of hydrochloric acid. It is then evaporated, filtered, mixed with Na Cl and KCl, and the mixture is charged into electrolytic cells. Graphite electrodes are used as anodes, and the walls of the tank act as cathodes at which magnesium particles form and float to the surface, where they are skimmed off. Chlorine gas is recycled to act on the magnesium hydrate in the first phase of the process.

Titanium is extracted of ores that contain the minerals ilmenite or rutile that are oxides of titanium. They are reduced in electric arc furnaces by the use of carbon or of aluminum. Rutile is converted first to TiC and further on, in chlorine atmosphere in a roasting furnace, to $Ti\ Cl_4$. This is then reduced with magnesium, processed by distillation and leaching, vacuum melted, alloyed, and cast. It can be further processed into plates and bars by hot rolling. It can be forged into various components, typically those used in military aircraft for strong and light structures.

3.10 POWDER METALLURGY

The powder metallurgy (PM) process has some similarities to casting and forging, and in certain applications (type A), it is used to produce parts that might otherwise also be produced by one or either of those two processes. In most (type B) applications, however, it is unique and delivers parts that may not be obtained by casting nor by forging.

As the name implies the starting material is not molten as in casting or solid as in forging; rather, it is fine-grained powder of one or more different constituents. The powder is compacted at room temperature into the desired shape and then heated and kept at the elevated temperature to produce bonds between the powder grains while between them a thin layer of another material is often applied as the binder. This is called sintering, and PM products are often also called sintered. The temperature may be high enough to cause melting of the binder but not of the basic grain, and the fundamental process by which the powder material becomes a strong granular structure is diffusion. The size and shape of the grains do not substantially change in the process, nor do the various constituents become chemical compounds.

For applications of type A PM may be chosen because under certain circumstances it offers *lower cost or a higher production rate.* Small parts such as gears, cams, pawls, and ratchets fall into this category. Dozens of small parts made by PM are used in a modern car. Otherwise, the process may offer *higher-quality material.* For instance, in the production of tools such as end mills or twist drills from high-speed steels, PM yields steels with fine and uniform grains leading to higher strength and superior durability of the cutting edge than cast steel, in which control of grain size in the solidification of the casting is much more difficult.

Type-B applications include the manufacture of parts in which the main constituents are refractory metals with *very high melting temperatures*: W (T_m = 3425°F) T_a (2980°C). They could hardly be cast because of the difficulties with achieving these temperatures and finding mold material that could withstand them. For example, consider sintered carbide cutting-tool inserts with a sample formulation of WC grains with Co binder material. Highly uniform structure is achievable that might be considered as a "designed" structure of a composite material. Actually, one of the first uses of PM was the development in the early 1900s of the method of making light-bulb filaments out of tungsten wire. Some parts for jet engines made from Co based or Ni-based superalloys are similarly produced by PM.

In other instances PM is used to produce porous parts such as filters and screens made of various metals and especially of stainless steels used in processing chemicals. Porous bushings made of bronze would be immersed in heated oil and impregnated by capillary action when they cool down. They can then be used as self-lubricating bearings.

In yet other applications PM is used to produce composite materials of metals with ceramics or other nonmetals. One such example is brushes for commutators of DC electric motors, which are made from powders of copper and graphite. The magnets for these motors are made of alloys such as $Fe_{14}Nb_2B$ or $SmCO_5$.

The PM process consists of producing the powders, mixing and blending them, compacting them into the "green" shape, and sintering.

3.10.1 The Powder

Powders are made of copper-based and iron-based materials, of stainless steel, aluminum alloys, nickel, tungsten, and cobalt-based alloys, and of metallic oxides and carbides.

The powders may be further ground in ball mills or crushers. It is very important to obtain uniform grains with a narrow distribution of sizes, which make it easier to achieve high density, compressibility, and green strength. Later on, powders are mixed with lubricants to ease the filling of the die and the compacting of the powder. Mostly the lubricants are volatilized or burned off in the sintering phase. Binder materials are mixed in to improve the final strength of the material.

The starting materials could be an extremely fine powder produced by deposition from the vapor phase (Zn), precipitation from a chemical solution (Cu, Ni), reduction of an oxide (Fe, Cu, Mo, W), or vapor of a compound (Ni, Fe), or electrolysis (Fe, Cu, Be). The metal may be reduced from its ores in a liquid form, refined, and then worked in the solid form. The solid is then converted into small particles, for example, by pounding until it disintegrates into small flakes, by machining fine chips, or even by chopping up fine wire. Alternatively, the stream is broken up (atomized) with powerful jets of water, air, or an inert gas into small (1–100-μm-diameter) particles. In special processes, the melt is broken up as it is formed, for example, by spinning an electrode. Some alloys may be produced by mixing powders of the constituent metals in the correct proportion; others are first alloyed by melting and then broken up into particles. Nickel and iron powders are often obtained by thermal decomposition of carbonyls.

The shape of powder particles, distribution of particle sizes, and surface conditions have a powerful effect on subsequent consolidation and sintering. Some metals (e.g., iron) are likely to be oxidized, but the oxide is readily reduced by a suitable atmosphere during sintering. Others, such as titanium, dissolve their own oxide and are thus reasonably suitable for powder processing. Still others are covered with a thin but very tenacious and persistent oxide film that greatly impairs the properties of the finished part, and these materials (typically those containing chromium and, in general, the high-temperature superalloys) must be treated by special techniques to keep oxygen content very low. Any contaminant that segregates on the surface is bound to create not only consolidation and sintering problems but will also greatly detract from the service properties of the material. Any remainders of a surface film automatically occupy the worst possible position on grain boundaries and act as crack initiators.

3.10.2 Compacting

Compacting is carried out by compressing the powders either at room temperature in a rigid mold or die in single-action, opposed-ram or multiple-action presses, or by cold or hot isostatic pressing in pressure vessels. The purpose is to achieve sufficient density of the material for successful sintering and sufficient green strength for safe mechanical handling of the compacts. In many materials densification is enhanced by plastic deformation of the particles.

A mold containing the negative of the component shape may be filled under gravity, giving a compact of low density and very low strength. Only with very careful handling can it be converted into a porous, sintered product (more likely, the mold itself will be heated to at least initiate sintering). Vibration of the mold during filling aids arrangement of the particulate and results in densities as high as two-thirds of the theoretical density, especially if the particles are of favorable shape and contain both larger- and smaller-size fractions so that interstices between the larger particles are filled by smaller ones.

A further increase in density is obtained by applying pressure to the powder. If the part shape is fairly simple and a die can be made of steel, high applied pressures are permissible. If the particulate can deform plastically, densities in excess of 90% of theoretical can be achieved. The effectiveness of pressing with a punch is limited, however, because particulate material does not transmit pressures as a continuous solid would. The pressure tapers off rapidly, giving compacts of higher density close to the punch and of diminishing density farther away. The situation obviously improves with a floating container or with several punches—that is, two counteracting punches compacting from the two ends of the die cavity. The logical extension of this concept is isostatic pressing, in which a deformable mold (reusable rubber or single-use metal) contains the particulate while hydrostatic (omnidirectional) pressure is applied by means of a hydraulic fluid inside a pressure vessel. This is called cold isostatic pressing. Hydraulic pressures up to 500 MPa may be used. Materials such as refractory metals and ceramics are compacted by hot isostatic pressing when contained in sheet metal molds of simple shapes and subject to pressures up to 300 MPa at temperatures to over 1500°C. In this latter process compacting and sintering occur simultaneously. Compacting methods other than pressing in rigid dies or isostatic pressing are also used, such as extrusion or rolling of powders.

A substantial barrier to attaining high green compact densities is the friction of individual particles relative to each other and also to the die wall. A thin film of lubricant, chosen so as not to interfere with subsequent sintering, can greatly improve the density of the compact. A small quantity of a lubricant, usually a stearate, is added to aid the densification of powders such as copper and iron that will be sintered with free access of the sintering atmosphere to the powder body. Iron powders are often mixed with varying quantities of carbon, copper, or nickel to impart better mechanical properties, aid sintering, and often also provide built-in lubrication for bearing applications.

For pressing in a rigid die, an accurately measured quantity of the powder is fed into the die and compacted with pressures ranging from 50 MPa for porous materials to 100–200 MPa for refractory materials and 300 up to 1000 MPa for brass, iron, and steel parts. Dies are usually of the true closed-die (trapped) configuration, as shown in Fig. 3.43. A punch penetrating from one side is suitable only for thin (low-height) parts *(a)*, although an action similar to opposed-punch pressing is obtained with a floating die *(b)*. Greatest control is achieved with two punches penetrating from opposite sides *(c)*, and special presses equipped with lower and upper rams are commonly used for thicker parts. In *(d)* the hydrostatic compaction is shown. It uses a flexible mold that is compressed by pressurized fluid or gas. In parts of greatly varying thicknesses (measured in the direction of punch movement), uniform density necessitates multiple punches guided within each other (see Fig. 3.44). Clearances between moving parts must be kept extremely small to prevent entry of powder. The dies are built of high-strength tool steel or, for larger production runs and severely abrasive conditions, of sintered tungsten carbide. Initial compaction may be accomplished by isostatic pressing or, less frequently, by slip casting.

3.10.3 Sintering

The green compact is sintered in an atmosphere chosen to provide a nonoxidizing, reducing, or, for steel, sometimes also a carburizing environment. With proper allowance for shrinkage, tolerances can be held to a 0.1–0.2 mm range on a 25-mm dimension.

The surface finish will be rougher than that of the compacting die because porosity is still significant: 4–10% depending on powder characteristics, compacting pressure, and sintering temperature and time. Particularly for bearing applications, density is often kept even lower so as to allow infiltration with a lubricating oil, a lubricating polymer

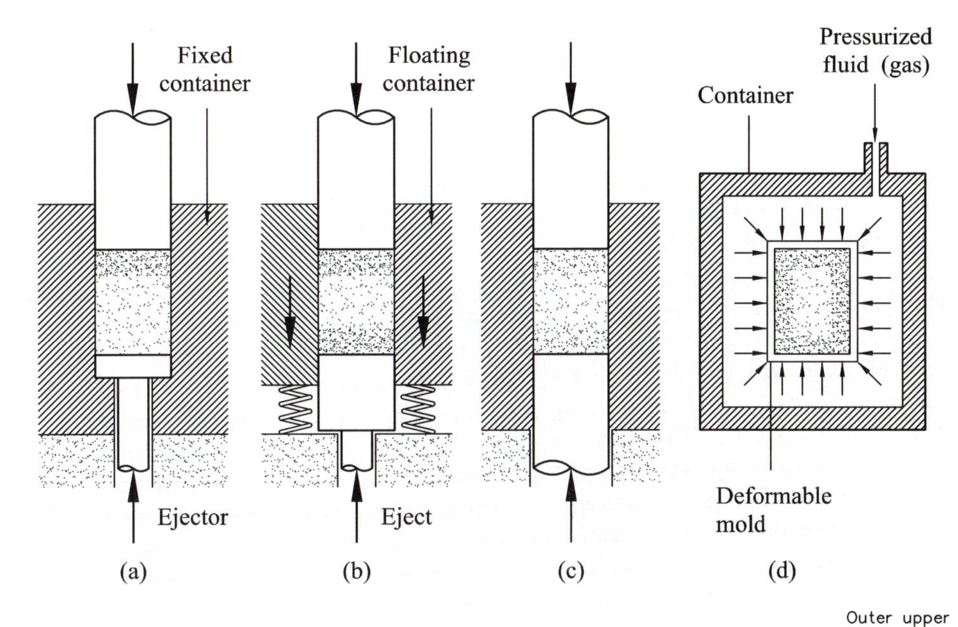

Figure 3.43
Methods of compacting particulate matter and the resulting density distribution: a) single-punch, fixed container; b) single-punch, floating container; c) counteracting punches; and d) hydrostatic (isostatic) compaction.

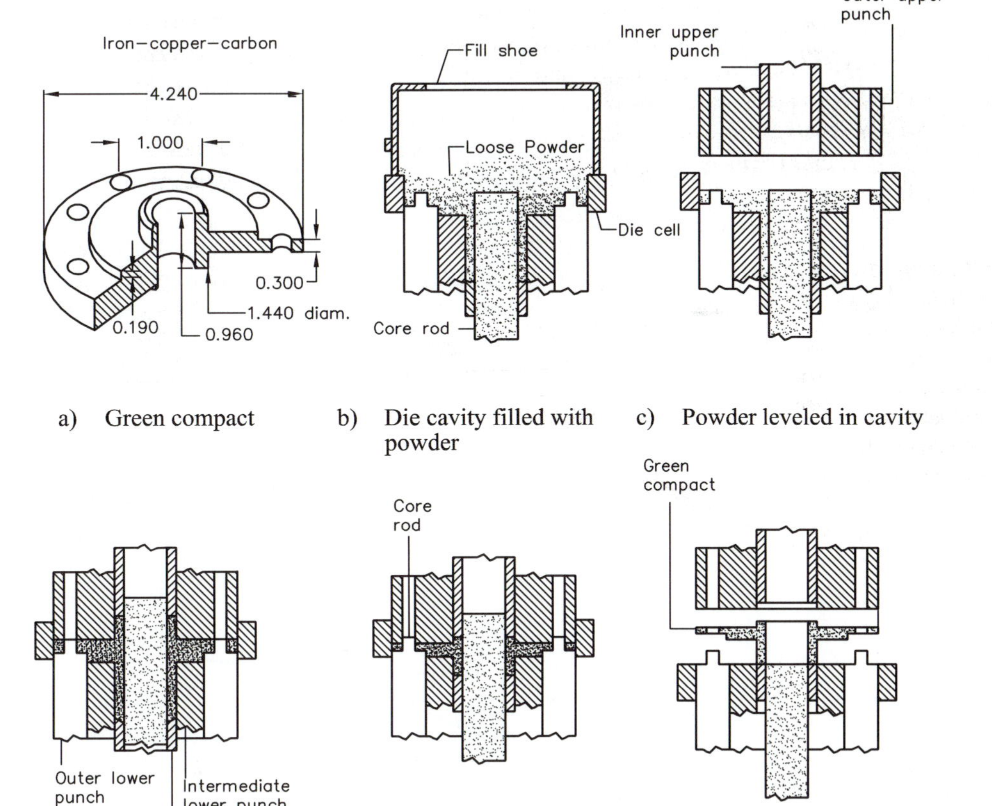

Figure 3.44
An example of a complex powder-pressing operation (dimensions in inches).

(such as PTFE), or a metal (such as Pb or Sn). The infiltrate is drawn into the sintered skeleton by capillary forces after it is heated above the melting point of the infiltrant. Copper infiltrated into iron increases strength and improves machinability.

Cold restriking (coining or sizing) of the sintered compact increases its density and improves dimensional tolerance to 0.025 mm. Further densification and strength improvement can be achieved by resintering the re-pressed compact (see Fig. 3.45) or by increasing sintering time (see Fig. 3.46).

The sintering process is carried out at temperatures of about 60–80% of the melting temperature of the basic matrix material and takes mostly 15–60 min. One constituent melts, as in the case of tungsten carbide tools, where the WC grains amount to 90–98% of weight and the rest is Co binder. Sintering is done in two steps. After pressing the compact at room temperature into the green state comes the presintering step. At about 700–800°C the binder grains soften, and the compact is pressed again to higher density than in the green state. Then it goes again into an oven where it is sintered at 1250–1350°C for about 30 minutes. The cobalt binder melts and fills the narrow spaces between the WC grains. The surface atoms of WC go by diffusion into solid solution in Co and form necks connecting the grains into a strong matrix. The structures and properties of sintered carbide inserts used for machining are discussed in more detail in Chapter 8, Section 8.3.

Most PM products, however, are sintered at temperatures below melting of all the constituents, and the consolidation of the compact is fully derived by solid-state diffusion. In any case the distributions of the particles, their sizes, and their shapes do not change much. Consequently, the composition and structure of the material can be rather accurately controlled.

Instead of the traditional pressing, sintering, re-pressing, and resintering sequence, the green compact may be preheated to the forging temperature and directly hot forged to close tolerances at full theoretical density. Such parts can possess the same properties (including toughness) as conventionally forged pieces and can be given shapes otherwise too complex to attain. For example, bevel gears can be forged to finish dimensions requiring only minimum surface finishing.

Compaction by rolling, followed by sintering and perhaps rerolling, is used both for manufacturing sheet and for cladding a solid base metal.

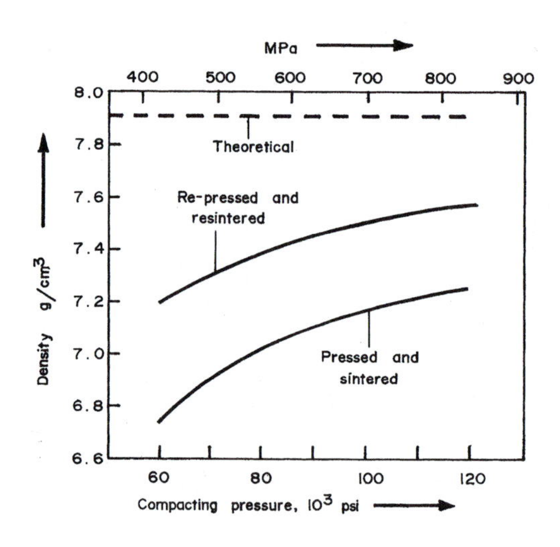

Figure 3.45
The effect of re-pressing at 700 MPa and resintering an electrolytic powder-iron compact (sintering 1120°C for 1 hr).

Figure 3.46
The process of sintering: a) changes in properties, and b) development of microstructure.

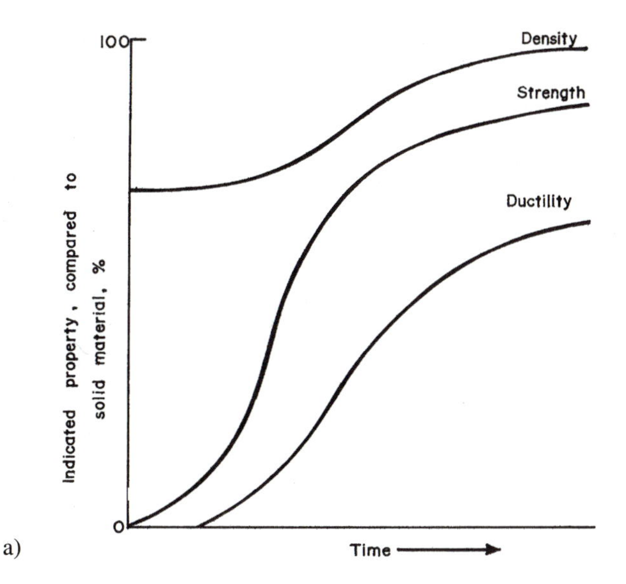

REFERENCES

1 McGannon, H. E., ed. *The Making, Shaping, and Treating of Steel.* 9th ed. United States Steel Corp., 1971.

2 Lankford, W. T., Jr.; N. L. Samways; R. F. Craven; and H. E. McGannon, eds. *The Making, Shaping, and Treating of Steel.* 10th ed. United States Steel Corp., 1985. Distributed by Association of Iron and Steel Engineers, Pittsburgh, PA.

3 Walton, C. A., and T. J. Opar, eds. *Iron Casting Handbook.* Iron Casting Society, 1981. Distributed by American Foundrymen's Society, Des Plaines, IL.

4 Steel Founders Society of America. *Steel Castings Handbook.* 6th ed. Steel Founders Society of America, 1995.

5 ASM International. *Metals Handbook.* Vol. 15. Casting. ASM International, 1988.

6 Schey, John A. *Introduction to Manufacturing Processes.* 2nd ed. McGraw-Hill, 1987.

7 Society of Manufacturing Engineers. *Tool and Manufacturing Engineers Handbook.* 4th ed. Vol. 2. Forming. Society of Manufacturing Engineers, 1984.

8 The Aluminum Association. *Aluminum: The 21st-Century Metal.* The Aluminum Association, 1994.

9 American Metals Society. *Aluminum.* Vol. 2. Design and Application. American Metals Society, 1967.

10 American Metals Society. *Aluminum.* Vol. 3. Fabrication and Finishing. American Metals Society, 1967.

11 The Aluminum Association. *Partnerships for the Future.* The Aluminum Association, 1996.

12 Klar, E., ed. *Metals Handbook.* Vol. 7. Powder Metallurgy. ASM International, 1984.

13 Hausner, H. H.; and M. K. Mal. *Handbook of Powder Metallurgy.* 2nd ed. Chemical Publishing, 1982.

14 German, R. M. *Powder Metallurgy Science.* 2nd ed. Metal Powder Industries Federation, 1994.

Illustration Sources

[AL/21] Reprinted with permission from *Aluminum, the 21st Century Metal*, 1994, The Aluminum Association, Washington, DC.

[AL/PF] Reprinted with permission by the Aluminum Association, Inc., Washington, DC, from *Aluminum Industry: Industry/Government, Partnerships for the Future*, 1996.

[ASM15] Reprinted with permission from *ASM Handbook, Vol. 15, Casting*, 1988, ASM International, Materials Park, OH 44073-0002.

[ICH] Reprinted from the *Iron Casting Handbook,* Iron Casting Society, 1981. With permission from American Foundrymen's Society, 505 State Street, Des Plaines, IL 60016-8399.

[SCH] Reprinted with permission from *Steel Castings Handbook,* 6th Ed., 1995, Steel Founders Society of America, 455 State Street, Des Plaines, IL 60016.

[SME2] Reprinted with permission from *Tool and Manufacturing Engineers Handbook,* 4th Ed., *Vol. 2, Forming,* Copyright 1984, Society of Manufacturing Engineers, Dearborn, MI.

[USS9] Reprinted from *The Making, Shaping, and Treating of Steel,* 9th Ed., United States Steel 1971. With permission from Association of Iron and Steel Engineers, Three Gateway Center, Suite 1900, Pittsburgh, PA 15222-1004.

[USS10] Reprinted from *The Making, Shaping, and Treating of Steel,* 10th Ed., United States Steel 1985. With permission from Association of Iron and Steel Engineers, Three Gateway Center, Suite 1900, Pittsburgh, PA 15222-1004.

QUESTIONS

Q3.1 Write down the stages in the production of steel sheet, starting from the iron ore. For each stage specify the form and substance of the material and specify the equipment used between these stages for each transformation.

Q3.2 Name the alternative processes applied to sheet steel exiting from the cold-rolling mill.

Q3.3 Name the four main components of the charge into the blast furnace.

Q3.4 Why is iron ore first converted in the blast furnace into pig iron (give its carbon content) and then, in other furnaces, into steel (give its carbon content)?

Q3.5 What is the fundamental chemical composition of iron ore? What is chemically the main reducing agent acting on the ore in the blast furnace?

Q3.6 How is coke produced? Name the by-products from coke ovens.

Q3.7 What is the role of limestone in the blast furnace?

Q3.8 Consider the three types of steel-making furnaces, (a) the open-hearth (OH), (b) the basic oxygen (BO), and (c) the electric arc (EA) furnaces. For each of them, state (1) sources of oxygen, (2) sources of heat, (3) type of charge, (4) average size (tons/cycle), (5) average time per cycle.

Q3.9 Explain the principle of heat regeneration as used in the blast furnace and in the open-hearth furnace.

Q3.10 Explain the difference between rimmed, semikilled, and killed steel ingots.

Q3.11 Refer to methods used to remelt steel in vacuum. What is the purpose? What is the purpose of vacuum casting of steel?

Q3.12 Explain the difference between slab, bloom, and billet.

Q3.13 What are the two possible paths from a BOF to a slab?

Q3.14 A two-high rolling stand is usually reversible if the product must undergo several passes. Which setup of a rolling stand eliminates the necessity of reversing the rotations of the rolls?

Q3.15 Why should the roll diameter be large for the rough hot-rolling passes and small for all the following passes, especially for the cold-rolling ones? In the latter case, what is the effect on the rolling force F, the rolling torque T, and the rolling power P?

Q3.16 What is the usual way of achieving smaller work-roll diameters, in which stand setup? Which design of a rolling mill brings the quest for the small work roll to the extreme?

Q3.17 Discuss the passes depicted in Fig. 3.25.

Q3.18 Explain the stress situation in the center of a billet squeezed in the rotary piercing mill that helps this method of making seamless pipes to work.

Q3.19 Name several casting processes in the expendable-mold category and in the permanent-mold category.

Q3.20 What is the purpose in a mold of the sprue, the runner, the gate, the riser, the core, the core print?

Q3.21 How is the sand mold designed to make it possible to take out the pattern? Which surface of the casting is determined by the mold and which by the core?

Q3.22 Which kind of machines are used to mechanize the making of a green sand mold?

Q3.23 Which expendable-pattern material is used for large castings and which for small, precision type castings?

Q3.24 Which mold materials are used for the casting of cast iron and steel, and which for aluminum or zinc?

Q3.25 Which materials are cast in hot-chamber die-casting machines, and which in the cold-chamber types? Why?

Q3.26 Name one method used in the mechanized filling of the cold chamber with molten metal.

Q3.27 Is it possible to produce end-milling tools made of high-alloy steel by investment casting?

Q3.28 What kind of ore is used for making aluminum, and by which process is it extracted? Compare the total yearly production of steel and aluminum in the United States.

Q3.29 Discuss the use of aluminum in airplanes, in automobiles, and in soda cans. In the last case what are the competitive materials? Which one seems to be gaining in this competition and why?

Q3.30 Sketch the process of forward and backward extrusion in the production of tubing, including the impact processes.

Q3.31 Name the main application fields of powder metallurgy and the reasons for its use.

Q3.32 Name the main stages in producing sintered carbide cutting inserts.

Q3.33 What kind of mold (die) material is used in cold isostatic compaction? (Can rubber be used? Try to compact a toy frog from iron powder.) What mold material is used in hot isostatic compaction?

PART 2

Traditional Processes

4 Metal-Forming Technology

This chapter is devoted to the practice of metal-forming operations. It is more descriptive than analytical, and it deals with the three main aspects of manufacturing technology: the features and conditions of the individual *operations,* the *machines,* and the *tools* used. Metal forming is an old art and trade. It has developed into a large number of specific operations, which are presented in two groups:

1. *Bulk* forming, mostly *hot* but also cold; all of it could also generally be called forging,
2. *Sheet-* (and plate-) metal forming, which is almost always done *cold.*

Automation is extremely important in metal forming. Most of the metal-forming operations have always included a high degree of automation; this is related to the use of sophisticated tooling in the form of complex die sets and the associated need for production in large quantities. There is no separate chapter on automation in metal forming, however, because many of the principles explained in Chapter 10 on automation in metal cutting also apply here. The various features of mass-production automation are discussed in the respective sections on forging and on sheet-metal forming. Because of the relative novelty of the applications of numerical control in metal forming, it is surveyed in a separate section in this chapter.

4.1 GENERAL OPERATING CONDITIONS, MACHINES, AND TOOLS

The operating conditions involve pressures, stress and strain distribution, strain rates and their distribution, temperatures, lubrication, and the corresponding friction. All the conditions have to be so arranged as to obtain, in the most efficient and economical way, a sound product of good quality, good accuracy and surface finish, without internal or superficial defects. First, some of the basic effects of hot and cold working on material properties are presented, especially with respect to grain-size control and some other special properties important in metal forming.

4.1.1 Hot Forming

Let us consider an ingot with its rather coarse and non-uniform grain structure, with voids (porosity) and nonmetallic inclusions. The ingot is subjected to the *first hot-forming* operations either by hot rolling or forging. These operations are essentially intended to *break down* the ingot structure. Diffusion processes are accelerated not only by the high temperatures but mainly by the reformation of grains. In this way the chemical nonhomogeneity of the casting is reduced. Blowholes and porosity are eliminated by the welding together of these cavities. Certain refinement of the grain is achieved. All this leads to increased toughness and ductility of the material.

Further changes are produced in subsequent hot forming that depend greatly on the combination of the degree of plastic deformation as expressed by the deformation ratio and of the temperature. The deformation ratio (forging ratio, rolling ratio) is the ratio A_0/A_1 where A_0 and A_1 are the areas perpendicular to the main direction of elongation, before and after the deformation, respectively. It is very important to carry out the hot work in the correct temperature range. The advantage of using higher temperatures, of course, is that forging is easier against lower yield strength. However, at temperatures that are too high the metal gets burned, and the corresponding damage cannot be repaired except by remelting.

At very high temperatures, steel *overheats.* Some of the effects arising are based on diffusion, so the duration of overheating is also important. The grains grow, and sulphur diffuses into the grain boundaries, forming compounds like FeS and MnS. These segregates prevent the formation of a fine-grained structure during further processing, and they affect the toughness and ductility of steel. These problems are not easy to remedy. However, they can be repaired by thorough forging and by special annealing treatments.

At temperatures close to the melting temperature, steel gets *burned.* Segregated regions of lower-melting-point material containing sulfides and phosphides occur at the grain boundaries and also inside of the grains. The work crumbles into pieces when deformed. This is called *hot shortness.* It is not possible to repair burned steel by mechanical work or heat treatment; it must be remelted.

Another effect of high temperatures used in forging is *oxidation* of the work surface. Primary oxidation occurs in the furnace. Several layers of oxides, Fe_2O_3, Fe_3O_4, and FeO, represent the scale that mostly crumbles away during subsequent forging. However, some of the scale is crushed to fine particles that combine with secondary oxides generated during forging. This secondary scale must subsequently be machined away, or, if hot forming is followed by cold forming, the scale is mostly removed by pickling before cold forming.

Having chosen an initial forging temperature that is high enough, but not so high as to cause overheating and burning, it is necessary to combine deformations with temperatures for proper *grain control.*

With a large degree of deformation, the grains are repeatedly crushed and refined during simultaneous recrystalization, even at high temperatures. Final stages of working, however, with a lower degree of deformation have to be carried out at lower temperatures. For instance, it is possible to rough-forge medium-carbon steel at 1200°C, but the finishing must be done between 1000°C and 900°C. If a complex shape of the forging requires higher finishing temperatures, then it is necessary to refine the grain of the forging in a normalizing treatment subsequent to forging.

Steel is best formable in the austenitic field. The temperature range for hot forming of carbon steels is indicated in Fig. 4.1. The upper limit is about 200–300°C below the solidus line, and the lower limit is above the line of the γ to α transformation. The finishing temperatures for hypereutectoid steels are lower. If their forming was finished above the S line (segregation of secondary cementite), continuous cementite structure would form on the boundaries of pearlite grains, and the steel would become very brittle. By continued forming below the S line, the cementite structure is broken into smaller particles, and the steel becomes tougher. Also, the lower temperatures minimize surface decarburization, which otherwise could be very pronounced in the high-carbon steels.

Grain size affects many properties of steels. Finer grain imparts higher yield strength and toughness and lowers internal strains and stresses. The variation of grain size during heating up and during forging is indicated in the diagram in Fig. 4.2 reproduced from [1]. Starting with size I and heating up there is no change until temperature A_1, point *a*. From then on until A_3, during the α–γ transformations the grain becomes finer (point *b*). During further temperature increase, in the austenitic field from *b* to *c*, where the maximum forging temperature is reached, the grain size increases considerably and continues to grow during the transfer of the part from the furnace into the forging machine, *c*–*d*. The first squeeze of the hot material produces finer grain and a slight increase of the work temperature, *d*–*e*. If forging was stopped now, the temperature and grain size would follow the line *f*–*g*–II. The final grain (II) would be coarser than the initial grain (I). However, if forging continues according to *f*–*h*–*i*–*j*–*k*–*m* and stops at a temperature close to A_3, the work then cools along *m*–*n*–III and ends up with a rather fine grain.

The effect of the final forging temperature on grain size is explained in another form in Fig. 4.3. The vertical scale represents temperature, and the widths of the individual columns represent grain size. The crosses represent forging. The columns illustrate the following:

(a) Cooling of an ingot from the teeming temperature to room temperature

(b) Heating up of the ingot for forging

(c), (d), (e) Forging finished at subsequently lower temperatures; in the last case this is at A_1 and results in the finest grain

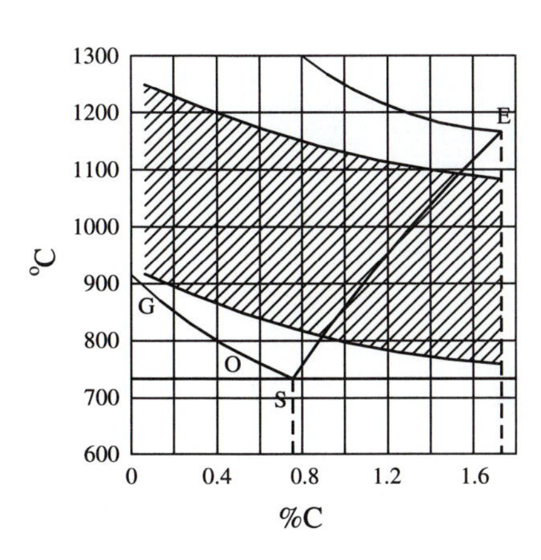

Figure 4.1
Range of hot-forming temperatures. [HASEK]

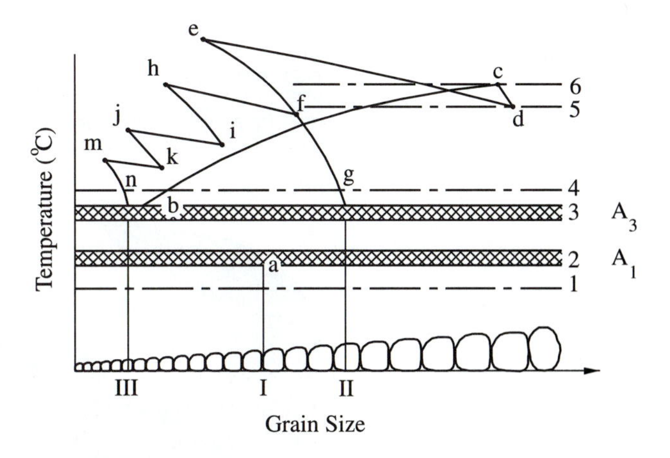

Figure 4.2
Variations of grain size during forging. [HASEK]

(f) Forging continued below A_1. This is eventually cold working, and there is no more grain refinement; the material strain-hardens.

In order to completely refine the original dendritic structure of the ingot, it is necessary, especially in open-die forging of large parts, to reach deformation degrees as expressed by forging ratios of about 5 to 6. During this deformation the material acquires a fibrous structure that leads to *anisotropy,* as expressed by the differences in *ductility, toughness,* and *fatigue strength* in the directions along and across the fibers. Achieving favorable directions of fibers is a significant consideration in the design of the individual steps in closed-die forging.

Impurities contained in the steel, consisting of sulphides, oxides, nitrides, phosphides, and of other types of inclusions, segregate along grain boundaries (see Fig. 4.4). During deformation, the grains and their boundaries with the segregated impurities elongate. While the grains reform and recrystallize, the segregates keep their elongated shapes. This process continues with the degree of deformation. The intensity of the fibers, or stringers, depends on the purity of the original ingot. Because of the inferior properties of the material in the directions across the fibers, it is important to design the forging process so as to result in fibers following the overall shape of the forging, as shown in Fig. 4.5.

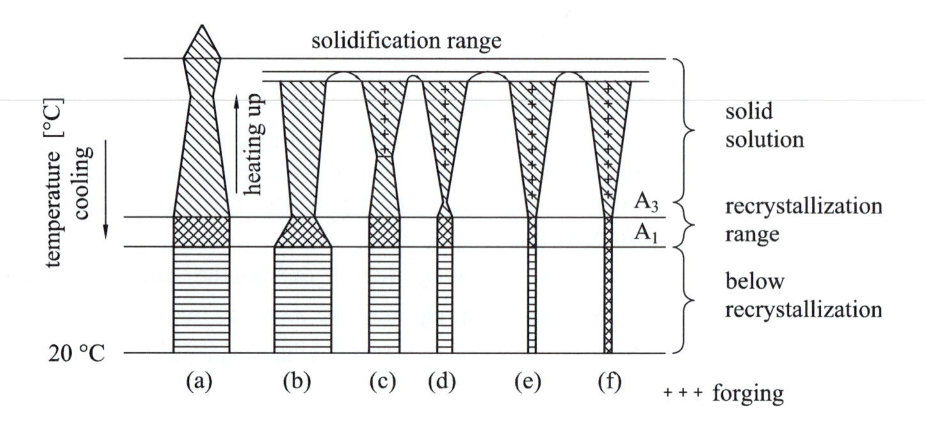

Figure 4.3
Effect of final forging temperature on grain size. [HASEK]

Figure 4.4
Generation of fibers of impurities in hot forming. [HASEK]

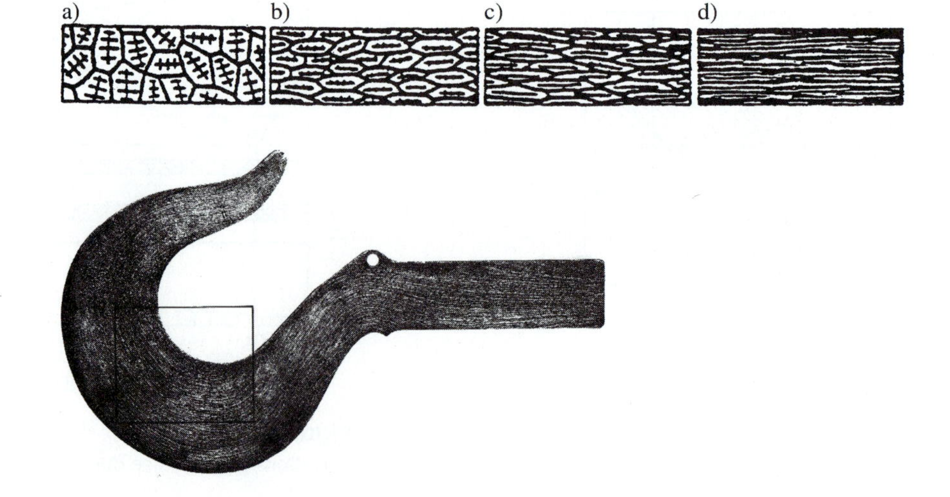

Figure 4.5
Streamlined fibers in a forging. [FIH]

4.1.2 Cold Work and Anneal Cycle

After primary hot forming, the structure and properties of the material undergo further changes during *heat treatment* or combinations of *cold work* and heat treatment. Primary cold work is essentially cold rolling and cold drawing. These *primary cold-forming operations* are usually followed by *annealing* before any of the *secondary cold-forming operations,* which again may be combined with heat treatment.

Figure 4.6
Lattice orientations from grain to grain in polycrystalline matter. [HASEK]

In order to understand the changes occurring in the material during cold work it is necessary to consider the nature of the plastic deformation of polycrystalline metals. The boundaries between grains are regions of disturbed lattice only a few atomic diameters wide. In general, the lattice orientation changes abruptly from grain to grain (see Fig. 4.6). The grain boundaries have high surface energy and therefore serve as preferential sites for solid-state reactions such as diffusion, phase transformation, and precipitation. They also contain higher concentrations of solute atoms than the interior of the grains.

If a single crystal is deformed, it happens usually on a single slip system. However, in polycrystalline material, due to the different orientation of grains, considerable differences arise in the deformation of neighboring grains and within each grain. Slip occurs on several systems in each grain and it occurs first in those grains in which the close-packed planes are located in the direction of maximum shear stress. Figure 4.7 is a diagram of a part of a sample that is being compressed. In each grain the slip planes are oriented differently with respect to the shear stress. After the slip occurs in the conveniently oriented grains, the deformation spreads to the other grains, too, while simultaneous lattice rotations lead to the formation of deformation bands. At this stage, the material strain-hardens considerably, due to a cellular substructure of high-density dislocation tangles. During further deformation, more and more lattice reorientation occurs as well as some lattice bending and fragmentation. At the deformation ratio of about 3 (25% area reduction), preferred orientations are easily discernible, and at a deformation ratio of 8 to 10 (70–90% area reduction), the preferred orientation is essentially complete. A *crystallographic fibrous structure* develops, which is different from the fibers of impurities that are generated in hot rolling and hot forging. The process of extensive cold work and of corresponding grain distortions is expressed in Fig. 4.8. Due to lattice rotations, the grains elongate in the directions of rolling or drawing, and they

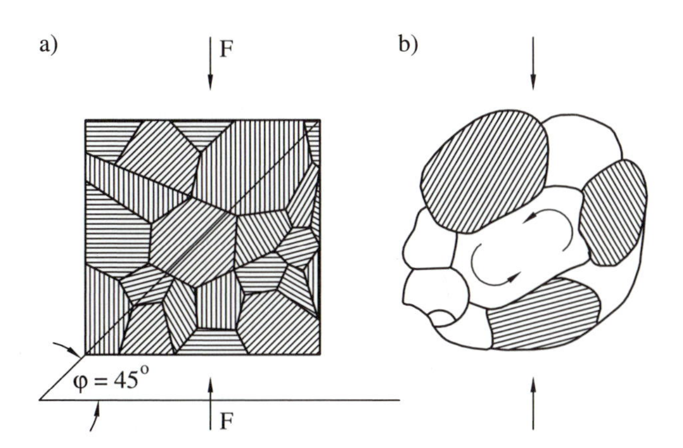

Figure 4.7
Deformation of a polycrystalline material. [HASEK]

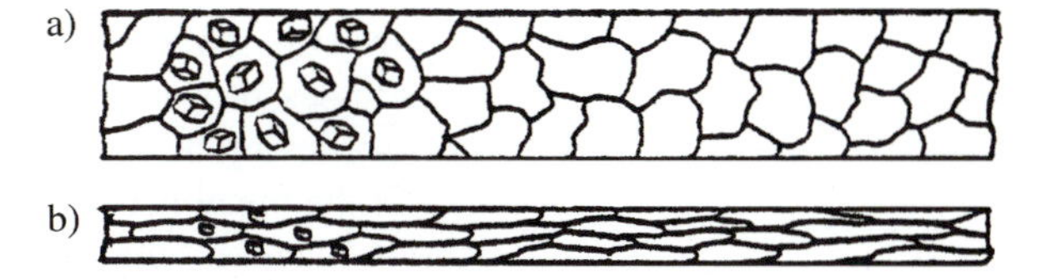

Figure 4.8
Elongation of crystals in cold forming; development of texture. [HASEK]

contain crystals with the preferred orientation. This kind of structure is called *texture,* and it can only be detected with X rays. It leads to *anisotropy* of material properties. This anisotropy is different from the anisotropy that arises in hot forming and is due to fibers of impurities. Now it includes anisotropy of plastic stress-strain relationships, whereas in hot forming it concerned only the limits of strains and failures.

The strongly textured material with the crystals of the individual grains aligned in parallel shows anisotropy similar to that of monocrystals of the metal concerned. Yield stress and componental strains along and across the direction of texture are different. One of the manifestations of this anisotropy is the formation of "ears," or non-uniform deformation in deep-drawn cups. An important parameter is the plastic strain ratio r of the width-over-thickness strain as measured in a tensile test, using a flat specimen with width w and thickness h:

$$\varepsilon_w = \ln\left(\frac{w_1}{w_0}\right),$$

$$\varepsilon_t = \ln\left(\frac{h_1}{h_0}\right),$$

$$r = \frac{\varepsilon_w}{\varepsilon_t}$$

The ratio r may be less or more than 1 depending on the crystallographic nature of the metal. A large value of r means high resistance to thinning in the thickness direction, normal to the plane of the sheet. This is very important in deep drawing. Most steels have values of r between 1 and 2; zinc has a very low value of 0.2 and a limit on the drawing ratio of about 1.9 (depth over diameter); and titanium has the highest value of 6, with a drawing-ratio limit of 3.

The *strain-hardening* coefficient n is an indicator of the resistance of the material to necking. If necking starts to occur in one spot, it is associated with an increased strain in that spot. This leads to a local increase of yield strength, which prevents further necking and spreading of the local deformation over a larger area. Correspondingly, in operations where necking arises as a problem, it is necessary to use material with a higher strain-hardening coefficient. In other operations where ductility is critical, a low strain-hardening coefficient is preferred.

Cold work is mostly combined with annealing and/or stress-relieving heat treatment. It is useful to recapitulate briefly the phenomena of recovery and recrystallization that occur during these treatments and to point out the associated aspects of grain-size control.

Recovery is achieved by keeping the material at a temperature not far below the recrystallization temperature: that is, in a range of from 0.3 to 0.5 $T_{m.}$ Dislocations move to form regular arrays. Grain boundaries do not change, and mechanical properties like strength, hardness, and ductility change only slightly. However, internal residual stresses produced by cold work are eliminated to a great extent. This is called stress-relieving heat treatment.

If the metal is heated up and kept for a period on the order of one hour above the recrystallization temperature, the grains distorted by cold work start to reform. New grains start to grow (see Fig. 4.9a) at nucleation sites, which are located at those points where the original crystal lattice was most distorted during cold work. *Recrystallization* can only occur in a metal that has been plastically deformed, and it depends strongly on the degree of that deformation. The higher the degree of cold work, the lower the recrystallization temperature and the greater the number of nucleation sites, leading to finer grain at the end of recrystallization. The progress of recrystallization is depicted in Fig. 4.9b and c. The new grains have a low density of dislocations, and they are stress-free. Recrystallization is completed when the boundaries of the new grains have all met, as shown in Fig. 4.9c, in which the original grains are indicated by dotted lines. If the metal is then kept longer at the recrystallization temperature, some of the grains start to grow by coalescence with neighboring grains. This grain-coarsening process, indicated in Fig. 4.10, becomes much more intensive if the temperature is increased above the recrystallization temperature.

The diagram shows one alternative of grain coalescence in which two grains unite to form one grain. In this process the lattice orientation of one of the grains must change to conform with that of the more stable grain. In other mechanisms one grain grows so as to progressively swallow up another grain, or several grains coalesce by progressive shifts of boundaries. Through recrystallization the ductility of the metal is fully restored, and all increase of strength and hardness obtained by cold work is eliminated. Through grain growth the toughness of the material deteriorates. However, if recrystallization is achieved by proper *annealing* or *normalization,* the grain may be efficiently refined, which leads to the best combination of strength, toughness, and ductility of the material.

The strong effect of the degree of cold work on recrystallization temperature and mainly on grain size can be expressed by recrystallization diagrams, which are a typical way to characterize the various kinds of metals. They indicate the coarsening of grain as a function of the preceding deformation and of the temperature of the subsequent heating up, for a certain time. One example of a recrystallization diagram is shown in Fig. 4.11. It shows that recrystallization begins at lower temperatures after a larger degree of cold work.

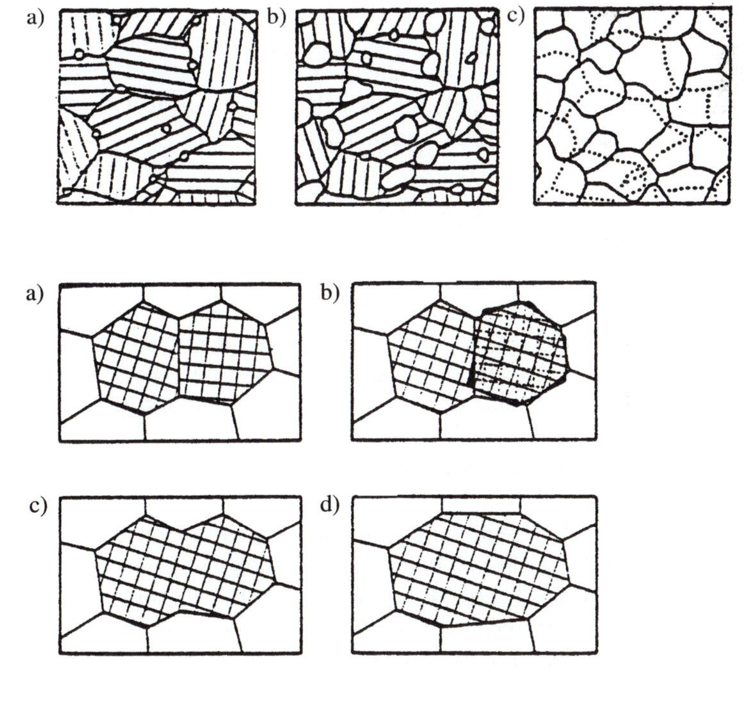

Figure 4.9
Progress of recrystallization: a) nucleation sites where original lattice has been most distorted during cold work; b) progress of recrystallization; c) completed recrystallization (dotted lines indicate original grains). [HASEK]

Figure 4.10
Grain coarsening by coalescence of two grains. [HASEK]

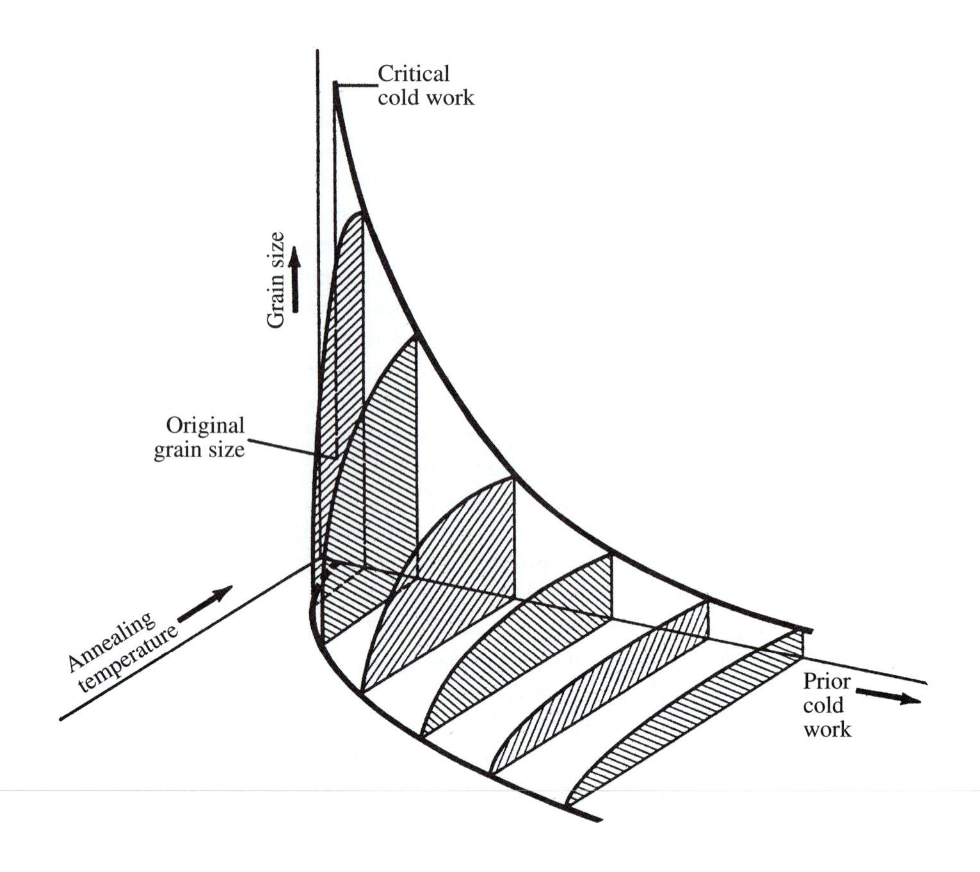

Figure 4.11
Effect of prior cold work on recrystallization. [SCHEY]

The crystallized grain size is much smaller after larger deformations because heavier work leads to the formation of more recrystallization nuclei. The grain size increases with temperature, especially after low amounts of cold work. If cold work was zero, no recrystallization occurs. There is a certain minimum deformation necessary, about 5% strain, which is called critical work, that leads to the formation of a small number of large new grains.

EXAMPLE 4.1 Combining Cold Work and Annealing for Desired Material Properties ▼

The effect of cold work on mechanical properties of a low-carbon steel is given in the graphs of Fig. 4.12, reproduced from [2]. The ultimate tensile strength *UTS,* the yield strength *YS,* and the maximum elongation e_m are plotted versus prior cold work as expressed by the area reduction,

$$q = \frac{A_0 - A_1}{A_0}$$

It is required to produce rod with diameter $d_2 = 6$ mm with $YS > 550$ MPa and maximum elongation $e_m > 20\%$. Available stock has diameter $d_0 = 9$ mm, and it is annealed. It would fail at cold work in excess of 60% area reduction. The work will be done in two steps with annealing in between. What will be the final diameter d_1 of the first step (i.e., the initial diameter of the second step)? Is it at all pos-

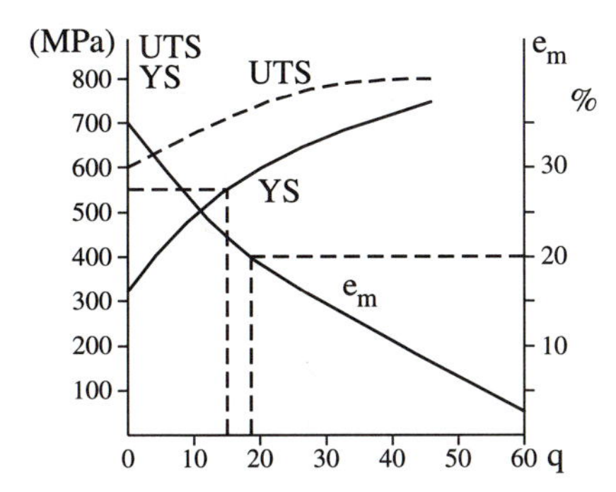

Figure 4.12
Strength and ductility as functions of cold work.

sible to do this work in two steps without exceeding the 60% reduction in the first step? What will be the area reduction q_1 in the first step?

Obviously, in order to achieve the required properties at the end of the second step the area reduction q_2 must be at least 15% to satisfy the strength requirement, and it should not exceed 19% in order to secure the required ductility. Let us choose:

$$q_2 = \frac{A_1 - A_2}{A_1} = 0.17,$$

or

$$1 - \frac{A_2}{A_1} = 0.17,$$

$$1 - \left(\frac{d_2}{d_1}\right)^2 = 0.17,$$

$$d_1^2 = \left(\frac{1}{0.83}\right) d_2^2 = 43.37 \text{ mm}^3$$

$$d_1 = 6.586 \text{ mm}$$

Checking on the area reduction in the first step:

$$q_1 = (A_o - A_1)A_o = 1 - \left(\frac{d_1}{d_0}\right)^2 = 0.456$$

This is less than the critical value of 0.6; therefore, the two-step operation is feasible. ▲

4.2 BASIC MACHINES FOR METAL FORMING

There are three basic classes of metal-forming machines: hammers, mechanical presses, and hydraulic presses. A variety of other machines are built on similar principles or utilizing the principles of rolling. They are described later in this chapter. First, however, the three basic types are discussed, starting with the diagrams in Fig. 4.13.

The three kinds of machines are differently constrained as regards their work. The hammers end their work when the kinetic *energy* of the ram is exhausted. The mechanical

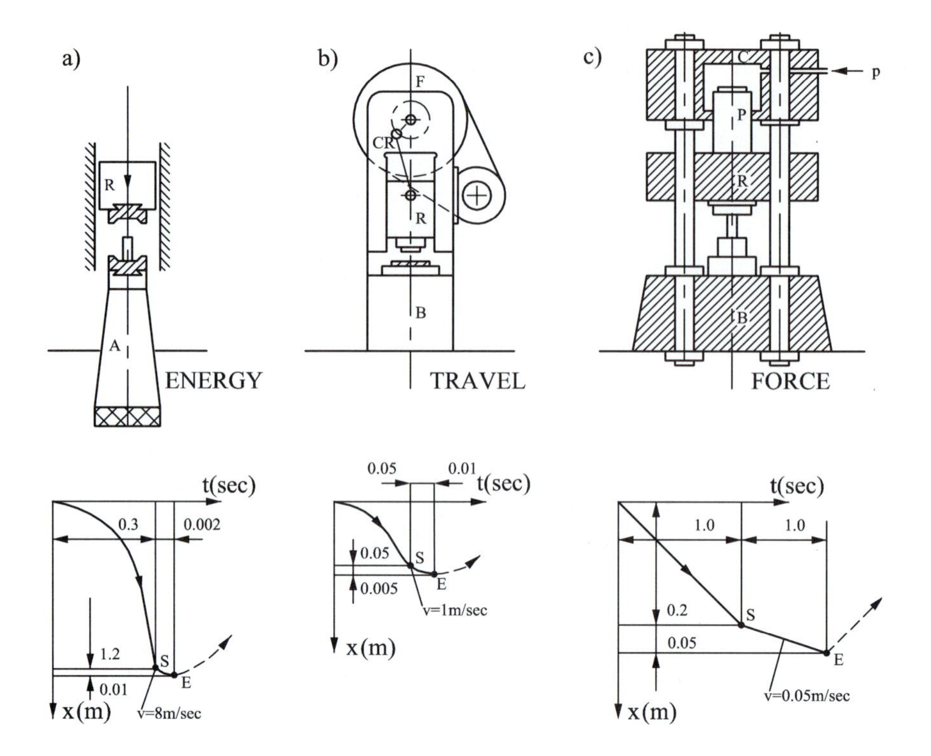

Figure 4.13
Basic types of forming machines: a) hammer; b) mechanical press; c) hydraulic press.

press, in which the ram is driven by a crankshaft, must reach the bottom *position* of its travel. The hydraulic press is constrained by the maximum *force* as obtained by the pressure p of the oil supplied to the work cylinder times the piston area. Its travel is reversed from an end switch, or else it will stop whenever the maximum force is exceeded.

The principle of the hammer is shown in Fig. 4.13a. The ram R with the upper half of the die is accelerated either simply by gravity or by gravity and force generated in an air-pressure or steam cylinder, and it acquires a certain velocity v and the associated kinetic energy before it strikes the work. Hammers are nominally rated by the weight of the ram. During the work the forming force times the forming travel consumes the kinetic energy of the ram. Some of this energy is also transferred to the anvil A and is consumed by the damped vibration of the anvil, which is supported flexibly against the ground. The anvil is made many times heavier than the ram; thus the energy lost in its motion is small compared to the energy used for forging. The bottom position of the ram motion is not exactly defined; it depends on the fall height and on the forging force. Hammers are used for both open-die and closed-die forging, the latter being prevalent. Work is usually done in a number of repeated blows.

A typical diagram of the travel x versus time t is shown in the lower part of the diagram, which shows particular parameters corresponding to a power drop hammer. Through a fall of 1.2 m the ram is accelerated to a final velocity of 8 m/sec. It decelerates to zero over a forging travel of 1 cm in 2 milliseconds. The "contact time" between the hot work and the dies is very short.

One variant of a crank (eccentric) press is shown in Fig. 4.13b. The crankshaft CR is driven from an electric motor via the flywheel F. The upper half of the die is on the ram R; the lower half is on the bed B. In the case of the hammer, the frame (not shown) practically does not feel the forming force at all, but here the frame, crankshaft, and connecting rod must transmit the forming force. This force is not supplied directly by the torque of the motor but mainly by the kinetic energy of the flywheel. In this way, the (average) power of the motor can be much smaller than the instantaneous power needed by the forming operation. The forming force could, in principle, rise to very high values. However, there are overload protection mechanisms in the drive, one of them being the clutch between the flywheel and the crankshaft. This clutch is primarily needed to start and stop (always in the top position) the motion. The press is used either intermittently, one revolution at a time, or continuously, if the stock is steadily (automatically) supplied. The (x, t) diagram shows that work is done close to the bottom position of the stroke, where velocity is decreasing to zero, and where a torque on the flywheel (or on the electric motor) translates into the largest available force on the ram (actually, at the end of the stroke this force would be infinite). Typically, with a stroke of 5 cm (0.05 m), work travel of 5 mm (0.005 m), and 600 strokes/min (10 strokes/sec), the striking velocity is 1 m/sec, and contact time is 10 msec, about five times more than for the hammer. Note that the (x, t) graphs for the three machines are not drawn in the same scale. A mechanical press is rated by the maximum force that it can safely take. This force is understood to be available at a specified distance before the bottom position of the ram.

The hydraulic press is shown in Fig. 4.13c. The force acting between the ram R and the bed B is produced in the hydraulic cylinder built into the crown C. Oil with pressure p drives the piston P. As for the crank press, the frame must be strong enough to take the forming force. This force is now well defined. The velocity of the ram can easily be controlled and maintained constant or arbitrarily varied. The lower diagram shows a faster approach and a slower forming velocity. The distinctive characteristic of the hydraulic press is its ability to supply full force over any length of travel. It is consequently uniquely suitable for deep drawing and extrusion operations. It can carry out in one stroke a forging operation that would require several strokes of a hammer. However, the contact times are much longer, and the forging dies would not be able to withstand such exposure except for easily forgeable materials like aluminum. A hydraulic press is rated by the maximum force for which it is designed, which is available over any part of the stroke.

Before going into a little more detail about the overall characteristics of the forming machines, let us review the force-stroke requirements of the basic forming operations, summarized in Fig. 4.14. Operations (*a*), (*b*), and (*c*) are bulk forming, and (*d*), (*e*), and (*f*) are sheet-metal forming. The scale for forces is arbitrary; in practice it depends on the size of the workpiece and on its material properties. In general, the bulk-forming operations require larger forces than sheet-metal forming. The individual diagrams are explained below:

(a) Unconstrained upsetting. The force increases with the area of the workpiece. Depending on the h/L or h/d ratio, nonhomogenous work or friction may play a role. The operation may be carried out in several steps. If it is done on a hammer, the work—the shaded area of the (F, X) diagram—will be equal for each blow if the same fall height is used.

(b) Closed-die forging. The force increases as the die fills, and there is a large increase of force when the flash is generated. Again, the total work is shown as being done in several steps.

(c) Backward and forward extrusion. The force is almost constant over the whole stroke. This kind of operation is mostly accomplished on hydraulic presses, although for smaller parts it can be done on hammers or mechanical presses as well.

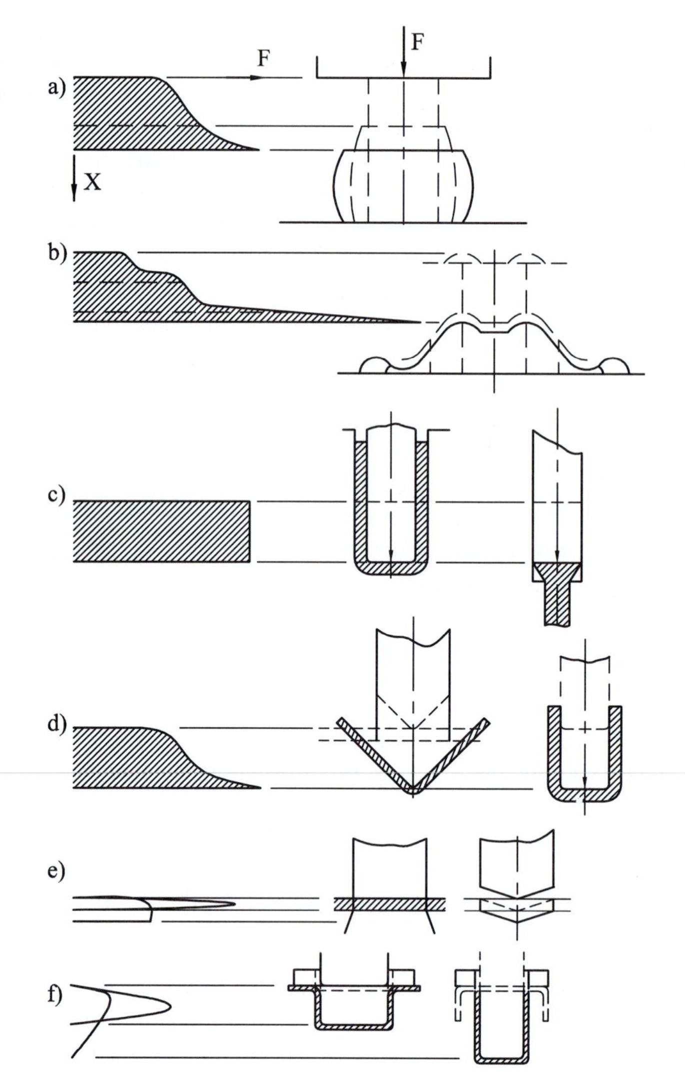

Figure 4.14
Force versus stroke for various forming operations: a) unconstrained upsetting; b) closed-die forging; c) backward and forward extrusion; d) bending; e) blanking, punching; f) deep drawing.

(d) Bending. The force increases over the stroke and especially toward its end for the hard contact and compressive yield intended to minimize springback.

(e) Blanking, punching. The force increases suddenly and reaches its peak at the penetration value, at which the material breaks; then the force drops suddenly. Lower force extended over a longer stroke is obtained when punching with "shear."

(f) Deep drawing is shown in two steps. In the second step the cup is reverse redrawn. During each step the force first steeply rises and then decreases during the stroke.

Summarizing, we see that operation (*e*) requires mostly a very short working travel, and the operations (*a*) and (*b*) may also involve short working travels. If these operations are performed on mechanical presses, the total stroke may be short, and the press may run rather fast. Short-stroke (30 mm) presses have recently been introduced with a maximum speed of up to 2000 strokes/min. Operations (*c*), (*d*), and (*f*) may require longer working travels, depending on the size of the workpiece. Examples are given in the later sections of this chapter.

Let us also briefly review the main variants of the three classes of basic forming machines. The hammers are summarized in the diagrams in Fig. 4.15. Diagram (*a*) shows the board drop hammer. The ram *R* is attached to the board *B*, which is squeezed between rollers *RL* and lifted. Then the roll lever releases the squeeze, and the ram with the upper die falls, driven by gravity, and strikes the lower die attached to an anvil cap on the anvil *A*. This hammer is intended for closed-die work. The frame *F*, which provides the guidance for the ram, is attached to the anvil and represents a part of the dead mass that supports the lower die. An alternative solution is indicated in Fig. 4.15b. The

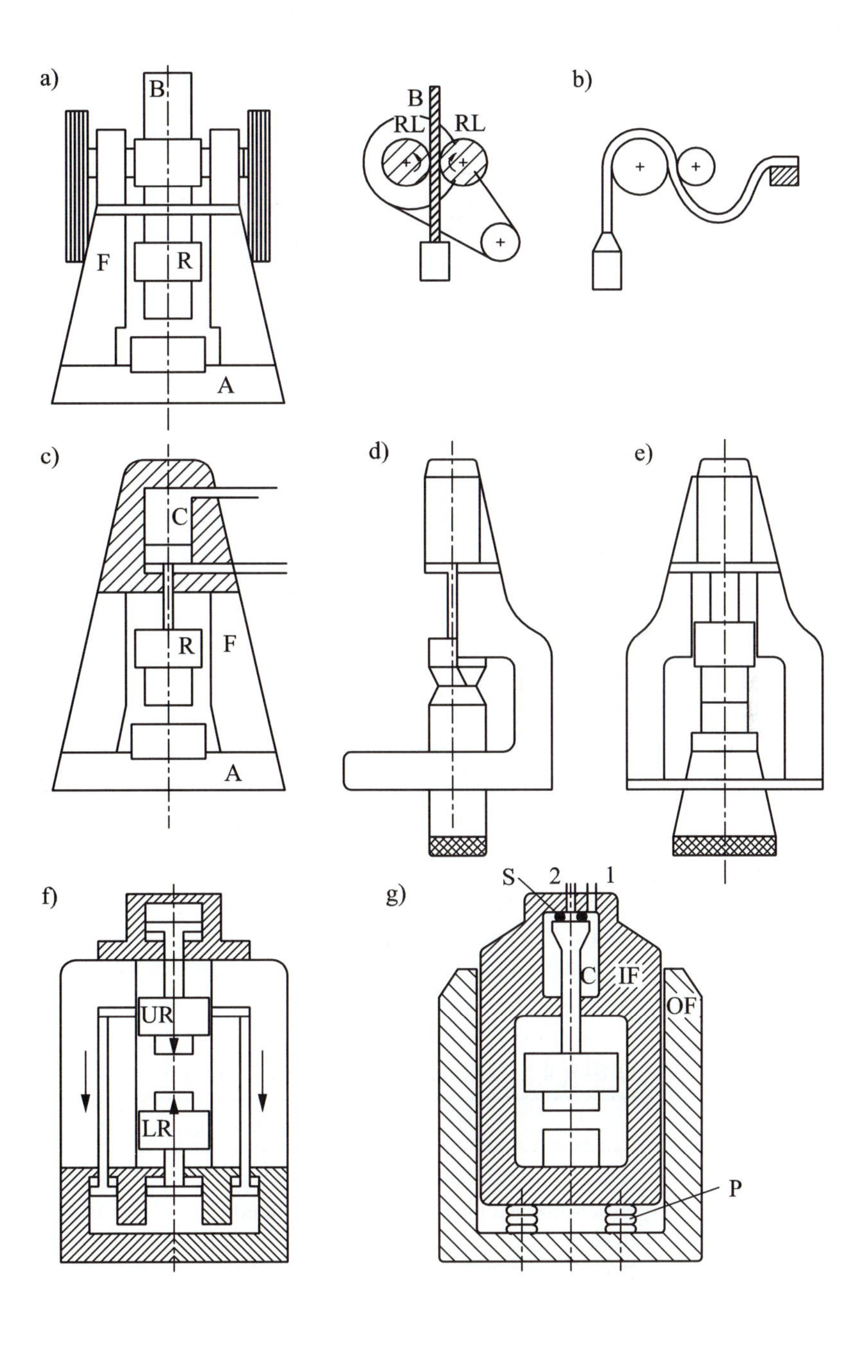

Figure 4.15
Various types of hammers.

ram is hung on a belt squeezed between two rollers. All the other hammers shown are power-drop hammers. The ram in (*c*) is accelerated not only by gravity but also by the expansion of steam or by air pressure in the cylinder *C*. This hammer also has an enclosed frame attached to the anvil and is intended for closed-die work. The hammers in (*d*) and (*e*) are intended for open-die work; their frames are single-sided and open in (*d*) and widely open and double-sided in (*e*). The operator thus has easier access for manipulating the stock as it is forged in successive blows between the hammer and the anvil. The anvil is not directly connected to the frame. It is rather massive and rests on wooden blocks that provide a partly elastic support against the ground.

Figure 4.15f shows a power-driven counterblow hammer. It has the upper ram *UR* and the lower ram *LR*. The upper ram is driven down by air pressure in the cylinder. Attached to it are piston rods of two pistons in hydraulic cylinders that drive oil under the piston that is attached to the lower ram, which moves upwards. The two rams and die halves have approximately the same mass, and they meet with equal and opposite velocities. No energy is transferred and lost to the frame and foundation.

Figure 4.15g shows one of various designs of high-energy-rate forming (HERF) machines. While the steam or air pressure in the conventional power-drop hammers operates with about 70 MPa maximum pressure, the gas used in the HERF machines is pressurized to about 1500 MPa. It is brought into the cylinder *C* through line 1. However, the piston is held upwards against the seal *S*. When high-pressure gas is fired under the seal through line 2, the piston moves and is driven down. The inner frame *IF* is simultaneously lifted inside of the outer frame *OF*, in which it is supported by pneumatic springs *P*. In this way very little impact is transmitted to the environment. The ram is subsequently lifted by hydraulic cylinders not shown in the sketch. The HERF machines supply high energies for the forging process that can only be reached in conventional power hammers of much larger sizes. However, the machine itself is more complex and usually does not offer short cycle times.

All hammers are cheaper than presses, and their frames are lighter because they do not transmit the forging force. On the other hand, mechanical presses perform more stroke cycles per minute, and hydraulic presses can deliver energies much beyond the capability of even the largest hammers. Table 4.1 gives the range of the basic parameters of hammers in use. Gravity-drop hammers are used to produce forgings of a few kilograms mass; power hammers produce forgings from 20 kg to several tons mass.

Mechanical presses exist in a great variety of sizes, tonnages, speeds, and designs, and with various modes of automation in the handling of workpieces. Perhaps the most significant distinction can be made between presses for forging (hot, bulk forming) and presses for sheet-metal forming. The former types are more compact, stronger, and more rigid because for the same size of work area they must develop much higher forces. The latter types are built in much larger table sizes for transfer die work. Presses for sheet-metal work are much more often provided with continuous feed of strip or

TABLE 4.1. Ratings of Forging Hammers

	Moving Mass (kg)	Energy at Strike (J)
Gravity-drop hammers	500–5,000	6,000–75,000
Power-drop hammers	500–18,000	18,000–600,000
HERF		500,000–5,000,000

with automated loading of blanks and unloading of parts. Because deep drawing may be included in the combined operations, they also often need blank-holding mechanisms. These either consist of rams driven from the main drive (double-acting presses) or of separate, pneumatically driven rams (cushions). Because of their larger spans, wider presses may be driven at several points. The lower loads make it possible to include in the drives mechanisms containing screws to adjust the "shut height," which allows easy setting for different dies used for a great variety of jobs. In a forging press the shut-height adjustment is done by means of a much more rigid and simpler wedge-type mechanism which, however, allows only a small adjustment range.

Several kinematic mechanisms are used in presses. Some of them are shown in Fig. 4.16. Diagram (a) shows the crank, which converts uniform rotation of the crankshaft into a reciprocating motion of the ram. The velocity of the ram varies almost harmonically; it becomes zero at the extremes of the stroke and is at its maximum near the middle of the stroke. Correspondingly, a constant torque on the crankshaft translates into large forces close to the bottom of the stroke where the work is done. At the very bottom the constant driving torque would deliver an infinite force on the ram. However, the maximum force permissible for the work is limited by the strength of the pins and bearings and of the crank. The driving force on the crankshaft as supplied through the belt and gear transmissions from the motor is not really decisive for the working capability of the ram. Work is supplied over a short part of one revolution by the kinetic energy of the flywheel, which slows down by up to 15%. Its speed and energy are

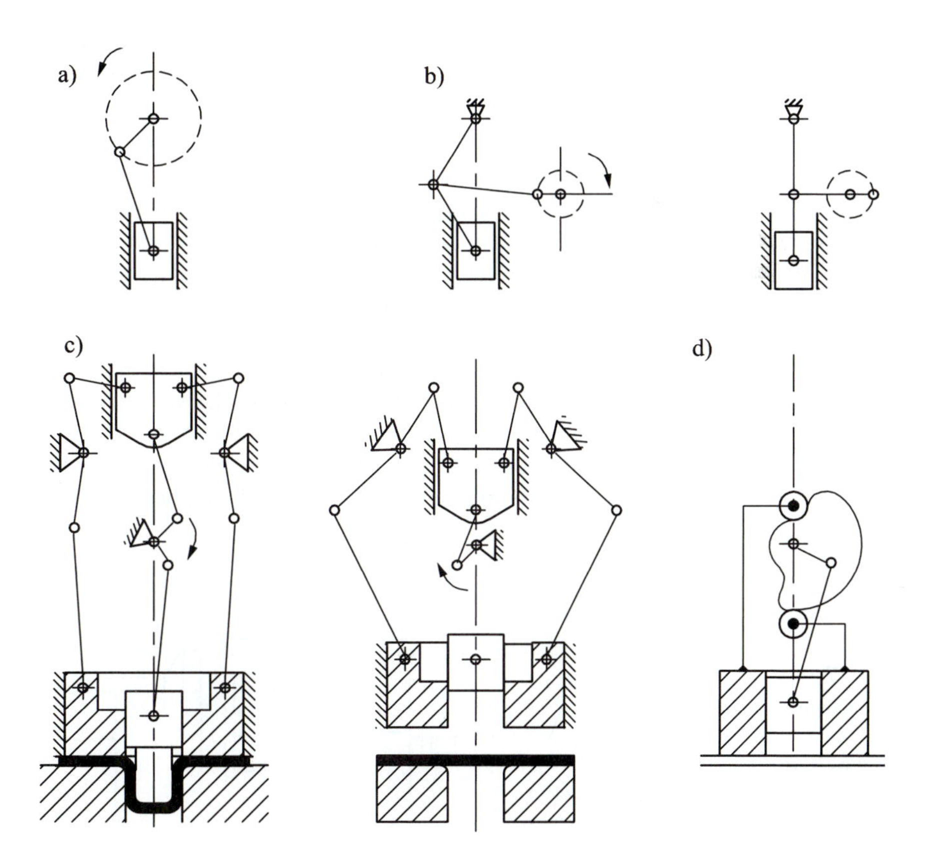

Figure 4.16
Kinematic mechanisms used in presses: a) the crank; b) a knuckle lever; c) toggle; d) cam mechanism.

restored by the driving motor over the idle part of the revolution. This is discussed more precisely in Example 4.2.

Figure 4.16b shows a knuckle-lever transmission, which increases the variation of ram velocity beyond that produced by the crank drive. The transmission ratio between the crankshaft and the ram is so modified that in the upper part of the stroke, shown at left, the knuckle transmits with the largest rate (about 1:2). This rate decreases to zero throughout the stroke toward the bottom position of the ram, shown at right. In this way very large forces are obtainable at the ram for small forces on the crank. This mechanism is used for clamping stock in upsetting machines.

A toggle mechanism is shown in Fig. 4.16c as applied to the hold-down blank-holder slide of a double-action press, the main action of the central ram being driven by a crankshaft. The position at left is close to the bottom extreme, while the position at right is close to the top of the blank-holder stroke. The mechanism as shown produces a long dwell at the bottom position.

For the purpose of the blankholder drive, a cam mechanism, as shown in Fig. 4.16d, is most suitable. Given the appropriate shape of the cam, the blankholder motion can be optimally adapted to the desired variation throughout the deep drawing cycle.

Of all the variety of sizes and forms of press frames, three are shown in Fig. 4.17. The open-back inclinable (OBI) blanking and punching press is shown in Fig. 4.17a. It provides for easy access for the strip feed and free fall of the punched-out parts through the opening in the frame into a bin. In (*b*) a stocky, strong and rigid, straight-side frame of a forging press is drawn in simplified contours. The frame is usually strengthened by four pre-stressed tie-rods. In (*c*) a very wide, three-column frame of a transfer press for sheet-metal forming is shown. The press has a single-point suspension drive in its left-hand section and a double-point suspension drive in the right-hand section. Many other kinds of mechanical press frames exist. Some of them are shown later in this chapter.

Hydraulic presses are built in two basic ways: the push-down and the pull-down. In the first, the cross head at the top of the frame contains the main cylinder and is stationary. The main piston pushes the ram down against the bed, which contains the return cylinders. The latter design incorporates a moving cylinder-frame assembly, with the stationary main cylinder attached at the bottom of the bed. The cross head with the upper die is pulled down against the lower die on the bed. The return cylinders are again supported by the bed. The pull-down design needs less head room. The hydraulic power is accommodated below floor level, and the length of piping between the pumps and the cylinder is reduced. The hydraulic drive exists also in two basic variants, which are shown in Fig. 4.18. Variant (*a*) incorporates direct supply of pressure oil from the pumps to the cylinder. Two parallel pumps are included. Pump 1 supplies a large flow of low-pressure oil, and pump 2 supplies a lower volume at high pressure. They

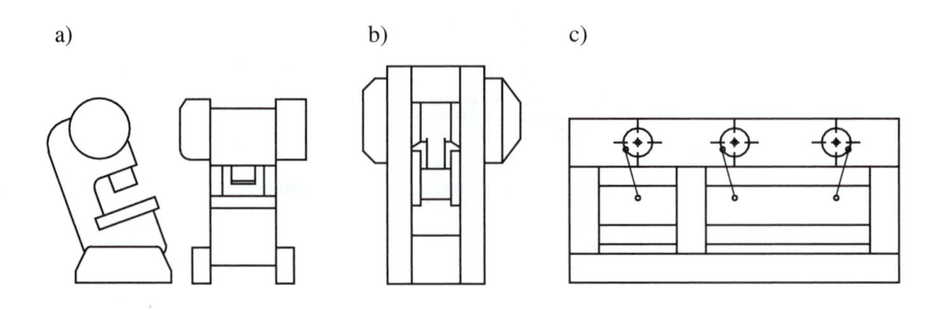

Figure 4.17
Various mechanical press frames: a) open-back inclinable; b) rigid, straight-side forging press, c) wide, three-column frame of a transfer press.

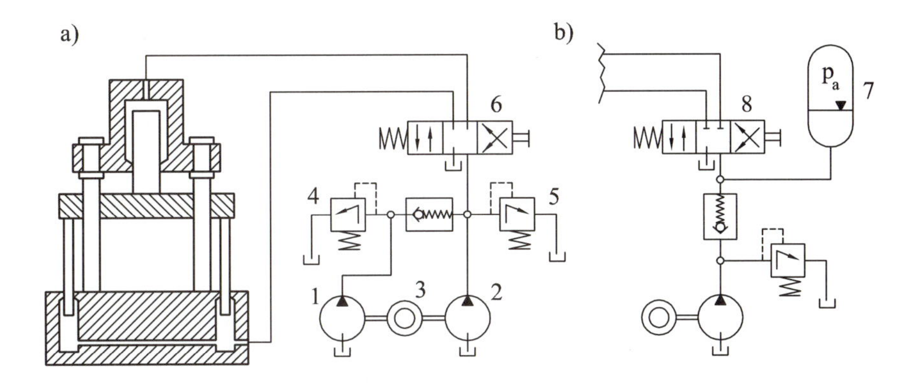

Figure 4.18
Hydraulic drives for presses.

are driven by the motor (3). At the initial, idle part of the stroke, low-pressure oil is supplied by both pumps. When the ram encounters the work load, the pressure in the system increases above the pressure set on the relief valve (4) of low-pressure pump 1. The supply of pump 1 is thus diverted back to the pump through relief valve 4. Relief valve 5 of pump 2 is set at the nominal working pressure. The directional valve (6) is operated from a microswitch set to signal the bottom end of the stroke, or it can be activated from relief valve 5 and switch over to return when the nominal pressure is exceeded. An alternative drive, shown in Fig. 4.18b, uses one pump and an accumulator (7). Energy in the accumulator is stored during the intermissions of the work of the press and whenever the required oil supply is less than the delivery rate of the pump. The drive of the pump needs only to have a power that is average for the press cycles. The work cycle is controlled by the directional valve (8). The characteristic parameters for mechanical and hydraulic presses are given in Table 4.2.

The data in Table 4.2 are general and typical; great variety exists within the general ranges indicated. Categories 1 and 2 are presses intended mainly for punching and blanking from strip. The larger presses in the category 3 are used for automotive and appliance sheet-metal pressing, and the transfer presses of category 4 are also used mainly in the automotive industries for simultaneous work on a number of stations. Correspondingly, the maximum load capacities in these categories range from 1 to 40 MN. The mechanical forging presses in category 5 are still heavier and range up to 80 MN. Hydraulic presses are built for still heavier loads. They are presented here in two categories. Heavy forging presses have been built with capacities up to 500 MN.

TABLE 4.2. Parameters of Presses

	Load Capacity	Strokes per Minute	Power (kW)
A. Mechanical presses			
1. Open-back inclinable	150–1250 kN	200–100	3–15
2. High-speed, straight-side	300–2000 kN	2000–200	
3. Larger straight-side	1–6 MN	100–20	10–60
4. Transfer presses	2–40 MN	50–10	
5. Forging presses	3–80 MN	100–30	20–500
B. Hydraulic presses			
1. Universal	4–25 MN		
2. Forging presses	2–500 MN		150–1000

EXAMPLE 4.2 Kinematics of a Crank Mechanism

The diagram of the crank mechanism is shown in Fig. 4.19. While the crank rotates from $\phi = 0$ to $\phi = 180°$, the ram moves from point A to B. The crank rotates around point C. It has radius r, and the length of the link (connecting rod, Pittman) is l. The motion of the ram is denoted x, and it is measured from the top position A. It is

$$x = l\cos\alpha - r\cos\phi - (l - r) \tag{4.1}$$

Because

$$u = r\sin\phi = l\sin\alpha,$$

it is,

$$\sin\alpha = \left(\frac{r}{l}\right)\sin\phi \tag{4.2}$$

and Eq. (4.1) transforms to

$$x = r(1 - \cos\phi) - l\left[1 - \left(\left(\frac{r}{l}\right)^2 \sin^2\phi\right)^{1/2}\right] \tag{4.3}$$

The angular velocity of the crank is assumed constant:

$$\phi = \omega t$$

and $\omega = 2\pi n$, where n is the speed of the crankshaft.

The velocity of the ram is obtained as follows:

$$v_x = \frac{dx}{dt} = r\omega\sin\phi - \frac{\omega\left(\frac{r}{l}\right)^2 \sin\phi\cos\phi}{\left[1 - \left(\frac{r}{l}\right)^2 \sin^2\phi\right]^{1/2}}$$

or,

$$v_x = \frac{dx}{dt} = r\omega\left\{\sin\phi - 0.5\left(\frac{r}{l}\right)\sin 2\phi \Big/ \left[1 - \left(\frac{r}{l}\right)^2 \sin^2\phi\right]^{1/2}\right\} \tag{4.4}$$

Assuming $r = 1$, $\omega = 1$, and $r/l = 0.3$, the expressions (4.2) and (4.3) are evaluated in a computer program and plotted in Fig. 4.20. The variations of both the displacement and the velocity are close to cosine and sine functions, respectively. Actually, for an infinite l, both would be purely harmonic. The displacement varies between 0 and $2r$ and the velocity between $+r\omega$ and $-r\omega$.

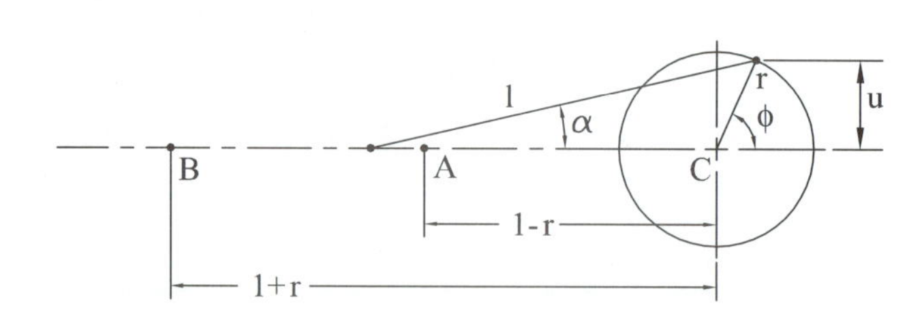

Figure 4.19
A crank mechanism.

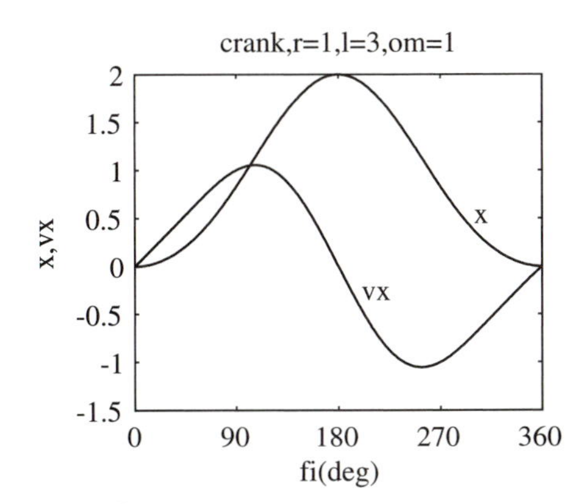

Figure 4.20 Displacement and velocity of the ram, from Example 4.2.

The ratio $\rho = v/r\omega$ expresses the transmission ratio between the crank and the ram, which determines the relationship between the torque T on the crankshaft and the force F available on the ram:

$$T\,d\phi = F\,dx$$

or, dividing by dt:

$$T\omega = Fv_z \tag{4.5}$$

For constant ω,

$$T = \frac{Fv_x}{\omega} \tag{4.6}$$

The velocity reaches its maximum positive and maximum negative values close to the middle of the stroke. Towards the bottom of the stroke, here at $x = 2$, the velocity goes to zero. This represents an infinite transmission ratio, and for any resisting force, the torque goes to zero at the bottom of the stroke. This applies to a completely rigid mechanism and a rigid frame. In reality, the system deforms, and the mechanism could be jammed.

For example, let us take a press with a stroke of 100 mm, which is rated at 600 kN. This usually means that the rated force is a maximum permissible force and can be utilized up to 6 mm from the bottom of the stroke. Assuming the same $r/l = 0.3$ ratio as used in the graph of Fig. 4.20, determine first the torque T_{lim} that would produce the maximum force of 600 kN at the corresponding position of $x = 94$ mm. Using Eqs. (4.3), (4.4), and (4.6), the value of

$$T_{\text{lim}} = 1.617\,e7 \text{ Nmm} = 1.617\,e4 \text{ Nm}$$

is obtained. Next, we ask which forces would produce the same torque at positions (a) 10 mm, and (b) 3 mm above the bottom of the stroke. The results are

(a) $x = 90$ mm, $F = 479$ kN

(b) $x = 97$ mm, $F = 836$ kN

In the latter case, the press would be overloaded because the force exceeds the rated force of 600 kN, although the clutch on the crankshaft would still transmit the necessary torque. It is necessary to carefully choose the forming jobs so as not to permit this kind of overload. Modern presses are equipped with sensors that measure the force acting between the dies and can thus check the load. In the case of overload, a collapsible member of the drive is released to protect all the other parts of the drive and the frame. ▲

EXAMPLE 4.3 The Crank-and-Toggle Mechanism ▼

The crank-and-toggle mechanism depicted in Fig. 4.21 is assumed to be driven by the crank mechanism of Example 4.2. The initial position of the toggle corresponds to $x = 0$, that is, to position A of the crank, and it is defined by the dimensions $a = 2$, $b = 6$. The length of the link is

$$l = (a^2 + b^2)^{1/2} = 6.325$$

At any given position it is

$$p = a - x$$

$$p^2 + q^2 = l^2$$

and

$$y = 2q - 2b = 2(l^2 - p^2)^{1/2} - 2b \tag{4.7}$$

To obtain the velocity of the ram, we differentiate (4.7):

$$2p\,\mathrm{d}p + 2q\,\mathrm{d}q = 0$$

$$p = a - x, \qquad \mathrm{d}p = -\mathrm{d}x$$

$$y = 2q - 2b, \qquad \mathrm{d}y = 2\mathrm{d}q$$

$$\mathrm{d}y = 2\mathrm{d}q = -2\left(\frac{p}{q}\right)\mathrm{d}p = 2\left(\frac{p}{q}\right)\mathrm{d}x$$

$$v_y = \frac{\mathrm{d}y}{\mathrm{d}t} = 2\left(\frac{p}{q}\right)v_x \tag{4.8}$$

The relation (4.7) is essentially between the position of the end point of the toggle and the end point of the crank mechanism driving it. The relation (4.8) is between the velocities of the two points.

In deriving these expressions a small error was neglected that arises because the end point of the connecting rod between the crank and the toggle moves in a circular arc instead of in a horizontal straight line, as was the case in deriving $x(\phi)$ in Example 4.2.

The displacement y and the velocity $\mathrm{d}y/\mathrm{d}t$ are both plotted versus the angle ϕ of rotation of the crankshaft in Fig. 4.22. Maximum velocity is reached in about a third of the downward stroke, and the approach to the bottom point is now much slower than in the crank mechanism alone. The transmission ratio between the crankshaft and the toggle end point is much smaller. Correspondingly, very large forces may be obtained in the very last portion of the motion. This is utilized, for example, in the clamping mechanism of an upsetting machine, which will be shown later.

Let us now check the transmission ratios in this last portion of the motion. The total stroke is 0.6491 for $x = 2.0$. The transmission ratios as expressed by the

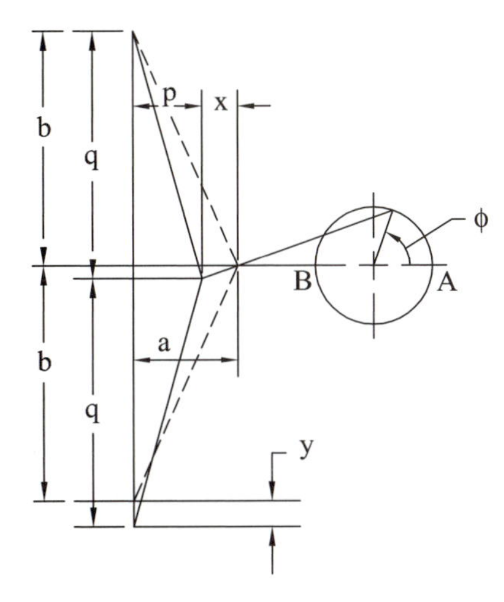

Figure 4.21
The crank-and-toggle mechanism.

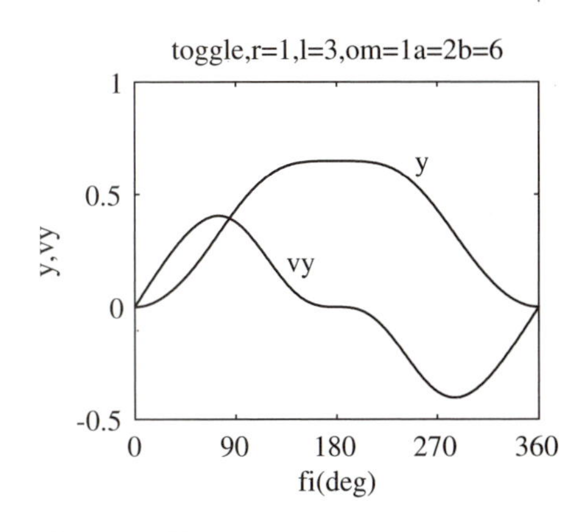

Figure 4.22
Displacement and velocity of the ram, from Example 4.3.

ratio $\rho_y = v_y/r\omega$ for the positions at 94% and 97% of stroke are compared with those of the crank mechanism only.

$x/x_{max}, y/y_{max}$	0.94	0.97
ρ_x	0.54	0.388
ρ_y	0.137	0.087

The transmission ratios for the crank and toggle are three to four times smaller than for the crank mechanism alone. ▲

EXAMPLE 4.4 Work Done in a Forging Hammer ▼

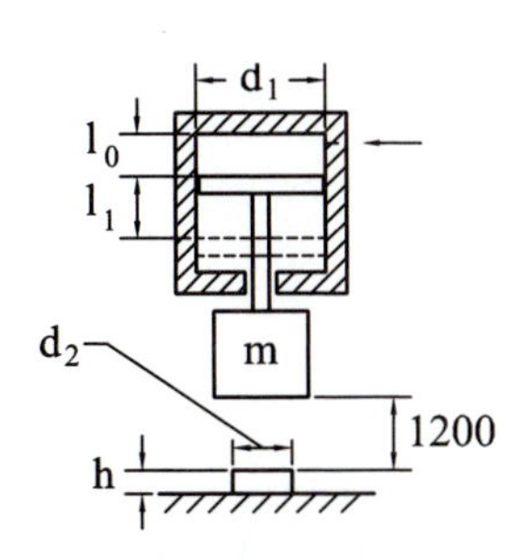

Figure 4.23
An air-pressure hammer upsetting a billet.

An air-pressure hammer is used to upset a billet, as shown in the diagram of Fig. 4.23. The air-cylinder diameter is $d_1 = 500$ mm, the initial volume height is $l_0 = 300$ mm, the stroke is $l_1 = 1200$ mm. The diameter of the blank is $d_2 = 180$ mm and its height is $h = 60$ mm. The yield stress of the hot steel blank is $Y = 200$ MPa. The mass of the hammer is $m = 2500$ kg, the initial air pressure is $p_0 = 700$ kPa. Determine the amount Δh of upset in one blow. We will assume adiabatic expansion of the air and neglect any inflow of air during the down stroke. We will further assume sticking friction between the hammer and the billet.

(a) Energy arising from the expansion of the pressurized air.

The (p, V) relationship is

$$pV^k = C, \qquad k = 1.4$$

$$E_a = \int_{V_0}^{V_1} \mathrm{p}\, dV = C \int_{V_0}^{V_1} \mathrm{V}^{-k}\, \mathrm{d}V = \frac{CV_1^{1-k} - CV_0^{1-k}}{1-k}$$

where $C = p_0 V_0^k = p_1 V_1^k$

and, correspondingly, $E_a = \dfrac{p_1 V_1 - p_0 V_0}{1 - k}$

where $p_0 = 7 \times 10^5 \text{ N/m}^2$

$$V_0 = \pi \times 0.25^2 \times 0.3 = 0.0589 \text{ m}^3$$

$$V_1 = \pi \times 0.25^2 \times 1.5 = 0.2945 \text{ m}^3$$

and $p_1 = p_0 (V_0/V_1)^{1.4} = 7.354 \times 10^4 \text{ N/m}^2$

and $E_a = \dfrac{-1.957 \times 10^4}{-0.4} = 4.893 \times 10^4 \text{ J}$

(b) Positional energy of the hammer.

$$E_p = mgl_1 = 2500 \times 9.81 \times 1.2 = 2.943 \times 10^4 \text{ J}$$

where g is the gravitational acceleration.
The total energy at strike is

$$E = E_a + E_p = 7.836 \times 10^4 J$$

in which the contribution by the pressurized air is almost two times that of gravity. The striking velocity is obtained from the kinetic energy:

$$E = \frac{mv^2}{2}$$

$$v = 7.92 \, m/\text{sec}$$

The forging force is obtained from the formula of Eq. (5.68):

$$F = Y\left(1 + 0.192\frac{d_2}{h}\right)\left(\frac{\pi d_2^2}{4}\right),$$

It is also $d_2/h = 3$ and, correspondingly, $F = 8.0209 \times 10^6$,

$$F \Delta h = E,$$

$$\Delta h = \frac{E}{F} = 0.0097 \text{ m} = 9.7 \text{ mm}$$

▲

EXAMPLE 4.5 Energy of a Hammer Strike Lost in the Motion of the Anvil ▼

The billet in open-die forging, or the lower half of the die with the billet in closed-die forging, is associated with the large mass of the anvil or of the hammer frame respectively. Let us denote the mass of the moving hammer m_1 and the mass of the anvil, or of the frame, m_2. The mass m_2 is supported by some kind of a flexible foundation, which may consist simply of wooden logs or of steel springs and dashpots. This mass will start to move as a result of the blow and will partially "soften the blow," or, more accurately, take a part of the striking energy.

Let us assume that the strike, in hot forging, can be taken as fully plastic and that the suspension of the anvil, apart from being elastic, also has an overcritical damping ratio.

Under these assumptions, first, the bodies m_1 and m_2 will move together after the blow, and their common kinetic energy equals that before the blow:

$$m_1 v_1 = (m_1 + m_2) v_2$$

$$v_2 = \frac{v_1 m_1}{m_1 + m_2}$$

Let us assume the ratio $\rho = m_2 / m_1 = 25$:

$$v_2 = \frac{v_1}{1 + \rho} = \frac{v_1}{26}$$

The kinetic energy before the blow is

$$E_1 = \frac{m_1 v_1^2}{2}$$

and after the blow it is

$$E_2 = (m_1 + m_2)\frac{v_2^2}{2} = (1 + \rho) m_1 \frac{1}{2}\left[\frac{v_1}{1 + \rho}\right]^2 = \frac{E_1}{1 + \rho} = \frac{E_1}{26}$$

The energy of $(m_1 + m_2)$ after the blow will not be recovered because of the overcritical damping of the suspension. The energy of the blow is

$$E = E_1 - E_2 = E_1\left[1 - \frac{1}{1 + \rho}\right] = E_1 \frac{\rho}{1 + \rho} = \frac{25E_1}{26} = 0.96E_1$$

Due to the motion of an anvil that is 25 times heavier than the hammer, 4% of the blow energy is lost. ▲

EXAMPLE 4.6 Slowdown of the Flywheel in a Punching Operation ▼

On a mechanical press, the work is done over a very small part of one revolution of the crankshaft. The power needed during the very short time of the forming operation is rather large. However, the average power over the cycle is low. Therefore, the drive is designed so that the energy for the actual work is supplied from the kinetic energy of the flywheel, the speed of which drops during the short time of the operation. The energy of the flywheel is recovered during the idle part of the cycle from the power of the electric motor. The link (rubber belts) between the flywheel and the electric motor is flexible enough to permit the slowdown, and the motor itself participates in the slowdown. In the following calculation we assume that all the work of the blanking operation is supplied from the energy of the flywheel.

The punching operation consists of simultaneous cutting of several holes with a total perimeter length of $l = 1$ m. The metal thickness is $h = 2$ mm. It is cold-worked carbon steel with a shear strength $S = 400$ MPa and penetration $p = 0.35$. The shearing force is

$$F = lhs = 1.0 \times 0.002 \times 4 \times 10^8 = 800 \text{ kN}$$

The shearing work is obtained by assuming shearing force constant over the depth of penetration:

$$W = Fhp = 8 \times 10^5 \times 2 \times 10^{-3} \times 0.35 = 560 \text{ J}$$

The flywheel has a diameter of $D = 0.75$ m, and width $b = 0.06$ m. It is made of steel with density $\rho = 7.8 \times 10^3$ kg/m^3 and rotates with a speed $n = 400$ rev/min. It is a simple disc, as shown in Fig. 4.24. Its moment of inertia is obtained from

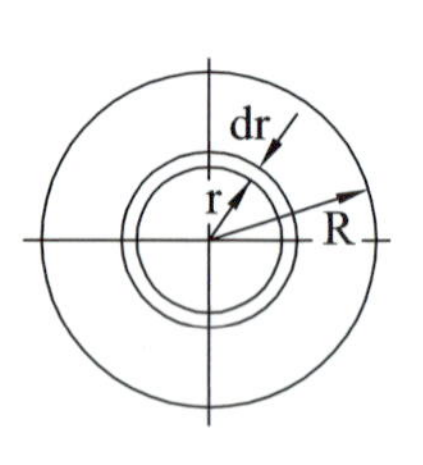

Figure 4.24
Determining the moment of inertia of a flywheel.

$$dJ = 2\pi r \, dr \, \rho b r^2 = 2\pi \rho b r^3 \, dr$$

$$J = \int_0^R dJ = \frac{2\pi \rho b R^4}{4}$$

In our instance, it is $J = 14.54$ kgm^2. Its speed before the operation is $\omega_1 = 2\pi n/60 = 41.9$ rad/sec; after the operation it will be ω_2:

$$\frac{J(\omega_1^2 - \omega_2^2)}{2} = W$$

$$\omega_1^2 - \omega_2^2 = \frac{2W}{J} = 2 \times \frac{560}{14.54} = 77 \text{ rad/sec}^2$$

$$\omega_2^2 = \omega_1^2 - 77 = 1679$$

$$\omega_2 = 40.97 = 0.978\,\omega_1$$

The flywheel has lost, through this operation, only $(100 - 97.8) = 2.2\%$ of its speed. The particular press, although it may not have a much larger force rating than the 800 kN needed in this operation, may be able to carry out work over longer strokes. It is usual to allow for about 5% loss of speed of the flywheel.

Even if it is of no practical consequence, it is interesting to calculate the instantaneous power during the punching operation. First, how long does the operation take? The press has a crank with a radius of 150 mm. Assuming, for simplicity, no deformations and an infinite connecting rod, and setting the operation for a 1-mm overrun of the punch over the bottom of the sheet, the operation starts 3 mm above bottom position and carries on for $ph = 0.35 \times 2 = 0.7$ mm, to 2.3 mm above bottom position (see Fig. 4.25). The angle ϕ is obtained as follows:

$$\phi_2 = \cos^{-1}\left(\frac{147.7}{150}\right) = 10.05°$$

$$\phi_1 = \cos^{-1}\left(\frac{147}{150}\right) = 11.48°$$

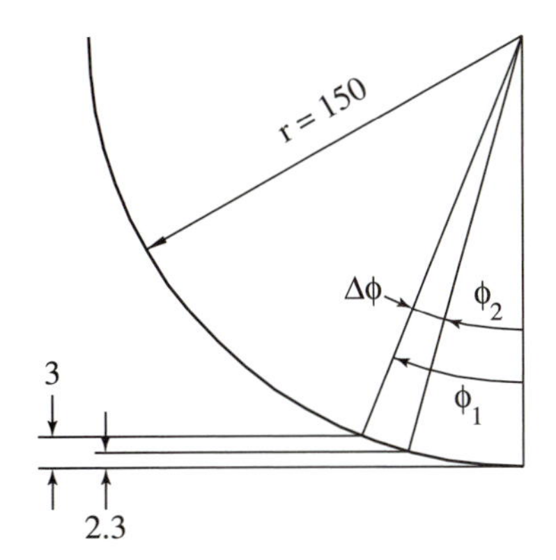

Figure 4.25
Geometry of the setup of the punching operations.

$$\Delta\phi = 1.43°$$

Neglecting the slowdown, the time of the operation is

$$\Delta t = \frac{\Delta\phi/360}{n/60} = 0.000596 \text{ sec}$$

The instantaneous power is

$$P = \frac{W}{\Delta t} = 940 \text{ kW}$$

The average power, which in this case could be the nominal power of the driving motor (for 100% efficiency of the transmission) is, however, only

$$P_{av} = \frac{Wn}{60} = 3.73 \text{ kW}$$

▲

4.3 FORGING

Forging is the name given in general to all the bulk-forming operations; most of them are carried out hot. They are usually distinguished according to the machine and tooling used. Only the actual forging equipment will be taken into account in the following discussions, although the heating and heat-treatment furnaces are also important parts of the equipment of a forging plant.

4.3.1 Open-Die Forging (ODF)

Open-die forging is distinguished by the fact that the metal is never completely confined or restrained, and that the dies used are rather simple and universal. All types of hammers or presses may be used in open-die forging. Forgings are made by ODF if (*a*) the forging is too large to be produced in closed dies; under open dies it is produced in many steps by forging only a part of it in each step, or (*b*) the quantity required is too small to justify the cost of complex closed dies.

The basic ODF operation is cogging, in which an ingot is rotated and axially moved under repeated squeezes between two flat dies. A photograph of this operation is shown in Fig. 4.26. A hydraulic press is used. The sequence of cogging operations from the ingot to a shaft is indicated in Fig. 4.27. These operations include sequences in which the forging is elongated and its cross section is reduced. This is called drawing. The various types of drawing operations are shown in the diagrams in Fig. 4.28. The basic principle of drawing between flat dies is in Fig. 4.28a. The dies are relatively narrow and have round edges. The narrow dies constrain the width of the forging and cause elongation in the direction of the stepwise feed, as indicated. The edges of the dies must be rounded because sharp edges might cause folds in the surface of the forging, which would later cause cracks. Similarly, if the central part only should be drawn, it is advis-

Figure 4.26
Cogging of an ingot into a shaft in a hydraulic press. [FIH]

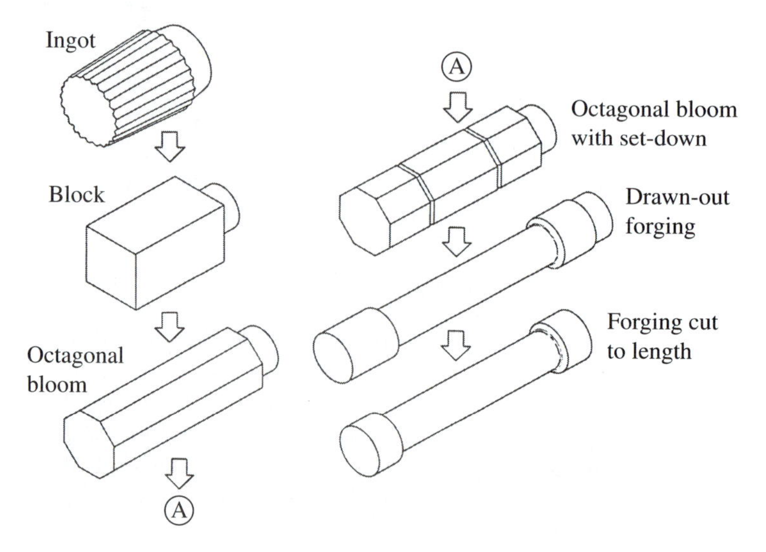

Figure 4.27
Sequence of ODF operations in cogging of an ingot. [ODFM]

able to first produce notches to prevent folding at the shoulders between the narrower middle portion and the ends. A forging may be drawn while also being rotated in steps and in this way it is symmetrically reduced in a round section as in (*c*) or in a square section as in (*d*). Diagram (*e*) shows the basic steps in producing a stepped shaft.

Other types of ODF operations are upsetting, punching, ring forging (Fig. 4.29), and ring expanding. In the upsetting operation a workpiece is compressed between flat dies: its height is reduced, and its cross section is enlarged. In ring expanding, a forg-

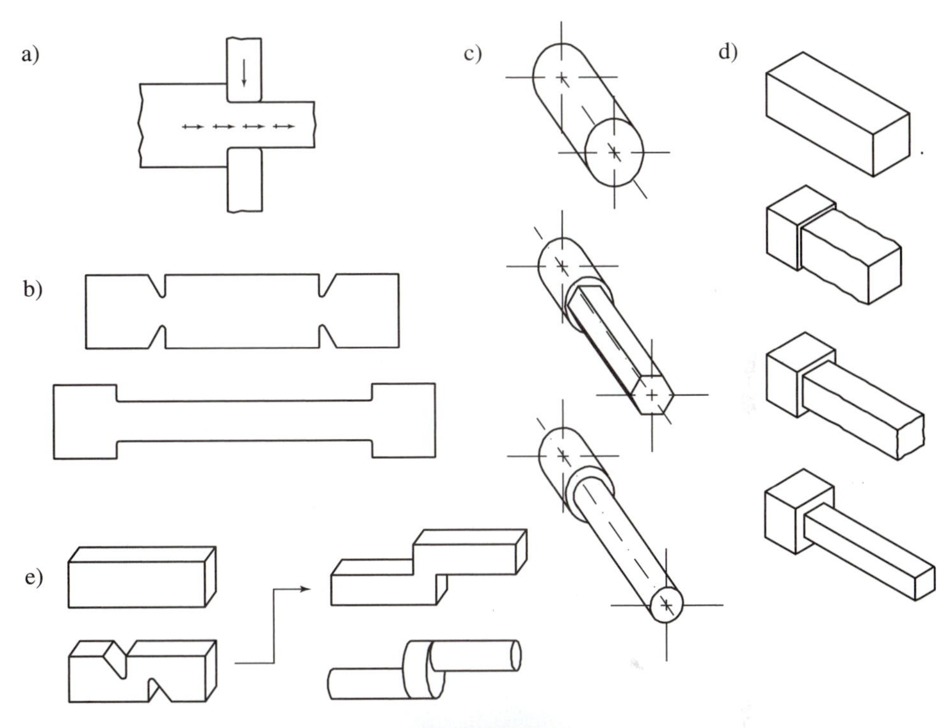

Figure 4.28
Examples of ODF drawing operations.

Figure 4.29
Ring forging. [ODFM]

ing that has first been upset and then punched is placed on a mandrel on a saddle, and its periphery is compressed in parts as it is stepwise rotated. The examples shown here involve rather large workpieces. All these operations may also be performed on a smaller scale; however, since the smaller workpieces may often be needed in larger quantities, they are usually made in closed dies.

4.3.2 Roll Forging

Roll forging is used to modify the shapes of heated bars by passing them in several steps between two driven rolls that rotate in opposite directions and have several form grooves. The principle of the process is explained in Fig. 4.30. The machine is attended by an operator who manipulates the stock and starts and stops the rotation of the rolls. First he feeds the stock between the open rolls (with die segments fully on the outside) against an end stop. Then the rolls start to rotate and push the stock back while forming it by the action of the die segments. The table of the machine has grooves on its top and they help to guide the stock between the rolls. The forming is done in several stages as the stock is passed through a number of grooves in the rolls (see Fig. 4.31). The roll-forging machines may have the shafts of the rolls supported at both ends for wide rolls incorporating a large number of grooves, with some additional ones on the outside rolls, or it may have just the overhang rolls, as shown in Fig. 4.32. These machines exist in a wide range of sizes with drives from 4 to 250 kW.

Roll forging serves two general areas of application. It may be itself the final forging operation. Most often, however, it is used as a preliminary operation for die forging to be carried out in a hammer or in a press. The preforming operation executed by rolling eliminates the need for too many impressions in the forging dies.

There are other types of roll forging, such as transverse rolling with round or straight dies, and skew rolling, which is used typically for the production of balls for ball bearings (see Fig. 4.33). The balls are subsequently finish-ground in special grinding machines.

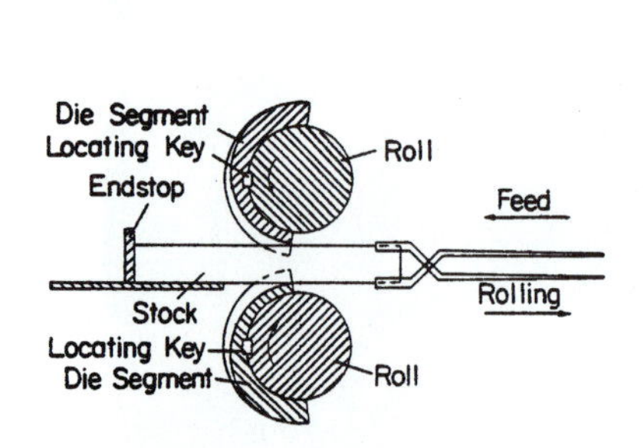

Figure 4.30
The principle of roll forging.
[SME2]

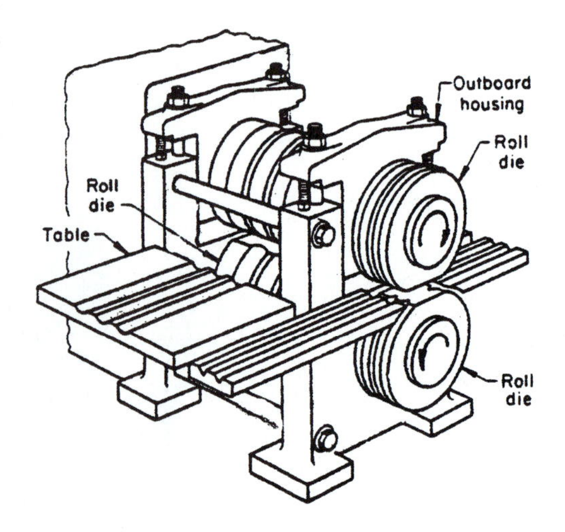

Figure 4.31
A roll-forging machine.
[ASM14]

Figure 4.32
Operating a roll-forging machine. [NATIONAL]

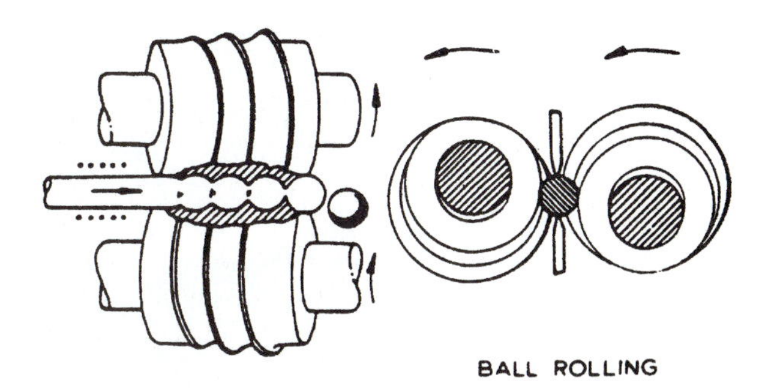

Figure 4.33
Skew rolling of balls. [HASEK]

4.3.3 Closed-Die Forging (CDF)

Closed-die forging is used to produce a wide variety of sizes of forgings from a wide variety of materials, often in low-volume, but mostly in high-volume production of parts with complex shapes and high requirements on structural integrity, strength, and toughness. The forgings may be made in various degrees of tolerance, correspondingly to what are called blocker, or conventional, or precision forgings with (respectively) a decreasing need for subsequent machining.

The largest user of forgings is the aerospace industry, which produces 32% of the total value of forgings in the United States. These represent jet engines and their parts as well as airframe parts. All these parts have to be as light and strong as possible; therefore, their shapes are very intricate. Some of these parts are shown in the following illustrations. A landing gear support beam made of a titanium alloy is shown in Fig. 4.34. It is six meters long and has a mass of 2000 kg. It was made on a hydraulic press with 500 MN capacity. Another large hydraulic press, also rated at 500 MN, is shown in Fig. 4.35. At the left-hand side are located banks of hydraulic accumulators.

The second largest user of forgings is the automotive industry, with a 20.5% share of the total forging business. Let us have a look at an example of a typical part. In general, a forging is made in a succession of operations that include several preforming stages followed by the blocking and finishing stages and, eventually, trimming of the flash. The distribution of the total work leads to better die life and improved quality of the forging. In the preforming stages the material, which originally is a rod, is redistributed along its length so as to roughly correspond to the needs of the final shape. Reducing one end is called drawing out; reducing in the center is called fullering; increasing the cross section of a part is called edging or rollering; and the material may also be bent and flattened. These operations may be carried out on hammers or presses in a corresponding succession of impressions in the die, but they are often most economically done by roll forging. When using hammers, usually several blows are used per impression, while in a press the stock is usually hit only once in each impression.

Figure 4.34
Jetliner main landing gear support beam forged from a titanium alloy. [FIH]

Figure 4.35
Hydraulic press with 500 MN capacity. [SCH]

Figure 4.36 shows the dies and the shapes in the individual stages for connecting rods which are made in pairs. The bar is first fullered to reduce the sections that will become the central connecting parts and then rollered, blocked, and finished. Trimming of the flash is done cold, separately on a press. In this case, all the forging operations are done in the single pair of die blocks, with manual transfer from impression to impression using tongs on the tonghold at the bottom end of the parts.

The design of forgings as well as the design of the preforms and of corresponding dies is a difficult task that requires specialized experience. Attempts are currently being made to use computers and to develop software for computer-aided design (CAD) that incorporates rules based on experience and on methods of computation of pressure and temperature loads in the dies. For illustration of some of the aspects of these considerations, examine the diagrams in the three following figures. Figure 4.37 illustrates the

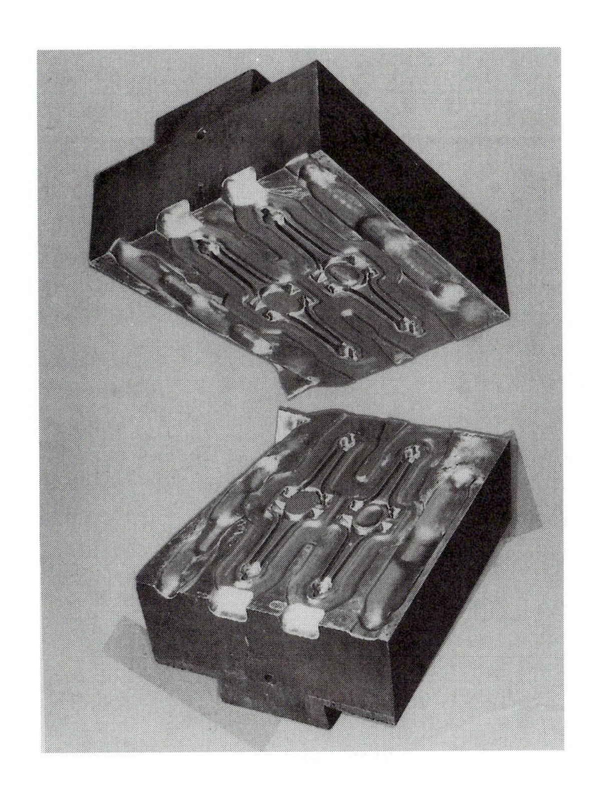

Figure 4.36
Dies and successive shapes in forging connecting rods. [FIH]

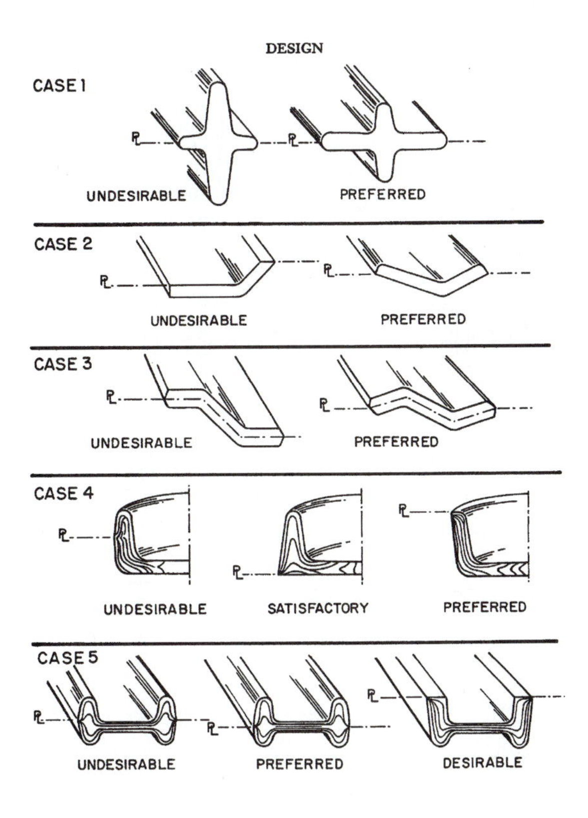

Figure 4.37
Illustrating the effects of location of the parting surface. [FIH]

choice of the parting surface of the die blocks, using simple, basic shapes of forgings. It shows undesirable and preferred parting-plane locations. Reasons for the preferred locations follow:

> *Case 1:* The preferred choice avoids deep impressions that might otherwise promote die breakage.
>
> *Cases 2 and 3:* The preferred choices avoid side thrust, which could cause the dies to shift sideways.
>
> *Case 4:* Preference is based on grain-flow considerations. The "satisfactory" location provides the least-expensive method of parting, since only one impression die is needed. The preferred location, however, produces the most desirable grain-flow pattern.
>
> *Case 5:* The choice in this case is also based on grain-flow considerations. However, the "desirable" location usually introduces manufacturing problems and is used only when grain-flow is an extremely critical factor in design. This, in turn, depends on the directional properties and cleanliness of the material being forged.

Methods for locating the parting plane on complex shapes are not usually as straightforward as these examples would indicate. However, most of the fundamentals involved are illustrated here.

In Fig. 4.38 the effect of webs and ribs on the increasing difficulty of forging is illustrated. Parts with thin and long sections are more difficult because the variations in shape produce more friction and temperature changes. Vertical projections such as ribs

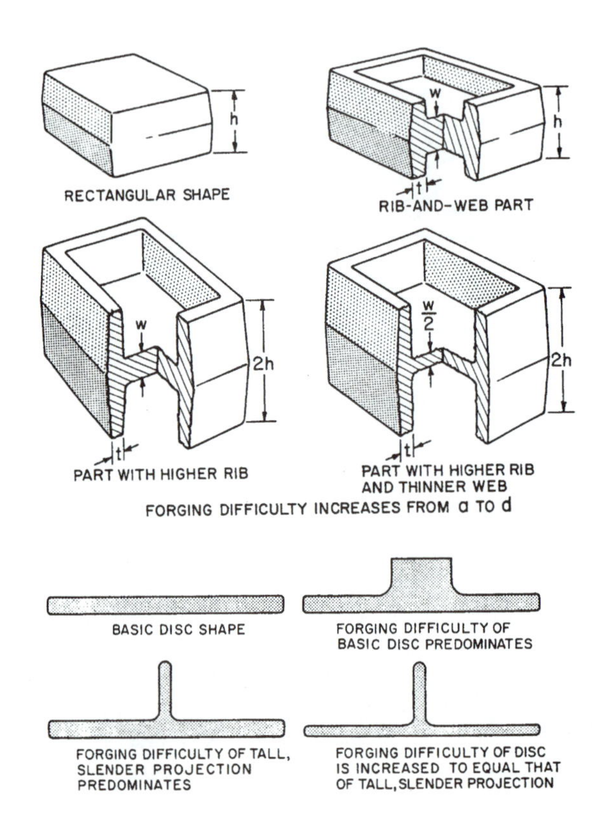

Figure 4.38
Effect of webs and ribs on difficulty of forging. [ASM14]

and bosses are harder to forge than lateral projections. Shapes with tall and slender vertical projections are most difficult to forge; they require more forging operations and thinner and wider flash channels, leading to higher forging forces.

Apart from the aspects of the loads on the die, it is also necessary to consider the possibility of defects like folds and laps in the product. There is a limit on the minimum section of the web between ribs or other projections. In forming the ribs, metal normally flows toward the center, causing the web to increase in thickness. If the web in the blocker forging is too thin it will buckle, causing laps in the web. Similarly, there are limits imposed on thin ribs. These and other aspects of forging engineering are discussed in much more detail in specialized books and handbooks.

4.3.4 Hot and Cold Upsetting

Hot upset forging is essentially a process of enlarging and reshaping the cross-sectional area of the end of a bar or tube of uniform, usually round, section. Originally, it was developed for heading of parts such as bolts. It is done on machines arranged horizontally, and it is a very efficient process that accounts for a substantial part of forging production. The products range from headed bolts or flanged shafts to cluster-gear blanks, and they may require a succession of operations including upsetting, extruding, piercing, and trimming. Figure 4.39 shows a display of a variety of parts produced by this method. The basic heading operation is shown in a diagram in Fig. 4.40. The hot end of the bar is inserted in the gripping die, against a stock gage. The die closes and holds the stock firmly against the axial force of the first-stage upsetting operation. The heading tool (the punch) moves axially against the die and upsets the end of the stock so as to fill the cavity of the die. The limits of the maximum length that can be upset successfully in a single stroke are set by the buckling of the unsupported end of the workpiece. In general, this cannot be longer than two to three diameters of the bar, depending on the guidance or support given to the die and on the squareness of the sheared end.

There exist two basic designs of hot upsetting machines: one is used almost exclusively in America, and the other one in Europe. The former is described here. Its design

Figure 4.39
Parts made on hot upsetting machines. [NATIONAL]

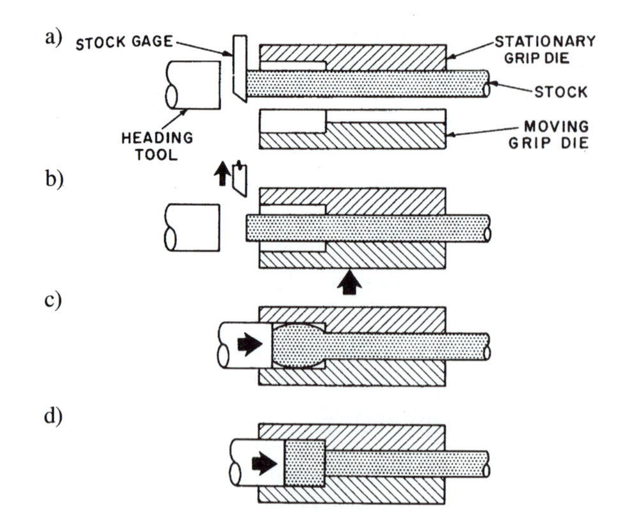

Figure 4.40
Basic upsetting operation. [FIH]

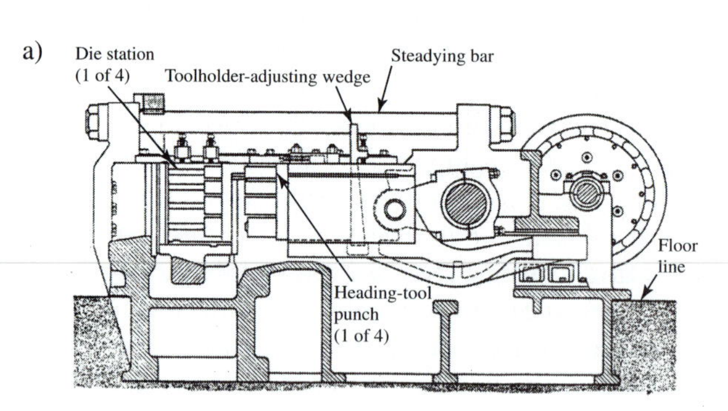

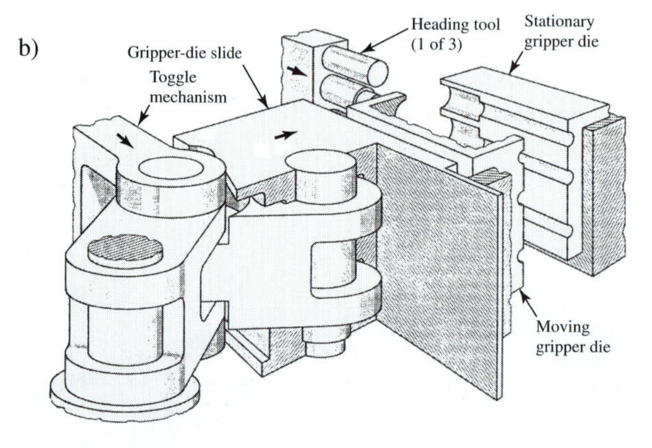

Figure 4.41
Drawings of the upsetting machine: a) cross section 216of an upsetting machine; b) detail of the toggle and gripper die. [ASM14]

is presented in Fig. 4.41. The motor drives—via a flywheel-pulley, clutch shaft, and gear transmission—the shaft carrying the eccentric that drives the heading slide with punches and carrying also a pair of cams driving the closing mechanism of the moving part of the gripping dies. These dies are split in a vertical plane, and several of them are arranged in this plane one below the other, opposite the corresponding number of punches. The bar of the stock being forged is normally held by the operator in tongs at its left-hand end. It is inserted during the open interval of the dies as well as moved from one die to the next lower one. The stroke of the gripper die is one of the basic parameters of the machine because it limits the maximum diameter of upset so that it can pass from one die to the next one. The closing and opening of the die is driven from the cams over the rollers, which are mounted in the cam slide, and further through the rocker and toggle. In the case of overload the linkage, which is held straight by the spring, collapses.

Die clamping is achieved independently from the heading action of the press. The nature and duration of the clamping force are different from the requirements of the heading force. The dies must be clamped before the first application of any heading force. If the working stroke of the header is long, clamping may need to occur very shortly after leaving top dead center. This motion is accomplished by the cam attached to the crankshaft, which has a rapid rise and then a dwell. The length of the heading stroke, which occurs in the forward direction after the die jaws have closed, is called stock gather: this is the actual working stroke of the machine. The dies remain clamped after the heading die has achieved maximum forward motion and begun retraction. The length of stroke rearward before the dies open is called hold-on. This is particularly useful if a deep piercing operation is being performed, since the work must be stripped from the heading tool. In very long operations, a draft angle may be required on the tool.

The tie rod strengthens the frame of the machine against the forging force. Notice the span of the guideways of the ram, which is guided in its front part as well as an extended rear part. The span is large enough to decrease the reaction forces on the guideway opposing a moment produced by the forging force not being in-line with the driving force of the connecting rod. This moment is largest when forging is done in the top or in the bottom die cavity. The detail in the bottom part of the illustration shows an open gripper die with three cavities and the head tools (punches). The machine is shown here embedded in the foundation. This is not always the practice; machines may be mounted on the floor and tied down with foundation screws. However, this connection must transmit substantial forces reacting to the inertia forces of the reciprocating motion of the ram. These forces may cause vibrations to propagate through the floor and disturb other machines, possibly of a precision type, in other parts of the shop.

Upset forging machines exist in a range of sizes that are primarily rated by the largest size of the upset diameter. Table 4.3 correlates the sizes with maximum forging forces *F,* average number of strokes per minute *S,* and motor power *P.* The forging force is obtained by estimating a yield pressure between 500 and 1000 N/mm^2 for forging alloy steels and difficult shapes. The larger value is applied to smaller sizes of upsets where the surface over volume ratio is larger. Commonly, carbon steel may require only 300 to 500 N/mm^2. The clamping force is 50 to 100% of the forging force.

TABLE 4.3 Principal Parameters of Hot Upsetting Machines

Rate Size (mm)	Forging Force F (MN)	Strokes S (strokes/min)	Power P (kW)
25	0.5	90	5
38	1	65	10
50	2	60	15
75	4	45	25
100	6	35	40
125	8	30	50
150	10	27	60
175	13	25	90
200	16	23	110
225	20	20	150

Most operations that are performed on upsetters require less than six stations in their die sets. This number corresponds mainly to automatic machines, where essentially redundant stations are an advantage for die life. Better control over material flow is also achieved because less material may be upset at each station, which reduces the possibility of cold shuts, laps, buckling, and other problems. Most operations require two to four stations, regardless of the complexity of the part produced. In most operations only one part is in the machine at a time. In manual machines this is the only possibility, and with an automatic machine, die-cooling requirements and heading force are limiting factors in this respect.

A photograph of a machine rated at 150 mm size is shown in Fig. 4.42. The heavy frame is cast of steel and reinforced by a longitudinal tie rod that partly supports the forging force as well by a transversal overhead beam helping to take up the die-clamping force.

An example of a four-pass operation in forging a cluster-gear blank is given in Fig. 4.43. In the first pass a flange is upset on the clamped bar. The workpiece is then transferred into the second die, in which it is further upset so as to create the final diameter of the central part and to increase the diameter of the left end. This end-diameter is further increased in the third pass, in which the piercing punch preforms the hole simultaneously. The hole is finished in the fourth pass, and the workpiece is severed from the bar stock. The drawing of the finished workpiece is shown at the bottom of the illustration. The production rate in this example was 80 pieces per hour. The die life was approximately 12,000 pieces, and the punches had a life of 4,000 to 6,000 pieces. The size of the forging machine used was rated at 100 mm.

Other types of hot upsetting machines used for forging of short parts are fully automated. They do not use gripper dies. Instead, the workpiece is first sheared from the bar stock and then transferred by a built-in transfer mechanism from station to station. These machines are used for the production of hexagonal nut blanks, ball bearing races, and gear blanks, and they are rather accurate and fast. On a large machine the maximum diameter of the parts is about 120 mm, and the maximum mass of the workpiece is about 3.2 kg. Machining allowance is about 0.5–1 mm per side. Hexagonal nuts as large as 60

Figure 4.42
Hot upsetting machine of a size rated at 150 mm. [NATIONAL]

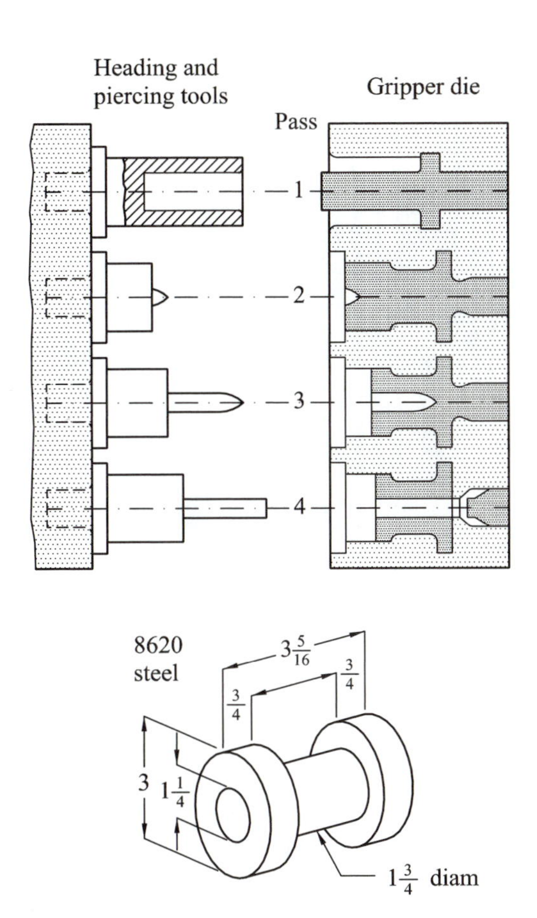

Figure 4.43
Forging a cluster-gear blank in a four-pass operation. [ASM14]

mm across the flats are made at rates of up to 3300 per hour. The machine is driven by a 300-kW motor. Dies are mounted in a stationary block, and each die has its own cam-controlled, adjustable kickout mechanism to push forgings out of the die and into the transfer fingers. Some part shapes are formed inside the punches. Timed knockouts can be provided on the punch slide to push such parts free of the punches.

A metal-forming operation called displacement is used. A round punch penetrates the metal, forcing it to flow from the hole outward into the die cavity. Since blank length before and after displacement is nearly the same, little extrusion occurs. Displaced metal flows into empty segments of the die cavity. When making a nut (see Fig. 4.44), the upset blank made in the first die is slightly smaller than the across-flats dimension of the second die. This helps keep the blank central so that displacement is uniform. As the punch drives into the center, metal flows outward to fill the die.

Since displacement uses metal from a central hole to fill die detail, most parts made on these machines have holes. The method works well for short parts because metal does not displace easily over long distances.

The first blow makes an upset, sometimes called a pancake because of its flattened shape. Hot shearing presents some conditions that the first upsetting blow can help. First, a better cutoff blank can be sheared if the length is greater than the diameter, preferably at least as long as the diameter. The first blow then converts this longer blank to a short, larger diameter suitable for displacement. Upsetting inside the first die allows the metal to move out against the die wall, giving a concentric blank with square ends.

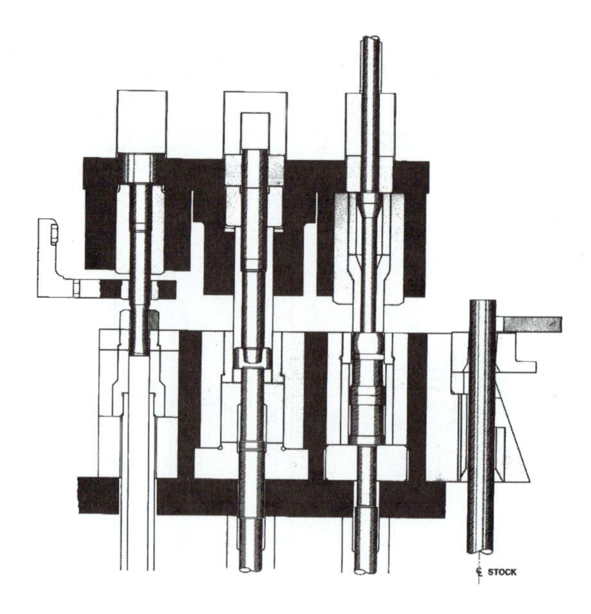

Figure 4.44
Automatic upsetting of nut blanks. [NATIONAL]

This works well when using induction heating, which creates little scale. When the bar has more scale, upsetting can be done at the face of the die allowing the scale to fall free and wash away. Preforming is done in the second die with controlled metal displacement. The part is formed to final shape and dimension with a web still left in the hole.

Web thickness should be as small as possible to minimize material waste, but thinner webs increase the forming pressure. Punchout of the web to produce the required through hole is done in the third die. A stripper removes completed forgings from the piercing punch, and the nut blanks fall into a discharge chute. The first and second dies contain knockout bars that move the blanks out of the die into the grippers of the transfer mechanism. A diagram of this mechanism is shown in Fig. 4.45. It consists of the upper and lower horizontal bars that move in a reciprocating motion in a stroke equal to the pitch between the dies, which are arranged in a horizontal plane in the forging machine.

The process described in the preceding paragraph can also be performed by *cold forming.* Many kinds of different parts are indeed formed cold. This offers the advantages of leaving no scale on the surface of the product; higher precision, so that often no further machining is necessary; and improved strength of the components. On the other hand, higher forging forces are needed, resulting in higher demands on the upsetter as well as on the dies. The process is automated starting from wire, rods, or sheared blanks. In the latter case the parts are supplied to the machine by a hopper magazine and feeder. Typical workpieces are nuts, bolts, spark plugs, piston pins, and so on. The forging machine used in this operation usually has six dies, each of them with a kick-out rod. The process normally follows the pattern of a coiled wire passing through feed and wire-straightening rolls on the way to the cut-off station, where a slug of proper length is sheared. From there on it is transferred by the properly adjusted fingers of the transfer mechanism from die to die in the sequence of forging operations.

Cold forging is also applied to relatively large parts—for example, the production of 1.25-in. (31.75-mm)-diameter bolts, 250 mm long with hexagonal heads. A bolt of this size has a mass of 2 kg. The machine cuts off, extrudes, heads, trims, points, and rolls the thread at a rate of 2280 finished parts per hour. Various cold-formed parts are shown in Fig. 4.46. They include stem pinions, front-wheel spindle, shafts, pinion gears, and other, mainly automotive parts.

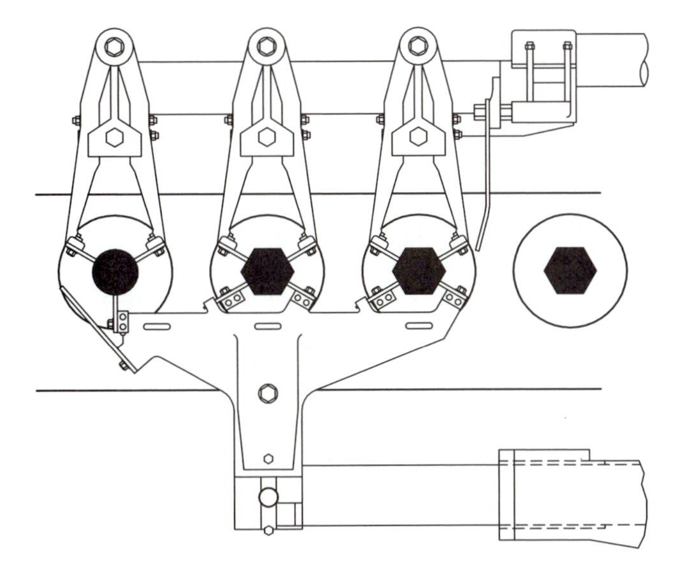

Figure 4.45
The transfer mechanism for nut blanks. [ASM14]

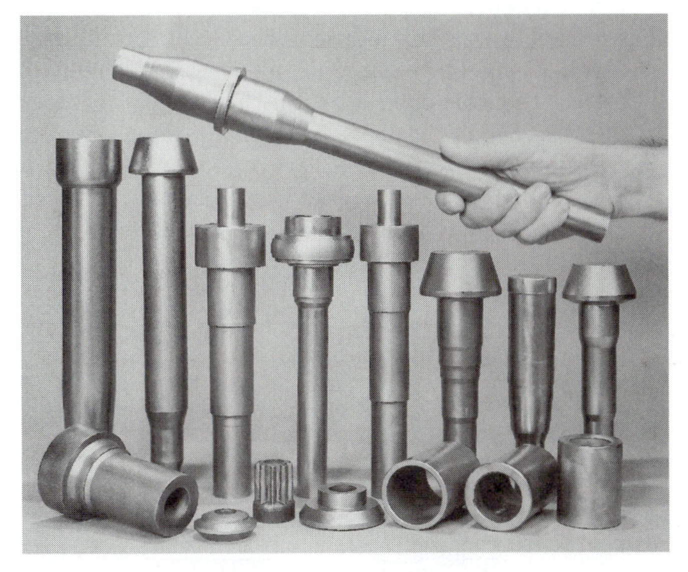

Figure 4.46
An assembly of various cold-formed parts. [NATIONAL]

4.3.5 Extrusion

Various types of extrusion operations are performed as steps in both hot and cold forging. Figure 4.47 shows some of the basic types of extrusion. In (*a*) and (*b*) we have examples of forward and backward solid extrusions, respectively. In these operations the cross section of the blank is reduced, and the length is increased. The extruded section may be round, or square, or of other shapes depending on the exit section of the die. Operation (*a*) is much more common than (*b*). However, in extruding hollow sections, as shown in (*c*) and (*d*), the backward process is quite common, and it is used to produce cups, containers, and tubes. A special type is called impact extrusion, and it is used to produce collapsible tubes from aluminum, tin, and zinc (toothpaste containers). Comparing the production of cups by drawing and extrusion, the latter process requires fewer steps and, correspondingly, less expensive tooling. Also, it can produce cups, cans, and shells with thick bottoms and flanges. On the other hand, the forces are larger. For shallow cups, drawing is preferable. In the forward process both the outer and inner contours of the section are generated. The outer one is determined by the shape of the die opening and the inner one by the form of the section of the mandrel. The forward and backward extrusion types may be combined, as shown in (*e*). In all these operations, when they are one of the steps on a forging press or on an upsetter, the length of the extruded parts is relatively small.

Long, profiled bars and other products are produced on special extrusion presses. Some specialized operations can be carried out, like sheathing of a cable (Fig. 4.47f), where the sheathing lead is introduced under pressure into the extrusion chamber. Cold extrusion may be carried out as hydrostatic extrusion (Fig. 4.47g). The extrusion chamber is filled with high-pressure oil, and there is no need of a ram. The advantage of this process is that there is no friction between the billet and the extrusion cylinder. The form of the billet is not important. In fact, it is possible to extrude wire from a coil inserted in the chamber. Its end is first shaped to fit the hole of the die and once it is inserted in the die the wire will uncoil automatically and extrude.

However, most of the work on special extrusion presses is the production of long beams and pipes with intricate sections, and it is done hot. Examples of sections that can be produced in this way are shown in Fig. 4.48. As shown, parts like gears may be extruded in the form of a long beam and cut off afterwards. Many of these section shapes cannot be produced in any other way than by extrusion. Materials most often extruded are aluminum and steel. The extrusion dies are made of sintered carbides, and the shaped holes in them are produced by EDM (electrical discharge machining). Extrusion presses exist with ratings up to 250 kN, and they can make parts up to 15 m long. Lubrication for the protection of the ram, chamber, and die is necessary. For hot-extruding steel, molten glass is used as a lubricant.

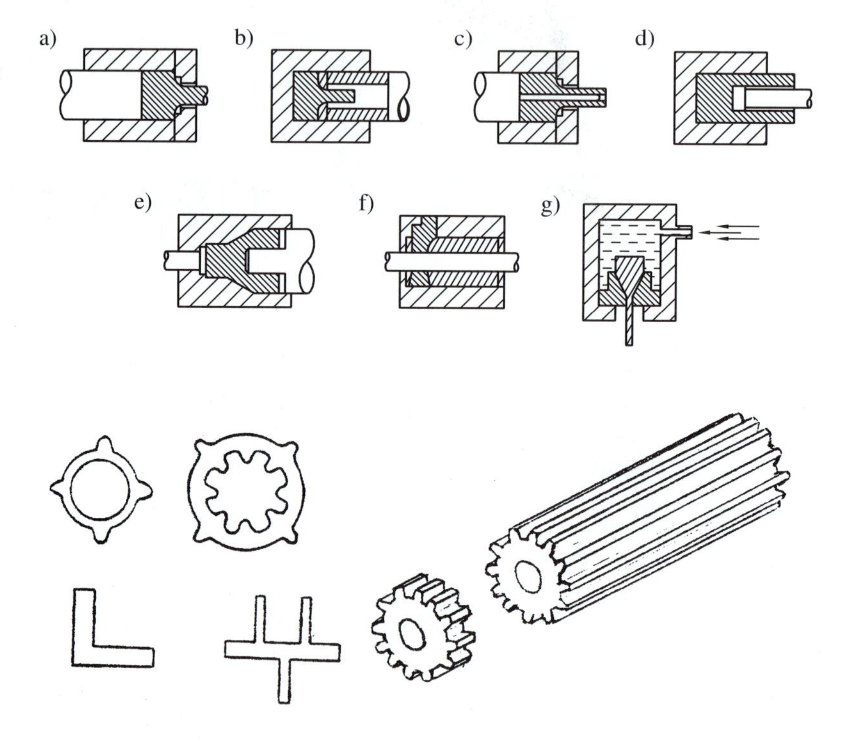

Figure 4.47
Various extrusion operations: a) forward solid extrusion; b) backward solid extrusion; c) forward hollow extrusion; d) backward hollow extrusion; e) forward and backward extrusion combined; f) sheathing of a cable; g) hydrostatic extrusion.

Figure 4.48
Extrusion press products.

4.3.6 Forgeability of Metals

Forgeability is a combination of two aspects: (1) the ease with which the material can be forged, expressed mainly by the flow stress and also by the ability to fill the die; and (2) the degree of deformation without failure (internal or surface cracks).

In general, the forgeability of metals increases with the forging temperature. However, in certain materials, the temperature should not exceed levels beyond which some adverse phenomena may occur: melting of a phase or fast grain growth. Fine-grain material has better forgeability. Materials with insoluble inclusions are brittle and exhibit low forgeability.

There are several types of standard tests for determining the forgeability of materials. The most commonly used test is the *upset test,* in which cylindrical specimens are upset in steps to an increasing degree (i.e., to a decreasing final height) until they start cracking radially or circumferentially. The test should be done so as to approach the conditions of practical application in the temperature of the specimen, the deformation rate, and the type of lubricant. In the *hot twist test* a round bar is inserted through a tubular furnace, heated to the test temperature, and twisted. The number of twist turns to failure is a relative measure of forgeability. Testing at different temperatures and strain rates establishes the best conditions for practical forging.

In Ref. [3] materials are ranked in forgeability as follows, in the order of increasing forging difficulty:

Aluminum alloys
Magnesium alloys
Copper alloys
Carbon and alloy steels
Martensitic stainless steels
Maraging steels
Austenitic stainless steels
Nickel alloys
Semi-austenitic PH stainless steels
Titanium alloys
Iron-base superalloys
Cobalt-base superalloys
Columbium alloys
Tantalum alloys
Molybdenum alloys
Nickel-base superalloys
Tungsten alloys
Beryllium

Comments regarding the forgeability of selected classes of materials follow.

Carbon and low-alloy steels are generally quite forgeable, except for grades that contain some insoluble compounds. The most common case of the latter type are free machining, resulphurized steels. Forgeability of steels improves as the deformation rate increases because the material heats up. On the contrary, the biggest obstacle to filling intricate dies is the cooling of the billet in its contact with the die. The forging temperature is generally chosen about 150°C below the melting temperature of the particular steel. Also the choice of lubricant is important. For deep dies, salt and sawdust are often used because they reduce sticking. For shallow dies in which there is considerable lateral flow, graphite suspensions are used.

Stainless steels are the most common class of alloy steels. The different classes of these steels differ in their forgeability. The forging characteristics of high-chromium, martensitic steels, series 400, are similar to carbon steels, but because of their higher strength they require 30–50% higher forging forces. Also, at high temperatures some of these alloys transform partly to delta ferrite, which reduces forgeability. Because they easily harden on cooling and might crack, they must be cooled down very slowly.

The austenitic steels, series 300, are more difficult to forge. They need higher forging pressures and exhibit strain-hardening even at high temperatures. Two to three times more forging energy is required than for carbon steels.

Aluminum alloys are the easiest to forge, for a number of reasons: they are ductile, can be forged in dies heated to the same temperatures as the workpiece (about 380 to 500°C), require low forging pressures, and do not develop scale when heated. The last characteristic, however, may cause the metal to stick to the die. To prevent this, the dies must be properly lubricated. Water-soluble soaps may be used for this purpose. It is necessary to take care to keep the temperature about 70°C below the melting point because the material becomes brittle at the onset of melting.

Titanium alloys are classified in three types: α, $\alpha\beta$, and β alloys. On heating, pure titanium transforms at about 885°C from a hexagonal, close-packed (α) to a body-centered, cubic structure (β). Various alloying elements stabilize one or the other of the two phases by either raising or lowering the transformation temperature. The various alloys differ in heat treatment and in the resulting strength and ductility. The transformation temperature is significant for the choice of the forging temperature. Above it, forging is easier because the beta phase is more ductile, and forging loads are lower. On the other hand this may lead to a coarse-grained structure; therefore, final forging operations are usually carried out in the alpha-beta range. In general, the forging loads are very sensitive to temperature and the recommended range must be strictly maintained. The flow stress increases strongly with strain rate. This means that more energy is required for hammer forging than for press forging. Hydraulic presses are preferable for large forgings. Because even low levels of oxygen, nitrogen, and hydrogen may have detrimental effects on the properties of titanium alloys, the heating is done in a vacuum or in an inert atmosphere. Some shops have had good results by coating the blanks with a protective layer of glass. Because of the sensitivity of flow stress to temperature drop, the dies for forging titanium alloys are always heated to about 260°C for hammers and mechanical presses and to 425°C for hydraulic presses, to minimize cooling in the contact of the forging with the die.

4.4 SHEET METAL FORMING

4.4.1 Basic Operations and Presses

A tremendous variety of parts are made of sheet metal, both in size and complexity. They range from small washers, clips, and brackets to long and narrow profiles for window frames; cabinets of appliances; arms and beams of automobile frames; the doors, roofs, and fenders of car bodies; and large containers and pressure vessels. These parts are usually produced in large quantities primarily on mechanical presses, sometimes on hydraulic presses, press brakes, and roll-forming machines. Substantial production expense goes into the necessary tooling, and a successful operation depends mostly on the ingenuity and skill used in the design of the various kinds of dies. Two-dimensional dies like those for punching and blanking operations are produced by form-grinding or EDM wire cutting, and the three-dimensional dies for press forming (e.g., of car body parts) are mostly precast and then finished by copy milling or NC milling.

The three basic types of operations that are diagrammed in Fig. 4.49 are cutting or shearing (blanking, punching or piercing, notching, trimming), bending, and drawing and stretching. In the cutting diagram (Fig. 4.49a), the action is limited to a line (the perimeter of the blank) on the plane of the sheet metal; one could call it a single-dimensional operation. In bending, the shape produced is essentially two-dimensional, because it is characterized by the cross section of the workpiece in any plane perpendicular to the longitudinal dimension of the bend. However, the strains arise in one direction of the plane of the sheet only: the one perpendicular to the longitudinal dimension at the beginning of the operation. The shape of the part obtained in drawing and stretching (Fig. 4.49c) is essentially three-dimensional. The strains produced arise in the two dimensions of the sheet metal. In deep drawing, one of these strains is compressive, and the other is tensile. In stretching, both strains are tensile.

In practice, actual operations are mostly combinations of the three basic processes and they are carried out in several stages using compound, combination or progressive dies. The terms *compound* and *combination* are used to define any particular station die in which two or more operations are completed during one press stroke. The former term is properly used if all these operations are of the cutting (shearing) type, and the latter term describes dies in which sheet-metal forming operations of various kinds are combined.

A progressive die performs a series of successive operations at two or more stations during each press stroke. The strip moves through all the stations to produce a complete part. The unwanted parts of the strip are successively cut off as the strip advances through the stations, but enough is left connected to each partially completed part to carry it through. The actual design of the die is often rather complicated, and it would not be easy to properly describe it in this text. It is discussed in detail in Refs. [1], [2], [5]. For illustration the successive shapes of a part and of the strip are shown in Fig. 4.50. The sequence of operations follows: (1) notch and pierce; (2) idle; (3) cupping; (4) redraw; (5) redraw; (6) redraw; (7) pierce holes 1/8″ and 3/16″; (8) notch; (9) coin; (10) blank; (11) completed workpiece.

The presses used are hydraulic or mechanical, single- or double- or triple-action, one- or two- or four-point suspension. The diagram of a double-action hydraulic press is shown in Fig. 4.51. It is a straight-side press (as opposed to C-frame). The frame consists of a bed, columns, and a crown. The columns are precompressed and strengthened by tie rods that connect the bed and crown with the columns. The main slide is driven down by the main hydraulic cylinder and piston and it carries the upper part of the die

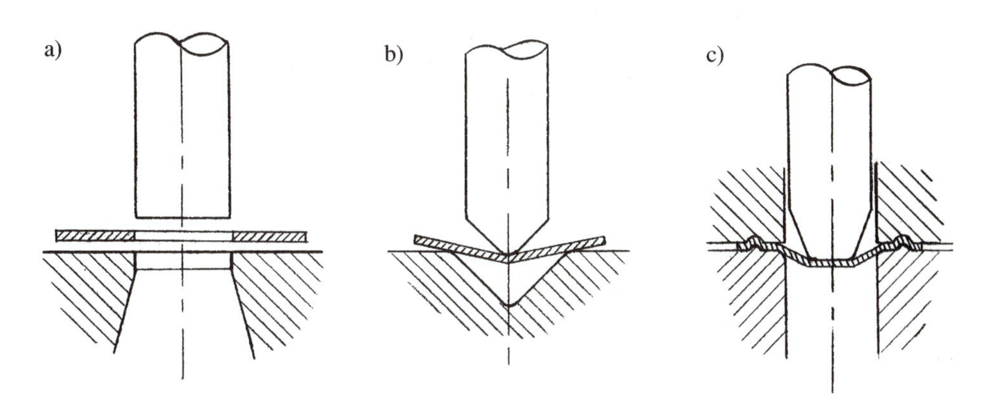

Figure 4.49
Three basic sheet-metal forming operations:
a) shearing; b) bending;
c) stretching.

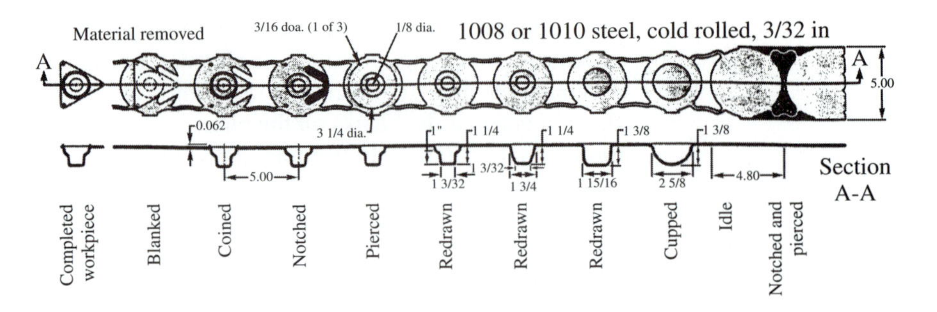

Figure 4.50
A mass-produced automotive spring seat in a progressive die. [ASM14]

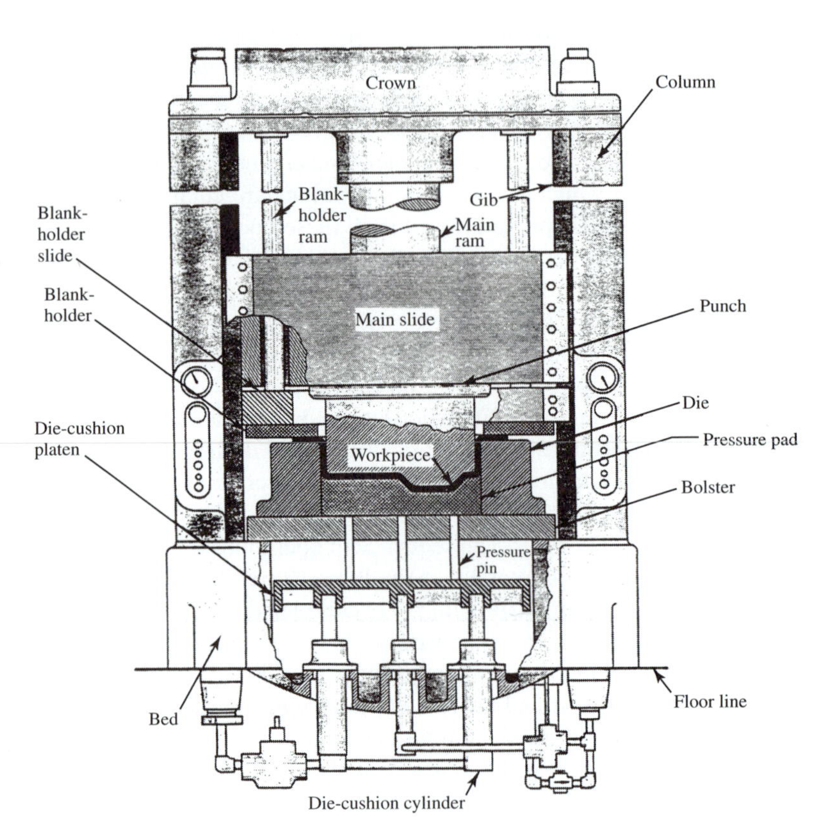

Figure 4.51
A double-action hydraulic press. [ASM14]

which is called the punch. The lower part is made of two parts, the die and the pressure pad. They sit on a bolster plate attached to the bed, which also guides the pressure pins of the die cushion. The cushion moves up to eject the workpiece; it is driven by hydraulic cylinders. The second action is the one of the blankholder slide, which is driven from auxiliary hydraulic cylinders and blankholder rams.

The design of a large-bed, single-action, four-point-suspension, eccentric-drive mechanical press is shown in Fig. 4.52. The drawing shows the drive shaft with the clutch and brake assembly, the eccentric gear, the connecting rods, and the slide. There are four eccentric gears and connecting rods. At the bottom of each is the slide-height adjustment mechanism. It is driven from an electric motor through work-wormwheel transmissions to a screw-and-nut transmission. It is used for adjusting the bottom position of the stroke for the proper mutual position of the upper and lower parts of the die. The air counterbalance cylinders support the weight of the slide and upper die and they

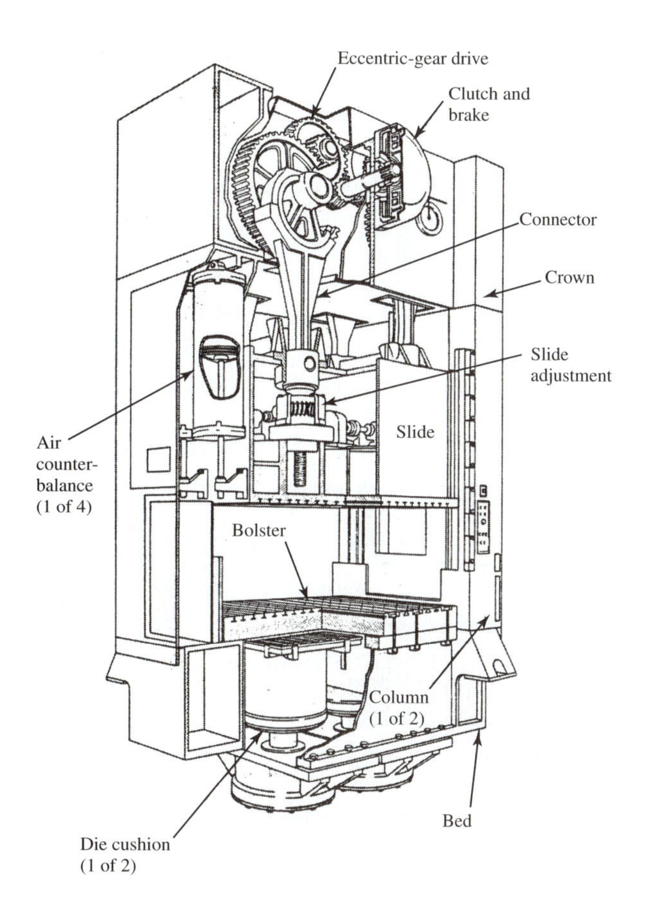

Figure 4.52. A single-action, four-point-suspension, eccentric-drive press. [ASM14]

help to smooth the motion of the slide and facilitate the adjustment. A bolster plate is attached to the bed, and under it are the two pneumatic cylinders of the cushion.

The die cushion in a single-action press is used for blankholding and stripping the finished part from the punch (see Fig. 4.53). The punch is on the lower die shoe, and the die is on the upper shoe. The cushion provides the force, through pressure pins and the pressure ring, to hold the blank in a deep drawing operation. In a double-action press, as shown in the lower part of the figure, this force is obtained from the blankholder slide. The punch is now the upper half of the die set, and the die is the lower part. Inside of the die moves the pressure pad, which is pressed against the bottom of the workpiece by the cushion. At the end of the operation the cushion ejects the workpiece from the die. The force produced by the cushion is usually 15–30% of the press capacity. The cushions are pneumatic or hydropneumatic.

Rather inexpensive tooling is applied to the production of various parts in processes that are generally denoted as rubber-pad forming. The rubber pad constitutes one half of the usual die set (mostly the female die part) and plays a universal role in that it adapts to different shapes of the counterpart. The disadvantage is its limited life, which, depending on the severity of the deformation, may not endure beyond about 20,000 pieces. Because of its universality, the process is suitable also for producing relatively small lots. There are several variants of the process. Three of them are diagrammed in Fig. 4.54. The oldest and simplest one is the Guerin process shown in (*a*). The rubber pad is either a solid block, or it consists of several slabs cemented together, and it is contained in a steel housing. The thickness of the pad must be about 30%

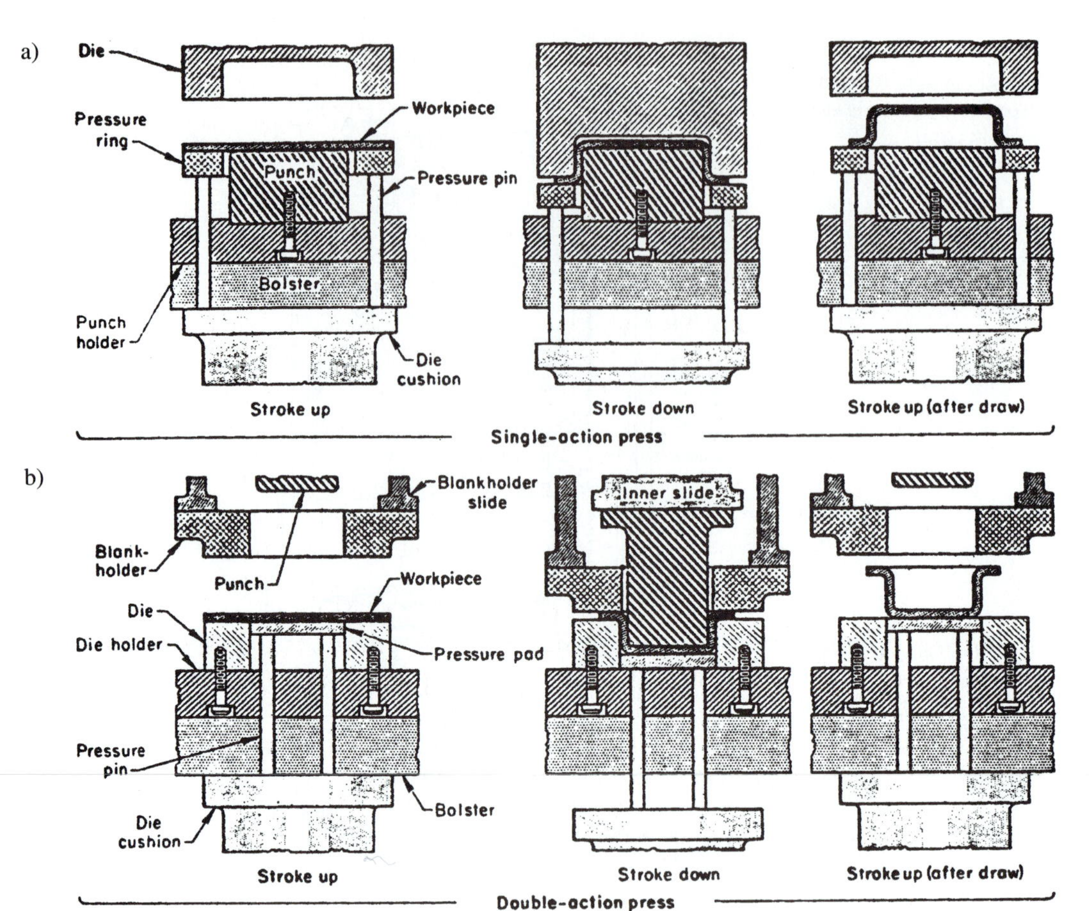

Figure 4.53 Use of cushions in single and double-action presses. [ASM14].

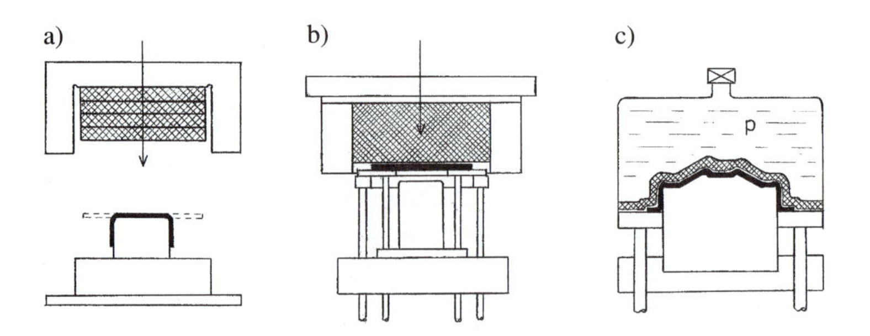

Figure 4.54 Rubber pad forming: a) Guerin process; b) Marform process; c) Rubber diaphragm process.

greater than the height of the form block. In the Marform process, sketch (*b*), a blankholder plate is added, which is pressed against the blank and rubber pad. Its role is to prevent wrinkling of the drawn part. In the rubber-diaphragm process, shown in (*c*), the pad is replaced by a rubber diaphragm backed up by pressurized oil. In this way a more uniform pressure distribution is obtained and more severe draws can be achieved than by the preceding two methods. In all these cases hydraulic presses are most often used to operate the die sets. Depending on the severity of forming, the sheet metals formed by these methods are aluminum alloys (2024-T4, 7075-W) up to 1.5–4 mm thick and austenitic stainless steels up to 0.5–1.5 mm thick.

4.4.2 Automation of Presswork

Automation is achieved by automating material loading and part unloading. For small parts, which are mostly made from a strip or from coil stock in composite and progressive dies, this means automating the stepwise feed of the strip. The devices used are either the so-called slide feeds or roll feeds. A slide feed consists of a slide that carries gripping rolls and moves reciprocatingly between stops. The reciprocating motion of the slide is mostly driven by a cam attached to the punch slide. The roll feed consists of roll that may be intermittently driven from the crankshaft of the press via a ratchet-type mechanism (see Fig. 4.55). Coil stock is generally handled as shown in Fig. 4.56. The coil is in this case put on a flat conveyor and passes through pinch rolls, and a set of straightening rolls, and the loop-control arm, finally entering the roll-feed mechanism. The coil may alternatively be carried by a motor-driven reel.

An example of a common type of a straight side automatic press is shown in Fig. 4.57. The drive bar of the roller feed is visible on the left-hand side of the press. The stock is fed right to left through openings in the sides of the press. These presses are built for capacities from 300 kN (30 t) to 2 MN (200 t), with bolster areas from 600 × 300 mm^2 to 1500 × 1100 mm^2, and speeds in the range from 1000 strokes per minute for the smaller presses to 500 strokes per minute for the large ones. An example of blanking and punching work is shown in Fig. 4.58. Both the rotor and stator laminations for an electric motor are made from the same coil stock. Special roller-cam feeds are used that maintain a reciprocating feed stroke to within 0.025 mm accuracy while feeding at a rate up to 120 m/min.

Larger parts are usually first blanked, and the blank subsequently passes through several presses for a number of successive tandem operations. A typical operation of this type, with manual handling of the blanks, is seen in Fig. 4.59. It is a part of a shop of the Truck and Coach Division of GMC. The shop consists of forty large, heavy-duty

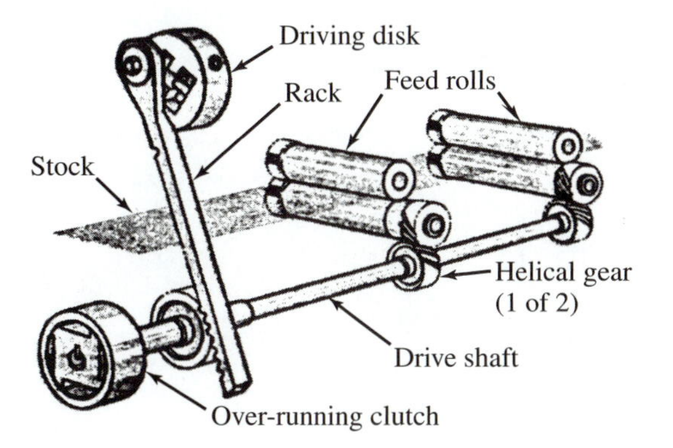

Figure 4.55
Roll feed of coil stock.
[SME2]

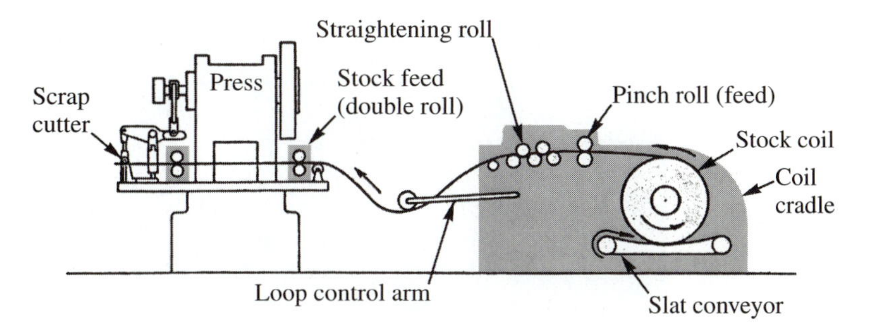

Figure 4.56
Power-operated coil cradle.
[SME2]

Figure 4.57
A straight-side mechanical press with automatic coil stock feed. [MINSTER]

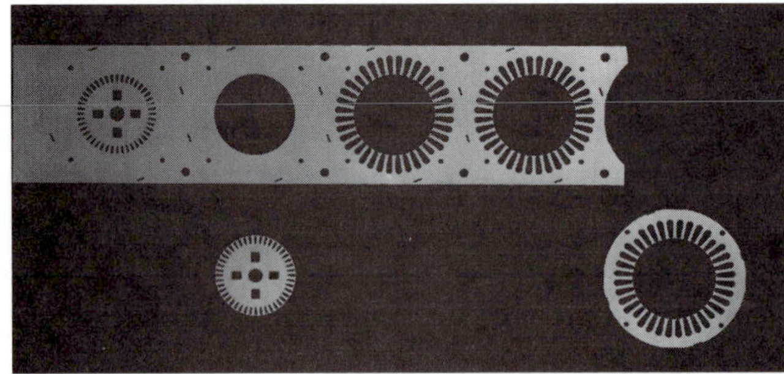

Figure 4.58
Blanking and punching laminations. [MINSTER]

Figure 4.59
A press shop in a truck manufacturing plant. [NIAGARA]

presses, mostly with 3000 kN capacity. Sixteen squaring shears are also located in this shop to supply the presses with the cut sheet metal. The material used is mild steel up to 5 mm thick. The presses have an eccentric shaft driven from both ends through double-gear transmissions. They have a stroke of 600 mm, a shutheight of 1200 mm, and a speed of 24 strokes per minute. The parts are manually loaded and unloaded, and they are transported between the presses on carts. As many as 300 die changes take place each day in this facility. Scrap produced in some of these operations is transported away through an underfloor handling system.

Automation of loading and unloading of parts in operations like those just described is obtained by using various kinds of devices. An example of an unloading device is a pneumatically driven swinging arm of the type shown in Fig. 4.60. Recently, the advantage of full automation has been obtained by using programmable robots. This is a flexible solution, because the robots can easily adapt to the frequent changes of the jobs.

Parts of intermediate size, especially those made of heavier sheet and requiring a larger number of successive operations, are made in large "transfer presses." There is lately a tendency to also form car body panels such as doors, inner and outer hood panels, and many others on transfer presses that are included in an automated system. A schematic of such a system is presented in Fig. 4.61. The process starts with the steel sheet in coils that are passed through leveling and straightening rolls and fed into the blanking press that produces the individual rectangles of sheet metal, stacks them onto powered carts that deliver them to an automatic storage and retrieval system (ASRS). From there they are retrieved and delivered on automatic guided vehicles (AGV) to the transfer presses (TP). The TP contains usually five stations, and a blank, or a pair of blanks in parallel, still attached or separate, undergoes five successive stamping operations, with a trimming die at the end. The five stations with a corresponding punch and

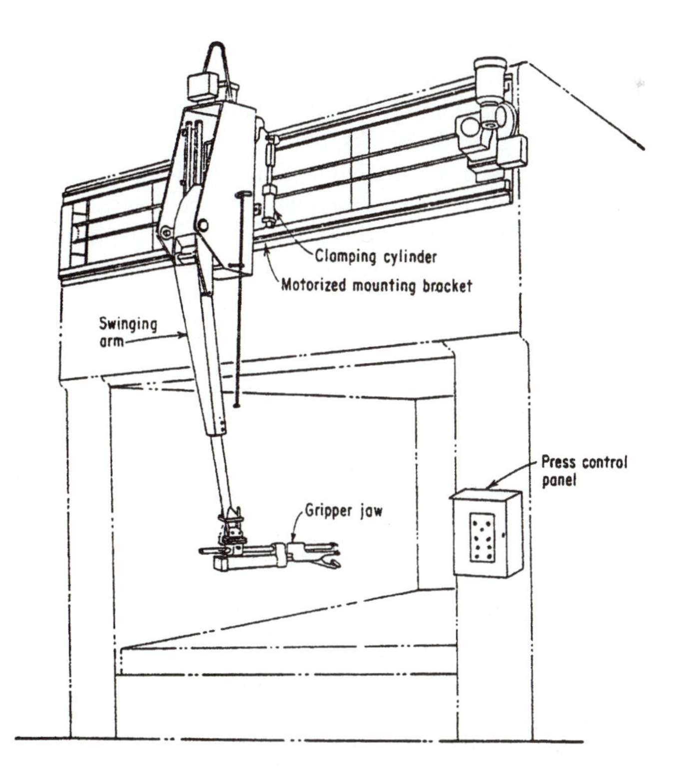

Figure 4.60
A pneumatically driven press-unloading arm. [SME2]

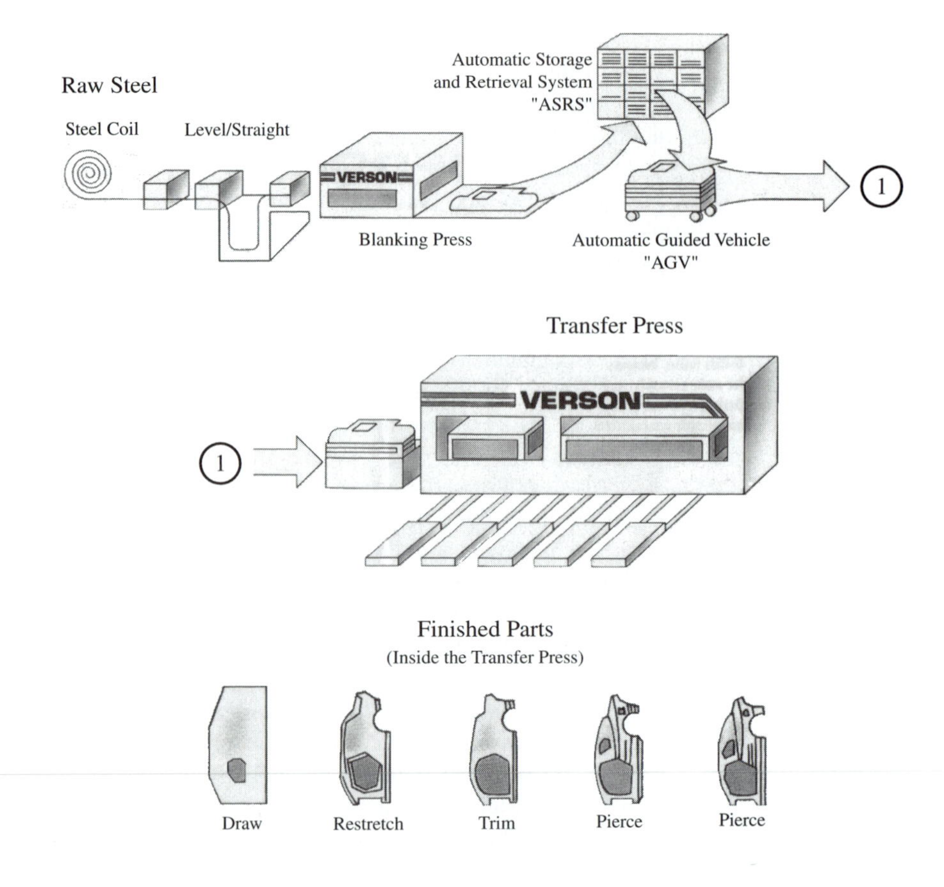

Figure 4.61
An automatic system involving a transfer press. [VERSON]

die in each are arranged along the whole width of the press, in line. The blanks are moved from station to station by a transfer system. One such system uses servomotors and computer control to drive the movement of crossbars with suction cups attached to them that grasp the blanks, lift them out of the die (250-mm stroke), move them on rails stretching over the whole length in transfer motions up to 2 m long, and deposit them into the next die. These motions are synchronized with the opening and closing of the dies. The press consists of two mechanical presses, each with 3300-ton (33-MN) capacity, one with two slides and one with three slides. The speed of the press is adjustable from 8 to 16 strokes/min. The bolsters carrying the dies are moved out and in at the front of the press. One of the users of the system confirms that a die-change takes only 5 min, permitting a very fast change of the parts being produced. The computer control of the transfer mechanism contributes substantially to the flexibility of the system.

Examples of the parts made in the system as double-attached mirror images include front and rear door inner panels. Door-hinge pillars are made in sets of four simultaneously. Parts made individually include front and rear floor pans, dash panels, roof, and inner and outer hood panels. The hood inner is said to be made at a rate of 16 parts/min. Parts that are too large for the transfer press go on the tandem line that was mentioned in reference to Fig. 4.59.

An overall view of a transfer press is shown in Fig. 4.62. It is an eccentric-gear, four-point-suspension, single-action, twin-drive link press. It is sized for 1800 tons, 9–25 strokes/min, and slide and bolster areas of 5500 mm by 2200 mm. It contains five stations, and the main drive motor power is 450 kW. The bolsters with the die sets are

Figure 4.62
A transfer press. The bolsters carrying the dies have been moved out. [VERSON]

shown in front of the press, on top of a common car, each with its own carriage. They are self propelled and driven by pneumatic motors.

Many small parts that are produced in very large quantities (at least 50,000) are made on small, fully automatic machines called *four-slide machines.* Very intricate parts can be made because of the combination of a press with a special forming device consisting of a central "kingpost" and four forming slides moving in and out in four directions perpendicular to the kingpost and to each other. The top of one such machine is shown in Fig. 4.63. Normally, this top sits on a bench-type bed. A chief engineer of a manufacturer of larger mechanical presses once called these machines "the sewing machines of the press industry." As small as these machines are, their output is very impressive. It is not possible to discuss their work here in any detail, but the basic

Figure 4.63
The top of a horizontal four-slide machine. [TORIN]

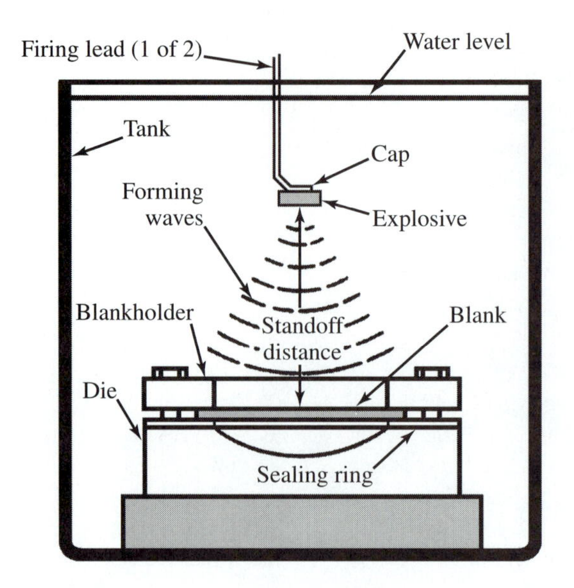

Figure 4.64
Unconfined system for explosive forming.

design of the machine may be seen from the photograph. It has four camshafts on the peripheries of all four sides. The strip arrives via a seven-roll straightener. The drive unit provides for the motion of all the camshafts as well as the stockfeed. Some of the cams visible in the picture are the left cam, the right cam, the front cam, and the cut-off cam. The rear cam is only partly visible, and the eccentric shaft of the press is hidden. The press has a capacity of 250 kN, and its three tie-rods are visible on the top. Punching, blanking, bending, or drawing dies are used in the press. The four-slide centerform head with a vertical kingpost is located to the right of the press.

On the other extreme of the forming equipment types is the technique of *explosive forming*. The main aspects of this extreme are the low production quantity, which may even consist of one workpiece only, and the size of the workpieces, which is often very large. There are two basic types of explosive forming operations, the confined and unconfined ones. We will describe an example of the latter type. The basic arrangement is illustrated in Fig. 4.64. The equipment consists of a tank filled, most commonly, with water. At its bottom is a base carrying the die. On the top of the die is the blankholder. Between the blank and the die is a ring sealing the water against the vacuum provided between the blank and the die. A charge of explosive is suspended over the center of the blank. When the charge is fired, shock waves travel through the water at the speed of sound and generate pressure over the blank. For illustration, a charge of 2 kg TNT generates a pressure of 70 N/mm^2 at a standoff distance of 1 m. The pressure draws the blank into the die. For multiple work, the dies are made of carbon steel or of alloy steel. However, for a small quantity of parts, the die may be made of concrete with an epoxy facing. Parts several meters in diameter can be made. The material of the part is often an aluminum alloy, and typical examples are covers and enclosures of aircraft engines or of rockets. However, workpieces made of stainless steel, titanium alloys, or heat-resistant alloys are also produced by explosive forming. In the cases of metals with low formability, a medium with higher density than water must be used. As an example of an ingenious way of solving the problem of forming a plate of tungsten 3.125 mm thick, a case was described in which molten aluminum was used as the medium. This provided the added advantage of the blank being heated to 670°C which made it a little more ductile. The explosive was protected in the molten aluminum by an insulated tube. The blank had a diameter of 112 mm and the die was made of alloy steel.

4.4.3 Press Brake Work

Bending operations may be carried out on presses as are the cutting and drawing operations, examples of which have already been described. However, if the parts are long, they are most often produced on press brakes in the case of small quantities, or by roll forming when the number is large.

Shear brakes and press brakes are machines similar in design, but the tooling part is different. Our discussion is limited to *press brakes* and their use for a great variety of bending operations on long parts. A typical press brake is shown in Fig. 4.65. This model is rated at 550 kN. Its manufacturer offers models with ratings from 120 to 1500 kN and with ram widths between 914 mm and 2540 mm. They are driven by hydraulic motors. An eccentric shaft and two connecting rods move the rather thin, wide, and high ram. Between the connecting rods and the ram are the ram-adjusting screws, which are used, along with digital indicators, to set the bottom position of the ram. Alternatively, press brakes may be purely mechanical, with a drive from an electromotor onto a clutch-brake-flywheel assembly, or they may be purely hydraulic, with hydraulic cylinders at the two ends of the ram. The sides of the frame are connected at the top by the crosshead containing the eccentric shaft and by the rather thin, wide, and high bed. Later, in Section 4.5 we will discuss the front and back stops, not shown in the present illustration, and the vertical stops. Now, let us illustrate the technology of press brake work by describing a selection of dies and corresponding bends, shown in Fig. 4.66.

The most basic are the V dies, which can be used in two different ways: for "air bending" as in (*a*), or for "bottoming" as in (*b*). In air bending the punch stops a certain distance above the bottom. The die set can be used for bending at any angle larger than 85°. The bottoming die bends the sheet metal at the angle of the die, which may be 90° or any other angle as shown in (*c*). The force is several times larger at the end of the bottoming stroke, and it is used for "coining," that is, for producing compressive stress at the outside of the bent sheet to minimize springback and prevent cracking. The "gooseneck" punches shown in (*d*) are used for return flanging in combination with standard female V dies. Then there are the hemming dies as shown in (*e*). In the sequence of forming a flat hem, the sheet is first placed in the die at an angle against the stop at the left outside of the die, position A, and in the first stroke a "V" bend is made. Subsequently, the metal is inserted in position B against the internal stop, and in the second stroke the hem is closed. Another

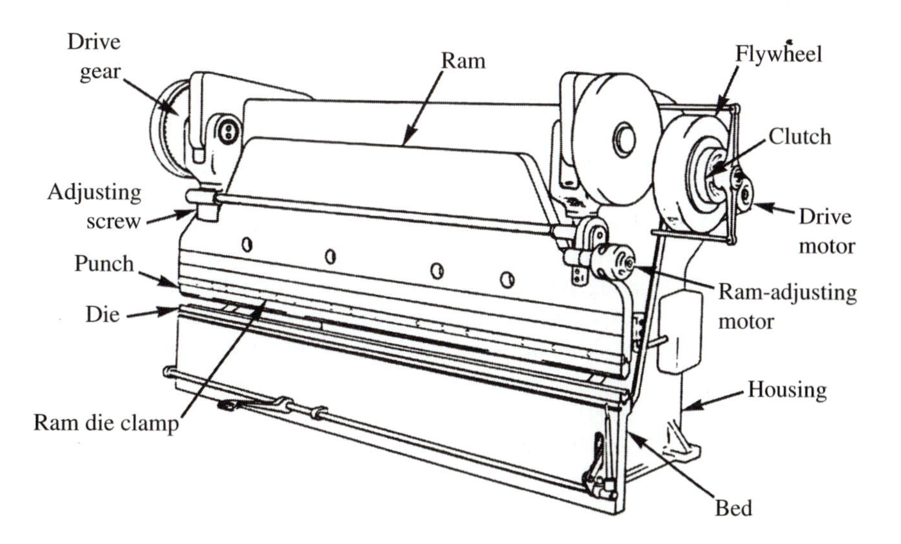

Figure 4.65
A typical mechanical press brake. [SME2]

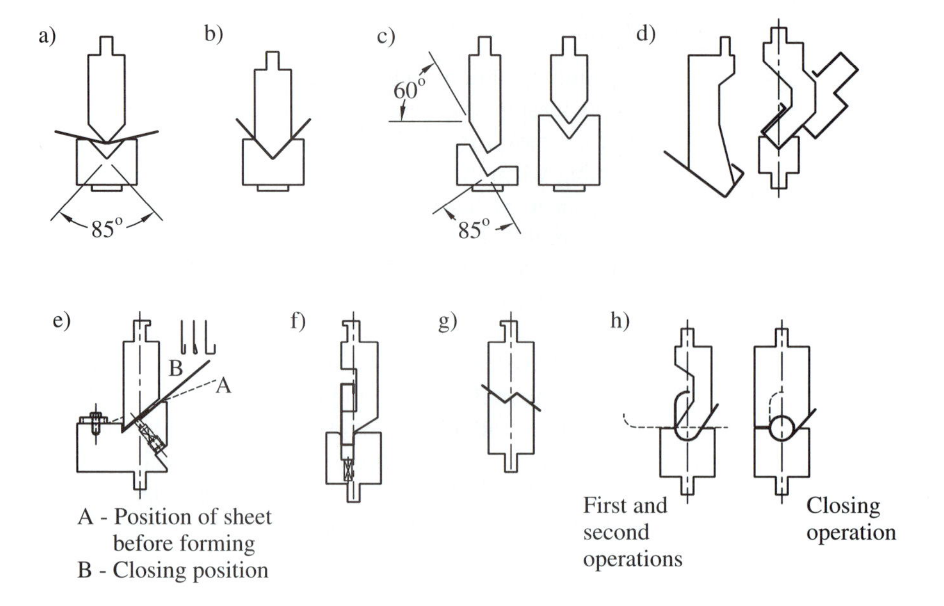

Figure 4.66
A selection of dies for press brake work: a, b, c) "air" and bottom bending; d) "gooseneck" die; e) hemming; f) channel punches.

type of a die set is the channel-forming die as shown in (*f*). Then there are the offset dies like that shown in (*g*) and curling dies like those shown in (*h*). In the latter two, die sets are used to perform three operations; the first and second are done in the set at the left, and the third one is done as shown in the die set at the right.

The examples given show that a great variety of operations can be performed on press brakes using standard dies. The handling of the sheet metal is done manually. Whenever a batch of workpieces has to go through several operations requiring different dies, they are passed through the press brake several times with a change of the die sets in between the individual passes. Obviously, the work is labor-intensive, and the production rate is not very high. Correspondingly, the batch sizes of press brake work are not large. Recently, the productivity of the work has improved by the introduction of numerical control, which eliminates the need for frequent die changes.

4.4.4 Cold Roll Forming

Long and narrow parts of sheet metal with intricate cross sections are preferably made by roll forming when the quantity of workpieces is large. The tooling, which consists of a number of roll sets, is almost always special-purpose and can only be used to produce the particular workpiece. This represents considerable expense, which can only be justified by the large number of identical workpieces required. On the other hand, the roll-forming machines are not very expensive. On an average, a roll-forming line will produce about 8,000 to 10,000 m of section in an 8-hour shift, based on an operating speed of 30 m/min and taking into account the delays for loading the strip coils, threading the mill, clearing away the finished product, and so on.

In comparison with press brake and press forming, it is estimated that roll forming is economically preferable if the production volume is at least in the order of 100,000 m annually.

Roll forming is used to make a variety of products, such as bicycle rims and fenders, door frames, auto trim, fence posts, nonwelded and welded tubing. A machine used to produce precision drawer slides for the furniture industry is shown in Fig. 4.67. It

Figure 4.67
A roll-forming machine used to produce precision drawer slides for the furniture industry. [YODER]

contains twenty-four stands and it is of the universal type in which one drive side can be used for rolling units with different span between axes and with a range of spindle lengths. The length of the drive shaft can be easily changed. These shafts have universal joints at their ends permitting the difference.

The roll-forming machine consists essentially of a line of roll-forming units, or stands, attached to a welded steel base. The individual stands are made up of the housing and two spindles with rolls and drives. The line may be working from precut lengths of sheet metal but mostly the initial material is coiled strip mounted on coil reel. Subsequently, it passes through the rolling line at the end of which is a "fly" shear and a run-out table. A fly shear is a crankshaft-driven press that has on its table a cut-off die that moves with the strip during the shearing action and then returns back to its starting position. The cut-off action is triggered either mechanically or electronically, when the predetermined length of the rolled section has approached the shear.

The profile is produced in successive rolling passes designed so as to provide smooth forming actions, which may combine bending and drawing. The number of passes depends on the material, its thickness, depth of section, and the complexity of its shape. Roll forming may be performed on prepunched stock or it may be combined and synchronized with perforating and dimpling. Roll forming is suitable for hot- and cold-rolled carbon steel, stainless and other alloy steels, copper, brass, zinc, and aluminum. The material may be prepainted or precoated, and it is also possible to combine different materials directly in the roll-forming process. For example, the designer may want to insert and lock a strip of wood, fabric, cord, felt, rubber, metal wire, or tubing into a roll-formed channel.

These products find applications in structural beams for high-rise construction, in roofing and siding, windows and doors, corrugated sheets, stainless steel skin for railroad cars, structural beams for automobile frames, and sections for metal furniture and appliance cabinets.

A selection of examples of roll-formed parts is shown in Fig. 4.68. In the upper right corner is the "double barrel shotgun" section made of a martensitic stainless steel

Figure 4.68
A variety of roll-formed products. [YODER]

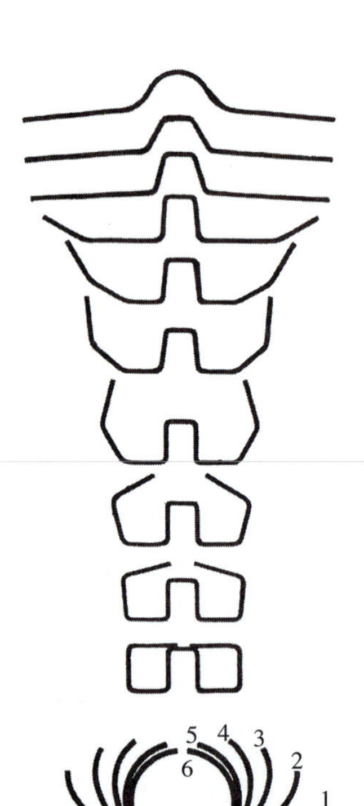

Figure 4.69
Examples of roll forming passes for a) the double channel, b) round tube. [YODER]

and used as trim on the bumper of an automobile. In the middle of the left-hand side is a "stringer" for an aircraft fuselage made of an aluminum alloy. Parts with various punch-outs and perforations are also included.

Two examples of roll-forming passes are given in Fig. 4.69. At the top is the double-channel profile shown as one of the parts in Fig. 4.68, and at the bottom is the process used in making round tubes.

The manufacture of tubes and pipes represents a significant part of the roll-forming industry. Tubes from about 12.5 mm diameter and larger can be made. Smaller diameters are mostly made by drawing. Tubes of various shapes are made; they may be round or rectangular.

Various welding methods are used as indicated in Fig. 4.70. These methods are discussed in Chapter 11. However, to mention them briefly, there is resistance welding with rotating copper electrodes, with AC low-frequency (360 Hz) or DC current, with low-voltage but high currents (20,000–70,000 A), radio frequency (450,000 Hz) induction or contact heating, and fusion welding using the various arc welding processes of gas metal arc, gas tungsten arc, plasma arc, and submerged arc, all of them successfully used and chosen according to the application.

Welding of the tube is followed by several sets of driven horizontal rolls and idle vertical rolls for final sizing of the tube, for weld-bead scarfing and potential reshaping into a rectangular shape.

4.4.5 Formability of Sheet Metals

Formability of sheet metal is expressed by its ability to deform without defects and without cracks. Among the defects are "stretcher marks." They often appear on parts made of low-carbon steel, which is the most common material in sheet-metal forming, on surfaces of the formed part that have undergone only a moderate strain. In order to avoid stretcher marks, it is necessary to use steel that has obtained a temper pass of about 1%

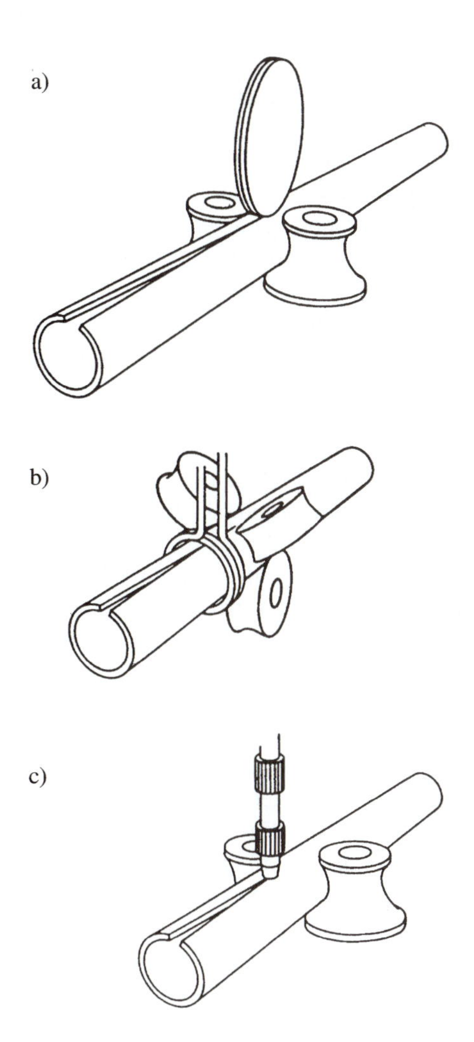

Figure 4.70
Welding processes used in welding tubes: a) resistance w., b) inductance w., c) arc welding. [YODER]

cold reduction after the final anneal. As mentioned in Section 2.5.2, steel may again become unsuitable due to aging. In this respect aluminum or titanium killed steel is preferable because these elements bind with nitrogen and prevent the aging effects.

Correspondingly, essentially two kinds of mild (low-carbon) steel are in use: rimmed and special killed steel. The latter, once given a temper pass, can be used for stamping, even after a longer period of storage. Rimmed steel must either be used soon after the temper pass, or if stretcher strains occurred due to aging, they can be eliminated by roller leveling preceding the stamping operation. Annealed steel, without a temper pass, cannot be roller-leveled because it would develop severe kinks that would show on the final product. Annealed steel, however, can be used for such stampings that will not be exposed and where stretcher marks can be tolerated.

Mechanical properties of the sheet metal affect formability in such a way that for larger strains, materials with high ductility and low yield stress are preferable. In these respects, killed steel is superior to rimmed steel. However, inferior surface properties and more surface defects can be expected from killed steel than from rimmed steel. Consequently, the former is only used for more severe forming operations.

The effect of grain size is such that a sheet with large grain size may exhibit a coarse, granulated surface after deep drawing; this is called "orange peel," and it is not acceptable on many stampings. On the other hand, fine-grained steel has high yield strength and hardness, and lower ductility, which is also not good for severe forming.

Several types of standard tests are used to assess formability. First, there is the tensile test, which is carried out on a standard specimen and results in the stress-strain diagram with the characteristic parameters: upper yield stress *UYS*, lower yield stress *LYS*, ultimate tensile strength *UTS*, uniform elongation e_u, and total elongation e_t. Furthermore, the value n of the strain-hardening exponent and the value r of the coefficient of normal anisotropy can be derived from the test data, and if the test is carried out at different speeds, the value m of the strain-rate sensitivity exponent can also be determined. High values of e_u, of the ratio *UTS/YS*, and of m are important for stretch forming, and a high value of r is important for deep drawing.

The most widely used test for ranking the formability of metals in a press shop is the hardness test, which is simple and fast. High hardness correlates, generally, with high *YS* and *UTS*, and low e_u. In general, metals with high hardness are more difficult to form, although there are a number of significant exceptions to this rule (formable high-strength steels, aluminum alloys). For most metals thicker than 1 mm, the Rockwell B scale is recommended; for those thinner than that, the Rockwell 30T is preferred.

The bulge test is carried out as shown in Fig. 4.71a. A circular sheet is clamped at the edge and deformed by hydraulic pressure into a dome. For an isotropic material, essentially uniform biaxial stress and strain exist over an appreciable region at the center of the diaphragm. Failure usually occurs in this region. The strains and stresses are evaluated by using extension-meters and measuring the radii of curvature and the pressure. Data similar to that obtained from the tensile test can be derived. There is no necking, because of the biaxial stretching, and therefore much larger uniform strains are obtained than in the tensile test. Correspondingly, data can be obtained (like the strain-hardening exponent n) under conditions that are much closer to the practical forming operations. This test with its complex instrumentation is carried out in a laboratory.

A much simpler biaxial stretch test is rather commonly used in press shops. It is the standard Olson test (Fig. 4.71b). The sheet metal is rigidly clamped in a blankholder. A spherical punch 22.2 mm in diameter is moved up to stretch the material. The height h to which the punch moves after initial contact, before the sheet metal fractures, is a measure of formability. In order to decrease the effect of friction in this test, a polyethylene sheet with oil is used between the punch and the sheet metal.

There are also other variants of stretch and bulge tests. It is possible to arrange tests so as to construct a formability limit diagram. In these tests small circles are first imprinted on the test sheet metal and their distortion around the fractured point indi-

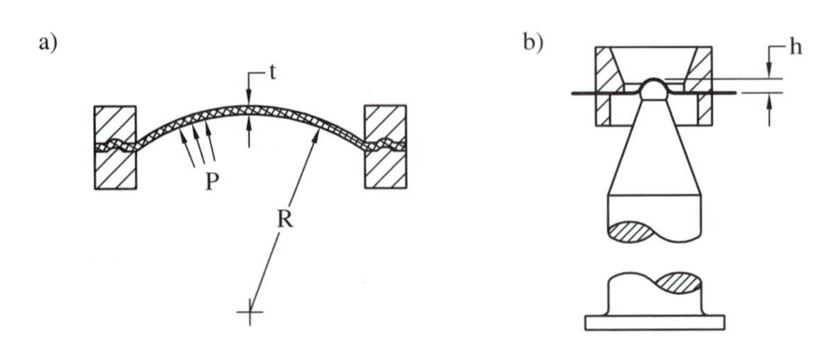

Figure 4.71
a) The bulge test.
b) The Olson test.

cates the magnitudes of the strains ε_1 and ε_2 in the two axes in the plane of the sheet. Typically, this diagram looks as shown in Fig. 4.72. The ε_2-axis represents plane strain such that while $\varepsilon_1 = 0$, the thickness of the sheet decreases, because—as discussed in Chapter 5—material maintains constant volume, and the sum of strains in the three axes has to be $\varepsilon_1 + \varepsilon_2 + \varepsilon_3 = 0$. It is seen that the strain ε_2 at fracture is smallest in this situation. It increases in the (+,+) quadrant, with both ε_1 and ε_2 tensile (the addition of stretch ε_1 prevents necking), and it increases still more in the (−,+) quadrant where ε_1 is compressive. Examples of corresponding situations are given: $\varepsilon_1 = \varepsilon_2 > 0$ is biaxial stretching; $\varepsilon_1 < 0 < \varepsilon_2$ corresponds to the strains in the flange of the material being drawn; and the plane strain to the strains in the walls of the cup being drawn where the punch prevents the circumference of the cup from shrinking ($\varepsilon_1 = 0$).

The second most common material in metal forming is stainless steel. In most cases, stainless steels work-harden quite strongly. While this property, on one hand, is favorable for resisting necking, it often leads to the exhaustion of ductility. This is best recognized by how closely the yield strength approaches the ultimate strength. In this respect the austenitic grades are superior because they retain a sufficient difference between *YS* and *UTS* even after larger amounts of cold work. Grade 301 has the best formability, and it can be used for a stretch of more than 35%. A comparison of the performance of the various grades of stainless steels is presented in Fig. 4.73, which gives the stress-strain curves and modes of failure in cup drawing.

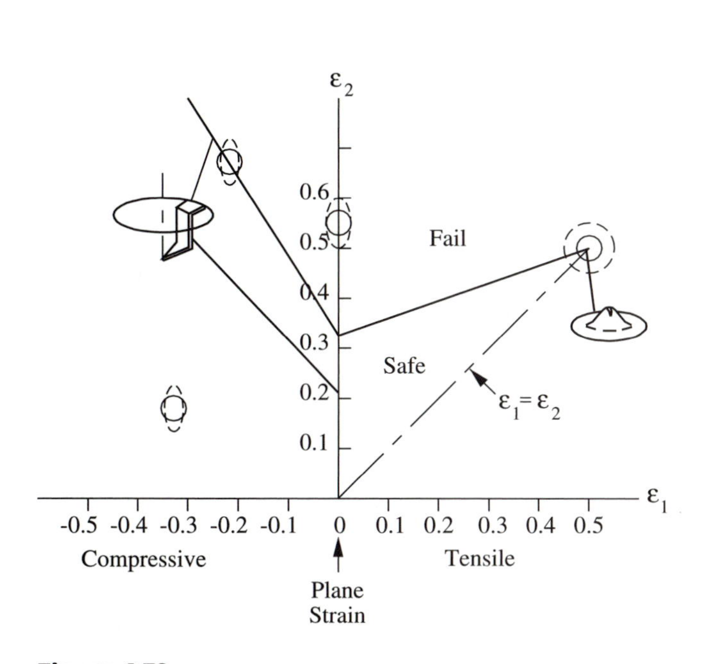

Figure 4.72
The formability limit diagram.

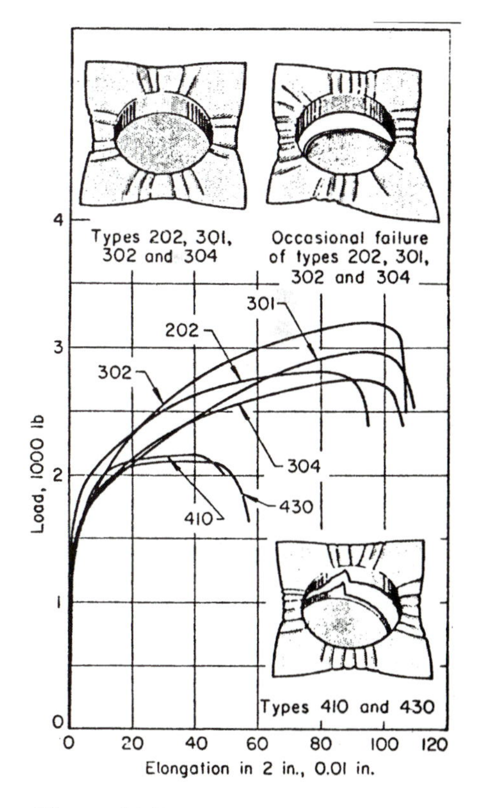

Figure 4.73
Formability of stainless steels. [ASM14]

4.5 NUMERICAL CONTROL (NC) IN METAL FORMING

Most metal-forming operations are based on very simple motions and complex tooling: a simple stroke of a press or of a hammer is combined with a die, or a sequence of dies. In closed-die forging and metal stamping the dies have complex, three-dimensional shapes; in punching and blanking they have complex, two-dimensional contours. In these instances numerical control (NC) is involved indirectly in the manufacture of the dies.

The forming operations just mentioned are such that the whole volume, or the whole periphery, are generated at once; they are global. These operations are very efficient and economical in the production of large quantities of the individual parts. Some metal-forming operations work in an incremental way. Thus, a shape involving complex bending may be generated in a number of successive steps, and a complex pattern of holes may be generated one by one. This would be the preference if the lot size is small and it is not economical to produce complex tools. Also in the bulk-forming operations (i.e., in forging), there are processes that have not been described in the preceding text, apart from open-die forging, in which the forged shape is generated successively in parts. As examples, the operations of rotary swaging and radial forging may be mentioned. The incremental forming machines involve complex motions and thus lend themselves to automation utilizing NC. Several examples are described.

4.5.1 Numerically Controlled (NC) Bending on a Press Brake

We have seen that multiple bends (e.g., like those in Fig. 4.66) are normally, on a manually operated press brake, produced in successive operations in such a way that each partial operation is produced on the whole batch of components in one given setting of the machine. Subsequently, the machine is reset and the next operation is produced on the whole batch, and so on. In contrast, on an NC press brake the settings for each operation are done automatically and very quickly. It is therefore much more efficient to carry out the whole sequence of operations on one part, in one loading into the machine, and then the next part, and so on. The settings are those of (1) the backstop and (2) the depth of penetration of the punch, which now always works in the "air-bending" way (see Fig. 4.74). The part is located against the backstop, which moves horizontally between the individual operations and determines by its position the location of the bend. The backstop is moved by means of leadscrews driven by servomotors, and its positioning is accomplished in a positional closed loop with encoders attached to the leadscrews. This servo does not differ much from those described in Chapter 10. The depth of penetration of the punch, which determines the angle of the bend, is set on mechanical press brakes by rotating the depth-setting screws on the ram; on a hydraulic press brake, it is set by using stops moved vertically by means of servo-driven screws. The NC program may include software that compensates automatically for springback. The operation shown in Fig. 4.74 is claimed to be carried out in 50 sec, in one handling of the part. The NC system of the machine has a memory facility for 500 bends, which would allow, for example, production of 100 different five-bend components.

Another type of bending machine works with a blankholder BH, and a counter blade CB, as well as with a bladeholder with an upper blade UB and a lower blade LB (see Fig. 4.75). The diagram shows three operations: the first two are down bends, and third is a reverse bend. Between the operations the blank holder releases the material, which is then shifted and clamped again.

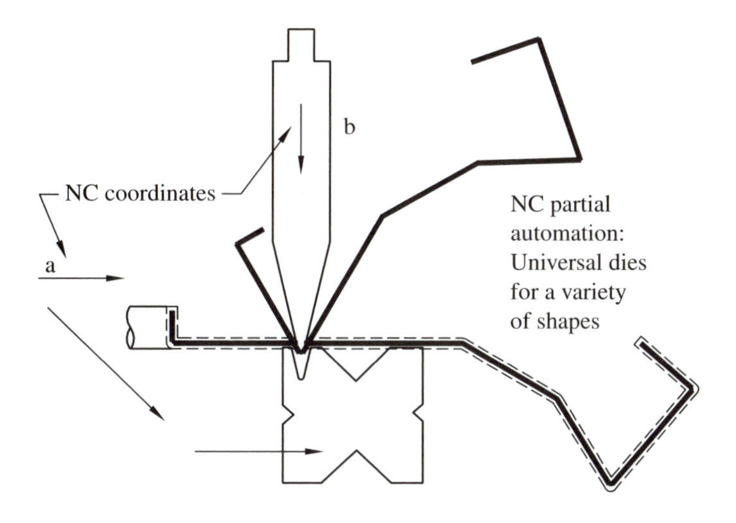

Figure 4.74
Bending on an NC press brake. Partial automation by using universal dies for a variety of shapes.

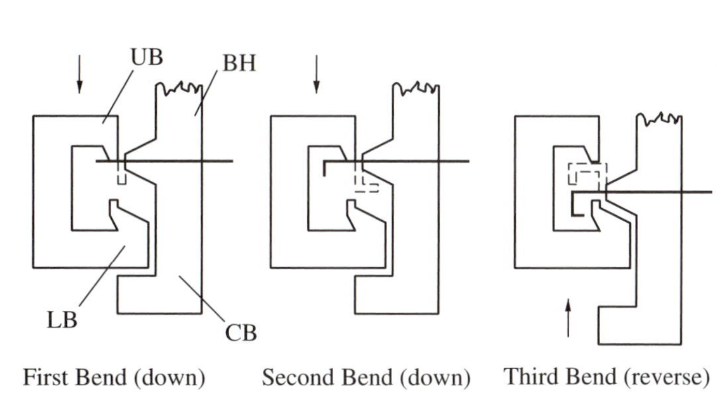

Figure 4.75
Bending on a panel bender.

4.5.2 NC Turret Punch Presses

Panels like the one in Fig. 4.76, when needed in small to medium lots, are most efficiently produced on NC turret punch presses. The illustrated part is an electrical panel. If needed in large lots they might be produced in a mechanical press with a complex die. In the NC press the holes are punched one after another using, in this case, eight different individual punching dies. Altogether, 198 hits are needed, and the time per piece is 3 min, 3 sec. The overall view of the press is shown in Fig. 4.77. At the sides of the top of the welded base the guide rails may be seen. A wide carriage moves on these rails in the coordinate Y in the direction to and from the turret head. Inside of this carriage is the leadscrew driving a head in the X coordinate, and this head carries the clamping fingers holding the sheet metal. In this way the material is moved in the (X, Y) coordinates between the individual hits and it is positioned with the hole to be made under the punch at the front of the large circular turret head. This head is just above the table; there is another one under the table that carries the punching dies. The diagram shows the turret drive motor and the chain transmissions to the upper and lower heads. At the top of the frame is the eccentric shaft, which is driven from the electric motor via belts onto the flywheel and from there via a clutch-and-brake system. The ram driven by the eccentric hits the tool that is positioned under it. The die set is illustrated in Fig. 4.78 and is designed for an easy and quick change. The punch

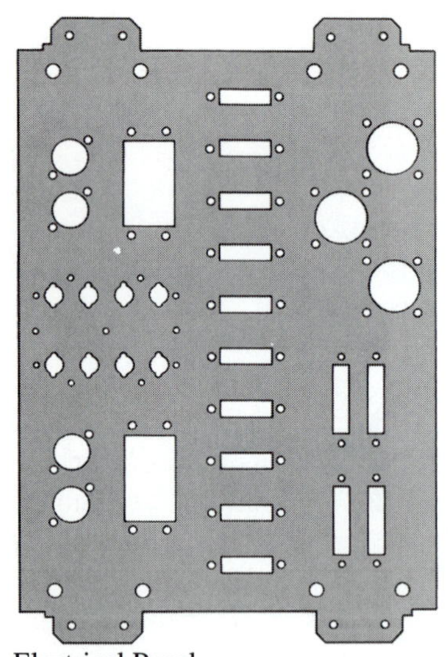

Figure 4.76
Typical part made on an NC turret punch press. [W&S]

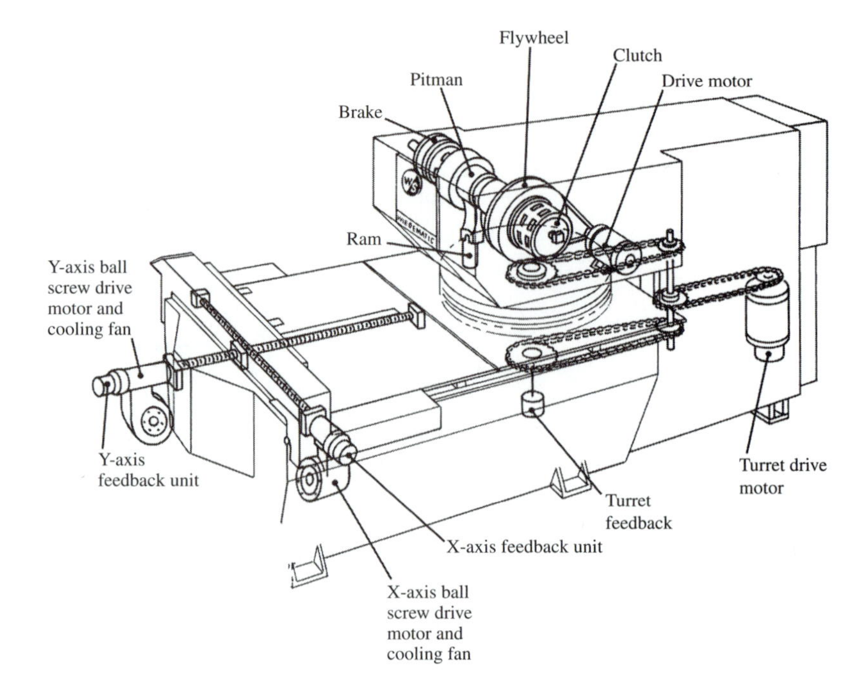

Figure 4.77
The kinematic diagram of the NC turret punch press. [W&S]

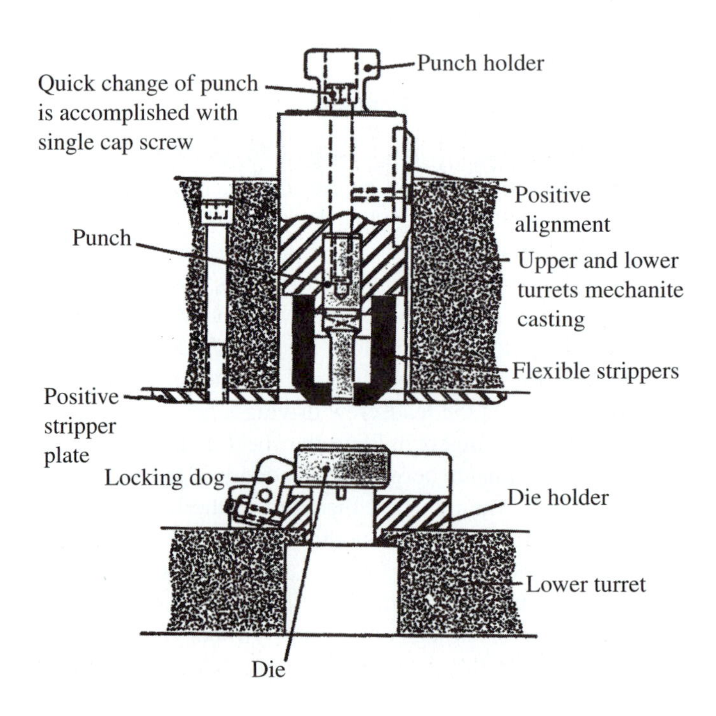

Figure 4.78
Quick-change dies for the NC turret punch press. [W&S]

is clamped in the punch holder outside of the machine. When needed, the holder with the punch and flexible stripper is inserted in one of the 24 upper turret head bores. Similarly the die is clamped in a standard die holder, which is easily inserted in one of the bores of the lower turret head. In this way complete tooling for a particular part can be quickly inserted. Once it calls the corresponding NC part program, the press is ready to produce a batch of the particular parts.

NC turret punch presses have become quite popular, and they are commonly used in many sheet-metal working shops with medium-size lot production. In the basic execution just described, the NC system works in an intermittent, point-to-point fashion. In some applications the NC provides continuous path control and the machine is equipped at one side with either a laser-beam cutting head or a plasma-arc cutting head. Panels may thus be produced with complex contours and/or hole shapes not limited to circular and square and other simple kinds produced by punching but also with much more complex shapes produced by laser-beam or plasma-arc cutting. An example of such a part is the rear cab panel for a tractor shown in Fig. 4.79.

Another type of a machine similar to the turret press is a shear. It has the same kind of X, Y table, but the punching head is replaced by one which carries a right-angle blade and moves it against a right-angle corner of the lower blade. The machine also contains hydraulic hold-down clamps. A computer program analyzes the total shearing requirements for a short period of time and selects the shear pattern using stock sizes resulting in the least amount of scrap (see Fig. 4.80). The stock sheets are automatically positioned and cut into the desired blanks.

Figure 4.79
A part made by plasma-arc contouring. [STRIPPIT]

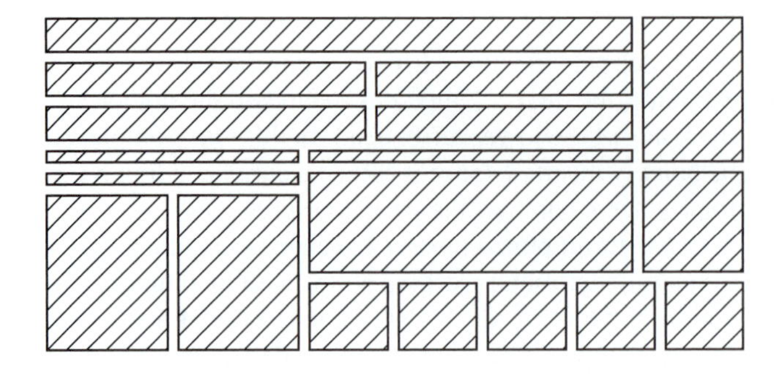

Figure 4.80
A layout of work for an NC shear.

REFERENCES

1 Hasek, V. *Zaklady tvareni kovu* (Fundamentals of Metal Forming; in Czech). Praha: SNTL, 1969.

2 Forging Industries Association. *Forging Industry Handbook.* Cleveland, OH: Forging Industries Association, 1970.

3 ASM International. *Metals Handbook*, Vol. 14, *Forming and Forging.* 9th ed. Materials Park, OH: ASM International, 1988.

4 Forging Industries Association. *Open-Die Forging Manual.* Cleveland, OH: Forging Industries Association, 1982.

5 Wick, C.; J. T. Benedict; and R. F. Veilleux, eds. *Tool and Manufacturing Engineers Handbook,* Vol. 2, *Forming.* Dearborn, MI: SME, 1984

6 Steel Founders Society of America. *Steel Castings Handbook.* 6th ed. Des Plaines, IL: Steel Founders Society of America, 1995.

Illustration Sources

[ASM14] Reprinted with permission from *Metals Handbook,* Vol. 14, *Forming and Forging,* 9th ed., 1988 ASM International, Materials Park, OH 44073-0002.

[FIH] Reprinted with permission from *Forging Industry Handbook,* Forging Industries Association, 55 Public Square, Suite 1121, Cleveland, OH 44113, 1970.

[HASEK] Reprinted from Hasek, V., *Zaklady Tvareni Kovu* (Fundamentals of Metal Forming, in Czech), SNTL, Praha, 1969, with permission of V. Hasek.

[ODFM] Reprinted with permission from *Open Die Forging Manual,* Forging Industries Association, Cleveland, OH, 1982.

[SCH] Reprinted with permission from *Steel Castings Handbook,* 6th ed., 1995, Steel Founders Society of America, 455 State Street, Des Plaines, IL 60016.

[SME2] Reprinted with permission from *Tool and Manufacturing Engineers Handbook,* Vol. 2. *Forming,* edited by C. Wick, J. T. Benedict, and R. F. Veilleux, SME, Dearborn, MI, 1984.

Industrial Contributors

[MESTA] Mesta Machine Co., PO Box 1466, Pittsburgh, PA 15230

[MINSTER] Minster Machine Co., Minster, OH 45865

[NATIONAL] National Machinery Co., Greenfield Street, Tiffin, OH 44883

[NIAGARA] Niagara Machine & Tool Works, PO Box 475, Buffalo, NY 14240

[STRIPPIT] Strippit Inc., 12975 Clarence Center Road, Akron, NY 14001-9902

[TORIN] Torin Corp., Torrington, CT

[VERSON] Verson Allsteel Press Co., 1355 East Ninety-third Street, Chicago, IL 60619

[W&S] Warner & Swasey, Wiedemann Div., 211 S. Gulph Road, King of Prussia, PA 19406

[YODER] Yoder Manufacturing, 26800 Richmond Road, Cleveland, OH 44146

QUESTIONS

Q4.1 Which range of temperatures is used in hot forming and which for cold forming?

Q4.2 Which improvements in material properties may be obtained in hot forming? Which deteriorations can occur, and under what circumstances?

Q4.3 Why should the hot-forming sequence be finished low over the recrystallization temperature?

Q4.4 Which changes of material properties result from cold forming? What is the beneficial effect of cold work on grain size in annealing and in normalizing treatment?

Q4.5 Name the three basic types of machines used in metal forming. Specify the differences in what determines the end of downward travel in each of them.

Q4.6 Name the methods used to lift the hammer mass in a drop hammer. What additional energy sources may be used to add to the potential energy of the hammer?

Q4.7 Use sketches to explain the differences between a crankshaft, an eccentric shaft, a knuckle drive, and a toggle mechanism.

Q4.8 Why does the pressurized oil supply in a hydraulic press work with essentially two pumps? How do these two pumps differ in flow volume and in pressure delivered? Which alternative oil supply may be used instead of one of the pumps?

Q4.9 Give typical parameters, force F (ton), strokes per minute n, and power P (kW) for (*a*) a high-speed mechanical press; (*b*) a forging eccentric press; (*c*) a large hydraulic forging press.

Q4.10 Explain the sequence of squeezing strokes in an ODF cogging operation that reduces the diameter of a shaft.

Q4.11 Explain the roll-forging process. Sketch the shape of a simple set of rolls used to reduce the diameter of a rod in a tapered form.

Q4.12 Name examples of parts and their materials that are made by closed-die (impression-die) forging in the aerospace and the automotive industries. Name basic rules for forging and die design intended to avoid difficulties.

Q4.13 Which is the basic, most common product made on an upsetting (heading) machine? What are the two distinct motions in such a machine and their roles? Which kinematic principles (go back to Question 4.7) are used in each of them?

Q4.14 Give the main parameters, force F (tons), strokes per minute n, and power P (kW) for a small, medium, and large upsetting machine.

Q4.15 Upsetting operations are commonly done in four steps. How is the part transferred between them? Name a mechanism mentioned in the text for transfer of short parts.

Q4.16 Sketch the following extrusion operations: forward and backward extrusion of rods and of tubes. Explain the difference between mechanical and hydrostatic extrusion.

Q4.17 Discuss briefly the differences in forgeability of aluminum, carbon steels, stainless steels, and titanium alloys, and describe the measures to use in each case for successful forging.

Q4.18 Name and sketch four basic sheet-metal forming operations. Explain the terms of compound and combination dies.

Q4.19 Explain the two actions in a double-action hydraulic press. Make a sketch of a simple case of using cushions in a single- and a double-action press.

Q4.20 Name three different devices used to automate press work, that is, to automate the transfer between successive forming operations performed on a workpiece.

Q4.21 Use sketches to indicate how a rectangular channel partially open on one of the longer sides would be formed on a press brake manually and also on a numerically controlled machine.

Q4.22 Which two different processes are combined on an NC turret press to cut standard as well as complex contoured openings?

5 Metal-Forming Mechanics

5.1 ELEMENTARY CONCEPTS

The purpose of metal forming is to transform the shape of the workpiece so as to obtain the desired dimensions and tolerances on one hand and the final properties of the workpiece material and surface quality on the other hand. With respect to this dual aim of form and material properties the latter is mostly understood as substantial improvement of material strength and/or toughness. In addition, we are requested to fulfill these aims in the most economical and effective way considering the most suitable tooling and machinery and the appropriate means of mechanization and automation.

At the start, there is usually the cast ingot, with a rather coarse material structure, large grains, voids, cavities, and inclusions, and considerable non-uniformity through its bulk. It is processed in the primary metalworking processes of hot rolling, hot forging, and hot extrusion. These are the hot, bulk-forming processes. The material assumes an intermediate and sometimes almost final form in the workpiece: its structure is refined, pores and cavities are largely eliminated, and impurities squeezed out into fibers. All this leads to increased toughness.

Subsequently, the blanks produced in the primary processes are further transformed in the secondary metalworking processes mostly into the final machine parts. These processes include cold forming, both bulk and sheet, such as cold rolling, cold extrusion, cold forging, cold drawing, sheet and plate bending, shearing, stretching, and deep drawing. In the bulk processes the surface-to-volume ratio and the shape both change, while in sheet forming the former aspect does not change. Expressed in a simplified way, one could say that in bulk forming the thickness of the part changes considerably, and in sheet and plate forming it does not. In cold forming the strength and hardness of the material can be substantially improved, although at the expense of ductility. By proper control of the deformation mode, of strains, strain rates, and temperatures throughout both primary and secondary metalworking, the final material properties may be varied over a wide range, and an optimum combination of strength and toughness for a given application of the part can be achieved.

Metal forming is, fundamentally, controlled plastic deformation of metals. It is possible thanks to the ability of metals to yield and deform plastically, often over a very large range of strains, before they fail by necking, cracking, and fracturing. The part must yield but not fail during the extent of the metal-forming operation. The conditions for successfully controlled yielding without failure are analyzed in metal-forming mechanics, with the following purposes in mind:

1. Determine the pressures, frictions, and temperatures as they affect the strength and life of the forming tools (dies).
2. Determine total forces and energies to specify; design and select the forming machines to use.
3. Assess whether plastic deformation without failure is feasible in the given operation.

In this text the analysis is kept on a rather elementary level, using a number of idealizing and simplifying assumptions. However, an attempt is made to explain and analyze all the significant aspects of metal-forming operations.

5.1.1 The Stress-Strain Diagram

The stress-strain diagram obtained in the tension test represents the simplest stress situation, uniaxial tension (see Fig. 5.1). In the graph (*a*) the engineering stress and strain are plotted showing the basic features of the elastic strain 0–1, the yield strength (*YS*) point 1, the ultimate strength (*UTS*) point 3, the necking portion 3–4 and the fracture point 4. The elongation e_{max} is a measure of ductility.

Recall from Section 2.2 that the engineering stress *S* and strain *e* are obtained by relating the load *F* and the deformed length *l* to the original area A_0 and to the original length l_0 of the specimen:

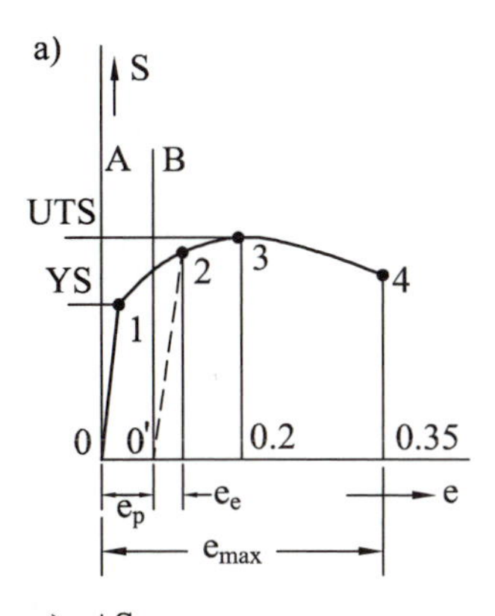

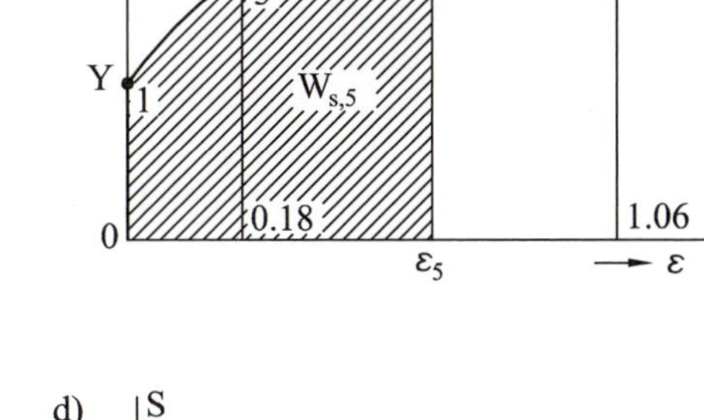

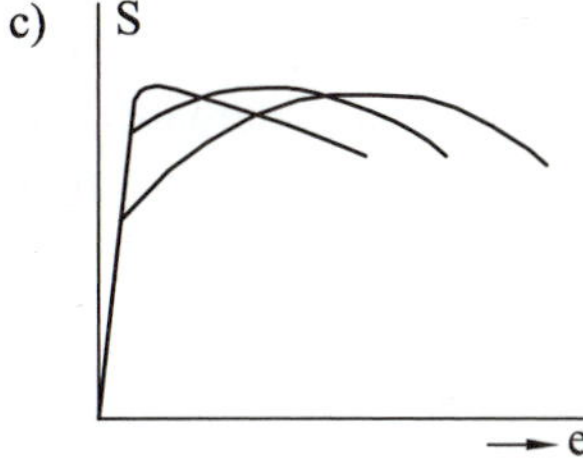

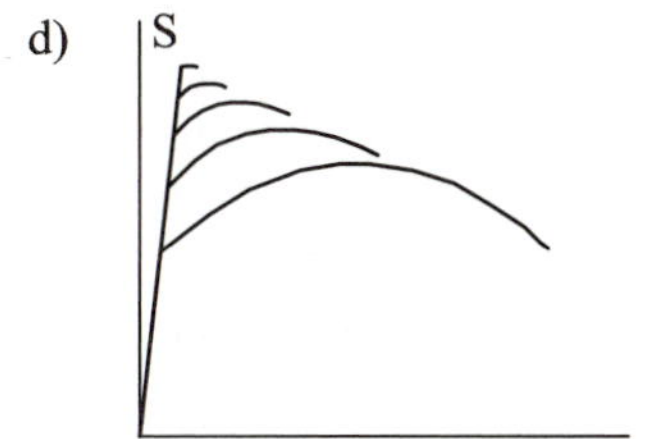

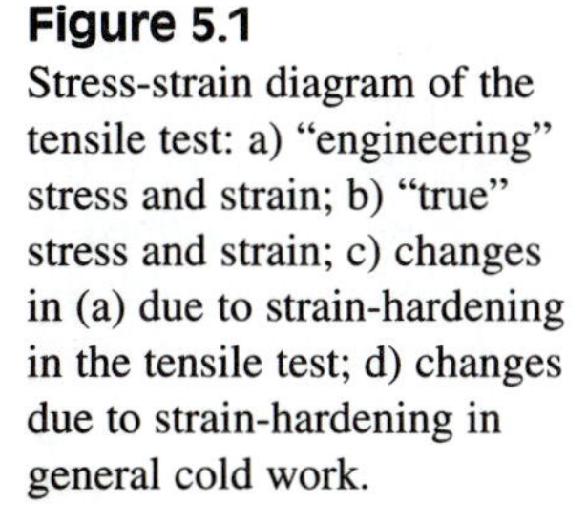
Figure 5.1
Stress-strain diagram of the tensile test: a) "engineering" stress and strain; b) "true" stress and strain; c) changes in (a) due to strain-hardening in the tensile test; d) changes due to strain-hardening in general cold work.

$$S = \frac{F}{A_0} \tag{5.1}$$

$$e = \frac{l - l_0}{l_0} \tag{5.2}$$

If loading is interrupted at point 2 and relieved down to point 0′, the elastic part of the strain e_e recovers, and its plastic part e_p remains (the scale of e_e in the figure is exaggerated as compared to e_p). The material "remembers" its elastic strain and is capable of retrieving it, but it does not remember the plastic strain. Let us look at the material obtained in the cycle 0–1–2–0′ as a product of operation A and start loading again, operation B. It is as if the stress-strain diagram is now plotted from a new origin 0′. When stress S_2 is again reached there will again be the corresponding elastic strain e_e but, so far, no plastic strain. The relationship between elastic stress and strain is conserved, but there is a completely new relationship between stress and plastic strain that follows the curve 2–3–4. Correspondingly, we will see that the *plastic "constitutive equations"* are written not between stress and strain but *between stress and strain increments.*

In most metal forming, plastic strains are much larger than elastic strains; thus, for simplification, in almost all subsequent analysis the *elastic part* of the stress-strain behavior *will be neglected.* This will not lead to significant errors in stresses, pressures, and forces. The only exception will be made in the analysis of *bending.*

The "necking" part of the engineering stress-strain diagram is a coarse distortion of reality because both the loading force and the deformation are almost solely affected by what is happening in the necked region of the tensile test specimen, while they are still related to the original total area and length of the specimen. Therefore, in all the subsequent text, we will use the true strains and stresses only. Correspondingly, Fig. 5.1b applies, which is obtained from the same test data as in (*a*), but it is a relationship between the true stress σ and the true strain ε. The parameters σ and ε were defined in Eqs. (2.3) and (2.5) as

$$\sigma = F/A \qquad \varepsilon = \ln (l/l_0)$$

In almost all of our analyses we will assume *homogeneous and isotropic material,* except for certain comments relating to *deep drawing.* For a homogeneous and isotropic material, plastic deformation occurs under *conservation of volume of the material*:

$$Al = A_0 l_0 = \text{const.}$$

Correspondingly, we have also

$$\varepsilon = \ln\left(\frac{l}{l_0}\right) = \ln\left(\frac{A_0}{A}\right) \tag{5.3}$$

The true stress–true strain diagram in Fig. 5.1b, from point 3 to point 4 concentrates on what is happening in the narrowed-down cross section A of the neck; therefore, ε is expressed using the right-hand alternative of Eq. (5.3). The strain ε_4 is the strain to fracture. It is necessary to realize that this particular fracture was obtained in the tension test, under nominally uniaxial tensile stressing and in association with the typical tensile necking feature, which involves triaxial tension. In other forming operations under various three-dimensional stress states and mainly in the presence of compressive hydrostatic stress, the strain to fracture will be different, mostly much larger.

Also, at high temperatures the strains without fracture, especially in compression, are almost unlimited.

Further on, we will express in proper form that the plastic deformation behavior as given by the (σ, ε) curve is the same in very different triaxial stress situations as long as they are represented by the "effective" values of σ and ε. On the other hand, the limit of the plastic flow, the ductile fracture, is different, and it is mainly associated with tensile stress values in particular regions of deformations.

The curves in Fig. 5.1a and b show an increase of σ with ε. This increase is associated with all cold-working operations; it is called *strain-hardening*. In relation to the tension test it is necessary to comment that it represents the amount of strain-hardening obtainable in the particular straining mode of the tension test. It would, for instance, seem when considering Fig. 5.1a that the yield strength *YS* as it is being increased by the steps of strain-hardening, as in diagram (*c*), will at the most reach the value of *UTS*. However, this is clearly a mistake due to the artificial concept of the "ultimate strength." As diagram (*b*) shows, the true maximum stress σ_{max} is much higher than *UTS*. Even the true stress σ_3 obtained at the engineering strain $e = 0.2$, that is, at true strain $\varepsilon = 0.182$, is 1.2 *UTS* (because the length is 1.2 times greater and the area is $A_0/1.2$). Straining to much larger strains than in a tensile test and throughout the whole volume of the material (unlike the local straining in the neck of the tensile specimen) can be obtained, for example, by cold rolling or cold extrusion. In such a case, if the tension test is repeated after the various amounts of cold work, the engineering stress-strain diagrams will develop, as shown in Fig. 5.1d. Both the yield strength *YS* and the ultimate tensile strength *UTS* increase. The former increases more than the latter, and they get closer to one another until entirely brittle behavior is reached at *YS* = *UTS* and zero elongation.

The area under the true stress–true strain curve is the specific plastic deformation work per unit volume. For instance, the area under the graph from $\varepsilon = 0$ to ε_5 is

$$W_{s,5} = \int_0^{\varepsilon_5} \sigma \, d\varepsilon = \int_{l_0}^{l_5} \left(\frac{F}{A}\right)\left(\frac{dl}{l}\right) = \frac{1}{Al}\int_{l_0}^{l_5} F \, dl = \frac{1}{Al}\int_{l_0}^{l_5} dW \qquad (5.4)$$

where $dW = F dl$ is the increment of work of the loading force over the distance dl, and Al is the strained volume; thus, the shaded area is plastic work per unit volume (specific work) for deformation between $\varepsilon = 0$ and ε_5. The yield strength is now, for simplicity, denoted Y instead of the *YS* of the engineering stress-strain diagram. If the "mean stress" for the part of the curve in diagram (*b*), between $\varepsilon = 0$ and ε_5, is denoted $\overline{Y}$, then the specific work for this plastic deformation is

$$W_{s,5} = \overline{Y}\varepsilon_5 = \overline{Y}\ln\left(\frac{l_5}{l_0}\right) = \overline{Y}\ln\left(\frac{A_0}{A_5}\right) \qquad (5.5)$$

The calculation of plastic work is useful in metalworking analysis.

The actual (σ, ε) diagram can be approximated in various ways by simple mathematical expressions that are graphically indicated in Fig. 5.2. Diagram (*a*) shows the simplest form:

$$\sigma = Y \qquad (5.6)$$

This is called rigid, perfectly plastic behavior. The former part of the denomination means "no elasticity"; the latter part means "no strain-hardening." This type of behavior corresponds to hot working where, however, Y will be a function of strain rate $\dot{\varepsilon}$. In diagram (*b*) a linear form of strain-hardening is used:

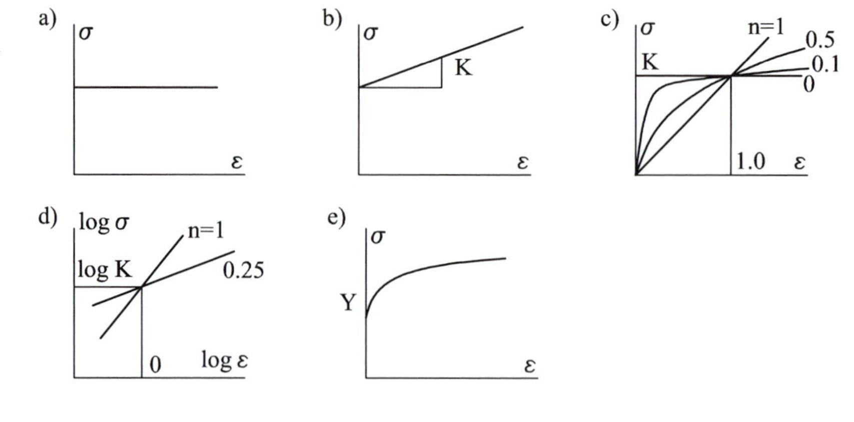

Figure 5.2 Various models of the stress-strain diagram: a) rigid perfectly plastic; b)–e) various forms of the strain-hardening effect.

$$\sigma = Y_0 + K\varepsilon \tag{5.7}$$

This may approximate cold working over a limited range where the slope of strain-hardening is $d\sigma/d\varepsilon = K$. Figure 5.2c shows a rather common form of the stress-strain formula:

$$\sigma = K\varepsilon^n \tag{5.8}$$

Two of the curves shown apply to extremes: $n = 0, \sigma = K$, which is the rigid, perfectly plastic case of Eq. (5.6), and $n = 1, \sigma = K\varepsilon$, which is the perfectly elastic case with K representing the module of elasticity. Most metals will behave in between the cases of $n = 0.1$ and $n = 0.5$. The parameter n is called the strain-hardening exponent. If Eq. (5.8) is plotted in $\log \sigma$, $\log \varepsilon$ coordinates (Fig. 5.2d), the slope of the corresponding line is $n = d(\log \sigma)/d(\log \varepsilon)$. Finally, a still closer approximation may be expressed as shown in graph (*e*):

$$\sigma = Y_0 + K\varepsilon^n \tag{5.9}$$

In the special case of $n = 1$, Eq. (5.9) reduces to Eq. (5.7).

In our text we will use the linear model expressed in (5.7) because it is simple enough, and yet it expresses the two main characteristics, an initial substantial yield strength Y_0 and its increase due to strain-hardening.

So far, the discussion has been based on the uniaxial tension test. In reality the state of stress is much more complicated. In courses on mechanics of solids, the student has already learned quite a lot about stresses in three dimensions. We will make use of stress analysis to derive the pressures and forces acting in some forming operations. However, this will be done mostly under simplified circumstances, assuming isotropic materials and assuming that the positions of the principal axes coincide with the axes of symmetry.

5.1.2 Stress in Three Dimensions

First, let us have a look at the three-dimensional stress situation oriented with respect to principal axes (Fig. 5.3). Here we have a cubic element of material where the directions X_1, X_2, X_3 are the directions of *principal axes*, perpendicular to each other and such that the sides of the cube are *principal planes*. There are no shear stresses on these planes. The normal stresses on these planes are $\sigma_1 > \sigma_2 > \sigma_3$; σ_1 is algebraically highest and σ_3 lowest. Tension is considered positive, compression negative. If two of the

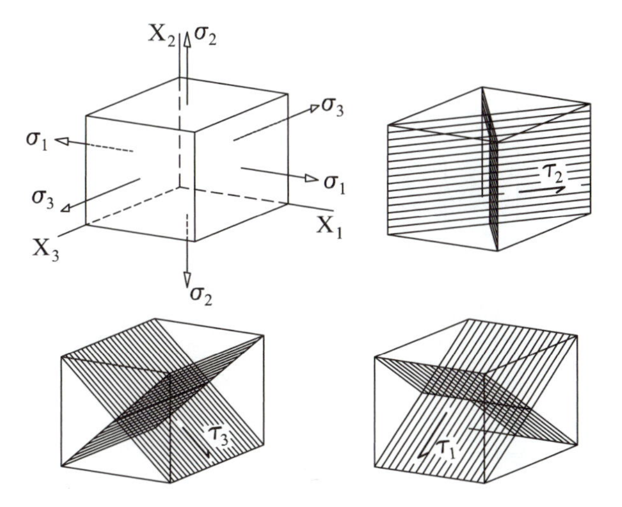

Figure 5.3
Three-dimensional stress. Normal stresses in principal axes and principal shear stresses.

principal stresses are equal, the state is called *cylindrical* stress. If all three are equal, it is the state of *hydrostatic*, or *spherical* stress.

All other planes than principal planes contain shear stresses. For all the planes passing through each of the principal axes, there are two orthogonal planes for which shear stresses are maximum. These are called *principal shear stresses*. They are located at 45° to the principal planes, and the stresses are

$$\tau_1 = \frac{\sigma_2 - \sigma_3}{2}$$

$$\tau_{max} = \tau_2 = \frac{\sigma_1 - \sigma_3}{2} \tag{5.10}$$

$$\tau_3 = \frac{\sigma_1 - \sigma_2}{2}$$

Because of the convention of subscripts to σ, the shear stress τ_2 is the largest of the three.

The stresses on all planes of the system as related to the principal stresses are easily obtained using Mohr's circles (see Fig. 5.4). In the system of the graph, its axes being those of normal stresses σ and of shear stresses τ, three circles are drawn located diametrally between points σ_1, σ_2, σ_3, with centers on the σ-axis. Each of the three circles represents stresses in planes passing through the X_1, X_2, X_3 axes, respectively. Stresses on all other planes correspond to points in the shaded area between the circles.

Let us look at some special stress states and comment on them with respect to the ratio of maximum shear stress to maximum tensile stress. Shear stress, as noted in the subsequent paragraph, is associated with plastic deformation without material failure. Tensile stress is associated with cracks and fractures. The ratio τ_{max}/σ_{max} then expresses, in a way, the ability to deform without fracture.

The various special cases are presented in Fig. 5.5. Case (*a*) is *uniaxial tension*. Maximum shear stress is found in planes at 45° to X_1 passing through X_2 as well as such planes passing through X_3, and it is $\tau_{max} = 0.5\sigma_1$. In case (*b*) of *uniaxial compression* with the compressive stress σ_3 (algebraically smallest) of the same magnitude as σ_1 in (*a*), the maximum shear stresses will be on analogous planes, as in (*a*), and they will be the same as in (*a*), $\tau_{max} = 0.5\sigma_3$. Plastic flow will be induced with the

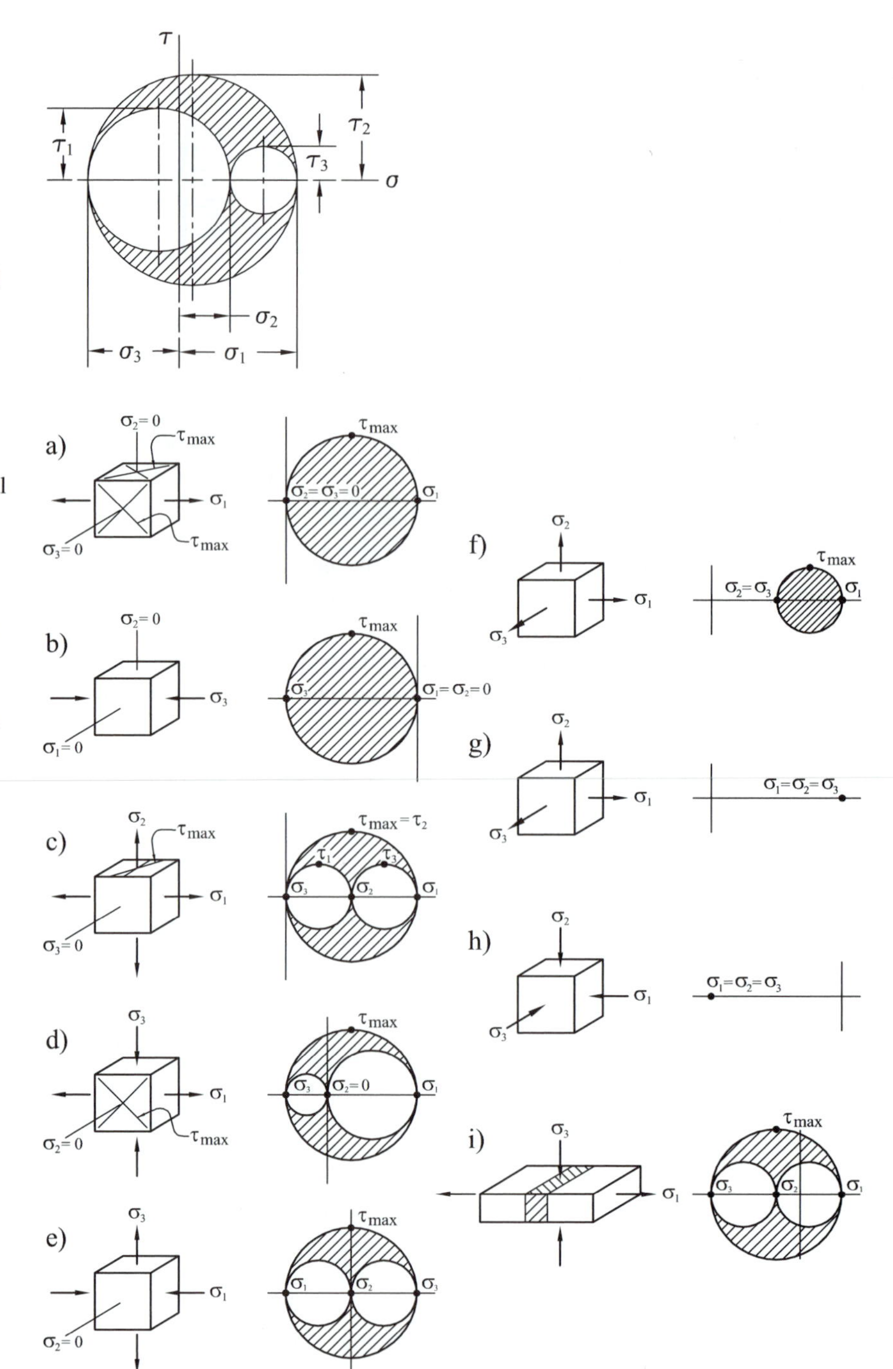

Figure 5.4
Mohr's circles for three-dimensional stress. Normal stresses are measured on the horizontal axis, shear stresses on the vertical axis. Each circle represents stresses in planes passing through one of the principal axes.

Figure 5.5
Various combinations of principal stresses: a) uniaxial tension; b) uniaxial compression; c) and d) plane-stress cases; e) special case of plane stress–pure shear; f) triaxial tension; g) uniform triaxial tension; h) hydrostatic compression; i) plane strain.

same intensity as in (*a*) but there will be no tendency to fracture because of zero tensile stresses. Cases (*c*) and (*d*) are called "plane-stress" situations because stresses exist in two dimensions.

In (*c*), in addition to tension σ_1, the specimen is also subjected to tension $\sigma_2 = \sigma_1/2$. The magnitude of σ_1 is the same as in (*a*). The maximum shear stress is again $\tau_{max} = \tau_2 = (\sigma_1 - \sigma_3)/2 = 0.5\sigma_1$. In the overall formability on one hand, and susceptibility to fracture on the other, this case does not differ much from (*a*). In (*d*), in addition to tension σ_1, there is compression $\sigma_3 = -\sigma_1/2$. Maximum shear stress is now $\tau_{max} = (\sigma_1 - \sigma_3)/2 = 3\sigma_1/4 = 0.75\sigma_1$. If this case was so stressed as to develop the same maximum shear stress τ_{max} as in cases (*a*), (*b*), and (*c*), where it was $\sigma_1 = 2\,\tau_{max}$, the corresponding maximum tensile stress would now be only $\sigma_1 = 4\tau_{max}/3 = 1.33\,\tau_{max}$. Obviously, in this case there is less tendency to fracture at the same intensity of plastic deformation.

This situation can be carried to the extreme in a plane stress such that $\sigma_3 = -\sigma_1$, and $\sigma_2 = 0$, case (*e*). One of the principal stresses is tensile, and the other is compressive, of the same magnitude. We will see later that this is a case of "pure shear" loading. There is no strain in the X_2 direction. The specimen contracts as much in X_1 as it extends in X_3. It is $\sigma_1 = \tau_{max}$, and there is still less tendency to fracture.

In the cases (*f*) through (*i*) stresses act in all principal axes. For the sake of brevity, we will not discuss them here.

5.1.3 Yielding: Plastic Deformation

It is generally recognized and accepted that the yielding of ductile materials is not affected by hydrostatic stresses. Pure hydrostatic stress, however large, does not cause yielding. Correspondingly, yielding is only affected by deviatory stresses, that is, by the differences between a complex stress state and the corresponding hydrostatic stress.

The two most widely stated criteria that have been developed empirically are those established by Tresca and by Von Mises.

The Tresca criterion is based on the maximum shear stress:

$$\tau_{max} = \frac{\sigma_1 - \sigma_3}{2} = \frac{Y}{2} = k \tag{5.11}$$

where Y is the yield stress obtained in the tensile test, and k is the shear-flow stress obtained in pure shear, where $\sigma_1 = -\sigma_3 = k$, and $\sigma_2 = 0$. Obviously, this leads to

$$k = \frac{Y}{2} \tag{5.12}$$

This criterion is rather simple, and it is a good approximation to experimental observations. However, the Von Mises criterion was found to be still closer to reality, and we will use this criterion in most of our exercises. It is also called the criterion of maximum distortion theory, and it combines all three deviatoric stresses in what is called the effective stress $\bar{\sigma}$. Yielding is obtained if

$$\bar{\sigma} = \sqrt{\frac{(\sigma_1 - \sigma_2)^2 + (\sigma_2 - \sigma_3)^2 + (\sigma_3 - \sigma_1)^2}{2}} = Y \tag{5.13}$$

or else,

$$(\sigma_1 - \sigma_2)^2 + (\sigma_2 - \sigma_3)^2 + (\sigma_3 - \sigma_1)^2 = 2Y^2 \tag{5.14}$$

where, again, Y is the tensile yield stress. Indeed, for uniaxial tension, $\sigma_2 = \sigma_3 = 0$, Eq. (5.14) leads to $\bar{\sigma} = Y$. For pure shear, $\sigma_1 = -\sigma_3 = k$, and $\sigma_2 = 0$ it is obtained as follows:

$$k = \frac{Y}{\sqrt{3}} = 0.577Y \tag{5.15}$$

Thus, the shear-flow stress k, according to this theory, is slightly larger than the one of Eq. (5.13).

Eq. (5.14) can also be rewritten, using Eq. (5.11) as follows:

$$\bar{\sigma} = \sqrt{2\,(\tau_3^2 + \tau_1^2 + \tau_2^2)} = Y \tag{5.16}$$

This shows that while the Tresca criterion uses the maximum shear stress τ_2, only the Von Mises criterion combines all the three shear stresses in a root mean square fashion.

Now we need to discuss the relations between stress and strain in plastic deformation. As mentioned before, the strains are not determined by the stresses; they depend on the entire history of loading. It is necessary to follow the strain increments through the changing stress situation and integrate them along the loading path. So, for instance, if a rod was first elongated by tension to a length $l_1 = 1.3 l_0$ and subsequently compressed to the original length, the total strain is

$$\varepsilon = \int_{l_0}^{l_1} \frac{dl}{l} + \int_{l_1}^{l_0} \left(-\frac{dl}{l}\right) = 2 \ln\left(\frac{l_1}{l_0}\right) = 0.52$$

This example illustrates that during all the changes of loading, it is just the increments of strain that are related to the stress situations. Accordingly, it is established that strain increments be proportional to the effective stresses. For the principal axes it is written as follows:

$$d\varepsilon_i = d\lambda\, \sigma_i \tag{5.17}$$

or, in more detail:

$$d\varepsilon_1 = \frac{2}{3} d\lambda\, [\sigma_1 - 0.5(\sigma_2 + \sigma_3)]$$

$$d\varepsilon_2 = \frac{2}{3} d\lambda\, [\sigma_2 - 0.5(\sigma_3 + \sigma_1)]$$

$$d\varepsilon_3 = \frac{2}{3} d\lambda\, [\sigma_3 - 0.5(\sigma_1 + \sigma_2)] \tag{5.18}$$

For the isotropic and homogeneous, ideally rigid, plastic material the incompressibility (constancy of volume) applies:

$$d\varepsilon_1 + d\varepsilon_2 + d\varepsilon_3 = 0 \tag{5.19}$$

In an analogous way as the effective stress (5.13), the effective strain $\bar{\varepsilon}$ may also be formulated:

$$d\bar{\varepsilon} = \frac{1}{3} \sqrt{2[(d\varepsilon_1 - d\varepsilon_2)^2 + (d\varepsilon_2 - d\varepsilon_3)^2 + (d\varepsilon_3 - d\varepsilon_1)^2]} \tag{5.20}$$

The coefficient of $\sqrt{2}/3$ is so chosen as to make it equal to the tensile strain $d\varepsilon$ in the uniaxial tension test.

Let us now look at how the yield criterion applies to some specialized stress-strain situations. For better illustration we will select a particular plastic stress-strain behavior. A very simple one is given in Eq. (5.7):

$$\bar{\sigma} = Y = Y_0 + K\bar{\varepsilon}$$

as it is now written out for effective stress and strain.

5.1.4 Special Cases of Yielding

Uniaxial Tension or Compression

This situation is represented in Fig. 5.6. A force F acts in direction X on an area A. For tension, it is

$$\sigma_1 = \sigma_x = \frac{F}{A}, \qquad \sigma_2 = 0, \sigma_3 = 0$$

The yield criterion, Eq. (5.14), reduces to $\sigma_1 = Y$ and the strains are, according to Eqs. (5.18):

$$\mathrm{d}\varepsilon_1 = \frac{2}{3}\mathrm{d}\lambda\sigma_1, \qquad \mathrm{d}\varepsilon_2 = -0.5\mathrm{d}\varepsilon_1, \qquad \mathrm{d}\varepsilon_3 = -0.5\mathrm{d}\varepsilon_1$$

If the particular yield stress of Eq. (5.7) is used, it is also

$$\varepsilon_1 = \varepsilon_x = \frac{Y - Y_0}{K}, \qquad \varepsilon_2 = \varepsilon_y = 0.5\varepsilon_1, \qquad \varepsilon_3 = \varepsilon_2 = -0.5\varepsilon_1$$

For compression, in an analogous way, it is

$$\sigma_3 = \sigma_x = -Y, \qquad \mathrm{d}\varepsilon_1 = \mathrm{d}\varepsilon_2 = -0.5\mathrm{d}\varepsilon_3$$

Plane Strain

In some situations the material is prevented from deforming in one of the principal axes. The most common such case arises in rolling and can be represented as shown in Fig. 5.7. A wide plate is compressed between anvils, the width b of which is much smaller than the width w of the plate. The material is prevented from spreading in direction Y by the material of the plate adjacent to the anvils. Correspondingly, the strain $\varepsilon_2 = 0$, and the material deforms in the directions X and Z only. We may assume tension σ_1 in the direction X. These stresses were depicted in Fig. 5.5i.

Using the second of Eqs. (5.18), it is

$$\mathrm{d}\varepsilon_2 = 0$$

$$\sigma_2 = 0.5(\sigma_1 + \sigma_3) \tag{5.21}$$

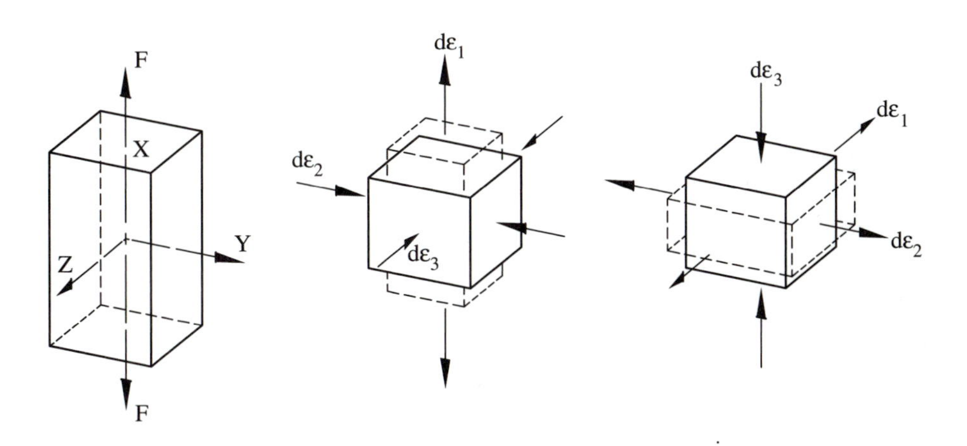

Figure 5.6
Strains in uniaxial tension and compression.

Correspondingly, Eq. (5.13) reduces to

$$\bar{\sigma} = \sqrt{\frac{(\sigma_1 - \sigma_3)^2 + \left(\frac{\sigma_1 - \sigma_3}{2}\right)^2 + \left(\frac{\sigma_1 - \sigma_3}{2}\right)^2}{2}} = Y$$

$$\bar{\sigma} = \sqrt{1.5\frac{(\sigma_1 - \sigma_3)^2}{2}} = Y \tag{5.22}$$

$$\sigma_1 - \sigma_3 = Y\sqrt{\frac{4}{3}} = 1.155Y$$

Plane Stress

In other situations stresses may occur in two principal axes only. These cases were depicted in diagrams (*c*), (*d*), and (*e*) of Fig. 5.5. In practice this applies often to sheet-metal forming and, as shown in Fig. 5.8, the stress σ_x being tensile, the stress σ_y may be either tensile or compressive. Let us consider the case with biaxial tension. The yield criterion for $\sigma_3 = 0$ reduces to

$$\bar{\sigma} = \sqrt{\frac{\sigma_1^2 + \sigma_2^2 + (\sigma_1 - \sigma_2)^2}{2}} = Y \tag{5.23}$$

Assuming a special case of $\sigma_1 = \sigma_2$, it is

$$\bar{\sigma} = \sigma_1 = Y$$

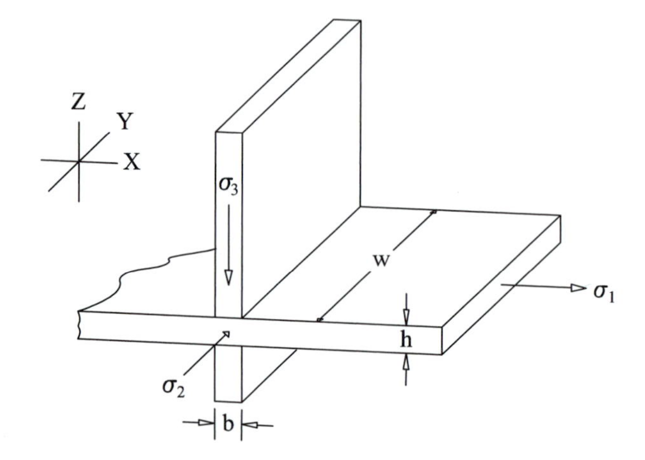

Figure 5.7
Plane-strain situation. In compressing over a long narrow strip across a wide sheet the surrounding material prevents sideways strain. Representative of rolling of wide sheet and plate.

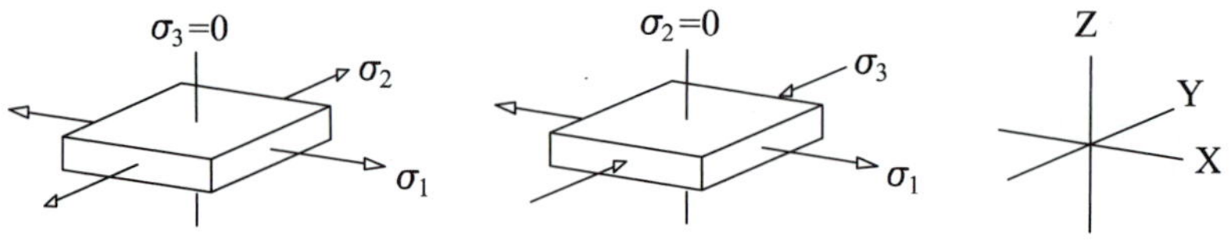

Figure 5.8
Plane-stress situation. Representative of sheet-metal forming.

The yielding is the same as in uniaxial tension. The strains are such that $d\varepsilon_1 = d\varepsilon_2$ (see Eq. [5.19]); consequently, $d\varepsilon_3 = -2d\varepsilon_1$.

Another significant special situation is the one of *axisymmetrical stresses*, which is discussed in the next section.

5.2 BULK FORMING: BASIC APPROACH—FORCES, PRESSURES

5.2.1 Wire Drawing: Work, Force, and Maximum Reduction Without Friction

Figure 5.9 shows the wire being drawn through a conical die. It is reduced from the initial diameter d_0 to the final diameter d_1. The drawing force is F_d, and the stress in the drawn wire is σ_d.

The total strain produced in this operation is

$$\varepsilon_1 = 2\ln\left(\frac{d_0}{d_1}\right)$$

Using the yield strength formula,

$$Y = Y_0 + K\varepsilon$$

the specific work is obtained as

$$W_s = Y_0\varepsilon_1 + \frac{K\varepsilon_1^2}{2} \tag{5.24}$$

Comparing the work done by the drawing force F_d over a length l_1 of the drawn wire with the work of the operation on the drawn volume $A_1 l_1$,

$$F_d l_1 = W_s A_1 l_1$$

the stress σ_d in the drawn wire is obtained:

$$\sigma_d = F_d/A_1 = W_s \tag{5.25}$$

The stress in the drawn wire is equal to the specific work and it increases with the reduction ratio $(d_0/d_1)^2 = A_0/A_1$, and there is, obviously, a limit to this ratio. This limit is reached when the stress σ_d reaches the yield stress Y_1 of the drawn wire. For a non-strain-hardening material, $K = 0$, $Y = Y_0$ it would be at the limit:

$$\sigma_d = Y_0, \qquad W_s = Y_0\varepsilon_1$$

$$\sigma_d = W_s, \qquad \varepsilon_1 = 1,$$

$$2\ln\left(\frac{d_0}{d_1}\right) = 1, \qquad \left(\frac{d_0}{d_1}\right)_{\lim} = 1.65$$

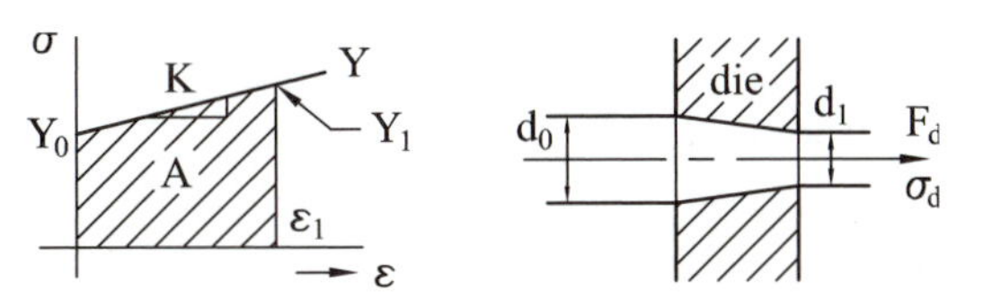

Figure 5.9
Simple analysis of wire drawing. Outlining the "work formula" for drawing stress.

For a strain-hardening material, the limit is obtained by equating $\sigma_{d_{\text{lim}}} = Y_0 + K\varepsilon_1$ and W_s as given in Eq. (5.24). If a larger reduction is required, it can be obtained in additional drawing stages. However, as the wire strain-hardens, it loses its ductility, and another limit will be reached when the wire exhausts all its ductility and breaks. The force and power needed for the operation with velocity v are easily obtained:

$$F_d = A_1\sigma_d, \qquad P = F_d v$$

5.2.2 Wire Drawing: Pressure on the Die, and Axisymmetric Yielding

For a small angle α of the die (see Fig. 5.10), we will assume that the stress state in the material being reduced is cylindrical, with one of the principal axes X_1 in the axis of the wire and the two other principal axes X_2, X_3 in two mutually perpendicular radial directions. It is $\sigma_1 = \sigma_{\text{ax}}, \sigma_2 = \sigma_3 = \sigma_{\text{rad}}$. The pressure p between the die and the work is shown for a slice of thickness dx. For a unit length of the circumference of this slice, the pressure p acts on a surface dx/cos α, and the radial component of this force is in balance with the radial force produced by the stress σ_2:

$$-\sigma_2\,\mathrm{d}x = p\left(\frac{\mathrm{d}x}{\cos\alpha}\right)\cos\alpha, \qquad -p = \sigma_2 \tag{5.26}$$

Although the pressure p is normally given as a positive number, we must treat σ_2 as negative according to the convention of the sign for compressive stress.

For cylindrical stress, $\sigma_2 = \sigma_3 = \sigma_{\text{rad}} = -p$, and $\sigma_1 = \sigma_{\text{ax}}$, the yield criterion (5.14),

$$(\sigma_1 - \sigma_2)^2 + (\sigma_2 - \sigma_3)^2 + (\sigma_3 - \sigma_1)^2 = 2Y^2$$

reduces to

$$2(\sigma_{\text{ax}} + p)^2 = 2Y^2, \qquad \sigma_{\text{ax}} + p = Y \tag{5.27}$$

where Y is the yield stress, and p is the pressure at the particular section.

At the entry plane A, the axial stress is zero because there is no force acting axially on the wire from the left; at A:

$$\sigma_{\text{ax}} = 0, \qquad p_0 = Y_0$$

At the exit the axial stress is the drawing stress; at B:

$$\sigma_{\text{ax}} = \sigma_d, \qquad p_1 = Y_1 - \sigma_d$$

In the extreme situation, for an extreme reduction, it will be

$$\sigma_d = Y_1, \qquad p_1 = 0$$

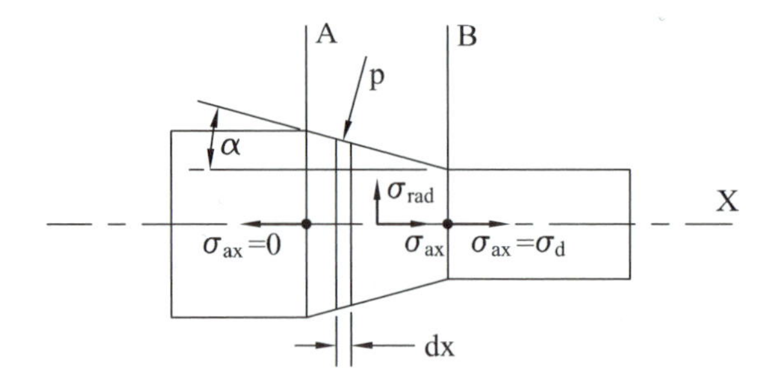

Figure 5.10
Diagram for deriving the expression for pressure on the die in wire drawing.

The pressure between the die and the drawn wire is at a maximum at the entry, where it equals the initial strength of the wire. It decreases toward the exit where for extreme reduction it will reach zero. This result is important for the die design. It does not have to be made of a material much stronger than that of the work, and it does not suffer at the exit where otherwise it might get damaged by having its particles drawn out. This makes the wire-drawing operation an easy one; wires were drawn a thousand years ago through dies made of stone.

5.2.3 Wire Drawing with Friction

The balance of forces in drawing with friction is illustrated in Fig. 5.11. In addition to forces shown in Fig. 5.10 the friction force $F_f = \mu\, p\, \pi\, D\, \mathrm{d}x / \cos\alpha$ must be considered in its axial component $F_f \cos\alpha$:

$$\frac{\pi}{4} D^2\sigma_x - \frac{\pi}{4}(D - \mathrm{d}D)^2(\sigma_x + \mathrm{d}\sigma_x) + p\frac{\mathrm{d}x}{\cos\alpha}\mu\pi D\cos\alpha$$

$$+p\frac{\mathrm{d}x}{\cos\alpha}\pi D \sin\alpha = 0$$

$\mathrm{d}D$ is taken as a positive number.

$$\frac{\pi}{4}[D^2\sigma_x - D^2\sigma_x + 2D\mathrm{d}D\sigma_x - D^2\mathrm{d}\sigma_x] + \pi D p(\mu + \tan\alpha)\mathrm{d}x = 0$$

Neglecting the second and higher powers of the differentials and expressing $\mathrm{d}x$ in terms of $\mathrm{d}D$, we have

$$\mathrm{d}D = 2\mathrm{d}x\tan\alpha \rightarrow \mathrm{d}x = \frac{\mathrm{d}D}{2\tan\alpha}$$

$$-\frac{D\mathrm{d}\sigma_x}{4} + \frac{\mathrm{d}D\sigma_x}{2} + \frac{p\mathrm{d}D}{2\tan\alpha}(\mu + \tan\alpha) = 0$$

$$\mathrm{d}\sigma_x = 2\frac{\mathrm{d}D}{D}[\sigma_x + p(1 + \mu\cot\alpha)]$$

Using the axisymmetric yield condition, $p = Y - \sigma_x$,

$$\mathrm{d}\sigma_x = 2\frac{\mathrm{d}D}{D}[\sigma_x + (Y - \sigma_x)(1 + \mu\cot\alpha)]$$

$$\mathrm{d}\sigma_x = 2\frac{\mathrm{d}D}{D}[Y(1 + \mu\cot\alpha) - \sigma_x\mu\cot\alpha] \tag{5.28}$$

The solution can be obtained by integrating this equation; an example follows.

EXAMPLE 5.1 Wire Drawing with Friction: Solution by Discrete Integration ▼

Use the expression for $\mathrm{d}\sigma_x$ in Eq. (5.28) and refer to Fig. 5.11. Write a simple computer program by proceeding in small steps. An example of the program listing is shown in Fig. 5.12. The results of the computation are plotted in Figs. 5.13 and 5.14.

$$L = \frac{D_0 - D_1}{2\tan\alpha} = 11.76$$

$$dD = \frac{D_0 - D_1}{30} = 0.167$$

$$dx = \frac{L}{30} = 0.392$$

$$Y = 300 + 320\varepsilon, \qquad D_0 = 10\text{ mm}, \qquad D_1 = 6.0\text{ mm}, \qquad \mu = 0.2$$

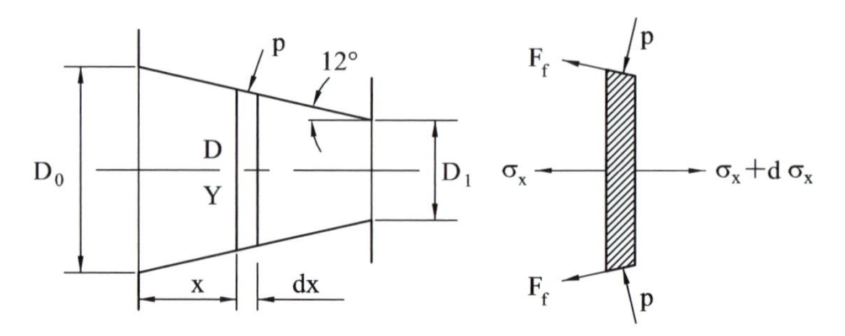

Figure 5.11
Balance of forces in drawing with friction.

```
function[S,Y,x,p]=wiredraw(Do,D1,Yo,K,mu)
% S is sigmax, mu is coefficient of friction, eps is epsilon
x(1)=0;S(1)=0;D=Do;Y(1)=Yo;p(1)=Yo;al=12*pi/180;
dD=(Do-D1)/200;dx=dD/(2*tan(al));
for n=1:200
% use Eq. (5.28)
   dS=2*(dD/D)*(Y(n)*(1+mu/tan(al))-S(n)*mu/tan(al));
   S(n+1)=S(n)+dS;
   D=D-dD;
   eps=2*log(Do/D);
   Y(n+1)=Yo+K*eps;
   p(n+1)=Y(n+1)-S(n+1);
   x(n+1)=x(n)+dx;
end
end
```

Figure 5.12
Listing of the Matlab program "wiredraw" of Example 5.1.

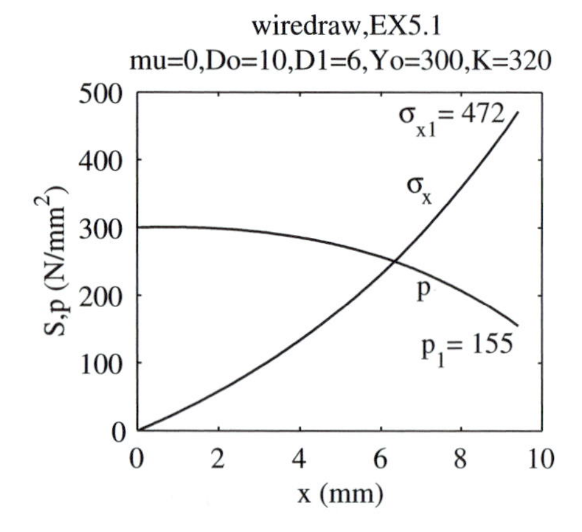

Figure 5.13
Plots of axial stress and of pressure on the die as results of Example 5.1, no friction.

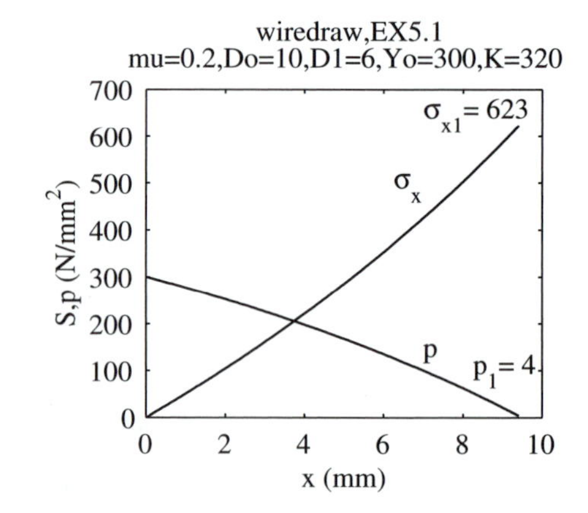

Figure 5.14
Plots of results of Example 5.1, with friction. Drawing stress is 30% higher; exit pressure on die drops to almost zero.

Start from $x = 0$, $\sigma_x = 0$, $D = D_0$, $Y = Y_0$, $p = Y_0$.
Proceed in 200 steps:

$$d\sigma_x = 2\frac{dD}{D}[Y(1 + \mu \cot \alpha) - \sigma_x \mu \cot \alpha] = \frac{0.33}{D}(1.94Y - 0.94\sigma_x)$$

$$\sigma_x = \sigma_x + d\sigma_x$$

$$D = D - dD$$

$$\varepsilon = 2 \ln\left(\frac{D_0}{D}\right)$$

$$Y = Y_0 + K\varepsilon$$

$$p = Y - \sigma_x$$

$$x = x + dx$$

Plot in one plot: σ_x versus x, p versus x, for (a) $\mu = 0$, (b) $\mu = 0.2$. Attach plots.
Write down the drawing force F_d:
Without friction ($\mu = 0$),

$$F_d = \sigma_d \times \frac{\pi D_1^2}{4} = 472 \times \frac{\pi}{4} \times 36 = 13{,}345 \text{ N}$$

With friction ($\mu = 0.2$)

$$F_d = 623 \times \frac{\pi}{4} \times 36 = 17{,}615 \text{ N}$$

You can see from Fig. 5.13 that there is still a rather high pressure on the die at the exit. This indicates that without friction, ($\mu = 0$), the drawing ratio could be increased. It is possible to use Eqs. (5.24) and (5.25), which apply for the frictionless case, and determine the value of the maximum possible drawing ratio such for which the drawing stress σ_d equals the yield stress Y_1 at the exit:

$$\sigma_d = Y_1$$

$$Y_0\varepsilon_1 + \frac{K\varepsilon_1^2}{2} = Y_0 + K\varepsilon_1$$

In our case,

$$300\varepsilon_1 + 160\varepsilon_1^2 = 300 + 320\varepsilon_1$$

$$160\varepsilon_1^2 - 20\varepsilon_1 - 300 = 0$$

$$(\varepsilon_1)_{1,2} = 1.433 \quad \text{and} \quad -1.308$$

Only the positive root is applicable, and

$$2 \ln\left(\frac{d_0}{d_1}\right) = 1.433$$

$$\left(\frac{d_0}{d_1}\right)_{\max} = 2.047$$

It would be possible to draw from $d_0 = 10$ mm down to $d_1 = 4.88$ mm. However, with friction coefficient $\mu = 0.2$ (see Fig. 5.14), the pressure p_1 is down to a very small value, and the ratio $d_0/d_1 = 10/6 = 1.667$ used in the example is just about the maximum possible. ▲

5.2.4 Extruding a Round Bar

This exercise will suffer more from our simplifications because in extrusion the die angle is usually quite large, and the assumption of the cylindrical stress with the principal axes axial and radial is only a rather coarse approximation of reality. Nevertheless it will indicate how much higher the pressures on the die are than in drawing.

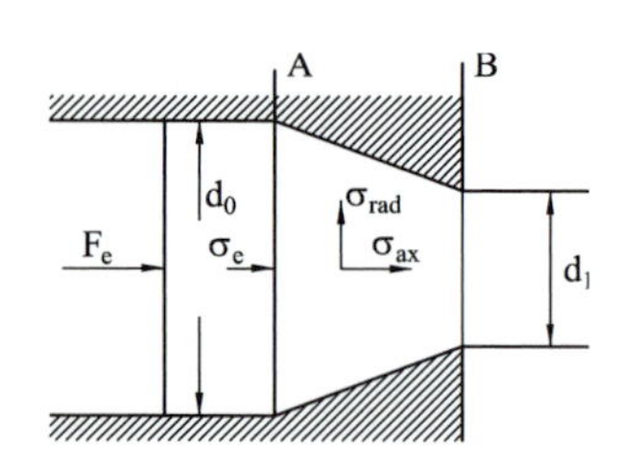

Figure 5.15
Diagram for stresses in extrusion.

The diagram in Fig. 5.15 outlines the situation. We will neglect the friction between the billet and the cylinder, so that the analysis will be closer to hydrostatic extrusion with extrusion pressure σ_e. The extrusion (pushing) force is F_e. In contrast to drawing, there is no simple limit to area reduction, because both the axial and radial stresses are compressive. Very large reductions are possible, and extrusion can be done both cold and hot.

Under the assumption of cylindrical stress, Eq. (5.27) applies just as it did in drawing. The extrusion pressure is again obtained from the work formula (5.5), but it is now negative:

$$\sigma_e = -W_s = -\bar{Y}\ln\left(\frac{d_0}{d_1}\right)^2 \tag{5.29}$$

and the pressure on the die at plane A is

$$\sigma_{ax} = \sigma_e, \qquad p_0 = Y_0 - \sigma_e = Y_0 + \bar{Y}\ln\left(\frac{d_0}{d_1}\right)^2 \tag{5.30}$$

At B, the exit, the axial stress is zero, and it is

$$\sigma_{ax} = 0, \qquad p_1 = Y_1 \tag{5.31}$$

For illustration, let us assume a non-strain-hardening material, $Y_0 = Y_1 = \bar{Y} = Y$, and a reduction ratio $d_0/d_1 = 4$. We have

$$\sigma_e = Y\ln 16 = 2.77Y \tag{5.32}$$

The pressure on the die at A is $p_0 = 3.77Y$, and at B it is $p_1 = Y$. Obviously the demands on the die material are much higher than in drawing. Therefore, much stronger material must be used for the die.

5.2.5 Rolling with Back and Forward Tension: Plane-Strain Yielding

This exercise presents a very simplified picture of rolling; as in the preceding exercises, we neglect friction, this time between the rolls and the work. However, friction actually plays a significant role. A more detailed analysis will be presented later.

With reference to Fig. 5.16a and b, the process will be simply regarded as compressing the strip of thickness h down to thickness h_1 over an area L wide across the strip of width w. The dimension L is the length of contact between the rolls and the work. The strip width w is many times greater than L. Thus the compressed material is bounded by the materials outside of L, and this situation can be considered as "plane strain," with deformations limited to the directions X and Z, and none occurring in Y. The principal axes are assumed to coincide with X, Y, and Z so that $\sigma_1 = \sigma_x$, $\sigma_2 = \sigma_y$, and $\sigma_3 = \sigma_z$. The yielding condition has been derived previously and was expressed in Eq. (5.22). It is

$$\sigma_1 - \sigma_3 = \frac{2Y}{3^{1/2}} = 1.15Y$$

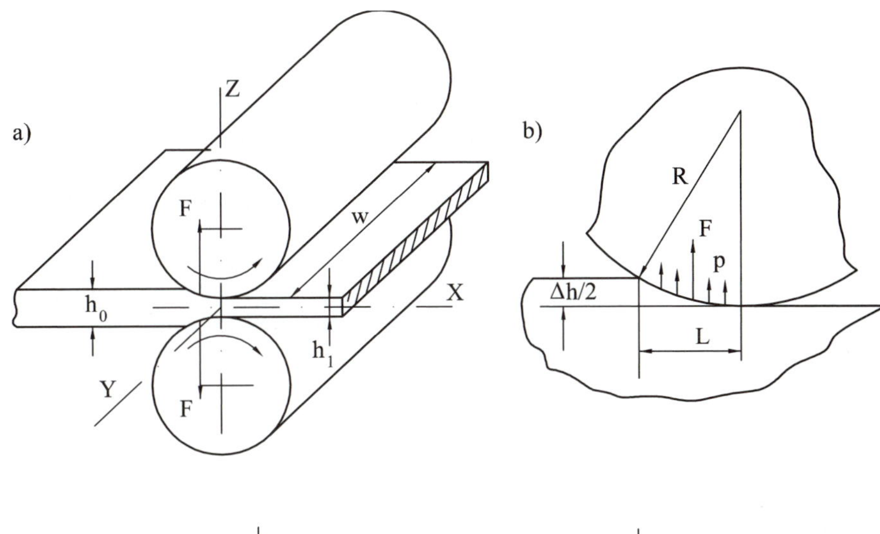

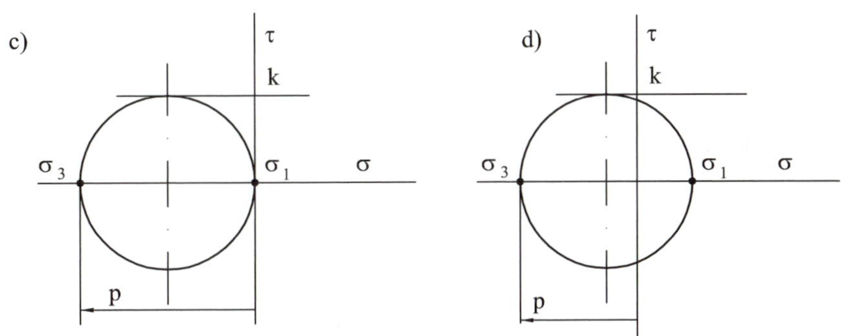

Figure 5.16
Simple analysis of rolling. The process is considered as one of compressing a wide strip on a narrow band of length L; see a), b). The Mohr's circles illustrate the cases c) without and d) with tension applied in the X-axis. Rolling pressure is reduced in the latter case.

For rolling without tension, as in hot rolling of slabs, the process is as described in the plane-strain compression test, with $\sigma_1 = 0$. The yield criterion (5.14) reduces to

$$\sigma_3 = 1.15Y \tag{5.33}$$

and the corresponding Mohr's circle is as shown in Fig. 5.16c. If, as is done in cold strip rolling where tension σ_1 is applied to the strip the necessary compressive stress decreases as indicated in Mohr's circle in Fig. 5.16d.

In order to obtain the rolling force and torque, the contact length L is obtained, as shown in diagram (*b*), and writing $\Delta h = h_0 - h_1$,

$$L^2 = \frac{\Delta h}{2}\left(2R - \frac{\Delta h}{2}\right)$$

and, for $\Delta h << 2R$, we may write, as a very good approximation,

$$L = \sqrt{R\,\Delta\,h} \tag{5.34}$$

Combining (5.33) and (5.34) gives us the rolling force;

$$F = Lw\sigma_3 = w(1.15Y - \sigma_1)\sqrt{R\,\Delta\,h} \tag{5.35}$$

Assuming that force F acts in the middle of the contact zone, the torque on one roll is obtained:

$$T = \frac{FL}{2} = w\,(1.15Y - \sigma_1)\frac{R\Delta h}{2}$$

The total power on the one stand is obtained by taking the torque T acting on each of the two rolls and assuming the peripheral roll velocity v:

$$P = 2T\omega = \frac{2Tv}{R} = w(1.15Y - \sigma_1)v\Delta h \tag{5.36}$$

The aim of the preceding exercises was to apply the yield criterion to simple stress situations and provide at the same time an elementary understanding of some of the metal-forming operations. In order to move a step further toward an analysis of metal forming that will be closer to reality, we must include the effects of friction and of nonhomogeneous deformation. This will be done in the following section.

5.3 BULK FORMING: EFFECTS OF REDUNDANT WORK AND FRICTION

For a proper analysis of the forming processes, it would be necessary to consider the distribution of stresses, strains, strain rates, and temperatures throughout the workpieces and to take into account the interplay between all these parameters and the yield stress of the material. Simultaneously the workability of the material must be considered. In bulk forming, the primary stresses are compressive. However, due to nonhomogeneous deformation, secondary tensile stresses develop that lead to the cracking of the workpiece. Recently, substantial progress has been made in metal-forming analysis through the application of finite-element computations while considering many of the mentioned effects. These methods, especially in combination with graphic terminals, are developing into a whole field of computer-aided design (CAD) for forming dies.

In our text we will try to arrive at good approximations of the above effects while using rather simple methods of analysis. We will consider the total work of the plastic deformation to consist of three parts:

1. W_h, the work for homogeneous deformation as it was considered in the exercises in Section 5.2
2. W_f, the frictional work
3. W_r, the redundant work due to internal distortions of the material

We begin by discussing the last of these.

5.3.1 Nonhomogeneous Deformation: Redundant Work

The redundant work is to be included because it increases the working stresses and forces. In addition, it introduces additional strain. This leads, for example, to an earlier exhaustion of ductility and limitations on total area reduction between annealing in wire drawing.

Let us illustrate nonhomogeneous deformation by analyzing the case of indenting a thick plate. First, let us consider a slab infinitely thick and very wide in both the horizontal dimensions, as indicated in Fig. 5.17. The tool is rather slender and compresses the surface of the slab along a contact length L, but it is very wide and extends over the whole workpiece width w. The deformation is obviously local, and it is feasible to rep-

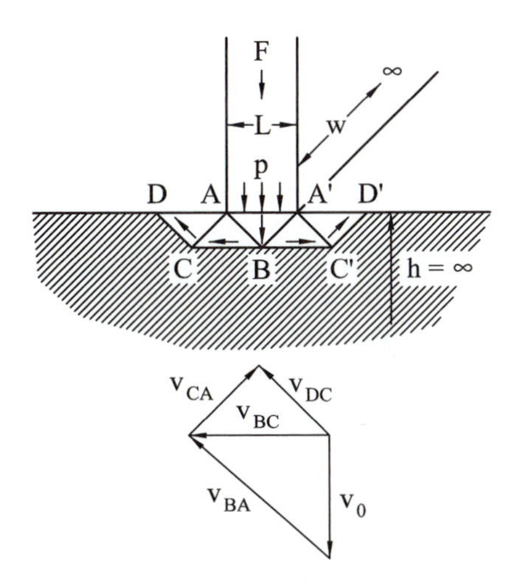

Figure 5.17
Indenting a thick plate. The narrow anvil pushes down a wedge ABA' that displaces material sideways and up. Indenting pressure is high because of all the redundant work.

resent it by the motions of nondeforming, triangular, wedge-type sections, as shown in the figure. The wedge under the punch moves down with velocity v, and displaces the two adjacent wedges horizontally with velocities v_{bc}. These, in turn, push the outer two wedges upwards so that they form bulges at the sides of the indentation. Work is done by shearing along boundaries *AB, BC, AC, CD*. The corresponding velocities are shown in the diagram at the bottom of the figure. The shearing force along any boundary, per unit width w, equals the shear yield stress k times the length of the boundary. The shearing velocities along these boundaries are

$$v_{ba} = v_0\sqrt{2}, \qquad v_{bc} = v_0, \qquad v_{ca} = \frac{v_0}{\sqrt{2}}, \qquad v_{dc} = \frac{v_0}{\sqrt{2}}$$

The total power to be delivered by the pressure p along AA' is the sum of all the shearing powers:

$$Lpv_0 = 2k\left(\frac{Lv_0\sqrt{2}}{\sqrt{2}} + Lv_0 + \frac{Lv_0}{2} + \frac{Lv_0}{2}\right)$$

where the terms in the parentheses correspond to boundaries *AB, BC, AC, CD* respectively, and each of them has to be counted twice. The expression simplifies to

$$p = 6k \tag{5.37}$$

or else, considering the relationship (5.15) between shear flow stress k and yield strength Y in tension,

$$p = \frac{6Y}{\sqrt{3}} = 3.46Y \tag{5.38}$$

The results of (5.37) and (5.38) are slightly on the higher side. This is due to the extreme simplification of the shearing deformation field presented in Fig. 5.17. The exact solution as presented in [6] leads to the values of $p = 5.14k = 2.97Y$.

Indeed, for the Brinell hardness test, which is well represented by this analysis, it is found in practice that for most steels there is a proven relationship:

$$UTS\,(\text{lb/in}^2) = 500\,BHN\,(\text{kg/mm}^2)$$

which if translated into equal units for Y and BHN results in

$$BHN\,(\text{N/mm}^2) = 2.84\ UTS\,(\text{N/mm}^2)$$

This is read as meaning that the indentation pressure in the hardness tests is 2.84 times the tensile strength, which, in turn, depending on the degree of strain-hardening, is about 10–20% higher than the yield stress. Thus practical experience agrees with the above analysis.

The analysis itself reveals that in the operation of indenting an infinitely thick slab, the compressive stress p between the tool and the workpiece is about three times higher than the yield stress of frictionless upsetting (compression) of a specimen, which, in turn, would be equal in magnitude to the tensile yield stress Y. This increased effort is due to the constraints on the locally (nonhomogeneously) deforming metal exercised by the surrounding thick, nondeformed slab.

The homogeneous deformation that is characterized by rectangular grid elements of the deforming material remaining rectangular could ideally be obtained by compressing an unconstrained specimen, as in Fig. 5.18a. The lack of constraint would have to be achieved not only by the specimen being compressed over its whole cross-sectional area (this is called upsetting) but also by frictionless contact between the tools and the specimen. This latter aspect is discussed later. Fairly homogeneous deformation is also obtained in indenting a plate with a thickness h that is small with respect to the contact length L (see Fig. 5.18b).

Actually, even up to $h/L = 1$ there is practically no redundant work (see Fig. 5.19). It is feasible to represent this case by two wedges meeting in the middle of the plate and pushing its two sides apart. The hodograph of the velocities along the boundaries is shown, and the lengths of the boundaries and the magnitudes of the velocities are

$$BC = CE = \frac{L}{\sqrt{2}}, \qquad v_{bc} = v_{ce} = \sqrt{2}\, v_0$$

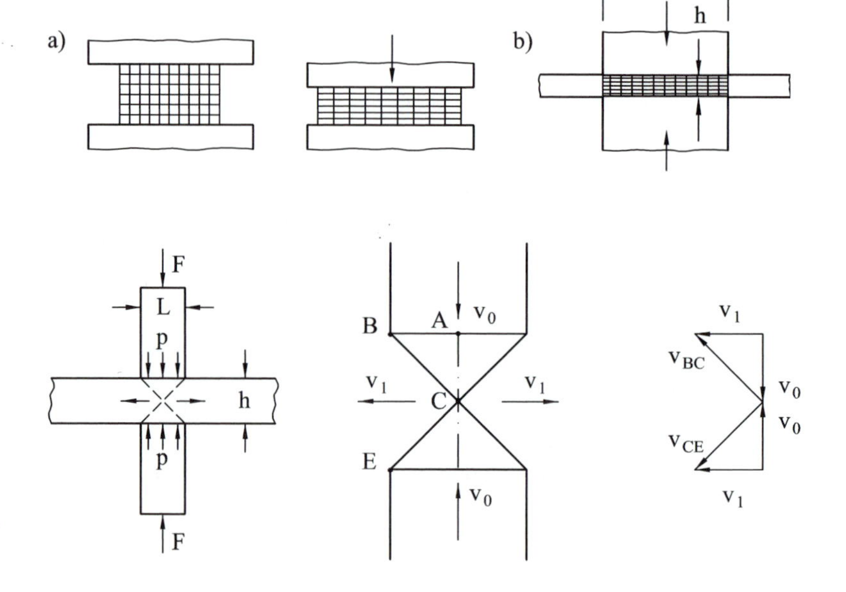

Figure 5.18
Homogeneous deformation in compression; a) unconstrained specimen with frictionless contact with the anvil; b) indenting of a thin plate.

Figure 5.19
Indenting at $h/L = 1$. There is no redundant work.

The power produced by the pressure p on both sides of the plate for one-half (left of the plane of symmetry S) of the action is

$$\frac{2pv_0L}{2} = 2v_0\sqrt{2}\frac{kL}{\sqrt{2}}, \qquad p = 2k = 1.15Y \tag{5.39}$$

This result is the same as in the solution of the "plane-strain" indenting case (5.22). The pressure p is still larger than in the simple upsetting case of Fig. 5.18a, and this is due to the constraints in the direction across the plate. However, there is no redundant work due to nonhomogeneous deformation.

The indenting operation has been analyzed by many authors. It is found that for thinner plates such that $h/L < 1$, there is no redundant work. For $h/L > 8.7$, the two deformation zones between the two indenters are entirely separated, and the amount of redundant work reaches the same level as for the infinitely thick slab. For thicknesses $1 < h/L < 8.7$, the compressive yield stress p is between $1.15Y$ and $2.97Y$.

A correction factor of redundant work Q_r can be introduced as a function of the ratio h/L such that the pressure p for yielding in indenting is

$$p = Q_rY \tag{5.40}$$

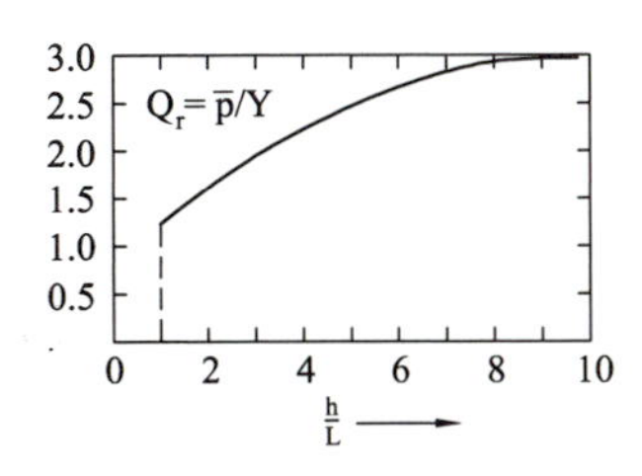

Figure 5.20
Correction factor for redundant work in plane-strain indenting. It increases with ratio $\Delta = h/L$.

The graph of Q_r versus h/L is given in Fig. 5.20.

With greater thickness, the two wedge-type zones under the indenters try to push the surrounding material apart, while there is a plastically undeformed zone between them. This generates "secondary tensile stresses" in the middle of the work, which either create cracks and fracture or, if they are not high enough for fracture, will leave residual stresses in the material after the operation.

Many metal-forming operations other than the indenting operation just described may be considered analogous to it as regards the mechanism of redundant work. Corresponding ratios h/L may be suitably formulated for them, as indicated in Fig. 5.21. In indentation forging, the ratio h/L is simply defined as in the preceding analy-

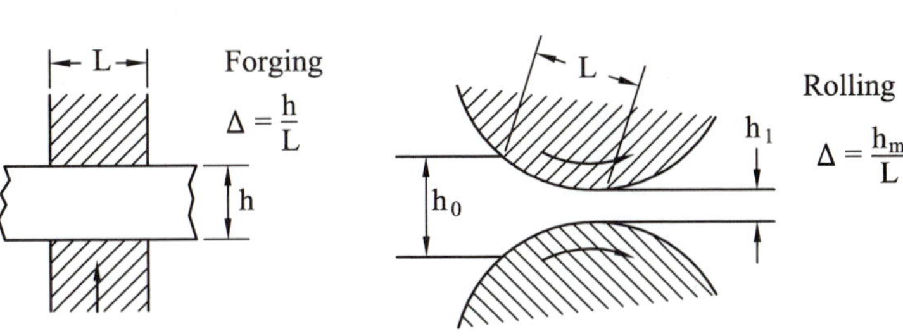

Figure 5.21
The characteristic ratio Δ for various operations in which pressure is applied over only a part of the work surface. Redundant work may be involved.

sis. For drawing or extrusion of flat parts being reduced from thickness h_0 to h_1 or of round parts reduced from diameter d_0 to d_1, the ratio of mean thickness to contact length is defined as

$$\Delta = \frac{h_0 + h_1}{2L} \quad or \quad \Delta = \frac{d_0 + d_1}{2L} \tag{5.41}$$

Respectively, it may be shown that the value of Δ depends on the reduction ratio d_0/d_1 as well as on the die half-angle α. In rolling, the mean thickness is again compared to the length of contact, $\Delta = h_m/L$.

Let us check briefly under which conditions the ratio Δ is greater than 1 in these operations, so that redundant work has to be considered. For drawing and extrusion, a brief discussion is based on Fig. 5.22, where the conditions are investigated for two die angles, $\alpha = 15°$ and $\alpha = 45°$. Let us determine the ratio $(d_0/d_1)_{\lim}$ for which $\Delta = 1$. It is

$$L = \frac{d_0 - d_1}{2 \sin \alpha}$$

and

$$\Delta = \frac{d_0 + d_1}{2L} = 1$$

Rewriting these two conditions, we have

$$d_0 - d_1 = 2L \sin \alpha,$$

$$d_0 + d_1 = 2L,$$

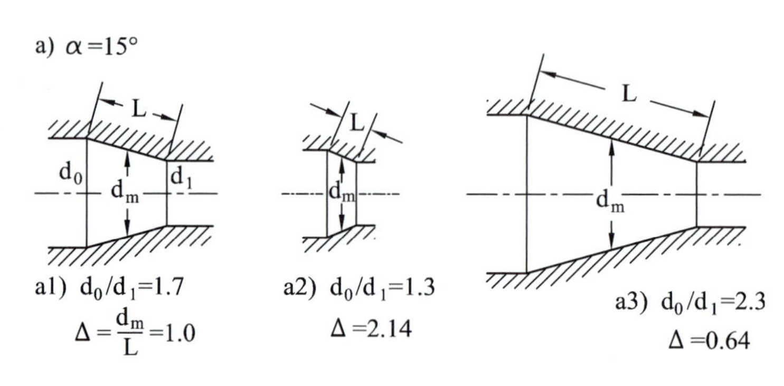

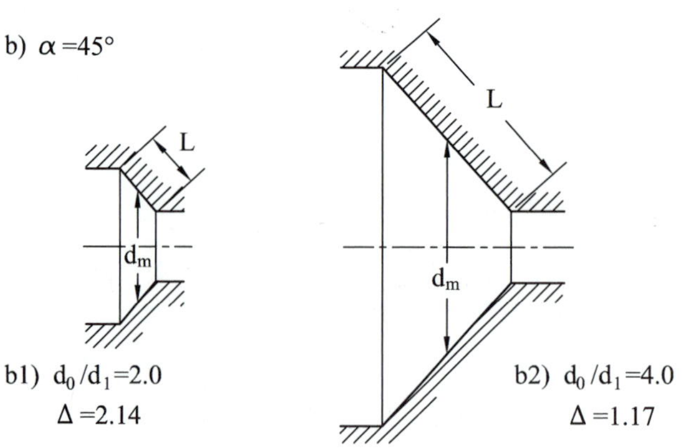

Figure 5.22
The ratios $\Delta = d_m/L$ in drawing and extrusion depend on the ratio d_0/d_1 and on the die angle. Conditions are shown under which $\Delta \geq 1.0$ (a1, a2, b1, b2) involving redundant work.

and

$$\Delta = \frac{\left(\dfrac{d_0}{d_1} + 1\right)\sin\alpha}{d_0/d_1 - 1} = 1 \tag{5.42}$$

Solving (5.42) for $\alpha = 15°$, we have $(d_0/d_1)_{\text{lim}} = 1.7$, and for $\alpha = 45°$, we have $(d_0/d_1)_{\text{lim}} = 5.9$.

It can be easily shown that in both cases $\Delta > 1$ for $d_0/d_1 < (d_0/d_1)_{\text{lim}}$. This is illustrated in the figure. In the cases of Fig. 5.22a, $\alpha = 15°$. In (*a1*), $d_0/d_1 = 1.7$ and $\Delta = 1.0$. In (*a2*), $d_0/d_1 = 1.3$ and $\Delta = 2.14$; this is a case of nonhomogeneous deformation and redundant work. In (*a3*), $d_0/d_1 = 2.3$ and $\Delta = 0.64$; in this case the deformation should be fairly homogeneous. Note that in cold drawing the maximum d_0/d_1 has been established as being about 1.65; therefore $\Delta > 1$. In extrusion d_0/d_1 is usually larger. However, the die angle is also made larger, and the cases shown in Fig. 5.22b, where $\alpha = 45°$ are more applicable. It was indicated above that for $\Delta < 1$, d_0/d_1 would have to be greater than 5.9. In both illustrated cases the reduction ratio is smaller than this value. In (*b1*), $d_0/d_1 = 2.0$ and $\Delta = 2.14$, and in (*b2*), $d_0/d_1 = 4.0$ and $\Delta = 1.17$. The latter case is close to homogeneous deformation. In general, the amount of redundant work increases with the die angle and decreases with increasing reduction ratio.

Another example of a strong distortion is presented in Fig. 5.23a. It was obtained experimentally by axisymmetric extrusion of superplastic Pb-Sn eutectic. The ratio $d_0/d_1 = 2.55$ and $\alpha = 45°$; correspondingly $\Delta = 1.62$. Figure 5.23b was obtained in axisymmetric extrusion of 1095 spheroidized steel at 300° C. Die angle $\alpha = 60°$, $d_0/d_1 = 1.625$, and $\Delta = 3.74$. The deformation is strongly nonhomogeneous.

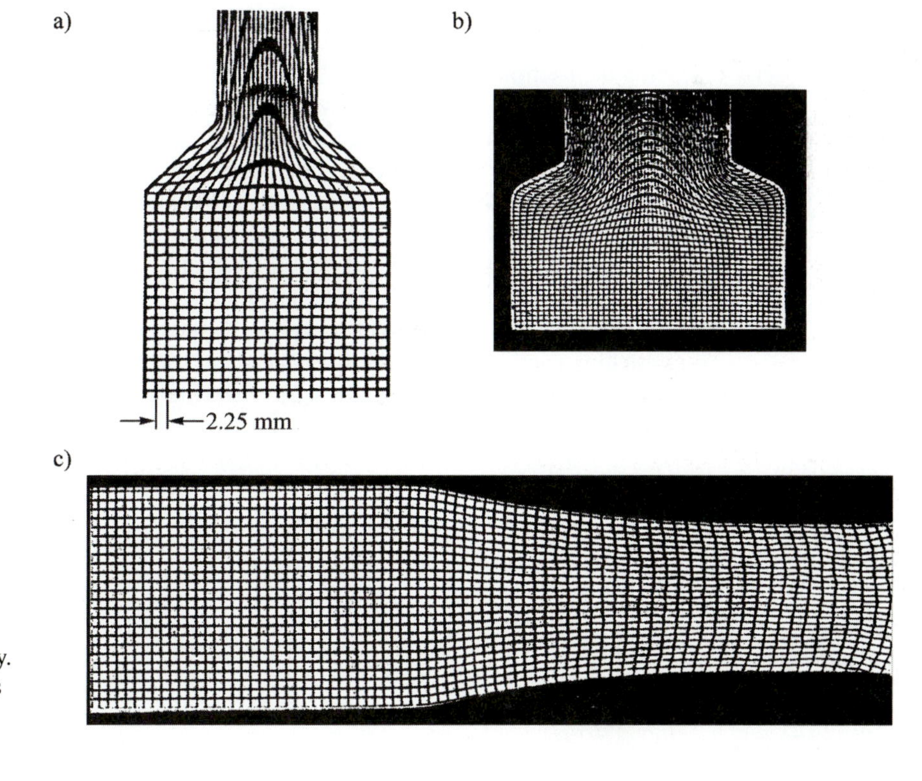

Figure 5.23
a), b) Nonhomogeneous deformation in extrusion demonstrated experimentally. In rolling, c), deformation is fairly uniform.

In cold rolling of thin strip the ratio Δ is usually rather small. Taking as an example a case of roll radius $R = 400$ mm and strip dimensions $h_0 = 2.5$ mm and $h_1 = 2.0$ mm, it is

$$L = (R\Delta h)^{1/2} = 14 \text{ mm}, \qquad h_m = 2.25 \text{ mm}, \qquad \Delta = 0.16.$$

Correspondingly, there will be very little redundant work, and deformation will be fairly homogeneous. An experimentally obtained case of evidence from Ref. [3] is shown in Fig. 5.23c. It corresponds to $\Delta = 0.55$ and shows that there is little grid distortion. In hot rolling the ratio Δ may be close to 1 or larger, and in some of those cases redundant work has to be considered.

5.3.2 The Effect of Friction in Plane Strain

Let us consider compressing a specimen in plane strain, such that it will not involve redundant work. This means that it is an indenting operation with a small ratio $\Delta = h/L$ (Fig. 5.24). The tools and work are considered very wide in the direction normal to the paper, and this leads to zero strain in that direction because the material at the sides of the tools constrains it. It was established in Eq. (5.22), that in these conditions the yield criterion is simplified to

$$\sigma_1 - \sigma_3 = 2k = 1.15Y \qquad \text{(reproduced Eq. 5.22)}$$

As the material is compressed, it spreads to the sides in the direction X. With respect to symmetry, the points in the axis Z do not move in X. It is sufficient to consider the half of the specimen to the left of Z. The coordinate x will be measured from the left hand edge to the tool. The friction between the tool and the work is acting against the spreading motion.

Two kinds of friction are considered. In cold forming, the Coulomb type of friction applies: the friction force is proportional to the pressure p at the interface involving the friction coefficient μ. The usual value of μ, assuming the usual means of lubrication may, for various work materials, be in the range $\mu = 0.06$ to 0.2. In hot forming there is more adhesion between the tool and the work, and the values of μ could be rather high. On the other hand, the shear flow of the hot material is rather low and the shearing force in the work material just at the interface often becomes smaller than the friction force would be. The material sticks to the tool and is sheared just under the contact layer. This is called "sticking friction." Considering Eq. (5.22) reproduced above, at the interface and relating k and $\sigma_3 = -p$, it is $k = (\sigma_1 + p)/2$. The stress σ_1 at the interface is never tensile, and so, at the most, $k = p/2$. This also means that the sticking friction sets in whenever the shear force is less than the friction would be, but in any case we never have $\mu > 0.5$.

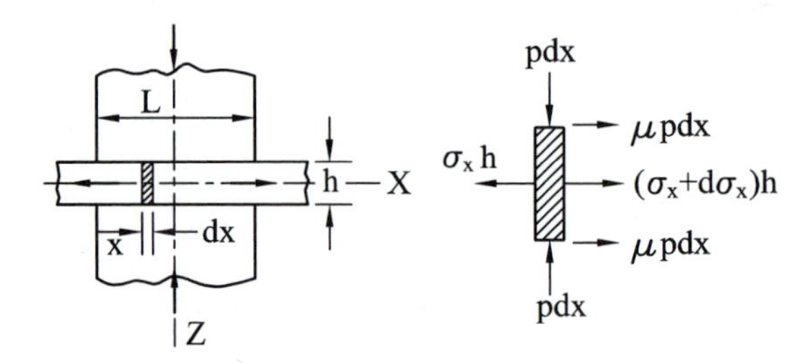

Figure 5.24
Deriving the effect of friction in plane-strain compression.

Dry Friction: Initial Derivation

The solution will first be carried out considering Coulomb friction as it would correspond to cold forging. Later on, after having dealt with sticking friction, we will see that this initial derivation is only valid within limits because also in cold forging the tangential stress at the interface as obtained by friction μp cannot exceed the value of the shear flow stress k. Proceeding from the left edge towards the center in the Fig. 5.24, the equilibrium of a slice dx at location x is considered as shown in the diagram. The forces acting on the slice of unit width w in the direction X are the normal pressures σ_x and $(\sigma_x + d\sigma_x)$ times the height h and the friction forces μp dx on both interfaces:

$$(\sigma_x + d\sigma_x)h - \sigma_x h + 2\mu p\, dx = 0$$

$$h\, d\sigma_x = -2\mu p\, dx \tag{5.43}$$

The pressure p and σ_x are effectively principal stresses σ_3 and σ_1, and for plane-strain yield we have

$$p + \sigma_x = 2k = 1.15Y \tag{5.44}$$

Then,

$$dp = -d\sigma_x \tag{5.45}$$

and, combining (5.43) with (5.45):

$$\frac{dp}{p} = 2\mu \frac{dx}{h} \tag{5.46}$$

The boundary condition at $x = 0$ is $\sigma_x = 0$, and from Eq. (5.44), $p = 1.15Y$. Integrating (5.46) with this condition gives

$$p = 2ke^{(2\mu x/h)} \tag{5.47}$$

The extremes of the interface pressure are

$$x = 0, \qquad p = 2k = 1.15Y$$

$$x = L/2, \qquad p_{max} = 1.15Ye^{\mu L/h} \tag{5.48}$$

For $\mu << 1$, Eq. (5.48) may be expanded and truncated and it is then

$$p_{max} = 2k\left(1 + \frac{\mu L}{h}\right)$$

The pressure has the form of the so called "friction hill." It is shown for $h/L = 0.2$ and various values of μ in Fig. 5.25a and for $\mu = 0.2$ and various values of $\Delta = h/L$ in Fig. 5.25b. The ratio Δ obviously plays an important role in the effect of friction on the interface pressure p. For large values of Δ, friction has very little influence. This is understood by realizing that the effect of friction is rather local to the layers close to the interface between tool and work. In a thick workpiece, while the surface layers are hindered in their spread by friction, the middle layers are not. For thin workpieces, however, the friction causes high peaks of interface pressure in the center of the compression zone. These high peaks are unrealistic, however, and a correction of their derivation is formulated in the section entitled "Limitation of Dry Friction," below.

It is interesting to note that the effect of Δ is here just the opposite of what it was on redundant work, which was largest for large values of Δ. For $\Delta \simeq 1$, neither friction nor redundant work are important and the simple analyses of Section 2.1 hold quite well.

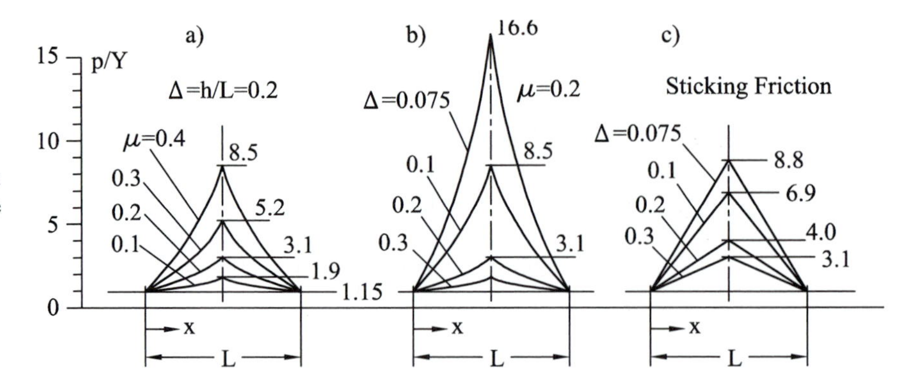

Figure 5.25
Interface pressure in plane strain: a), b) dry friction; c) sticking friction. Some parts of graphs a) and b) are not valid. This happens when the friction stress exceeds the shear flow stress; see Figs. 5.26 and 5.27.

The mean pressure p_{av} is obtained by integrating (5.47) from $x = 0$ to $x = L/2$ and dividing by $L/2$:

$$p_{av} = \frac{2}{L}\int_0^{L/2} p(x)\,dx = \frac{2kh}{\mu L}(e^{\mu L/h} - 1) = 1.15Y\frac{h}{\mu L}(e^{\mu L/h} - 1) \quad (5.49)$$

and the total force F is obtained by multiplying p_{av} by the area $A = Lw$:

$$F = p_{av}Lw \quad (5.50)$$

Sticking Friction

A similar exercise will now be carried out for the case of sticking friction corresponding to hot forging. The friction forces $\mu p dx$ in Fig. 5.24 are now replaced by shear forces $k dx$, and the equilibrium of forces acting on the slice dx is written

$$h d\sigma_x = -2k\,dx \quad (5.51)$$

Again, $dp = -d\sigma_x$,

$$h dp = 2k\,dx$$

$$\frac{dp}{2k} = \frac{dx}{h} \quad (5.52)$$

$$p = \frac{2kx}{h} + C$$

At $x = 0$, $\sigma_x = 0$, and $p = 2k$, which leads to $C = 2k$, and

$$p = 2k(1 + x/h) \quad (5.53)$$

At the extremes the interface pressure is

$$x = 0, \qquad p = 2k = 1.15Y$$

$$x = \frac{L}{2}, \qquad p_{max} = 1.15Y\left(1 + \frac{0.5L}{h}\right) \quad (5.54)$$

The values of $p/2k$ for various values of Δ are plotted in Fig. 5.25c. The peak pressure is now higher than the one of Coulomb friction, and $\mu = 0.2$ for larger values of Δ. For the very small values of $\Delta = 0.1, 0.075$, the very sharp peaks of the graph in Fig. 5.26b do not occur. However, in reality for those sharp peaks the friction force in the center of the compression zone would have exceeded the shear flow stress, and they would have been flattened also in cold forging.

The mean pressure is now

$$p_{av} = 1.15Y\left(1 + \frac{0.25L}{h}\right) \tag{5.55}$$

Limitation of Dry Friction: Corrected Values of Pressures

Returning to Eq. (5.46), for dry friction we have

$$\frac{dp}{dx} = \frac{2\mu}{h}p \tag{5.56}$$

Comparing this with Eq. (5.52), for sticking friction, expressing the condition of shear flow at the contact surface, we have

$$\frac{dp}{dx} = \frac{2k}{h} \tag{5.57}$$

Obviously, the slope of the graphs in Fig. 5.25 is limited by the shear flow to the value of (5.57), and those parts of the graphs (*a*) and (*b*) *where the slope is higher* are not valid. As the slope of these graphs increases with increasing x, once it reaches the value of Eq. (5.57) the plot of (p, x) from there on proceeds under a constant slope, as would correspond to sticking friction. Such a limit point is obtained by equating (5.56) and (5.57):

$$\frac{2\mu}{h}p = \frac{2k}{h} \tag{5.58}$$

or

$$p_{\lim} = \frac{k}{\mu} \tag{5.59}$$

The initial slope of the graph of p versus x under the condition of dry friction, at $x = 0$ is obtained from the initial value of $p_0 = 2k = 1.15Y$, as formulated in (5.22) for $\sigma_1 = 0$.

$$\frac{dp}{dx_{x=0}} = \frac{4\mu k}{h} \tag{5.60}$$

For example, for $\mu = 0.2$, the initial slope is

$$\left[\frac{dp}{dx}\right]_{x=0} = \frac{0.8k}{h} \tag{5.61}$$

The slope will reach the value of Eq. (5.57) at a position x such that the condition (5.59) is satisfied; see Eq. (5.47):

$$p = 2ke^{2\mu x/h} = \frac{k}{\mu}$$

$$\frac{2\mu}{h}x_{\lim} = \ln\left[\frac{1}{2\mu}\right]$$

$$x_{\lim} = \frac{h}{2\mu}\ln\frac{1}{2\mu}$$

Thus, for the case of $\mu = 0.2$,

$$x_{\lim} = 2.5h \ln 2.5 = 2.29h \tag{5.62}$$

The exponential form of the (p, x) graph extends as far as this point; from there on it continues as a straight line at the slope of $2k/h$. The dry friction exponential form would extend over the whole contact for $x_{\lim} = L/2$ if it is

$$\frac{L}{2} \leq \frac{h}{2\mu} \ln\left(\frac{1}{2\mu}\right) \tag{5.63}$$

Thus, for the case of $\mu = 0.2$,

$$\frac{h}{L} > 0.218$$

For combinations of h/L *and* μ *not satisfying the condition of Eq. (5.63), the graphs of Fig. 5.25a and b are not valid.*

The graph of p versus x is then a combination of the forms (*a*), (*b*), and (*c*) of Fig. 5.25. The two parameters of interest, that of p_{max} and, most important, p_{av}, the latter determining the total force between the punch and the work, are best obtained by running a simple computer program.

EXAMPLE 5.2 Determine p_{max} and p_{av} for Plane-Strain Compression Under the Combined Conditions of Dry Friction and the Friction Shear Flow ▼

The listing of the computer program "frict" is given in Fig. 5.26. The variables obtained from this program are the pressure p as it would arise if there were no limit on the friction force, *the pressure* ps *resulting from the combination of dry friction and friction shear flow*, and the corresponding average values *pav* and *psav*. The program can be run for various input values of the coefficient of dry friction μ and of the work thickness h, expressed by the ratio h/L. The pressures are expressed in ratios versus yield strength Y. The shear flow stress is $k = 0.577Y$.

An example is plotted out in Fig. 5.27 for $\mu = 0.2$ and $h/L = 0.1$. Under these conditions the dry friction acts between $0 < x/L < 0.23$, and for $x/L > 0.23$ the shear flow sets in. The peak pressure reaches $5.97Y$, and the average pressure is $3.284Y$.

This program can be used as a subroutine in determining the upsetting force. The effort is simplified by the use of the graph generated by the program "nfrict" listed in Fig. 5.28; sample output is presented in Fig. 5.29. This program is identical with the one in Fig. 5.26 except that it runs for five values of the coefficient of friction μ and for eight cases of h/L. The result is plotted as *psav* versus h/L, for the individual values of μ. In order to use it for determining the compressive force, the value of the "friction correction coefficient $Q_f = psav/Y$" is found by interpolation from the graph. The compressive force F is obtained as

$$F = Q_f Y A = Q_f Y Lw \tag{5.64}$$

where w is the width of the workpiece, and L is the length, in the direction X of the compressed area.

It is also of interest to look at the increase of the compressive stress σ_x, from the value of $\sigma_x = 0$ at $x = 0$ towards the middle of the indented length L. This is

Figure 5.26
Matlab listing "frict" for Example 5.2, to compute pressures in plane-strain compression with dry friction limited by shear flow stress (partly dry, partly sticking friction).

```
function[p,ps,x,pav,psav,sig,sigs]=frict(mu,h)
% p,ps are pressures relative to yield strength, Y=1;
%sig, sigs are stresses in X, relative to Y=1;
% p,sig apply to dry friction unlimited,
% ps, sigs apply to sticking friction setting in if p>k/mu,
% h is work thickness relative to L=1; h should be read as h/L;
% pa, psa are obtained by integrating p,ps versus x; when divided
% by their number over L/2 (100)they determine the average
% pressures ps, psa
% initial values at x=0 are:
sig(1)=0;sigs(1)=0;Y=1;p(1)=1.15*Y;L=1;dx=0.005*L;x(1)=0;
k=0.577*Y;ps(1)=p(1);pa=p(1);psa=ps(1);
pa=p(1)/Y;psa=p(1)/Y;
% run the loop for x from 0 to L/2=100dx=0.5:
for n=1:100
x(n+1)=x(n)+dx;
% use Eq. (5.56)
dp=2*mu*dx*p(n)/h;
% integrate dp
p(n+1)=p(n)+dp;
% use Eq. (5.44)
sig(n+1)=1.15*Y-p(n+1);
% check if dry friction holds
if p(n)<=k/mu
ps(n+1)=p(n+1);
sigs(n+1)=sig(n+1);
else
% use Eq. (5.52)
dps=2*k*dx/h;
ps(n+1)=ps(n)+dps;
sigs(n+1)=1.15*Y-ps(n+1);
end
% keep on adding individual pressures p, ps to update their sum
pa=pa+p(n+1);
psa=psa+ps(n+1);
end
pav=pa/100;
psav=psa/100;
end
```

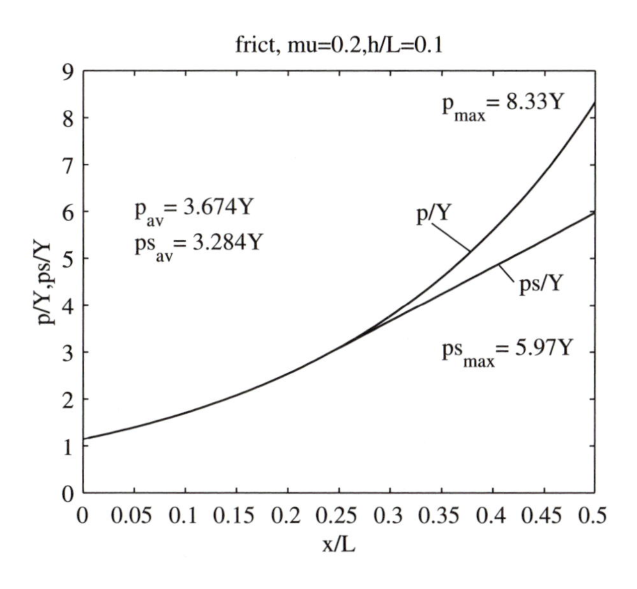

Figure 5.27
Results of Example 5.2. The pressure *ps* between the anvil and the work for plane-strain compression resulting from the limitation of the action of dry friction (pressure *p*) by the onset of shear flow stress at the interface and the resulting sticking friction participation. It is seen how the sharp pressure peaks obtained in Fig. 5.25 a) and b), where shear flow was disregarded, are reduced.

Figure 5.28
Matlab listing "nfrict" for producing the graph of Fig. 5.29 that can be used for finding the correction coefficient Q_f for plane-strain compression with a combination of dry and sticking friction. It is obtained by multiple runs of the program in Fig. 5.26 for various values of *h/L* and of friction coefficient.

```
function[pav,psav,mu,h]=nfrict
% mu is coefficient of friction
% this program computes pav,psav for mu=0.1,0.15,0.2,0.25
% and for h/L= 0.025 to 0.2 in steps of 0.025
% p,ps are pressures relative to yield strength, Y=1;
% sig, sigs are stresses in X, relative to Y=1;
% p,sig apply to dry friction unlimited,
% ps, sigs apply to sticking friction taking over if p>k/mu,
% h is work thickness relative to L=1; h should be read as h/L;
% pa, psa are variables integrating p,ps versus x;
% start outer loop for various values of mu:
for i=1:5
mu(i)=i*0.05;
% start the next inner loop for various values of h
for j=1:8
h(j)=j*0.025;
%initial values at x=0:
sig=0;sigs=0;Y=1;p=1.15*Y;L=1;dx=0.005*L;x=0;
k=0.577*Y;ps=p;pa=p/Y;psa=ps/Y;
% run the innermost loop for x from 0 to L/2=100dx=0.5,equivalent
% to program "frict.m"
for n=1:100
dp=2*mu(i)*dx*p/h(j);
p=p+dp;
if p<=k/mu(i)
ps=p;
else
dps=2*k*dx/h(j);
ps=ps+dps;
end
pa=pa+p;
psa=psa+ps;
end
pav(i,j)=pa/100;
psav(i,j)=psa/100;
end
end
% plot(h,psav)
```

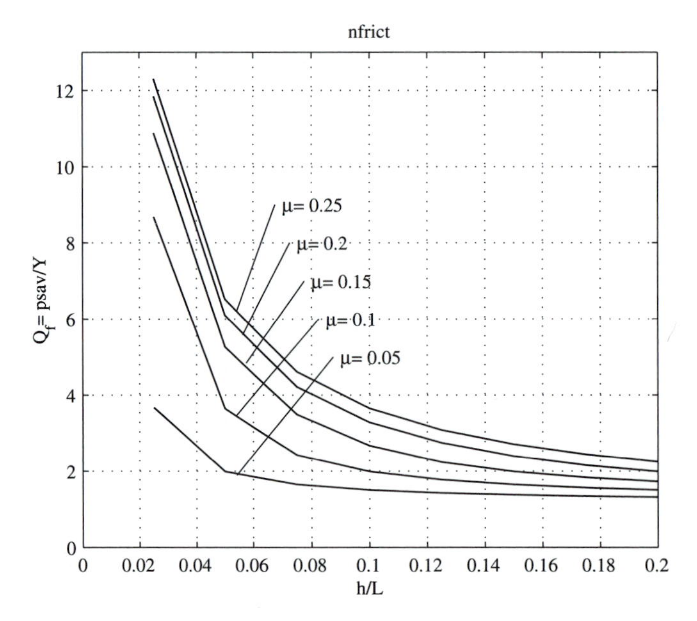

Fig. 5.29
Average pressure in plane-strain compression under combined action of dry friction and shear flow stress. The graph was obtained using program *nfrict*.

obtained from the program "frict," and presented in Fig.5.30, which shows that for $\mu = 0.2$ and $h/L = 0.1$ it reaches the value of almost $5Y$. ▲

5.3.3 Effect of Friction in Upsetting a Cylindrical Workpiece

In an upsetting operation the tool platens overlap the workpiece, as indicated in Fig. 5.31a. Instead of plane strain, the situation involves cylindrical stresses. To simplify, we will assume that throughout the deformation cylindrical sections will remain cylindrical. If an element of a tubular slice is taken out of the specimen so that it corresponds to an angle $\Delta\phi$, one can, in an approximation, associate the principal axes with the radial, circumferential, and axial directions. The corresponding principal stresses are then denoted σ_r, σ_c, σ_a (see Fig. 5.31b). For the equilibrium in the radial direction, for a unit thickness in the direction Z, the radial force ΔF_r,

$$\Delta F_r = \sigma_r \Delta\phi\,(r + \Delta r - r) = \sigma_r\,\Delta\phi\,\Delta r$$

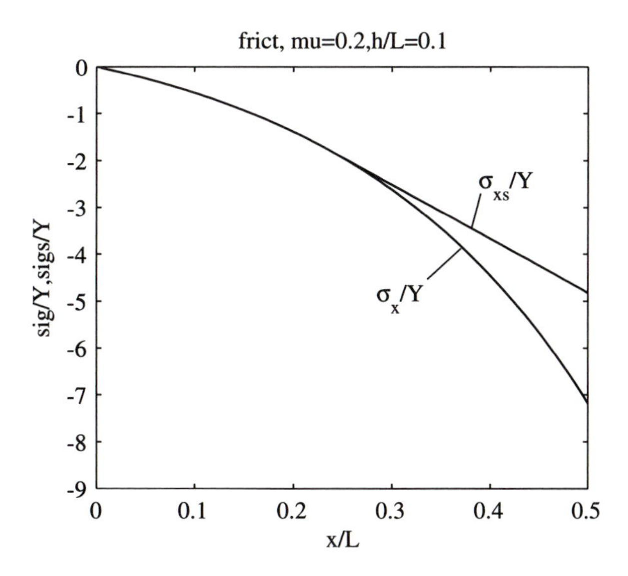

Figure 5.30. Longitudinal stress in plane–strain indenting for unlimited dry friction, σ_{xu}, and for combined dry friction and friction shear σ_{xs}. The peak of the compressive stress parallel with the face of the anvil has been reduced.

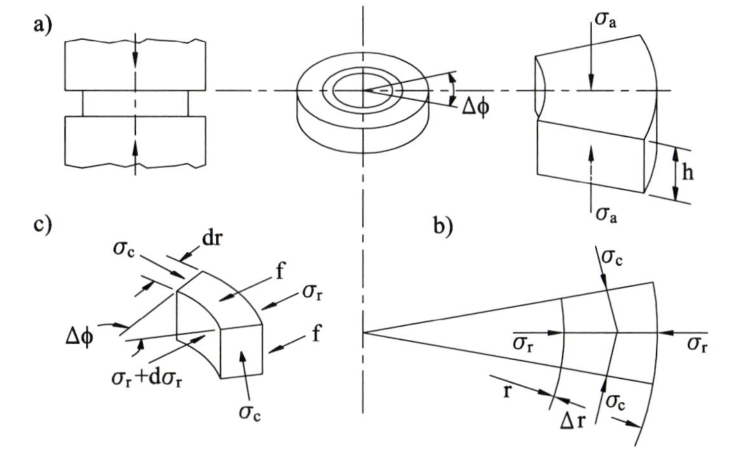

Figure 5.31 Balance of forces used in deriving the effect of friction in upsetting a cylinder: a) the set-up, b) an annulus of a wedge is taken out with axial stress σ_a, radial stress σ_r, and circumferential stress σ_c; for radial force balance it is shown that $\sigma_c = \sigma_r$, c) frictional forces f must be included.

is counteracted by the radial components of the circumferential stresses acting on the sides of the element:

$$\sigma_r \, \Delta\phi\Delta r = \frac{2\sigma_c \, \Delta r\Delta\phi}{2}$$

$$\sigma_r = \sigma_c$$

This shows that the circumferential stress is equal to the radial stress, and the yield criterion

$$(\sigma_1 - \sigma_2)^2 + (\sigma_2 - \sigma_3)^2 + (\sigma_3 - \sigma_1)^2 = 2Y^2$$

reduces to

$$\sigma_a - \sigma_r = Y \tag{5.65}$$

and it is

$$d\sigma_r = d\sigma_a = -dp \tag{5.66}$$

where p is the interface pressure.

Dry Friction: Initial Derivation

At this stage we introduce frictional stress at the faces of the element as shown in Fig. 5.31c. First, let it be Coulomb friction, as in cold upsetting:

$$f_c = \mu p \tag{5.67}$$

The radial equilibrium is now influenced by the addition of the frictional forces due to which there is also a radial gradient $d\sigma_r/dr$ of the radial stress. The circumferential stress equals the mean radial stress:

$$h\Delta\phi \, [(\sigma_r + d\sigma_r)r - (r + dr)\sigma_r + \sigma_c dr] = 2f_c dr\left(r + \frac{dr}{2}\right)\Delta\phi$$

$$r d\sigma_r - \sigma_r \, dr + \sigma_c \, dr = 2f_c r\frac{dr}{h}$$

Now, applying Equations (5.66) and (5.67):

$$-dp = 2\mu p\frac{dr}{h}$$

$$\frac{dp}{p} = \frac{-2\mu}{h}dr \tag{5.68}$$

Integrating (5.68) and applying the boundary condition based on Eq. (5.65), at $r = R$, $\sigma_r = 0$, and $p = Y$, we have

$$p = Ye^{2\mu(R-r)/h} \tag{5.69}$$

At the extremes it is

$$r = R, \qquad p = Y, \quad \text{and} \quad r = 0, \qquad p = Ye^{2\mu R/h} \tag{5.70}$$

$$p_{av} = \frac{1}{\pi R^2}\int_0^R p \, 2\pi r dr = \frac{2Y}{R^2} e^{2\mu R/h} \int_0^R re^{-2\mu R/h} \, dr \tag{5.71}$$

Evaluating the integral gives

$$\int_0^R re^{-(2\mu R/h)}\,dr = \left[\frac{e^{-(2\mu R/h)}}{-2\mu/h}\left(r + \frac{h}{2\mu}\right)\right]_0^R$$

$$= -\frac{h}{2\mu}\left[e^{-(2\mu R/h)}\left(R + \frac{h}{2\mu}\right) - \frac{h}{2\mu}\right]$$

$$= \frac{h^2}{4\mu^2}\left[1 - e^{-(\mu D/h)}\left(1 + \frac{\mu/D}{h}\right)\right]$$

Returning to Eq. (5.71), we have

$$p_{av} = \frac{2Y}{R^2}e^{\mu D/h}\frac{h^2}{4\mu^2}\left[1 - e^{-\mu D/h}\left(1 + \frac{\mu \mathrm{D}}{\mathrm{h}}\right)\right]$$

$$= 2Y\left(\frac{h}{\mu D}\right)^2\left[e^{\mu(D/h)} - 1 - \frac{\mu D}{h}\right] \tag{5.72}$$

Sticking Friction: Friction Shear Flow

For sticking friction (hot forging), the friction stress is replaced by the shear stress k:

$$f_h = k \tag{5.73}$$

and Eq. (5.68) is replaced by

$$dp = -\frac{2k}{h}dr \tag{5.74}$$

This leads to

$$p = \frac{2k}{h}(R - r) + Y = Y\left[1 + \frac{1.15(R - r)}{h}\right] \tag{5.75}$$

$$p_{av} = \frac{2\pi}{\pi R^2}\int_0^R pr\,dr = \frac{2Y}{R^2}\int_0^R\left[\left(1 + 1.15\frac{R}{h}\right)r - \frac{1.15}{h}r^2\right]dr$$

$$= \frac{2Y}{R^2}\left|\left(1 + 1.15\frac{R}{h}\right)\frac{r^2}{2} - \frac{1.15}{h}\frac{r^3}{3}\right|_0^R$$

$$= 2Y\left[\frac{1}{2}\left(1 + 1.15\frac{R}{h}\right) - \frac{1.15}{h}\frac{R}{3}\right]$$

$$= 2Y\left[\frac{1}{2} + \left(\frac{1.15}{2} - \frac{1.15}{3}\right)\frac{R}{h}\right]$$

$$= Y\left[1 + 1.15\left(1 - \frac{2}{3}\right)\frac{R}{h}\right]$$

$$= Y\left[1 + 0.3833\frac{R}{h}\right] \tag{5.76}$$

Upsetting a Cylindrical Part: Combination of Dry Friction and Friction Shear Flow

As in the case of dry friction in plane-strain compression (indenting), in cylindrical upsetting the action of the dry friction is limited by the onset of shear stress flow once the friction stress μp exceeds the shear flow stress k; see Eq. (5.58). Referring to Eq. (5.68),

$$\frac{dp}{dr} = \frac{-2\mu}{h} p \tag{5.77}$$

For sticking friction, according to Eq. (5.74),

$$\frac{dp}{dr} = \frac{-2k}{h} \tag{5.78}$$

These two equations are analogous to (5.56) and (5.57) obtained for indenting. Correspondingly, the "friction hill" in round upsetting, may be plotted as a graph of p versus r, starting from $r = R$ to $r = 0$, using variable $x = R - r$ and will be identical with the graph in Fig. 5.27 as regards p and ps. With the transformation $x = R - r$, Equation (5.69) is indeed identical with (5.47). However, the values of pav and $psav$ and thus also of the resulting force are different, because the integration is now carried out according to Eq. (5.71).

The computer program "frict" of Fig. 5.26 must be modified accordingly, as presented in Fig. 5.32 under the name "cylfrict." For the same inputs of $\mu = 0.2$ and $h = 0.1D$, the graph in Fig. 5.33 of ps versus $x = D/2 - r$ is similar to the graph in Fig. 5.27 obtained for indenting, except that for $x = 0$, we have $p = Y$ and not $1.15Y$. The average pressures are much smaller because now, although the pressure increases about as much towards the center of the circular surface as it did toward the middle of the rectangular surface in indenting, the annular surface at which the increasing pressure acts decreases in proportion to the decreasing radius. The value of $psav = 2.104Y$ as opposed to $psav = 3.146$ is the result of this effect. It is also obvious that shear flow stress has less significance because it is concentrated on the inner part of the round surface, with a lesser area. This is recognized by comparing the value of $psav = 2.104Y$ resulting from the combination of dry friction and shear friction with the value of $pav = 2.153Y$ obtained from Eq. (5.72) considering dry friction only, for the same set of inputs $\mu = 0.2$ and $h/D = 0.1$.

Finally, program "ncylfric" in Fig. 5.34 is analogous to the one in Fig. 5.28, this time for a cylindrical workpiece. The graph in Fig.5.35 is analogous to the one in Fig. 5.29, and it can help in determining the average pressure $psav$ for various combinations of μ and h/D. It expresses the correction coefficient Q_f to take account the effect of dry and shear flow friction on the upsetting force F:

$$F = Q_f YA = Q_f Y\pi \frac{D^2}{4} \tag{5.79}$$

The use of the program "ncylfric" and of the graph in Fig. 5.35 is illustrated in the following example.

EXAMPLE 5.3 Using a Gravity Hammer for Cold Upsetting of a Cylindrical Workpiece ▼

A hammer with a mass $m = 200$ kg falls from a height $H = 4$ m on the workpiece. Initial dimensions are $D = 120$ mm, $h = 20$ mm. Determine the amount of com-

Figure 5.32
Matlab listing "cylfrict" for upsetting cylindrical parts, for combination of dry friction and friction shear.

```
function[p,ps,x,pav,psav]=cylfrict(mu,h)
% p,ps are pressures relative to yield strength, Y=1;
% p applies to dry friction unlimited,
% ps applies to sticking friction setting-in if p>k/mu,
% h is work thickness relative to D=1; h should be read as h/D;
% variable x=R-r is used instead of r to be able to plot pressures
% versus a variable that increases toward the center
% F, Fs are forces obtained by integrating p,ps versus x over annular areas;
% initial values at x=0 are:
Y=1;p(1)=Y;D=1;dx=0.005*D;x(1)=0;
k=0.577*Y;ps(1)=p(1);F=0;Fs=0;
% run the loop for x from 0 to D/2=0.5 (i.e. for r from R to zero)
for n=1:100
x(n+1)=x(n)+dx;
r=D/2-x(n);
% use Eq. (5.68), recall that dx=-dr
dp=2*mu*dx*p(n)/h;
p(n+1)=p(n)+dp;
% check if dry friction holds
if p(n)<=k/mu
ps(n+1)=p(n+1);
else
% use Eq. (5.74)
dps=2*k*dx/h;
ps(n+1)=ps(n)+dps;
end
% force increment over an annulus at r with thickness dx:
dF=p(n)*2*pi*(r-dx/2)*dx;
% keep adding force increments
F=F+dF;
dFs=ps(n)*2*pi*(r-dx/2)*dx;
Fs=Fs+dFs;
end
% average pressures are obtained
% by dividing by the area with diameter D
pav=F/(pi*D^2/4);
psav=Fs/(pi*D^2/4);
```

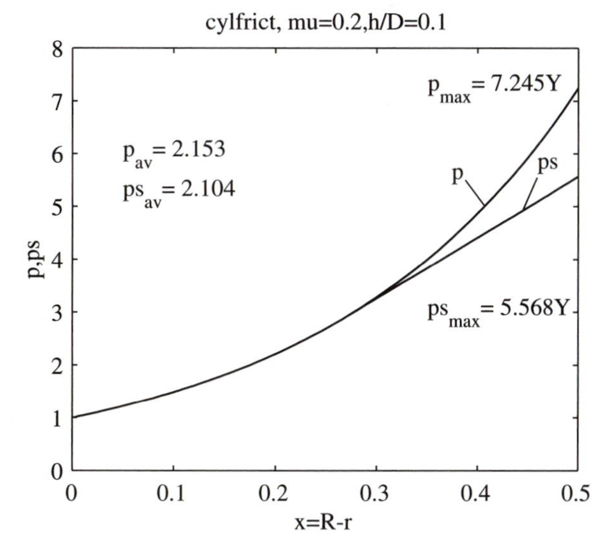

Figure 5.33
Solution of Example 5.3. Pressure p for dry friction only and ps for the combination of dry friction and friction shear stress versus the radius r.

Figure 5.34
Matlab listing "ncylfric" for multiple computing of pressures in cylindrical upsetting.

```
function[pav,psav,mu,h]=ncylfric
% this program computes average pressures in upsetting
% a cylindrical workpiece, for mu=0.05,0.1,0.15,0.2,0.25
% and for h/D=0.025 to 0.2 in steps of 0.025;
% p,ps are pressures relative to yield strength, Y=1;
% p applies to dry friction unlimited,
% ps applies to sticking friction setting in if p>k/mu,
% h is work thickness relative to D=1; h should be read as h/D;
% F, Fs are forces obtained by integrating p,ps versus x=R-r over
% annular areas;
% start the outer loop for the various values of mu
for i=1:5
mu(i)=i*0.05;
% start the next inner loop for various values of h
for j=1:8
h(j)=j*0.025;
%initial values at x=0:
Y=1;p(1)=Y;D=1;dx=0.005*D;x(1)=0;
k=0.577*Y;ps(1)=p(1);F=0;Fs=0;
% run the loop for x from 0 to D/2=100dx=0.5; it is equivalent
% to the program "cylfrict.m"
for n=1:100
x(n+1)=x(n)+dx;
r=D/2-x(n);
dp=2*mu(i)*dx*p(n)/h(j);
p(n+1)=p(n)+dp;
if p(n)<=k/mu(i)
ps(n+1)=p(n+1);
else
dps=2*k*dx/h(j);
ps(n+1)=ps(n)+dps;
end
dF=p(n)*2*pi*(r-dx/2)*dx;
F=F+dF;
dFs=ps(n)*2*pi*(r-dx/2)*dx;
Fs=Fs+dFs;
end
pav(i,j)=F/(pi*D^2/4);
psav(i,j)=Fs/(pi*D^2/4);
end
end
% plot(h,psav)
```

pression from the first blow, h_1. Several blows will follow, and the height of the workpiece will be reduced to $h_2 = 12$ mm. What will be the diameter D_2? Determine the amount of compression Δh_2 from the next blow. The yield strength of the material is $Y = Y_0 + K\varepsilon$, $Y_0 = 300$ N/mm^2, $K = 250$ N/mm^2. The coefficient of dry friction is $\mu = 0.2$.

1. $Y_0 = 300$ N/mm^2, $h/D = 0.167$, $\mu = 0.2$. Running program "cylfrict," Fig. 5.32, yields

$$psav = 1.542Y = 1.543 \times 300 = 462.6 \text{ N/mm}^2$$

Instead of running "cylfrict," we can use the graph of Fig. 5.35. For $\mu = 0.2$ and $h/D = 0.167$, the point is found as marked by the x, and the value of $Q_f \approx 1.54$ is found for the same results as above.

The upsetting force is then

$$F_1 = \pi \times \frac{120^2}{4} \times 462.6 = 5.2319e^6 \text{ N}$$

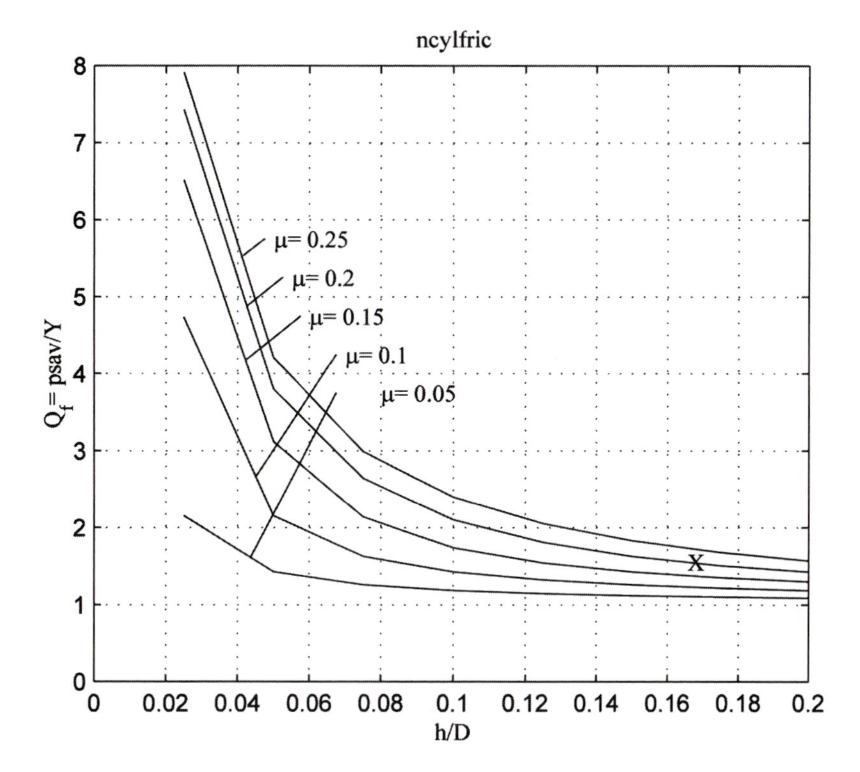

Figure 5.35
Graph for average pressures in cylindrical upsetting under a combination of dry friction and shear flow stress at the interface for various values of *h/D* and of friction coefficient.

The energy in the hammer is

$$E = mg\,H = 200 \times 9.81 \times 4 = 7{,}848 \text{ Nm}$$

$$F_1\,\Delta h_1 = E$$

$$\Delta h_1 = \frac{E}{F_1} = 0.0015\text{m} = 15 \text{ mm}$$

2. For $h_2 = 12$ mm and for constant work volume,

$$h_2\,D_2^{\,2} = h_1\,D_1^{\,2}$$

$$D_2 = D_1\sqrt{\frac{h_1}{h_2}} = 154.92 \text{ mm}$$

$$h/D = 0.077$$

The strain between h_1 and h_2 is $\varepsilon = \ln(h_1/h_2) = 0.511$
The yield strength is $Y_1 = 300 + 0.511 \times 250 = 427.7 \text{ N/mm}^2$
Running "cylfrict" for $\mu = 0.2$ and $h/D = 0.077$ results in

$$psav = 2.585\text{Y} = 1105.6 \text{ N/mm}^2$$

$$\text{F}_2 = psav \times \frac{D_2^{\,2}}{4} = 20.841\text{e}^6 \text{ N}$$

$$\Delta h_2 = E/F_2 = 0.0037\text{m} = 0.37 \text{ mm}$$

A simple program could be written that would follow *h, D, h/D*, and *Y*, and include "cylfrict" as a subroutine to determine the succession of Δh values in the whole operation. ▲

5.3.4 Summary of the Effects of Friction and Redundant Work

We have seen that in both the effects of friction and redundant work on pressures and forces in cold and hot forging the characteristic ratio Δ plays an important role. Redundant work was discussed in relation to the "indentation" type of compression, where the compressive action is applied to a part of the surface only and the surrounding material constrains the deformation. This includes most of the forming operations and applies not only to the actual indenting (open-die forging) but also to drawing, extrusion, and rolling. Redundant work becomes important for $h/L >> 1$. The effect of friction at the tool/work interface applies to both the indenting and upsetting type of operations, as long as the ratio Δ is small, $\Delta = h/L < 1$.

Figure 5.36 summarizes both effects in showing the coefficients Q_r (Eq. 5.40) and Q_f (Eq. 5.64) for $\mu = 0.1$ and $\mu = 0.5$, as functions of Δ for plane-strain compression. Note that at $\Delta \simeq 1$ both the effects are negligible. In this case the operations are most efficient. For cases where $\Delta < 1$ it is sufficient to consider friction alone, and for $\Delta >> 1$, for constrained operations (indenting type), redundant work must be considered.

In reference to Fig. 5.21 we indicated that redundant work as derived for indenting compression can also be applied to other operations like drawing, extrusion, and rolling if the characteristic ratio Δ is suitably formulated. Correspondingly, we can also use the correction coefficient Q_r as given in the graph of Fig. 5.20 for these operations and combine it with the stress and force formulas derived in Section 5.2 in the way that we expressed the forging pressure by the formula (5.40).

However, the effect of friction as it was derived for plane strain and axisymmetrical compression is not simply applicable to some of the other operations shown in Fig. 5.21, especially not to drawing and extrusion. Neither of these operations involves a point of no sliding between tool and workpiece that would correspond to the middle points in Figs. 5.24 and 5.31. The pressure between the work and the die varies strongly from one end to the other even without friction, as was shown in Equations (5.30) and (5.31). The angle of the die plays a role of its own apart from affecting the characteristic ratio Δ. For these reasons the role of friction in drawing, extrusion, and rolling has been investigated separately.

There are many practical implications of the effects described in the preceding sections. For instance, in indenting a deep material, as in using a punch for piercing as indicated in Fig. 5.37a, the pressure on the end of the punch will be approximately $3Y_w$

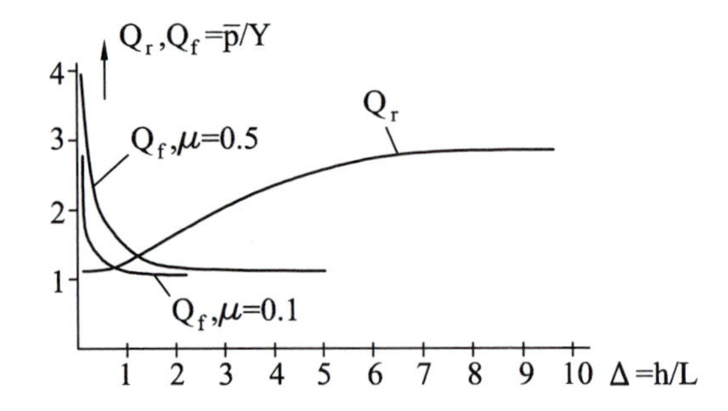

Figure 5.36
Both correction coefficients, Q_f for friction and Q_r for redundant work as functions of $\Delta = h/L$, for plane-strain compression.

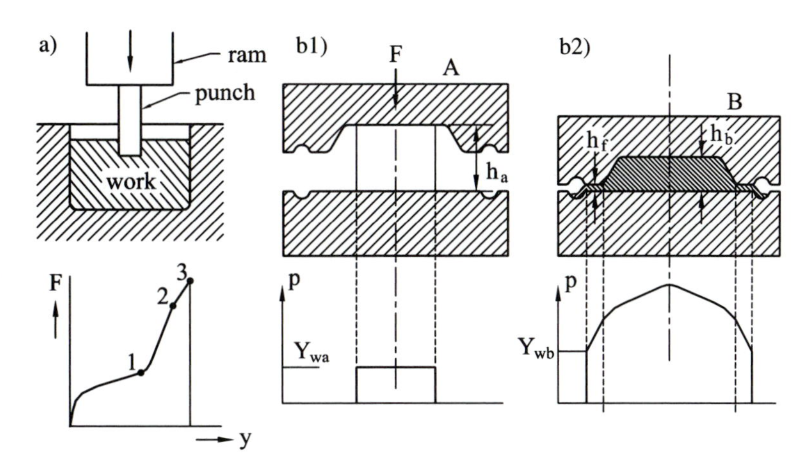

Figure 5.37
Pressures in a) punching and b) die forging. Diagrams b1) and b2) show the start and end positions, *A* and *B* of the upper die motion. The diagram c) indicates the variation of force *F* with travel *y* of the upper die. The effect of friction on the pressure between die and work increases as *h/D* is decreasing.

(see discussion following Eq. (5.38)), where Y_w is the yield strength of the work material. This means that the material of the tool must have a yield strength $Y_t > 3Y_w$. This is not difficult to achieve in hot working, but for cold working the punch must be made from a very strong, hardened tool steel. On the other hand, the ram that drives the punch can be made of common steel not much stronger than that of the workpiece because it would not yield under pressure $p < 3Y_p$, where Y_p is the yield strength of the platen. It is, of course, also easy to arrange for the top end of the tool to have a larger diameter than its working bottom end.

For closed-die forging the diagram of a simple case is given in Fig. 5.37b1 and b2. At the beginning of the particular operation shown, with the upper half of the die in position *A*, we have unconstrained upsetting with a very high ratio *h*/*d*. There is practically no effect from friction, and because there is no constraint there is little redundant work. The pressure between die and work is Y_{wa}, which is the yield strength of the work material as it corresponds to the given temperature and strain rate.

In Fig. 5.37c the force versus downward travel *y* is plotted. As the upper die descends (point 1 on this graph), the flash starts to form. In position *B* (Fig. 5.37b2) the die is filled, and flash continues to be squeezed out. At this stage the pressure distribution is as shown in the graph below the diagram. At the outer diameter of the flash, the pressure is Y_{wb}, which is the yield strength of the material as it corresponds to the now lower temperature and higher strain rate in the flash. The pressure increases inward rather steeply due to friction because of the rather low value of h_f/e at the flash area. From the entrance of the flash area towards the center of the diameter *d* of the forging, the pressure increases further due to friction, while corresponding to the ratio h_b/d, which is higher than that of the flash area. Because of the larger value of Δ the pressure increases less steeply towards the center (stages 1–2 of Fig. 5.37c). There is still an additional movement during which the flash is further squeezed, and the force reaches its peak at 3. The peak force may be 5–10 times higher than the force in the "free" forging stage before the material started to flow in the flash area.

It is also interesting to review, with reference to Fig. 5.21, the operations of drawing and extrusion. Drawing of wire and rod is a cold-forming operation. It is carried out in dies with small angles $\alpha = 7°–15°$. The maximum reduction ratio per pass is limited by the strength of the drawn wire, and it is about $(d_0/d_1)_{max} = 1.6$. For operations close to this ratio (see Fig. 5.22a1), the characteristic ratio $d_m/L \simeq 1$; thus, it is a rather efficient operation where there is little effect of redundant work. In lesser reductions with $\alpha = 15°$ the ratio Δ would be larger, and redundant work would occur. However, this can be remedied by using a smaller angle α. One can conclude that cold drawing can be made very efficient by properly adjusting the die angle α with respect to the reduction ratio.

In extrusion no other limit is imposed on the reduction ratio than the strength of the die. For good efficiency at large reductions it is obviously necessary to increase the die angle. Extrusion-area ratios used in practice may be 20:1 (i.e., $d_0/d_1 = \sqrt{20} = 4.5$) for harder alloys and up to 100:1 ($d_0/d_1 = 10$) for aluminum. In these cases the ratio $d_m/L = 1$ would be obtained for $\alpha = 40°$ and $\alpha = 55°$ respectively (see Fig. 5.22).

Due to the high pressure both between the work and the container and the work and the die it is important to use efficient lubrication (e.g., graphite in cold extrusion of steel, glass in hot extrusion of steel). Often a 90° die angle is used, in which case a dead zone of the work material forms that acts as a die with the optimum angle. Between the dead zone and the work, shear sliding occurs. This method is often used in unlubricated hot extrusion of aluminum.

5.3.5 Force and Neutral Point in Cold Rolling

In this exercise we analyze the pressures between the roll and the strip while considering friction, strain-hardening, and the effects of back and forward tensions. For the geometry of the case, refer to Fig. 5.38b. The roll radius is R. This is the "actual radius," which may differ from the "nominal radius" because the roll flattens elastically during contact. This flattening may be considerable in rolling thin sheet with small reductions. However, for most cold rolling it is not substantial.

The starting thickness at the entry is h_s, and the exit thickness is h_e. Somewhere between the entry and the exit is the "neutral point" N, at which there is no slip, and

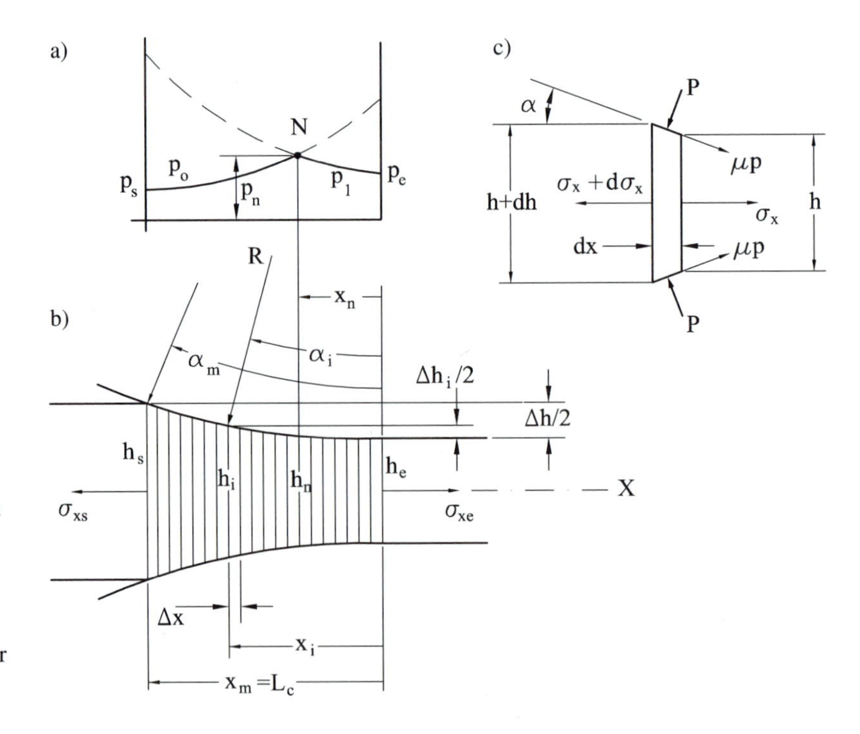

Figure 5.38
Pressures in rolling:
a) pressures p will be obtained in the computation by starting from both ends and meeting in the neutral point N; b) shows the geometry of the computation; and c) is the basis for the force balance equation.

the velocity of the strip v is equal to the peripheral velocity of the roll v_r. The width of the strip is considerably greater than its thickness. The typical values will be $h_s = 2.5$ mm, $h_e = 2$ mm, $R = 375$ mm, and the width $w = 1000$ mm.

This is clearly a case of plane-strain yielding:

$$\sigma_1 - \sigma_3 = 2k = 1.15Y$$

Considering strain-hardening, we will use a formula of the following type:

$$Y = Y_0 + K\varepsilon = Y_0 + K \ln \frac{h_s}{h} \tag{5.80}$$

For particular values to use in an example we will choose

$$Y = 400 \text{ N/mm}^2, \qquad K = 300 \text{ N/mm}^2 \tag{5.81}$$

Considering the balance of forces on a slice with thickness dx, the vertical components of all the forces acting on the slice from both the rolls eliminate each other. The balance in the direction X will depend on whether the slice is located before or behind the neutral point N in the direction of strip motion. From entry to N, the periphery of the rolls moves faster than the strip; there is slippage between the two, and the friction F_f acts on the strip in the direction of its motion. Between N and the exit the strip moves faster than the roll, and the friction force acts against the motion of the strip. The velocity v of the strip varies from v_s to v_e, and at any location it is given by

$$v = \frac{v_n h_n}{h} \tag{5.82}$$

where $v_n = v_r$ is the velocity at the neutral point equal to roll peripheral velocity, and h_n is the thickness at the neutral point.

The forces involved in the balance are (see Fig. 5.38c) listed below:

(a) The contact pressure force. The contact pressure p acts on an area dx/cos α, and its horizontal component, from both rolls, is

$$Fp = 2p\left(\frac{\mathrm{d}x}{\cos\alpha}\right)\sin\alpha = 2p\mathrm{d}x \tan\alpha \tag{5.83}$$

(b) The friction forces (on both rolls):

$$F_f = 2\mu p\left(\frac{\mathrm{d}x}{\cos\alpha}\right)\cos\alpha = 2\mu p\ \mathrm{d}x \tag{5.84}$$

where μ is the coefficient of friction. This force is positive (acts in the $+x$ direction) for slices between N and the exit, and it is negative for slices between entry and N.

(c) The normal forces: acting on the right face,

$$F_{xr} = -h\sigma_x$$

and on the left face,

$$F_{xl} = (h + \mathrm{d}h)(\sigma_x + \mathrm{d}\sigma_x) \tag{5.85}$$

Considering the yield criterion, we find

$$\sigma_x + p = 1.15Y, \qquad \sigma_x = 1.15Y - p, \qquad \mathrm{d}\sigma_x = -\mathrm{d}p \tag{5.86}$$

and considering the geometry, we have

$$\alpha = \sin^{-1}\left(\frac{x}{R}\right) \tag{5.87}$$

$$h = h_e + 2R(1 - \cos \alpha) \tag{5.88}$$

$$dh = 2\, dx \tan \alpha \tag{5.89}$$

The equilibrium of the forces F_p, F_f, F_{xr}, F_{x1}, for the part between N and exit, is written under consideration of Eq. (5.89) as

$$2(p\, dx \tan \alpha + \mu p dx) + (h + dh)(\sigma_x + d\sigma_x) - h\sigma_x = 0$$

$$2(p\, dx \tan \alpha + \mu) + h\, d\sigma_x + \sigma_x\, dh = 0$$

(the term $d\sigma\, dh$ was neglected as a second-order differential), and finally:

$$dp = 2\, dx \frac{1.15Y \tan \alpha + \mu p}{h} + 1.15dY \tag{5.90}$$

For the part between the entry and N, the friction force has the opposite direction; it would be

$$dp = 2\, dx \frac{1.15Y \tan \alpha - \mu p}{h} + 1.15dY$$

However, in the computing routine, we will be proceeding in this part from left to right, and the increment dx will thus have to be considered negative. If it is taken just as a positive number, the balance equation will apply in the following form:

$$dp = 2dx \frac{\mu p - 1.15Y \tan \alpha}{h} + 1.15dY \tag{5.91}$$

where dY has to be considered properly both in (5.90) and (5.91): for (5.90) proceeding right to left, with x increasing, Y is decreasing, and dY will automatically become negative, while in (5.91) proceeding in the direction of decreasing x (decreasing h), the value of Y (see Eq. 5.80) increases, and dY is positive.

In the computer program the pressure p is incremented from both ends of the roll-work contact (see Fig. 5.38a), starting from the pressures at the ends. The pressure p_1 starts from the exit pressure:

$$p_1 = 1.15Y_e - \sigma_{xe} \tag{5.92}$$

where Y_e is obtained from Eq. (5.80) for $h = h_e$, and σ_{xe} is the normal stress of the forward tension in the strip. The pressure p_1 assumes the direction of friction against the strip motion and is obtained from Eq. (5.90). It applies for the part of the contact after the neutral point.

The pressure p_0 starts from the entry ("start") pressure:

$$p_0 = 1.15Y_s - \sigma_{xs} \tag{5.93}$$

where Y_s is obtained from Eq. (5.80) as $Y_s = Y_0$, and σ_{xs} is the normal stress of the "back tension" in the strip. The pressure p_0 assumes the direction of friction driving the strip motion and is obtained from Eq. (5.91). It applies for the part of the contact before the neutral point.

By computing both p_1 (x) and p_0 (x), the point is found where they cross, which is the neutral point with the coordinate x_n:

$$p_1(x_n) = p_0(x_n) = p_n$$

Obviously, only the solid parts of the p_1 and p_0 lines apply and not the dashed-line parts. The point where the two lines meet is the neutral point. Correspondingly, the pressure p is obtained as follows:

$$\text{if } p_0 < p_1, \qquad p = p_0$$

$$\text{if } p_1 < p_0, \qquad p = p_1 \tag{5.94}$$

and the force between a roll and the strip is obtained by integrating:

$$dF = p\,dx$$

over the length of contact.

The solution is obtained by computation in the following steps:

1. Input data
   w, h_s, h_e, R, v_r, for the geometry of the case
   Y_0, K, μ, for work material properties and friction
   σ_{xs}, σ_{xe}, for the back and forward tensions applied
   Δx as the parameter for discretizing the situation

2. Derived parameters

$$\Delta h_m = h_s - h_e$$

$$\alpha_m = \cos^{-1}\left[\frac{(R - \Delta h_m/2)}{R}\right]$$

$$x_m = L_c = R \sin \alpha_m$$

$$m = \text{integer}\left(\frac{x_m}{\Delta x}\right)$$

$$Y_e = Y_0 + K \ln\left(\frac{h_s}{h_e}\right) \tag{5.80}$$

$$p_s = 1.15Y_0 - \sigma_{xs}, \tag{5.93}$$

$$p_e = 1.15Y_e - \sigma_{xe} \tag{5.92}$$

3. Compute $p_{0,i}$ by repeating the following loop for $i = 1$ to m:

$$x_i = i\,\Delta x$$

$$\alpha_i = \sin^{-1}\left(\frac{x_i}{R}\right)$$

$$\Delta h_i = 2R\,(1 - \cos \alpha_i)$$

$$h_i = h_e + \Delta h_i$$

$$Y_1 = Y_0 + K \ln\left(\frac{h_s}{h_i}\right), \tag{5.80}$$

(a) for $i = 1$

$$\Delta Y_i = Y_i - Y_e$$

$$\Delta p_{1,i} = 2\Delta x(1.15 Y_i \tan \alpha_i + \mu p_e)/h_i + 1.15\Delta Y_i \tag{5.90}$$

$$p_{1,i} = p_e + \Delta p_{1,i}$$

(b) for $i > 1$

$$\Delta Y_i = Y_i - Y_{i-1}$$

$$\Delta p_{1,i} = 2\,\Delta x \frac{1.15\, Y_i \tan \alpha_i + \mu p_{1,i-1}}{h_i} + 1.15\,\Delta\, Y_i \tag{5.90}$$

$$p_{1,i} = p_{1,i-1} + \Delta p_{1,i}$$

4. Compute $p_{0,i}$ by repeating the following loop for $i = 1$ to m:

$$k = m + 1 - i$$

for $i = 1$, $\quad p_{0,m} = p_s$

$$\Delta p_{0,k} = 2\,\Delta x \frac{\mu p_{0,k+1} - 1.15\, Y_k \tan \alpha_k}{h_k} - 1.15\,\Delta Y_k$$

In this expression the sign ΔY is reversed as compared to the one established in Step 3 because now we are progressing in a sequence of descending subscripts.

for $i > 1$, $\quad p_{0,k} = p_{0,k+1} + \Delta p_{0,k}$

5. Find the neutral point, and establish p, F, v_s, v_e.
   (a) If $p_{0,1} \leq p_e$, $n = 1$, $p_n = p_e$, $x_n = 0$.
   (b) If (a) is not true, n equals the smallest i for which $p_{1,i} > p_{0,i}$
   for $i = 1$ to $(n - 1)$, $p_i = p_{1,i}$
   for $i = n$ to m, $p_i = p_{0,i}$

$$F = w\Delta x \sum_1^m p_i$$

$$v_s = v_r \frac{h_n}{h_s}$$

$$v_e = v_r \frac{h_n}{h_e}$$

6. Plot p versus x.

EXAMPLE 5.4 Compute the Pressures and Force in Rolling ▼

The listing of the Matlab program used in this example follows the above sequence of steps 1–5 and is presented in Fig. 5.39. Execute program "rollfrc" for the following input data:

$h_s = 2.5$ mm, $h_e = 2.0$ mm, $R = 375$ mm, $w = 1000$ mm,
$Y_0 = 400$ N/mm^2, $K = 300$ N/mm^2, $\mu = 0.1$, $\Delta x = 0.1$ mm

Figure 5.39
Matlab listing "rollfrc" for computing forces in cold rolling, Example 5.4.

```
function[x,po,p1,p,M,N,F]=rollfrc(he)
% hs and he are start and exit thicknesses, R is roll radius,
% w is strip width, mu is coefficient of friction, dist is distance
% between stands, sigs and sige are interstand tensile stresses at start and end
% al is alpha, L is contact length=M*dx, N is the value of subscript n
% at neutral point, m is the subscript denoting each of the four different
% sets of interstand tensions
hs=2.5;R=375;w=1000;Y0=400;K=300;mu=0.1;dx=0.1;
dt=0.00025;dist=2000;E=2e5;he=2;
del=0.25;als=acos((R-del)/R);L=R*sin(als);
M=round(L/dx);
Ye=Y0+K*log(hs/he);x(M+1)=L;
for m=1:4
if m==1
sigs=0;sige=0;
end
if m==2
sigs=0;sige=160;
end
if m==3
sigs=80;sige=0;
end
if m==4
sigs=160;sige=160;
end
ps=1.15*Y0-sigs;pe=1.15*Ye-sige;
p1(1,m)=pe;po(M+1,m)=ps;
% run the loop from exit towards start
for n=1:M
x(n)=(n-1)*dx;
y(n)=(M-n+1)*dx;
al=asin(x(n)/R);
dlh=2*R*(1-cos(al));
h=he+dlh;
Y=Y0+K*log(hs/h);
if n<2
dlY=Y-Ye;
else
% Ypr is previous Y
dlY=Y-Ypr;
end
Ypr=Y;
% use Eq. (5.90)
dp1=2*dx*(1.15*Y*tan(al)+mu*p1(n,m))/h+1.15*dlY;
p1(n+1,m)=p1(n,m)+dp1;
end
% run the loop from start towards exit
for i=1:M
k=M+1-i;
x(k)=(k-1)*dx;
al=asin(x(k)/R);
dlh=2*R*(1-cos(al));
h=he+dlh;
Y=Y0+K*log(hs/h);
if i<2
dlY=Y-Y0;
else
% Ybf is Y before
dlY=Y-Ybf;
end
Ybf=Y;
dpo=2*dx*(-1.15*Y*tan(al)+mu*po(k+1,m))/h+1.15*dlY;
po(k,m)=po(k+1,m)+dpo;
end
```

(continued)

Figure 5.39
(continued)

```
F(m)=0;
% determine the position N of the neutral point to decide which of
% p1 or po to accept as p
for n=1:(M+1)
dlta=p1(n,m)-po(n,m);
if dlta<0
p(n,m)=p1(n,m);
r=n;
N(m)=max(r);
else
p(n,m)=po(n,m);
end
F(m)=F(m)+p(n,m)*dx*w;
end
end
%plot(x,p)
```

and

(a) $\sigma_{xs} = 0$, $\sigma_{xe} = 0$
(b) $\sigma_{xs} = 0$, $\sigma_{xe} = 160$ N/mm^2
(c) $\sigma_{xs} = 80$ N/mm^2, $\sigma_{xe} = 0$
(d) $\sigma_{xs} = 160$ N/mm^2, $\sigma_{xe} = 160$ N/mm^2

The results of the computations are plotted in Fig. 5.40. The forces obtained are

(a) $F = 9.32e^6$ N (c) $F = 8.26e^6$ N
(b) $F = 7.92e^6$ N (d) $F = 6.20e^6$ N

The effects of the interstand tensions are clearly demonstrated. By applying tension both upstream and downstream, the force decreased by a third. The neutral point remained at almost the same location. Further decrease of the rolling force could obviously be obtained by using larger back and forward interstand tensions. ▲

5.3.6 Material Failure in Bulk Forming

Material failure in bulk forming is due to excessive shear along the so called "shear bands" or else to secondary tensile stresses that arise in certain zones of the workpiece in

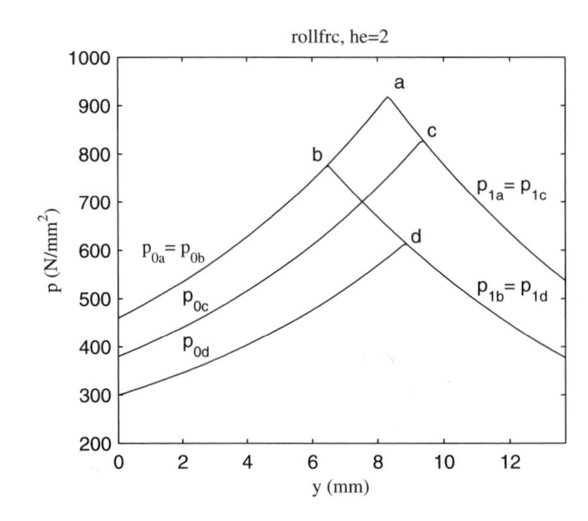

Figure 5.40
Results of Example 5.4. Pressures in rolling under various combinations of interstand tensions. The extremes are a) no interstand tension is applied and d) tensile stress of 160 MPa applied both at entry and exit. Peak pressure dropped 1.7 times and the total force dropped 1.5 times. Higher force reduction could be obtained with higher interstand tension.

which voids start to coalesce into increasing cavities and eventually form cracks. The voids and cavities may be in the material from the preceding casting operation, or they may arise around impurities and hard inclusions and, in hot forming, at stress concentrations along grain boundaries. The latter is especially likely in coarse-grained material. Therefore, for example, forging is first done to crush and refine the grains before proceeding to produce the intricate final shapes. In cold forming, fractures may occur due to strain-hardening up to a complete exhaustion of ductility of the material.

In upsetting the two dangerous phenomena are the formation of shear bands (Fig. 5.41a) and surface cracking due to tensile stresses arising from barreling of the work (Fig. 5.41b). The shear bands (shown by dashed lines) result when friction constrains the deformation of the metal at the contact with the dies. Under these contacts "dead zones" form. At their boundaries very large shear strains occur that are conducive to the development of larger cavities and voids. The shear bands are not very detrimental in cold forging, where they are diffuse due to work hardening. Improvements in lubrication usually lead to the elimination of the problem. The shear bands may become a serious problem in hot forging. Barreling and the associated cracking are also linked with the friction at the dies and the radial flow occurring in the middle part of the specimen. Whenever this problem arises, it can be solved either by improved lubrication or by redistributing the sequence of the successive steps of the entire deformation.

Another phenomenon is "center bursting" (see Fig. 5.41c), due to high secondary tensile stresses in nonhomogeneous compression (high h/L ratio) of circular sections between flat platens. This condition is usefully employed in hot seamless tube making (the Mannessman process), where a round bar is rolled between crossed cylinders. The contact between the rolls and the workpiece is consequently very narrow leading to a very high h/L ratio. This initiates a cavity in the center that is expanded by axially pushing the work against a mandrell.

The highly nonhomogeneous deformation that can occur in extrusion leads to tensile stresses in the center of the extruded bar. Characteristics cavities occur, as shown in Fig. 5.41d. They are called chevron-type cavities. When they are encountered, it is

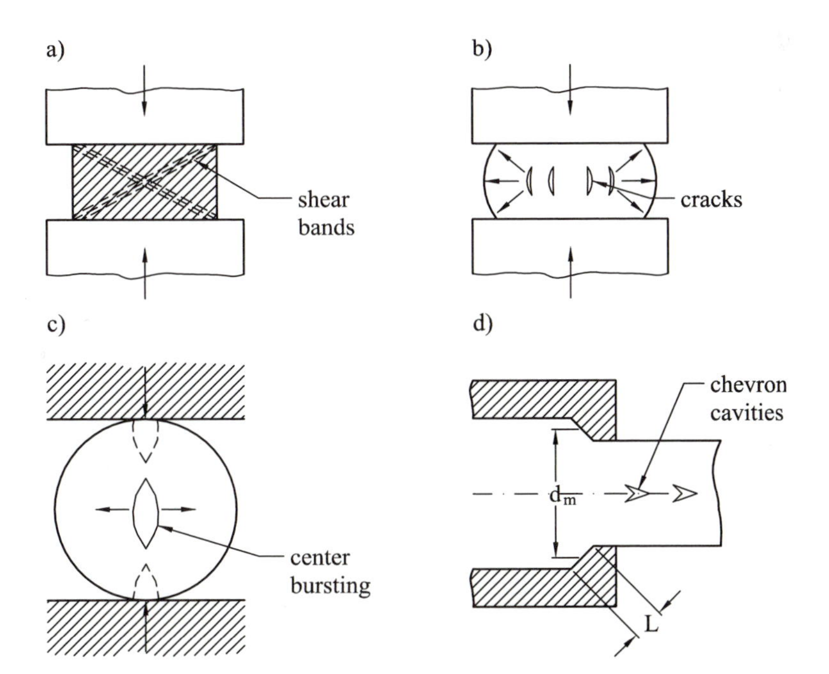

Figure 5.41
Failures in bulk forming: a) shear bands; b) surface cracking due to barreling; c) center bursting; d) chevron cavities in extrusion.

necessary to change the extrusion ratio or the die angle so as to reduce the nonhomogeneity of the deformation.

Concern with the integrity of the product is one of the most important factors in the design of the individual stages of bulk-forming operations.

5.4 ANALYSIS OF PLATE- AND SHEET-METAL FORMING

In the plate- and sheet-metal forming operations the surface area and the thickness of the work do not change significantly at all. The changes are limited, on one hand, to cutting the area (shearing, punching, blanking, etc.) and, on the other hand, to out-of-plane form changes (bending, stretching, deep drawing). These various operations are presented here first in their simplest form and with simplified formulas for the forces involved. Later on, a more detailed analysis of some selected aspects of these operations is carried out.

5.4.1 Simplified Analysis

The *shearing* operations are well represented by the operation of *punching* (see Fig. 5.42). It is used to cut out openings in sheet metal and plates. The perimeter o of such openings may be round or of any other shape. The tools used are called the punch and the die, representing the male and female parts. The opening in the die is greater than the periphery of the punch by a small clearance. The process of plastic deformation here is shear along the perimeter o and through the thickness h of the plate. If the shear flow strength of the material is k, the force F must increase to the value

$$F = koh \tag{5.95}$$

to produce the shearing action. The force versus stroke y diagram is shown in the figure. A shearing fracture occurs after a certain amount of shear strain, which in practical terms is measured by the "penetration" p of the punch. The penetration p is expressed

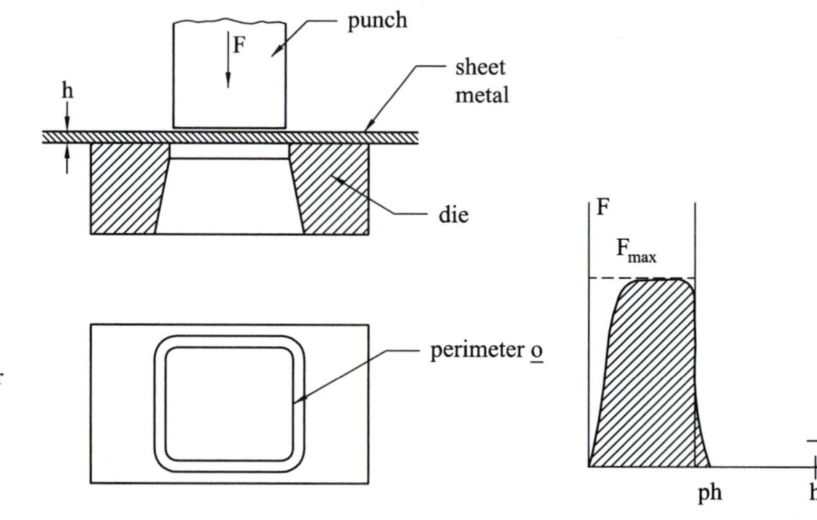

Figure 5.42
Force in punching is proportional to the perimeter of the work and the thickness of the sheet. Force drops to zero at partial penetration.

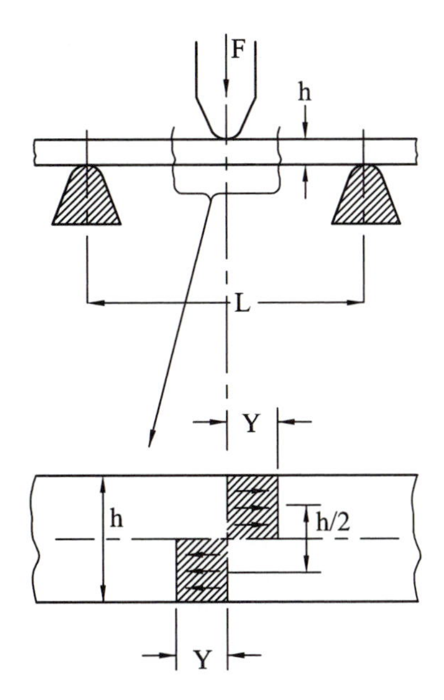

Figure 5.43
Force in bending. External moment equivalent to moment of internal stresses. Fully plastic case is shown.

as a fraction of the thickness h. Once this amount of plastic deformation is obtained, the plate fractures along the perimeter of the punch and the force drops to zero. For a simplified evaluation, and assuming a safety factor, the force diagram will be represented as shown by the dashed line indicating a sudden increase of the force at the beginning and its constant value $F = F_{max}$ over the penetration part of the travel with a sudden drop to zero at fracture. Correspondingly, the work expended in the operation is

$$W = Fph \tag{5.96}$$

In *bending*, (Fig. 5.43), the purpose is to produce yielding all across the thickness of the plate in those sections that should be bent. In the simplest approach a one-dimensional stress is assumed with compression above the neutral axis and tension below it, and a non-strain-hardening material that yields both in tension and compression at constant stress Y. With plate width w and thickness h, the moment of the stresses in the section C is

$$M_s = Y\frac{h}{2}\cdot\frac{h}{2}w = Y\frac{h^2}{4}w$$

and the moment from the external force F, or else from the reactions $F/2$, is

$$M_f = \frac{F}{2}\cdot\frac{L}{2} = \frac{FL}{4}$$

Writing out the balance of the internal and external moments gives the bending force:

$$\frac{FL}{4} = \frac{wYh^2}{4}$$

$$F = \frac{wYh^2}{L} \tag{5.97}$$

In *stretching* (Fig. 5.44) the sheet metal is required to permanently conform to a shape where most of the bending is very light, with large radii. Correspondingly, if the sheet was just pressed (e.g., by means of a conforming punch) against the die, the stresses caused by this bending would remain elastic and not exceed the yield stress at all, or not through much of the material thickness. The sheet would almost completely spring back to its original flatness. Therefore, forces F are applied at the edges of the sheet that produce tensile yield stress in the material. Actually, at section A on the underside of the sheet there will be some small compressive stress due to the bending σ_b, and the tension σ_t from forces F must be high enough to produce tensile yielding, $\sigma_t = Y_0 + \sigma_b$. This tensile stress is transmitted through section C where the material first yields at Y_0 and then strain-hardens to $Y_1 = Y_0 + \sigma_b$. This simple presentation neglects the effect of friction between the sheet and the die. However, at the first approximation, the stretching force must be slightly higher than

$$F = CY_0hw \tag{5.98}$$

where the factor C may have a value of about $C = 1.2$–1.5 to include both the strain-hardening and the friction, h is the sheet thickness, and w is its width.

In *deep drawing* (Fig. 5.45), which is used for production of cup or can-type workpieces, the sheet metal is drawn by the punch into the die from the original flat blank. For instance, the blank may be a round piece of sheet metal with diameter d_0, and the drawn cup will be cylindrical and have diameter d_1 and height L. The thickness h of the sheet will not significantly change during this operation; it will be approximately

$$\frac{\pi d_0^2}{4} = \frac{\pi d_1^2}{4} + \pi d_1 L \tag{5.99}$$

The blank being drawn is compressed by a "hold-down" ring (HD) so as to prevent the wrinkling that would otherwise occur due to circumferential compression as the diameter of the periphery of the blank is shrinking progressively from d_0 to d_1.

The deformation in this operation consists of the just-mentioned plane shrinking on the top surface of the die, of bending 90° around the radius r_1, unbending 90° into the cylindrical wall, and bending 90° around the radius r_2. The last part concerns only a small amount of the material in the vicinity of the diameter d_1.

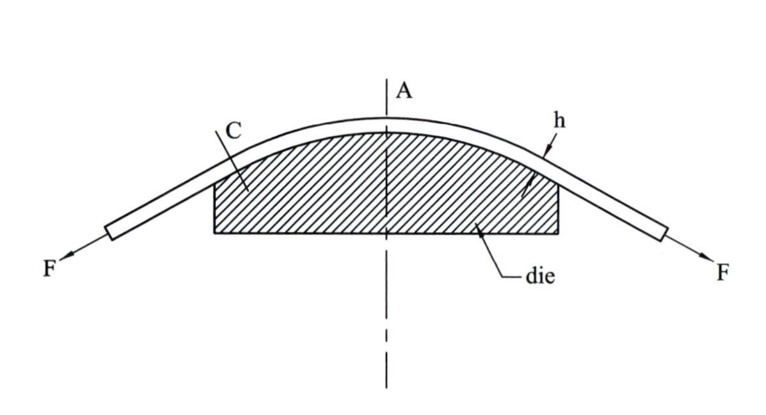

Figure 5.44
Stretch forming. Tensile stresses are added to those of bending to cause permanent strain also over large curvatures.

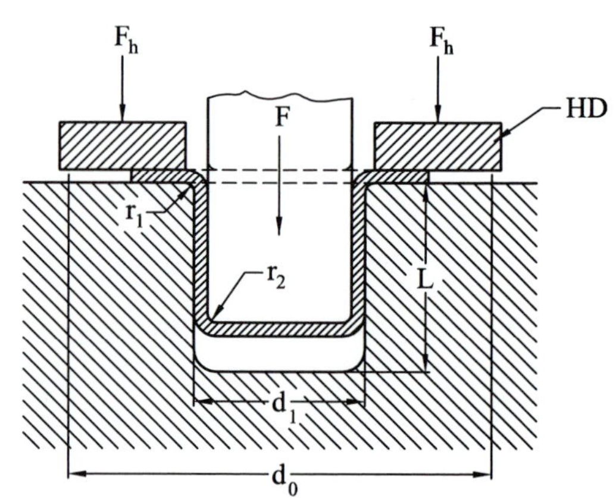

Figure 5.45
Deep drawing. Compression in the diminishing flange is driven by the tensile stress in the cup wall.

In the simplest approach, we may reason fairly realistically that the limit on the drawing force F is imposed by the tensile yield strength of the cylindrical wall of the cup.

Correspondingly,

$$F = Y\pi d_1 h \tag{5.100}$$

From this condition a limit is derived for the ratio d_0/d_1, or as expressed in another form, for the ratio of L/d_1. This will be discussed later. In the following we analyze some aspects of the various operations in more detail.

5.4.2 Elastic and Plastic Bending

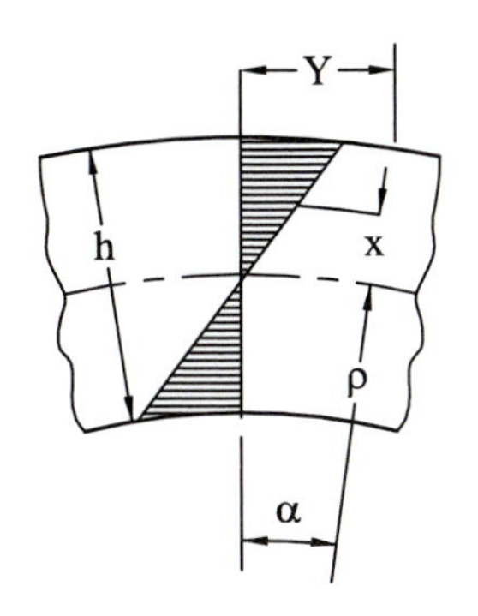

Figure 5.46
Elastic bending. Both stress and strain proportional to x.

In the preceding we very simply assumed that in bending the yield stress is reached through the whole section of the plate. Strictly speaking, this is not possible; there will always be a zone, even if very small, close to the neutral axis where the yield stress will not be attained. Let us consider several stages of severity of bending as they are characterized by the bend radius ρ (see Fig. 5.46). Let us take the case of the limit elastic deformation where the stress, tensile on the outside and compressive on the inside, just reaches the value Y on these outer surfaces.

Strain at the distance x from the neutral axis is obtained if a length of the plate is considered as it corresponds to an angle α; at the neutral axis this length is $l = \rho\alpha$, and it was the same at the level x before bending. Now, it is $l_x = (\rho + x)\alpha$.

Correspondingly,

$$e_x = \frac{\Delta l}{l} = \frac{(\rho + x)\alpha - \rho\alpha}{\rho\alpha} = \frac{x}{\rho} \tag{5.101}$$

and $\varepsilon = \ln(1 + e)$ and, approximately, $\varepsilon = e$.
At the extreme fibers it is $x = h/2$ and $e = h/(2\rho)$, and the stress is

$$\sigma_{\max} = Y = eE = \frac{hE}{2\rho} \tag{5.102}$$

where E is the modulus of elasticity.

The strain and stress are distributed linearly from $\sigma = 0$ at the neutral axis to $\sigma_{\max}$ at the surfaces. The moment produced by these stresses is

$$M = 2Y\frac{h/2}{2}\cdot\frac{2}{3}\cdot\frac{h}{2}w = \frac{Yh^2w}{6} \tag{5.103}$$

Combining Equations (5.102) and (5.103) by eliminating Y from both of them, we obtain the relationship between the radius ρ and the bending moment M, which applies to the extreme elastic bend:

$$\frac{1}{\rho} = \frac{M}{EI}, \qquad \text{where } I = \frac{wh^3}{12} \tag{5.104}$$

Figure 5.47a shows a diagram of stress distribution in a section of a bent plate, where over a part of the section the strain is plastic. A rigid, perfectly plastic material is assumed with no strain-hardening. Correspondingly, the stress in the plastic part is constant and equal to Y. The layer with thickness a around the neutral axis is in elastic strain and the stress in the part varies linearly from zero at the neutral axis to Y at the distance $a/2$. The strain remains linear through the whole cross section. The elastic strain according to formula (5.101) is $a/(2\rho)$, and the corresponding stress is

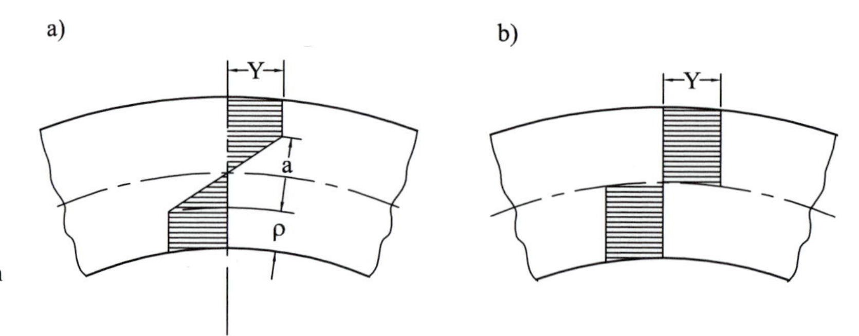

Figure 5.47
Elastic and plastic bending: a) a layer with thickness a remains elastic; b) the theoretical extreme case is shown of fully plastic situation. In reality, this is not possible because of the fracture limit of tensile strain on the outer surface.

$$\sigma_{a/2} = e_{a/2}E = \frac{Ea/2}{\rho} = Y$$

and the corresponding curvature is

$$\frac{1}{\rho} = \frac{2Y}{Ea} \tag{5.105}$$

The moment from the elastic part is (see Eq. 5.103)

$$M_e = \frac{Ya^2w}{6}$$

and the moment from the plastic part is

$$M_p = 2Yw\left(\frac{a}{2} + \frac{h-a}{4}\right)\left(\frac{h-a}{2}\right) = Yw(h^2 - a^2)/4$$

The total moment is

$$M = M_e + M_p = Yw\left(\frac{h^2}{4} - \frac{a^2}{12}\right) \tag{5.106}$$

In the limit case, as shown in Fig. 5.47b, $M = Yh^2w/4$ and, theoretically, $a = 0$, and $\rho = 0$. But this cannot even be approached, because long before this happens the strain at the surface of the beam exceeds the maximum elongation to fracture. There is always an elastic layer inside even if very small.

For steel it is usual to take the yield point at a strain of $e = 0.002$. Accordingly, the elastic layer is determined by $e = (a/2)/\rho = 0.002$, or $a = 0.004\rho$.

If, for example, a steel beam can take $e_{\max} = 0.3$, it will be possible to bend it to the smallest radius:

$$\rho_{\min} = \frac{h}{2 \times 0.3} = 1.6h$$

and the corresponding elastic layer will be $a = 0.0064h$.

The bent beam, if it is not very wide, will narrow down at the outside and widen at the inside to compensate for the circumferential elongation and compression respectively (see Fig. 5.48). If the beam is wide these lateral strains cannot occur except at the ends of the section (see Fig. 5.48b). For the major part of the section, $\varepsilon_y = 0$; in order to achieve this, a lateral stress σ_y is generated that is compressive on the inside and tensile on the outside. Radial stress is neglected, $\sigma_2 = 0$, and it is, for elastic bending,

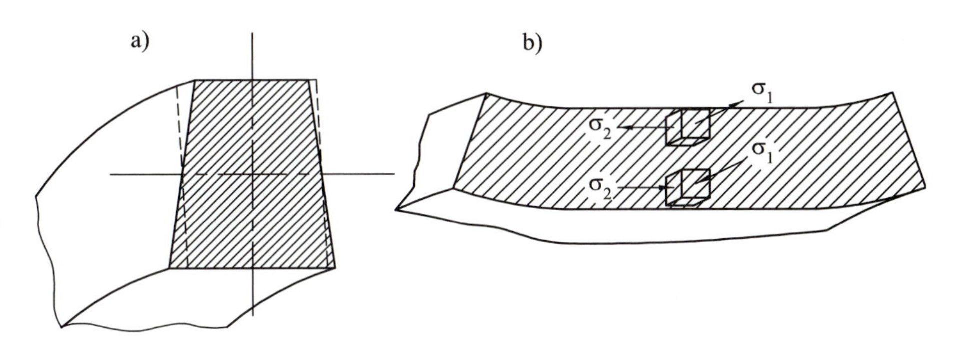

Figure 5.48
Distortions in bending; a) in a narrow beam the width decreases on the outside and increases on the inside. In b), a wide beam is shown with lateral stress that affects the circumferential stress.

$$\varepsilon_y = 0 = \frac{1}{E}[\sigma_y - \nu(\sigma_2 + \sigma_x)]$$

where ν is Poisson's ratio. For $\sigma_2 = 0$, $\sigma_y = \nu\sigma_x$ and $\sigma_x = [E/(1 - \nu^2)]\varepsilon_1$. This shows that in a wide beam the circumferential stress is $1/(1 - \nu^2)$ times larger than for narrow beams, and also the elastic bending moment is greater in the same ratio; the curvature in elastic bending is

$$\frac{1}{\rho} = (1 - \nu^2)\frac{M}{EI} \tag{5.107}$$

For severe bending to small radii, the radial stress should not be neglected. The corresponding solution, not discussed here, leads to a shift of the neutral axis.

Let us now analyze the more realistic case of a *strain-hardening material*. The stress-strain diagram of the material determines the stress distribution in the cross section of the beam. Recalling that in a beam bent to a radius ρ the strain is $\varepsilon = \ln(1 + x/\rho)$ the (σ, e) graph may be plotted as a $(\sigma, x/\rho)$ graph with σ_{max} corresponding to $\varepsilon = \ln(1 + h/2\rho)$, as in Fig. 5.49. It is possible to derive the relationship between the radius ρ and the bending moment M. If the beam is bent by a total angle α the bending moment does external work:

$$W_{ext} = M\alpha \tag{5.108}$$

The internal stresses do such work that at the distance x from the neutral axis, the force is

$$F = \sigma(x)w\,dx$$

and the displacement is $dl = x\alpha$. The width of the beam is w.

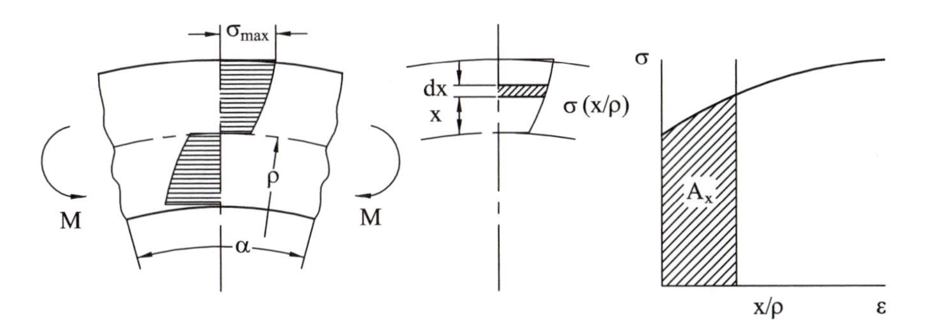

Figure 5.49
Bending of strain-hardening material.

The differential work done in this layer is

$$dW_{\text{int}} = F\,dl = \sigma(x)w\,dx\,x\alpha$$

and the total internal work is

$$W_{\text{int}} = 2\int_{x=0}^{h/2} dW_{\text{int}} = 2w\alpha\int_{0}^{h/2} \sigma(x)x\,dx$$

and considering that $\varepsilon \simeq x/\rho$, and $x = \rho\varepsilon$:

$$W_{\text{int}} = 2w\alpha\int_{\varepsilon=0}^{h/2\rho} = \sigma(\varepsilon)\rho\varepsilon\,d\rho\varepsilon = 2w\alpha\rho^2\int_{\varepsilon=0}^{h/(2\rho)} \sigma(\varepsilon)\varepsilon d\varepsilon \tag{5.109}$$

Comparing W_{ext} (5.108) and W_{int} (5.109) the expression for the moment is obtained:

$$M = 2w\rho^2\int_{\varepsilon=0}^{h/2\rho} = 2w\rho^2 A(h/2\rho) \tag{5.110}$$

The function $A(\varepsilon) = \int_0^{\varepsilon} \sigma(\varepsilon)\varepsilon\,d\varepsilon$ can be derived from the stress-strain diagram by plotting the area of the stress-strain diagram as function of ε. Let us explain this by assuming a simple form of strain-hardening:

$$\sigma = Y_0 + K\varepsilon \tag{5.111}$$

and referring to Fig. 5.50a, b:

$$A(\varepsilon) = \int_0^{\varepsilon} (Y_0 + K\varepsilon)\varepsilon\,d\varepsilon = \int_0^{\varepsilon} (Y_0\varepsilon + K\varepsilon^2)d\varepsilon = \frac{Y_0\varepsilon^2}{2} + \frac{K\varepsilon^3}{3}$$

With this function the relationship between bending moment M and the radius ρ is

$$M = 2w\rho^2\left[\frac{Y_0}{2}\left(\frac{h}{2\rho}\right)^2 + \frac{K}{3}\left(\frac{h}{2\rho}\right)^3\right],$$

$$M = w\left(\frac{Y_0h^2}{4} + \frac{Kh^3}{12\rho}\right) \tag{5.112}$$

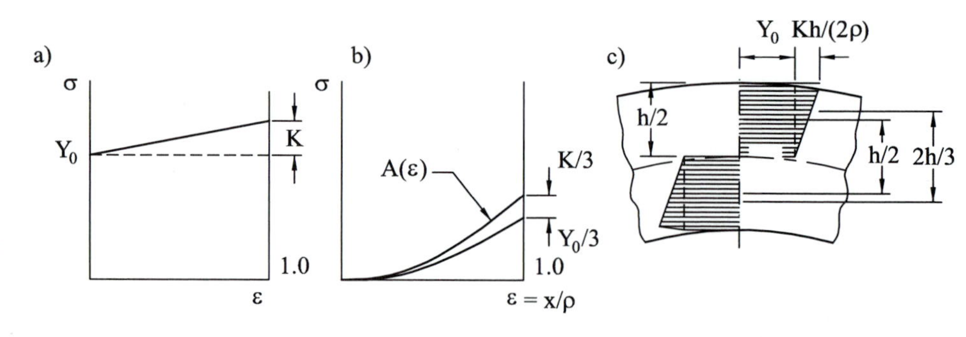

Figure 5.50
Deriving the formula for bending radius: a) the assumed stress-strain diagram; b) the area A (work) of the stress-strain plot as function of distance x from neutral axis; c) the stress distribution.

To verify this result, let us consider Fig. 5.50c. The stress distribution is as shown. The resisting moment consists of two terms:

$$M = w\left(Y_0 \frac{h}{2}\frac{h}{2} + \frac{Kh}{2\rho}\frac{h}{2}\frac{1}{2}\frac{2h}{3}\right)$$

$$M = w\left(\frac{Y_0 h^2}{4} + \frac{Kh^3}{12\rho}\right)$$

which is identical with (5.112).

EXAMPLE 5.5 Determine the Shape of the Deformation of a Plate Loaded in the Middle Between Supports ▼

A plate is loaded as indicated in Fig. 5.51. It is $w = 50$ mm wide and $h = 20$ mm thick, and it is supported over a span of $L = 480$ mm and loaded in the middle by a force $F = 10{,}000$ N. Its yielding properties are characterized by the equation $Y = Y_0 + K\varepsilon$, where $Y_0 = 200$ N/mm^2 and $K = 150$ N/mm^2. Determine the depth of deflection d and the shape of the deformed plate. Neglect elastic deformations.

The moment acting at a distance x from the middle of the plate is

$$M = 5000\,(240 - x)(\text{Nmm}) \tag{5.113}$$

Using Eq. (5.112), we have

$$5000\,(240 - x) = 50\left(200 \times 20^2/4 + 150 \times \frac{20^3}{12\rho}\right),$$

$$1200 - 5x = 1000 + \frac{5000}{\rho} \tag{5.114}$$

$$\frac{1}{\rho} = 0.04 - 0.001x$$

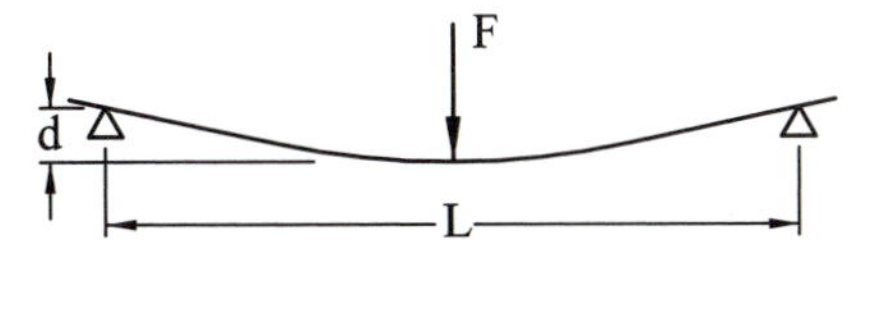

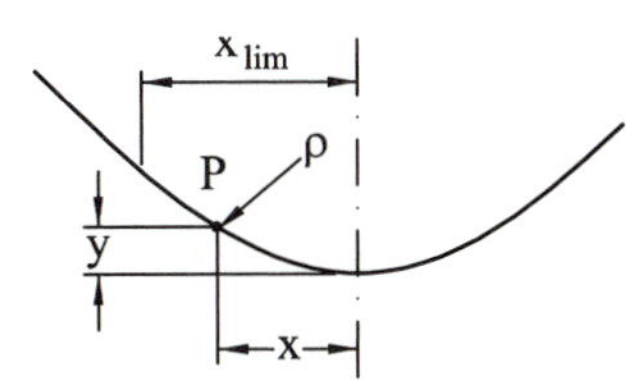

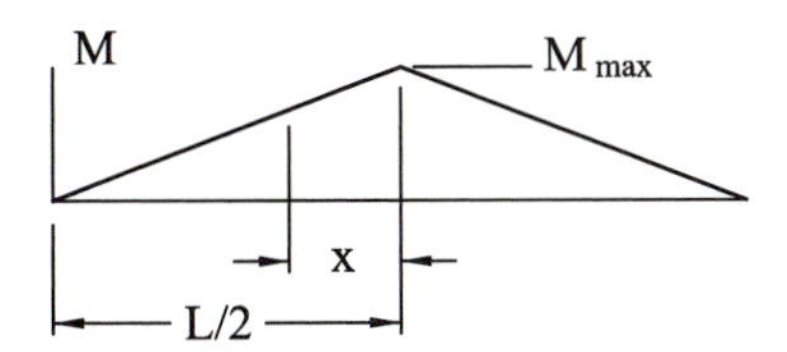

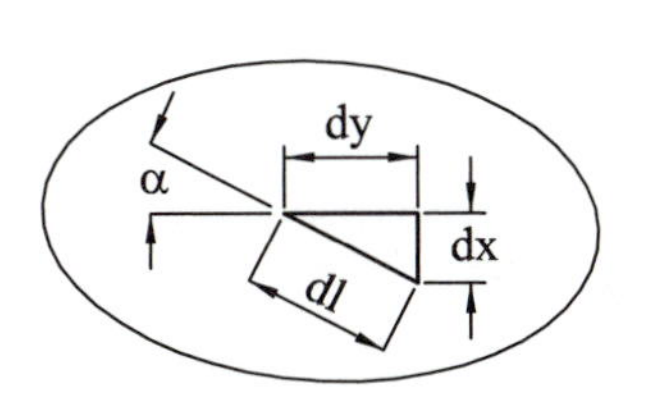

Figure 5.51 Moment distribution and geometry of the plastically bent shape for the case of Example 5.5.

The beam will be curved in its central part until the distance x_{lim}, wherefrom it will remain straight (we are neglecting elastic strains). The value of x_{lim} is obtained for $\rho = \infty$:

$$\frac{1}{\rho} = 0 = 0.04 - 0.001x,$$

$$x_{lim} = 40 \text{ mm}$$

The curvature is related to the change of the angle α:

$$\frac{1}{\rho} = \frac{d\alpha}{dl} = 0.04 - 0.001x$$

Further it is

$$dl = \frac{dx}{\cos\alpha}$$

leading to

$$\frac{d\alpha}{(dx/\cos\alpha)} = 0.04 - 0.001x,$$

$$\cos\alpha\, d\alpha = (0.04 - 0.001x)\, dx$$

Integrating, we have

$$\int_0^\alpha \cos\alpha\, d\alpha = \int_0^x (0.04 - 0.001x)\, dx \quad \text{for } x = 0, \alpha = 0$$

$$\sin\alpha = 0.04x - 0.0005x^2 \tag{5.115}$$

Eq. (5.115) defines the shape of the deflection curve. This shape can be computed by proceeding in small steps of Δx in a loop, for the range $x = 0$ to $x_{lim} = 40$ mm.

$$x \leftarrow x + \Delta x$$

$$\alpha = \sin^{-1}(0.04x - 0.0005x^2)$$

$$\Delta y = \Delta x \tan\alpha$$

$$y \leftarrow y + \Delta y$$

The values of x, y, and α can be printed out and the plot (y, x) produced. Table 5.1 gives the results in coarse steps of $\Delta x = 2.5$ mm. The values of α for $x = x_{lim}$ determine the slopes of the straight lines for the range 40 mm $< x <$ 240 mm.

TABLE 5.1 The Coordinates of the Deflection Curve

x	2.5	5	7.5	10	12.5	15	17.5	20	22.5	25	27.5	30	32.5	35	37.5	40
y	0.24	0.72	1.42	2.36	3.52	4.92	6.55	8.40	10.6	12.9	15.5	18.3	21.4	21.6	24.9	28.2
α	5.6	10.8	15.8	20.5	24.9	29.2	33.2	36.9	40.3	43.4	46.2	48.6	50.5	51.9	52.8	53.1

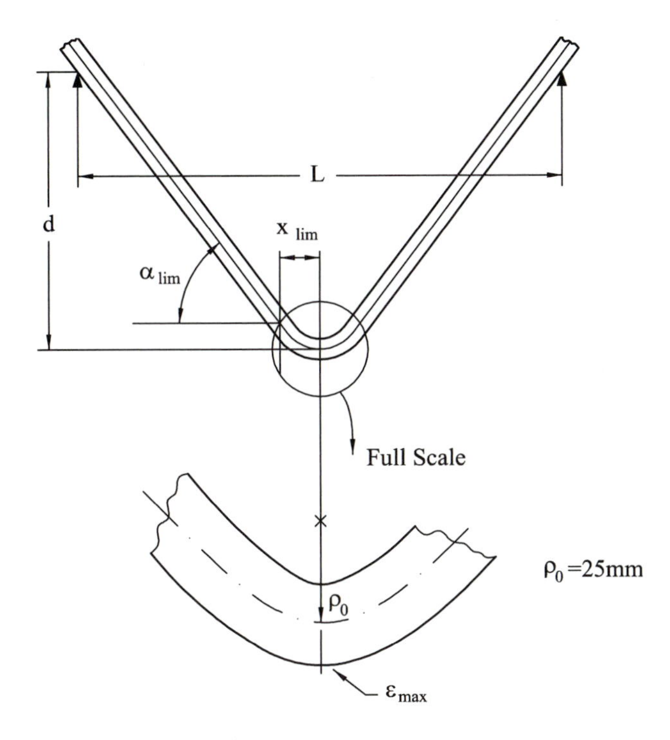

Figure 5.52
Results of the Example 5.5.

The result of the computation is graphically shown in Fig. 5.52. The radius in the middle of the span is given by setting $x = 0$ in equation (5.114):

$$\frac{1}{\rho} = 0.04$$

$$\rho = 25 \text{ mm}$$

The maximum strain at the outer surface under the load is

$$e = \frac{h}{2\rho} = \frac{20}{50} = 0.4$$

The angle $\alpha_{\text{lim}} = 53.1°$, and the depth of deflection d is

$$d = \left(\frac{L}{2} - x_{\text{lim}}\right) \tan \alpha_{\text{lim}} + y_{\text{lim}} = 294.6 \text{ mm}$$

▲

5.4.3 Residual Stresses

It was shown in Fig. 5.46 and explained in the accompanying text that there is always an elastic core in the plastically deformed material. However, apart from that, a part of the stress in the whole section is elastic. This applies to the whole cross section of thickness h, including the part between the elastic layer a and the outer surfaces. This fact has been obscured so far in the preceding analysis because the behavior of the material was assumed "plastic, rigid" and described by Eq. (5.111). In reality the stress and strain in each layer dx (see Fig. 5.49) have passed through an elastic range. Once the bending force is removed, an "elastic recovery" occurs, which is essentially analogous to the recovery described in the case of unloading in a tensile test in Fig. 5.1.

However, in the case of bending where the stress and strain distribution varies through the section, the elastic recovery is more complicated than in a tensile test. In explaining it we will, for simplicity, assume a non-strain-hardening material. Still referring to Fig. 5.1 and now also Fig. 5.53, we may consider the mechanism of recovering from a situation where large plastic strain (plus some elastic strain) was produced by a certain stress created by an external moment. This moment will be canceled by an equal moment corresponding to purely elastic strain. The difference of the two will be equivalent to the removal of the loading moment. Figure 5.53a shows the distributions of both the plastic and an opposite elastic stress such that their moments are equal:

$$\frac{wYh^2}{4} = w\sigma_{\text{elmax}}\left(\frac{h}{2}\right)\left(\frac{1}{2}\right)\left(\frac{2h}{3}\right)$$

$$\frac{wYh^2}{4} = \sigma_{\text{elmax}}\frac{wh^2}{6} \tag{5.116}$$

$$\sigma_{\text{elmax}} = 1.5Y$$

The sum of the original plastic stress and the opposite elastic stress is shown as the difference between the plastic stress and positive elastic stress in Fig. 5.53b, and it represents the "residual stress." It is compressive in the outer fibers and tensile in the inner fibers outside of the neutral axis, while inside of the neutral axis it is arranged in the opposite way.

In the case of partly elastic and partly plastic loading, as shown in Fig. 5.47a, the residual stresses are obtained on unloading, as shown in Fig. 5.53c.

Springback in Plane Stress (Narrow Beams)

The diagram in Fig. 5.54a illustrates the stress distribution assuming a non-strain-hardening material. Equations (5.105) and (5.106) apply, and they are reproduced here:

$$\frac{1}{\rho} = \frac{2Y}{Ea}$$

$$M = Yw\left(\frac{h^2}{4} - \frac{a^2}{12}\right)$$

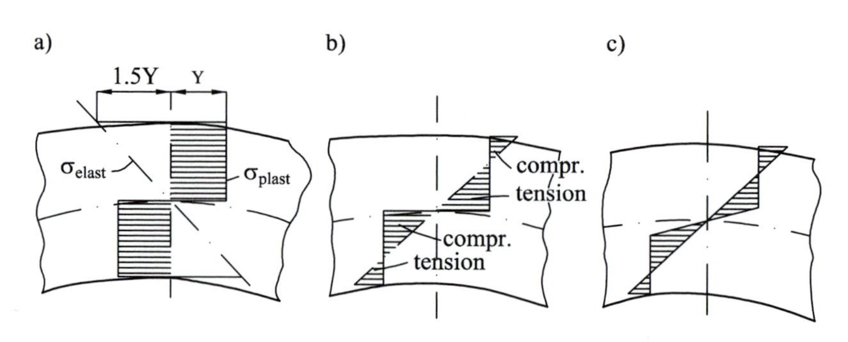

Figure 5.53
Residual stresses in bending: a) an elastic stress moment is equated to the moment of the fully plastic stress; b) the former relieves the latter to obtain the distribution of the residual stress. In c), a graph analogous to (b) is shown for a case with an initial elastic layer of Fig. 5.47a.

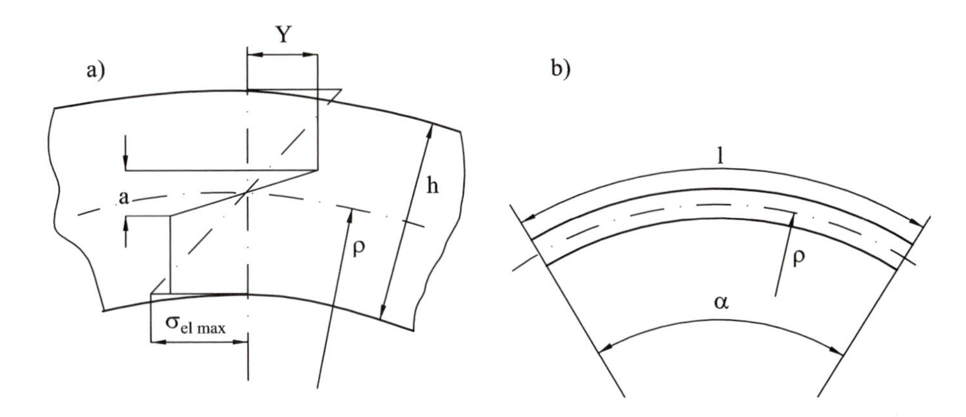

Figure 5.54
Springback in bending of a non-strain-hardening material: a) the initial stress distribution; b) relating the bend angle to the radius.

Combining the two equations, the moment for the deflection of the beam is a function of ρ:

$$M = w\left(\frac{Yh^2}{4} - \frac{\rho^2 Y^3}{3E^2}\right) \tag{5.117}$$

On unloading, an elastic stress moment of the same magnitude occurs which, on its own, would produce the curvature:

$$\frac{1}{\rho_{\text{el}}} = \frac{M}{EI} = \frac{12M}{Ewh^3} \tag{5.118}$$

This "elastically produced curvature" is the *springback*. The final curvature is the difference of the total and the elastic curvatures:

$$\frac{1}{\rho_{\text{f}}} = \frac{1}{\rho} - \frac{1}{\rho_{\text{el}}} \tag{5.119}$$

and

$$\frac{\rho}{\rho_{\text{f}}} = 1 - \frac{\rho}{\rho_{\text{el}}} \tag{5.120}$$

Equating the moments of Eqs. (5.117) and (5.118),

$$\left(\frac{w}{12}\right)\left(3Yh^2 - \frac{4\rho^2 Y^3}{E^2}\right) = \frac{w}{12} \cdot \frac{Eh^3}{\rho_{el}}$$

Multiplying both sides by ρ/Eh^3,

$$\frac{3Y\rho}{Eh} - 4\left(\frac{Y\rho}{Eh}\right)^3 = \frac{\rho}{\rho_{\text{el}}} \tag{5.121}$$

and combining (5.120) and (5.121), we obtain the ratio of the radius ρ produced under load and of the final radius ρ_{f} that is left after the removal of the load:

$$\frac{\rho}{\rho_{\text{f}}} = 1 + 4\left(\frac{Y\rho}{Eh}\right)^3 - 3\left(\frac{Y\rho}{Eh}\right) \tag{5.122}$$

For example, if $Y = 2 \times 10^2$ N/mm^2 and $E = 2 \times 10^5$ N/mm^2, then

$$\frac{\rho}{\rho_f} = 1 + 4 \times 10^{-9}\left(\frac{\rho}{h}\right)^3 - 3 \times 10^{-3}\left(\frac{\rho}{h}\right)$$

For $\rho/h = 3$, $\rho/\rho_f = 0.991$, $\rho/h = 10$, and $\rho/\rho_f = 0.97$. For severe bending, $\rho/h = 3$, and the springback is 0.9%, while for less severe bending to a radius of 10 h, the springback is 3%.

If a certain length l of the beam was bent to a radius ρ (see Fig. 5.54b), the total angle α was $\alpha = l/\rho$. Since the length of the neutral fiber does not change during bending, angle α is directly proportional to the curvature, and

$$\frac{\alpha_f}{\alpha} = \frac{\rho}{\rho_f} \tag{5.123}$$

For example, if a steel beam with $Y = 200$ N/mm^2 must be bent to 90° with a radius $\rho = 5h$,

$$\frac{\alpha_f}{\alpha} = \frac{\rho}{\rho_f} = 1 + 4 \times 10^{-9} \times 5^3 - 3 \times 10^{-3} \times 5 = 0.985 \tag{5.124}$$

and the die must be made to bend to 90° × 0.985 = 88.65°. After springback the angle α will equal the required 90°.

5.4.4 Failures and Limitations in Bending

In bending a solid plate or beam with a rectangular section, the limit (i.e., the smallest bending radius) is basically determined by the maximum possible tensile strain on the outer surface, that is, by the ductility of the material. The ductility depends on the amount of strain-hardening and on the presence of voids and microscopic or larger cracks in the surface of the material.

The processing of the sheet, plate, or beam preceding the bending operation is of great importance. For sheet metal and plate, the anisotropy produced in cold rolling has a great effect (see Fig. 5.55a). The material can stand higher strain if bending is done with tension in the direction of previous rolling (bending across rolling) rather than perpendicular to it (bending axis along rolling).

Material which has been sheared to width before bending has been strain-hardened at its sides and may also have cracks that originated in the shearing process (see Fig. 5.55b). This leads to a decrease of the ultimate strain. In any case, the edge with shearing burr should either be rounded off before severe bending or it must be located on the inside of the bend.

The maximum strain at the outer fibers is

$$e_{max} = \frac{h}{2\rho}$$

a)

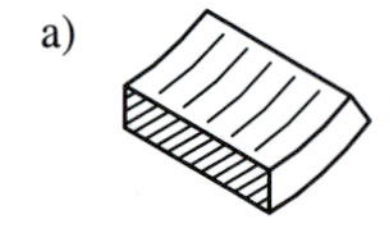

b)

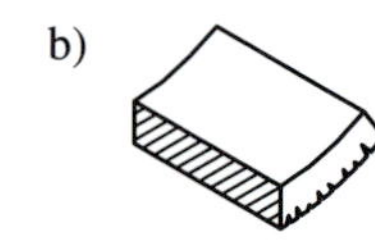

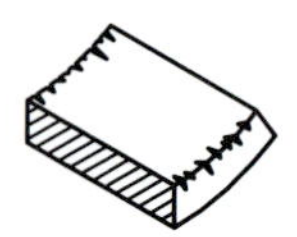

Figure 5.55
Cracks in bending a) rolled, b) sheared material. In a), bending about an axis perpendicular and parallel to the rolling direction is illustrated. In b), the sheared edges may have burrs or cracks; these have to be rounded off or at least located on the inside of the bend.

and should not exceed the strain corresponding to area reduction q at fracture in the tensile test:

$$q = \frac{A_0 - A_f}{A_0} = 1 - \frac{A_f}{A_0} = 1 - \frac{l_0}{l_f}$$

The limit strain can be evaluated as

$$e_{\text{lim}} = \frac{l_f - l_0}{l_0} = \frac{l_f}{l_0} - 1 = \frac{q}{1 - q} \tag{5.125}$$

and, correspondingly:

$$\rho_{\text{min}} = \frac{h}{2e_{\text{lim}}}$$

The extreme, that is, minimum, values of ρ/h as established by practical experience are listed in Table 5.2 for some commonly used materials, for bending across and along the direction in which they were rolled.

In bending profiles and in flanging (see Fig. 5.56), tearing may occur on the outer surface and wrinkling on the inner surface. Cross sections of profiles become distorted. Some of these limitations may be prevented by the use of form blocks.

5.4.5 Stretch Forming and Deep Drawing

Typical press forming operations like those carried out in car body production consist of combinations of stretch forming and deep drawing. Other instances, such as the production of cans or pots and pans, usually involve deep drawing. In all these operations the sheet metal is deformed in two directions in its plane. In stretch forming (see Fig. 5.57a), the "draw beads" cause the deformation to be tensile in both directions, whereas in deep drawing (Fig. 5.57b), the metal is stretched in one direction and shrunk in the other.

We will concentrate on an analytical exercise of deep drawing. Let us first describe all the typical features of this process before suitable simplifications are formulated for the analysis. As shown in Fig. 5.58 the tooling consists of three parts: the die D, the punch P, and the blankholder B. Both the punch and the die are designed with substantial radii on their perimeters. In this operation, as the simplest case, a

TABLE 5.2. Smallest Radii in Bending (ρ_{min}/h)

		Annealed		Hard	
Material	UTS (MPa)	Across	Along	Across	Along
Steels	280–450	0.2	0.6	0.6	1.2
	420–500	0.6	1.2	1.5	3.0
	500–600	1.0	2.0	2.0	4.0
	550–700	1.5	3.0	3.0	5.0
Copper	210	0	0.2	0.8	1.5
Aluminum	200	0	0.2	0.3	0.8
Hard Al alloys	400	1.0	1.5	3.0	4.0

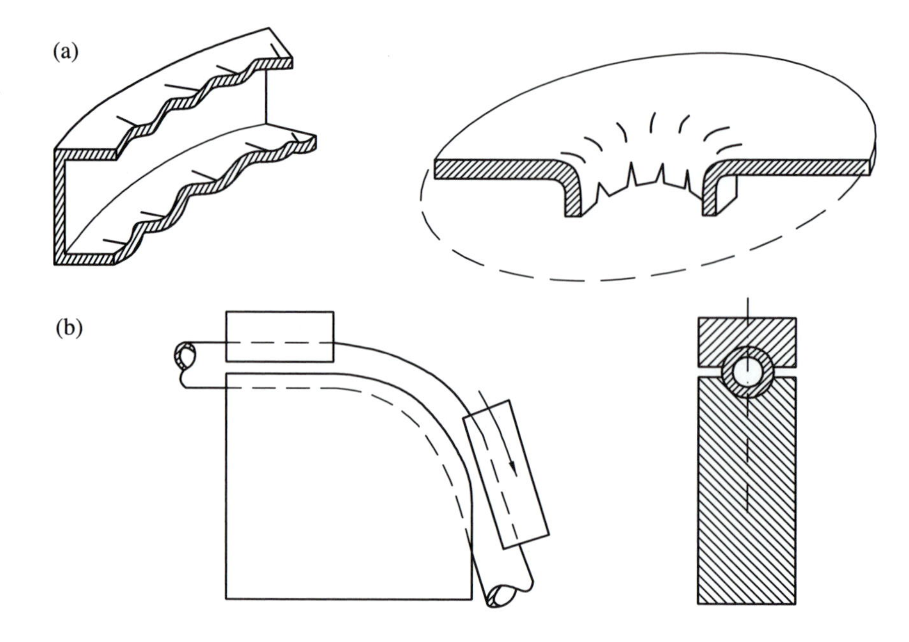

Figure 5.56
a) Cracks and wrinkles in bending. b) Form blocks for pipe bending.

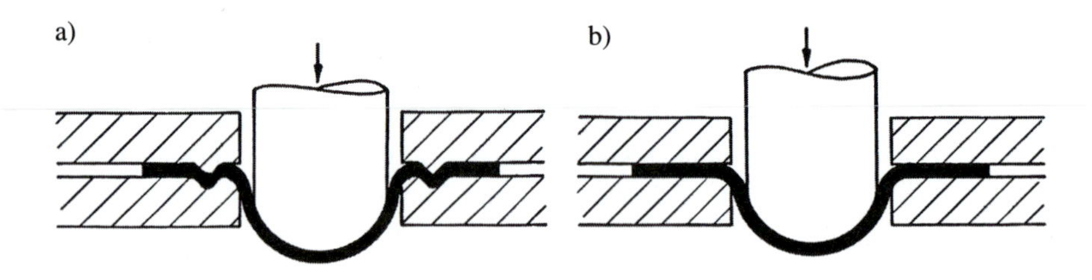

Figure 5.57
a) Stretch drawing; b) deep drawing.

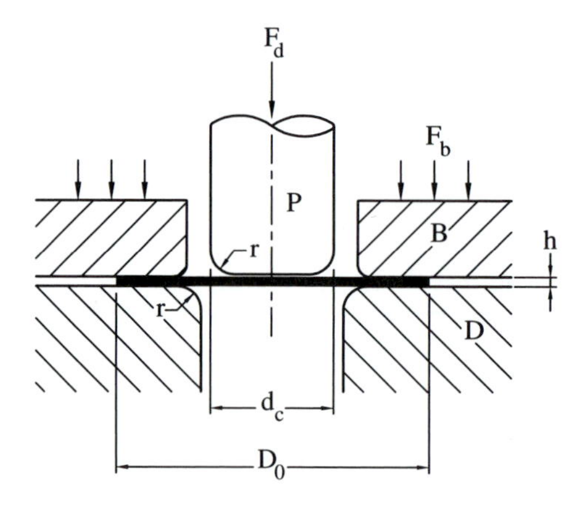

Figure 5.58
The geometry of deep drawing. Shown is punch *P*, die *D*, and blankholder *B*.

round blank with diameter D_0 is drawn into a cup with diameter d_c. Obviously, most of the process consists of shrinking the flange, and the extreme of this action concerns the outer diameter D_0 which will become the rim of the cup with diameter d_c. The ratio D_0/d_c is called the drawing ratio. The metal in the flange has a tendency to wrinkle. Therefore, it is necessary to use the blankholder. A blankholding force F_b resulting in a blankholding pressure of 2.5 MPa is generally found satisfactory for deep drawing of mild steel (with yield strength in the range 400–800 MPa). This pressure does not influence the drawing force F_d in any substantial way.

Consider the various stages of the operation as shown in Fig. 5.59. In the initial stage the outer annular zone X is in contact with the die (the flange), the inner circular zone Z is in contact with the punch and the intermediate annular zone Y is not in contact with either the punch or the die. The process in the zone X, which diminishes as drawing proceeds, is one of radial drawing. It is decisive for the drawing force and is analyzed in the following exercises.

Material in zone Z is radially stretched. However, the stretching is prevented, to some extent, by the friction between the punch head and the material. At the circumference of this zone the metal is stretched and bent around the punch radius. This may lead to a local neck. This is where cracks occur if the drawing ratio exceeds the limit for a given material. The value of the limit drawing ratio is analyzed later on.

At the outer periphery of the zone Y, the metal, in the initial stage of drawing, is stretched and bent over the radius of the die. This constitutes another local thinned region where failure is possible. The middle of zone Y is not significantly thinned because of circumferential compression.

The thinning of the two above-mentioned bands, and possibly the cracking, occurs during the initial stage (a) of the operation in which the drawing force is maximum. During the cup formation, stage (b), there is tensile axial stress in the cup wall. The strain situation in this wall is one of plane strain because the punch body prevents circumferential shrinking of the wall. The axial tensile stress provides the force for the radial drawing in zone X.

Although there is local thinning in zones Y and Z, and thickening of the material in zone X, especially at the outer radius, the average thickness of the wall material does not change. This is the simplifying assumption in the following analytical exercises. This assumption permits us to express the depth l of the drawn cup as follows:

$$\frac{\pi D_0^2 h}{4} = \frac{\pi d_c^2 h}{4} + \pi d_c h l$$

$$D_0^2 - d_c^2 = 4d_c^2\left(\frac{l}{d_c}\right) \tag{5.126}$$

$$\frac{l}{d_c} = \frac{D_0^2/d_0^2 - 1}{4}$$

Eq. (5.126) relates the depth-to-diameter ratio to the drawing ratio. For example, for a drawing ratio $D_0/d_c = 2$, the depth-to-diameter ratio of the cup is $l/d_c = 0.75$. These

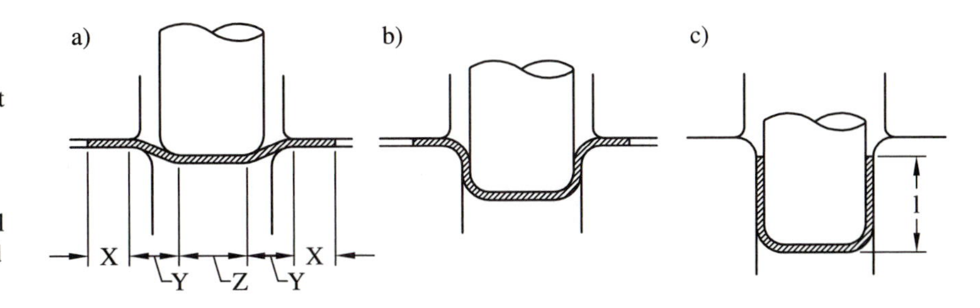

Figure 5.59
Stages of deep drawing. In a) zone X is in contact with the die and zone Z with the punch while zone Y does not contact either. Zone X is in the state of radial drawing. In the intermediate stage of b) the axial stress in the wall prevails. In c) the completed draw is shown.

are the common limits of a single deep drawing operation. Deeper draws (i.e., larger diameter reductions) can be achieved by multiple drawing in which the once-drawn cup is repeatedly redrawn.

5.4.6 Pure Radial Drawing of a Non-Strain-Hardening Material

This exercise is rather artificial because most of the materials used for deep drawing exhibit considerable strain-hardening. Nevertheless, it permits some basic insight into the process and forms a basis for the next exercise, in which we consider strain-hardening of a selected type.

In pure radial drawing, any segment of the flange can be considered as a flat wedge (see Fig. 5.60a). The drawing stress σ_d at the final radius of the flange, which is the cup radius r_c, will vary as the outer radius r_0 of the flange decreases continuously from the initial radius R_0 to r_c. At an intermediate stage characterized by the value of r_o, it can be determined that any initial intermediate radius R has decreased to a radius r, which is obtained from the equality of the shaded areas in Figs. 5.60b and c:

$$R_0^2 - R^2 = r_0^2 - r^2$$

and the ratio R/r is obtained as

$$\frac{R}{r} = \left(\frac{R_0^2 - r_0^2}{r^2} + 1\right)^{1/2} \tag{5.127}$$

The true strain at the radius r is then,

$$\varepsilon = \ln\left(\frac{R}{r}\right) \tag{5.128}$$

if an assumption of no change in the thickness h of the sheet is accepted. This assumption, of $d\varepsilon_2 = 0$, expresses a situation of plane strain.

The equilibrium of forces acting on a strip of width dr at radius r (see Fig. 5.61) is obtained as

$$\alpha h[\sigma_1(r + dr) - (\sigma_1 + d\sigma_1)r] = \sigma_3 dr\alpha h \tag{5.129}$$

where the last term is explained in the inset vector diagram in the figure. Eq. (5.129) can be simplified:

$$(\sigma_1 - \sigma_3)dr - r\,d\sigma_1 = 0 \tag{5.130}$$

If now the plane-strain yield criterion is applied:

$$\sigma_1 - \sigma_3 = 1.155Y$$

and we obtain

$$d\sigma_1 = 1.155Y\,dr/r \tag{5.131}$$

The drawing stress σ_d would be

$$\sigma_d = \int_{r_0}^{r_c} d\sigma_1$$

provided that all the work is homogeneous. Under the circumstances, there is always some redundant work. The corresponding correction factor Q_r can be derived using slip-line analysis, although this method is not discussed in the present text. With refer-

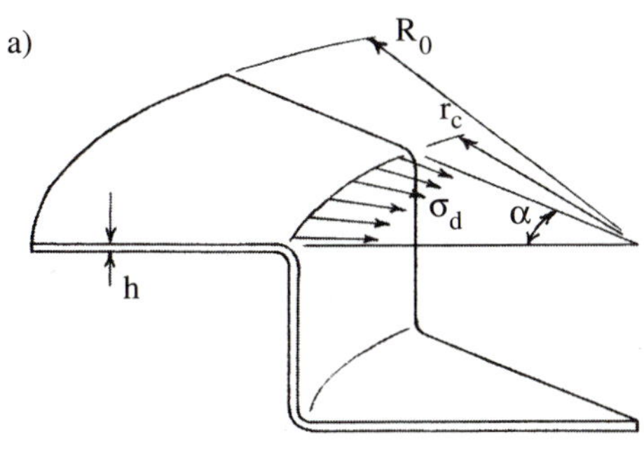

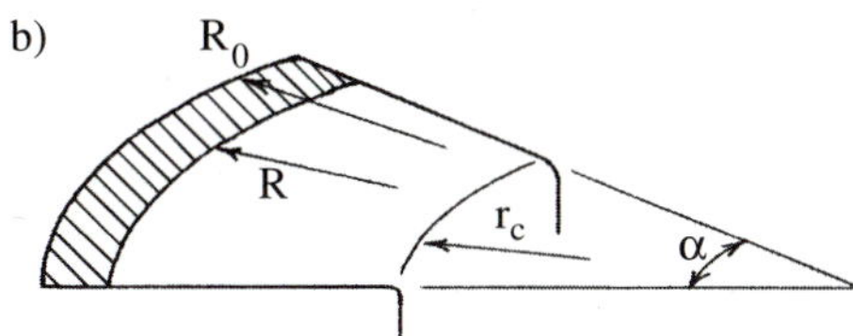

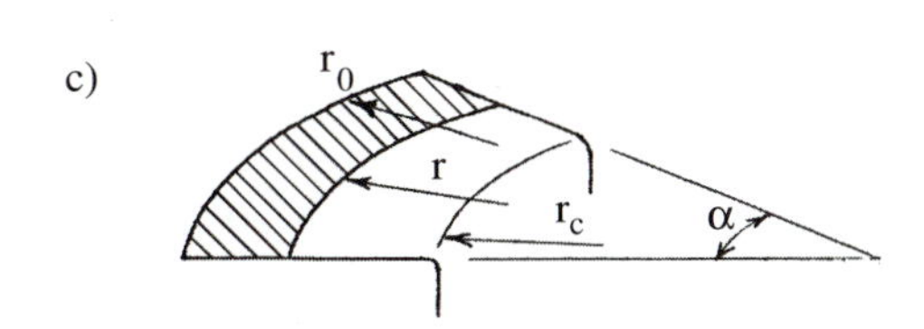

Figure 5.60
Analysis of radial drawing: a) the initial situation; b) the initial annular strip (shaded) changes into the intermediate one in c). This defines the strain and leads to the derivation of radial stress.

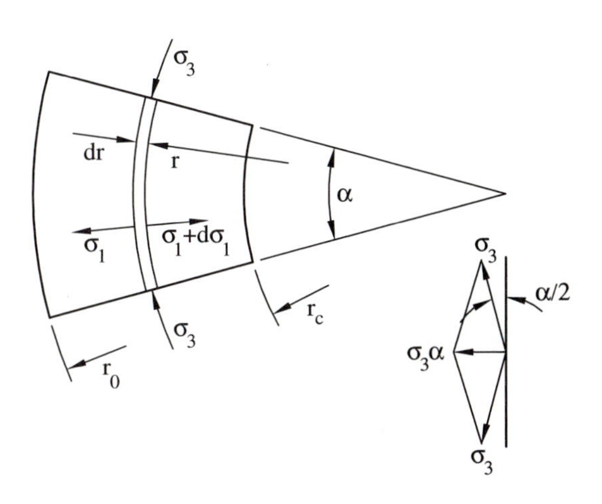

Figure 5.61
Balance of forces on an element of a radial drawing operation.

ence for example to [6], it is, for $R_0/r_c \simeq 2$, $Q_r = 1.3$, and this has to be included in the expression for the drawing stress, which finally becomes:

$$\sigma_d = Q_r \int_{r_0}^{r_c} d\sigma_1 = 1.5 \int_{r_0}^{r_c} Y \frac{dr}{r} \tag{5.132}$$

The simplest solution is obtained for the assumption of a *non-strain-hardening* material, $Y =$ const. The solution of Eq. (5.132) is then

$$\sigma_d = 1.5Y \int_{r_c}^{r_0} \frac{dr}{r} = 1.5Y \ln\left(\frac{r_0}{r_c}\right) \tag{5.133}$$

The largest drawing stress occurs at the very beginning, for $r_0 = R_0$,

$$\sigma_{\text{dmax}} = 1.5Y \ln\left(\frac{R_0}{r_c}\right) \tag{5.134}$$

The limit situation occurs when $\sigma_{\text{dmax}} = Y$, when the material stretched between the die and the cup starts to tear:

$$1.5Y \ln\left(\frac{R_0}{r_c}\right) = Y$$

and this gives

$$\left(\frac{R_0}{r_c}\right)_{\text{lim}} = 1.95 \tag{5.135}$$

This applies to tearing at the critical point at the bottom of the cup wall. If the failure is considered in the cup wall where the deformations correspond to plane strain (because the punch does not permit any circumferential shrinking), it is,

$$1.5Y \ln\left(\frac{R_0}{r_c}\right) = 1.155Y$$

and the corresponding limit for the drawing ratio is

$$\left(\frac{R_0}{r_c}\right)_{\text{lim}} = 2.16 \tag{5.136}$$

The radial stress σ_1 in the flange is expressed as a function of the radius r, for a given stage of drawing characterized by the decrease of R_0 to r_0, as follows:

$$\sigma_1 = \int_{r_0}^{r} \mathrm{d}\sigma_1 = 1.5Y \ln\left(\frac{r_0}{r}\right) \tag{5.137}$$

and it increases from zero at the radius r_0 to a maximum of $\sigma_1 = \sigma_\text{d}$ at $r = r_c$. Choosing $Y = 500$ MPa, the variation of σ_1 as a function of r is graphically expressed in Fig. 5.62 for two stages of drawing from $R_0 = 2$ to $r_c = 1$. The upper line corresponds to the beginning when $r_0 = R_0$, and the lower line corresponds to an intermediate stage of $r_0 = 1.5$. The variation of the drawing stress σ_d during the drawing of this cup, as the flange radius varies from R_0 to r_0, is expressed in Fig. 5.63. The drawing stress, and the drawing force F_d,

$$F_d = 2\pi r_c \sigma_\text{d} h \tag{5.138}$$

are maximum at the beginning and drop to zero at the end of the operation.

5.4.7 Radial Drawing of a Strain-Hardening Material

Let us consider deep drawing of steel 1008 with the following stress-strain formula of the type of Eq. (5.8):

$$Y = 657\, \varepsilon^{0.24}\ \text{(MPa)} \tag{5.139}$$

Assume that the sheet ready for deep drawing has already undergone a strain of $\varepsilon = 0.1$. Correspondingly, we have

$$Y = 378 + 657\, \varepsilon^{0.24}\ \text{(MPa)}$$

where, according to Eq. (5.128),

$$\varepsilon = \ln\left(\frac{R}{r}\right)$$

and, using Eq. (5.127) for strain, we have

$$\varepsilon = 0.5 \ln\left(\frac{R_0^2 - r_0^2}{r^2} + 1\right)$$

and

$$Y = 378 + 657\left[0.5 \ln\left(\frac{R_0^2 - r_0^2}{r^2} + 1\right)\right]^{0.24}$$

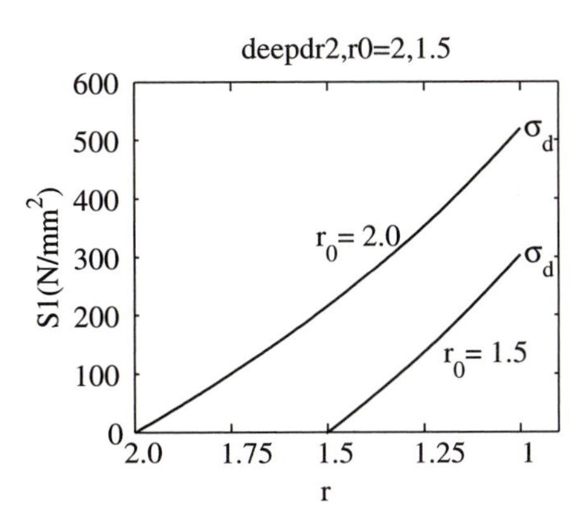

Figure 5.62
Variation of the radial stress σ_1 in the flange for two stages of the operation.

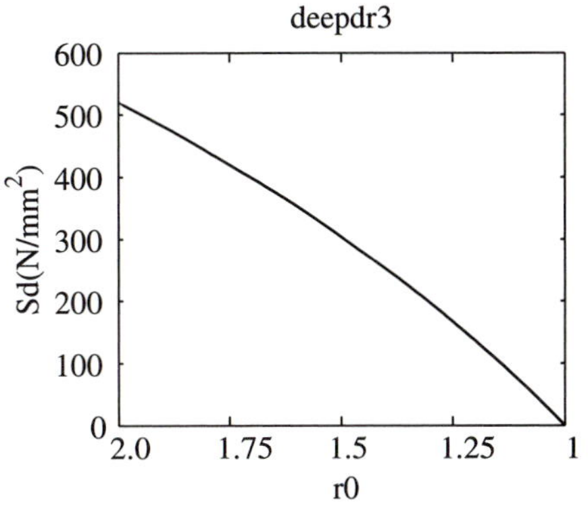

Figure 5.63
Variation of the drawing stress during radial drawing of non-strain-hardening material.

or

$$Y = 378 + 556\left[\ln\left(\frac{R_0^2 - r_0^2}{r^2} + 1\right)\right]^{0.24} \tag{5.140}$$

During the drawing operation, as the radius r_0 decreases from R_0 to r_c, the material in the flange strain-hardens. This is illustrated graphically in Fig. 5.64, where the strain ε is plotted versus the radius r in a number of stages of drawing from $R_0 = 2.0$ to $r_c = 1.0$, for $r_0 = 2.0$, 1.75, 1.5, 1.1 and, finally 1.01. The corresponding values of the yield strength at the points $r = r_0$ and $r = r_c$ are written in the diagram.

We see that the initial value of $Y = 378$ MPa increases to $Y = 980$ MPa at the end of the operation. This will certainly affect the drawing force. There will be two opposite effects: the reduction ratio r_0/r_c decreases, and this will lead to a decrease of σ_d; strain hardening will tend to cause an increase of σ_d, and there may be a maximum

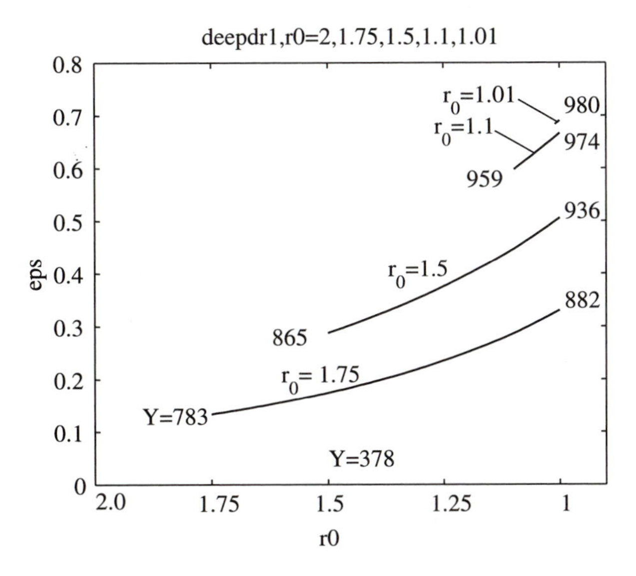

Figure 5.64
Strain-hardening during radial drawing. Yield strength Y is plotted versus radius r for a number of successive stages determined by the decreasing outer radius r_0 of the circular blank. The material hardens considerably.

somewhere between $r_0 = R_0$ and $r_0 = r_c$. Equation (5.132), in which Y is a function of r as well as of the lower limit r_0 of integration as expressed in Eq. (5.140), leads to

$$\sigma_d = 1.5 \int_{r_0}^{r_c} Y \frac{dr}{r} = \sigma_A + \sigma_B \tag{5.141}$$

where

$$\sigma_A = 567 \int_{r_0}^{r_c} \frac{dr}{r} \tag{5.142}$$

and

$$\sigma_B = 834 \int_{r_0}^{r_c} \left[\ln\left(\frac{R_0^2 - r_0^2}{r^2} + 1 \right) \right]^{0.24} \frac{dr}{r} \tag{5.143}$$

The part σ_A is easily integrated as

$$\sigma_A = 567 \ln\left(\frac{r_0}{r_c} \right) \tag{5.144}$$

The part σ_B may be integrated numerically:

$$\sigma_B = 834 \int_{r_0}^{r_c} y(r)\, dr \tag{5.145}$$

where

$$y(r) = \frac{\{ \ln[(R_0^2 - r_0^2)/r^2 + 1] \}^{0.24}}{r} \tag{5.146}$$

The integrand $y\,(r)$ which expresses $d\sigma_B$ has the form shown in Fig. 5.65, which illustrates the stage of $r_0 = 1.5$. We may use the simplest integration procedure by incrementing r in steps of Δr and replacing the integral in Eq. (5.145) by the sum

$$\sigma_B = 834 \sum_{1}^{N} (y_n + y_{n+1}) \frac{\Delta r}{2} \tag{5.147}$$

where, for brevity, we used the notation $y_n = y(r_n)$, and we have,

$$r_1 = r_0, \qquad r_n = r_0 + n\Delta r \tag{5.148}$$

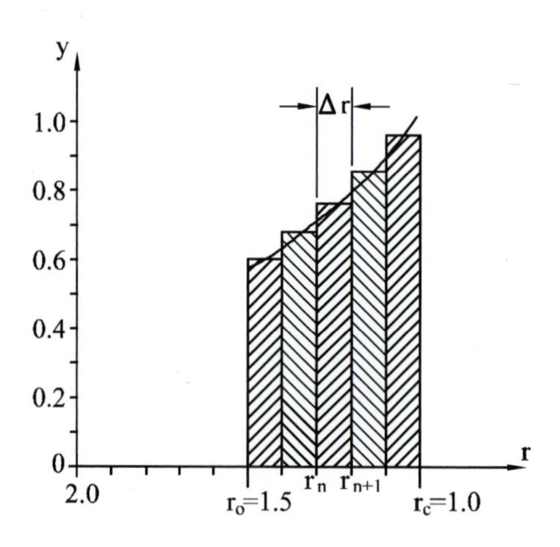

Figure 5.65
Method of integrating the strain-hardening part of the radial stress.

and

$$\Delta r = \frac{r_0 - r_c}{N} \tag{5.149}$$

where N is an integer.

The computer program for determining σ_d as it varies during drawing will be carried out for a chosen number of values of r_0 decreasing from R_0 to r_c in steps of Δr_0. In each step the part σ_B will be integrated numerically in an internal loop in which r is changing between r_0 and r_c in steps of Δr. These latter steps have to be small enough to minimize the error due to the discrete evaluation of the integral. In order to always have an integer number of steps in the integration, the step Δr_0 must also be an integer multiple of Δr.

An algorithm for the computation follows:

1. Input data: R_0, r_c, N, Δr, and parameters of Eq. (5.139).
2. Define $(r_0)_n = R_0 - (n - 1)\,\Delta r_0$, $\Delta r_0 = (R_0 - r_c)/N$
3. Carry out steps 4, 5, 6, for $n = 1$ to N.
4. Set initially $\sigma_B = 0$ and carry out the following loop, from $r = (r_0)_n$ until $r = r_c + \Delta r$ [See Eq. (5.145)].

$$y_1 = \frac{\{\ln([R_0^2 - (r_0)_n^2]/r^2 + 1)\}^{0.24}}{r}$$

$$p = r + \Delta r$$

$$y_2 = \frac{\{\ln([R_0^2 - (r_0)_n^2]/p^2 + 1\ )\}^{0.24}}{p}$$

$$d\sigma_B = \frac{834}{2}\Delta r\,(y_1 + y_2)$$

$$\sigma_B = \sigma_B + d\sigma_B$$

$$r = r - \Delta r$$

5. Make $(\sigma_B)_n = \sigma_B$ and compute σ_A from Eq. (5.142):

$$(\sigma_A)_n = 567\ \ln\left(\frac{(r_0)_n}{r_c}\right)$$

6. $(\sigma_d)_n = (\sigma_A)_n + (\sigma_B)_n$
7. $(\sigma_d)_{N+1} = 0$, $(r_0)_{N+1} = r_c$
8. Plot $(\sigma_d)_n$ versus $(r_0)_n$, for $n = 1$ to $N + 1$.

EXAMPLE 5.6. Compute the Variation of the Radial Drawing Stress for a Strain-Hardening Material ▼

Input data: $R_0 = 2.0$, $r_c = 1.0$, $N = 20$, and $\Delta r = 0.01$. The yield stress of the material is given by Eq. (5.139). The computation follows the algorithm presented above. The result is given in Fig. 5.66.

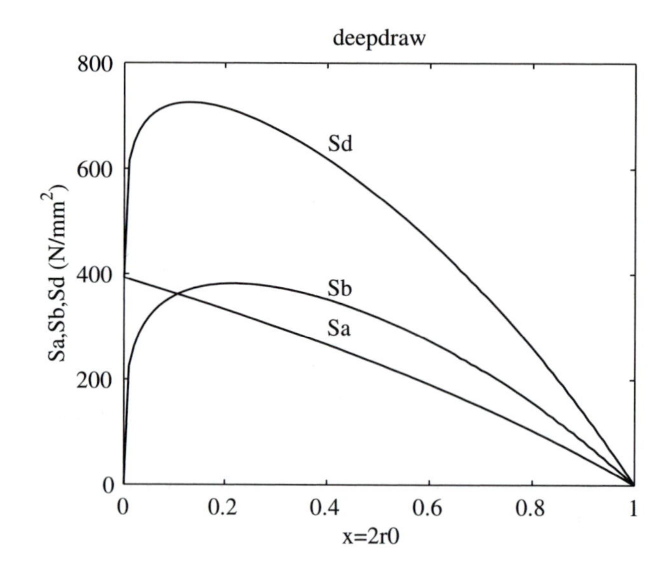

Figure 5.66
Result of Example 5.6. Variation of the drawing stress in radial drawing of a strain-hardening material versus the size of the decreasing outer radius.

It is seen that at the beginning of the operation, due to the term σ_B dependent on strain hardening, the drawing force first steeply increases. It reaches a maximum at about $r_0 = 1.85$ and then it decreases progressively. Quite naturally, for $r_0 = r_c$ it drops to zero. ▲

5.5 CHATTER IN COLD ROLLING

In multistand, tandem cold rolling, when attempts are made to increase the rolling speed, violent vibrations often occur that are clearly identified as of the "self-excited" type (see Chapter 9 for more details about another case of self-excited vibrations, that of chatter in machining). These vibrations in rolling are also referred to as "chatter."

5.5.1 A Simple Rolling Chatter Theory

The theory presented here is based on deriving the condition for the limit of stability of the simple case, as illustrated in Fig. 5.67. Here, only the vibration of stand 2 is considered; the preceding, nonvibrating stand 1 is just for the purpose of establishing input tension for stand 2 in its incoming strip. It is also assumed, for further simplification, that roll speed is constant and the neutral (no-slip) point in stand 2 is in the exit plane (in the plane passing through the axes of the rolls). Subsequent computer simulation of a multistand mill in which the above simplifications have been removed has shown that the simple case of Fig. 5.67 nevertheless contains all the pertinent features of the phenomenon.

Two basic ideas presented in Ref. [7] are essential to the theory:

1. Vibration of the rolls produces periodically variable tension between stands that affects the rolling force in such a phase as to represent negative damping and self-excite the vibration.

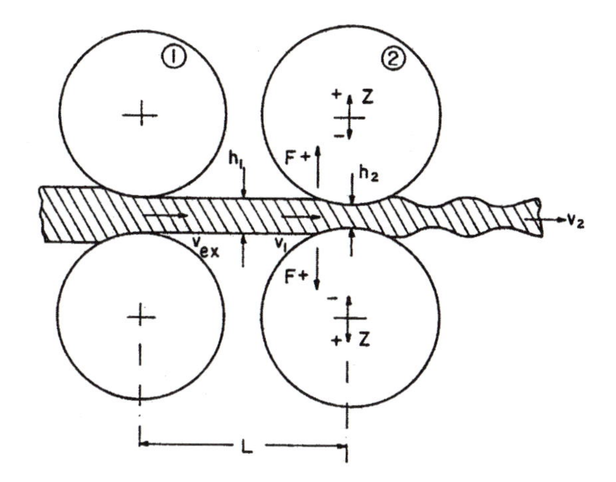

Figure 5.67
Simplified case of a single-stand vibration in cold rolling of sheet metal. The entering, constant-thickness sheet exits undulated as result of the vibrating rolls. Exit speed is practically constant and equal to roll peripheral speed. Hence, the sheet is sucked in with varying speed, causing variable entry tension and affecting rolling force with a 90° phase shift and acting as negative damping.

2. Periodic variation of the contact length at the exit of the strip produces a variable component of the rolling force which represents positive damping. After formulating these two effects, we obtain the condition of the limit of stability by checking which one of the two prevails.

In Fig. 5.67 the input thickness h_1 is constant. The exit-strip thickness h_2 varies,

$$h_2 = h_{2m} + 2Z \sin \omega t \tag{5.150}$$

where Z is the amplitude of vibration, ω is its frequency (rad/sec) and h_{2m} is the mean thickness.

It is assumed that the neutral point on stand 2 is at the exit. It is accepted that the roll speed does not vary. This corresponds well with observations. The frequency of chatter is 120–140 Hz, while the natural frequencies of torsional drive vibrations are much lower, and the drive cannot follow the high-speed vibration. Consequently, the exit speed v_2 is equal to the roll peripheral speed and is constant.

From the equation of constant mass throughput,

$$h_1 v_1 = h_2 v_2$$

it follows correspondingly that the input velocity must vary:

$$v_1 = \frac{v_2}{h_1} h_2 = \frac{v_2}{h_1} (h_{2m} + 2Z \sin \omega t)$$

The input speed has a mean and variable component:

$$v_1 = \frac{v_2 h_{2m}}{h_1} + \frac{2v_2\, Z \sin \omega t}{h_1} = v_{1m} + v_{1var}$$

The exit speed from the first stand v_{ex} is constant and is equal to v_{1m}.

The variation of the input speed v_1 produces a variable "stretch" between stands 1 and 2. This stretch ΔL (as an elongation) is the integral of the variable speed

$$\Delta \mathrm{L} = \int^t v_{1var}\, \mathrm{d}t = \frac{2Zv_2}{h_1} \int \sin \omega\, \mathrm{d}t = -\frac{2Zv_2 \cos \omega t}{h_1 \omega} \tag{5.151}$$

and it produces a variable component of the tension between stands 1 and 2:

$$\sigma_{12var} = \frac{\Delta L}{L} E = \frac{2Zv_2 E}{L\, h_1\, \omega} \cos \omega t \tag{5.152}$$

The interstand tension σ affects the rolling force in the following manner:

$$F = (S - \sigma)\, W\sqrt{Rr} \tag{5.153}$$

where S is the yield stress, R is the roll radius, W is the strip width, and $r = (h_1 - h_2)$ is the reduction in strip thickness.

Due to the variable tension component (5.152), the rolling force will have a variable component of the first kind, F_{1var}.

$$F_{1var} = -\sigma_{12var}\, W\sqrt{Rr_m} = \frac{2WZ\, v_2\, E\sqrt{Rr_m}}{L\, h_1\, \omega} \cos \omega t \tag{5.154}$$

A cosine function (5.154) precedes the sine function of the vibration (5.150), and thus the harmonic force F_{1var} "leads" the harmonic vibration z by 90°. Such a force represents negative damping and provides self-excitation.

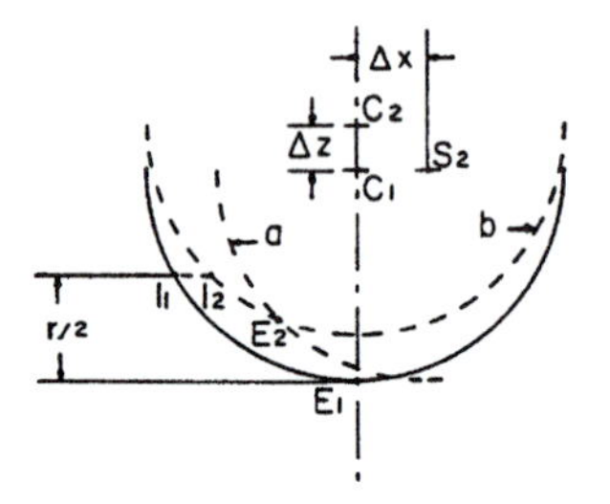

Figure 5.68
Shift of intersects due to vibrations.

The rolling force has also two other variable components F_{2var} and F_{3var}, which are both due to the variation of the contact length ℓ_c as it is produced by the vibration z. Let us look at Fig.5.68 which, in an exaggerated scale, represents the situation in two subsequent instants during the upward vibratory motion of the roll. The solid lines represent the roll and strip at instant 1. The contact profile extends from the input point I_1 to the exit point E_1. The center of the roll is C_1. After a short time the roll has moved upward by the amount Δz with center at C_2, and its periphery is now indicated by the dashed line b. The strip moves to the right by the amount Δx, and its previous surface has now the center S_2 and is indicated by the dashed line a. The contact length extends now from the input point I_2 to the exit point E_2, so it is shorter from both sides. (From E_2 to the right the previously rolled surface is intact).

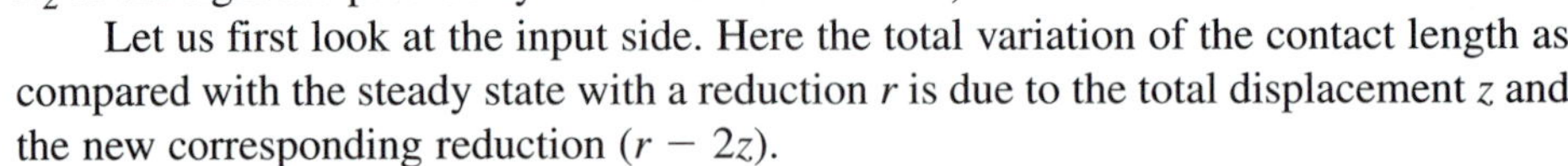
Let us first look at the input side. Here the total variation of the contact length as compared with the steady state with a reduction r is due to the total displacement z and the new corresponding reduction $(r - 2z)$.

The corresponding variable component of the rolling force is

$$F_{2var} = W(S - \sigma)\left(\sqrt{R(r - 2z)} - \sqrt{Rr}\right)$$

for $z << r$, it is possible to use the binomial series and approximate:

$$F_{2var} = -W(S - \sigma)\, z\sqrt{\frac{R}{r}}$$

This component of force variation is proportional to the vibrational displacement z and opposes it the same way as a spring would. Thus, the ratio

$$\frac{d\,F_{2var}}{dz} = W(S - \sigma)\sqrt{\frac{R}{r}} \tag{5.155}$$

represents the "rolling contact stiffness." Therefore, it does not provide negative or positive damping but is added to the elastic forces of the stand and increases the frequency of its natural vibrations.

As an example, let us use formula (5.155) to estimate the rolling stiffness for $W = 1$ m, $(S - \sigma) = 70 \times 10^7$ N/m^2, $R = 0.4$ m, and $r = 0.0005$ m. The rolling contact stiffness $= 1.97 \times 10^{10}$ N/m. Now, let us consider the variation of the contact length on the exit side. In Fig. 5.68, the exit point was shown to move left and to shorten the contact length. This will produce another variation of the rolling force, F_{3var}.

In Fig. 5.69, the circumstances are presented in a very exaggerated scale. In diagram (a) the exit intersect between the roll and the strip surface is shown assuming no vertical motion of the roll. During an instant Δt the rolled surface has moved to the right by Δx.

The intersect representing the exit point of the rolling contact is located $\Delta x/2$ to the right of the roll centerline. In diagram (*b*) the roll has moved upward by Δz, and the strip surface moved to the right by Δx. The intersect is obviously on the line of symmetry between the centers C_2 and S_2.

The change in contact length $\Delta \ell_c$ due to the two incremental motions shown in (*a*) and (*b*) is

$$\Delta \ell_c = \frac{-R\Delta z}{\Delta x} \tag{5.156}$$

In Fig. 5.69c the situation shows a downward motion of the roll Δz and a motion of the strip to the right Δx. The change in contact length is

$$\Delta \ell_c = \frac{R\Delta z}{\Delta x}$$

Obviously equation (5.156) applies if the proper sign of Δz is used. As it is, since $\Delta x = v_2\, \Delta t$, we have

$$\Delta \ell_c = \frac{R}{v_2} \frac{\Delta z}{\Delta t}$$

The variable force component can be expressed by using the force formula

$$F_{3var} = (S - \sigma)\, W\, \Delta \ell_c$$

and for the limit $\Delta t \rightarrow dt$, we have

$$F_{3var} = -\frac{(S-\sigma)WR}{v_2}\frac{dz}{dt} = -\frac{(S-\sigma)WR\, Z\omega \cos \omega t}{v_2} \tag{5.157}$$

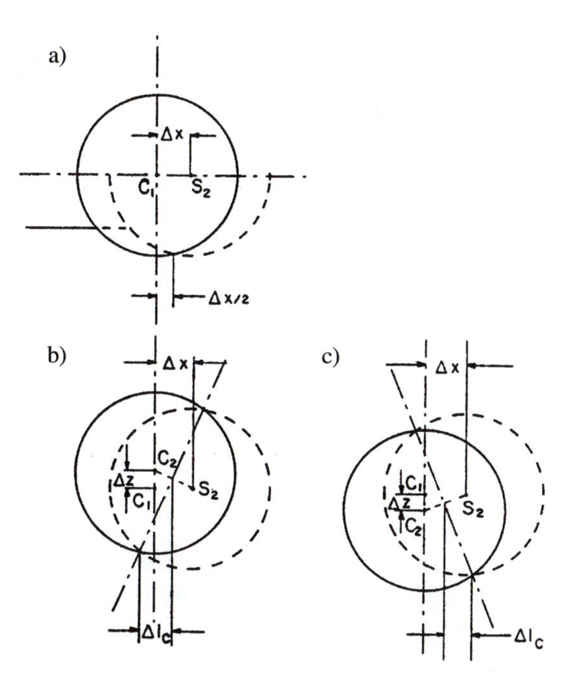

Figure 5.69
Variation of exit contact length; a) no vertical motion, the rolled surface moves right by Δx; b) vertical motion of plus Δz is added; c) vertical motion of minus Δz is added.

This force is now proportional to the velocity of the vibration, and it "lags" the vibrational displacement by 90°. It represents positive damping and opposes self-excitation. There is, of course, also another positive damping inherent in the structure of the rolling stand. The "limit of stability" is reached if the negative damping F_{1var} is equal to the positive damping F_{3var} plus the structural damping.

$$F_{stdamp} = -c\,Z\,\omega\cos\omega t$$

where c is the damping coefficient.

If we neglect the structural damping, we may obtain a condition for the limit of stability that will be on the "safe" side: instability (self-excited vibration) will arise if

$$\frac{2\,WZ\,v_2\,E}{L\,h_1\,\omega}\sqrt{Rr_m}\cos\omega t > \frac{(S-\sigma)WR\,\omega\,Z}{v_2}\cos\omega t$$

or

$$\frac{2E\,v_2^2}{(S-\sigma)L\,h_1\,\omega_2}\sqrt{\frac{r_m}{R}} > 1 \tag{5.158}$$

The influences are such that in order to stabilize, it is necessary to change the individual parameters as follows:

Young's modulus E	decrease
Yield strength of strip $(S - \sigma_m)$	increase
Distance between stands L	increase
Input thickness h_1	increase
Very strongly (second power)	
Exit speed (roll speed) v_2	decrease
Natural frequency ω	increase
Strip width W (the frequency increases with strip width)	increase
Lighter influences (0.5 power)	
Reduction r_m	decrease
Roll radius R	increase

Some of these influences (rolling speed, strip width, input thickness) check extremely well with experience.

Also quantitatively, if we take an example:

$h_1 = 0.0022$ m,	$(S - \sigma) = 56 \times 10^7$ N/m^2	$f_n = 125$ Hz
$h_2 = 0.0017$ m,	$E = 2.1 \times 10^{11}$ N/m^2	$\omega = 785$ rad/sec
$r = 0.0005$ m,	$L = 3.2$ m	
$v_{2m} = 11$ m/sec,	$R = 0.4$ m	

With these parameters, the left side of expression (5.158) has a value of 0.74. The case is stable but not far from the limit of stability.

This is, of course, a very simplified evaluation; however, it explains rather well the effects of the individual parameters. A more detailed analysis could be carried out by means of a simulation program. The results of such an exercise are presented in Ref. [7].

REFERENCES

1 ASM International. *ASM Handbook*. Vol. 14, *Forming and Forging*. Materials Park, OH: ASM International, 1988.

2 Altan, T., S.-I. Oh, and H. Gegel. *Metal Forming: Fundamentals and Applications*. Materials Park, OH: ASM International, 1983.

3 Avitzur, B. *Handbook of Metal Forming Processes*. New York: John Wiley & Sons, 1983.

4 Cotrell, A. H. *The Mechanical Properties of Matter*. New York: John Wiley & Sons, 1964.

5 Hosford, W. F., and R. M. Caddell. *Metal Forming Mechanics and Metallurgy*. Englewood Cliffs, NJ: Prentice Hall, 1983.

6 Johnson, W., and P. B. Mellor. *Engineering Plasticity*. New York: Van Nostrand Reinholt, 1973.

7 Lange, K., ed., *Handbook of Metal Forming*. New York: McGraw-Hill, 1985.

8 Tlusty, J.; G. Chandra; S. Critchley; and D. Paton. "Chatter in Cold Rolling." CIRP Annals 31, no. 1 (1982).

9 Schey, J. A. Introduction to Manufacturing Processes, 2nd Ed. New York: McGraw-Hill, 1987.

QUESTIONS

Q5.1 Which parameters of the stress-strain diagram express "ductility" and "toughness" of the materials?

Q5.2 **(a)** What are the approximate temperature ranges in which hot and cold working are done?

(b) Why is the finishing stage of a hot-forming operation done just above the recrystallization temperature?

Q5.3 **(a)** Why do the yield strength *YS* and ultimate tensile strength *UTS* increase with increasing amounts of cold work?

(b) How can the effects of cold work be eliminated?

Q5.4 How are the material properties and the microstructure of a material affected in a hot-forming operation?

Q5.5 Compare and contrast the process characteristics and the resulting material properties for the operations of hot and cold work.

Q5.6 Which factor limits the reduction ratio of a drawing operation? Neglecting friction and redundant work, determine the maximum area reduction obtainable in drawing. What limits the maximum reduction ratio in extrusion?

Q5.7 What is the yielding condition for a metal sheet in a cold-rolling operation? With the use of Mohr's circle, explain why it is advantageous to apply a back and forward tension to the strip.

Q5.8 **(a)** In an indenting process, for which value of ratio h/L (workpiece thickness/indentor length) is there (1) no redundant work; (2) the same amount of redundant work as for an infinitely thick slab. What are the values of the redundant work correcting factor Q_r for the above cases?

(b) For which values of $\Delta = h/L$ does the effect of friction become significant in the plane-strain compressing of a workpiece?

Q5.9 **(a)** In a hot-rolling operation, what condition must be satisfied if the workpiece is required to enter the rolls unaided? If this is the limiting condition, what is the maximum reduction obtainable?

(b) What is the "neutral point" in a rolling operation, and in which directions do the frictional forces act before and after this point?

Q5.10 How are the amounts of redundant work and friction work affected by the reduction ratio D_0/D_1 in a drawing operation? What effect does die angle have on these factors?

Q5.11 Make some sketches showing the most common types of material failures in bulk-forming operations. Briefly describe the causes of these failures.

Q5.12 Draw the stress distribution resulting from a bending moment being applied to a beam. Consider four cases and determine the expression for the value of the moment in each case.

(a) Elastic bending
(b) Fully plastic bending
(c) Elastic and plastic bending
(d) Plastic bending of a strain-hardening material

Q5.13 What factor limits the smallest radius to which a plate or sheet can be bent? How is this limit affected by the previous processing of the sheet?

Q5.14 What is the basic difference, in terms of strains, between stretching and drawing? What is the primary purpose of the blankholder in a deep drawing operation?

PROBLEMS

P5.1 *Stress-strain diagram*

(a) Refer to Fig. P5.1a. Additional data follows: Diameter at fracture $d_f = 15$ mm; initial diameter $d_0 = 20$ mm; $e_A = 0$; $e_B = 0.25$; $e_C = 0.38$; $S_A = 250$; $S_B = 430$; and $S_C = 400$ N/mm^2. Determine true stress σ and true strain ε at B and C and plot the (σ, ε) graph. Connect Y_0 and $(\sigma_C, \varepsilon_C)$ by a straight, dashed line, and determine the coefficient K in the yield stress equation (5.9). Calculate specific work to fracture $W_{s,f}$.

(b) Draw and label the engineering stress-strain diagram for carbon steel using the following data obtained in a tensile test: Young's modulus $E = 200$ GPa; yield strength $YS = 280$ MPa; ultimate tensile strength $UTS = 480$ MPa; stress at fracture 380 MPa; and elongation $e_{max} = 28\%$.
Indicate graphically how the shape of the stress-strain diagram changes with increasing amounts of cold work applied to the material.

(c) The engineering load-strain diagram shown in Fig. P5.1c was obtained from a tensile test on a specimen having an initial diameter of 12.0 mm. The diameter of the fractured neck at point 2 was 6.25 mm.

(i) Determine the true stress σ (MPa) and true strain ε at points 1 and 2.

(ii) Determine the value of the parameters n and K of the relationship $\sigma = K\varepsilon^n$.

(d) A tensile test was performed on a steel alloy specimen having initial dimensions 15 × 2 mm and a gage length of 100 mm. The yield strength was found to be 370 MPa at a 0.2% offset, and a modulus of elasticity was 205 GPa.

(i) Determine the engineering stress and strain if a 10-kN load is applied. What is the gauge length during the application of the load, and after the load is removed?

(ii) What load would be required to produce a stress equal to the yield strength? What is the gauge length during and after the application of this load?

(e) The following load-extension data was obtained from a tensile test on an aluminum specimen.

Load (N)	*Extension (mm)*
257	0.0991
512	0.1969
600	0.3938
690	0.5907
756	0.7876
845	1.181
970	1.772
1067	2.560
1179	3.938
1290	6.892
1379	11.81
1401	19.45

The specimen had initial dimensions of $l_0 = 50.8$ mm, $w_0 = 12.73$ mm, and $t_0 = 1.00$ mm. The gauge length at failure was $l_f = 70.25$ mm.

(i) Calculate the engineering stress $S = F/A_0$ and the engineering strain $e = (l - l_0)/l_0$. Draw the stress-strain diagram and determine the yield strength YS, the ultimate tensile strength UTS, the elongation e_{max}, and the modulus of elasticity E using a 0.2% offset.

(ii) Determine the true stress σ and true strain ε, and draw the stress-strain diagram.

(iii) Determine the parameters K and n for the relationship $\sigma = K\varepsilon^n$ by plotting log σ vs. log ε of the points in the plastic range. The strain hardening relationship is then represented by the straight line: $\log \sigma = \log K + n \log \varepsilon$.

(f) The effect of cold work on the ultimate tensile strength UTS and elongation e_{max} for 70% Cu, 30% Zn brass is shown in Fig. P5.1f. The area reduction is $r = [(A_0 - A_1)/A_0] \times 100\%$.

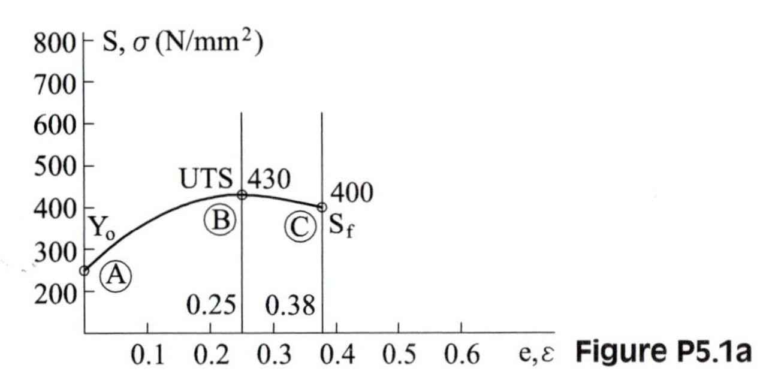

Figure P5.1a

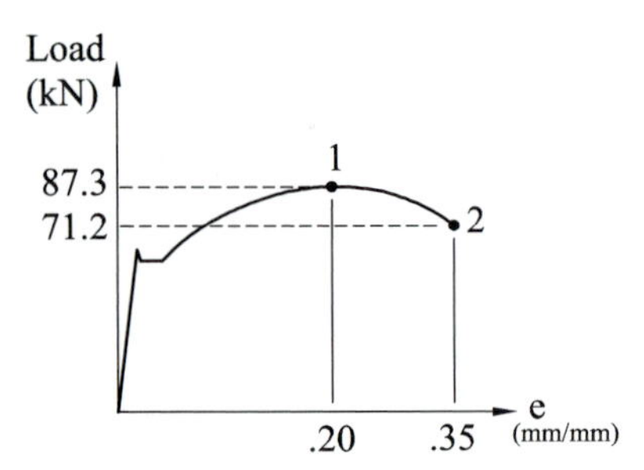

Figure P5.1c

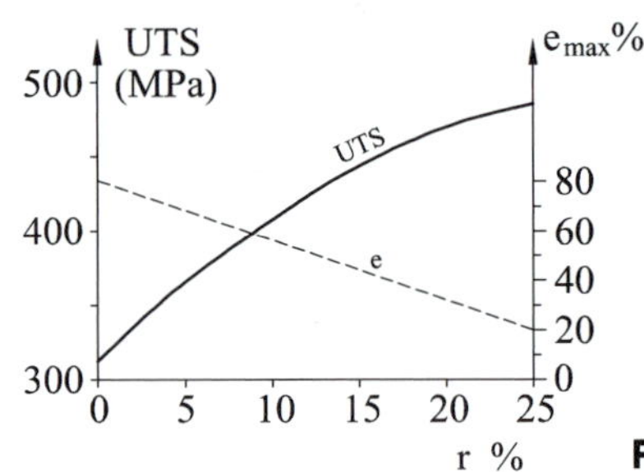

Figure P5.1f

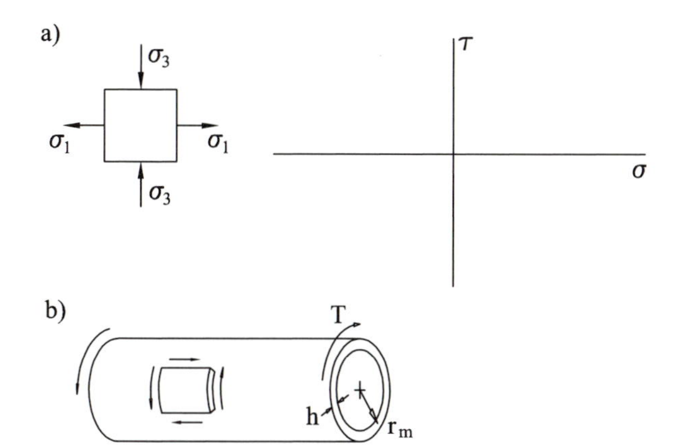

Figure P5.2

A bar of 5.25 mm diameter is to be produced with $UTS > 410$ MPa and elongation > 30%. The available stock is annealed brass with a diameter of 8.95 mm. The material will fail if the amount of cold work exceeds 50% area reduction. The work will be done in a number of steps with annealing in between. What will be the final diameter of the first step, which is the same as the initial diameter of the second step? Describe the sequence of steps needed to produce the required bar.

P5.2 *Yielding: pure shear*

(a) See Fig. P5.2. $\sigma_3 = -\sigma_1$, $\sigma_2 = 0$. Draw Mohr's circle. Indicate by dashed lines the location of planes with maximum shear stress $\tau_{1,3} = \tau_{max}$. Determine τ_{max}.
What are the values of normal stresses σ in those planes? Write out the yield criterion and express the value of σ_1 at yield and for $Y = 300$ N/mm^2.

(b) Pure shear in torsion. We are given the following data: $Y = 350$ N/mm^2; $d_{outer} = 98$ mm; $d_{inner} = 90$ mm; $r_m = 47$ mm; and $h = 4$ mm. Determine the torque T (Nm) for yielding (see Fig. P5.2b).

P5.3 *Simplified calculation of cold rolling.* Refer to Fig. P5.3. Use average yield strength $Y = 400$ N/mm^2, $h_0 = 2$ mm, $h_1 = 1.4$ mm, $v_0 = 10$ m/sec, $\sigma_x = 180$ N/mm^2, width $w = 1000$ mm, and $R = 250$ mm. Determine v_1 (m/sec), the length of the compression zone L (mm), the compressive force F (N), the torque on each roll T (Nm), and the total power P (kW) (use average $v = (v_0 + v_1)/2$) and the temperature increase ΔT (°C) of the work ($\rho c = 3.7$ N/mm^2°C).

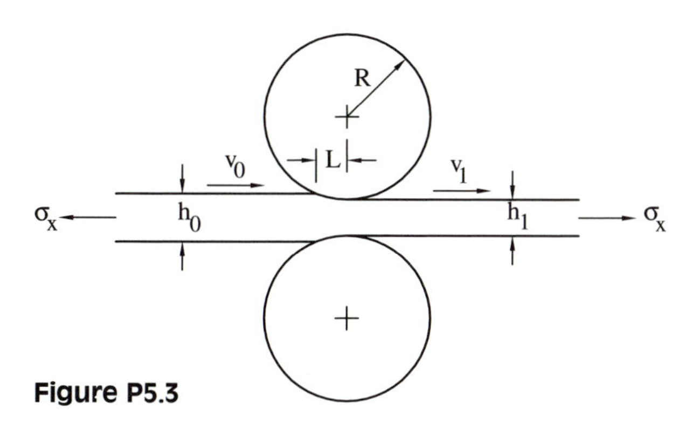

Figure P5.3

P5.4 *Forming: stress-strain diagram; yield criterion.*

(a) In a stress-strain test the specimen, with an initial diameter $d_0 = 20$ mm fractured with a diameter $d_f = 12$ mm. The engineering stress at fracture was determined as $S = 350$ N/mm^2. Determine the true strain ε_f and true stress σ_f (N/mm^2) at fracture.

(b) In a rolling operation of steel (refer to Fig. P5.4), the average yield stress is $Y_{av} = 300$ N/mm^2. Interstand tension causing stress $\sigma_x = 150$ N/mm^2 is provided. The length of contact is $L = 16$ mm, and the width w of the strip is 1000 mm. Determine the rolling force F. Write out the applicable yield criterion. Determine the compressive stress σ_3 (N/mm^2) and the force F (N).

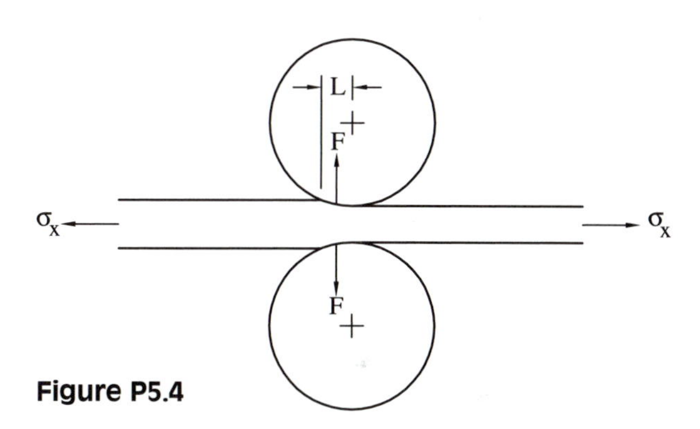

Figure P5.4

P5.5 *Wire drawing: neglect friction.* Refer to Fig. P5.5. $Y = Y_0 + K\varepsilon = 280 + 320\varepsilon$ (N/mm^2), $d_0 = 11$ mm, $d_1 = 6$ mm, $v_d = 4$ m/sec, $\rho c = 3.7$ (N/mm^2°C). Determine ε_1, Y_0 (N/mm^2), Y_1 (N/mm^2), F_d (N), the drawing power P (kW), the increase of the work temperature, and the maximum possible ratio of (d_0/d_1).

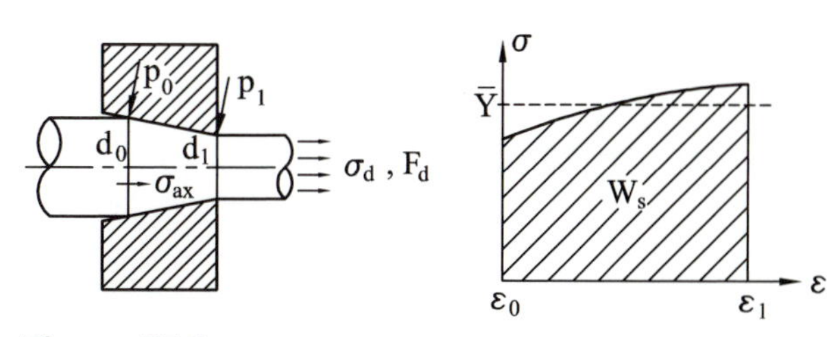

Figure P5.5

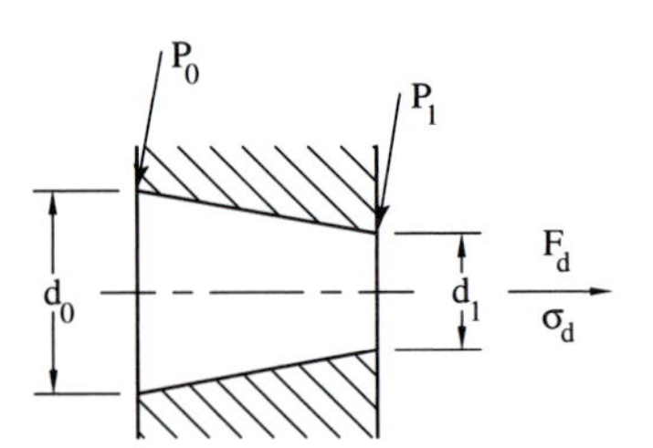

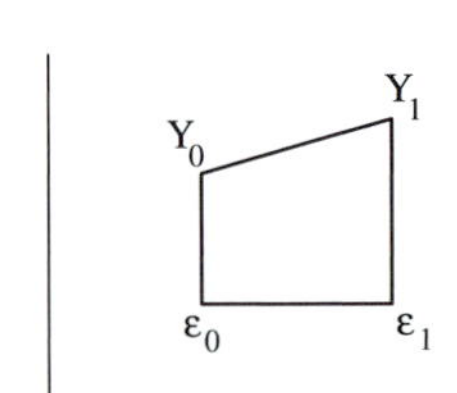

Figure P5.6

P5.6 *Wire drawing: neglect friction—strain-hardening material.* Refer to Fig. P5.6.
Given $Y = Y_0 + K\varepsilon = 250 + 300\varepsilon$ N/mm^2, $d_0 = 6.5$ mm, $d_1 = 4$ mm.

(a) Determine ε_1, Y_1 (N/mm^2), σ_d (N/mm^2), F_d (N).

(b) Determine the pressure p_0 and p_1. Write out the yield criterion for the axisymmetric case as a relationship between σ_x and p.

P5.7 *Rod drawing: extrusion.* Refer to Fig. P5.5 and Fig. P5.6.

(a) Drawing from $d_0 = 18$ mm to $d_1 = 10$ mm. The yield strength of the material is $Y = 200 + 150\varepsilon$ (N/mm^2). Determine ε_1, yield strength Y_1 (N/mm^2), drawing stress σ_d (N/mm^2), pressures between work and die p_0 (N/mm^2), p_1 (N/mm^2).

(b) Extrusion: same geometry as in (*a*) and the same material. Determine σ_{ex} (N/mm^2), pressure between work and die, $p_0 =$ (N/mm^2), p_1 (N/mm^2).

P5.8 *Wire drawing with friction.* Refer to Fig. P5.8.
Use the expression for $d\sigma_x$ in Eq. (5.28). Write a simple computer program by proceeding in small steps.

(a) Use the following: $Y = 250 + 300\varepsilon$, $D_0 = 15$ mm, $D_1 = 9.0$ mm, and $\mu = 0.2$. Start from $x = 0$, $\sigma_x = 0$, $D = D_0$, $Y = Y_0$, $p = Y_0$. Proceed in 100 steps:

$d\sigma_x = ?$
$\sigma_x = \sigma_x + d\sigma_x$
$D = D - dD$
$\varepsilon = ?$
$Y = ?$
$p = ?$
$x = x + dx$

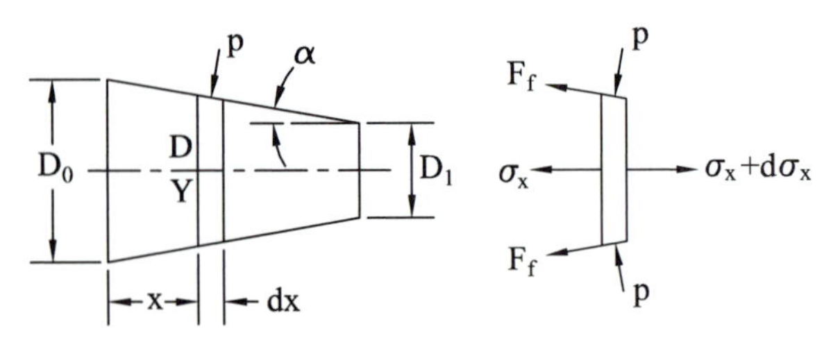

Figure P5.8

Plot in one plot: σ_x versus x, and p versus x.
For (1) $\mu = 0$ and (2) $\mu = 0.2$, attach plots. Write down the drawing force F_d:
Without friction ($\mu = 0$), $F_d =$
With friction ($\mu = 0.2$), $F_d =$

(b) Use the following: $Y = 280 + 320\varepsilon$, $D_0 = 11$ mm, $D_1 = 6.0$ mm, and $\mu = 0.2$. Carry out the same tasks as in (*a*).

P5.9 *Cold upsetting on a hammer.* See Fig. P5.9.
A round workpiece is being upset by multiple blows of a hammer. The hammer, which has mass $M = 400$ kg, falls from a height of 5 m. The material properties follow: $Y = 260 + 280\varepsilon$ (N/mm^2) and the friction coefficient between the hammer and the work is $\mu = 0.225$. The initial dimensions are $d_0 = 120$ mm and $h_0 = 25$ mm.

(a) Determine the intermediate parameters and the amount of height reduction Δh_0 (mm) in the first blow: Y_0 (N/mm^2), $p_{0,av}$ (N/mm^2), F_0 (N), Δh_0 (mm).

(b) After several more hits, the height is reduced to $h_1 =$ 14 mm. What will be the diameter d_1, and the reduction Δh in the next blow? Determine d_1 (mm), Y_1 (N/mm^2), $p_{1,av}$ (N/mm^2), F_1 (N), Δh_1 (mm)

This exercise is similar to Ex. 5.3. Use "cylfrict" and cross check with the graph 5.35. Determine the limit value of the *h/D* ratio for which the Coulomb friction would apply over the whole surface.

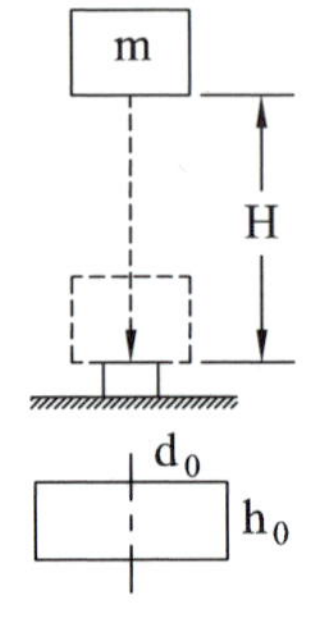

Figure P5.9

P5.10 *Cold upsetting on a hammer.* Refer to Fig. P5.9.
A round workpiece is forged by repeated blows of a hammer. The hammer mass is $M = 200$ kg, and it falls from a height of $H = 4$ m. The material properties are $Y = 300 + 250\varepsilon$ (N/mm^2) and the friction coefficient is $\mu = 0.18$. The initial dimensions are $d_0 = 120$ mm and $h_0 = 20$ mm. After several hits the height is reduced to $h = 9$ mm. What will be the height reduction Δh in the next blow? Determine Y, p_{av}, force F, and Δh in this blow. Use the program "cylfrict."

P5.11 *Forming: upsetting a cylindrical specimen, with sticking friction.* See Fig. P5.9.
The initial diameter of the workpiece is $d_0 = 100$ mm, and the initial height is 30 mm. The yield strength is $Y = 200$ N/mm^2. A hammer is used with mass $M = 400$ kg and fall height $H = 4$m. Determine the height h_1 and h_2 after the first and second blows.

First, convert the hammer energy E (Nm) to (N/mm). Determine average pressure for the first blow p_{av} (N/mm^2), force F_1 (N), compression Δh_1 (mm), height h_1 (mm), height h_2 (mm), the diameter after the first blow d_1 (mm), and the compression Δh_2 (mm).

P5.12 *Hot forging: sticking friction.* See Fig. P5.9. Neglect strain rate effect and assume constant $Y = 150$ N/mm^2. The initial diameter of the cylindrical workpiece is $d_0 = 150$ mm, height $h_0 = 60$ mm. The mass of the hammer is $M = 600$ kg, and height $H = 4$ m. After several hits the height has decreased to $h = 40$ mm. Determine the amount Δh in the next hit.

P5.13 *Cold upsetting a cylindrical workpiece by repeated hammer blows.* See Fig. P5.13.

(a) Use the following: initial diameter $d_0 = 110$ mm, $h_0 = 40$ mm, coefficient of friction $\mu = 0.2$, and $Y = 260 + 300\varepsilon$ (N/mm^2). Apply $m = 10$ successive blows. Write a simple program. In each step use the initial d_n and h_n, and use "cylfrict" to determine p_{av}, the force F, the corresponding Δh, then h_{n+1} and d_{n+1}, and so on. Plot d, h, and F versus m. Attach plots.

(b) Same as (a) but hot forging, with sticking friction. Use the following: $d_0 = 210$ mm, $h_0 = 50$ mm, and constant yield stress $Y = 120$ N/mm^2. Again, plot d, h, and F versus m. Attach plots.

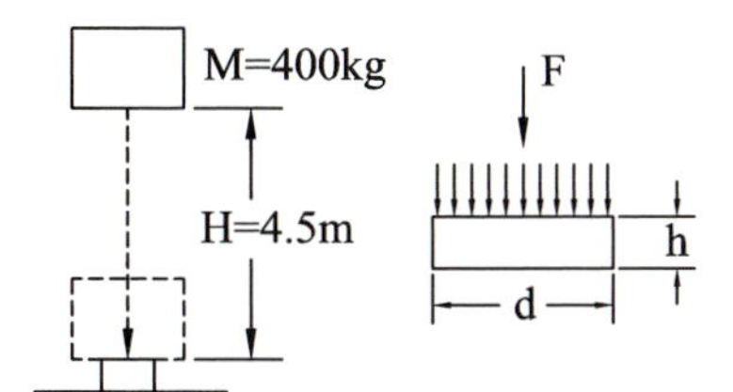

Figure P5.13

P5.14 *Hot upsetting, considering strain rate effect.* (A reworking of Example 4-12 of Schey [9].)
A billet of steel 1045, $d_0 = 50$ mm and $h_0 = 50$ mm is upset down to height $h_1 = 10$ mm, with velocity $v = 80$ mm/s. The yield strength formula which considers the strain rate effect is used:

$$Y = C\dot{\varepsilon}^m, \text{ where } \dot{\varepsilon} = \frac{v}{h}$$

and parameters C, m are $C = 105$ N/mm^2, $m = 0.11$. Sticking friction is considered, $p_{av} = Y(1 + 0.19\,D/h)$. Write a simple program to plot F versus h. Proceed in steps of $\Delta h = 1$ mm and then compute

$$W = \sum F dh$$

Initial values: $d = d_0$, $h = h_0$, $W = 0$.
The strain rate ε as a time derivative of ε is called *deps*.

For $n = 1{:}\ 40$
$deps = v/h_n$
$Y = C \times (deps)^m$
$F_n = (1+0.19\ d/h) \times Y \times \pi d^2/4$
$dW = F \times dh.$
$h_{n+1} = h_n - dh$
$d = d_0\ \text{sqrt}\ (h_0/h_{n+1})$
$W = W + dW$

Plot (h_n, F_n). Evaluate W.

P5.15 *Redundant work; friction effects; plane-strain compression.* Refer to Fig. P5.15.
For all cases: $Y = 420$ N/mm^2, $\mu = 0.2$. Determine force F (N): $w = 1000$ mm, $L = 15$ mm, $h = 60$ mm.
Case 1: F (N) = ?
Case 2: (2a) $F_t = 0$; (2b) $F_t = 300{,}000$ N; $w = 1000$ mm, $L = 15$ mm, $h = 1.4$ mm. Determine p_A, p_B, F for (2a) and (2b).

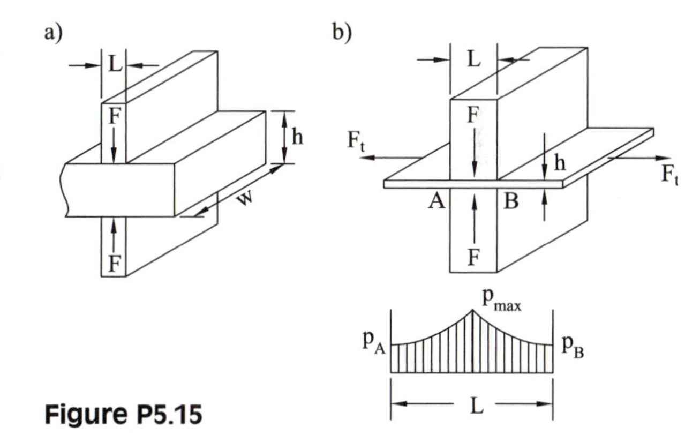

Figure P5.15

P5.16 *Metal forming; effect of friction, and of redundant work.* Refer to Fig. P5.16.
Where applicable, the coefficient of friction is $\mu = 0.22$, $Y = 400$ N/mm^2, $L = 12$ mm, and $w = 900$ mm. Determine force F.

(a) $h = 40$ mm, F (N) =
(b) $h = 40$ mm, F (N) =
(c) $h = 2$ mm, F (N) =

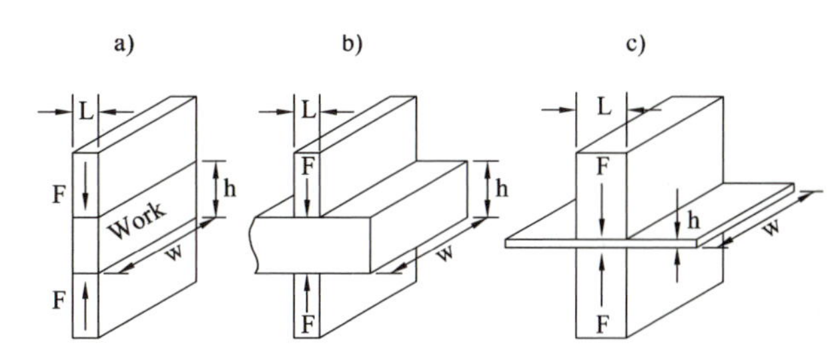

Figure P5.16

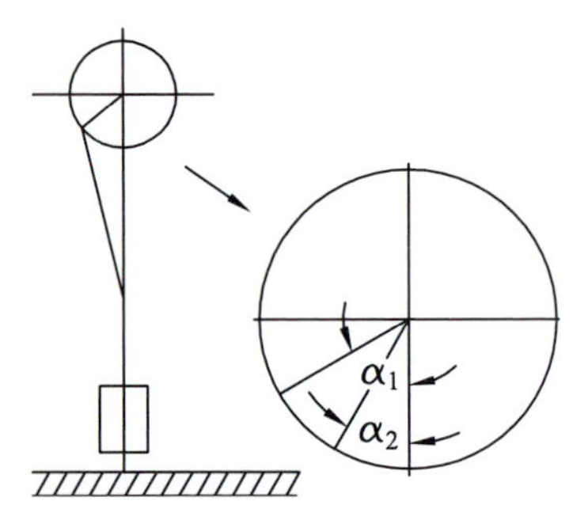

Figure P5.17

P5.17 *Press work.* Refer to Fig. P5.17.
A press is to punch 5 round holes, $d = 25$ mm. Shear flow stress $k = 150$ N/mm^2, penetration $p = 0.28$, crank radius $R = 80$ mm, and sheet metal thickness $h = 2.5$ mm. The bottom of the stroke is flush with the underside of the sheet metal. At which angle α_1 does the work start, and at which angle α_2 does it end? Determine punching force F (N). Also determine the work W (Nm), the average power per revolution, at $n = 300$ rpm, P_{av} (kW), the time of the actual punching operation t_p (sec), and the instantaneous power during the punching work, P_{inst} (kW).

P5.18 *Punching operation on a mechanical press: punching a set of holes.* Refer to Fig. P5.18.
The total perimeter of the holes is $o = 550$ mm, the sheet-metal thickness is $h = 2$ mm, shear flow strength $k = 150$ N/mm^2, and penetration factor $p = 0.27$.

(a) Calculate the total force F (assuming all punch faces in one plane, all holes punched simultaneously) and the total work W (Nm).

(b) Assume crank radius $r = 60$ mm, and an infinitely long connecting rod. In the bottom position the punch will pass through the sheet metal by a clearance $e = 0.5$ mm. This will mean that the work is done through an angle α of the crank rotation. Determine α. Determine the radius R (m) of a steel flywheel (solid disk 0.08 m wide; specific mass $\rho = 7.8 \times 10^3$ kg/m^3), rotating at $n = 50$ rpm such that it will slow down through the operation by 10%. Then, neglecting the slowdown, determine the instantaneous average power through the punching operation P_{inst} (kW) and the average power over one cycle P_{av} (kW).

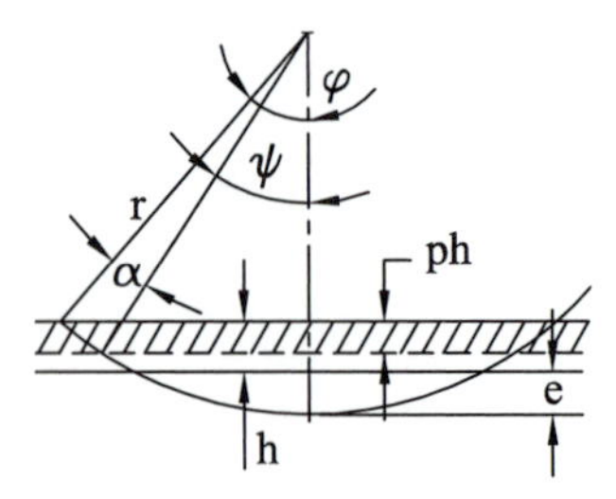

Figure P5.18

P5.19 *A blanking operation.*
A blanking operation is used to punch a 6.0-cm-diameter hole in a 3-mm-thick piece of sheet metal having a shear flow strength k of 200 MPa.

(a) What is the force F (kN) required to shear the plate?

(b) If the penetration is $p = 0.45$, determine work W (kJ) expended during the operation. Make a sketch of the force variation during the stroke of the punch.

(c) The punch is driven by a flywheel that has a mass moment of inertia $J = 10$ kgm^2 and a speed of 1.5 rev/sec. Determine the kinetic energy of the flywheel E (Nm) before and after a single punching operation. What is the final speed n (rev/sec) and the percent slowdown of the flywheel?

P5.20 *Cold rolling; effect of friction.* Refer to Section 5.3.5.
Evaluate the pressure between the roll and the strip of a cold-rolling operation while considering friction, strain-hardening, and the effects of forward and backward tension. Use the approach of and the values of Example 5.5 to make a graph of pressure variation on the rolls between the entry and the exit planes. Also determine the total force F (kN) on each roll, and the location of the neutral point. Repeat the computation using a higher friction coefficient $\mu = 0.15$.

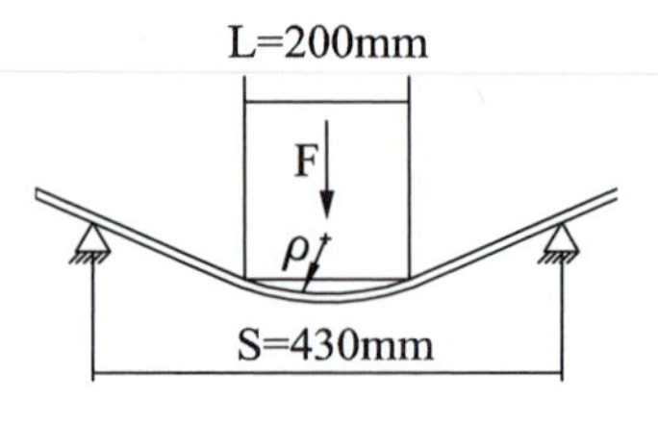

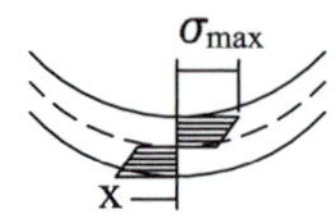

Figure P5.21

P5.21 *Plastic bending with strain-hardening.* Refer to Fig. P5.21.
Plate thickness $h = 12$ mm, $Y = 220 + 300\varepsilon$, $\rho = 225$ mm, and $w = 100$ mm. Determine the force F (N) and maximum stress σ_{max}.

P5.22 *Bending of non-strain-hardening material; springback.* Refer to Figs. 5.53 and 5.54.
A plate of thickness $h = 5$ mm and width $w = 80$ mm is to be bent to a radius of curvature of $\rho = 40$ mm. Determine the die angle α_d that is required to produce an angle of $\alpha = 60°$ in the plate after springback. The material is non-strain-hardening with a yield strength of $Y = 350$ MPa and Young's modulus $E = 200$ GPa. Make a sketch of the cross section of the plate showing the size of the elastic and plastic portions of the stress distribution before springback has occurred.

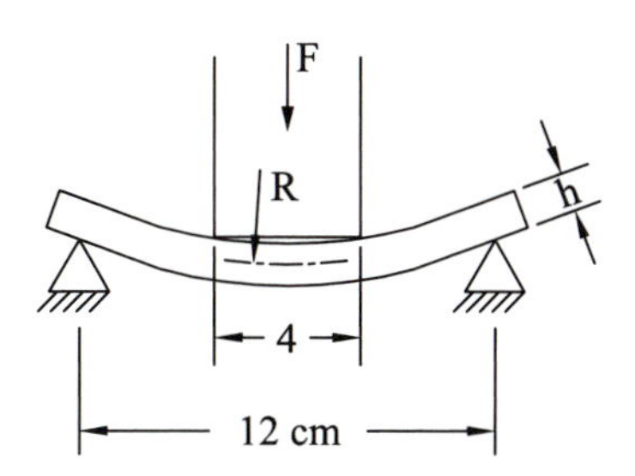

Figure P5.23

P5.23 *Plastic and elastic bending with strain-hardening.* Refer to Example 5.5 and Figs. P5.23, 5.51, and 5.52. Determine the force F (kN) that must be applied to the punch to bend a plate having a width $w = 50$ mm and thickness $h = 10$ mm.

Bend the plate to a radius of 35 mm, measured to the neutral fiber at the midpoint of the thickness. Consider a strain-hardening material characterized by the equation $Y = Y_0 + K\varepsilon$, where $Y_0 = 200$ N/mm^2 and $K = 150$ N/mm^2. Determine the dimension $x_{\lim}$ and the angle $\alpha_{\lim}$ at the point on the beam where the curved section meets the straight section (neglecting elastic deformations). Make a sketch of the deflected beam showing the areas of constant curvature, changing curvature, and no curvature.

P5.24 *Plastic bending of strain-hardening steel.* Refer to Fig. P5.24.
The dimensions are $w = 80$ mm, $h = 18$ mm, $F = 5550$ N, and material yield strength is $Y = 300 + 350\varepsilon$ N/mm^2. The modulus of elasticity is $E = 2.1\text{e}5$ N/mm^2.

(a) Determine the limit distance x_1 beyond which the beam remains straight.

(b) Determine the radius $\rho_{\min}$ at the root of the beam and the angle α of the slope of the straight part.

(c) Determine the maximum stress $\sigma_{\max}$ at the root of the beam, and the final radius ρ_f at the root after the removal of the load.

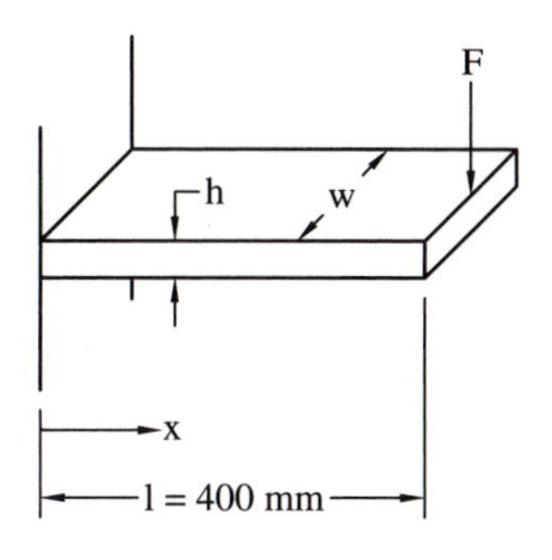

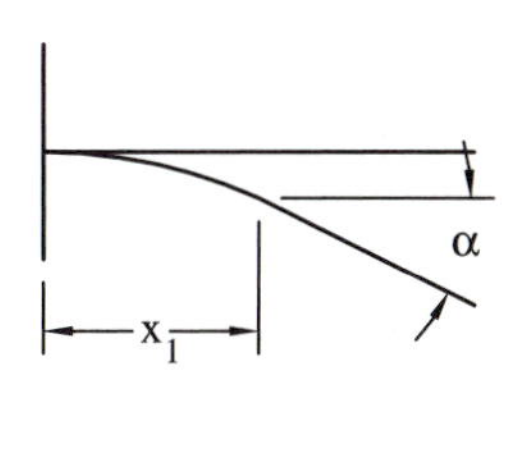

Figure P5.24

P5.25 *Radial drawing without strain-hardening*
A cylindrical cup of thickness $h = 1.0$ mm is being formed in a drawing operation as shown in Fig. 5.60. The material is non-strain-hardening, with a yield strength of $Y = 450$ MPa, and a redundant work correction factor $Q_f = 1.3$.

(a) Determine the radial stress σ_1 (MPa) in the flange at four radii: $r = R_0 = 10$ cm, $r = 8.5$ cm, $r = 6.5$ cm, and $r = r_c = 5$ cm for the initial position of the blank: $r_0 = R_0$.

(b) Determine the drawing stress σ_d (MPa) at the position $r = r_c = 5$ cm, and the drawing force F_d (kN) required as the punch moves downward from its initial position to four subsequent positions of the blank: $r_0 = 9$, 8, 7, and 6 cm.

P5.26 *Radial drawing.* Refer to Figs. 5.58–5.66.
Provide solutions for **(a)** a non-strain-hardening material $Y = 310$ (N/mm^2) and for **(b)** a strain-hardening material $Y = 310 + 450\varepsilon^{0.3}$ (MPa). Given: $D_0 = 140$ mm, d_c = (cup diameter) = 80 mm, $h = 1$ mm. Neglect friction between sheet metal and the die. Determine the drawing force on the punch for one instant in the draw when the initial diameter D_0 has been reduced to $d_0 = 120$ mm, for both (*a*) and (*b*).

P5.27 *Radial drawing with strain-hardening*
Write a computer program to determine the variation of radial drawing stress for a strain-hardening material. Use the approach of Section 5.4.7, Example 5.6, and Fig. 5.66. The material characteristics are $Y = 100 + 500\varepsilon^{0.2}$ (MPa). Use the dimensions $R_0 = 2$ and $r_c = 1$. Plot the variation of stresses $\sigma_d = \sigma_A + \sigma_B$ as the value of r_0 changes from R_0 to r_c.

6 Processing of Polymers

6.1 INTRODUCTION: PROPERTIES USED IN PROCESSING

This chapter deals with the processes used to manufacture products made from the many types of plastics and rubbers, including composites with polymer matrix. The words *polymers* and *plastics* are generally used interchangably, with emphasis on the latter as applied to the man-made polymers as distinct from those found in nature. Chapter 2 presented a brief discussion of the nature of polymers and of their useful properties, such as strength and stiffness, and optical and electrical characteristics. Many books exist with information about the design and use of such products; see Refs. [1], [2], [3]. A great number of plastics exist with a wide range of the various properties.

The use of polymers, in the three basic classes of thermoplastics, thermosets, and elastomers, has been steadily increasing for the past fifty years. Presently they play an important role in household materials for the family, and their consumption is approaching that of aluminum or even steel. As a group they possess unique properties of low weight, corrosion resistance, resistance to both acid and basic environments, and electrical and thermal insulating capability. They find successful application as parts of automobiles, as packaging materials, in electrical and electronic components, household articles, utensils, tubing, and foamed products. Most of them are easily recycled.

In this chapter we are concerned only with the properties involved in the *processing* of plastics, such as enthalpy (specific heat) and thermal conductivity as they affect both the initial plastification stage and also the final cooling stage, and viscosity in the range of temperatures used in processing, as it affects the flow through the dies in extrusion and through the mold cavities in transfer and injection molding. These materials, even with their great variety, are all processed under similar conditions. Thus we can discuss the equipment and the processes of mixing, plastifying, shaping, curing, and cooling by using examples of a limited number of materials.

All of these processes are strongly affected by two main characteristics of polymers: (1) the rather high viscosity at the shaping and processing temperatures, and

(2) the rather low thermal conductivity. While the viscosity of water is 0.001 Nsec/m^2 (Pas) at 20°C, that of oils at room temperature is 0.1–1.0 Pas, for dough it is 300 Pas, and for molten glass it is 100–10,000. For most plastics, it is between 100 and 1000 Pas in the 100°–300°C temperature range used in extrusion or injection molding. We will review later on the notion that viscosity is the coefficient of proportionality between shear rate and shear stress in viscous flow. For fast processing such as extrusion and injection molding, where shear rates are of the order of 10^2 to 10^5, rather high pressures in the range of 20–200 MPa are needed to produce the flow. Injection-molding machines deliver pressures up to 200 MPa (27,500 psi) and range in sizes from 100 to 10,000 tons of die-clamping force.

The thermal conductivity of polymers is in the range of only 0.1–0.35 W/°C·m as compared to glass with 0.84, carbon steel with 47, and copper with 390 W/°C·m. Consequently, it is difficult both to heat and to cool plastics by conduction from the walls into the material. Any significant heating or cooling effect takes 17 min to reach a depth of 100 mm and 10 sec to reach a depth of 1 mm [1]. Therefore, a significant role in heating the material in an extruder screw is played by the viscous heating in the material itself due to its mixing and shearing in the screw channels. The longest part of the injection-molding cycle of a thermoplastic is the cooling stage. While injecting may take 1 sec, cooling may take as long as 10–20 sec. Therefore, injection-molded parts are preferably made with thin walls. In extrusion, the cooling of the parts is done either in air for films, or in water for films and pipes; again, thin walls help. To provide at least a basic understanding of such problems and of the tasks in developing the processes and the machinery, simple calculations of flows of plastics are included in Section 6.3.

The overall management of heat plays an important role in a plastics-processing plant using injection-molding machines. It is necessary to provide the energy to heat the material, for plasticizing or melting it, for producing the flow through a die or into a mold, and for cooling the molded parts. A plant may use twenty machines, each running at 50 kW that they convert to heat and then use 50 l/min chilled water at 30°C with 5°C temperature rise through the heat exchanger to remove the heat. This amounts to a total of 1000 kW and 1000 l/min water flow with a central chiller using an additional 70-kW compressor and 15 kW of pump power.

6.2 SUMMARY OF SELECTED POLYMERS

Some basic characteristics and functional properties of polymers were briefly discussed in Section 2.5. We will now summarize the characteristics of the most commonly encountered polymers. They are listed in Table 6.1 where the tonnage of annual consumption is given for each material, for the year of 1993, in the groups of thermoplastics, thermosets, and fibers. The total amounts to 34.8 million tons. This compares with about 100 million tons for the total annual production of steel, which has seven to eight times greater density. Obviously, the volume of plastics use is higher than that of steel.

Generally, polymer molecules consist of many repeating units called monomers or simply *mers*. Many natural polymers exist, but we will be dealing with synthetic polymers called *plastics*. The various plastics differ in the composition and structure of the basic building blocks. They may consist of just one single type, called *homopolymers*, or of combinations of several types, known as *copolymers*. Those with two different

TABLE 6.1 U.S. Sales of Most Common Polymers (millions of metric tons)

Thermosets	Sales	Thermoplastics	Sales	Fibers	Sales
Synthetic rubbers		LDPE	6	Polyester	1.6
Styrene-butadiene	0.9	PVC and copolymers	4.7	Nylon	1.2
Polybutadiene	0.5	HDPE	4.8	Others	1.0
Ethylene-propylene	0.3	Polypropylene	4.1		
Nitrile	0.1	Polystyrene	2.5		
Polyurethanes	1.4	Polyesters	1.3		
Phenols	1.4	ABS	0.6		
Urea formaldehydes	0.8	Nylons	0.3		
Polyesters	0.6	Acrylates	0.3		
Epoxy	0.2	Polycarbonates	0.3		
		Polyurethanes	0.1		
		Acetals	0.1		

components are *binary* (e.g., a mixture of polyethylene and polypropylene), and those with three are ternary or *terpolymers*, such as the well-known plastic ABS (acrylonitrile-butadiene-styrene). Another significant feature is the number n of repeating mers in the molecule. The average value of n is expressed as the *degree of polymerization* (*DP*); higher *DP* indicates higher strength of the material in its use but also higher viscosity in its processing. In another way, the *DP* affects the *molecular weight* (*MW*) which is obtained as *DP* times the molecular weight of the mers. (Molecular weight is the weight in grams of 6.02×10^{23} molecules.) Thus, for the ethylene mer, C_2H_4, the molecular weight is 28; for the polypropylene mer, C_3H_6, it is 42; for the polystyrene mer, C_8H_8, it is 104, and so on. The typical sizes of some polymer molecules follow: for polyethylene (PE), $DP = 10{,}000$ and $MW = 280{,}000$; for polystyrene, $DP = 3{,}000$; for polycarbonate, $DP = 200$. Actually PE exists in low density LDPE ($DP = 1{,}000$ to $2{,}000$), high-density HDPE ($DP = 5{,}000$ to $10{,}000$), and also as UHMWPE (ultra-high-molecular-weight PE) with $DP = 150{,}000$ and $MW = 4$ million.

Polymers exist in three types of structures: *linear*, *branched*, and *network* (see Fig. 2.27). In the linear structure the chain continues sequentially (see Fig. 6.1) either in a smooth fashion, as in Fig. 6.1a, or with pendant groups that can be located in certain positions in the *isotactic* manner (Fig. 6.1b), the *syndiotactic* manner, as in Fig. 6.1c, or the *atactic* manner (Fig. 6.1d). The double-strand backbone (Fig. 6.1e) of aromatic polymers that contains benzene rings is called *ladder structure*. The double bond makes these polymers highly temperature-resistant. The chains are normally not straight lines but twisted like spaghetti, with side branches. Some polymers can be made free of branches such as linear LDPE (LLDPE) or polytetrafluorethylene (PTFE). Lightly branched and lightly cross-linked structures are typical of elastomers. A tight network of cross-linking that develops during the curing process characterizes thermosetting polymers, which explains why thermosets cannot be reverted to the viscous state on reheating. The structure of copolymers depends also on the sequence of the individual components, such as *alternating* (ABABAB), or *block* (AAAABBBAAABBBB), or *random*. One of the components may constitute branches off the chain of the other component, and this is called the *graft* copolymer.

The cohesion of the molecules of a polymer is due to secondary intermolecular forces. These forces are much weaker than the primary covalent bonds that hold the

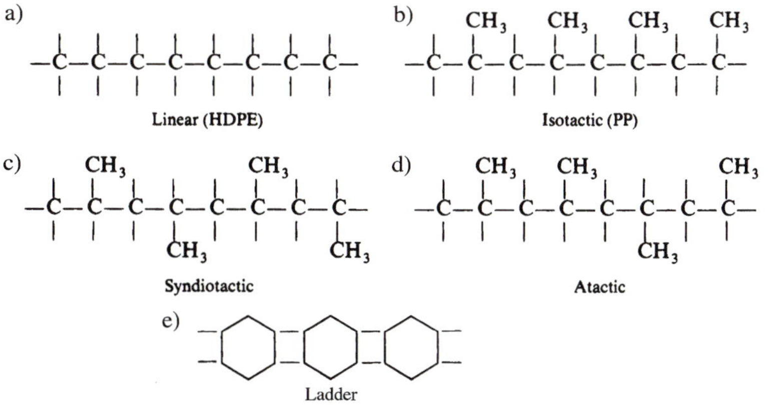

Figure 6.1
Different types of ordering (a, b, c) of pendant groups along the linear backbone chain *(a)* of polymers. Aromatic polymers contain benzene rings providing ladder structure *(d)*. [SCHEY]

atoms in the mer molecule together. The latter are due to electrons shared by the atoms, and the energy needed to break 1 mol (6.022×10^{23} bonds) is of the order of 300–800 kJ/mol. Most of the secondary bonds are of the very weak van der Waals type (2–10 kJ/mol). The dipole bonds in polar molecules are stronger (6–12 kJ/mol) and the hydrogen bond that acts between H and O, N or F is even stronger (15–30 kJ/mol). In any case, the secondary bonds are easily weakened at higher temperatures, which is why the molecules of the plastic slide over each other in the viscoplastic flow.

Polymers are produced by one of two different methods, the addition type and the stepwise polymerization. The first one consists of using a chemical catalyst to open the carbon double bond that is characteristic for many monomers, see Fig. 6.2. A typical example is ethylene. The one released internal bond with its unpaired electron attaches very actively to another such monomer and very fast long chains are generated. Most of the polymers produced by the addition process are thermoplastic but some become thermosets such as polystyrene.

In step polymerization two monomers join to create a new type of a molecule, see Fig. 6.3, and these newly created monomers gradually combine in many chains of varying length. In this process often another compound, mostly water, is discharged. Both thermoplastics and thermosetting polymers are created by this method. Polyesters exist in both classes; the common kind PET is thermoplastic. Thermoplastics are also polyamids and polycarbonates. The silicones and polyurethanes are thermosets.

Finally (as illustrated in Fig. 2.27), polymers may be either *amorphous* when the long molecular chains are entangled in an entirely random fashion that is often likened to a bowl of spaghetti, or there may be regions in the bulk of the plastic where regular, ordered geometry develops, for instance, where the molecule repeatedly folds into parallel straight sections. The geometric order is evocative, but never as perfect as that of the crystal structure of metals. However, we speak about the *crystallinity* of such polymers and about the degree of crystallinity as the portion of the total volume in which this order develops. This may be 40%, 60%, and as high as 95%, but never 100%. Crystallization develops on cooling from the melt, with slower cooling producing greater crystallization.

Only polymers with simple structure, such as linear polymers, can crystallize. Those with side branches and cross links, and also copolymers, especially the random and graft type, do not crystallize. As for the linear structures, those that are isotactic crystallize readily; the syndiotactic ones crystallize less easily, and the atactic ones

Figure 6.2
Common monomers used in addition polymerization.
[SME8]

Monomer	Structure	Monomer	Structure
Butadiene	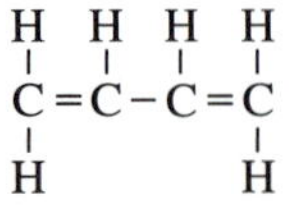 H H H H C=C–C=C H H	Acrylonitrile	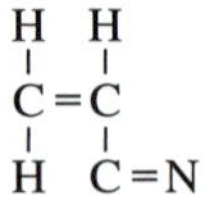 H H C=C H C≡N
Isoprene	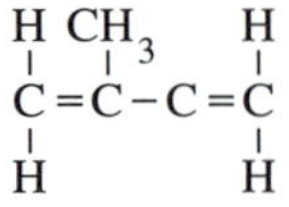 H CH_3 H C=C–C=C H H	Propylene	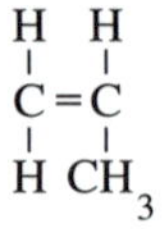 H H C=C H CH_3
Styrene	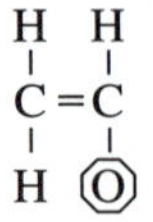 H H C=C H ⌬	Vinyl chloride	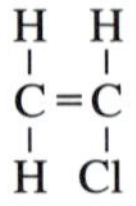 H H C=C H Cl
Ethylene	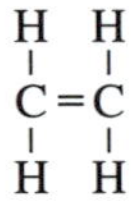 H H C=C H H	Acrylonitrile	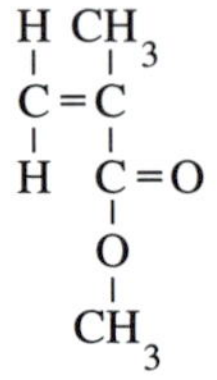 H CH_3 C=C H C=O O CH_3
Tetrafluoroethylene	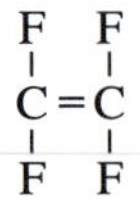 F F C=C F F	Propylene oxide	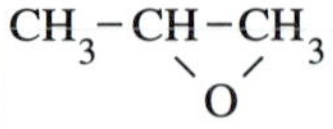 CH_3–CH–CH_3 \\ / O

almost never crystallize. Crystallic polymers are stronger, have a higher modulus of elasticity, and need higher temperatures for processing. Amorphous polymers in their rigid form are transparent; the crystalline types are not.

The properties of the various polymers can be enhanced by the use of *additives* and *fillers*. There are many kinds of these substances, and they are widely used. Additives include antioxidants and UV light stabilizers, plasticizers, flame retardants, colorants, lubricants, cross-linking agents, foaming agents, fungicides, pesticides, and odorants. Fillers can improve thermal, electrical, and mechanical properties and are often added to reduce cost without appreciably degrading the properties of the polymer.

Many polymers are adversely affected by oxidation and/or exposure to UV light, both of which cause destruction of the chain links and loss of strength of the material. A specific reaction occurs in Cl-containing polymers such as PVC, where a release of HCl produces the destructive effect. Antioxidants and UV stabilizers are used in all PE and PVC plastics.

Plasticizers are made of high-molecular-weight fluids that are mixed with the polymer. They depress the glass-transition temperature T_g in order to lower the effective viscosity at a given temperature or else to permit a lower processing temperature at which a level of viscosity appropriate for the flow of the polymer in a particular process is achieved. The most common application is in processing PVC.

The use of *flame retardants* is important because all plastics normally burn rather easily, and some of them produce toxic fumes. Various agents act in various ways by

Figure 6.3
Monomers used in stepwise polymerization. [SME8]

Reaction	Result	
~C(=O)-OH+H-O~ Carboxylic Acid + Alcohol ~C(=O)-Cl+H-O~ Acid Chloride + Alcohol	~C(=O)-O~ Polyester	$[(CH_2)_2$-O-C(=O)-⟨O⟩-C(=O)-O$]_n$ Poly(ethylene terephthalate)-PET (Trade names: Mylar, Dacron)
~C(=O)-OH+H_2N~ Carboxylic Acid + Primary Amine Acid Chloride + Primary Amine	~C(=O)-N(H)~ Polyamide	$[(CH_2)_6$-O(H)-C(=O)-$(CH_2)_4$-C(=O)-N(H)$]_n$ Poly(hexamethylene adipate) (Nylon-66)
~OH+ Cl-C(=O)-Cl Alcohol + Phosgene	~O-C(=O)-O~ Polycarbonate	[⟨O⟩-C(CH_3)(CH_3)-⟨O⟩-O-C(=O)-O$]_n$ Bisphenol A polycarbonate (Trade Name: Lexan)
Cl-Si(R_1)(R_2)-Cl+ H_2O Chlorosilane + Water	~Si(R_1)(R_2)~O~ Polysiloxane	[Si(CH_3)(CH_3)- O$]_A$ Poly(dimethyl siloxane)
~NCO+ H_2O~ Isocyanate + Primary Amine ~CHO+ 2~ NH_2 Aldehyde + 2 Primary Amines	~N(H)-C(=O)-N(H)~ Polyurea	[N(H)-C(=O)-N(H)-$CH_2]_n$ Urea-formaldehyde resins*

* This material is a cross-linked polymer in its commercial form. Hydrogen abstraction from the nitrogen and subsequent reaction forms the basis for cross-linking in these systems.

either increasing the combustion temperature, or producing gases that do not support burning and impede flame propagation.

Plastics are not painted; instead many *colors* are achieved by using organic dyes or mixing the plastic powder with organic and nonorganic *pigments*. Pigments are also used to create opacity in otherwise translucent materials. Dyes in solvents are used to produce transparent plastics in various colors.

Lubricants are added to facilitate the flow of plastics in molds and dies. They can be sprayed on the mold surface to help the release of the product from the die.

Fillers such as wood paste or various mineral powders may be added to a polymer to reduce its price. Other fillers are used to enhance electrical or thermal conductivity, or to reduce shrinkage. Furthermore, many fillers are added to improve the mechanical properties of the plastic. These may be in the form of powders, or flakes, or most effectively fibers. In all instances it is important that materials be selected that are wetted by the polymer so as to adhere to it, or else special coupling agents must be used. The powders could be minerals such as clay, quartz, or metallic oxides. An example of a flaky filler material is mica. Many different fibers are used in materials called composites, and a separate section is devoted to them. The impact strength of glassy polymers can be improved by adding small rubber particles to their matrix. This technique is commonly applied to PVC and to polymethyl acrylate.

With all these various structural characteristics in mind, let us briefly discuss the individual polymers listed in Table 6.1.

6.2.1 Thermoplastics

Thermoplastics are rigid below a temperature designated as the glass-transition temperature T_g, which for some of them is far below room temperature. For others, it is above room temperature or even above the boiling point of water. Above T_g they become viscoelastic, and their viscosity decreases with temperature. The transition to liquid state is gradual. On cooling they resume their properties and become rigid below T_g. This cycle can be repeated unless they overheat and degrade. This means that they can be recycled, reprocessed, and reshaped. It is common to classify them into different groups according to end use as commodity, intermediate, engineering, advanced, and key engineering and specialty plastics. Their mechanical properties and some other special properties improve from the commodity to the engineering and advanced grades, but their cost increases accordingly. The most popular commodity grades, polyethylene (PE), polypropylene (PP), polystyrene (PS), and polyvinyl chloride (PVC), represent two thirds of all plastics used, and they are inexpensive.

Polyethylene (PE)

The three types of polyethylene—the low-density (LDPE), linear low-density (LLDPE), and high-density (HDPE)—are synthesized by different types of addition polymerization. The first one contains a large number of side chains, the second does not have long branches, and the third contains essentially no branches. Correspondingly, the degree of crystallinity increases in order from the first to the second and the third type, and the mechanical properties also improve. The UHMWPE type of the HDPE is considered an engineering grade. Polyethylenes have good chemical and electrical properties, but they need antioxidants and UV stabilizers. Most of the LDPE is used as film for wrapping, bags for food, and coating for wires. HDPE is used in injection and blow molding for production of bottles and other containers.

Polyvinyl Chloride (PVC)

Rigid PVC finds many uses in the housing industry for pipes, gutters, window frames, and also in automobile exteriors for molding. It can be made softer and pliable by adding various amounts of plasticizer and is then used in car interiors for instrument panels, seats, and headrest covering, and for films and bottles, wire insulation, garden hose, raincoats, and so on. It needs UV stabilizers to prevent degradation in sunlight, and it needs heat stabilizers both for its processing and its use.

Polypropylene (PP)

PP is almost always used in the isotactic form. It is similar to HDPE, but it is a little stronger, which means that it softens at higher temperature. The applications are for films, bottles, and containers, and for a variety of auto parts, including heating and air-conditioning ducts.

Polystyrene (PS)

PS is transparent and brittle. It can be made tougher by the addition of rubber particles. It can be used in copolymers in combination with other polymers, but most of it is used as linear, amorphous homopolymer. In the foamed form it is used for cups and trays and as loose-fill, protective packaging material, as well as in wall and roof insulation slabs.

Polyesters

Polyesters may be made either thermoplastic or thermosetting. The most common thermoplastic polyester is polyethylene tere-phtalate (PET). It is amorphous and transparent, and is used for soda bottles, as film, and as fibers in the garment industries. It is known also under the trade names Mylar, used as film for transparencies, and Dacron, used as textile fiber.

Acrylonitrile-Butadiene-Styrene (ABS)

ABS is a terpolymer, as indicated by the three initial mers, but it is obtained as a copolymer of two binary copolymers, of a hard glassy styrene-acrylonitrile and styrene-butadiene rubber. Its properties can be varied by varying the proportions of the monomers. It is used for housings of consumer electronics and communication equipment, and as instrument and door panels in automobiles and in appliances.

Polyamide (PA)

The common type of PA is known as *nylon 66*. It is an engineering plastic—strong, tough, and abrasion-resistant. It retains good mechanical properties to temperatures over 100°C. It is used for machine parts such as bushings and gears and in various automobile parts.

Most of it is used as fiber in the textile and tire industries.

Acrylic

The most important of the acrylics is the polymethal methacrylate (PMMA) known as *plexiglas*. It is hard, brittle, and transparent, and it replaces glass in many uses such as optical lenses, enclosures, and windows. In these applications it is less shatter-prone than glass. Other acrylics are used as glue, dental protheses, and contact lenses.

Polycarbonate (PC)

PC is transparent, rigid, and tough. It is used as safety glass (e.g., Lexan). It is rigid up to 150°C, and it is flame-retardant and a good electrical insulator.

Cellulose Acetate (CA)

Cellulose is a natural fiber, a constituent of wood and cotton. One of the synthetic forms is cellulose acetate used as film and mainly as a textile fiber. A derivative of it is the textile fiber viscous rayon.

6.2.2 Thermosets

Thermosets must first be melted for shaping and then cured to harden. This is mostly done in two steps. Final curing is done by one of two methods that produce the heavy cross-linking of the polymer molecular chains: (1) radiation or (2) chemically induced cross-linkage. The latter method uses heat alone or in combination with pressure, or a catalyst, or the mixing of two chemicals. The former method provides more precise control in an application such as electronic circuits or as curing of a tooth filling.

Phenolic Resins (PR)

Phenolics are produced from a reaction of phenol or cresol with an aldehyde such as formaldehyde. They are manufactured in a two-stage process. Partially formulated polymer is shipped to the final processing plant. Phenol formaldehyde was the first polymer commercially made under the name of Bakelite. It is made with a variety of additives such as wood flour, glass, cotton fibers, minerals, accelerators, and colorants. It is hard and heat-resistant. It is used for knobs and handles of cookware, for electrical boxes and contactor bodies, for table tops, and as a bonding base for grinding wheels.

Urea Formaldehydes (UF)

UF is produced by the reaction of formaldehyde with urea, made into powder, and then shipped for processing by transfer molding at high temperature of about 150°C and at pressures on the order of 30–50 MPa. It can also be processed by injection molding. It is used for bottle caps and electrical housings. Melamine formaldehyde is a similar polymer with somewhat better heat resistance. It is sold under the name of Formica for use on furniture and counter tops.

Polyesters

In linear form polyesters are thermoplastic. They can be used as prepolymers, mixed with additives, and cured by heat. They are used as matrix for composites for pipes, boats, tanks, tubs, and automobile parts.

Epoxies

Epoxies are available either as two-component systems capable of curing at room temperature or as single-component resins curing at elevated temperatures. They are strong, resistant to temperature and chemicals, and because of good adhesion are often used for bonding mechanical parts. Their major use is in composites such as the fiberglass-epoxy systems. They are used for encapsulation of electronic parts. In a mixture with sand, glass, and marble they make high-quality concrete for swimming pool decks, or for wash basin tops.

6.2.3 Elastomers

Elastomers, or rubbers, are not as heavily cross-linked as thermosets; therefore, they retain a high degree of elasticity. However, like thermosets and unlike thermoplastics, they cannot be reverted to the plastic state once cured. They are amorphous, and to be rubbery they must be used above T_g. Most rubbers serve down to −50 to −60°C; silicone rubbers serve as low as −90°C. Below T_g they are hard and brittle. Most synthetic rubbers are random copolymers. Elastomers are capable of very large elastic elongation, 500–700%.

Natural Rubber

The cross-linkage of rubber is called vulcanization, a process invented by Charles Goodyear in 1884. Crude natural rubber is essentially polyisoprene, and it is a sticky substance. Goodyear made it into a useful product by heating it with sulphur, which produced the necessary cross-linking. It was later discovered to be the most useful material for bicycle tires and later as the best material for automobile tires, for which it is mixed with carbon black as an essential additive.

Synthetic Rubbers

Although natural rubber is still used for high-quality tires, much greater use, for tires and other applications, such as belts and hoses, is made of synthetic rubbers of several kinds. *Styrene-butadiene rubber (SBR)* is a copolymer of butadiene (C_4H_8). Other synthetic rubbers are *ethylene-propylene-diene* terpolymer, used for wire and cable insulation; *chloroprene* rubber, which is resistant to ozone, heat, and flame, and is used for hoses and conveyors; and *nitrile rubber*, which is a copolymer of butadiene and acrylonitrile and is used for gasoline hoses and for shoes.

6.3 THERMAL PROPERTIES: VISCOSITY

The structural changes that occur during temperature changes differ among amorphous, semicrystalline, and highly crystalline polymers. The diagrams in Fig. 6.4 indicate characteristic changes of (*a*) specific volume and (*b*) the pressure versus strain rate exponent m, of elongation, and of viscosity η versus temperature. Below the temperature T_g, called the glass-transition temperature, the polymers are rather stiff elastic, with short-term modulus of elasticity between 500 and 3500 MPa. (Long-term behavior is affected by the inherent viscoelasticity and creep). Some are rather brittle, some have modest up to rather high ductility. The percentage of elongation at break ranges between 3 and 400 for thermoplastics and up to 700 for elastomers. In the amorphous polymers the T_g temperature is rather well defined, and through an increase of 10–20°C the material starts to be rubbery, and viscoelastic. Its viscosity decreases gradually with temperature. For a particular polymer, the value of T_g depends on how fast it is reached on cooling. This transition is less pronounced for the partially crystalline materials that remain predominantly rigid-elastic, diminishing as the degree of crystallinity increases. On the other hand, these materials exhibit a rather sudden solid-liquid transformation at the melting temperature T_m. This change is very gradual with the amorphous materials. The latter do not possess any latent heat of the solid-liquid transformation while the crystalline plastics do (see Fig. 6.5). Above T_m polymers are essentially viscous liquids. Shaping of polymers involves viscous flow and is carried out in the higher range of the spread between T_g and T_m for amorphous materials, and just above T_m for

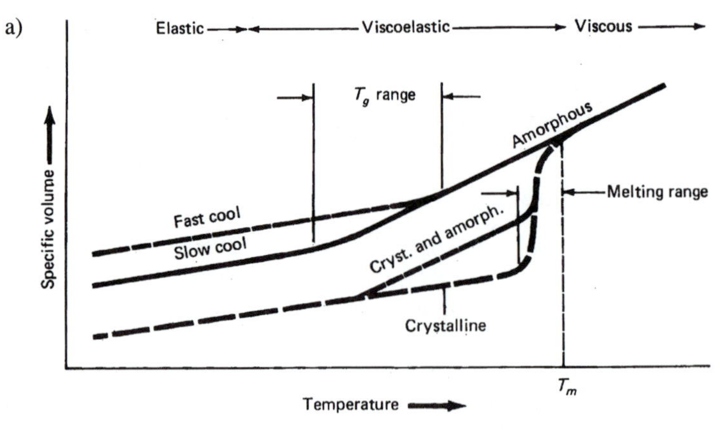

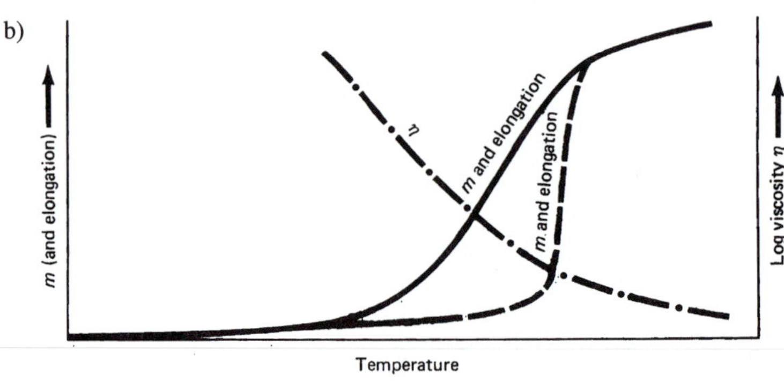

Figure 6.4
Characteristic changes of a) specific volume, and b) viscosity and strain rate sensitivity, with temperature reflect structural changes. [SCHEY]

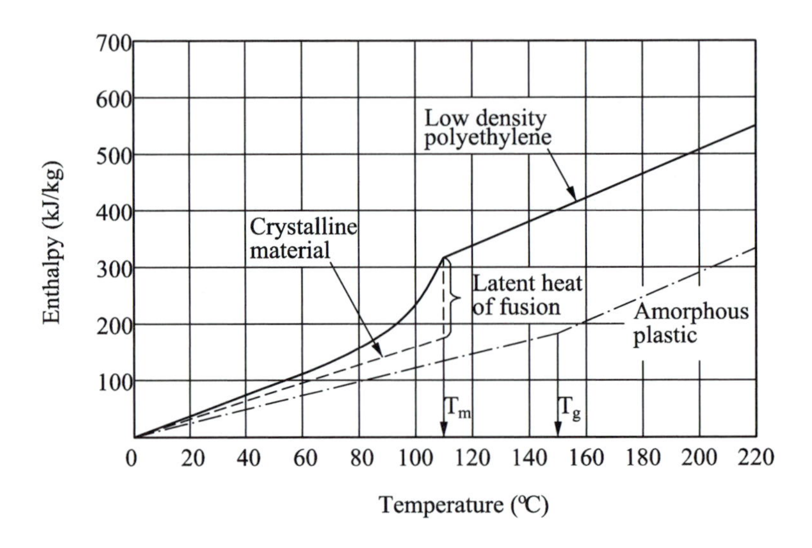

Figure 6.5
Enthalpy variation with temperature. [CRAWFORD]

partially crystalline ones. The values of T_g, T_m, latent heat of fusion H_f, processing temperatures T_p, and total heat to process H_p are listed in Table 6.2. Column 2 of this table indicates whether the material is amorphous (*am*) or, if it is crystalline, it gives the percentage of crystallinity.

It may be observed that T_g is far below room temperature for the crystalline thermoplastic materials polyethylene (PE), isotactic polypropylene (PP), and the acetal polyoxymethylene (POM). However, this is not of any real significance because the crystalline materials do not appreciably soften above T_g until they reach the melting temperature T_m. The situation is very different with rubbers. We have already mentioned that the elastomers have a very low T_g, of the order of -50 to -60°C. These materials are highly elastic at commonly encountered temperatures. The amorphous plastics in Table 6.2 serve below T_g at the common temperatures of use, up to the range of 70–150°C, and they are rigid and often brittle. Above T_g they soften; their viscosity decreases with temperature. Thus, the processing temperatures for amorphous materials may be chosen slightly below T_m especially for the slower processes such as compression molding. For extrusion and injection molding, the processing temperatures T_p listed in Table 6.2 are slightly below or above T_m for the crystalline materials. At higher processing temperatures, flow is easier, but the material may be damaged by the heat, and even if it is not, cooling takes longer. Correspondingly, the choice of T_p is a compromise between these aspects; the temperatures given in the table are shown over a range, leaving the final choice to the practitioner. Typical values of viscosity η (Pas) are given applicable at the processing temperature and a shear rate of 1000/sec. Other useful thermal properties of polymers are listed in Table 6.3. The dimensions of the parameters are thermal conductivity k (J/m·sec.°C), specific heat c_p (kJ/kg·°C), latent heat of fusion H_f (kJ/kg), total heat to process H_p (kJ/kg), specific mass ρ (g/cm^3), diffusivity α (cm^2/sec).

The most important parameter for the extrusion, blown film, and injection-molding processes is the *viscosity* of the material. Viscosity is defined in the relationship between shear stress τ and shear rate γ. In the "Newtonian" fluids such as water, viscosity at a given temperature is a constant. It can be demonstrated in the example of *drag flow* (see Fig. 6.6). Consider a viscous liquid under simple shear between two parallel plates

TABLE 6.2 Processing Parameters of Thermoplastics

Material	Percent cryst/am	T_g	T_m	T_p	η	n
Polyethylene						
LDPE	55	−100	120	160–240	65	0.35
LLDPE			125			0.6
HDPE	92	−115	130	200–282	240	0.5
Polyvinyl chloride (PVC)	am	80	212	160–210	80	0.3
Polypropylene (PP)						
Isotactic	high	−15	175	200–300	75	0.35
Polystyrene (PS)						
Homopolymer	am	100	240	180–260	220	0.3
Polyester						
PET	am	70	270			0.6
ABS	am	115		180–240	210	0.25
Polyamide, nylon 6.6	high	55	260	260–290	100	0.75
Acrylic (PMMA)	am	105	200	180–250	440	0.25
Polycarbonate (PC)	am	150	230	280–310	225	0.7
Polyacetal (POM)	75	−50	180	185–240	140	
Cellulose acetate (CA)	am	105	306	300–320		
Polytetrafluoroethylene (PTFE)	95	125	327			

TABLE 6.3 Thermal Properties of Polymers

	k	c_p	H_{fusion}	H_p	ρ	α
LDPE	0.24	2.25	150	540	0.92	0.0013
HDPE	0.25	2.30	209	720	0.96	
PVC	0.21	1.10		180	1.40	
PP	0.15	2.10	100	250	0.91	
PS	0.12	1.20		200	1.06	0.00065
PET	0.29	1.55		430	1.35	
ABS	0.25	1.45		300	1.02	
PA, nylon 6.6	0.24	2.15	130	570	1.14	
Acrylic (PMMA)	0.20	1.45		300	1.18	
PC	0.19	1.40		350	1.2	
POM		1.45	163	465	1.42	
CA		1.5		195	1.3	

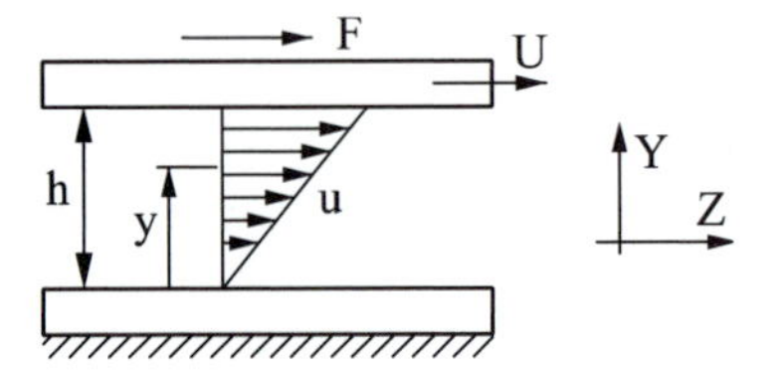

Figure 6.6
Drag flow demonstrates the concept of viscosity relating shear stress to shear strain rate.

separated by thickness h. The width w of the plates, seen here in the direction into the plane of the paper is assumed large, and we can neglect the effect of the side walls. The top plate is moving in direction Z with velocity U. The shear strain rate is

$$\dot{\gamma} = \frac{U}{h} \tag{6.1}$$

and the shear stress τ is generated as

$$\tau = \eta\,\dot{\gamma} = \frac{F}{A} = \frac{F}{wh} \tag{6.2}$$

The velocity u increases linearly from $u = 0$ at $y = 0$ to $u = U$ at $y = h$

$$u = \dot{\gamma}\,y \tag{6.3}$$

The average velocity is

$$u_{\text{av}} = \frac{U}{2} \tag{6.4}$$

and the total flow is

$$Q = \frac{Uwh}{2} \tag{6.5}$$

The shear rate $\dot{\gamma}$ and shear stress τ are both constant over the cross section of the flow.

6.3.1 Newtonian Flow in a Rectangular Channel (Slit)

The flow will be analyzed as produced by a pressure difference ΔP over a length L of the channel (see Fig. 6.7). It is assumed that there is no slip at the walls; the melt is incompressible; the flow is steady, laminar; and end and side effects are negligible. The balance of forces on an element of the fluid with dimensions dy, dz, and w is

$$w\,\mathrm{d}y\,\mathrm{d}p = w\mathrm{d}z\,\mathrm{d}\tau \tag{6.6}$$

$$\frac{\mathrm{d}p}{\mathrm{d}z} = \text{const} = \frac{\Delta P}{L} = C$$

$$\mathrm{d}\tau = C\,\mathrm{d}y \tag{6.7}$$

$$\tau = Cy + C_1$$

At $y = 0$, $\tau = 0$, $C_1 = 0$

$$\frac{\mathrm{d}u}{\mathrm{d}y} = \dot{\gamma} = \frac{\tau}{\eta} = \frac{C}{\eta}y$$

$$u = \frac{C}{2\eta}y^2 + C_2 \tag{6.8}$$

At $y = h/2$, $u = 0$, $C_2 = -Ch^2/8\eta$,

$$u = \frac{C}{2\eta}\left(y^2 - \frac{h^2}{4}\right) = \frac{\Delta P}{2\eta L}\left(y^2 - \frac{h^2}{4}\right) \tag{6.9}$$

At $y = 0$,

$$u_{\max} = -\frac{\Delta P h^2}{8\eta L} \tag{6.10}$$

The average velocity u_{av} is obtained as follows:

$$Q = 2w\int_0^{h/2} u\,\mathrm{d}y = 2w\int_0^{h/2} \frac{\Delta P}{2\eta L}\left(y^2 - \frac{h^2}{4}\right)\mathrm{d}y$$

$$= \frac{\Delta P w}{\eta L}\left|\frac{y^3}{3} - \frac{h^2}{4}y\right|_0^{h/2} = -\frac{\Delta P w h^3}{12\eta L} \tag{6.11}$$

$$u_{av} = \frac{Q}{wh} = -\frac{\Delta P h^2}{12\eta L} \tag{6.12}$$

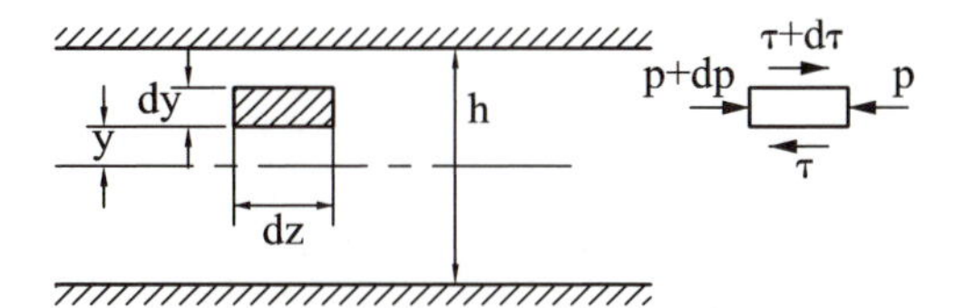

Figure 6.7
Flow in a flat rectangular channel.

The distribution of velocity, shear rate, viscosity, and shear stress is obtained as follows: at $y = 0$:

$$u = u_{max}$$

At $y = h/2$:

$$u = 0$$

$$u = u_{max}\left[1 - \left(\frac{2y}{h}\right)^2\right]$$

In between:

$$\tau = \eta\dot{\gamma} \tag{6.13}$$

$$\dot{\gamma} = \frac{du}{dy} = u_{max}\frac{8y}{h^2}$$

At $y = 0$, $\dot{\gamma} = 0$; at $y = \frac{h}{2}$, $\dot{\gamma} = -u_{max}\frac{4}{h^2}$. (6.14)

$$\dot{\gamma}_{av} = -\frac{2u_{max}}{h} \tag{6.15}$$

Both Q and u come out negative, which is just to indicate that the direction of the flow is opposite that of the pressure gradient. We may as well take them as positive.

Polymers do not behave as Newtonian fluids. Their viscosity under isothermal conditions decreases with shear rate. This results from the effect of shear on disentanglement of the molecular chains. This "shear thinning" behavior is commonly modeled using the so-called Power Law:

$$\eta = K\left(\frac{\dot{\gamma}}{\dot{\gamma}_0}\right)^{n-1} \tag{6.16}$$

where γ_0 is a reference shear rate that may well be taken as 1. Then we have

$$\eta = K\dot{\gamma}^{n-1} \tag{6.17}$$

and

$$\tau = \eta\dot{\gamma} = K\dot{\gamma}^n \tag{6.18}$$

The values of the exponent n are given in Table 6.2. For an average value, $n = 0.4$, we have

$$\eta = K\dot{\gamma}^{-0.6}, \qquad \tau = K\dot{\gamma}^{0.4} \tag{6.19}$$

This is graphically expressed in Fig. 6.8, where the relationship between strain rate and shear stress is plotted for a Newtonian fluid and for a shear-thinning fluid. The power law is a simplification of reality. It is a reasonable approximation over a certain range of stresses or of strain rates. Over a large range this relationship may be obtained experimentally.

Furthermore, viscosity is also strongly dependent on temperature. It decreases with temperature according to an exponential relationship expressed by the William-Landel-Ferry (WLF) equation:

$$\log_{10}\left(\frac{\eta T}{\eta T_g}\right) = \frac{-17.44(T - T_g)}{51.6 + (T - T_g)} \tag{6.20}$$

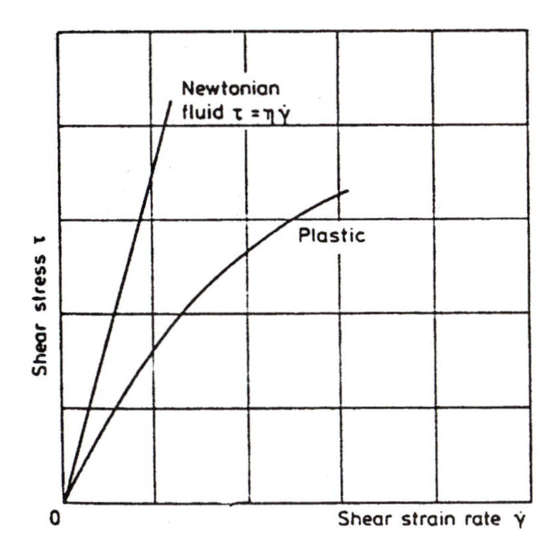

Figure 6.8
Relationship between shear stress and shear rate. [CRAWFORD]

For example, the graph in Fig. 6.9 expresses the temperature effect on viscosity as obtained from experimental data [5], for a shear rate of 1000/sec. The values of "apparent viscosity" η in Table 6.2 apply also for a shear rate of 1000/sec and for the processing temperatures T_p in the middle of the given range.

Local viscosity is also dependent on local pressure. Obviously, with the complex influences of material properties, shear rate, temperature, pressure, and local shear stress, the computation of the flows involved in injecting the melt through a sprue and a system of runners and gates into a mold cavity with all the corresponding pressure and temperature changes is a rather difficult task. However, extensive research has been devoted to this task, and finite difference and finite element computer programs currently exist that deal with these tasks and help in the design of a mold that will be completely filled and in which the melt will be properly compacted, cooled, and solidified.

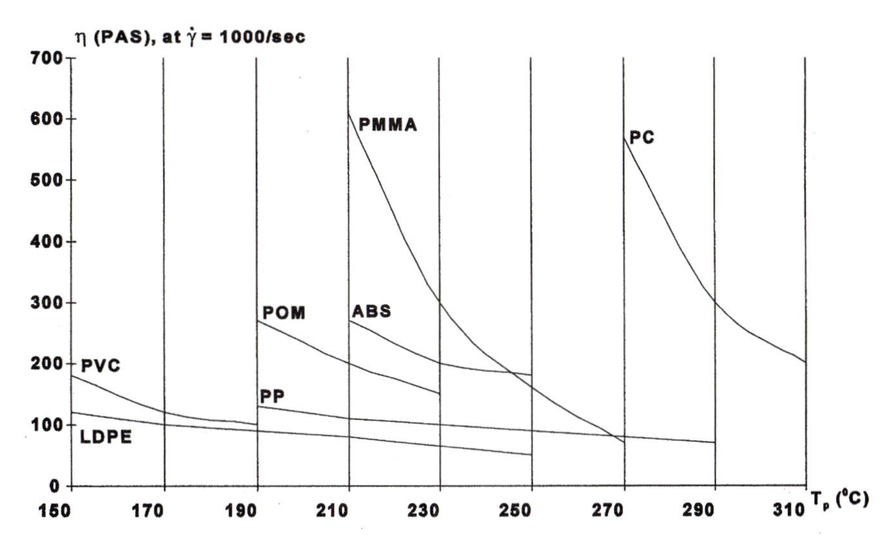

Figure 6.9
Strong effect of temperature on viscosity for selected plastics.

This software also deals with the problems of residual stresses and geometric distortions of the moldings.

Nevertheless, useful lessons can be learned by calculations under simplified assumptions, and many such exercises have been carried out in the literature; see [1], [2], [5], and [7]. In our text some of the simpler tasks will be carried out. As in other texts, constant temperature will be assumed for flow calculations. This is called *isothermal flow*. The effect of pressure on viscosity will be neglected. At this stage, we will review only the cases of *drag flow* and *flat channel flow*, by considering the *non-Newtonian, power-law fluid.*

In the case of drag flow, the distribution of the main flow variables is the same as for the Newtonian liquid. The velocity distribution remains linear, and the shear rate and shear stress are constant over the cross section of the channel. To express the force F the "apparent viscosity" corresponding to the given velocity U and shear rate $\dot{\gamma} = U/h$ must be used.

6.3.2 Non-Newtonian, Power-Law Flow in a Flat Channel

Refer again to Fig. 6.7 and to Eq (6.6), which is rewritten here:

$$w\,\mathrm{d}y\,\mathrm{d}p = w\,\mathrm{d}\tau\,\mathrm{d}z \tag{6.21}$$

$$\frac{\mathrm{d}\tau}{\mathrm{d}y} = \frac{\mathrm{d}p}{\mathrm{d}z} = \frac{\Delta P}{L} = C \tag{6.22}$$

$$\tau = Cy$$

Using

$$\tau = K\dot{\gamma}^n = K\left(\frac{\mathrm{d}u}{\mathrm{d}y}\right)^n = Cy$$

$$\frac{\mathrm{d}u}{\mathrm{d}y} = \left(\frac{C}{K}\right)^{1/n} y^{1/n} \tag{6.23}$$

$$u = \left(\frac{C}{K}\right)^{1/n} \int_0^y y^{1/n}\,\mathrm{d}y = \left(\frac{C}{K}\right)^{1/n} \frac{n}{n+1} y^{(n+1)/n} + C_1$$

At $y = h/2$, $u = 0$, and

$$C_1 = -\left(\frac{C}{K}\right)^{1/n} \frac{n}{n+1} \left(\frac{h}{2}\right)^{(n+1)/n}$$

$$u = \left(\frac{C}{K}\right)^{1/n} \frac{n}{n+1} \left[y^{(n+1/n)} - \left(\frac{h}{2}\right)^{(n+1/n)}\right] \tag{6.24}$$

$$= \frac{n}{n+1}\left(\frac{\Delta P}{LK}\right)^{1/n}\left(\frac{h}{2}\right)^{(n+1)/n}\left[1 - \left(\frac{2y}{h}\right)^{(n+1)/n}\right]$$

$$\text{At } y = 0,\ u_0 = u_{\max} = \frac{n}{n+1}\left(\frac{\Delta P}{LK}\right)^{1/n}\left(\frac{h}{2}\right)^{(n+1)/n} \tag{6.25}$$

The expression of Eq. (6.24) can be made more compact by using Eq. (6.25):

$$u = u_0\left[1 - \left(\frac{2y}{h}\right)^{(n+1)/n}\right] \tag{6.26}$$

The flow Q is obtained as follows:

$$\begin{aligned} Q &= 2\,wu_0\int_0^{h/2}\left(1 - \left(\frac{2y}{h}\right)^{(n+1)/n}\right)\mathrm{dy} \\ &= 2wu_0\left| y - \frac{n}{2n+1}\left(\frac{2}{h}\right)^{(n+1)/n} y^{(2n+1)/n}\right|_0^{h/2} \\ &= 2\,wu_0\left| y\left[1 - \frac{n}{2n+1}\left(\frac{2y}{h}\right)^{(n+1)/n}\right]\right|_0^{h/2} \\ &= whu_0\frac{n+1}{2n+1} = w\left(\frac{\Delta P}{KL}\right)^{1/n}\frac{2n}{2n+1}\left(\frac{h}{2}\right)^{(2n+1)/n} \end{aligned} \tag{6.27}$$

The average velocity u_{av} is

$$u_{av} = \frac{Q}{wh} = u_0\frac{n+1}{2n+1} = \left(\frac{\Delta P}{KL}\right)^{1/n}\frac{n}{2n+1}\left(\frac{h}{2}\right)^{(n+1)/n} \tag{6.28}$$

Choosing a common value of $n = 0.4$, the distribution for velocity, shear rate, viscosity, and shear stress along the Y-axis can be evaluated as

$$u_0 = u_{max} = 0.286\left(\frac{\Delta P}{KL}\right)^{2.5}\left(\frac{h}{2}\right)^{3.5} \tag{6.29}$$

$$u_{av} = 0.778\,u_{max} \tag{6.30}$$

$$u = u_0\left[1 - \left(\frac{2y}{h}\right)^{3.5}\right] \tag{6.31}$$

$$\dot{\gamma} = \frac{\mathrm{du}}{\mathrm{dy}} = 3.5\,u_0\left(\frac{2}{h}\right)^{3.5} y^{2.5} \tag{6.32}$$

$$\eta = K\dot{\gamma}^{n-1} = K\left[3.5\,u_0\left(\frac{2}{h}\right)^{3.5}\right]^{-0.6} y^{-1.5} \tag{6.33}$$

EXAMPLE 6.1 Newtonian and Non-Newtonian Flow in a Flat Channel ▼

Given: Rectangular cross-section channel $h = 2$ mm, $w = 15$ mm, $L = 50$ mm, flow $Q = 60\ \mathrm{cm^3/sec} = 6\ \mathrm{e}{-5}\ \mathrm{m^3/sec}$.

(a) Newtonian flow: apparent viscosity $\eta = 100$ Pas

$$\tau = \eta\dot{\gamma} = K\dot{\gamma}^n = K\left[3.5\,u_0\left(\frac{2}{h}\right)^{3.5}\right]^{0.4} y \tag{6.34}$$

$$\text{Eq (6.11): } \Delta P = \frac{12\eta LQ}{wh^3} = \frac{12 \times 100 \times 0.05 \times 6\,\text{e} - 5}{0.015 \times 0.002^3} = 30 \text{ MPa}$$

$$\text{Eq (6.12): The average velocity is} \qquad \text{u}_{\text{av}} = \frac{6\,\text{e} - 5}{0.015 \times 0.002} = 2 \text{ m/sec}$$

Eq (6.10): Maximum velocity is $u_{\max} = 3\text{m/sec}$
Eq (6.14): The average shear rate is $\dot{\gamma}_{\text{av}} = 3000/\text{sec}$
Eq (6.15): The average shear stress is $\tau_{\text{av}} = 300000$ Pa

In the case of a Newtonian fluid, the viscosity is not dependent on shear rate. Next we investigate a non-Newtonian flow in the same channel, and its viscosity will be referred to the value of 100 Pas at $\dot{\gamma} = 3000/\text{sec}$ as above.

(b) Flow of a non-Newtonian, Power-Law fluid

Viscosity is used as defined in Eq (6.16), and the value of the exponent $n = 0.4$ is applied.

$$\eta = K\left(\frac{\dot{\gamma}}{\dot{\gamma}_0}\right)^{-0.6} = 100\left(\frac{\dot{\gamma}}{3000}\right)^{-0.6} = 12198\dot{\gamma}^{-0.6} \tag{6.35}$$

This equation yields $\eta = 100$ Pas for $\dot{\gamma} = 3000/\text{sec}$ as in *(a)*. The constant $K = 12198$.

(b1) What ΔP is needed for the same flow of $Q = 6\,\text{e}{-5}\ \text{m}^3/\text{sec}$, $u_{\text{av}} = 2$ m/sec?

$$\text{Eq (6.28): } u_{\text{av}} = 0.778 \times 0.286\left(\frac{\Delta P}{KL}\right)^{2.5}\left(\frac{h}{2}\right)^{3.5} = 2 \text{ m/sec}$$

where $K = 12198$, $L = 0.05$ m, and $h = 0.002$ m.

$$(\Delta P)^{2.5} = (KL)^{2.5}\left(\frac{2}{h}\right)^{3.5} \times 8.991$$

$$\Delta P = 23.269 \text{ MPa}$$

The necessary pressure is $23.269/30 = 0.776$ of that needed for the Newtonian fluid.

(b2) What flow Q is obtained if the same $\Delta P = 30$ MPa is applied as in *(a)*?

$$u_{\text{av}} = 0.778 \times 0.286\left(\frac{30\text{ e6}}{12198 \times 0.05}\right)^{2.5}\left(\frac{0.002}{2}\right)^{3.5}$$

$$= 3.775 \text{ m/sec}$$

$$Q = 3.775 \times 0.002 \times 0.015 = 11.33\,\text{e} - 5 \text{ m}^3/\text{sec}$$

The flow is $11.33/6 = 1.89$ times higher for the Power-Law fluid when using the same viscosity law, that is, assuming the same polymer considered in *(a)* in a simplified mathematical version of Newton's law.

Using expressions (6.9) and (6.13) on one hand and (6.29), (6.31), and (6.32) on the other hand, the distributions of velocity u and shear rate $\dot{\gamma}$ are expressed in the listing of the program "chanflow" in Fig. 6.10. Thus, for the Newtonian flow,

Figure 6.10
The listing of a Matlab program for computing distribution of velocity and of shear rate.

```
function[y,ua,ub,gdota,gdotb]=chanflow
% Evaluating expressions of EX.6.1, refer to Fig.6.7.
% ua is Newtonian and ub non Newtonian flow, gdota and gdotb are shear rates
for n=1:100
y(n)=-0.001+2e-5*n;
ua(n)=3e6*(1e-6-y(n)^2);
ub(n)=4.854*(1-31.623e9*abs(y(n))^3.5);
gdota(n)=6e6*abs(y(n));
gdotb(n)=16.989*31.623e9*abs(y(n))^2.5;
end
end
```

$$u = \frac{\Delta P h^2}{8\eta L}\left[1 - \left(\frac{2y}{h}\right)^2\right] = 3\,(1 - 1\text{ e}6\,y^2)$$

$$\dot{\gamma} = u_{\max}\frac{8y}{h^2} = 6\text{ e}6y$$

and for the Power-Law flow,

$$u = u_{\max}\left[1 - \left(\frac{2y}{h}\right)^{3.5}\right] = 4.854\,(1 - 31.623\text{ e}9\,y^{3.5})$$

$$\dot{\gamma} = 16.989 \times 31.623\text{ e}9 \times y^{2.5}$$

The results are plotted in Fig. 6.11 for the flow velocities and in Fig. 6.12 for the shear rates. The shear rates close to the walls are much higher in case *(b)*. This causes the drop in viscosity and the resultant higher flow than in case *(a)*. In case *(b)* the velocity is almost constant for the central half of the channel and drops down towards the walls. ▲

EXAMPLE 6.2 Viscous Heat Generation ▼

Calculate the heat generated in the flow considered in Example 6.1(b2). What temperature increase would it cause if it were uniformly distributed?

Obviously the heat generation in that flow is very non-uniform. Denoting power as W, the local power loss in a unit volume is expressed as

$$\mathrm{d}W = \tau\dot{\gamma} = \eta\dot{\gamma}^2 = K\dot{\gamma}^{n+1}$$

for $n = 0.4$, $\mathrm{d}W = K\dot{\gamma}^{1.4}$

It has been shown in Fig. 6.11 that $\dot{\gamma}$ is highest close to the walls, and that is where most of the power loss occurs. The heat spreads through the flow by conduction and also passes through the channel walls. The latter part is substantial because of the proximity of the heat generation there. Computation of the temperatures throughout the flow is rather complex, and it is to be done by using the FD and FE programs mentioned above. However, it will be good for a general orientation to carry out the simple task outlined above.

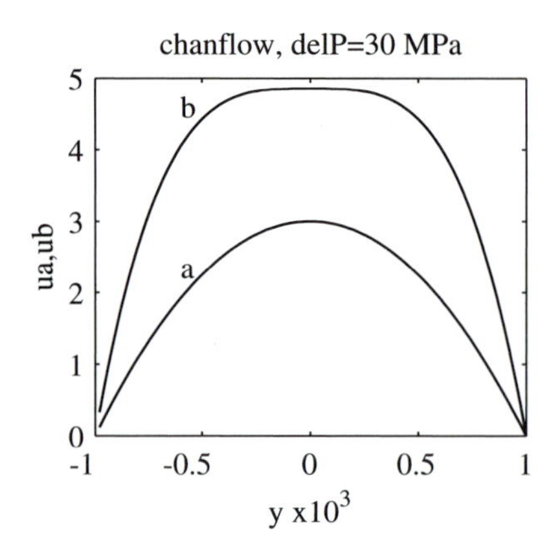

Figure 6.11
Flow velocities in a flat channel. a) Newtonian flow; b) Power Law flow for the same pressure drop. Channel height is $h = 0.002$ m.

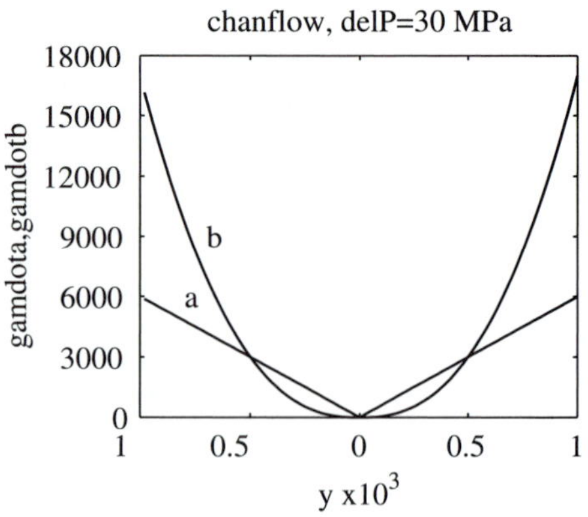

Figure 6.12
Shear rate in flat-channel flow: a) Newtonian flow; b) Power Law flow for the same pressure drop. Viscosities are equal at $y = \pm h/4$. Channel height is $h = 0.002$ m.

The total power loss in the flow of $Q = 11.33\,\mathrm{e}{-5}\ \mathrm{m^3/sec}$ through the channel with area $0.002 \times 0.015 = 3\,\mathrm{e}{-5}\ \mathrm{m^2}$, with average velocity $u_{av} = 3.775$ m/sec as produced by the pressure drop $\Delta P = 30$ MPa is

$$W = \Delta P \times Q = 11.33\ \mathrm{e} - 5 \times 30\ \mathrm{e}6\ \mathrm{Nm/sec} = 3399\mathrm{W}$$

The temperature increase ΔT due to this power loss is:

$$\Delta T = \frac{W}{Q\rho c} = \frac{\Delta P}{\rho c}$$

The specific heat is found in Table 6.3. If we use LDPE for this example, we have

$$\rho = 0.92\ \mathrm{g/cm^3} = 920\ \mathrm{kg/m^3}$$

$$c = 2.3\ \mathrm{kJ/kg^\circ C}$$

and then

$$\Delta T = \frac{30\ \mathrm{e}6}{920 \times 2.3\ \mathrm{e}3} = 14.2^\circ C$$

This is quite a substantial temperature increase and, according to Fig. 6.9, it might cause a decrease of viscosity from 100 Pas to about 89 Pas. However, most of the heat may be conducted away through the walls of the channel. ▲

EXAMPLE 6.3 Is the Flow in Example 6.1 (b2) Laminar? ▼

The Reynolds number for a flow in a flat channel with thickness h and average velocity u_{av} is obtained as

$$Re = \frac{u_{av}h\rho}{\eta} = \frac{3.775 \times 0.002 \times 920}{\eta} = \frac{6.946}{\eta}$$

The viscosity $\eta = K\dot{\gamma}^{n-1}$, and $K = 12198$,
From the graph of Fig. 6.12, the average shear rate is $\dot{\gamma} = 3000/\text{sec}$.
Accordingly $\eta = 12198 \times 3000^{-0.6} = 100$ Pas.

$$Re = \frac{6.946}{100}\frac{\text{m·m·kg·m}^2}{\text{sec·m}^3\text{·N·sec}} = 0.0695\frac{\text{m·m·N·sec}^2\text{·m}^2}{\text{sec·m}^3\text{·m·N·sec}}$$

The dimensions check Re is dimensionless. Flow is known to become turbulent for $Re >$ (2,000 to 10,000). Here Re is so low that the flow is solidly laminar. ▲

6.3.3 Flow in a Tube

The derivations for flow in a tube are similar to those presented in 6.3.1 for the flat channel. It was found that especially those for the Power-Law fluid were rather lengthy. Therefore, we will skip many of the steps and explain only the initial formulation for the Newtonian flow and the final formulas.

Newtonian Flow

Refer to Fig. 6.13. The balance of forces on an annular element dr thick and dz wide is

$$\pi[(r + \text{d}r)^2 - \pi r^2]\text{d}p = 2\pi[(r + \text{d}r)(\tau + \text{d}\tau) - r\tau]\text{d}z$$

where the left side is the pressure force and the right is the shear force. Neglecting second-order differentials, we have

$$2\pi r \,\text{d}r \,\text{d}p = 2\pi \,(\tau \,\text{d}r + r \,\text{d}\tau) \,\text{d}z$$

$$\frac{\text{d}p}{\text{d}z} = \frac{\tau \,\text{d}r + r \,\text{d}\tau}{r\text{d}r}$$

where dp/dz = $\Delta P/L = C$, and the numerator is the differential d (τr).
Then we have

$$\frac{\text{d}\tau}{\text{d}r} = C - \frac{\tau}{r} \tag{6.36}$$

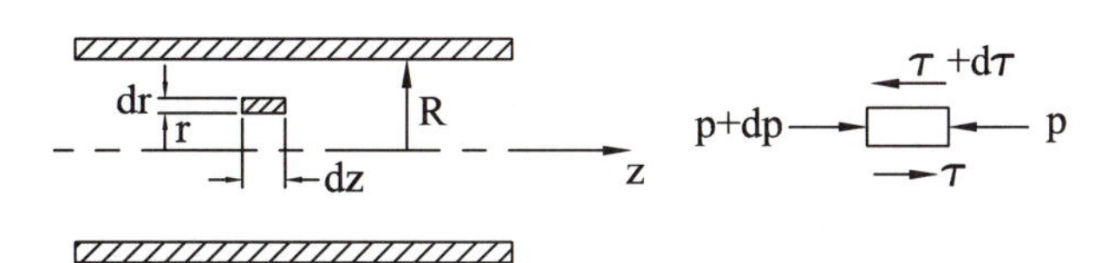

Figure 6.13
Conditions for calculating flow in a tube.

which is solved as

$$\tau = \frac{C}{2} r = \frac{\Delta P}{2L} r \tag{6.37}$$

The shear stress is zero in the center (see Fig. 6.13), and it is maximum at the wall, $r = R$.

$$\tau_{\max} = \frac{\Delta P R}{2L} \tag{6.38}$$

We also have $\tau = \eta\, \dot{\gamma} = \eta\, \mathrm{d}u/\mathrm{d}r$,

$$\dot{\gamma} = \frac{\mathrm{d}u}{\mathrm{d}r} = \frac{\Delta P r}{2L\eta} \tag{6.39}$$

and so on. Finally,

$$u = \frac{\Delta P}{4\eta L}(r^2 - R^2) \tag{6.40}$$

$$\text{at } r = 0,\ u_0 = u_{\max} = \frac{\Delta P R^2}{4\eta L} \tag{6.41}$$

The velocity, taken as positive, is maximum at the center and varies in a parabolic way. The shear strain rate is zero at the center and reaches a maximum at the wall:

$$\dot{\gamma}_{\max} = \frac{\Delta P R}{2\eta L} \tag{6.42}$$

The flow Q is obtained as

$$Q = \int_0^R 2\pi r u\, \mathrm{d}r = \frac{\pi \Delta P R^4}{8\eta L} \tag{6.43}$$

It is interesting to note the strong effect of the radius: the flow is proportional to its fourth power.

Non-Newtonian, Power-Law Flow

Let us start from Eq (6.37), rewritten here:

$$\tau = \frac{\Delta P}{2L} r \tag{6.44}$$

and Eq. (6.39):

$$\dot{\gamma} = \frac{\mathrm{d}u}{\mathrm{d}r} = \frac{\Delta P r}{2L\eta}$$

$$\eta = K\dot{\gamma}^{n-1} \tag{6.45}$$

$$\left(\frac{\mathrm{d}u}{\mathrm{d}r}\right)^n = \frac{\Delta P}{2LK}$$

$$\frac{\mathrm{d}u}{\mathrm{d}r} = \left(\frac{\Delta P}{2LK}\right)^{1/n} r^{1/n} \tag{6.46}$$

Integrating (6.46) between the limits $r = 0, R$, where $u = 0$ at $r = R$, yields the following expressions:

$$u = \frac{n}{n+1}\left(\frac{\Delta P}{2LK}\right)^{1/n} R^{(n+1)/n}\left[1 - \left(\frac{r}{R}\right)^{(n+1)/n}\right] \tag{6.47}$$

At $r = 0$,

$$u_0 = u_{\max} = \frac{n}{n+1}\left(\frac{\Delta P}{2LK}\right)^{1/n} R^{(n+1)/n} \tag{6.48}$$

$$Q = \frac{n+1}{3n+1}\pi R^2 u_0 = \pi\frac{n}{3n+1}\left(\frac{\Delta P}{2LK}\right)^{1/n} R^{(3n+1)/n} \tag{6.49}$$

For $n = 0.4$,

$$Q = 0.571\left(\frac{\Delta P}{2LK}\right)^{2.5} R^{5.5} \tag{6.50}$$

As with the Newtonian formula, here there is also a very strong effect of the radius of the tube on the flow; it increases to the 5.5 power of R.

6.4 PROCESSING METHODS AND OPERATIONS

6.4.1 General Considerations

A number of different processing methods are used. The choice of a particular process is dictated primarily by the desired form of the product: film, sheet, a profiled bar, tube, or a bulky shape such as a cup, bucket, car bumper, or chair. The quantity required also influences the choice of process. In most instances large quantities are being made and either *extrusion* (EX) or *injection molding* (IM) is chosen, respectively, for long and for bulky parts. For smaller quantities, *compression molding* (CM) and *transfer molding* (TM) are used. Some materials are processed by *casting*. In all these methods the material is input in the form of particles such as granules, or scrap, either cut up, or chopped up. Finally, some products are best made, either in small or large quantities, by *thermoforming* (TF) where the material is input in the form of a film or sheet, which may have been preproduced by extrusion.

The actual conditions and details of the operations depend on the type of the plastic, and while most of them can be formed by either main method, there are advantages and disadvantages, especially with respect to whether the material is a thermoplast or thermoset, amorphous or crystalline. H. Rees in [8] has formulated the corresponding considerations. One of these refers to the dwell at the melting point, whether in heating or cooling, and to the latent heat of fusion, which is one of the basic characteristics of the crystalline materials, such as PE, PP, nylon, and PET, and is not found in amorphous plastics such as PS, PC, or PVC. This affects the behavior of the material in the extruder screw as it is heated up and also in the mold as it cools down.

The next concern is that of the danger of thermal degradation of the plastic. This depends on the combination of time and temperature of the exposure. For each material the corresponding characteristic has been determined, and it is a different kind for thermoplasts (Fig. 6.14a) and for thermosets (Fig. 6.14b). In *(a)* the form of the graph is a hyperbola indicating that degradation depends essentially on the product of the temperature T and time t for which the plastic was held at that temperature. The higher the T the shorter the permissible t. For thermosets there is a zone. Processing is done in the area to the left of the dotted zone. Thereafter, the material is cured (cross-linking is developed) under conditions within this zone. To the right of the zone the thermoset degrades, or "scorches." Practically, this means that it is processed, for instance by IM, so that for the time of its heating up in the screw it will reach temperature just below the curing temperature. Then it must be injected into a mold kept at higher temperature but can only be held there long enough to set properly; it must be ejected safely in time before it degrades. Some polymers are removed from the mold as soon as they are strong enough for it. Full cross-linking is obtained subsequently in an oven. Materials that cross-link fast may have to be injected in a cold mold and held there for a complete heating and cooling cycle. Injection molding of thermosets is much more difficult than of thermoplasts, and the temperature in the screw must be carefully controlled. Actually, they are there mainly for the mixing and transportation; there is no compression due to the screw rotation. Pressure is developed by the final axial injecting motion only. Therefore, screw extruders are not suitable for thermosets.

The fundamental difference in processing of the two types of materials is that while both must be heated to lower their viscosity for forming, thermoplasts must then be cooled to stiffen, and thermosets must first be heated even more to stiffen and then allowed to cool.

6.4.2 Casting

In casting, the liquid material is poured into the mold and settles by gravity. For thermoplastics the monomers or prepolymers are cast, and full polymerization is achieved in the mold (e.g., for machine parts made of nylon). PMMA sheet is cast as monomer between plates or between endless steel belts where it is heated for several hours to polymerize. PVC is cast as a plastisol, which is a suspension of PVC particles in a plasticizer. Many other plastisols are cast through multihole dies to produce fibers. Thermosets are not cross-linked until they are in the mold.

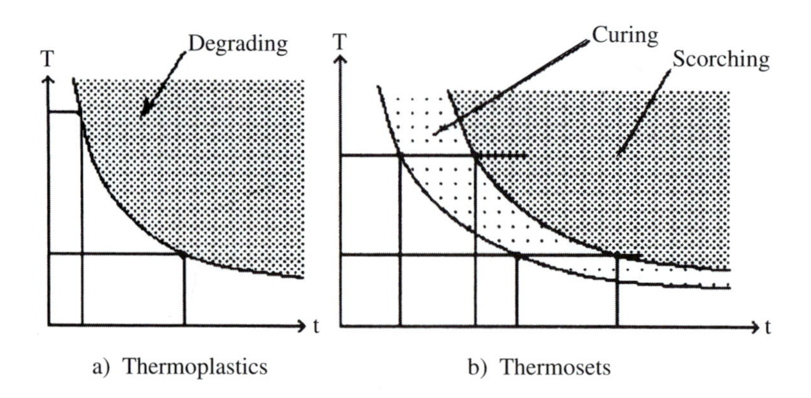

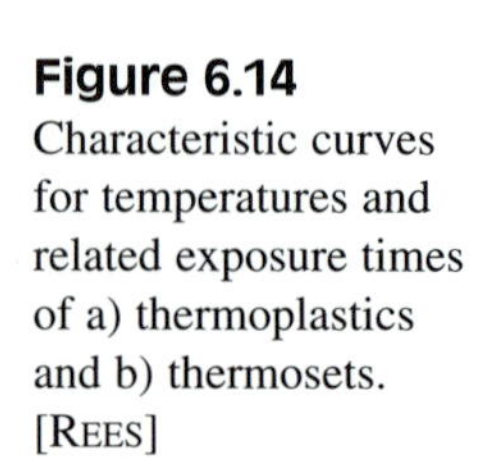

Figure 6.14
Characteristic curves for temperatures and related exposure times of a) thermoplastics and b) thermosets. [REES]

6.4.3 Compression (CM) and Transfer Molding (TM)

Both compression and transfer molding are used to make piecewise parts, mostly "bulky," but also disks for musical recording, which do not fit this general shape classification, but can be considered the exception. Briefly, the parts are made in molds.

In CM the premeasured amount of the material that may have been prepared in a mixer and precompressed, preheated, and prepolymerized into a pellet (cake) is placed in an open mold (see Fig. 6.15). It is then compressed by the downward movement of the punch. It is not possible to achieve an exact fill of the mold, so a slight surplus of the blank material is used that is then squeezed out as "flash." The flash must be removed later in a separate operation by breaking or machining it off.

The CM process is used mostly for thermosets but also for thermoplastics that contain large amounts of filler material (e.g., composites). These materials do not flow well in the fast injection-molding process. Thermoset material is preheated just below the curing temperature; it is then heated in the mold as required for setting. Typical products are tires and dinnerware. For thermoplastics, the preheated material is put in a cooler mold. As soon as it becomes rigid enough to avoid distortion, it is ejected and cools off outside of the mold. Typical products are phonograph records.

In the TM process the material is loaded into a heated pot from which it is forced through connecting channels into a separate closed mold or usually several separate closed molds (see Fig. 6.16). No flash is produced on the molded part, but the channels consisting of the central sprue, the runners leading to the individual mold, and narrowed-down gates at the entry to each mold are discarded after the mold parts are ejected. They are easily broken off at the gates. The separation of the preheating process and the molding process makes it possible to control separately the temperatures in the two phases, so TM is very suitable for thermosets, because it separates the precure and the final setting. The multiple-mold arrangement speeds up production.

Transfer molding was the precursor of injection molding, in which the central pot and plunger are replaced by the rotating and axially reciprocating screw. The screw itself, however, without the injecting motion, is the main element of the extrusion process, which is discussed first.

6.4.4 Extrusion

The extruder shown in Fig. 6.17 consists of a barrel with a cylindrical bore in which the screw rotates, the drive of the screw, a hopper from which the granulated material is fed by gravity into the end zone of the screw, a breaker plate and screen pack in front of the screw, and a die attached to the front end. On the outer surface of the barrel are heater bands. This is the most common design. Other designs have coaxial twin screws rotating in the same direction or in opposite directions. Here only the single-screw extruder is discussed. The screw consists of three sections: the feed section with a

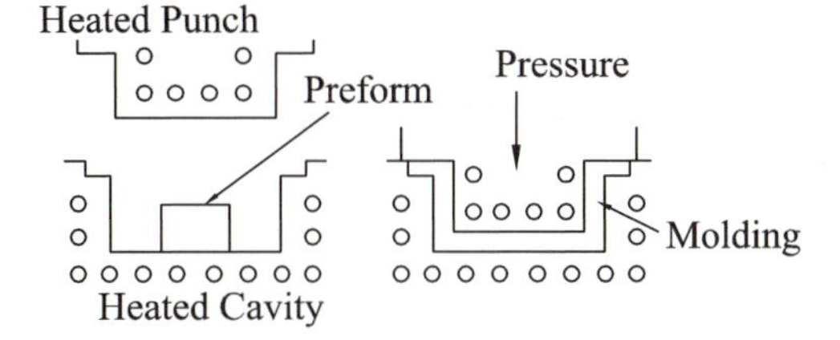

Figure 6.15
Principle of compression molding. [CRAWFORD]

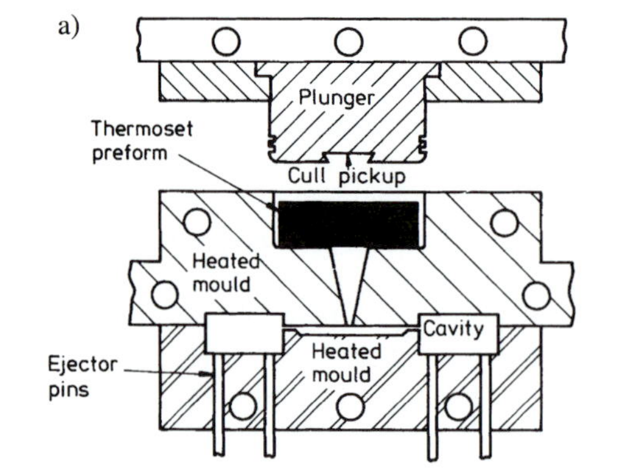

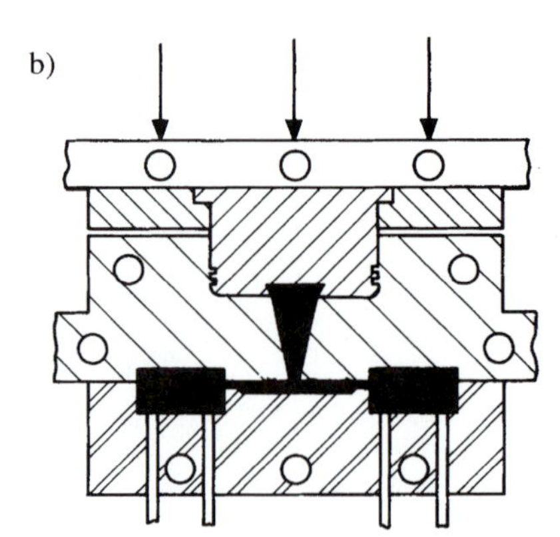

Figure 6.16
Transfer molding of thermosetting materials: a) preforming position; b) material forced into cavities. [CRAWFORD]

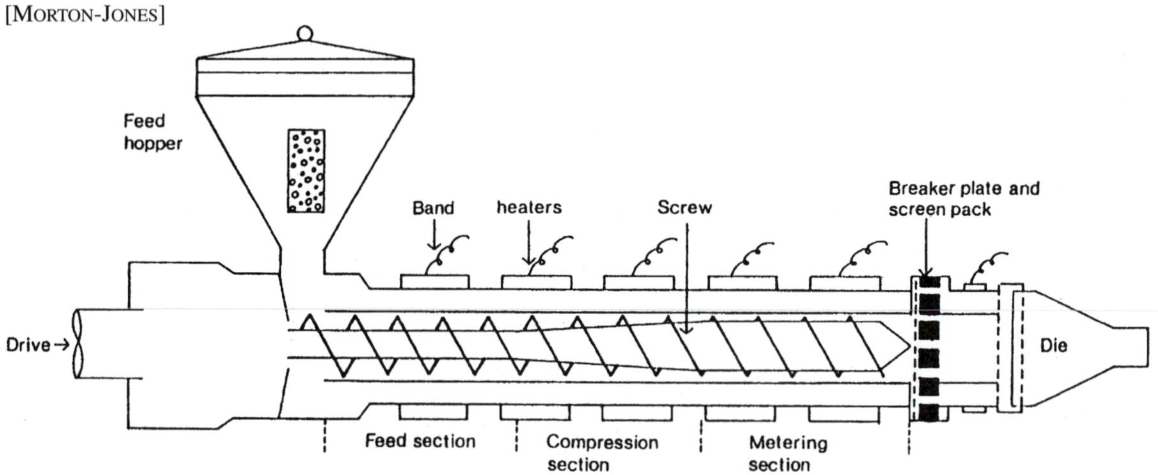

Figure 6.17
Main features of a single-screw extruder. [MORTON-JONES]

cylindrical core, the compression section in which the core is conical and therefore the depth of the channel between the screw flights decreases, and the metering section. The compression section is also called the plasticating zone because in it the granular material melts completely. The basic processes occurring are shown in Fig. 6.18. In the feed zone the granular material is compacted into a solid mass and some parts of it start to melt due to the heat generated by the mixing of the granules and the motion of the mass in the channel as well as friction with the screw and the barrel. In the plastication zone the melting progresses, the pressure builds up, and heating continues. In the metering zone the material is liquid and internal heating is due to the viscous flow. Apart from the internal heat sources, heat is added from the outside by the electric heater bands. In spite of the poor thermal conductivity of polymers, the external heating is efficient because of the hectic cross-flow in the channel (see Fig. 6.19) and also because the channel at the end of the middle zone and in the metering zone is only about 3–5 mm deep. The common L/D ratio of the screw is 20:1 or 24:1, but it may be as low as 12:1 and as high as 42:1. The ratio of the channel depth between the feed zone and the melting zone is 2.5:1 to 4:1. The screw as depicted in Fig. 6.18 with the three sections about equally long is called the PE screw. For materials other than polyethylene, the compression section may be shorter for materials that melt suddenly, such as nylon, and much longer for materials that soften very gradually, such as PVC.

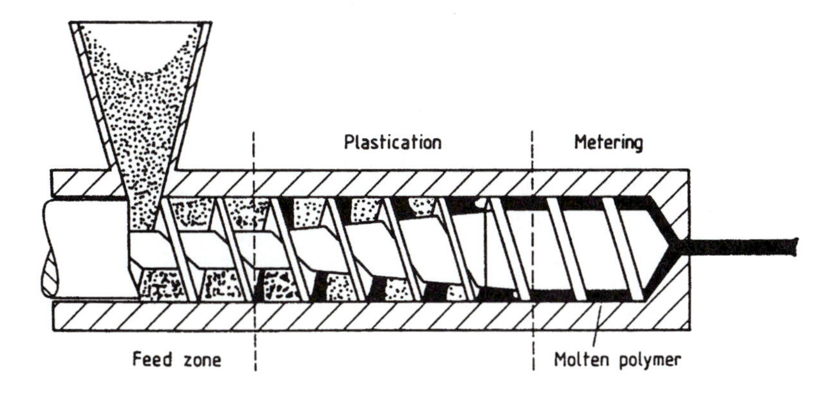

Figure 6.18
Single-screw plasticating extruder. [AGASSANT]

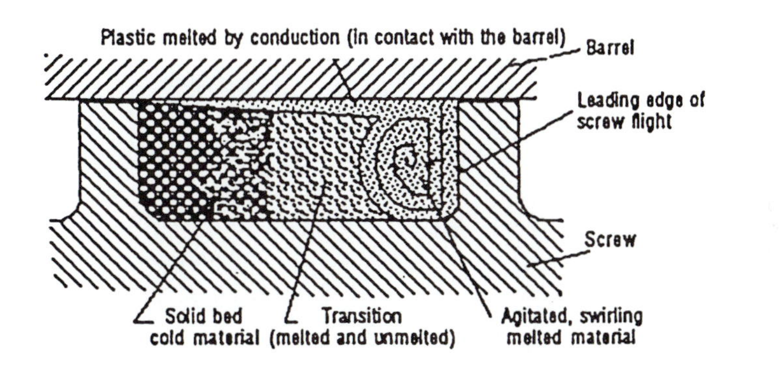

Figure 6.19
Melting of plastic within one screw flight. [REES]

Just in front of the die is the fine screen pack pressed against the much stronger breaker plate with coarser holes. The screens would normally have a mesh to filter down to 120 μm and even less for some applications. It is absolutely necessary to remove all foreign, often metallic, particles that may damage the die or be embedded in the plastic product. The role of the much thicker breaker plate is to support the screens and also to remove the spiral "memory" of the material. The screens are periodically replaced.

The role of the extrusion screw is to mix and compact the solid charge, to melt it, expel gases, heat the melt to a uniform temperature, and to compress the melt to provide the pressure essential to producing the flow of the material through the die. The pressures that can be generated are of the order of 10–20 MPa. Data from one manufacturer show the output rates for a range of sizes, from screw diameter 64 mm to 114 mm and up to 203 mm, for different polymers; for example, for LDPE they are 210–260; 695–850; 2200–1580 kg/hr, respectively.

By using various special dies, the extruder can be used to manufacture a great variety of products. The design of the die must take into consideration some general features of polymer behavior. One of these is the swell, as shown in Fig. 6.20 for a rod and a pipe. The swell is due to a combination of the elastic recovery and the change in velocity of the material close to the die walls. We have seen that in a round channel, the maximum velocity of flow is in the center, and there is none at the wall. As the extruded rod exits, the velocity, after a transient, becomes uniform through the section. This generates tensile stresses that may cause "sharkskin" surface or "bambooing" (Fig. 6.20c). Similarly, if a square or a T profile is desired, the die cross section must be different, as

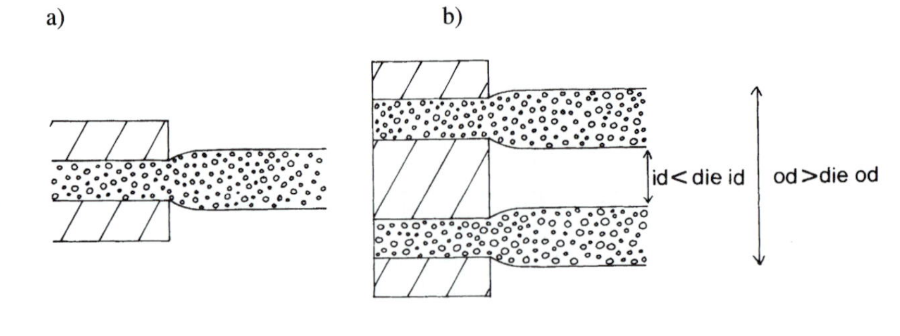

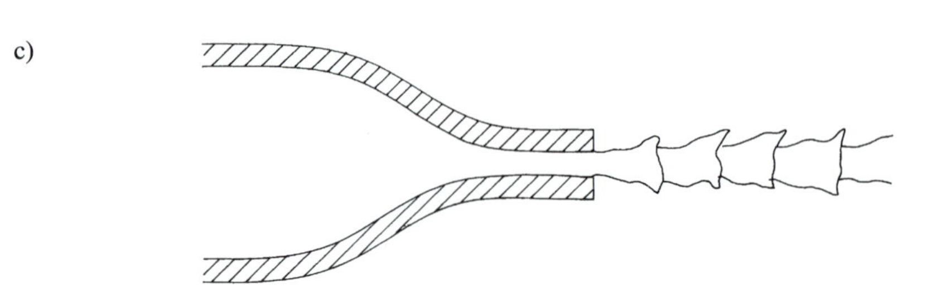

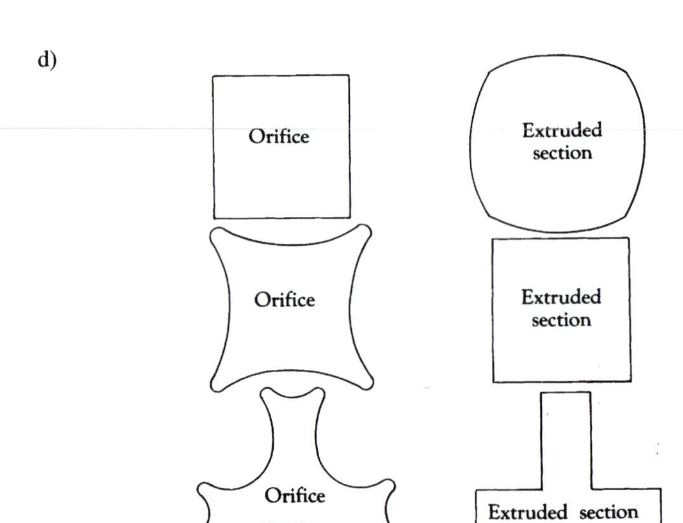

Figure 6.20
Special features of polymer behavior: a), b) swell; c) surface fracture ("bambooing"); d) profile distortion.
[MORTON-JONES 4.28, 4.29], [RICHARDSON]

shown in Fig. 6.20d. In a die used for pipe extrusion the central mandrel is supported by spider legs that disrupt the flow of the polymer. To minimize the effect on the profile of the extrusion, the mandrel is tapered gradually (Fig. 6.21a). Furthermore, in order to produce the desired inner and outer diameters with acceptable accuracy the pipe must pass through internal and external sizing operations.

Two specialty dies are shown in Fig. 6.22. In *(a)* is a tubing die, in *(b)* is a wire-coating die. The latter design is a cross-head where the wire runs horizontally straight through the die, and the polymer runs vertically down from the extruder and into a cavity that distributes it through a tubular die into a cylindrical channel around the wire. This arrangement is used for a large range of wire diameters from telephone wire with $d = 1$ mm to submarine cable with $d = 150$ mm. An example of a telephone-wire pro-

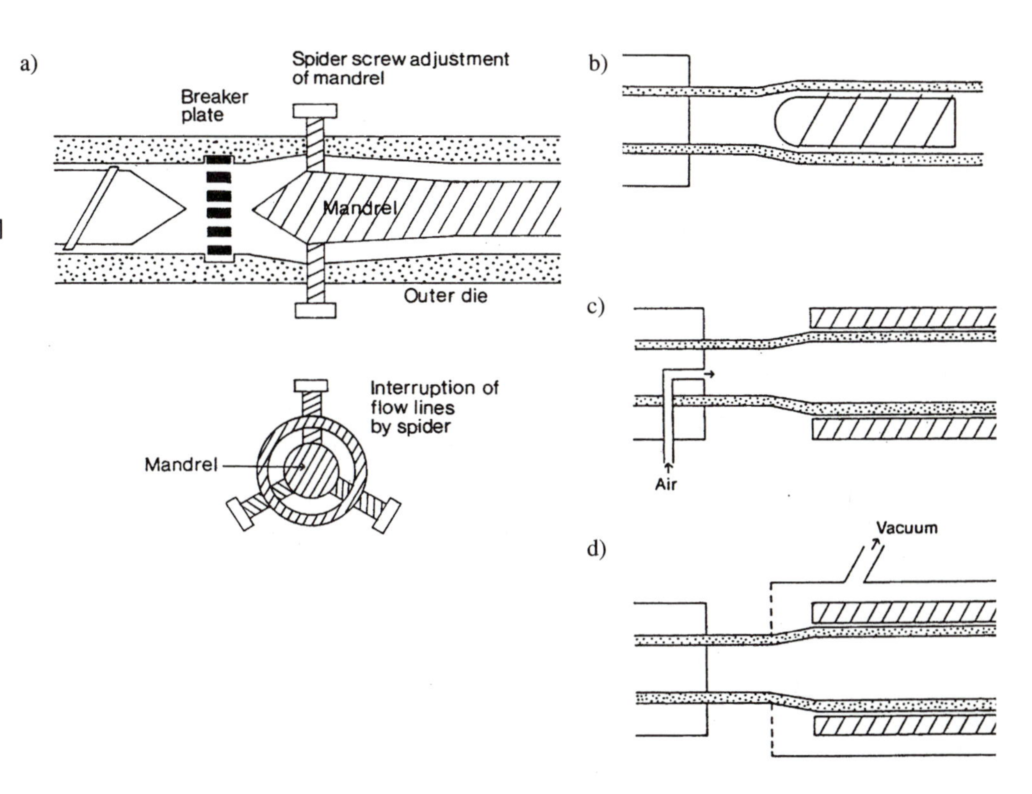

Figure 6.21
Extruding a pipe: a) the die; b) internal sizing mandrel; c) external sizing using air pressure; d) external sizing using vacuum. [MORTON-JONES]

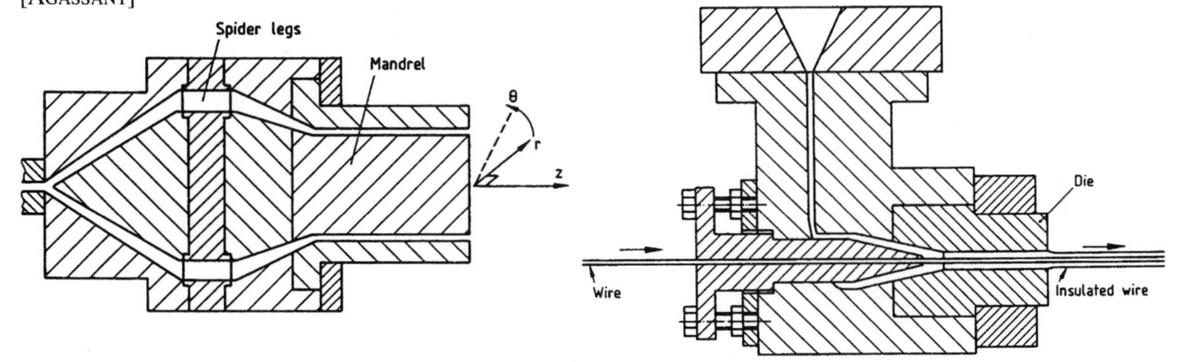

Figure 6.22
Extrusion dies: a) tubing die; b) wire-coating die. [AGASSANT]

duction line is shown in Fig. 6.23. It combines a wire-drawing operation with the wire-coating operation. Many different polymers are used for wire coating; for one kind, a double coating of nylon over PVC is used.

Sheet is extruded through a slit die commonly called the "coat hanger" die (see Fig. 6.24). The polymer pumped by the extruder screw flows first through a widening manifold *a* that decreases in depth as it widens, then through a narrow slit *b*, past a straining bar *c*, a relaxation channel *d*, and the distribution lips *e*. The design strives to achieve uniform thickness and temperature over the width of the sheet being extruded.

An example of a sheet production line is shown in Fig. 6.25. There is the barrel *1*, screen changer *2*, die head *3*, calender *4*, calender heating control *5*, pulling and edge-trimming unit *6*, cross cutter *7*, and suction-type sheet stacker *8*. At the end is a cooling

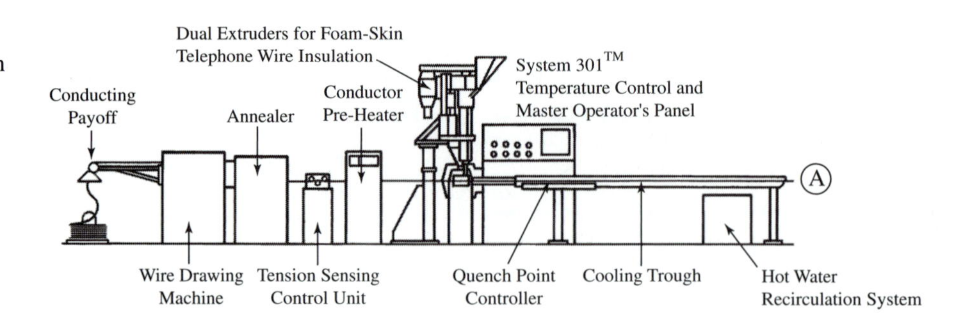

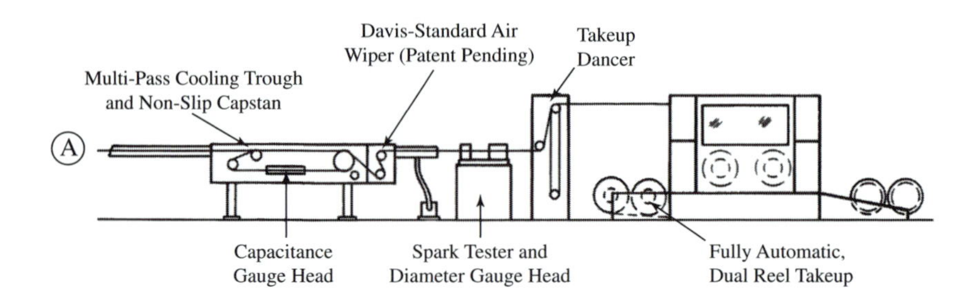

Figure 6.23
A telephone wire production line. [DAVID STANDARD CO]

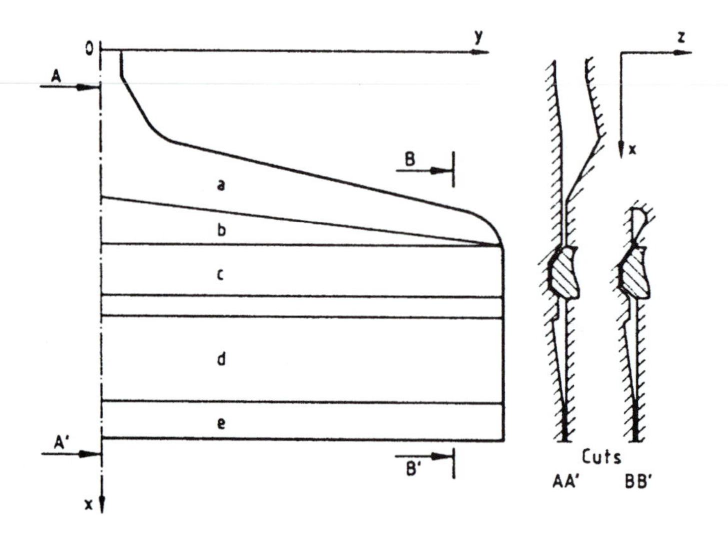

Figure 6.24
A "coat hanger" sheeting die. Only one half of the die is shown. [AGASSANT]

table. The die head is mostly of multiple type for co-extruded multilayer sheet. This can be of several types, either with an adhesive between layers or nonadhering peelable sheets. Materials processed are PS, PE, PP, PMMA, ABS, and rigid and flexible PVC. The thickness of the sheet ranges from 2 to 12 mm, and exit speed ranges from 0.7 to 7 m/min.

The most common way of producing thin sheet, film, and plastic bags is *film blowing* (Fig. 6.26). The molten plastic flows through an annular tubing die of a special design for preventing turbulence. Air is blown inside of the tube to produce a bubble that is inflated, and the plastic is cooled by a jet on the outside of it. It is drawn upwards by a pair of rollers. As the polyethylene cools it starts to crystallize at a freeze

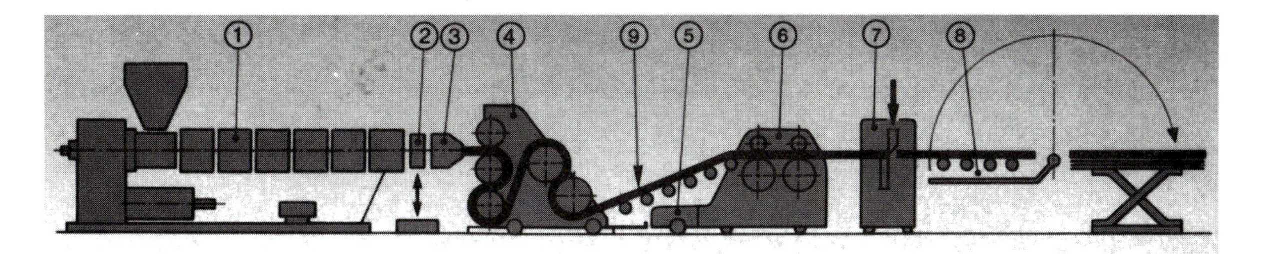

Figure 6.25
A sheet production line. [McNeil Akron Co]

Figure 6.26
Diagram of blown film process. [Morton-Jones]

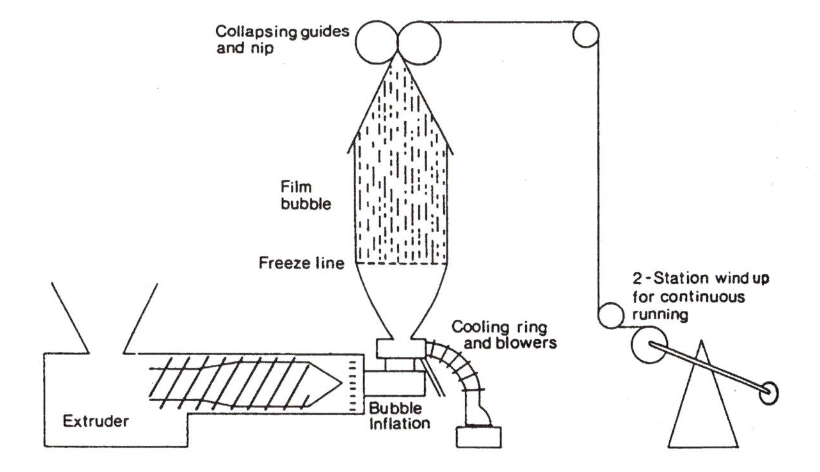

line and becomes less transparent. In this process a biaxial tension stress is induced rather than the shear stress of other extrusion operations. The air pressure promotes the circumferential stress and the pull at the rollers determines the longitudinal stress. The tensile strain in LDPE produces an opposite effect than shear strains that cause shear thinning; it stiffens the polymer, and this stabilizes the film-blowing process. The produced film can be wound on drums or gussetted and cut to length for plastic bags.

An example of a machine for production of 500-mm-wide, layflat T-shirt bag tubing at rates up to 115 kg/hr is shown in Fig. 6.27. The film is wound onto 1-m-diameter rolls; a semiautomatic cut-off system provides for scrap-free roll changes. A pneumatic lift mechanism facilitates unloading finished rolls.

The *extrusion stretch blow-molding* process is used to manufacture bottles. The process is analogous to the traditional method of producing glass bottles. A semimolten tube is prepared, called the parison, which is clamped between two halves of the bottle mold and inflated to conform to the surface of the mold, which is kept cold to freeze the shape of the bottle. A competitive process is *injection stretch blow molding*, in which the parison is first prepared by injection molding and then reheated for the stretch blow forming. This alternative is gaining more acceptance, especially for making bottles for carbonated drinks from polyethyleneteraphtalate. The principle of the process is shown in Fig. 6.28. The extruder head is directed vertically down; it produces the parison and stops. The parison sags under its weight and swells on the exit

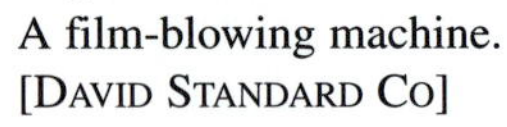

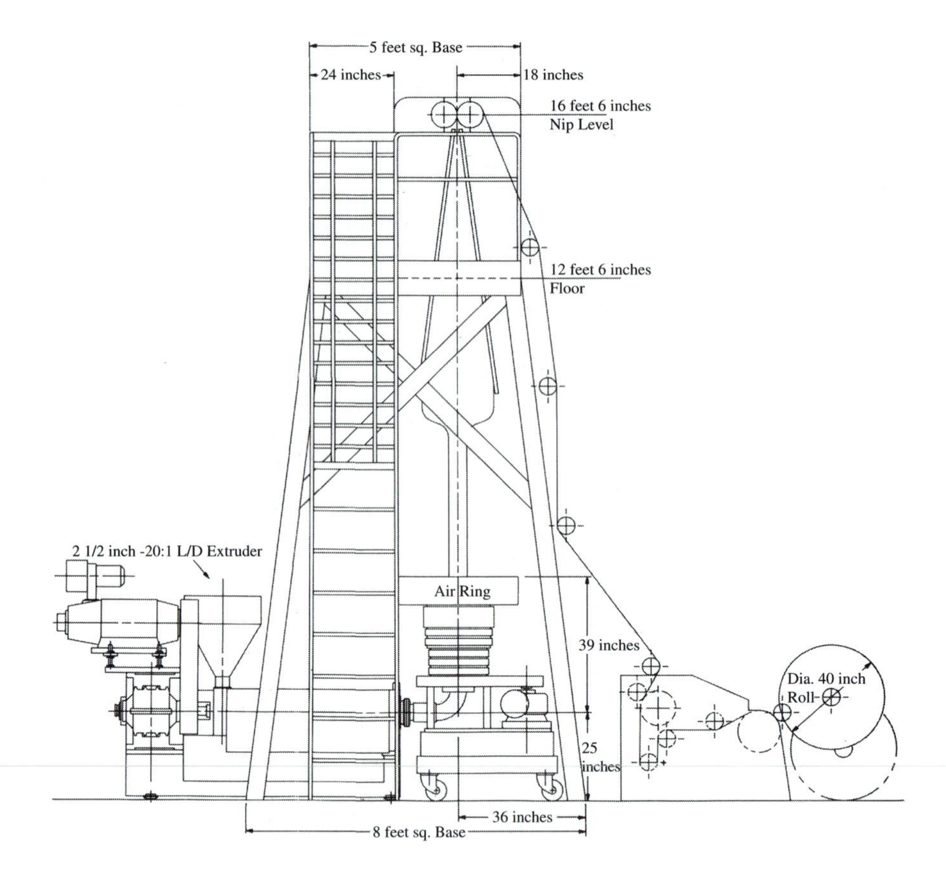

Figure 6.27
A film-blowing machine.
[DAVID STANDARD CO]

from the die. Subsequently it is inflated to create a preform. It is then transferred to another mold in which it is first stretched longitudinally and then inflated. These two steps create biaxial tension that is beneficial for the crystallinity and strength of the product. Finally, the mold opens and the product is ejected.

6.4.5 Injection Molding (IM)

The IM machine consists of the injection unit, the clamp unit, the heating and cooling system, and the drives and controls. The injection unit consists of the barrel and screw very similar to those of the extruder, except that the screw rotation is the principal motion only in one part of the cycle, and it is used to mix, compact, plasticize, and heat the material and build pressure in it. This pressure build-up is needed to provide a dense melt, but it is not essential for the injection process. Although the pressure in this stage usually reaches the 10–20 MPa that are the working values for the extruder, there are instances (e.g., processing thermosets) in which the compression ratio of the screw is 1:1. In the injecting stage the screw is driven axially by a piston to generate the working pressure, which is set to 150–250 MPa. This is about ten times more than in extrusion, and the injecting phase normally takes only a few seconds. The combination of the rotary and translative drives is achieved so that the translation is connected with the screw by means of a thrust bearing based commonly on a pair of tapered roller bearings, and the rotary drive permits the translative motion. This may be done as schemat-

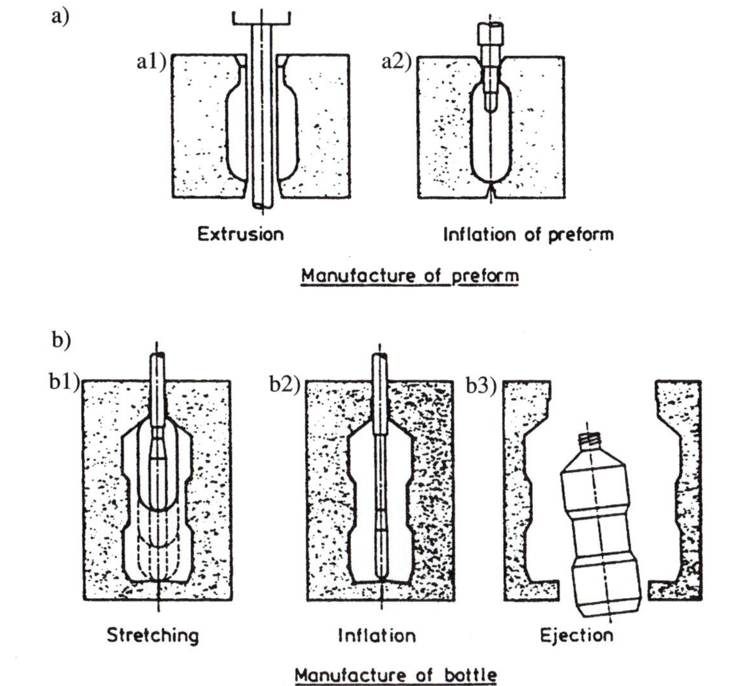

Figure 6.28 Extrusion stretch blow-molding process. [CRAWFORD]

ically indicated in Fig. 6.29, where a very wide driving pinion meshes with a narrow gear mounted on the screw. Another solution uses a spline shaft/spline bushing coupling that transmits rotation while allowing relative axial motion between the driving motor and the screw. This design is used in the machine shown in Fig. 6.33. The polymer is injected into a die contained in the clamp unit, discussed later. First, the cycle of the screw motions is explained in Fig. 6.30.

The holding phase *(a)* starts right after the injection. The screw remains in the forward position under the holding pressure, which may be about half of the injection pressure. It does not rotate but can still move ahead to supply material and compensate for shrinkage while the molding cools down. The screw remains in the holding mode until the gates freeze and no additional material can be injected. Then the screw starts to rotate and begins to move back against a rather low pressure that has at the same time been set in the driving cylinder, stage *(b)*. This restarts the plasticizing function of the screw. In the phase *(c)* the mold has closed again, and the screw moves forward in the injecting motion. In *(d)* everything is back to start phase *(a)* again.

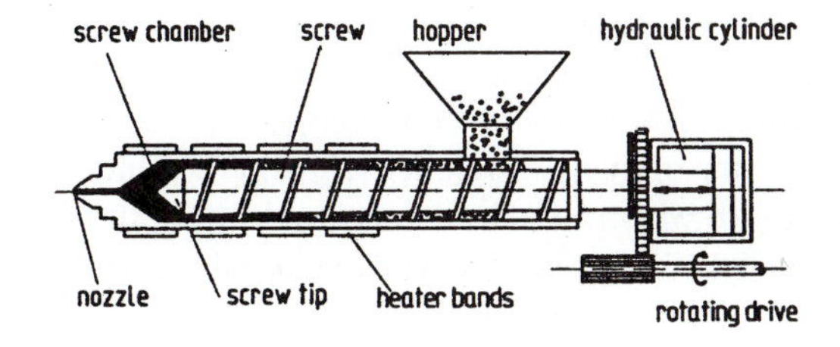

Figure 6.29 Screw plasticating and injection units. [POTSCH]

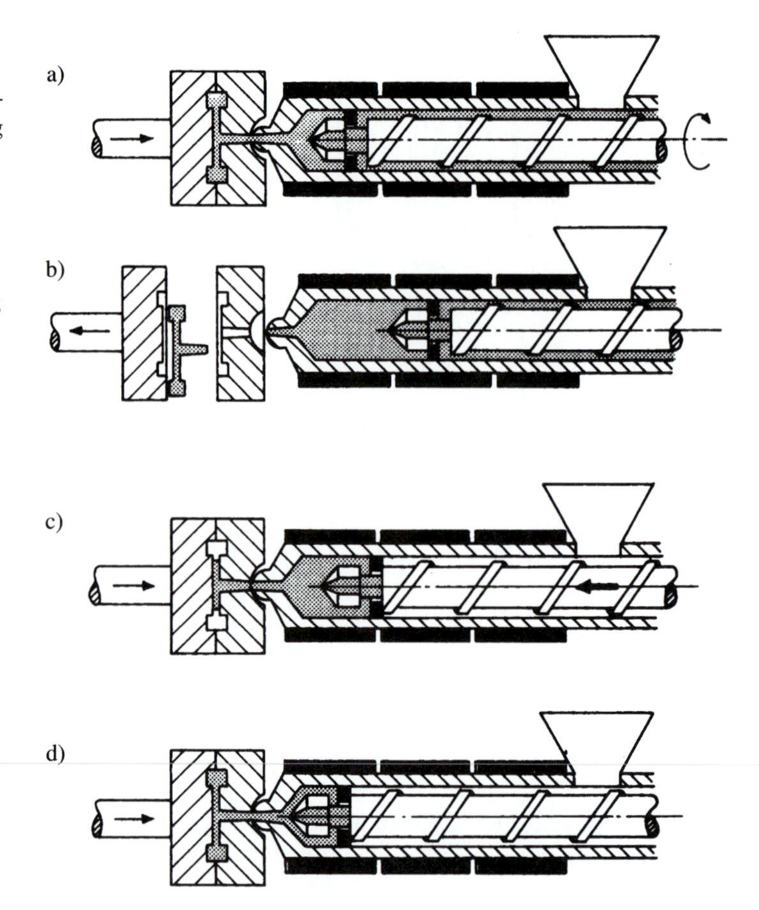

Figure 6.30
Typical cycle in reciprocating screw injection-molding machine: a) holding pressure after injection until gates freeze; b) screw rotates, moves back under low pressure, and starts plasticizing; c) mold closed, screw moves ahead to inject; d) injection completed, back to (a). [CRAWFORD]

A typical time distribution for the injection-molding cycle is presented in Fig. 6.31. The graph starts with *1*, the closing of the mold [phase *(c)* in Fig. 6.30]. Phase *2* is the advance of the nozzle to the sprue bushing of the mold. Phase *3* is the actual injection, which may last from a fraction of a second to several seconds. From here on starts the cooling phase. Phase *4* is holding, which lasts until the gates freeze. Next, in phase *5* the screw moves back while plasticating, which continues in phase *6*. Phase *7* is the time needed to open the mold and take out the part. Between phases *6* and *7* the machine waits for the part to cool down to removal temperature, when the part is rigid enough to be safely ejected. Obviously, the cycle time is dictated primarily by the cooling time; the injection itself is a small fraction of it. During the injection molding of thermosets, the die must remain closed until the material has set sufficiently for safe removal. Shorter injection time requires higher pressure. Long injection time may lead to narrowing down of passages and an increase of injection pressure as well. There is an optimum both as regards pressure and also temperature, which may increase with fast flow and decrease with a longer injecting period.

The connection between the barrel and the die is controlled by a check valve; Fig 6.32a shows one of many different designs. When the screw is injecting or holding, the valve is closed; when it is plasticizing and filling the front space while moving back, the valve is open. With very viscous materials, such as rigid PVC, there is no check

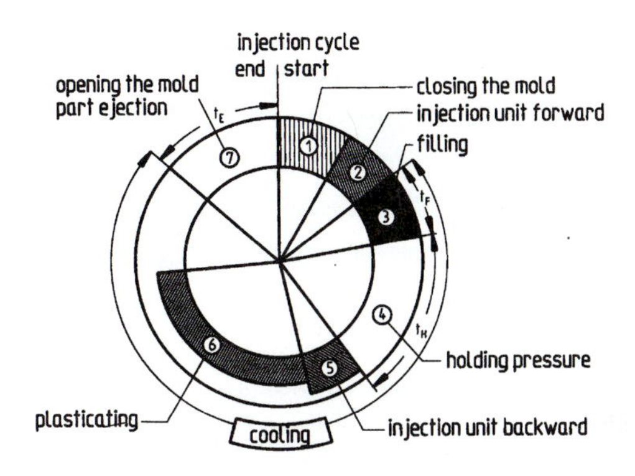

Figure 6.31
Typical time distribution in the injection molding cycle. Part 3 is the actual injecting phase. [POTSCH]

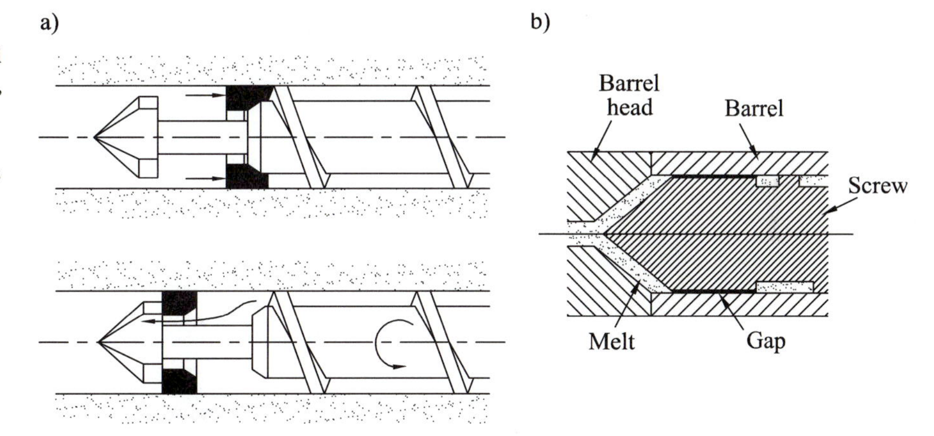

Figure 6.32
a) A check valve. It is closed during injection and holding, and open when plasticizing and filling the space in front of screw. b) Plain screwtip is used for very viscous materials. [CRAWFORD], [REES]

valve. This is called the plain screw tip, shown in Fig. 6.32b. Between the screw and barrel head there is a free passage through a small gap. When injecting, the plastic is pushed forward fast so that the small amount of material that manages to flow back through the gap is negligible.

An example of the whole machine is shown in Fig. 6.33, which is the overall view. The injection unit is on the right, and the clamping unit is on the left. Clamping of the two halves of the die is hydraulic. The clamping cylinder with thick walls to withstand the high pressures is on the left, and its piston, called here the ram, carries on its right end the moving platen, which is also designed for high rigidity. The "core" half of the die is attached to this platen. The "cavity" half of the die is attached to the stationary platen. Returning to the right side, at the very end is a radial piston hydraulic motor which provides the rotary drive to the screw. The funnel-shaped hopper is visible on top of the injection unit, and left of this housing is the barrel. The whole injection unit is mounted on a slide and can be moved to the right in order to change the screw or clean the system.

The detail of the clamping unit is shown in Fig. 6.34a. It shows the massive cylinder *A* and ram *B*, which carries the movable platen *H* with the conical core half of the die. The cavity half of the die is attached to the fixed platen *G*. The clamping force is contained by tie rods *I* between the body of the main cylinder and the fixed platen. The hydraulic oil is partly held in the tank above the machine so that it can keep filling the

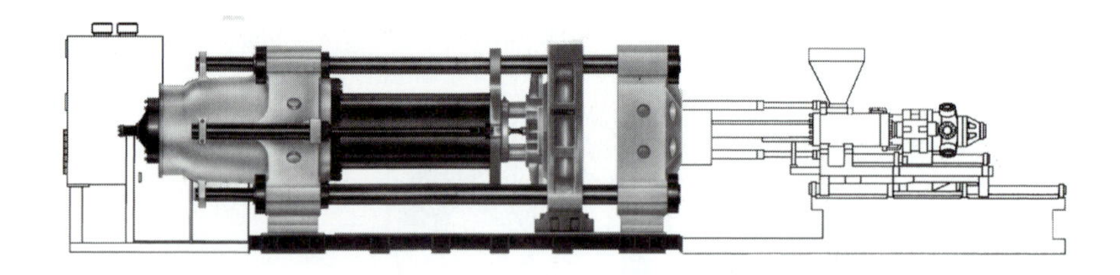

Figure 6.33
An injection-molding machine, overall view. [CINCINNATI MACHINE]

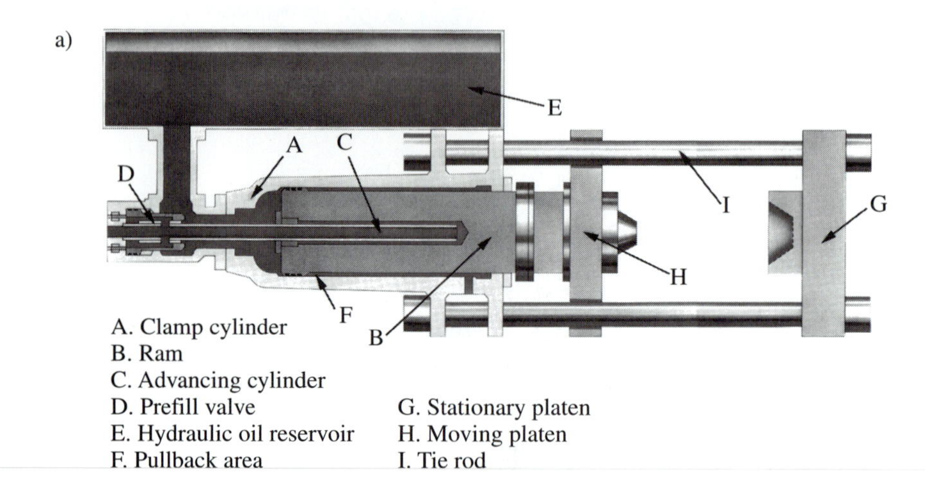

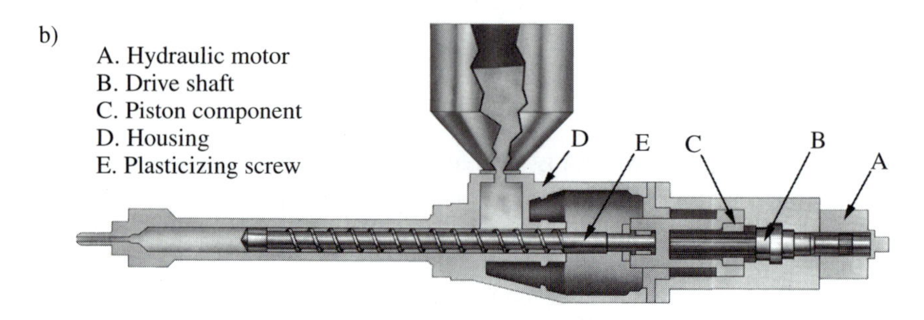

Figure 6.34
a) The hydraulic die-clamping unit; b) the injection unit. [CINCINNATI MACHINE]

main cylinder when the ram is being moved to the right in the rapid approach part of the cycle, during which the core die is moved ahead to close the die after part ejection, for which the die had to be wide open. The rapid approach motion is driven by oil pumped through the advancing cylinder *C*. The corresponding pump is low-pressure, high-delivery type. The valve *D* between the top tank drain and cylinder *A* is closed during this phase. Just before the two halves of the die touch, the passage from the upper tank is closed off. This is accomplished by the hollow body enclosing the valve *D* moving to the right until its chamfered front presses against the opposing seat. A low-volume, low-pressure pump drives the ram as far as a contact between the die halves. A high-pressure pump is then turned on to rapidly build up the clamping force. A pressure switch signals the injection unit to start injection in advance to shorten overall time. At the end of the cycle the motions act in reverse.

Some other machines use mechanical clamping that incorporates various forms of toggle linkage, with various degrees of complexity. The basics of the toggle linkage were shown in Chapter 4, Fig. 4.21.

The injection side of the machine is shown in Fig. 6.34b. At the right is the rotary hydraulic motor *A* driving shaft *B*, which is hollow with internal splines. The external splines of the right end of the screw engage with the splines of the shaft *B*. In this way the rotary connection remains active while the screw mounted in tapered roller bearings in the left face of the hollow piston *C* is driven axially by the piston during injection. The hopper is seen at the top. No check valve is shown at the front of the screw in this simplified picture.

IM machines are marked by the magnitude of the clamping force in tons; they are built in sizes between 20 and 10,000 tons. Another characteristic parameter is the screw diameter, available from 18 to 120 mm. The injection pressure is mostly about 200 MPa. Another parameter is the volume of the shot, which ranges between 20 and 6000 cm^3. This latter number is actually 6 liters (almost two gallons). This illustrates the scale of the operation, just for the purpose of making buckets, bumpers, or plastic furniture with high production rates (Fig. 6.35). A greater variety of parts, large and small, is assembled in Fig. 6.36. They include automotive components such as wheel covers, fan wheel, instrument panel, light covers and lenses, as well as bottles, cups, and audio and video cassettes.

Here are some other interesting numbers for one make of a 1400-ton machine: screw diameter, 100 mm; screw stroke, 440 mm; screw speed, max 100 rpm; 13,921 Nm torque on the screw at 17 MPa plasticizing pressure at the head of the screw; barrel heater capacity, 58 kW; heat exchanger water use, 151 l/min. The given torque and screw speed result in a power of 146 kW, which is capable of providing about 2.5 times more internal heating than the 58 kW of the heater bands.

In an alternative design of the IM machine, the plasticizing function is continuous and separate from the injecting function (see Fig. 6.37). At the top is an extruder unit,

Figure 6.35
Examples of large injection-molded parts. [HUSKY]

Figure 6.36
Examples of various injection-molded parts. [HUSKY]

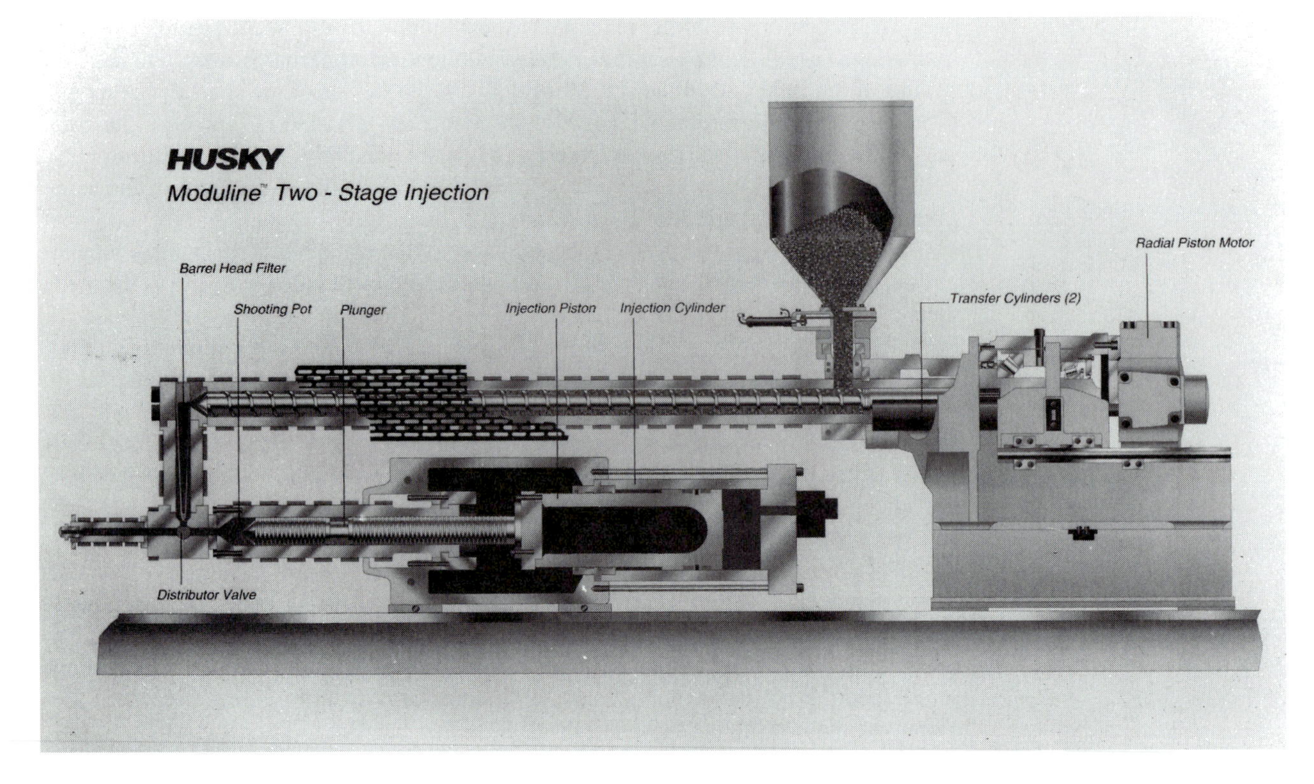

Figure 6.37
Two-stage injection. The plasticating screw unit is at the top; the injection unit at the bottom. [HUSKY]

and it supplies, through a distribution valve, the melt into the shooting piston of the injection plunger. This machine can produce up to 40% more than the standard reciprocating screw machine with the same size of the plasticating unit. However, it is more complicated and more expensive than the much more popular standard machine. Many plastics manufacturers prefer to cover additional production capacity by simply installing more of the standard machines.

The most important part of the IM operation is the mold. Its design is decisive for the quantity and quality of the operation, and it is entrusted to very experienced engineers because it represents the largest part of the production expense. Computer programs have been developed to simulate the filling of the mold, the progression of the solidification, pressure distribution, welding line generation, shrinkage of the part, residual stresses, and distortions of the part. Even so, as good as these programs are, they are just aids in the creative work of the die designer. In this text we cannot explain even the basic factors and parameters affecting a good mold design. It is only possible to outline the simplest forms and types of the injection-molding molds.

The very basic type of the mold is a two-part mold (Fig 6.38). It consists of the "cavity" part on the right to be attached to the stationary platen, and the "core" part on the left to be attached to the movable platen. The locating ring centers the die in the platen so that the sprue bushing becomes coaxial with the nozzle of the injection unit. The core is aligned with the cavity by guide pins. The injecting channels, the sprue, the runners, and the gates are built into the cavity part. Shown is a die with two, perhaps four mold cavities. Dies are commonly built with multiple cavities. The usual numbers of them in a rectangular layout are 2, 4, 6, 8, 12, 16, 24, 32, 48, 96, and 128. Typical layouts are shown in Fig. 6.39. Returning for now to the mold features shown in Fig.

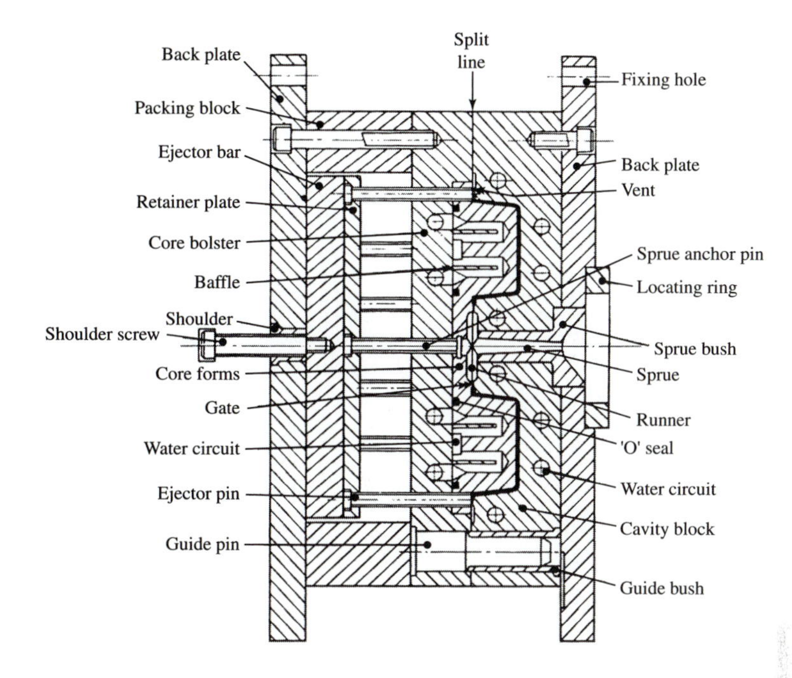

Figure 6.38
A two-part injection mold. [CRAWFORD]

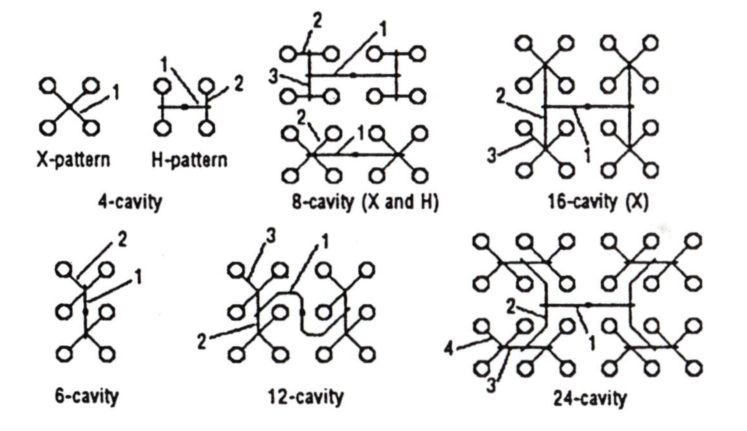

Figure 6.39
Typical layouts of different multicavity mold arrangements (1, 2, and 3 refer to the order in which the material reaches each runner branch in the mold). [REES]

6.38, there are cooling channels in both parts, and there is an ejector bar or plate that carries ejector pins and the sprue anchor pin. The retainer plate with the ejector pins is moved by the shoulder screw hitting a stop when the mold opens, and the pins eject the molding. The sprue anchor pulls the sprue away from the nozzle together with the core. The gates are narrowed-down passages that break when the sprue and runner system is separated from the molding. This is, however, an additional, perhaps manual operation, and all these parts are scrap (see Fig. 6.40a).

In a three-plate mold the runner system is in a separate parting plane than the basic parting plane of the mold. As a result the sprue and runner system separates automatically when the mold opens, as illustrated in Fig. 6.40b. Still, the channeling system becomes scrap. This is eliminated in a mold with a "hot runner" system (see Fig. 6.41). The sprue and runners are insulated or even heated, and they remain hot and molten and will become a part of the polymer injected into the mold in the next cycle.

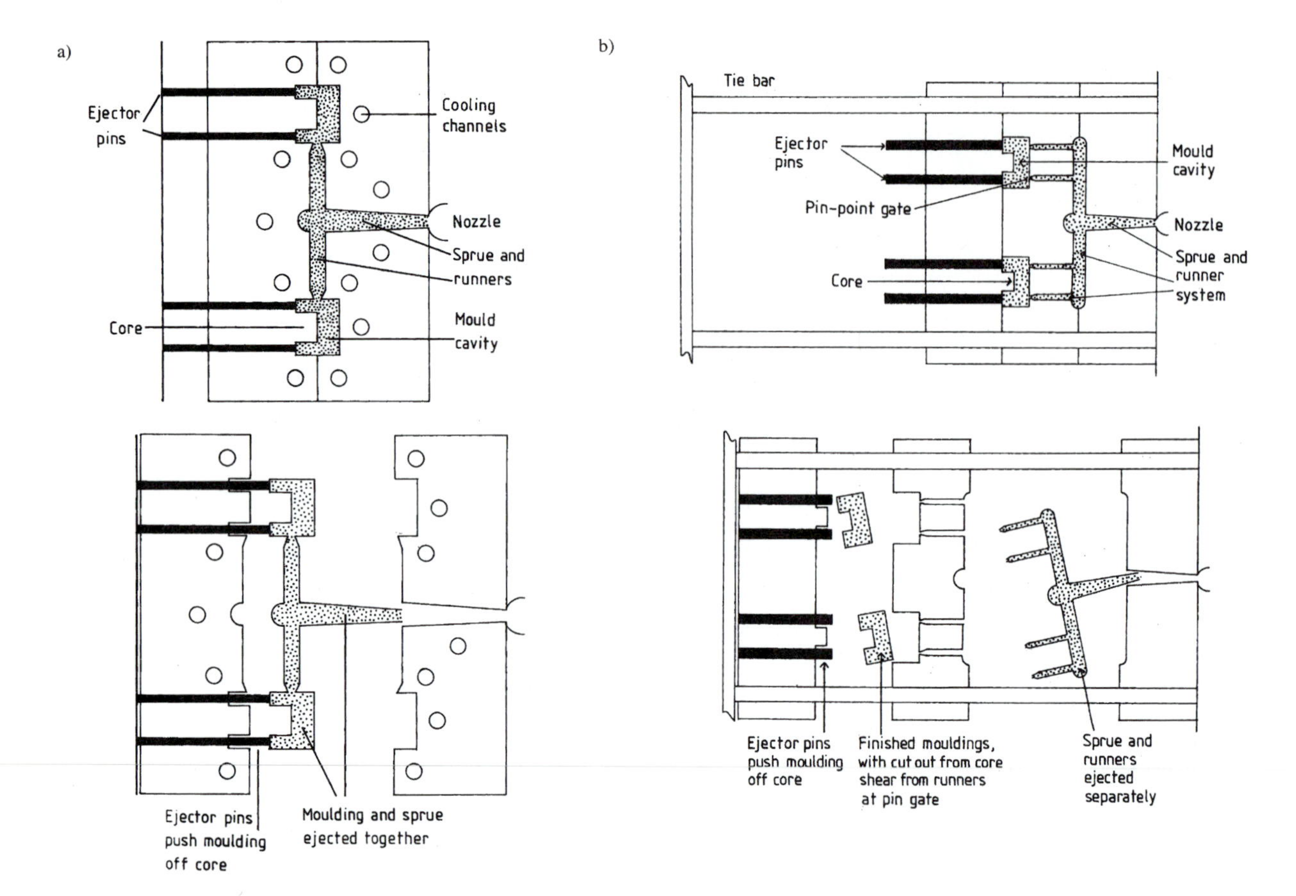

Figure 6.40
a) Two-plate mold, closed and opened. The sprue and runner system is ejected together with the part and has to be separated in an additional operation.
b) The principle of the three-plate mold. Sprue and runners separate automatically as mold opens. [MORTON-JONES]

We conclude the necessarily brief discussion of IM operations by mentioning the equipment used to mechanize and automate the handling of the molds for their insertion and removal. Figure 6.42 shows an automatic crane for the change of rather large molds, and Fig. 6.43 shows a robotic arm used for the removal of large moldings (in this case, a car bumper that is being taken out). Robotic systems are used when the molding cannot be ejected without external assistance, when the molding must be protected against falling, or when the handling operation involves also additional work, such as cutting off the runner system, or positioning inserts. Five-axis, numerically controlled robots are commonly available for these tasks.

Manufacturing plants specializing in injection molding often have large installations of ten, twenty, or more machines. They run on installed electric power on the order of 5000 kW, have to supply 3000 l/min chilled water, and must solve complex logistics of materials handling. They may also include die design and manufacture or else subcontract this significant task to specialized firms.

6.4.6 Thermoforming

In thermoforming a film or sheet of a thermoplastic material is heated above T_g and then shaped into or onto a female or male half die that is kept at a selected temperature to cool the product and stiffen it for permanent shape. The shaping may be done by one of several methods illustrated in Fig. 6.44. Method *(a)* uses vacuum to draw the sheet

Figure 6.41
Layout of hot runner mold. Sprue and runners do not become scrap. They are kept molten and used in the next injection. [CRAWFORD]

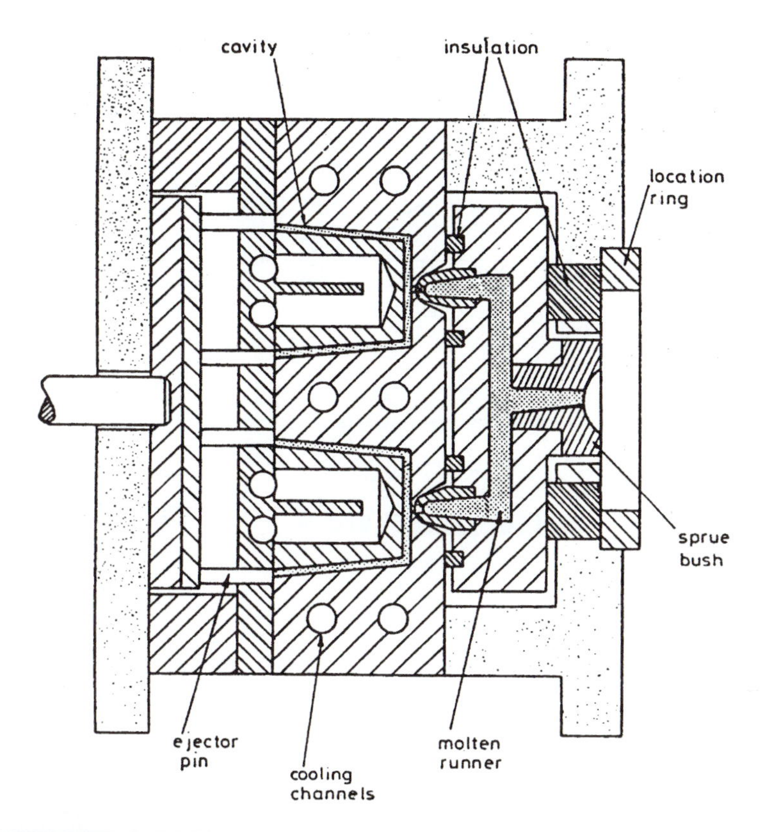

Figure 6.42
An automated crane for mold changing. [HUSKY]

Figure 6.43
A robot for removing moldings. (Example: car bumper). [HUSKY]

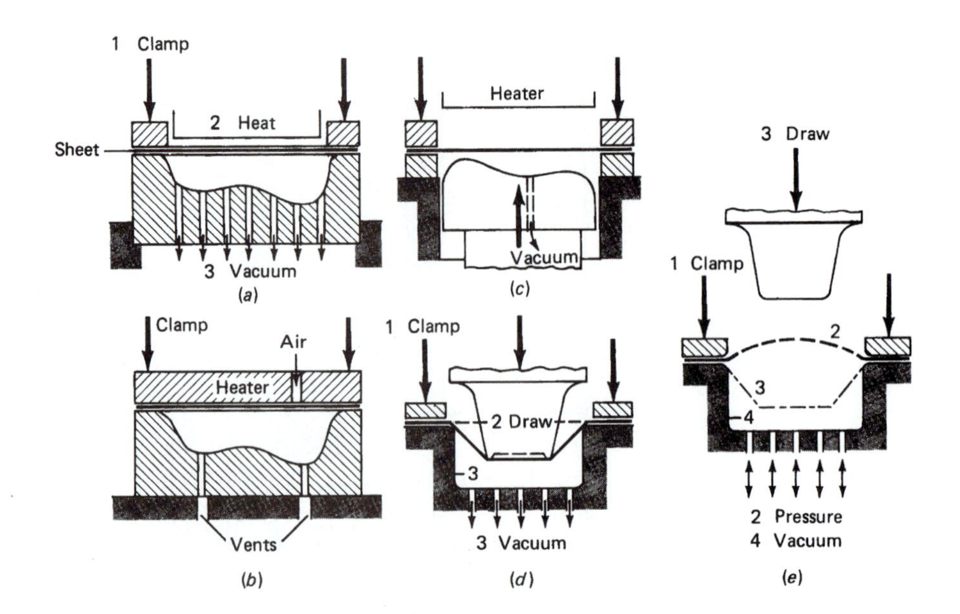

Figure 6.44
Various methods of thermoforming: a) vacuum; b) pressure; c) drape-vacuum; d) plug assist; e) pressure-bubble plug-assist methods. [SCHEY]

into a female die. Atmospheric pressure at about 100 kPa does the work. In the method *(b)* a pressure between 200 and 1000 kPa is used to press the sheet into a female die. In both cases the pressures are relatively low. Rather inexpensive equipment is used, and dies can be made of wood, plaster, fiberglass, or aluminum. The latter has the advantage of easy temperature control for consistency of the product and may be used for large production quantities with a long die life. Method *(c)* uses vacuum again, but the sheet is draped over a male die. Method *(d)* combines vacuum forming and stretching by means of a plug. It provides better uniformity of thickness of the walls over the entire shape of the product. The best thickness control is achieved by first blowing the sheet into a dome shape, then reversing the deformation by using the plug, and finally using the vacuum to produce the final shape, as in method *(e)*.

The blank material, the sheet or film, is often produced in an extruder, and the two operations may be combined in a line without much cooling between them. Multilayer products are also made in this way.

Small parts with intricate shapes and wall thickness less than 0.25 mm can be made in large quantities with very high production rates. On the other hand, very large parts (e.g., jacuzzi tubs) can also be made economically by this process.

6.5 ANALYSIS OF THE PLASTICATING SCREW

In analyzing the conveying of polymers in the single-screw extruder, we find a substantial difference in the transportation mechanism at the back of the screw, in the feeding section, where the material is solid granular, and in the metering section, where the material is liquid.

The approach to the treatment of the conveying process in the feed section generally assumes that the granules have already been compacted into a solid mass. Two extreme situations are considered, depending on whether the friction of the material in contact with the barrel prevails over the friction on the surface of the core and on the sides of the flights of the screw or vice versa. In the extreme, if friction existed on the barrel only and none on the screw, all the material would move ahead by the full axial motion of the flanks of the screw, and an efficiency of 100% for material transport would be achieved. In the opposite case, with friction on the screw and none on the barrel, the material would just simply rotate with the screw without any axial transport movement. With equal friction on barrel and screw it is a toss. Rather sophisticated models have been developed by many authors in an attempt to explain the clear empirical experience of the fact that the screw works. Those who first tried the process have accepted this fact unaware of the theoretical complexities. We will present a simplified summary of some of the explanations.

The conveying of the viscous liquid in the metering section is rather easy to analyze. There are two competing mechanisms. The *drag* flow described in Section 6.3.1 tends to move the material ahead in the channel between the screw and the barrel, and the average velocity of this flow is approximately half of the peripheral velocity of the surface of the screw core measured in the tangential direction down the helical direction. If the helix angle is ϕ, this velocity is

$$v_d = \pi D n \cos \phi$$

where D is the screw core diameter, and n is the rotational speed in rpm.

Assuming, for illustration, a commonly used angle $\phi = 17.66$ deg, so that the pitch of the screw is equal to its diameter, then the width of the channel is

$$w = (D - e) \cos \phi$$

where e is the thickness of the flight of the screw. The flow down the channel is then

$$Q_d = \frac{1}{2} v_d \, w \, h = C \, D^2 \, n$$

where h is the channel depth that is the clearance between the screw core and the barrel, and the flow is proportional to the square of the screw diameter and to the rotational speed n, independent of the viscosity of the material and independent of the pressure.

However, there is a *counterflow due to the pressure gradient* that is being built up. This counterflow can be evaluated as flow in a rectangular channel, which was analyzed in Section 6.3.2, and the non-Newtonian flow Q was expressed by Eq. (6.27), which for a typical value of $n = 0.4$ comes out as

$$Q_p = w \left(\frac{\Delta P}{KL} \right)^{2.5} \frac{0.8}{1.8} \left(\frac{h}{2} \right)^{4.5}$$

This counter flow increases with the 2.5th power of the pressure gradient and with the 4.5th power of the channel depth. The effective flow is the difference of the drag and pressure flows:

$$Q = Q_d - Q_p$$

Obviously, the efficiency of the conveyance in the metering section strongly depends on the channel depth h, and for a small value of h it can be very high.

6.6 PROCESSING OF POLYMER-BASED COMPOSITES

Traditionally, the best-known polymer-based composite material is made of glass fibers in a polyester matrix. However, there are many more fiber-matrix combinations in use, and the composites may also be classified according to the form in which the reinforcing fiber is introduced in the matrix. Both thermoplastic and thermoset materials are used. Of the latter, apart from the polyester, the epoxies are commonly used in combination with glass, kevlar, or carbon fiber.

6.6.1 Preforms

Preforms are generally used for processing of composites. The powders or granules of the polymer are premixed and preheated. They may be prepared in the form of *bulk-molding compounds* (BMC). Another preform is the *prepreg* in a sheet or tape form. *Sheet-molding compounds* (SMC) are often called mats. They are precombined sheets of resin, reinforcement, fillers, and fabrics, with an upper and lower carrier film. An example of a thick molding compound (TMC) is shown in Fig. 6.45. The fibers are made in various combinations of profiles, sizes, and forms.

The strongest composites are made from continuous filaments, which may be gathered in bundles. Others are woven into various cloth and tape orientations. Carbon fibers are available in continuous lengths up to 13,000 m in bundles (tows) of 1,000–160,000 filaments of 7.5 micrometer diameter. Short fibers, flakes, and particulates are randomly dispersed in the matrix. Fibrous reinforcements are supplied as *rovings* (continuous filaments in strands or yarns), *chopped strands* (cut from rovings in lengths 3 to 50 mm), *mats* in various widths and lengths, *woven rovings*, and *tapes*. A comparison of two types of woven and knitted rovings is shown in Fig. 6.46.

There are also *laminates*, illustrated in Fig. 6.47 a) and b). In the sandwich form the core may be a honeycomb, waffle, or corrugated structures as well as foam.

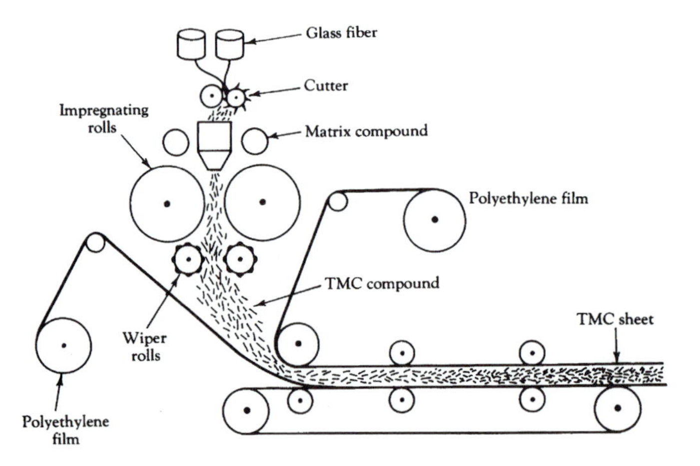

Figure 6.45
Thick molding compound (TMC). [RICHARDSON]

Figure 6.46
Woven and knitted rovings.
[RICHARDSON]

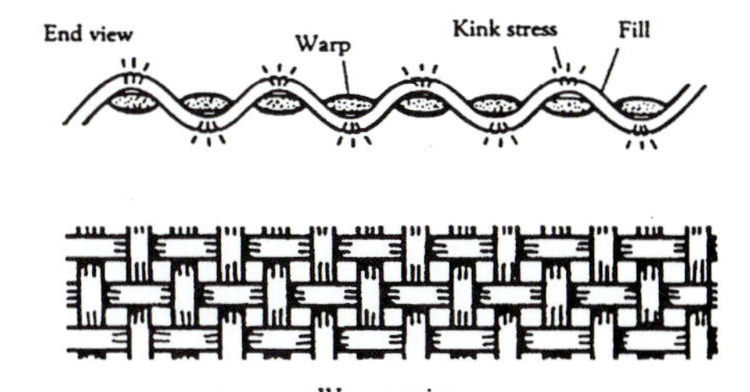

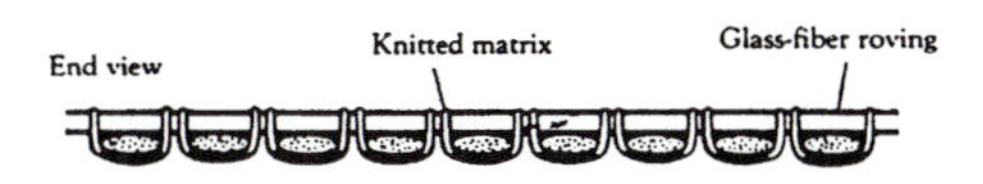

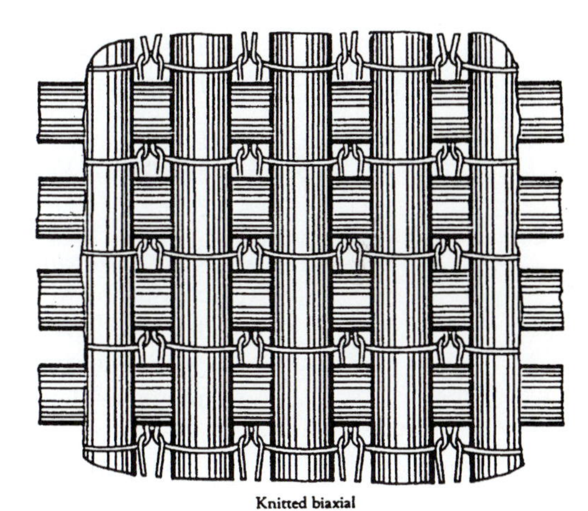

Figure 6.47
a) Laminate piles tailored to meet specific requirements of composite design. b) Two distinct classes of laminar composites. c) Various types of sandwich cores.
[RICHARDSON]

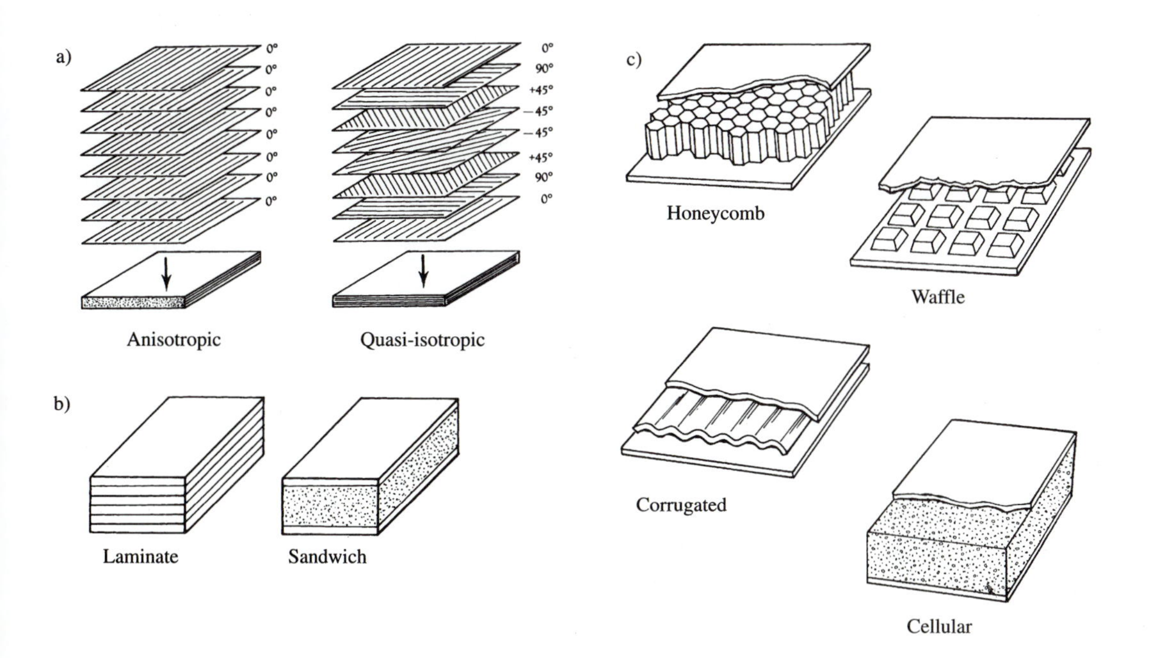

6.6.2 Hand Lay-up and Spray-up Molding

The simplest processing technique is the *hand lay-up* process (see Fig. 6.48). A one-sided, male or female form is used. A surface skin layer called the gel coat is first applied to the surface of the mold followed by the application of several layers of composite material. Usually the materials cure at room temperature although surface heaters may be used as well. The process is used in the production of large boats, swimming pools, shower stalls, septic tanks, and aircraft parts. Polyesters and epoxies are used in the gel coating, resulting in a smooth surface. The resin may be mixed in a bucket and applied by a brush, or squeegee, cloth, mat, and other forms of reinforcements can be used.

An alternative process that may have a higher degree of mechanization is the *spray-up molding* technique (see Fig. 6.49). The chopped fiber is sprayed together with the resin and catalyst, and it is manually spread in thin layers of about 1.5 mm, repeatedly, if desired. The process may be further mechanized using special machines that control the motion of the spreading roller.

6.6.3 Filament Winding

A rather sophisticated technique used for the production of strong and rigid structures is *filament winding*. It is used in the manufacture of aircraft parts, wind-turbine blades, pressure vessels, rocket-engine cases, helicopter blades, and other applications where the directional control of the filament reinforcement leads to the best mechanical properties. The strength and stiffness of the parts may by comparable to those made of aluminum or steel, but they are much lighter.

There are two basic alternatives of the process: wet and dry winding. In the former the filament and the resin are applied separately and simultaneously; in the latter prepregs are used, mostly in the form of tapes.

There are many different kinematics of the filament-winding process. Three of these are selected and shown in Fig. 6.50. These three are known as the helical winder, the normal-axial winder, and the braid-wrap winder. The filament-wound shell is pro-

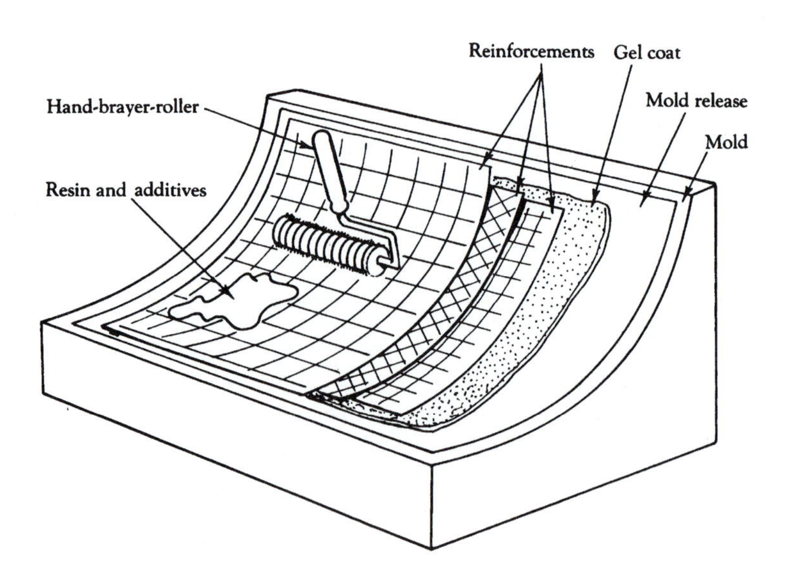

Figure 6.48
Concept of hand lay-up process. [RICHARDSON]

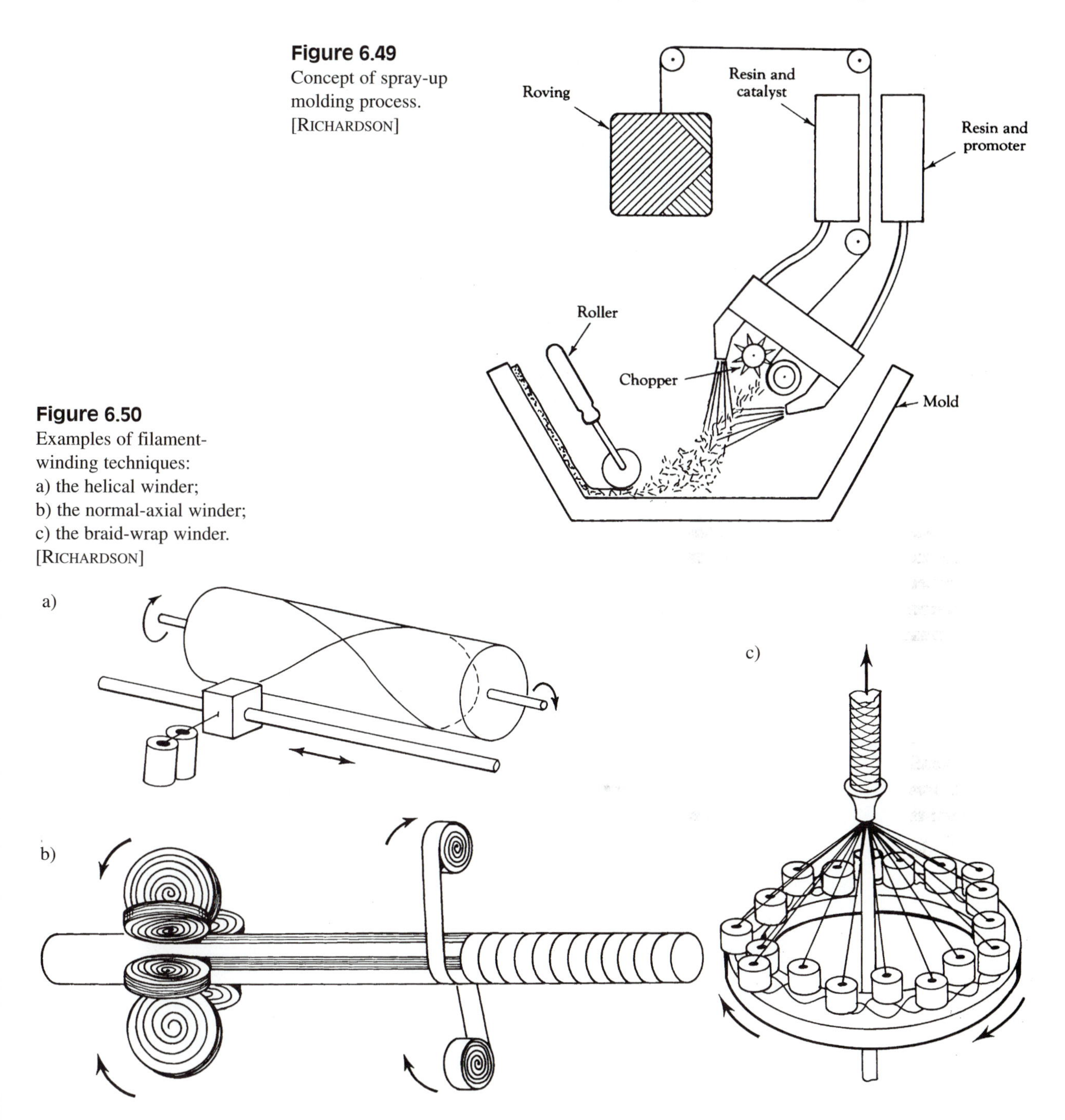

Figure 6.49
Concept of spray-up molding process. [RICHARDSON]

Figure 6.50
Examples of filament-winding techniques:
a) the helical winder;
b) the normal-axial winder;
c) the braid-wrap winder.
[RICHARDSON]

duced over a mandrel. Most mandrels are intended to be washed out, melted out, broken out, or collapsed after the shell has been finished. Constructions made of wood, cardboard, plaster, and low-melting-point plastics are used.

Large multicoordinate NC tape-laying machines are on the market. An example of one of these, a fiber-placement machine, is shown in Fig. 6.51. (For principles of numerical control, see Chapter 10.) The machine has seven NC axes: the rotation of the

Figure 6.51
An NC filament-placement machine. [CINCINNATI MACHINE]

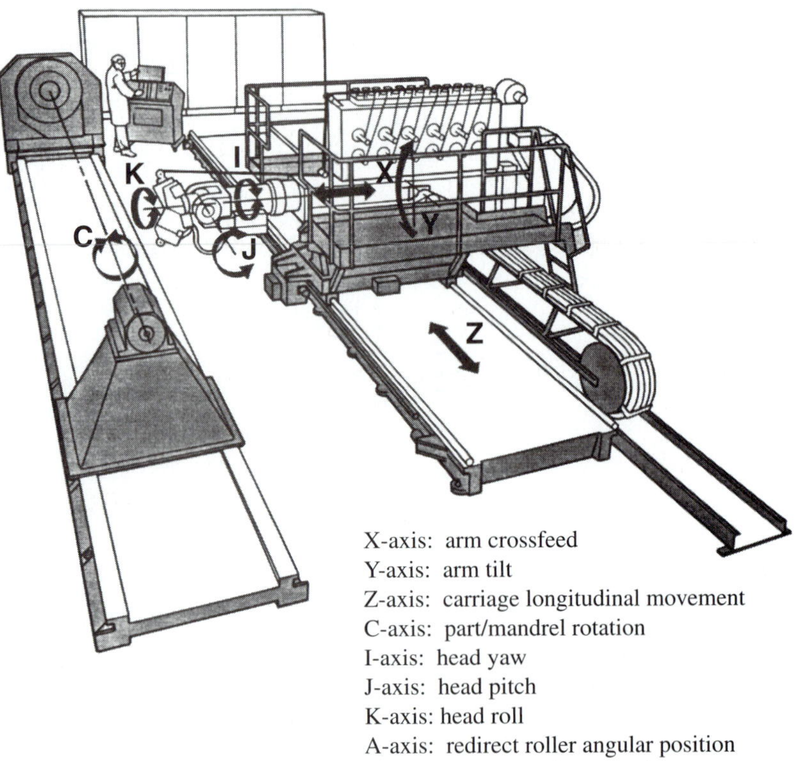

Figure 6.52
The kinematics of the filament-placement machine. [CINCINNATI MACHINE]

mandrel carrying the part being made; the longitudinal travel, in parallel with the axis of the mandrel, of the saddle-carrying the ram; the rotation, on the saddle, of the base of the ram around a horizontal axis parallel with the axis of the mandrel; extension of the ram towards the rotating part, and three rotations of the fiber-placement head and a rotation of the ram around its axis (see Fig. 6.52). Two additional controlled motions are used for restarting the tow. Although Fig. 6.51 shows an operator at the control console, the actual fiber-placement operation is fully automated under computer control.

This includes functions additional to the tow lay-up, such as adjusting tow tension, varying the bandwidth in-process, automatically starting, stopping, cutting, clamping, and restarting individual tows as needed to place window and door openings in the aircraft fuselage (shown further on in Fig. 6.54). The fiber-placement head functions are shown in Fig. 6.53. Up to 24 tows, each up to 3.18 mm wide, are collimated into a fiber band. The tows are delivered from an air-conditioned creel in which there are bidirectional tensioning devices to ensure that the carbon fiber/epoxy material is consistently tensioned. The fiber band is applied to the part by means of a compaction roller while it is heated by a stream of hot air to achieve the right plasticity. An example shown in Fig. 6.54 of a product made on this machine is the fuselage of the Raytheon Aircraft's Premier I executive jet airplane. The wall of the structure is made as a sandwich with inner and outer carbon fiber/epoxy layers over a middle layer of honeycomb material. In its production the inner carbon fiber prepregs are the first to be wound on the mandrel. Operators then hand place the honeycomb layer, and the machine then overlays the outer carbon fiber tows. A final layer of fabric, woven with metal filaments to provide lightning strike protection, is then applied. The fuselage is finally pressure- and heat-cured in a large autoclave.

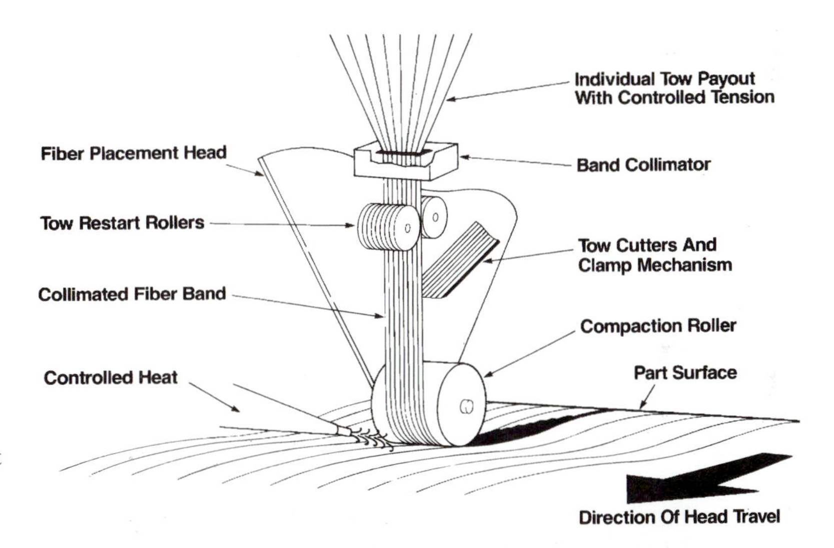

Figure 6.53
The functions of the filament placement head. [CINCINNATI MACHINE]

Figure 6.54
Fuselage of the Raytheon Aircraft Premier I airplane produced on the machine of Fig. 6.51. [CINCINNATI MACHINE]

The control system of the machine is rather advanced. For instance, the programmer does not have to determine the motions of the individual axes. Instead, the path of the center of the compaction roller is determined in world (cartesian) coordinates, and its orientation is also expressed in this system; the controller then automatically carries out the coordinate transformation and drives the individual axes accordingly. The CAD/CAM process of programming makes use of finite element (FE) software to determine optimum strength by varying the orientation of fibers. Nominal laydown rate is up to 15 m/min. The application force of the head ranges up to 500 N. The machine is used for a variety of composite parts, symmetrical, asymmetrical, rotational or not, such as engine ducts, blades, struts, propellers, fuselages, "C" channels, nozzle cones, and tapered casings. The advantages of this technology are (1) the use of FE computation to obtain high strength, which is made possible by the flexibility of the control of the motions, permitting optimum configuration of the fibers throughout the structure, and (2) the possibility of achieving a one-piece construction instead of an assembly of several simpler parts that would result from hand lay-up.

6.6.4 Pultrusion

Pultrusion is a continuous process that is the opposite of extrusion, in which the product was pushed through a die. In pultrusion, it is pulled through a die. Both processes are analogous to wire drawing and extrusion of metals as they were described and discussed in Chapters 4 and 5. In pultrusion various continuous forms of prepregs are pulled through a bath of resin, and a die (mold), and a curing oven; see Fig. 6.55. The process is used for the production of automotive springs, airfoils, skis, golf shafts, tennis rackets, bars, rods, and tubes.

Many other processes are used in the manufacture of parts from composites. However, this brief review is about as much as befits the overall concept of our book. A number of more detailed texts are referred to in the list of references at the end of this chapter.

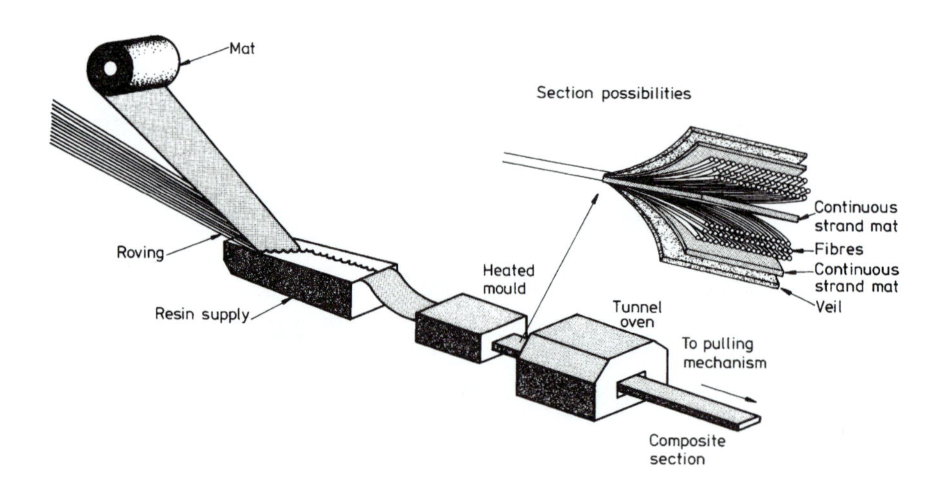

Figure 6.55
The pultrusion process.
[CRAWFORD]

REFERENCES

1 Agassant, J. F.; P. Avenas; J. Sergent; and P. J. Carreau. *Polymer Processing: Principles and Modeling.* Munich: Hanser Publishers, 1991. Distributed by Oxford University Press, New York.

2 Crawford, R. J. *Plastics Engineering.* 2nd ed. Pergamon Press, 1987.

3 Lubin, G., ed. *Handbook of Composites.* Van Nostrand Reinhold, 1982.

4 Kennedy, P. *Flow Analysis of Injection Molds.* Munich: Hanser Publishers, 1995. Distributed by Hanser/Gardner Publishers, Cincinnati.

5 Morton-Jones, D. H. *Polymer Processing.* London: Chapman and Hall, 1989.

6 Potsch, G.; and W. Michaeli. *Injection Molding: An Introduction.* Munich: Hanser Publishers, 1995. Distributed by Hanser/Gardner Publishers, Cincinnati.

7 Rauwendahl, C. *Polymer Extrusion.* 3rd ed. Munich: Hanser Publishers, 1994. Distributed by Hanser/Gardner Publishers, Cincinnati.

8 Rees, H. *Understanding Injection Molding Technology.* Munich: Hanser Publishers, 1994. Distributed by Hanser/Gardner Publishers, Cincinnati.

9 Rees, H. *Mold Engineering.* Munich: Hanser Publishers, 1995. Distributed by Hanser/Gardner Publishers, Cincinnati.

10 Richardson, T. *Composites: A Design Guide.* New York: Industrial Press, 1987.

11 Strong, A. B. *Fundamentals of Composites Manufacturing: Materials, Methods, and Applications.* Dearborn, MI: Society of Manufacturing Engineers, 1989.

12 Mitchell, P., ed. *Tool and Manufacturing Engineers Handbook.* Vol. 8. *Plastic Parts Manufacturing.* 4th ed., Dearborn, MI: Society of Manufacturing Engineers, 1996.

Illustration Sources

[AGASSANT] Reprinted with permission from J. F. Agassant, P. Avenas, J. Sergent, P. J. Carreau, *Polymer Processing: Principles and Modeling*, Hanser/Gardner Publications, 6915 Valley Ave., Cincinnati, OH 45244-3029, 1991.

[CRAWFORD] Reprinted from R. J. Crawford, *Plastics Engineering*, 2nd ed., Pergamon Press, 1987, with permission of Heineman Publishers, Oxford.

[MORTON-JONES] Reprinted from D. H. Morton-Jones, *Polymer Processing*, Chapman and Hall, London, 1989, with permission of Chapman and Hall, Andover, Hampshire, England.

[POTSCH] Reprinted with permission from G. Potsch and W. Michaeli, *Injection Molding: An Introduction*, Hanser/Gardner Publications, 6915 Valley Ave., Cincinnati, OH 45244-3029, 1995.

[REES] Reprinted with permission from H. Rees, *Understanding Injection Molding Technology*, Hanser/Gardner Publications, 6915 Valley Ave., Cincinnati, OH 45244-3029, 1994.

[RICHARDSON] Reprinted from J. Richardson, *Composites: A Design Guide*, Industrial Press, Inc., New York, 1987, with permission of Industrial Press, Inc.

[SME8] Reprinted with permission from: *Tool and Manufacturing Engineers Handbook*, Vol. 8, Plastic Parts Manufacturing, 4th ed., 1996.

[SCHEY] Reprinted from: J. A. Schey, *Introduction to Manufacturing Processes*, 2nd ed., McGraw-Hill Book Co. 1987 with permission of the McGraw-Hill Co.

Industrial Contributors

[CINCINNATI MACHINE] Cincinnati Machine, a Unova Company, 4701 Marburg Ave, Cincinnati, OH 45209-1025

[DAVID STANDARD CO] David Standard Co., Pawcatuck, CT

[HUSKY] Husky Injection Molding Systems Ltd., 500 Queen Street S., Bolton Ontario, Canada, L7E 5S5

[MCNEIL AKRON CO] McNeil Akron Co, Akron, OH

QUESTIONS

Q6.1 Which two properties of polymers in general are most decisive in choosing the methods for processing them into usable shapes?

Q6.2 What is the main conceptual difference in processing thermoplastics versus thermosets? How does it affect the processes of extrusion and of injection molding? What is the difference in the design of the injection-molding screw?

Q6.3 From Table 6.2 select two polymers with low and two with high viscosity (given for a recommended processing temperature) and the corresponding influential parameter from Table 6.3 to assess the effects determined in Question 6.1.

Q6.4 Which processes start with the polymer in the form of granules, and which start with sheet or film?

Q6.5 Which polymers from Table 6.2 are viscoelastic at room temperature? Which are transparent?

Q6.6 Describe the extrusion dies (make sketches) for wire coating and for sheet forming. What special process is used for making thin sheet (film)?

Q6.7 Characterize "die swell" in the extrusion of pipes. How are internal and external diameters of a desired size obtained?

Q6.8 Name the two opposing parts of an injection-molding machine and the types of drives used in them.

Q6.9 Compare injection molding with die casting, pointing out similarities and differences.

Q6.10 Give the main parameters of a middle-size injection-molding machine: injection pressure, clamping force, barrel-heating power, power for plastication, cooling-water flow capacity.

Q6.11 Explain the function of the check valve in an IM machine. For which material is a plain screwtip used instead?

Q6.12 Explain the difference in the function of a two-part and a three-part mold as regards ejection of the molding and of the sprue and runners. What is the purpose of gates?

Q6.13 Is it common to inject a single part or multiple parts? How many?

Q6.14 Under what circumstances would a particular part preferably be made by either transfer or injection molding?

Q6.15 Which part of the IM cycle is longer, injection or holding? About how many times longer?

Q6.16 Which devices are used in larger IM machines for unloading the molding and for changing the die sets?

Q6.17 Name examples of products made by extrusion, by blow molding, by film blowing, and by injection molding.

Q6.18 Name the various forms of prepregs for composite-part processing. Name and explain various processes for composites.

Q6.19 Which is the most common composite you would use to make yourself a small boat? Which process would you use? Which process would you use to make ten such boats per month?

Q6.20 Describe three different systems of filament winding.

Q6.21 Which composite material is used for aircraft skins and also for aircraft shells? What kind of machine is best suitable for this kind of production? What kind of control system is used?

PROBLEMS

P6.1 *Effect of channel dimensions on polymer flow.* Assume **(a)** a flat channel with height $h = 2$ mm and **(b)** a round channel with radius $R = 1$ mm and any given pressure drop at any given length and a particular polymer. Consider (*1*) Newtonian flow, and (*2*) non-Newtonian, Power-Law flow with $n = 0.4$. In doubling the channel height or radius, respectively, without changing pressure or channel length, how many times will flow rate Q increase in the individual cases (a1), (a2), (b1), and (b2).

P6.2 *Polymer flow characteristics.* How would the graph with the coordinates of Fig.6.9 look for the material used in Example 6.2?

P6.3 *Flow through a round channel.* In a way analogous to Figs 6.11, 6.12, and 6.13 write a Matlab computer program and produce plots of velocity distribution and of strain rate distribution for **(a)** a Newtonian fluid with viscosity 100 Pas at a strain rate of 3000/sec, and **(b)** a non-Newtonian fluid with viscosity characterized by parameters $K = 12{,}198$ and $n = 0.4$. See Eqs. (6.39), (6.40), and (6.47) for flow in a round channel with $R = 2$ mm, length $L = 50$ mm, and pressure drop $\Delta P = 30$ MPa. Plot u and $\dot{\gamma}$ versus r.

P6.4 *Flow through a round channel.* Calculate pressure drop ΔP (MPa) for a flow of $Q = 30$ cm^3/sec for both cases **(a)** and **(b)** of Problem 6.3.

P6.5 *Viscous flow power generation.* At the end of Example 6.2 it was stated that because most of the heat generated by the flow is conducted away through the channel walls as it is generated close to the walls, cooling is efficient, and the temperature increase of the polymer is less than calculated by considering the average velocity and average strain rate. Referring to the graphs in Figs. 6.12 and 6.13, and to the underlying calculations, determine the local power ΔW (W) generated in a small volume of 1 mm^3 located at **(a)** $y = 0.2$ mm and **(b)** $y = 0.95$ mm for both the Newtonian and non-Newtonian fluids. Furthermore, by assuming each of the two volumes is insulated from the surrounding fluid and using the same values of specific heat and density as in Example 6.2, calculate the local temperature increase ΔT for the two locations and the two types of fluids.

7 Cutting Technology

7.1 INTRODUCTION

The cutting operations, often referred to as "chip removing," represent the largest class of manufacturing activities in engineering production, especially in the production of all kinds of machines. Mostly, they involve metal cutting, but they are applied to machining of nonmetallic materials as well. Metal cutting is a traditional manufacturing process, and its development, together with the development of metal-cutting machine tools, marked the general development of mechanical industries. Cutting is a universal operation and is used to produce products of all sizes, from small parts like shafts for watches to very large parts like the frames of rolling mills, housings of turbines, and road-building machinery. It is capable of high machining rates using powers of the order of 100 KW and involving cutting forces of the order of 50,000 N. Generally, it produces parts with tolerances on the order of 25 μm. Finish machining achieves accuracies specified by errors as small as 1 μm on parts ranging in size from 0.1 to 1 m.

The significance of metal cutting may be illustrated by a simple estimate of its annual value in the United States. There are 1.87 million machine tools, see Ref [6], with an average labor cost and overhead of about $45/hour. Following the reasoning presented by Metcut Res. Ass. of Cincinnati we consider one operator per machine and 8 hours/day as an average; in some instances one operator looks after several machines and many machines work for more than one shift a day. Taking the average number of 250 work days per year, the total value of the machining operations is $1.87 \times 10^6 \times 45 \times 8 \times 250 =$ $168,000,000,000 annually ($168 billion).

Cutting operations are essential to almost any engineering production and there is hardly any shop in existence that does not contain a lathe, a drill, and a grinder. Metal-cutting machine tools in their overall variety represent the ultimate in sophistication and in the application of computer technology to production equipment. In Chapter 10, we will use examples of machine tools to explain the basics of numerical control and of flexible manufacturing systems, which are applicable also to other classes of production machinery.

This chapter presents a brief description of the cutting processes, tools, and nonautomated machine tools and introduces the elementary empirical formulas for cutting forces and powers. The analytical treatment is presented in Chapter 8. Cutting processes and machines are described in a number of texts, several of which are listed in the References at the end of this chapter.

7.2 SINGLE-POINT TOOL OPERATIONS

The single-point tool operations are turning, boring, planing, and shaping. The tools involved act with a single cutting edge. The photograph in Fig. 7.1 shows a chip being cut in a turning operation. The tool is a triangular carbide insert mechanically clamped in a toolholder. The tool moves parallel with the axis of the cylindrical workpiece, from the right to the left. Marks of the feed per revolution of the workpiece are seen both on the surface previously machined which is presently being cut (work surface, left of the tool) and on the just-generated surface (machined surface, right of the tool). The chip is curled and breaks away in short pieces. This is one of the most desirable chip forms because it is easily cleared away from the working space of the machine. Otherwise a long, ribbonlike chip might be obtained that would clog the working space. Thus, the main function is to cut the chip from the workpiece, but it is also important to curl and break it in a suitable form.

A typical single-point tool operation is cylindrical outer turning of a shaft, as shown in Fig. 7.2. In this operation the tool T_1 moves parallel with the axis of rotation of the workpiece and reduces its diameter from d_1 to d_2. The workpiece has the rotational speed n (rev/sec). In general practice, rotational spindle speeds are usually given in rpm, and we use this occasionally, too. The work surface is denoted *WS*, and the machined surface *MS*. The surface that is actually produced by the cutting edge of the tool may be called the transient surface *TS*, or the surface being cut, *CS*. The periph-

Figure 7.1
A chip being cut and curled in a turning operation. [SANDVIK]

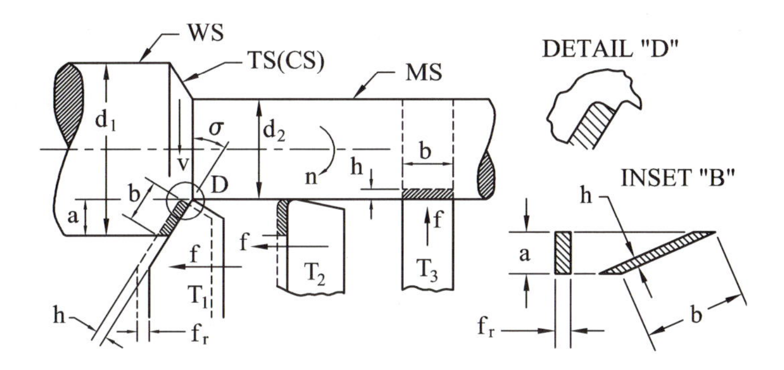

Figure 7.2
The geometry of cylindrical turning.

eral speed v (m/sec) of the workpiece is called the *cutting speed.* Obviously, along the cutting edge of the tool the cutting speed varies from $v = \pi d_1 n$ to $v = \pi d_2 n$, and it is practical to use the mean value $v_m = \pi n(d_1 + d_2)/2$. The tool moves with such a feed rate that it travels the feed f_r per one revolution of the workpiece. The *depth of cut* is $a = (d_1 - d_2)/2$. The shaded area is the chip area $A = af_r$, or also $A = bh$, where b is the *chip width,* and h is the *chip thickness.* The cutting edge of the tool is inclined with respect to the direction normal to the feed motion by the *side-cutting-edge angle* (SCEA) σ. Thus, $a = b \cos \sigma$, and $f_r = h/\cos\sigma$.

A side-cutting tool like T_2 could be used for this operation. Indeed, in most cases this type of tool, with $\sigma = 0$ is used with the advantage that it works against square shoulders between d_1 and d_2. In this case $a = b$, and $f_r = h$. On the other hand, with a larger value of σ, the cutting load for a given chip area A is distributed over a longer cutting edge $b > a$, and with a lesser load per unit edge length corresponding to a lesser chip thickness $h < f_r$. This is shown in the inset "*B*" of Fig. 7.2.

A *plunge-cut* operation with tool T_3 is different because the feed direction is radial instead of longitudinal. There is no sense in talking about a depth of cut a. Instead, the width of cut is equal to chip width b, and the feed per revolution $f_r = h$.

7.2.1 Metal Removal Rate: Cutting Force

The *metal removal rate* (*MRR*) Q is obtained as the product of chip area and cutting speed:

$$Q = Av = bhv = af_r v \tag{7.1}$$

For simplicity: a, b, h, and f_r will usually be given in (mm) and v in (m/min). Then

$$Q(cm^3/\text{min}) = a\,(\text{mm})\,f_r\,(\text{mm})\;v\,(\text{m/min}) \tag{7.2}$$

Using the units given in Eq. (7.2), numbers of reasonably small decimal orders are obtained for all the parameters involved.

EXAMPLE 7.1 Metal Removal Rate in Turning ▼

We are given the following data: depth of cut a = 6mm, feed per revolution f_r = 0.25 mm, cutting speed v = 150 m/min, and Q = 225 cm^3/min. The cutting force components associated with the production of the chip are discussed in some detail in Chapter 8. At this stage it is sufficient to say that in the first approximation the main force component is tangential to the cut surface at the point of cutting and as such falls into the direction of the cutting speed v; therefore, it is the power-producing component. This component F_t is proportional to chip area A:

$$F_t = K_s A \tag{7.3}$$

In this relationship K_s is the "specific force," and it is primarily determined by the material being cut. For example, for cutting steels it is on the order of $K_s = 2000$ N/mm^2.

Taking Equation 7.3 and multiplying both sides by v,

$$F_t v = K_s A v,$$

we obtain, with respect to Eq (7.2),

$$P(\text{W}) = K_s(\text{N/mm}^2) \times \frac{Q(\text{cm}^3/\text{min})}{60} \tag{7.4}$$

where the factor 60 applies if, as is customary, the cutting speed is input in Eq. (7.2) in m/min. We have now obtained a very practical relationship between *MRR* and the power needed to drive the operation. Again, typically, for the units used here, the specific power for steel is about 2000/60 = 33 W/(cm^3/min). The values of K_s are listed in Table 7.1 for selected workpiece materials. ▲

Tool Geometry

The geometry of the single-point cutting tool is characterized by, apart from the SCEA, a number of other angles and by the nose radius. Some surfaces may have multiple facets, and in combination with the chip-curling grooves, the geometry may be quite complex. The single-point tools (turning, boring, planing) usually consist of a cutting insert mechanically clamped to a tool holder. The insert is made of the cutting material (e.g., commonly of sintered carbide), and the holder is made of steel. The basic geometry, which generally corresponds to the American Standard ANSI B94.5-1966, is illustrated in Fig. 7.3. There are also other systems of defining the tool geometry that differ from this one, but we will not deal with them.

TABLE 7.1 Specific Force and Power for Selected Workpiece Materials

Material	*UTS*	K_s	Material	*UTS*	K_s
Gray cast iron *HBN* 200		1500	Ni-based Inconel X	1450	3500
Carbon steel 1020 *N*	400	2100	Ni-based Udimet 500	1500	3550
Carbon steel 1035 *N*	500	2300	Co-based L605	1250	3700
Carbon steel 1045 *N*	650	2600			
Stainless steel 302	700	2700	Ti (Al6,V4)	1350	2000
Alloy steel 4140 *H*	900	2800			
Alloy steel 5140 *H*	950	2800	Al 7075-T6	530	850

Note: K_s is given in N/mm^2, or in W · sec/cm^3, *UTS* is in N/mm^2, *N* means normalized, *H* means heat-treated.

Figure 7.3
Geometry of a turning tool.

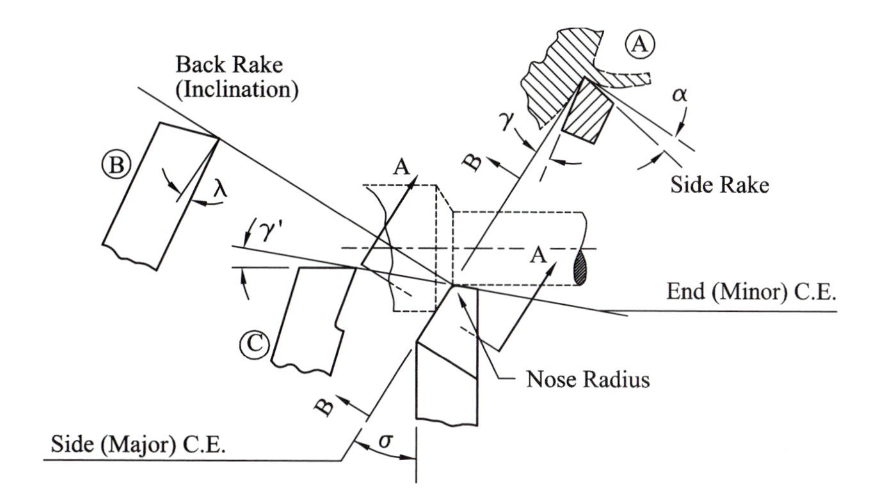

As is shown in the detail of Fig. 7.2, the tool cuts with a large part of the side cutting edge (main cutting edge, or simply cutting edge), with the nose radius, and with a small part of the end cutting edge (secondary cutting edge), depending on the magnitude of f_r.

The geometry as shown in Fig. 7.3 is oriented with respect to the base of the tool (assumed horizontal) and to the side (major) cutting edge. The most important angle is the *(side) rake angle* α; it is measured in a vertical plane A perpendicular to the base and perpendicular to the side cutting edge. It can practically be considered as the active rake angle. As shown in the diagram, it is taken as *positive.* If the top of the insert (the rake face) is sloping down towards the edge, the rake angle is *negative.* In the same plane we see the *(side) relief angle* γ. Both these angles define cutting by the main part of the cutting edge. The *back rake angle* λ, which is also called the *inclination angle*, is measured in a plane B perpendicular to the base (vertical) and passing through the main cutting edge. It can also be positive or negative (sloping down to the nose). In the former case an impact in interrupted cutting is taken by the tip of the tool; the chip is curled away from the machined surface. In the latter case these two aspects are opposite. The view C is along the end (minor) cutting edge, and the angle γ' is the end relief angle.

7.2.2 The Tools

A variety of tool holders are used in external turning operations, some of which are shown in Fig. 7.4. The "turning" operation indicates generating an outer cylindrical surface, and various tool holders may be used. Style *A* with a round insert is suitable for heavy roughing because it does not have the concentration of load on a tip as the triangular and square inserts do, and it can be rotated several times before the whole cutting edge is used. Style *A* with a triangular insert has the advantage that it can turn against square shoulders. All the other styles *B, C, D,* and *E* offer a range of the side-cutting-edge angle (SCEA) from 15° to 45°; correspondingly, the cutting load is spread over longer cutting edges. For the chamfering operation, a couple of style-*D* tools are shown to generate chamfers simultaneously on one inner and one outer diameter, in one longitudinal plunge cut. The operation called "trace" involves turning under a

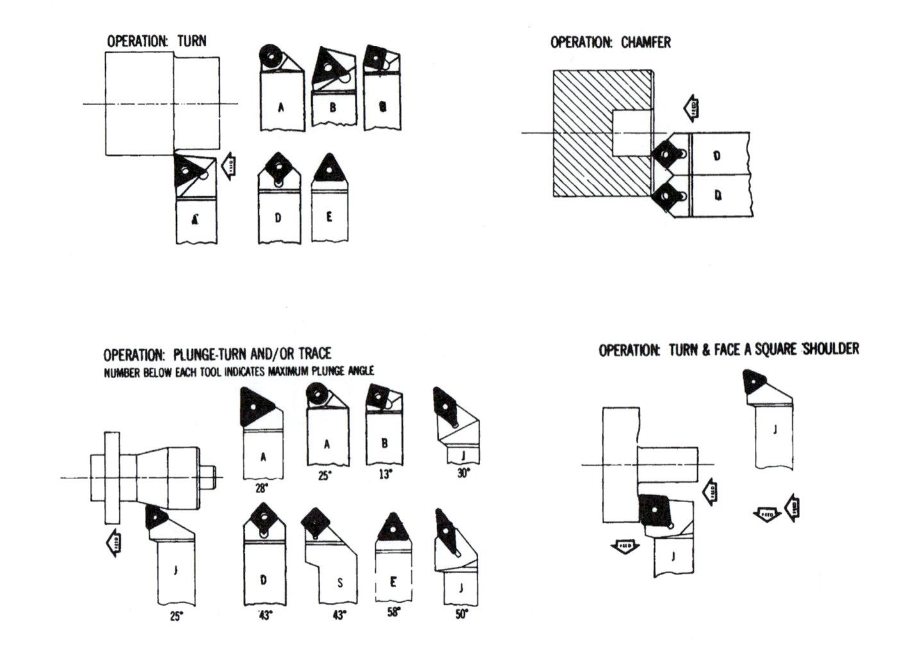

Figure 7.4
Some external turning operations and corresponding tools.

combination of both the X and Z motions, producing either a conical or shaped (form) surface. With a slope inward as shown, the slope angle is limited by the type of the holder and insert. Apart from the styles used for the turning operation, additional styles are used, like I and especially the J styles, which are most commonly used in copy-turning lathes and in numerical-control (NC) lathes for turning workpieces with complex contours. Finally, the facing operation is included in the figure with J-style holders such that the edge that produces the face is slightly inclined with respect to the face, so that the tip of the tool generates the finished surface. Apart from outer turning, internal operations are performed in the variously shaped holes or cavities of rotational workpieces and these are carried out, on one hand, by various drilling tools to be mentioned in Section 7.2 and, on the other hand, by "internal turning" tools, which are usually called boring tools. Some of these operations are shown in Fig. 7.5. Boring operations in which the workpiece rotates and the tool feeds in should be distinguished from other boring operations in which the tool rotates; these are performed on boring machines to be mentioned later.

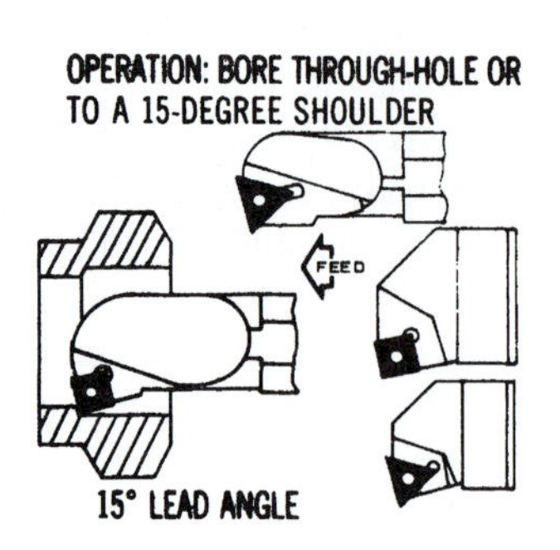

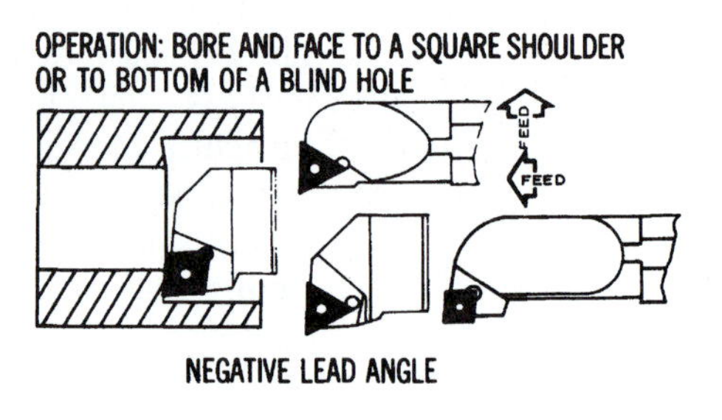

Figure 7.5
Some internal turning (boring) operations and corresponding tools.

The three operations shown are distinguished by the value of the "lead" angle, which is another name for the SCEA. In case *(a)* it is possible to use a square insert for through hole boring. In case *(b)* the tool moves first longitudinally in the *Z*-axis to bore the cylindrical surface of the hole and then radially inward to produce the bottom face of the bore. In case *(c)* the square shoulder is produced at the end of the longitudinal motion by virtue of SCEA = 0.

An assembly of turning and boring tools with inserts is shown in Fig. 7.6. They are intended for various operations: external radius turning, grooving, parting, internal turning of reliefs, internal threading, ring grooving, and the turning operations indicated in Figs. 7.4 and 7.5. They also include a boring bar with interchangeable heads, a solid boring bar, and a boring bar with an adjustable head. The inserts exist also in a great variety of shapes and sizes, as illustrated in Fig. 7.7. This variety is justified; each insert finds a particular application for which it is best suitable. It is beyond our scope to discuss this in detail; refer to the catalogues and handbooks published by the manufacturers of tools as needed.

The details of two-insert clamping systems are shown in the following two illustrations. Figure 7.8 shows the "classical" system, which uses a plain type of the insert on top of which there is a chip-breaker platelet. Both these parts are pressed by a clamp onto a support shim and against the walls of a pocket in the toolholder. Figure 7.9 shows an alternative clamping mechanism that employs a pivoted lever in an insert with a central hole. The lever presses the insert into the pocket. No separate chip breaker is used because the insert incorporates a chip-breaking groove.

So far we have shown tools with cutting inserts clamped in holders. Their main advantage is that when a cutting edge is worn down, the insert can be taken out and indexed for a fresh cutting edge. After all its usable cutting edges have been worn, the insert is discarded. The insert-changing time is short, and positional repeatability may

Figure 7.6
Turning and boring tools with inserts. [SANDVIK]

Fig 7.7
Inserts for turning and boring tools. [SANDVIK]

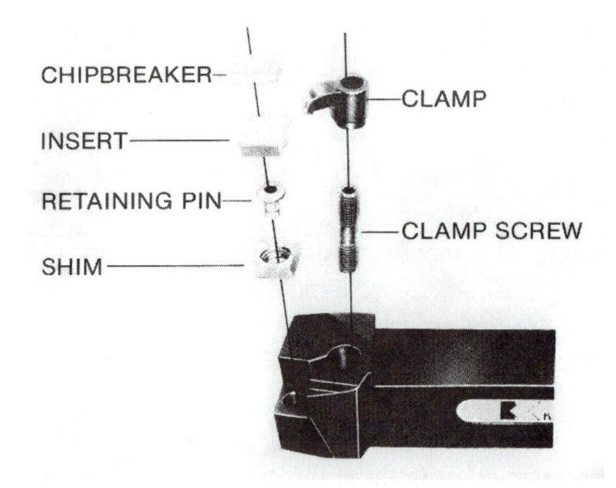

Figure 7.8
Clamping of the insert with an external clamp and separate chip-breaker platelet. [KENNAMETAL]

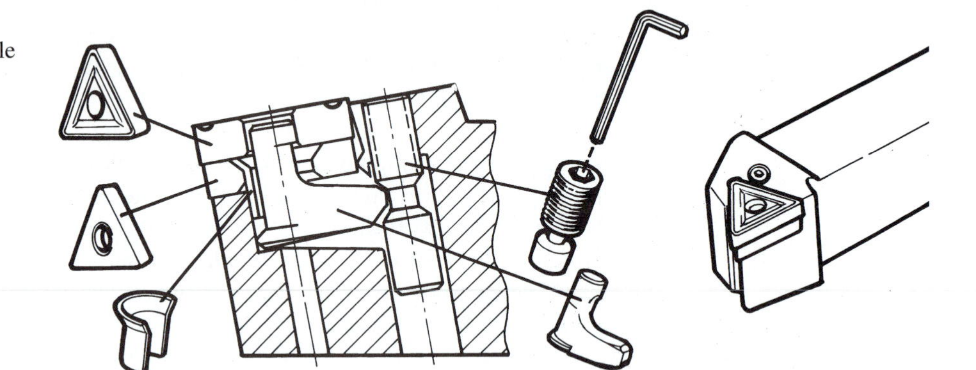

Figure 7.9
Clamping in the central hole of the insert. [SANDVIK]

be rather high. In some tools the inserts are brazed to the tool holders, or the tool is made of a single piece of high-speed steel. Such tools must be reground when worn. Because of the simplicity of the single-point tools and of their inserts, the indexable insert types are most common.

All the inserts may be categorized as either "negative" or "positive" (see Fig. 7.10). The faces of a negative insert are all square to each other. It must be located in a holder with negative inclination, which, for a plain insert (Fig. 7.10a), results in a negative rake angle and provides for a positive relief (clearance) angle. The side faces of a positive insert (Fig. 7.10b) enclose with the top face an angle smaller than 90°. The

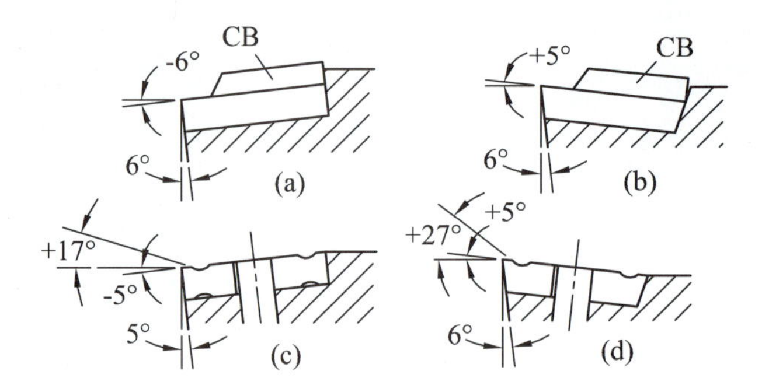

Figure 7.10
Positive and negative inserts, with chip breakers and with chip-forming grooves.

insert can be positioned under a positive rake angle and still permit positive clearance on the flank. The actual values of angles given in the diagrams here are tentative; other values are possible. The negative insert can be turned over and thus has twice as many usable cutting edges as the positive one. For example there are eight edges available on a square negative insert and only four on a square positive insert. The plain inserts in *(a)* and *(b)* are shown with chip breakers CB on their tops. The inserts with holes and form grooves as shown in *(c)* and *(d)* are also made in the basic negative and positive shapes; correspondingly, the type shown in *(c)* can be turned over, while that shown in *(d)* cannot. However, the effective rake angles depend mainly on the shape of the groove, while there are still negative or positive narrow "lands" at the cutting edges, respectively. This brief discussion does not, by far, explain the variety of the geometries of inserts, which is continuously being developed so as to result in the optimum formation of chips in cutting many different workpiece materials under different cutting conditions. It is a special field in itself, and a tooling engineer can obtain the detailed information from the tool suppliers.

In order to summarize our rather introductory discussion of turning and boring tools, let us consider in Fig. 7.11 an example of a workpiece as presented in [5], enumerate the various operations carried out on it from its two ends, and list the corresponding tool types:

1. Drill a short hole.
2. Turn face.
3. Recess an outer surface of this end of workpiece.
4. Plunge-turn a deep groove.
5. Turn a form groove with 45° sides. The tool has SCEA 35° and is moved along a complex path to produce the desired shape of the groove.
6. Cut a relief in the inner corner of the recess.

Between 6 and 7 the workpiece is turned over for machining at its second end.

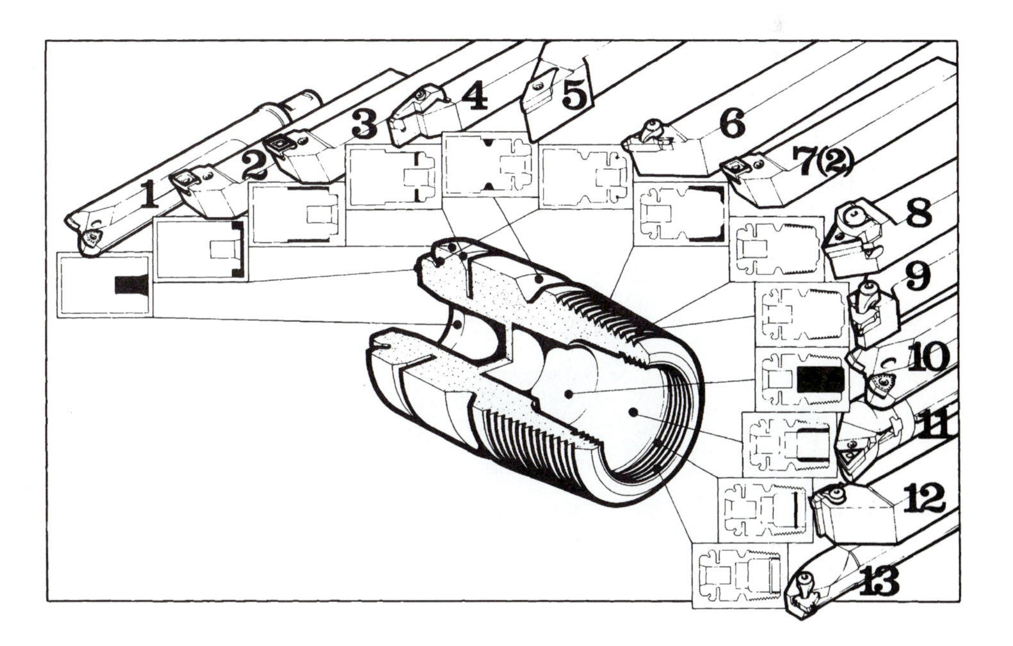

Figure 7.11
Example of a complex turned workpiece and the thirteen different tools and operations needed. [SANDVIK]

7. Turn a taper. The same tool is used as in 2.
8. Rough-turn a tapered thread.
9. Finish the thread.
10. Drill a short hole.
11. Bore the hole to a larger and more accurate diameter.
12. Cut an internal groove.
13. Cut a short thread in the end of the hole.

7.2.3 The Machine Tools

The machine tools for the operations of turning, boring, planing, and shaping exist in a number of basic, manually operated types and in a wide range of sizes with masses from 0.50 to 100 t. From the basic types a great variety of types have been further developed in the application of the various principles and methods of automation. These are discussed in Chapter 10.

In this chapter we describe some of the basic types of manually operated machine tools. The most widely used and known of all machine-tool types is the center lathe. A photograph of a typical model is presented in Fig. 7.12. Under the bed and between the legs are the trays for the collection of the cutting fluid and of the chips. On top of the bed at left is the headstock and at the right the tailstock. The headstock carries the spindle with a chuck at its front end. A long workpiece (not shown) is clamped in the chuck and supported at its right end by the center. A short workpiece would be clamped in the chuck only, in "overhang." The spindle is driven from a motor inside the base via gear transmissions in the headstock. The tool is clamped in the tool post, which sits on top of the upper carriage. This carriage is mounted on the front end of the cross slide, which itself moves transversely on the saddle, which moves longitudinally on the guideways of the bed. The upper carriage is shown in a position in which it would move longitudinally but it can be swivelled so as to move in any direction between the transversal and longitudinal ones. The motions of the cross slide and of the saddle are driven by the feed bar or by the leadscrew. The various levers on the headstock are used to shift the gears in the spindle drive and set the spindle speeds. The levers on the feedbox are used to select the feed rate (or the thread pitch). The handwheels and levers on

Figure 7.12
A center lathe. [MAZAK]

the apron are used to either manually drive or select the mechanical drive for the saddle and the carriage. The individual subgroups of the lathe are indicated again in the diagrammatic drawing of Fig. 7.13 in which, however, the upper carriage is left out. The two basic feed motions are denoted X for the transversal (cross) and Z for the longitudinal. Their combined action moves the tool in a horizontal plane, and its path determines the contour of the workpiece. Due to the rotation of the workpiece all the cross sections of the workpiece perpendicular to its axis are circular. The drawing also shows a stop and stop drum, which may be used to automatically end the Z motion.

By the basic kinematics of the lathe, the rotary "main" motion of the spindle is driven from the motor via belts and the main gearbox. The gearbox contains several shafts that shift gears. By engaging the various combinations of gears, various spindle speed steps are obtained. The number of these steps on a universal lathe may be as many as 22. The different spindle speeds are needed to obtain various cutting speeds for the various workpiece materials, and for a range of diameters. The feed motions are derived from the spindle via the feed gearbox which, again, contains shafts with gears to combine and engage to provide different values of feed per spindle revolution. From the feed box the drive normally goes via the feed bar into the apron. In the apron it passes through an overload clutch, which disengages when the saddle runs against a stop, and further through clutches choosing longitudinal (Z) or transversal (X) motions. The Z motion is finally driven by a pinion mounted in the apron (saddle) that engages with a long rack mounted on the bed. The X motion is finally driven by a leadscrew and nut. The Z motion can alternatively be also driven by a leadscrew and nut, but this is only done when cutting threads.

Bear in mind that many actions are performed manually by the operator who uses this kind of a machine tool. These actions are discussed systematically in Chapter 10 with regard to the subject of automation. However, it is useful to make this point now in conjunction with the description of this basic machine tool. Fig. 7.11 has shown that the total work on a workpiece may involve a variety of operations requiring various tools, various spindle speeds and feed rates, and frequent changes in the direcuons of the feed motion, as well as frequent adjustment of the end points of the tool travel. The following is a summary list of the manual actions:

1. Load the workpiece into the machine; unload it.
2. Clamp the workpiece in the chuck and center; unclamp.
3. Load, set, clamp, and change tools.
4. Set and change speeds and feeds.

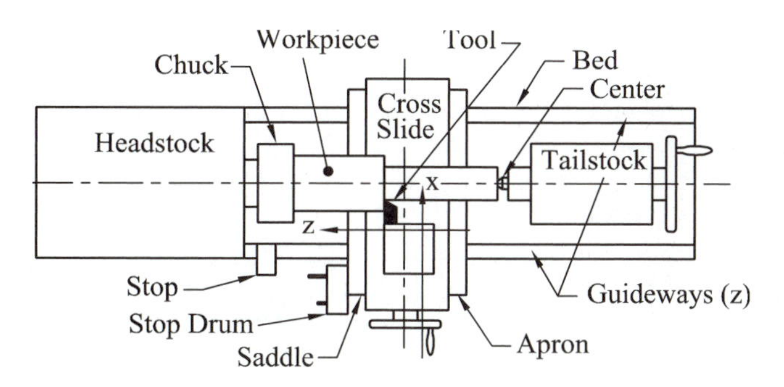

Figure 7.13
Main components of a center lathe: workpiece is clamped in chuck and tailstock center; spindle rotation provides main cutting motion and power; tool is fed in X and Z motions.

5. Control the path of the tool relative to the workpiece so as to obtain the desired form and dimensions of the workpiece.
6. Carry out measurements necessary for 5.
7. Perform additional actions like clearing chips away, switch coolant on and off, and so on.

Automating machine tools means automating these actions. The two groups of actions that occupy most of an operator's time are steps 3, 5, and 6. The illustrations in Fig. 7.11 show clearly that with so many tools used, changing and setting them may be a lengthy process. For the purpose of mechanizing this process, there has been in existence for over one hundred years a separate class of lathes, the turret lathes. One particular model of a ram-type turret lathe is shown in Fig. 7.14. This one has a hexagonal turret head with a vertical axis. Other types may have turret heads with a horizontal axis and a different number of turret head positions. In this example up to six different tools are preset in the head, and they are prepared for action by moving the turret head slide back, which simultaneously causes it to index to the next position. Each of the six tools may even be a group of several tools.

We conclude our discussion of lathes by mentioning that center lathes exist in sizes characterized by the "swing over bed" D_{max} and maximum length between centers L_{max}. The available ranges are from D_{max} = 0.1 to 4 m, L_{max} = 0.5 to 20 m. An example of a particular kind of heavy lathe is the roll lathe for machining of rolls for rolling mills. Lathes for short workpieces of larger diameters exist either as chucking lathes or disc turning lathes with T-shaped beds, but mainly as vertical lathes, which are commonly called "vertical boring mills." A medium-size (swing 1.6 m) lathe of this type is shown in Fig. 7.15. The vertical spindle is mounted in the base of the machine

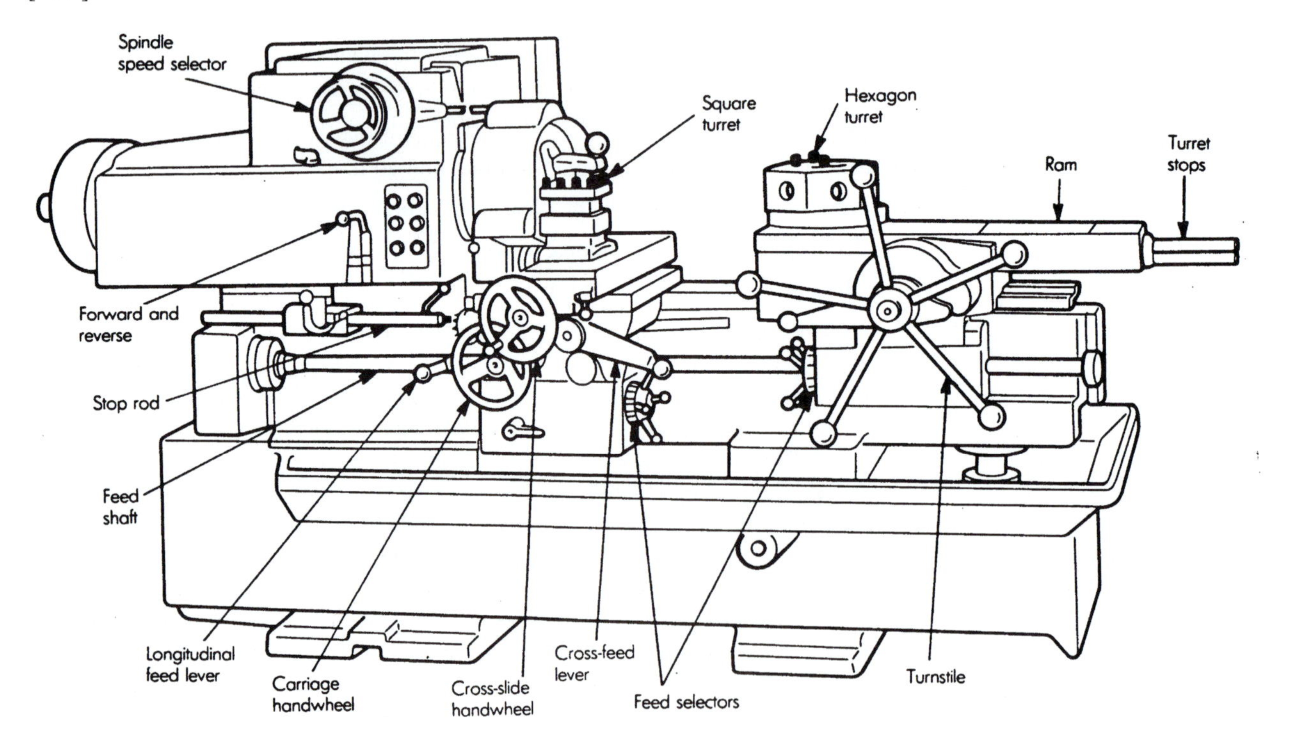

Figure 7.14
Ram-type turret lathe. Tools, mostly center-cutting (drills, reamers, taps), will be clamped in the six sides of the main turret head and brought into action sequentially, by indexing the head. Other tools for outer turning will be clamped in the square tool post on top of the compound slide. [SME1]

Figure 7.15
A medium-size vertical turret lathe (boring mill). The workpieces, mostly barrel-type, are clamped on the rotary table mounted in the base. The portal carries a traverse with two saddles carrying a vertical ram each with tools at their lower ends, to perform internal and external turning by feeding down vertically. One of the rams is equipped with a turret head. [OMNITRADE]

Figure 7.16
A large boring mill. In addition to two vertical rams, it is equipped with a horizontal slide for horizontal motions and turning on the outside of the workpieces. [OMNITRADE]

and it carries a round table that replaces the chuck or faceplate of a center lathe. The horizontal plane of the table makes it much easier to load and clamp the heavy workpiece. The model shown has two vertical rams in separate saddles moving horizontally (P and U coordinates) on the cross-rail. The cross-rail is vertically adjustable on the vertical guideways on the column. Both rams can carry out, independently, vertical feed motions. The ram on the right side carries a turret head and is intended mainly for internal turning (boring). The ram on the left side can be tilted under an angle for conical turning. An additional slide on the right side moves radially to the workpiece and is used for plunge cutting and grooving. In other designs, the vertical turret lathes may have just one ram and a fixed rail. In Fig. 7.16 a large vertical boring mill is shown with a portal-type frame, a vertically adjustable cross-rail, and two rams. These may be used in parallel to divide a large cut onto two tools, one at the end of each ram. The maximum diameter to be turned on this machine is 5 m. The base of this machine with the drive of the table and with its mounting is under the floor level. Machines of this and larger sizes are used in the manufacture of turbines and electrical generators.

Other types of machine tools for single-point tool operations are boring, planing, and shaping machines. Although internal turning is also called boring, there is a separate class of boring machines that have the boring bar clamped in a boring spindle that rotates while the workpiece is stationary. In other types, both the workpiece and the tool may rotate. Boring machines are mostly specialized either as part of a complex sequence of operations in mass production, where they may exist as boring stations in a transfer line, or as, for example, deep boring machines for final machining of hydraulic cylinders or gun barrels. In general, however, the boring function is one of several on a more universal drilling, boring, and milling machine to be described later on.

7.3 DRILLING AND ALLIED OPERATIONS

Drilling and related operations include drilling with twist drills, spade drills, gun drills, and drills that have carbide inserts, and with tools for counterboring, reaming, and tapping.

The twist drill is the most common cutting tool. Hundreds of twist drills are used for drilling holes with diameters from 0.25 to 50 mm. They are mostly made of high-speed steel in one piece. Some of the common tools directly associated with the twist drill are illustrated in Fig. 7.17. The twist drill has two cutting edges and two flutes to provide space for the clearing away of chips. Drilling does not produce an accurate and smooth hole. Counterboring enlarges the diameter of the hole at its beginning and countersinking produces a tapered enlargement of the starting part of the hole. Reaming is used to enlarge the hole by a very small amount for obtaining close tolerances and good surface finish. The basic geometry of the twist drill is given in Fig. 7.18. There are two different parts of the cutting edge: the chisel edge and the lips. As seen in Sections A-A, the rake angle α on the lips decreases from the outer diameter toward the center, while the clearance angle γ increases. The clearance angle shown is measured with respect to a conical surface passing through the lip. The effective clearance angle should be measured with respect to the helical surface produced in the workpiece as the drill rotates and advances axially. The helix angle of this surface increases with the feed per revolution.

The back face is produced by grinding while rotating and axially advancing the drill. Obviously the advance per revolution during grinding must be greater than the feed per revolution when using the drill in order to avoid rubbing the back face against the generated helical bottom of the hole, that is, in order to secure positive clearance angle γ all along the lip.

As shown in Section B-B, the chisel edge is associated with a very large negative rake angle α_3. No actual chip forming occurs at the chisel edge. The action there is more like indentation. Correspondingly the thrust force generated on the chisel edge is rather large. There exist various forms of grinding the center of the tip so as to improve the action at the chisel edge and decrease the thrust force.

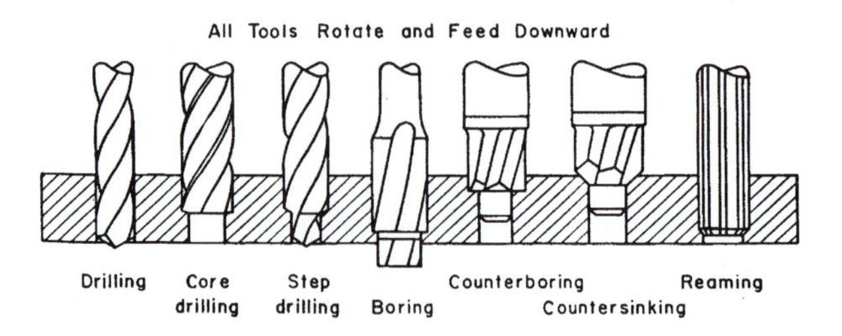

Figure 7.17
Tools for drilling and allied operations.

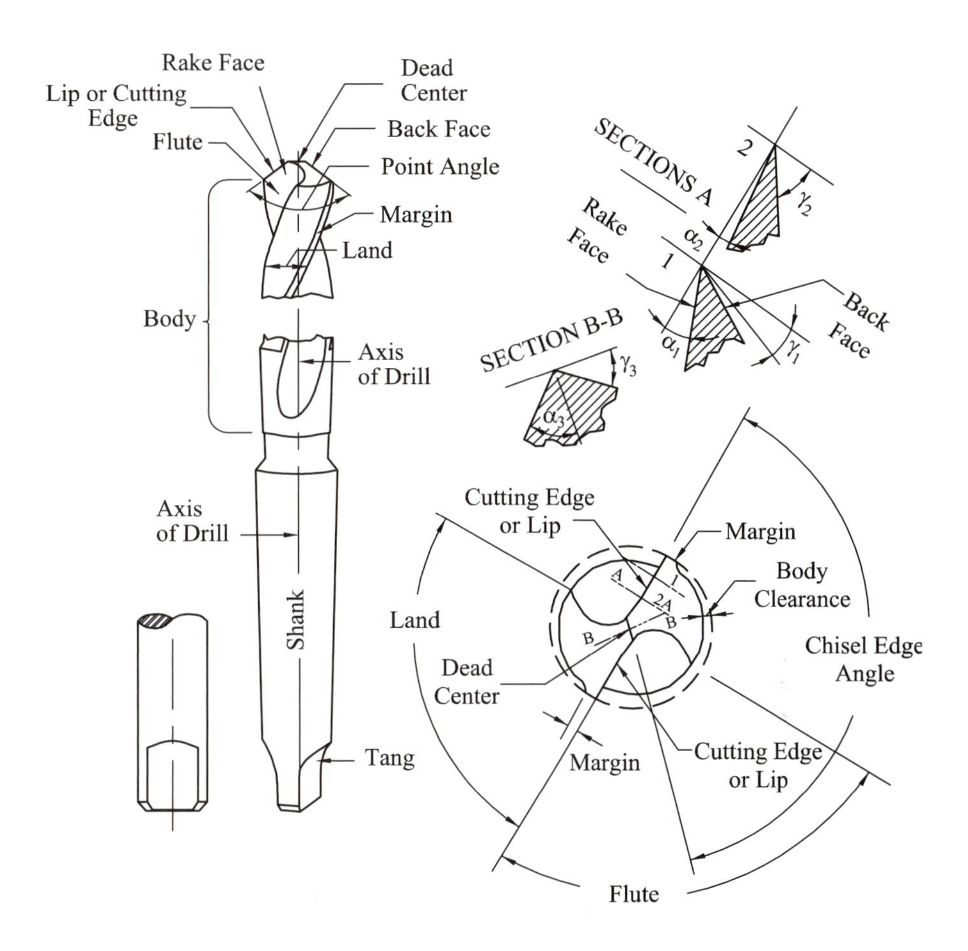

Figure 7.18
The geometry of a twist drill.

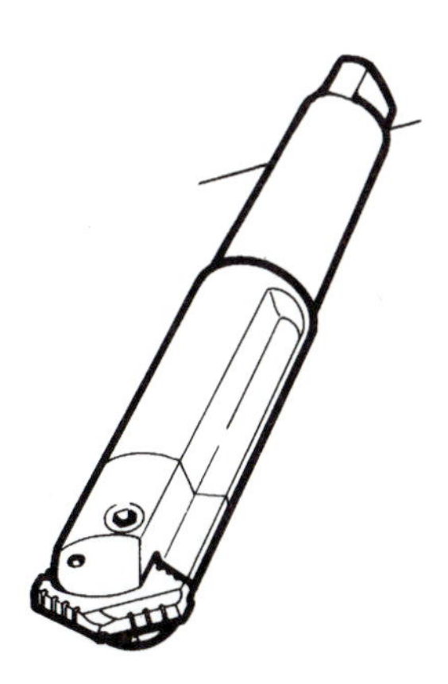

Figure 7.19
The spade drill. [SANDVIK]

Drilling of larger holes from the solid may be done by using *spade drills* (see Fig. 7.19). They consist of a holder and an interchangeable blade. The blade is commonly made of high-speed steel and can be reground when worn. The usual range of diameters drilled with spade drills is 25 mm to 150 mm. The geometry of the cutting edges is similar to that of a twist drill, with the chisel edge in the center. The lip cutting edges have serrations that break the chip for easier removal. The rake face is ground so as to obtain good curl on the chips.

Tools with indexable, mechanically clamped, or brazed carbide inserts for hole-making operations are basically of three kinds (see Fig. 7.20): *solid drilling, trepanning*, and *counterboring*. The distinction between them is obvious. The trepanning tool does much less work than solid drilling, and it leaves a core. For large dimensions, the lower power and thrust force are of considerable advantage, but the core is often difficult to handle when the tool penetrates through the material.

Carbide insert drills for short and not-too-large holes are designed with inserts so that one of them cuts in the center and the other on the periphery; the resulting forces are balanced so as not to produce any significant radial component. For larger and longer holes, the tools contain multiple inserts and use two additional features: guiding pads to support the drill and guide it for producing a straight hole, and channels for the supply and outflow of cutting fluid to flush out chips. Details of the geometry of a drilling head with brazed inserts are shown in Fig. 7.21. The central cutting edge replaces the chisel edge of a twist drill. The three cutting inserts have geometries that balance out the radial

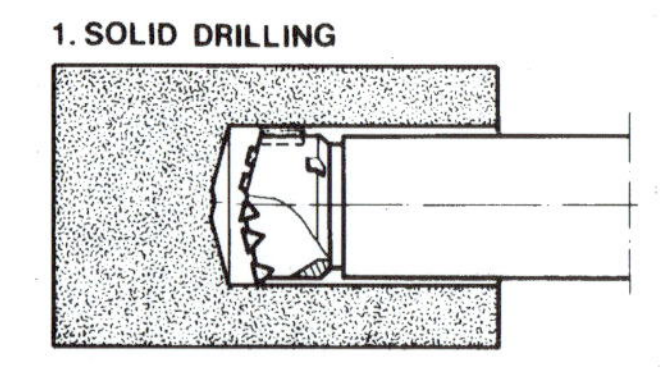

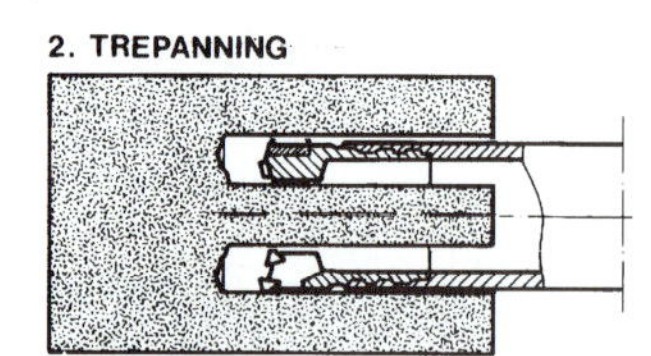

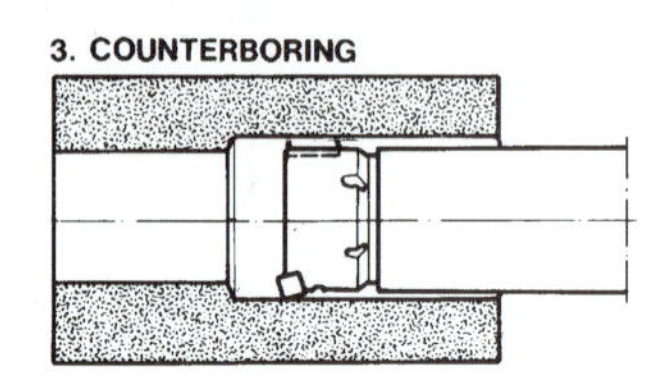

Figure 7.20
Operations with carbide-tipped drills. [SANDVIK]

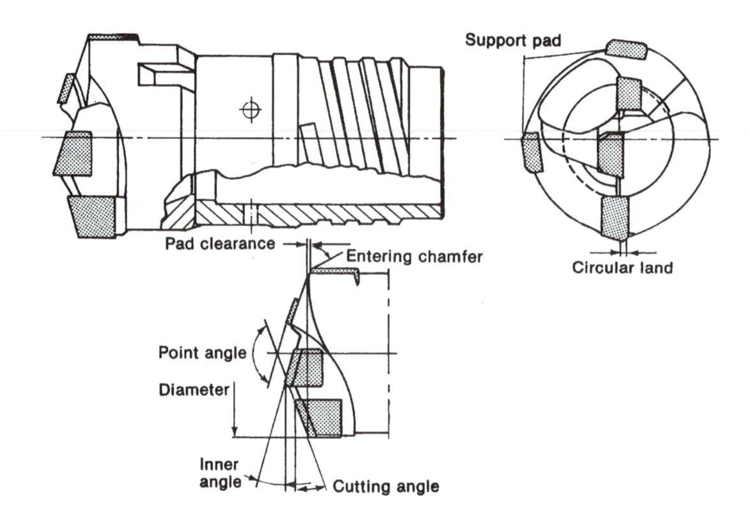

Figure 7.21
Geometry of a drilling head. [SANDVIK]

force components. This cannot be achieved completely, but the two support pads and the circular land on the outer cutting insert guide the tool safely. The hollow body of the head permits the flow of the coolant that flushes the chips. The drill head is attached to a holder consisting of an outer tube and an inner tube. The coolant is supplied through the outer tube and flushes the chips away through the inner tube.

Reamers are used to finish a hole for size and surface finish. The size can correspond to a variety of "fits" of holes within a close tolerance range. The reamer removes usually only between 0.1 and 0.75 mm of the diameter and produces a tolerance of about ±0.001 mm. Details of the geometry of one reamer type are shown in Fig. 7.22. Reamers are generally cylindrical or conical; they have two or more flutes, either parallel to the axis or in a right- or left-hand helix; and they are used as hand or machine tools. They are either of a fixed diameter ground to the exact dimension or they have adjustable blades.

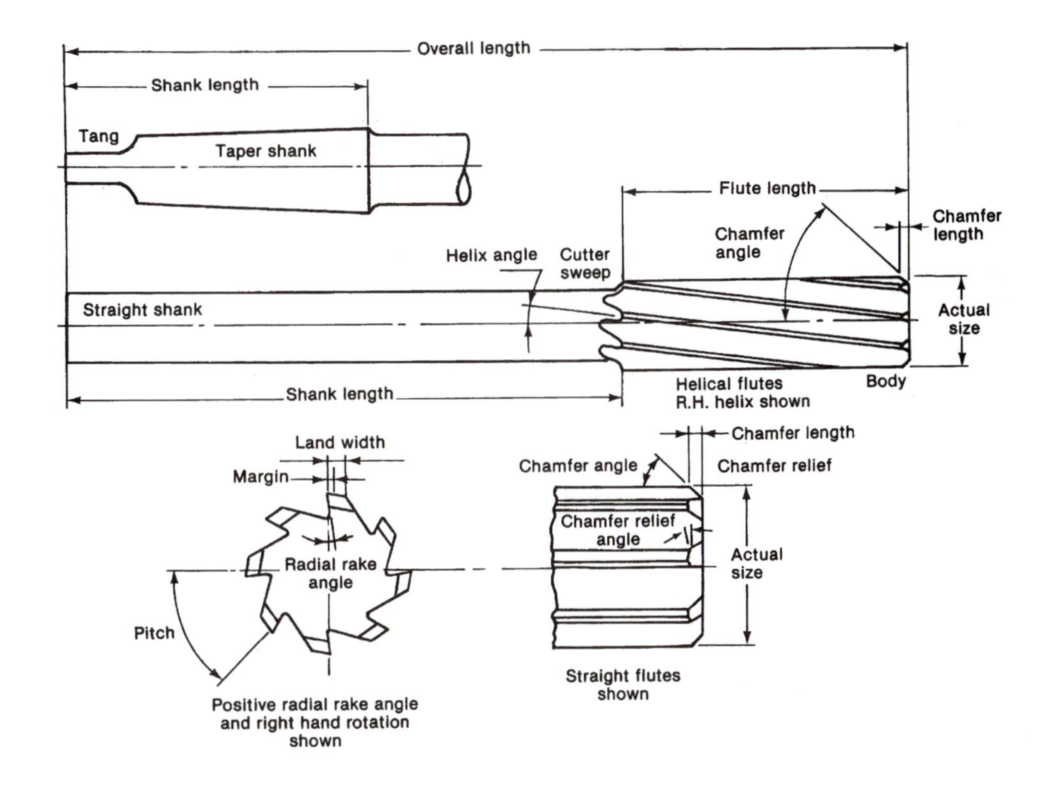

Figure 7.22
Geometry of reamers.

Finally, because many holes must contain threads, we need to mention, among the tools associated with drilling, the *taps*. The geometry of taps is rather complex. The cutting edges have the form of a thread interrupted by three or four flutes that provide room for chips. The flanks of the threads are relieved back from the cutting edge inwards toward the axis of the tool. Several threads at the tip are ground with a conical chamfer to ease the beginning of the cut and to guide the tap into the predrilled hole.

7.3.1 Metal Removal Rate: Force, Torque, and Power

The metal removal rate *(MRR)* of a solid drill is

$$Q = \frac{\pi d^2}{4000} n f_r \qquad Q\,(\text{cm}^3/\text{sec}),\ d\,(\text{mm}),\ f_r\,(\text{mm}),\ n\,(\text{rev/sec}) \tag{7.5}$$

where d is the hole diameter, n is the rotational speed, and f_r is the feed per revolution. The cutting force components are indicated in Fig. 7.23, assuming a drill with two cutting edges (lips). The chip area for each edge is $A = f_t d/2$, where $f_t = f_r/2$ is the feed per tooth. The normal force component F_n and the tangential component F_t in the direction of the cutting speed v are analogous to those in turning.

Correspondingly,

$$F_t = \frac{K_s d f_t}{2} = \frac{K_s d f_r}{4} \qquad F_t\,(\text{N}),\ K_s\,(\text{N/mm}^2),\ d\,(\text{mm}),\ f_r\,(\text{mm}) \tag{7.6}$$

The torque is

$$T = \frac{F_t d}{2} = \frac{K_s d^2 f_r}{8000} \qquad T\,(\text{Nm}),\ K_s\,(\text{N/mm}^2),\ d\,(\text{mm}),\ f_r\,(\text{mm}) \tag{7.7}$$

Figure 7.23
Forces on a drill.

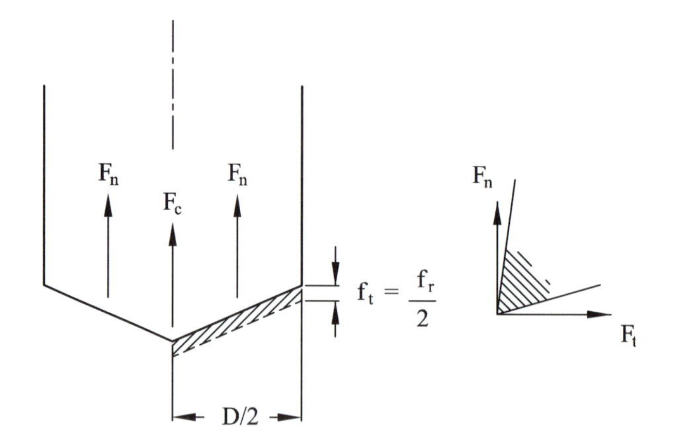

and the power is

$$P = T\omega = 2\pi T n = \pi \frac{K_s d^2 f_r n}{4000}$$

$$P\ (\mathrm{W}),\ K_s\ (\mathrm{N/mm^2}),\ d\ (\mathrm{mm}),\ f_r\ (\mathrm{mm}),\ n(\mathrm{rev/sec}) \tag{7.8}$$

By comparing Eq. (7.8) with Eq. (7.5), we find simply

$$P = \frac{K_s Q}{60} \qquad P\ (\mathrm{W}),\ K_s\ (\mathrm{W\ sec/cm^3}),\ Q\ (\mathrm{cm^3/sec}) \tag{7.9}$$

which is identical with Eq. (7.4) and where K_s has the meaning of specific power.

The *thrust (axial) force* F_a necessary to feed the drill into the cut is the sum of the two normal forces F_n and of the central force F_c on the chisel edge. The force F_c depends strongly on the length of the chisel edge, but for the common condition an empirical formula can be used that assesses the total thrust to be about one half of the sum of the tangential cutting forces F_t; correspondingly,

$$F_a = 0.5(2F_t) = \frac{K_s d f_r}{4} \qquad F_a(\mathrm{N}),\ K_s\ (\mathrm{N/mm^2}),\ f_r\ (\mathrm{mm}) \tag{7.10}$$

The values from Table 7.1 can be used for K_s.

7.3.2 Drilling Machines

An *upright drill press* is a common machine in small shops, tool rooms, and maintenance shops for job-type production. A particular model of this machine is shown in Fig. 7.24. Its table can be adjusted on the column for the height of the workpiece. Spindle speeds can be selected according to the tool type and size. The spindle is axially movable in a quill, and the axial feed is hand-driven with the hand acting at the end of the feed lever. The workpiece is held in a vise, the position of which is manually set so as to position the location of the hole to be machined under the spindle. For solid drilling, this position may, in the simplest case, be marked by an indentation made beforehand with an indenting punch.

Figure 7.24
An upright drill press.
[ROCKWELL]

Larger workpieces that cannot be accommodated on an upright drill press, cannot be easily moved by hand, and cannot, when drilling larger holes, be held by hand are drilled on a radial drilling machine, which exists in a range of sizes. A medium-size machine is shown in Fig. 7.25. It has a base plate with T-slots to clamp the fixture in which the workpiece is clamped. The radial arm is guided on a cylindrical column and can be moved up and down as well as swung around the column. A headstock moves radially on the guideways of the arm. The up-down motion of the arm is driven mechanically for height adjustment. The swing of the arm and the travel of the headstock may be adjusted by hand to position the spindle above the location of the hole to be drilled. Subsequently, it is locked in this position. The axial feed motion of the quill is mechanically driven and is derived from the spindle drive via a feed box, as in the case of the lathe. In this way the required feed per revolution in drilling and the required pitch in tapping are obtained. The mechanical feed of the quill may also provide for rapid approach and withdrawal, or else these are done manually on the release of the mechanical feed. In tapping the return motion is obtained by reversing the spindle with the feed drive kept engaged.

Obviously, of all the manual actions that the operator carries out, the most difficult one is the positioning of the spindle relative to the workpiece for the proper location of the hole. If the location is specified to a close tolerance, this cannot be accomplished manually without additional aids. Traditionally, such an aid, especially in the production of larger series of workpieces, has been the jig, which is a fixture in which the workpiece is clamped in a precise position against locating surfaces, and which contains bushings to guide the drilling tools. A drawing of a simple jig for drilling the holes in a disc is shown in Fig. 7.26.

A jig may be, of course, an expensive device, depending on its complexity. A special class of machine tools that has been in existence for a long time are known as "jig-boring machines." They have a rigid frame and accurately positioned *X*, *Y* coordinate motions to drill and bore holes in jigs with precise positions. The lengths of these coordinate motions are controlled either from highly accurate leadscrews or from optical

Figure 7.25
A radial drilling machine.
[GIDDINGS & LEWIS]

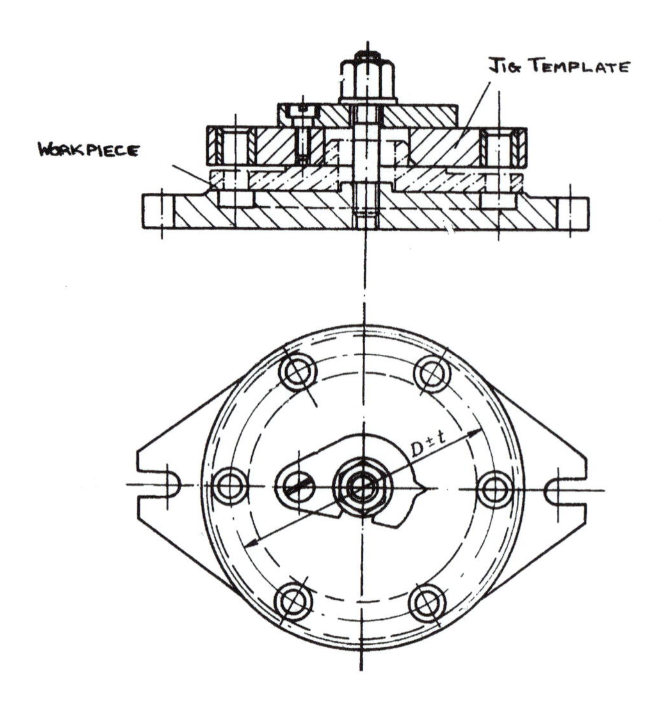

Figure 7.26
A simple jig. Workpiece is clamped under the jig template with bushings that provide for positioning and guidance of the drills.

systems reading accurate graduated scales. A jig is economical for production in medium-size series of workpieces. Flexible automation of the positioning of workpieces on drilling and boring machines, so that it is economical for small-lot production, is achieved by numerical control, which is discussed in Chapter 10.

For mass production, it is economical to build special, multispindle drilling machines or stations as part of a transfer line. The headstock carries a number of spindles that are mounted in positions corresponding to those of the holes on the workpiece, and the speeds of these spindles correspond to the sizes of the drills. The headstock carries the motor and contains gear transmissions driving all the spindles. Feed rate has to be compromised because the feed motion is common for all the tools. The headstock moves on a column unit, and its motion is derived from a drive that works in a cycle of rapid approach, working feed, and rapid withdrawal. Typically, a transfer line for drilling, reaming, and tapping of holes in an engine block of an automobile may contain 20–40 such stations.

7.4 MULTIPOINT TOOL OPERATIONS: MILLING

In milling the main cutting motion is the rotation of a multitoothed cutter that machines a workpiece that performs translative (mostly rectilinear) feed motions. The geometry of milling operations is presented in Fig. 7.27. Two basic modes are shown: *(a)* face milling, and *(b)* peripheral milling. These names refer to the fact that it is the face and periphery respectively of the cutter that generate the machined surface *S*. However, if we look at the cutting action of the face-milling tooth in the detail *D*, we see that the nose of the tooth generates the surface *S*, but the chip is mainly produced by the main cutting edge set at the SCEA σ; thus, most of the cutting is actually performed by the periphery of the cutter. Correspondingly, most aspects of chip formation, especially the character of the variation of chip thickness, are common to both the basic modes of milling,

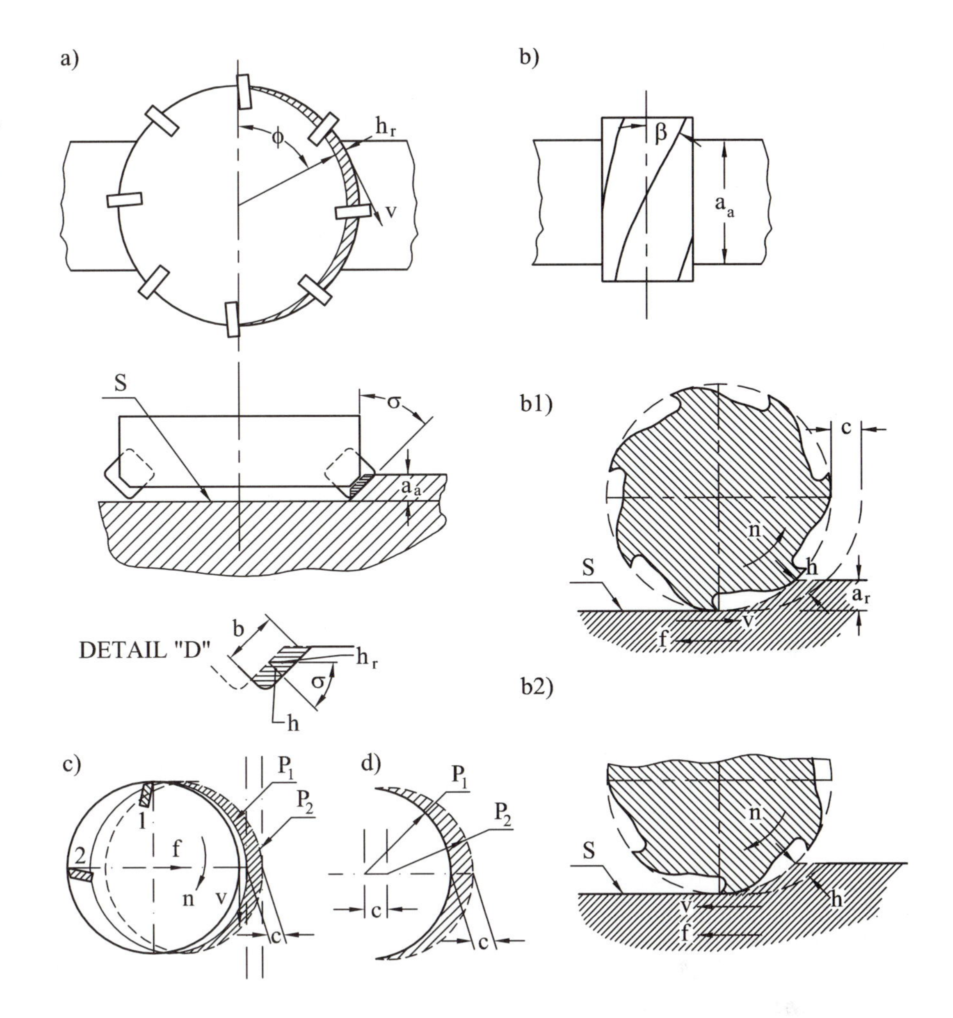

Figure 7.27 Geometry of milling operations: a) face milling; b) slab milling; b1) up-milling, b2) down-milling; c) cycloidal motion of the cutting edges may be approximated by circular arcs moved in the feed direction by the amount of the chip load c.

and they relate to the peripheral cutting action. The relative motion of the cutting edge with respect to the workpiece is a sum of the rotation of the cutter with speed n (rev/sec) and of translation with *feed rate* f (mm/sec). The feed is carried out either by the cutter or by the workpiece. The peripheral speed of the cutter is the cutting speed v (m/sec). The path resulting from the rotation and translation is a cycloid as shown in diagram *(c)*, in which the paths P_1 and P_2 are shown of the teeth 1 and 2 of cutter C. The material removed by tooth 2 is the difference between P_1 and P_2 on the forward side, as shown by the shaded area. However, because the *feed per tooth* f_t is very small if compared with the diameter of the cutter (e.g., $f_t = 0.2$ mm, $d = 25$ to 250 mm; the ratio is about 100 up to 1000), the cycloid is very "tight" and the cutting paths of the teeth can, with good approximation, be considered as circular arcs that are mutually shifted by f_t. The feed per tooth is commonly called the "chip load" c, and it is

$$c = f_t = \frac{f}{nm} \tag{7.11}$$

where m is the number teeth of the cutter.

In *face milling* the depth of the layer removed from the workpiece is the axial depth of cut a_a, and the width of the workpiece is the radial depth of cut a_r. The chip thickness varies with the angle ϕ of cutter rotation, as indicated by the shaded areas representing the cut taken by one tooth. The width of chip is b, and the chip thickness is h. In the radial section normal to the cutter axis, the chip thickness is seen: $h_r = h/\cos\sigma$. For $\sigma = 0$, $h_r = h$.

The type of *peripheral milling* shown in diagram *(b)* is called plain milling, and its geometry is analogous to that of face milling if referred to the cutter axis. The section normal to the cutter axis shown at the bottom of the diagram corresponds to that at the top of diagram *(a)*. However, when referred to the workpiece, the geometry is reversed. The thickness of the layer removed from the workpiece is now the radial depth of cut a_r, and the width of the workpiece is the axial depth of cut a_a (width of cut). The cutter teeth are helical; the helix angle is β. The angle β has an entirely different role than the angle σ had in the face milling case; actually the SCEA σ for a cylindrical peripheral cutter is $\sigma = 0$. The helix angle has a similar significance as the back rake (inclination) angle λ of a single-point tool in Fig. 7.3. Correspondingly, in the diagram *(b)* the radial chip thickness is also the actual chip thickness, $h_r = h$. The chip width, however, is now $b = a_a/\cos\beta$. Depending on the relation between the direction of rotation of the cutter (of the cutting speed v) and the direction of feed f, one distinguishes between *up-milling (b1)* and *down-milling (b2)*. In the former case the chip thickness is zero at the start of the cut and increases through the cut. In the latter case it is the reverse.

A very common type of peripheral milling is *end milling.* Photographs of a face-milling cut and of an end-milling cut are given in Fig. 7.28 on the left, top and bottom, respectively. An end mill is a cutter of a smaller diameter (usually between 5-mm and 30-mm diameter) clamped in overhang, and its length is several times its diameter. There is not much of a fundamental difference between end and face milling, except that in the face milling, the result of the operation is the horizontal surface S left behind the cutter (perpendicular to the cutter axis), and in end milling the principal result is the machined side S of the workpiece (parallel to the cutter axis), although at the same time a face is created in this cut. The operation shown is up-milling.

The operation shown at the top right results in two slots made by the two cutters in the form of discs that are carried by an arbor clamped at its left end by the spindle and supported at its right end by two bearings mounted on two round, horizontal overarms (only the front one is visible). The bottoms of the slots are created by the peripheries of the cutters, and the sides of the slots are created by the edges on the faces of the cutters. The operation at bottom right is called square-shoulder face-milling. The cutter used is a face-milling cutter with the angle $\sigma = 0$. Obviously, the only difference between the two bottom pictures is in the length-to-diameter ratio of the cutters (end and face mills, respectively).

The geometry of end milling is shown again in the drawing of Fig. 7.29. The three cases shown are *(a)* up-milling, $a_r = d/2$; *(b)* down-milling, $a_r = d/2$; and *(c)* slotting, $a_r = d$. The ratio a_r/d is called "the radial immersion," and cases *(a)* and *(b)* are called half-immersion; case *(c)* is called full immersion. Slotting can be considered as the sum of the half-immersion up- and down-milling cases.

7.4.1 Mean Chip Thickness, MRR, and Power

Let us now express the variation of chip thickness h_r, (see Fig. 7.30). The paths of two subsequent teeth are represented by two circles mutually shifted in the direction f by the chip load c. The position of the tooth concerned is denoted by the angle ϕ measured from the point in which an up-milling cut would start. The width a_r of the workpiece is located so that the tooth starts to cut at the angle ϕ_s and ends at the angle ϕ_e. In this sense the angles associated with the operations indicated in Fig. 7.29 would be:

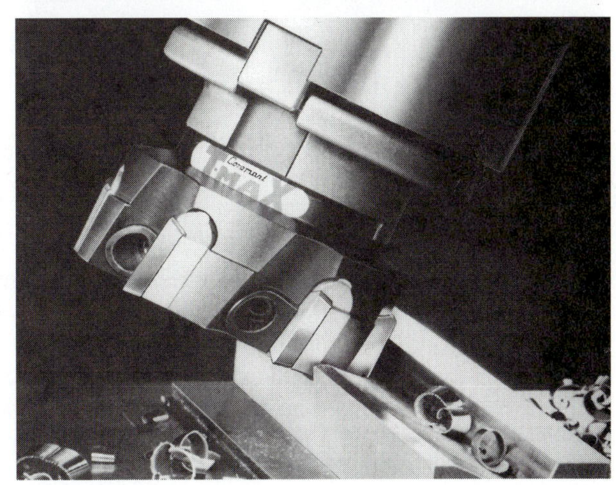

Figure 7.28
Various types of milling operations; top left: face milling, top right: slotting with disc type cutters, left bottom: end milling, right bottom: square shoulder milling. [SANDVIK]

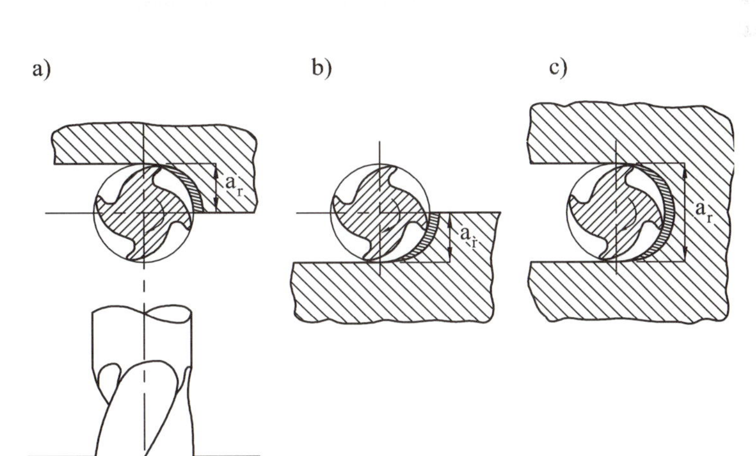

Figure 7.29
Geometry of end milling: a) half immersion up-; b) half immersion down-; c) full immersion (slot) milling.

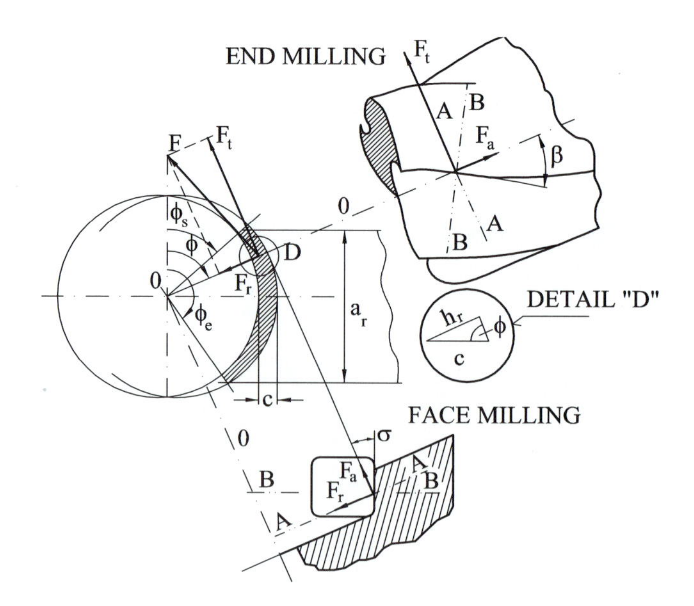

Figure 7.30
Geometry of chip thickness on a helical end mill and on a face mill with side cutting edge angle σ. Radial F_r, tangential F_t, and axial F_a force components and chip thickness as measured in plane A normal to tool axis or B normal to cutting edge are shown.

(a) $\phi_s = 0$, $\phi_e = 90°$; *(b)* $\phi_s = 90°$, $\phi_e = 180°$; *(c)* $\phi_s = 0$, $\phi_e = 180°$. Detail D shows that

$$h_r = c \sin \phi \tag{7.12}$$

The thickness h_r is measured in the radial plane A, which is normal to the axis O of the cutter. The cutting force acts in a plane B normal to the cutting edge, which in face milling is inclined by σ to plane A; in end milling, it is inclined by β to plane A. We will look at the projections of the force components into plane A, which are the tangential component F_t and the radial component F_r. We will commonly assume $F_r = 0.3\, F_t$. These two together compose the force F; there is also an axial component $F_a = F_r \tan \sigma$ in face milling and $F_a = F_t \tan \beta$ in peripheral milling. Thus, we have established constant ratios between the force components, and we can limit further discussion to the force F_t which has the direction of the peripheral speed of the cutter (the cutting speed v) and corresponds to the force F_t in Eq. (7.3).

We must realize that on a helical tooth of a peripheral cutter the angle ϕ determining the tooth position varies along the tooth (edge) length. Correspondingly the directions of the force components F_t and F_r vary along the edge length, and this affects the resulting force acting on the cutter. We analyze this further in Chapter 9.

Presently, we are interested in the effect of the force on power. In this respect every incremental force F_t generated on an incremental length of a helical edge has always the direction of the cutting speed v, which also varies along the edge. Therefore, the incremental powers $dP = dFv$, being scalars, can simply be summed.

Moreover, we are now interested in the *mean chip thickness* h_m, in the *mean metal removal rate* (MRR_m), and in the *mean power*. This should be understood as follows. The chip thickness on every tooth varies according to Eq. (7.12). The number of teeth engaged simultaneously in cutting also varies as the individual teeth enter and leave the angle of engagement (ϕ_s, ϕ_e) in face milling and as the helical teeth in peripheral milling are engaged over a continuously varying length of the helix.

Correspondingly, the total chip area $A_{c,tot}$ being cut at an instant varies periodically with the "tooth period" $T_t = 1/nm$, and the instantaneous $MRR = (A_{c,tot}\, v)$ varies with the same period, and the total force acting on the cutter and the instantaneous

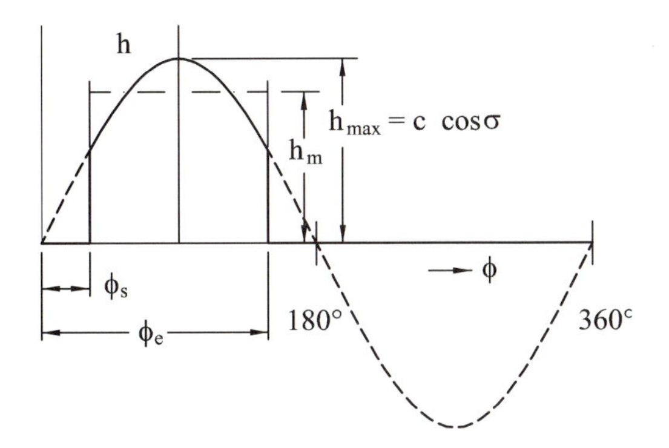

Figure 7.31
Chip thickness variation.

power vary with the same period. However, the total volume of chips per unit time (i.e., the *MRR*, on average) must be equal to the area of the layer being removed from the workpiece A_w times the velocity of the workpiece, which is the feed rate *f*. The mean metal removal rate MRR_m is

$$MRR_m = Q = 0.001 a_r a_a f \qquad Q\,(\text{cm}^3/\text{min}),\ a_r\,(\text{mm}),\ a_a\,(\text{mm}),\ f\,(\text{mm/min}) \tag{7.13}$$

Although at any instant various parts of the layer A_w are being removed by the individual teeth at different rates, the average as given by Eq. (7.13) holds.

Referring to Eqs. (7.3) and (7.4) of Section 7.1, we can determine the mean power:

$$P = \frac{K_s Q}{60} \qquad P\,(\text{W}),\ K_s\,(\text{W sec/cm}^3),\ Q\,(\text{cm}^3/\text{min}) \tag{7.14}$$

The formula of Eq. (7.14) is identical with (7.4), but the value K_s actually depends also on the mean chip thickness. This will be explained in Chapter 8. The K_s values are found in Table 7.1 It remains to establish the parameter h_m.

The variation of h_r during the cut was expressed in Eq. (7.12) as a sine function. Graphically the variation of h_r on one tooth and also of $h = h_r \cos \sigma$, is expressed in Fig. 7.31. It is $h = c \cos \sigma \sin \phi$ for $\phi_s < \phi < \phi_e$ and $h = 0$ over the rest of the cutter revolution. Over the cutting range, the mean value h_m is

$$\begin{aligned} h_m &= \left(\frac{c \cos \sigma}{\phi_e - \phi_s}\right) \int_{\phi_s}^{\phi_e} \sin \phi \, d\phi \\ &= c \cos \sigma \frac{\cos \phi_s - \cos \phi_e}{\phi_e - \phi_s} \end{aligned} \tag{7.15}$$

where ϕ_e and ϕ_s are given in (rad).

EXAMPLE 7.2 Mean Chip Thickness in Milling ▼

For a half-immersion, up-milling cut with $c = 0.2$ mm, and $\sigma = 0°$,

$$\phi_s = 0°, \qquad \phi_e = 90°$$

$$h_m = c\,(1 - 0)/(0.5\,\pi - 0) = c2/\pi = 0.64c = 0.127 \text{ mm}.$$ ▲

EXAMPLE. 7.3 Power in Face Milling ▼

Determine the mean power for a face-milling operation, as shown in Fig. 7.32: cutter diameter $d = 150$ mm, $m = 8$ teeth, $a_a = 8$ mm, $\sigma = 30°$, $c = 0.2$ mm, workpiece material is steel 1045, cutting speed is $v = 150$ m/min. In Table 7.1 we find $K_s = 2600$ W sec/cm^3.

Spindle speed is $n = v/\pi d = 318$ rpm.

Feed rate $f = cmn = 0.2 \times 8 \times 318 = 509$ mm/min.

$Q = 0.001 \times 80 \times 8 \times 509 = 326$ cm^3/min.

$P = Q \times K_s / 60 = 14{,}127$ W $= 14.127$ kW ▲

7.4.2 Design of Milling Cutters

Face-milling cutters and the disc-type slotting cutters are most commonly constructed as bodies with mechanically clamped carbide inserts. Some of the end mills are also made in that way. A photograph of a group of such cutters and of various inserts is shown in Fig. 7.33. A variety of insert shapes and a variety of mechanisms for clamping inserts in the cutter body are available. The mechanisms must ensure safe clamping to hold the insert and withstand the impacts of entry into and exit out of the cut, provide a fast and easy way of changing the inserts, a minimal radial and axial run-out, and allow sufficient space for chips. It is beyond the scope of our text to discuss the many existing and continuously developing solutions to these problems. As a basic illustration, one simple design is shown in Fig. 7.34. The cutter body (1) is made of hardened alloy steel. Large chip pockets (2) are provided in front of each insert to allow free chip formation. The insert (5), which in this case may be a plain one without a hole, is clamped by means of the wedge (3) made of hardened steel, and it is radially located against the rest button (4). Apart from a choice of the SCEA, the geometry of a face mill can be one of three basic types, as shown in Fig. 7.35. According to the sign

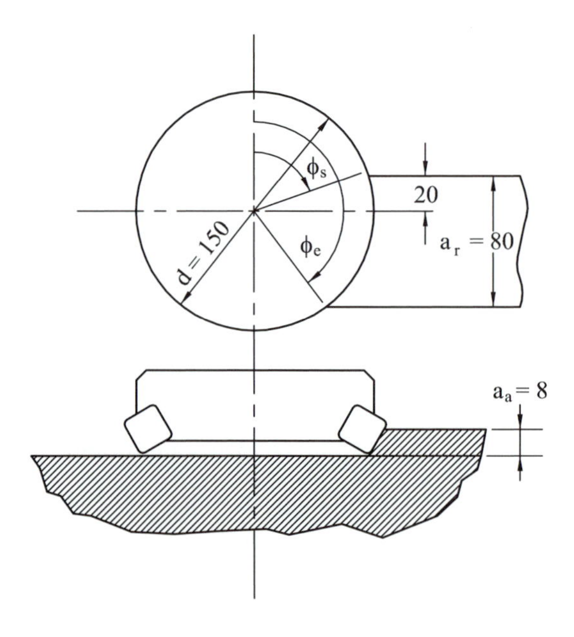

Figure 7.32
Geometry of the case in Example 4.3.

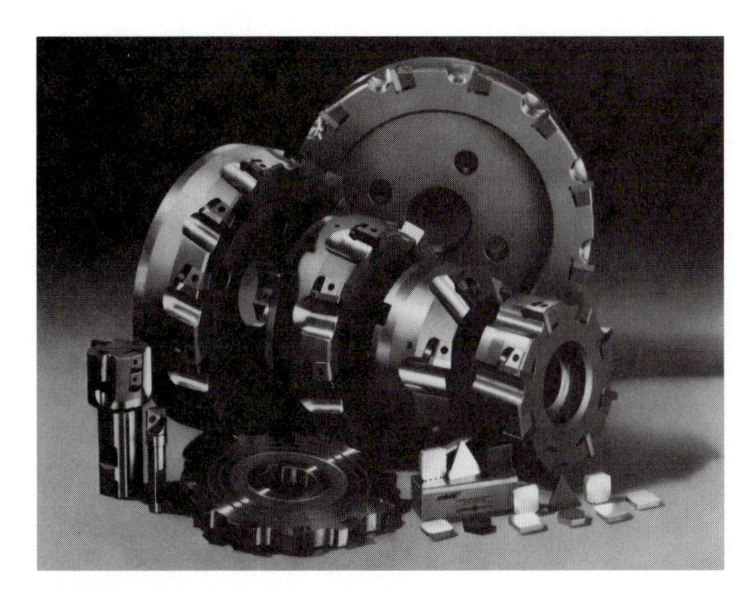

Figure 7.33
Various types of milling cutters and inserts. [VALENITE]

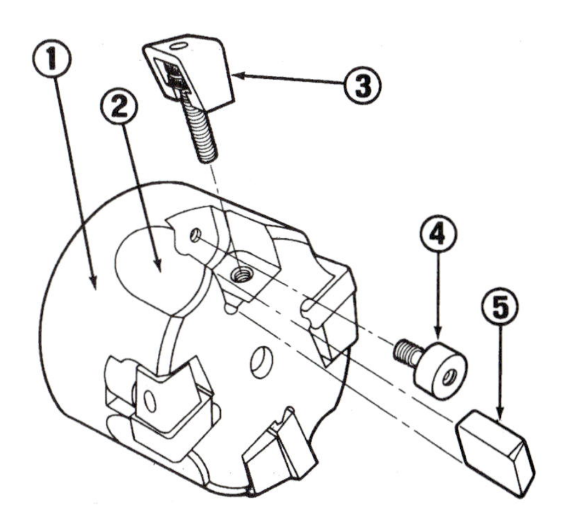

Figure 7.34
One method of clamping inserts in the milling cutter body. Wedge *3* provides the clamping force. Axial position of the insert can be adjusted by the stop *4*.

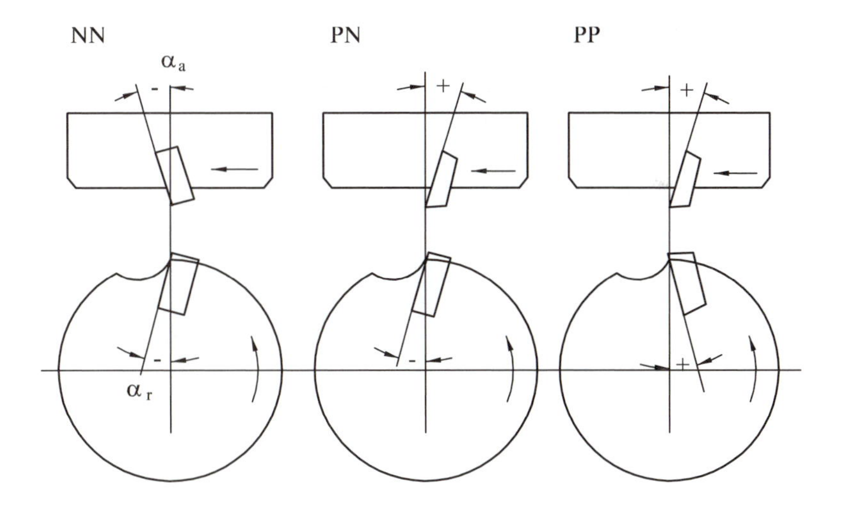

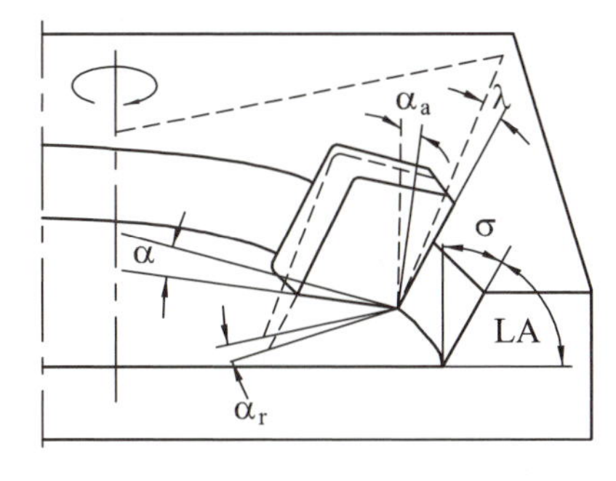

α, EFFECTIVE RAKE ANGLE
α_r, RADIAL RAKE ANGLE
α_a, AXIAL RAKE ANGLE
λ, INCLINATION ANGLE
σ, SIDE CUTTING EDGE ANGLE, COMPLEMENT OF LEAD ANGLE, LA

Figure 7.35
Basic geometries of face-milling cutters: *NN* (negative axial, negative radial); *PN* (positive axial, negative radial); *PP* (positive axial and radial).

of the axial rake α_a and of the radial rake α_r angles one can have cutters that are double negative (*NN*), positive negative (*PN*), or double positive (*PP*). The bottom diagram shows the general situation and the geometry of the cutting edge. The effective rake angle α is measured in a plane perpendicular to the cutting edge. Whilst *NN* is strongest against the entry impact, and *PP* gives the lowest specific forces and best throw-out of chips, the design of *PN* is a compromise with good entry conditions as well as good clearance of chips. Only the *NN* design permits the use of "negative inserts" with double the number of available cutting edges (see also Fig. 7.10).

End mills are most often made of high-speed steel, and they either have plain continuous cutting edges or edges that are serrated or sinusoidally wavy (see Fig. 7.36). The same picture shows a shell end mill and two form-milling cutters. The advantage of serrated and undulated edges is that an otherwise wide chip is divided into a number of narrow chips that are easily cleared from the cutting space. Some end mills possess the ability to cut also with axial feed as a drill does. They must have a corresponding form of edges on their end face. Another common type of end mill is the "ball-nosed" one. It is suitable for machining curved, "sculptured" surfaces like those on turbine blades or some airfoils. A particular design of an end mill with mechanically clamped carbide inserts with serrated edges is shown in Fig. 7.37. Another advantage

Figure 7.36
High-speed steel milling cutters. [TRW]

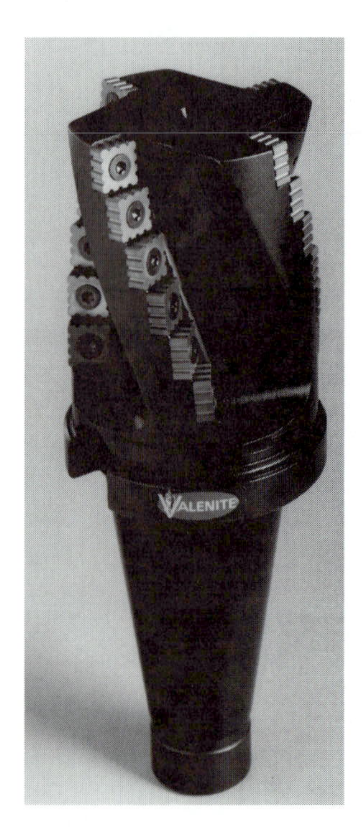

Figure 7.37
An end mill with serrated inserts. Serrations separate a wide chip into strips easy to clear away and they may also have a vibration-stabilizing effect. [VALENITE]

of the serrations and undulated edges is that they increase the stability of milling against chatter vibrations, as explained in Chapter 9.

Three basic ways of clamping milling cutters are shown in Fig. 7.38. The mounting of a face mill directly onto the nose of the spindle as in *(a)* gives the shortest overhang and maximum stiffness on the cutter teeth. It permits the most powerful milling without chatter. The mounting on a mandrel with a tapered shank clamped into the internal taper of the spindle (Fig. 7.38b) is less rigid than *(a)* but permits quick, automatic tool change, and it is the type most frequently used on NC machining centers. Finally, end mills and cutters with cylindrical shanks are often mounted in a cylindrical hole, either as shown in Fig. 7.38c or clamped in adapters with collets. Recently, especially in applications of high-speed milling (see Section 9.9), the cylindrical shank of the cutter is shrink-fit into the undersized hole of the holder. The shrink-fitting is accomplished by heating the holder in the coil of a high-frequency, induction-type electric furnace before inserting the cutter and letting the assembly cool. Heating is also used to release the tool from the holder.

7.4.3 Milling Machines

For smaller workpieces, the classical, manually operated milling machine has traditionally been the knee-type milling machine with either a vertical or horizontal spindle. The knee, which can be moved vertically on the guideways on the column, can be

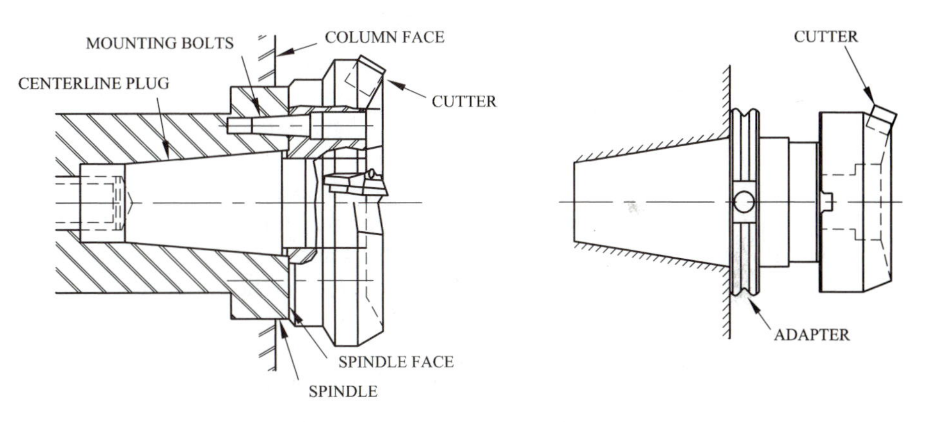

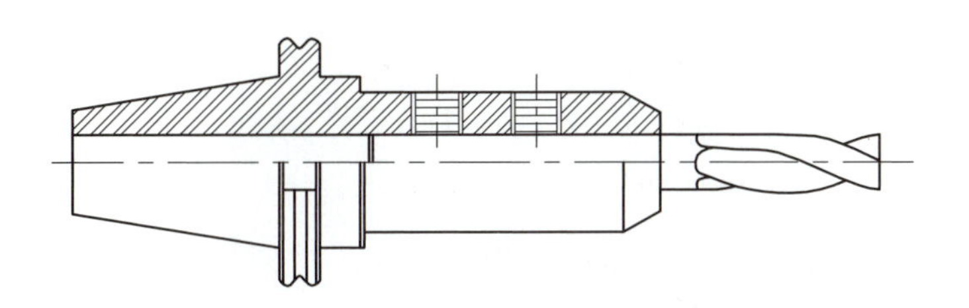

Figure 7.38
Various ways of clamping milling cutters. a) Large face mills are clamped directly on the spindle face. b) Smaller face mills (shell mills) are clamped on an adapter that itself is held in the internal taper of the spindle. c) End mill is shown clamped by set screws in a holder. Alternatively, it may be held in a collet-type holder or, for best accuracy and stiffness, shrink-fit in an accurate hole in the holder.

adjusted in Z according to the height of the workpiece, so that the machined surface is at a level convenient for the operator. The knee carries the saddle that executes the Y motion, and on top of it the table that executes the X motion. While Z is just an adjusting motion, X and Y are driven with working feed motions as well as with rapid traverse. The spindle is mounted in the column, which also contains the motor and gearbox for the spindle drive. Milling cutters can either be clamped in overhang in the spindle or on an arbor that is supported by bearings in brackets attached to the overarm. Feeds are driven from a separate electric motor attached to the knee and kinematically distributed to X, Y, Z. Thus, there is no kinematic relationship between feed rates and spindle speeds, as there is on a lathe. The handwheels, cranks, and push buttons that control the direction and speed of feed motions are located on the front of the knee. The "universal" operation of this machine, which permits the guideways of the table to be swiveled in the horizontal plane, is typically used for milling flutes in drills or in milling cutters with helical teeth; the workpiece is clamped on a mandrel in a "dividing head" for indexing of the individual flutes and supported by a tailstock.

Once the controls of a milling machine became more sophisticated, including electromagnetic clutches for engaging and disengaging the feed motions, or even separate servomotors for each feed motion, the "bed-type" horizontal and vertical milling machines began replacing the knee-type models, ensuring more rigidity and accuracy of the position of the table. A vertical-spindle, bed-type milling machine is shown in Fig. 7.39. It is intended for face and end milling of small- and medium-size parts. The saddle moves in Y, the table in X, and the Z motion, which can be used also as a working motion for drilling and boring, is executed by the headstock, which moves on the vertical guideways of the column. The horizontal spindle alternative is preferred for machining a box type workpiece clamped on a rotary table attached to the top of the longitudinal table. This arrangement permits access to the workpiece from many sides; it is the most popular configuration for NC machining centers and is described in Chapter 10.

Large and very large workpieces are machined on planer-milling machines, which exist in a variety of configurations as single-column or portal types. For instance, the portal-type machine may carry four headstocks; one on each side on the vertical guideways on the columns and two on the horizontal guideways of the cross-rail. The motion

Figure 7.39
Vertical spindle bed-type milling machine. Such machines mostly replaced the previously popular knee-type machines. They provide better rigidity of the table and less influence of the moving workpiece weight on accuracy. [OMNITRADE]

of the table on the guideways of the bed is the most frequently used feed motion. However, the headstocks can also move individually on their guideways, and their quills move axially.

Drilling, Boring, and Milling Machines

All the three types of operations are often combined in one machine, the most common type of which is the table-type horizontal boring machine illustrated in Fig. 7.40. The table of the machine moves horizontally in the X coordinate on top of a saddle, which itself is guided on the bed and executes the motion W. The headstock moves vertically on the column in the Y coordinate. In the headstock is mounted an assembly of a hollow spindle, within which the boring spindle can move axially in and out in the Z coordinate. Light milling cutters, drills, and boring bars are clamped into the boring spindle. Heavier cutters are clamped directly onto the flange of the hollow spindle. For long holes, the boring spindle can carry a long boring bar supported in a bearing, the housing of which can move vertically on the counter-column, and its position is synchronized with the Y position of the spindle. Essentially, boring is executed by feeding the saddle in the W coordinate, while the Z coordinate is used as a setting motion to adjust for the depth of the bore and may be used as the working motion for drilling and tapping. The floor-type horizontal boring and milling machine is used for machining large workpieces. In the example shown in Fig. 7.41 such a machine is performing a

Figure 7.40
Table-type horizontal boring machine intended for boring of long bores or of bores in opposing walls of box-type workpieces. It includes an outer bearing housing opposite the main spindle. This bearing is movable vertically on the outer column. A long boring bar (not shown) inserted in the spindle passes through the workpiece and is supported by the outer bearing. The workpiece is mounted on the table that is carried by a saddle for the longitudinal feed. [OMNITRADE]

Figure 7.41
A floor-type horizontal boring and milling machine. The heavy workpiece is aligned on a floor plate. The column travels horizontally along the workpiece on a bed also mounted on the floor. A headstock travels vertically on the column and carries a ram that includes the spindle. The ram moves in and out towards the workpiece. [SKODA]

Figure 7.42
Another floor-type horizontal boring and milling machine. It is shown drilling holes in the flange of the housing of a rotary furnace. The workpiece is clamped on the floor plate. In other applications a vertical axis rotary table may be mounted on the floor plate so as to provide access to a workpiece from many angles. [MITSUBISHI]

face-milling operation on a part of a large frame. The workpiece is clamped to a large floor plate. The base of the column moves in X on a bed alongside the floor plate and below its level. A heavy headstock with a platform for the operator moves horizontally (Z) in a saddle, which moves vertically (Y) on a large column. Another floor-type boring and milling machine is shown in Fig. 7.42; it is engaged in end-contour milling and in drilling of a large ore concentrator part. Such machines use a large variety of tooling accessories, such as extension supports to increase the stiffness of milling cutters for milling inside of workpieces, angle heads for face milling of horizontal surfaces and of vertical surfaces that are parallel to the spindle, surfacing heads for single-point tools used in machining the faces of flanges around bores, and so on. They also use coordinate and rotary tables to be placed on the floor plate so as to enhance the versatility of the operations. Machine tools like these typically have X-travels of 5–30 m, vertical Y-travels of 2–6 m, and spindle-drive powers of 30–60 kW. They are mostly equipped with numerical control.

7.5 BROACHING

Like milling, broaching is an operation in which several (often many) cutting edges act simultaneously. However, the cutting motion is not rotary but rectilinear. One might consider broaching as milling with an infinite cutter radius. However, there is a substantial difference in that there is no feed motion. Instead, the teeth on the tool are so arranged that they increase in height, penetrating farther and farther into the workpiece. The machined surface is the inverse of the profile of the tooth. The tool is called the broach, and it is usually rather long and often accommodates as many teeth as are necessary to complete the machining in one pass of the tool over the machined surface.

The original and most frequent purpose of broaching is to machine unround holes: holes with keyslots, holes with multiple slots to match spline shafts (e.g., for shifting gears in gearboxes). These operations can hardly be carried out economically by any other method. However, because of the speed of the operation (carried out in one pass) and the cost of the mostly special-purpose tool, it is an operation well suited to mass production of surfaces that are straight in at least one direction. It is used for machining all kinds of internal and external surfaces, from simple flat ones to internal and external gears.

An essentially round broach is shown in Fig. 7.43. It is intended to be pulled through a hole premachined to the "root diameter." The radial offset (step) of the roughing teeth is larger than that of the semifinishing teeth. The step on the roughing teeth may typically be 0.1–0.2 mm, and on the semi-finishing teeth, 0.02–0.04 mm. The finishing teeth have equal height. They start to wear from the front, so that the last ones are preserved to keep the correct size of the hole for a rather long time. The cutting speeds used are relatively low, 15–25 m/min for high-speed steel broaches and 60–100 m/min for carbide broaches. This leads to rather long tool lives, which are economically necessary because the regrinding of a broach is costly. The broach shown in the figure is the pull type. The shank length must be sufficient to permit the broach to pass through the workpiece and be attached to the puller before the roughing teeth engage the workpiece. The front pilot aligns the broach in the blank hole. Because the complete operation is mostly finished in a single pass of the broach and the geometry of the operation is simplified for the single motion, the auxiliary times are minimized, especially on the double-acting machines. This makes broaching very efficient and well suited to mass production.

The step per tooth determines the chip thickness; the profile on the periphery of the broach determines the shape of the hole machined and the total chip width. More details

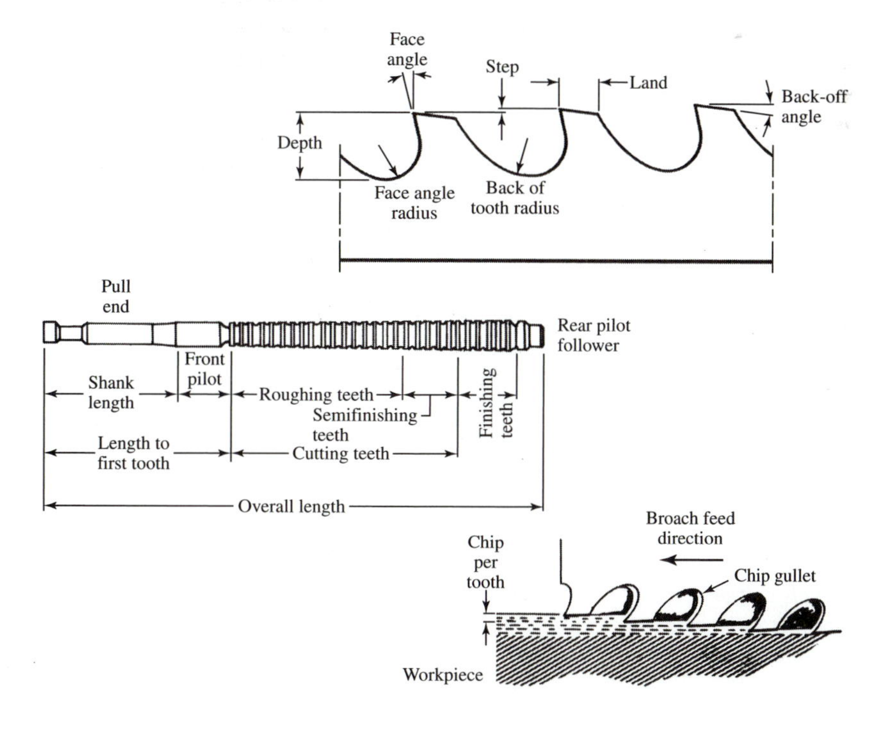

Figure 7.43
The geometry of a broach.
[NATIONAL BROACH]

of the geometry of the teeth of broaches are shown in Fig. 7.44. In the diagram a) details of chipbreakers on a flat and on a round broach are presented. They relieve the total load by splitting the heavy chips in the roughing sections of the tool. The diagram b) shows the keyway broach, the multiple spline broach, and the helical spline broach.

Broaching is also suitable for mass production of gears. Figure 7.45 shows the tool and the workpiece. The workpiece is an internal, helical automobile transmission gear. Figure 7.46 shows a "pot broach," which is used to produce an external helical gear. A variety of automotive transmission gears are produced by broaching in a single-pass operation that eliminates the previously used finishing method of shaving. The teeth of these broaches are so designed as to assure concentricity of the internal and external gear diameters, respectively, with the pitch diameter of the gears. In broaching helical gears, the workpiece is rotated in synchronization with the motion of the broach.

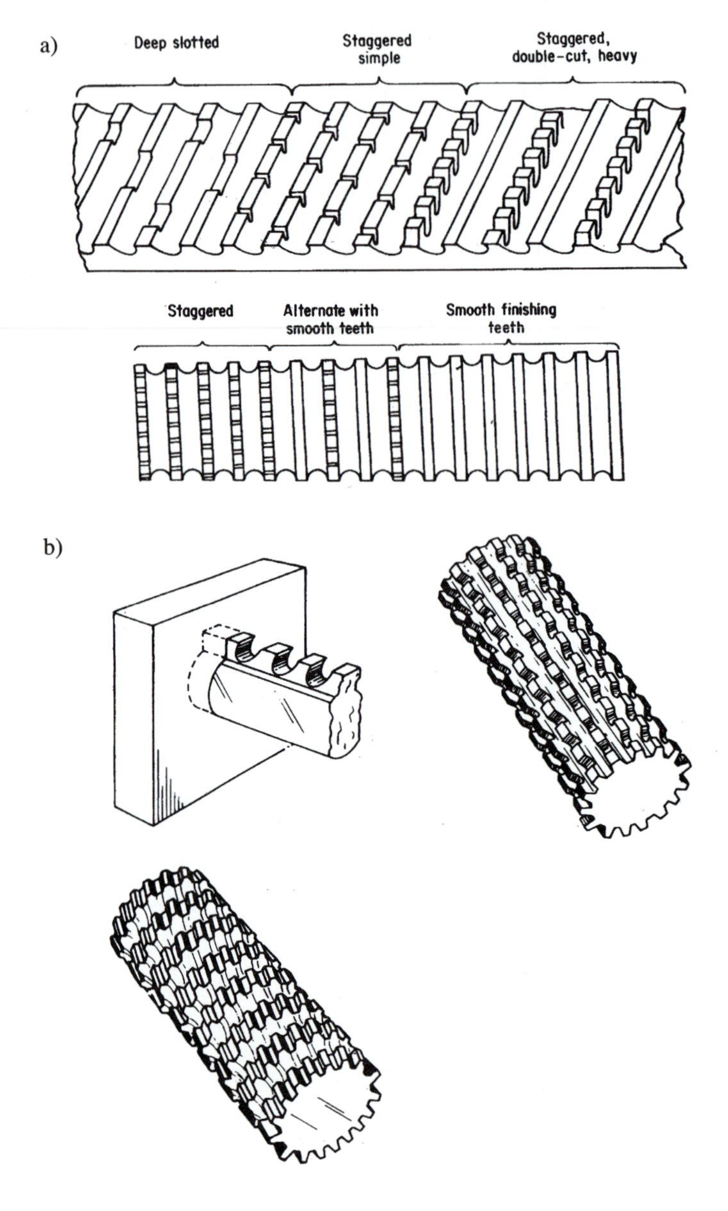

Figure 7.44
Top: Chipbreakers on flat and round broaches. Bottom: Keyway and spline broaches. [NATIONAL BROACH]

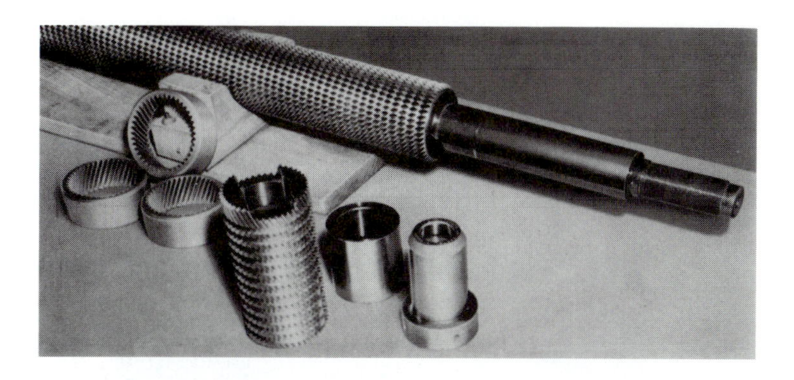

Figure 7.45
The tools and workpieces: internal gears. [National Broach]

Figure 7.46
The tool (a pot broach) and workpiece: an external gear. [National Broach]

Broaching machines are very simple because the geometry of the broach completely determines the whole operation; the machine just has to supply one linear cutting motion. In external broaching, large forces arise between the tool and the workpiece in directions perpendicular to the cutting motion. The machine must be rigid enough not only to sustain these forces but also to prevent chatter vibrations, which are prone to arise considering the large accumulated chip width.

Broaching machines may be classified into the pull and push types, or specialized types (e.g., machines for which the workpieces move on a chain while the tool is stationary), and these are either vertical or horizontal. The vertical machines occupy much less floor space, but their stroke is limited to about 1.5 m. For longer strokes, horizontal machines are used. The cutting motion is generally driven by a hydraulic cylinder.

As an example, a horizontal-type internal broaching machine is shown in Fig. 7.47. It shows the machine with a spline broach and the workpiece, which is a bevel gear with a splined hole. The machine can be equipped with automatic part loading and unloading and with automatic tool handling. As an example of part handling, consider small gears that have splines broached in their holes. Each part passes through a chute into the loading position, and a loader pushes it into the broaching position. Then the broach puller engages the broach, and the part is machined. The completed part drops into the unloading chute, the broach is returned, and the cycle is repeated. The complete cycle takes 8 seconds. Machines of this type are built in various sizes, for maximum pulling force from 40 kN to 1.5 MN.

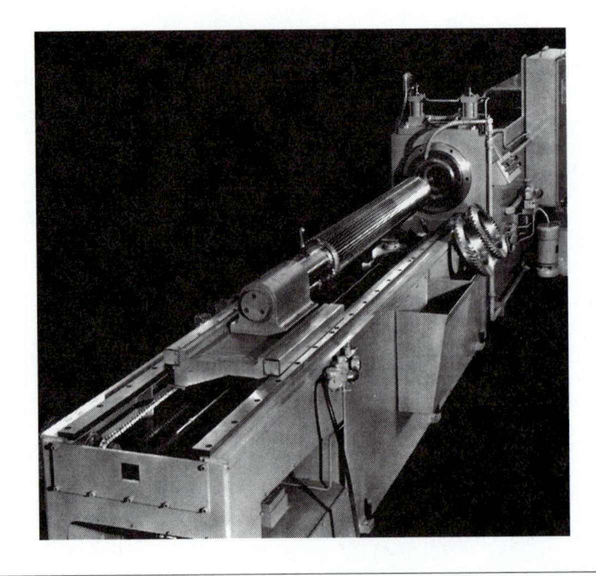

Figure 7.47
A horizontal internal broaching machine. Illustrated is a round broach for production of multiple splines in the holes of bevel gears (shown at the side). These gears will be clamped on the face of the housing opposing the broach. [NATIONAL BROACH]

REFERENCES

1 Boothroyd, G. *Fundamentals of Metal Machining and Machine Tools.* Washington, DC: Scripta Book Co., 1975 and McGraw-Hill Book Co., New York.

2 Boothroyd, G.; and W. A. Knight. *Fundamentals of Metal Machining and Machine Tools.* New York: Marcel Dekker, 1989.

3 Society of Manufacturing Engineers. *Tool and Manufacturing Engineers Handbook.* Vol. 1. *Machining*. 4th ed. Dearborn, MI: Society of Manufacturing Engineers, 1983.

4 ASM International. *Metals Handbook.* Vol. 16. *Machining.* Materials Park, OH: ASM International, 1989.

5 *Modern Metal Cutting.* Fair Lawn, NJ: Sandvik Coromant, 1980.

6 *Economic Handbook of the Machine Tool Industry*, 1993/94. Association for Manufacturing Technology, McLean, VA.

Illustration Sources

[SME1] Reprinted with permission from *Tool and Manufacturing Engineers Handbook,* 4th ed., vol. 1, *Machining*, by permission of the Society of Manufacturing Engineers, Dearborn, MI.

Industrial Contributors

[DETROIT BROACH] Detroit Broach and Machine, 950 S. Rochester Road, Rochester, MI 48063.

[GIDDINGS & LEWIS] Giddings and Lewis Co., 142 Doty Street, Fond Du Lac, WI 54935.

[KENNAMETAL] Kennametal Inc., Latrobe, PA 15650.

[MAZAK] Mazak Machine Tools, 1, Norifune, Oguchi-Cho, Niwa-Gun, Aichi-Ken, Japan.

[MILACRON] Milacron Inc., 4701 Marburg Ave., Cincinnati, OH 45209-1025.

[MITSUBISHI] Mitsubishi Heavy Industries, Ltd., 5-1, Marunouchi 2 chome, Chiyoda-ku, Tokyo, Japan.

[NATIONAL BROACH] National Broach and Machine Co., 5600 St. Jean Ave., Detroit, MI.

[OMNITRADE], [SKODA] Omnitrade Machinery, 78 Torlake Crescent, Toronto, Ontario, Canada M8Z 1B8.

[ROCKWELL] Rockwell Int. of Canada, Power Tool Div., 40 Wellington West, Guelph, Ontario, Canada N1H 6M7.

[SANDVIK] Sandvik Coromant, 1702 Nevins Road, Fair Lawn, NJ 07410.

[TRW] TRW, Putnam Tool Div., 2981 Charlevoix Ave., Detroit, MI 48207.

[VALENITE] Valenite Inc., 31700 Research Park Drive, P.O. Box 9636, Madison Hts., MI 48071-9636.

QUESTIONS

Q7.1. Consider the geometry of turning. What is the significance of the side-cutting-edge angle σ? Write out the relationship between feed per revolution f_r and chip thickness h.

Q7.2. Write out the formula for metal removal rate in turning. What is the machining time t_m for turning a cylindrical workpiece of diameter d, with spindle speed n, and feed f_r.

Q7.3. Write out the relationship between cutting speed v, diameter of workpiece d, and spindle speed n.

Q7.4. Write out a formula for the tangential cutting force F_t in turning. How does it depend on chip thickness? What is the "specific force" K_s, and how is it related to "specific power" P_s?

Q7.5. Discuss the significance of the rake angle α and of the relief angle γ on a turning tool.

Q7.6. Sketch at least five different types of turning tools, using different types of inserts and side cutting edge angles.

Q7.7. Why does a carbide turning tool consist of a holder and an insert? How is the insert attached to the holder?

Q7.8. What is the practical difference in the usage of negative and positive inserts? When is each kind used? Make a sketch of the two types.

Q7.9. Sketch the kinematics of a lathe. What is the transmission ratio set between the spindle and the leadscrew for making a thread with a pitch of 0.2 mm if the leadscrew pitch is 5 mm? What ratio is set for turning with feed $f_r = 0.2$ mm if the feed rack and pinion have $DP = 8$ and the pinion has 12 teeth? In the latter case a deviation from nominal feed of $\pm 1\%$ is permitted.

Q7.10. Name the functions to be carried out by the operator to accomplish a turning operation.

Q7.11. Which of the manual operating functions is automated on a turret lathe?

Q7.12. Make a sketch of the frame (bed, columns, cross-rail, ram, table) of a vertical turret lathe (boring mill).

Q7.13. Name and sketch the various tools used in machining simple (small) holes.

Q7.14. How do the rake angle α, and the cutting speed v, vary along the edge of a twist drill? Give the approximate value of the rake angle on the chisel edge.

Q7.15. Make sketches of a spade drill, of a carbide-tipped drill for short holes, and of a trepanning drill.

Q7.16. What kind of inserts are located on a carbide-tipped drilling head, apart from those which form the cutting edges?

Q7.17. What is the necessary additional feature of drilling tools for deep hole drilling?

Q7.18. What is the role of reamers, and what are the basic types?

Q7.19. Write out the formula for the metal removal rate of a solid drill, and the formula for the torque.

Q7.20. Name and sketch the basic types of drilling machines.

Q7.21. How is the operating function of locating the hole for drilling made easier, or eliminated, in a medium-series production? in mass production? How is it automated for small-lot production?

Q7.22. Make sketches indicating the following operations: face milling, cylindrical peripheral milling, end milling.

Q7.23. Make a sketch of chip thickness variation in milling and derive the formula for the mean chip thickness.

Q7.24. Write out the formula for metal removal rate and for power in a milling operation. How will you find the applicable "specific power" P_s?

Q7.25. Make a sketch showing the combinations of positive and negative radial and axial rake angles on milling cutters.

Q7.26. Sketch the various ways of clamping a milling cutter. Which method is used on NC machining centers?

Q7.27. Name the basic types of milling machines and, in a simple sketch, indicate their kinematics.

Q7.28. Make sketches of a table-type and a floor-type horizontal milling and boring machine, and indicate their axes of motions.

Q7.29. Name typical applications of broaching.

PROBLEMS

P7.1. A turning operation such as the one shown in Fig. 7.2 has the following operating parameters:

$d_1 = 75$ mm	Length $= 80$ mm
$a = 5$ mm	Cutting speed $v_m = 150$ m/min
$\sigma = 0°$	$f_r = 0.3$ mm
	Specific force $K_s = 2000$ N/mm^2

Determine the following:

(a) Tangential force F_t (N)
(b) Metal removal rate Q (cm^3/min)
(c) Power P (kW)
(d) Machining time t_m (min)

P7.2. A shaft of 1035 steel is turned down to a diameter of 80 mm from 100 mm. The length of the shaft is 100 mm, and the side cutting edge angle σ is 0°. The cutting speed $v = 120$ m/min with a feed rate $f_r = 0.25$ mm.

Determine the following:
(a) Chip area A (mm^2)
(b) Metal removal rate Q (cm^3/min)
(c) Specific force K_s, using Table 7.1.
(d) Power P (kW)
(e) Machining time t_m (min)

P7.3. (a) What is a drill jig, and when is it used?
(b) How is the drilling of multiple holes done in large-series production?
(c) A multispindle machine is to drill holes of diameters 5, 10, and 25 mm. The cutting speed in each case should be 30 m/min. What are the corresponding spindle speeds in rev/min?
(d) If the feed per revolution is proportional to the drill diameter, $f_r = 0.02d$, what is the feed f (mm/min) required by each of the drills?
(e) Assuming the specific force $K_s = 2000$ N/mm^2, calculate the torque and power required for each drill.

P7.4. A milling operation is being performed similar to the one shown in Fig. 7.32. The feed/tooth $c = 0.2$ mm, and the start and end angles are $\phi_s = 30°$, and $\phi_e = 160°$.
(a) Determine the average chip thickness h_{av}.
(b) Indicate in a graph of h vs. ϕ, the variation of h on one tooth over two revolutions of the cutter.

P7.5. Graphically express the variation of torque in an end mill. In a graph of torque vs. angle ϕ, plot the torque of three subsequent teeth and find the maximum total torque on the end mill. The cutter has 8 straight teeth and a diameter of 100 mm. The workpiece engages the cutter from $\phi_s = 45°$ to $\phi_e = 150°$, and the axial depth of cut $a_a = 20$ mm. Assume a specific force of $K_s = 2000$ N/mm^2.

P7.6. A four-fluted end mill of 40 mm diameter is performing an up-milling operation as shown in Fig. 7.29a, into a workpiece of 1035 steel. The axial depth of cut $a_a = 20$ mm, and the radial depth of cut $a_r = 25$ mm. The cutting velocity $v = 45$ m/min, with a feed tooth of $f_t = 0.25$ mm.

Determine the following:
(a) Spindle speed n (rev/min)
(b) Feed rate f (mm/min)
(c) Metal removal rate Q (cm^3/min)
(d) Average chip thickness h_{av} (mm)
(e) Specific power K_s (W/cm^3), from Table 7.1
(f) Power P (kW)

P7.7. A face-milling operation is being performed similar to that of Fig. 7.32. A 300-mm-diameter cutter with 16 teeth is cutting a 200-mm-wide strip of 1035 material, which is centrally located under the cutter axis. The operating parameters are $a_a = 5$ mm, $v = 150$ m/min, and $c = 0.25$ mm.

Determine the following:
(a) Spindle speed n (rev/min)
(b) Feed rate f (mm/min)
(c) Metal removal rate Q (cm^3/min)
(d) Average chip thickness h_{av} (mm)
(e) Specific power K_s (W/cm^3/sec)
(f) Power P (kW)

P7.8. A broaching operation must produce a slot 8 mm wide by 3 mm deep by 60 mm long in a workpiece of 1035 steel. The broach has a step/tooth of $s_t = 0.075$ mm.
(a) How many teeth will the broach have?
(b) If the pitch is 12.5 mm, how long will the broach be?
(c) How does the number of teeth cutting simultaneously vary?
(d) Determine the specific force K_s, and use the maximum number of teeth engaged simultaneously to find the total broaching force.
(e) If the cutting speed is 20 m/min, what is the power required?

8 Cutting Mechanics

The central problems of the art of metal cutting are *tool wear* and *tool breakage.* These phenomena limit the cutting speed and feed rate, and consequently the *metal removal rates* that can economically be used in machining various workpiece materials. The role of the production engineer is to select the most suitable tools and the economically *optimum cutting speeds and feeds.* The decisive factor in the increase of metal removal rates is the development of *tool materials.* Another consideration involves possible modifications of the workpiece materials to improve their *machinability.* The analytical work in this chapter focuses on the following topics:

1. *Temperature field in the chip and tool.* In cutting, rather high temperatures (500–1000°C) are generated at the interface between the chip and the tool. These together with interface pressures on the order of 1000–3000 MPa have a strong effect on tool wear. We will analyze the influence of depth of cut, feed rate, and cutting speed, as well as of the mechanical and thermal properties of the workpiece on these temperatures. Thus we will gain insight into the requirements for tool materials and an understanding of the different degrees of machinability of materials. Our analysis also leads into a useful exercise in writing a finite difference program for heat transfer computation.
2. *Discussion of cutting tool materials.* The requirements imposed on these materials can be briefly stated as high hot hardness, toughness, and lack of chemical affinity to the workpiece materials. Historically, the increase of metal removal rates has progressed with the development of tool materials. The main classes described here are high-speed steels, sintered carbides, ceramics, borazon, and diamond. We will discuss their individual suitability for various tool designs and for machining different materials.
3. *Tool wear.* Basic forms and mechanisms of tool wear are discussed, and a classic formula is presented of the relationship among feed rate, cutting speed, and tool life. Using this formula, optimum feed and speed are derived, first for a

single tool/single pass operation and, subsequently, for multi tool–multi workpiece operations, such as they are encountered on multi spindle automatic lathes and on transfer lines.

4. *Tool breakage.* Fundamental features of chipping and breakage of the cutting edge are presented for both continuous and interrupted cutting. Stress analysis is used to explain how, in interrupted cuts, breakage occurs during the exit from the cut. This is then used to show the advantage of down milling over up milling.

No detailed discussion of the chip formation process for the purpose of determining the cutting forces is included. The analysis is very complex, and no good, practical solutions exist yet. However, a great deal of empirical data have been collected over almost a century of tool life and cutting-force measurements, and they fully satisfy the accuracy with which forces must be predicted. Once again, the central problem is tool wear and breakage, not the exact cutting force.

We will discuss all these aspects of metal cutting. At the beginning, we need some understanding of the process of producing the chip. This is called the mechanics of metal cutting. First, however, we will explain the concepts of metal removal rate and cutting force using the example of turning as one of the simplest metal-cutting operations.

8.1 THE CUTTING FORCE

The cutting force acting on the tool is generated by the actively engaged part of the cutting edge, which is drawn with a thick line in Fig. 8.1; it includes the main cutting edge, the nose radius, and a small part of the secondary cutting edge. The direction of the force depends on the ratio of the components of the edge, on the size of the radius with respect to feed f_r, and on the side-cutting-edge angle σ. The force F is split into components: *feed force* F_f, which determines the direct load on the feed drive, the *radial*

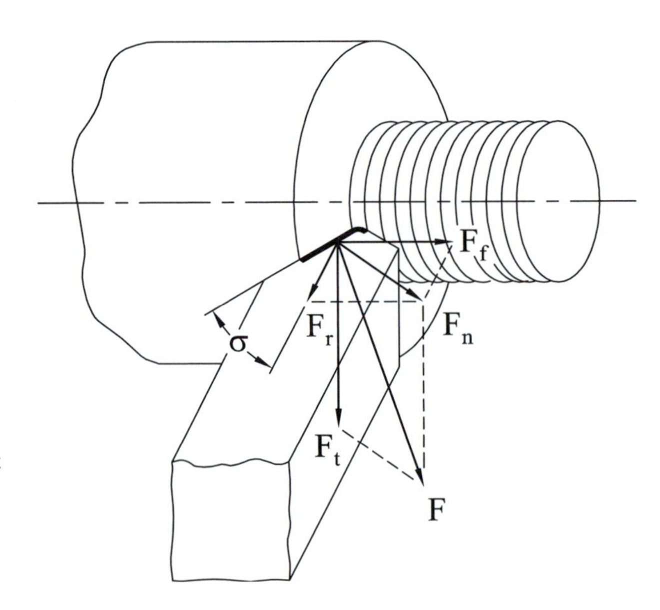

Figure 8.1
The cutting force components in turning. The force F is resolved in the plane normal to the main cutting edge into normal component F_n and tangential component F_t. The normal force F_n itself is further split into the feed force F_f and the radial force F_r.

component F_r, which is decisive for the deflections affecting the accuracy of the machined surface, and the *tangential force* F_t, which has the direction of the cutting speed v and determines the cutting power $P = Fv$. It is tangential to the cut surface. The components F_f and F_r may be combined to give the force F_n, which is normal to the cutting edge.

The magnitude and direction of the cutting force could be determined by an analysis of the chip formation process, as mentioned in the following text. However, for most practical purposes the cutting force is commonly determined by using empirical formulas and data based on numerous cutting tests carried out in many laboratories. This is the approach we take here. For most purposes, it is not necessary to know the force very accurately. Most often the magnitude of the force F_t is required to estimate the power necessary for the operation and to ensure that the tool holder will not break; neither instance requires accuracy better than about 10%. In other cases, such as those of deflection and its effect on accuracy and of self-excited vibrations, it is more the type of relationship between force and deflection that is of interest, rather than accurate data about the magnitude of the force.

In our first approximation the tangential cutting force is considered to be directly proportional to the chip area A:

$$F_t = K_s A = K_s bh = K_s a f_r \tag{8.1}$$

where, as it was shown in Fig. 7.2, b is chip width, h is chip thickness, a is depth of cut, f_r is feed per revolution, and the subscript of the constant K_s means "specific," that is, per unit chip area. Conveniently, the dimensions to use are F_t (N), b (mm), h (mm), A (mm^2), and K_s (N/mm^2).

Multiplying both sides of (8.1) by the cutting speed:

$$\begin{aligned} F_t v &= K_s A v \\ P &= K_s Q \end{aligned} \tag{8.2}$$

If we use the dimension v (m/s), the other parameters are cutting power P (W), metal removal rate Q (cm^3/sec), and K_s (W · sec/cm^3).

Thus, depending on the application, the same number K_s has a different meaning: in (8.1), K_s (N/mm^2) is the *specific force;* in (8.2), K_s [W/(cm^3/sec)] is the *specific power.*

Within the first approximation K_s is a workpiece-material constant. Its values for selected materials are given in Table 7.1. Furthermore, in a first approximation it is common to estimate:

$$F_n = 0.3\, F_t \tag{8.3}$$

The values of K_s in Table 7.1 apply to an average cutting operation, assuming chip thickness of about 0.2 mm and a tool geometry and a cutting speed common for each of the materials. The values of K_s also depend on the state of the material, that is, its heat treatment and amount of previous cold work. In order to include these effects, the ultimate tensile strength is given for which the values of K_s were measured.

In a closer approximation it is recognized that the cutting force is less than proportional to the chip thickness. It is found that the specific force and specific power are greater for thin chips than for thick chips. Some researchers maintain that the relationship between the force F_t and chip thickness is like that in Fig. 8.2a:

$$F_t = b(K_0 + K_1 h) \tag{8.4}$$

This means that the tangential force is proportional to chip width b and as chip thickness h decreases to zero there remains a threshold force F_{t0}. From there on, it

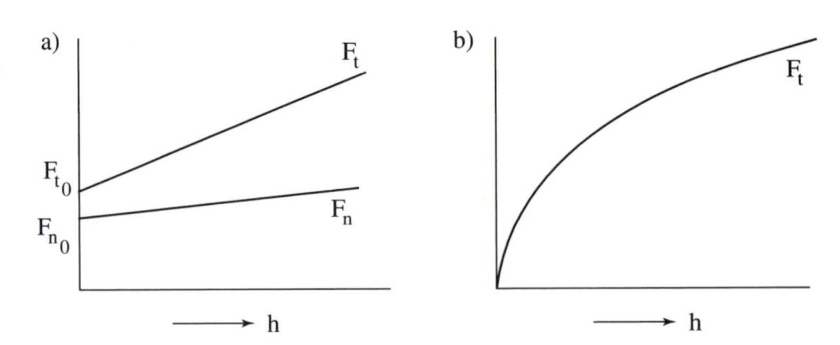

Figure 8.2
Relationship between cutting force and chip thickness as expressed by two models: a) linear; b) power type. In both cases it is acknowledged that force is less than proportional to chip thickness, but for many purposes simple proportionality is satisfactorily used.

increases in proportion to h. A similar relationship applies to the normal force F_n, where the threshold effect is still stronger. All this has to do with the finite sharpness of the cutting edge; as the chip thickness drops below the edge radius there is more rubbing than cutting. Others prefer a relationship like that in Fig. 8.2b, which is expressed as

$$F_t = Kbh^{(1-c)} \tag{8.5}$$

where c is small; typically, we may have $c = 0.2$. Formally, one may still use the concept of specific force, or specific power, K_s which, however, becomes a function of h:

$$F_t = Kh^{-c}\,bh = K_s bh, \tag{8.6}$$

where

$$K_s = Kh^{-c} \tag{8.7}$$

In further, closer approximations, not to be discussed at this stage, the cutting force depends on the geometry of the tool and on the cutting speed. So far, the values of K_s obtained from Table 7.1 should be considered as average; they apply for the usual tool geometry and cutting speed used for cutting the respective materials.

8.1.1 Chip Generation

In general, the chip is being produced by the main cutting edge, the nose, and a small part of the secondary cutting edge of the tool. Our analysis is limited to the action of the main cutting edge only, as seen in a section by a plane *A-A,* which is perpendicular to the edge (see Fig. 8.3.) Thus the process is taken as two-dimensional only. This is called "orthogonal cutting." The picture contains the tool wedge with the associated rake angle α and the relief angle γ. The workpiece is passing the cutting edge with cutting speed v, and the chip is produced by removing a layer of thickness h from the workpiece. In a simplified representation the workpiece material is plastically deformed by a shearing action in a "shear plane" to become the chip. The chip slides on the rake face of the tool along the contact area and then curls away.

The transformation of the workpiece layer into the chip is often studied by means of an experimental technique using a "Quick-Stop" device, as shown in Fig. 8.4. It is essentially a pivoted tool holder supported by a shear pin which, at an instant of cutting, is broken, for example, by "shooting the tool off." In this way the cutting action is suddenly interrupted, and the chip being generated is "frozen" on the workpiece. Subsequently it can be cut off and prepared as a metallographic sample of the kind shown in Fig. 8.5. It was etched and polished after it was obtained so as to show the structure of the material. This particular sample was obtained in cutting low carbon "free machining" steel with inclusions of manganese sulfide, which are visible in the

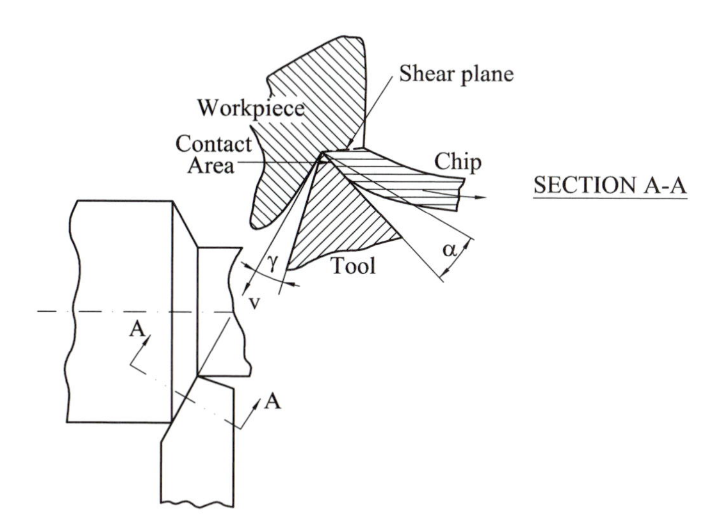

Figure 8.3
Orthogonal cutting: main cutting edge only is considered, neglecting cutting at tool nose. For a wide chip, this is a good approximation. The shear plane and the chip-tool friction zone in the contact area are pointed out.

Figure 8.4
The Quick-Stop experiment. The gun is used to break a shear pin supporting the tool and quickly interrupt the cut to preserve the chip.

structure. The shear plane is indicated by a broken line drawn on the photograph. It clearly separates the undisturbed structure of the workpiece material, which arrives from the right. In place of the tool is the empty black space on the left side of the picture, and the chip moves upwards from the shear plane.

Many materials, mainly most of the steels, provide a continuous chip like the one shown in Fig. 8.5. Sometimes, this kind of chip runs off almost straight like a long ribbon and gets entangled around the toolholder or other parts of the working space. Cutting must be interrupted, and even then it takes considerable time to clear the chip away. This may become an especially grave problem in automatic lathes with turrets and in multispindle lathes. It is therefore necessary to form the chip into tight curls and break it in short pieces. This is achieved by the use of chip breakers and of chip-forming grooves. Some materials are cut in the form of segmented chips, such as cast iron, high alloy steels, and titanium alloys.

Segmented chips are the easiest to remove from the working space of the machine. For machining of gray cast iron, no chip breakers are necessary. However, the surface

Figure 8.5
Quick-Stop metallographic sample. The shear plane is marked by a broken line. It separates clearly the chip from the workpiece material. The laminations in the chip are not parallel with the shear plane.

quality is not as good as in cutting low-alloy steel with high cutting speed. In the extreme, cast iron chips may create problems if they occur almost as dust, and they may clog the passages through which they are supposed to be carried away; therefore no coolant is usually used in cast iron machining.

Another phenomenon that is important in the chip formation process is a "built-up edge" (BUE), which consists of highly strained material adhering to the tool tip. A photograph of a Quick-Stop sample of a case in which a large BUE was generated is shown in Fig. 8.6. In the picture the tool is again missing; instead there is black space at the left hand side of the picture. However, the BUE has come off the tool and is shown between the chip root and the machined surface of the workpiece.

In general, the chip sliding over the rake face of the tool is in a very intimate contact with the tool. The normal pressures between the chip and the tool are very high, typically on the order of 1000 N/mm^2–2000 N/mm^2 in cutting steels, and the temperatures at the contact are high, typically 600°–1000°C for cutting steels, depending mainly on the cutting speed. Under these conditions, there is hardly any actual sliding and a layer of the chip material adheres to the tool. The bulk of the chip moves while there is a shear flow in this layer.

The diagrams in Fig. 8.7 indicate the changes in the chip-tool contact with the change of cutting speed, typically for cutting medium-carbon steel. For a rather low cutting speed v = 0.5 m/sec, a well-developed BUE exists, and parts of it move away with both the workpiece and the chip (Fig. 8.7a). In this case the temperature on the tool face may reach 500°C. Parts of strain-hardened workpiece material adhere to the tip of the tool and act as a projection of the tool nose. In this way, the BUE makes cutting easier because it provides a highly positive rake angle and protects the cutting edge. However, it is not very stable; parts of it are always breaking away and get

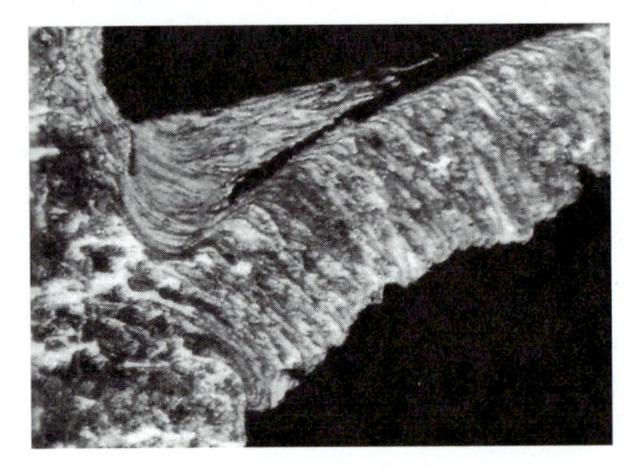

Figure 8.6
A Quick-Stop specimen with built-up edge. The BUE has separated from the tool and adhered to the chip.

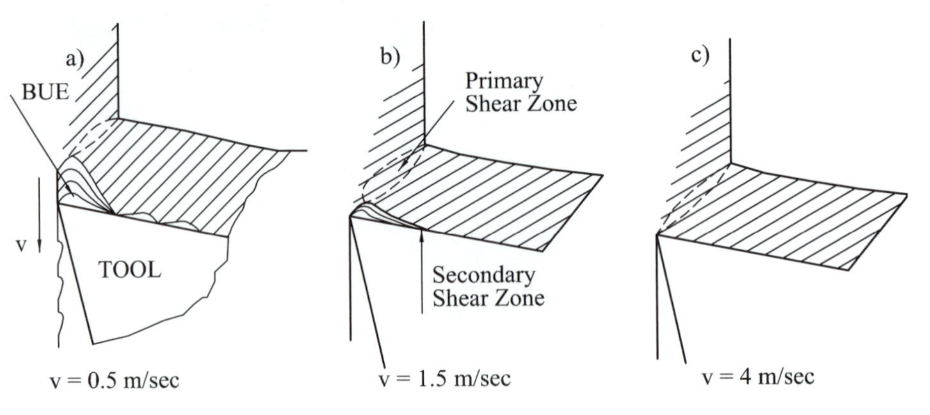

Figure 8.7
Typical variations of the secondary shear zone with cutting speed.

squeezed between the flank of the tool and the generated work surface, which degrades the surface finish of the machined surface. They may also take away particles of the cutting edge and thus increase the rate of tool wear. In Fig. 8.7b with $v = 1.5$ m/sec the temperature on the tool face increases to about 750°C. The BUE has almost disappeared, and there is a thin layer of chip material on the tool face within which there is a steep gradient of velocity change and a shear flow. The zone in which the workpiece material is transformed into the chip is called the primary shear zone, and the layer of the shear flow on the rake face is called the secondary shear zone. In both these zones plastic transformation of the workpiece material takes place, and heat is generated. The structural pattern in the chip is indicated by the shading lines in the diagrams. At high cutting speeds, as in *(c)*, where the temperature on the tool face may reach 1100°C, the secondary shear zone becomes very thin and may not even exist as a continuous layer. Similar changes occur between the flank of the tool and the machined surface. Correspondingly, with the increase of cutting speed the surface finish of the machined surface improves, and the underside of the chip becomes smoother and shinier. The cutting force is primarily a function of the chip area, but it also varies with cutting speed. It is high for very low cutting speeds. It drops to a minimum at a speed at which there is the largest BUE, which provides a very sharp extension of the tool nose. It increases again as the BUE diminishes and reaches a maximum at a speed at which the BUE disappears. From there on it decreases with the increase of cutting speed, at a diminishing rate. These variations are not very significant. For most practical purposes they may be neglected.

The cutting force depends also on the geometry of the tool, especially on the value of the side rake angle α. A change of α by 1 degree plus or minus causes a decrease or increase, respectively, of the tangential force F_t by about 1.5% and of the radial and feed components F_r, F_f by about 5%. The values of K_s in Table 7.1 correspond to the angles commonly used for machining the respective materials.

The model of chip formation is shown in the diagram of Fig. 8.8a, which represents "orthogonal cutting"; that is, it shows a section by a plane perpendicular to the cutting edge through the cutting process carried out on the straight main cutting edge. The effects of cutting on the tool nose are neglected. This process is identical all along the main cutting edge, and correspondingly the magnitudes of the cutting-force components are proportional to the length of the cutting edge or else to the width of the chip. A wedge of the tool opposes the incoming material by its top face, the "rake face," which is inclined with respect to the normal N cut surface by the rake angle α. The tip of the tool wedge is the intersection of the plane of the section (the plane of the paper in the figure) with the cutting edge. It is assumed that the plastic transformation of the workpiece layer being removed and having thickness h_1 into the chip is concentrated in the shear plane S, which is inclined to the direction of the cutting speed v by the shear-plane angle ϕ. The dimension h_1 is called the undeformed chip thickness, and the orthogonal cutting process extends into the plane of the paper by the chip width b (not shown). The chip has the deformed chip thickness h_2, which is always greater than h_1. The ratio $r = h_1/h_2$ is called the chip ratio.

In reality the chip is generated not in the (infinitely) thin plane S but in a shear zone of certain thickness. The shear plane is an abstraction. The actual chip-generation process starts with a stress field upstream of the shear zone. The material passing through the shear zone undergoes plastic strain, which affects its yielding, which, on the other hand, is also affected by the temperature increase and strain rates in the material. The stress field in front of the tool is also influenced by the processes between the tool and the sliding chip. A satisfactory analysis of the process has not yet been

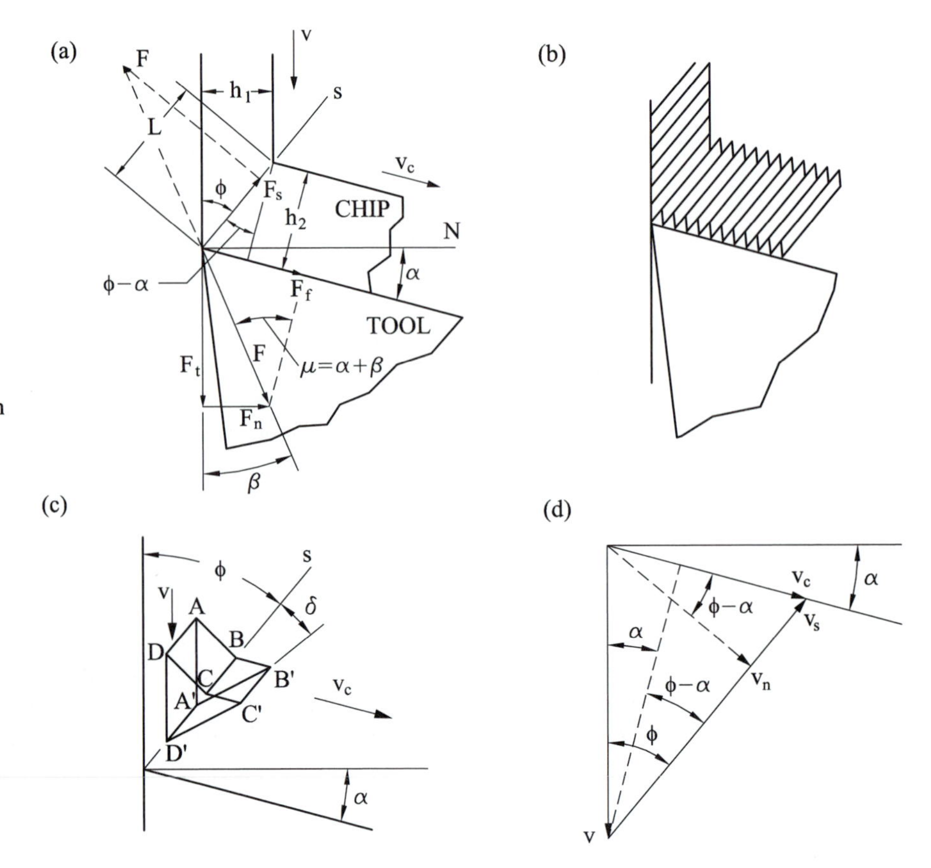

Figure 8.8
Geometric relationships in the model of chip formation. a) Cutting force components and their transformation into the shear and friction forces. Shear angle ϕ, friction angle μ, and tool rake angle α govern the geometry of the process. b) The "deck of cards" model. c) Explaining the slope of the laminations in the chip. d) The hodograph of the cutting, shearing, and chip-sliding velocities.

developed, so far, it is not possible to determine the shear-plane angle ϕ and the cutting force analytically to a good degree of approximation. Instead, both ϕ and F are usually measured experimentally. The angle ϕ can either be obtained from a Quick-Stop test sample like the one in Fig. 8.4, or it can be calculated from the chip ratio r obtained by measuring the chip thickness h_2.

$$h_1/h_2 = r \tag{8.8}$$

The calculation proceeds as follows. The shear-plane length L is obtained both from h_1 and h_2:

$$L = \frac{h_1}{\sin\phi} = \frac{h_2}{\cos(\phi - \alpha)}$$

$$r = \frac{\sin\phi}{\cos\phi\cos\alpha + \sin\phi\sin\alpha} \tag{8.9}$$

$$\phi = \tan^{-1}\frac{r\cos\alpha}{(1 - r\sin\alpha)}$$

The cutting force is determined by either its components F_t and F_n or by its magnitude and the angle β.

The chip-generating process is often visualized by the "deck of cards" model as shown in Fig. 8.8b, which was proposed in 1937 by Piispanen. The workpiece layer to be removed is represented by slices that slip on one another like a deck of cards as they arrive in the shear plane.

The texture of the chip does not, however, show these layers as it is seen in the photograph Fig 8.5 and explained in the diagram of Fig. 8.8c. Let us take a square element *ABCD* of the workpiece material just arriving with its side *BC* into the shear plane *S*. When the side *AD* arrives with velocity v into the shear plane to points A', D', points *B* and *C* will have moved with the chip velocity v_c into points B', C'. The original square is now strongly distorted, and its longest axis indicates the characteristic direction of the structure of the chip, which is inclined by an angle δ to the shear plane. This angle has no significance for the chip formation process; it is mentioned here for the sole purpose of understanding why the chip texture does not correspond directly to the shear angle ϕ. However, the shear plane itself is clearly discernible in Fig. 8.5 as it separates the structure of the workpiece material from that of the chip. The motion of the chip along the rake face of the tool is considered in our model as simple sliding that involves friction. This sliding replaces the actual shearing process in the secondary deformation zone. The chip velocity v_c is obtained from the mass continuity:

$$\begin{aligned} h_1 v &= h_2 v_c \\ v_c &= rv \end{aligned} \tag{8.10}$$

It is now possible to determine the forces, velocities, and powers of both the primary (in the shear plane; let us call this one shearing) and the secondary (friction) deformation processes. The force acting on the workpiece is equal and opposite to the force F acting on the tool. Its component falling into the shear plane is the shearing force F_s:

$$F_s = F \cos(\beta + \phi) \tag{8.11}$$

The force F can be decomposed into a component normal to the rake face and one tangential to the rake face, and the latter is the friction force F_f opposing the sliding of the chip:

$$F_f = F \sin(\alpha + \beta) \tag{8.12a}$$

The shearing velocity v_s is obtained from the hodograph in Fig. 8.8d. The vector of the cutting speed v of the layer cut from the workpiece transforms into the vector of the chip velocity v_c by adding the vector of the shearing velocity v_s.

$$v_s = \frac{v \cos \alpha}{\cos(\phi - \alpha)} \tag{8.12b}$$

The chip-forming process is very complex because the stress field between the rake face of the tool and the shear plane is affected by the temperature field and by the strain rate field in this area, all of which affect the elastic and plastic properties of the material. Furthermore the boundary of the sliding zone on the rake face of the tool is strongly affected by the temperature at this contact and by the sliding speed. However, in an approximation in which the material properties of the chip were the same as those of the workpiece material, it would be possible to see the compressive force F as the one producing the shearing process and, correspondingly, assume that the angle $(\phi + \beta)$ between F and F_s has to be 45°. The angle β depends on the rake angle α and on the friction angle $\mu = \alpha + \beta = \sin^{-1}(F_f/F)$. By this simple reasoning the shear-plane angle should be:

$$\phi = 45° - \beta \tag{8.13}$$

where $\beta = \mu - \alpha$. The feel for this approach is obtained by imagining a tool with zero rake angle and no friction between chip and tool, which would lead to $\phi = 45°$.

However, it is obvious that the seemingly simple expression for ϕ depends on the ability (or lack of it) to judge the "friction" process between the chip and the tool and to assess the value of μ.

The shearing power that is consumed and dissipated in the shear plane is

$$P_s = F_s v_s \tag{8.14a}$$

and the friction power consumed and dissipated in the contact of the chip with the rake face of the tool is

$$P_f = F_f v_c = F \sin(\alpha + \beta) r v \tag{8.14b}$$

It was explained above that with the increase of cutting speed, the temperature in the chip-tool contact increases, and thus the "friction" decreases. This leads generally to an increase of the shear-plane angle ϕ and to an increase of the chip ratio r, which, however, is always less than 1. An increased angle ϕ means a shorter shear-plane length L and a smaller area Lb, and thus, for a given shear stress of the material, a decrease in the cutting force.

It is useful to mention the plastic strain and strain rate involved in the shearing process. With reference to Fig. 8.8d, the strain γ may be obtained as the ratio of the shearing velocity v_s and the velocity v_n normal to the shear plane:

$$v_n = v \sin \phi,$$

$$\gamma = \frac{v_s}{v_n} = \frac{\cos \alpha / \cos(\phi - \alpha)}{\sin \phi} \tag{8.15}$$

$$= \frac{\cos \alpha}{\sin \phi \cos (\phi - \alpha)}$$

The strain rate could be evaluated if one knew how long it took for an element of the workpiece material passing through the shear zone to become an element of the chip; let us denote this time increment by Δt. The strain rate is then

$$\dot{\gamma} = \frac{\gamma}{\Delta t} \tag{8.16}$$

For the idealized "shear-plane" model, Δt is zero and $\dot{\gamma}$ is infinite. However, if a finite thickness d of the "shear zone" is used, we have

$$\Delta t = \frac{d}{v_n}$$

and the strain rate can be established.

EXAMPLE 8.1 Shearing and Friction Power in Chip Formation ▼

Machining steel 1035, chip thickness $h_1 = 0.2$ mm, chip width $b = 6$ mm, cutting speed $v = 3$ m/sec. The rake angle $\alpha = 10°$. Shear-plane angle was measured as $\phi = 28°$. Tangential force is obtained with reference to Table 7.1: $K_s = 2300$ N/mm^2; thus we have $F_t = bh_1K_s = 2760$ N. The normal component is taken as $F_n = 0.3\, F_t = 828$ N. The angle β is obtained as $\beta = \tan^{-1} 0.3 = 16.7°$. The force $F = 2881$ N.

Shear plane length $L = h_1/\sin \phi = 0.426$ mm

Shearing force $F_s = F \cos (\beta + \phi) = 2881 \cos 44.7° = 2048$ N

Shearing stress $\tau = F_s/Lb = 764$ N/mm^2

Deformed chip thickness $h_2 = L \cos 18° = 0.41$ mm

Chip ratio $r = h_1/h_2 = 0.49$

Chip velocity $v_c = rv = 1.48$ m/sec

Shearing velocity $v_s = v \cos 10°/\cos 18° = 3.1$ m/sec

Shearing power $P_s = F_s v_s = 6363$ W

Friction force $F_f = 2881 \sin 27° = 1294$ N

Friction power $P_f = F_f v_c = 1917$ W

Total power $P = F_t v = 2760 \times 3 = 8280$ W

Also $P = P_s + P_f = 8280$ W; the two last lines check well against each other. The metal removal rate $Q = 0.2 \times 6 \times 3 = 3.6\ \text{cm}^3/\text{sec}$. Specific power $P_{sp} = P/Q = 2300\ \text{W}/(\text{cm}^3/\text{sec})$; this value checks exactly with the value of the specific force $K_s = 2300\ \text{N/mm}^2$. ▲

EXAMPLE 8.2 Strain and Strain Rate in Chip Formation ▼

Let us take the same case as in Example 8.1 and estimate the thickness of the shear zone $d = 0.02$ mm.

$$\gamma = \cos 10°/(\sin 28° \cos 18°) = 2.2$$

$$v_n = v \sin \phi = 3 \sin 28° = 1.41\ \text{m/sec}$$

$$\Delta t = d/v_n = 2 \times 10^{-5}/1.41 = 1.41 \times 10^{-5}\ \text{sec}$$

$$\dot{\gamma} = \gamma/\Delta t = 1.56 \times 10^5\ \text{sec}^{-1}$$

Both the strain γ and the strain rate $\dot{\gamma}$ encountered in metal cutting are very high. ▲

8.1.2 Simplified Formulations

For the calculations of the thermal fields in chip and tool, we will simplify the geometry of the cut and the force formula. In practice, depending on the combination of workpiece and tool materials, positive or negative rake angles are used (see Fig. 8.9). Mostly they are chosen in the range of $\pm 8°$. In the following calculations we will simply keep $\alpha = 0°$. Then we can draw the diagram of forces as shown in Fig. 8.10. The cutting force F, under angle β with respect to the cutting speed v, is split into the tangential F_T and normal F_N components. The latter appears as the friction force F_f between the chip and the tool. The chip is formed by shear in the shear plane at angle ϕ with respect to the cutting speed v, $\phi = \tan^{-1}(h_1/h_2)$ where h_2/h_1 is the plastic strain of the chip. Letting the force F act on the work in the way opposite to the action on the tool, we obtain the shear component F_s of the force by projecting F onto the shear plane. The chip is in contact with the tool over the length L_c. The shear velocity v_s is obtained from the hodograph at the lower right of Fig. 8.10, in which the cutting speed v changes into the chip speed v_c.

The tangential force F_T is derived from the chip cross section using the specific force K_s:

$$F_T = K_s\, b\, h_1 \tag{8.17}$$

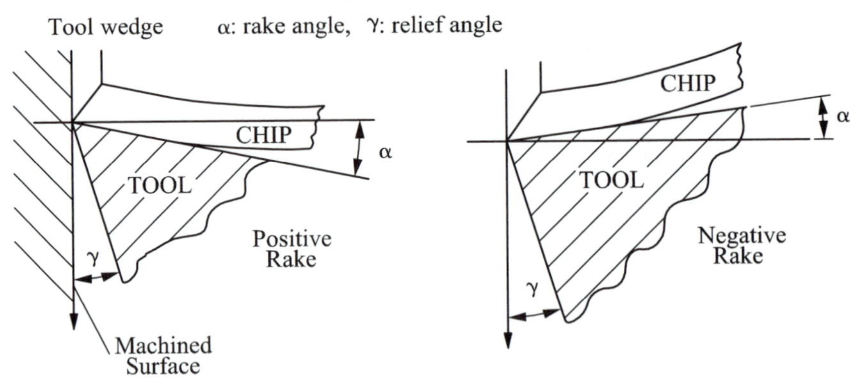

Figure 8.9
Positive and negative rake angles. The former allows easier cutting and lower forces; the latter provides a stronger tool wedge for cutting harder work materials, especially in interrupted cuts (e.g., milling).

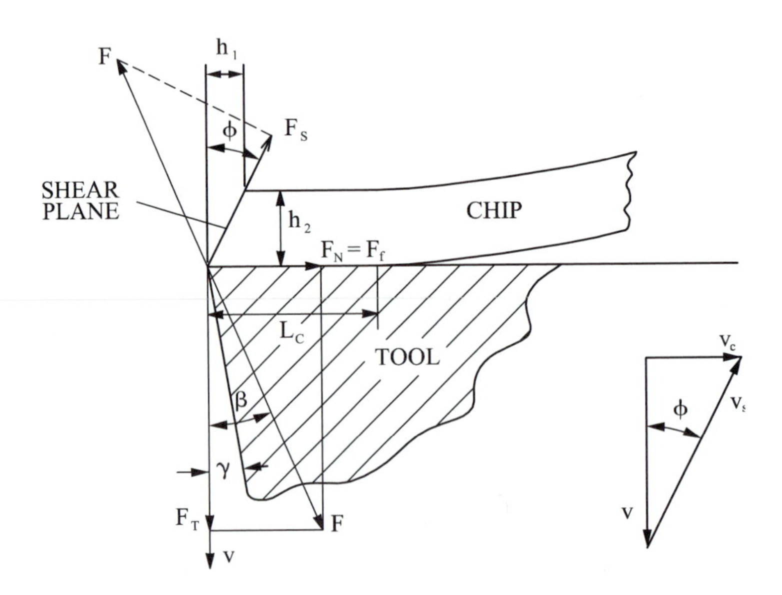

Figure 8.10
Shear and friction forces, and velocities for the simple case of zero rake angle.

where h_1 is the uncut chip thickness and b is the chip width. We assume knowledge of the angles β and ϕ. It is now possible to determine the shearing force, velocity, and power and the friction, force, velocity, and power.

$$F_s = F\cos(\phi + \beta) \qquad v_s = \frac{v}{\cos\phi}$$

The shearing power is

$$P_s = F_s v_s$$

$$F_f = F_T \tan\beta \tag{8.18}$$

$$v_f = v_c = v\tan\phi$$

The friction power is

$$P_f = F_f v_c \tag{8.19}$$

A different approach in which the forces and angles are derived from the shear flow stress is used in problem P8.10.

8.2 TEMPERATURE FIELD IN THE CHIP AND IN THE TOOL

The temperature at the contact between the chip and the tool and, in general, in the tool tip is one of the most decisive factors for tool wear rate, and it plays an important role in the machinability of various workpiece materials. It is affected by both the mechanical and thermal properties of the work material and the cutting conditions, especially by the cutting speed v and by the chip thickness h.

A considerable amount of research work has been devoted to the analysis of temperature fields in the chip and in the tool. These fields can be measured, and various ways have been developed to do so. To mention just a few of the most common ones, the "average temperature in the cut" is measured by using the tool-workpiece contact as a thermocouple; infrared radiation is used to measure temperature at various points on the chip and tool. Temperatures may also be assessed by determining metallurgical changes in the microstructure of a high-speed steel tool. Methods of computing the temperature fields in the chip and the tool have been highly developed. The finite-element method in particular gives rather accurate results.

We will use rather transparent finite-difference computations under simplifying assumptions. Nevertheless, the errors due to these assumptions will be held within very reasonable limits. Our purpose, however, is to provide an insight into the role of the most significant factors as well as to offer a useful exercise. Simple calculations as well as computations using the finite-difference method applied to simplified cases of chip formation are presented in what follows. We use a method that is a numerical modification of an approach used by Boothroyd [3] in his dissertation. He transformed the two-dimensional steady-state problem into a single-dimensional transient one. This provides a computational advantage.

8.2.1 Shear Plane Temperature

It is assumed that shear stress is constant (uniform) over the whole shear-plane area. Correspondingly, heat generation is uniform, and the work material when passing through the shear zone is immediately and uniformly heated up. Heat is generated within the material, and in passing through the shear plane the temperature of the work increases from room temperature $T_r = 20°C$ to shear-plane temperature T_s.

The shearing power represents work (heat) in unit time, and the temperature increase $(T_s - T_r)$ is obtained by dividing it by the heat capacity of the material passing through the shear zone in unit time, which is the heat capacity of the metal removal rate Q:

$$(T_s - T_r) = \frac{P_s}{Q \rho c}$$

where ρ is specific mass, and c is specific heat. These parameters are expressed in convenient units as follows:

P_s is shearing power (mW = Nmm/sec)
b is chip width (mm)
v is cutting speed (mm/sec)
h (mm) is chip thickness
ρ is specific mass (kg/m^3)
c is specific heat (Nm/(kg · °C))
ΔT is temperature increase (°C)

The product (ρc), which represents "specific heat per unit volume," may be given in the following units:

$$\rho c\left(\frac{\mathrm{N}}{\mathrm{mm}^2\cdot{}^\circ\mathrm{C}}\right) = \rho\left(\frac{\mathrm{kg}}{\mathrm{m}\cdot\mathrm{mm}^2}\right) c\left(\frac{Nm}{kg\cdot{}^\circ C}\right)$$

$$= 1\times 10^{-6}\,\rho\left(\frac{\mathrm{kg}}{\mathrm{m}^3}\right) c\left(\frac{\mathrm{Nm}}{\mathrm{kg}\cdot{}^\circ\mathrm{C}}\right)$$

In all our calculations the first of the above alternatives will be used for ρc. In this case we have:

$$(T_s - T_r)\,({}^\circ\mathrm{C}) = \frac{P_s\,(\mathrm{Nmm/sec})}{[bh\,(\mathrm{mm}^2)\;v\,(\mathrm{mm/sec})\;\rho c\,(\mathrm{N/(mm^{2\circ}C)})]} \tag{8.20}$$

The values of the various mechanical and thermal parameters are given for selected workpiece materials in Table 8.1. The materials are characterized by their *UTS* values.

Question: How will T_s depend on changes of h and v?

In the first approximation, the cutting force depends on the undeformed chip dimensions only; it does not change with cutting speed. In that same first approximation the shear-plane angle ϕ does not depend on cutting speed either. Under these assumptions, we have:

TABLE 8.1 Mechanical and Thermal Properties of Selected Workpiece Materials

No.	Material	*UTS*	K_s	k	α	T_m	ρc
1	Cast iron BHN		1500	43	12	1220	3.7
2	Steel 1020 N	400	2100	43	12	1520	3.7
3	Steel 1035 N	500	2250	43	12	1500	3.7
4	Steel 1045 N	650	2650	43	12	1490	3.7
5	Stainless steel 302	700	2700	15	4.4	1425	3.6
6	Alloy steel 4140 H	900	2800	38	10	1510	3.7
7	Alloy steel 5140 H	950	2800	40	11	1500	3.7
8	Ni-based AISI 688 (Inconel X)	1350	3300	12	3.2	1370	3.7
9	Ni-based AISI 684 (Udimet 500)	1400	3400	12	3.2	1370	3.7
10	Co-based AISI 670 (L605)	1050	3300	10	2.8	1370	3.6
11	Ti (6Al, 4V)	1350	2000	7	2.6	1600	2.7
12	Al 7075-T6	530	850	140	60	540	2.3

Note: Column heads represent the following:
UTS, ultimate tensile strength, N/mm^2 (MPa)
K_s, specific force, N/mm^2
k, thermal conductivity, N/(sec · °C)
$\alpha = k/(\rho c)$, thermal diffusivity, mm^2/sec
T_m, melting temperature, °C
(ρc), specific heat per volume, N/(mm^2·°C)
T_s, shear plane temperature, °C

$$(T_s - T_r) = \frac{P_s}{bhv\rho c}$$

$$P_s = F_s v_s = \frac{F_t}{\cos\beta}\frac{\cos(\beta+\phi)\,v}{\cos\phi}$$

Let us denote

$$D = \frac{\cos(\beta+\phi)}{\cos\beta\cos\phi}$$

Then

$$P_S = F_T vD = K_S bhvD$$

$$T_s - T_r = \frac{K_s bhvD}{bhv\,\rho c} = \frac{K_s D}{\rho c} \tag{8.21}$$

where K_s (N/mm^2), ρc (N/mm$^2 \cdot$ °C)

Thus, in the first approximation, the shear-plane temperature does not depend on the chip dimensions b and h, nor on the cutting speed v. It is proportional to the specific force K_s, and inversely proportional to specific mass ρ and specific heat c. In this way, T_s is determined by the mechanical and thermal properties of the workpiece only, and it applies even to very slow cutting. This perhaps surprising result is easy to explain. While the power spent in the shear plane increases in proportion with speed and chip thickness, the flow rate of material through the plane that gets heated by that power increases in the same proportions.

In reality, because K_s slightly decreases with chip thickness h, and cutting force F further decreases with cutting speed v, the shear-plane temperature slightly decreases with the increase of both h and v.

Accepting the simplification of the above first approximation, let us compare the shear-plane temperatures of various workpiece materials. We will use the values of the mechanical and thermal parameters given in Table 8.1, and choose for all the materials the following average values of the chip formation parameters: $\alpha = 0°$, $\beta = 17°$, $\phi = 28°$, $r = 0.5$. The constant D then becomes $D = 0.8$ and, for $T_r = 20$°C, the resulting temperatures T_s in the shear plane of the various workpieces are graphically expressed in Fig. 8.11. They also represent, in a simplified way, the temperatures between tool and workpiece at very low cutting speeds.

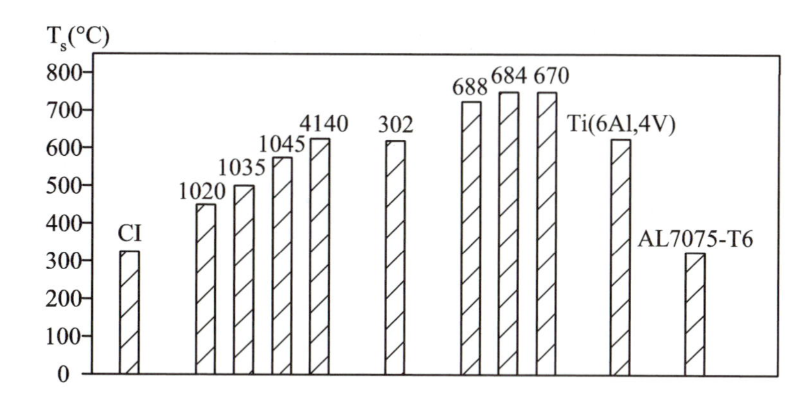

Figure 8.11
Common shear-plane temperatures for various workpiece materials.

The friction power is generated in the contact between the chip and the tool (see Fig. 8.12). The distribution of this power follows the distribution of the normal pressure. In the numerical treatment we divide the chip into elements both in the direction X of the face of the tool and in the direction Y of the shear plane. The subscript for the former division is denoted k, and it starts with $k = 1$ on the cutting edge and amounts to KK at the end of the contact length L_c. The pressure on the tool has been studied experimentally by Yellowley and found to be maximum and constant over the distance $h_1/2$ from the cutting edge and to decrease linearly between this point and the end of contact.

We can derive the friction power acting over the area $b\Delta x$ of one slice. Assume that the length of contact is proportional to the undeformed chip thickness

$$L_c = mh_1 \tag{8.22}$$

Then

$$\frac{h_1}{2} = \frac{L_c}{2m} = \frac{KK}{2m}\Delta x$$

and

$$p_{\max}\left[\frac{KK}{2m} + \frac{1}{2}KK\left(1 - \frac{1}{2m}\right)\right] = P_f \tag{8.23}$$

and for $k \leq KK/2m$, $p_k = p_{\max}$.

For

$$\frac{KK}{2m} < k \leq KK, \qquad p_k = p_{\max} - \Delta p\left(k - \frac{KK}{2m}\right) \tag{8.24}$$

where

$$\Delta p = \frac{p_{\max}}{KK\,(1 - 1/2m)} \tag{8.25}$$

for $k > KK$, $p_k = 0$.

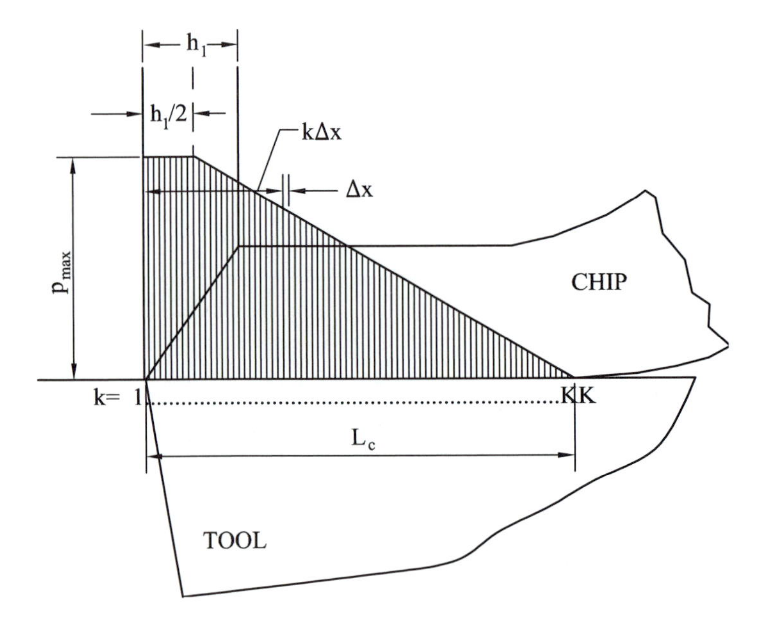

Figure 8.12
Distribution of friction power over the rake face of the tool. It is assumed constant from the cutting edge through half of undeformed chip thickness and decreases linearly to the end of the contact zone (AFTER YELLOWLY).

So, for instance, if $m = 4$, $KK = 400$,

$$p_{\max}\left(\frac{KK}{8} + \frac{1}{2} \times \frac{7}{8} KK\right) = P_f \tag{8.26}$$

$$\begin{aligned} p_{\max} &= \frac{P_f}{50 + 175} = \frac{P_f}{225} \\ \Delta p &= \frac{p_{\max}}{350} = \frac{P_f}{7875} \end{aligned} \tag{8.27}$$

$$\begin{aligned} p_k &= p_{\max} \qquad \text{for } k = 1, 50 \\ p_k &= p_{\max} - \Delta p\,(k - 50), \qquad k = 50, 400 \end{aligned} \tag{8.28}$$

8.2.2 Computing the Temperature Field

Phase 1: Neglecting Heat Escaping Through the Tool

It is assumed that the heat convected from the chip into the surrounding air is completely neglected, and no heat is conducted away through the tool. The former assumption is quite realistic, while with the latter a more significant error is introduced that will be evaluated and corrected in Phase 2. Both these assumptions together mean that all heat is conducted away by the chip.

We will consider chip area ABCD as shown in Fig. 8.13. The boundary conditions are such that along the shear plane AB the temperature is T_s, along AC and BD the chip is insulated, and along AC over the contact length L_c there is power (heat) input p due to the chip-tool friction. The friction force and, correspondingly, the heating power p are distributed as shown: it is maximum and constant from the cutting edge A over half the undeformed chip thickness h_1, and thereafter it decreases linearly towards the end of the contact. The heat spreads in the chip by conduction in the direction Y and by conduction and mass transfer (chip motion) in the direction X.

The whole field of the chip is divided in incremental slices in both directions. Those in the direction X are subscripted k, from 1 to KK over the contact length and further, up to a total of KKK. The slices in the direction Y are subscripted j, from 1 to 20. In this way, the whole field is divided in 20 times KKK elements. Applying the method of finite differences, we assume that the temperature is constant over every element and changes discretely from element to element.

It has been shown by Boothroyd that the mass transfer of heat in X is much more powerful than conduction. The latter can be neglected; it is sufficient to consider the former mode only. Correspondingly, we will consider heat conducted in Y and heat moved by mass transfer in X. This is symbolically expressed by assuming that every vertical slice moves in direction X with the velocity of the chip v_c while being insulated, in the direction X, against the neighboring slices, as indicated in Fig. 8.13b for the kth slice. Instead of considering a *steady-state, two-dimensional* (X, Y) heat transfer problem, we formulate *one-dimensional* (Y), *transient* case. We will follow one vertical slice as it moves, in discrete increments of time, through positions 1 to KKK. Over the contact length, from 1 to KK, it obtains at its lower end heat input p_k and the heat spreads in this slice in the direction Y, by conduction. Between KK and KKK there is no more heat input. The temperature of every element $T_{j,k}$ varies through the steps 1 to KKK. Under our assumption of no convection of heat out of the chip, the temperature in such a slice,

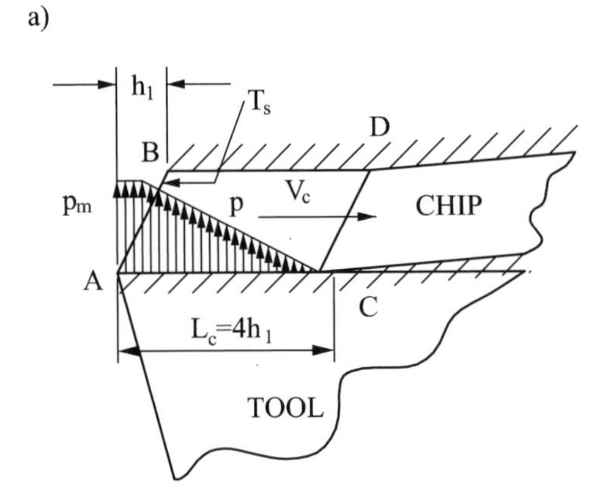

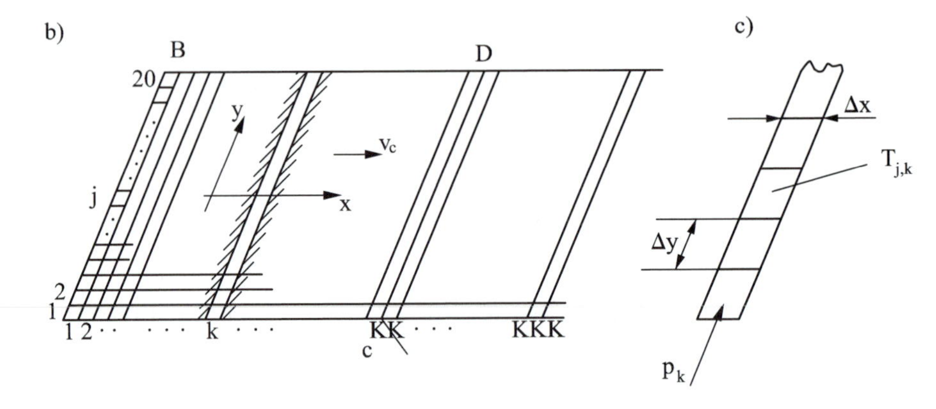

Figure 8.13
Finite-difference formulation for temperatures in the chip. Convection of heat out of chip is neglected. A uniform shear-plane temperature is assumed on the left boundary. At this stage also heat conducted into the tool is neglected. Friction heat is pumped into the chip at the chip-tool interface. Heat spreads by conduction in Y and by mass transfer with velocity v_c as well as conduction in X. However, mass transfer is much more powerful than conduction in X which is, therefore, left out—this is indicated by showing the kth vertical slice insulated on its sides. In this way, the process is then expressed by a single dimensional (direction Y) transient field in a slice Δx thick moving in X. The temperature profile in the Y direction in the slice as it moves in the steps k expresses the steady state XY temperature field.

far away from the end of the contact zone, would approach a uniform, constant value. For practical purposes, we choose the following:

$$KKK = \frac{3\,KK}{2}.$$

The shear-plane temperature is determined from Eq. (8.20). It will further be accepted that, on average, for most our demonstrations,

$$L_c = 4h_1$$

The total heat flow over the area of contact length L_c times chip width b equals the friction power P_f, Eq. (8.19).

The heat flux distribution is as shown in Fig. 8.12, and it is expressed in Eqs. (8.23), (8.24), and (8.25) in the form of power injected into each slice. The initial state of the thermal field in the chip is such that the slice of $k = 1$ has the following uniform shear-plane temperature:

$$j = 1 \text{ to } 20, \qquad T_{j,1} = T_s \tag{8.29}$$

At every instant k the lowest element, $j = 1$, receives heat input p_k and loses heat by conduction to the next higher element, $j = 2$. The surplus of heat is used to increase the temperature over the time instant Δt from $T_{1,k}$ to $T_{1,k+1}$:

$$p_k - \frac{T_{1,k} - T_{2,k}}{\Delta y}\Delta xbk = b\Delta x\,\Delta y\,\rho c\,\frac{T_{1,k+1} - T_{1,k}}{\Delta t} \tag{8.30}$$

where the units used are p_k (N · mm/sec), T (°C), b (mm), Δx (mm), Δy (mm), ρc [N/(mm^2 · °C)], Δt(sec), k [N/(sec. °C)] is thermal conductivity, and

$$\Delta t = \frac{\Delta x}{v_c} = \frac{L_c}{KKv_c} \tag{8.31}$$

The heat conducted per time step Δt from $j = 1$ to $j = 2$ is the second term in the bracket on the left side of Eq. (8.30). It is proportional to the temperature gradient $(T_{1,k} - T_{2,k})/\Delta y$, to the area $A = \Delta xb$, and to the thermal conductivity k.

The right side is the heat capacity of the element (which is the product of its volume $b\,\Delta x\,\Delta y$ and of the specific mass ρ and specific heat c multiplied by the temperature increase $(T_{1,k+1} - T_{1,k})$ over time Δt.

For the elements $j = 2$ to 19, instead of the heat input p_k the heat flow from the next lower element is used:

$$\left(\frac{T_{j-1,k} - T_{j,k}}{\Delta y} - \frac{T_{j,k} - T_{j+1,k}}{\Delta y}\right)\Delta xbk = b\Delta x\Delta y\rho c\frac{T_{j,k+1} - T_{j,k}}{\Delta t} \tag{8.32}$$

For $j = 20$, the second term in the brackets on the left side of Eq. (8.32) is missing because heat is neither conducted out of this element nor convected out of the chip.

Eqs. (8.30) to (8.32) can be used to express the new temperatures at time $(k + 1)$:

$$j = 1, \qquad T_{1,k+1} = \left[\frac{p_k}{b\Delta x\,\Delta y\,\rho c} - \frac{(T_{1,k} - T_{2,k})\,\alpha}{(\Delta y)^2}\right]\Delta t + T_{1,k} \tag{8.33}$$

$$j = 2 \text{ to } 19, \qquad T_{j,k+1} = (T_{j-1,k+1} + T_{j+1,k} - 2T_{j,k})\frac{\alpha\,\Delta t}{(\Delta y)^2} + T_{j,k} \tag{8.34}$$

$$j = 20, \qquad T_{20,k+1} = (T_{19,k+1} - T_{20,k})\frac{\alpha\Delta t}{(\Delta y)^2} + T_{20,k} \tag{8.35}$$

where $\alpha = k/\rho c$ is thermal diffusivity as given in Table 8.1. In Eqs (8.34) and (8.35) the first term on the right side is subscripted $(k + 1)$ and not k as might have been expected. This means that we are updating temperatures as soon as the new values are available, which improves the convergence and stability of the computation.

The time increment Δt, Eq. (8.31), must not be chosen too large because then the computation in discrete steps would not converge. This condition leads to a minimum value of KK for computational stability. We will recommend values of KK for the individual examples.

Phase 2: Correcting for Heat Escaping Through the Tool

The temperatures $T_{j,k}$, $j = 1, 20$, $k = 1, KKK$ will be obtained by using a corresponding computer program. First, however, *Phase 2* of the program is explained, in which we determine the heat flow (power) escaping through the tool and subsequently introduce a correction for it.

It is assumed that heat escaping into the surrounding air is neglected. A greater part of the heat is carried away by the chip, and a smaller part is conducted away through the tool; how much is to be established in the exercise. First, we will derive the corresponding equations.

The shape of the tool and the thermal field in it are strongly simplified (see Fig. 8.14). The general set up is shown in diagram *(a)*. Heat is generated first in the shear plane *S* and then in the chip/tool contact. Most of it is taken away by the chip, but some of it flows through the tool to the tool holder, and into the machine structure. Obviously, some of the heat is convected and radiated out of the tool to the surrounding air, and we simply assume that the tool far away from the cutting zone is already at room temperature. We further assume that the isotherms in the tool are, more or less, lines of equal distance from the cutting zone. Therefore, we will represent the tool as a wedge with the chip/tool contact at the flat top, as in Fig. 8.14b. It will be divided into slices with thickness Δz_i, width w_i, and depth (into the paper) *b*. The thermal field is assumed single-dimensional in *z*, while every slice has a constant temperature throughout. The sides are assumed isolated and the bottom is at room temperature T_r. From the point of view of the tool the temperature along the contact length L_c is also constant. It will be taken as equal to the average temperature along L_c. Thus we formulate a *steady-state, single-dimensional* problem. The temperature T_{cav} is expressed as

$$T_{cav} = \frac{1}{KK} \sum_{1}^{KK} T_{1,k} \tag{8.36}$$

The tool consists of two parts: part *A* made of sintered carbide, and part *B* of steel. The dimensions are defined as: $L_c = 4h_1$, $A = 6$ mm, $B = 14$ mm, and the slopes of the tool sides are 45°. Consider altogether 40 layers with thickness $\Delta z = 0.5$ mm.

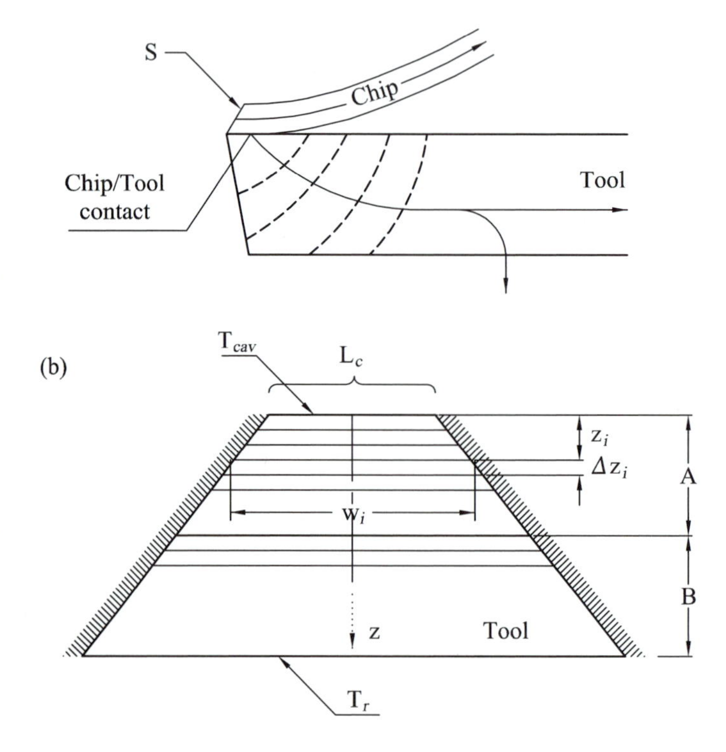

Figure 8.14
Diagram of the finite-difference formulation for the tool (single-dimensional, steady-state process). This heat flow from the chip-tool interface is used to correct the first phase of the computation in which the conduction into tool was neglected. Convection from the tool sides is neglected. At the top of the tool wedge is the contact with the chip; the average temperature at this contact is used on this boundary, and room temperature is used at the bottom end.

The properties of sintered carbide are taken as follows:

Thermal conductivity $k_a = 70$ N/(sec · °C)

Specific heat is $c = 200$ Nm/(kg · °C)

Specific mass ρ is in the range 12,000 to 15,400 kg/m^3

However, we will not need ρ and c. The thermal conductivity of the steel shank is $k_b = 43$ N/(sec · °C).

The temperatures T_i of the individual slices do not change (steady state). All heat entering a slice leaves it. The power flowing through the tool is P_t and it is the same for each slice (see Fig. 8.15). The first step goes from the top of the tool (the interface between chip and tool) to the middle of the top slice where temperature T_1 is located. The mean width is

$$w_1 = L_c + \frac{\Delta z}{2} \tag{8.37}$$

For all the following slices, we step from the middle of one to the middle of the next one: the distance is Δz, and the mean width is

$$w_i = L_c + 2(i - 1)\,\Delta z, \qquad i = 2, 40 \tag{8.38}$$

For the last slice, the step goes from T_{40} to T_r over a distance of $\Delta z/2$:

$$w_{41} = L_c + 79.5\Delta z \tag{8.39}$$

Then we have

$$\begin{aligned}
P_t &= \frac{T_{cav} - T_1}{\Delta z/2} \times w_1\, bk_a, & R_1 &= \frac{\Delta z/2}{w_1} \\
P_t &= \frac{T_1 - T_2}{\Delta z} \times w_2\, bk_a, & R_2 &= \frac{\Delta z}{w_2} \\
P_t &= \frac{T_i - T_{i+1}}{\Delta z} \times w_i\, bk_i, & R_{i+1} &= \frac{\Delta z}{w_{i+1}} \\
P_t &= \frac{T_{41} - T_r}{\Delta z/2} \times w_{41}\, bk_b, & R_{41} &= \frac{\Delta z/2}{w_{41}}
\end{aligned} \tag{8.40}$$

Now, expressing the individual temperatures in terms of T_{cav},

$$T_1 = T_{cav} - \frac{P_t}{bk_a} R_1$$

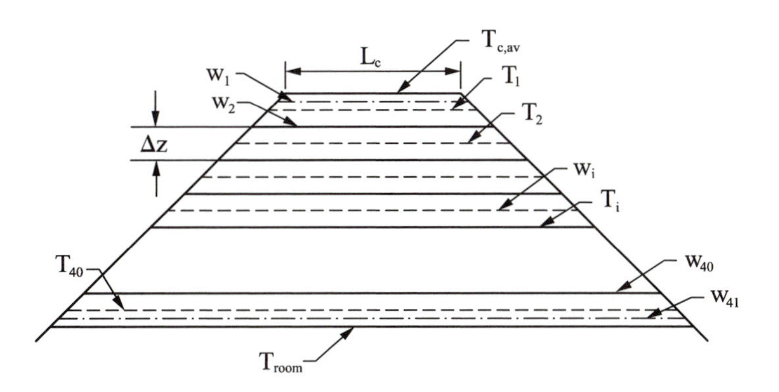

Figure 8.15 Specification of the thermal field in the tool as steady-state, single-dimensional. Heat flow is constant from slice to slice across the stepwise increasing width w.

$$T_2 = T_1 - \frac{P_t}{bk_a} R_2 = T_{cav} - \frac{P_t}{b}\left(\frac{R_1}{k_a} + \frac{R_2}{k_a}\right) \tag{8.41}$$

$$T_{i+1} = T_i - \frac{P_t}{bk_i} R_{i+1} = T_{cav} - \frac{P_t}{b} \sum_{1}^{i+1} \frac{R_i}{k_i}$$

Denoting $r = \sum_{1}^{41} \frac{R_i}{k_i}$, (8.42)

$$T_r = T_{cav} - \frac{P_t r}{b}$$

and

$$P_t = (T_{cav} - T_r)\frac{b}{r} \tag{8.43}$$

The "thermal resistance" r is determined by the dimensions and thermal conductivities of the tool. We can now formulate the computer program, which consists of two sections.

1. Provide input data: h_1 (mm), h_2 (mm), b (mm), L_c (mm), β (deg), K_s (N/mm^2), v (mm/sec), *KK, KKK, con* (N/sec · °C), k_a, k_b, (ρc) (N/(mm^2 · °C)), α (mm^2/sec), and Δz (mm). So as not to confuse the count k with thermal conductivity, the latter is now called *con.*
2. Derive the parameters of the case v_c (mm/sec), ϕ (deg), v_s (mm/sec), P_s (mW = N · mm/sec), T_s (°C), P_f (N·mm/sec), r (°C · sec/N).

   First Run ($i = 1$): Assume that no heat escapes through the tool and all friction power goes into the chip.
3. *Phase 1:* Determine the temperature field $T_{j,k}$: for $j = 1, 20$; $k = 1, KKK$, Eqs. (8.33), (8.34), and (8.35). Determine the average temperature T_{cav} in the contact between chip and tool, Eq. (8.36).
4. *Phase 2:* With T_{cav} as input determine the power P_t (n), passing through the tool, Eq. (8.43). The index i numbers the runs.
5. *Correct* for the power escaping through the tool for the next run. In the formula for p_{max}, Eq. (8.23) and also in (8.26) the friction power will now be replaced by a smaller one; let's call it P_c. For $i = 1$, $P_c = P_f$ and then: $P_c(i + 1) = P_f - P_t(i)$. Check for the level of convergence. Establish the value of an error as the relative difference in the average temperature in the chip-tool contact in two subsequent runs:

$$err = \frac{T_{cav}(i) - T_{cav}(i-1)}{T_{cav}(i)}$$

   If $err < 0.02$, end. Else, increment i and go back to step 3.

Once the error check has been satisfied, print out the temperature field in the chip and plot out the temperatures on the tool face. Evaluate the part of friction power that escapes through the tool, $e = P_t(i)/P_f$.

EXAMPLE 8.3 Compute Temperature Fields for a Particular Case ▼

Compute the temperature fields for machining steel 1035 with a cutting speed of $v = 120$ m/min $= 2$ m/sec. The geometry of the cut is given with reference to Fig.

8.10 as $h_1 = 0.2$ mm; $\phi = 28°$; $\beta = 20°$; and $L_c = 0.8$ mm. Discretize for $j = 1, 20$; $KK = 400$, $KKK = 600$; $z = 1, 41$; and $\Delta z = 0.5$. The tool is made of one material only with a different thermal conductivity than the workpiece. Discuss the result.

The listing of the corresponding program chptmp1 written in Matlab is in Fig. 8.16. The parameters of the machined material are taken from Table 8.1. The shearing temperature T_s is evaluated according to Eq. (8.20). The friction power P_f is evaluated using Eq. (8.19). The thermal resistance of the tool is obtained using Eq. (8.42). The computation of the temperature field in the chip is derived from the power $P_c = P_f - P_t$, where P_t is the power escaping through the tool, Eq. (8.43), which at the beginning, for $i = 1$, is set at zero; consequently the parameter $e = P_t/P_f$ is also zero. The initial value of the "error" is set large, $err = 1.0$, and the initial value of the average temperature T_{cav} in the chip-tool contact is set low, equal to the shear-plane temperature, $T_{cav} = T_s$.

```
function[Td,T,Tc,p,x,y,Pt,e,err,Tcav]=chptmp1(v)
% Refer to EX.8.3, v(m/sec)is cutting speed
% con is thermal coductivity of workpiece, cont is that of tool
% roc is specific heat per volume, fi is shear angle, Ts is shear
% plane temperature, Eq (8.21), vc is chip velocity, Pf is friction power, alp is
% thermal diffusivity of workpiece, Tr is room temperature
Ks=2250;h1=0.2;con=43;cont=70;roc=3.7;KK=400;
fi=28*pi/180;beta=20*pi/180;h2=h1/tan(fi);b=10;KKK=fix(KK*1.5);
Ts=Ks*cos(fi+beta)/(roc*cos(fi)*cos(beta))+20;vc=v*tan(fi);
Pf=Ks*b*h1*tan(beta)*vc*1000;alp=con/roc;Tr=20;
Lc=4*h1;dx=Lc/KK;dy=h2/20;dt=dx/(1000*vc);dz=0.5;
% Initialize the shear plane temperature
for j=1:20
T(j,1)=Ts;
y(j)=j*dy;
end
% calculate the thermal resistance r of the tool, Eq. (8.42)
% cont is thermal conductivity of the tool
w(1)=Lc+dz/2;R(1)=dz/(2*w(1));
for l=2:40
w(l)=Lc+2*(l-1)*dz;
R(l)=dz/w(l);
end
w(41)=Lc+79.5*dz;R(41)=dz/(2*w(41));
r=0;
for l=1:41
r=r+R(l)/cont;
end
i=1;Tcav(1)=Ts;Pt(1)=0;e(1)=0;err(1)=1.0;
% Start the outer loop indexed by i; it provides correction
% for heat Pt escaping through the tool; it is the "while" loop;
% err is defined in the last line of this loop, Pf is friction power,
% Pc is power entering the chip, Pt is power passing through tool
while abs(err(i))>0.02
Pc=Pf-Pt(i);
% with input Pc calculate pmax, Eq.(8.23)
for k=1:(KKK-1)
pmax=Pc*16/(9*KK);
dp=pmax*8/(KK*7);
if k<=(KK/8)
p(k)=pmax;
elseif k>KK/8 & k<=KK
p(k)=pmax-dp*(k-KK/8);
```

Figure 8.16 Matlab program for computing the temperature field in chip and tool.

Figure 8.16
(continued)

```
else
p(k)=0;
end
% Start the computation of the thermal field in the chip, Eq.(8.33)-(8.35)
T(1,k+1)=(p(k)/(b*dx*dy*roc)-(T(1,k)-T(2,k))*alp/dy^2)*dt+T(1,k);
for j=2:19
T(j,k+1)=(T(j-1,k+1)+T(j+1,k)-2*T(j,k))*alp*dt/dy^2+T(j,k);
end
T(20,k+1)=(T(19,k+1)-T(20,k))*alp*dt/dy^2+T(20,k);
end
% This ends the computation of the field in the chip
% Calculate the temperatures Tc in the chip tool contact and
% later, the average temperature Tcav
sum=0;
for k=1:KK
x(k)=(k-1)*dx;
Tc(k)=T(1,k);
sum=sum+Tc(k);
end
% For printing reduce the size of the T matrix to the Td(m,n) matrix
for m=1:10
for n=1:50
Td(m,n)=round(T(2*m-1,12*n-11));
end
end
i=i+1
Tcav(i)=sum/KK;
Pt(i)=(Tcav(i)-Tr)*b/r;
% e is the fraction of the friction power passing into the tool
e(i)=Pt(i)/Pf;
err(i)=(Tcav(i)-Tcav(i-1))/Tcav(i);
if i>8,break,end
end
% This is the end of the while loop; if err<0.02 the
% computation ends
% print: e,err,Tcav,Td'; plot; mesh(T); plot: contour(T,s);
% s is the vector of the contour temperatures, e.g. s=[550:50:900]
```

The computation of the temperatures in the chip proceeds in loops marked by incrementing the subscript *k*. It starts with determining the distribution of the friction power *p*(*k*) injected into the chip from the friction contact, using Eqs. (8.23), (8.24), and (8.25). Then the temperatures in the *k*th slice are updated using Eqs. (8.33) through (8.35). Once the whole field in the chip is determined, the average temperature T_{cav} in the chip-tool contact and the power escaping through the tool P_t are calculated. The evaluation parameters *e* and *err* are also calculated. The computation then returns to the beginning of the program for the thermal field in the chip as marked by the condition "while," and carries it out with a new value of P_c. The repetitions are counted using subscript *i*, until the error no longer satisfies the condition stated at "while" for the absolute value of the error. Then the program ends.

The results of the computation may be inspected in several ways. Figure 8.17 shows how the values of the share coefficient *e*, of the error *err*, and of the average contact temperature T_{cav} converged in the four repetitions of the computation of temperatures $T_{j,k}$. It is seen that the convergence was rather fast and the error fell below the limit of 0.02 after four repeats. By then the ratio $e = P_t/P_f$ indicating the share of the heat escaping through the tool settled on 16.8% and the average contact temperature settled on 835.34 °C.

Figure 8.17
Printout of the evaluation parameters *e, err,* T_{cav} for the four repetitions of the program needed to achieve acceptable convergence. Thermal field temperature printout; shear plane is at the top.

```
Tcav =  510.42,   900.42,   829.65,   835.34
err =    1.0000,    0.4331,   -0.0853,    0.0068
e =   0,    0.1815,    0.1669,    0.1680
 A=Td'
  510  510  510  510  510  510  510  510  510  510
  591  516  511  510  510  510  510  510  510  510
  645  532  512  511  510  510  510  510  510  510
  689  551  517  511  510  510  510  510  510  510
  725  572  523  512  511  510  510  510  510  510
  757  593  531  514  511  510  510  510  510  510
  783  613  540  517  512  511  510  510  510  510
  805  632  551  521  513  511  510  510  510  510
  824  650  561  525  514  511  511  510  510  510
  840  667  572  531  516  512  511  510  510  510
  854  682  583  536  518  512  511  511  510  510
  866  696  593  542  521  513  511  511  510  510
  876  709  603  548  524  514  511  511  510  510
  884  721  613  554  527  516  512  511  511  510
  891  732  623  561  530  517  512  511  511  510
  896  741  632  567  534  519  513  511  511  510
  900  750  641  573  537  521  514  512  511  511
  902  758  649  580  541  523  515  512  511  511
  904  765  657  586  545  525  516  512  511  511
  904  771  664  592  549  527  517  513  511  511
  904  776  671  598  553  530  518  513  511  511
  902  780  677  603  558  532  520  514  512  511
  900  783  683  609  562  535  521  515  512  511
  896  786  688  614  566  537  523  515  512  511
  892  788  693  619  570  540  524  516  513  511
  887  789  697  624  574  543  526  517  513  512
  881  790  701  628  577  546  527  518  514  512
  874  790  704  632  581  548  529  519  514  512
  866  789  707  636  585  551  531  520  515  513
  858  788  709  640  588  554  533  521  515  513
  849  786  711  643  592  557  535  522  516  513
  839  783  712  646  595  559  537  524  517  514
  829  780  713  649  598  562  539  525  518  514
  818  776  713  651  601  564  540  526  518  515
  807  772  713  654  604  567  542  527  519  515
  798  767  713  655  606  569  544  529  520  516
  790  762  712  657  609  572  546  530  521  517
  783  758  711  658  611  574  548  532  522  517
  777  754  710  659  613  576  550  533  523  518
  771  749  708  660  615  578  552  534  524  519
  765  746  707  660  616  580  554  536  525  520
  760  742  705  661  618  582  555  537  526  521
  756  738  704  661  619  584  557  539  527  522
  751  735  702  661  620  585  559  540  528  522
  747  732  700  661  621  587  560  541  530  523
  743  728  699  661  622  589  562  543  531  524
  739  725  697  661  623  590  563  544  532  525
  736  723  695  660  624  591  565  546  533  526
  733  720  694  660  625  592  566  547  534  528
  729  717  692  660  625  594  568  548  535  529
```

Furthermore, in the same figure the whole final thermal field of the chip is printed out. However, in order to fit it on a page the variables T have been replaced by T_d such that T_d equals every second T in the j count and every twelfth T in the k count. So, instead of 600 × 20 numbers, we only have 50 × 10 to print.

The values of T_d obtained in the final i run are printed, rounded off to three-digit integers. The orientation of the printout is such that the top row is the $k = 1$ row, that is, the shear plane. The first column is the $j = 1$ column, which is the chip-tool contact, at least for the first 33 rows that extend over the contact length L_c. The remaining 17 numbers in the column apply to the underside of the chip after it separates from the tool.

Another form of the same information is given in Fig. 8.18, where we see the isotherms of the field in the chip. This graph is obtained in Matlab by invoking the function *"contour."* This time the orientation is such that the x-axis is the $j = 1$ contact line, and the y-axis is the $k = 1$, the shear plane. The lower left corner point is the cutting edge. The vertical axis is the $T = 510°$ isotherm. Next is the 550° line, and the next lines go by increments of 50°. The scale on the x-axis is expressed in the values of k. It would be converted into x (mm) by $k \times dx$, where $dx = 0.002$ mm. The hottest point is enclosed by the $T = 900°$ line, at $k = 225$ (i.e., $x = 0.45$ mm). At $k = 400$ we have the end of the chip-tool contact. It is possible to understand how the heat from the contact line spreads into the chip as it moves right. The friction power intensity is being reduced with increasing distance x from the cutting edge. At first, then, the temperature on the contact increases as the chip slides over the contact, but from $k = 250$ on, the heat spreads away in y direction faster than it is produced by friction, and the temperatures start to get lower. This trend is even stronger after $k = 400$, when there is no longer any input into the chip from friction.

Finally, the most compact information is the plot in Fig. 8.19 of the contact temperatures T_c versus the count k from the cutting edge. These are the temperatures in the first column of Fig. 8.17 and those on the x-axis in Fig. 8.18, but the horizontal range is different. While all $T(1, k)$ for $k = 1$ to 600 are used in Fig. 8.17 and 8.18, here only the range $k = 1$ to 400 of the contact $L_c = 0.8$ mm is used. The graph starts at the cutting edge, $x = 0$ with $T(1, 1) = T_s = 510°C$ and reaches the maximum of 933°C at $k = 225$, i.e., $x = 0.45$ mm.

The graph in Fig. 8.19 obviously provides the most important information because the temperatures on the rake face of the tool are most decisive for the wear phenomena on both the rake face and the flank of the tool. Computing these temperatures gives an excellent insight into the effects of material properties and of cutting conditions on the machinability involved in an operation, as illustrated in the following examples. ▲

EXAMPLE 8.4 Compute Temperature Fields for Four Different Materials ▼

Compute the temperatures between chip and tool for four different materials:

1. Carbon steel 1035
2. Stainless steel 302
3. Titanium alloy Ti (6Al, 4V)
4. Aluminum Al 7075-T6

Use the same geometry of the cut (h, β, ϕ) as in Ex. 8.3 for all the cases. The mechanical and thermal properties of the materials are given in Table 8.1. First, use the same cutting speed of $v = 2$m/s for each material.

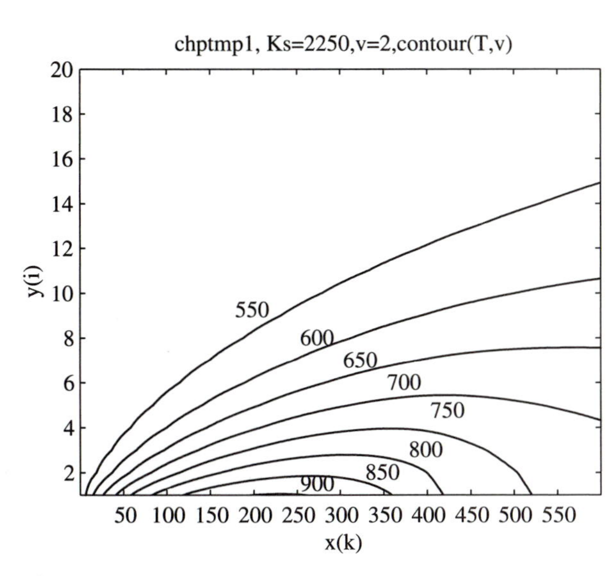

Figure 8.18
The thermal field expressed by isotherms. Shear plane is on the left vertical axis. The horizontal axis represents the bottom face of the chip. The scale of it is expressed in the k counts of steps of $\Delta x = 0.002$ mm. The chip-tool contact extends from $k = 1$ (cutting edge) as far as $k = KK = 400$ (0.8 mm). From there to $k = KKK = 600$ the chip curls away from the tool. The friction heat input is constant for $k = 1, 50$, and then it decreases and stops completely at the tool contact end. The temperature at the chip-tool interface increases from the cutting edge where it has shear plane temperature and reaches a maximum around $k = 250$ and then it decreases as the friction heat input decreases and the heat spreads out towards the top surface of the chip.

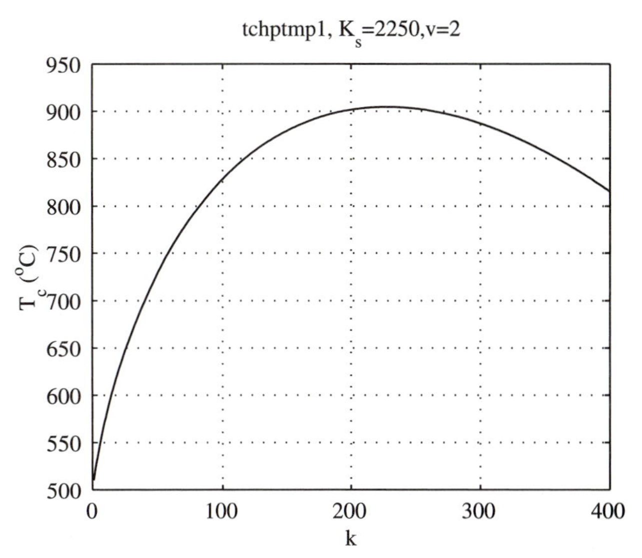

Figure 8.19
Plot of the temperatures along the chip-tool contact. This is just the $y = 0$ boundary of the XY field of Fig. 8.18, but it contains all the essential information about temperature affecting the tool and its wear.

A computer program was written, similar to the one in Fig. 8.16, to compute all four cases and plot the results in one graph, which is shown in Fig. 8.20. Note that while Al gives very low temperatures, Ti and stainless steel 302 machine very hot, and there is a substantial rise of their temperature over the shear-plane temperature, which does not occur as much for steel 1035. By checking with Table 8.1, we see that the low temperature for Al is due both to its low strength, as expressed by the specific cutting force K_s, and its very high thermal conductivity k and thermal diffusivity α. Carbon steel 1035 is stronger and has lower conductivity. The stainless steel is also stronger, and its conductivity is about three times less than that of the carbon steel. The titanium alloy has about the same strength as 1035, but six times lower conductivity. While in aluminum the friction heat spreads quickly into the chip, in titanium the heat remains confined to a thin layer at the chip-tool contact; there is less volume to heat. This explains why titanium has the highest temperature.

Obviously, it is not appropriate to machine all these materials with the same cutting speed. Therefore, we compute again and use different cutting speeds:

1. 1035: $v = 2$ m/s
2. 302: $v = 0.75$ m/s
3. Ti: $v = 0.5$ m/s
4. Al: $v = 10$ m/s

The results are shown in Fig. 8.21. The temperatures for the stainless steel and titanium drop from about 1300 and 1400°C down to about 970°C and 900°C. The increase of the maximum temperature over the shear-plane temperature is much smaller. For Al, in spite of increasing the cutting speed five times, the peak temperature is still low enough for any cutting-tool material.

These computations explain the large differences in the machinability of various materials. It is true that other factors strongly affect tool wear, such as the chemical affinity between work and tool materials, and this is explained later on. Nevertheless, the temperature in the cut generally has the strongest effect both because the diffusion rate, which plays an important role in tool degradation, increases with the fourth power of temperature, and because the tool strength and hardness decrease with temperature. For these reasons, we have placed the thermal analysis at the beginning of the discussion of cutting mechanics.

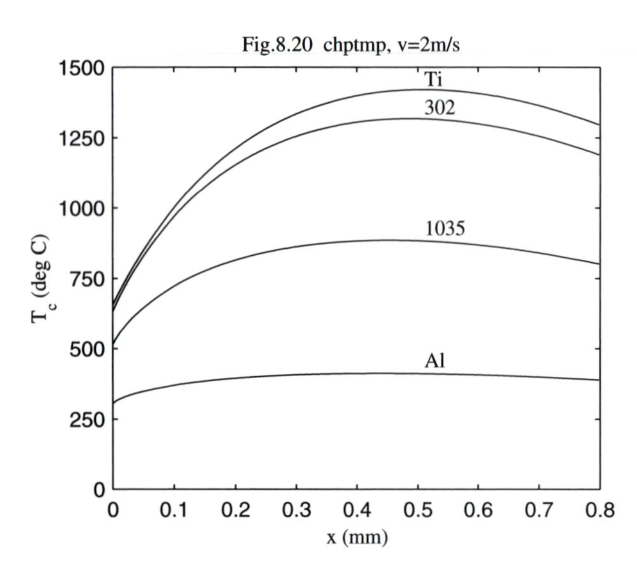

Figure 8.20
Temperatures along the chip-tool contact for four different materials cut with the same speed of $v = 2$ m/s: titanium alloy Ti 6Al4V, stainless steel 302, carbon steel 1035, aluminum alloy Al 7075-T6. For titanium and stainless steel the temperatures rise very high, which would cause fast tool wear.

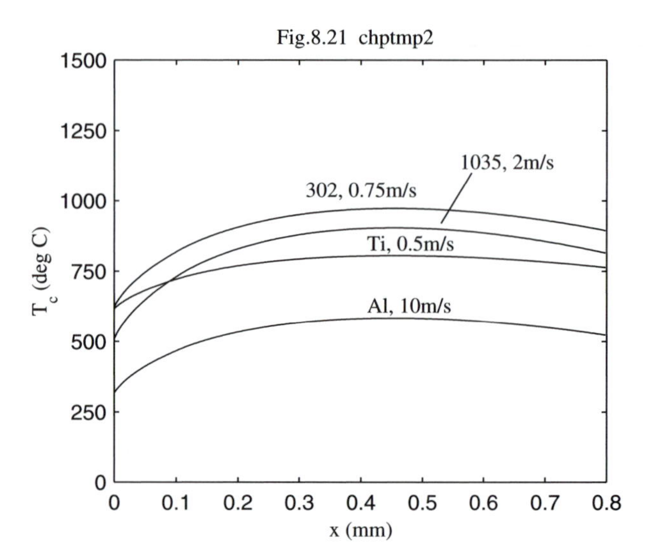

Figure 8.21
Temperatures along the chip-tool contact for the four different materials of Fig. 8.20 cut with different, appropriate speeds. To obtain acceptable temperatures, it was necessary to lower cutting speed by a factor of 4 for Ti and of 2.6 for 302. For aluminum it was possible to increase speed by a factor of 5 and still cut rather cool. Big differences in "machinability" of the various materials are due to both mechanical and thermal properties.

The computations carried out in Examples 8.3 and 8.4 were based on the following simplified assumptions:

- No heat escapes from the shear plane into the workpiece.
- Heat conduction can be neglected as compared to heat mass transfer in the direction of the chip motion.
- It is acceptable to use average values, equal for all materials, for shear-plane angle, rake-face friction coefficient, and chip/tool contact length, and to consider them to be unaffected by cutting speed.
- Thermal parameters of materials are assumed as they apply at room temperature, and their change with temperature is neglected.

In spite of these simplifications the temperature fields obtained are not significantly different from those obtained by much more complex computations and measured by various authors. The error is estimated not to exceed 20% at the most, and it is much less in many cases.

Most important, these computations permit us to recognize the effects of the following individual parameters on the maximum temperature at the rake face of the tool:

The specific force K_s, to which shear-plane temperature T_s and friction heat input into the chip and tool are proportional

The specific heat per volume (ρc) to which T_s is inversely proportional

The thermal diffusivity $\alpha = k/(\rho c)$: the higher it is, the lower the increase beyond T_s of temperatures, including T_{max}

The cutting speed v: T_{max}/T_s increases with v, and this effect is as strong as that of α, but inverse to it. ▲

8.3 CUTTING-TOOL MATERIALS

The ability to increase metal removal rates depends primarily on the development of cutting-tool materials. The metal removal rate, $MRR = bhv$ [see Eq. (7.1)], can be increased by increasing either the width of chip b (and, consequently the depth of cut a), or the chip thickness h, or the cutting speed v. The chip width (depth of cut) is limited by the allowance (depth of stock to remove) or by the static or dynamic flexibility of the machine tool. The increase of chip thickness leads to tool breakage or to faster tool wear, and the increase of cutting speed increases strongly the tool wear rate (mainly because of the increase of the chip/tool interface temperature). The tool wear rate (or, in the inverse formulation, the tool life) as it depends on h and v plays the basic role in the economy of machining. Thus, tool materials that permit larger chip thickness and higher cutting speeds for a given tool wear rate make higher metal removal rates economical. Simply put, the productivity of metal cutting depends on the durability of the small tool tip. Through the history of metal cutting, increased productivity has depended primarily on discoveries of new tool materials and on their improvements.

Although we can only understand fully the requirements imposed on tool materials after discussing the various modes of tool wear and tool failure in Section 8.4, we can see from the presentations given so far about forces acting on the tool tip and temperatures in the tool tip that the tool material must possess the following characteristics:

1. *High hot hardness.* It is obvious that a tool can penetrate the workpiece and form a chip only if it is harder than the workpiece, at the rather high temperatures existing in the cut. A tool will permit higher cutting speeds if it retains high hardness at higher temperatures.
2. *High strength and toughness.* The stresses in the tool tip are very high, as explained in more detail later. Both shear and tensile stresses are involved as well as, quite often, impacts. Tools that lack toughness and are very brittle are subject to various forms of breakage.
3. *Resistance to chemical interaction with workpiece material, to oxidation, and to corrosion.* At the high temperatures at the chip/tool/workpiece contacts, chemical interaction in the form of diffusion can lead to rapid erosion and abrasion of the tool. Therefore, combinations of tool/workpiece materials that have high mutual chemical affinity must be avoided. Tool materials are preferable that are as widely inert as possible. Similarly, because all tools are exposed in some parts of the tool tip to air and often also to a coolant, a high degree of resistance to oxidation and to corrosion is required.

Aspects 1 and 2 of hardness and toughness are contrary to each other because the nature of materials is such that high hardness usually combines with low toughness and vice versa, so it is mainly a matter of obtaining the best possible combination. Hot hardness, however, is the primary requirement in tool materials, and this characteristic is discussed first. The development of cutting-tool materials proceeded historically in steps characterized by new inventions. The periods in between these steps were devoted to gradual improvements and refinements. Until the beginning of the twentieth century, cutting tools were made of high-carbon steels, or lightly alloyed "tool steels." These materials no longer have any practical significance. All other basic kinds of cutting-tool materials, representing individual milestones in the developments of the twentieth century, still exist side by side, and each has its particular field of application. These milestones can be approximately dated as follows:

High-speed steels	
Tungsten-based	1900
Molybdenum-based	1930–1940
Sintered carbides	
"Straight" grades, WC + Co	1927
Steel-cutting grades, WC, TiC, TaC + Co	1931
Ceramics	1950–present
CBN, Synthetic diamond	
Cubic boron nitride, polycrystalline diamond	1960–present
Coated carbides	1970–present

Some of these developments led, on average, to a general increase of cutting speeds for most workpiece materials, especially steels. The dramatic improvements obtained are well illustrated by the graph in Fig. 8.22, which is reproduced from Sandvik [13]. It gives the (average economical) machining time for a steel shaft of

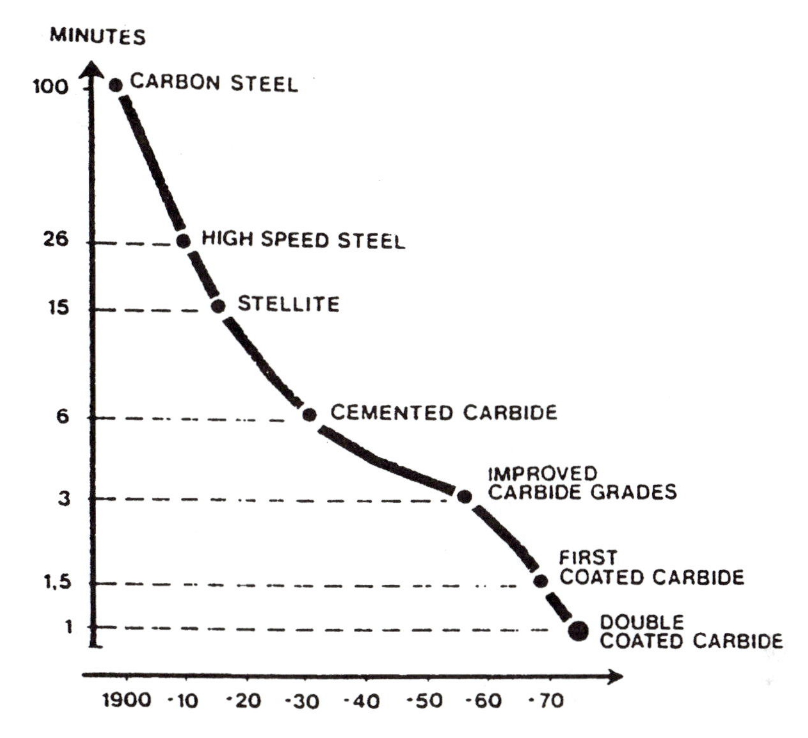

Figure 8.22
Change in cutting time over the past ninety years for machining a 100-mm-diameter, 500-mm-long steel shaft with different cutting materials. The 100-fold decrease expresses the crucial significance of the development of tool materials for the productivity of machining. [SANDVIK]

diameter 100 mm and length 500 mm obtainable with different cutting materials, with the dates of their development.

In other instances these developments have made it possible to machine new materials that are more difficult to machine. This is especially true for the "super high-speed steels" and for the CBN and polycrystalline diamond. Even if such tool materials do not find wide application, they are crucial for machining special and important products like turbines and structural parts of supersonic aircraft and spacecraft.

The overall usage of the individual materials is presented in Table 8.2. The total value of cutting tools sold in the United States in 1997 was $1.916 billion (May 1998, Tooling and Production: www.toolingandproduction.com). From Table 8.2 it is obvious that the two most important kinds of tool materials are high-speed steels (HSS) and sintered carbides (SC). The latter class is more efficient in producing chips. However, the fields of application of HSS and SC differ mainly by the types of tools, due to the

TABLE 8.2 **Usage of Different Tool Materials (1997)**

	Percentage of Total Purchased Value	Percentage of Total Chips Produced
High-speed steels	65	28
Sintered carbide	33	68
Ceramics	2	4
Diamond	<1	<1
Other	<<1	<<1

different ways of manufacturing them. HSS tools are mostly made by machining (turning, milling) the tool in an annealed state when it is relatively soft and subsequently hardening it, which is then followed just by sharpening (grinding). SC tools are pressed in molds and sintered, and they cannot be machined at all except for grinding and polishing. The molds are expensive and difficult to make in complex shapes. Sintered carbides are therefore mostly made as simple inserts mechanically clamped (or, less often, brazed) to steel holders. They are thus used for tools with rather simple shapes like turning and face-milling tools. HSS tools have complex forms and are often of such small size that mechanical clamping of inserts is impossible: for example, twist drills, reamers, taps, broaches, gear-cutting tools. Before presenting more detailed discussion of the individual classes of tool materials, we give some general parameters of the tool materials and their constituents.

Some indication of potential constituents of tool materials may be obtained from their hardness at room temperature and their melting temperature T_m. The higher these two parameters, the better the chance for success. Most of the potential constituents are carbides, oxides, or nitrites of metals. The room-temperature hardness HV (N/mm^2) and melting temperature T_m are given for a selection of tool-material constituents in Table 8.3. The most important property, the hot hardness, is expressed, for the main tool materials, in the diagram of Fig. 8.23, where hardness is plotted versus temperature. Sintered carbides are indicated as a wide band. At its top are the finishing steel-cutting grades, and at its bottom the roughing grades with high Co content. The aluminum oxide represents both the ceramic cutting tools and grains used in grinding-wheels. The silicon carbide is used for grinding-wheel grains only. The cubic boron nitride is used both for cutting tools and grinding wheels. Diamond is outside of the range of the graph; its room temperature hardness is 85,000 N/mm^2.

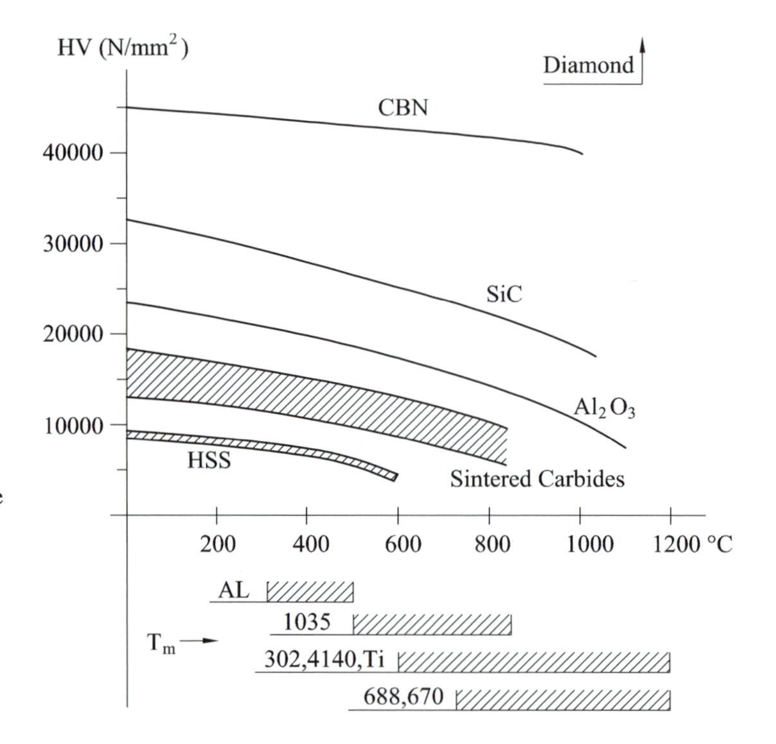

Figure 8.23
Hardness versus temperature for basic classes of tool materials as compared with melting temperatures of workpiece materials. Sintered carbides and HSS are still used for most machining tasks.

TABLE 8.3 Hardness and Melting Temperature of Tool-Material Constituents

	HV(N/mm^2)	T_m(°C)
Fe_3C	11,500	1,500
Cr_3C_2	18,000	1,900
TaC	18,000	3,800
WC	21,000	2,700
Al_2O_3	24,000	
V_4C_3	25,000	2,500
ZrC	26,000	3,500
HfC	27,000	3,900
W_2C	28,000	2,600
VC	29,500	2,800
TiC	32,000	3,200
BN	45,000	3,200

In the same diagram the melting temperatures T_m are indicated for machining several workpiece materials. They represent the range of maximum chip/tool interface temperatures for a range of cutting speeds from $v = 0$ to the usually applicable cutting speeds. It would seem that HSS cannot be used to machine (e.g., drill) the Ni-based 688 and Co-based 670 alloys. However, with HSS tools a built-up edge will form, acting as an extension of the tool into a large positive rake angle, and this produces a lower cutting force. Also, a coolant is used that, at the rather low cutting speeds used with HSS, will sufficiently penetrate between the chip and tool rake face, lowering the friction and the corresponding T_m. Therefore, in practice, some of the HSS grades with the highest hardness will be able to cut these alloys. Now the individual classes of tool materials are discussed in more detail.

8.3.1 High-Speed Steels

High-speed steels were invented by two Americans. One of them, F. W. Taylor, was a production engineer who is considered to have founded the science of metal cutting and who undertook an extraordinary, systematic metal-cutting testing program. The other one, M. White, was a metallurgist. Their major contribution was not primarily the composition of the steel but rather the special heat treatment, which was novel at that time. On one hand they raised the hardening temperature just below the melting point of steel at approximately 1250°C instead of the then-usual 750°C. Tempering was also done at higher temperatures than usual. These steels permitted a dramatic increase of cutting speed for machining steel from the 6 m/min for carbon steel tools to 30 m/min.

The high-speed steel of Taylor and White had the following composition: C, 0.67; W, 18.9; Cr, 5.5; Mn, 0.1; V, 0.3; Fe, balance, which was very close to the presently popular grade T1.

The typical heat-treatment cycle for HSS is given in Fig. 8.24. With carbon or low-alloy steels, the rather high hardening temperature, at 3, would lead to very coarse-grained structure and to corresponding brittleness of the steel. In the HSS the alloying elements W, Cr, and V form very strongly bonded carbides that tend to dissolve as temperature is raised. However, even up to the melting point some of them remain intact and prevent the grains of steel from growing.

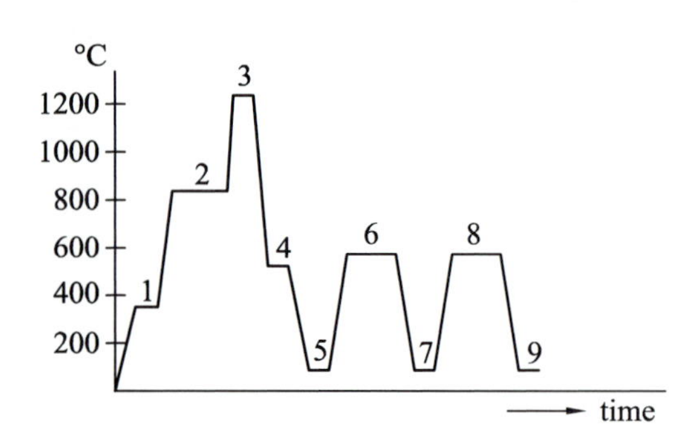

Figure 8.24
Heat-treatment cycle for high-speed steels: 1, 2—preheating; 3—heating 25 min/1260 °C; 4—quenching in salt bath; 5—cooling in air to about 50°C; 6—first tempering 2 hours /590 °C; 7—cooling in air to room temperature; 8—second tempering; 9—cooling to room temperature. The strong carbide-forming alloying elements prevent grain coarsening at the high hardening temperature. Repeated tempering results in very fine particles in a martensite matrix.

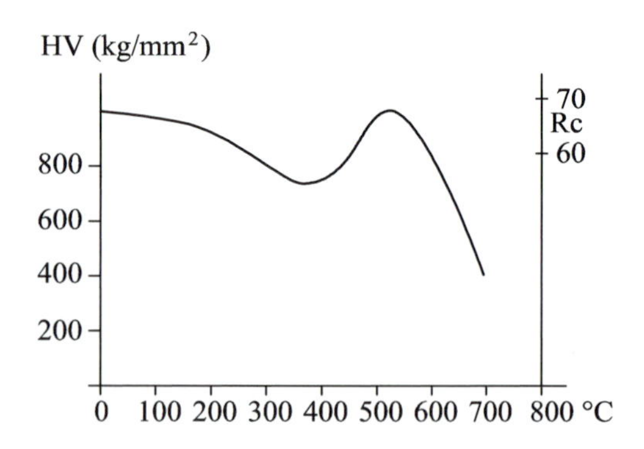

Figure 8.25
Hardness as function of tempering temperature for HSS. The particular feature of hardness peak achieved by tempering at 560°C is discussed in the text.

After quenching and formation of martensite, these carbide particles remain in the martensitic matrix as very hard grains. While the carbon and low-alloy steels soften with higher tempering temperatures; this is not the case with HSS. The diagram in Fig. 8.25 shows the hardness of HSS as a function of the tempering temperature. As temperature increases, the hardness first decreases and then increases and reaches a maximum after tempering at about 560°C. This "secondary hardening" is caused by the formation within the martensite of very small particles, about 0.01 μm in diameter, which precipitate throughout the structure. Another effect of tempering consists of converting residual austenite to martensite, and this is achieved more completely be repeating the tempering cycle. If the HSS is heated above 600°C, it softens, and its hardness can only be restored by another complete hardening cycle.

The HSSs exist nowadays in a great variety of compositions. The alloying elements used and their effects are briefly recapitulated below.

W, Mo — In the 1930s and during the World War II when tungsten was scarce, it was found that molybdenum could be used in its place. Correspondingly the two basic classes of HSSs are denoted T and M depending on whether their main alloying element is tungsten or molybdenum. Both form carbides, which are essential for secondary hardening. Without any Mo, about 18% of W is used. Without any W about 8% of Mo is used. Otherwise they may be combined using Mo in a ratio of 4/9 to replace W.

V — Vanadium forms mostly the carbide VC, which is rather stable and dissolves very little during the heating part of the hardening cycle. VC is very hard and contributes substantially to the wear resistance of the steel. On the other hand it makes the steel difficult to grind. The amount of V may be up to 5% but is usually 1 or 2%.

Cr — A chromium content of about 4% is used in all HSSs. Chromium forms carbides that dissolve fully during hardening. It improves both the primary and secondary hardenability.

Co — The main effect of cobalt is to increase hot hardness, to raise the temperature at which hardness starts to fall. It does not form carbides but it is dissolved in the metal and stabilizes precipitated carbide particles.

All HSS materials are usually classified in four groups A to D as presented in Table 8.4, which includes the composition and basic properties of selected grades. The properties of wear resistance, toughness, hardness, and cost are rated from 1 (low) to 10 (high).

Group A: Conventional High-Speed Steels

Typical grades among the conventional HSSs are the T1, as mentioned above, and the M2, the latter being the most commonly used grade of HSS. It is used for drills, reamers, end mills, broaches, saws, and hobs and for machining carbon and low-alloy steels up to 375 BHN hardness, cast steels and cast irons up to 255 BHN, and for nonferrous metals.

Group B: Conventional High-Speed Steels with Cobalt Added

Steels with cobalt added have less toughness but greater hot hardness. They are used for drills and end mills in machining stainless steels, Ti-alloys and Ni- and Co-based alloys.

Group C: High-Vanadium High-Speed Steels

The high-vanadium HSSs have high wear resistance due to the very hard carbide of vanadium. On the other hand, they are difficult to grind with ceramic grinding wheels; electrolytic grinding or grinding with diamond-impregnated wheels must be used to sharpen them. They are used for machining steels where long tool life is required, as

TABLE 8.4 Compositions and Properties of Selected High-Speed Steels

Steel Class and Grade	C	W	Mo	Cr	V	Co	Wear Resistance	Toughness	Hardness	Cost
A. Conventional										
T1	0.75	18.0	—	4.0	1.0	—	4	8	5	5
Mi	0.80	1.75	8.5	3.75	1.15	—	4	10	5	3
M2	0.85	6.0	5.0	4.0	2.0	—	5	10	5	3
B. Cobalt added										
M33	0.88	1.75	9.5	3.75	1.15	8.25	5	5	8	5
T5	0.8	18.0	—	4.25	2.0	8.0	5	4	8	6
T6	0.8	20.0	—	4.5	1.75	12.0	5	2	9	8
C. High vanadium										
M3	1.05	6.0	5.0	4.0	2.4	—	6	6	6	4
M4	1.3	5.5	4.5	4.0	4.0	—	9	6	6	4
T15	1.5	12.0	—	4.5	5.0	5.0	10	9	9	6
D. High-hardness Co steels										
M42	1.1	1.5	9.5	3.75	1.15	8.25	6	9	9	5
M44	1.15	5.25	6.5	4.25	2.0	12.0	6	3	10	6

on single-spindle or multiple-spindle automatic lathes, especially for form tools, or else for machining of refractory metals.

Group D: High-Hardness Cobalt Steels (M-40 types)

The high-hardness cobalt steels have the highest hot hardness; for example, compare the hardness at 540°C of the following three steels:

M42	6550 HV (N/mm^2) or	58 R_c
T15	5750 HV	54 R_c
M2	5250 HV	51 R_c

They are used for drills and end mills in machining heat-treated steels, like 4340 at 40–55 R_c, Ti-alloys, Ni- and Co-based alloys.

Apart from chemical composition and heat treatment, the properties of HSS are affected favorably by the uniformity of carbide distribution and the size and uniformity of carbide particles; they are affected unfavorably by oxidation or decarburization as they depend on the manufacture and heat treatment of the steel. High-speed steels are usually produced from cast ingots that are subjected to forging or rolling and/or drawing into bars of various sizes. These manufacturing processes and the heat treatment of tools must be carried out so as to obtain fine and uniform carbide grains, uniformly distributed, and so as to avoid oxidation and loss of carbon. Fine and uniform structures are obtained by electroslag refining. The best structures are obtained in HSS tools manufactured by powder metallurgy. These steels are also easiest to grind.

8.3.2 Sintered Carbides

Sintered carbides consist of hard carbide grains bonded together by a ductile metal. The size of the carbide grains is between 0.5 and 5 μm, and they constitute 85–97% (weight percent) of the material. The binder metal exists as a thin layer with 0.4 to 1 μm thickness between the carbide grains. Thus, the hard carbide particles are the principal constituents, whereas in HSS they constitute only about 10–25% of the material.

The two main classes of sintered carbides are

1. The "straight" grades, consisting of the carbide WC and Co as binder.
2. The "complex" or "steel-cutting" grades. The binding metal is again Co, but the carbide grains are partly WC and partly complex WC-TiC-TaC carbides. The TiC and TaC are entirely soluble in each other, and they can dissolve up to 70% by weight of WC.

Apart from the two basic classes, there is also a group of sintered carbides with no WC content:

3. The "straight TiC" grades, in which the carbide phase is TiC and the binder is Ni, Mo. This group was never very significant, and it became even less so with the introduction of coated carbides.

The structure of a medium-grained straight grade with 10% Co is shown in Fig. 8.26. The angular grains of WC are seen surrounded by the lighter-appearing Co. The structure of a complex grade with 75% WC, 16% (Ti, Ta, W) C, and 9% Co is shown in Fig. 8.27. The complex WC-TiC-TaC grains are globular in shape and are easily distinguished from angular, lighter gray WC grains.

Figure 8.26
Structure of medium-grained 90WC/10Co grade, sintered at 1450°C,1500X. Best strength and toughness is attainable with finer and more uniform grain. [KENNAMETAL]

Figure 8.27
Structure of 75WC/16 (Ta, Ti,W)C/9Co, sintered at 1500°C,1500X. [KENNAMETAL]

The properties of both basic classes of sintered carbides depend on

Co content
Grain size
Uniformity of grain size
Thickness of bonding layer
Carbon balance

The decisive criteria for judging the individual grades can only be based on the actual performance of the tool in a given machining operation. However, there are some nominal properties that can be measured in simple tests. The most commonly used parameters follow:

- Hardness, measured in the Vickers scale (N/mm^2) or in the R_A scale
- Transverse rupture strength (TRS), in N/mm^2

The hardness of sintered carbide expresses its ability to cut hard materials, and it is significant for wear resistance at low cutting speeds but not at high cutting speeds, which are associated with high temperatures in the cut.

The TRS is obtained as a result of a bending test that is standardized by ISO 3327. A ground specimen with rectangular cross section, *H* high and *W* wide (see Fig. 8.28) is supported by two cylindrical rests and loaded by a sphere until fracture occurs. The tensile bending stress at fracture is calculated as $\sigma(\text{TRS}) = 3FL/WH^2$. This result expresses, indirectly, the tensile stress of the material. Using the Weibull formula of the statistical theory of strength, it is given as $UTS \approx 0.6\,\sigma(\text{TRS})$. The TRS is considered also a measure of the toughness of sintered carbides.

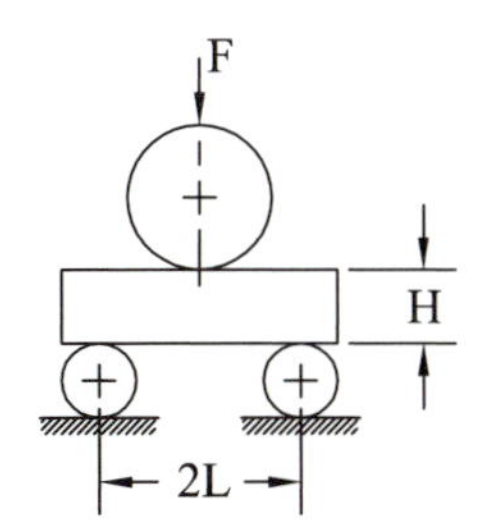

Figure 8.28
The standard transverse rupture strength test for sintered carbides.

The "straight" WC-Co grades were the first sintered carbides made. They possess a unique combination of hardness, strength, and toughness. This is due not only to the excellent mechanical properties of the WC carbide but mainly to its ability to form

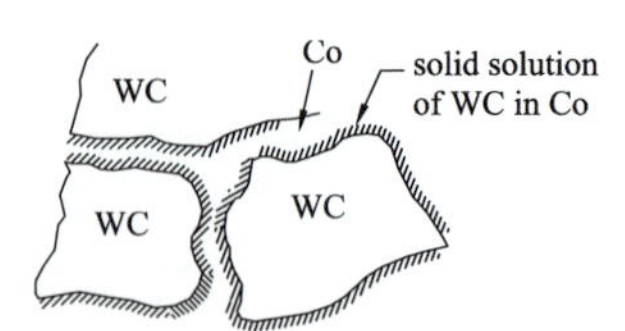

Figure 8.29
Detail of the WC/Co microstructure. Tungsten carbide grains are bound by a thin layer of cobalt; at the boundaries the carbide is dissolved in the binder, which improves the strength.

strong bonds with Co. Figure 8.29 shows that the layer of the bond between the carbide grains contains solid solution of WC in Co, which is richer in WC the closer it is to the boundary. The carbides TiC and TaC do not dissolve in Co and, therefore, could not provide a strong structure with this binder.

The carbide WC not only dissolves easily in Co, it is also easily soluble in Fe, and the two have a strong affinity that leads to strong diffusivity. The "straight grades" cannot be used for machining steel; they fail at high cutting speeds because the chip erodes a crater on the rake face of the tool. This erosion is based on diffusion. It was not until the "complex" grades were invented, in which the WC-TiC-TaC carbides resist erosion by steel chips, that sintered carbides could be used to machine steel. Since steel was the most common workpiece material, this was a significant development. The following numbers, quoted from [13], serve to illustrate the aspect of chemical affinity:

Carbide	Solubility in Fe at 1250°C
WC	7%
TiC	<0.5%
50/50 TiC/WC solid solution	0.5%
TaC	0.5%

The straight grades are used for machining cast iron, which is sufficiently saturated by C and does not dissolve WC, and for machining nonferrous metals as well as nonmetallic materials. On the other hand, for similar reasons of chemical affinity, sintered carbides containing TiC cannot be used to machine Ti-alloys; straight grades must be used instead.

Both the TRS and hardness depend on the content of Co and also on grain size. The binder is ductile and it imparts toughness to the structure. With increasing Co content, the TRS increases for both "coarse" grains (CG) and "fine" grains (FG), a distinction only recently established: "coarse" is the average common grain size (e.g., 3 μm), and fine relates to the modern sintered carbides with grain sizes below 1 μm and with very high uniformity of grain size and of the thickness of the Co layer. Actually, for the common grades with grain sizes between 1 and 5 μm the maximum strength is obtained with grain size about 3 μm. It is interesting to note the high level of strength, 2000–3000 N/mm^2 of the SC (compare with steels with *UTS* in the range of 250–1500 N/mm^2), which underscores our previous statement of the extraordinary combination of strength and hardness.

Hardness generally decreases both with the increase of Co content and with the increase of grain size. For both the straight and complex compositions the strength increases with Co content. The strength at room temperature of the complex grades is much lower than that of the straight grades, the more so the higher the TiC content. This is due to the superior bond between WC and Co and poor bonding between TiC and Co. However, the strength of the 50/50 TiC/WC composition actually increases with temperature and reaches its maximum at 800°C, when it is roughly equal to that of the straight grades. In general, the straight grades are superior at lower cutting speeds, while the complex grades are useful at high cutting speeds.

The quality (TRS, hardness, wear resistance) of carbides depends further on the carbon balance. The amount of C in the composition should be exactly what is necessary for binding all W, Ti, and Ta in carbides. Excessive C will exist as graphite, and insufficient C leads to the formation of W_2C and of a brittle phase in the structure. In both these cases the strength of the structure is strongly affected. The carbon balance as well as other aspects like uniformity of grain size, density, and so on, depend on the quality of the manufacturing process of the sintered carbide. This manufacturing process consists of the following steps.

1. **Production of W, TiO_2, and Ta_2O_3 powders.** In this stage, it is important to produce uniform grains of the desired size.
2. **Production of WC and WC/TiC/TaC.** This consists of blending and milling the powders produced in step 1 with carbon black and heating in furnaces, mostly in a hydrogen protective atmosphere. Temperatures for WC are between 1400 and 2650°C (the temperature must be high enough to give full carburization within an acceptable time but not so high as to induce grain growth). For WC/TiC/TaC, it is about 2100°C. In this stage, it is critical to reach the correct content of C.
3. **Production of grade powders.** Milling and mixing the powders of step 2 with Co powder so as to obtain uniform coating of carbide grains with Co. This takes days or even weeks.
4. **Pressing.** The powders are compressed in dies to the desired form and shape of the carbide tool (insert). Allowance must be made for shrinkage during sintering. Powders do not "flow" easily and, in order to obtain good density and proper fill of the die, it is necessary to use a lubricant, usually paraffin wax. Pressures of about 100–200 N/mm^2 are used in cold pressing. The dies must be strong, with small clearances, and they are expensive. Cold isostatic pressing, hot pressing, and hot isostatic pressing, respectively, lead to improved quality.
5. **Lubricant removal and presintering.** This is done continuously in furnaces with protective atmosphere and temperatures increasing up to 700°C.
6. **Sintering.** During sintering, the compact shrinks by between 18% and 26% linearly, and virtually all porosity is eliminated. Sintering is carried out in the liquid phase of the binder, that is, above the melting temperature of Co (1320°C). The processes occurring in this stage are (1) rearrangement of carbide particles into dense packing under the influence of the surface tension of the liquid; (2) solution and reprecipitation of WC in and out of the liquid; (3) conversion of the binder from a pure metal to a pseudo-binary eutectic; and (4) diffusion, coalescence, and welding of carbide particles. Sintering is done in furnaces with hydrogen atmosphere for straight WC and low TiC grades or in vacuum furnaces for the complex grades, at temperatures between 1350°C (high Co) and 1650°C (low Co).

Because of the complexity of the manufacturing process, high-quality SC tools are only obtainable from manufacturers who have had extensive experience. Each of them offers a large variety of grades of proprietary compositions. All these grades fit loosely into the existing classification systems. Two of these, the ISO and the American system, are presented in Table 8.5 in an abbreviated form. For the complete ISO standard, refer to ISO 513:1975(E). These standards classify the SC grades according to the applications. The nominal compositions and hardnesses have been included in Table 8.6 with reference to [12].

The developments of the recent decade have led to a general improvement of the combination of strength and hardness through the improvement of the uniformity of

TABLE 8.5 Classification of Sintered Carbides

Designation Categories		Work					Machining
ISO	American	material	WC	Co	TiC	TaC	Operations
01	C8		64	3	25	8	Finishing
10	C7	Steels	76	6	12	6	
P 20							Roughing
30	C6		82	8	8	2	
40	C5		70	12	6	12	Interrupted
50							Cutting
10		Steels,					Finishing
M 20		cast iron,					
30		nonferrous					
40		metals					Roughing
01	C4	Cast iron,	97	3	—	—	Finishing
K 10	C3	nonferrous	96	4	—	—	
20	C2	metals,	94	6	—	—	
30	C1	nonmetallic materials	94	6	—	—	Roughing

grain size, uniformity of the Co layer, the introduction of micrograin carbides, improvement of the quality and uniformity of the manufacturing process, and the use of cold and hot isostatic compaction. However, the most important factor was the development of coated carbides.

Coated Carbides

We have seen that the strong and wear-resistant straight WC-Co grades cannot be used for machining steel because of the chemical affinity between WC and Fe. To overcome this problem the complex WC-TiC-TaC grades must be used, but these have less strength and wear resistance. Another solution was found to provide a thin protective coating over the base carbide material, the "substrate." Such coatings were first made of solid TiC (not combined with Co).

The main problem in coating is to obtain a bond between the coating layer and the substrate that is strong enough to withstand the stresses due to the different thermal coefficient of expansion of the two materials and to the temperature gradient between them. It was soon found that the coating layer must be rather thin, 3–5 μm; otherwise it would crack and peel off. The substrate must be strong, and its composition has to offer good bonding with the coating. The coating is generally carried out by chemical vapor deposition (CVD). To coat the inserts with TiC they are heated to a temperature of about 1000°C in an atmosphere of hydrogen containing vapors of titanium tetrachloride ($TiCl_4$). The chloride decomposes at the tool surface to deposit a thin layer of Ti that is carburized by methane included in the atmosphere. The techniques of carbide coating are still developing, and various compounds are successfully used. Although TiC is the one most commonly used, there are other popular coatings, such as TiN and $Al_2 0_3$, and they may all be combined.

Titanium carbide TiC provides good bonding with the substrate and acts as a diffusion barrier between the substrate and the steel chip. Titanium nitride TiN gives the lowest coefficient of friction with a steel chip, which is especially important at lower cutting speeds because it diminishes the tendency to form a built-up edge. Aluminum oxide Al_2O_3 has good chemical stability and wear resistance at high cutting speeds.

A microphotograph of the Coromant grade GC 310 is shown in Fig. 8.30. The top coating layer of Al_2O_3 is 1 μm thick and is applied over the basic coating layer of TiC, which is 6 μm thick. It is recommended for light to medium rough milling of cast iron at high cutting speeds. Another grade, GC 415, is triple-coated with the basic coat of TiC, an intermediate coat of Al_2O_3, and finally a very thin layer of TiN. It is intended for light and medium roughing of both steels and cast irons at low and high cutting speeds.

In general, coated carbides are more universal than the uncoated ones, and they find a wider range of applications. This is a great advantage because it reduces the necessary inventory of carbide inserts. They give about two times longer tool life and in some applications permit the use of much higher cutting speeds. They have now been widely accepted and represent almost half of all carbide inserts in use.

However, coated inserts are not suitable for finishing operations with fine feeds because they must have a less sharp cutting edge (larger radius of rounding of the edge); otherwise, the coating layers would peel off at the edge. They are also not suitable for heavy roughing, especially of castings with hard scale, because under such conditions the coating is destroyed in a short time. Nevertheless, as a whole, the coated carbides represent a significant step in the development of tool materials.

8.3.3 Ceramic Tools

The basic composition of ceramic cutting inserts is Al_2O_3 sintered mostly under hot isostatic compaction without any bonding agent. This tool material is more brittle than sintered carbides but very inert chemically, and it has high abrasion resistance. It cannot be used for heavy cuts or on work-hardening materials like stainless steels, and it has a chemical affinity to aluminum alloys. The best applications are for finishing and medium roughing of various kinds of cast irons. Characteristic workpieces are brake drums, brake discs, flywheels, clutch pressure plates, cylinder liners, and so on. In these applications it permits rather high cutting speeds.

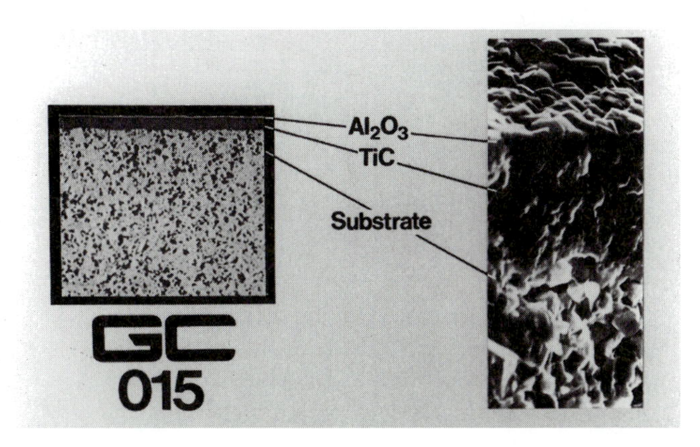

Figure 8.30 Structure of the coating of a sintered carbide. Several thin layers of different compositions are commonly applied so as to improve their cohesion in spite of different thermal expansion coefficients and to provide diffusion barriers. [SANDVIK]

Another significant ceramic material is silicon nitride Si_3N_4, which is used for machining cast iron at rather high cutting speeds of 500 to 1000 m/min. It is both hard and tough. It has been used successfully in an automotive engine factory in the form of inserts in large face-milling cutters for machining cast iron engine blocks and heads. This operation involves interrupted cutting as the heads move over the various bores and openings on the faces of the workpieces. The corresponding transfer line contains multiple spindles and is a rather massive application of this ceramic.

8.3.4 Borazon and Polycrystalline Diamond

These two materials are manufactured in a similar way, by superpressure hot sintering of a layer of cubic boron nitride (CBN; General Electric trade name Borazon) or of synthetic diamond particles (polycrystalline diamond, PCD) respectively, on top of sintered carbide inserts. However, solid CBN or PCD inserts are also coming into use. The individual crystal particles are from 1 to 100 μm in size, depending upon the application. These are the hardest of all materials, as is shown in Fig. 8.32. They are, of course, rather brittle; therefore they only make strong roughing tools when applied to the strong and tougher base of a WC-Co sintered carbide insert body.

They differ in their resistance to high cutting temperatures, as shown in Fig. 8.31. This graph plots the yield strength of the CBN and of diamond and compares it with that of tungsten carbide and with the more difficult to machine steel 4340 and with the most difficult to machine Ni-based alloys, Inconel and René 95. Note that the vertical scale of the diagram is logarithmic. At room temperature the *YS* of diamond and CBN is, respectively, 6 and 3 times higher than that of WC, which is itself 5 times higher than that of heat-treated alloy steel. The most important temperatures, however, are those at which the individual cutting materials succumb to chemical reactivity, oxidation, and disintegration. This limits the cutting ability of diamond and of sintered carbides to about 700°C and 1000°C, while CBN can take up to 1200°C. This makes the CBN suitable for machining not only very hard materials like chilled iron but also the "superalloys," which generate high cutting temperatures.

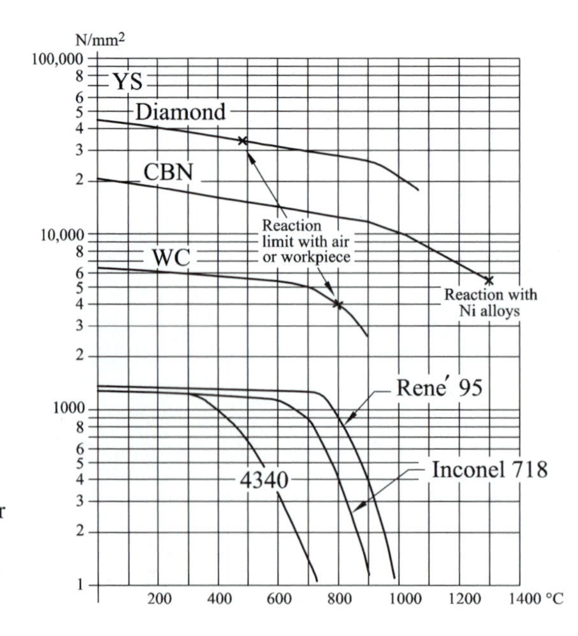

Figure 8.31
Yield strength versus temperature of CBN and diamond compared with WC and some workpiece materials. These two materials are used in special machining applications: CBN for hard steel and chilled iron, as well as superalloys, and diamond for Al-Si alloys containing hard SiO_2 grains and for grinding sintered carbides.

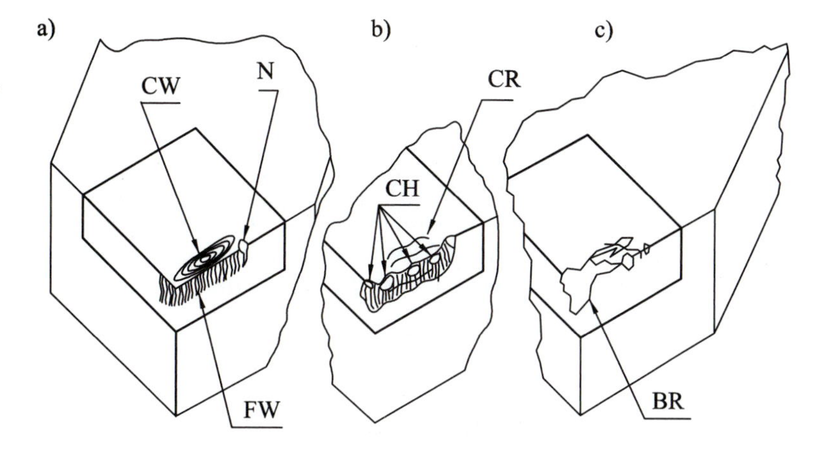

Figure 8.32
Various forms of tool damage. a) The continuously growing features: flank wear, crater wear, and the notch. b) The stochastically but gradually developing features: cracks and chippings of the edge. c) The random massive breakage.

Chilled cast iron is rather hard because it consists essentially of grains of cementite (Fe_3C). It may have a hardness of $68R_c$, which is about 10,000 HV (N/mm^2) and equals the hardness of hardened steel. It can be machined with sintered carbides only with great difficulty; the cutting speed for carbide tools would be about 25 m/min, and the temperature on the tool face would reach about 650°C. With CBN, a cutting speed of 120 m/min can be used with temperatures in the cut of about 800°C. Metal removal rate is about 5 times higher. For Co-based alloys, the cutting speeds with carbides are about 30 m/min for Inconel 718 and about 15 m/min for René 95. With CBN tools cutting speeds up to 6 times higher can be used.

These examples illustrate the application fields of CBN tools. On one hand, they are used for turning operations on hardened steel or chilled iron components (e.g., the rolls used in rolling mills) or even on components made of sintered carbides. In these instances turning with CBN tools replaces much more efficiently the traditional grinding operations. The other field is machining of Ni-based and Co-based heat-resistant alloys.

The tools incorporating polycrystalline diamond are used quite differently. Figure 8.31 shows that diamond cannot stand temperatures higher than about 650°C; thus, it cannot be used to machine materials developing high cutting temperatures. Moreover, it has chemical affinity to steel, which is "carbon hungry" and would very quickly destroy a diamond tool.

Applications for diamond tools thus exclude ferrous metals and are most successful on materials that are very hard or contain very hard particles and do not develop high cutting temperatures. Such materials have high values of thermal diffusivity α (see Table 8.1). The best examples of such materials are aluminum alloys with high silicon content. These alloys are widely used for automotive cylinder blocks and pistons and other parts because of their good castability and wear resistance. They are hypereutectic Al-Si alloys (e.g., up to 18% Si), and contain primary silicon, mostly as $Si\ O_2$, which abrades HSS and SC tools. Polycrystalline diamond tools are very successful in this application in that they allow cutting speeds of $v = 500$ m/min with tool lives of over 30 min. Another notable application is in machining (turning, milling, grinding) of sintered carbide parts. The value of α for the WC-Co material is 35 mm^2/sec, which is about 3 times that of steels and about half that of aluminum. Another special application is for grinding CBN tools. In this case the diamond is the only tool material applicable. Diamond is also used to machine Cu-alloys (e.g., commutators of electric motors), bronzes, plastic, and composite materials (glass laminates, carbon-fiber composites, boron-fiber-Al composites).

Finally, we should note that both CBN and diamond tools are very expensive and cannot be used for common machining where other, less expensive materials perform well.

Apart from polycrystalline diamond tools, the *single-crystal diamond tools* are also used in special applications. A typical example is in fine turning of elliptical mirrors made of aluminum or bronze (for space applications). In this case the main advantage of the diamond tool is the very high surface finish as well as very long tool life, which guarantees that there is no effect of tool wear on accuracy of form of the machined surface.

8.4 TOOL WEAR: CHOICE OF CUTTING CONDITIONS, MACHINABILITY OF MATERIALS

Various forms of damage of the cutting tool develop during cutting. The various features of tool wear, cracks, and tool breakage are indicated in Fig. 8.32. In sketch *(a)* the "regular" wear features are shown: flank wear FW, crater wear CW, notching N. These forms of wear grow rather uniformly with cutting time. The basic feature is flank wear, which is found on all tools, in all operations. Sketch *(b)* shows features that do not occur in all operations and could often be avoided by proper choice of tool material and geometry, and of cutting conditions (speed and feed). They are: cracks CR, which can be transversal or parallel to the cutting edge, and chipping CH. Cracks may occur soon after the start of the operation or after a certain time of cutting, but they do not grow uniformly. Chipping, which consists of small parts of the cutting edge breaking out, develops and grows with cutting time but this growth is rather erratic and not well repeatable. Sketch *(c)* shows massive breakage BR of a chunk of the nose of the tool. This form of damage has very little systematic relationship to cutting time. It occurs suddenly either soon after the start of the operation or after some time. The probability of this breakage increases with the chip load c.

The distinction between the regular features *(a)* of gradual wear and the irregular or catastrophic features in *(b)* and *(c)* will be clearly maintained, and the two classes of tool damage are discussed separately under the general headings of "wear" and "breakage," respectively.

8.4.1 Tool Wear

The three forms of tool wear are recapitulated in Fig. 8.33, which also gives their quantitative measures. Flank wear develops on the flank of the main cutting edge over the length that equals the chip width b and also on the flank of the tool tip and of the minor cutting edge, all along the part of the tool edge that is engaged in cutting. The size of flank wear is expressed by the average flank wear width FWW_{av} and/or its maximum FWW_{max}. Internationally, flank wear width is often denoted as VB (from the German *Verschleissmarken Breite*).

Crater wear develops on the rake face of the tool and can be expressed by measuring the crater depth CD (internationally, KT: *Krater Tiefe*). Crater develops only in some cutting operations, depending on the tool/workpiece material combinations and on the cutting speed used.

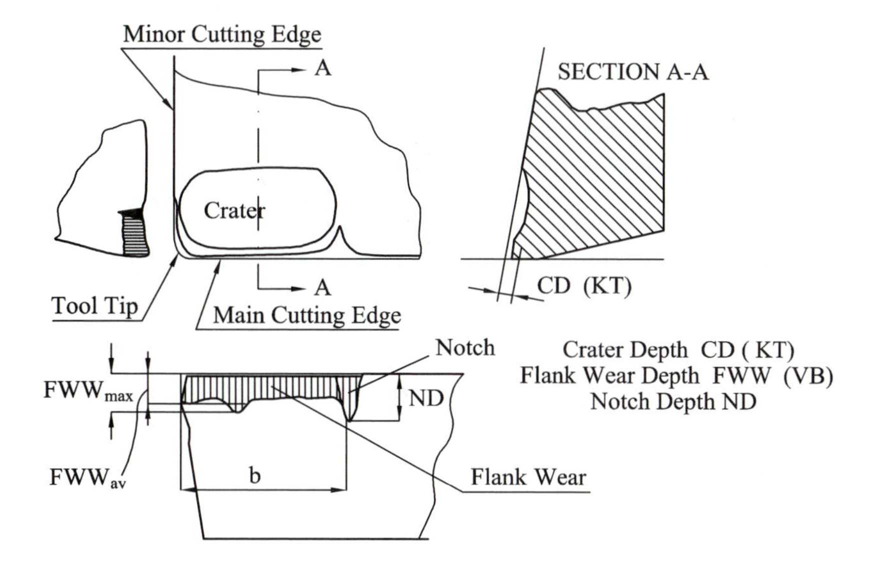

Figure 8.33
Parameters used to measure tool wear features.

At both ends of the cutting edge engagement, but not in all cases, a notch develops. It is primarily an oxidation phenomenon, and this is why it is generated at a point of the edge that is rather hot but not protected from the access of air by the chip and by the machined surface. Photographs of the various tool wear forms are given in Fig. 8.34. In the two views the crater, flank wear, and the notch are seen. These photographs were made under a tool-room microscope with a moderate magnification of about 10×.

Any of the three forms of wear (flank wear, crater, notch) may grow so much as to become the limiting factor of the particular cutting operation, that is, to determine the end of tool life. The criteria for the end of tool life as expressed by the maximum permissible value of *FWW* or *CD* or *ND* may differ depending on the type of the tool and the type of the operation.

High-speed steel tools are reground after tool life ends. If any form of wear is too great, too much of the tool material on the tool flank or on its rake face or both must be ground away. Therefore, only a small *FWW* or *CD* can be permitted so as to obtain a

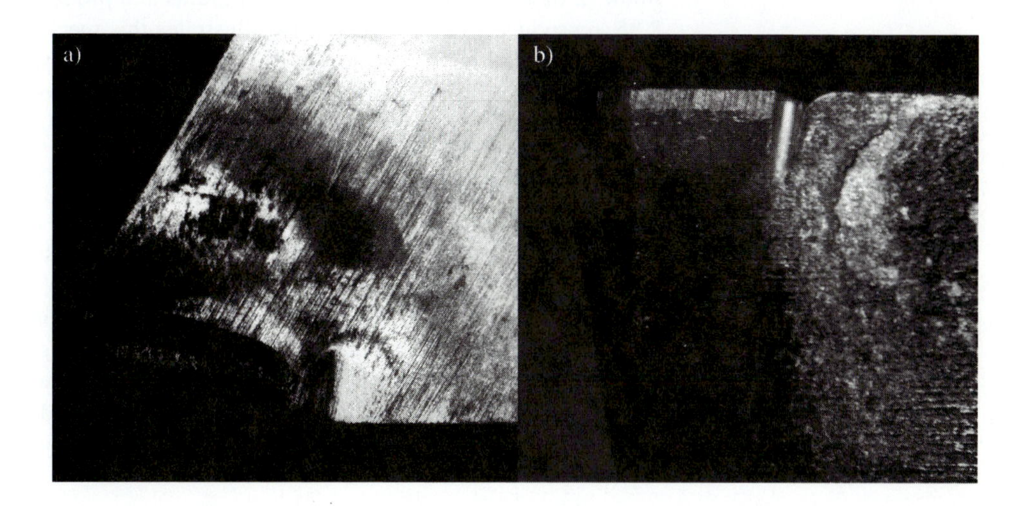

Figure 8.34
Photographs of a) crater and notch; b) flank wear and notch.

reasonable and economical number of regrinds. Sintered carbide tips are not reground but indexed for the next available edge, and after all edges have been worn out the insert is discarded (tools with "throw-away" tips). The tool-life criterion is then not the aspect of regrinding but of the ability of the tool to perform satisfactorily. If flank wear width is too large, the friction between the tool and the machined surface increases, and wear rate accelerates rapidly, indicating the end of tool life. In finishing operations the decisive aspect of wear is an increase of an outer dimension of the workpiece (decrease of an inner dimension), and the limit is reached when this affects the tolerance of the dimension. If a crater grows too close to the edge, it weakens the edge of the tool, which may suddenly break. Tool-life end must be declared safely before this stage is reached.

There are several different causes and mechanisms of tool wear. Friction on the rake face and on the flank of the tool occurs under intimate contact of freshly created surfaces of the workpiece material with practically no presence of air. The coolant penetrates into this contact only at very low cutting speeds. The pressure in the contact is at least equal to the yield stress of the workpiece material. The temperatures in the contact are high and may reach the melting temperature of one of the materials in the contact, most often that of the workpiece material.

The various mechanisms that contribute to the wear process are

1. Mechanical overload causing microbreakages (attrition)
2. Abrasion
3. Adhesion
4. Diffusion
5. Oxidation

Attrition

The grains of the various components of the tool material hold together at grain boundaries. Those on the rake face and on the flank are supported on at least half of their surface. However, those on the cutting edge are held on perhaps only a quarter of their surface and can therefore be rather easily broken out, embedded in the machined surface and in the underside of the chip, and dragged over the tool surface. Some of them may then break out other grains and produce a kind of chain effect. Fig. 8.35, reproduced from Vieregge [14] illustrates this mechanism. Diagrams *(a)* and *(b)* represent the initial and an advanced stage of the process, and diagram *(c)* shows the form of the worn tool. Attrition is the most significant mode of flank wear.

Abrasion

Abrasion is the commonly known wear process in which a harder material scratches a softer material over which it is sliding under normal pressure. This mechanism is significant for tool wear only in those instances where the workpiece material is very hard or contains hard particles: cast iron with grains of cementite, various metals containing hard inclusions like hypereutectic aluminum with SiC grains, steel killed with aluminum and containing Al_2O_3, and so on. The machined surface is cooler than the tool flank, and it may happen that the tool material is softened more than some of the constituents of the workpiece materials, which creates the conditions for abrasion.

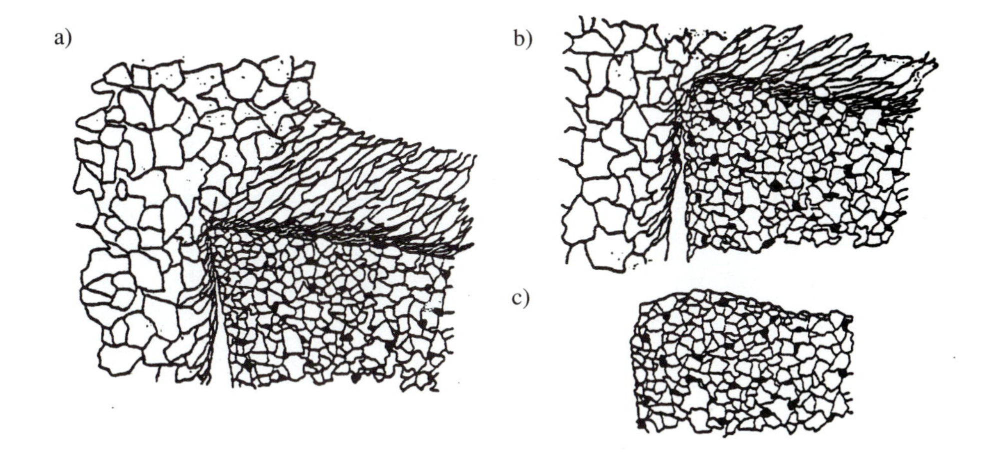

Figure 8.35 Attrition tool wear mechanism. (AFTER VIEREGGE)

Adhesion

In the conditions of the intimate contact between the tool and the freshly created surfaces on the workpiece and on the underside of the chip, welding of the workpiece surface and of the chip to the tool can often be observed. The extreme case is the built-up edge, which is formed in the low and middle speed range. Layers of workpiece material welded to the tool are found in ductile materials like in ferritic and austenitic steels, titanium alloys, and nickel-based alloys. The welded layers and points are periodically sheared away. This mechanism contributes to flank wear as well as to the formation of the crater.

Diffusion

Diffusion is an important mechanism and plays a significant role at higher cutting speeds in some workpiece/tool material combinations. The diffusion rate, that is, the amount of atoms of a material penetrating into another material, depends on the affinity of the two, very strongly on temperature, and on the gradient of concentration of the penetrating atoms in the solvent material. The latter aspect is very special in cutting, because the chip material that absorbs atoms of the tool material is continuously being carried away, and all the time new, virgin, unsaturated material is always arriving. The diffusion rate is highest at the point of highest temperature. This is about in the middle of the chip contact length, and that is where the crater becomes deepest. Diffusion is the most significant factor in crater wear, but it also participates in flank wear. Diffusion rate increases with cutting speed as the temperature at the contacts on the rake face and on the flank increases.

In cutting steel, diffusion occurs between the chip and the iron in the HSS and the Co in the SC, as well as between the carbon-hungry steel and carbides like WC and SiC (the latter in grinding wheels; correspondingly, Al_2O_3 grinding grains must be used on steel), between steel and the carbon of diamond, between titanium alloys and the TiC in carbides, and so on (see Fig. 8.36).

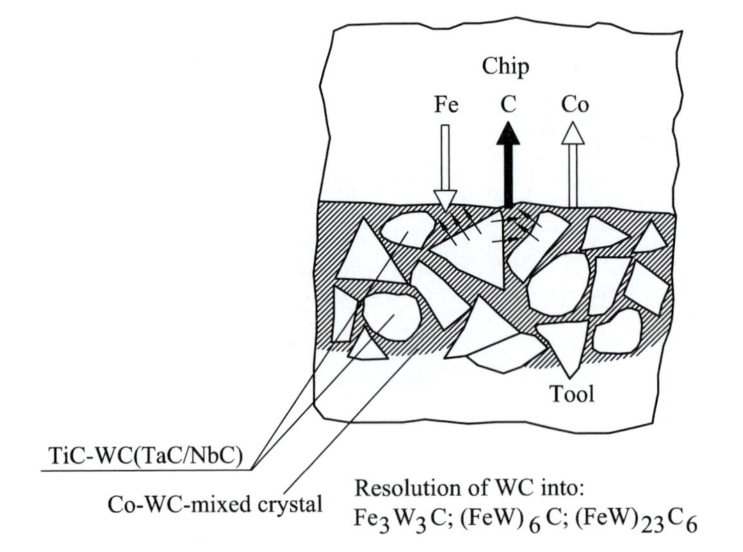

Figure 8.36
Diffusion processes between chip and a complex carbide tool. (AFTER KOENIG)

8.4.2 Tool Wear Rate and Tool Life

All the various wear mechanisms described above increase in intensity with increasing load on the cutting edge and with increasing temperature on the rake face and on the flank of the tool. These parameters depend in turn on the strength (shear-flow stress), on thermal properties of the workpiece material, and on the chip thickness *h,* and cutting speed *v.* Chip thickness is proportional to feed per revolution, or to feed per tooth. What is said here about flank-wear width *FWW* applies as well to crater depth *CD* or notch depth *ND* (as long as these two occur). All three forms of tool wear increase with cutting time and depend on cutting conditions.

Flank wear typically increases with the time of cutting, as shown in Fig. 8.37. At the beginning, Phase I, there is an initial faster increase that is followed by a steady increase in proportion to cutting time, Phase II. When the wear reaches a certain size, it will accelerate and may lead to a sudden failure of the edge, Phase III. As an approximation to actual wear, we can accept what is expressed by the dashed line indicating wear proportional to time:

$$FWW = r_w t \tag{8.44}$$

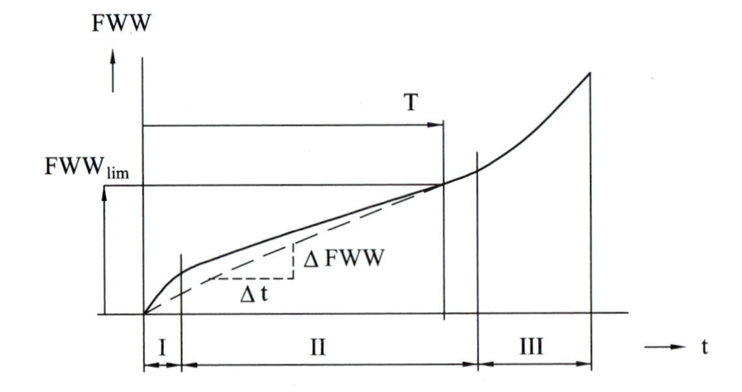

Figure 8.37
Development of flank-wear width with time; definition of tool life *T* as the time when a limit value of flank wear is reached.

where $r_w = \Delta FWW/\Delta t$ is the rate of wear. A certain value of wear may be chosen as the permissible limit FWW_{lim}. The time at which this limit is reached is called the tool life T.

The wear rate depends, for a given work/tool material combination, on cutting speed v and chip thickness h. Within a practical range of v and h, this relationship is usually expressed in algebraic form:

$$r_w = C_r v^p h^q \tag{8.45}$$

The exponent p on v is rather high, between 2 and 6, indicating the significant influence of v, which is due to its effect on the temperatures of the tool. The exponent q on h is usually between 1.5 and 3 and is due to the influence of h on the load on the tool. Combining (8.44) and (8.45) we have

$$FWW = C_r v^p h^q t$$

and, for $t = T$, we have

$$FWW_{\text{lim}} = C_r v^p h^q T$$

or, combining FWW_{lim} and C_r,

$$v^p h^q T = C^* \tag{8.46}$$

Equation (8.46) is the tool life equation in a form very similar to that chosen already by F. W. Taylor in 1905, except that his formulation involved a relationship between v and T only. Considering that in the turning operation, chip thickness is related, for a given side-cutting-edge angle σ, to the feed per revolution f_r, we have

$$h = f_r \cos \sigma$$

We could also write

$$v^p f_r^q T = C \tag{8.47}$$

where $C = C^*/(\cos \sigma)^2$

Equation (8.47) is the usual form of the tool life equation expressing the relationship between cutting speed v, feed f_r, and tool life T, and it is established considering a particular value of the angle σ. However, Eq. (8.46) is more fundamental. It is common to obtain the parameters in Eq. (8.47) experimentally and express them graphically on log-log paper as a relationship between two of the three variables involved.

EXAMPLE 8.5 Determine the Parameters of the Tool Life Equation ▼

Tool life tests were carried out by turning normalized steel 1045 using carbide grade P20 with the following results. (These values are mean values from a number of repeated tests. We do not discuss here the scatter and the probabilities; the situation is simplified as a deterministic case.)

	v (m/min)	f_r (mm)	T (min)
1.	100	0.2	80
2.	200	0.2	10
3.	200	0.1	40

The units used here for v, f_r, and T are the most commonly used ones in workshop practice. Assuming the tool life equation in the form of Eq. (8.47), we have

$$\left(\frac{v_2}{v_1}\right)^p = \frac{T_1}{T_2}, \qquad 2^p = 8, \qquad p = 3$$

$$\left(\frac{f_{r2}}{f_{r3}}\right)^q = \frac{T_3}{T_2}, \qquad 2^q = 4, \qquad q = 2$$

$$C = 100^3 \times 0.2^2 \times 80 = 3.2 \times 10^6$$

The corresponding tool life equation is

$$v^3 f_r^2 T = 3.2 \times 10^6 \tag{8.48}$$

This equation is expressed in three forms: (1) in the graph of Fig. 8.38 in coordinates (T, v); (2) in Fig. 8.39 in coordinates (T, f_r); and (3) in Fig. 8.40 in coordinates (v, f_r), while the remaining variable is always used as a parameter. It is, of course,

$$\log T = -3 \log v - 2 \log f_r + 6 + \log 3.2$$

and

$$\log v = -\frac{2}{3} \log f_r - \frac{1}{3} \log T + 2 + \frac{\log 3.2}{3}$$

Correspondingly, in the three graphs all three relationships are represented, in the log-log scales, as straight lines with slopes: (1) -3; (2) -2; (3) $-2/3$. In all the three graphs the point A is indicated: $v = 200$ m/min, $f_r = 0.2$ mm, and $T = 10$ min.

The graphs of Figs. 8.38 and 8.39 show the intensity of the effects of cutting speed and feed (chip thickness) on tool life. The graph of 8.40 is interesting from another point of view. Recalling the equation for the metal removal rate, $Q = af_r v$,

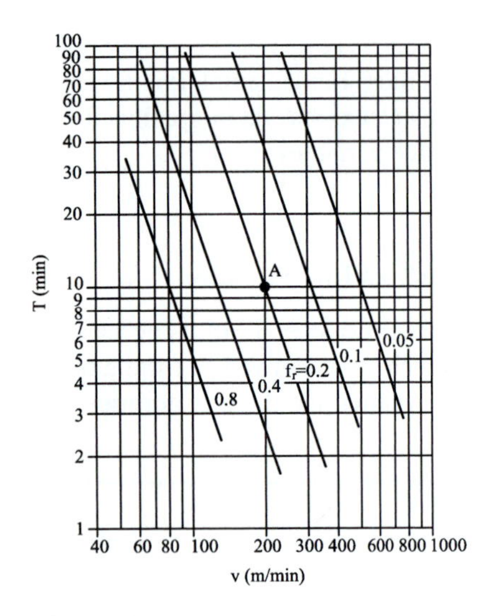

Figure 8.38
Taylor-type graph of tool life versus cutting speed.

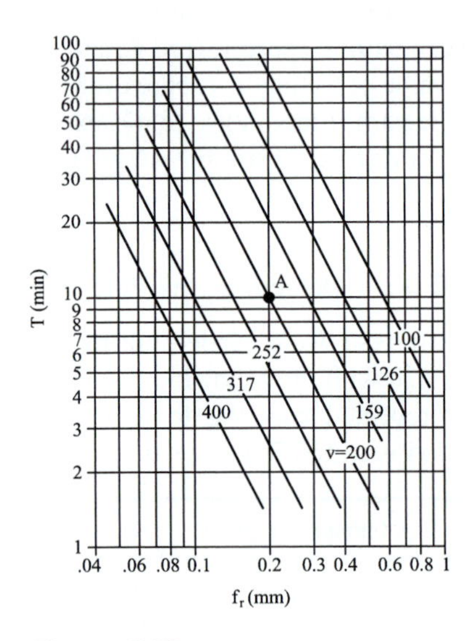

Figure 8.39
Taylor-type graph of tool life versus feed per revolution.

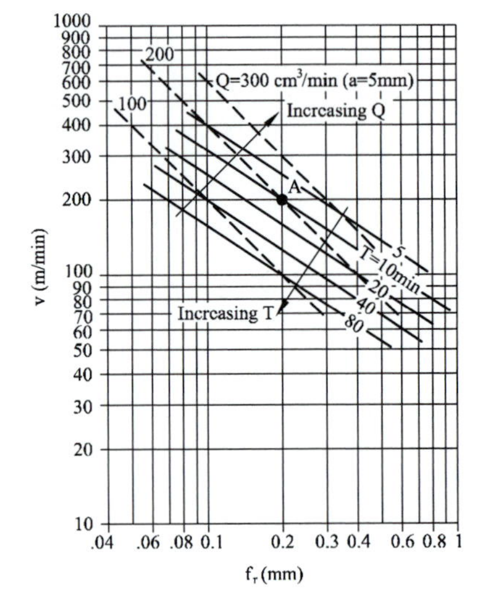

Figure 8.40
Lines of constant tool life and of constant metal removal rate as functions of cutting speed and of feed.

it is possible to include in this diagram the lines for constant metal removal rates for a given depth of cut. For example, choosing $a = 5$ mm yields

$$Q\,(\mathrm{cm^3/min}) = 5 \times f_r(\mathrm{mm}) \times v(\mathrm{m/min})$$

Three such lines, for $Q = 100, 200, 300\ \mathrm{cm^3/min}$ are entered in the graph as broken lines. They have a slope of (-1). The directions are indicated in which metal removal rate Q increases for the $Q =$ const lines and in which the tool life T increases for $T =$ const lines.

8.4.3 Optimizing Cutting Speed and Feed in a Single-Cut Operation: Taylor-Type Tool Life Equation

This exercise is done for a single-tool, single-pass turning operation. The cutting conditions can be optimized from two different points of view:

1. Minimum cost of the operation
2. Minimum time of the operation

The first criterion is most common and basic to all manufacturing. Criterion 2 is applied if a faster delivery represents an advantage that is greater than the slightly higher cost.

First, we analyze *the criterion of minimum cost per part* C_p. The cost per part C_p of an operation consists of three parts:

$$C_p = C_{\mathrm{fix}} + C_m + C_t$$

where C_{fix} is fixed cost unrelated to cutting speed v and feed f_r, such as the cost of the material of the part and cost of all the auxiliary operations, C_m is the cost of machining, and C_t is the cost related to tool wear and to tool changing.

The component C_{fix} has no effect on optimizing v and f_r. The two other components have opposite trends (see Fig. 8.41). With increasing speed v or increasing feed f_r, the machining time and hence C_m decrease; tool wear, however, accelerates, and it is necessary to change the tool more often and pay more often for the changing as well as for the worn edge; the cost C_t thus increases. The sum $C_m + C_t$ shows a minimum M.

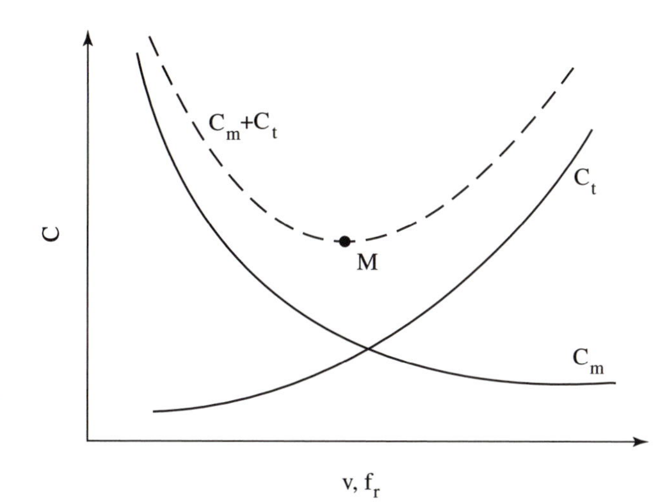

Figure 8.41
Effect of cutting speed v and feed f_r on the cost per part of components due to machining time and due to tool cost. The parameters for achieving minimum cost are determined using the tool life equation.

The two basic terms involved are the machining time t_m and tool life T, both functions of v and f_r:

$$t_m = VOL/MRR \tag{8.49}$$

where VOL is the volume to be removed from a part, and MRR is metal removal rate. The volume VOL depends on the type of operation; for cylindrical turning, see Fig. 8.42.

$$VOL = \pi D_m aL$$

The metal removal rate is

$$MRR = vaf_r$$

The machining time is then

$$t_m = \frac{\pi D_m L}{vf_r} = \frac{A}{vf_r} \tag{8.50}$$

where the constant A represents the dimensional details of the case and also consolidates the following dimensions: t_m (min), v (m/min), and f_r (mm).

The tool life equation of the Taylor type (8.47) will be used to express

$$T = Cv^{-p} f_r^{-q} \tag{8.51}$$

The cost per part can now be written as follows:

$$C_p = C_{\text{fix}} + t_m r_m + \frac{(t_{tch} r_m + C_{te})\, t_m}{T} \tag{8.52}$$

where

r_m is the rate for using the machine tool, (\$/min)

t_{tch} is the tool changing time, (min)

C_{te} is the cost per tool edge

and

$$C_{tr} = t_{tch} r_m + C_{te} \tag{8.53}$$

where C_{tr} is tool-related cost and is expended once per tool life T, that is per n parts: $n = T/t_m$. Correspondingly, the tool-related cost per part is

$$C_t = \frac{C_{tr}}{n} = \frac{C_{tr} t_m}{T} \tag{8.54}$$

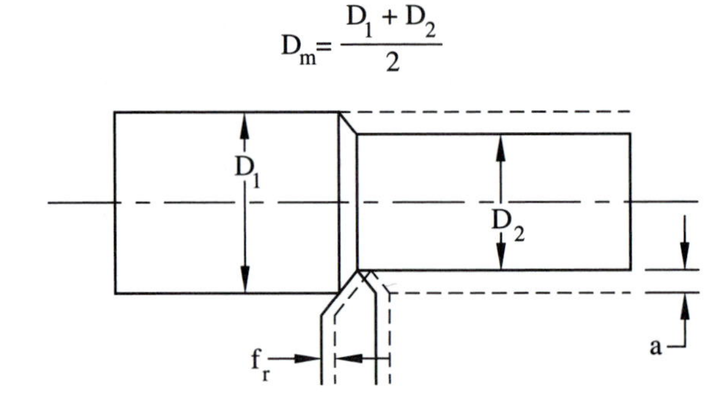

Figure 8.42
Volume removed in a single-pass, cylindrical turning operation.

Combining Eq. (8.52) with Eqs. (8.53) and (8.54),

$$C_p = \frac{Ar_m}{vf_r} + \left(\frac{C_{tr}A}{vf_r}\right)\left(\frac{v^p f_r^q}{C}\right) \tag{8.55}$$

$$C_p = A\left(\frac{r_m}{vf_r} + \frac{C_{tr}v^{p-1}f_r^{q-1}}{C}\right)$$

In order to determine the values of v_{opt} and f_{ropt} for which C_p is minimum, partial derivatives of Eq. (8.55) with respect to the two variables should equal zero:

$$\frac{\partial C_p}{\partial v} = A\left[\frac{-r_m}{v^2 f_r} + \frac{(p-1)\,C_{tr}v^{p-2}f_r^{q-1}}{C}\right] = 0 \tag{8.56}$$

$$\frac{\partial C_p}{\partial f_r} = A\left[\frac{-r_m}{vf_r^2} + (q-1)\frac{C_{tr}v^{p-1}f_r^{q-2}}{C}\right] = 0 \tag{8.57}$$

$$v^p f_r^q = \frac{Cr_m}{(p-1)C_{tr}} \tag{8.58}$$

$$v^p f_r^q = Cr_m/[(q-1)C_{tr}] \tag{8.59}$$

Considering the form of Eqs. (8.58) and (8.59), it is obvious that they cannot be used to give unique finite values of v and f_r. Correspondingly, they represent, in the field of variables v and f_r of the graph in Fig. 8.40, two lines of constant tool life. Such lines are parallel to each other and do not intersect except in infinity. However, the graph gives a good insight into this optimization problem.

First, assume the task to optimize v and f_r for a particular value of tool life T (e.g., $T = 10$ min). For various combinations (v, f_r) the corresponding metal removal rate varies; this is recognized by intersecting the various lines of $Q = \text{const}$. For a particular tool life T, the cost component C_t of Eq. (8.52) decreases with t_m, and the cost component C_m also decreases with t_m; that is, they both decrease with increasing metal removal rate Q. In which direction should we move on a line $T = \text{const}$ to increase Q? The answer is obvious from Fig. 8.40: moving to the right, towards larger values of f_r and lower values of v, the values of Q increase all the time. Thus the optimum is obtained for $f_r = \infty$. An infinite feed rate is obviously not possible; there are constraints on the maximum feed that can be used: in a finishing cut, it is the required surface finish, and in a roughing cut, it is the load on the cutting edge that would lead to its breakage. The maximum cutting force that the tool holder can take also puts a limit on feed, so that there is a particular permissible value $f_{r\max}$. This is then the optimum value of f_r.

The reason why it is preferable to increase the feed and decrease the speed and not vice versa is that although both affect MRR equally, v affects tool wear more than f_r because $p > q$, and the slope of the $T = \text{const}$ line in Fig. 8.40 is smaller (in absolute value) than that of the $Q = \text{const}$ line: $|-2/3| < |-1|$. If we had $p < q$, the situation would be reversed, and the optimum would be reached for maximum cutting speed. However, $p > q$ applies in most cases.

Once the optimum feed is selected as $f_{\max}$, Eq. (8.58) can be used to determine the optimum value of v.

$$v_{\text{opt}} = \left[\frac{Cr_m}{(p-1)C_{tr}f_{r\max}^q}\right]^{1/p} \tag{8.60}$$

and the optimum tool life is

$$T_{\text{opt}} = \frac{(p - 1)C_{tr}}{r_m} = (p - 1)t_{\text{teq}} \tag{8.61}$$

Recall that C is the tool life equation constant, p is the exponent on v in the tool life equation, C_{tr} is the tool-related cost of Eq. (8.53), and r_m is the machine rate. The ratio (C_{tr}/r_m) can be interpreted as the "tool cost equivalent time" t_{teq}, that is, the time of using the machine that would cost as much as tool changing and using a new tool edge. Expression (8.61) is easy to remember.

Minimum Production Time per Part

As mentioned above, alternatively, one may want to optimize with respect to this criterion. The production time per part t_p is

$$t_p = t_{\text{fix}} + t_m + t_{tch} \tag{8.62}$$

where t_{fix} is time that is independent of v and f_r and includes all the auxiliary times (loading, etc.), t_m is the machining time, and t_{tch} is the tool-changing time. In looking for the optimum values of v and f_r under this criterion one would proceed in a way analogous to that used for the minimum cost criterion.

8.4.4 Optimizing Cutting Speed and Feed in a Single-Cut Operation: Tool Life Equation Non-Taylor-Type

The Taylor-type tool life equation of Eq. (8.47) is a simplification of reality. The straight lines in the graph of Fig. 8.40 should be replaced, in a better approximation of reality, by curves of the type shown in Fig. 8.43. The difference between this graph and Fig. 8.40 is that the T = const curves are convex. This signifies that increasing v and f_r towards more extreme values of each, respectively, affects the tool life at an accelerated rate. For example, at the left-hand end of these curves, feeds are light and speeds high; any small change of speed has to be compensated by a much greater change of feed than in the middle of the field.

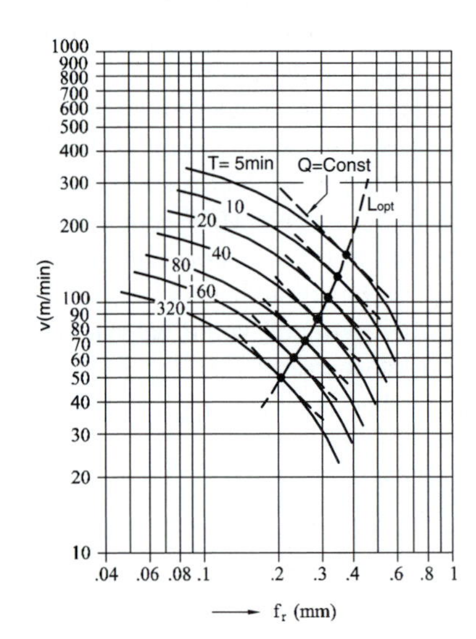

Figure 8.43
The v, f, T relationship for non-Taylor-type tool equations. The locus of optima follows constant removal rate tangent points to the constant tool life lines.

To optimize, we will assume that the T = const curves have been obtained experimentally. We will not try to express them by a mathematical formula and will proceed on the basis of the graphical input.

The "local" optimum for every T = const curve is obviously the one where a Q = const line with a (-1) slope is tangent to it. This determines the point of maximum *MRR* for the particular tool life. For a given tool life, indeed, the point of maximum *MRR* gives the shortest cutting time and, correspondingly, the lowest machining cost and the most parts per tool change. Such points on all the available T = const curves provide the locus L_{opt} on which the overall optimum will be located.

Using the expression for the cost per part C_p,

$$C_p = C_{\text{fix}} + t_m r_m + (t_{\text{tch}} r_m + C_{te}) \frac{t_m}{T}$$

$$C_p = C_{\text{fix}} + A r_m/(v f_r) + \left(\frac{A C_{tr}}{v f_r}\right)\left(\frac{1}{T}\right) \tag{8.63}$$

$$C_p = C_{\text{fix}} + A\left(\frac{r_m T + C_{tr}}{v f_r T}\right)$$

We are looking for the minimum of the term in the parentheses:

$$B = \frac{r_m T + C_{tr}}{v f_r T} \tag{8.64}$$

The parameters r_m and C_{tr} are available for the particular operation, and the combinations of v, f_r, T can be read off the diagram of Fig. 8.43 for every known point on the locus L_{opt}. By evaluating B for all these points, we find that the one for which B is minimum indicates the optimum $(v, f_r)_{\text{opt}}$ combination.

8.4.5 Optimizing Speeds and Feeds for a Multitool Operation: Tool Life Equation of the Taylor Type

In some instances many tools may be engaged simultaneously on one or several identical workpieces. The extreme example is the transfer line (see Fig. 10.13), where hundreds of tools may work simultaneously on dozens of workpieces (stations). An intermediate case is a multispindle lathe (see Fig. 10.10). On an 8-spindle lathe, for example, a complex machining operation is divided into eight suboperations, each of them carried out in one of the eight positions. The spindles are arranged in a drum and indexed through the eight positions, step by step. The time between subsequent indexes is the cycle time t_c. In a single cycle the eight suboperations are carried out, one of them being the final one. In this way, one workpiece (part) is finished in every cycle. At each station several tools work simultaneously. All together, there are n tools; in our example $n = 24$ tools working on the 8 spindles.

Let us subscript the tools by $i = 1$ to n and the spindles by $j = 1$ to 8. In Fig. 8.44 the diagram shows two spindles with two tools on each. Each tool works on a particular diameter D_i over a length of cut L_i. In the case of tool 4 the mean diameter was taken, for simplicity (this operation could be handled in a more complex way). The spindle speeds are n_j (rpm).

Quite generally, the total number of variables v_i and f_{ri} to determine is $2n$. However, the following constraints reduce the number of independent variables to one only:

1. There is a maximum permissible feed $f_{ri\max}$ for each tool.

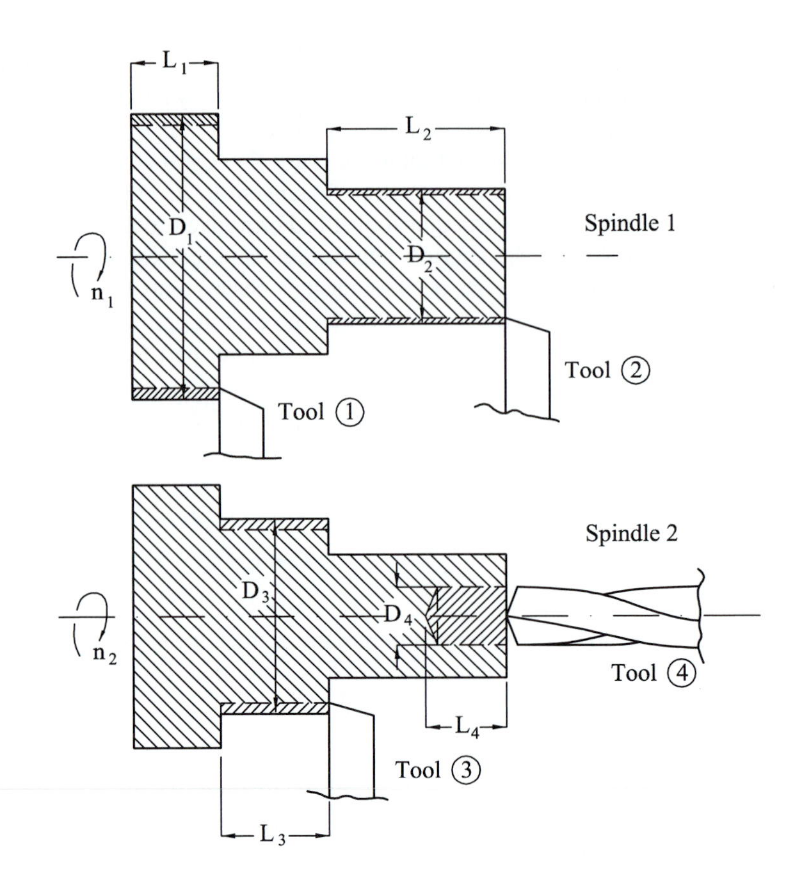

Figure 8.44
Part of a multispindle operation. Two spindles are shown with two tools machining in each of the two stations. Speeds and feeds for all tools can be optimized using the constraint dictating that all tools must work for full cycle time.

2. Cutting speeds $v_{i,j}$ for equal j, that is, of all tools working on the same spindle, are related to the spindle speed n_j: for $j = k$,

$$v_{ik} = \pi D_{ik} n_k \tag{8.65}$$

3. Provided that the exponents in the tool life equation are $p > 1$ and $q > 1$ (which is almost always the case), the rule applies that it is most economical if every tool works for the whole cycle time. We will now justify this rule. Let us assume that the cycle time is t_c, and

   (a) the tool (i) which has to machine a rather short length L_i is cutting for only a part of the cycle time, let us say for only $t_c/2$. Let us assume $p = 3$ and $q = 2$. As it is, the tool wear increases through the cycle by

$$\Delta FWW_a = \frac{r_{wa} t_c}{2}$$

   In order to make the tool work for the whole cycle time we do either of the following:

   (b) decrease the cutting speed to one half; the tool wear rate will then be [refer to Eq. (8.45)]

$$r_{wb} = \frac{r_{wa}}{2^p} = \frac{r_{wa}}{8}$$

$$\Delta FWW_b = \left(\frac{r_{wa}}{8}\right) t_c = \frac{\Delta FWW_a}{4}, \text{ or}$$

(c) decrease the feed rate to one half; the tool wear rate will then be

$$r_{wc} = \frac{r_{wa}}{2^q} = \frac{r_{wa}}{4}$$

$$\Delta FWW_c = \left(\frac{r_{wa}}{4}\right)t_c = \frac{\Delta FWW_a}{2}$$

In cases (*b*) and (*c*) in which the tool worked for the whole cycle time, the tool wear increment was less than in the case (*a*), where it worked for part of the cycle time only. This is understandable, because tool wear is proportional to cutting time, which is itself inversely proportional to v or f_r. However, the tool-wear rate is more than proportional to v and f_r.

The number of variables is reduced by applying constraints 2 and 3. Provided that $p > q$, we will again obtain an optimum for maximum possible feeds within the constraint 1.

There may be different strategies for changing tools. The basic strategy is such that every tool is changed whenever its tool life ends. Alternative strategies consist of changing several tools at once whenever one of them arrives at its tool-life end, even if the others could still work. This latter strategy is based on the condition that tool-changing time for a group of tools would be less than the sum of the individual tool-changing times. Here, we discuss only the basic strategy.

The cost per part equation is

$$C_p = C_{\text{fix}} + t_c r_m + \sum_{1}^{n} (t_{tch,i}\, r_m + C_{te,i}) \frac{t_c}{T_i} \tag{8.66}$$

In keeping with the constraint 3 above, the machining time t_m of Eq. (8.52) is replaced here for all tools by the cycle time t_c.

The parameters r_m (machine rate), $t_{tch,i}$ (tool-changing time) and $C_{te,i}$ (cost per tool edge or per tool regrind) are supposed to be known for each tool. Also the tool life equation:

$$T_i = \frac{C_i}{(v_i^{pi} f_r^{qi})}$$

is supposed to be known for every tool.

In minimizing C_p it is usually convenient to first determine all feeds by applying constraints. Next, the constraints 2 and 3 are applied to express all cutting speeds by one spindle speed n_k only. ▲

EXAMPLE 8.6 Optimize Speeds and Feeds for a Multitool Operation According to Fig. 8.44 ▼

The strategy of changing each tool individually will be considered. We are given the following:

$D_1 = 112$ mm, $D_2 = 60$ mm, $D_3 = 86$ mm, $D_4 = 28$ mm,
$L_1 = 40$ mm, $L_2 = 80$ mm, $L_3 = 50$ mm, $L_4 = 30$ mm

The tool-related costs are identical for tools 1 to 3:

$(t_{tch})_{1,2,3} = 5$ min, $(C_{te})_{1,2,3} = \$2.00$

For the twist drill it is

$t_{tch} = 3$ min, $C_{te} = \$5.00$

The tool life equation for tools 1, 2, 3:

$$v^3 f_r^2 T = 3.2 \times 10^6, \qquad v\,(\text{m/min}), f_r\ (\text{mm}), T(\text{min})$$

and for the drill, tool no. 4:

$$v^5 f_r^2 T = 2 \times 10^7$$

Limits on feed:

$(f_{r\max})_{1,2,3} = 0.4$ mm, $f_{r\max,4} = 0.3$ mm

Execute the following:

1. Feeds on spindle 1, machining time:

$t_{m1} = L_1/(n_1 f_{r1}) = 40/(n_1 f_{r1}), f_{r1} \leqq 0.4$
$t_{m2} = L_2/(n_1 f_{r2}) = 80/(n_i f_{r2}), f_{r2} \leqq 0.4$
$t_{m1} = t_{m2} = t_c;\ 40/f_{r1} = 80/f_{r2}, f_{r1} = 0.5\,f_{r2}; f_{r2} = 0.4$
$f_{r1} = 0.2$

Feeds on spindle 2.

$t_{m3} = L_3/(n_2 f_{r3}) = 50/(n_2 f_{r3}), f_{r3} \leqq 0.4$
$t_{m4} = L_4/(n_2 f_{r4}) = 30/(n_2 f_{r4}), f_{r4} \leqq 0.3$
$t_{m3} = t_{m4} = t_c = 50/(f_{r3}) = 30/f_{r4}$
$f_{r4} = 0.6\,f_{r3}\ \ f_{r3} = 0.4$
$f_{r4} = 0.24$

All the feeds have now been determined: $f_{r1} = 0.2$ mm, $f_{r2} = 0.4$ mm, $f_{r3} = 0.4$ mm, and $f_{r4} = 0.24$ mm

2. Speeds on spindle 1 are $v_1 = \pi \times 0.112 \times n_1$, and $v_2 = \pi \times 0.060 \times n_1$.

Speeds on spindle 2 are $v_3 = \pi \times 0.086 \times n_2$, and $v_4 = \pi \times 0.028 \times n_2$.

Applying constraint 3, we have

$t_{m1} = t_{m3}$
$L_1/(0.2n_1) = 50/(0.4n_2)$
$10n_1 = 16n_2$
$n_2 = 0.625n_1$

All the cutting speeds can now be expressed by means of spindle speed n_1:

$v_1 = \pi\ 0.112n_1 = 0.3519\ n_1$
$v_2 = \pi\ 0.060n_1 = 0.1885\ n_1$
$v_3 = 0.625\ \pi\ 0.086\ n_1 = 0.1689\ n_1$
$v_4 = 0.625\ \pi\ 0.028\ n_1 = 0.0550\ n_1$

The cost equation (8.66) can now be written as a function of the single variable n_1.

$$C_p = C_{\text{fix}} + t_c r_m + (C_{tr})_{1,2,3}\,(1/T_1 + 1/T_2 + 1/T_3)\,t_c + C_{tr,4} t_c/T_4$$

So we have

$t_c = t_{m1} = L_1/(f_{r1} n_1) = 40/(0.2n_1) = 200/n_1$

The machine rate is given as

$r_m = 1.0$ \$/min

$(C_{tr})_{1,2,3} = t_{tch} r_m + C_{te} = 5 \times 1.0 + 2.0 = 7.0$

$C_{tr,4} = 3 \times 1.0 + 5.0 = 8.0$

From the tool life equations we obtain:

$$1/T_1 = v_1^3 f_{r1}^2/(3.2 \times 10^6) = (0.3519\, n_1)^3(0.2)^2/(3.2 \times 10^6)$$

$$= 5.447 \times 10^{-10} n_1^3$$

$$1/T_2 = (0.1885 n_1)^3\,(0.4)^2/(3.2 \times 10^6) = 3.349 \times 10^{-10} n_1^3$$

$$1/T_3 = (0.1689 n_1)^3\,(0.4)^2/(3.2 \times 10^6) = 2.409 \times 10^{-10} n_1^3$$

$$1/T_4 = (0.055 n_1)^3(0.24)^2/(2 \times 10^7) = 1.450 \times 10^{-15} n_1^5$$

$$C_p = C_{\text{fix}} + 200/n_1 + (7 \times 200 \times 1.121 \times 10^{-9} n_1^3)/n_1 +$$

$$+ (8 \times 200 \times 1.450 \times 10^{-15} n_1^5)/n_1 \qquad (8.67)$$

For finding optimum n_1 the derivative of C_p with respect to n_1 is set to zero:

$$\frac{dC_p}{dn_1} = \frac{-200}{n_1^2} + 3.1374 \times 10^{-6} n_1 + 9.2768 \times 10^{-12} n_1^3 = 0,$$

$$9.2768 \times 10^{-12} n_1^5 + 3.1374 \times 10^{-6} n_1^3 - 200 = 0 \qquad (8.68)$$

The only real positive root of Eq. (8.68) is $n_1 = 359$ rpm, which is the optimum speed of spindle no. 1. This gives the following:

Cutting speeds

$v_1 = 126$ m/min.
$v_2 = 68$ m/min.
$v_3 = 61$ m/min.
$v_4 = 20$ m/min.

Tool lives

$T_1 = 40$ min.
$T_2 = 65$ min.
$T_3 = 90$ min.
$T_4 = 116$ min.

Cost per part

$$C_p = C_{\text{fix}} + C_m + C_{t1} + C_{t2} + C_{t3} + C_{t4}$$

$$C_p = C_{\text{fix}} + 0.557 + 0.0975 + 0.060 + 0.0433 + 0.0384,$$

$$C_p = C_{\text{fix}} + C_m + C_t = C_{\text{fix}} + 0.557 + 0.239 = C_{\text{fix}} + 0.796\ (\$)$$ ▲

EXAMPLE 8.7 How Does the Number of Tools Cutting Simultaneously Affect Optimum Speed and Optimum Tool Life? ▼

The answer to this question depends, of course, on the combined conditions of all of the tools that are working simultaneously. However, in order to understand the basic character of this influence we may simply assume that all the tools working simultaneously carry out identical tasks and that they all have the same limit on feed and follow the same tool life equation.

For a single-tool operation, the optimum cutting speed and optimum tool life were expressed by Eqs. (8.60) and (8.61).

Now, we turn our attention to the cost equation (8.66) which, with our assumption of N identical simultaneous operations becomes

$$C_p = C_{\text{fix}} + t_c r_m + NC_{tr}\left(\frac{t_c}{T}\right) \tag{8.69}$$

where $T = C/(v^p f_r^q)$ is the identical tool life for all N tools.

We also have $t_c = A/vf_r$, where A is a constant depending on the volume to remove, and $f_r = f_{r\max}$.

Correspondingly,

$$C_p = C_{\text{fix}} + \frac{Ar_m}{vf_{r\max}} + NC_{tr}\left(\frac{A}{C}\right)v^{p-1}f_{r\max}^{q-1}$$

For an optimum,

$$\frac{dC_p}{dv} = \frac{-Ar_m}{v^2 f_{r\max}} + (p-1)NC_{tr}\left(\frac{A}{C}\right)v^{p-2}f_{r\max}^{q-1} = 0$$

$$v_{\text{opt}} = \left[\frac{Cr_m}{(p-1)NC_{tr}f_{\max}^q}\right]^{1/p} \tag{8.70}$$

and the optimum tool life is

$$T_{\text{opt}} = \frac{(p-1)NC_{tr}}{r_m} = (p-1)Nt_{teq} \tag{8.71}$$

Comparing Eq. (8.70) with Eq. (8.60) and (8.71) with (8.61), we find that

$$(v_{\text{opt}})_n = \left(\frac{1}{N}\right)^{1/p}(v_{\text{opt}})_1 \tag{8.72}$$

and

$$(T_{\text{opt}})_n = N(T_{\text{opt}})_1 \tag{8.73}$$

where the subscripts n and 1 denote simultaneous operation with N tools and a single tool, respectively. In multitool operations the optimum economy is obtained at lower cutting speeds and at N times longer tool lives.

To indicate, if $p = 3$ and $N = 10$, then

$$(v_{\text{opt}})_{10} = 0.46(v_{\text{opt}})_1$$

and

$$(T_{\text{opt}})_{10} = 10(T_{\text{opt}})_1$$

Referring to Example 8.6, if the case of a single operation was identical with that of tool no. 1, the optimum cutting speed and tool life would be, using Eqs. (8.60) and (8.61), respectively,

$$v_{\text{opt}} = \left[3.2 \times 10^6 \times \frac{1.0}{(2 \times 7 \times 0.2^2)}\right]^{1/3} = 179 \text{ m/min}$$

instead of 126 m/min, and

$$T_{\text{opt}} = \frac{2 \times 7}{1.0} = 14 \text{ min}$$

instead of the recommended 40 min. In that example the four tools were not identical; therefore, the ratios of v_{opt} and T_{opt} for four and one tools do not exactly correspond to Eqs. (8.72) and (8.73), but they are close.

The result of this analysis can be explained by pointing out that the increased significance of the cost related to tool changing and the consequent slowdown of cutting speeds in multitool operations is the result of the condition that whenever any one of the tools is changed, all the other tools must stop working too.

If, on the contrary, the work is arranged on separate machines or on stations separated by buffers, then the change of a tool on one machine (station) does not require stopping the tools on the other machines (stations), each of them will thus be treated as a single-tool operation and will, optimally, produce faster. ▲

8.4.6 General Conclusions for the Choice of Cutting Speeds and Feeds

Detailed calculations of optimum feeds and speeds are important for operations that are used repeatedly many times, in large lot and mass productions. If only one or a few pieces of a particular workpiece is needed the requirement of a minimum cost of such an operation is obviously less significant. However, even in small-lot variable machining productions, it is necessary to use good practice, to save tools in general and maintain, on average, an efficient machining shop. It is advisable to follow general rules and make good use of available handbooks.

The procedures presented in the preceding examples were included here primarily for the purpose of understanding the role of the individual parameters r_m, t_{tch}, and C_{te} that characterize the machine tool and the tool, and of the parameters p, q, and C that are involved in the tool life equation. However, the latter parameters are not readily available for the many possible combinations of work/tool materials. Moreover, many other factors affect the outcome of the operation, such as special conditions of the work material or of its surface and many details of the tool geometry, as well as the part configuration and the speed, power, and stiffness characteristics of the machine tool. Coolant, when it is used, may also play a significant role. Nevertheless, there are some basic rules, which will now be recapitulated.

The choice of the feed f_r is dictated (1) in a finishing operation by the requirements of surface finish or of the accuracy of form of the workpiece (see Chapter 9) or (2) in a roughing operation essentially by the strength of the tool edge (see discussion later in this chapter). Then the cutting speed must be chosen, and in this respect Eqs. (8.60) and (8.61) are most useful, especially the latter, because it is very simple to recall and understand.

Through the years extensive handbooks of cutting data have been compiled. One of the most comprehensive ones is the handbook [17]. Many large companies have their own handbooks. These handbooks contain recommendations of tool material, feeds, and cutting speeds for the various machining operations performed on various workpiece materials. These recommendations are nominal, for an average operation, and serve only as initial guidelines; they should be modified to optimize a particular operation, which may be carried out repeatedly in a given workshop; once optimized, its parameters are stored for further use. The conditions recommended in the manual will lead, on average, to some standard tool life (e.g., 60 min or 30 min).

Computerized machining data banks now exist where many detailed conditions of successfully executed operations are stored. The computer can match the operation for which cutting data are required with a similar one in the data bank and produce the established and verified machining conditions.

TABLE 8.6 Parameters Used in Cutting Data Banks

Workpiece and Cutting Data	Machine Tool Data	Cutting Tool Data
Material no.		Price of insert
Surface condition	Machine identification	Tool holder depreciation
Coolant	Power	No. of cutting edges
Tool grade	Machine type	Operator's rate
Workpiece hardness	Spindle speed range	Machine overhead rate
Workpiece diameter	Feed range	Tool-changing time
Depth of cut	Type of fixture	
Surface finish		
Tool geometry		
Tool life required		
Tolerance required		

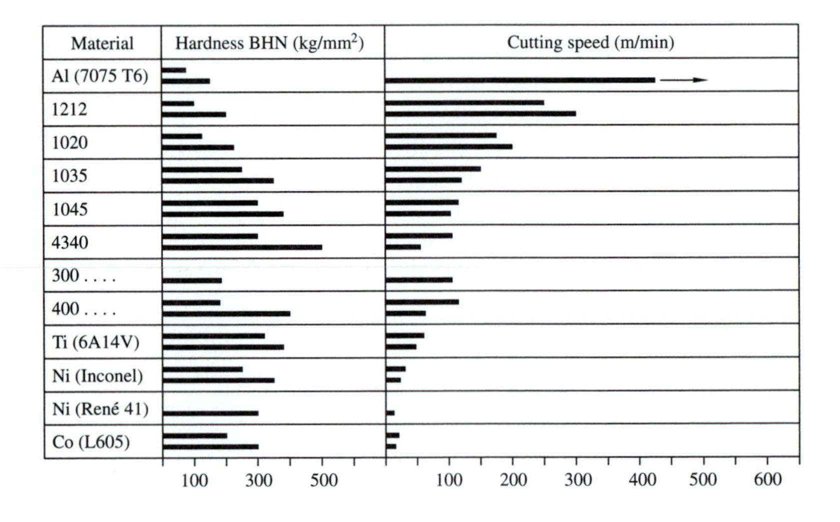

Figure 8.45 Recommended common cutting speeds for selected workpiece materials. On average, these result in tool lives between 30 and 60 minutes when the appropriate tool types are used.

Typically data entries into such systems may include the information shown in Table 8.6.

As a simplified example and for general information, Fig. 8.45 presents recommended cutting speeds for medium rough turning of selected materials using the best suitable uncoated carbide grades. These speeds would, on average, lead to 30–60-min tool life. Students should remember some of these numbers, which are perhaps as important as the strengths of selected steels, which most students undoubtedly have already memorized. Professional engineers graduated from a Mechanical Engineering program should be able to guess machining conditions for most common steels within an order of magnitude.

Returning now to the matter of basic rules and Eq. (8.61), we have

$$T_{\text{opt}} = (p - 1)t_{teq} \qquad t_{teq} = t_{tch} + \frac{C_{te}}{r_m}$$

This indicates that

1. For higher values of p, tool life should be chosen longer. This is understandable because high p means a large increase in tool life for a small decrease of v; we lose very little on machining time but gain a lot in tool life. The average values

of p are $p = 5$ to 6 for HSS and 3 to 4 for carbide tools. Correspondingly, cutting speed should be chosen so as to give tool life about 5.5 and 3.5 times longer, respectively, than the "tool equivalent time" t_{teq}.

2. The tool equivalent time increases with increasing tool-changing time and with increasing cost per tool edge.

The tool-changing time t_{tch} may be as short as 4 sec on an NC machining center and as long as 5 min on a manually operated machine tool or on an automatic machine tool with manual tool change.

The second term, C_{te}/r_m, is twice as large for positive inserts (3 to 4 edges) as for negative inserts (6 to 8 edges). Typically, it is about 0.4 to 1.0 min for an expensive machining center and 5 min for a simple, manually operated machine.

Thus, there will be two extremes:

1. $t_{teq} = 0.066$ min + \$0.8/(\$1.0/min) = 0.866 min, for an expensive machining center with automatic tool change; tool lives as short as 4 min. might be economical. However, this may require frequently changing the tools in the tool magazine, and this is done manually. Therefore, tool lives of (actual machining time) 15 min are more characteristic.
2. $t_{teq} = 5$ min + \$1.0/(\$0.2/min) = 10 min, for an inexpensive (perhaps old and well depreciated), manually operated machine tool with manual tool change and a lengthy tool setting for an accurate position. The economical tool life will be on the order of 40–60 min. With $p = 3.5$, this means about 1.5 to 2 times slower cutting speeds than in case 1.
3. Finally, it is important to recall the result of Example 8.6, which has shown that for N simultaneous multitool operations without buffers between stations (e.g., the 8-spindle automatic lathe) the most economical tool life is N times larger than for a single-tool operations. Thus, for example, in a 24-tool operation on an 8-spindle automatic lathe, one may be considering tool lives corresponding to one or two shifts of work. This also points out the importance of dividing operations with many tools working simultaneously (e.g., transfer lines), into sections separated by buffers.

8.5 TOOL BREAKAGE: WEAR AND BREAKAGE IN MILLING

We have noted, with reference to Fig. 8.32, that the damage to the tool may take the form of cracks, chipping, and massive breakage. The latter two forms of breakage occur in continuous cutting (e.g., turning) with heavy feeds, and all these "irregular wear" phenomena are associated, also for less heavy feeds, with interrupted cutting. The latter is represented mainly by milling or by the turning of workpieces with slots or otherwise interrupted cross sections. Even the most common form of wear, flank wear, has a different character in milling than in continuous cutting. First we discuss breakage under heavy loads in continuous cutting and then all the special wear phenomena associated with interrupted cutting.

8.5.1 Breakage in Continuous Cutting

The illustrations in Fig. 8.46 obtained in tests described by Massood [10] are characteristic examples of breakage in continuous cutting. In this summary of six tests carried

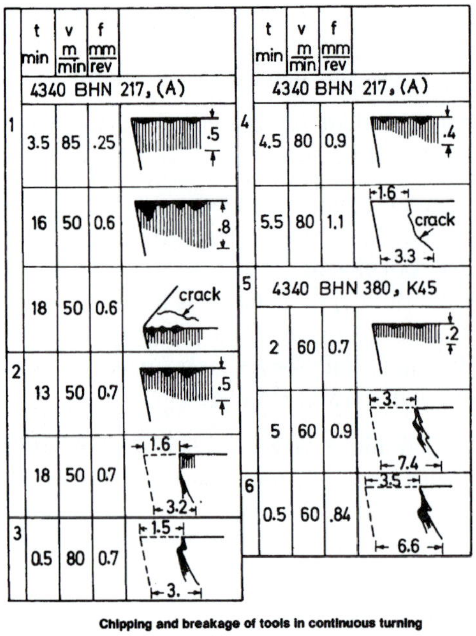

Figure 8.46
Damage of tool edge observed in turning with heavy feeds.

	t min	v m/min	f mm/rev	
	4340 BHN 217, (A)			
1	3.5	85	.25	.5
	16	50	0.6	.8
	18	50	0.6	crack
2	13	50	0.7	.5
	18	50	0.7	1.6, 3.2
3	0.5	80	0.7	1.5, 3.

	t min	v m/min	f mm/rev	
	4340 BHN 217, (A)			
4	4.5	80	0.9	.4
	5.5	80	1.1	1.6, crack, 3.3
5	4340 BHN 380, K45			
	2	60	0.7	.2
	5	60	0.9	3., 7.4
6	0.5	60	.84	3.5, 6.6

Chipping and breakage of tools in continuous turning

out in turning steel 4340 with carbides of the grades C6 and C7, the parameters given are the time of cutting, the cutting speed, and the feed per revolution f_r. The values of f_r ranged from 0.25 to 1.1 mm, which for the side-cutting-edge angle $\sigma = 30°$ used corresponds to thicknesses h in the range of 0.22–0.95 mm, and these represent rather heavy loads per unit length of cutting edge. These loads are about 600–2667 N/mm. In all the tests we can observe flank wear (the shaded areas), chipping (the black smaller areas on the cutting edge), cracks (as indicated), and, at the heaviest feeds of $f_r = 0.7$ to 1.1 mm, complete massive breakage of the tool tip. The cutting speeds used were not very high and did not produce very high temperatures in the cut. We will distinguish between the chipping of the edge and the breakage of the tool tip.

These two modes of breakage in general are shown separately in Figs. 8.47 and 8.48, respectively, in two views in each case. Microphotographs of the fractured surfaces have revealed that the broken surface shows the carbide grains with sharp edges, while the chipped surface shows the grains covered by a layer of cobalt. Correspondingly, breakage is of a brittle nature and occurs on the surfaces of or across the carbide grains, while chipping is of a ductile nature and occurs by shear flow in the cobalt layer. It is important to discuss the distribution of stresses in the tool under the highly localized load exerted by the cutting operation. Only mechanical stresses will be considered; thermal stresses are disregarded because in continuous cutting the latter do not cause problems.

Results of finite-element computations of the two principal normal stresses and of maximum shear stresses are given in Fig. 8.49. The load distribution assumed in the computation is shown in the diagram *(a)*. It consists of normal and tangential stresses on the rake face and on the flank. Chip thickness $h = 0.25$ mm is assumed, and chip contact length on the rake face is 1 mm. This is the length from the cutting edge included in diagrams *(a)*, *(b)*, and *(d)*. The compressive load on the rake face is taken as a maximum of 1400 N/mm^2 and is constant over a distance of 0.125 mm; then it decreases linearly to zero at the end of the contact length. The tangential load τ, which is due to

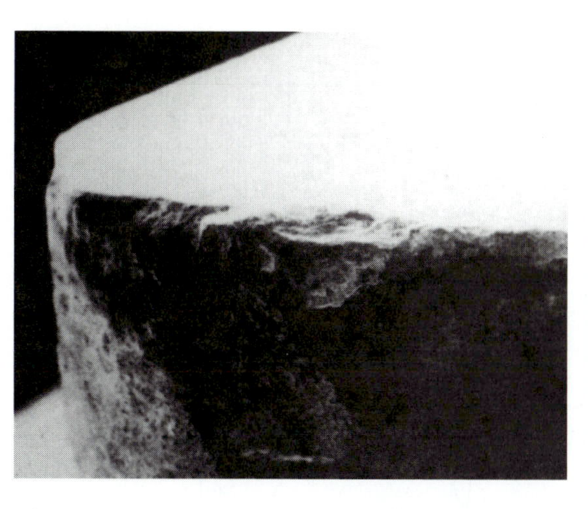

Figure 8.47
Chipping of the cutting edge: view of the flank side (left); view onto the rake face of the tool (right).

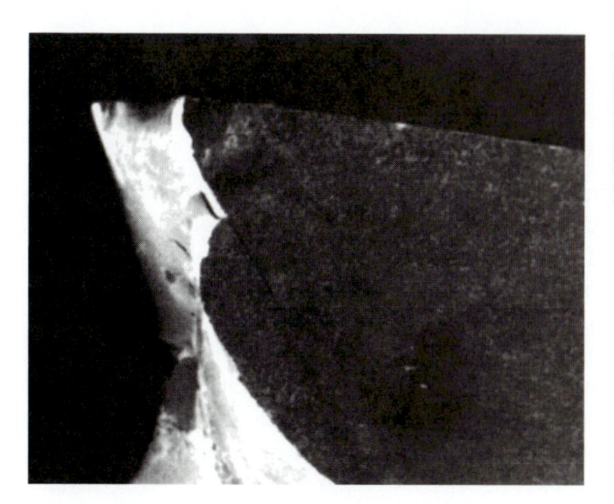

Figure 8.48
Breakage of the tool tip seen from the flank side (left) and looking at the tool tip (right).

chip/tool friction, is also distributed trapezoidally. Both the compressive and tangential loads on the flank decrease in proportion with the distance from the cutting edge.

The diagram of Fig. 8.49c shows the lines of constant maximum principal stresses; the positive values indicate tension, and the negative values, compression. The small area at the tip of the tool is reproduced in an enlarged scale in the diagram *(a)*. It is seen that close to the cutting edge there are no tensile stresses. As seen in *(c)*, tension starts at about the end of the contact length, increases to a maximum of 276 N/mm^2 on the rake face at a distance from the cutting edge equal to about twice the contact length, and then decreases and starts to slowly increase again in proportion to the distance from the cutting edge. At heavier feeds (at loads about 3 to 4 times the one used in the computation) one can assume $(\sigma_1)_{max} \simeq 1000$ N/mm^2 at the top tip, and the brittle breakage occurs as shown in Fig. 8.48.

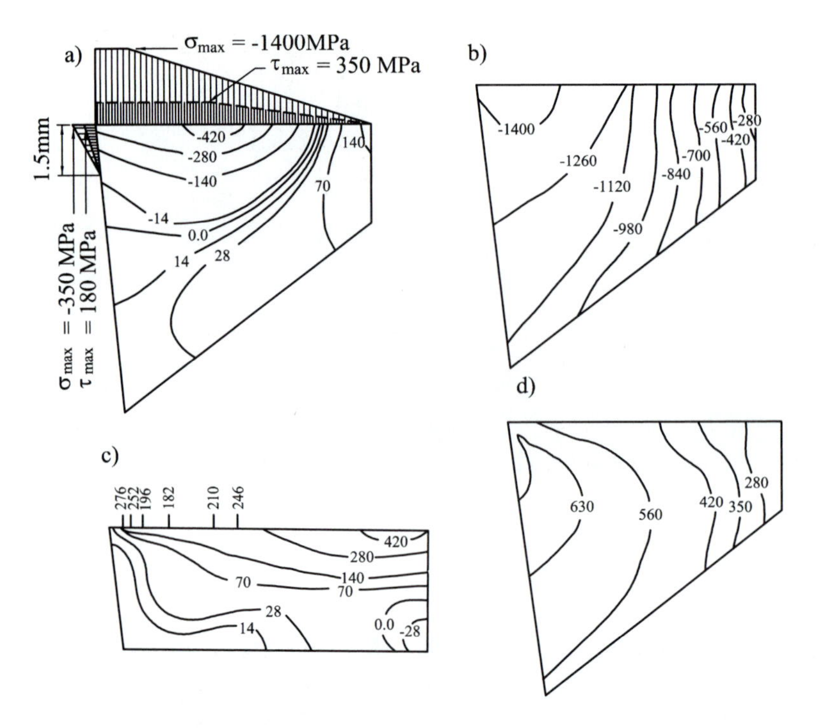

Figure 8.49
Stress distribution in the tool: a), c) maximum principal stresses; b) minimum principal stresses; d) shear stresses. Chipping is ductile failure and is caused by high shear stresses close to the cutting edge. Breakage is brittle failure and is caused by tensile stress reaching a maximum at a short distance from the edge. (Ref. [10]).

In Fig. 8.49b the minimum principal normal stresses (i.e., practically, the larger compressive [negative] stresses) are given. These reach their maxima close to the cutting edge. Diagram *(d)* gives the maximum shear stresses, and these, again, are highest at the cutting edge. They are high enough even at the more moderate feeds to produce the ductile fractures of the "chipping" kind.

To summarize, although chipping and breakage are different kinds of fracture, the former being ductile and arising from shear stresses and the latter being more of a brittle nature and caused by tensile stresses, they both result from heavy loads on the cutting edge. These heavy loads are associated with strong and hard workpiece materials and great chip thicknesses.

To prevent these fractures, stronger types of tool materials are recommended. For sintered carbides this means grades with higher Co content. Furthermore, wherever possible it is recommended to use a larger side-cutting-edge angle σ; in turning chilled iron rolls, it is not unusual (see Fig. 8.50) to use $\sigma = 87°$, where a feed as heavy as $f_r = 6$ mm/rev results in chip thickness of only $h = 0.3$ mm. Otherwise, in

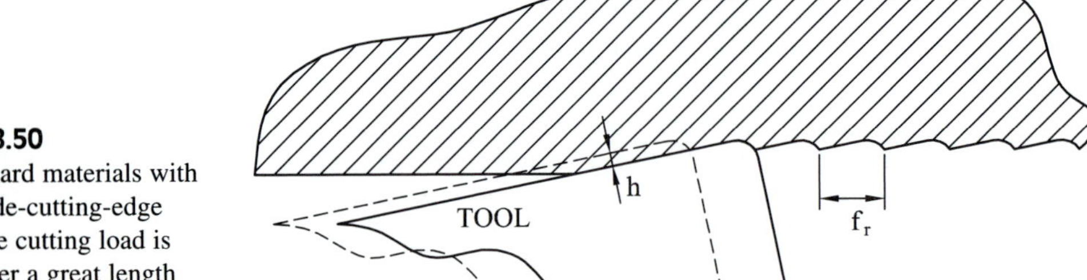

Figure 8.50
Turning hard materials with a large side-cutting-edge angle. The cutting load is spread over a great length of the edge.

most operations, it is necessary to use lighter feeds in order to make sure that breakage does not occur. On automatic, unattended machine tools such tool breakage might lead to the breakage of the tool post or to the loosening of the workpiece from the chuck and serious damage to the machine.

8.5.2 Tool Wear and Breakage in Interrupted Cutting

Under interrupted cutting we will essentially consider milling. In general, in many interrupted cutting operations tool chipping and breakage occur at about half or less of the chip thicknesses at which they are found in continuous cutting. This has to do with special stress distributions during the entry into the cut and, mainly, during the exit from the cut.

In milling the entry of a tooth into the cut will start at a point on the rake face of the tool. In the first instants of cutting, the load of starting the chip formation concentrates around this point. Depending on the geometry of the tool and on the entry angle ε_1 (see Fig. 8.51), the point of the first contact can be *S*—tip of the tool; *T*—on the cutting edge, away from the tip; *V*—on the minor cutting edge, away from the tip; *U*—on the rake face, away from the cutting edges. The most unfavorable case is *S*.

Such a case may happen with a negative entry angle ε_1, while it is avoided with a positive entry angle ε_1. The angle ε_1 is measured between the radius of the tool and the entry side of the workpiece. It is taken as positive if the outside of the workpiece is also on the outside of the cutter. With a negative-negative geometry of the cutter, the *S*-contact is safely eliminated. The entry damage plays a significant role in milling hard materials. For the majority of more ductile and less hard materials, the exit damage is more decisive.

The mechanism of the exit from the cut is best illustrated in Fig. 8.52, which is reproduced from Pekelharing [4] and was based on a high-speed film. The sequence of events proceeds from the bottom to the top. A tool is approaching the end of the workpiece, shown as moving upward and removing a layer with thickness $h = 0.25$ mm. In *(a)* it is 0.9 mm from the end side of the workpiece. The chip is still being formed along the regular shear zone, denoted as the "positive shear zone." The chip/tool contact length is about $3h$. Because the cut is so close to the end side of the workpiece, the stress distribution in front of the tool changes: instead of maximum shear stresses concentrating along the usual shear zone, they form in a different location called here the

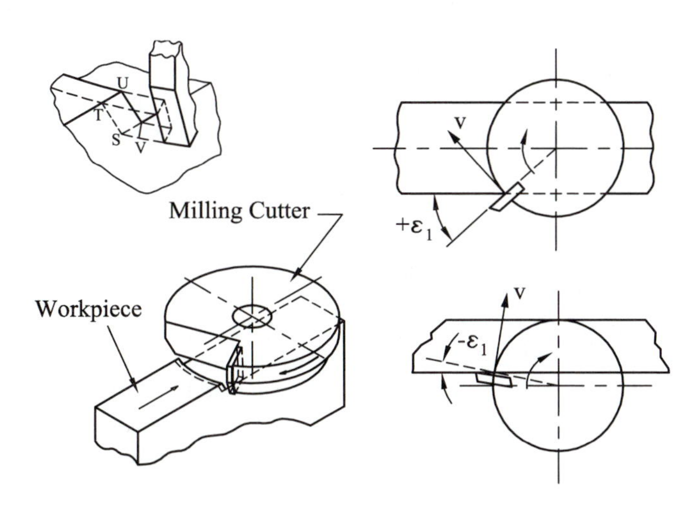

Figure 8.51
Point of first contact of a milling cutter tooth at entry into the cut. The four alternative contacts are marked as *V, U, T, S*. At the bottom the most stressful tip contact (*S*) is shown.

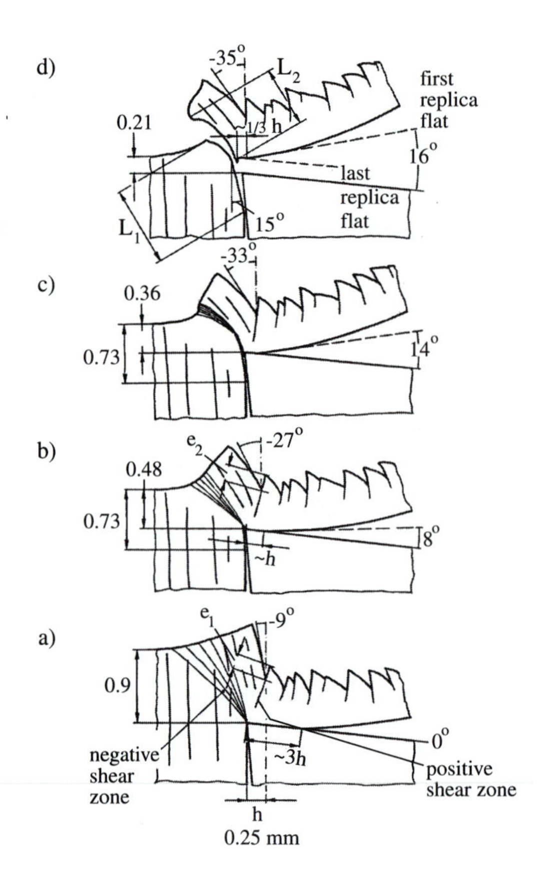

Figure 8.52
Formation of the negative shear zone at the exit from the cut (after Pekelharing). In positions from A to D the tool is approaching (moving upwards) the flat end, square to the tool velocity, of the workpiece. The shear zone is rotating to the left, eventually reversing the direction of the slide over the rake face of the tool and causing tensile stresses in the tool tip. The chip ends up with a "foot," and the tool edge chips off.

"negative shear zone." Negative is meant in the sense that its angle is measured to the left of the cutting direction instead of to the right, as for the positive shear zone. The area in front of the negative shear zone is becoming what will be a "foot" of the chip, with the end side of the workpiece bulging out. An angle of 9° is associated with this incipient foot. In *(b),* the tool is 0.48 mm distant from the workpiece end. The regular shear zone is not active anymore, and all the material cut off the workpiece goes into the foot, which is rotated by 27°. The chip/tool contact is only about *h.* This process continues in *(c),* where the distance from end is 0.36 mm, and *(d),* distance 0.21 mm. The chip/tool contact diminishes, and the chip with the foot separates from the workpiece along the negative shear zone leaving a certain amount of chamfer and burr at the workpiece end. The foot formation is more or less pronounced, depending on the properties of the workpiece material; it increases with ductility.

During the final stages of this process the loading on the tool changes, and the tangential load on its rake face (friction) may even reverse its direction. Increased tensile stresses arise in the tool, and the shear stresses at the edge increase. This leads to both chipping and breakage of the tool at chip thicknesses *h,* much smaller than in continuous cutting.

One way to prevent or at least diminish chipping and breakage is to strengthen the tool edge by giving it a chamfer or a highly negative land (see Fig. 8.53). The two edge-form modifications are essentially the same, except that the land is preformed on the insert, while chamfer is usually ground additionally. The stress distribution at the modified edge has never been properly investigated because it is not known what kind of pressure acts on the chamfer or land. A part of the chip being formed is partly arrested

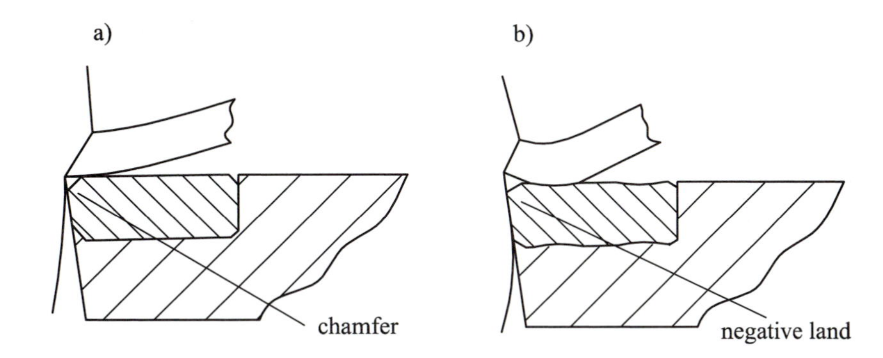

Figure 8.53
a) A chamfer and b) a negative land. Both are intended to strengthen the cutting edge.

on them like a built-up edge, but for the usual size of a chamfer, which is about 1/4 of the undeformed chip thickness, the shear plane is essentially created similarly as with a sharp tool. Even without knowing the proper stress distribution, experience shows that chamfers and negative lands are effective in diminishing the tendency to chipping. The best size and angle of the chamfer has to be determined experimentally for a given work/tool material combination and for the given type of operation with its particular form of interrupted cutting.

The disadvantage of chamfers, especially those made at steep angles, is that they lead to an increased rate of flank wear and also to larger radial forces, and consequently to larger deflections (less accuracy) and a greater tendency to chatter.

Another more important solution is to arrange for the best possible entry and especially the best possible exit conditions. The process involving the negative shear zone and chipping and breakage was described with reference to Fig. 8.52, where the end face of the workpiece was square to the direction of the cutting motion. Often this is not so, and the exit angle ε_2, as shown in Fig. 8.55, is either less or more than 90°. The angle ε_2 is measured between the direction of the cutting motion outward of the cut and the end face of the workpiece below the cut. Various configurations of peripheral milling (end or face milling), both up- and down-milling, are shown with the exit angle ε_2 indicated.

First, let us look at Fig. 8.54, which is also reproduced from Pekelharing [4]. It gives the number of cuts before chipping of the cutting edge for a particular test operation, for different values of the exit angle ε_2. The conditions of these tests follow: work material 1045, $v = 150$ m/min, $h = 0.25$ μm, carbide grade P10. It may be seen that $\varepsilon_2 = 90°$ to 105° is worst, and as ε_2 becomes less than 90°, there is less tendency to chipping and for $\varepsilon_2 < 30°$ there is none. Similarly, beyond $\varepsilon_2 = 120°$, cutting is safe against breakage, but a large burr is created. Correspondingly, of the four operations shown in Fig. 8.55, the "half-immersion" up-milling in case *(b)* is the worst arrangement. The quarter-immersion up-milling with $\varepsilon_2 = 60°$ is considerably better. Figure 8.55c, with $\varepsilon_2 = 120°$ is safe against breakage, and (*d*), down-milling with $\varepsilon_2 = 0°$ and with zero chip thickness at the exit is the safest of all.

8.5.3 Flank Wear in Milling

The interrupted way of cutting by milling also affects strongly the mechanism of flank wear. Tlusty and Orady [11] have observed that the amount of flank wear in all the three operations shown in Fig. 8.56 is roughly the same after the same total length of travel with all the cutting conditions (cutting speed, feed per tooth) equal except the radial depth of cut. Now we will not consider chipping, which under the given conditions may or may not occur; instead, we concentrate on flank wear. In spite of the

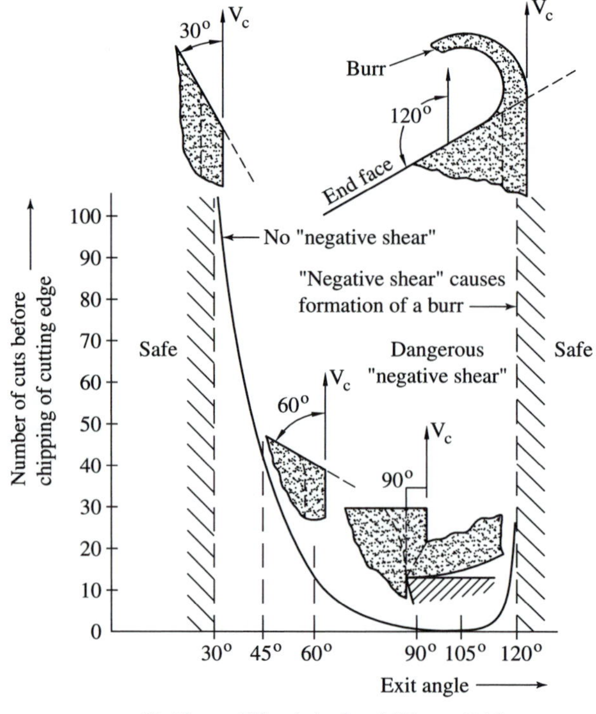

Figure 8.54
Safe and dangerous exit angles as expressed by the number of cuts made before chipping of the edge occurred in the experiment.

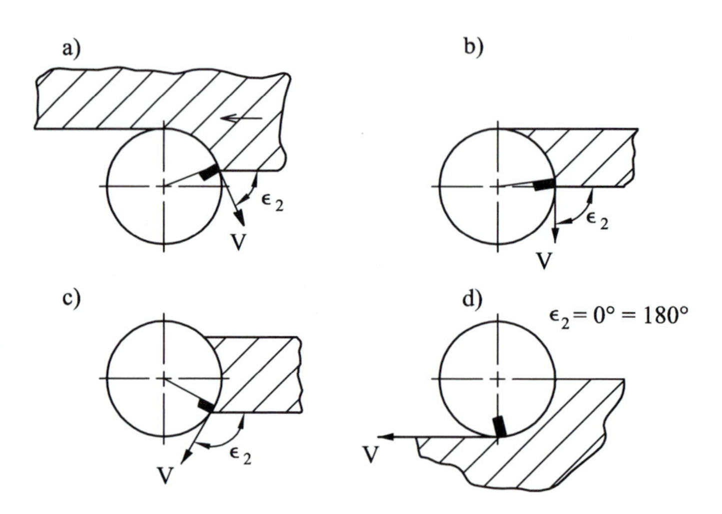

Figure 8.55
Exit angles in various arrangements of milling. a) Up-milling, quarter immersion; b) up-milling, half-immersion—cut starts with a thin chip but exits with a thick chip (large exit force); c) down-milling, quarter immersion; d) down-milling, half immersion—cut starts with a thick chip but ends with a thin chip (zero exit force). Down-milling is much safer from chipping than up-milling.

strongly different amounts of material removed, the three operations end up with the same flank wear width *FWW.* The metal removal rates are different in the ratio of the radial depth of cut to the cutter diameter, *a/d: (a)* 1/4, *(b)* 1/2, *(c)* 1/1. Also the total length of travel of each edge in the cut is strongly different, being proportional to the angle of engagement, which is *(a)* 60°, *(b)* 90°, *(c)* 180°.

It has been shown [11] that the decisive factor for flank wear in milling is the variation of thermal stresses on the flank and on the cutting edge, which is due to thermal cycling. Figure 8.57 shows the variation of temperature on the rake face of the cutter teeth, as computed for the three operations of Fig. 8.56. At the beginning of the cut the temperature rises and stabilizes rapidly, and after the exit from the cut it drops and stabilizes rapidly. The differences between the maximum and minimum temperatures on the tool are about the same for the three operations. The severity of thermal cycling is about the same. In each of the thermal cycles, during the heating-up period the surface of the tool is hotter than the inside and, correspondingly, compressive stresses develop on the surface of the tool. During the cooling-off period, the surface becomes cooler than the subsurface, and tensile stresses occur. During each such cycle this variation of stresses causes some amount of grains to ease out of the surface. The total wear is then proportional to the number of cycles which, for the three operations of Fig. 8.56, is the same for the same length of travel.

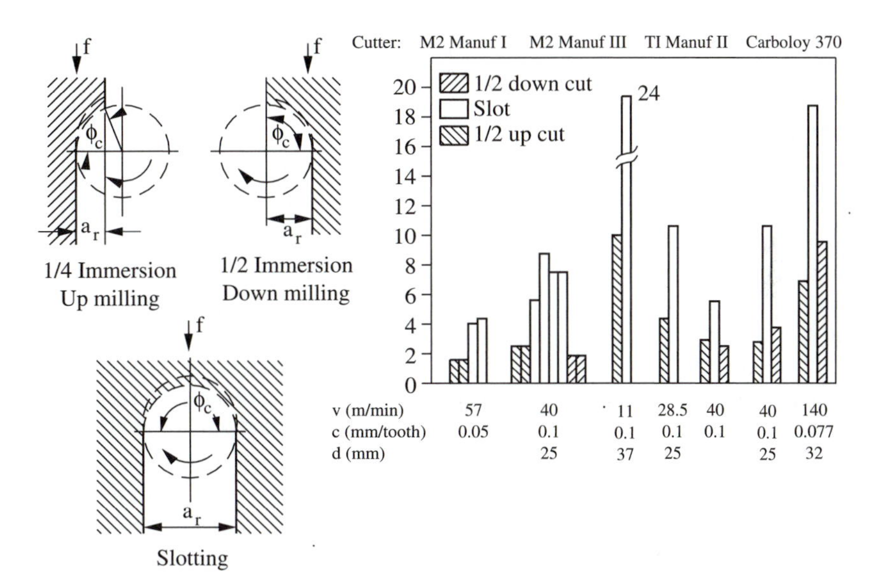

Figure 8.56
Peripheral milling with various radial depths of cut. Tool life in slotting is about twice that of the half-immersion cuts.

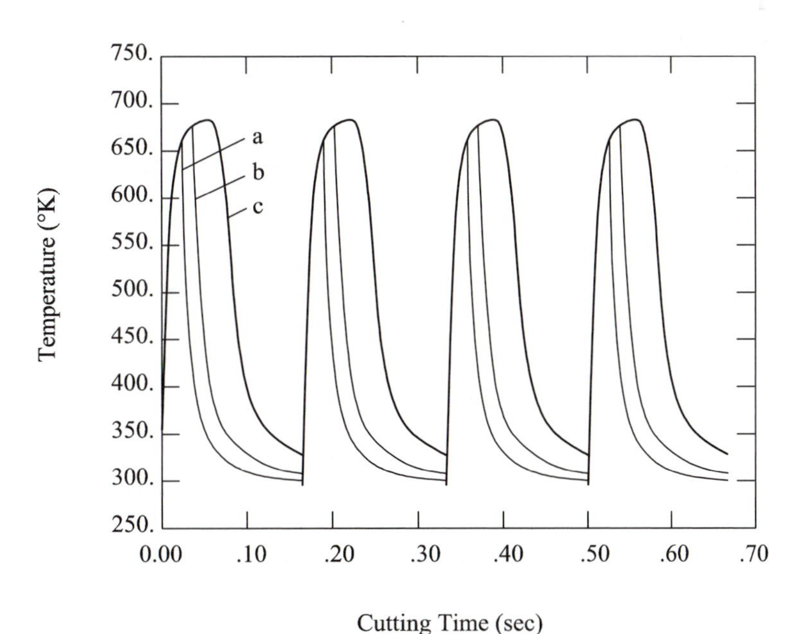

Figure 8.57
Variation of temperature on the rake face and on the flank of a milling cutter for the three types of operations given in Fig. 8.56. First four cycles are shown for *(a)* quarter, *(b)* half and *(c)* full radial immersion. Temperature rises very fast during the cut and drops fast also as the tool returns through the noncutting part of the revolution. Severe thermal cycling causes tool wear.

Naturally, the wear rate depends strongly on cutting speed, because its value affects the maximum temperatures developed and thus the severity of thermal stress cycling. One important lesson is learned from the explanations given here. There is no sense, from the point of view of tool life, in dividing the total radial depth of cut into several passes. There may be reasons for one roughing and one finishing cut, but they relate to the accuracy of form of the machined surface, which is explained in Chapter 9. The mechanism of flank wear in milling as explained here should help in the development of tool materials for milling cutters. They should be less sensitive to temperature stresses. This is true especially in the microstructural sense with respect to temperature and stress gradients between the individual grains of the tool material. So far, research has not yet progressed enough to give the actual practical recommendations for such favorable microstructures. However, further research may lead to important developments in this direction.

REFERENCES

1 Armarego, E. J. A., and R. H. Brown. *The Machining of Metals.* Engelwood Cliffs, NJ: Prentice Hall, 1969.

2 Boothroyd, G., and W. A. Knight. *Fundamentals of Metal Machining and Machine Tools.* New York: Marcel Dekker, 1989.

3 Boothroyd, G. "Temperatures in Orthogonal Metal Cutting." *Proc. IME* 177 (1963): 789.

4 Pekelharing, A. J. "The Exit Failure in Interrupted Cutting," CIRP Annals, Vol. 27, No. 1, 1978.

5 Merchant, M. E. "Mechanics of the Metal Cutting Process." *J. Appl. Phys.* 16, no. 5 (1945): 267, and no. 6: 318.

6 Oxley, P. B. L. *Mechanics of Machining: An Analytical Approach to Assessing Machinability.* New York: John Wiley & Sons, 1989.

7 Shaw, M. C. *Metal Cutting Principles.* New York: Oxford University Press, 1984.

8 Stephenson, D., and J. S. Agapiou. *Metal Cutting: Theory and Practice.* New York: Marcel Dekker, 1996.

9 Taylor, F. W. "On the Art of Cutting Metals." *Trans. of ASME* 28 (1906): 31.

10 Tlusty, J., and Z. Masood. "Chipping and Breakage of Carbide Tools." *ASME J. of Eng. for Ind.* 100 (Nov. 1980): 403–412.

11 Tlusty, J., and E. Orady. "Effect of Thermal Cycling on Tool Wear in Milling." 9th NAMRC Conf., Penn. State U., May 1981.

12 Society of Manufacturing Engineers. *Tool and Manufacturing Engineers Handbook.* Vol. 1. *Machining.* 4th ed. Dearborn, MI. 1983. Society of Manufacturing Engineers.

13 Sandvik Coromant. *Modern Metal Cutting.* Sandvik Coromant, Fair Lawn, NJ, 1996.

14 Vieregge, G., *Zerspanung der Eisenwerkstoffe,* 2nd ed., Verlag Stahleisen, Düsseldorf, 1970.

15 Trent, E. M., *Metal Cutting.* Butterworth & Co, London, 1978.

16 Am. Soc. of Tool and Manufacturing Engineers, *Cutting Tool Material Selection,* H. J. Swinehart Ed., 1968.

17 Machinability Data Center, *Machining Data Handbook,* 3rd ed., MetCut, Cincinnati, 1980.

Industrial Contributors

[SANDVIK] Sandvik Coromant, 1702 Nevins Road, Fair Lawn, NJ 07410; Modern Metal Cutting.

[KENNAMETAL] Kennametal Inc., Latrobe, PA 15650.

QUESTIONS

Q8.1 **(a)** What is a "built-up edge," and under what conditions does it occur?

(b) How does it affect the cutting operation and the tool wear rate?

Q8.2 How does the cutting force vary with cutting speed? What are the reasons for this variation?

Q8.3 Make a sketch of the model of chip formation showing the shear plane and the vector of the cutting force. Show the geometry of deriving the force component F_s in the shear plane and the friction component F_f on the rake face of the tool.

Q8.4 In a metal-cutting operation, which properties of the workpiece material and which cutting conditions affect the following?

(a) The shear-plane temperature. Write out the formula for T_s.

(b) The maximum temperature at the chip-tool interface. Which cutting parameters and workpiece material properties affect shear-plane temperature T_s and maximum temperature on the face of the tool T_m? Explain how. Fill out the table, using the following: *I*, increased parameter increases temperature; *D*, decreases; *N* has negligible influence.

Parameter	T_s	T_m
h		
b		
v		
K_s		
(ρc)		
k		

Q8.5 What is the effect of cutting speed and feed on **(a)** shear-plane temperature and **(b)** temperature on the rake face of the tool? Give a brief, approximate explanation of these effects.

Q8.6 Sort workpiece materials according to temperature on the face of the cutting tool when using the appropriate, commonly accepted cutting speeds for each material (fill in numbers 1 lowest, 4 highest).

Stainless Steel ( ), Aluminum ( ), Titanium ( ), Carbon Steel ( ).

Q8.7 Write out the finite-difference equations that apply to heat transfer in one slice of a chip sliding on the tool face. State the boundary conditions and the heat transfer modes for the whole chip area.

Q8.8 Make a sketch indicating the general form of isotherms in the chip and in the tool.

Q8.9 What are the basic requirements for cutting tool materials?

Q8.10 Name the main classes of tool materials in order of increasing hardness.

Q8.11 **(a)** Briefly describe the heat-treatment cycle for high-speed steels.

(b) Which are the two basic classes of high-speed steels, and what are their compositions?

Q8.12 Compare the difference in chemical composition, and applications in metal cutting, of the two main classes of sintered carbide tools.

Q8.13 What is the effect of Co content on hardness and strength of sintered carbides? Describe the Co contents for roughing and for finishing grades.

Q8.14 How does the transverse rupture strength of a sintered carbide change with the percentage of TiC in the mixture? Under what conditions is a high TiC content advantageous?

Q8.15 Name the tool materials most likely to be used for **(a)** a twist drill, and **(b)** a turning tool.

Q8.16 Describe the manufacturing process of sintered carbide tools.

Q8.17 How are coated carbides made? What is their main advantage over noncoated carbides?

Q8.18 What are the basic compositions and major applications of ceramic tools?

Q8.19 What are the differences in high-temperature cutting ability and in the types of applications for cubic boron nitride and polycrystalline diamond?

Q8.20 **(a)** Make sketches showing the major features of tool wear and tool breakage.

(b) Describe the mechanisms of the various types of tool wear.

Q8.21 **(a)** Write out the "Taylor-type" equation for tool life T as a function of cutting speed v and feed f_r.

(b) Determine the parameters p, q, and C of the Taylor-type tool life equation $v^p f^q T = C$ from the following test results:

Test No.	v (m/min)	f (mm)	T (min)
1	100	0.1	80
2	100	0.2	20
3	200	0.1	10

Q8.22 **(a)** Draw the diagram of (log v, log f_r) showing the lines of constant tool life and constant metal removal rate. Which way should we move along a line of constant tool life in order to maximize metal removal rate?

(b) Why is the result of **(a)** dependent on the condition that $p > q$?

(c) What constraints are usually applied to the maximum permissible feed rate?

Q8.23 Write the equation expressing cost per part as function of cutting speed v and feed f_r.

Q8.24 Explain the basic procedure of optimizing v and f_r for a non-Taylor-type tool life equation.

Q8.25 In a multitool operation, why is it advantageous to have all of the tools working for the entire cycle time?

Q8.26 Is the optimum cutting speed of an operation that is a part of a multitool operation higher or lower than for the same operation carried out as a single-tool operation? Why?

Q8.27 Give the typical cutting speed used for HSS and for sintered carbide machining of steels 1020, 4340, Ti alloys, and Al alloys.

Q8.28 **(a)** Explain the differences between chipping and breakage of tools in terms of the mechanism of fracture and material properties.

(b) Make a sketch of a tool bit, showing the location of the maximum tensile and maximum shear stresses.

(c) What methods can be employed to prevent tool fractures?

Q8.29 **(a)** With the aid of sketches, explain why an exit angle of 90° would contribute to the tendency for chipping or breaking to occur in a cutting operation. Which exit angles should be used to prevent this tendency?

(b) What are the disadvantages of using a chamfered tool or a tool with a negative land when trying to reduce the possibility of chipping or breaking?

Q8.30 What is the basic factor in flank wear of milling cutters? Is there any advantage in terms of tool life, in dividing the total radial depth of cut of a milling operation into several smaller passes? Explain.

Q8.31 Which of the three milling operations shown in Fig. Q8.31 is most susceptible and which is least susceptible to tooth breakage?

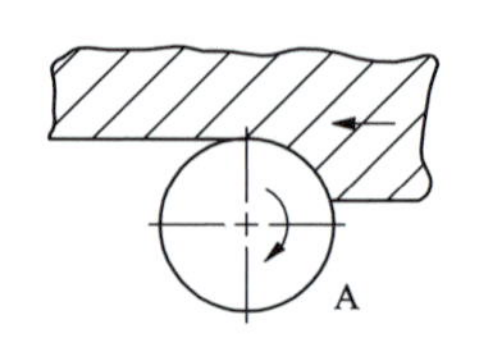

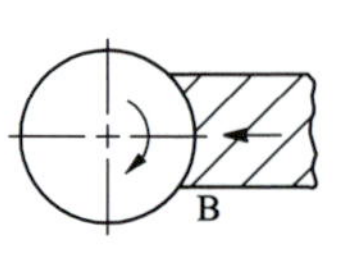

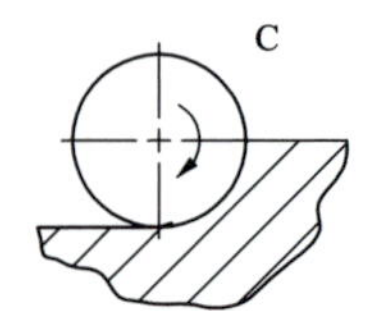

Figure Q8.31

PROBLEMS

A. Chip Formation: Simple Force, Power, and Temperature Calculations

P8.1 Refer to Section 8.1.2. The cutting operation shown in Fig. P8.1 has the following characteristics: chip width $b = 5$ mm, $\phi = 28°$, rake angle $\alpha = 0°$, and cutting speed $v = 3$ m/s. Determine:

(a) The deformed chip thickness, h_2

(b) The shearing velocity v_s and the chip velocity v_c. Sketch the hodograph.

(c) The shearing force F_s and shearing power P_s.

(d) The friction force F_f and friction power P_f.

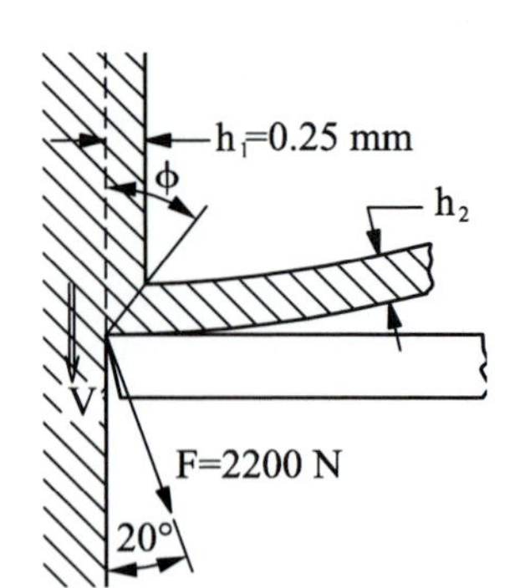

Figure P8.1

P8.2 A material of density $\rho = 7800$ kg/m^3 and specific heat $c = 470$ Nm/(kg°C), is being machined. Refer to Section 8.1.1. and Fig. 8.8a. The cutting speed is 3 m/s, and the resultant force F is 2400 N at an angle $\beta = 20°$. The other parameters are $h_1 = 0.2$ mm, $h_2 = 0.4$ mm, chip width $b = 6$ mm, rake angle $\alpha = 10°$.

(a) Determine the shear-plane angle ϕ.

(b) Calculate the shear force F_s, shear velocity v_s, and shear power P_s.

(c) Determine the shear-plane temperature assuming an ambient temperature of 20°C.

B. Temperatures in the Chip: Simple Calculations Without Using Computer Programming

P8.3 Consider temperature in the chip, neglecting heat conducted through the tool. Refer to Section 8.2.2 and Fig. P8.3.

(a) Chip width $b = 8$ mm, $h_1 = 0.12$ mm, $v = 3000$ mm/sec, $K_s = 2000$ N/mm^2, $L_c = 1.0$ mm, $\phi = 30°$, $\beta = 20°$, $T_{room} = 20°$C, $\Delta x = 0.002$ mm, $\Delta y = 0.01$ mm, $k = 40$ N/(sec · °C), $(\rho c) = 3.6$N/(mm^2 · °C). Determine:

Shear-plane temperature: v_s (mm/sec), F_s (N), P_s (mW), T_s (°C),

Friction power distribution: F_f (N), v_c (mm/sec), P_f (mW), p_{max} (mW):

Start of the potential computation:

Determine $T_{1,1}$, $T_{2,1}$, $T_{3,1}$, $T_{1,2}$, $T_{2,2}$, $T_{3,2}$

(b) Chip width $b = 10$ mm, $h_1 = 0.15$ mm, $v = 2500$ mm/sec, $K_s = 2000$ N/mm^2, $L_c = 1.0$ mm, $\phi = 30°$, $\beta = 20°$, $T_{room} = 20°C$, $\Delta x = 0.002$ mm, $\Delta y = 0.01$ mm, $k = 40$ N/(sec · °C), $(\rho c) = 3.6$ N/(mm^2 · °C).

(c) $b = 10$ mm, $h_1 = 0.1$ mm, $h_2 = 0.2$ mm, $\Delta x = 0.002$ mm, $y = 0.01$ mm, $K_s = 1800$ N/mm^2, $v = 2500$ mm/sec, $k = 24.5$ N/sec · °C, $(\rho c) = 3.5$ N/(mm^2 · °C), $\phi = 26.57°$, $\beta = 20°$, $T_{room} = 20°C$.

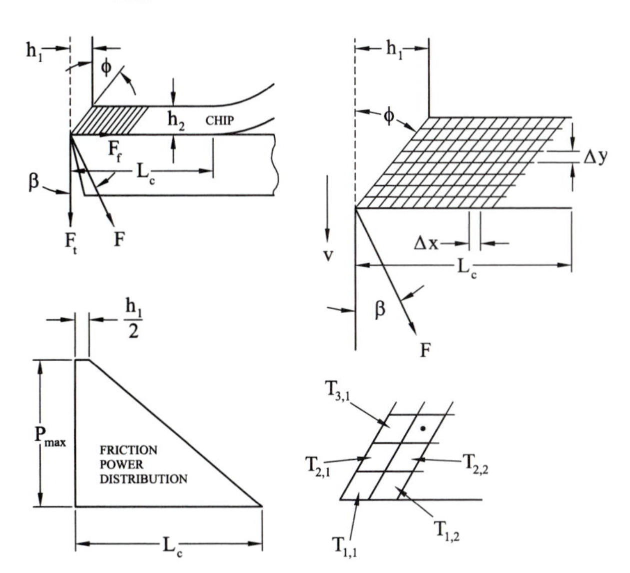

Figure P8.3

P8.4 Consider temperature in the chip; neglect heat conducted through the tool. Refer to Section 8.2.2 and Fig. P8.3. Machining steel 1035, $k = 43$ N/sec · °C, $\alpha = 12$ mm^2/sec, $\rho c = 3.7$ N/mm^2 · °C, $h_1 = 0.2$ mm, $b = 10$ mm, $L_c = 0.8$ mm, $v_c = 1.5$ m/sec, $\Delta x = 0.0025$ mm, $\Delta y = 0.02$ mm, $\beta = 20°$, $\phi = 25°$.

The following values have been precomputed: Shear-plane temperature $T_s = 510°C$, friction power $P_f = 2.07 \times 10^6$ N mm/sec (mW).

(a) Determine p_{max}.

(b) Determine the initial temperatures in the thermal field: $T_{1,1}$, $T_{2,1}$, $T_{3,1}$, $T_{1,2}$, $T_{2,2}$.

P8.5 Consider temperatures in the chip; neglect heat through the tool. Refer to Section 8.2.2 and Figs. P8.5 and 8.10. Cutting steel $K_s = 2000$ N/mm^2, $(\rho c) = 3.7$ N/(mm^2 · °C), $h_1 = 0.25$ mm, $b = 8$ mm, $\beta = 25°$, $\phi = 30°$, $\alpha = 0°$, $v = 3$ m/sec. Determine cutting force F(N), shearing force F_s (N), shearing velocity v_s (mm/sec), shearing power P_s (Nmm/sec), and shear-plane temperature T_s (°C), assuming $T_{room} = 20°C$. Thermal diffusivity $\alpha = 12$ mm^2/sec. Determine chip velocity v_c, friction force F_f, and friction power P_f (Nmm/sec). Element dimensions: $\Delta x = 0.0025$ min, $\Delta y = 0.022$ mm. Determine time step Δt. In the course of computation, we find: $T_{1,165} = 779.2°C$, $T_{2,165} = 706.0$ °C, $T_{3,165} = 670.1$ °C, and $T_{1,166} = 781.0$ °C. Determine $T_{2,166}$.

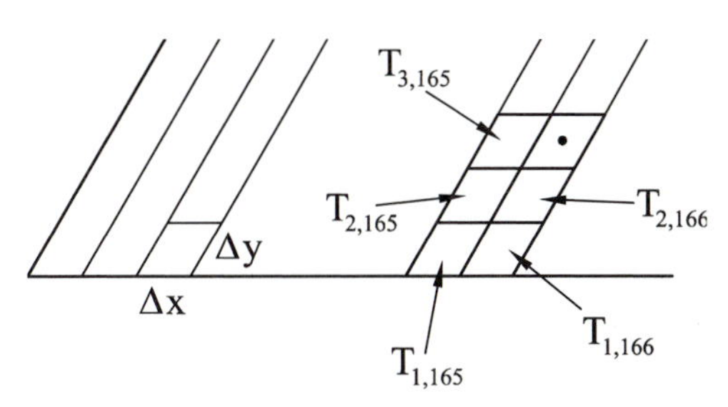

Figure P8.5

P8.6 Understanding the computation of temperatures in the chip and the tool. See Fig. P8.6. Cutting mild steel, $K_s = 1600$ N/mm^2, $\rho c = 3.7$ (N/mm^2 · °C), chip width $b = 8$ mm, $h_1 = 0.12$ mm, $h_2 = 0.2$ mm, contact length, $L_c = 4h_1 = 0.48$ mm, force angle $\beta = 18°$, cutting speed $v = 3000$ mm/sec.

(a) Determine tangential force F_t, cutting force F (N), shear angle ϕ, shearing velocity v_s (mm/sec), shearing power P_s (mW), shear-plane temperature T_s ($T_r = 20°C$), chip velocity v_c (mm/sec), friction force F_f (N), and friction power P_f (mW).

(b) Assuming initially that no power escapes through the tool and $KK = 400$ slices over L_c (see Fig P8.3) determine: p_{max} (mW), and the power p_{150} (mW).

(c) Assume thermal conductivity of the tool $k_{tool} = 45$ N/sec · °C, $\Delta z = 1$mm. Assume further that 5% of P_f escapes through the tool, and determine the temperature drop $\Delta T = T_{cav} - T_1$.

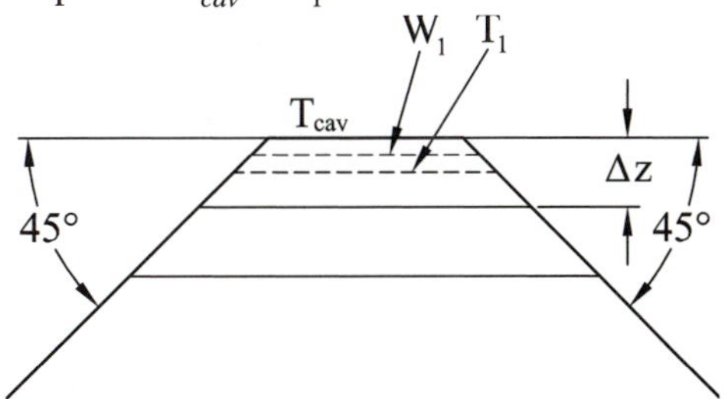

Figure P8.6

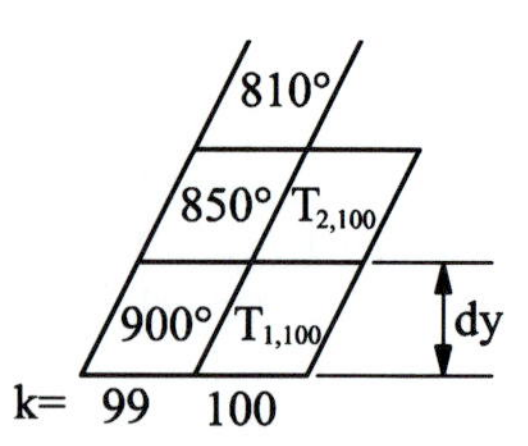

Figure P8.7

P8.7 Temperature field in the chip. Refer to Figs. P8.3 and P8.7.

(a) Cutting steel 1045, K_s = 2650 N/mm^2, ρc = 3.7 N/mm^2 · °C, k = 43(N/sec · °C), α = 12 (mm^2/sec), L_c = 0.6 mm, cutting speed v = 3 m/sec, v_c = 1.5 m/sec, β = 25°, h_1 = 0.15 mm, chip width b = 6 mm, KK = 400, dy = 0.015 mm.

Determine: the tangential force F_t (N), the friction power P_f (Nmm/sec =mW): Assuming the same power distribution as in P8.3, determine the power p_{100} entering the slice of k = 100. Assuming that the temperature distribution at the lower end of the slice in position k=99 is as shown, determine the temperatures $T_{1,100}$ and $T_{2,100}$.

C. Computations of the Temperature Field in Chip and Tool

P8.8a See Fig. P8.8. Compute the temperature field in the chip and in the tool for different workpiece materials. Adjust cutting speeds.

The geometry of the cut: h_1 = 0.2 mm, h_2 = 2 h_1, L_c = 4 h_1, b = 10 mm, cutting force angle β = 20°. The finite difference grid in the chip and tool: $\Delta y = h_2/20$, $\Delta x = l_c/400$, Δz = 0.5 mm. The tool is of one material only with thermal conductivity k = 40 N/(sec · °C). The grid in the tool extends for j = 1, 20; for k: KK = 400, KKK = 600, α = 0°. Follow the algorithm as outlined in Section 8.2.2.

1. Print out the value of $e = P_t(i)/P_f$, of *err* and of T_{cav} for each step of repetition.
2. Plot the temperature along chip/tool contact in the form of Figs. 8.20 and 8.21, that is, all $T_{1,k}$ versus x.

Do all the above for:

i) Equal cutting speed for all materials, v = 1.5 m/sec

	K_s	k	ρc	α
a) Carbon steel	2300	43	3.8	11.3
b) Ti alloy	2100	7	2.7	2.6
c) Al alloy	800	140	2.3	61
	N/mm^2	N/(sec · °C)	N/(mm^2 · °C)	mm^2/sec

ii) Different speeds: v(m/sec)

(a) Carbon steel: 2.5 (b) Ti alloy: 0.75 (c) Al alloy: 10

P8.8b Same as Problem 8.8a, but h_1 = 0.1 mm, b = 10 mm, β = 17°, $\Delta x = L_c/500$, Δz = 0.4, tool conductivity k = 45 N/(sec · °C), KK = 500, and KKK = 700.

i) Equal cutting speed for all materials v = 1 m/sec

ii) Different speeds: carbon steel v = 3 m/sec, Ti alloy v = 0.5 m/sec, Al alloy v = 15 m/sec.

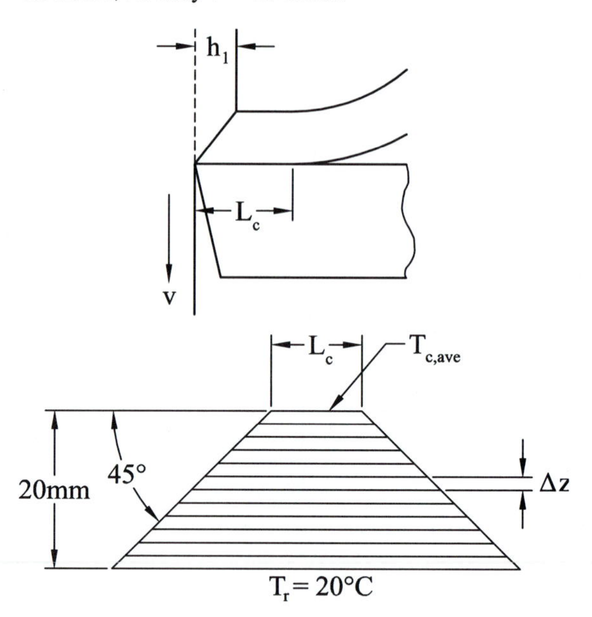

Figure P8.8

P8.9 Temperature field in chip and tool. Effect of cutting speed v and of chip thickness h. Assume the same geometry of the cut, that is, angles β and ϕ as in Problem 8.8a, and tool with thermal conductivity k = 40 N/(sec · °C), the contact length L_c = 4 h_1. Discretize for j = 20, KK = 400, KKK = 600.

(a) Investigate the effect of cutting speed v for the carbon steel and for the titanium alloy in three steps: steel v = 2.5, 3.75, 6.25 m/sec, Ti alloy v = 0.5, 0.75, 1.5 m/sec, using h_1 = 0.1 mm, b = 10 mm.

(b) Choosing the middle speed v of each of the three above, increase the chip thickness to h_1 = 0.25 mm. Compare so as to see this effect. Plot the temperatures along the chip/tool contact versus x, for k = 1, KK for (1) all the three speeds and two materials combinations in one plot and (2) all the four (one speed and two chip thicknesses for each material) combinations in another plot. Plot isotherms in the chip for steel, v = 3.75 m/sec, h_1 = 0.1 mm, and h_1 = 0.25 mm.

P.8.10 Investigate machining Ti alloys with a high rake angle, thin chips, and high cutting speed. Refer to Fig. P8.10.

(a) Take the case of Example 8.4 with v = 0.5 m/sec, h_1 = 0.2 mm, α = 0°, β = 22°, ϕ = 23°, b = 10 mm as reference *(a)*.

In order to properly include the effect of change in the rake angle and the corresponding changes of the shear angle ϕ and of the length L_c of the shear plane, the cutting force F will be derived from the shearing force F_s, which itself is obtained from the shear flow stress $\tau_s = 800$ N/mm^2:

$$F_s = \tau_s L_s b$$

From this the value of the friction force F_f is further derived. Refer to Fig. 8.8a and Eq. (8.13).

(b) First change the rake angle to $\alpha = 12°$. This will affect also the other angles of the cut in the way that has been discussed and expressed in Eq. (8.13). Assume friction angle $\mu = 22°$; this results in $\beta = 10°$ and $\phi = 35°$. Assume $L_c = 4h_1$.

(c) Add the next change, $h_1 = 0.05$ mm.

(d) Now, see if you could afford to triple the speed to $v = 1.5$ m/sec and still keep the peak temperature within acceptable limits. Plot $T_{1,k}$ (1,*KK*) versus x for all four cases in one plot.

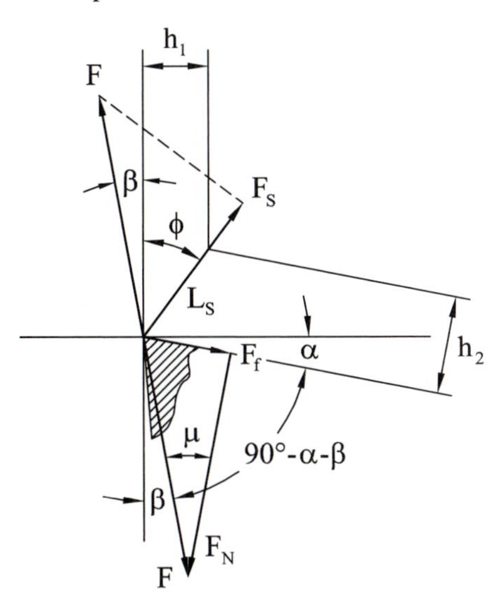

Figure P8.10

D. Single-Tool, Single-Pass Turning

P8.11 Metal removal rate, power in turning. Refer to Fig. P8.11.

A simple, single-pass, cylindrical turning operation is considered. Workpiece material is 1035 steel, $K_s = 2000$ (N/mm^2), length $L = 350$ mm, mean diameter $d_m = 92$ mm. Cutting speed $v = 120$ m/min, feed per revolution $f_r = 0.25$ mm, depth of cut $a = 5$ mm.

Determine tangential force F_T (N), power P (W), metal removal rate Q (cm^3/min), and time of machining t_m (min).

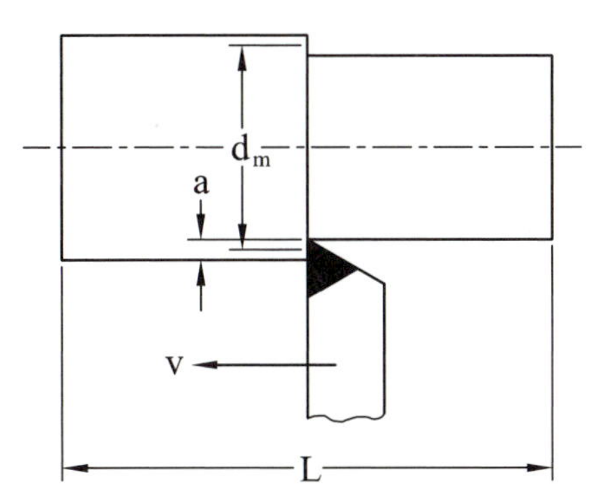

Figure P8.11

P8.12 Power, *MRR*, cost in a turning operation. Refer to Fig. P8.11.

Mean diameter $d_m = 75$ mm, length $L = 200$ mm, depth of cut $a = 5$ mm, feed per revolution $f_r = 0.25$ mm. Specific force $K_s = 2000$ N/mm^2, cutting speed $v = 150$ m/min. Tool life equation $v^3 f_r^{\,2} T = 8 \times 10^6$ [v (m/min), f_r (mm), T(min)], $C_{\text{tool edge}} = \$4.0$, $t_{\text{tool change}} = 8$ min, machine rate $r =$ \$0.4/mm. Determine the power consumed (kW), machining time t_m, cost per part *C/p*.

P8.13 Optimum speeds, feeds, cost. Use Fig. P8.11.

A single-tool, single-pass turning operation has the following tool life equation: $v^{3.5} f_r^{\,2.5} T = 15.24 \times 10^6$; v (m/min), f_r (mm), T (min). The rate for using the machine is $r_m =$ 0.5/min, the tool-changing time is $T_{tch} = 5$ min, and the cost per tool edge is $C_{te} = \$2.50$, $d = 80$ mm, $L = 400$ mm.

(a) The feed f_r is limited by the maximum permissible cutting force of $F_{t,\max} = 2516$ N. If the cutting force is determined by $F_t = 1400\, b f_r$ and $b = 5$ mm, what is the maximum feed?

(b) Express the machining time t_m as a function of v and determine the optimum cutting speed v_{opt} (m/min).

(c) What is the corresponding machining time t_m and the minimum cost per part C_p?

P8.14 Single-pass turning. Minimum cost. Use Fig. P8.11.

Determining metal removal rate, power, average chip temperature, cost per part, and optimizing speed.

Mean diameter $d_m = 80$ mm, length of cut $L = 250$ mm, depth of cut $a = 5$ mm, feed per revolution $f_r = 0.25$ mm. Cutting steel, $K_s = 2000$ (N/mm^2) $(\rho c) = 3.7$ N/(mm$^2 \cdot$ °C). The tool life equation applies: $v^3 f_r^2 T = 1 \times 10^7$ [v(m/min), f_r (mm), T (min)] and tool-change time $t_{tch} = 7$ min, cost per tool edge is $C_{t,e} = \$3.0$, the machine rate is $r = \$0.65$/min. Determine cost per part C_p, n_{opt}, and $C_{p,\text{opt}}$. Also determine: tangential force F_t (N), metal removal rate *MRR* (cm^3/min), power consumed P (kW), and, assuming all heat carried off by the chip, the final (average) chip temperature T_c (°C).

E. Turning with Two Tools on One Spindle. Refer to Fig. P8.15

P8.15 Metal cutting: tool life, cost, forces, power. Do not try to optimize spindle speed n; we will select it and simply calculate the cycle time, cost/part, and so on. Select $n = 750$ rpm.

(a) $d_1 = 50$ mm, $d_2 = 75$ mm, $l_1 = 100$ mm, $l_2 = 50$ mm, $v^3 f_r^2 T = 7 \times 10^7$, [$v$(m/min), f_r(mm), T(min)], $f_{r\max} = 0.25$ mm, depth of cut $a_1 = 5$ mm, $a_2 = 4$ mm, specific force $K_s = 1.800$ N/mm^2, machine rate $r = \$0.8$/min, tool-change time $t_{tch} = 6$ min, $C_{\text{tool edge}} = \$9.00$.

Calculate v_1 (m/min), f_{r1} (mm), tool life T_1 (min), v_2, f_{r2}, T_2, cycle time t_c (min), cost per part C_p (\$). Further: cutting forces F_1 (N), F_2 (N), total torque T (Nm), total MRR (cm^3/min), power needed P(kW).

(b) $d_1 = 60$ mm, $d_2 = 80$ mm, $l_1 = 105$ mm, $l_2 = 60$ mm, $v^3 f_r^2 T = 7 \times 10^6$, [$v$(m/min), f_r(mm), T(min)], $f_{r\max} = 0.3$ mm, and also depth of cut $a_1 = 5$ mm, $a_2 = 4$ mm, specific force $K_s = 2000$ (N/mm^2), machine rate $r = \$0.5$/min, tool-change time, 5 min, $C_{\text{tool edge}} = \$5$. Calculate the same parameters as in (*a*). Select $n = 500$ rpm.

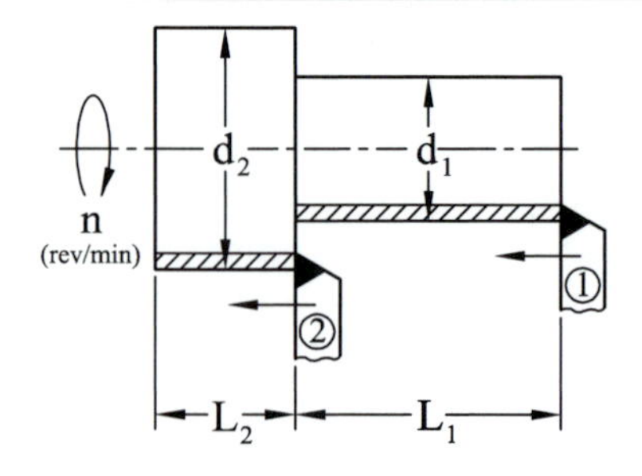

Figure P.8.15

P8.16 Optimum spindle speed for minimum cost. Refer to Fig. P8.15.

(a) $d_1 = 55$ mm, $d_2 = 80$ mm, $l_1 = 80$ mm, $l_2 = 40$ mm, $v^3 f_r^2 T = 5 \times 10^7$, ($v$ (m/min), f_r (mm), T (min)), machine rate $r = \$2$/min, tool change time 5 min, $C_{\text{tool edge}} = \$8$, $f_{r\max} = 0.4$ mm. Determine: f_{r1} (mm), f_{r2} (mm). Determine optimum spindle speed n_{opt} (rev/min), cost per part C_p (\$), tool lives T_1, T_2 (min). Carry out the same tasks in (*b*) and (*c*).

(b) $d_1 = 60$ mm, $l_1 = 80$ mm, $d_2 = 100$ mm, $l_2 = 40$ mm. For both tools: $f_{r\max} = 0.25$ mm, $t_{tch} = 5$ min, $C_{te} = \$7$, machine rate $r = \$1.5$/min, $v^{3.5} f_r^{2.5} T = 5 \times 10^7$ (v/mm), f_r (mm), T (min)).

(c) $d_1 = 40$ mm, $d_2 = 50$ mm, $l_1 = 80$ mm, $l_2 = 40$ mm, $f_{r\max} = 0.3$ mm/rev, $t_{tch} = 5$ min, $C_{te} = \$5$, $r = \$0.8$/min, $v^{3.5} f_r^{2.5} T = 1 \times 10^7$.

F. Optimizing Spindle Speeds in a Multitool, Multispindle Turning Operation

P8.17 See Fig. P8.17.

(a) $d_1 = 80$, $l_1 = 40$, $d_2 = 60$, $l_2 = 80$, $d_3 = 70$, $l_3 = 20$, $d_4 = 55$, $l_4 = 40$, all in (mm). Maximum permissible feed on all four tools $f_{r\max} = 0.3$ mm, tool life equation for all tools $v^3 f_r^2 T = 2 \times 10^7$, $t_{tch} = 6$ min, $C_{te} = \$4.0$, machine rate $r = 0.8$ \$/min.

Optimize spindle speeds n_a and n_b for minimum C/part. Determine n_a, n_b; cutting speeds v_1, v_2, v_3, v_4; cycle time t_c; tool lives T_1, T_2, T_3, T_4, and cost per part C_p for the optimum conditions.

(b) $d_1 = 100$, $l_1 = 50$, $d_2 = 55$, $l_2 = 88$, $d_3 = 110$, $l_3 = 47$, $d_4 = 50$, $l_4 = 95$, all in (mm), maximum feed on tools 1 and 3 is 0.25 mm and on tools 2 and 4 it is 0.30 mm, tool life equation is the same for all four tools; $v^4 f_r^3 T = 3.24 \times 10^8$, $t_{tch} = 7.5$ min, $C_{te} = \$ 8$, machine rate $= \$1.25$. Determine optimum spindle speeds, cutting speeds, tool lives, and optimum cost per part.

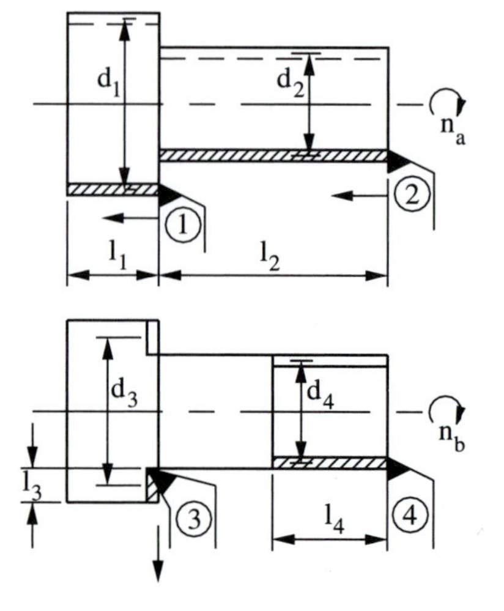

Figure P.8.17

PART 3

Machine Tools

9 Design of Machine Tools: Drives and Structures

9.1 GENERAL DESCRIPTION OF MACHINE-TOOL DESIGN

In this chapter we discuss the various criteria and computations used as the basis for designing machine tools. The problems related to controls and drives of the coordinate motions are treated in Chapter 10, which deals with automation. The main subject of this chapter, apart from a brief discussion of main drives, is structures. The design of machine-tool structures is a unique exercise in mechanical engineering. Unlike most other kinds of machinery (i.e., excavators, turbines, ships, airplanes), where most of the parts are dimensioned with respect to some criterion of strength, the overwhelming concern regarding the parts of machine-tool structures is for static and dynamic deflections, and the corresponding criteria of rigidity must be applied. In terms of strength, these structures are grossly overdimensioned. In general, a machine tool can be subdivided into three groups of parts:

1. *The structure,* which consists of stationary bodies (e.g., beds, columns, portals) and moving bodies (e.g., tables, slides, saddles) that carry out the coordinate motions, as well as moving bodies that carry out the main cutting motion (this is in most instances the spindle), and guideways and bearings that represent the movable joints between the moving bodies and the stationary bodies. The whole structure must ensure high accuracy of the motions and high rigidity, both static and dynamic, with respect to forces acting on the structure—primarily the cutting forces.
2. *The drives,* which provide the torque, or force, and speed of the motions; taken together these are the power of the motions. They can be classified into "main drives" providing the cutting power, and feed drives. They consist of motors and of transmissions.
3. *The controls,* which take care of switching on and off the individual motions in the desired directions, and of controlling their speeds, individually or in a coordinated way so as to shape the path of the tool relative to the workpiece.

Two documents have been generated that deal with the specification, testing, and analysis of the various machine-tool characteristics, grouped under slightly different headings of accuracy, mechanics, and controls. They consider both the design and the use of machine tools. In 1969–71, Tlusty, with a small group at the University of Manchester Institute of Science and Technology, in cooperation with researchers from Germany, Czechoslovakia, Sweden, and the United States, formulated a large report called *Specifications and Tests of Machine Tools* [1]. Problems of accuracy and of dynamics and generally of structures of machine tools are discussed in Koenigsberger and Tlusty [2]. In 1978–80, the Lawrence Livermore Laboratory with support from the U.S. Air Force organized a large international group of experts to prepare four volumes of a report of the Machine Tool Task Force. The mechanical aspects are dealt with in Tlusty [3]. These documents provide the framework of machine-tool science. Professor M. Weck of the Technical University in Aachen, Germany has written a four-volume book, *Handbook of Machine Tools,* that has been published in English [4]. It is still the best comprehensive text. For the design of components of high-precision machine tools, Slocum [5] has published a thorough book, *Precision Machine Design.* Another excellent text on high precision that originated at Prof. Weck's institute is [6]. There is a basic introductory text about machine tools by Boothroyd and Knight [7]. This is a selected list of significant general texts. References on individual, specialized topics are given later, but without attempting to establish a detailed bibliography.

Let us now comment on the three machine-tool subgroups enumerated above. The controls and some of the drives (servodrives of feeds) are discussed in Chapter 10, especially in the section on numerical control. The topic of main drives would include gear transmissions, steplessly variable electric motors, and to a lesser extent, hydraulic drives. All three topics are fairly specialized and are usually discussed in other courses: gear transmissions in Kinematics and Design of Machine Parts; electric motors in a corresponding course of Electrical Engineering; and hydraulic drives in Fluid Mechanics. For NC programming refer to Zeid [18]. Main drives are discussed in this text only from the point of view of establishing the torque and power versus speed requirements so that we can properly specify the design of motors and transmissions. This text focuses especially on the subgroup of machine-tool structures. The simultaneous requirements of extremely high accuracy, as expressed by deviations on the order of micrometers, and of rather large forces acting on these structures lead to the critical importance of rigidity, of stiffness. Stresses (strength criteria), which are of prime concern in most other kinds of machinery, remain very small. Guideways and bearings are considered parts of the structure and, again, the combination of high accuracy and high stiffness is the primary characteristic required of them.

As an example and illustration the parts of a model of a machining center shown in Fig. 9.1 will be described. In the center of the picture the circular shape of the front end of the spindle housing strikes the eye, along with the dark, slender, cylindrical shape of the tool holder and of the end milling cutter sticking out of the face of the spindle. The front end of the end mill is behind the workpiece being machined. This workpiece is clamped on the side facing the tool of a dark, boxlike fixture sitting on the top of a square table. On the side facing the viewer, the one away from the tool, another workpiece is clamped. It has a rectangular box shape, with its longer dimension horizontal, and two round openings on its front. This workpiece has either already been machined or awaits machining after the one being machined is completed. The table carrying the towerlike fixture can obviously be indexed by 90 degrees around its vertical axis so as to bring up to four faces in sequence towards the engagement with the spindle. The indexing table is mounted in a housing attached to the floor and also to the front of the main bed of the machine. The bed carries a big saddle with its left end carrying telescoping guideway covers that have semidark, shiny, flat tops. Under these covers are the left ends of X direction guideways attached to the bed underneath. Just below the covers, on the floor level, the black nut of one mounting bolt attaching the bed to the floor is visible, just right of the corner post of the white fence. At the center top of the saddle are two guideways in the form of bars with rectangular sections laid in the direction parallel with the spindle axis; this direction is denoted Z. It is horizontal and square to the X direction. The base of a vertical column slides on the Z guideways. The spindle housing moves vertically in direction Y on the front of the column. The lower ends of the two corresponding guideways are visible below the spindle head. On the left (looking towards the machine) side of the column is a tool magazine shown in black. Various tools can be discerned sticking out of the magazine. The magazine is covered by a protective screen. Just at the bottom right corner of the magazine is the tool-changing arm. Its top is empty, ready to accept the current tool from the spindle, and its lower end holds another tool that will be swapped with the current one. For the exchange, the arm will move out and swing to the side of the spindle. All the three linear coordinate motions are driven by electric servomotors via recirculating ball screws and nuts. The Y servomotor is visible as a white, vertical, cylindrical body on the top of the column.

The three basic structural bodies of this machine, the bed (longer in this case), the saddle, and the column are shown in Fig. 9.2. They are made of cast iron. They could also be made of steel plates welded into the required shapes that would differ to some extent from the cast ones as befits different production processes. No reasons exist to choose one over the other from a performance point of view, the choice is dictated by

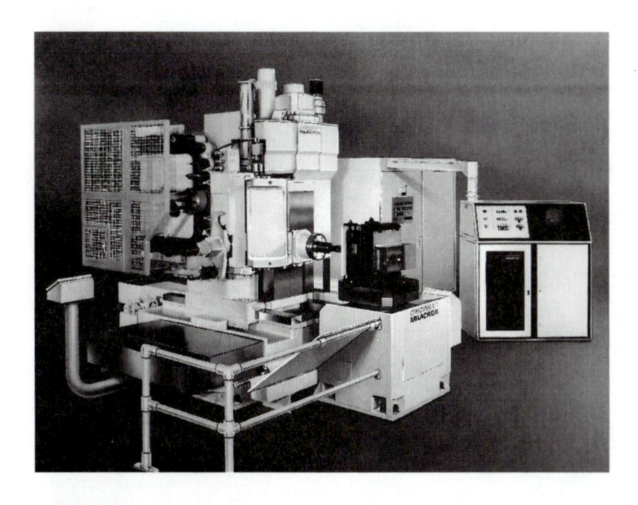

Figure 9.1
A horizontal machining center with an indexing table carrying a fixture with two workpieces clamped back to back. [MILACRON]

Figure 9.2
The bed and column of a horizontal machining center of the type of Fig. 9.1. All the coordinate motions are compounded on top of each other. The headstock (not shown) rides vertically (Y) on the column guideways. The column travels (Z) on the guideways on top of the saddle that moves (X) on the bed. [MILACRON]

economy, depending on which of the two processes is better available in a given plant or in a given region. The guideways are clearly distinguished. They are of the plain sliding type and are made of hardened and ground steel. The mating surfaces on the sliding bodies are often coated with a plastic material that gives low and uniform friction characteristics and has long life. Semi-dry friction exists at the rather low sliding velocities, and the combination of plastic against hard steel eliminates scoring of the guideway surface if any contamination occurs, and it also eliminates weldments that would arise between steel and steel under lack of lubricant. Lately, it is becoming common to use rolling-type guideways with packs of recirculating balls or rollers riding on races ground into rails attached to the basic bodies. This is the prevailing solution for high-speed machines. On large machines it is common to use hydrostatic guideways with flat guideways like those in Fig. 9.2 and pockets arranged in the mating surfaces of the sliding bodies. Pressurized oil is pumped into these pockets and supports the sliding body without contact with the guideway. Opposing pockets assure stiffness of the sliding connection, and oil is fed in the individual pockets through restrictors so as to make them independent of each other. Although the hydraulic system makes this type of guideway more complicated and more expensive, the advantages—theoretically no wear and low friction—often outweigh the extra cost. The long leadscrew driving the table in the X motion is visible, as is the servomotor mounted on its left end.

A sketch of a similar arrangement of a machining center is presented in Fig. 9.3. In this case, however, the bed has a T shape, with the bar of that letter in the X direction and its stem in the Z direction. The X table carries a round indexing table in its middle, and it also has fixed T-slotted surfaces at the sides of the rotary table, for clamping longer workpieces that do not need indexing. The tool-changing arm is mounted above the spindle. In front of the drum-type tool magazine is an auxiliary station that prepares the tool to be grabbed by the arm, and it also accepts the tool to be

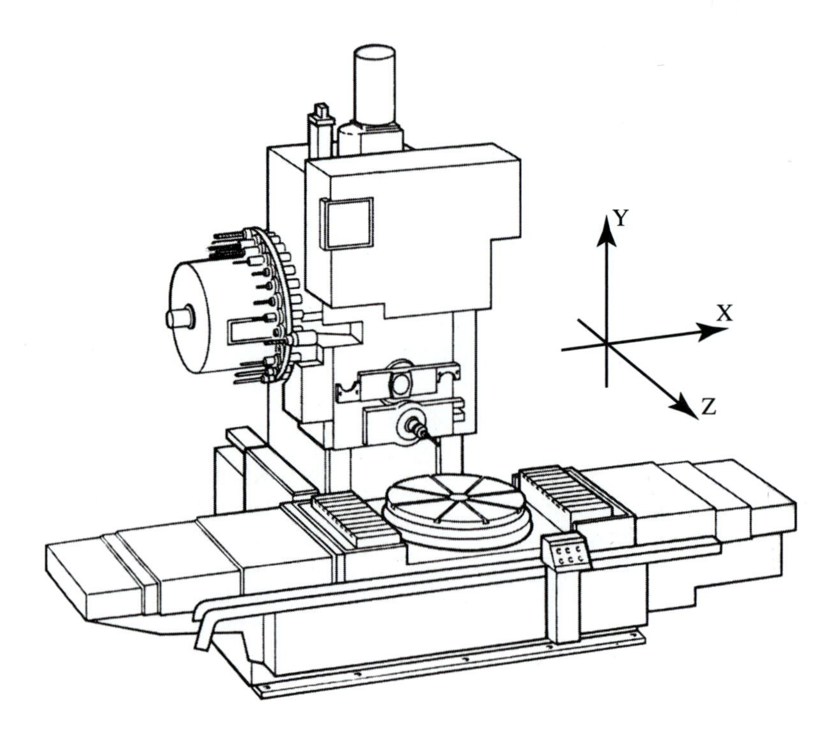

Figure 9.3
A horizontal machining center with a T-type bed. The X motion of the table is separate from the Z motion of the column.

returned to the magazine. The structure of this type of machining center can be configured in variations shown in Fig. 9.4. Note that the T-type bed consists of two parts bolted together. The guideways G_z and G_y are exposed in these diagrams, but the G_x guideways are hidden under the table and under the guideway covers.

The most important part of the structure, the "heart" of the machine tool, is the spindle. The primary requirement for spindle rotation is accuracy: deviations from true and accurate rotation are usually required not to exceed the order of 1 μm. The spindle mounting must permit a rather high speed of rotation, without play and with the maximum possible rigidity. Cylindrical roller or tapered roller bearings are often used for rigidity. They are mounted with a preload, and there is a danger that due to thermal expansion, because of the heat generated even with little friction at high speed, the inner ring will expand more than the outer ring, and the preload would increase too much and destroy the bearing. Careful lubrication and possibly internal cooling must be used to control these effects. External cooling around the bearings is used to minimize thermal deformations of the headstock and consequent loss of accuracy. An example of the insides of the headstock of a machining center is shown in Fig. 9.5. It has an 18.6 kW drive and spindle speeds range from 4 to 2400 rpm.

The transmissions provide three shift-gear speed ranges, within which stepless speed variation is obtained from the DC electric motor. The spindle is mounted in a double-row cylindrical roller bearing in the front and in a pair of back-to-back angular contact ball bearings in the rear that provide both radial and axial support. Inside of the spindle is a draw bar for clamping the tool holder. It is actuated by a battery of Belville washer springs. For tool release, these springs are compressed from a hydraulic cylinder arranged coaxially behind the spindle.

More recently, especially for high-speed machine tools, a wide range of spindle-motor speed regulation is obtained by more advanced controls without using any gear transmissions in the spindle drive. Over most of the speed range, constant maximum torque is available, and the demand on this torque determines the size of the motor.

Just as the spindle and its mounting take care of the "main," (i.e., cutting) motion between tool and workpiece, the system of guideways is basic for the "feed" motions

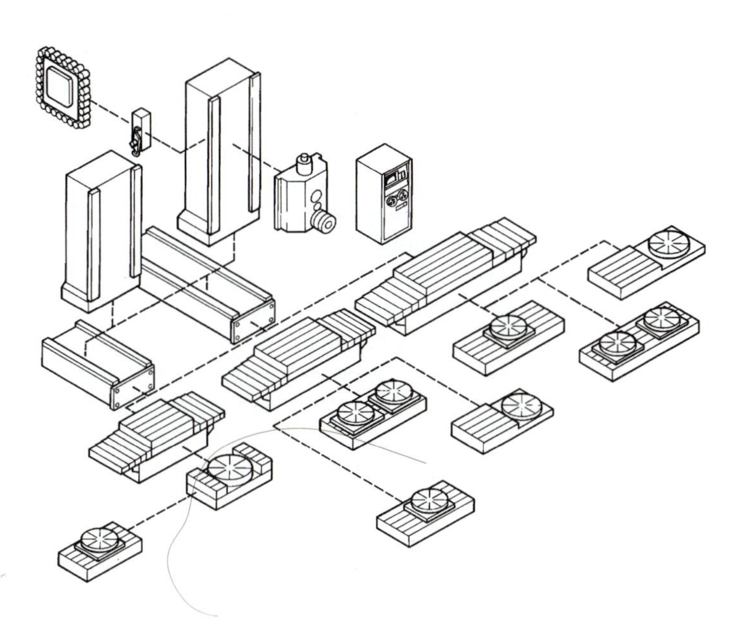

Figure 9.4
Various configurations of the structure of a machining center. Shown are columns, T-bed parts, tables, and rotary tables. In front of the column is the headstock containing the spindle. On its side are the tool magazine and the tool-changing arm. [MILACRON]

Figure 9.5
Headstock of a machining center. Spindle is mounted in roller (front) and ball (back) bearings. It carries a face mill and it is driven via a triple-range gear transmission. Combination of stepless motor speed variation with gear transmission provides a broad speed range needed for machining wide variety of materials. [KEARNEY & TRECKER]

by which the form of the machined surfaces is determined. The feed, or coordinate, motions are comparatively slow. While the peripheral velocity in a spindle bearing usually reaches the order of 100–2000 m/min, the feed rates are mostly in the order of 0.1–5 m/min. The most important requirements here are for accuracy and rigidity as well as longevity of accurate service. The aspects of accuracy are discussed in more detail in Section 9.2; just for orientation, however, a common standard for the order of deviations from straightness of motion is 10 μm over a travel of 1 m. As regards rigidity, special care is necessary in the design of the guideways because, inevitably, they are the weak links in the structure. It is impossible to provide for mutual motion of two bodies and still have as stiff a connection in the movable joint as if it were a part of a solid body.

There are three different fundamental types of guideways: plain, rolling type, and hydrostatic. The first kind is the most common and consists of planar surfaces with a sliding freedom in one direction. Most often one side is made of hardened steel or of hardened cast iron, and the other side of gray cast iron or special plastic. Semi-dry lubrication is provided, and the necessary play between the pair is limited to about 10 μm, unless it is eliminated by forces acting always in one direction, as it is, for example, when a heavy table moves horizontally on a bed.

The rolling-type guideways consist most often of special units with recirculating rollers between the two bodies. Hydrostatic guideways have pockets created in one of the bodies that are supplied with oil under pressure. These pockets are so arranged that they straddle a guideway with forces on the two sides opposing each other. In this arrangement the sliding body is centered on the guideway with enough internal stiffness to oppose external forces and prevent metal-to-metal contact between the two bodies. Hydrostatic guideways are more complex and more expensive than the other two types because they contain a rather sophisticated hydraulic system and because of the need to collect and recirculate the oil flowing out of the pockets. Therefore, they are used only occasionally, on very large machine tools.

Machine tools exist in a great variety of configurations, some of which were mentioned in Chapter 7; others are described in Chapter 10. However, all of them consist of subgroups similar to those described here with reference to machining centers. Constraints on forces, speeds, powers, accuracy, and static and dynamic stiffness can be formulated in a general manner and applied to the design of particular machine tools. In this chapter, we explain and derive this general theoretical background.

At the beginning of the development of a machine-tool type, it is necessary to determine the types and sizes of workpieces to be machined, their materials, the types of tools to be used, expected metal removal rates, and tolerances to be achieved. The maximum size of the workpiece will dictate the lengths of travels in the individual coordinate axes and therefore the lengths of beds and columns and the overall size of the machine. In order to determine the basic dimensions of the sections of the main structural bodies and their shapes, one must use computational criteria derived from the requirements of accuracy and of metal removal rates. The art of successful design consists in skillfully using these criteria so as to satisfy the requirements with minimum cost of building the machine. This mostly approximately translates into satisfying the requirements with minimum weight of the structures. The following discussion develops the theoretical background for the specification of the characteristics of main drives, for specifications and tests of accuracy, and for static and dynamic stiffness of structures.

Precision Engineering

In the 1980s, a new class of machine tools emerged, born out of the need for very high precision in the aerospace industry on one hand and in computer engineering on the other. In the former case, a typical example of high-precision parts are large mirrors for satellites. In the latter case, super precision is needed in the production of VLSI chips as well as of the discs of hard drives.

In the preceding text, we mentioned accuracies characterized by errors down to 25 μm for the general population of machine tools and to 1 μm for high-precision types. In the new super-precision class, a typical representative is the numerically controlled diamond turning lathe (and other specialized lathes or grinding machines). The beds of these machines are made of granite whose surface is lapped to an overall flatness of 0.0001 in (2.5 μm), with hydrostatic guideways based on rails made of alumina ground and lapped to a flatness of 1 μm. Lengths of motion are controlled by feedback from linear scales with resolution of 10 nanometers (0.01 μm). Special care is devoted to the compensation and elimination of thermal deformations in these machines; for instance, the whole work space of a diamond turning lathe is showered by oil the temperature of which is kept within 0.1 degree Celsius. The above-mentioned scales are available as made of Zerodur, a material with a thermal expansion coefficient lower than 1E-7. For a very detailed treatment of these problems, refer to the Slocum [5] and Weck and Hartel [6].

9.2 SPECIFYING THE CHARACTERISTICS OF MAIN DRIVES

The drive of the cutting motion is designated as the main drive, in distinction from feed drives. It is the drive of a spindle carrying the workpiece in turning and the drive of the spindle carrying the tool in milling, drilling, and boring. With the exception of single-purpose machine tools used in mass production, every machine tool is used for a variety of machining operations, requiring the spindle to rotate at various speeds. We will now discuss the required range of spindle speeds and the corresponding torque and power. This could be done both for lathes and for milling machines; however, because there is a complete analogy between the two, most of our reasoning uses the lathe for illustration.

The analogy between the lathe and the milling machine is based on the analogy between the workpiece diameter for the former and milling cutter diameter for the latter. The spindle speed n is obtained from the cutting speed v as

$$n = \frac{v}{\pi d} \tag{9.1}$$

in compatible units, or,

$$n = \frac{1000\,v}{\pi d}$$

if the usual units of v (m/min), d (mm), n (rpm) are used; d signifies the workpiece diameter on the lathe and the cutter diameter on the milling machine.

On a lathe a certain range of workpiece diameters has to be machined, say, from 25 to 250 mm. On the milling machine a range of milling cutter diameters is used in the various operations. Again, it could be that this range will be from 25 to 250 mm. In general, let us assume the required range from d_{min} to d_{max}. From now on we will talk about them as being workpiece diameters and describe turning, but remember that a complete analogy exists for milling.

In Chapter 8 the choice of the cutting speed v was explained and how the optimum speed depends on the workpiece material, the tool material, the feed rate used (roughing or finishing), and on some other economic parameters like the cost of using the machine, the tool-changing time, tool cost, and so on. Assume for the moment that only one cutting speed v is considered. Then, the required range of spindle speeds would be from

$$n = \frac{v}{\pi d_{max}}$$

to

$$n = \frac{v}{\pi d_{min}}$$

and it would require a ratio of speeds:

$$R_d = \frac{d_{max}}{d_{min}}$$

However, there is a variety of workpiece materials to machine, a variety of feed rates, and a variety of tool grades, and for every workpiece diameter one may want to use a whole range of cutting speeds from v_{min} to v_{max}. For instance this may be from $v =$ 15 m/min in roughing tough steel with HSS tools to $v = 300$ m/min in finishing mild steel with carbide tools, or to $v = 600$ m/min in finishing cast iron with silicon nitride tools, or to $v = 1000$ m/min for machining aluminum with carbide tools. Thus, assuming just a single diameter d of the workpiece, it would be necessary to vary spindle speeds from

$$n = \frac{v_{min}}{\pi d}$$

to

$$n = \frac{v_{max}}{\pi d}$$

and this would require a ratio of speeds:

$$R_v = v_{max}/v_{min}$$

If both a range of diameters and a range of cutting speeds are realistically considered, then a range of spindle speeds is needed from

$$n_{min} = \frac{v_{min}}{\pi d_{max}}$$

to

$$n_{max} = \frac{v_{max}}{\pi d_{min}} \tag{9.2}$$

with a total speed ratio of

$$R = R_d R_v \tag{9.3}$$

This ratio might be very large. Taking the examples of diameter range $R_d = 10$ and speed range $R_v = 20$, a spindle-speed range $R = 200$ is obtained, which is quite common. It is of course mainly the top speed n_{max} that is the real difficulty, although with fully geared transmissions and constant-speed motors, the number of shafts and gears in the gearbox increases with the increase of R.

Let us now discuss the correlation of power and spindle speed. We are going to establish a plot in log scales: (log P, log n), as shown in Fig. 9.6. Let us first consider rough machining a particular material, say steel 1045, over a range of diameters $R_d = d_{max}/d_{min}$. Let us assume a certain maximum depth of cut a_{max} and maximum feed f_{max} resulting in a certain maximum force F_{max} and a certain optimum cutting speed v_1 corresponding to the best suitable grade of carbide tool. Irrespective of the diameter, a constant power $P_1 = F_{max}\, v_1$ is required over the range of $n_{11} = v_1/(\pi d_{max})$ to $n_{12} = v_1/(\pi d_{min})$. This is represented by the line *1, 2*. Machining the same size of chip, with the same force F_{max}, but with a lower cutting speed v, will require less power $P = F_{max}\, v$ and correspondingly a lower spindle speed. Considering diameter d_{max}, the power decreases with spindle speed along the line T_1 between points *1* and *3*. It is a line of constant torque:

$$T_1 = F_{max}\frac{d_{max}}{2} = \frac{P}{2\pi n} \tag{9.4}$$

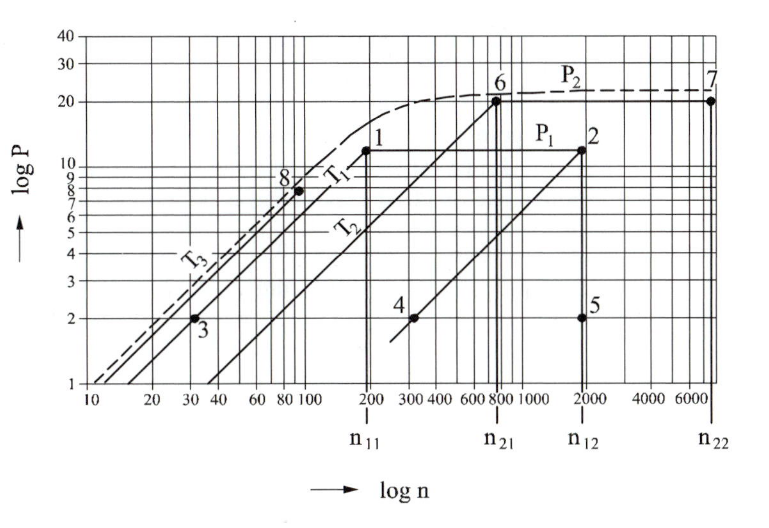

Figure 9.6
The power versus spindle-sped diagram. Left is the constant torque part and right is the constat power range. The torque, power, an spindle-speed parameters of the diagram depend on the range of sizes of the workpieces (on the lathe) or cutters (on milling machines), an on the workpiece materials determining cutting speeds.

Equation (9.4) is a straight line under 45° because

$$P = 2\pi T_1 n$$
$$\log P = \log (2\pi T_1) + \log n \tag{9.5}$$

Similarly, machining with the same F_{max} the diameter d_{min} at any speed v follows the line $T' = F_{max} d_{min}/2$ between points 2 and 4. All machining of any diameter between d_{max} and d_{min} with F_{max} is enclosed between the lines T_1, P_1. On the other hand, machining using the same cutting speed v but for smaller chip areas and with smaller forces is enclosed between the lines n_{11}, P_1, n_{12}. Thus, machining of steel 1045 will all be enclosed in the area of the diagram between the lines T_1, P_1, n_{12}, that is between the lines connecting points 3, 1, 2, and 5.

Let us now consider machining an easier-to-machine material, for example, aluminum, and use the same maximum depth of cut a_{max} and feed f_{max}. The force will be about 2.5 times lower, but the optimum speed may be 4 times higher. For d_{max}, the torque will be $T_2 = T_1/2.5$, but the power will be $4/2.5 = 1.6$ times higher, $P_2 = 1.6\ P_1$, and these parameters determine the point 6 at $n_{21} = 4n_{11}$. From point *6* the constant power line extends to point 7, corresponding to d_{min} and $n_{22} = n_{21}\, d_{max}/d_{min}$.

On the contrary, machining a tougher material, like heat-treated 4340, may give 1.25 times the force (torque T_8) but require lowering the cutting speed by a factor of 2. This would determine the corner point 8 of the area corresponding to this material. Assuming this to be the toughest material to regularly machine, the line T_3 is the characteristic for all speeds below n_{31}. The dashed line is the resulting characteristic to be used for the design of the main drive of this machine. Its shape roughly corresponds to the generally available characteristics of steplessly variable electric motors that deliver constant power over a certain part at the top of the speed range and constant torque in the lower part of the range.

EXAMPLE 9.1 Drive Characteristics ▼

Consider machining steel 4140, steel 1020, and aluminum alloy 7075-T6. Use Table 7.1 to obtain corresponding specific forces. Assume a range of workpiece diameters from $d_{min} = 30$ mm to $d_{max} = 300$ mm. Use Fig. 8.45 to find the recommended roughing speeds. Consider maximum chip cross section $6 \times 0.4\ mm^2$. Follow the reasoning presented above in constructing the diagram in Fig. 9.6 and determine the values of spindle speed *n*, power *P*, and torque *T* corresponding to the points 1, 2, 6, 7, and 8 of the diagram.

From Table 7.1 we find the values of K_s, and for the above maximum chip area, we obtain the following maximum forces:

Al: $K_s = 850\ N/mm^2$, $F_{max} = 2040$ N
1020: $K_s = 2100\ N/mm^2$, $F_{max} = 5040$ N
4140: $K_s = 2800\ N/mm^2$, $F_{max} = 6720$ N.

The roughing cutting speeds follow:

Al: $v = 400$ m/min
1020: $v = 175$ m/min
4140: $v = 60$ m/min

Thus we have the following:

Point	n (rpm)	P (kW)	T (N/m)
1	186	14.7	756.0
2	1860	14.7	75.6
6	425	13.6	306.0
7	4250	13.6	30.6
8	64	6.7	1008.0

▲

Let us now briefly discuss the expected future development of the requirements for main drives.

The development of tools and corresponding increases of cutting speed call, first of all, for an increase in spindle speeds and powers; see Fig. 9.7. If *1-2-3* was the original characteristic for machining a certain workpiece material over a certain range of diameters with a certain chip section, and if a new tool permits doubling the cutting speed, the characteristic *1-4-5* arises. If in addition the new tool material is also tougher and permits larger feeds, the characteristic *6-7-8* arises. The vertical shift represents an increase of both torque and power over the same spindle-speed range. Further, such an increase may be due to better stability of the machine, permitting a larger depth of cut, or it may be due to the introduction of a chip-breakage sensor, and so on.

The demand is for universal machine tools. Typically the aircraft and spacecraft industries use large, multi-axis NC machine tools and, for the sake of flexibility in production control, require that these machines should machine a wide range of materials from Ti-alloys to high-strength steels and Al-alloys. This kind of requirement applies to most batch manufacturing, mostly in a smaller range of workpiece-tool material combinations. Thus, for example, it may be required to use ceramic tools on cast iron, or carbide tools on tough steels. In another instance the materials may range from mild steels to heat-resistant alloys; thus, spindle drives and bearings would have to span from high-torque, low-speed applications to high-speed, high-power applications. The bearings must have large diameters to provide for high stiffness of spindles and must also be capable of rotating at high speeds.

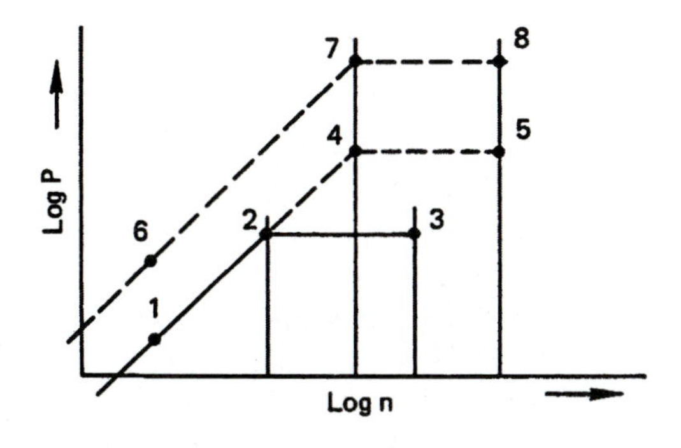

Figure 9.7
Further development of drive characteristics (follow explanations in the text).

9.3 ACCURACY OF MACHINE TOOLS

In order to establish a proper concept of machine-tool accuracy, reference must be made to workpiece accuracy. The two concepts are not simply related. Workpiece accuracy is specified by errors of dimensions and errors of form of surfaces bounding the workpiece. Machine-tool accuracy, as explained in the following text, is essentially based on the accuracy of machining motions.

Bear in mind, however, that machine-tool accuracy is only one of several inputs to workpiece accuracy (see Fig. 9.8.) There, on the part of the machine tool, "geometric accuracy" is mentioned. This is a concept that was historically established when in the 1930s Professor Schlesinger in Berlin undertook the task of preparing accuracy specifications and tests for the acceptance of machine tools on the part of both manufacturer and customer. These tests, which have been adopted as a basis of many national and international standards, were conceived as nonmachining tests; instead, they concentrated on measuring the accuracy of parts of the machine-tool structure by means of levels and dial gages. Since then a lot of work has been devoted to the development of machine tool metrology. The following section introduces the essence of the present state of this art.

Returning to Fig. 9.8, it is sufficient to state that the concept of geometric accuracy includes those effects on workpiece accuracy that derive from machine-tool motions, without machining. These effects depend mainly on the accuracy with which the machine-tool structure was manufactured. Figure 9.8 also indicates that, apart from accuracy in manufacturing the structure, geometric accuracy is influenced by thermal deformations of the structure and by weight deformations. Thermal deformations arise from internal heat sources like the friction in spindle bearings or energy losses in hydraulic circuits. Heat is also generated in the cutting process, and it may affect both the structure and the workpiece. This part of the effect is shown to bypass the block of geometric accuracy, which was defined as excluding machining. Weight deformations are discussed separately in Section 9.3.2 on static deformations. They depend on the stiffness of the structure.

Clamping accuracy is shown as a separate influence. It involves both the design of the clamping devices (chucks, vises, fixtures [pallets]) and the flexibility of the

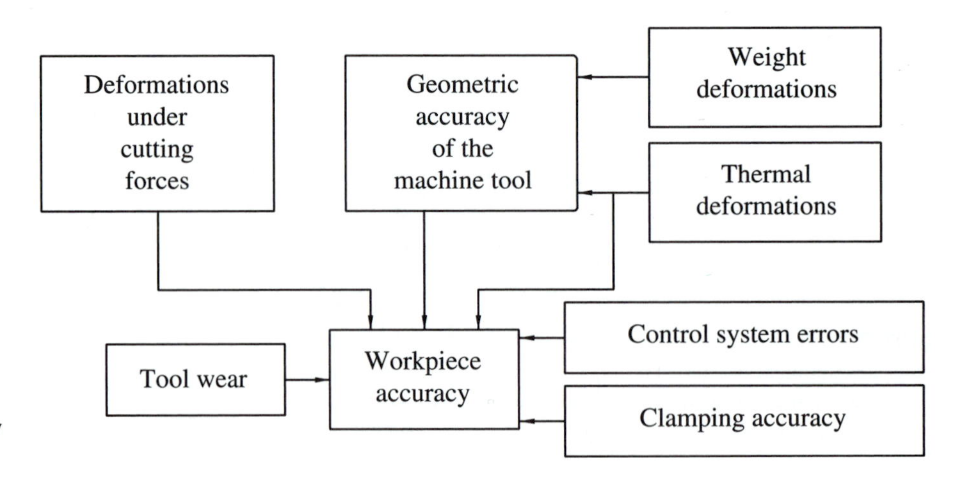

Figure 9.8
Factors influencing accuracy of the workpiece.

workpiece. Also separately shown are errors due to dynamics and to steady-state errors arising in the servomechanisms and, generally, in the control system. This applies essentially to numerical control, and all these errors are discussed in Chapter 10.

The cutting process affects workpiece accuracy in two ways. First, the cutting forces produce deformations of the machine-tool structure, and these cause errors of dimension and of form in the workpiece. This problem is analyzed in Section 9.3.3. Second, tool wear is generated in cutting and affects the dimensions of the machined surfaces. This effect can be limited by in-process or post-process gaging of the workpieces, by manual or automatic corrections of the setting of tools, or by shifting the origin of coordinates in an NC system.

9.3.1 Geometric Accuracy: Machine-Tool Metrology

It was mentioned above that Schlesinger established a system of measurements defining the geometric accuracy of machine tools sixty years ago. This system consisted of specifications of accuracy for various surfaces of the machine-tool structure: those that carry the workpiece and the tool (tables, spindle flanges, etc.) and those that carry the moving parts (guideways, spindle rotation). Typical measurements of which such a system consists are flatness of a table, straightness of a guideway, and mutual squareness of X and Y guideways. This system is closely related to the manufacture of the structure, but it is not easy to determine how the measured deviations will affect the accuracy of the workpiece.

In further development of machine-tool metrology (see Tlusty [8]), a different approach was adopted that is based on the measurement of errors of the machining motions (the coordinate motions and spindle rotation) in such a way as to directly relate errors in these motions to errors in the workpiece. The techniques of measuring spindle-rotation accuracy are too specialized to discuss here; they are well described in [9]. However, we will discuss the measurement of coordinate motions. One very important event in the development of these kinematic methods was the introduction of a well-developed and easy-to-use laser interferometer, which is an excellent tool for measuring errors of motions, while it is much less suitable for measuring errors of surfaces, and useless for the static measurement of distances.

The modern metrology of machine tools is based on a principle called *master part tracing*. It can be formulated as follows: Replace the tool by a gage and replace the workpiece by a master part. A master part is an ideally accurate workpiece. Move the gage along the surfaces of the master part with the same motions as if machining them with a tool. Any variations in the signal from the gage represent errors that would occur on the workpiece. Obviously, if the coordinate motions were perfectly accurate, a gage clamped in the spindle and sliding on an ideal workpiece would not show any signal variation. But how do we prepare an ideally accurate workpiece that would incorporate all kinds of surfaces, representing all kinds of workpieces? And how do we choose the motions of the gage so as to represent all kinds of machining motions? The master part is represented in its elements by straightedges or laser beams strategically located, and the individual measurements are so chosen as to eventually, in their sum, express conformity with tolerance requirements in the whole working zone of the machine tool.

Let us consider a horizontal boring machine, or a machining center, in the configuration shown in Fig. 9.9. The workspace is above the table, the coordinate motion X is executed by the table, Y is executed by the headstock traveling vertically on the column, and Z is carried out by the saddle. Let us select three lines in the work space:

A, B, and C. These will be the reference lines, and the tests must be arranged so as to have them exactly square to each other (as accurately as possible; much more accurately than the degree of accuracy tested). These lines are considered parts of the Master Part. Along each line, measurements of three *translative errors* will be made. These measurements are explained in Fig. 9.10 for the motion *X* along the line *A*. In Fig. 9.10a two gages are shown attached to the spindle (in place of the tool) sliding on a straightedge (with very accurately straight surfaces) located in the place of line *A* (of the workpiece). If the table of the machine is accepted as reference, the straightedge *A* will normally be aligned to be parallel with the surface of the table in direction *Y* and parallel with the side of a T-slot, in direction *Z*. The deviations (variations of the gage

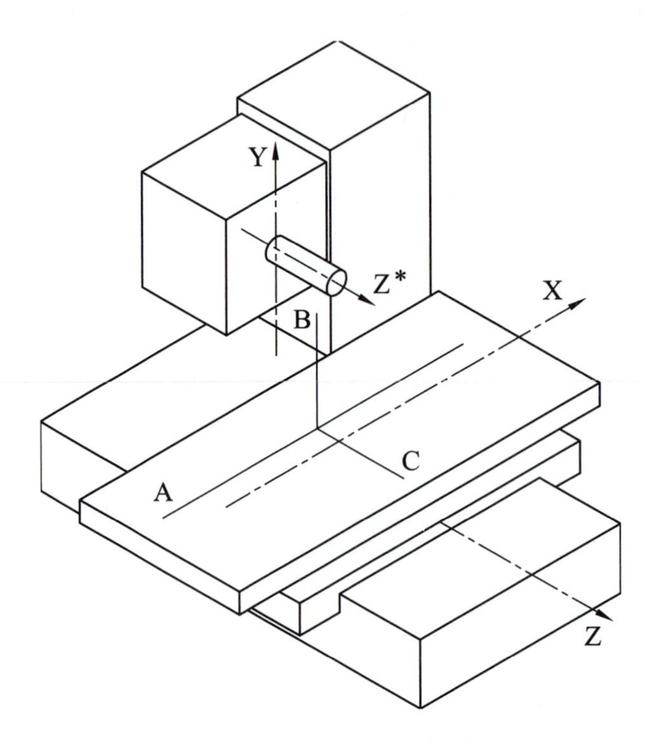

Figure 9.9
Coordinate axes and workspace of a machining center. The *X* and *Z* motions are compounded, the *Y* motion is separate. The workspace is above the table. Geometric accuracy of the machine is related to the coordinate motions throughout the workspace.

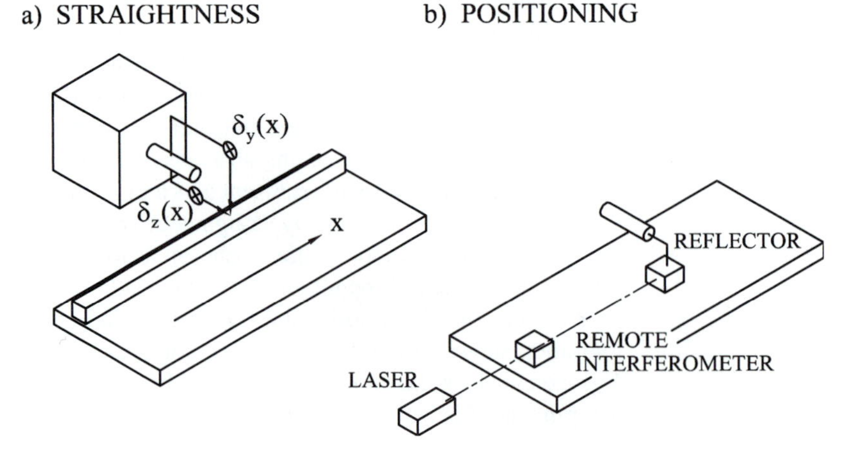

Figure 9.10
Three measurements of translative errors associated with a motion along a straight line: two straightness measurements and one of positioning.

signals) obtained in these measurements are denoted δ with a subscript denoting the direction of the deviation and with the moving coordinate in parentheses. So we have

- Error in direction Y of straightness of motion X: $\delta_y\ (x)$,
- Error in direction Z of straightness of motion X: $\delta_z\ (x)$, both along the line A.

Again, it is obvious that if the two gages did not show any signal variation during the test, the motion would be perfectly straight, and a line machined on the workpiece as part of a surface would be accurately straight.

The third measurement deals with deviations in direction X, during the motion X. This means that it deals with the accuracy of the lengths of the motion from any point to any other point; that is, it deals with the accuracy of distances between points along the line X. The result is called positioning error in the direction X: $\delta_x\ (x)$. Currently, this measurement is commonly done using the laser interferometer. This is indicated in Fig. 9.10b, and it is explained in more detail later.

An idealized example of a record of a positioning measurement is presented in Fig.9.11a. It is so located that the gage (laser interferometer) was zeroed at $x = 0$. The error δ_x is plotted versus the x coordinate of the motion. A positive error means longer motion, a negative error shorter motion. It is seen that the total motion, from $x = 0$ to $x = x_3$, was found to be longer than desired by $\delta_x\ (x_3)$. It can be seen that the relative error between points x_1 and x_2 is the difference $\delta_x\ (x_{1,2}) = \delta_x\ (x_2) - \delta_x\ (x_1)$ by which the distance between these two points is longer than it should be. A more realistic example of the positioning error is shown in Fig. 9.11b. It is a record obtained during both the $+X$ and $-X$ (return) motions. It contains a slowly varying error as in *(a)*, and also a dead zone DZ as the difference between errors in the two directions, and a

a)

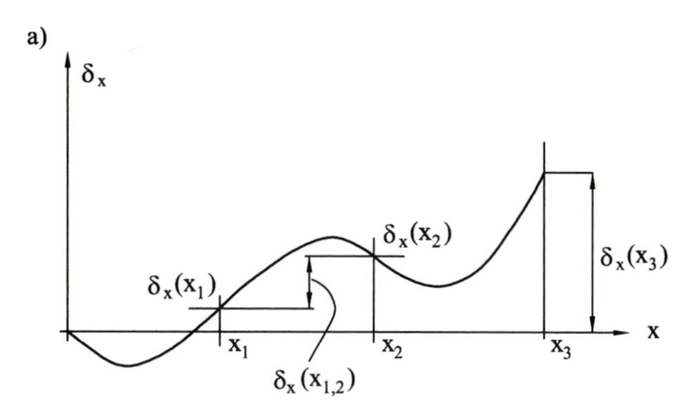

b)

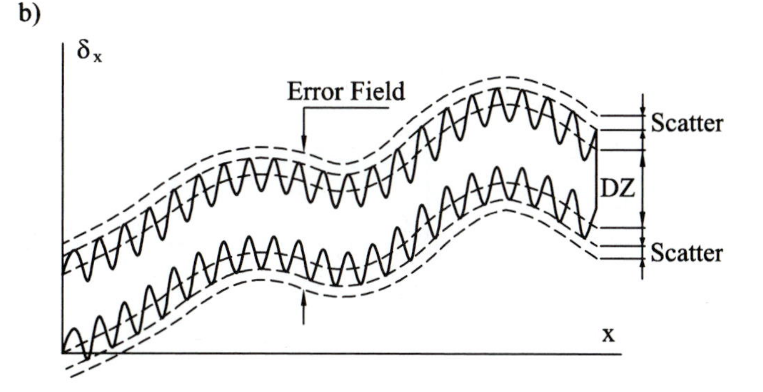

Figure 9.11
Model of a record of a positioning error: a) average error obtained in a single direction of the *X* motion, b) the graph includes also the reversal of motion effect and the periodic error component, as well as scatter. Positive error means travel was longer; negative error means travel was shorter than commanded.

periodic error (due, e.g., to an axial run-out of the thrust bearing of the lead screw), and scatter obtained in repeated measurements. The top and bottom envelopes of these errors represent the error field.

Similarly an example of a record of an error obtained in a straightness measurement is presented in Fig. 9.12, which shows the traces *a* and *b* obtained during the $+X$ and $-X$ motions, respectively. Let us look at the curve *a,* which corresponds to the $+X$ direction of motion. From the machine-tool builder's point of view, it can be considered as consisting of a "straightness of motion error" e_s (the minimum distance of two parallel lines enclosing the curve) and an "error of direction" e_d (the slope of these lines). In the case of $\delta_y(x)$, e_d would be the error of parallelism of the table surface with motion *X;* for $\delta_z(x)$, it would be the error of parallelism of the T-slot with motion *X.* However, from the user's point of view, that is, from the point of view of the accuracy of the workpiece, these distinctions have no sense at all, and the functions $\delta_y(x)$ or $\delta_z(x)$ must be simply considered as errors of location of points of the workpiece. Later we will discuss how to evaluate them. Still, it is useful to realize right now that the difference $[\delta_y(x_1) - \delta_y(x_2)]$ between the errors of points 1 and 2 located at coordinates (x_1, y_A) and (x_2, y_A) is the *y*-component error of the distance of the two points (which in this case ideally should be zero).

The three component measurements of two straightnesses and one positioning could be done along any line in the working space of the machine. In practice, they are done along three lines like the lines *A, B,* and *C* in Fig. 9.9. Mostly, however, the locations of these three lines will be selected so that they pass through the center of the working space.

The errors measured are evaluated by comparing them with specified tolerances. This is done in a way that was introduced by the National Machine Tool Builders Association. It uses the "tolerance template," which is reproduced in Fig. 9.13a. There, the positioning error found (including scatter) is plotted in coordinates *x* (length of motion in *X*) and δ_x (error of *x*). The rule requires that a doubled-sided template *T,* if shifted along the error field, should always be able to include the whole of it. The tolerance template is shaped to allow for constant error Δ_x along a length *b* and for an additional error increasing in proportion to the distance extending beyond *b.* The template is free to be shifted in the directions of both coordinate axes of the figure while remaining parallel to itself. This is significant in that distance *x* and error δ_x are taken as relative only between any two points. In Fig. 9.13b a simplification of this rule is suggested in the form of a template where distance *b* is diminished to zero. This is com-

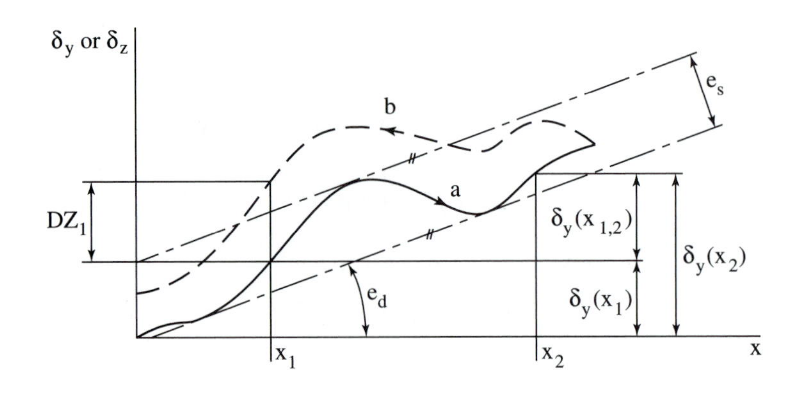

Figure 9.12
Characteristic features of a record of straightness-of-motion error. Included is error of direction (e.g., nonsquareness to another axis).

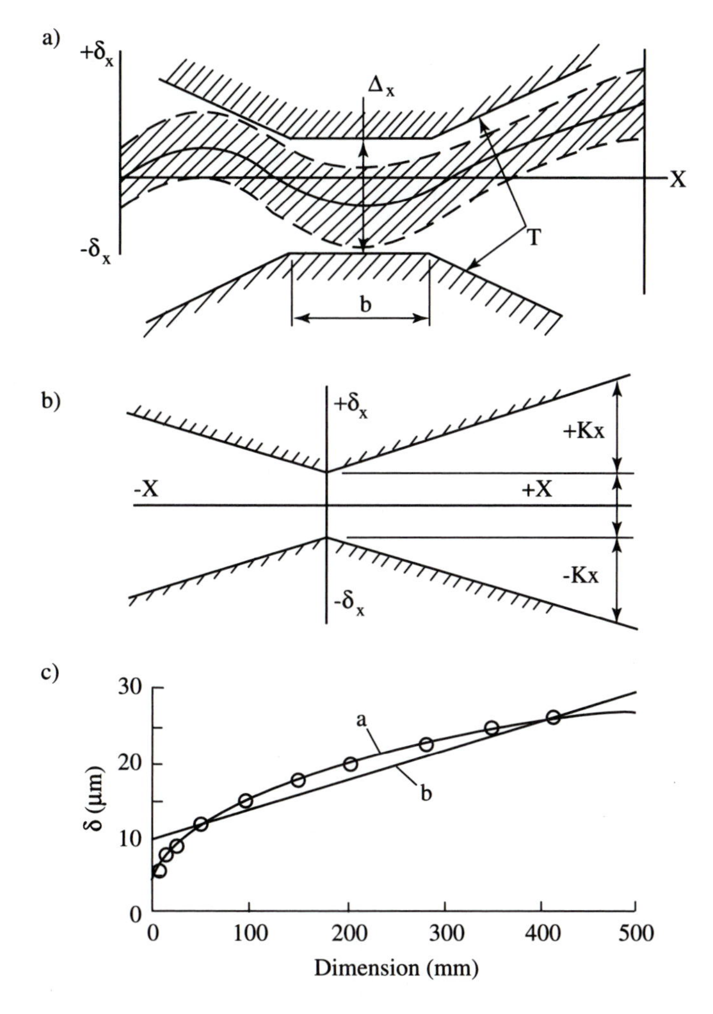

Figure 9.13
The tolerance template:
a) the error field is enclosed by the "bow tie" template;
b) the simplified form of the template and its components; c) relationship to dimensional tolerance.

parable to the ISO system of tolerances of dimensions. Figure 9.13c expresses graphically the IT 5 class tolerance δ as a function of the dimension D (curve a). This function is nonlinear. However, it has two fundamental features: the tolerance does not decrease to zero when dimension D approaches zero; the tolerance increases with the dimension. Line b represents a linearization of the curve a. The tolerance template of the form given in Fig. 9.13b is the interpretation of the rule expressed by line b in Fig. 9.13c if both the tolerance and the dimension are taken in the relative way.

The mathematical expression of the template form depicted in Fig. 9.13b is

$$|\,\delta_x(x_1) - \delta_x(x_2)\,| < A_{xx} + K_{xx}\,|\,x_1 - x_2\,| \tag{9.6}$$

where x_1 and x_2 are the coordinates of any two points along a positioning path in the direction X, and $\delta_x(x)$ is the error in the direction X of the positions x. Equation (9.6) can be read, "The absolute value of the error of the distance of two points (whether the distance is greater or smaller is not relevant) may be equal to or greater than a constant value A_{xx} plus another value proportional by K_{xx} to the distance (while it is irrelevant in which direction the distance is taken as positive)."

The lateral errors resulting from the straightness measurement have, from the point of view of workpiece accuracy, exactly the same significance as the positioning error. Furthermore, the character of the lateral error field is similar to that of the positioning error field. Therefore, the NMTBA tolerance rule may be extended using the notation A_{yx}, A_{zx} and K_{yx}, K_{zx} for the lateral constants:

$$
\begin{aligned}
|\delta_y(x_1) - \delta_y(x_2)| &< A_{yx} + K_{yx}|x_1 - x_2| \\
|\delta_z(x_1) - \delta_z(x_2)| &< A_{zx} + K_{zx}|x_1 - x_2|
\end{aligned}
\tag{9.7}
$$

These rules may also be graphically expressed by the "bow tie" template. The interpretation of the rules is such that (see Fig. 9.14) when moving in the X direction from point x_1 to point x_2 we end up within a box symmetrically located around the ideal location of x_2, and its sides a, b, and c are obtained by Eqs. (9.6) and (9.7), respectively, all of them having terms proportional to the distance of points x_1, x_2.

$$
\begin{aligned}
a &= A_{xx} + 2K_{xx}|x_1 - x_2| \\
b &= A_{yx} + 2K_{yx}|x_1 - x_2| \\
c &= A_{zx} + 2K_{zx}|x_1 - x_2|
\end{aligned}
\tag{9.8}
$$

The values of A and K constants have to be given in the specification of the machine. Typically, these tolerances may be $A = 10\ \mu\text{m}$, $K = 10\ \mu\text{m/m}$ for high-quality machine tools.

So far, we have only considered the three translative errors along one line of measurement in each coordinate. However, if the errors were measured, say during the motion X along other lines parallel with A, like lines A' and A'' in Fig. 9.15, the results

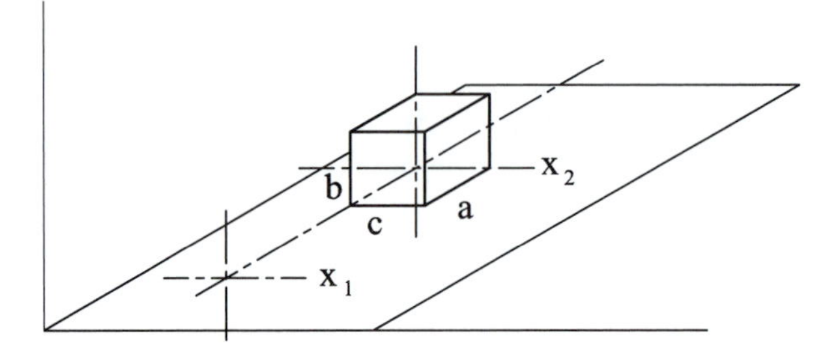

Figure 9.14
The tolerance rule applied to three dimensions. Component errors of the position of the end of the path $x_1 - x_2$. In this case of a motion in one coordinate, these end errors result from the three translative errors associated with that coordinate. In general all three coordinates get involved.

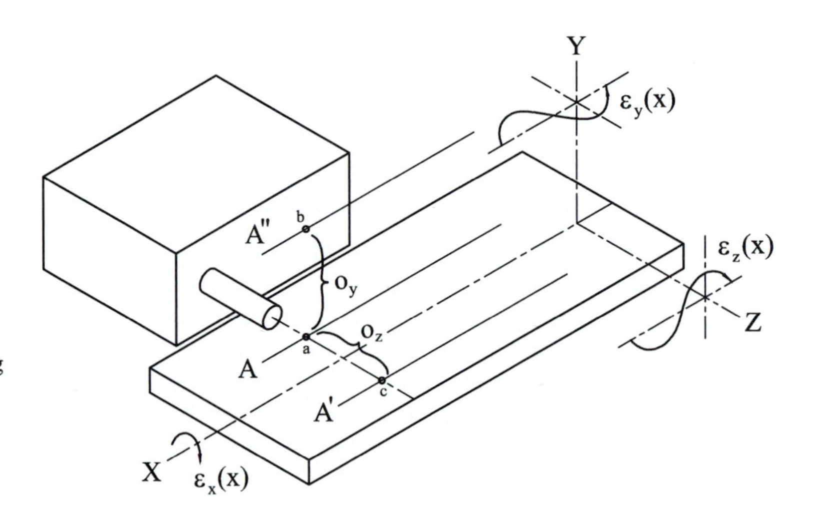

Figure 9.15
Angular errors and their effects on translative errors. Depicted are the roll, pitch, and yaw of the table moving in X and the offsets of lines in the workspace through which the angular errors affect translative errors.

would be different. This is so because apart from the three above-mentioned types of translative errors associated with the motion of the table, there are also three types of angular errors arising: the roll, the yaw, and the pitch. The table does not remain parallel to itself during the X motion but carries out the three small angular motions. We will denote them by ε with a subscript indicating the axis around which it rotates and the axis of motion in parentheses. Thus, as shown in Fig. 9.15, we have,

Roll: $\varepsilon_x(x)$ Yaw: $\varepsilon_y(x)$ Pitch: $\varepsilon_z(x)$

These angular error motions cause differences in the translative errors measured at offsets. There is an offset o_y between lines A and A'' and an offset o_z between lines A and A'. Obviously, we have

$$\begin{aligned} \delta_x(x)_{A'} &= \delta_x(x)_A + o_z\, \varepsilon_y(x) \\ \delta_y(x)_{A'} &= \delta_y(x)_A + o_z\, \varepsilon_x(x) \\ \delta_x(x)_{A''} &= \delta_x(x)_A + o_y\, \varepsilon_z(x) \\ \delta_z(x)_{A''} &= \delta_z(x)_A + o_z\, \varepsilon_x(x) \end{aligned} \tag{9.9}$$

It is necessary to measure the three translative errors along one line during each coordinate motion and also the three angular errors of the body carrying out each coordinate motion. It is then possible to calculate the translative errors along any line in the working space by using equations of the type (9.9). If this is done for lines located at the extremes of the working space—that is, for lines having mutually extreme offsets—and if each of these errors satisfies the tolerance template, then any error measured along any lines in the workspace will satisfy the tolerance.

Let us illustrate this statement by taking positioning measurements at lines A and A''; they are related by the first and third of Eqs. (9.9). If another measurement was made along a line located between A and A'', with an offset a from A, the error on this line would be,

$$\delta_x(x) = \delta_x(x)_A + [\delta_x(x)_{A''} - \delta_x(x)_A]\,\frac{a}{o_y} \tag{9.10}$$

This means that this error is obtained by linear interpolation between the errors at the extreme locations, and if each of those fits into the tolerance limit, this error will too.

EXAMPLE 9.2 Determining Errors at Offsets ▼

This example is an illustration of the application of relationships like those expressed in Eqs. (9.9). The records of errors used are very simplified in order to make it easy to see the effects. Let us refer to Fig. 9.16a, which shows two lines A_1 and A_2 that are located in parallel with the coordinate motion X and are offset by the distance b in the direction Y. The error of pitch $\varepsilon_z(x)$ is shown as being positive as a rotation around axis Z such that it would increase $(x_1)_2$ with respect to $(x_1)_1$. It is assumed that the positioning error along A_1 was measured as given in Fig. 9.16b and that the pitch error associated with the coordinate motion X of the table was measured as shown in diagram *(c)*. Its dimension is (μm/m), and since the offset is $b = 1$ m, the combination of the two is a simple addition resulting in the positioning error along A_2 as shown in graph *(d)*. ▲

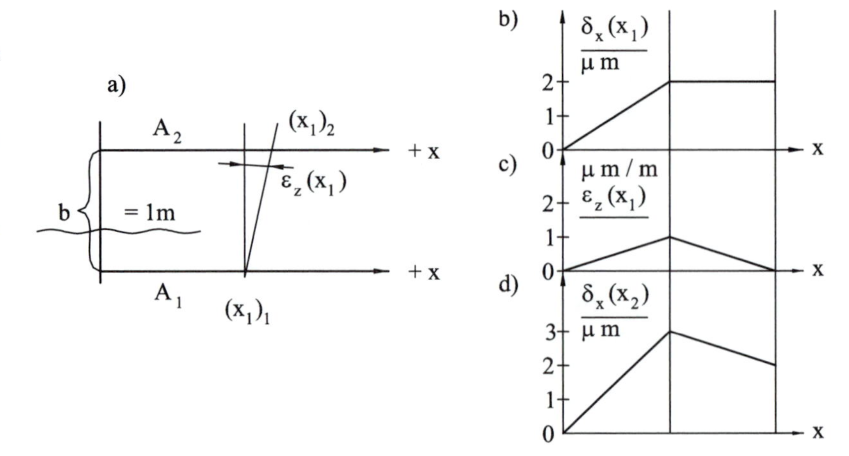

Figure 9.16
Determining the effect of the pitch error on positioning error. Error along line A_1 shown in b) when combined with pitch error given in c) results in error along A_2 shown in d), given the offset b = 1m.

In Fig. 9.10 we saw that the measurements of straightness of motions can be made by using straightedges and gages sliding on them. These gages are usually electronic and give a continuous signal that can be recorded on a chart so as to directly obtain graphs like that in Fig. 9.12. Straightness measurements can also be made using the laser interferometer, although its primary use is for the positioning measurements. The angular errors like roll $\varepsilon_x(x)$ and pitch ε_z (x) can be made by using precision levels, preferably electronic levels based on the principle of the pendulum, which also give a continuous electric signal. Errors of pitch and yaw, however, are usually measured with the use of the laser interferometer, and only roll has to be measured with levels or some other more complex means.

The various uses of the laser interferometer are shown in Fig. 9.17. In *(a)* the positioning measurement is explained in some detail. A He-Ne laser generates two frequencies of light f_1 and f_2. In the first beam-splitter, a part of the light is directed away as a reference signal; it is converted as a beat frequency $f = f_2 - f_1 = 2$ MHZ into voltage in the first photodetector. The light which passes through arrives at the second beam splitter where the frequency f_2 is diverted to a fixed corner cube reflector, and the frequency f_1 continues as a beam to be reflected by the moving external corner cube. The laser with the first beam-splitter is located in a housing mounted outside of the machine on a tripod.

The second beam splitter with the fixed reflector is a unit called the remote interferometer, which would be attached to the table of the machine. The external reflector is attached to the spindle. In this case actually the remote interferometer moves and the external reflector is stationary. However, what is important is that there is relative motion between the two. The beam returning from the fixed reflector still has frequency f_2, but the beam returning from the external reflector has frequency f_1 modified by the motion. The two returning beams are combined again in the splitter, and their sum arrives at the second photodetector, where it is converted into the Doppler signal $f_d = (f_1 \pm \Delta f_1)$, where the plus sign applies to the direction of motion shortening the distance traveled and the minus sign applies to travel in the opposite direction. The wavelength of the light in this beam is about 0.3 μm; thus, a motion of about 0.15 μm between the remote interferometer and the external reflector (the beam gets shorter or longer by twice the motion) is recognized as one fringe pulse between the Doppler signal and the reference.

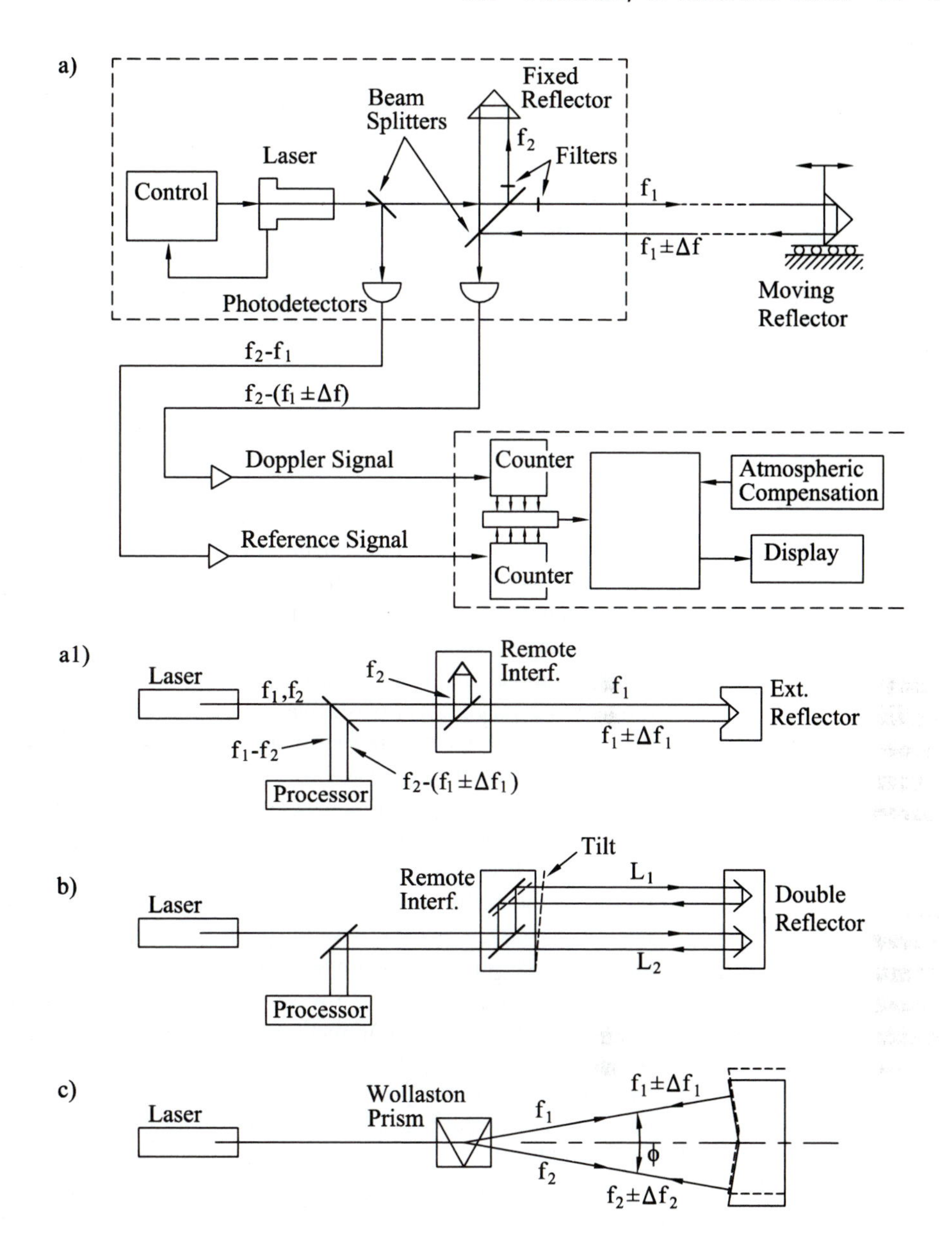

Figure 9.17 Measuring the errors of a) positioning, b) pitch, and c) straightness using the laser interferometer; a1) is a simplified version of diagram a).

Counting these pulses establishes the distance traveled. The accuracy of this measurement is in practice about ± 1 μm/m, and measurements at velocities up to 18 m/min can be made. Figure 9.17a is reproduced in a simplified way in diagram *(a1)*. It shows that, essentially, the interference is obtained between a beam of constant length that reflects from the fixed internal prism and a beam of varying length that reflects from the external prism.

The measurement of pitch is done according to Fig. 9.17b. There are two external corner cube reflectors located one above another. They may be attached to the spindle. The beam-splitter and a mirror are located in the remote interferometer, and interference is obtained between the beams directed to the two reflectors. If the body of the interferometer, which is attached to the moving table, tilts, a difference is generated between lengths L_1 and L_2 of the two paths, which is a measure of the tilt. For the measurement of yaw, the two reflectors are arranged side by side in a horizontal plane.

The measurement of straightness of motion is arranged as shown in Fig. 9.17c. The beam from the laser passes through a Wollaston prism that is attached to the spindle. The beam is split in two with a very small angle ϕ included between them. They are reflected, respectively, from two mirrors that are mutually inclined by the same angle ϕ. These mirrors are attached to the table, and their axis of symmetry represents the straightedge. If the table moves sideways in an error of straightness of motion, the length of the upper beam increases, and that of the lower beam decreases. An interference between the two beams provides for detecting this difference as a measure of the sideways displacement of the table.

9.3.2 Weight Deformations

Changes of the form of the structure due to variations in the weight of workpieces and due to the moving weights of traveling headstocks, tables, columns, and so on, must be kept within the limits dictated by the required geometric accuracy of the structure.

Various methods are employed to compensate for these deformations, to counterbalance the weights, or to minimize their effects. However, ultimately, there is always a remaining requirement of stiffness of the corresponding parts of the structure, which becomes one of the criteria for the design of the structure. Several examples that illustrate these points are given below.

First, look at Fig. 9.18a. The cross-rail of a portal-type milling machine deforms under the weight of the headstock. This deformation is largest when the headstock is in the middle. It consists of both bending and torsion, and it affects the straightness of motion of the milling cutter over the table. If the headstock is at one side during the vertical displacement of the cross-rail, the lead screw on this side will be more elongated, and the cross-rail will be clamped with a deviation from the horizontal that is indicated by the dashed line. Figure 9.18c shows that with two headstocks, the resulting deformation depends on the combination of the positions of both of them. The deformation of the cross-arm of a single-column structure like the one shown in *(b)* may be still more significant than in the portal type. Figure 9.18d shows that a heavy workpiece on the table of a horizontal boring machine may deform the table, saddle, and bed. This deformation will affect the positioning in the X direction high above the table, because it directly produces a considerable pitch error associated with the table motion.

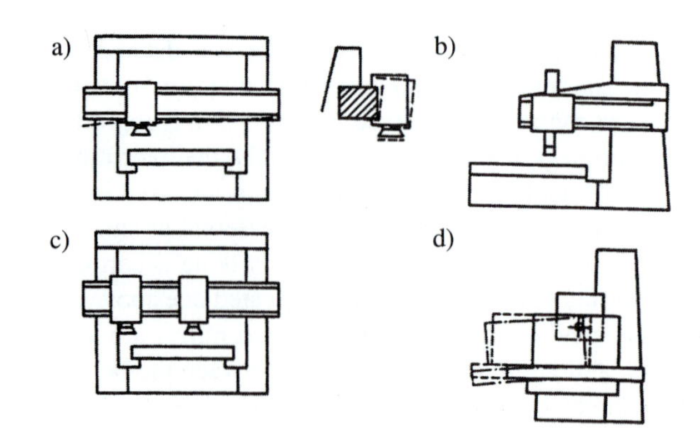

Figure 9.18
Examples of weight deformations: a) effect of headstock weight on the cross-rail of a portal-type milling machine; b) effects of two headstocks vary, depending on their individual and combined positions; c) a single-column structure; d) tilt of the table of a horizontal boring machine.

Various methods are employed to compensate for or counteract these deformations. Some of them are shown in Fig. 9.19, applied to the cross-rail of a portal-type machine. Diagram *(a)* indicates that the guideways may be preshaped with a deviation from straightness that will just be eliminated by deformation under the weight of the traveling headstock of Fig. 9.18a. It is not possible to properly preshape for the two headstocks of Fig. 9.18c. The method indicated in Fig. 9.19b uses an additional traverse above the cross-rail with connections between the two so that both are deformed, while the deformation of the cross-rail is such that it will just be compensated for by deformation under the weight of the headstock. The traverse need not be rigid and may be light and inexpensive. In the method shown in *(c)*, a carriage rides on the traverse and carries a lever with a weight that counterbalances the weight of the headstock. Thus the guideways on the cross-rail, which determine the straightness of the headstock motion, are not deformed under its weight at all.

These examples indicate clearly that although it is often possible to compensate for or alleviate the weight deformations, it is generally necessary to design the responsible parts of the structure with sufficient stiffness so as to keep the errors due to weight deformations within limits arising from the rather tight tolerances imposed on the geometric accuracy of the machine tools.

9.3.3 Deformations Under Cutting Forces

The cutting force produces deformation between tool and workpiece. If this deformation remained constant throughout the machining operation a workpiece of accurate form could be obtained. However, it varies, and its variation causes errors of form of the workpiece.

The deformation x depends on the cutting force F and on the stiffness k_a between tool and workpiece:

$$x = \frac{F}{k_a} \tag{9.11}$$

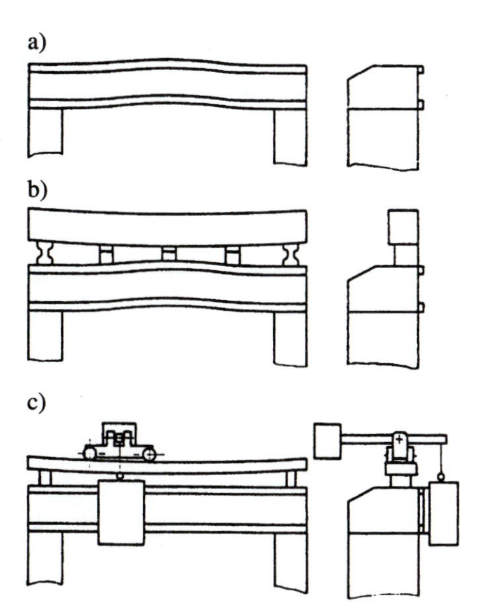

Figure 9.19
Compensating for and eliminating the deformation of a cross-rail: a) preshaped guideways; b) deforming the guideway from a separate traverse; c) headstock weight is counterbalanced from a separate traverse.

The definition of the stiffness k_a is given later. First, we can see that a variation of x is caused by either of the following conditions:

Type A: Variation in the cutting force F due to the varying depth of cut

Type B: Variation of stiffness k_a along the tool path

Often both types of variations a and b occur simultaneously; however, we will consider them separately. The two effects can then be superimposed.

Type A: Deformations Due to Variation of the Cutting Force

The blank of a workpiece entering a machining operation does not have an exact geometric form. The sections of a cylinder will not be exactly round; longitudinally it may be barrel-shaped or conical, and due to inaccurate clamping it will be rotating with a "run-out" e (see Fig. 9.20a). Similarly, a surface to be face-milled will not initially be perfectly flat. Or it may be that in one turning cut the tool path is so programmed that the initial diameter d will be reduced to two different diameters d_1 and d_2 (Fig. 9.20b).

For all the above examples, the cutting force varies during the cut. Consequently, the deformation between tool and workpiece also varies, and initial form errors will be "copied" onto the machined surface. In Fig. 9.20c the cutting force, F, is a function of the depth of cut a and of feed per revolution f_r. Commonly this relationship is expressed as

$$F = K_s f_r\, a$$

The coefficient of proportionality, r_a, between depth of cut a and force F is called the "cutting stiffness" with respect to depth of cut:

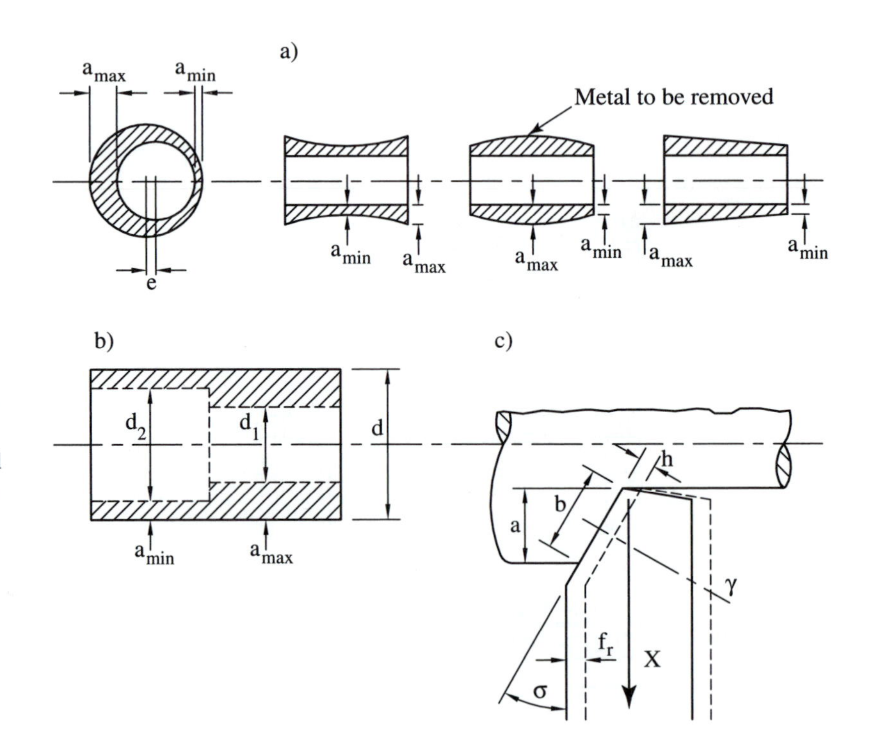

Figure 9.20
Cutting force varies due to variation of depth of cut: a) the blank of a cylindrical workpiece may be rotating off center ("run-out"), and it may be concave, barrel-shaped, or conical; b) machining a cylinder to two different depths, c) expressing the cutting force as function of the depth of cut a.

$$r_a = \frac{F}{a} = K_s f_r \tag{9.12}$$

and it can be used to express the cutting force as

$$F = r_a a \tag{9.13}$$

The cutting force has a direction that depends on the angle σ. It lies in a plane γ that is approximately perpendicular to the cutting edge, and it has a larger tangential component. This force causes a deflection of the system. The component x of this deflection is normal to the surface left behind the cut. It is this deflection component that affects the dimension, and hence the geometric form, of the machined workpiece.

Therefore, a particular stiffness k_a related to the Type A effect now being discussed is defined as a ratio between a force F acting in the direction of the cutting force and a deflection in the direction x, as expressed by Eq. (9.11).

The ratio of the cutting stiffness r_a to the machine stiffness k_a has a special significance. We will denote it μ:

$$\mu = \frac{r_a}{k_a} \tag{9.14}$$

Let us now consider, as an example, a longitudinal cut as shown in Fig. 9.21a.

The nominal depth of cut varies between a_{min} and a_{max}. The difference between these two values can be denoted as the maximum initial error of form of the workpiece Δ:

$$\Delta = a_{max} - a_{min} \tag{9.15}$$

The deflection between the tool and the workpiece varies correspondingly between x_{min} and x_{max}. The actual depths of cut a' are obtained as the difference between the nominal depths a and deflection x.

$$\begin{aligned} a'_{max} &= a_{max} - x_{max} \\ a'_{min} &= a_{min} - x_{min} \end{aligned} \tag{9.16}$$

Using Eq. (9.13) to express the force, one has to apply the values a':

$$\begin{aligned} F_{max} &= r_a a'_{max} \\ F_{min} &= r_a a'_{min} \end{aligned} \tag{9.17}$$

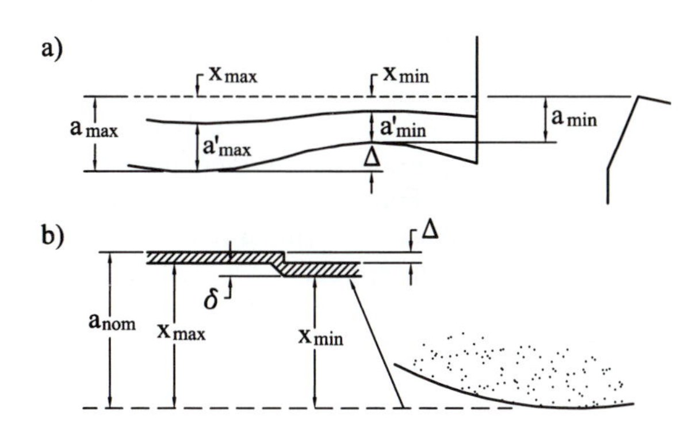

Figure 9.21 Copying the error of form of workpiece: a) turning a workpiece with varying depth of cut a, deflection x varies accordingly; b) grinding of a step of size Δ; the error δ left after the grinding pass is still large because of a high ratio of cutting stiffness over machine stiffness.

and the deflections are obtained by means of the machine stiffness k_a:

$$x_{max} = \frac{F_{max}}{k_a} = \mu\, a'_{max}$$
$$x_{min} = \frac{F_{min}}{k_a} = \mu\, a'_{min} \tag{9.18}$$

If there were no deflections, the machined surface would have no form error. As it is, there will be an error of form δ left after the cut which equals the maximum variation of the deflections:

$$\delta = x_{max} - x_{min} \tag{9.19}$$

Inserting Eq. (9.18) in Eq. (9.16), we have

$$a'_{max}\,(1 + \mu) = a_{max}$$
$$a'_{min}\,(1 + \mu) = a_{min}$$

Further, using Eq. (9.18), we have

$$x_{max}\frac{(1 + \mu)}{\mu} = a_{max}$$
$$x_{min}\frac{(1 + \mu)}{\mu} = a_{min} \tag{9.20}$$

Subtracting the second Eq. (9.20) from the first one yields

$$\delta = \frac{\Delta\,\mu}{1 + \mu} \tag{9.21}$$

Eq. (9.21) relates the error of form after the operation to the error of form before the operation. The coefficient of proportionality between the two could be called the rate of copying i of the workpiece form error,

$$i = \frac{\mu}{1 + \mu} \tag{9.22}$$

The rate of error-copying depends on the ratio μ of cutting stiffness r_a over machine stiffness k_a. Later we will see that this ratio is rather small for most cutting operations. Typically, for turning or boring, it may be $\mu = 0.01$. In that case it could be written

$$\mu \ll 1, \quad i \simeq \mu \tag{9.23}$$

Eq. (9.12) shows that r_a depends on the feed. Correspondingly, there will be a value μ_r for roughing larger than that for finishing μ_f, and after one roughing and one finishing cut, the rate of error-copying decreases to,

$$i = \mu_r\,\mu_f \tag{9.24}$$

Even if the value of μ remained the same in both cuts, we would have $i = \mu^2$ after two cuts, $i = \mu^3$ after three cuts, and so on. The initial form error decreases geometrically with the number of cuts. For the above-mentioned value $\mu = 0.01$, an initial error of form of 5 mm will decrease to 50 μm after one cut, to 0.5 μm after two cuts.

On the other hand, in grinding operations the cutting stiffness is rather high. A value of $\mu = 20$ is not unusual. Then we have

$$i = \frac{\delta}{\Delta} = \frac{\mu}{1 + \mu} = 0.95 \tag{9.25}$$

This is shown in Fig. 9.21b, where we can see that after one pass of grinding, the initial form error diminishes very little; many passes are needed to obtain the required accuracy of form.

Machine Stiffness k_a and Direction Orientation

Let us now analyze the character and typical values of the machine stiffness k_a, especially of the directional orientation involved. We have explained that the particular stiffness denoted as k_a relates the cutting force to the X component of the relative displacement between tool and workpiece. The direction X is normal to the surface left behind the tool (created by the tool).

Machine stiffness depends on the configuration of the machine tool and on the position of the cutting process in the machine tool structure. As this position changes during the tool motion, stiffness k_a varies during the cutting operation. The effect of variation in k_a is considered in the next section. For periodically varying forces, the stiffness k_a is considered to be a "dynamic" stiffness ratio between force and deflection; however, in the current discussion the variation of cutting force is slow, and k_a can be considered a "static" stiffness.

The directional orientation of the cutting process in the machine is involved in the following ways:

(a) The structure is usually most flexible in a certain direction; dynamically, the system only deforms in certain "modes." Depending on the direction of cutting, the projection of the force in this sensitive direction may be strong.

(b) The deflection of the system has to be projected into the "sensitive" direction X to be effective. The direction normal to the surface left behind the tool is indicated as sensitive.

Let us illustrate this by using the example of a lathe (Fig. 9.22a). The cutting force F acts between the tool and the workpiece (Fig. 9.22b) in a plane perpendicular to the cutting edge, which itself has a side-cutting-edge angle σ. The decisive direction X of the deformation affects the dimension and form of the workpiece and is perpendicular to the surface left behind the tool. The flexible direction of the displacement of the workpiece is any radial direction. Force F has a component F' in the radial plane (Fig. 9.22c) and another component F'' in the direction X, which produces the displacement x_w. The action of the force F' on the tool is depicted in Fig. 9.22d. Depending on the design of the slides and of the toolholder, the flexible direction of the deformation of this part may be direction T. Force F' is projected into F''', which produces the displacement δ_t with a component x_t in the direction X. In the diagram this displacement is shown to be "negative," that is, into the workpiece. Most often it will be positive—out of the workpiece. The resulting displacement will be the sum.

$$x = x_w + x_t \tag{9.26}$$

Obviously, the angle σ plays an important role, and for $\sigma = 0$ there will be almost no deflection x. In reality there are also some smaller flexibilities in directions other than those indicated, and even with $\sigma = 0$, the force has an X component due to the nose radius of the tool and therefore a small x-deflection. Note, however, that the tool deflection does not depend only on the X component of the force.

The lengthy example of the lathe should have indicated the importance of the directional orientation. We will not discuss in detail the directional orientation and the corresponding value of machine stiffness k_a for all types of machine tools, but we mention a few other examples. In a vertical milling machine used for face milling, the horizontal direction is the flexible direction most sensitive to the bending of the column, the twist of the attachment between headstock and column, and the bending of the spindle. Thus, the flexible direction is almost perpendicular to the sensitive vertical direction. In face milling on a horizontal spindle, floor-type machine, the flexibility of the column (at its base) may cause significant deflections in the sensitive direction, which is normal to the cut surface.

Figure 9.23a shows the cases of a boring bar and boring spindle. The effect of the side-cutting-edge angle is here similar to the case of the lathe. The main concern is the inevitably low stiffness of the boring bar. In addition, this case often involves dynamic

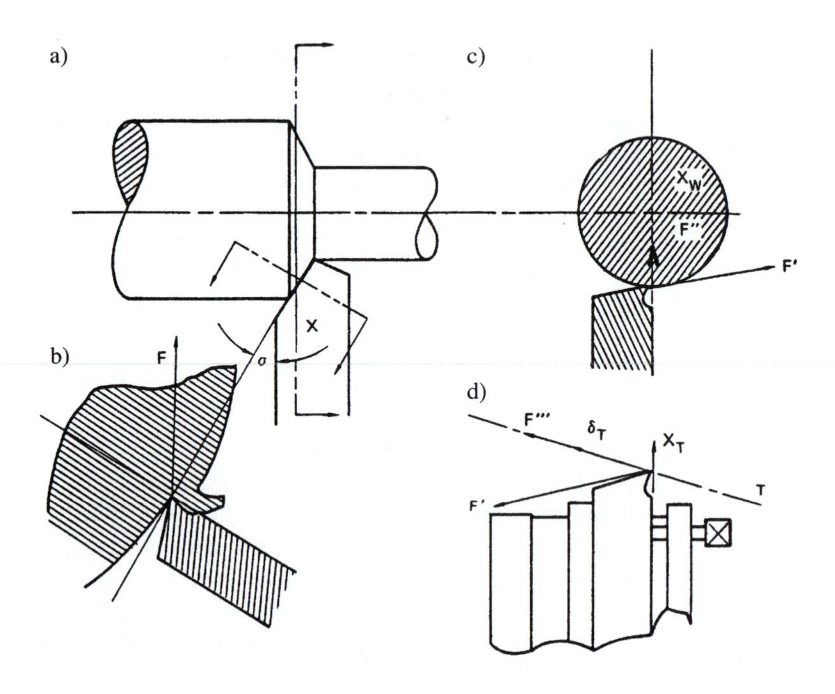

Figure 9.22
Direction of forces and deformations on a lathe: a) top view; b) section perpendicular to the cutting edge; c) section perpendicular to axis of rotation; d) force on the tool and deflection of the tool post; it may cave in.

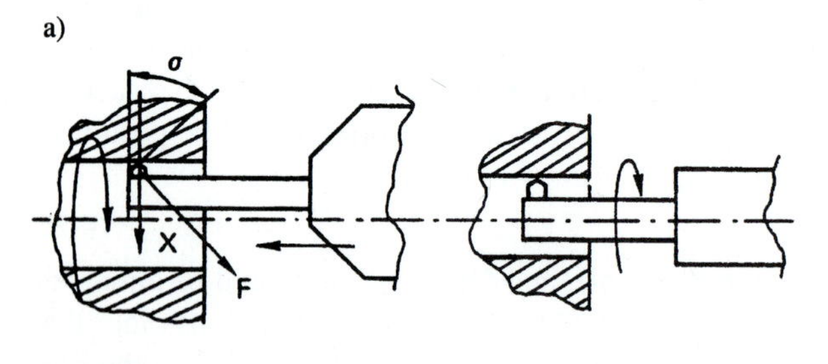

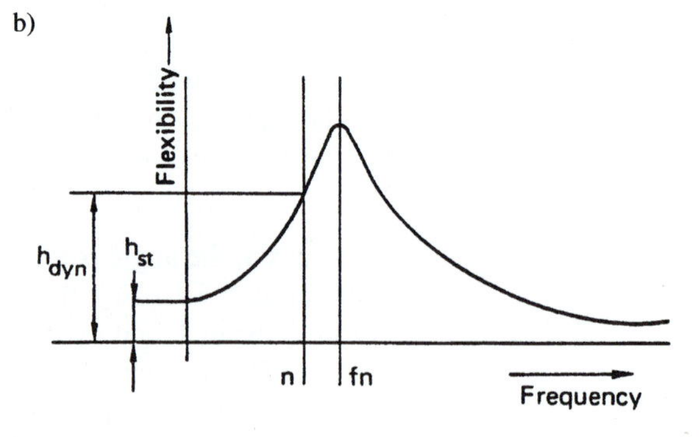

Figure 9.23
Directional orientation and dynamic effect in boring: a) effect of side-cutting-edge angle σ; b) effect of the frequency of the form error in relation to natural frequency of the boring bar.

flexibility. If a boring spindle produces a periodic force, say a variation of force once per revolution, and its frequency n is close to the natural frequency f_n of the bar, then the dynamic flexibility h_{dyn} may be several times higher than the static flexibility h_{st} and must be considered. This is expressed by the displacement versus speed characteristic in Fig. 9.23b. However, for a small side-cutting-edge angle σ, the radial force component in the flexible direction perpendicular to the boring bar axis is small, $F_r = F \sin \sigma$, and any deflection in the sensitive direction is then mainly due to the radial force generated by the radius of the tool tip only.

In some operations, like boring or end milling, the values of μ are large enough to make the problem of form error-copying significant. These are illustrated in the following examples.

EXAMPLE 9.3 Copying of Form Error of Workpiece ▼

Turning

Let us assume turning of steel 1045 in one roughing cut, with feed $f_r = 0.25$ mm/rev, and one finishing cut, with feed $f_r = 0.1$ mm/rev. The side-cutting-edge angle is $\sigma = 45°$. The initial form error is $\Delta = 4$ mm.

The value of the cutting stiffness depends on the value of the specific force K_s, found in Table 7.1:

$$K_s = 2600 \text{ N/mm}^2$$

For the roughing cut: $r_{a,r} = 2600 \times 0.25 = 650$ N/mm.

For the finishing cut: $r_{a,f} = 2600 \times 0.1 = 260$ N/mm.

The stiffness k_a is difficult to determine. As it was explained in Fig. 9.22, this stiffness might even become infinite if the tool "caves in" as much as the workpiece deflects. Without any further analysis, we will give a value based on experience. Taking into account the given side-cutting-edge angle, we shall use

$$k_a = 25{,}000 \text{ N/mm}$$

Then $\mu = r_a/k_a$ is, for roughing, $\mu_r = 0.026$, and for finishing, $\mu_r = 0.010$. The total ratio is

$$\frac{\delta}{\Delta} = \mu_r \, \mu_f = 0.00027$$

The final error will be $\delta = 1.0\ \mu$m, which is practically negligible.

Boring

Let us assume a boring bar with diameter $d = 50$ mm and length $l = 250$ mm, made of steel with $E = 2 \times 10^5$ N/mm^2 as the modulus of elasticity. Consider a side-cutting-edge angle $\sigma = 45°$, a roughing feed rate $f_r = 0.2$ mm/rev, and a finishing feed rate $f_f = 0.1$ mm/rev. The initial error is $\Delta = 4$ mm. How many finishing cuts will be necessary to obtain $\delta < 1\ \mu$m? Neglect any dynamic effects.

The radial stiffness at the end of the boring bar, under the assumption of an infinitely rigid clamping, is

$$k_r = \frac{EI}{l^3} = \frac{3E\pi d^4}{64 l^3} = 1.18 \times 10^4 \text{ N/mm}$$

Taking into account the flexibility of the clamping of the bar, we will consider only 60% of this value. Assuming the radial component of the cutting force $F_r = 0.3\,F_t$, then the oriented stiffness at the end of the bar (radial deflection due to cutting force) is

$$k_a = \frac{k_r}{0.3 \cos \sigma} = 4.7 k_r$$

$$= 4.7 \times 0.6 \times 1.18 \times 10^4 = 3.3 \times 10^4 \text{ N/mm}$$

The cutting stiffness is

$$r_{a,r} = K_s f_r = 2600 \times 0.2 = 520 \text{ N/mm}$$

$$r_{a,f} = K_s f_f = 2600 \times 0.1 = 260 \text{ N/mm}$$

and $\mu_r = 0.016$, and $\mu_f = 0.008$. After one roughing and one finishing cut, we have

$$\delta = \Delta\, \mu_r\, \mu_f = 0.5\ \mu m$$

One finishing pass is sufficient. ▲

Type B: Deformations Due to Variation of Flexibility

During a cutting pass the cutting force varies and often also the flexibility between the tool and workpiece varies. We have decided to deal with the two effects separately. They can always be superimposed. The former effect was discussed in the preceding section. Now, it is assumed that the workpiece has no initial form error and, essentially, the cutting force remains constant throughout the cut. However, it is recognized that in some operations the flexibility h_a (the inverse of stiffness: $h_a = 1/k_a$) between the tool and workpiece varies quite significantly. A typical example is a boring operation with a rotating tool and nonrotating workpiece, as shown in Fig. 9.24. Overhang boring with a long and slender boring spindle is shown in *(a)*; throughout the operation the spindle remains maximally extended and the feed motion is executed by the table. In *(b)* the same operation is carried out in a different mode: the table is stationary, and the spindle moves from a small extension to the largest one. In *(a)* the flexibility h_a remains practically constant, and in *(b)* it varies from h_{min} to h_{max}. Similarly, the turning operation on a vertical boring mill is carried out by the feed of the ram. There is a significant variation of flexibility from h_{min} to h_{max}. In external cylindrical grinding of a long shaft, the deflections on the workpiece are due partly to the deflections of the centers supporting it at the ends and partly due to the workpiece itself. The tailstock center is more flexible. The spindle center deflects due to the reaction to the grinding force and also due to the force acting on the driving pin. Correspondingly, this deflection between the grinding wheel and the ground shaft varies during one revolution of the workpiece and also along its length.

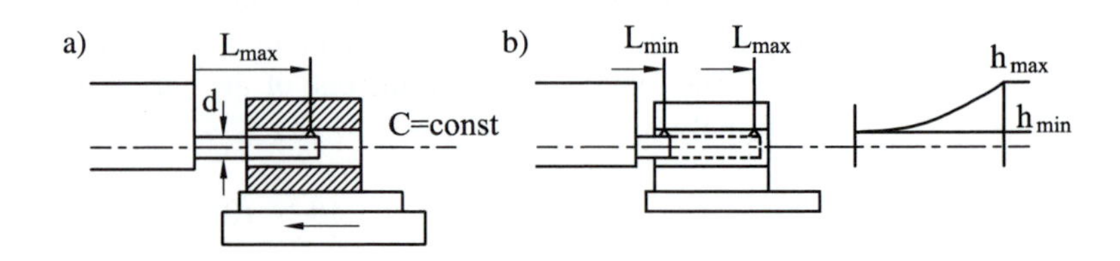

Figure 9.24
Variation of flexibility in boring: a) using table feed, flexibility is constant; b) extending the boring spindle, flexibility varies strongly.

The error of form after the machining operation is determined as the difference between the maximum and minimum deflections during the operation:

$$\delta = F(h_{max} - h_{min}) \tag{9.27}$$

Obviously, this error depends almost solely on the last, finishing pass, and it does not matter how many finishing passes are made. This type of error may be quite large.

To illustrate, let us consider a boring spindle with a diameter of 63 mm extending from $L_{min} = 50$ mm to $L_{max} = 350$ mm. The direct flexibility at the end of the spindle varies between 1.4×10^{-5} mm/N and 9.0×10^{-5} mm/N. Assuming a finishing depth of cut $a = 1$ mm and a finishing feed $f_r = 0.1$ mm/rev, a radial force F_r that produces the radial deflection x in combination with the direct flexibilities will be about 100 N. The resulting error of cylindricity would be

$$\delta = 100 \times 7.6 \times 10^{-5}\ \text{mm} = 7.6\ \mu\text{m}$$

Compare this with an error that would be caused by copying an initial out-of-roundness of 2 mm with the spindle at its maximum extension. The cutting stiffness is $r_a = 100$ N/mm, the spindle stiffness is $k_a = 1/(9 \times 10^{-5}) = 1.1 \times 10^4$ N/mm, and the ratio μ is $\mu = 0.009$. After one pass, the original error of 2 mm decreases to 18 μm, and after two passes to only 0.16 μm. This is 47 times less than the error due to the variation of the spindle flexibility. This example illustrates how important it is to do boring by the method shown in *(a)* rather than the one shown in *(b)*. In general, it is necessary to avoid situations in which large variations of flexibility arise.

9.4 REVIEW OF FUNDAMENTALS OF MECHANICAL VIBRATIONS

9.4.1 Vibrations: Natural, Forced, Self-excited

It is important to distinguish the three classes of mechanical vibrations as given in the title of this section: natural, forced, and self-excited vibrations.

Natural, or *free, vibrations* occur (see Fig. 9.25a) if a vibratory system, represented in the diagram by a simple combination of a spring, a dashpot, and a mass, is first deflected from its equilibrium and then left free to move. The result is a motion with an *amplitude* decaying, and a *frequency* equal to the natural damped frequency f_d of the system.

There is *no external force* acting on the system. The differential equation of this motion is

$$m\ddot{x} + c\dot{x} + kx = 0 \tag{9.28}$$

and it has a solution depending on the initial conditions. For an initial displacement x_0 and no initial velocity, this motion is

$$x = x_0 e^{-\lambda t} \cos(2\pi f_d t + \phi) \tag{9.29}$$

where the exponent λ, the frequency f_d, and the phase shift ϕ depend on the parameters of the system: the mass *m,* the damping coefficient *c,* and the spring stiffness *k.* Natural vibrations have little practical significance because of their short, transient character.

Forced vibrations are carried out if there is an *external, periodic force F* acting on the system:

$$F = F_0 \cos 2\pi ft$$

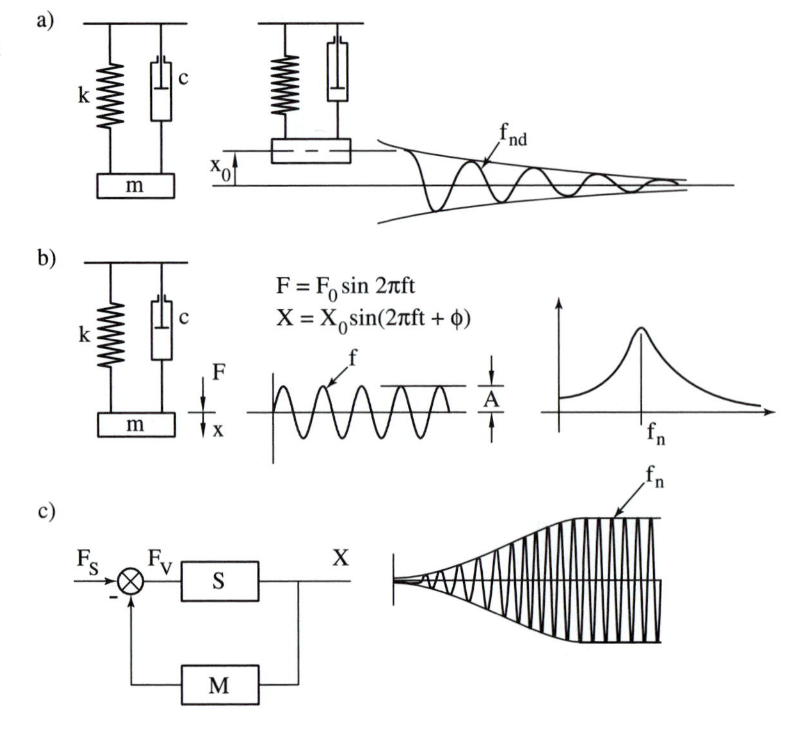

Figure 9.25
a) Free vibration: no external force, frequency natural, amplitude decays. b) Forced vibration: frequency that of the force, amplitude steady, dependent on the ratio of force frequency to natural frequency. c) Self-excited vibration: no external periodic force, frequency close to natural, amplitude increases to saturation.

The resulting motion, after an initial transitory period in which there is also a natural vibration present, reaches a steady state in which (1) the *amplitude A* is constant, and (2) the *frequency* of vibration f is *equal* to the frequency f of the exciting force.

The amplitude A of the vibration depends on the amplitude F_0 of the force, on the stiffness k and damping c of the system, and on the ratio f/f_n of the exciting frequency over the natural frequency of the system. For $f = f_n$, a *resonance* is obtained at which the amplitude of vibrations is maximum (see Fig. 9.25b). More details about this are given in Section 9.4.2.

Forced vibrations are present in all kinds of machinery where periodic forces are generated. These are mostly due to unbalanced rotating shafts or to reciprocating motions. Usually they are significant only in the cases of resonance, and the problem may be solved by changing the natural frequency so as to avoid resonance.

In machining, the only practically significant cases of forced vibrations exist in finish and fine machining like fine boring and grinding, where they produce waviness of the machined surface. Again, this is a substantial problem only if there is a resonance. For instance, if the speed of the grinding spindle (rev/sec) is equal to a natural frequency f_n (Hz) of an important structural mode of the grinding machine, then the ground surface will show waviness due to even a slight imbalance of the grinding wheel.

In milling, forced vibrations are excited by the periodic component of the cutting force. Consequently, they are synchronized with the periodic generation of the machined surface and they do not cause waviness of that surface. However, they affect the location of the machined surface so that it becomes overcut or undercut. In end milling with cutters with helical teeth, the forced vibrations cause nonflatness of the machined surface as measured in the direction parallel with the cutter axis, as discussed in Section 9.5.

Self-excited vibrations develop due to a built-in mechanism (providing for a closed-loop relationship) in the system. This mechanism is capable of modulating a steady, nonperiodic external energy source and generating a periodic force through the vibration of the system such that it sustains the vibration; hence, it is self-exciting.

As a basic example, consider a clock. There is a steady source of energy: the pulling weight. The machinery of the clock contains the so-called escape mechanism, which transforms the steady force on a toothed wheel into a periodic force sustaining the periodic motion of the pendulum. If the pendulum is stopped, the periodic force ceases. Thus, the characteristic features of self-excited vibrations are (1) the *amplitude increases,* until it stabilizes on a constant value (due to a nonlinearity in the system); (2) the *frequency of the vibration is equal to or close to the natural frequency of the system;* (3) there is *no independent, external periodic force;* (4) there is a steady energy source from which the system derives a periodic force through its vibration.

In distinction to forced vibrations, which exist whenever there is a periodic force acting on the system and where the relevant question concerns the amplitude of the vibration, self-excited vibrations either exist or they do not. If they occur, they will grow until some nonlinearity in the system provides a limit.

Self-excited vibrations are very common. Let us give a few examples from sound generation: playing the violin or flute, whistling, and so on. In the first case a bow is drawn steadily along a vibrating string which (through the "negative" friction-velocity characteristic) derives a periodic force component sustaining its vibration; the frequency is equal to the natural frequency of the string. In wind instruments, the steady air flow is modulated by a vibrating element that sustains its own vibration. In servomechanisms and, in general, in systems with feedback, growing vibrations are often generated; it is said that the system becomes unstable. Then it is mostly a question of *gain* whether the system behaves in a *stable* or *unstable* fashion. Of decisive interest are the *conditions for the limit of stability.*

These characteristics are symbolically expressed in Fig. 9.25c. First, the mutual interdependence is shown in the closed loop of vibration X and the variable force F_v. There is an external steady force F_s which is modulated into the variable force F_v acting on the system S and producing its vibration X. This vibration, through a mechanism M, which is particular to every case, produces a force F that causes the modulation of F_s. The oscillations may or may not be sustained, depending on the gain in the feedback loop. If the gain is larger than a limit value, the vibration will grow until the phenomenon saturates and the amplitude of the vibration stabilizes.

We are not going to discuss the conditions for stability in general. However, we will indicate the mechanisms of self-excitation in metal cutting in Section 9.6 and derive the condition of the corresponding limit of stability. First, however, it is necessary to recapitulate a little more of the formulations of mechanical vibrations.

9.4.2 Harmonic Variables

In order to deal efficiently with the following exercises concerned with vibrations, we will first recapitulate the mathematics of expressing harmonic variables. The two basic harmonic variables, as shown in Fig. 9.26, are

$$x_1 = A \cos \omega t, \qquad x_2 = A \sin \omega t \tag{9.30}$$

where A is the amplitude of the motion, $\omega = 2\pi f$ is the "circular frequency" (rad/sec), and f is the frequency of the motion as expressed in Hz (cycles/second). It is obvious that the two are identical except that x_2 is back-shifted in time against x_1 by a time

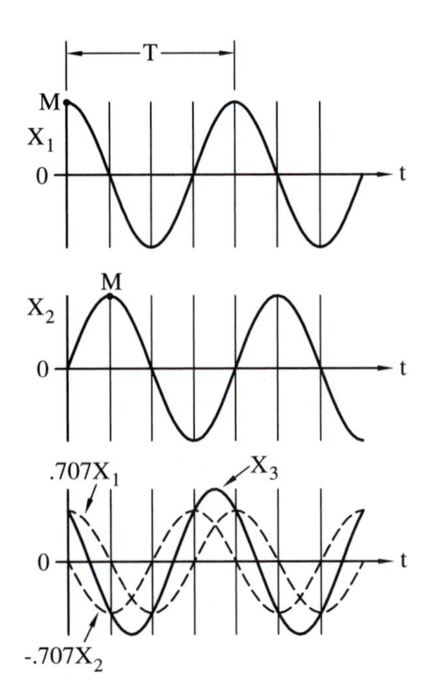

Figure 9.26
Harmonic variables in the time domain: cosine, sine, and their combination.

corresponding to a quarter of the period T. Thus, for example, the positive peak M (the maximum) is found on x_1 earlier by $T/4$ than on x_2. The functions of sine and cosine are periodic over 2π, or in degrees, over 360°. Correspondingly, the period $T = 1/f$ is such that

$$\omega T = 2\pi$$
$$T = \frac{2\pi}{\omega}, \qquad \omega = 2\pi f \tag{9.31}$$

The time shift ϕ between x_1 and x_2 is expressed as a quarter of the period:

$$\phi = \frac{\pi}{2}$$

and

$$\sin \omega t = \cos\left(\omega t - \frac{\pi}{2}\right) \tag{9.32}$$

It is also said that the sine lags the cosine by 90°. A harmonic function may be a sine or a cosine, or a mixture of the two, depending on its state at $t = 0$, and depending on when the timing began. Thus, a function shown in x_3, for which recording begins an eighth of a period ($\phi = 45°$) later than the cosine function, will be

$$x_3 = A \cos(\omega t + \phi) = A(\cos \omega t \cos \phi - \sin \omega t \sin \phi)$$
$$= A \cos \phi \cos \omega t - A \sin \phi \sin \omega t$$

and for $\phi = 45°$, $\sin \phi = \cos \phi = 0.707$,

$$x_3 = 0.707A(\cos \omega t - \sin \omega t) = 0.707(x_1 - x_2)$$

For another phase shift ϕ, a harmonic function would be a sum of sine and cosine functions of the same frequency but of different amplitudes.

All this is simple to follow by using vectorial representation in the "complex plane" (see Fig. 9.27a). A harmonic function will be represented by a vector of length A rotating with angular velocity ω:

$$X = Ae^{j\omega t} = A(\cos \omega t + j \sin \omega t).$$

The sine component is taken as imaginary and the cosine component as the real part of the vector. The actual value of the function at any instant is its real part, the cosine component. However, the complex presentation permits us to express the phase relationships. If in Fig. 9.27b the timing begins with the vector in an initial position indicated by a phase angle ϕ, which determines the "complex amplitude" X, we have,

$$Ae^{j(\omega t + \phi)} = Ae^{j\phi}e^{j\omega t} = Xe^{j\omega t}$$

where $X = Ae^{j\phi}$.

Figure 9.27c shows that a sine function is such that the initial position of the vector is X_1 ($\phi = -\pi/2$); a cosine function starts at X_2 ($\phi = 0$), and a function with $\phi = \pi/4$ starts at X_3.

With these representations, it is usual to generally express a harmonic function as

$$x = Xe^{j\omega t} \tag{9.33}$$

where the complex amplitude X is itself a vector:

$$X = Ae^{j\phi} \tag{9.34}$$

Thus we have $X_1 = Ae^{-j\pi/2}$, $X_2 = A$, and $X_3 = Ae^{j\pi/4}$.

9.4.3 Basics of Vibrations: Transfer Function of a System with a Single Degree of Freedom

The transfer function (*TF*) is a concept in forced vibrations; it expresses the relationship between the periodic force F and the vibration X it produces. It is a very descriptive characteristic of a vibratory system; thus it will be used in dealing with self-excited vibrations as well.

First, we consider the *TF* of the simplest vibratory system, such as that shown in Fig. 9.25b. It consists of a mass m attached to a spring with stiffness k that is fixed at the other end. In parallel with the spring there is a dashpot (damping element) with a damping coefficient c. The force produced by the dashpot is proportional to the velocity of the motion (viscous damping). A harmonically variable force F is acting on the mass of the system. Let us write

$$F = F_o e^{j\omega t} \tag{9.35}$$

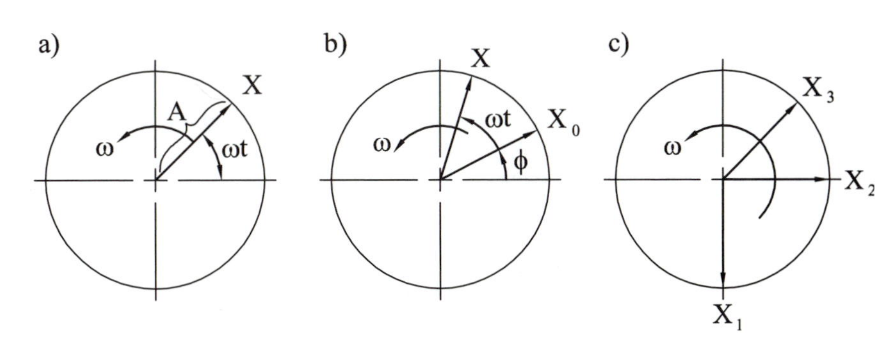

Figure 9.27 Harmonic variables as vectors in the complex plane. a) Vector X of size A rotates with angular speed ω. b) Initial position of vector X determined by phase angle ϕ. c) The sine vector starts at X_1, the cosine vector at X_2; X_3 is a combination of both.

where F_o is the amplitude and ω the frequency of the force. The differential equation of motion of the system expresses the balance of forces acting on mass m:

$$m\ddot{x} + c\dot{x} + kx = F_o e^{j\omega t} \tag{9.36}$$

Neglecting a transient solution, the steady-state solution has the form

$$x = Xe^{j\omega t} \tag{9.37}$$

which gives $\dot{x} = j\omega x$, and $\ddot{x} = -\omega^2 x$; then Eq. (9.36) solves for the complex amplitude X,

$$X = \frac{F_o}{k - m\omega^2 + jc\omega}$$

and the transfer function G, being the ratio of the output amplitude X over the input amplitude F_o:

$$G = \frac{X}{F_o} = \frac{1}{k - m\omega^2 + jc\omega} \tag{9.38}$$

Introducing the following notation,

$k/m = \omega_n^2$, which is the square of the natural frequency of the system;

$c/2\sqrt{km} = \zeta$, which is the damping ratio of the system;

we have

$$G = \frac{X}{F_o} = \frac{1/k}{1 - \omega^2/\omega_n^2 + 2j\zeta\omega/\omega_n} \tag{9.39}$$

In Eq. (9.39) the TF is expressed as a function of the ratio of the frequency ω of the exciting force over the natural frequency ω_n of the system. Instead of the circular frequencies, we could use the frequencies in Hz:

$$G = \frac{X}{F_o} = \frac{1/k}{1 - (f/f_n)^2 + 2j\zeta f/f_n} \tag{9.40}$$

The TF is a ratio of the vectors of the two complex amplitudes. It is customary to choose the vector of the force amplitude as real (like the vector X_2 in Fig. 9.27) and to determine the phase ϕ by which the vector X lags behind the vector F_o.

For brevity of expression, let us denote $p = f/f_n$ and rewrite Eq. (9.40):

$$G = \frac{X}{F_o} = \frac{1/k}{1 - p^2 + 2j\zeta p} \tag{9.41}$$

One form of expressing the TF is by specifying the magnitude $|G|$ and the phase ϕ as functions of (p):

$$|G| = \frac{1/k}{\sqrt{(1 - p^2)^2 + 4\zeta^2 p^2}} \tag{9.42}$$

$$\phi = \text{atan2}\,\frac{-2\zeta p}{1 - p^2} \tag{9.43}$$

Another form of the TF is obtained using the real Re[G] and imaginary Im[G] parts as functions of (p):

$$\text{Re}[G] = \frac{(1/k)\,(1 - p^2)}{(1 - p^2)^2 + 4\zeta^2 p^2} \tag{9.44}$$

$$\text{Im}[G] = \frac{-\,(1/k)\,(2\zeta p)}{(1 - p^2)^2 + 4\zeta^2 p^2} \tag{9.45}$$

Equations (9.42), (9.44), and (9.45) are expressed by the graphs in Fig. 9.28. Damping ratios are marked by the letter z.

The TF can also be expressed by plotting the vector G in the complex plane; see the graph in Fig. 9.29. The frequency ratio (f/f_n) is now noted in the individual points along the graph. For any frequency, the length of the vector $|G|$ and its phase ϕ can be seen as well as its real and imaginary parts.

Special features of the TF are obtained at several frequencies:

1. At $(f/f_n) = 0$, the value of the TF is $G = 1/k$. It is purely real, and it is the static flexibility.
2. At $(f/f_n) = 1$, the phase shift is exactly $\phi = \pi/2$. This is called the resonance. The value of G is purely imaginary, and it is very close to being the maximum negative (the minimum) imaginary. Its value is

$$G = \frac{-j}{2k\zeta} \tag{9.46}$$

3. The maximum of the magnitude $|G|$ occurs close to resonance; to be precise, it occurs at

$$\frac{f}{f_n} = \sqrt{(1 - \zeta^2)}$$

and it is approximately,

$$|G|_{\max} \simeq 1/2k\zeta$$

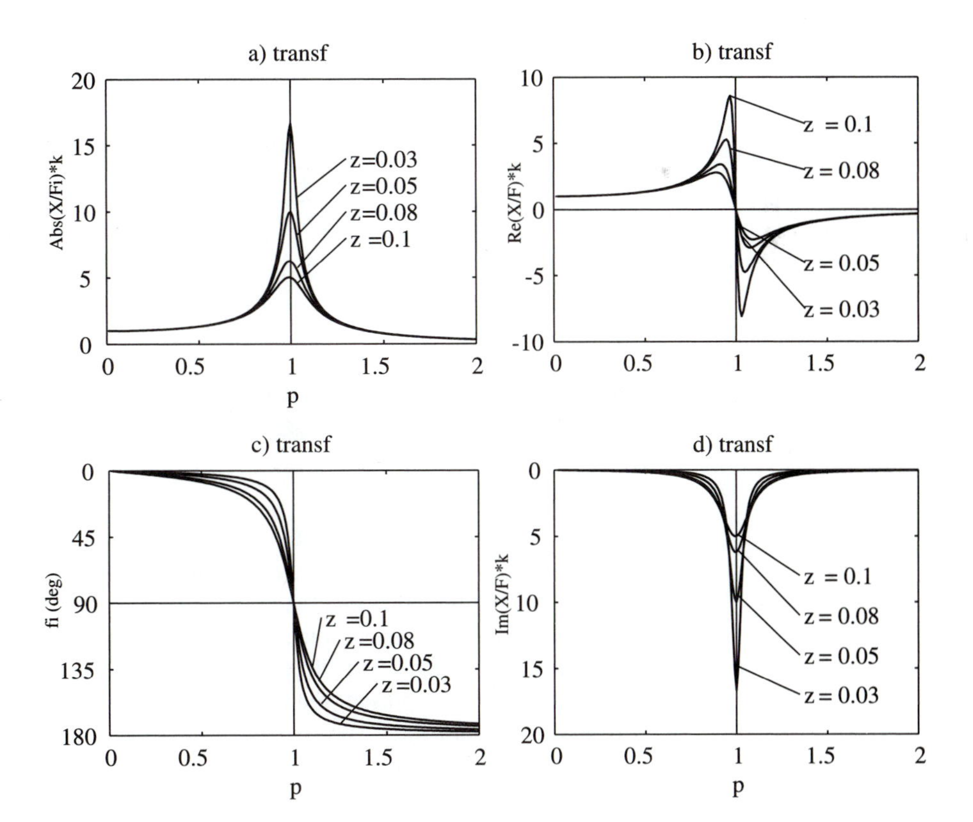

Figure 9.28 Single-degree-of-freedom transfer function expresses the ratio of harmonic vibration amplitude over the harmonic force amplitude, as a function of force frequency: a), c) magnitude and phase; b) real; and d) imaginary parts of the transfer function.

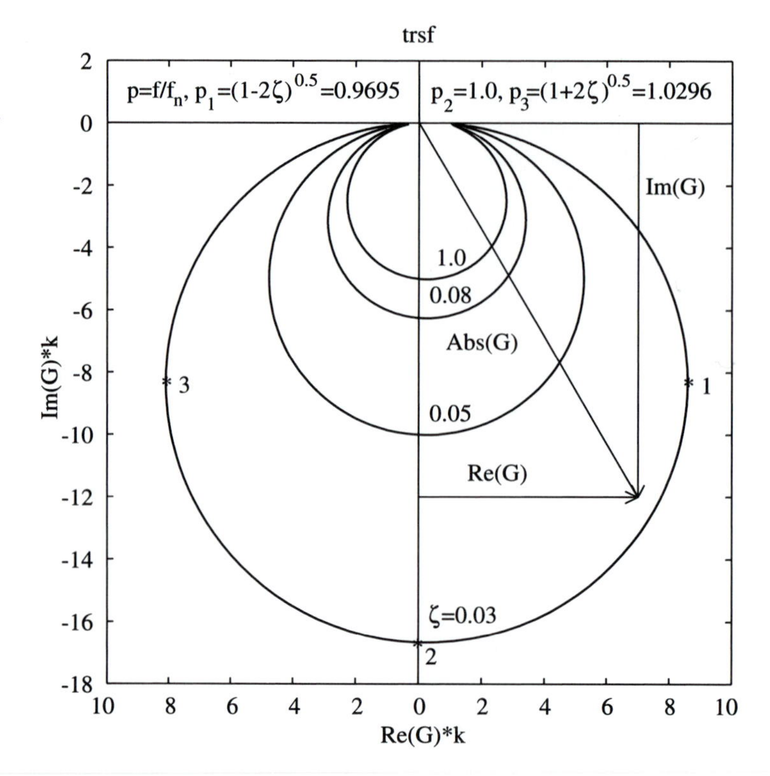

Figure 9.29
The transfer function defined in the complex plane. Frequency (or its ratio p over natural frequency) must be marked along the periphery of the graph. Because of lack of space, only points 1, 2, and 3 are marked in the figure, and the corresponding p values are given in the top legend.

4. The real part Re[G] has a special significance for the limit of stability of chatter. It has the value of $G = (1/k)$ at $(f/f_n) = 0$, the value of Re[G] $= 0$ at $(f/f_n) = 1$, and it has two extremes (see Fig. 9.28). These occur at frequencies $(f/f_n) = (1 \pm 2\,\zeta)^{1/2}$, or approximately $f/f_n = 1 \pm \zeta$, and their values are

$$\mathrm{Re}[G]_{\max} = \frac{1}{4k\zeta(1-\zeta)}$$
$$\mathrm{Re}[G]_{\min} = \frac{1}{4k\zeta(1+\zeta)} \tag{9.47}$$

9.4.4 Transfer Functions of a Selected System with Two Degrees of Freedom: Uncoupled Modes in Two Directions

We often encounter, at least in an approximation, a system where a mass can vibrate simultaneously in two mutually perpendicular directions in a plane; yet these two modes of vibrations are, internally in the system, not coupled. Physically, such a system is represented by a bar with a rectangular cross section that is fixed at one end and carries a mass at the other end (see Fig. 9.30). The bar is considered massless, and the center of gravity of the mass coincides with the centroid of the section of the bar, which is located on the Z-axis. The directions X_1 and X_2 are the principal axes of the section of the bar, and they represent the directions of the maximum and minimum transversal stiffness at the mass m:

$$k_1 = \frac{E\,a^3 b}{4l^3}$$

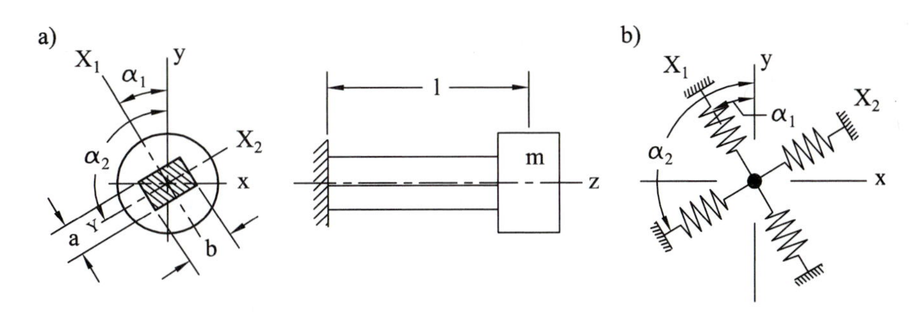

Figure 9.30
a) Two mutually perpendicular degrees of freedom: system of a bar with mass at its end. The two motions are uncoupled if the center of mass is in the centroid axis of the bar. b) A spring-mass model of the system.

$$k_2 = \frac{E\,ab^3}{4l^3}$$

where E is the modulus of elasticity. The system can also be represented as shown in Fig. 9.30b, as a mass suspended in the (X, Y) plane on two perpendicular springs in the directions X_1, X_2.

If a harmonic force acts in the direction X_1 it produces forced vibrations in this direction. These vibrations are governed by the corresponding TF involving mass m and stiffness k_1. In an analogous way, a force acting in the direction X_2 produces vibrations x_2 according to a TF involving m and k_2.

If a harmonic force acts in a direction different from X_I or X_2, say in the direction Y, it delivers components both in X_1 and X_2, and vibrations are produced simultaneously in both directions with the same frequency f of the exciting force. The amplitudes and phase shifts in the two directions are, however, different because of the different natural frequencies $\omega_{n1} = (k_1/m)^{1/2}$ and $\omega_{n2} = (k_2/m)^{1/2}$ and the correspondingly different ratios (f/f_{nl}) and (f/f_{n2}).

EXAMPLE 9.4 Forced Vibration of a System with Two Uncoupled Modes in a Plane ▼

Let us have the following conditions:

The directions of the modes, $\alpha_1 = 30°$, $\alpha_2 = 120°$.

The direction of the force is Y; its amplitude is $F_o = 500$ N and its frequency is $f = 450$ Hz.

The dimensions of the bar are $a = 60$ mm, $b = 70$ mm, $l = 300$ mm.

The mass is $m = 4$ kg $= 4$ Nsec^2/m.

The damping in the bar is such that the damping ratios in both directions are $\zeta_1 = \zeta_2 = 0.05$.

The bar is made of steel with a modulus of elasticity $E = 2 \times 10^5$ N/mm. Determine the amplitudes A_1, A_2 and phase shifts ϕ_1, ϕ_2 of the two vibrations, with respect to the force. What is the shape of the resulting motion of the mass m in the plane (X, Y)?

The stiffnesses are

$$k_1 = \frac{Ea^3b}{4l^3} = 2.8 \times 10^4 \text{ N/mm} = 2.8 \times 10^7 \text{ N/m}$$

$$k_2 = \frac{Eab^3}{4l^3} = 3.811 \times 10^4 \text{ N/mm} = 3.811 \times 10^7 \text{ N/m}$$

The natural frequencies are

$$f_{n1} = \frac{1}{2\pi}\sqrt{\frac{k_1}{m}} = \frac{1}{2\pi}\left(\frac{2.8 \times 10^7}{4}\right)^{1/2} = 421 \text{ Hz}$$

$$f_{n2} = \frac{1}{2\pi}\sqrt{\frac{k_2}{m}} = \frac{1}{2\pi}\left(\frac{3.811 \times 10^7}{4}\right)^{1/2} = 491 \text{ Hz}$$

The amplitudes of the force components are

$$F_{01} = F_0 \cos 30° = 0.866\, F_0 = 433 \text{ N}$$

$$F_{02} = F_0 \cos 120° = -0.5\, F_0 = -250 \text{ N}$$

Using Eqs. (9.42) and (9.43), we have

$$f/f_{n1} = \frac{450}{421} = 1.069$$

$$A_1 = F_{01}\,|G_1|$$

$$= \frac{433}{2.8 \times 10^4}[(1 - 1.069^2)^2 + (2 \times 0.05 \times 1.069)^2]^{-1/2} = 0.087 \text{ mm}$$

$$f/f_{n2} = \frac{450}{491} = 0.961$$

$$A_2 = F_{02}\,|G_2|$$

$$= \frac{-250}{3.811 \times 10^4}[(1 - 0.916^2)^2 + (2 \times 0.05 \times 0.916)^2]^{-1/2} = -0.016\text{mm}$$

$$\phi_1 = -143.17° = -2.5 \text{ rad}$$

$$\phi_2 = -29.65° = -0.52 \text{ rad}$$

If the force is $F = 500 \cos 900\,\pi t$, then $x_1 = 0.016 \cos(900\,\pi t - 2.5)$, and $x_2 = -0.087 \cos(900\,\pi t - 0.52)$

It is generally known that a sum of two mutually perpendicular vibrations with the same frequency, unequal amplitudes, and a mutual phase shift is an ellipse. Special cases are (1) amplitudes equal and mutual phase 90° give a circle; (2) any amplitudes and mutual phase 0° give a straight line. ▲

EXAMPLE 9.5 Determine the Transfer Function (*Y/F*) for a System with Two Uncoupled Modes in a Plane ▼

Let us take the same system as in Example 9.4. This time, however, the harmonic force acts in a direction *F* that is inclined by 70° to the direction *Y*. We are interested in the component of the resulting vibrations in the direction *Y* (see Fig.

9.31). We will measure the directions X_1 and X_2 from Y in the counterclockwise direction as α_1 and α_2. We have $\alpha_1 = 30°$, and $\alpha_2 = 120°$. The ratio of the complex amplitudes of this vibration over the force as a function of the frequency f of the force will give the TF that we are interested in. This TF will be called a "cross" TF because the directions of the exciting force and of the vibration of interest do not coincide; if they do, it is called a "direct" TF. This example differs from the previous one in that in Example 9.4 only one force frequency was considered, whereas now we are looking for the entire transfer function.

The force F excites each of the vibrations by a component falling into the corresponding direction:

$$F_1 = F\cos(\alpha_1 - \beta) = F\cos 40° = 0.766\,F \tag{9.48}$$

$$F_2 = F\cos(\alpha_2 - \beta) = F\cos 50° = 0.6428\,F \tag{9.49}$$

a)

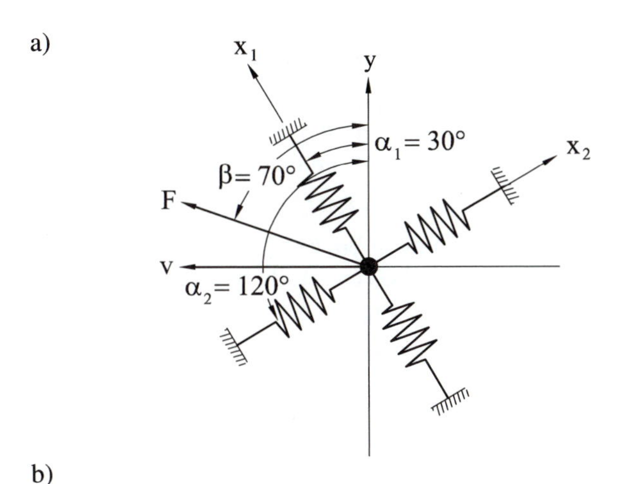

b)

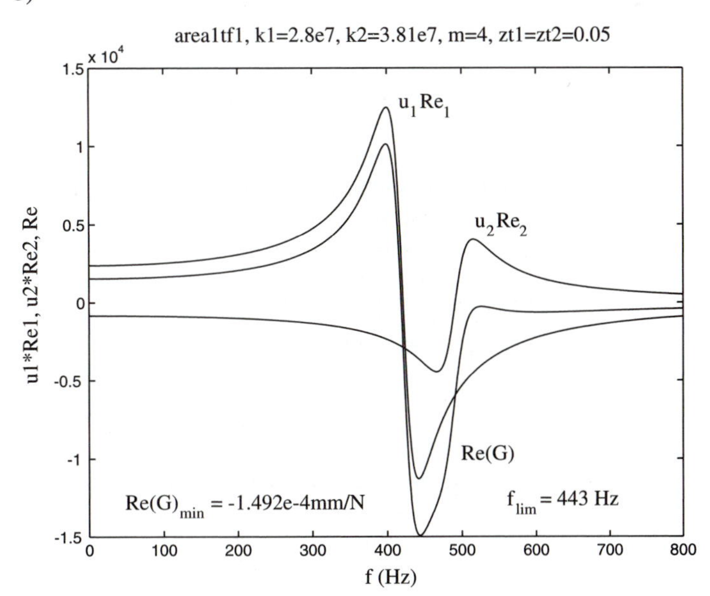

Figure 9.31
a) A force acting on the uncoupled two-degree-of-freedom system. We are interested in vibration Y that has a different direction than the exciting force F. b) The oriented transfer function Y/F of the boring bar.

The direct transfer functions in the directions X_I and X_2 are defined as ratios of vibrations in each of the directions over the forces acting in these directions:

$$G_1 = \frac{X_1}{F_1} \qquad G_2 = \frac{X_2}{F_2} \tag{9.50}$$

The resulting vibration is obtained as the sum of vibrations X_1 and X_2, each projected into the direction Y:

$$Y = X_1 \cos \alpha_1 + X_2 \cos \alpha_2 \tag{9.51}$$

By combining Eqs. (9.48), (9.49), (9.50), and (9.51), we have

$$\begin{aligned} Y &= F_1 G_1 \cos \alpha_1 + F_2 G_2 \cos \alpha_2 \\ &= F\left[G_1 \cos \alpha_1 \cos(\alpha_1 - \beta) + G_2 \cos \alpha_2 \cos(\alpha_2 - \beta)\right] \end{aligned} \tag{9.52}$$

Let us denote $u_i = \cos \alpha_i (\cos \alpha_i - \beta)$ and call it the directional factor:

$$G = \frac{Y}{F} = G_1 u_1 + G_2 u_2 \tag{9.53}$$

The resulting cross TF is obtained as the sum of the individual direct TFs multiplied by the corresponding directional factors.

The terms in Eq. (9.53) are complex, and they have to be added correspondingly. However, if we are interested in the real part of G only, then we have

$$\text{Re}(G) = u_1 \text{Re}(G_1) + u_2 \text{Re}(G_2) \tag{9.54}$$

and the summing in Eq. (9.54) is a simple arithmetic adding of real numbers. The functions G_1 and G_2 can easily be constructed from Eq. (9.40) and from Eq. (9.44) and according to Fig. 9.28. This is now done for the parameters of our example:

$$u_1 = \cos 30° \cos 40° = 0.6634$$

$$u_2 = \cos 120° \cos 50° = -0.321$$

The static flexibilities and the peaks of u_1 Re(G_1) and u_2 Re(G_2) are

$$\frac{u_1}{k_1} = 2.37 \times 10^{-5} \text{ mm/N}$$

$$\frac{u_2}{k_2} = -8.42 \times 10^{-6} \text{ mm/N}$$

The natural frequencies are $f_{n1} = 421$ Hz and $f_{n2} = 491$ Hz. The peaks of the two TFs will be found at frequencies of $(1 \pm \zeta) f_n$, that is, at (1 ± 0.05) 421 Hz = 400 Hz and 442 Hz for G_1, and (1 ± 0.05) 491 Hz = 466 Hz and 515.5 Hz for G_2. The values of the peaks follow

$$u_1 \text{Re}(G_1)_{\text{max}} = 12.5 \times 10^{-5} \text{ mm/N}$$

$$u_1 \text{Re}(G_1)_{\text{min}} = -11.3 \times 10^{-5} \text{ mm/N}$$

$$u_2 \text{Re}(G_2)_{\text{max}} = -4.4 \times 10^{-5} \text{ mm/N}$$

$$u_2 \text{Re}(G_2)_{\text{min}} = 4.0 \times 10^{-5} \text{ mm/N}$$

Because of the negative value of the directional factor u_2, the maximum of G_2 becomes the minimum of $u_2 G_2$ and vice versa for the minimum.

The two functions u_1 Re(G_1) and u_2 Re(G_2) are drawn in Fig. 9.31 as well as their sum, the "oriented cross transfer function" Re(G) for the given system. ▲

9.5 FORCES AND FORCED VIBRATIONS IN MILLING

9.5.1 Accuracy of End Milling: Straight Teeth, Static Deflection

The subject of accuracy in end milling is interesting and significant for several reasons. First, end mills are rather flexible, and cutting stiffness is high. Second, end milling is a very significant operation in the aerospace industries. Third, it involves a special aspect of the periodically variable force and of indirect relationship between deflections and accuracy. Let us analyze the last aspect first. We will start with a simplified treatment, assuming cutters with straight teeth and assuming an instantaneous response of deflection to force (the static approach). The indirect nature of the effect of deflections on the accuracy of the machined surface will be pointed out.

Figure 9.32 shows simplified pictures of end milling cutters with straight teeth. The end mill and its shank and support are flexible. In up-milling, cases *(a)* to *(c)*, the end mill deflects mostly towards the workpiece. Every tooth starts with a very small force at point *A* and reaches the maximum force at point *B*. Every tooth generates the final surface *S* while at point *A*. In the case of a two-fluted cutter as shown in *(a)*, there is practically no deflection when the tooth is at *A*. In *(b)*, when the tooth is at *B*, there is a large deflection but there is no tooth at *A* to imprint this deflection on the generated surface. However, the four-fluted cutter shown in *(c)*, if it cuts in a radial depth of cut at least equal to the cutter radius, will transfer the deflection to the machined surface. In down-milling, case *(d)*, the situation is similar, but the deflection is away from the workpiece. In reality, the teeth are not straight but helical, and the relationship between forces, deflections, and the form of the machined surface is more complicated.

From this explanation it follows that for cutters with straight teeth, two-fluted cutters are preferable to four-fluted cutters. The complex relationship between force, deflection, and accuracy of the machined surface has, among other aspects, this very

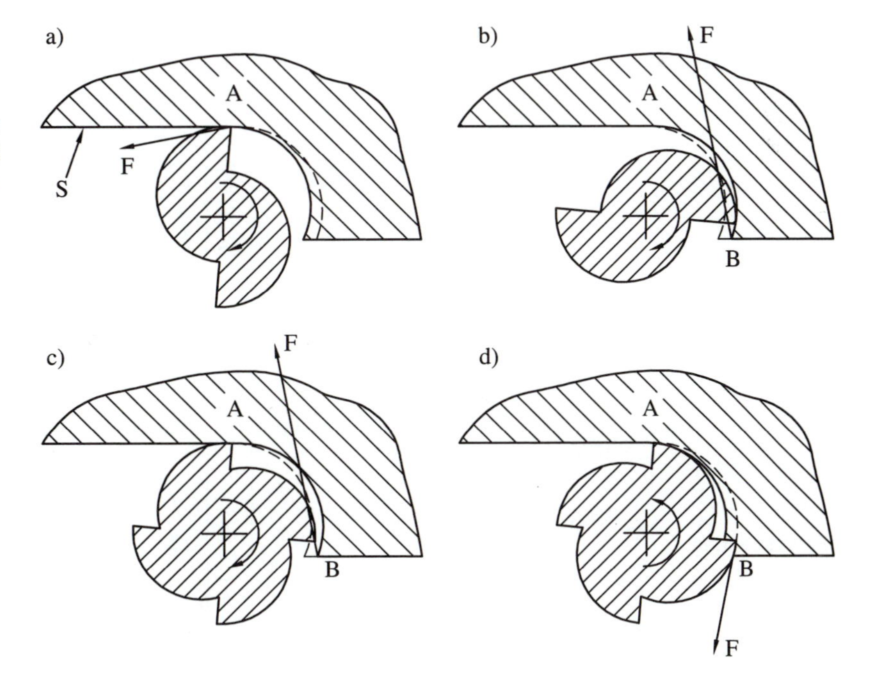

Figure 9.32
Forces on end mills with straight teeth and their effect on accuracy of the machined surface. a) Tooth in position *A* determines surface *S* but does not produce force. b) For a two-fluted cutter, the force may be supplied by tooth in *B*, but there is at that instant no tooth in *A* to imprint the deflection on surface *S*. c) and d) For the four-fluted cutter, if the radial depth of cut is at least equal to cutter radius, tooth in *B* produces force and deflection and tooth in *A* imprints it on *S*, both for c) up-milling and d) down-milling.

unexpected one: even if one can observe substantial periodic deflection on a two-fluted cutter, the machined surface obtained is perfectly accurate and unaffected by these deflections. This is, of course, true only in the "static" approach, which may be accepted for low spindle speeds when the tooth-passing frequency is well below the natural frequency of the vibratory system on the tool. Furthermore, most end mills have helical teeth, and the analysis is a little more complicated.

9.5.2 The Dynamics: Forced Vibrations, Straight Teeth

At this stage, we retain the simplification of assuming straight teeth on the cutter. This is approximately the case of low axial depth of cut within which the effect of a helix on the peripheral position of the cutting edge is small. The cutting force variation has briefly been mentioned briefly in Chapter 7, Fig. 7.31. Let's recapitulate it. Using the simple approach of Chapter 7, Eq. (7.3), where the tangential force F_t is proportional to the chip cross section and the normal component $F_n = 0.3\ F_t$, (see Fig. 9.33), and taking the tool path as a circular arc, tooth after tooth this path moves in X by the feed per tooth (chip load) c, tooth 1 in position ϕ_1 encounters chip thickness $h_1 = c \sin \phi$ and a force F_{t1}:

$$F_{t1} = K_s bc \sin \phi_1 \tag{9.55}$$

If, for instance, this is the case of up-milling, the tooth engages over the arc of cut, $\phi_s < \phi_1 < \phi_e$, where, in this case the starting angle $\phi_s = 0°$ and the end angle $\phi_e = 65°$, as shown in the sketch. Tooth 2 has in the meantime moved out of this engagement. In the case of slotting, $\phi_e = 180°$, and tooth 2 encounters the force

$$F_{t2} = K_s bc \sin \phi_2$$

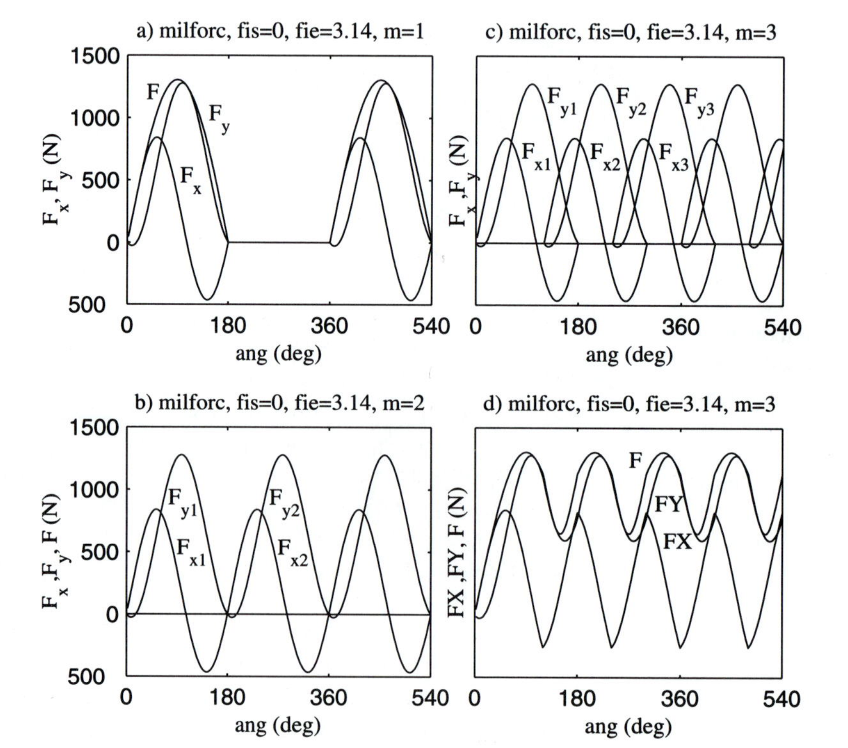

Figure 9.33
Formulation of forces acting on a tooth of a cutter, as functions of angle ϕ. Tangential force $F_{t\,i}$ and normal force $F_{n\,i}$ $(i = 1,2)$ can be resolved for each tooth into axes X and Y and added in F_x and F_y.

where $\phi_2 = \phi_1 + 90°$, for a cutter with four teeth. In these equations K_s is the specific force, and b is the axial width of cut (the depth of cut). The forces acting on the teeth can be added and reflected into the F_x and F_y components in the tool axis:

$$F_x = \sum_{i=1}^{m} F_{ti} \cos \phi_i - F_{ni} \sin \phi_i$$

$$= K_s bc \sum_{i=1}^{m} \sin \phi_i \cos \phi_i + 0.3 \sin^2 \phi_i \tag{9.56}$$

$$= \frac{K_s bc}{2} \sum_{i=1}^{m} \sin 2\,\phi_i - 0.3 + 0.3 \cos 2\,\phi_i$$

$$F_y = \sum_{i=1}^{m} F_{ti}\, sin\, \phi_i - 0.3\, F_{ti} \cos \phi_i$$

$$= K_s bc \sum_{i=1}^{m} \sin^2\phi_i - 0.3 \sin \phi_i \cos \phi_i \tag{9.57}$$

$$= \frac{K_s bc}{2} \sum_{i=1}^{m} 1 - \cos(2\phi_i) - 0.3 \sin(2\phi_i)$$

where m is the number of teeth on the cutter, and every force component is counted only within the limits of $\phi_s < \phi_i < \phi_e$.

Both force components are periodic in 2 ϕ, where $\phi = 2\pi(n/60)t$, with n (rpm) the spindle speed. Because of the summation for the individual teeth, which pass with circular frequency $2\pi\, mn/60$, this tooth frequency is the dominant periodicity of the force.

Except for the slotting case where $\phi_s = 0$ and $\phi_e = 180°$, the functions F_x and F_y contain abrupt drops in the magnitude of the individual components at $\phi_i = \phi_e$. This produces strong harmonics over the basic tooth frequency. Let's examine a few examples.

EXAMPLE 9.6 Variation of F_x and F_y Forces on Cutters with Straight Teeth ▼

A computer program is presented in Fig. 9.34 to be used for the computation of forces on cutters with straight teeth. In the examples shown in Fig. 9.35, forces F_x and F_y are shown for the individual teeth i of a cutter with m teeth over a span of 1.5 revolutions of the cutter. For all of them, the start of the cut is at $\phi_s = 0$; that is, we deal here with up-milling, and the arc of cut spans over 180°, the so-called slotting cut. The plots were obtained from the computer program "milforc" written in Matlab.

Note that the program loops in 270 steps subscripted n with the cutter rotating by *dfi* = 2° between each step. Inside of these loops there is internal looping in steps subscripted *i,* and there are as many of these loops as there are teeth, that is, *m.* Equations (9.56) and (9.57) are used to calculate the forces; however, the forces are first checked against the angle *al* to see whether it is within the range of cut (*fis, fie*). The angle *al* is derived from the leading angle *fi (n)* and steps back in every internal loop by the pitch angle of the teeth $2\,\pi/m$. Outside of the *i* loops the forces on all the *m* teeth are summed in every *n*th position to determine the component forces *FX* and *FY* and, eventually, the resulting force *F.*

The use of the program is demonstrated in several plots for cutters with various numbers of teeth and various cutting engagements. The first case, Fig. 9.35a, one of slotting, shows the forces F_x, F_y for a cutter with a single tooth. The horizontal axis of the graph is the angle *fi* of rotation, over a range of 3π. The forces

Figure 9.34
Matlab program "milforc" for computing cutting forces on milling cutters with straight teeth.

```
function[ang,fi,Fx,Fy,FX,FY,F]=milforc(fis,fie,m)
dfi=2*pi/180;Ks=2000;b=5;c=0.25;FX(1)=0;FY(1)=0;F(1)=0;
Fx=zeros(270,m);Fy=zeros(270,m);
% run the program for one and one-half revolutions (ang =1-540 deg)
% m is the number of teeth on the cutter, i is the subscript for
% moving from tooth to tooth, fi(n) is the angular position of the leading tooth
% and al is the angular position for every tooth in turn
for n=1:270
 sumx=0;sumy=0;
 for i=1:m
   fi(n)=n*dfi;
   ang(n)=fi(n)*180/pi;
  al=fi(n)-(i-1)*2*pi/m;
  % check if tooth is outside of the arc of cut; if so forces are zero
  if al<fis
  elseif al>fie & al<(2*pi+fis)
  elseif al>fie+(2*pi)
   Fx(n,i)=0;
   Fy(n,i)=0;
else
  % determine forces, use Eq (9.57)
  Fx(n,i)=(Ks*b*c/2)*(sin(al)*cos(al)+0.3*sin(al)^2);
  Fy(n,i)=(Ks*b*c/2)*(sin(al)^2-0.3*sin(al)*cos(al));
  end
  % add forces Fx and Fy for all the m teeth as they are obtained
  % in the loops of i=1:m
  sumx=sumx+Fx(n,i);
  sumy=sumy+Fy(n,i);
 end
 FX(n)=sumx;
 FY(n)=sumy;
 F(n)=sqrt(FX(n)^2+FY(n)^2);
end
% plot(ang,Fx,ang,Fy)
% plot(ang,FX,ang,FY,ang,F)
```

exist for $0 < fi < \pi$ and $2\pi < fi < 3\pi$; otherwise they are zero. The shapes are harmonic functions in 2ϕ; overall the basic period is once per revolution. In this basic frequency, the signals are nonharmonic because of the gap. In Fig. 9.35b the forces are plotted for a cutter with $m = 2$ teeth. Apart from DC shifts the forces are harmonic, 2 times per revolution. The gap between π and 2π that existed with the single-toothed cutter is now filled with the forces on the second tooth. Both the component forces and the resultant force are shown. Forces for a cutter with 3 teeth are plotted in Fig. 9.35c. This is the picture of 9.35a with two more graphs of the same shape shifted by 120° ($2\pi/3$). The dominant frequency is 3 times per revolution. Over some ranges two teeth overlap. The resulting forces (sums of the forces on the individual teeth) are presented as *FX, FY,* and *F* in Fig. 9.35d. The tooth frequency $f_t = 3n/60$ is clearly distinguished. Strong DC shifts are seen. Forces F_x and F_y for the individual teeth of a cutter with $m = 4$ teeth in slotting are shown in Fig. 9.36a. However, once the forces on the two simultaneously engaging teeth are added, nonperiodic constant forces are obtained (see Fig. 9.36b). Steady state of the computation is not reached until $\phi = \pi/2$. This perhaps unexpected result of constant forces is explained later in Example 9.11. An example of a cutter with three teeth but cutting over only an arc of 60° (*fie* = 60°) is given in Fig. 9.36c. The force variation is now strongly nonharmonic. Let us also look at the case of a cutter with four teeth in a cut over 60° engagement, *fie* = 60°, Fig 9.36d. The resulting forces are again strongly nonharmonic. ▲

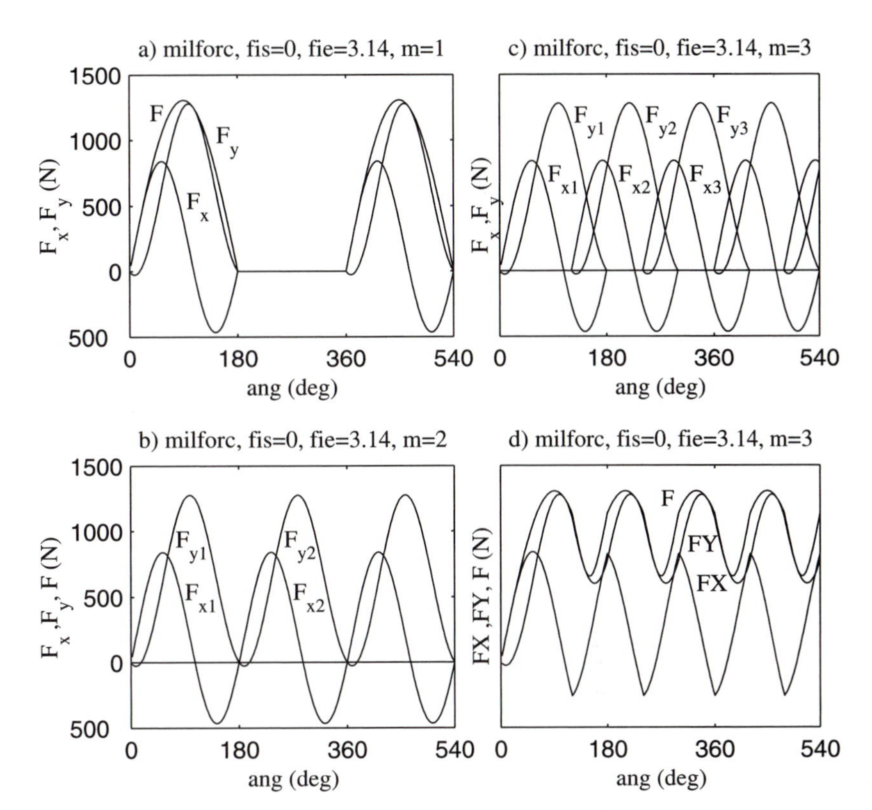

Figure 9.35 Milling forces on cutters with straight teeth. a) Component forces for a cutter with a single tooth, for 1.5 revolutions (3π) of the cutter, in slotting. b) Component forces for a cutter with 2 teeth in slotting. c) Component forces for a cutter with 3 teeth. d) Sums of F_x and F_y forces from all the 3 teeth and the resultant force F for a cutter with three teeth.

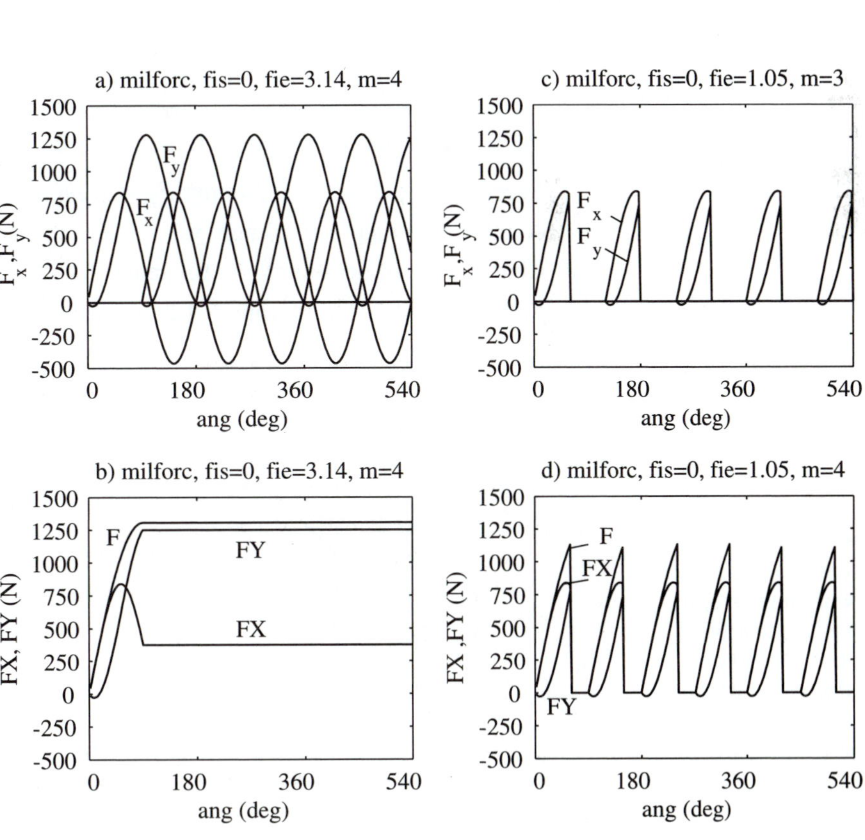

Figure 9.36 Milling forces on cutters with straight teeth. a) Slotting with a four-toothed cutter, individual forces F_x and F_y, b) summary forces FX, FY, F. At steady state, the forces are constant, without variation. c) Forces F_x and F_y for a cutter with 3 teeth, up-milling over engagement angle of 60°. d) Summary forces FX, FY, F for a four-toothed cutter, up-milling over an engagement of 60°. In c) and d), forces are strongly nonharmonic.

9.5.3 Forced Vibrations and Their Imprint as Error of Location of the Machined Surface

Refer to Section 9.4, where the basic notions of vibrations were reviewed. It is important to know that a harmonic force produces harmonic deformations, vibrations, of the same frequency as that of the force. The amplitude of the vibrations depends not only on the amplitude of the force and on the stiffness of the vibratory system but also on the ratio of the frequency of the force over the natural frequency of the system. If the two frequencies are equal, we have the case of "resonance," and vibration amplitude is maximum. Also, we should recall that the displacements are phase-shifted behind the force by an angle that depends solely on the ratio of the two frequencies. For more detail, see Section 9.4.3.

In the preceding text, Section 9.3.3, Eqs. (9.15) to (9.22), we learned that there is a mutual effect between cutting force and deflection because the deflection affects the chip thickness, which affects the force. This was also illustrated in Example 9.3. This mutual effect is called regeneration of the deformations imprinted on the surface and is the basic cause of chatter; it is discussed later in Section 9.6. Regeneration plays a minor role in forced vibration in milling, and it is not considered at this stage. Forces will be expressed by Eqs. (9.55), (9.56), and (9.57) and will be assumed to be unaffected by the displacements. The problem of the error of the machined surface will be illustrated by means of examples. First, simple examples will be worked out without using the computer.

EXAMPLE 9.7 Slotting with a Two-Fluted Cutter: Resonant Vibration ▼

A two-fluted cutter with straight teeth is machining a slot producing the surfaces S_A and S_B whenever a tooth passes through point A and point B (Fig. 9.37). We need to determine the displacement y of the cutter at these instants as caused by the Y component of the cutting force F_y. Given:

Specific force $K_s = 1000$ N/mm^2, $F_T = K_s bh$, $F_N = 0.3\ F_T$

Axial depth of cut $b = 10$ mm

Chip load (feed per tooth) $c = 0.1$ mm

Spindle speed $n = 7200$ rpm $= 120$ rev/sec

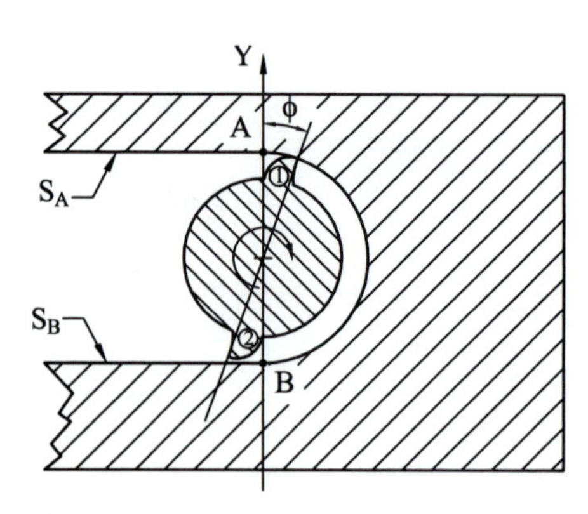

Figure 9.37 Slotting with a two-fluted cutter with straight teeth. A simple case to solve in Example 9.7 for forces and deflections (vibrations) and the resulting error of location of the machined surfaces S.

Write F_y as a function of time using Eq. (9.57) and taking into account the action of both teeth,

$$F_y = K_s bc \sin\phi\,(\sin\phi - 0.3\cos\phi)$$

Express F_y as a function of (2ϕ). It will be a sum of a DC term and two harmonic terms.

$$F_y = \frac{K_s bc}{2}(1 - \cos 2\phi - 0.3\sin 2\phi)$$

$$F_y = 500\,(1 - \cos 2\phi - 0.3\sin 2\phi)$$

$$F_1 = -500\cos 2\phi \text{ N}$$

$$F_2 = -150\sin 2\phi \text{ N}$$

$$\text{DC} = 500 \text{ N}$$

Plot by hand the three components of the force $F_{y,}$ as in Fig. 9.38a.

Produce the effects of the three force components separately. Assume that the natural frequency f_n, of the vibratory system on the tool is exactly the same as that of the force, $f_n = f = 240$ Hz; the tool is running *at resonance.* With stiffness $k = 1000$ N/mm and damping ratio $\zeta = 0.04$, what are the amplitudes A_1, A_2 (mm) and phases ϕ_1, ϕ_2, of the vibrations behind the force? As shown in Section 9.4.3, phase shift at resonance is $-90°$. The vibration amplitudes are $Y_1 = F_1/(2k\zeta) = 6.25$ mm and $Y_2 = F_2/(2k\zeta) = 1.875$ mm. The vibrations will have the same frequency as the force, $f = 2 \times 120 = 240$ Hz. A 90° phase shift converts $\cos\omega t$ into $\sin\omega t$ and $\sin\omega t$ into $-\cos\omega t$. What is the value of the DC component of the deflection?

Therefore $y_1 = -Y_1\sin(2\phi)$, and $y_2 = Y_2\cos\phi$.

The value of the DC component is DC $= 500/1000 = 0.5$mm, and the values of y_1 and y_2 at $\phi = 0$ are $y_{1,0} = 0$, $y_{2,0} = 1.875$ mm. Plot out the deflections as functions of time and indicate by an asterisk the value of the error of location of surfaces S_A and $S_{B,}$ (Fig. 9.38b).

$$\delta = 0.5 + 0 + 1.875 = 2.375 \text{ mm}$$

The straight teeth pass through points A and B at $\phi = 0$ and $\phi = 180°$, that is, at $t = 0$ and $t = 0.00425$ sec and after any other half a revolution. These instants are

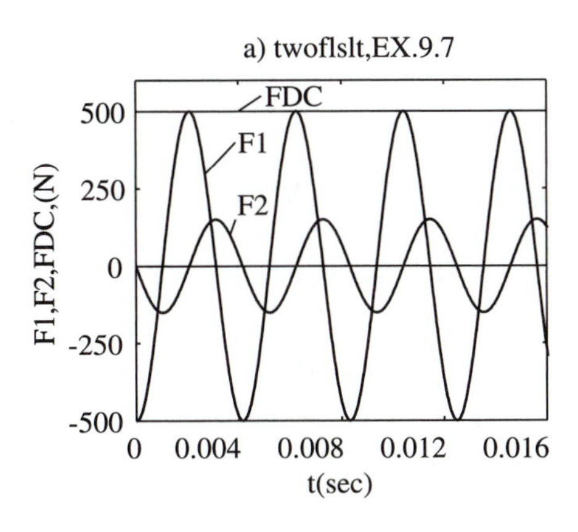

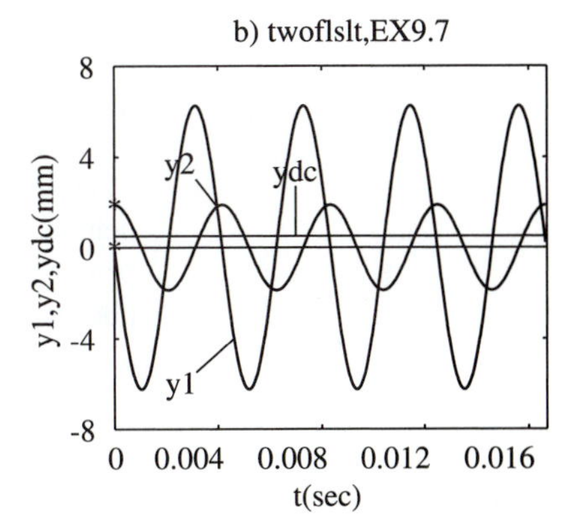

Figure 9.38 Milling forces and vibrations and error of surface location. Slotting with a two-fluted cutter. a) Force components in Example 9.7 *at a resonant speed.* b) Vibration components. They are lagging by 90° behind the forces. The asterisks on the plots at $t = 0$ indicate the errors of location of the milled surface.

marked at the $t = 0$ axis in the graph and show the three deflection components that determine the error of location of the machined surface. The surface S_A is "overcut" by 2.375 mm. ▲

EXAMPLE 9.8 Slotting with a Two-Fluted Cutter, Nonresonant Conditions ▼

Assume the same natural frequency of the system, $f_n = 310$ Hz, and the spindle speed $n = 8400$ rpm $= 140$ rev/sec; the tooth frequency is $f_t = 280$ Hz; the periodic force is exciting the system below resonance.

Axial depth of cut $b = 10$ mm

Specific force $K_s = 1000$ N/mm

Chip load $c = 0.1$ mm

Stiffness on the tool $k = 1000$ N/mm

The components of the cutting force F_y are expressed in the same way as in Example 9.7, except for the frequency:

$F_{y1} = -500 \cos(2\pi \times 280\, t)$ N

$F_{y2} = -150 \sin(2\pi \times 280\, t)$ N

DC $= 500$ N

and the force graph is the same as Fig. 9.38a except for the time scale.

The phase shift of the vibration behind the forces is obtained using Eq (9.41):

$$\phi = \text{atan2}\left(\frac{-2\zeta p}{1 - p^2}\right)$$

where $p = f_t/f_n$. Thus

$$p = \frac{280}{310} = 0.9032 \qquad \phi = -38.11°$$

and the ratio of the vibration amplitude A to the force amplitude is obtained from Eq (9.42) as

$$\frac{A}{F} = \frac{1/k}{\sqrt{(1 - p^2)^2 + 4\zeta^2 p^2}} = 0.00427$$

Thus we obtain $A_1 = 2.135$ mm, and $A_2 = 0.64$ mm.

The forces and vibration components are shown in Fig. 9.39, where they are represented by rotating vectors, the projection of which onto the real axis determines the instantaneous values. At time $t = 0$ a positive cosine function is on the positive real axis, and a positive sine function is on the negative imaginary axis. Thus, F_1 (negative cosine) is on the negative real axis, and Y_1 is 38.11° lagging behind it. F_2 (negative sine) is on the positive imaginary axis, and Y_2 is lagging behind by 38.11°. The projections of Y_1 and Y_2 on the real axis determine the displacements y_1 and y_2 at $t = 0$:

$$y_1 = -2.135 \cos 38.11° = -1.68$$

$$y_2 = 0.64 \sin 38.11° = +0.395$$

and the error δ is obtained as

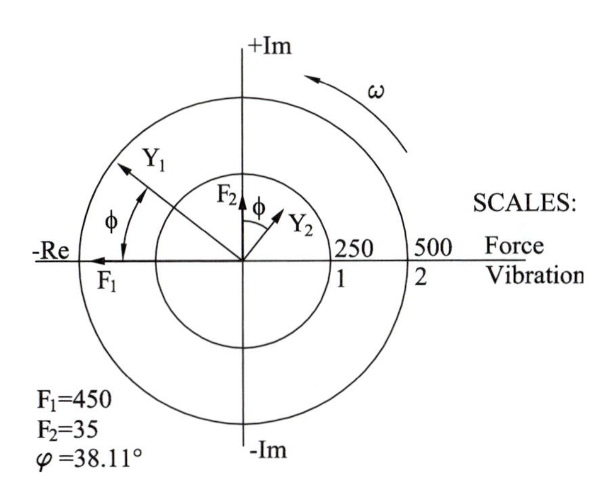

Figure 9.39
Milling forces and vibrations and error of surface location. Slotting with a two-fluted cutter. Forces and vibrations as rotating vectors in Example 9.8 *at nonresonant conditions.* Both the amplitudes and phase shifts of the vibrations depend on the ratio of force frequency over the natural frequency.

$$\delta = 0.5 - 1.68 + 0.395 = -0.785 \text{ mm}$$

which is the amount by which the surface is "undercut."

The next example is such that it can best be solved by writing and running a computer simulation program because the force function is strongly nonharmonic. ▲

EXAMPLE 9.9 Up-Milling with a Four-Fluted Cutter with Straight Teeth: Only One Tooth in Cut; Forces and Deflections ▼

This is not a slotting case; the cutter engagement angle is less than 90°: The parameters of the case follow:

Axial depth of cut $b = 10$ mm

Chip load $c = 0.1$ mm

Stiffness on cutter $k = 4 \times 10^6$ N/m

Mass $m = 0.88$ kg

Damping coefficient $c = 150$ N/(m/sec)

Spindle speed $n = 5000$ rpm

Cutter diameter $d = 20$ mm

Radial depth of cut $a = 3$ mm

Number of teeth $m = 4$

Specific force $K_s = 750$ N/mm^2

$\phi_s = 0°$, $\phi_e = 45.57°$; see Fig. 9.40

Determine the following:

$$\text{Natural frequency:} \quad f_n = \frac{1}{2\pi}\sqrt{\frac{k}{m}} = 0.159\sqrt{4.54 \times 10^6} = 339 \text{ Hz}$$

$$\text{Tooth frequency:} \ f_t = \frac{4n}{60} = 333.33 \text{ Hz}$$

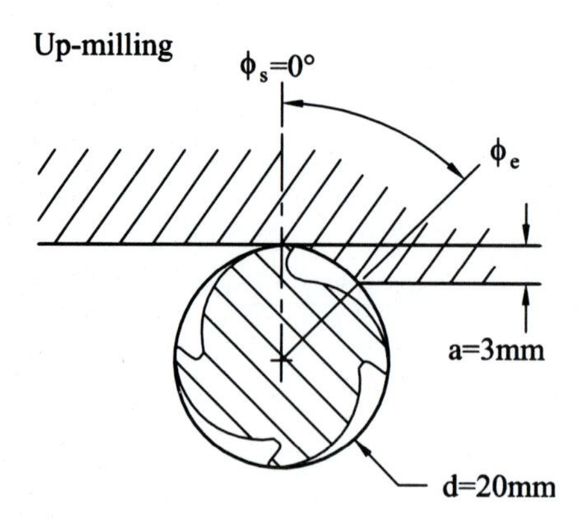

Figure 9.40
Up-milling with a four-fluted cutter as discussed in Example 9.9. The radial immersion is 3/20.

$$\text{Damping ratio: } \zeta = \frac{c}{2\sqrt{km}} = \frac{150}{2\sqrt{3.52 \times 10^6}} = 0.0399$$

given that $n = 5000$ rpm $= 83.33$ rev/sec, $T_{rev} = 0.012$ sec, $dt = 5e{-}5$ sec; that yields 240 steps per revolution.

The process will be simulated by proceeding in small time steps.

$$\phi_s = 0° \qquad \phi_e = 45.573°$$

Write the formula for F_y, omitting regeneration, and plot F_y versus time for one revolution of the cutter.

$$F_y = K_s bc(\sin^2\phi - 0.3 \sin\phi \cos\phi) = 750\,(\sin^2\phi - 0.3 \sin\phi \cos\phi)$$

$$\phi = \frac{2\pi\, n}{60} t = 523.6\, t$$

The computer program will follow the rotation of the cutter in 240 steps/revolution; $dfi = 360/240 = 1.5°$ (see Fig. 9.41). For the angle of engagement, $(\phi_e - \phi_s) = 45.57°$ there will be $K = \text{int}\,(45.573/1.5) = 30$ steps, and $M = 60$ steps per tooth period of 90°. There will never be more than one tooth in the cut. This simplifies the writing of the program. The force is computed over one tooth period; after that it is just being repeated:

$$fi(n) = (n - 1)dfi$$

$n = 1, 31,\;$ F is obtained from the formula (9.57)

$$n = 32, 61, \quad F = 0$$

$$n > 61, \quad F(n) = F(n - M + 1)$$

The program runs for 3000 steps, that is, 12.5 revolutions. During several initial tooth periods, vibrations start to develop and then reach the steady state in which the displacement y at an instant of $t{=}ndt$ is determined, where n is an integer multiple of the number of the 60-steps-per-tooth period plus 1. The next part of the program uses the equation of motion for the vibratory system:

$$m\ddot{y} + c\dot{y} + ky = F$$

in such a way that, for selected initial values of $y = 0$, $\dot{y} = 0$, the initial acceleration is expressed as

Figure 9.41
Matlab listing of the program "upmill" for forces and displacements in milling with cutters with straight teeth.

```
function[F,y,t,T,FF,Y]=upmill(sp)
%This program carries out EX.9.9. It determines the Y component of the
%force denoted just F, and the y displacement for a cut with a 3/20 radial
%immersion. The variables F,y,t are generated for the whole run of 12.5
%revolutions. For the purpose of plotting them for the last 2.5 revolutions
%of the fully developed vibration they are renamed FF,Y,T.
%Input parameters are:
dt=(60/sp)/240;dfi=pi/120;m=0.88;k=4e6;y(1)=0;dy=0;c=150;t(1)=0;
K=30;M=60;
%sp(rev/min) is spindle speed, m(kg) is mass, k(N/m) is stiffness,
%c(N/(m/sec)) is damping coefficient of the vibratory system, dy is
%first and ddy second derivative of y.
%Simulation is run for 3,000 steps, i.e. 12.5 revolutions at sp=5,000 rpm
%and 240 steps per revolution dfi=2pi/240,dt=5e-5sec and total time is
%0.150sec. With 4 teeth on the cutter this counts for M=60 steps per tooth
%and with exit angle 45.57 deg K=30 steps per cutting engagement
for n=1:3000
%Determine force F
fi=(n-1)*dfi;
if n<=K
F(n)=750*(sin(fi)^2-0.3*sin(fi)*cos(fi));
end
if n>K & n<=M
F(n)=0;
end
%force repeats per every M steps
if n>M
F(n)=F(n-M);
end
end
for n=1:2999
%Derive vibrational displacement y
ddy=(F(n)-c*dy-k*y(n))/m;
dy=dy+ddy*dt;
y(n+1)=y(n)+dy*dt;
t(n+1)=t(n)+dt;
end
% Waiting for vibration to settle look for force FF and displacement
% Y in the last 2.5 revolutions, i.e.the last 600 steps.
for i=1:600
T(i)=t(2400+i);
FF(i)=F(2400+i);
Y(i)=y(2400+i);
end
%plot(T,FF),(T,Y)
```

$$\ddot{y} = \frac{F - c\dot{y} - ky}{m}$$

and it is twice integrated into the next velocity and next displacement:

$$\dot{y} = \dot{y} + \ddot{y}\,\mathrm{d}t$$

$$y(n + 1) = y\,(n)\; + \dot{y}\,\mathrm{d}t$$

For the last 600 steps new variables *T, FF,* and *Y* replace *t, F, y* just for the purpose of plotting this part of well-established vibrations and determining the error δ. The plots of the force *FF* and of vibrations *y* for the whole time and in the initial period of 0.03 sec, as well as plots of vibrations *Y* in the last 2.5 revs are in Fig. 9.42a, b, c. At $n = 3001$ the value of y determines the error $\delta = -0.13$mm. ▲

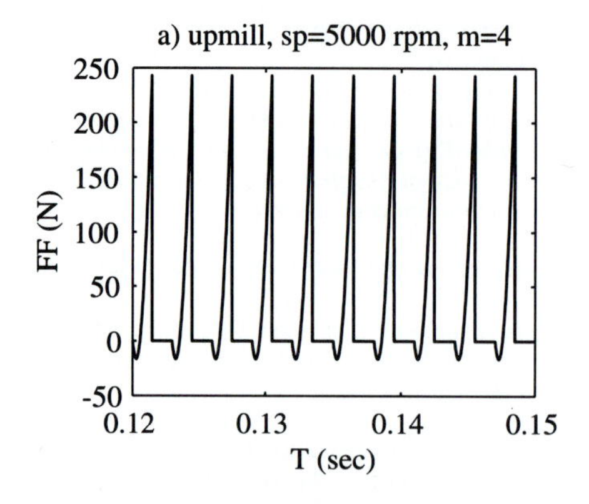

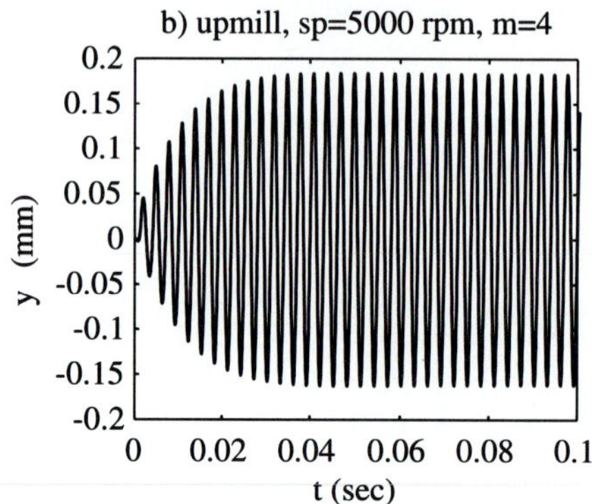

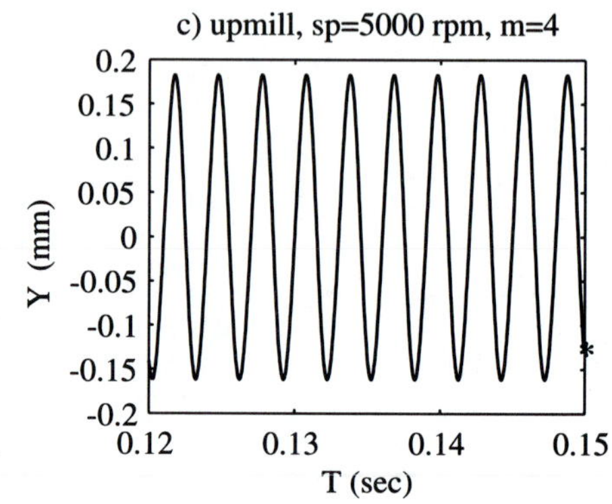

Figure 9.42
Force, vibration, and error of surface location in Example 9.9. A cutter with 4 teeth engaged over less than 90° of rotation. a) The force *FF* in the *Y* direction; b) vibration *y* over the whole of 0.15 sec; c) vibration *Y* in steady state. Error of surface location is marked by the asterisk at 0.15 sec.

9.5.4 Forces on End Mills with Helical Teeth

In this exercise all explanations and illustrations are based on an end mill with helical smooth edges, but they apply as well to any other type of a milling cutter. The knowledge of the cutting force variation is important for its effects on forced vibrations and on the accuracy of the machined surface.

The analysis is based on the diagrams in Fig. 9.43. Diagram *(a)* shows an element of a cutter over a workpiece with radial depth (width) a with the angle of tooth engagement ϕ measured from the position 0 and the cutting engagement spread over $\phi_s < \phi < \phi_e$. The cutting angle is $\phi_c = \phi_e - \phi_s$. The axial depth of cut is b. The direction of the feed motion f of the workpiece against the rotating cutter is as indicated, and feed per tooth is c. An element of a tooth cuts chip thickness h, which according to Eq. (7.12) is $h = c \sin \phi$.

Figure 9.43c is a view of the "unrolled" circumference of the cutter. The axial depth of cut b spreads from level A to level B. Every helical tooth edge becomes on unrolling a straight line under the helix angle β. All the edges move with peripheral velocity v (cutting speed). Their positions in the diagram are expressed simply (symbolically) by means of corresponding angles ϕ; properly they should be measured as peripheral distances ($r\phi$), where r is the radius of the cutter. The engagement of a tooth spreads over the angle ψ, which is obtained as

$$\psi = \frac{b \tan \beta}{r} \tag{9.58}$$

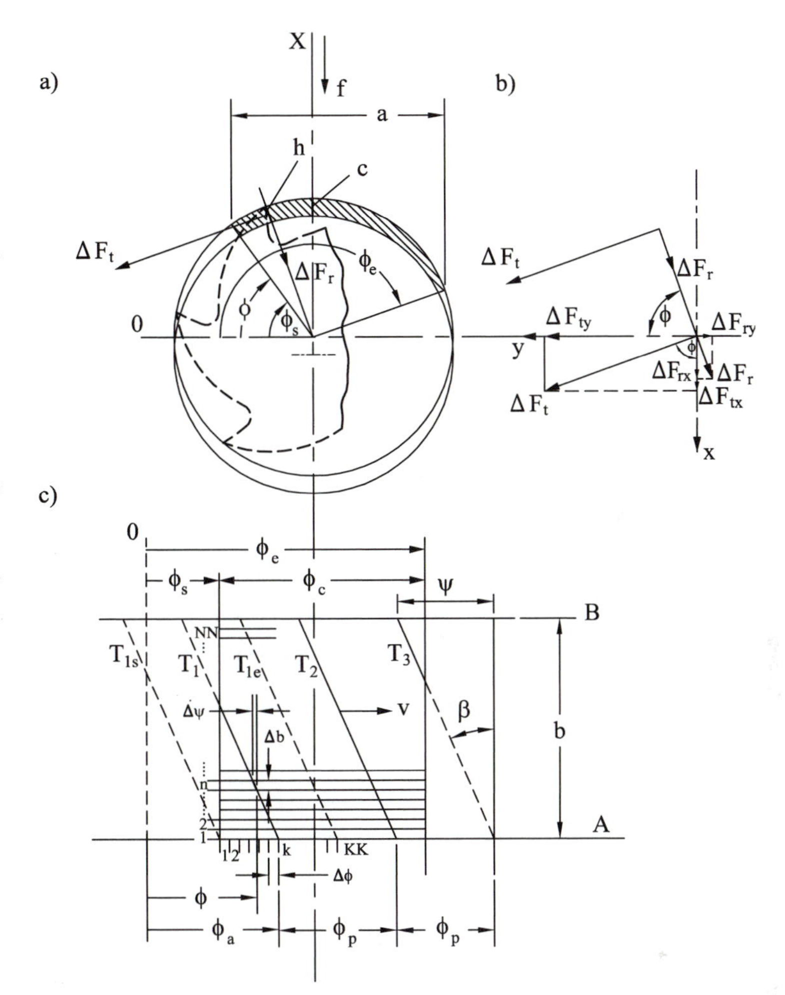

Figure 9.43 The diagram for the computation of the milling force for cutters with helical teeth. a) One of the horizontal slices with thickness Δa. b) Elements of the forces generated in this slice. c) The cylindrical periphery of the cutter is unrolled onto the plane perpendicular to the direction X of the feed f. The helical edges unroll into straight lines inclined under helix angle β. The sequence of the analysis goes in loops: 1) up along the edge in slices on tooth T_1, for $n = 1, NN$, adding forces ΔF_x and ΔF_y; 2) from the first tooth to the others, for $m = 1, M$; 3) rotating the cutter by moving the leading edge in steps $\Delta\phi$ for $k = 1, KK$ where $KK\,\Delta\phi = \phi_p$, the pitch angle.

However, the actual cutting engagement of any tooth exists only over the part that is located between ϕ_s and ϕ_e. In our case we have three teeth T_1, T_2, and T_3 cutting simultaneously; the cutting parts of their edges are drawn in solid lines; the rest are dashed lines.

The variation of the cutting forces is obviously periodic within a "tooth period," which corresponds to a rotation by a pitch angle ϕ_p. Thus, we may start with the leading point of tooth T_1 at level A just entering the cut at $\phi_a = \phi_s$ and examine the motion of this tooth until $\phi_a = \phi_s + \phi_p$. The tooth T_1 will move through this range from the position shown as T_{1s} to the position T_{1e}. The teeth T_2 and T_3 move accordingly being always by ϕ_p and $2\phi_p$ ahead, respectively. The motion of the teeth through the range of ϕ_p is incremental by $\Delta\phi$ with the count k, and there are KK steps. Every edge is considered incrementally with count n in a total of NN elements, each spreading over $\Delta\psi$ and having a height Δb.

The computer program according to the algorithm in Fig. 9.44 proceeds so that in every kth step the rotation stops, and force components are computed in the slices n

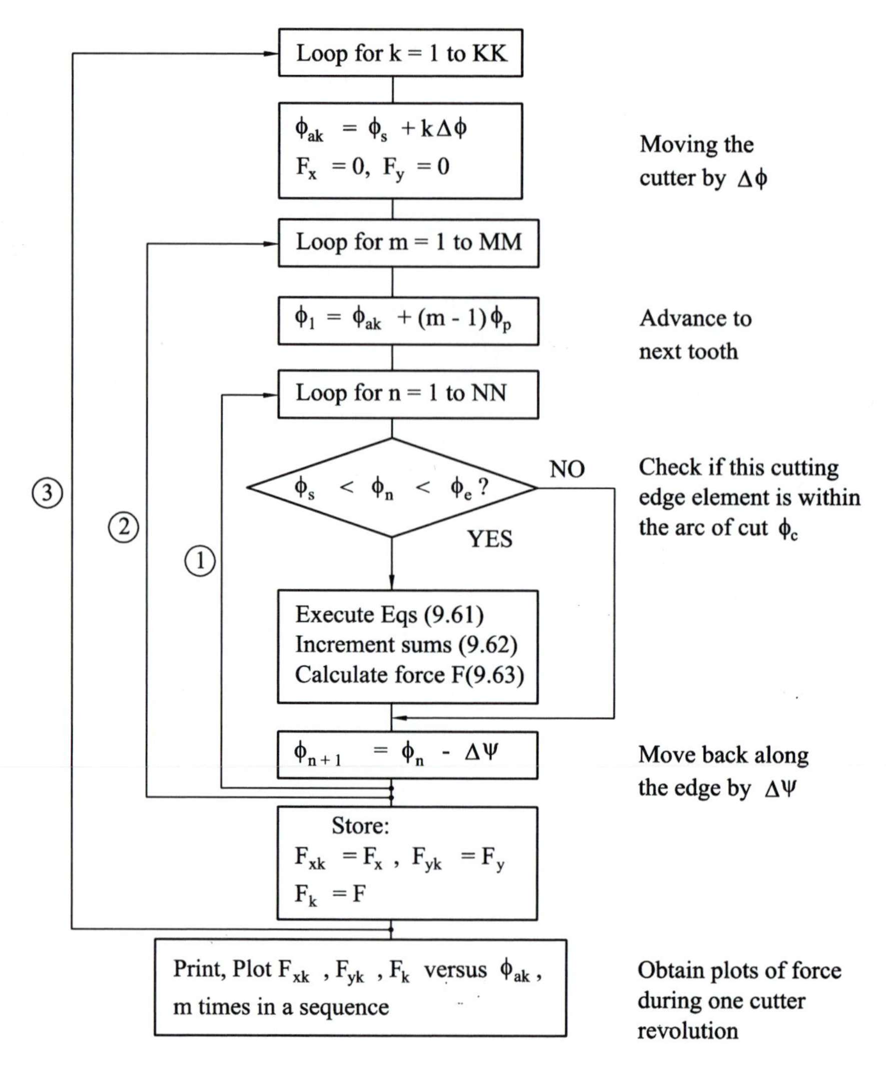

Figure 9.44 The algorithm for computing the milling force for cutters with helical teeth.

Loop 1 : Calculating and summing the incremental forces along one edge, in one position of the cutter.

Loop 2 : Repeating 1 over all edges engaged in the cut

Loop 3 : Repeating 1 and 2 for successive incremental positions of the cutter over one tooth period ϕ_p.

through the cutter and on the teeth T_1, T_2, T_3, and the F_x and F_y are summed respectively; then the cutter rotates by $\Delta\phi$ to position $k + 1$, and so on.

The element of the tangential force ΔF_t is proportional to the corresponding elemental chip area.

$$\Delta F_t = K_s\, \Delta b\, h = K_s\, \Delta b\, c \sin\phi = C \sin\phi \tag{9.59}$$

where $C = K_s\, \Delta b\, c$

The magnitude of the normal component of the force is

$$\Delta F_n = 0.3\, \Delta F_t \tag{9.60}$$

In order to obtain the total force acting on the cutter axis at any instant, that is, at any angular position of the cutter, it is necessary to add the elemental forces acting along the engaged parts of all the cutting edges. These elemental forces have different directions. Therefore, as shown in Fig. 9.43b, the forces ΔF_t and ΔF_r are moved into the axis of the cutter and decomposed into components falling into fixed axes X and Y. Then we have

$$\begin{aligned} \Delta F_{tx} &= \Delta F_t \cos\phi = C \sin\phi \cos\phi \\ \Delta F_{ty} &= \Delta F_t \sin\phi = C \sin^2\phi \\ \Delta F_{nx} &= 0.3\,\Delta F_t \sin\phi = 0.3\,\Delta F_{ty} \\ \Delta F_{ny} &= -0.3\,\Delta F_t \cos\phi = -0.3\,\Delta F_{tx} \end{aligned} \tag{9.61}$$

and

$$\begin{aligned} F_x &= \Sigma(\Delta F_{tx} + \Delta F_{nx}) \\ F_y &= \Sigma(\Delta F_{ty} + \Delta F_{ny}) \end{aligned} \tag{9.62}$$

$$F = (F_x^2 + F_y^2)^{1/2} \tag{9.63}$$

The summations in (9.62) are carried out over all elements engaged in cutting on all the teeth in the given position of the cutter.

The number m of edges that may cut simultaneously is, at the most, as many as will enter into the span of the cutting engagement $\phi_c = \phi_e - \phi_s$ plus the spread ψ of an edge:

$$m = integer\left(\frac{\phi_c + \psi}{\phi_p}\right) + 1 \tag{9.64}$$

The force, as obtained in (9.63), can be plotted versus the angle ϕ of cutter rotation, for example, with reference to the leading point of tooth T_1 (i.e., versus angle ϕ_a). This provides one period of force variation for $\phi_s < \phi_a < \phi_s + \phi_p$. Examples of such plots are shown later where the graph over the tooth period is repeated several times (e.g., as many times as there are teeth on the cutter).

The computer program starts with the input data: cutter radius r, number of teeth m, helix angle β, axial depth of cut a_a, feed per tooth c, angles ϕ_s and ϕ_e of the start and end of the cutting arc, the specific force K_s. Further, the discretization parameters of the computation are selected as $\Delta\phi$, NN, and the derived parameters are computed:

$$\psi[\text{Eq. (9.58)}]: \phi_p = \frac{2\pi}{m}, \qquad M[Eq.\ (9.64)],\ \Delta b = a_b/\text{NN},$$

$$C\,[\text{Eq. (9.59)}]: KK = integer\left(\frac{\phi_p}{\Delta\phi}\right), \qquad \Delta\Psi = \frac{\Psi}{NN}$$

EXAMPLE 9.10 Forces in Milling with Helical Teeth ▼

The examples of results of force computations are presented for an end-milling cutter with a radius $r = 15$ mm and four teeth, $m = 4$, helix angle $\beta = 30°$, feed per tooth $c = 0.1$ mm, and specific cutting force $K_s = 2000$ N/mm^2. The individual cases are illustrated in Fig. 9.45 as they differ in the radial depth of cut a and the axial depth of cut b, and they are denoted C_{mn} where the subscript m refers to

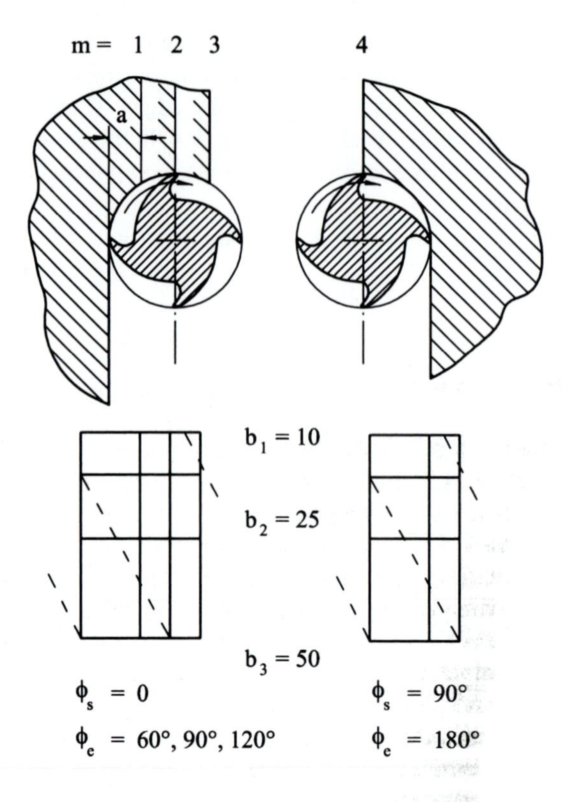

Figure 9.45
The cases of force computations of Example 9.10. They differ in radial immersions *a* as marked by the number *m* and in depths of cut *b*.

the value of *a* and the subscript *n* to that of *b*. The cases with $m = 1, 2, 3$ could be considered up-milling with 1/4, 1/2, 3/4 immersion respectively and the one with $m = 4$ is down-milling 1/2 immersion. The values of ϕ_s, ϕ_e, and *b*, and of the number *M* of teeth engaged simultaneously are given in the following tables.

m	1	2	3	4
ϕ_s	0	0	0	90°
ϕ_e	60°	90°	120°	180°

n	1	2	3
b	10	25	50 mm

Values of M

n/m	1	2	3	4
1	1	2	2	2
2	2	2	2	2
3	2	3	3	3

The plots of the magnitude of the resulting force *F* are presented in Fig. 9.46. One could also calculate and plot the variation of the components F_x and F_y and/or of the direction α of the force with respect to the *X* axis, $\alpha = \tan^{-1}(F_y/F_x)$.

Fig. 9.46 is so arranged that cases with the same arc of cutting are assembled together in each of the four diagrams; the three graphs in each diagram correspond to the three axial depths of cut $b = 10$, 25, and 50 mm respectively. It may be seen that the variable component of the force is about the same in diagrams *(a)*, *(b)*, and *(d)*, and it is almost equal for $b = 10$ and 50 mm, while it is about 50% larger for

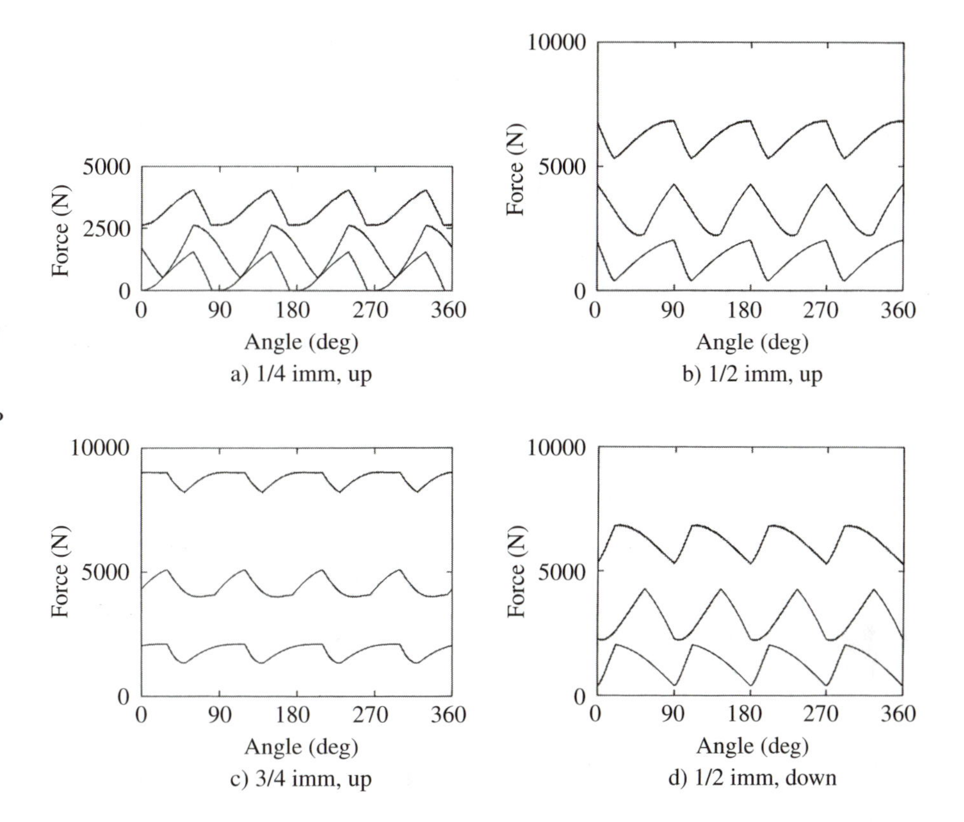

Figure 9.46
Plots of milling forces of Example 9.10 for a four-fluted cutter. In each graph the three traces correspond to the three values of depth of cut, b = 10, 25, 50 mm. The graphs differ by radial immersion (imm) as expressed by the ratio of the radial depth to the cutter diameter. As the immersion increases, the DC level of the force increases. The PTP values of the variable part of the force do not change much with the depth of cut; they increase from the 1/4 imm through 1/2 imm but diminish for the 3/4 imm and would disappear completely for slotting.

b = 25 mm. It decreases substantially in diagram *(c)*, for the 3/4 immersion. The static component (the mean level) of the force increases both with the cutting arc ϕ_c and with the axial depth of cut b. ▲

EXAMPLE 9.11 Prove that the Cutting Force on a Four-Fluted Cutter in Slotting Is Constant ▼

The drawing in Fig. 9.47 represents a section through a four-fluted cutter in a slotting operation; we will consider a slice with axial thickness Δz. At whichever axial level this slice is taken there will be two edges in the cut between $\phi_s = 0°$ and $\phi_e = 180°$: one at the angle ϕ_1 and the other one at the angle $\phi_2 = \phi_1 + 90°$. The magnitudes of the forces on the two edges will be

$$\Delta F_{t1} = K_s\,\Delta b\,h_1 \qquad \Delta F_{t2} = K_s\,\Delta b\,h_2$$

$$\Delta F_{r1} = 0.3\,\Delta F_{t1} \qquad \Delta F_{r2} = 0.3\,\Delta F_{t2}$$

and the forces will have directions as shown in the drawing. F_t is the tangential and F_r is the radial component of the force. The values of h are $h_1 = c \sin \phi_1$, and $h_2 = c \sin \phi_2$.

The forces can be decomposed according to Eqs. (9.61). Denoting $C = K_s\,\Delta bc$, we have

$$\Delta F_x = \Delta F_{t1x} + \Delta F_{t2x} + \Delta F_{r1x} + \Delta F_{r2x}$$

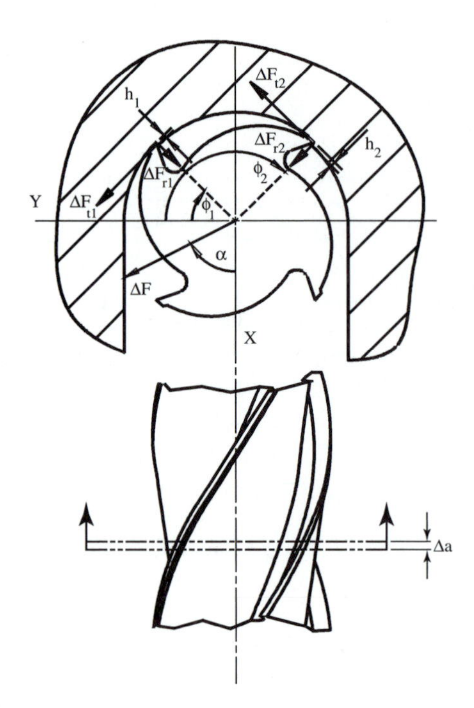

Figure 9.47
Diagram for forces in slotting with a four-fluted cutter.

$$\Delta F_x = C(\sin\phi_1 \cos\phi_1 + \sin\phi_2 \cos\phi_2 + 0.3\sin^2\phi_1 + 0.3\sin^2\phi_2)$$

For $\phi_2 = \phi_1 + 90°$, we have

$$\sin\phi_2 = \cos\phi_1, \cos\phi_2 = -\sin\phi_1$$

Hence,

$$\Delta F_x = C(\sin\phi_1\cos\phi_1 - \sin\phi_1\cos\phi_1 + 0.3\sin^2\phi_1 + 0.3\cos^2\phi_1) = 0.3C$$

$$\begin{aligned}\Delta F_y &= \Delta F_{t1y} + \Delta F_{t2y} + \Delta F_{r1y} + \Delta F_{r2y} \\ &= C(\sin^2\phi_1 + \sin^2\phi_2 - 0.3\sin\phi_1\cos\phi_1 - 0.3\sin\phi_2\cos\phi_2) \\ &= C(\sin^2\phi_1 + \cos^2\phi_1 - 0.3\sin\phi_1\cos\phi_1 + 0.3\sin\phi_1\cos\phi_1) \\ &= C \qquad (9.65)\end{aligned}$$

Result: both the ΔF_x and ΔF_y components remain constant. Consequently, the resulting force on the slice is constant and has a constant direction α:

$$\Delta F = C(1 + 0.09)^{1/2} = 1.044\,C$$

$$\alpha = \tan^{-1}\frac{F_y}{F_x} = \tan^{-1} 3.333 = 73.3° \qquad (9.66)$$

All the ΔF_x and ΔF_y on all the slices are equal and constant, and the resulting force dF on each slice has the same value and direction as given by Eq. (9.66). The total force is constant and has a constant direction:

$$F = 1.044\,NNC \qquad \alpha = 73.3°$$

where NN is the number of slices. ▲

EXAMPLE 9.12 Show that at Certain Axial Depths of Cut the Milling Force Is Constant

Consider Fig. 9.48. A cutter is engaged over a cutting angle ϕ_c, whatever its value. In diagram *(a),* the periphery of the cut is unrolled, and we see that in this instance the axial depth of cut b has been determined such that the pitch angle of the teeth, $\phi_p = 360°/m$, where m is the number of teeth, is equal to the edge-spread angle ψ:

$$b \tan \phi = r\psi = r\,\phi_p \tag{9.67}$$

The leading point of the tooth engaged in cutting (solid line) aligns with the trailing edge of the preceding tooth. The part drawn with a solid line of one cutting edge is so spread that if we divide the cutting engagement angle ϕ_c in small increments $\Delta\phi_c$ at an angle $\phi_n = n\Delta\phi_c$, then each of these, for any n, is filled by the presence of one cutting edge. All of the corresponding chip thicknesses h and force increments ΔF are represented once. If this cutter moves to another position, as shown in *(b),* where parts of two teeth are engaged, the situation is the same, and every $\Delta\phi_c$ is filled by one edge increment only. The cutting force in *(b)* is the same as in *(a);* indeed, it does not vary with the cutter rotation at all.

If on the contrary, the axial depth of cut is smaller, as shown in *(c),* there is always a part of the angle ϕ_c that is not occupied by a cutting edge, and this "gap" moves through the varying chip thickness. Correspondingly, the cutting force varies with cutter rotation.

If the axial depth of cut is greater than that given by Eq. (9.67), as in graph *(d),* there is a part of ϕ_c that is occupied by two edges, and this part moves through the varying chip thickness, and the cutting force varies again.

The constant force applies to the axial depth of cut as given by Eq. (9.67) and to axial depths that are integer multiples of that one. ▲

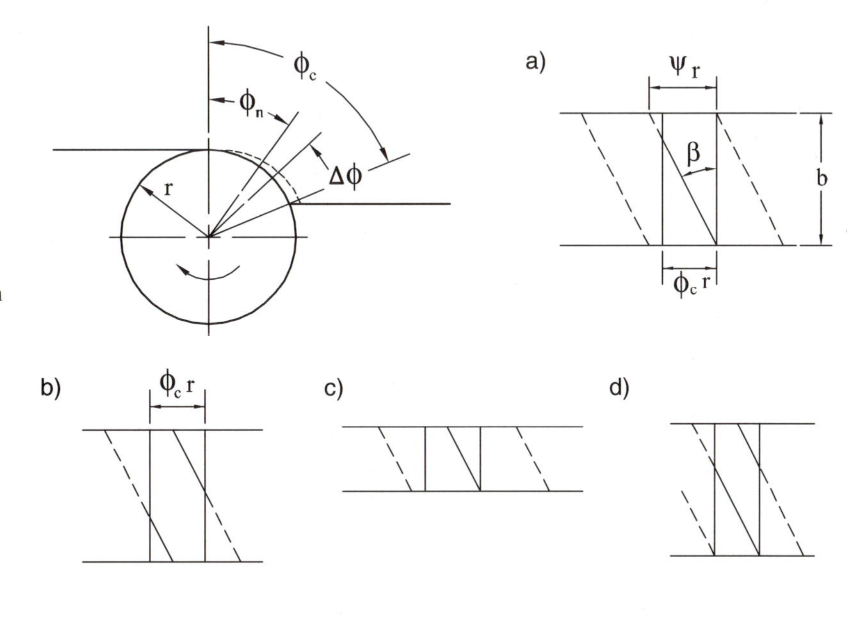

Figure 9.48
Showing that for a particular depth b, force is constant. In a) and b) depth of cut is such that the tooth spread ψ is equal to the tooth pitch ϕ_p. At any rotational position of the tooth, one and only one cutting point is found. That is not so in c), where spread is smaller than pitch, and in d), where spread is larger than pitch.

9.5.5 Errors of Surface Produced by End Mills with Helical Teeth: Static Deflections

Let us consider Fig. 9.49a, which deals with a case of up-milling. We will limit ourselves to up-milling. The analysis of down-milling would be analogous. The position of a point of the cutting edge is defined by the angle measured from the beginning of the cutting arc at the line *0*, which represents the contact of the cylindrical envelope of the end mill with the generated surface *S*. The line *0* is obtained as the intersection of a plane passing through the axis of the end mill and perpendicular to the direction of feed *X*, with the surface *S*. Every point of the cutting edge cuts from $\phi = 0$ to the exit angle $\phi = \phi_e$.

In diagram *(b)* the periphery of the cutter and the cut surface are unrolled. All the cutter teeth appear as straight lines inclined at the helix angle β. In the same way as in Fig. 9.43, the pitch angle of the teeth is denoted ϕ_p, and the span angle over which one edge extends for the given axial depth of cut b is denoted ψ. The radius of the cutter is r. Thus we have

$$r\psi = b \tan \beta$$
$$\psi = b \tan \frac{\beta}{r} \tag{9.68}$$

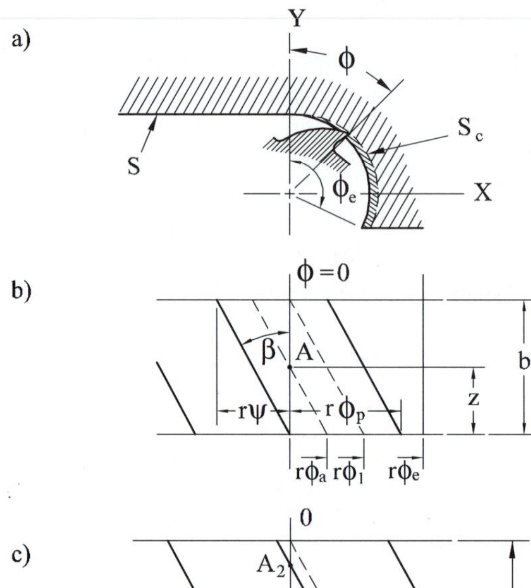

c)

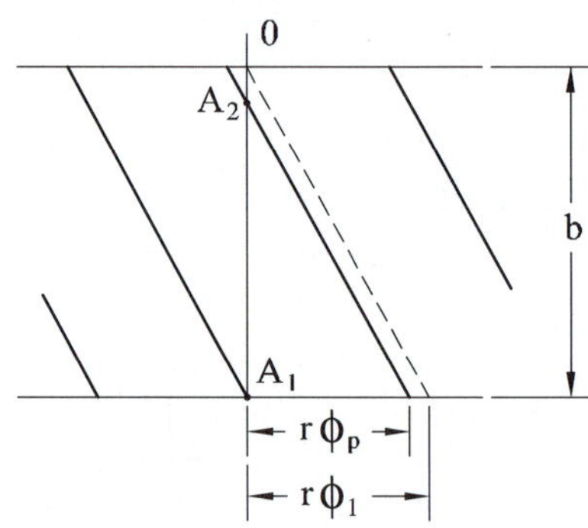

Figure 9.49
Analysis of the effect of force on accuracy of surface machined with end mills with helical teeth. a) Section through the cutter at any level. b) Depth of cut is small. Tooth climbs up on line *A* before next tooth starts at the bottom. c) At a larger depth, two teeth pass through *A* simultaneously.

Diagram *(b)* depicts a case in which $\psi < \phi_p$. One cutting edge is shown with its leading point exactly on the line *0*. As this edge progresses into the cut so that its leading point is at the angle ϕ_a, it will be generating the surface *S* on the line *0* in the point *A* at the height *z:*

$$z = \frac{r\,\phi_a}{\tan\beta} \tag{9.69}$$

It is obvious that the surface *S* is generated on the line *0* (which moves with the feed motion of the cutter) point after point and as the cutting edge passes, its generating point climbs from the bottom upwards, through the length *b.* This is repeated for each tooth.

Section 9.4.3 explained that the cutting force variation is periodic per one tooth period. The deflection *y* in the direction *Y,* which is perpendicular to the feed motion, is produced by the F_y component of the cutting force, expressed by Eqs. (9.56) and (9.57). A computer program was described in Fig. 9.44 for the computation of F_y as function of the angle ϕ_a, which was measured from the line *0* in the same way as indicated in Fig. 9.43. The computation and, if required, plotting of the cutting force starts at $\phi_a = 0$ and is carried out for ϕ_a increasing in small steps until $\phi_a = \phi_p$, whereupon the force repeats itself.

In a "static" approach it is possible to calculate the deflection *y:*

$$y = \frac{F_y}{k_a} = F_y h \tag{9.70}$$

where *h* is the flexibility at the cutting edge of the end mill. As F_y is obtained as a function of ϕ_a, Eq. (9.70) establishes *y* also as a function of ϕ_a. For every instantaneous position of the cutting edge defined by ϕ_a, there is a point *A* where the deflection *y* imprints itself on the surface *S* being generated. The position *z* of this point is established as a function of ϕ_a in Eq. (9.69). Eventually, we may eliminate the variable ϕ_a and establish *y* as function of *z.* In this way we obtain the profile of the error of the machined surface *S.*

Figure 9.49b shows that for a small axial depth of cut *b,* the cutting edge moves over an angle $\phi_1 < \phi_p$ when it reaches the top of line *0.* This means that deflections over a time span lesser than the tooth period are generating the machine surface. The extreme of straight teeth, that is, $\beta = 0$, produces the case of Fig. 9.37a, where we had $\phi_1 = 0$, and the generation of the surface was limited to a small fraction of the cutter rotation. The condition $\phi_1 < \phi_p$ arises if

$$b < b_p \tag{9.71}$$

where

$$b_p = \frac{r\,\phi_p}{\tan\beta} \tag{9.72}$$

is the axial depth of cut at which the span angle ψ of one cutting edge is $\psi = \phi_p$.

If, on the contrary $b > b_p$ (see Fig. 9.49c), then the surface *S* is generated in the line *0* simultaneously at two points, A_1, A_2, located on two cutting edges.

In an analogous way, once the force is being computed using the procedure developed in Section 9.5.4, the dynamic case of vibrations *y* can be resolved in the same way as it was in Example 9.10, and it can be expressed as function of the coordinate *Z* along the line 0. Thus the imprint of vibrations *y* on the machined surface can be determined.

Such a complete exercise is not included in this text. Instead reference is made to publications [10] and [11]. Professional software is available that includes this kind of exercise in simulation programs and also deals with regenerative force-deflection problems leading to chatter vibrations; refer to [12]. The procedure of the complete computations of the nonregenerative imprints of vibrations on the end-milled surface is outlined in Fig. 9.50. An assumed variation of force is plotted in *(a)*. It is assumed that this force excites vibration as plotted in *(b)*. It is shown as a single harmonic component with the same frequency as is the fundamental periodicity of the force, and it is phase-shifted by ϕ. The end mill with four helical teeth is drawn in *(c)*. Every tooth cuts through a surface starting at line *A-A* and ending at line *B-B*. The tooth starts to cut in point *1* at the base of the *A* line and it moves as shown by the successive helical lines, of which the one corresponding to the instantaneous position of the tooth shown in the cross section of the cutter is drawn as a solid line. However, only the point *5* where this line intersects line *A* contributes at this instant to the generation of the surface *S*. As the tooth rotates it keeps imprinting the vibrations through the succession of points *1, 2, . . . 6*. During the time it takes the tooth to climb line *A* from point *1* to *6* it vibrates, and these vibrations produce the surface profile error indicated in *(d)*. The computation has to follow these individual transformations.

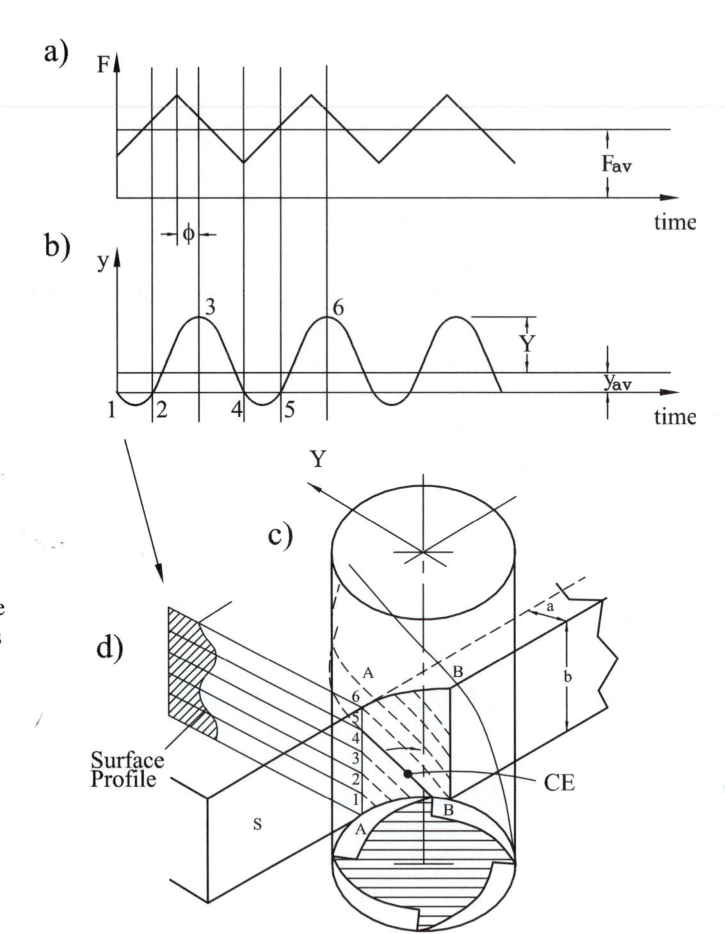

Figure 9.50
Surface error in end milling with helical teeth, the non-regenerative approach. Force F_y, vibrations y, and profiles of machined surface. To illustrate: a) assume this as the force variation and b) this as the corresponding vibration of the tool in direction *Y* shown in c). As the cutting edge *CE* rotates, it climbs on line *A-A* and imprints the vibration on surface *S* as shown in d).

9.6 CHATTER IN METAL CUTTING

9.6.1 General Features

Chatter is a self-excited type of vibration that occurs in metal cutting if the chip width is too large with respect to the dynamic stiffness of the system. Under such conditions these vibrations start and quickly grow. The cutting force becomes periodically variable, reaching considerable amplitudes, the machined surface becomes undulated, and the chip thickness varies in the extreme so much that it becomes dissected. Chatter is easily recognized by the noise associated with these vibrations, by the chatter marks on the cut surface, and by the appearance of the chips (see Fig. 9.51). Machining with chatter—unstable cutting—is mostly unacceptable because of the chatter marks on the machined surface and because the large peak values of the variable cutting force might

Figure 9.51
Chatter marks and chips. Top left: end-milled, peripheral, and bottom surfaces; right: turned surface; bottom left: face-milled periphery; right: chips are dissected by the vibration. [DISSERTATION OF F. ISMAIL]

cause breakage of the tool or of some other part of the machine. Correspondingly, the chip width and also the metal removal rate must be kept below the limit at which chatter occurs. In this respect, chatter is often the factor limiting metal removal rate below a rate which would normally correspond to the available power and torque on the spindle. Extensive work has been published on the various aspects of chatter in machining. Selected sources are in references [2], [13], [14], [15], and [16].

9.6.2 Mechanisms of Self-Excitation in Metal Cutting

The most significant cutting parameter, which is decisive for the generation of chatter, is the *width of cut* (*width of chip*) b (see Fig. 9.52). For sufficiently small chip widths, cutting is stable, without chatter. By increasing b chatter starts to occur at a certain width b_{lim} and becomes more energetic for all values of $b > b_{\text{lim}}$. The value of b_{lim} depends on the dynamic characteristics of the structure, on the workpiece material, cutting speed and feed, and on the geometry of the tool. In milling, the cumulative chip width b_{cum} has to be considered, which is the sum of the chip widths of all the teeth cutting simultaneously.

There are two main sources of self-excitation in metal cutting:

1. Mode coupling
2. Regeneration of waviness

Mode coupling is a mechanism of self-excitation that can only be associated with situations where the relative vibration between the tool and the workpiece can exist simultaneously in at least two directions in the plane of the orthogonal cut. This is symbolically expressed in Fig. 9.53, where the tool is shown attached to a mass suspended on two sets of springs perpendicular to each other. Simultaneous vibration in the directions X_1 and X_2 with the same frequency and a phase shift between the two results in an elliptical motion as shown. The workpiece may be moving with a steady cutting speed v. Assume that the tool moves on the elliptical path in the direction of the arrows. The cutting force F has the direction as shown. For the part of the periodic motion of the tool from A to B the force acts against this motion and takes energy away. During the motion from B to A the force drives the tool and imparts energy to its motion. Because motion $B \rightarrow A$ is located deeper in the cut the force F is larger than

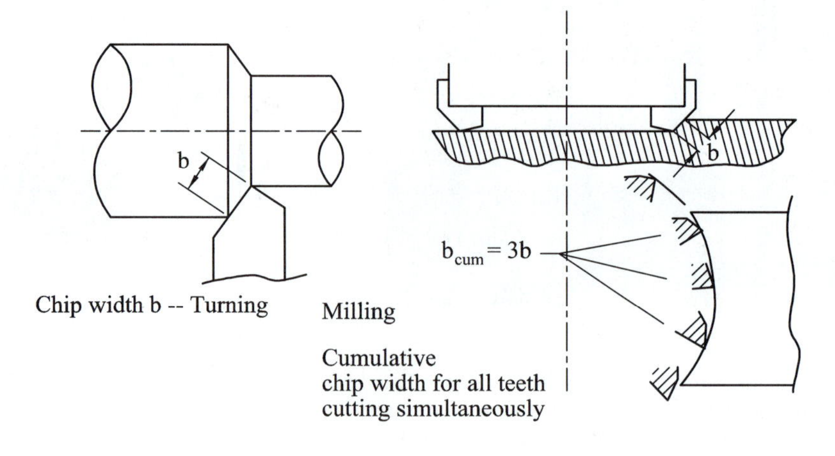

Figure 9.52
The chip width b as shown for turning and for milling is the decisive parameter for chatter.

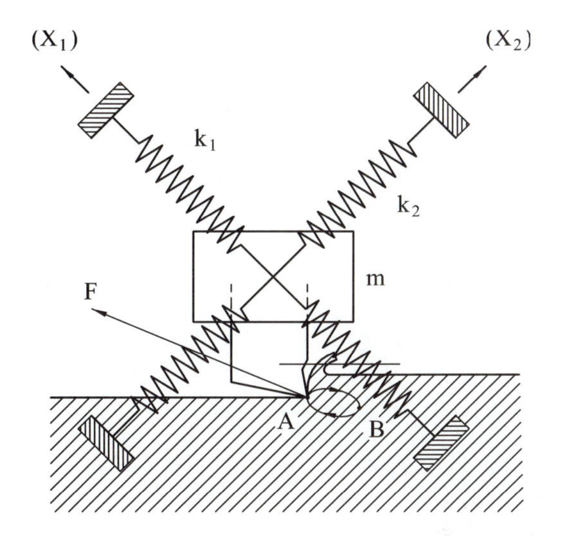

Figure 9.53
Mechanism of mode coupling in a system with two mutually perpendicular degrees of freedom. The two vibrations are phase-shifted and combine in an elliptical motion that is driven by force F in a bigger depth (bigger force) than when moving against the (smaller) force.

during the motion $A \rightarrow B$, and the energy delivered by the force F to the periodic motion in section $B \rightarrow A$ is larger than the energy taken away in the section $A \rightarrow B$. Periodically, there is surplus of energy sustaining the vibrations against damping losses. For various directional orientations of the system, its ability to derive energy in this way varies. This is illustrated in Section 9.6.4.

Regeneration of waviness (Fig. 9.54) is possible because in almost all machining operations the tool removes the chip from a surface that was produced by the tool in the preceding pass, that is, the surface produced in turning during the preceding revolution or, in milling, by the preceding tooth of the cutter. If there is relative vibration between tool and workpiece waviness is generated on the cut surface. The tool in the next pass (next revolution in turning, next tooth in milling) encounters a wavy surface and removes a chip with periodically variable thickness. The cutting force is periodically variable. This produces vibrations and, depending on conditions derived further on, these vibrations may be at least as large as in the preceding pass. The newly created surface is again wavy and in this way the waviness is continually regenerated.

Regeneration of waviness is influenced by the geometry of the operation, which imposes a constraint on the phasing between undulation produced in subsequent passes. Only in a planing, or shaping, operation is there no such constraint; see Fig. 9.54(a). Between the individual passes, the tool exits from the cut, and its vibrations decay. At the beginning of the next cut its vibration may freely adjust its phase for maximum regeneration. In turning, as in diagram *(b)*, where for simplicity of presentation a plunge cut is depicted, the phase ε between subsequent undulations is determined by a relationship between spindle speed n and frequency of chatter f. The number of waves between subsequent cuts is

$$N + \frac{\varepsilon}{2\pi} = \frac{f}{n} \tag{9.73}$$

where f is frequency in Hz, and n is spindle speed in rev/sec. N is the largest possible integer such that $\varepsilon/(2\pi) < 1$. In other words, there are N full waves and a fraction $\varepsilon/(2\pi)$ of a wave on the circumference of the workpiece. Depending on ε, the chip thickness variation $(Y_0 - Y)$, as shown in the inset, can be either zero for $\varepsilon = 0$ or maximum for

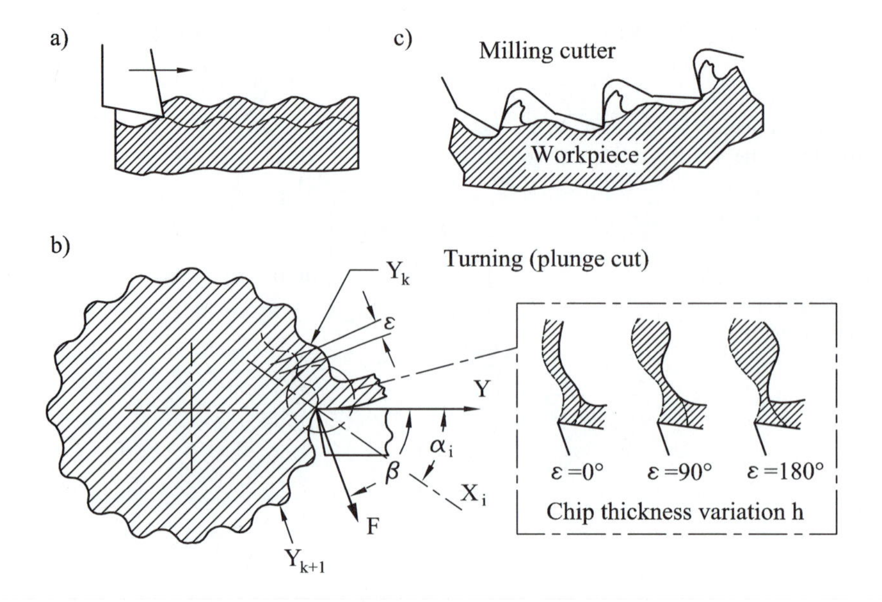

Figure 9.54
Regeneration of surface waviness: a) shaping; b) turning; c) milling. The chip thickness variation depends on the phase between vibrations of two subsequent passes.

$\varepsilon = \pi$. Obviously, there would be no chatter with $\varepsilon = 0$ because no periodic self-excitation force would be generated. It will be shown later that for maximum self-excitation, the phasing is close to $\varepsilon = -\pi/2$. If the geometric constraint just described led to a strongly different ε, the system would adjust frequency to achieve the most favorable phasing. In turning, even a small change of frequency leads to a large change of ε; the change of the number N by 1 produces a full 2π change of ε. If, for example, there are $N = 50$ waves on the circumference of the workpiece, a change of 2% in frequency produces a 2π change of ε. Correspondingly, the geometric constraint on phase ε has little effect on the limit of stability in turning. It should be noted that 2% change of spindle speed has the same effect on ε.

In milling (Fig. 9.54c) the geometric constraint is much more important. Regeneration of waviness on the surface is not done in subsequent spindle revolutions as in turning but by subsequent teeth of the cutter. Eq. (9.61) is now changed to

$$N + \frac{\varepsilon}{2\pi} = \frac{f}{nm} \tag{9.74}$$

where m is the number of teeth of the cutter. It may easily be that the number N is 2 or 3; that is, there are two and a fraction (three and a fraction) waves between subsequent teeth. Now, a change of ε by 2π, for $N = 2$, requires a 50% change of frequency or of cutting speed. The geometric condition has an important influence on stability, and this has a great practical significance to be discussed later on.

At this stage, no geometric constraint on the phasing of subsequent undulations in regenerative chatter is considered. This is equivalent to saying that we will be dealing with chatter as it would develop under the most favorable phasing, which in practice can always occur at some spindle speed in milling and which is almost always obtained in turning. In other words, it is a realistic simplification for turning, and it is the worst possible case in milling. It is sometimes called the borderline, or "critical" stability. The expressions for the limit of stability for regenerative and nonregenerative (mode coupling) borderline chatter will now be derived.

9.6.3 The Condition for the Limit of Stability of Chatter

The theory of chatter presented here is based on a number of simplifications which, however, have been proven not to substantially alter the most important effects on the limit of stability as they are found experimentally. These simplifications follow:

(a) The vibratory system of the machine is linear.

(b) The direction of the variable component of the cutting force is constant.

(c) The variable component of the cutting force depends only on vibration in the direction of the normal to the cut surface (Y).

(d) The value of the variable component of the cutting force varies proportionally and instantaneously with the variation of chip thickness.

(e) The frequency of the vibration and the mutual phase shift of undulations in subsequent overlapping cuts are not influenced by the relationship of wavelength to the length of cut; this assumption corresponds to an infinite length for every cut or, practically, to planing.

Assumption *(e)* is practically valid in planing, boring, and turning. In milling the phasing of undulations produced by subsequent cutter teeth may sometimes have a significant effect on stability. This is discussed again later.

The condition for the limit of stability is derived in a way published by Tlusty and Polacek [13]. It is discussed with reference to the diagram in Fig. 9.55. The structure of the machine tool is depicted as a frame with the workpiece at one end and the tool at the opposite end. There is the relative cutting motion between the two, in the direction of the cutting speed v. The structure is a vibratory system that is characterized by the individual modes of vibration, each of which represents a freedom of the relative motion between the tool and workpiece and has a particular direction. Directions X_k and X_j of two such modes are indicated. The vibration component, which is normal to

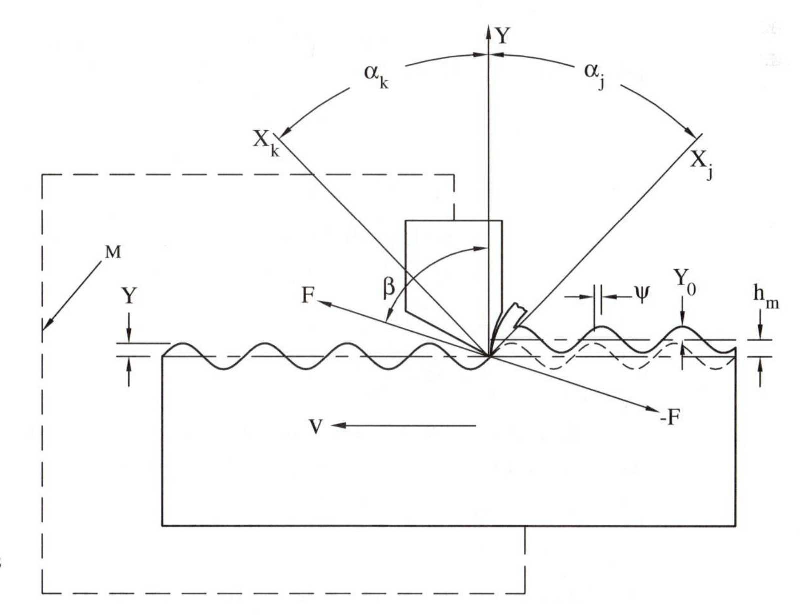

Figure 9.55
The regeneration diagram relating force, surface waviness, and vibration. Vibration Y between tool and workpiece as opposing points in the machine structure causes the tool to cut into the wavy surface Y_0 of the preceding pass; the variable chip thickness between the two causes variation of the force F that, in a feedback fashion, causes vibration Y.

the cut surface, produces undulations with an amplitude Y_0 in one cut, and in the subsequent cut this component has amplitude Y. These amplitudes express both the magnitude and the mutual phase shift of the vibrations, as well as of the undulations, in the two subsequent passes of cutting. The direction of the cutting force F is inclined by the angle β from Y, and the directions of the individual modes of vibration are measured from Y by corresponding values of angles α_i.

The process of self-excitation is a closed-loop one in which the vibrations cause a force variation, and the variable force in turn produces vibrations. The *force depends on vibrations* in (at least) two subsequent passes:

$$F = K_s bh,$$

where the chip thickness h consists of a steady-state value h_m (the mean chip thickness), which is equal to the down-feed per pass; and of the variations Y_0 at the top, the undulation of the surface from the previous pass; and of the variations Y at the bottom due to the vibration in the current cut:

$$h = h_m + (Y_0 - Y)e^{j\omega t} \tag{9.75}$$

Correspondingly, the force also has a mean component F_m and a variable component. Because we are considering a linear system, we may neglect the mean components and write the relationship between the *amplitude F of the variable force and the amplitude of the chip thickness variation:*

$$F = K_s b\,(Y_0 - Y) \tag{9.76}$$

where b is the chip width, and $(Y_0 - Y)$ is the chip thickness variation.

The feedback relationship of vibrations caused by this force is, in general:

$$Y = FG(\omega) \tag{9.77}$$

where $G(\omega)$ is the oriented transfer function of the system in the same sense used in Example 9.5 and Fig. 9.31; that is, it is a ratio of the complex amplitude of the Y component of all the X vibrations over the complex amplitude of a force acting in the direction F, as a function of frequency ω; both the vibrations and the force are taken as relative between the tool and the workpiece.

The oriented TF is obtained as a sum of all the direct TFs of the modes G_i multiplied by the directional factors u_i:

$$u_i = \cos\alpha_i \cos(\alpha_i - \beta)$$

$$G = \sum_1^i u_i G_i \tag{9.78}$$

Let us now combine Eqs. (9.76) and (9.77) so as to eliminate the force:

$$Y = K_s bG\,(Y_0 - Y)$$

After modification,

$$\frac{Y_0}{Y} = \frac{1/(K_s b) + G}{G} \tag{9.79}$$

The condition for the limit of stability may be formulated so that vibrations do not decay nor increase from pass to pass, or so that the magnitudes $|\bar{Y}_0|$ and $|Y|$ are equal:

$$\left|\frac{Y_0}{Y}\right| = 1 \tag{9.80}$$

Combining Eqs. (9.79) and (9.80), we obtain

$$\left|\frac{1}{K_s b} + G\right| = |G| \tag{9.81}$$

which expresses the equality of the absolute values of two complex numbers (the function $G(\omega)$ is complex, while $(K_s b)$ is real). This condition then has two parts:

$$\text{Im}(G) = \text{Im}(G)$$

which is obvious, and

$$\frac{1}{K_s b} + \text{Re}(G) = \pm \text{Re}(G),$$

here the + sign leads to $b = \infty$; the − sign gives

$$\frac{1}{K_s b} = -2\,\text{Re}(G)$$

which is the actual condition for the limit of stability. We can express the limit value of the chip width as follows:

$$b_{\text{lim}} = \frac{-1}{2K_s\,\text{Re}(G)} \tag{9.82}$$

The chip width b is a positive number. Eq. (9.82) can therefore only be satisfied for the negative part of the function $\text{Re}[G(\omega)]$. Moreover, of all the values b that satisfy Eq. (9.82), there is a minimum one, the smallest chip width at which chatter can occur. This is the actual critical limit of stability, and it corresponds to the largest negative value of $\text{Re}(G)$, to the minimum $\text{Re}(G)_{\text{min}}$:

$$b_{\text{lim,cr}} = \frac{-1}{2K_s\,\text{Re}(G)_{\text{min}}} \tag{9.83}$$

For chip widths $b < b_{\text{lim}}$ cutting is stable: there is no self-excited vibration. For $b > b_{\text{lim}}$ chatter will occur and grow. In practice, because of nonlinearities in the phenomenon, [see later, Eq. (9.96)], the amplitude of chatter will stabilize at a finite value.

Equation (9.83) has great practical significance. It is used to analyze and design structures with maximum stability at minimum weight. It offers a clearly defined criterion for the dynamic properties of machine-tool structures. Examples of the "modal analysis" of machine tools are discussed later.

9.6.4 Analyzing Stability of a Boring Bar

For now, let us look again at the example of the simple vibratory system with two uncoupled modes in a plane, as embodied by a rectangular bar with a mass at an end. Let us first consider the system as specified in Example 9.4 and Fig. 9.30. In Example 9.5, its oriented transfer function (OTF) was established between the cutting force F and the normal Y to the cut, as shown in the sketch of Fig. 9.31(a). The real part of this OTF was plotted in Fig. 9.31(b). Let us assume that there is a tool attached to the mass of the system (Fig. 9.56a) that is cutting a material with specific force $K_s = 1800$ N/mm^2. How wide will the tool be at the limit of stability?

Consider the OTF in Fig. 9.31. Its minimum, that is, its largest negative value is $\text{Re}(G)_{\text{min}} = -14.9 \times 10^{-5}$ mm/N, and it occurs at the frequency $f_{\text{lim}} = 443$ Hz. Using Eq. (9.83), we establish that

$$b_{\text{lim,cr}} = \frac{-1}{2K_s\,\text{Re}(G)_{\text{min}}} = 1.86 \text{ mm}$$

For chip width $b > b_{\text{lim}}$, chatter will occur with frequency f_{lim}.

Let us now change the orientation of the bar by rotating its section by 90° so as to reach the situation shown in Fig. 9.56b. The two modes with transfer functions G_1 and

G_2 have exchanged their places. We will still call the direction with the lower stiffness X_1; it is now associated with an angle $\alpha = 120°$, and the value of the directional factor is $u_1 = -0.321$. The factor $u_2 = 0.6634$ is assigned to the direction X_2 of the higher stiffness. The OTF is then obtained as

$$G = u_1 \operatorname{Re}(G_2) + u_2 \operatorname{Re}(G_1) \tag{9.84}$$

where the functions $\operatorname{Re}(G_1)$ and $\operatorname{Re}(G_2)$ are the same as before. The new OTF is plotted in Fig. 9.57; we can see that not only is the more flexible function now multiplied by a smaller directional factor, but because this factor is negative, it is now the lower

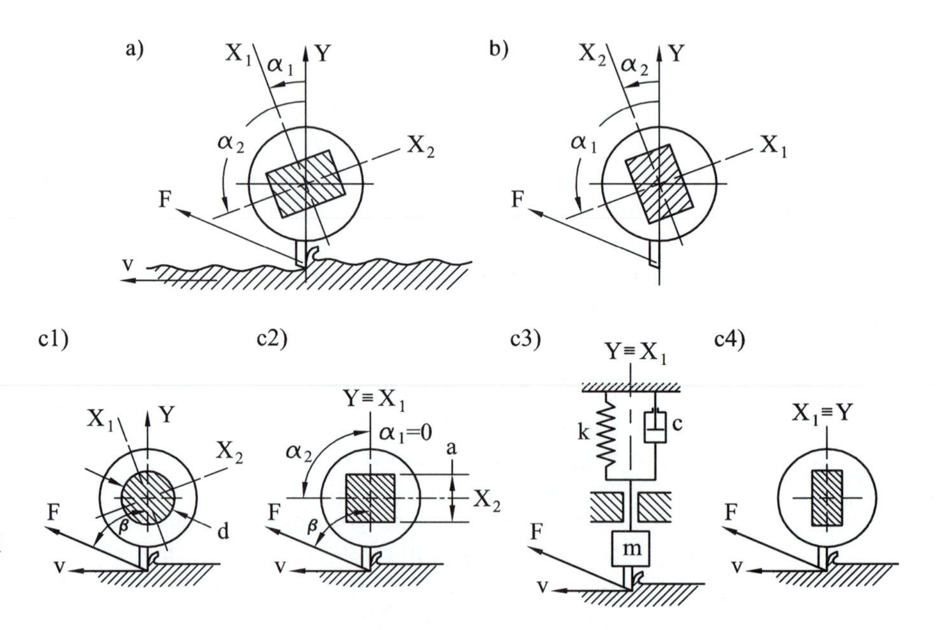

Figure 9.56
Various orientations of the boring bar. a) The orientation of a rectangular bar defined in Example 9.4 (and in Fig. 9.31a), with the OTF $\operatorname{Re}(G)_{\min}$ presented in Fig. 9.31b. b) The bar is rotated by 90° with respect to tool location. Now the direction X_2 of the larger stiffness halves the angle between normal Y and cutting force F. c) Orientations leading to an action as of a single-degree-of-freedom system (SDOF): c1) round section, c2) square section, c3) model of an SDOF, c4) a rectangular bar with $X_1 - Y$.

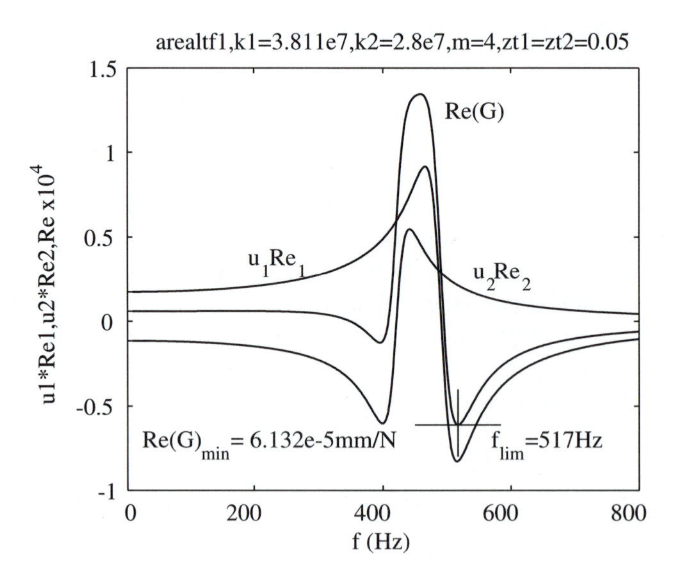

Figure 9.57
The OTF for the rectangular bar in orientation (b) of Fig. 9.56. The absolute value of $\operatorname{Re}(G)_{\min}$ is about two times smaller than in orientation (a).

frequency function that is reversed, and the summing of the two functions is completely different than before.

The minimum of the OTF has a value of $\mathrm{Re}(G)_{\min} = 6.1 \times 10^{-5}$ mm/N, and it occurs at $f_{\lim} = 517$ Hz.

The limit of stability is

$$b_{\lim,\mathrm{cr}} = \frac{-1}{2K_s \, \mathrm{Re}(G)_{\min}} = 4.62 \text{ mm}$$

In this case the limit width of the chip is 2.5 times larger than for the orientation of Fig. 9.56a.

The comparison of the stability of the two orientations of this bar shows that for the same weight of the structure, we can obtain large differences in stability depending on the mutual relationships and orientation of modes. Proper analysis allows us to optimize the design of a structure for optimum stability.

Let us now look at the stability of a bar with a square or round section and refer to Fig. 9.56c. If two mutually perpendicular directions X_1 and X_2 are chosen, the stiffnesses will be the same in both these directions, however they have been chosen. A round section, as in *(c1)*, does not have any particular orientation; thus it is preferable to choose the two directions so that one of them coincides with the cutting speed v and the other with the normal Y to the cut surface. Similarly, a square section *(c2)* may be oriented in this way. In deriving the OTF we find that the directional factor of the mode parallel with v, the mode X_2 is $u_2 = \cos \alpha_2 \cos (\alpha_2 - \beta) = 0$ because $\alpha_2 = 90°$. It is actually obvious that vibration X_2 cannot participate in self-excitation because it does not modulate the chip thickness (it does not provide any component in Y).

Consequently, for the case of chatter, these systems may be viewed as having a single degree of freedom, as modeled in *(c3)*. Any rectangular section bar with this particular orientation will also act as a single-degree-of-freedom (SDOF) system of *(c3)*. In this sense, *(c4)* is equivalent to *(c2)* if it has the same stiffness in the $X_1 = Y$ direction.

Let us take the case of a bar with a round section, with diameter d such that the section area is equal to that of the two previous examples based on the specification in Example 9.4. The length of the bar is again $l = 300$ mm, and the mass at its end is $m = 4$ kg. The damping ratio is chosen again as $\zeta = 0.05$. Thus we have

$$\frac{\pi d^2}{4} = ab = 60 \times 70 \qquad d = 73.13 \text{ mm}$$

The moment of inertia of the section is

$$I = \frac{\pi d^4}{64} = 1.40375 \text{ mm}^4$$

and the stiffness at the end of the bar is

$$k = \frac{3EI}{l^3} = 3.2666 \times 10^4 \text{ N/mm}$$

The limit of stability for the round bar is obtained by considering the transfer function $\mathrm{Re}(G)$ of the single applicable degree of freedom, formula (9.62), and the directional factor,

$$u = \cos \alpha \cos (\beta - \alpha), \qquad \alpha = 0$$

$$u = \cos \beta = \cos 70° = 0.342$$

It is not necessary to construct the whole OTF. It is sufficient to express the value of the minimum point, from formula (9.46):

$$u\,\mathrm{Re}(G)_{\min} = \frac{u}{4k\zeta\,(1+\zeta)} = 4.985 \times 10^{-5}\ \mathrm{mm/N},$$

$$b_{\lim} = \frac{-1}{2K_s u\,\mathrm{Re}(G)_{\min}} = 5.57\ \mathrm{mm}$$

The round bar gives a still higher stability, 1.2 times more, than the rectangular bar in the orientation of Fig. 9.56b. However, if we orient the rectangular bar as in (c4) so that only its mode with higher stiffness of $k_2 = 3.81 \times 10^4$ N/mm applies, which is higher than the stiffness $k = 3.267 \times 10^4$ N/mm of the round bar, it will give the highest stability:

$$b_{cr} = \frac{3.811}{3.2666}\,5.57 = 6.5\ \mathrm{mm}$$

The directional orientation of the bar is obviously very important. The values of $b_{\lim}$ have been calculated for various orientations. It was found that maximum stability of $b_{\lim} = 6.5$ mm corresponding to the orientation $\alpha_1 = 78°$ is 4.6 times higher than the minimum of $b_{\lim} = 1.8$ mm to $\alpha_1 = 35°$.

9.6.5 Another Way of Deriving the Limit of Stability, Using the Nyquist Criterion

We now examine another way of deriving the expression (9.82) that is based on the classical approach of the control systems theory. It also gives an insight into the role of the phase shift between undulations of subsequent cuts. Refer to Fig. 9.55, which involves the OTF G of the vibratory system and the generation of a variable cutting force F as a result of both the undulated surface from which a chip is being cut and the vibration of the tool in the cut. This situation can be expressed by the block diagram in Fig. 9.58. The variable force F_v excites the structure. Vibration, relative between tool and workpiece, is produced through the corresponding transfer function $G(\omega)$ of the structure. The difference between Y and vibration Y_0 of the preceding pass, which produced the waviness of the surface from which the cut is taken and which was delayed by τ_r plus the feed per pass H_m, is the average chip thickness H. The delay time τ_r is the time between the two cutting passes (e.g., the time of a revolution in turning). The time delay for a vector rotating with frequency ω results in a phase shift $\varepsilon = \tau_r\omega$, and it is represented by the factor $e^{-j\varepsilon}$. When multiplied by $K_s b$, it produces the force F, which, when compared with the mean force F_m, leaves the variable force F_v that excites the structure. The diagram includes implicitly the relationships expressed in Eqs. (9.76) and (9.77).

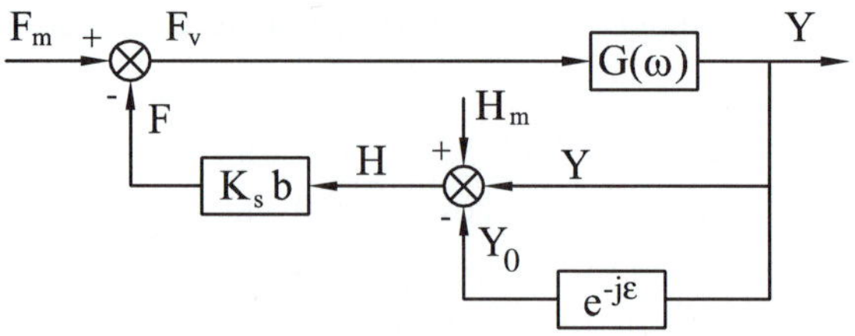

Figure 9.58
Block diagram of vibration in regenerative cutting. It is used in the control system approach in deriving limit of chatter stability. Difference between vibrations Y and Y_0, the latter delayed by τ_r (time between two passes or revolutions in turning), is multiplied by $K_s b$ to result in cutting force that is fed back. The time delay is expressed in the frequency domain by negative phaseshift $\varepsilon = \tau_r\omega$.

At the *limit of stability* any vibration would remain constant without decaying or increasing. The absolute values of amplitudes of vibrations in subsequent passes are equal:

$$|Y| = |Y_0| \tag{9.85}$$

This will be obtained, according to the Nyquist criterion if the open-loop transfer function has the value -1:

$$K_s\, b\, G(\omega)\,(1 - e^{-j\varepsilon}) = -1 \tag{9.86}$$

The value of the chip width b at the limit of stability is

$$b_{\lim} = \frac{-1}{K_s\, G(1 - e^{-j\varepsilon})} \tag{9.87}$$

As the coefficient K_s was chosen to be real, and b is obviously always real, the condition (9.87) is only satisfied if

$$G(1 - e^{j\varepsilon}) = G - Ge^{-j\varepsilon} \tag{9.88}$$

is real. Because $e^{-j\varepsilon}$ is a unit vector, we have

$$|G| = |Ge^{-j\varepsilon}| \tag{9.89}$$

The two vectors of Eq. (9.89) are depicted in Fig. 9.59. The force F is taken as real and scaled to a unit value. The transfer function G of the system is plotted in the complex plane as in Fig. 9.29, except that this one represents a system with two degrees of freedom, like the one in Fig. 9.53. The vector G drawn between the origin of coordinates and point 1 corresponds to a particular frequency of the force F. The vector $Ge^{-j\varepsilon}$ has to be equal in magnitude, and the difference $(G - Ge^{-j\varepsilon})$ must be, according to Eq. (9.88), real; therefore, in the diagram it is parallel with the real axis. We then see that

$$G - Ge^{-j\varepsilon} = 2\mathrm{Re}(G) \tag{9.90}$$

and this equation combined with (9.85) leads to

$$b_{\lim} = \frac{-1}{2K_s\,\mathrm{Re}(G)}$$

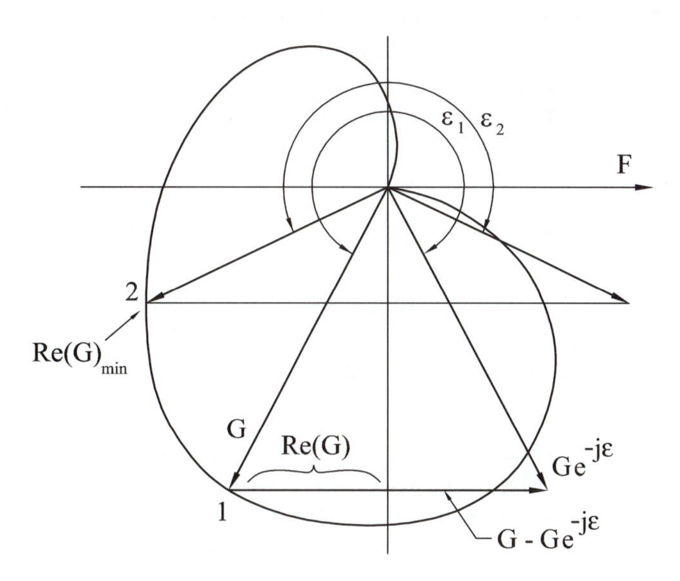

Figure 9.59
The vectors of force and of two subsequent undulations at the limit of stability. The diagram shows the phase angle ε between Y and Y_0 for two different speeds. The chip thickness variation $(Y - Y_0)$ is a real vector.

which is identical with Equation (9.82) obtained before. The vector G drawn to point 1 corresponds to a particular value of the phase shift ε between undulations in subsequent cuts, and there is a corresponding frequency of vibration that corresponds to point 1 on the transfer function. The smallest value of b_{lim} will correspond to the largest value of Re(G), which is found at point 2. This gives the critical value of b_{lim} as expressed in Eq. (9.83):

$$b_{cr} = \frac{-1}{2K_s \operatorname{Re}(G)_{\min}}$$

However, this critical case can only occur if the phase shift ε_2 is possible. This condition may become important for the stability of milling, and we discuss this later in reference to Fig. 9.74.

9.6.6 Time Domain Simulation of Chatter in Turning

The development of chatter can be well illustrated and understood by using digital simulation. This approach not only permits good insight into the behavior of the vibrating system but makes it possible to correctly account for the basic nonlinearity of the process and the cross effects of teeth of milling cutters. The aspect of nonlinearity is explained as we proceed. For now, let us indicate that it is mainly due to the fact that when vibration grows larger, the tool jumps out of the cut for a part of the vibrational period, and the cutting force disappears for this time instant. The cross effects in milling arise from the different directions of the normals to the cut surface at each tooth. For example, a vibration that is tangential to the cutting of one tooth and does not, therefore, modulate its chip thickness, may be most efficient in modulating the chip thickness for a tooth 90° away on the cutter circumference for which it is normal to the cut surface, and vice versa.

The model used for simulation of turning is shown in Fig. 9.60. A system with two mutually perpendicular degrees of freedom X_1 and X_2 is used. A tool is attached to the mass of the system; it cuts an undulated surface and leaves another undulated surface behind. The normal to the cut surface is denoted Y.

The simulation is carried out by proceeding in small time steps dt and following the motion of the system as it is determined by the differential equations of balances of forces in the direction X_1 and X_2, starting from some initial conditions. The count of the time steps is denoted n, which is also the subscript of selected variables at a given time.

Subsequent passes of cutting follow one another after a period T, which corresponds to one revolution of the workpiece and contains i time steps so that

$$t_n = n\mathrm{d}t \tag{9.91}$$

is the time from the beginning of the simulation, and

$$T = i\mathrm{d}t \tag{9.92}$$

is the time period between subsequent passes.

The average chip thickness is h_{av}. The instantaneous chip thickness is usually taken as the difference of the tool positions in two subsequent passes:

$$h = h_{\text{av}} + y_{n-i} - y_n \tag{9.93}$$

but, as vibrations grow more than one preceding cut may be involved. Looking at the bottom of Fig. 9.60, at any time t_n we may check which of the cuts preceding the present cut reached lowest in the workpiece material and denote such position $y_{\min}$:

$y_{\min}$ is the lowest of $y_{n-i} + h_{\text{av}}$, $y_{n-2i} + 2h_{\text{av}}$, $y_{n-3i} + 3h_{\text{av}}$, and so on.

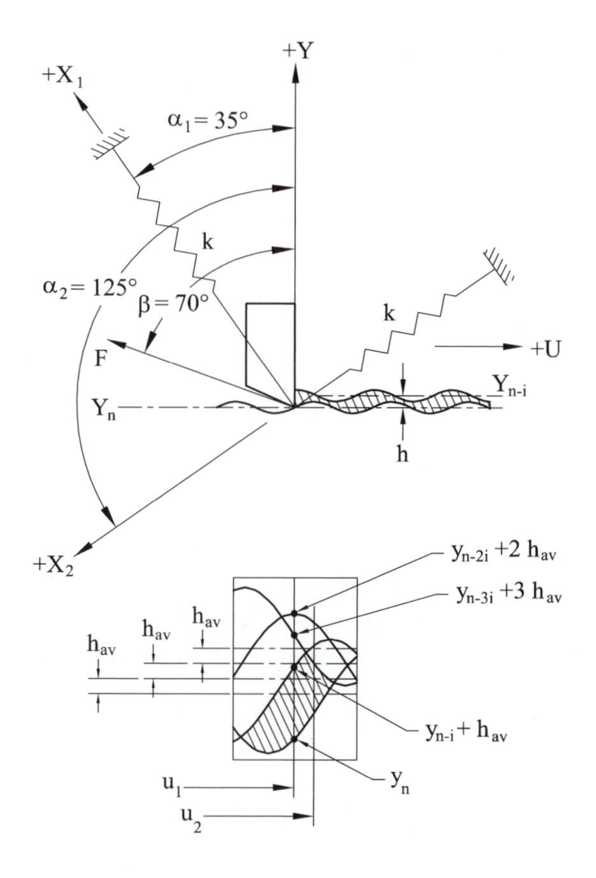

Figure 9.60
Model for simulation of turning. In the *n*th pass the vibrating tool cuts into surface waviness created in preceding passes.

The instantaneous chip thickness is:

$$h = y_{\min} - y_n \tag{9.94}$$

and it is this equation that applies rather than Eq. (9.93). The material removed in the pass y_n is indicated by the shaded area, and chip thickness at the distance u_1 is $h = y_{n-i} + h_{\text{av}} - y_n$, but at distance u_2 it is $h = y_{n-3i} + 3\,h_{\text{av}} - y_n$. The cutting force at the instant t_n is proportional to chip thickness h and to chip width b:

$$F_n = K_s bh \tag{9.95}$$

but, if

$$F_n < 0, \quad F_n = 0 \tag{9.96}$$

Equation (9.96) expresses the basic nonlinearity. The force is proportional to chip thickness h, but if the tool moves above the surface, this is formally recognized by obtaining $h < 0$, $F_n < 0$. This should not be accepted; instead, the force should be set to zero: The cutting force is never negative.

The force excites vibration X_1 and X_2 by its components:

$$\begin{aligned} F_1 &= F_n \cos(\beta - \alpha_1) \\ F_2 &= F_n \cos(\beta - \alpha_2) \end{aligned} \tag{9.97}$$

The forces excite vibrations according to

$$m_j\ddot{x}_j + c_j\dot{x}_j + k_jx_j = F_j \quad j = 1, 2 \tag{9.98}$$

In the simulation, in each step n the acceleration $\ddot{x}_{jn}$ is determined from (9.98), and the displacement $x_{j,n+1}$ is obtained by double integration. Thus, for example, for x_1:

$$\ddot{x}_{1,n} = \frac{F_{1,n} - c_1\dot{x}_{1,n} - k_1 x_{1,n}}{m_1} \tag{9.99}$$

$$\begin{aligned} \dot{x}_{1,n+1} &= \dot{x}_{1,n} + \ddot{x}_{1,n}\,\mathrm{d}t \\ x_{1,n+1} &= x_{1,n} + \dot{x}_{1,n+1}\,\mathrm{d}t \end{aligned} \tag{9.100}$$

We obtain $x_{2,n+1}$ in a similar fashion. It is then possible to express the tool motion in the direction Y:

$$y_n = x_{1,n}\cos\alpha_1 + x_{2,n}\cos\alpha_2 \tag{9.101}$$

The system is linear as long as the tool does not leave the cut. Therefore, for incipient vibration, Eq. (9.83) for the basic limit of stability width of chip b_{lim} applies:

$$b_{\text{lim}} = \frac{-1}{2K_s\mathrm{Re}(G)_{\text{min}}}$$

where $\mathrm{Re}(G)_{\text{min}}$ is the real part of the OTF between vibration y and force F, and most favorable phasing is considered between vibrations in subsequent cuts. This latter condition is practically fulfilled if the number of waves between subsequent cuts is large.

The simulation is performed in a computer program by reading in the following:

Input data: k_1, c_1, k_2, c_2, m, α_1, α_2, β, h_{av}, K_s, b, $\mathrm{d}t$, i, and the total time of simulation $t_{\text{tot}} = M\mathrm{d}t$

Initial conditions: $t_1 = 0$, $x_{1,1} = 0$, $x_{2,1} = 0$, $\dot{x}_1 = 0$, $\dot{x}_2 = 0$

Then carry out the computation in loops from $n = 1$, to $n = M$, according to equations (9.91) and (9.94) through (9.101).

For the first revolution, for $1 < n < i + 1$, the force will obviously not be expressed by the formula (9.95) using formulas (9.93) and (9.94), because the chip is being removed from a non-undulated surface. Therefore we have, for $n < i + 1$,

$$h = h_{\text{av}} - y_n \tag{9.102}$$

Then, the computer program proceeds formulating a loop in a sequence: Eq. (9.93) or Eq. (9.94), Eq. (9.95), Eq. (9.96); this establishes the applicable force; further, the component forces are obtained from Eq. (9.97); displacement x_1 is determined using Eqs. (9.99) and displacement x_2 is determined in an analogous sentence of second and first derivatives and finally the displacement itself; then, y is obtained using Eq. (9.101); return to the start of the loop. The loop is repeated in small time steps $\mathrm{d}t$, altogether M times. So, simply,

```
for n = 1:M
if n < (i + 1)
F(n) = K_s b (h_av - y(n));
else
F(n) = K_s b (h_av - y(n) + y(n - i));
end
if F(n) < 0
F(n) = 0;
end
F_1 = F(n) * cos (beta - alpha_1);
F_2 = F(n) * cos (beta - alpha_2);
xdd_1 = (F_1(n) - c_1*xd_1 - k_1*x_1(n))/m_1;
xd_1 = xd_1 + xdd_1*dt;
x_1(n+1) = x_1(n) + xd_1*dt;
```

```
% insert here analogous three lines for x2
y(n) = x1(n)*cos(alpha1) + x2*cos(alpha2);
end
```

As a result of the computation, it is interesting to plot (y_n, t_n) for $0 < t < t_m$ or over selected periods and, for selected cycles, to plot (y_n, u_n). The latter would demonstrate the elliptical motion of the tool, to be compared with Fig. 9.53.

EXAMPLE 9.13 Simulation of Chatter in Turning ▼

Write a simulation program for a system according to Fig. 9.60 for the following parameters:

$k_1 = 4.0 \times 10^7$ N/m, $k_2 = 5.6 \times 10^7$ N/m

$c_1 = 5320$ N sec/m, $c_2 = 7480$ N sec/m

$m = 100$ kg

$\alpha_1 = 30°$, $\alpha_2 = 120°$, $\beta = 70°$

$h_{av} = 0.10$ mm

$dt = 0.0001$ sec,

$K_s = 2000$ N/mm^2

$i = 1000$, $M = 5000$

The OTF of this case is shown in Fig. 9.61. The value of Re $(G)_{min} = 1.226$ e-4 mm/N, and the critical frequency is 105 Hz.

In the manner demonstrated in Section 9.6.4, it could be determined that the limit chip width for this system is

$$b_{lim} = 2.04 \text{ mm}$$

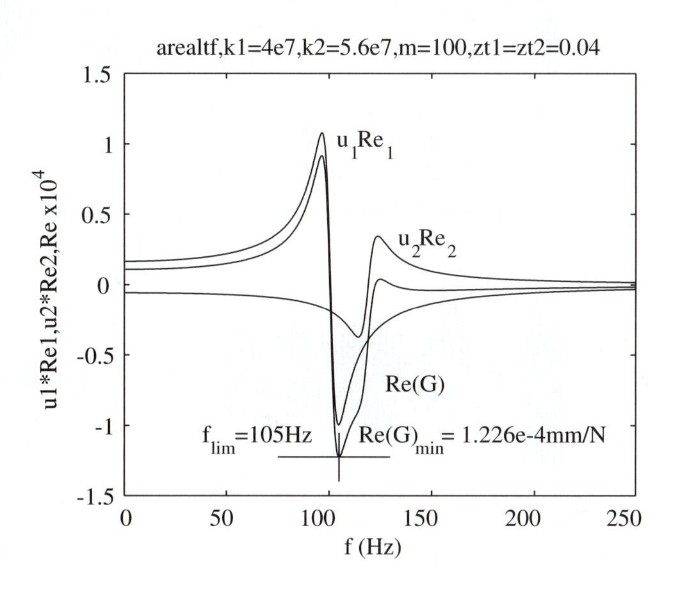

Figure 9.61
The OTF for the system defined in Example 9.13.

Carry out the simulation as advised in Eqs. (9.91) and (9.94) through (9.102), for the following:

(a) $b = 0.8\, b_{\text{lim}} = 1.63$ mm
(b) $b = b_{\text{lim}} = 2.04$ mm
(c) $b = 1.2\, b_{\text{lim}} = 2.45$ mm
(d) $b = 1.5\, b_{\text{lim}} = 3.06$ mm

Plot y versus t. Note: For simplicity, you may disregard Eq. (9.94) and just use Eq. (9.93) instead. The results of the computations are presented in Fig. 9.62. At the beginning, in all these cases, the system meets a step input in the cut with chip thickness $h_{\text{av}} = 0.10$ mm. This produces the first event of predominantly natural vibration which decays to almost zero. The vibration is centered at a steady deflection of

$$y_m = K_s b h_{\text{av}} \left(\frac{u_1}{k_1} + \frac{u_2}{k_2} \right) \tag{9.103}$$

where $u_i = \cos \alpha_i \cos (\alpha_i - \beta)$; thus we have

$$y_m = 2000 \times 0.10 \left(\frac{0.663}{4} \times 10^4 - \frac{0.321}{5.6} \times 10^4 \right) b$$

$$= 0.00216\, b$$

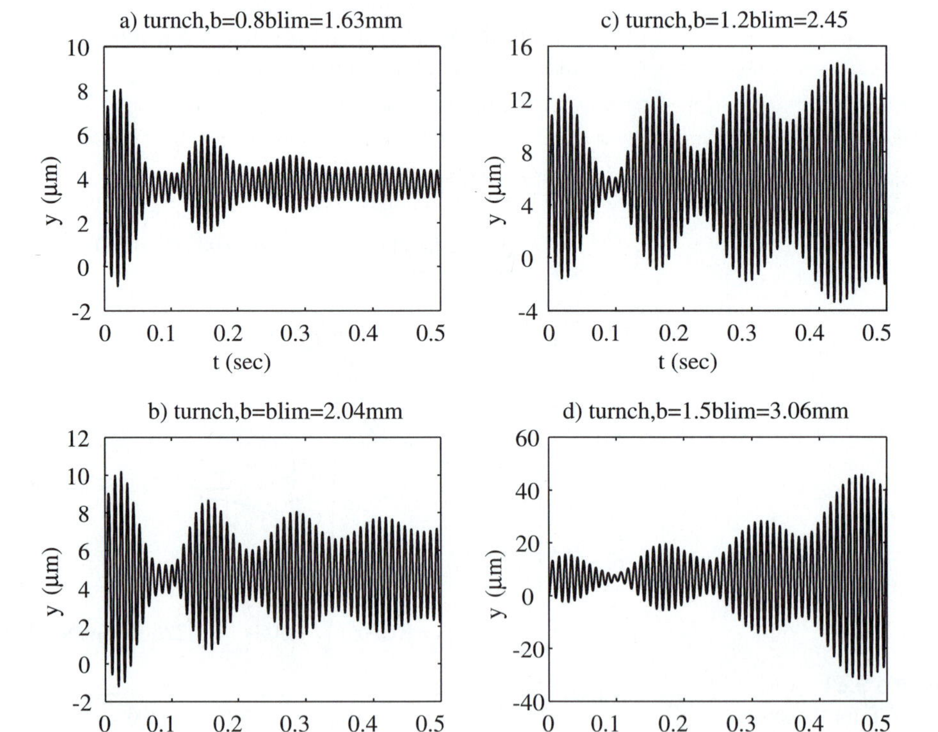

Figure 9.62
Results of Example 9.13. a) Simulated chatter for $b = 0.8\, b_{\text{lim}}$; b) simulated chatter for $b = b_{\text{lim}}$; c) simulated chatter for $b = 1.2\, b_{\text{lim}}$; d) simulated chatter for $b = 1.5\, b_{\text{lim}}$.

For the individual cases:

(a) $y_m = 0.0035$ mm $= 3.5\ \mu$m

(b) $y_m = 0.0044$ mm $\simeq 4.3\ \mu$m

(c) $y_m = 0.0053$ mm $\simeq 5.3\ \mu$m

(d) $y_m = 0.0066$ mm $\simeq 6.6\ \mu$m

One revolution of the workpiece takes $idt = 0.1$ sec. Each diagram corresponds to five revolutions. During the first revolution the principle of regeneration of waviness did not work because cutting was done from a non-undulated surface. However, some self-excitation is possible due to mode coupling. In all the four cases the vibrations are regenerated in the second, third, and subsequent revolutions. However, in case *(a),* where $b = 0.8\ b_{\text{lim}}$, the vibrations are decreasing revolution after revolution. This case is stable. In case *(b)* where $b = b_{\text{lim}}$, vibrations are becoming more and more uniform and, after a greater number of revolutions they would develop a constant amplitude. The case is at the limit of stability. The magnitude of vibrations during the first five revolutions is, on average, about 6 μm peak-to-peak (PTP). In case *(c)* with $b = 1.2\ b_{\text{lim}}$, vibrations start to grow already during the first revolutions, and in the fifth revolution their average magnitude is about 16 μm, PTP. In case *(d)* with $b = 1.5\ b_{\text{lim}}$, the growth of vibrations is very fast. In the fourth revolution they reach an average value of about 70 μm, PTP, and they grow still further.

The amplitude of vibrations in Fig. 9.62d is growing larger than the average chip thickness $h_{\text{av}} = 100\ \mu$m. At $t = 0.5$ sec it has already reached almost 35 μm. In order to follow the development, the simulation of case *(d)* was continued for another 0.5 sec. The result is shown in Fig. 9.63. After about eight revolutions (after 0.8 sec) the amplitude reached about 125 μm and did not grow any more. The reason is that for parts of the vibrational cycles the tool is moving in the air and no longer generates self-excitation energy. This is evident in the plot of the force F in graph *(b)*. For parts of the cycle the force drops to zero. This is the basic nonlinearity of self-excited vibrations in machining. ▲

9.6.7 Chatter in Milling

Chatter in milling is more complex than in turning because the system of forces on the individual teeth rotates with respect to the directions of flexibilities of the vibratory

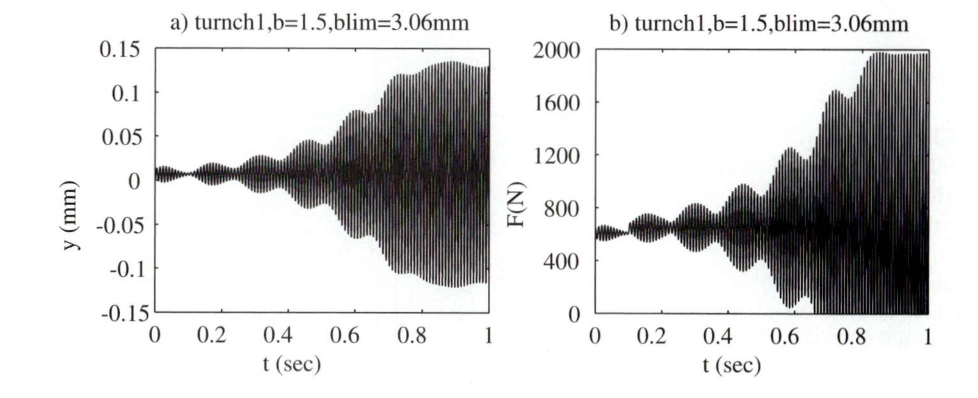

Figure 9.63 Example 9.13. Simulated chatter for $b = 1.5\ b_{\text{lim}}$ for a double time period. a) Vibration y; b) force F_y: saturation is indicated by the truncations $F_y = 0$.

system. These are usually related to the nonrotating part of the structure. We will not present a comprehensive case; instead, we present an example of a particular case. Let us consider a milling cutter with 8 teeth (see Fig. 9.64) in a slotting cut, that is, with each tooth cutting through a 180° engagement. There are 4 teeth cutting simultaneously. The spindle is mounted in a housing with one vibrational degree of freedom in each of the structural axes *X* and *Y*. The tool rotates with *N* rpm, and the workpiece feeds with feed rate *f*, chip load *c* per tooth in the direction $-X$.

A tooth at the angular position ϕ produces a tangential force F_T and a normal force F_N. The force F_T is proportional to the chip cross section:

$$F_T = k_s\, b\, h$$

where *b* is the axial width of cut, and

$$h = c \sin\phi - z(t) + z\,(t - T)$$

where $z(t)$ is the current instantaneous vibrational displacement of the cutting edge in the radial direction and $z(t - T)$ is the displacement of the preceding tooth, when it was at the current position, that is, by the time *T* back, where *T* is the period of the passage of the teeth: $T = 60/Nm$, where $m = 8$ is the number of teeth, and *N* is spindle speed (rpm). In the computer program of Fig. 9.65 this displacement is called z_{old}. The displacements *z* result from displacements *x* and *y*:

$$z = x \sin\phi - y \cos\phi$$

The program runs in time steps d*t*, with 320 steps per revolution, that is, 40 steps between teeth, the cutter rotating by $dfi = \pi/160$. In the first part the displacements *z* on all 4 teeth and instantaneous chip thicknesses *h* are established. The factor 1000 is used because *x* and *y* have been obtained in meters, but the specific force will use *z* and *h* in millimeters. Then the forces *F1* through *F4* are obtained while a check is made whether a tooth really cuts at a particular instant, $h > 0$, $F > 0$ or whether it is jumping out of the cut, $h < 0$, $F = 0$. Further on, the component forces F_x and F_y are obtained and accelerations $\ddot{x}$ and $\ddot{y}$ calculated and integrated into velocities and further integrated into the next displacements $x(n + 1)$, $y(n + 1)$.

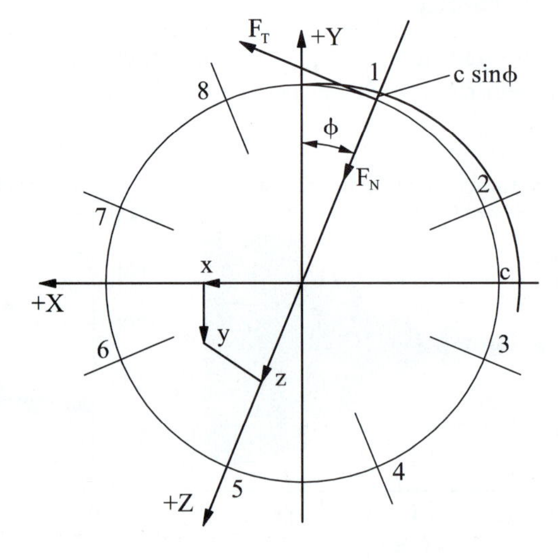

Figure 9.64
Milling cutter with 8 teeth in a slotting cut. The cutter rotates clockwise and feeds in the $-X$ direction. The stability of this case of milling is analyzed in Section 9.6.7.

Figure 9.65
Matlab listing "millchatt" for milling chatter for the case of Fig. 9.64.

```
function[zold1,zold2,zold3,zold4,x,y,Fx,Fy,t]=milchatt(b,N,M)
mx=4;kx=7e7;cx=535;my=4;ky=1e8;cy=737;Ks=1500;c=0.1;
x(1)=0;dx=0;y(1)=0;dy=0;i=0;dfi=pi/160;dt=60/(320*N);
%This is simulation of chatter in milling.Refer to Fig.9.64. b is depth of cut(mm),
%N is spindle speed (rev/min),M is total number of computational loops,
%c is chip load(mm),m=8 is number of teeth on the cutter;
%there are 40 steps per tooth, 320 steps per revolution
%because No of teeth is multiple of 4 and it is a slotting case there
%is no periodicity per tooth. There are a maximum 4 teeth simultaneously
%in the cut. fi1 through fi4 is the angle of rotation of individual teeth
for n=1:M
t(n)=(n-1)*dt;
fi1=i*dfi;
fi2=pi/4+i*dfi;
fi3=pi/2+i*dfi;
fi4=3*pi/4+i*dfi;
i=i+1;
%after 40 steps the positions of the teeth repeat
if i>39
i=0;
end
%radial displacement z on every tooth is obtained as the sum
%of projections of the displacements x and y on the tooth radius
% the factor 1000 is used to obtain z in mm while x and y are in m
z1=1000*(x(n)*sin(fi1)-y(n)*cos(fi1));
z2=1000*(x(n)*sin(fi2)-y(n)*cos(fi2));
z3=1000*(x(n)*sin(fi3)-y(n)*cos(fi3));
z4=1000*(x(n)*sin(fi4)-y(n)*cos(fi4));
if n<=40
% formulating chip thicknesses
h1=c*sin(fi1)-z1;zold1(n)=z1;
h2=c*sin(fi2)-z2;zold2(n)=z2;
h3=c*sin(fi3)-z3;zold3(n)=z3;
h4=c*sin(fi4)-z4;zold4(n)=z4;
else
h1=c*sin(fi1)-z1+zold1(n-40);
h2=c*sin(fi2)-z2+zold2(n-40);
h3=c*sin(fi3)-z3+zold3(n-40);
h4=c*sin(fi4)-z4+zold4(n-40);
end
% checking if tooth is in cut and not displaced (h>1)above the
% surface left by the previous tooth; the z coordinate of the surface
%zold being left behind the tooth is determined for both instances of
% h either positive or negative; forces on teeth are determined
if h1>0
F1=Ks*b*h1;
zold1(n)=z1;
else
F1=0;
zold1(n)=z1+h1;
end
if h2>0
F2=Ks*b*h2;
zold2(n)=z2;
else
F2=0;
zold2(n)=z2+h2;
end
if h3>0
F3=Ks*b*h3;
zold3(n)=z3;
else
F3=0;
```

Figure 9.65
(continued)

```
zold3(n)=z3+h3;
end
if h4>0
F4=Ks*b*h4;
zold4(n)=z4;
else
F4=0;
zold4(n)=z4+h4;
end
%Fx and Fy components are separately added for all teeth
Fx(n)=F1*(cos(fi1)+0.3*sin(fi1))+F2*(cos(fi2)+0.3*sin(fi2));
Fx(n)=Fx(n)+F3*(cos(fi3)+0.3*sin(fi3))+F4*(cos(fi4)+0.3*sin(fi4));
Fy(n)=F1*(sin(fi1)-0.3*cos(fi1))+F2*(sin(fi2)-0.3*cos(fi2));
Fy(n)=Fy(n)+F3*(sin(fi3)-0.3*cos(fi3))+F4*(sin(fi4)-0.3*cos(fi4));
%the accelerations ddx and ddy are produced by forces Fx and Fy and
%displacements x and y are obtained by double integration
ddx=(Fx(n)-cx*dx-kx*x(n))/mx;
dx=dx+ddx*dt;
x(n+1)=x(n)+dx*dt;
ddy=(Fy(n)-cy*dy-ky*y(n))/my;
dy=dy+ddy*dt;
y(n+1)=y(n)+dy*dt;
end
%There have been M+1 values of x and y generated and only M for t,Fx,Fy;
%for plotting purpose add the (M+1) values of the latter
Fx(M+1)=Fx(M);
Fy(M+1)=Fy(M);
t(M+1)=M*dt;
% plot(t,x),(t,y),(t,Fx),(t,Fy)
```

The program will be run for various values of the depth of cut b at $N = 3000$ rpm. In order to get an estimate of the chatter limit b_{lim}, the rotating system is represented by the stationary mean position of the middle tooth, that is, at $\phi = 90°$. The corresponding directional factors are $u_x = \cos 70° = 0.342$ and $u_y = 0$. The structural dynamic parameters are

$$k_x = 7e7 \text{ N/m} \qquad m_x = 4 \text{ kg} \qquad c_x = 535 \text{ N/(m/s)} \qquad \zeta = 0.016$$

$$k_y = 1e8 \text{ N/m} \qquad m_y = 4 \text{ kg} \qquad c_y = 737 \text{ N/(m/s)} \qquad \zeta = 0.018$$

Specific force is $K_s = 1500 \text{ N/mm}^2$. In the very simplified way of a "mean" stationary system in which the Y mode does not participate, and using formula (9.83), we have

$$b_{\text{lim}} = \frac{4k \times \zeta \times 2}{2\mu_x \times K_s m} = 1.09 \times 10^{-3} m = 1.09 \text{ mm}$$

In this evaluation the factor $m/2$, the number of teeth cutting simultaneously, was used in the denominator. This very approximate method gives us a starting value, say $b = 1$ mm, for running the program. The resulting vibration is shown in Fig. 9.66a, and it is a stable case. Using $b = 2$ mm, the result in Fig. 9.66b is obtained, which is clearly unstable. In this case the approximate simple method gave a reasonable result. However, depending on the differences in the X and Y vibrational modes, and especially for milling operations other than slotting, it is necessary to use the simulation computer program to obtain a reasonably accurate solution.

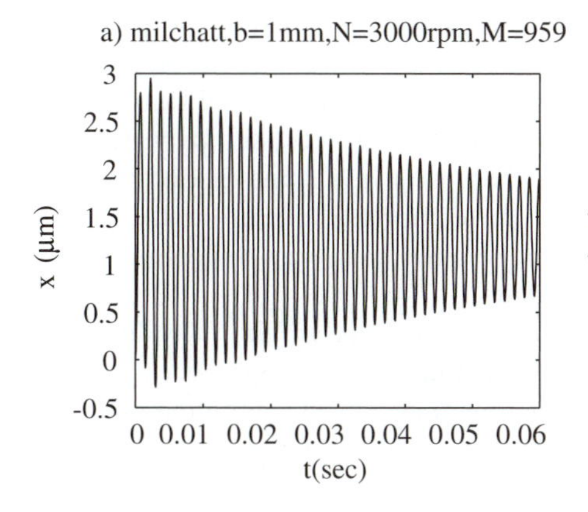

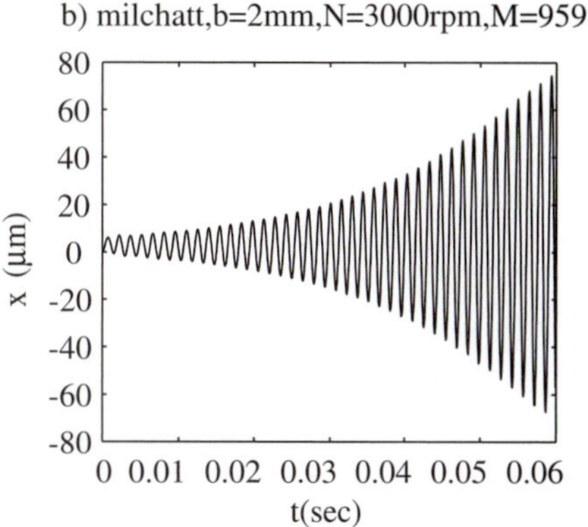

Figure 9.66
a) Milling chatter simulation for $b = 1$ mm. The cut is stable. b) Milling chatter simulation for $b = 2$ mm. The cut is unstable.

9.7 DESIGNING MACHINE-TOOL STRUCTURES FOR HIGH STABILITY

In the preceding text we have explained the nature of chatter in machining and, in a quantitative way, determined and analyzed the stability of well-specified systems with two mutually perpendicular uncoupled degrees of freedom representing the structure of a machine tool that carries a tool against a rigid workpiece. This was certainly a simplification, but it illustrated well all the basic features of a structure that are involved in determining its degree of stability. Let us summarize these main features.

Degree of Stability

The *degree of stability* of a given structure exhibiting modes of vibration between tool and workpiece can be expressed by the value of the *limit width of chip, b_{lim}*, at which chatter just starts to occur.

Thus, stability of the machine tool is measured by the most decisive parameter of the cutting process. The other cutting conditions, such as material of the workpiece (its specific cutting force K_s), feed rate (average chip thickness h_{av}), cutting speed v, tool geometry, and degree of tool wear, also play a role but not as systematic and strong as the chip width. The effects of these parameters are briefly discussed later. The brief statement above may be qualified by saying that b_{lim} is a measure of stability, provided that all the other cutting conditions are standardized.

Oriented Transfer Function

The degree of stability of a machine-tool structure is determined by the minimum value $\text{Re}(G)_{\text{min}}$ of the oriented transfer function (OTF) between the tool and the workpiece. This is obtained as a sum of the $\text{Re}(G)_i$ functions of the individual "modes of vibration" of the structure, each multiplied by its directional factor u_i. The directional factor depends on the direction of vibration in the mode and on two directions related to the cutting process in the machine tool: (1) the direction of the normal to the cut surface Y and (2) the direction of the cutting force F.

From this it follows that even for the same machine tool and the same configuration of its parts, *stability may vary with the orientation of the cutting process.* Thus on a lathe, most of the flexibility is found on the workpiece in the plane *AA* perpendicular to the workpiece axis. It is obvious that the angle σ of the tool plays an important role in influencing the values of the directional factors u_i that include the factor $\sin^2 \sigma$. For a side-cutting tool (SC) with $\sigma = 0°$, $\sin^2 \sigma = 0$, cutting is essentially almost infinitely stable. In reality there is always some flexibility, and also the nose radius of the tool creates a radial force component; therefore, the stability of side cutting is not infinite, but it is high. On the other hand, in plunge cutting we have $\sigma = 90°$, $\sin^2 \sigma = 1$, and stability is very low.

In face milling (also in end milling) the directions of the force F and of the normal to the cut Y when taken as average over the arc of cut may vary in a wide range depending on the direction of feed and on the mutual positions of the cutter and of the workpiece (up- or down-milling). Correspondingly, stability of these various cases may differ substantially.

Combination of Orientations

Once again, the degree of stability is determined by the Oriented Transfer Function. The comparison of stability of two systems with reversed modes, as established in Figs. 9.31 and 9.57, has shown how strongly the value $\mathrm{Re}(G)_{\min}$ is influenced by the *combination of differently oriented modes.* This applies especially for mutually perpendicular modes with close natural frequencies.

However, it is most important to realize that the dynamic flexibilities of the individual modes (the magnitudes of the peaks in the real parts of their transfer functions) are *inversely proportional to the product* $(k_i\zeta_i)$. Whatever the combination of orientations, the stability of a structure is increased by increasing the *stiffness* and *damping* of either all of the modes, or more suitably, of the decisive modes. Looking back at Fig. 9.32, we see that the value of $\mathrm{Re}(G)_{\min}$ is most influenced by the mode with natural frequency 421 Hz (the more flexible mode), and in Fig. 9.57 it is most influenced by the mode with the natural frequency 491 Hz (the stiffer mode). This illustrates what is meant by the decisive mode. As a simple rule we could say: If we test a machine and chatter occurs, then usually the *decisive mode* of the structure is the one with natural frequency closest to the chatter frequency. It is recommended to increase its stiffness or damping, or both.

The stiffness of a mode is increased by increasing the sectional moments of inertia of the machine part that represents the spring of the mode. The damping of a mode is most efficiently increased by attaching a "tuned damper" to a suitable point of the structure. These measures are illustrated in the following text. Having summarized the basic factors affecting stability of a structure, let us now discuss some of the corresponding design features.

In many instances a single part of the structure is decisive for the $\mathrm{Re}(G)_{\min}$ value between the tool and the workpiece, and most often it is a boring bar or the spindle. These parts might, in the fundamental simplification, be considered with a single degree of freedom, as explained for a round bar. In such a case, considering side-cutting-edge angle σ (and taking the cutting-force angle $\beta = 70°$) the value of $b_{\lim}$ would be

$$b_{\lim} = \frac{1}{2K_s \,\mathrm{Re}(G)_{\min}}$$

and with $\mathrm{Re}(G)_{\min} \simeq \sin^2 \sigma \cos 70°/(4k\zeta)$,

$$b_{\text{lim}} = \frac{4k\zeta}{2K_s \sin^2 \sigma \cos 70°} \tag{9.104}$$

Let us now, for a quick orientative assessment of sample cases, choose $K_s = 2000$ N/mm^2, which is typical for cutting steel, and use a very common value of the damping ratio $\zeta = 0.035$. The formula (9.104) then becomes, for this purpose and expressing stiffness k in N/mm,

$$b_{\text{lim}}(\text{mm}) = 1.0 \times 10^{-4} \frac{k}{\sin^2 \sigma} \tag{9.105}$$

We shall consider four cases of "overhang" machining (see Fig. 9.67):

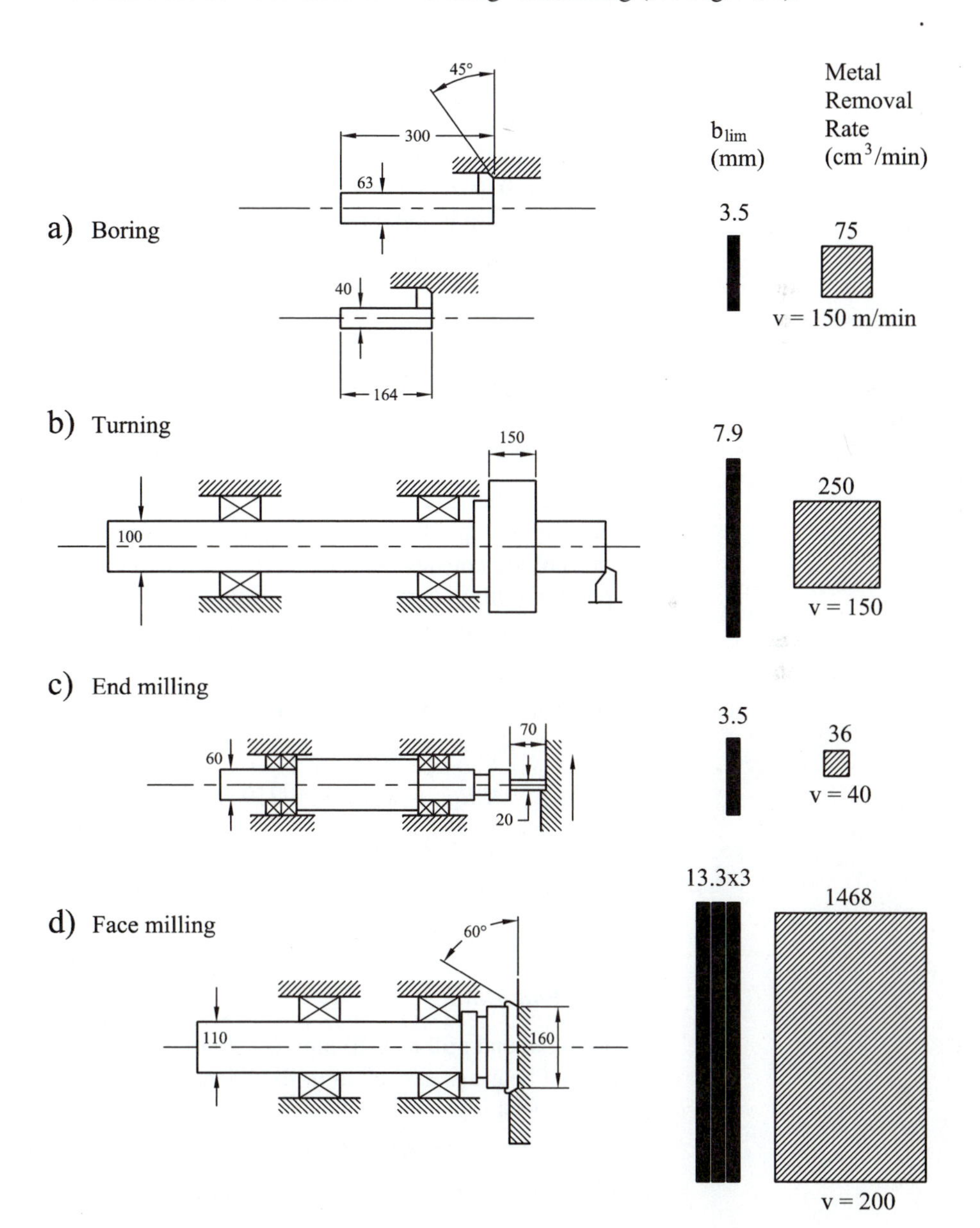

Figure 9.67 Four cases of overhang machining and the corresponding limit depth of cut b_{lim} and the limit for metal removal rate to achieve stable machining of 1035 steel.

Case 1: *Boring*

The stiffness at the end of a boring bar with circular section is

$$k = \frac{3E\,\pi d^4}{64 l^3}$$

Considering steel with $E = 2 \times 10^5$ N/mm^2, we have

$$k = 2.95 \times 10^4 \frac{d^4}{l^3} \text{ (N/mm)}$$

The two illustrated boring bars, $d = 63$ mm, $l = 300$ mm, and $d = 40$ mm, $l =$ 164 mm, have the same stiffness:

$$k = 1.71 \times 10^4 \text{ N/mm}$$

Using formula (9.105) with $\sigma = 45°$, we have

$$b_{\text{lim}} = 3.5 \text{ mm}$$

This can be further evaluated by taking a common roughing feed rate $f_r = 0.2$ mm/rev, depth of cut $a = b \cos \sigma = 2.5$ mm, a common cutting speed $v = 150$ m/min, and expressing the limit metal removal rate as follows:

$$MRR_{\text{lim}} = a_{\text{lim}} f_r v = 75 \text{ cm}^3/\text{min}$$

Assuming specific power $P_s = 2300$ Wsec/cm^3 $= 38$ Wmin/cm^3, the limit of chatter-free power is

$$P_{\text{lim}} = 2.875 \text{ kW}$$

Case 2: *Turning*

Consider overhang turning at the end of a workpiece that is 1.5 times longer than the diameter of the spindle bearings. Bearings of a very stiff type ($d = 100$ mm, double-roller type) are considered with a very good and optimized spindle design. Inevitably, the point of cutting is rather distant from the front bearing. From practical experience, it is known that in this case a typical value of the stiffness at the point of cut is

$$k = 4 \times 10^4 \text{ N/mm}$$

Using formula (9.105), with $\sigma = 45°$, we have

$$b_{\text{lim}} = 7.9 \text{ mm}$$

$$MRR_{\text{lim}} = a_{\text{lim}} f_r v = 250 \text{ cm}^3/\text{min},$$

$$P_{\text{lim}} = 9.5 \text{ kW}$$

The cutting conditions are chosen as $f_r = 0.3$ mm, $a_{\text{lim}} = b_{\text{lim}} \cos \sigma = 5.6$ mm, and v $= 150$ m/min.

Case 3: *End Milling*

A case of an end mill $d = 20$ mm, $l = 70$ mm is considered using a spindle with angular contact bearings $d = 60$ mm. Experience shows that in this case the stiffness at the end of the tool is

$$k = 3.5 \times 10^4 \text{ N/mm}$$

In this case $\sigma = 90°$, and according to Eq. (9.83),

$$b_{\text{lim}} = 3.5 \text{ mm}$$

For a cutter with $m = 4$ teeth, feed per tooth $c = 0.2$ mm, $v = 40$ m/min, and radial depth of cut $a = d$ (slotting), $b_{lim} = 3.5$ mm, spindle speed is $n = v/(\pi d) = 637$ rpm, and the feed rate is $f = nmc = 509$ mm/min.

$$MRR_{lim} = f\,b_{lim}d = 36\ \text{cm}^3/\text{min}$$

$$P_{lim} = 1.37\ \text{kW}$$

Case 4: ***Face Milling***

Consider a spindle with double-roller bearings, with $d = 110$ mm, in a very compact design with a minimum distance of the cutter teeth from the front bearing. In this case we can expect a stiffness of

$$k = 3 \times 10^5\ \text{N/mm}$$

For $\sigma = 60°$, we have

$$b_{lim} = 40\ \text{mm}$$

This value, however, must be taken as cumulative over all teeth cutting simultaneously.

Assume a cutter of $d = 160$ mm, with 16 teeth cutting a width of cut $B = 80$ mm with feed per tooth $c = 0.25$ mm, and cutting speed $v = 200$ m/min. The spindle speed will be $n = v/(\pi d) = 398$ rpm, with feed rate $f = nmc = 1590$ mm/min. The number of teeth cutting simultaneously will be $m = 3$, and the depth of cut $a_{lim} = b_{lim} \sin \sigma/3 = 11.5$ mm.

$$MRR_{lim} = fb_{lim}a = 1468\ \text{cm}^3/\text{min}$$

$$P_{lim} = 55.8\ \text{kW}$$

These four examples offer an illustration of how the output of the metal-cutting operations is limited by the criterion of the limit of stability against chatter. They show also the rather wide range of the degrees of stability, whether they are expressed by the b_{lim} values or by the MRR_{lim} or P_{lim} values. Compare the cases of turning and face milling, where the spindles are almost identical but there is a difference between MRR_{lim} values in a ratio of 1468/250 = 5.87/1, because the cutting at the face mill is done very close to the front spindle bearing, while there is a large overhang in the turning operation.

The four simple examples were treated as SDOF cases. In each of them, in reality, there will be two mutually perpendicular modes of vibration with different natural frequencies due to the effects of the asymmetry of the spindle housings, and stability may vary with the orientation of each case.

In practice machine-tool structures exhibit a number of more or less prominent modes of vibration that must be taken into account as they influence the OTF. Each of these modes has a natural frequency f_n, a static stiffness k, a damping ratio ζ, and a direction of vibration between the tool and the workpiece, and with each of these modes a particular *mode shape* is associated. The mode shape is determined by the vector of the ratios of vibrations at various points of the structure and it indicates which parts of the structure act more like springs and which more like masses in this mode.

The mode shapes and the OTF depend on the distribution of stiffness, damping, and mass throughout the structure. They could, ideally, be computed from the drawings of the machine using either the finite-element method or, in a more approximative way, using a model of the structure consisting of "lumped" springs, dampers, and masses.

Practically, such a computation cannot be accomplished with the desired degree of accuracy because the structure of the machine tool is rather complex and because it is extremely difficult to compute local deformations around joints and guideways. However, it is often sufficient to neglect most of the structure and concentrate on a subgroup like that of a spindle, which can be reasonably well computed.

In most instances one considers an existing machine or its prototype and measures its structural characteristics. Subsequently the model of the machine as an assemblage of lumped springs, dampers, and masses can be identified from these measurements. This model can then be used for computing results of various design changes.

To illustrate such a procedure, Fig. 9.68 shows a sketch of a light milling machine, with the five most prominent mode shapes of this machine plotted as measured. The 22 Hz mode is the rocking of the machine on the floor; the 75 Hz mode is represented by the twisting of the ram, and the head with the motor is its inertia; the 125 Hz mode is characterized by the rotation of the ram in the flexible turret at the top of the column; the 350 Hz is a higher mode related to the 75 Hz mode, with the top of the head (including the motor) vibrating in counterphase to the lower part of the head; and the 570 Hz mode is the higher mode related to the 125 Hz mode, with prominent flexibility at the base of the U-shaped end of the ram. Two transfer functions (TF) of the machine are shown in Fig. 9.69.

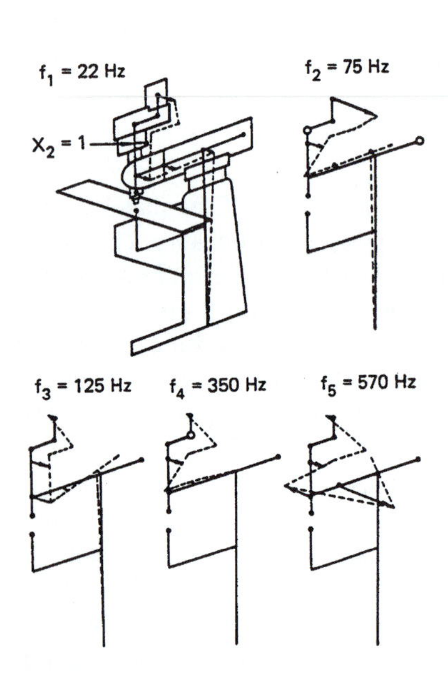

Figure 9.68
Mode shapes of a toolroom-type milling machine. Displacements of the individual points are related to $x_2 = 1$. It is possible to estimate where the "localized springs" are in each mode: (1) rocking of machine on floor; (2) rotation of overarm (twist) in its base on top of column; (3) rotation of overarm in its base around vertical axis and swinging of the top mass (motor) on the twisting transmission housing; (4) twist of overarm and swinging of motor; (5) all the three rotations mentioned in (2), (3), and (4) plus bending of the U-type frame holding the spindle housing.

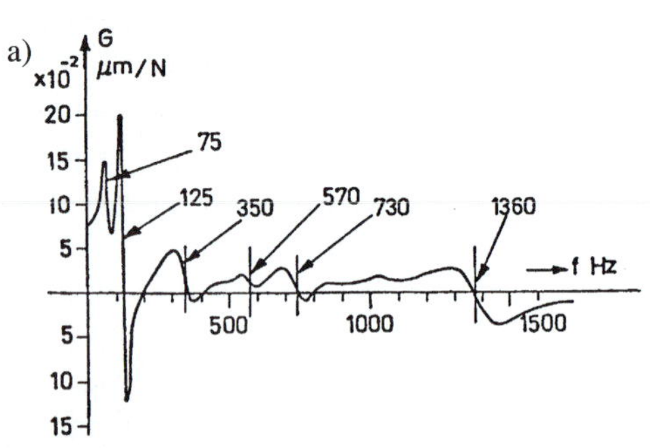

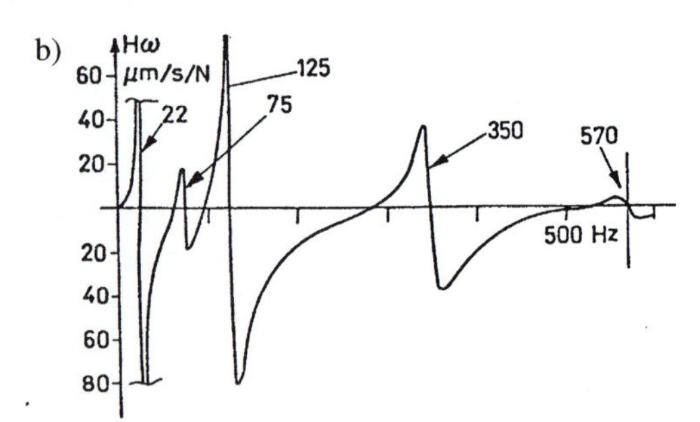

Figure 9.69
Transfer functions of the milling machine.
a) Relative displacements between spindle end and table. b) Absolute motions (velocities) on point x_2. The latter identifies frequencies and stiffnesses of individual modes.

The TF in graph *(a)* is relative between tool and workpiece (the real part), and it indicates that the 125 Hz mode is most responsible for chatter on this machine. One could immediately recommend stiffening this mode and thus increase the stability of the machine. This could be done by more properly and more abundantly dimensioning the turret at the top of the column and the guideways in which the overarm is clamped. Alternatively, the damping of this mode could be increased by attaching a "tuned damper" to the top of the headstock. Such a damper would consist of a spring, a mass, and a dashpot; it would be tuned to 125 Hz. It would participate strongly in the vibration of this mode and dissipate its energy.

Graph *(b)* is an absolute TF on the point x_2 of Fig. 9.68, and it serves to determine the modal parameters of the structure. A lumped spring-mass model of the machine is shown in Fig. 9.70. The individual concentrated masses are indicated by black circles. The double lines represent rigid links. Five springs are denoted by flexibilities γ_1 to γ_5. There are five basic coordinates x_1 to x_5 of the model, corresponding to the five degrees of freedom, as expressed by the five modes visible in the transfer function in Fig. 9.69b. In a project described in Tlusty [15] the "local" parameters of this model were identified. Subsequently various changes were tried, so as to minimize the negative peak of the TF in Fig. 9.69a. It is interesting to note that the most effective stabilizing measure was found to be a redistribution of the masses on the headstock, which affected very favorably the stiffness of the 125 Hz mode at the point of the tool.

Another example of a measure to increase stability of a machine-tool structure is diagrammatically shown in Fig. 9.71. This shows the case of a vertical turning and boring machine, where the tool head is located at the end of a long and slender ram. The sectional dimensions of the ram are limited by the requirement that it must pass into a long bore in the workpiece of a certain minimum diameter. On one hand the section of the ram should be designed for optimum orientation according to the rules explained in connection with Fig. 9.56. On the other hand, the most efficient decrease of the $\text{Re}(G)_{\text{min}}$ value at the end of the ram is achieved by a significant increase of the damping ratio of the ram system. A four- to six-fold increase of stability is possible by incorporating a tuned damper in the end of the ram.

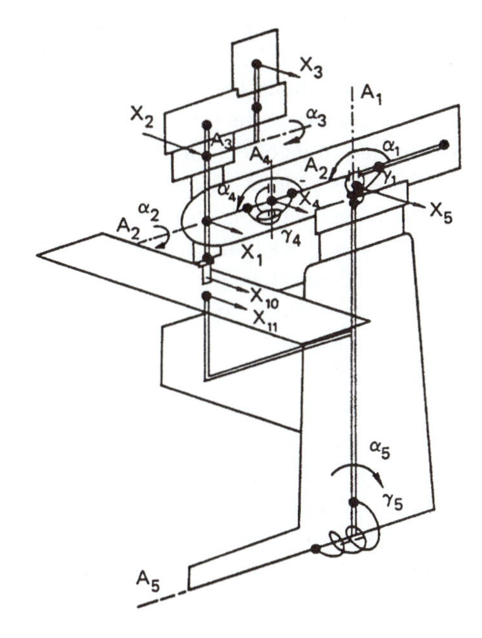

Figure 9.70
Lumped spring-mass model of the milling machine. Concentrated masses are marked by black circles. All springs are torsional. Measured mode shapes and transfer functions make it possible to identify stiffnesses of the springs and magnitudes of the masses (for the latter, additional constraints are used).

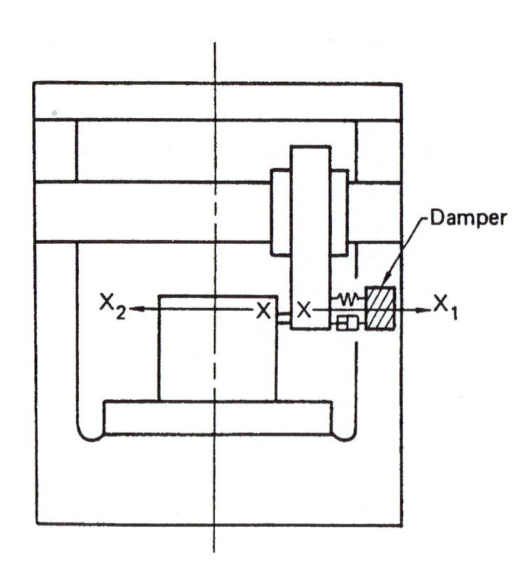

Figure 9.71
Attaching a tuned damper to the ram of a vertical boring mill. Natural frequency of added spring-mass system is tuned equal to that of the most flexible mode at end of ram.

9.8 EFFECT OF CUTTING CONDITIONS ON STABILITY

The effect of the chip width *b,* and correspondingly of the directly related depth of cut *a* has been emphasized in the preceding text as being primary and direct. For a given vibration and, consequently, variation in chip thickness, the feedback provided by the generated variable force is proportional to *b.* The chip width represents the "gain" in the closed-loop of self-excitation. With sufficiently small *b,* cutting is always stable; as *b* increases, a limit value b_{lim} is reached beyond which chatter arises.

We shall now briefly discuss the effects of the other cutting conditions: workpiece material, feed, cutting speed, and tool geometry. The effect of the *workpiece material* is essentially expressed by the specific force K_s. This is clearly understood from formula (9.82). In general, material *A,* which has 1.5 times higher K_s than material *B,* will give 1.5 times lower value of b_{lim}. (We may refer to Table 7.1 in Chapter 7 in which the values of K_s are given.)

The effect of *feed f* is actually the effect of the average chip thickness h_{av}. There is a direct proportionality between the two that was expressed in Fig. 7.2. The effect of *h* on b_{lim} is mainly related to its effect on the value of K_s. This effect was formulated in the Eq. (7.9), which shows that K_s decreases with the increase of *h.* Indeed, practical experience shows that in operations like turning and boring, stability improves with an increase of feed. Often b_{lim} is smaller for small feeds, and operations with typically thin chips (e.g., hobbing of gears) are especially prone to chatter. This effect, however, does not apply equally to milling operations, where chatter usually increases with increased feed rate. As an explanation, one should realize that in most milling operations every chip either starts from zero thickness (up-milling) or ends with zero thickness (down-milling). There are, however, instances where a thin-chip/low-feed effect on lowering stability is seen; if feed is stopped, then for a moment chatter occurs as the deflection of the cutter gradually vanishes and the chip thickness goes gradually to zero. In summary, the effect of feed may be both stabilizing (turning) and destabilizing (milling), but it is mostly not very strong. However, once a case becomes unstable, the amplitude at which the chatter vibration stabilizes is larger for higher feedrates because the saturation effect of the teeth jumping out of the cut sets in at the thicker chips.

Tool geometry does not have a strong effect on chatter. It could be assumed to be mainly due to the change in the value of K_s or also perhaps to a change of the direction of the cutting force and a corresponding change in the directional factor u. In this sense tools with negative rake angles may increase the tendency to chatter compared to those with positive rake angles, because the component of force into the direction Y of the normal to the cut surface is larger. However, these effects are weak.

Cutting speed v affects stability in two very different ways. The first of them is often connected with the "damping in the cutting process." Equation (9.76) states that the variable component of the cutting force is proportional to the variation of chip thickness, and it was assumed that the coefficient or proportionality ($K_s b$) was a real number, meaning that there was no phase shift between the two variations. However, it has been shown that there may actually be a phase shift in the generation of the variable force, which signifies that there is damping generated in the vibratory motion of the tool in its action of chip formation. This damping is significant at very low cutting speeds, typically below 25 m/min and mainly at speeds below 10 m/min. The increase of stability at the very low cutting speeds is very significant. The diagram in Fig. 9.72 is typical for machining carbon steels. It expresses the variation of the value of the limit width of chip b_{lim} as a function of cutting speed, for systems with three different natural frequencies: 200, 500, and 1000 Hz. It shows that, for example, for the latter two, chatter will practically never occur at speeds below 25 m/min, and for the low frequency system, at speeds below 10 m/min. This very strong stabilizing effect is utilized in practice. Some operations that are inherently difficult from the point of view of chatter because of the very wide chips involved can only be carried out at very low cutting speeds. As an example we quote broaching of "Christmas-tree slots" in turbine discs. The perimeter of the broach tooth edge is typically 70 mm wide, and with two or three teeth engaged simultaneously the total chip width is about 200 mm. On the other hand, the fixture holding the disc does not have very high stiffness. In spite of these severe conditions the cut is chatterfree at the low cutting speed of about 8 m/min.

An explanation for the high damping in the cutting process at low cutting speeds combined with high frequencies is offered in Fig. 9.73. It shows the tool moving on a

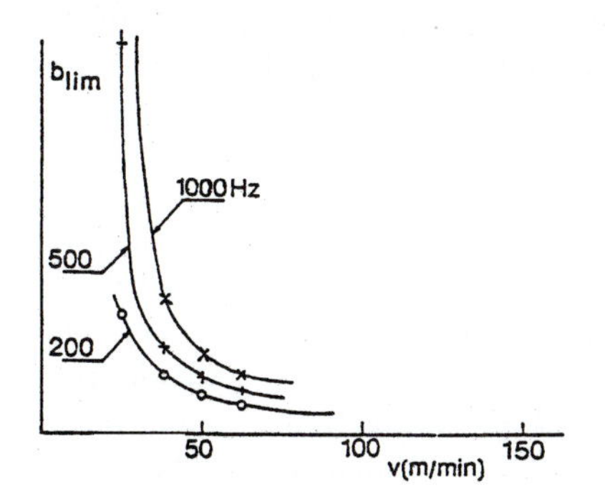

Figure 9.72
Typical effect of low cutting speed on stability, as established in an end-milling experiment. Due to "process damping," stability increases with decreasing speed, in the low-speed range.

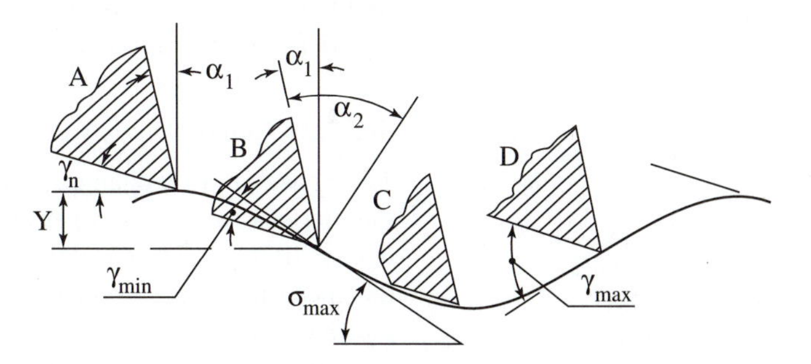

Figure 9.73
Source of damping in the cutting process. Variation of the actual relief angle γ of a tool cutting while vibrating causes a variable force component 90° out of phase with vibration: the extremes of γ occur halfway between those of vibratory displacements.

sine wave when removing a chip. In positions *A* and *C,* top and bottom, the actual clearance angle at the flank of the tool is equal to the nominal value γ_n. In the middle of the way down, at position *B,* the direction of the cut is at the largest slope σ_{max}, and the actual clearance of the flank $\gamma_{min} = \gamma_n - \sigma_{max}$ is very small or even negative. This leads to an increase of the thrust component of the cutting force. On the contrary, during the upward motion, in position *D,* the actual clearance angle is maximum, and the thrust force decreases.

Correspondingly, during the half cycle in which the velocity of the vibratory motion is downward (from *A* to *C*) it is opposed by a higher force than the force that drives the tool upwards (from *C* to *A*). This variation of the thrust force is in opposite phase with velocity (90° out of phase with displacement), and it represents damping. This damping is larger for shorter waves because they have steeper slopes. The wavelength *w* of the undulations on the surface is obtained as

$$w = \frac{v}{f}$$

Shorter waves are obtained at low cutting speeds and high frequencies.

Another kind of effect of cutting speed *v* on stability is the effect of spindle speed *n,* which is significant in high-speed milling. It has been indicated already in Fig. 9.54, where it was shown how the phasing with which undulations in subsequent cuts meet affects the chip-thickness variation. In Section 9.6.5 Fig. 9.59 illustrated how the value of Re(*G*) depends on the phase angle ε between the two vectors *Y* and Y_0. Using this diagram, it is possible, for a given system with a given OTF, to derive the variation of b_{lim} with the variation of spindle speed *n* and the corresponding variation of the phase angle ε.

The phase angle ε is obtained by calculating the number *p* of waves between subsequent teeth of a milling cutter. This number is obtained by dividing the number of waves per second *f* by the number of teeth per second *(mn),* where *n* is spindle speed in revolutions/second and *m* is the number of cutter teeth:

$$p = \mathrm{N} + \frac{\varepsilon}{2\pi} = \frac{f}{mn} \tag{9.106}$$

The number *p* consists of an integer N and the fraction $\varepsilon/2\pi$. We will not carry out here the computation establishing the mentioned relationship and its effect on stability. An intuitive approach is now used to establish the concept of the "lobing diagram" instead of a detailed derivation.

Stability Lobes

The phase angle ε changes with spindle speed. In turning there usually is a large *N* number of chatter waves per circumference of the workpiece, $p = f_c/n$, where f_c is the chatter frequency that is close to the natural frequency of the system f_n, and *n* is spindle speed in rev/sec. Typically, the natural frequency may reside in the system where the mass of the workpiece combines with the stiffness of its clamping in the chuck. Suppose it is $f_n = 200$ Hz and that in turning a steel workpiece with cutting speed $v =$ 150 m/min on a diameter of 200 mm, the spindle speed is $n = v/(\pi d) = 3.98$ rev/sec, with $p = 50.25$. To obtain the same phasing, a change of speed to obtain $p = 49.25$ should occur of 1/50.25 = 0.0199 (i.e., about 2%). That is, 2% of change in chatter frequency would bring the case into the same level of stability. The regeneration mechanism will adapt easily to such a change. This means that changing spindle speed would affect the level of stability very little; it would always approximately correspond to the worst phasing, as expressed by Eq. (9.83).

In milling, however, regeneration occurs not revolution after revolution, as in turning, but tooth after tooth of the rotating cutter. In the equation for phase between subsequent undulations, spindle speed n is replaced by the product $(m\ n)$ of number of teeth m and spindle speed. The spindle speeds are higher than in turning because in the equation for cutting speed $v = \pi\ d\ n$, d is now the cutter diameter, which is often much smaller than the workpiece diameter in turning. Instead of the d = 200 mm used above for turning, we use a face mill with d = 100 mm or an end mill with d = 19 mm. Taking the case of the end mill with m = 4 teeth, the tooth frequency $f_t = m\ n = v/(\pi d)$ = 167.53 Hz is obtained for cutting speed of v = 150 m/min in cutting steel. In cutting aluminum, we use v = 600 m/min or more, with n = 167.53 rev/sec, which gives tooth frequency f_t = 670.13 Hz. Let us assume natural frequency on the end mill f_n = 880 Hz. The number of waves between teeth is now p = 880/670.13 = 1.31. For the same phase (i.e., the same fractional part of p), for example, in order to get p = 2.31, the spindle speed would have to change by 76%—quite a drastic change. Within that range, stability would strongly change. This illustrates that in milling, especially in high-speed milling, a change of spindle speed may be a very effective means for increasing the limit depth of cut and, correspondingly, the metal removal rate (*MRR*).

This effect is recapitulated in the diagram in Fig. 9.74. At the top it is shown how in milling the waves that are cut into the surface during chatter vibrations by a tooth get recut by the subsequent tooth. At the spindle speed in *(a1)* several waves and a fraction occur between the two teeth. In *(a2)* the spindle speed is lower. Frequency of wave generation has not changed, because it is still chatter frequency that is close to the natural frequency of the tool. Correspondingly, a larger number of waves occur between two subsequent teeth. In *(a3)*, on the contrary, spindle speed has been increased, and only one and a small fraction of a wave occur. In the middle row of the figure, in *(b1)* there is exactly one wave between teeth. Although vibration is assumed, the waves produced by two subsequent teeth are in phase, and no chip thickness variation results; thus there is no force variation either. The vibration is not reexcited and will die out; this is a stable case. In *(b2)*, with one and a half waves between teeth, for the same vibration amplitude, a variation in chip thickness with twice the vibration amplitude occurs, resulting in a large force variation that excites further vibration. This may have been the result of a 50% increase of spindle speed. This is not exactly the worst phasing; as shown in Fig. 9.59 the worst phase is about 270°, that is, about 1.75 waves between teeth. Once these circumstances are properly analyzed as, for instance in Tlusty [15], a diagram such as the one at the bottom of Fig. 9.74 is obtained. Since its publication in Tobias [20], it has been called the "lobing diagram." The vertical coordinate is the ratio $q = b_{lim}/b_{cr}$, where b_{cr} is the lowest b_{lim} obtained for the phasing most favorable for chatter generation, and it occurs repeatedly through the variation of spindle speed. The horizontal scale expresses the value of the number $p = m\ n/f_n = f_t/f_n$, which is the ratio of the tooth frequency over the natural frequency of the system. The value b_{cr} is the highest depth of cut at which cutting is stable, and no chatter occurs at any speed. It depends on the stiffness and damping between tool and workpiece and on the K_s value for the workpiece material in a way similar to that expressed by Eq. (9.83), except that the resulting value of b is now shared by all the teeth cutting simultaneously, and the directional orientation factor is a more complex concept because it is periodically variable. An explicit formula of the type in Eq. (9.83) does not exist for milling; it can only be approximated by using "average directional factors" corresponding to the mean position of each tooth. Realistic solutions are obtained by simulations of the kind shown in Section 9.6.7.

The individual "lobes" in the diagram each correspond to a different integer N in Eq. (9.74). The practical interpretation of the graph is to consider the envelope of all the lobes as the boundary between the stable field below that envelope and the chatter field (shaded), above the envelope. The upturn of the stability boundary at the left end of the horizontal scale is the effect of process damping. At the high-speed end, at the right, gaps of increased stability occur. The highest stability, permitting the highest value of stable depth of cut is obtained with the spindle speed at which the tooth frequency equals the natural frequency of the system. The graph has been derived for an SDOF system. The system is stable at a resonance of the periodic force with the tooth frequency and the system natural frequency. Forced vibrations will be at their highest, which is mostly not serious. Actually, in slotting with a cutter with four teeth, there is no periodic force once per tooth; see Example 9.11 associated with Fig. 9.47. The forced resonant vibrations repeat per tooth and create a condition shown in diagram *(b1)* of Fig. 9.74.

It may be seen that the diagram has the form of "lobes of stability" that repeatedly go down to $q = 1.0$ and, in between, rise to peaks of stability. These are found close to values of $p = 1(N + 1)$ where N is the integer part of the number of waves between cutter teeth. For low spindle speeds, N is large, and the peaks of stability are close to each other and not very high. This means that for many waves between subsequent cuts—as is always the case in turning and in general milling—the variation of stability with spindle speed is insignificant. But as the spindle speed approaches the values $n = 0.5 f_n /m$, and mainly $n = f_n/m$ (one wave between subsequent teeth), substantial stability increase may be achieved by selecting exactly the right speed. The numbers N of waves between teeth are marked at the individual lobes of the diagram.

In practice, this all means that in cases of milling where the number of chatter undulations between cutter teeth may become low, it is important to have steplessly variable spindle speed so as to be able to select the most stable speed. Such cases are those of high-speed milling of aluminum or of milling of cast iron with silicone nitride tools and also carbide face milling on large machine tools like floor-type horizontal boring machines, where the natural frequencies of the structure may be low.

In reality several prominent vibratory modes exist between tool and workpiece, and the lobing diagram contains several sets of lobes. However, usually one mode is dominant and one stability gap is highest, indicating the spindle speed for deepest stable cut.

Commercially available software exists that can be used for very efficient simulation runs such as those in Figs 9.66a and b. Moreover, such runs can be automatically programmed over ranges of spindle speeds and depths of cut, resulting in comprehensive graphs of vibration and force amplitudes over a whole field of combinations of the cutting conditions of axial and radial depths of cut, and spindle speeds. These "maps" can be used for NC programming of optimum operations. In order to establish these maps, however, it is necessary to measure the dynamic characteristics (transfer functions) on all the tools to be used in the particular NC program. An alternative method for finding the most stable speeds is the use of a control system that incorporates a microphone to capture and process the sound of the milling operation so as to determine, in real time, the chatter frequency and calculate the best speed, which gives a tooth frequency equal to the chatter frequency. This is based on the fact that the chatter frequency is very close to the natural frequency of the dominant vibratory mode. If this approximation is not fully satisfactory a second run is performed in which chatter

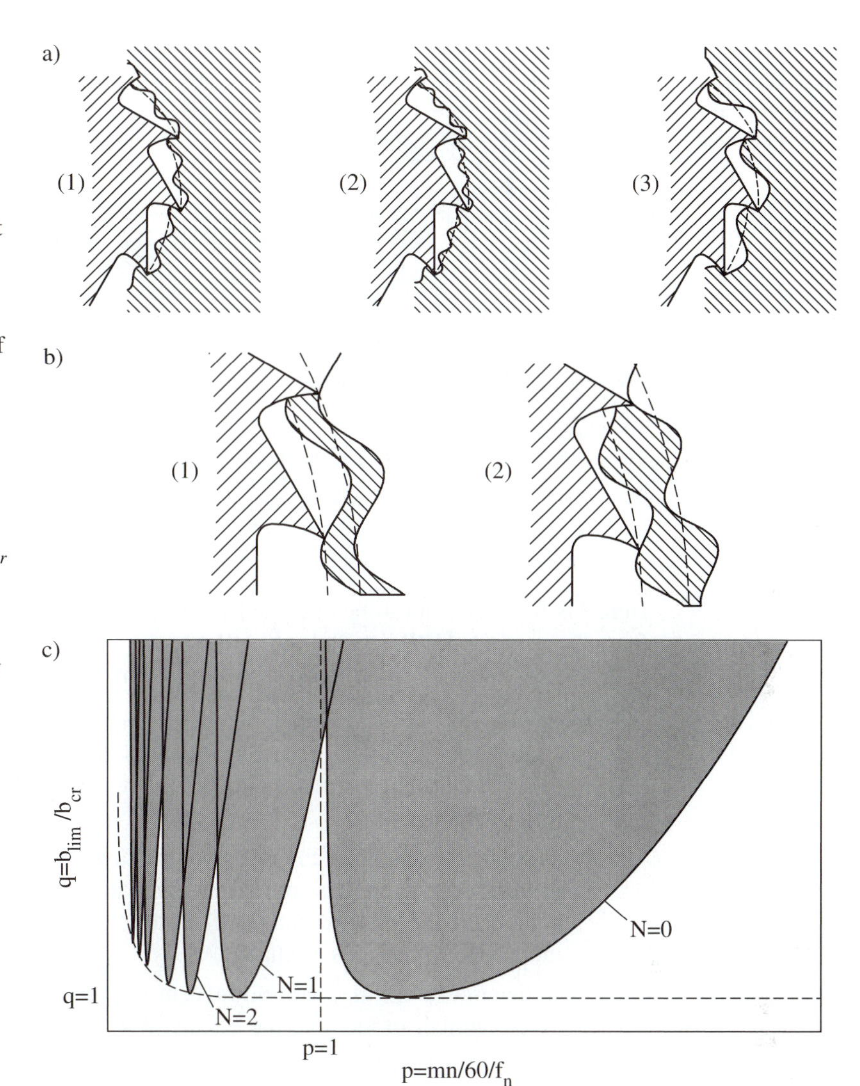

Figure 9.74 Stability lobes diagram. a) At different speeds the number of waves between teeth varies. b1) A single wave between teeth; the waves created by subsequent teeth align; the chip thickness and cutting force do not vary: no chatter feedback. b2) One and a half waves between teeth cause maximum chip-thickness variation. c) The "lobing diagram" for milling expresses the ratio of the limit for stable depth of cut b over the "critical" value b_{cr} versus tooth frequency $p = mn$ (number of cutter teeth times spindle speed rev/sec) divided by natural frequency f_n of the system. The diagram is drawn for a single-degree-of-freedom system.The phasing effect varies with speed resulting in the higher speed range in increasing stability peaks (increasing stable depth of cut *b)* between the lobes. The highest of the peaks occurs when $p = 1$, i.e., if tooth frequency equals natural frequency of the system.

frequency becomes still closer to the natural frequency. This automatic system is described in [18].

Let us now briefly summarize the effects of all the cutting parameters on chatter while referring to Fig. 9.75 for the notations. A face mill and an end mill are shown, the first in a down-milling cut and the latter in an up-milling cut. Spindle speed is marked n (rpm), cutter diameter is d (mm), peripheral speed (cutting speed) is v (m/min), radial depth of cut (width of cut) is a (mm), axial depth of cut (depth of cut) is b (mm), chip load (feed per tooth) is c (mm), the number of teeth on the cutter is m. The units given here are commonly used, but they do not always balance out, and multipliers must sometimes be used. The following relationships determine the combined parameters, and in them the proper multipliers are used. However, they are often left out, and it is up to the user to balance out the dimensions:

$$\frac{v = \pi\, d\, n}{1000} \qquad f\,(\text{mm/min}) = m\, n\, c \qquad MRR\,(\text{cm}^3/\text{min}) = a\, b\, f \tag{9.107}$$

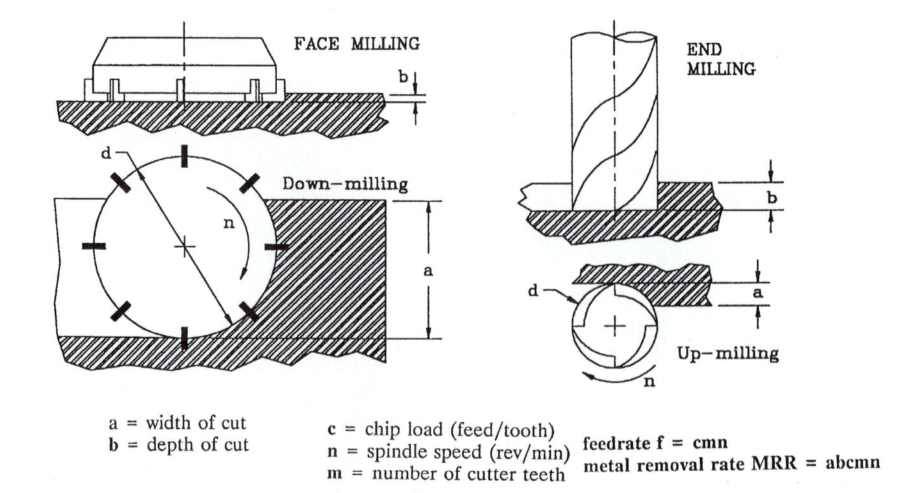

Figure 9.75
Milling process parameters and their notations. Radial depth (width) is a and axial depth is b.

Here f is feed rate, not to be confused with frequency; the distinction must be judged from the context. *MRR* is metal removal rate.

The effects of the individual cutting parameters on chatter are summarized as follows:

1. Chip width b in turning; axial depth of cut b in milling: These express the "gain" in the self-excitation process; increasing b always leads from stable cutting to chatter.
2. Number of teeth m on the milling cutter and the radial immersion $i = a/d$: the product $(m\ i)$ expresses approximately twice the number of teeth cutting simultaneously; the product $(m\ i\ b/2)$ is the "cumulative chip width" in milling; increasing $(m\ i)$ leads to chatter.
3. In general in milling, we find (except for very small radial depth a) that for a given number of teeth m, the effects of a and b are interchangeable. Thus the product $(ab)_{\text{lim}}$, determines the limit of stability. It also means that for a given spindle speed and chip load, this product indicates a constant limit *MRR* independent of the combination of radial and axial depths of cut.
4. Chip thickness h, feed per revolution f_r in turning, chip load c in milling: these have very little effect on the limit of stability b_{lim}. Once the cut is unstable, however, higher h $(f_r,\ c)$ leads to higher amplitude of saturated chatter; see Fig. 9.63.
5. Cutting speed v: At low cutting speeds process damping suppresses chatter. This leads to higher values of b_{lim}. It is actually the wavelength $w = v/f_c$ of chatter undulations on the surface that is decisive. Here f_c is chatter frequency. For about $w < 2$ mm, process damping starts to play an increasing role as v decreases. Obviously, this condition is reached at higher speeds for higher natural frequencies (e.g., for end mills) and at very low speeds on large machines with low natural frequencies.
6. Spindle speed n: Its effects are seen first as stated in (5) above, because of its direct proportionality with cutting speed. In addition, in milling at high speeds, generally for $f_t = mn/60 > (f_n/4)$, the stability lobes start to play an increasing role. By correct choice of n the value of b_{lim} may be substantially increased.

Special Milling Cutters

It is also possible to improve stability against chatter by using special milling cutters that are designed to disturb the regeneration of waviness on the cut surface, which is the main mechanism of self-excitation. One such design uses nonuniform tooth spacing. As indicated in Fig. 9.76a, if the pitches p_{12} and p_{23} of subsequent cutter teeth are not equal, then the phase angles of subsequent undulations ε_{12} and ε_{23} are not equal, and they cannot both adjust simultaneously to the optimum phasing for regeneration. Another design is shown in diagram *(b),* which uses teeth with alternating helix. The principle of the action of this design is indicated in the diagram *(c),* which shows waves left by a tooth that would have zero helix angle. The following tooth in the cut has a steep helix angle and instead of reproducing the preceding waves it cuts across them; the average chip thickness variation is thus very small. The cutters with unequal tooth spacing and those with alternating helix are not very common, but they are used in instances where other ways of achieving stable cutting are not possible.

The type of special cutter design shown in Fig. 9.77, which either has interrupted or undulated edges, is becoming quite common and serves two purposes. One is to divide a wide chip into narrower and shorter chips that do not clog the flutes of the cut-

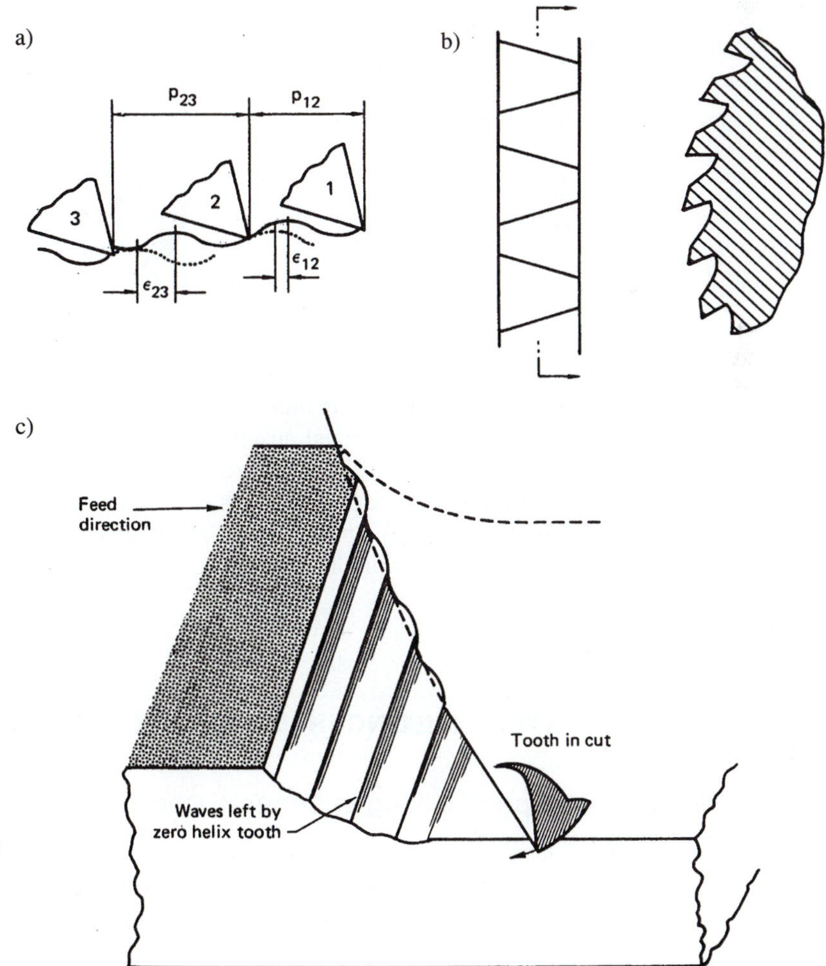

Figure 9.76
Cutters with a) unequal spacing and b) with alternating helix. In c) it is shown that the oblique edge of the tooth in the cut cannot reproduce waves created by a preceding straight edge, thus increasing stability against regenerative chatter.

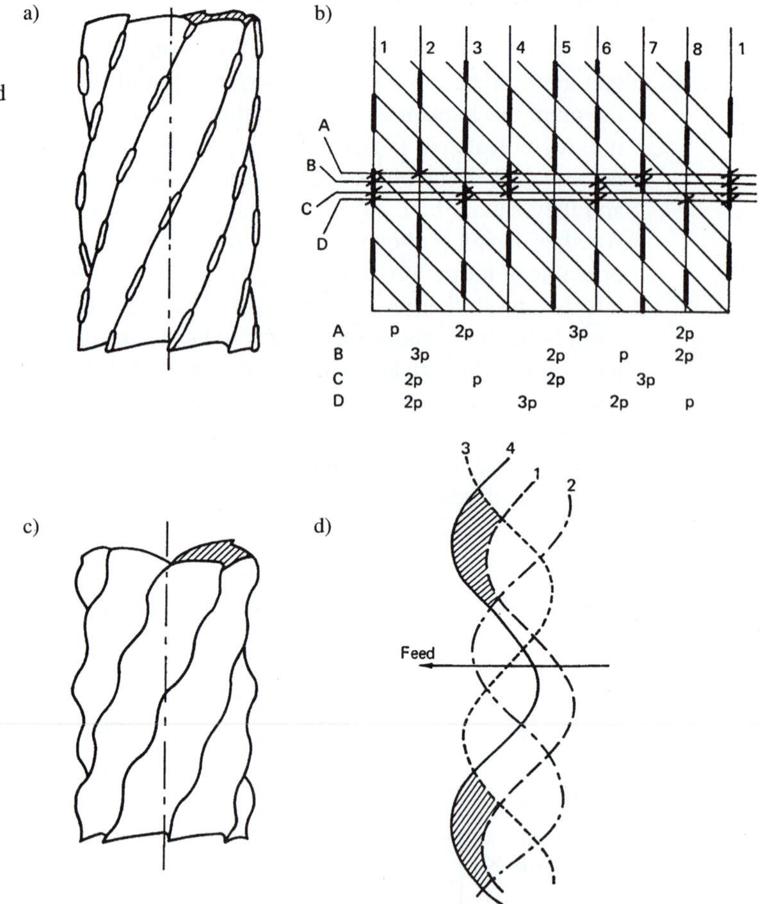

Figure 9.77
Cutters with a) serrated and b) undulated edges. Effects of nonuniform pitch are generated at the various height levels, creating a stabilizing effect.

ter and are easily cleared away. The other purpose is to achieve increased stability of cutting, based on the fact that in individual sections perpendicular to the cutter axis, the cutter is found to have unequal spacing. This is shown in Fig. 9.77b, which shows an unrolled periphery of the cutter. The active parts of the cutting edges are shown by thick lines. Between them are the serrations arranged helically. The sizes of spacings—*p* (single), *2p* (double pitch), *3p* (triple pitch)—are given for sections A, B, C, and D. Similar effects are obtained with the undulated edges as shown in Figs. 9.77c and 9.77d.

9.9 CASE STUDY: HIGH-SPEED MILLING (HSM) MACHINE FOR ALUMINUM AIRCRAFT PARTS

9.9.1 High-Speed Milling in General: Operations with a Lack of Stiffness

First, let us have a brief lesson in the use of words in the practical engineering world. A new technique, a new method, or a new device that gains significance in a particular stage of technology development gets a name that helps to spread its application, and this name does not always have a well-defined meaning. In electronics one started with integrated circuits (ICs), then large scale ICs (LSICs), then very large scale ICs (VLSICs), and so on, without clearly defining the ranges. We speak about superfinishing (honing with a special form of motion), about superalloys (see Chapter 2). In Chapter 8, the story was briefly told of how two engineers at the beginning of the twentieth century invented a special formulation and a special method of heat treatment for a class of alloy steels used for cutting tools that have permitted a substantial increase, two to three times, of cutting speeds as compared with the then commonly used high-carbon steels. They called them high-speed steels. This denotation has stuck and survived the era, and thirty years later, when another new class of tools, sintered carbides, permitted another jump in cutting speed, that former class remained denoted as HSS steels. In the 1980s a movement started to further increase cutting speeds, partly by introducing new types of tool materials such as CBN (cubic boron nitride) for machining hard steels and chilled irons, Si_3 N_4 (silicon nitride), for three to four times more cutting speed on cast iron than with the carbides, but primarily and especially in the application to end milling of aluminum by developing spindles that could rotate much faster while retaining relatively good stiffness. This movement is known as high-speed machining (HSM) and is especially important in milling: see Tlusty [15], [17].

Another characteristic feature of HSM is not at all indicated by the name of the technique. In most instances HSM is applied to such machining operations where there is an *inherent lack of stiffness* in the system, primarily on the part of the tool. In the preceding section we demonstrated that in those instances the depth of cut is limited by chatter vibrations, and in milling also the width of cut is limited by chatter. Consequently, increase of the speed remains as the only parameter available to achieve higher metal removal rates. It is then necessary to use tool materials that permit higher speeds. Especially in the milling of aluminum, which is easily machinable (see Section 8.2), it is necessary to develop spindles that can rotate faster. In all the applications it is necessary to expand the limits of speed as much as possible. Typically, spindles with rolling-bearing-bore diameter of 100 mm rotate up to 15,000 rpm, and those with bearing-bore diameter of 50 mm rotate up to 40,000 rpm.

Let us illustrate how often the circumstances of an inherent lack of stiffness arise. One such case is milling of cast iron stamping dies (Fig 9.78). To reach into the cavity of the female counterpart of the depicted male die, various spindle attachments of the extension type and of the right-angle/extension type are used. Three of these are outlined at the bottom of the figure. These are housings that carry inside shafts in ball bearings that act as the effective spindles into which the tool holders for face or end mills are clamped. The right-angle attachments also contain the bevel gear transmissions. All these attachments would, in turn, be clamped on the face of the main spindle housing of the machine and be driven from the main robust spindle with bearings of 140 mm bore diameter. Thus the main spindle is just a part of the drive, but the stiffness of the tool is determined by the housings and spindles of the attachments, which are long and slender. Inevitably, the stiffness of the tools is rather limited. The attachments must be carefully designed to provide maximum stiffness within these limitations. HSM is then the only way to obtain increased *MRR*s. Another case of an inherent lack of stiffness is associated with a popular type of a horizontal spindle milling machine shown in Fig. 9.79. In this case the headstock moves vertically on the column of the machine. A hollow spindle is rotationally mounted in the headstock, and a round bar is mounted in a sliding fashion in the spindle. This bar, typically of 125 mm diameter, carries the tool such as the face-milling cutter shown in the drawing. It may extend typically up to 750

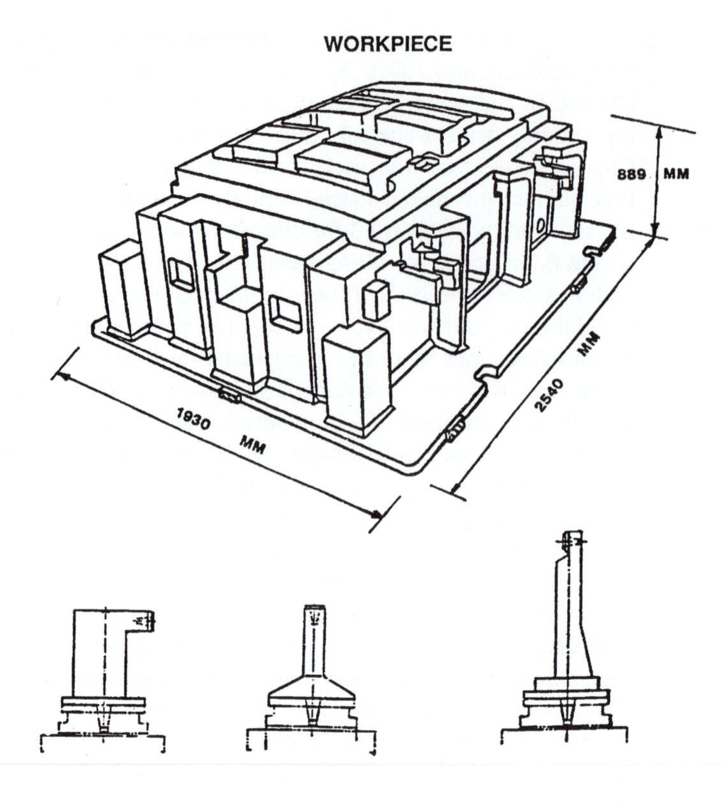

Figure 9.78
Spindle attachments for milling of stamping dies. In order to reach into the female counterpart of the die shown at top, long and slender spindle attachments shown at bottom are needed, but they lack in stiffness.

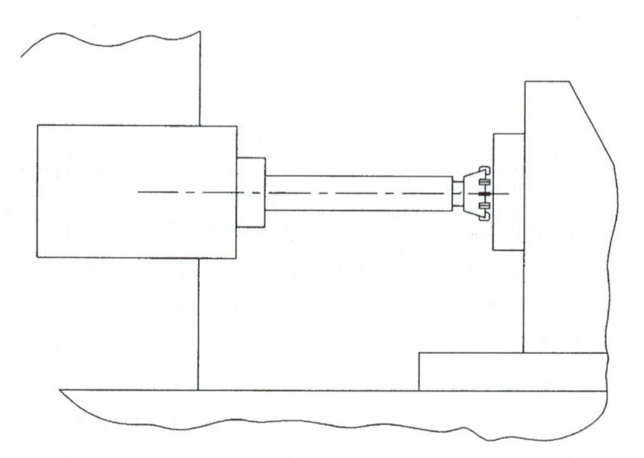

Figure 9.79
Horizontal machining center with an extendable spindle. This permits good reach of spindle end into workpiece, but at longer extensions, stiffness of the cutter is low.

mm. This design concept has the advantage of permitting the tool to reach surfaces at various levels of depth in a complex cast workpiece while bypassing the frame of the fixture in which the workpiece is clamped, but it suffers from the unavoidable limitation of stiffness on the tool, due to the long and relatively slender spindle/bar. Again, HSM appears to be a good way to iprove *MRR*s.

Many other examples involve end milling where the stiffness limitation is due to the flexibility of the tool itself, which is often a slender rod with flutes; the diameters range between 6 and 50 mm with length/diameter ratios between 4 and 8. The most common case is milling of thin-walled and pocketed aluminum structures such as the one shown in Fig. 9.80. All parts of an airplane must be as light as possible, and those that are part of the fuselage, wings, and other load-bearing components must also be strong and stiff. These combined requirements are best satisfied by monolithic (one-piece, integral) design that is obtained from a solid forging by end milling out 90–98% of the material in the form of pockets with thin walls and floors. The stiffness/weight requirement is especially stringent in military aircraft that are capable of accelerations on the order of 7 g. While taking off vertically, every 1 kg of mass is felt as 7 kg. Commercial aircraft parts may currently have ribs typically 75 mm high and 5 mm thick, requiring end mills with diameters of 19, 25, and 38 mm. Parts of fighter planes have ribs up to 100 mm high and down to 1 mm or even 0.5 mm thick with small pocket corner radii; end mills with 8–19-mm diameters and lengths up to 110 mm must be applied. End milling of aircraft parts is perhaps the largest milling operation in the world. Many machine shops containing dozens, even hundreds of NC milling machines, many of them very large and with multiple spindles, are engaged in the milling of aluminum. The annual value of these operations in the U.S. aerospace industry may be estimated at $10 billion. The importance of HSM has been recognized, and this technique is being readily adopted.

Another example of end milling where HSM is applied is the milling of scrolls for scroll pumps, either those made of aluminum and milled with carbide cutters or those made of cast iron and milled with CBN cutters. In both instances end mills with 12-mm diameter 60 mm long are used, with spindle speeds of 40,000 rpm. Milling of fan blades on blisks (blades and disc in one piece) and also of impellers for jet engines involves carbide end mills machining Ti alloys. The tools are rather slender and long so as to reach in between the blades; Fig. 9.81 shows the case of an impeller. It is being machined on a 5-axis NC machining center, and abundant coolant is seen flowing over

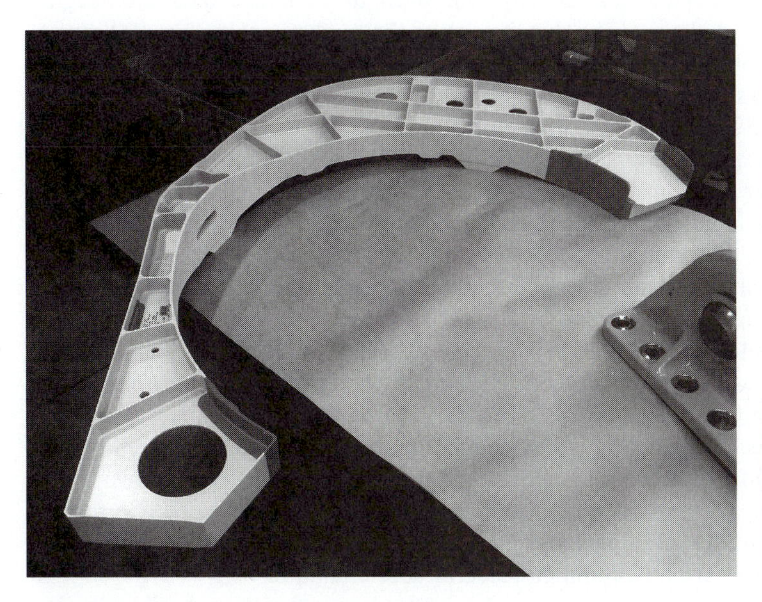

Figure 9.80
A typical aircraft aluminum part with pockets end-milled from a solid forging. A stiff, strong, and light structure is created. Long and slender end mills with inevitably low stiffness must be used. The process is very challenging from the point of view of chatter vibrations.

Figure 9.81
Machining of a jet engine impeller made of titanium. Very long and slender end mills must be used to reach between the blades—another case of chatter challenge.

the workpiece. In the case of the blisks, the tools are very long, 250 mm, with $d = 32$ mm at the root and tapering down to a ball nose with $d = 12$ mm. It is very difficult to use HSM on titanium because of its very low thermal conductivity and the consequent high temperature on the tool. Two different methods are used: "point" milling, where only the ball nose is engaged in cutting, and "flank" milling, in which about 100-mm width of the cutter edge is engaged. The latter is only applicable when the airfoil of the blade has a shape consisting of straight lines. A relatively moderate HSM is possible for point milling only while using very low depth and width of cut and so keeping the temperature in the cut at levels that the tungsten carbide tool can tolerate. The rather wide flank cuts are so demanding from the point of view of chatter stability that they can only be performed at very low speeds so as to stay in the regime of "process damping."

9.9.2 Developing an HSM Machine for Aluminum Aircraft Parts

In the following the development of the HSM machine at the Machine Tool Research Center at the University of Florida will be discussed. Similar machines have been developed in industry, partly in cooperation with the University of Florida. The basic principles applied in this development are valid generally.

Specification of the Main Characteristics of the Machine

Most aircraft structural parts are made of aluminum alloys because of their outstanding strength-over-weight ratio. An example of such an alloy is 7075-T6 (see Chapter 2, Section 2.2). It is supplied in the form of forged plates up to 100 mm thick. Hence, most of these parts are flat and, as already mentioned, they are machined into integral, pocketed, thin-walled shapes. The dimensions of the pockets are within 100 mm × 100–400 mm. It follows that (1) the workpieces are light; (2) the feed motion stops often in the corners of the pockets, demanding high decelerations and accelerations; and (3) basically a horizontal spindle is used to enable an easier evacuation of chips from the pockets than would be the case with a vertical spindle pointing downward into the pocket. Furthermore, as mentioned in the preceding paragraph, slender, small-diameter end mills are used. This fact combined with the well-established fact that the easy machinability of aluminum means that tool wear is trivial, especially if carbide tools are used, means that the fastest possible spindle speed should be used. Fast spindle speeds lead to fast feed rates and high accelerations.

It is shown below that a 36,000-rpm (max) spindle was developed, with a 10-Nm torque through the whole speed range, delivering 36 kW at top speed. Assuming a four-fluted cutter and a chip load $c = 0.2$ mm, the maximum feed rate is $f = nmc = 30$ m/min $= 0.5$ m/sec. Considering a small but common pocket size of 100 mm × 100 mm, it is clearly difficult to accelerate to the top feed rate on the way from one corner to the next. In Fig.9.82 the velocity of motion over the 100-mm path is plotted for two cases with constant acceleration/constant velocity/constant deceleration: *(a)* with a drive capable of the currently common value of acceleration of 0.2g = 2 m/sec^2 and *(b)* with a drive delivering acceleration of 2g = 20 m/sec^2. Also shown is the distribution of the three motion phases over the path. In case *(a)* full velocity of 0.5 m/sec is not achieved; for half the distance the slide is accelerating, and for the other half it is decelerating. In case *(b)* full velocity is reached in 25 msec, and 87.5% of the distance is traveled with full velocity. The total time from corner to corner is *(a)* 447 msec and *(b)* 225 msec. This shows that the requirement of high acceleration is critical, and a goal of achieving $a = 2$g has been specified.

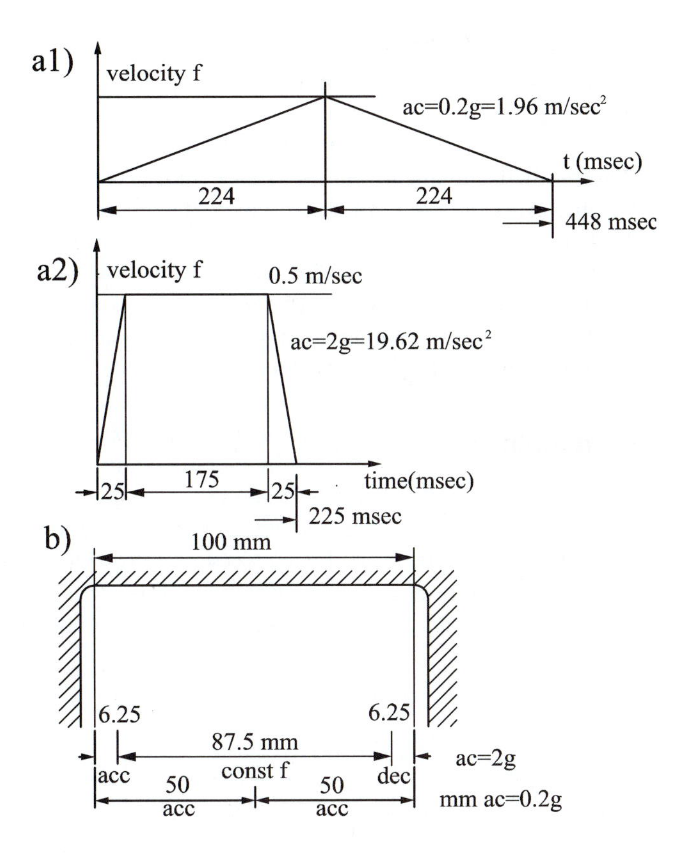

Figure 9.82
Velocity variation along the path of one 100-mm side of a pocket. In order to reach high feed rate within the limits of the short travel between corners, high acceleration of 2g is needed. Pass time is just half of the one obtained with 0.2g.

High-Speed, Highly Stable Spindle

Bear in mind that the most important requirement on the spindle/tool-holder/tool (SHT) system is to achieve the highest possible stiffness at the end of the tool for maximum stability against chatter. It is true that the tools to be used are rather flexible, with diameters between 12.5 mm and 25 mm and length-over-diameter ratios of 4 to 8. However, they are clamped in overhang to the spindle shaft mounted in ball bearings, and bending of the spindle contributes strongly to the deflection at the tool end. The tool holder also plays an important role in that its length extends the overhang between the front bearing of the spindle and the tool end.

Let us first quickly dispose of the tool-holder problem. Several types of tool clamping have been considered. The final solution came in the form of very short holders with just a plain hole into which the tool is shrink-fitted. A special induction-type furnace is supplied that has a heating coil into which the tool and tool holder are inserted, both for clamping and releasing the tool. Each of these operations takes less than 10 sec.

From the beginning, the spindle was designed so that the drive is not transmitted by a belt; rather, the rotor of the electric motor is mounted directly on the spindle shaft, between the bearings. Preloaded, angular-contact ball bearings were selected. The crucial problem was the choice of the spindle-shaft diameter, which is also the bore diameter for the bearings. The larger the bearing diameter, the higher the rolling velocity, spins and skids of the balls, and the centrifugal force on the balls. In spindle design the difficulties with the proper and robust running ability of the bearings and of their life are traditionally associated with the so-called *DN* number, which is the product of the mean bearing diameter *D* (mm) and the spindle speed *N* (rpm). For angular-contact bearings, the current practice did not recommend more than *DN* = 750,000. Here, the target spindle speed of 36,000 rpm and the desired bearing-bore diameter of 60 mm (bearing mean diameter: 77 mm) call for *DN* = 2,776,000. Also other parameters became overwhelming. So, for instance, the aluminum bars in the squirrel cage of the induction type electromotor, at a diameter of 80 mm, are subject to a centrifugal acceleration $a_c = r\omega^2 = 0.04 \times (2\,\pi n/60)^2 = 568{,}489\ \text{m/sec}^2 = 58{,}000$ g. Indeed, in the first prototype the rotor bars started to extrude through the slits in the rotor laminations.

Why a 60-mm spindle shaft diameter? It was necessary to balance the dynamic stiffness of the SHT system. For many diameter–length combinations of tools the mode shapes, modal stiffnesses, and natural frequencies have been computed for various spindle dimensions. An example of such a computation is presented in Fig. 9.83 for a d = 19 mm, HSS tool, of 65 mm free length clamped in a Weldon-type holder, in a 60-mm-diameter spindle. The first 5 modes of natural vibrations are shown. Modal frequencies and stiffnesses are marked. In this case the lowest modal stiffness of 0.96 e7 N/m belongs to mode 3, with frequency of 1846 Hz. Apart from the HSS tools, solid carbide tools were also considered, as well as other types of holders; the designers finally settled on the shrink-fit type. Without going into any more details, it was found

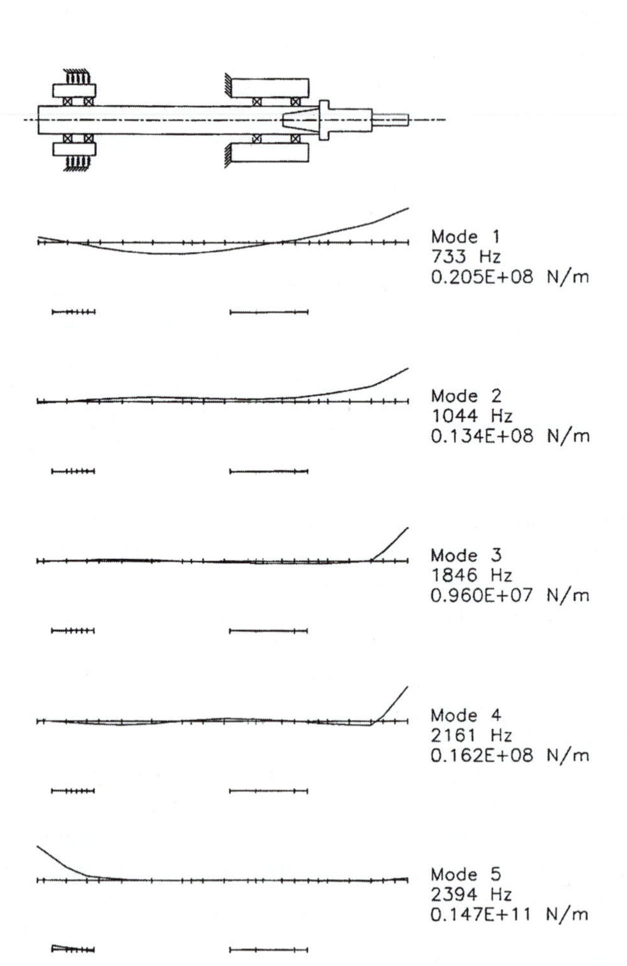

Figure 9.83
Typical spindle/holder/tool mode shapes of a high-speed spindle carrying a typically slender end mill. Natural frequencies and modal stiffnesses are marked for each mode. In this case, the "tool-bending" mode 3 has only half the stiffness of the "spindle-bending" mode 1. This kind of analysis helps in designing optimum spindle and holder.

that the 60-mm spindle diameter provides optimum dynamics for the SHT system for most of the tools. The design of the spindle incorporates two pairs of low-contact-angle, hybrid (silicone nitride balls on steel races) bearings. The ceramic balls have 0.4 the mass of steel balls. This alleviates the problem of centrifugal forces on the balls and the associated skidding. A special lubrication system is used that delivers reliably just about one drop of oil every 5 min, thus minimizing viscous energy losses so as to keep the heat generation in the bearings at a very low level. The rear pair of bearings is axially preloaded by means of springs against the front pair. It is very important to keep the preload constant at all speeds. This is assured by having the rear bearings mounted in a bushing that is inserted in a roller cage, thus eliminating any axial friction and any danger of losing the floating capability of these bearings by their getting stuck in their seats. The spindle proved to give a very satisfactory performance.

The Configuration of the Axes of the Machine

The size of the machine was chosen so as to accommodate a selected medium range of parts within a space of 750 mm × 750 mm × 200 mm. It should be mentioned that milling machines for aircraft parts exist also with travels in the long coordinate up to 10 or even 20 m. The common configuration of a medium-size machine with a horizontal spindle is shown in Fig. 9.84. The spindle head moves vertically in the *Y* direction on the column. A saddle moves horizontally on the bed in the *Z* direction parallel with the mean position of the spindle axis. On the top of the saddle are the guideways for the motion of the table in the direction *X* that is horizontal and perpendicular to *Z*. The top of the table is horizontal. The flat aircraft parts could not be machined by the horizontal spindle when clamped directly on the table. It is necessary to attach an angle plate to the top of the table and clamp the parts onto the face of the angle plate. In other instances instead of the angle plate an indexing rotary table with a horizontal top is put on the *XZ* table, and a tower with a square cross section is mounted onto the rotary table. This tower is commonly called the tombstone. Flat parts are clamped to all the four faces of the tombstone and are brought in sequence against the spindle for

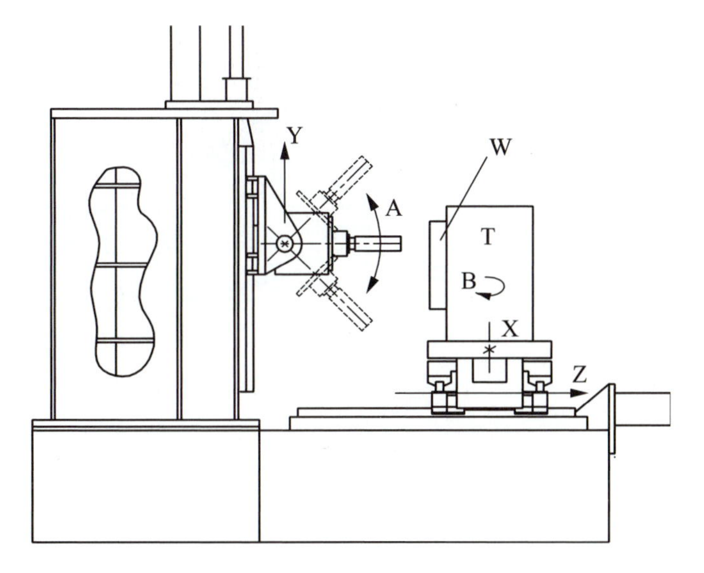

Figure 9.84
Conventional, horizontal-spindle machining center. A heavy "tombstone *T*" must be used to support the workpiece *W* opposite the spindle. This "dead" mass limits the achievable acceleration.

machining by indexing the rotary table. The angle plate, and especially the tombstone, are much heavier than the workpiece; therefore in this configuration both X and Z coordinate motions must accelerate these additional "dead" masses. The rotary table may be built for continuous motion in the rotary coordinate B and the spindle may be controlled in the rotary coordinate A.

A different configuration has been conceived. It was decided to locate the table in a vertical plane opposing the spindle as is shown in Fig. 9.85. The table, which actually can be made very light just to provide clamping points for a palette carrying the flat workpiece, moves in vertical direction Y on guideways provided on the front of a saddle that moves horizontally in X. The spindle head is mounted on a slide that moves horizontally in the Z direction. All guideways are of the recirculating roller type to minimize friction and suppress the dead zone. The X and Y feeds are driven from rotary DC electric servomotors via ball screws and nuts. Because the X must accelerate the compound masses of the X saddle and of the Y table, it is driven by two motors and leadscrews in parallel. The Z motion is driven by two parallel, linear, electric servomotors. A trend may be observed towards replacing rotary servomotors by linear ones in general, especially for longer travels. For shorter travels, as in our machine, the linear motors eliminate the use of the leadscrew/nut transmissions, and for long travels, they replace the classical rack-and-pinion pairs. On top of the Z slide is a frame rotating in the vertical B-axis and providing bearings for rotation of the spindle housing around the horizontal A-axis. The machine is numerically controlled in five axes. Considerable effort was spent on the controls of the positional servos in order to achieve the desired goal of 2g accelerations under stable conditions in the feedback loops. The problems encountered in such tasks and their solutions are dealt with in Chapter 10.

The design described so far considered just the basic and bare machine without looking at tasks such as the covers of guideways and the enclosure of the machine to contain the coolant and the chips, chip transport, and so on. Figure 9.86 is a photograph of the bare machine in a three-axis configuration. In the completed machine the work-

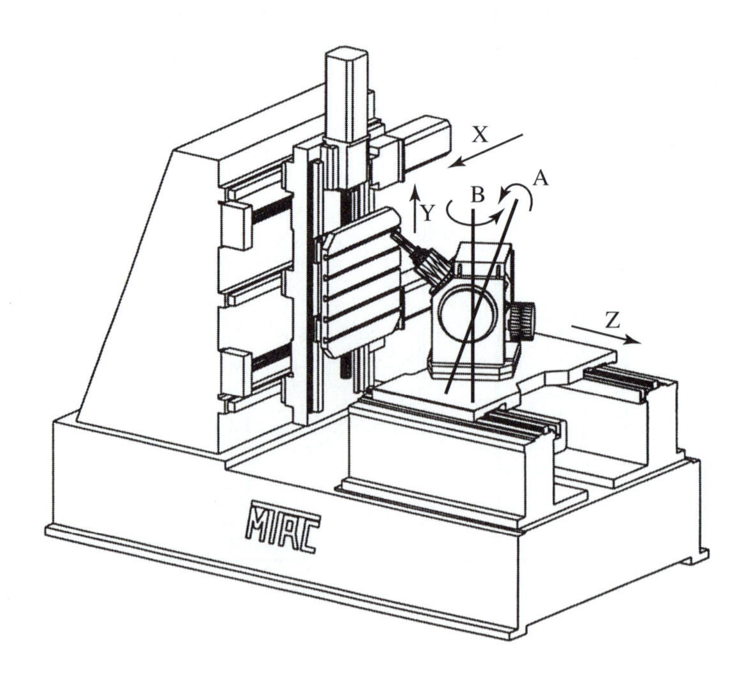

Figure 9.85
The design concept of the UF high-speed milling machine. The worktable moves in a vertical X-Y plane. No extra support masses are necessary. A five-axis $XYZAB$ design is presented.

Figure 9.86
The UF high-speed milling machine in a three-axis *XYZ* configuration. The spindle is seen from its rear end in the front part of the picture. The dual leadscrews and triple roller guideways of the *X*-axis and the vertical leadscrew and double guideways of the *Y*-axis are seen in the back, located on the vertical plane of the column. The enclosures of the machine are removed for visibility of the features.

space is fully enclosed for several reasons: it contains the chips, which are taken out by a chip transporter that is not shown; it protects the operator against flying broken tool tips (rare) and against exposure to the lubricating mist supplied to the cutting edges and against the noise of the high-speed spindle. A hoist may be used that picks up a pallet with a workpiece from a table located next to the machine (not shown) and moves it to the top of the table, where spring-operated clamps attach it to the table. When the part is finished, the clamps are released by means of hydraulic cylinders mounted on the face of the column. Another important device is a tool magazine and automatic tool-changing mechanism, which may be located under the ceiling of the enclosure, above the headstock.

Some aircraft parts have to be machined in more than three axes. In another application, that of milling injection molding and forging dies, there is a strong need for two additional rotary NC axes. The design of the *AB* head that converted our machine into a five-axis machine was shown in Fig. 9.85. The rotation about the vertical axis *B* is performed on circular guideways on the *Z*-axis saddle, and it carries a frame that houses the horizontal rotary axis *A*.

The *AB* head invited analysis of how much it contributed to the flexibilities at the tool end. Finite-element computations have revealed and measurements confirmed that indeed this head has several vibrational modes that are only about 3 and 5 times stiffer than the spindle/holder/tool modes. However, these modes have rather low frequencies, between 90 and 120 Hz. As long as the machine is used for high-speed machining, at spindle speeds above 12,000 rpm and using tools with at least two teeth, resulting in tooth-passing frequencies above 400 Hz, these cases would be located at least four times above the highest "lobe of stability" in the diagram of Fig. 9.74. Therefore the stiffnesses of the modes of the *AB* head are safely above the stability limits of the SHT system.

Once this machine became functional in three axes, it was challenged by one of the leading machine tool companies to test-produce a very difficult part with pockets 100 mm deep, walls 0.5 mm thick, and floors 1 mm thick, with an opening in one of the floors, to be milled out of a solid block of aluminum. A photograph of the part is in Fig. 9.87. The part has been made repeatedly in many copies, with superb finish and

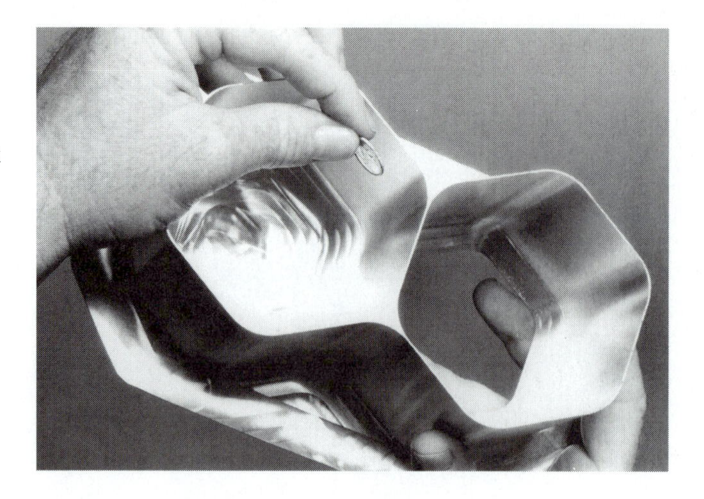

Figure 9.87
Photograph of an extremely thin-walled component milled on the machine of Fig. 9.86 out of a solid block of aluminum. Wall thickness is 0.5 mm, and the walls are 100 mm deep. High stiffness and strength over weight ratio is achieved.

accuracy, in an overall machining time five times shorter than when the part is made conventionally, that is, without high-speed milling. The techniques of HSM, while still in an initial stage of wide acceptance, have already been proven in leading aircraft manufacturing plants. The design exercise described here shows that in new developments it is necessary to identify and solve the problems particular to each task.

REFERENCES

1 Tlusty, J., and F. Koenigsberger. *Specifications and Tests of Machine Tools.* Manchester: UMIST, April, 1970.

2 Koenigsberger, F., and J. Tlusty. *Structures of Machine Tools.* Oxford: Pergamon Press, 1970.

3 Tlusty, J. *Criteria and Methods in Structural Analysis.* Machine Tool Task Force, Lawrence Livermore National Laboratory, SME, October 1980.

4 Weck, M., *Handbook of Machine Tools.* Vol. 1. *Types of Machines, Forms of Construction, and Applications.* Vol. 2. *Construction and Mathematical Analysis.* Vol. 3. *Automation and Controls.* Vol. 4. *Metrological Analysis and Performance Tests.* Wiley Heyden Ltd., 1980.

5 Slocum, A. H. *Precision Machine Design.* Englewood Cliffs, NJ: Prentice-Hall, 1992.

6 Weck, M., and R. Hartel. "Ultraprecision in Manufacturing Engineering." *Proceedings of the Intnl. Congress for Ultraprecision Technology.* Aachen: Springer Verlag Berlin, May 1988.

7 Boothroyd, G., and W. A. Knight. *Fundamentals of Metal Machining and Machine Tools.* New York: Marcel Dekker, 1989.

8 Tlusty, J. *A Specific Approach to Accuracy Testing.* Machine Tool Task Force, Lawrence Livermore National Laboratory, SME, 1980.

9 "Axes of Rotation," ANSI/ASME B89.3.4–1985, American Society of Mechanical Engineers, New York.

10 DeVor, R. E.; W. A. Kline; and W. J. Zdeblick, "A Mechanistic Model for the Force System in End Milling with Application to Machining Airframe Structures." *Proc. NAMRC* 8 (1980).

11 Smith, S., and J. Tlusty. "An Overview of Modeling and Simulation of the Milling Process." *ASME J. Eng. for Industry* 113 (1991): 169–75.

12 MILSIM software for simulation of milling forces, vibrations, and surface profile. Manufacturing Laboratories, Inc., Gainesville, FL.

13 Tlusty, J. and M. Polacek. "The Stability of the Machine Tool Against Self Excited Vibration in Machining." Prod. Eng. Res. Conf. ASME, Pittsburgh, 1963.

14 H. E. Merritt. "Theory of Self-Excited Machine Tool Chatter." *ASME J. for Eng. in Industry* 87 (1965): 447–54.

15 Tlusty, J. "Machine Dynamics." Chapter 4 in *Handbook of High-Speed Machining Technology.* ed. R. I. King. New York: Chapman and Hall, 1985.

16 Tlusty, J. and F. Ismail. "Special Aspects of Chatter in Milling." *Trans. ASME, J. of Vib, Stress, and Reliability in Des.* 105 (Jan. 1983): 24–32.

17 Tlusty, J. "Dynamics of High-Speed Milling." Symp. on High-Speed Machining, WAM, ASME, 1984.

18 Delio, T.; S. Smith; and J. Tlusty. "Use of Audio Signals for Chatter Detection and Control." ASME, *J. of Eng. for Ind.,* 114 May 1992: 146–57.

19 Zeid, I. *CAD/CAM Theory and Practice.* New York: McGraw-Hill, 1991.

20 Tobias, S. A. *Machine Tool Vibrations.* London: Blackie & Son, 1965.

Illustration Sources

[KEARNEY & TRECKER] Kearney & Trecker Co., Milwaukee, WI 53214.

[MILACRON] Milacron Inc., 4701 Marberg Ave., Cincinnati, OH 45209–1025.

QUESTIONS

Q9.1 Name typical parts of a machine tool in the three basic groups.

Q9.2 Make a sketch of a typical mounting of a machine tool spindle. What are the special requirements to satisfy?

Q9.3 Name the three basic types of guideways in machine tools.

Q9.4 How is the total speed range of a machine-tool spindle determined?

Q9.5 For which workpiece material is the requirement of power in the main drive highest?

Q9.6 Name the factors affecting the accuracy of the workpiece. How is geometric accuracy of machine tools defined?

Q9.7 What principle governs modern metrology of machine tools, and how is it applied?

Q9.8 What are the three translative error measurements made along a line in the workspace of a machine tool?

Q9.9 What are the three angular errors associated with the motion of a body? Make a sketch defining them for the motion of a table of a horizontal machining center.

Q9.10 Write out the equations expressing the effect of pitch on positioning error in the workspace of a machining center and the effect of roll on straightness errors. Make illustrative sketches.

Q9.11 Make sketches illustrating the principle of using a laser interferometer for measurements of the positioning error, of the straightness-of-motion error, and of pitch error.

Q9.12 Make sketches illustrating three examples of weight deformations. Show how the effect of the weight of a headstock on a cross-rail can be eliminated by counterbalancing.

Q9.13 What are the two independent causes of the variation of deformation between tool and workpiece.

Q9.14 Derive the formula for the rate of copying of a form error of a workpiece.

Q9.15 What are the detrimental effects of chatter? How is chatter recognized? Which of the resulting parameters of the machining process is limited?

Q9.16 Distinguish between natural, forced, and self-excited vibrations as regards frequency, amplitude, and external variable force. Give examples of self-excited vibrations.

Q9.17 What are the two basic harmonic functions? By how much are they mutually phase-shifted? How is a harmonic variable represented in the complex plane? What is its complex amplitude?

Q9.18 Plot the approximate shapes of the graphs of the three kinds of representations of the transfer function of a single-degree-of-freedom system.

Q9.19 A harmonic force acts at the end of a bar with rectangular section and mass at the end. What is the shape of the motion of the mass in the plane perpendicular to the axis of the bar?

Q9.20 Explain the concept of directional factors involved in the oriented transfer function of a boring bar. Indicate the directions involved: of the modes, of the cutting force, of the normal to the cut surface. Write the formula for the directional factors u.

Q9.21 Which cutting parameter is most decisive for stability of cutting? Show it in sketches for turning, face milling, and end milling.

Q9.22 Explain the two mechanisms of self-excitation in metal cutting.

Q9.23 Which characteristic of the vibratory system is decisive for the value of the width of chip at the limit of stability against chatter? How is this characteristic expressed for a single-degree-of-freedom system as a function of its stiffness k, damping ratio ζ, and directional factor u? Write the formula for b_{lim}.

Q9.24 How and to what extent does the stability of a boring bar with rectangular section vary with its directional orientation? Make an explanatory sketch and an approximate polar plot of stability.

Q9.25 How is the oriented transfer function obtained for a round bar with diameter d and length l, with damping ratio ζ? Express the corresponding b_{lim}.

Q9.26 How is a measure of stability of a machine tool obtained and tested?

Q9.27 Indicate variations of directional orientation of the cutting process in the machine-tool structure for turning and for face milling.

Q9.28 How is the decisive (for chatter) mode of a machine tool recognized? How is stability of this machine tool increased?

Q9.29 For overhang turning, and for face milling, which geometric parameter of the spindle-holder (chuck)/workpiece (tool) system is most important for stability of machining?

Q9.30 Which is the most efficient way of increasing the damping of a structural mode? Give an example of an indicated use of this method.

Q9.31 How is the effect of workpiece material on chatter explained?

Q9.32 Discuss the effect of feed on chatter.

Q9.33 Discuss damping in the cutting process and the effect of cutting speed.

Q9.34 Discuss the stability lobes expressing the effect of spindle-speed variation on stability of milling. Draw a typical diagram.

Q9.35 Which kinds of special milling cutters can be used for increased stability?

PROBLEMS

P9.1 Determine the drive characteristics and plot the (P, n) envelope diagram, similar to Fig. 9.6, for $d_{min} = 10$ mm, $d_{max} = 150$ mm, chip section $A = 2$ mm^2, machining of cast iron and of steels 1020 to 4310.

P9.2 *Geometric accuracy of a machine tool.* Figure P9.2 gives the results of measurements $\delta_x(x)$, $\delta_y(x)$, $\varepsilon_x(x)$, $\varepsilon_y(x)$ as obtained along line A. Determine errors $\delta_x(x)$ and $\delta_y(x)$ along line B and draw them in the same graphs as $\delta_x(x)$ and $\delta_y(x)$ along A. Use the signs of motions and of errors as shown.

P9.3 *Geometric accuracy of a machine tool.* Assume that the measurements shown in Fig. P9.3 were made along line A. Determine the errors $\delta_x(x)$, $\delta_y(x)$, and $\delta_z(x)$ along line B and plot them in the corresponding graphs. Use the signs of motions and of errors as shown.

P9.4 *Copying of form error.* Consider the turning operation shown in Fig. P.9.4. The tool has a lead angle of 45°. The force F is generated in the plane A-A, perpendicular to the cutting edge. In this plane the ratio of the normal and tangential forces is $F_N/F_T = 0.35$. The flexibility between the tool and workpiece is concentrated in the direction X normal to the axis of the workpiece and located in the horizontal plane (the plane of the top of the tool); its inverse is the stiffness $k_m = F_x/x = 5 \times 10^4$ N/mm.

The initial error of roundness $\Delta = a_{max} - a_{min} = 1$ mm. Using $K_s = 2000$ N/mm^2, roughing feed $f_r = 0.25$ mm, and finishing feed $f_f = 0.1$ mm, what will be the magnitudes of the error δ_r after the roughing and δ_f after the finishing cut?

P9.5 *Error of position of the machined surface in end milling* (see Fig. P9.5). Radial depth of cut $a = d/2$, chip load (feed per tooth) $c = 0.2$ mm, axial depth of cut $b = 10$ mm. Assume that deflection follows the force instantaneously, without phase shift, like a static case, and the flexibility of the tool is $1/k = 2\text{e}{-5}$ mm/N. Specific force $K_s = 1500$ N/mm^2,

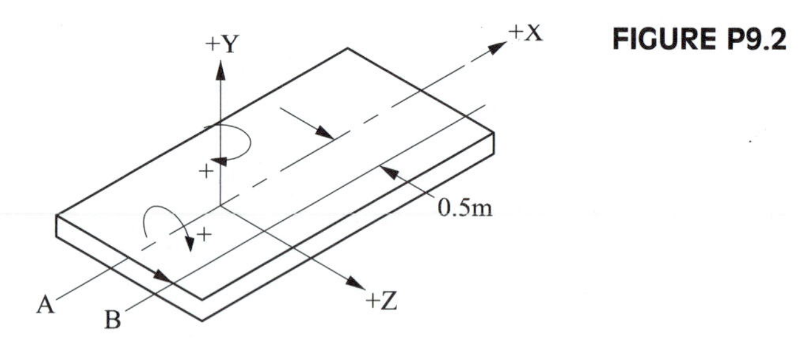

FIGURE P9.2

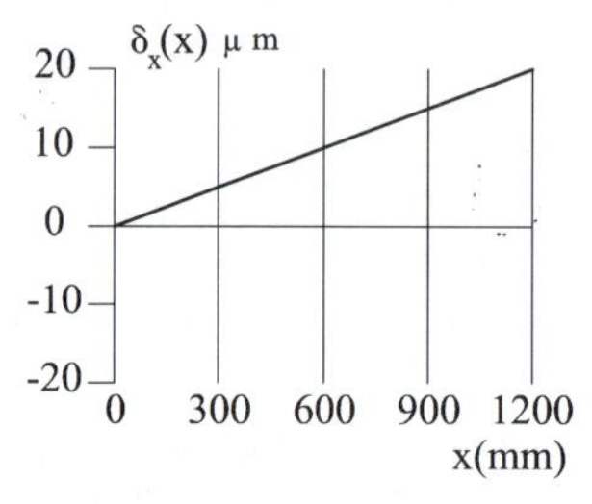

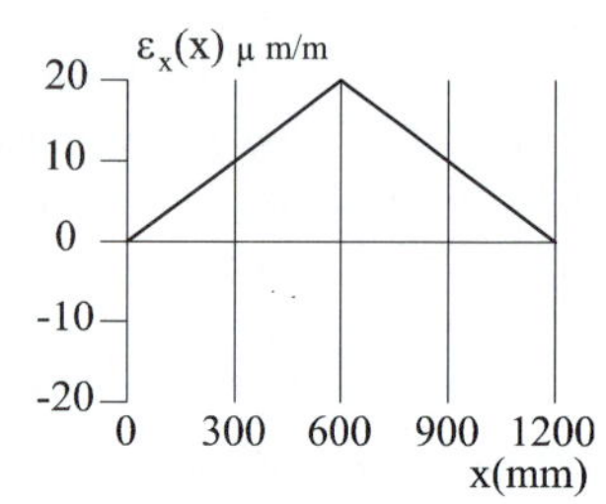

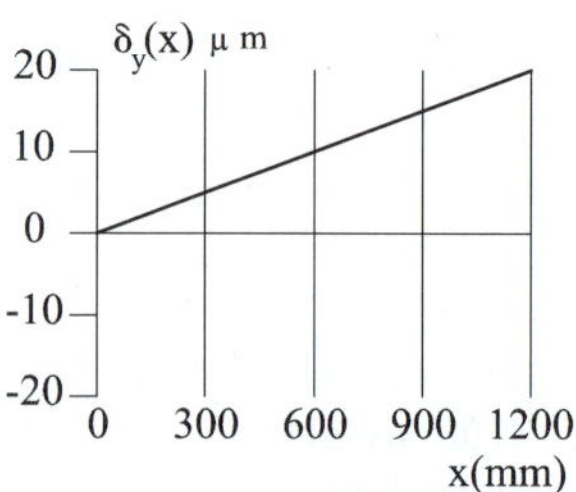

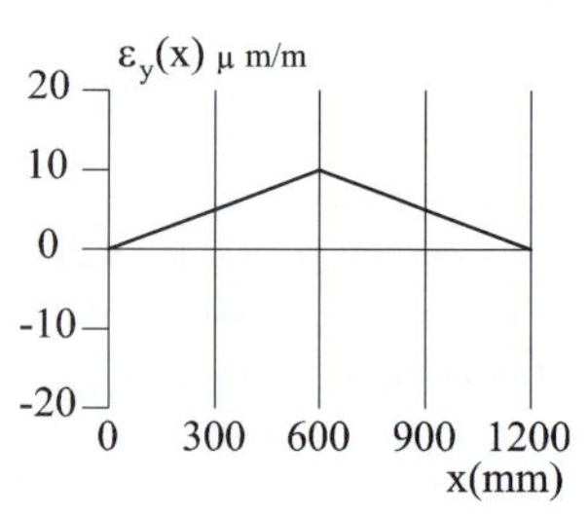

$F_N = 0.3\, F_T$. Determine the tangential force F_T, the force F_y normal to the machined surface S, and the error of position of the machined surface δ.

P9.6 *Error of position of milled surface: slotting with a two-fluted cutter.* This is an exercise like those in Examples 9.7 and 9.8. The F_y force component consists of harmonic terms.

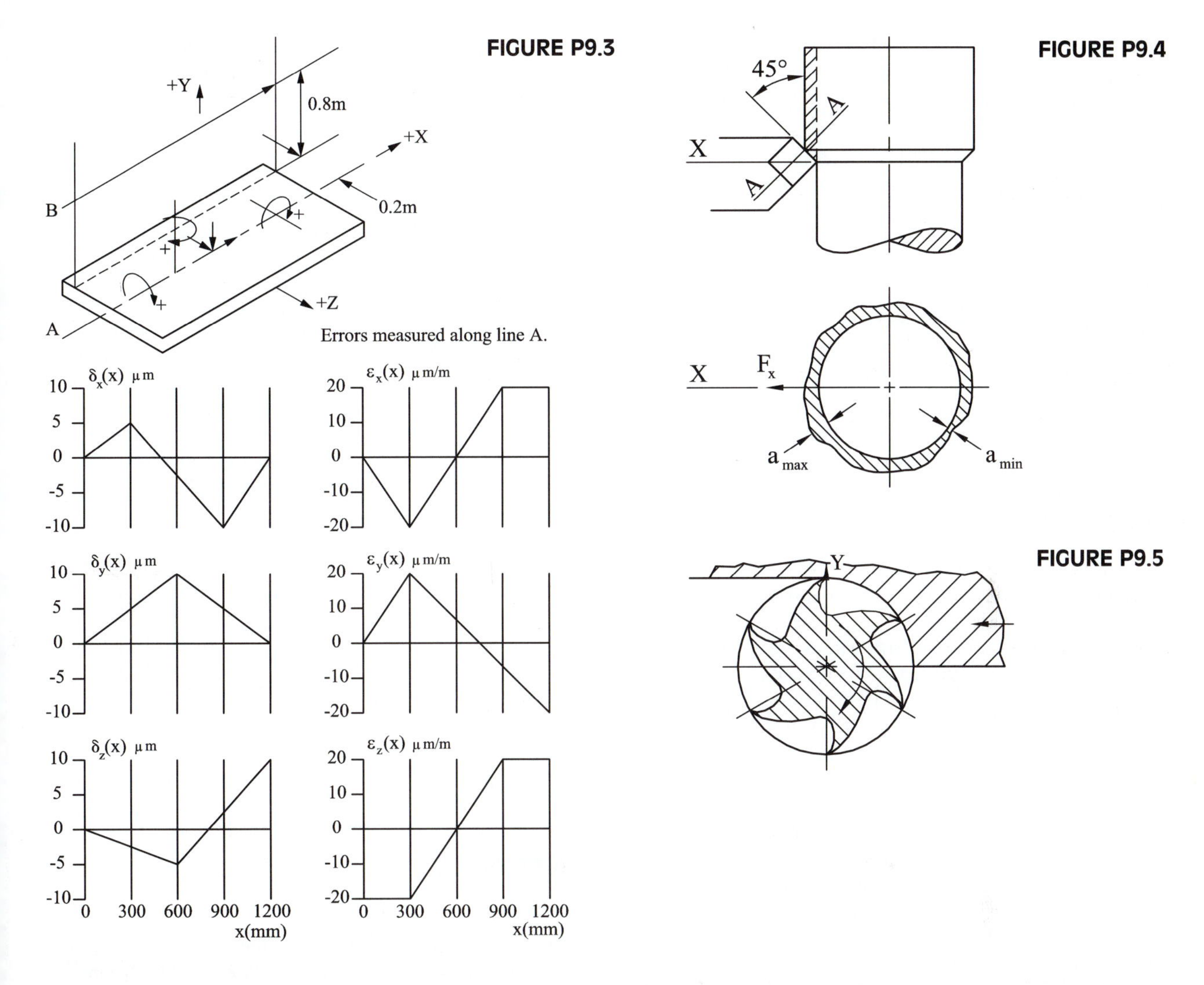

Thus it is possible to calculate the error of position of the machined surface without resorting to time-domain simulation.

In Example 9.7 spindle speed was $n = 7200$ rpm $= 120$ rev/sec. Correspondingly, the angle of rotation was $\phi = 2\pi \times 120 \times t$, and the frequency of the force components was 240 Hz. Different spindle speeds and dynamic parameters are used here. Refer to Figs. 9.38 and 9.39.

The dynamic parameters of the tool in direction Y are $k_y =$ 2000 N/mm, natural frequency $f_n = 250$ Hz, and damping ratio $\zeta = 0.05$. Use three different speeds such that they result in under-resonant, resonant, and above-resonant conditions: (1) $n = 6900$ rpm, (2) $n = 7500$ rpm, (3) $n = 8400$ rpm. Determine amplitudes A_1, A_2, and DC of the three components of the deflection y and their phase shifts ϕ. Plot diagrams such as in Fig. 9.39 and evaluate the error δ for all the three cases.

P9.7 *Milling force, vibration, error of position of machined surface: cutter with straight teeth.* Consider **(a)** up-milling and **(b)** down-milling for a cutter with 4 straight teeth as shown in Fig. P9.7. We have the following: axial depth of cut $b = 10$ mm, chip load $c = 0.1$ mm, stiffness of cutter $k = 4.8 \times 10^6$ N/m, mass $m = 1.0$ kg, damping coefficient $c = 150$ N/(m/sec), spindle speeds are (1) $N = 4800$ rpm, and (2) $N = 6000$ rpm.

Determine the following: natural frequency f_n (Hz), tooth frequency f_t (Hz), and damping ratio ζ. Determine the starting and ending angles of cut ϕ_s and ϕ_e. Write the formula for F_y (use $K_s = 750$ N/mm^2), omitting regeneration, and plot F_y versus time for one revolution of the cutter for *(a)* and *(b)*.

Write a simulation program using 200 steps per revolution that will formulate force F_y and vibration y. Plot y versus t for the initial vibration transient over the first 12 revolutions and then for the last two revolutions (fully developed vibration) for *(a1)*, *(a2)*, *(b1)*, and *(b2)*. In each case plot F_y vs. t and y vs. t as two separate graphs, but place them on the same page, one directly above the other. Mark on the y versus t plot of the

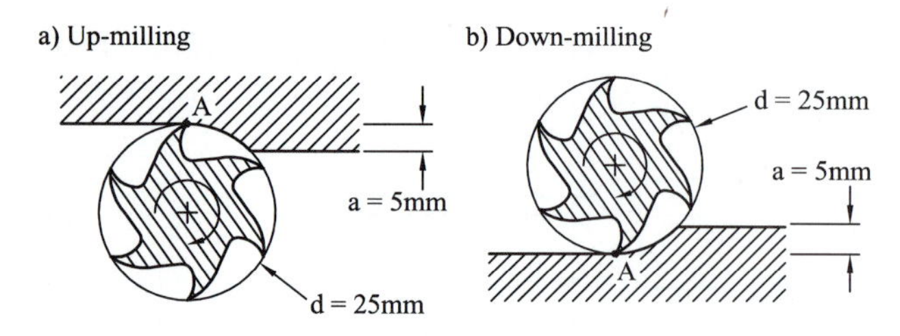

FIGURE P9.7

fully developed vibration the point of passage through the point A generating the surface, and write on it the value of error e.

P9.8 *Error of position of an end-milled surface: cutter with straight teeth.* This is an extension of Problem 9.7, concentrating on the *up-milling case* and on the *role of spindle speed.*

We have a four-fluted cutter with an engagement of $\phi_s = 0°$, $\phi_e = 45°$, axial depth of cut $b = 10$ mm, chip load $c = 0.1$ mm, stiffness on the cutter $k = 4e^6$ N/m, mass $m = 0.8765$ kg natural frequency $f_n = 340$ Hz, damping ratio $\zeta = 0.04$ [$c = 150$ N/(m/sec)].

Run simulations at four different speeds sp specified below. Use 200 steps per revolution, $M = \text{fix}((\pi/2)/d\phi)$ is the number of steps per tooth period (90°). Cutting engagement extends over $K = M/2$ steps. Plot F and y over the total of 12 revolutions and then for the period of the last two revolutions. On the latter record mark the passage of a tool through point A of the cut and measure the error of position of the cut.

The cutting force is strongly nonharmonic and will have strong higher harmonics. The speeds as specified will produce the first and second harmonics close to and below and above resonance, and the vibration will correspondingly contain a mixture of those frequencies. For the speeds (a) to (d), the tooth frequency f_t and its second harmonic will be distributed as indicated:

(a) $sp = 2400$ rpm, $f_t = 160$ Hz, $2f_t = 320$ Hz
(b) $sp = 3180$ rpm, $f_t = 212$ Hz, $2f_t = 424$ Hz
(c) $sp = 4800$ rpm, $f_t = 320$ Hz, $2f_t = 640$ Hz
(d) $sp = 5700$ rpm, $f_t = 380$ Hz, $2f_t = 760$ Hz

The approximate distribution of these excitations on the response function of the system is shown in Fig. P9.8.

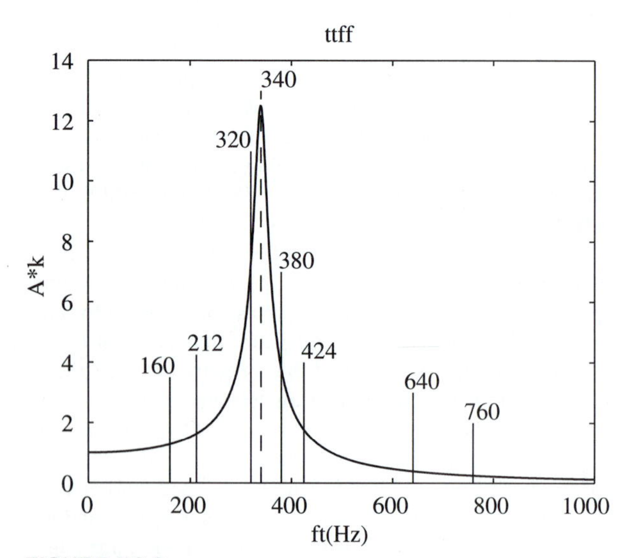

FIGURE P9.8

P9.9 *Milling force, deflection, surface error.* The parameters of the case are cutter diameter $d = 25$ mm, $m = 4$ teeth, helix angle $\beta = 30°$ up-milling. Consider two depths of cut: **(a)** $b = 22.67$ mm, **(b)** $b = 51$ mm. Simplified graphs of one period of the F_y force are given in Fig. P9.9. The vertical scale is (N). The flexibility on the cutter is 1 mm /4000 N. Plot (by hand) the corresponding surface profiles.

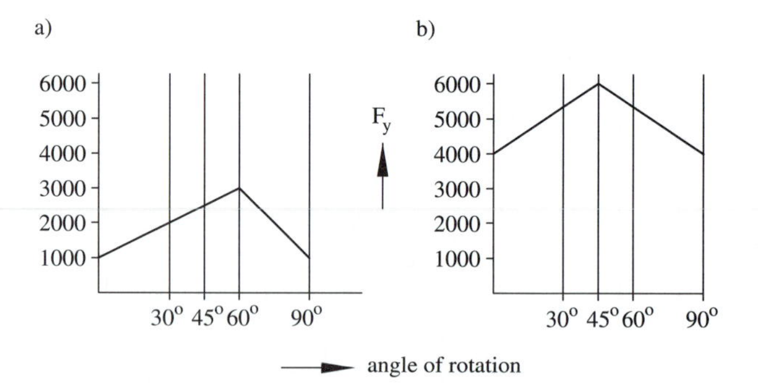

FIGURE P9.9

P9.10 *Limit of chatter stability.* Assume a single-degree-of-freedom system as shown in Fig. P9.10; $K_s = 1200$ N/mm^2 = 1.2×10^9 N/m^2, $k = 2 \times 10^7$ N/m, $\zeta = 0.04$, $m = 20$ kg. Determine the chatter frequency f_c (Hz) and the limit chip width b_{lim} (mm).

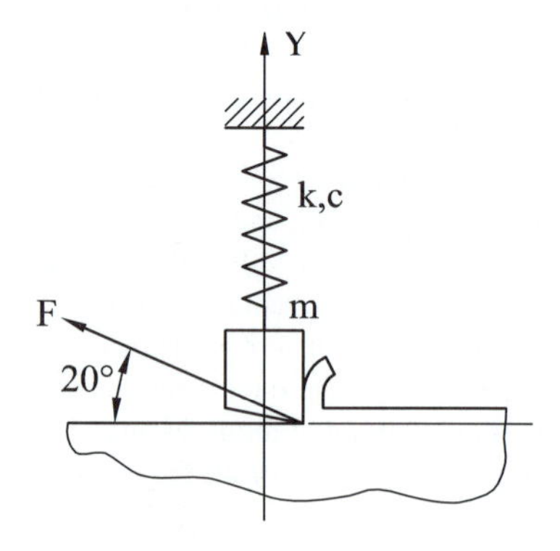

FIGURE P9.10

P9.11 *Chatter vibrations, two degrees of freedom (see Fig. P9.11).* Calculate the minimum of the real oriented transfer function $\text{Re}\,[G]_{or} = u_1 \text{Re}\,[G_1] + u_2 \text{Re}\,[G_2]$ and calculate the critical limit width of chip b_{lim}, assuming workpiece material with $K_s = 2000$ N/mm^2. The parameters are chosen so as to make this calculation very easy; it is a very special case: $k_1 = 4.3604e6$ N/m, $m = 50$ kg, $f_{n1} = 47$ Hz, $\zeta_1 = 0.06$, $k_2 = 5.5447e6$ N/m, $f_{n2} = 53$ Hz, $\zeta_2 = 0.06$. Determine the directional factors u_1, u_2. At which frequency f_1 will u_1 Re $[G_1]$ have its negative peak? At which frequency f_2 will u_2 Re $[G_2]$ have its negative peak (positive peak of Re $[G_2]$)? Draw preliminary sketches of the two Re $[G]$-oriented functions.

Make an approximate sketch of u_1 Re $[G_1]$ + u_2 Re $[G_2]$. Determine the magnitude of the sum of the minima of both real transfer functions. Determine Re $[G_{or}]_{min}$ and b_{lim}.

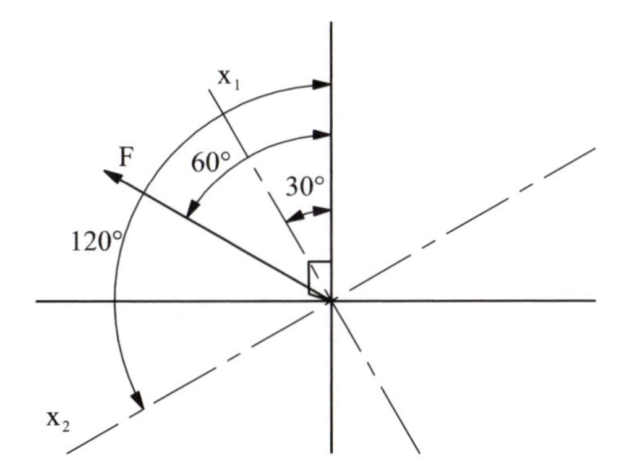

FIGURE P9.11

P9.12 *Chatter in turning (see Fig. P9.12).*

$\alpha_1 = 35°$, $\alpha_2 = 125°$, $\beta = 70°$

$K_s = 2000$ N/mm^2

$\zeta_1 = \zeta_2 = 0.04$

$h_m = 0.1$ mm

For any one of the four specifications *(a)* through *(d)* given below, carry out Tasks 1 and 2.

Task 1. Determine the value of directional factors u_1 and u_2 and plot out u_1 Re $[G_1]$, u_2 Re$[G_2]$ and Re $[G] = u_1$ Re $[G_1]$ + u_2 Re $[G_2]$ over the range 0–400 Hz. Use $\Delta f = 1$ Hz. Calculate $b_{lim,cr}$ (mm).

Task 2. Write a simulation program and plot the force component *Fy* and vibration *y* for six revolutions of the workpiece. Choose dt = 0.0002 sec, spindle speed 600 rpm; this will result in $i = 500$ computation steps per workpiece revolution. Run the simulations for $b1 = 0.8\, b_{lim}$, $b2 = 1.2\, b_{lim}$, $b3 = 1.5\, b_{lim}$, and $b4 = 2.0\, b_{lim}$.

The four parameter combinations are given below.

	k_1 (N/m)	k_2 (N/m)	m (kg)
(a)	4e7	6e7	50
(b)	6e7	4e7	50
(c)	3e7	8e7	30
(d)	8e7	3e7	30

Plot vibrations y versus t and force F versus t.

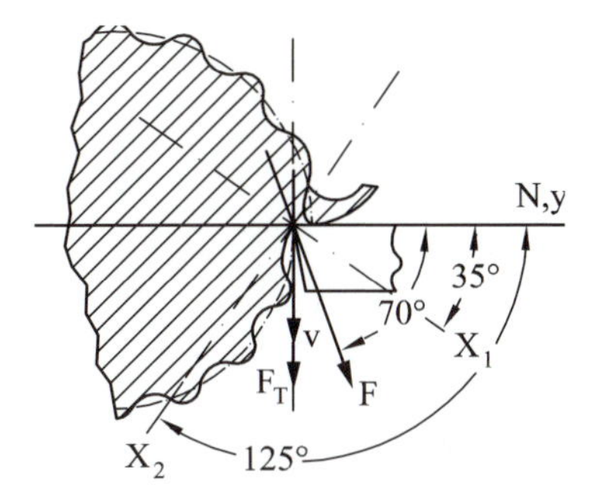

FIGURE P9.12

P9.13 *Chatter in internal plunge turning (boring) (see Fig. P9.13).* We are given $m = 5$ kg, $a = 60$ mm, $b = 67$ mm, $l = 240$ mm, $E = 2e5$ N/mm^2, damping ratio in both x_1 and x_2 is $\zeta = 0.05$, $K_s = 1500$ N/mm^2, $\beta = 72°$.

Determine b_{lim} for the cases **(a)** and **(b),** as shown in the figure.

Take case *(a)* only and use workpiece rotational speed of 1200 rpm and feed per revolution $f_r = 0.15$ mm. Simulate the vibrations for *(a1)* $b = 0.7\, b_{lim}$, *(a2)* $b = 1.6\, b_{lim}$. Choose dt = 0.0002 and total simulation time 1 sec. Plot out the results.

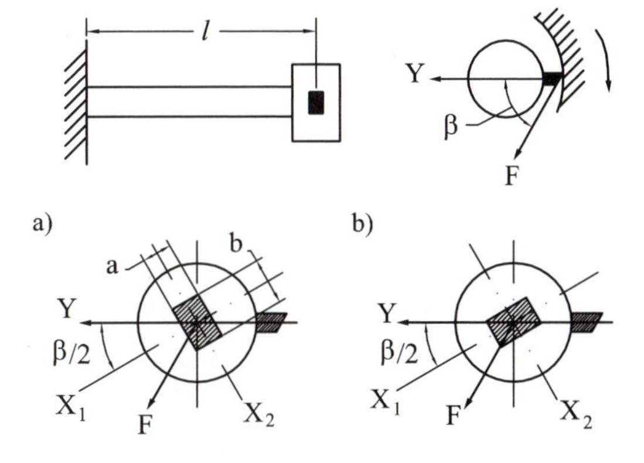

FIGURE P9.13

P9.14 *Chatter in milling.* Modify the case illustrated in Fig. 9.64. Instead of slotting, assume:

(a) Up-milling with radial immersion $a/d = 0.3$

(b) Down-milling with $a/d = 0.3$

The tool feeds in direction $-X$. The parameters of the vibratory system are the same as in Section 9.6.7.

1. First, determine the approximate value of b_{lim} in a similar way as done for turning in Section 9.6.4 by considering the cutter frozen with a tooth in the middle of the arc of cut, deriving the corresponding directional factors u_x and u_y, and evaluating Re $(G_{or})_{\text{min}}$.
2. Modify the program listed in Fig. 9.65 to accommodate the different radial immersions and the fact that only one tooth is cutting at a time, and run simulations for $b = 0.5\, b_{\text{lim}}$ and for $b = 1.8\, b_{\text{lim}}$. Use spindle speed $N =$ 3,000 rpm and $M = 1{,}280$ total computation steps. Plot x, y, F_x, F_y.

10 Automation

This chapter deals with the important topic of the automation of individual metal-cutting machine tools, and also of groups of machine tools. The distinction between "hard," or "rigid," and "flexible" automation is explained; the former is used in large-series and mass production and the latter in medium- and small-lot production. Flexible automation is then discussed in more detail, in the form of numerical control (NC), adaptive control (AC), flexible manufacturing systems (FMS) and unmanned machining centers (UMC). The last part of the chapter includes the control of industrial robots. While the principles and methods of NC are illustrated in applications to metal-cutting machine tools, where NC originated and has been widely developed, and to robots, the knowledge derived is easily applied to other types of production machinery used for sheet-metal forming, forging, electrical-discharge machining, wire cutting, oxygen flame and plasma arc cutting, laser machining, and so on.

In particular, a large part of the chapter is devoted to the control system aspects of coordinate motion drives of machine tools and of robots, in the form of positional servomechanisms. The steady-state and transient errors in the execution of the commanded path are determined while limiting the discussion to simultaneous motions in two coordinates only. However, this is fully sufficient to derive all the pertinent characteristics and to explain the ways and methods used to minimize the errors. First, the control systems theory is briefly reviewed as it applies to a servo described by a second-order system. Beyond that, an essential problem in machine tools and robotic drives is dealt with: the flexibility and inertia of the transmissions and the potential of unstable behavior. Improvements obtainable by applying suitable feedback corrections or feedforward compensations are investigated. The presentation is held to a relatively simple level of treatment of linear systems by using Laplace transform, Bode plots, and the Nyquist criterion, as well as by using time-domain simulations for both linear and nonlinear cases.

Automation in mass production is about one hundred years old. However, it was not until 1950–1960 when NC as the most flexible form of automation was introduced, making it possible to economically automate the small-batch productions that represent a large majority of engineering manufacturing.

The graph in Fig. 10.1, which gives the number of NC machine tools in the United States over the period of 1970–1990, shows that it was not until recently that such machine tools gained wide acceptance. The reasons for this will be explained later, but on one hand, it took twenty years to bring the development of these machines to the high degree of performance, accuracy, and reliability that they now possess, and on the other hand, it took about the same length of time for the manufacturing industries to change their organization of production and acquire the personnel necessary to use these machines successfully.

NC machine tools now represent about 15% of the almost 2 million machine tools in the U.S. metalworking industries. However, we may estimate that they deliver about 60% of all production. The hard automation used in mass production furnishes about 15–20% of generated value. These numbers indicate the extraordinary significance of automation.

Automation is the main driver of the increase in productivity. Of the roughly five-fold general increase of productivity in U.S. manufacturing since the 1950s, about one-third was due to better tools, while two-thirds was achieved by improved machine tools, mainly by automation.

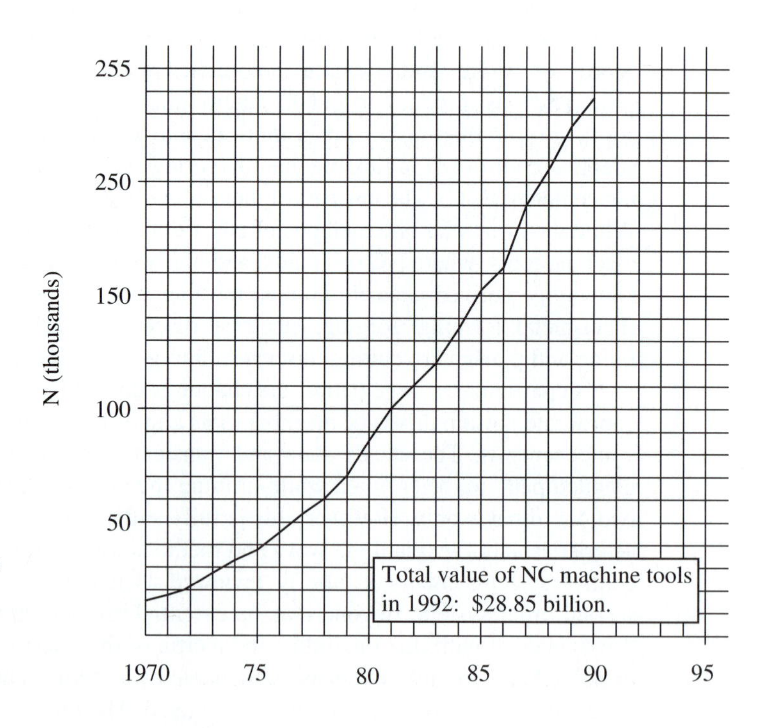

FIGURE 10.1
NC machine-tool cumulative installations in U.S. industries. These machines have dramatically changed machining technology.

10.1 AUTOMATION OF MACHINE TOOLS

Automation is concerned with all the actions that must be performed by the operator on a manually operated machine tool. These are the actions that occupy the noncutting part of the operation of a machine tool. They may be referred to as auxiliary actions, as distinguished from the main action, the actual machining. These actions are listed in Table 10.1.

Let us now comment on the individual entries in the table. In order to automate each of these actions, it is necessary to incorporate devices that were not included in the manually operated machine. Examples of such devices are given in the table.

Action 1: Loading and Unloading the Workpieces

Workpieces come in all sizes. Automated loading is used on small- and medium-size parts. In bar machines the loading is automated by simply shifting the bar by the corresponding length, and this motion is usually derived from a cam. Short, cylindrical workpieces may be loaded simply from a chute-type magazine down to between the centers of, for example, a grinder. Various designs of pneumatically operated loading arms are used on shaftlike workpieces. Robots may be used for loading NC lathes. On NC machining centers, workpiece loading is accomplished by pallet-shuttle devices.

Action 2: Clamping the Workpiece

A bar shift is usually associated with automatic opening and closing of the collet clamping the bar. For "chucking work," mechanized chucks are used, and for box-type workpieces, automatic clamping of fixtures or pallets carrying the workpiece or simply mechanized vises are used.

Action 3: Loading, Setting, Clamping, and Changing Tools

This is one of the most important groups of actions that in manually operated machining occupies a large part of the noncutting time. An average operation requires several different tools, between approximately 4 and 20 in turning, and often many more in

Table 10.1 **Actions to Automate and Devices to Use**

Action	Devices
1. Load, unload workpiece	Bar feed; magazine and chute; loading arms; robots; pallet shuttles
2. Clamp workpiece.	Mechanized collets, chucks, vises
3. Load, set, clamp, and change tools.	Turret head; tool magazine and changer
4. Set and change feeds and speeds.	Hydraulic gear shift
	Hydraulic clutches
	Steplessly variable drives
5. Control relative tool/workpiece path.	Cams; dead stops; copying; numerical control
6. Measure.	In-process gaging
	Post-process gaging
7. Other.	Coolant on, off; clearing of chips

milling, boring, and drilling operations. Each of the tools must be inserted, clamped, and set. Setting means either locating the tool in an accurate relation to the tool holder or else performing a series of manually operated motions of the slide, combined with measurements. Because of the significance of tool changing, this was one of the first actions to be automated and the resulting machine tool, the turret lathe (see Fig. 7.14), has been in existence for more than one hundred years. In a turret the various tools are preset, and they are changed in the machining cycle by simply indexing the turret head. Another solution is a magazine and tool changer, as described later.

Action 4: Set and Change Speeds and Feeds

On a manually operated machine, changing speeds and feeds is accomplished by actuating levers that shift the gears for the various transmission ratios. In automatic machines the gearshift is often replaced by the use of hydraulic clutches, but steplessly variable motors or transmissions are most often used for both the spindle and feed drives. Such spindle drives are mostly DC electric motors, but AC induction-type motors with frequency converters are also used.

Action 5: Control the Relative Tool/Workpiece Path

This group of actions is most important and central to the whole automatic system of the machine tool. The proper definition of these actions follows: Those actions that control the relative motion between tool and workpiece so as to obtain the desired form and dimensions of the part. The design solution for automating these actions influences the mode of control of most of the other actions. Thus, if tool slides are driven by cams, then bar shift, clamping, turret head indexing, and so on are also derived from cams. The coordination of the various motions and actions is simply obtained by having all the cams on the same camshaft. Another type of control of simple slide-motion cycles is based on dead-stops. In this case the feed drives are usually accomplished by means of hydraulic cylinders. Other types of control are copying and numerical control. This group of actions is discussed in more detail later.

Gains and Costs of Automation

Let us categorize the gains and the costs of automation. The gains are obtained by reducing the time of the manufacturing operation performed by the machine tool as well as by labor savings when one operator can supervise more than one machine tool. In other words: more parts per hour are obtained from one machine tool, possibly with also less cost of labor directly involved at the machine tools. These gains may be categorized as follows.

A. Savings of Auxiliary Time

1. Time saved on some auxiliary actions that are performed faster automatically than by hand (e.g., performing a tool change by indexing a turret head).
2. Overall time savings on other auxiliary operations that may not be faster when automated (e.g. loading a shaftlike workpiece in a lathe); however, by eliminating the fatigue factor of the operator, more such operations will be performed over a longer time period than when done manually.

B. Savings in Machining Time

Basically, the cutting operation is not different, whether carried out on an automatic machine or on a manually operated one. However, on an automatic machine more than one slide can work at the same time, and work can be done simultaneously on more than one workpiece. This cannot be accomplished manually. Therefore, substantial savings in the total machining time per part may be obtained by simultaneous action of several slides and simultaneous action on several workpieces.

C. Savings of Labor Cost

One operator may be able to supervise more than one automatic machine tool. This may be accomplished quite often, for example, in shops with automatic bar lathes, single or multiple spindles, in the flexible manufacturing systems, and ultimately in unmanned machining systems.

D. Savings Due to Better Utilization of Machine Tools

This type of saving is not really due to any other characteristic of the automatic machine tool itself except that the duration of an automatic operation is well predictable. For this reason and also because of the higher value of the work, the organization and supply of work to the automatic machine tool is usually superior to that of a manual one and this leads to better utilization.

E. Savings Due to Improvement of Quality

The automatic machine usually delivers much more uniform quality and a lesser percentage of scrap.

All of these savings are obtained for some cost of automated production additional to that of manual production. In order to categorize the components of this additional cost, let us first characterize the periodic nature of the discrete manufacturing processes (see Fig. 10.2). On a particular automatic machine tool batches of various parts requiring particular machining operations will be processed. These are indicated as the x's, the circles, the crosses, and so on. The individual types of parts reoccur periodically, and there will be altogether n_i batches made of a particular part i, during its life, with m_i parts in each batch. The life of a part, or of the whole product to which it belongs,

Work on a particular machine tool.

[SU_1] x x x x x x x x x x [SU_2] o o o o o o o o o o o o [SU_3] + + + + + + · · · · · · · ·

· [SU_1] x x x x x x x x x x · · · · · · · · · ·

· · · · · · · · [SU_3] + + + + + + · · · · · · · · · · · · · ·

FIGURE 10.2
Diagram of batch production. Batches of various parts (the x's, circles, crosses, etc.) are recurrently processed on a particular machine. That affects the repeated use of special tooling, but the setup (SU) cost has to be expended for every batch.

is the period over which the product is being repeatedly produced until it is replaced by a new model. The parts are made in several batches per year instead of the whole yearly amount at once, partly because of the savings on storage, and mainly so as not to have capital frozen in these goods over long periods between the manufacture and the sales. Over the life of the particular machine tool, the whole sum of all the batches of all the various parts will be produced. With respect to this picture, we can categorize the *cost of automation* according to three basic components:

Increased Initial Acquisition Cost of the Machine, C_m

The cost of an automatic machine tool is higher than that of a manual one because of all the additional devices and systems. This additional cost will be depreciated against the gains obtained on all the N_m operations carried out during the life of the machine:

$$N_m = \Sigma\, m_i n_i$$

As the number N_m is large, the cost C_m is not the most decisive component.

The Cost of Special Tooling, C_{st}

Special tooling consists of all the special devices to be used for automating a particular operation *i*. It involves the cost of special fixtures, form tools, cams (in cam-controlled machines), templates (in copying machines), NC programs (in NC machines), and so on. Such special tooling can be used repeatedly whenever a batch of workpieces *i* is processed. This cost can be depreciated against the gains obtained on all the N_{st} parts of the *i* kind:

$$N_{st} = m_i n_i$$

This may still be a substantial number.

Setup Cost, C_{su}

Setup cost has to be expended at the beginning of every batch, as indicated by setups SU_1, SU_2, SU_3 in Fig. 10.2. Setting up consists of inserting the corresponding cams, or setting the corresponding stops, or calling the corresponding NC program, and of setting all the necessary tools. Setup cost can be depreciated only over the number N_c of parts in a batch:

$$N_{c,i} = m_i$$

This number may be small: 1–10; medium: 10–500; or large: over 500. These limits are rather arbitrary and depend mainly on the size and complexity of the part; for example, 1000 may be considered a medium or a large lot size. Obviously, for the automated production of large lots (large series), it is quite economical to accept a substantial setup cost C_{su}. However, for small lot sizes the setup cost is critical. Correspondingly, it is mainly the setup cost C_{su} that is decisive for whether one speaks about *hard,* or *rigid,* automation—large C_{su}, good for large-series manufacturing—and *flexible* automation, in which it is easy to set up the machine for another automatic cycle; it is suitable also for small-lot (batch) manufacturing.

10.1.1 Rigid and Flexible Automation

We have now arrived at acknowledging the significance of the setup cost for the degree of flexibility of automation. We have also learned that the most decisive actions to automate are No. 3 and No. 5 of Table 10.1 (tool setting and changing and controlling the tool path). Thus, in hard automation, which is used in large-series productions, we usually find a combination of complex tools, multiple slides, and simple motions, while in flexible automation, which is used in small-lot productions, simple tools are combined with complex motions. These principles are illustrated in Fig. 10.3. In the drawings *(1a)* and *(1b),* machining of a multidiameter shaft is presented.

In *(1a)* six tools are employed that are preset, three and three, in a rather complex configuration on tool plates located on the front and rear slides of the machine, respectively. Each of the slides performs a simple rectangular cycle of motions: 1-2 machining, 2-3-4- rapid return, 4-1 approach for a new cycle. These simple cycles require only one accurate end position of each slide in each of the two directions of motion, that is, position 1 in *X* and position 2 in *Z*. This is usually accomplished by using a hydraulic cylinder to drive the slide against a "dead stop" in each of the two motions.

In *(1b)* the same shaft is machined with one tool moving in a more complex path: 1-2-3-4 machining, 5-6 return, 6-1 approach for a new cycle. This motion cycle requires three accurate end positions in both the *X* and *Z* axes. In this cycle part *A* of the workpiece is finished. After the whole batch is machined in this way, the workpieces may be turned over and passed again through the machine where the same tool on a different path will machine part *B* of the workpiece. The more complex path (which may often include tapered or circular sections) will be based either on a copy system or on numerical control.

Obviously, setting up the six tools of *(1a)* and the four dead stops in their respective precise positions is a rather lengthy procedure. The cycle is "rigid." This method

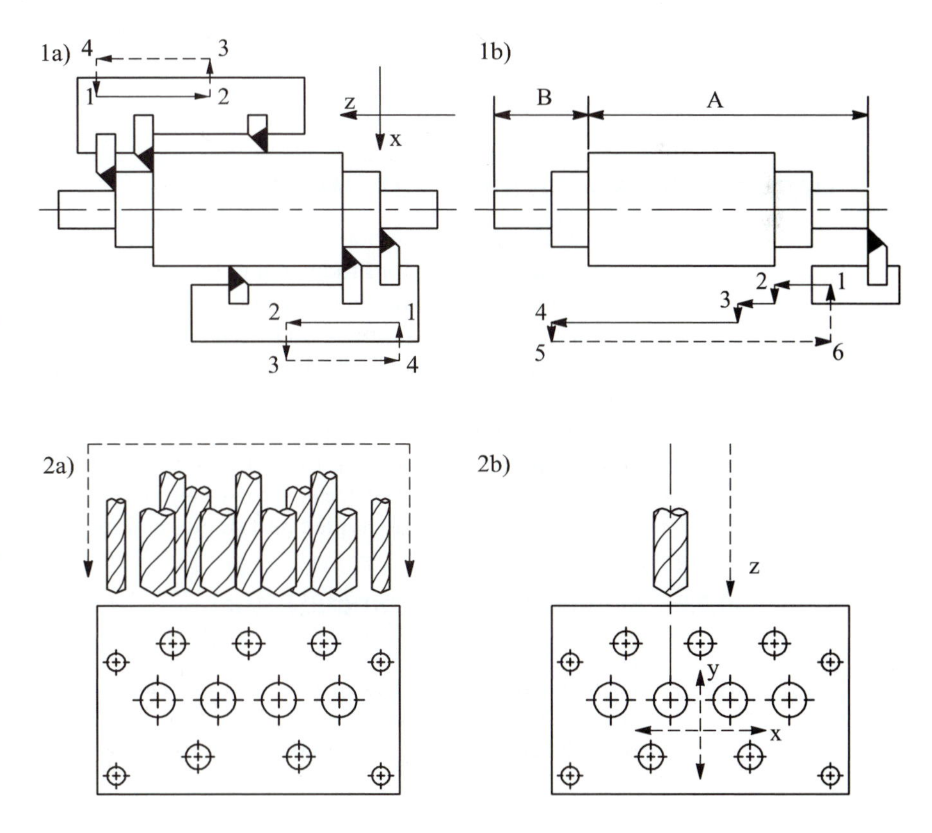

FIGURE 10.3
Rigid (a) and flexible (b) automation. 1) An example of turning a shaft, a) on a lathe with rigid front and rear rectangular motion cycles. Setting all the tools and end stops is costly; b) on an NC lathe. Although now only one tool works at a time, resulting in a longer machining operation, setting up for a cycle means just calling a different part program, resulting in great savings in SU cost. 2) An example of drilling a pattern of holes, a) using a multi-spindle drilling head. This is fast, but the cost of the special head is high. b) On an NC machine only one tool works at a time, but the machine can easily be used for another pattern of holes.

can only be economically used for a larger batch of workpieces. On the other hand, savings in machining time are obtained thanks to the simultaneous action of several tools. In a certain part of the cycle all six tools will be cutting. Setting up the single tool in *(1b)* is very easy, especially when using the "zero shift" function of NC. The motion cycles *A* and *B* are simply called from the computer memory. This method is very "flexible," and it can economically be used for small lots of workpieces. The machining time will inevitably be longer than in *(a)* because there is only one tool. Thus, method *(a)* is preferable if the size of the batch permits it. However, method *(b)* is the only way of automating small-lot production.

The examples of drilling a pattern of holes in a plate, as shown in diagrams *(2a)* and *(2b)*, are based on the same principles as the preceding ones, now applied to a different machining operation. For large-series production of these plates, a multispindle drilling head may be made and used as shown in *(2a)* for simultaneous drilling of all the holes. Again, we have here a complex set of tools combined with a very simple working motion. For a small lot, the price of the multispindle head could not be justified. Automation can be achieved by drilling the holes one by one and moving the table in between the two coordinates *X* and *Y*, using numerical control. The three different sizes of drills will either be preset in a turret head, or changed manually, or changed automatically from a magazine.

Obviously, the distinction between rigid and flexible automation is not absolute. The degree of rigidity or flexibility of automation is judged mainly by the difficulty (or ease, on the other hand) of setting up the machine for a particular automatic cycle. There is quite a transition between the most rigid, special-purpose machine tools and the most flexible, universal machine tools. However, if a division must be made between rigid and flexible types of automation, the latter is best limited to NC machine tools, while those involving cam or dead-stop controls are considered rigid.

10.2 MACHINE TOOLS WITH RIGID AUTOMATION

Once again, there are various degrees of "rigidity." The title of this section could have been "Machine Tools for Production in Large Series," and the words *large series* could mean a range from batches of several hundred to several thousand workpieces up to what is called "mass production." The latter expression is applied to production processes and machine tools where an operation on one machine tool (or one production station) does not vary: the same operation is repeated all the time, for days, months, years. We may classify all these machine tools into two large groups: those for turning operations producing rotational parts, and those for operations carried out on plate and box-type workpieces, like drilling, boring, and milling. The automatic turning machines can be subdivided into single- and multiple-spindle types.

10.2.1 Single-Spindle Automatic Lathes

There are several traditional types of the single-spindle automatic lathe. On the very small end, they are used in the manufacture of shafts for watches, and the required short motions are driven from cams. The corresponding automatic lathe is called the "Swiss type." It is of a special, unconventional design that originated in the 1880s and has survived until today, although its main application area, the manufacture of tiny shafts and axles for watches, has shrunk considerably due to the introduction of digital watches. The parts are made out of bar stock that must have an accurate and smooth diameter, preferably obtained by centerless grinding. The bar passes through a hollow spindle and moves coaxially with the spindle housing illustrated in Fig. 10.4a. The tools, of which only one, located vertically, is shown, move in directions perpendicular to the workpiece axis. The bar of the blank is guided and rotates with a very small play in a bushing shown in black. The tools act right there at the bushing. The arrangement of the tools is shown in Fig. 10.4b. The two horizontal tools are mounted on a "rocker" that pivots around a point just below the workpiece. The upper three slides move in linear guideways by being pushed against springs. The motions of all five tools are derived by a system of levers from individual, disc-type cams, all mounted on the same camshaft, which also carries the cam that moves the headstock axially. In this way all the motions are synchronized.

a)

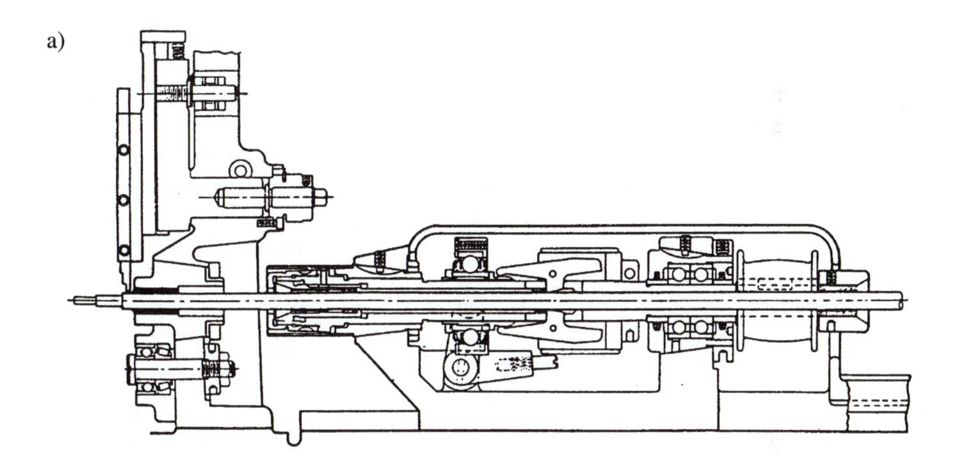

b)

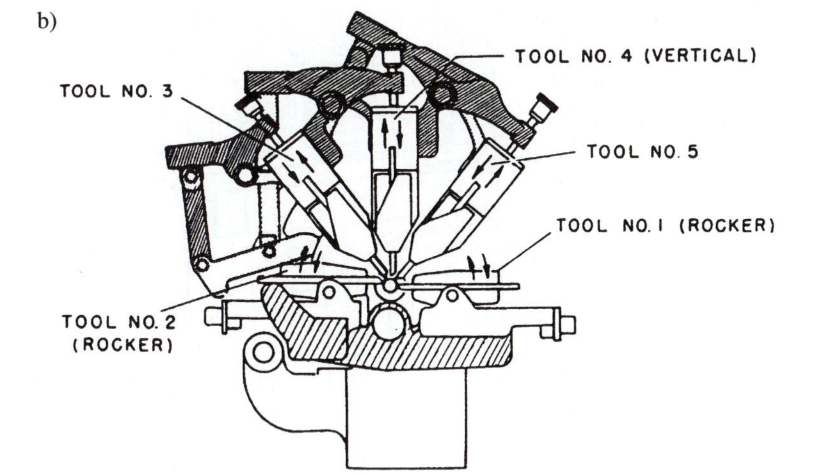

FIGURE 10.4
Swiss-type automatic lathe. a) Section through spindle head. b) Arrangement of tool slides. Tool motions are driven and controlled from a camshaft via levers that have to be set. SU is costly. (REPRINTED FROM R. A. LINDBERG, "PROCESSES AND MATERIALS OF MANUFACTURE," 2ND ED. [BOSTON: ALLYN AND BACON, 1977]

The small- to medium-size machines for bar diameters in the range 10–50 mm are also based on cam control and cam drive of tool slides. A very common type is often called the screw automatic (because one of its early uses was for manufacturing screw bolts). It is equipped with a turret head slide and with several transversal slides. An overall view of a model of this type of machine is shown in Fig. 10.5. The turret head indexes a horizontal axis into six positions. The camshaft carries the cams of the cross slides and is linked with the lead cam, which drives the turret slide. The clamping of the bar in a collet in the spindle and the bar feed are operated by cams on the left end of the camshaft. The driveshaft handwheel is used when setting the tools. A typical part for a five-cycle machine, the sequence of operations, the tools used, and cams for the motions of the two cross slides and for the longitudinal motion of the turret slide are shown in Fig. 10.6.

The examples given here support comments and statements that can be made in summary. It is obvious that mechanical cam control leads to a compact machine design and permits the manufacture of rather complex parts in very short cycles. Such machines survive and are still successfully used in mass production of very small and small parts. The trend in their development is based on a certain amount of programmable devices intended to permit faster setup times. However, the fundamental feature remains of dividing the total operation into the actions of many tools, some of which work simultaneously.

Single-spindle automatic lathes for larger parts are also built with turret heads and multitool slides, but because of the longer motions and higher cutting forces, cam drives are not used; instead, hydraulic cylinders or motors, or electric motors, are used to drive the feed motions. In the "rigid" automation mode, these motions are rather simple, with adjustable end switches. An example of such a machine is shown in Fig. 10.7.

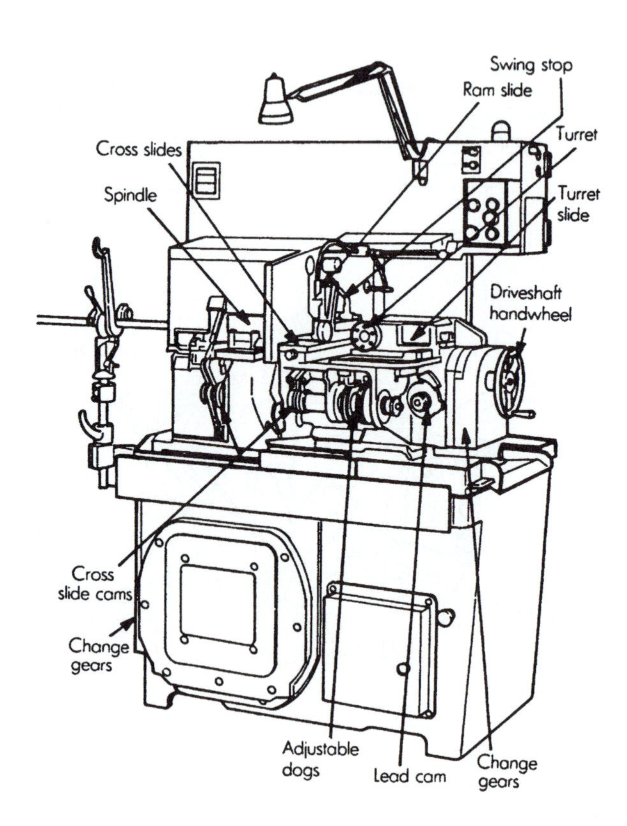

FIGURE 10.5
Single-spindle bar automatic lathe (screw-type automatic). Both the turret head and cross slide are driven from a set of cams. SU is costly. [Brown and Sharpe]

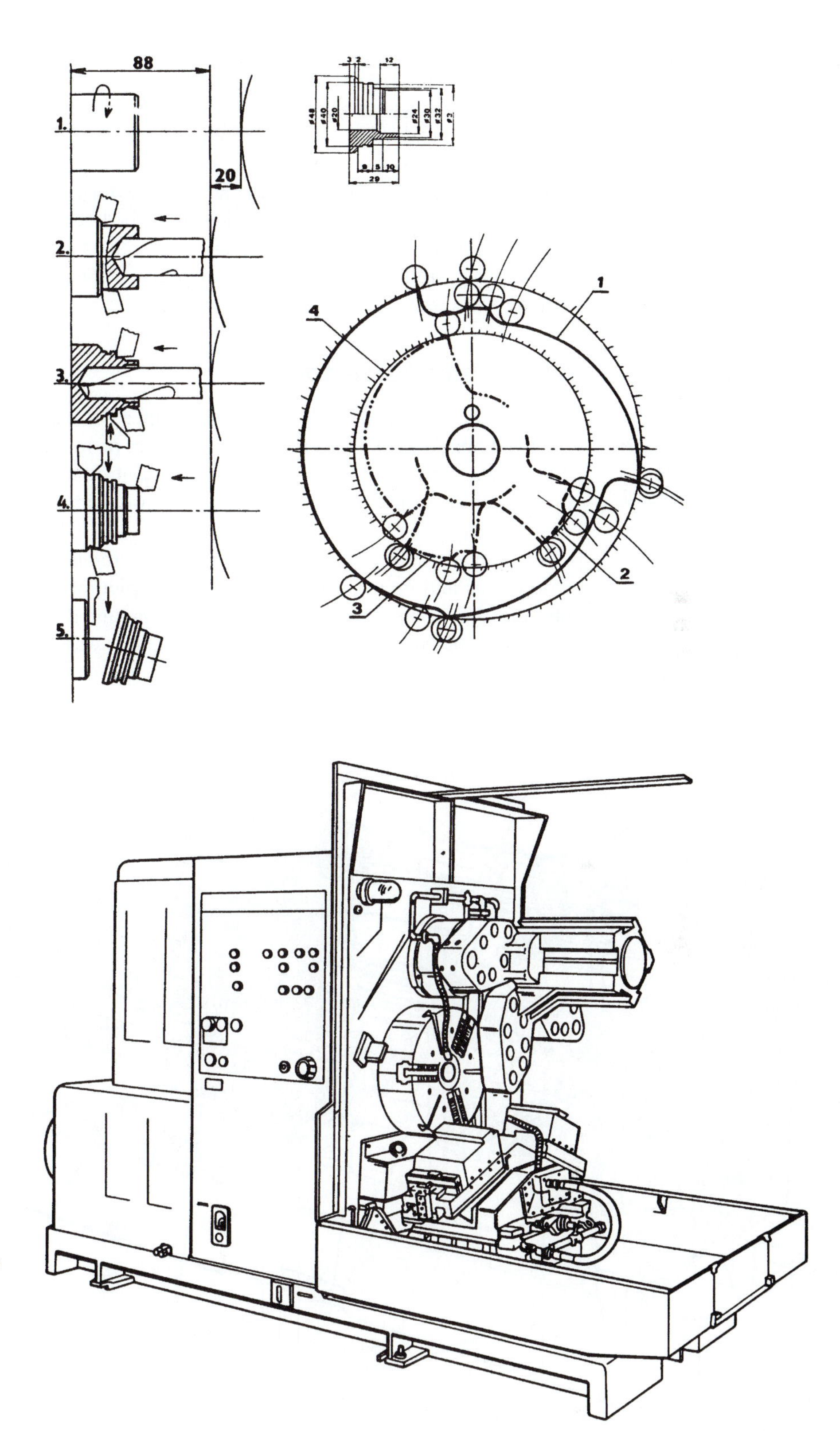

FIGURE 10.6
Example of work and cams on a screw-type automatic. Cams are special for each particular part, and all tools have to be set in their positions. SU is costly. [OMNITRADE]

FIGURE 10.7
Large, single-spindle bar automatic lathe. The motions of the hexagonal end slide and of the cross slides are controlled either by cams or by a hydraulic system. An NC version also exists. [WARNER & SWASEY]

It is intended for chuck work: machining both externally and internally short rotational parts of medium-size diameters, up to 300 mm. Machines for larger chuck work usually use vertical spindles, that is, rotary tables. The machine depicted here is equipped with a turret head in the form of a five-sided box as the end of a large-diameter round tube sliding in and out of the box-type column of the machine. This motion is parallel with the work spindle. The head carries multiple tool holders attachable to its five sides. Additionally, there are two cross slides on the bed below the spindle. Each of these can carry a set of tools. The complex machining cycle is created by the combination of various turning, drilling, and boring tools moving coaxially with the spindle, and outer turning tools moving in the two cross directions.

10.2.2 Multispindle Automatic Lathes

Multispindle automatic lathes are made either for bar or chuck work, with spindles arranged horizontally or vertically. The number of spindles n may be four, six, or eight. The whole complex operation to be carried out on a part is divided into n suboperations, each executed in one of the n tool stations and each taking the same time, the cycle time t_c.

An example of a six-spindle ($n = 6$) bar automatic lathe is shown in Fig. 10.8. No workpieces and no tools are included in this picture so as not to clutter the view and obscure the orientation. The spindles are mounted in a barrel-type carrier that is indexed by 60° in the work-cycle intervals. In this way the rather complex turning oper-

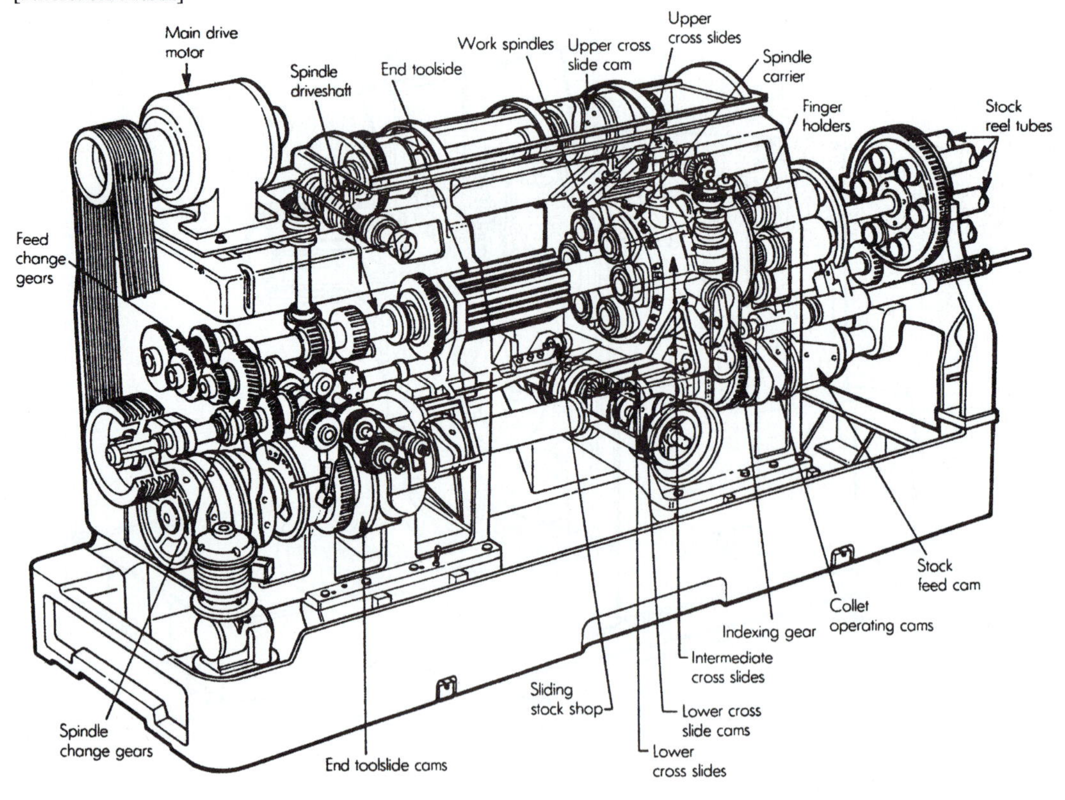

FIGURE 10.8
A six-spindle bar automatic lathe. Motions of cross slides and of the hexagonal end slide are driven from drum cams. A very productive machine but SU of the many tools and mechanical lever transmissions for a particular part may take an eight-hour shift, and special form tools must be made. [NATIONAL ACME]

ation is divided into six simpler ones, yet each of them employs two or three tools. Each part passes through all six positions. Six parts are machined simultaneously, so that one part is completed for each work cycle time. The bars of the stock blanks (not shown) are clamped in collets in the work spindles and extend back into the stock reel tubes. The feed motions of the central hexagonal tool slide and of six cross slides are mechanically driven from the various drum-type cams. The closing and opening of the collets and feeding of the bar stock in the first station is also derived from a drum-type cam. The rotation of the spindles is provided from the main electric motor via change gears (to set the desired rotational speed), and further via a drive shaft that passes through the center of the end (central) slide and finally onto gears, one on each spindle. The rotation of the upper and the lower camshafts is also derived from the spindle drive shaft, via other sets of gears.

The diagram in Fig. 10.9 shows the central hexagon and the six cross slides in a diagrammatic representation. There may be one or more tools attached to each side of the hexagon and to each cross slide. The two upper and the two lower cross slides have T-slots on their top surfaces for clamping the tools. The two cross slides in the middle positions are drawn in this illustration with wedges that are used to accurately set the longitudinal position of the tools. Various other devices exist for clamping and adjusting the tools. Overall, there will be 12–30 tools working simultaneously. There are many accessories available for additional operations, such as cross drilling and tapping.

The system of cams and levers is successfully employed to feed simultaneously all the various tools in a synchronized manner. However, the forces and velocities in the system must be kept at a rather low level in order to limit deflections and at the same time overcome friction in the guideways. All this is "old-fashioned" compared to the electrical servodrives used in modern automatic machine tools, including those of the rigid type such as transfer lines. Setting and clamping all the tools in rather exact positions, and changing the gears and the lever transmissions, is an extensive operation that usually takes one or even two eight-hour shifts. In order to more than cover the cost of the setup from savings obtained by automating the operations and by the simultaneous work of many tools, a rather large quantity of the particular workpiece must be produced for each setup.

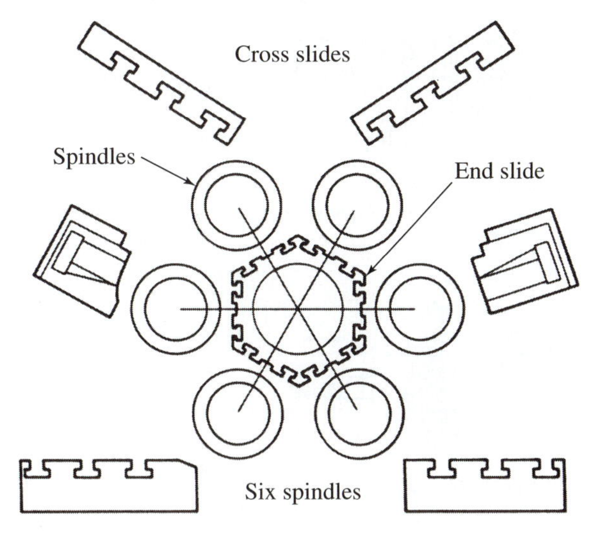

FIGURE 10.9
The arrangement of tool slides on the six-spindle bar automatic. Each spindle is served by a cross slide and shares the action of the hexagonal end slide. [National Acme]

An example of operations performed on an eight-spindle bar machine in producing spark plugs is given in Fig. 10.10. The stock is hexagonal bars, the cycle time is 4.5 sec, and gross production (without any downtime) is 800 parts per hour. The tools used are drills, reamers, form tools for plunge cutting, knurling tools, and a cutoff tool. The majority of tools used are made of HSS. Workpiece material is free-machining, low-

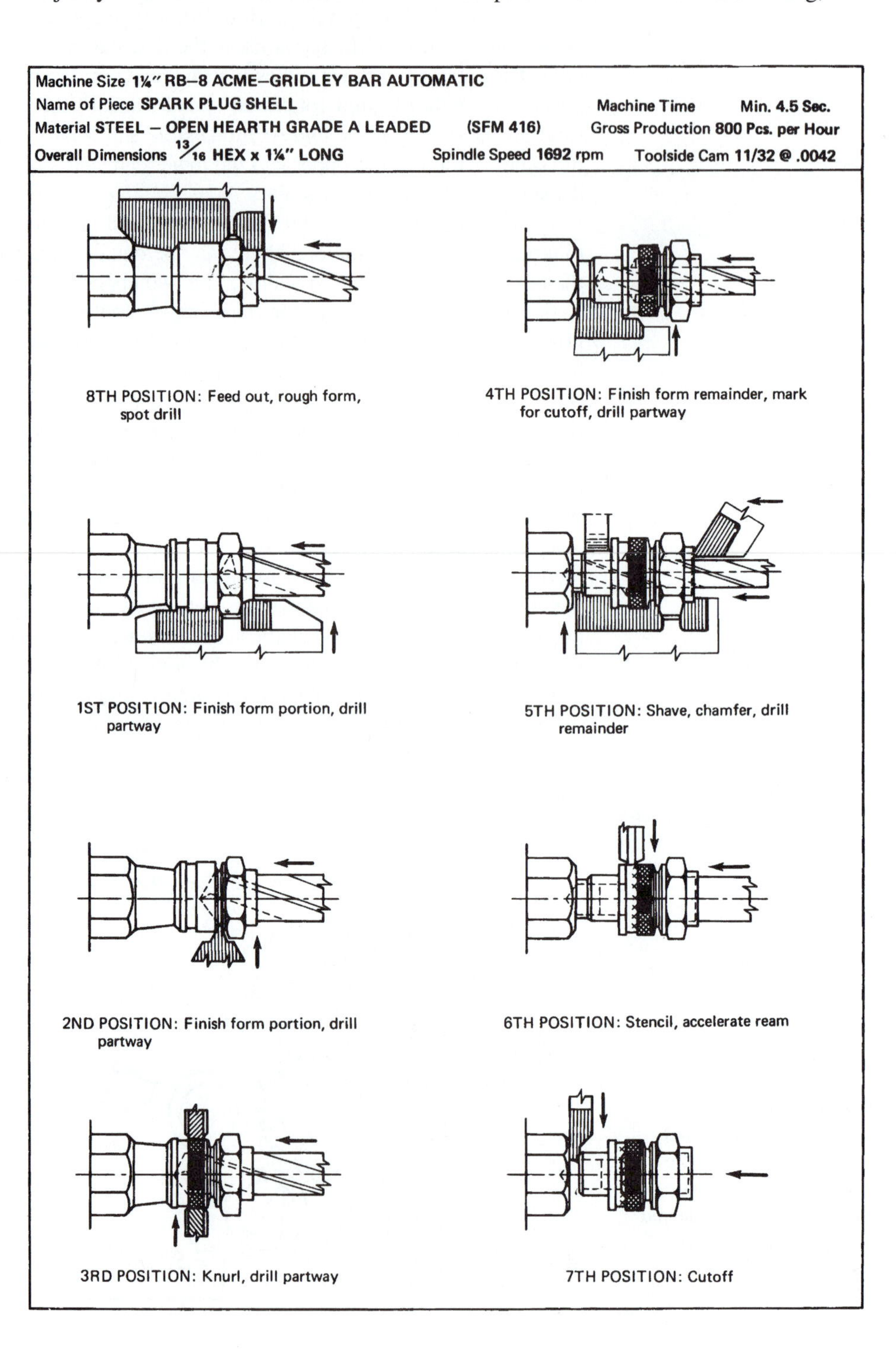

FIGURE 10.10 Example of work on an eight-spindle bar automatic. Notice the number of expensive, special form tools. Each part passes through the eight stations for successive operations. Many tools work at the same time. One part per index of the spindle drum is completed. [NATIONAL ACME]

carbon steel, permitting long tool life at relatively high cutting speeds. Typically a machine of this size has spindle speeds of up to 2000 rpm and a total power of 15 kW.

An eight-spindle chucking automatic lathe in a vertical configuration is shown in Fig. 10.11. The eight vertical slides are guided on the central column and may carry various sets of turning tools or other auxiliary units for drilling or tapping. The spindles are mounted vertically in a round table that indexes in a rotary motion around the column. The feed motions are not derived from cams; each slide has its own hydraulic or electromotor-leadscrew drive mechanism. The spindle speeds are also selected individually. Workpieces with diameters up to 700 mm can be machined. Operations may be set up in a more flexible and easier way than on the cam-controlled machine. This machine can alternatively be equipped with numerical control for the most flexible automation.

10.2.3 Dial-Index Machines and Transfer Lines

Dial-index machines are used in large-series and mass production of workpieces in the form of boxes, plates, housings, levers, and so on. The operations performed are mainly various types of drilling, reaming, tapping, boring, and milling. The complete, rather complex set of operations is distributed to a number of stations. In each station one or more tools work in a cycle and all the suboperations are carried out simultaneously on one workpiece in each station. Each workpiece is stationary during the operations and is moved to the next station between the cycles. Thus, each workpiece passes subsequently through all the stations, and once per cycle one workpiece is completed, and one workpiece enters the line.

For smaller operations that are divided among a small number of stations (essentially between two and ten), the machine is based on a rotary indexing table with the stations arranged on its periphery and the workpieces clamped in fixtures or pallets on

FIGURE 10.11
Vertical, eight-spindle automatic lathe. Similar in principle to the machine in Fig. 10.10. Vertical spindles are mounted in a drum in the base. It can accommodate large-diameter parts. [BULLARD]

the table. This arrangement is called a dial-index machine. The operations performed consist of drilling and reaming a number of holes.

For large operations the stations are arranged in lines with an overall shape of a U or a rectangle. Some of the main features of these transfer lines are described in the following particular example. A typical workpiece is shown in Fig. 10.12. It is an automobile transmission case. The picture shows surfaces that were milled or bored, and holes drilled, tapped, reamed, and bored. The corresponding transfer line is shown in Figs. 10.13 and 10.14. It consists of 84 stations arranged in a rectangular line pattern. The operations performed are face milling, planetary milling, fly cutting, rough boring, trepanning, hollow milling, finish boring, spot facing, drilling, chamfering, core drilling, reaming, gun drilling, gun reaming, grooving, tap drilling, tap milling, wire brushing, and counterboring. The production rate is 200 parts per hour. The workpieces are attached to pallets that are moved from station to station by hydraulically activated cycloidal drives and positioned and clamped accurately for processing in each station.

FIGURE 10.12
The transmission case to be machined on a transfer line. Notice the complex scope of the machining cycle, involving face milling of mounting surfaces, boring of a number of large and precise openings, and drilling, reaming, and tapping of many holes. [F. J. LAMB]

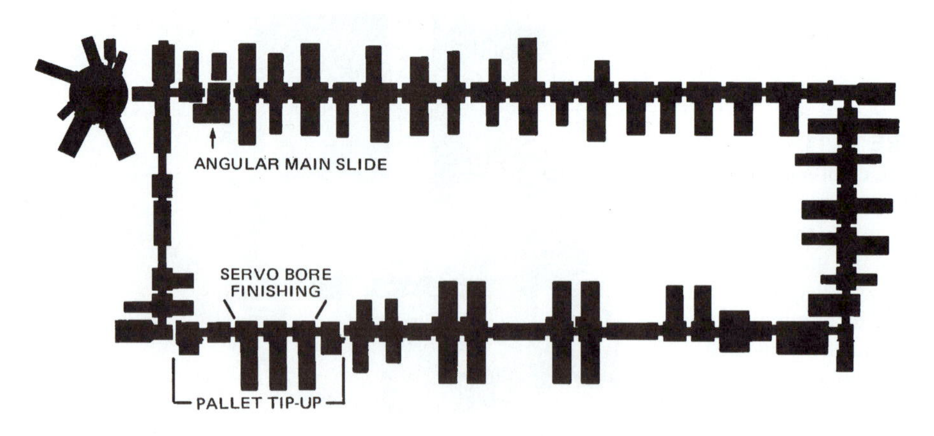

FIGURE 10.13
Outline of the transmission case transfer line. Parts on pallets are moved from station to station. In each one, slides with drilling, boring, or milling heads are attached from one or both sides. This is a massive, multitool, multiworkpiece, simultaneous operation. [F. J. LAMB]

FIGURE 10.14 Photograph of the transmission case transfer line. It is built to produce the particular workpiece shown in Fig. 10.12. The complexity and high cost of this piece of equipment is apparent. [F. J. LAMB]

In most of the stations machining is carried out from both sides of the line by heads moved either in a motion perpendicular to the line or on compound slides permitting two- or even three-dimensional motion cycles. Most of the heads carry multiple spindles for simultaneous drilling of a pattern of holes or several milling spindles. Considering the size and complexity of this piece of machinery and the corresponding investment, it is understandable that it may only be economically used for mass production by repeatedly performing for a long period of time the set of operations for which it was designed. The savings obtained thanks to the automatic operation and, primarily, the simultaneous action of many tools on 84 workpieces are multiplied in this particular case by the 1.2 million parts per year (assuming three-shift work), multiplied by the number of years of the life of the particular part design. Thus, the economic aspect favors a long design life and works against design change.

If the part design is changed, a new transfer line is needed. In order to minimize its cost, the machinery is designed to use as many standard components as possible: unit beds, pedestals, slides, tables, feed drives, and so on. Even so, current trends in the automobile business tend to more frequent changes of design. This applies not only to the styling of the body parts but also, to a lesser extent, the engine and drive components. In the former case, this calls for efficient and flexible manufacture of stamping dies; the latter case calls for more flexibility in the transfer line design. This problem is currently the focus of much work; however, it is not easy to solve.

Figure 10.3 shows that flexibility is commonly obtained by sacrificing the simultaneous action of many tools. So, for instance, instead of a multispindle head drilling eight holes at the same time, a single spindle would have to drill a hole faster and move rapidly from one to the other. Often we hear the requirement of one second from hole to hole. It is certainly a fascinating task, but within the scope of our text it is not possible to elaborate on this problem in any more detail.

10.3 NUMERICALLY CONTROLLED MACHINE TOOLS

10.3.1 Basic Operation

For numerically controlled (NC) machine tools, the program for the cycle of the relative tool/work motions is expressed by a sequence of numbers. Apart from this basic function of motion control, the numerical program also contains other commands like those for spindle speeds, tool changes, and so on. The development of NC machine tools started from a task supported by the U.S. Air Force and originally assigned to MIT, where the first model of an NC milling machine was demonstrated in 1952. The Air Force, which recognized the need for such machine tools in the machining of complex jet aircraft parts, sponsored the development of large NC milling machines at several machine-tool manufacturing companies. Since then this most flexible type of automatic machine tool has undergone tremendous development, and hundreds of models of NC machine tools are currently produced in all the industrially advanced countries of the world.

The basic characteristics of NC machine tools are explained in the diagrams of Figs. 10.15 to 10.19. In all these cases the discussion is limited to 2 dimensions, but in general, it could involve 3, 4 or 5 coordinates (often called *axes*).

Figure 10.15 shows that the process of numerical control starts from the drawing and the associated geometry of the machining operation. A simple case is shown of contour end milling. The path of the center of the cutter is equidistant from the contour of the workpiece by the radius of the cutter. It will be shown later that it is not always necessary to have exactly the same cutter radius because the NC controls incorporate what is known as "cutter radius compensation," which automatically modifies the path of the center of the cutter. The path in the example consists of three circular and three straight sections that will be called segments. The process sheet is the next step. It includes the details of the geometry of the sections of the workpiece contour and the tool path strategy, (it may be that the cut will be made in two passes or even in a more complex way, depending on the form and dimensions of the initial blank), and on the data about the machining operation, like spindle speed and feed rate. These data are then processed in the "programming" process.

This process is presently performed on CAD/CAM systems; the part is designed using CAD software, and the computer holds all the details of the part geometry. This definition is used for semi-automatic generation of the tool path using a corresponding CAM software. The tool path is defined in a standard NC code that is then input to the NC controller of the machine. The controller generates the actual commands to the servodrives on the individual axes of the machine tool.

The control command will consist of increments Δx, Δy of the desired tool path that are issued at regular time intervals. The typical value of such time interval may be 1 to 5 msec. The command generation is actually an interpolation of tool path data between the initial and end points of each segment; for most NC machine tools, it is limited to linear and circular interpolations. The controller also issues commands to other functions of the machine: spindle speed, tool change, and so on. The commanded path increments go to the positional servomechanisms X, Y of the machine, which include the servomotors and transmissions driving the saddle and the table.

The form of the command for linear interpolation is illustrated in Fig. 10.16. A motion is required between the starting point S with $x_s = 20$, $y_s = 20$ and the end point E with $x_e = 60$, $y_e = 50$; these dimensions are in mm (see Fig. 10.16a). Assume the

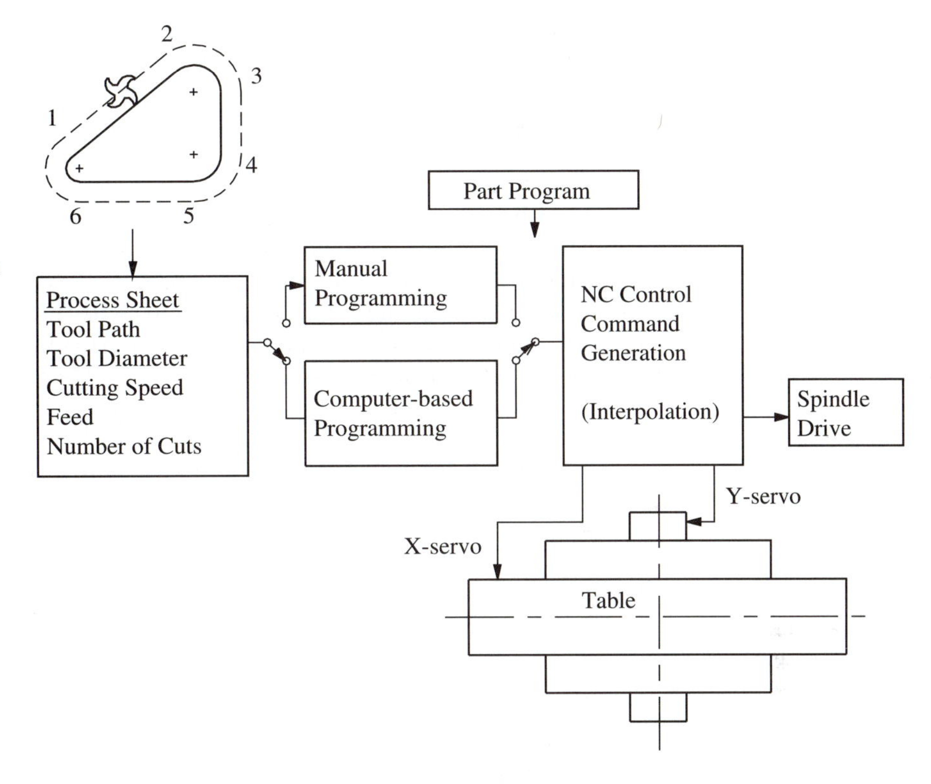

FIGURE 10.15 Information processing in numerical control, from the drawing of the part through the decisions on the machining parameters and tool path to the actual path generation. This process is mostly computer-aided, from the definition of the part in CAD (design) to CAM (manufacturing). Various software packages exist.

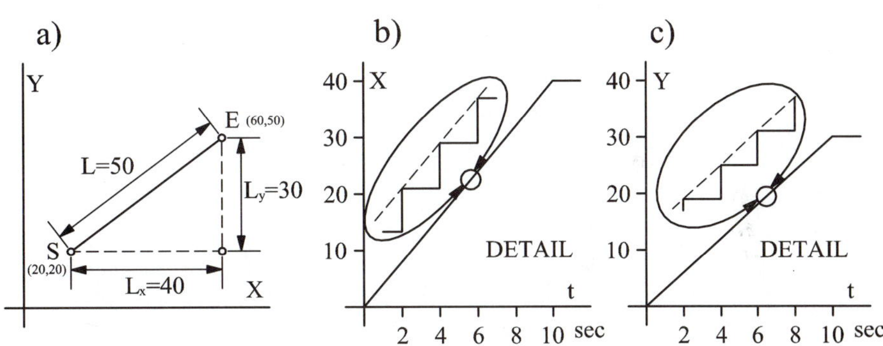

FIGURE 10.16 Linear interpolation commands. Constant motion increments per command period are issued in each participating coordinate.

time of the command-generation cycle $\Delta t = 0.005$ sec, and the commanded velocity of motion (feed rate) $f = 5$ mm/sec. The total motions are $L_x = 40$ mm in X and $L_y = 30$ mm in Y; the length of the motion between points S and E is $L = 50$ mm.

The control as well as the positional servos work digitally with numbers that are understood to be expressed in basic length units (BLUs). Let us suppose that in our case such a basic length unit is $BLU = 0.001$ mm.

The total time for this motion is $T = L/f = 50$ mm/(5mm/sec) $= 10$ sec, and the component velocities are $v_x = 40$ mm/10 sec $= 4$ mm/sec, and $v_y = 30$ mm/10 sec $=$ 3 mm/sec. Correspondingly, the command will be issued every 5 msec in increments of Δx, Δy represented by the following numbers:

$$\Delta x = 4000 \text{ BLU/sec} \times 0.005 \text{ sec} = 20 \text{ (BLU)}$$

$$\Delta y = 3000 \text{ BLU/sec} \times 0.005 \text{ sec} = 15 \text{ (BLU)}$$

The total numbers issued will be counted up to 40,000 in X and to 30,000 in Y.

The motion commands X and Y versus time are shown in Fig. 10.16b. The increments $\Delta x = 20$ and $\Delta y = 15$ are issued discretely, as "chunks" of the travel commands per every $\Delta t = 5$ msec, as shown in the "Details." In reality, however, the servos cannot execute these sudden jerks, and the actual motions are rather smooth in both axes. In any case, the geometric relationship of diagram (*a*) would hold even if the incremental commands could be executed as suddenly as they were issued.

The circular interpolation pattern is presented in Fig. 10.17. An example of a semicircular segment of a path from point S, $x_s = 10$, $y_s = 10$ to point E, $x_e = 70$, $y_e = 10$ is given in diagram (*a*). Let us again assume BLU $= 0.001$ mm, a peripheral feed rate of $f = 5$ mm/sec, and the command interval $\Delta t = 0.005$ sec. This would result in increments Δs on the tangent of the path:

$$\Delta s = f\Delta t = 5000 \text{ BLU/sec} \times 0.005 \text{ sec} = 25 \text{ (BLU)}$$

The component velocities f_x, f_y vary all the time:

$$f_x = f \sin \phi, \qquad f_y = f \cos \phi$$

Figures 10.17b and c show the commanded travels in X and Y as functions of time. In the bottom part of Fig. 10.17, magnified details of diagram are given for points A, B, C on the path. The increments, in BLUs, in the two axes at these points are at A, $\phi = 22.5°$,

$$\Delta x = f\Delta t \sin \phi = 25 \sin \phi = 9.5670858$$

This has to be truncated to integer BLUs and the accumulated remainder added whenever it reaches 1. Correspondingly, we have the following sequences around point A:

$$\Delta x = 9, 10, 9, 10, \ldots$$
$$\Delta y = f\Delta t \cos \phi = 25 \cos \phi = 23.0969883$$
$$\Delta y = 23, 23, 23, \ldots$$

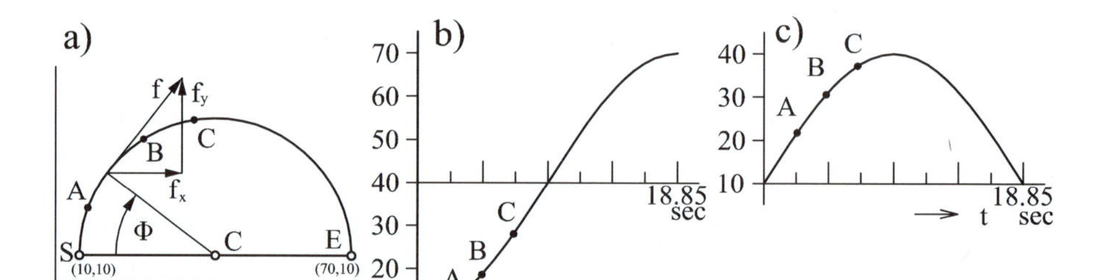

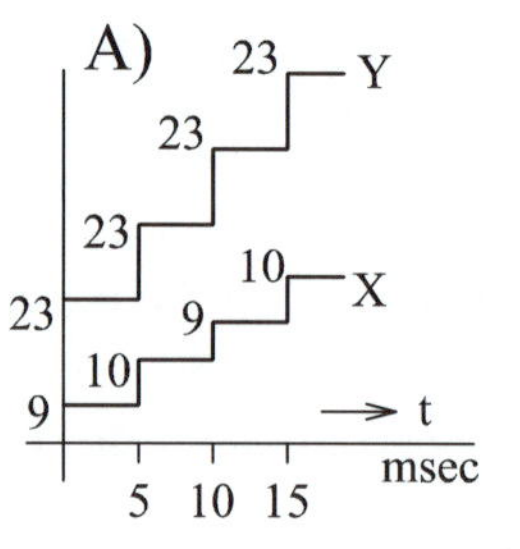

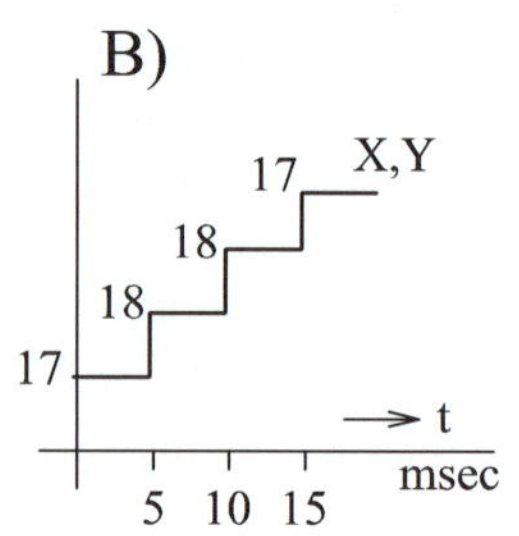

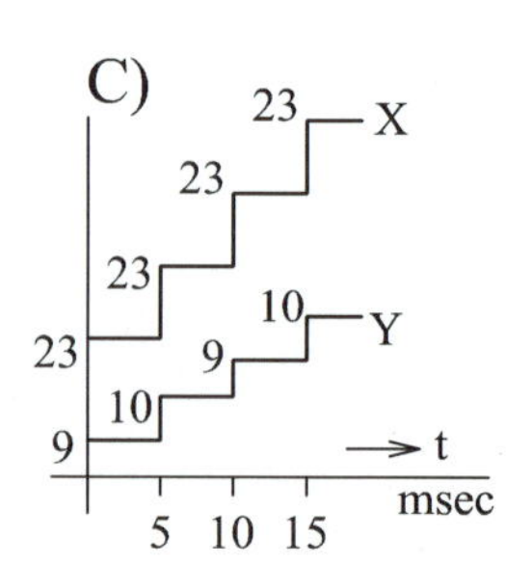

FIGURE 10.17
Circular interpolation commands. The increments in individual coordinates vary accordingly.

At B, $\phi = 45°$, the component velocities are equal, $f_x = f_y = f \sin \phi = f \cos \phi$,

$\Delta x = \Delta y = 25 \sin \phi = 17.6776695$, and the sequence is:

$\Delta x = \Delta y = 17, 18, 18, 17, 18, \ldots$
At C, $\phi = 67.5°$,

$\Delta x = 23, 23, 23, \ldots$

$\Delta y = 9, 10, 9, 10, \ldots$

The commands generated by the interpolator are executed by the positional servomechanisms. The diagram of one such servo is shown in Fig. 10.18. The increments Δx from the interpolator are being accumulated in the counter x_{com}. Depending on the direction of the commanded motion X, the numbers Δx may be positive or negative, and x_{com} is correspondingly increasing or decreasing. Another counter x_{act} accumulates the feedback pulses Δx_{act}. The difference of the two registers ($x_{\text{com}} - x_{\text{act}}$), as obtained in the positional discriminator PD, is the error of position e_x, and it is interpreted by the servo as a command to move. So far, the functions just described may be executed as a software routine in the microprocessor. The error e_x is then output through a digital/analog converter and becomes the control signal for the velocity servo. In the velocity discriminator VD this command signal is compared with the tachogenerator feedback, and the velocity error e_v is fed into the power amplifier with gain G_a, and further into the servomotor. The servomotor drives the table of the machine through a leadscrew and nut transmission. This transmission is of the "recirculating ball" type. It has very high efficiency, and the nut consists of two halves preloaded axially against each other so as to eliminate any backlash. High-precision gears transmit the rotation of the leadscrew to a tachogenerator TG for velocity feedback and also to an encoder EN for positional feedback. Alternatively both can be attached directly to the motor shaft. The encoder emits one pulse for every BLU of the table motion. Correspondingly, the positional feedback has the form of a train of pulses; their frequency corresponds to the velocity of the motion, and their total number to the total distance traveled.

It is obvious that in this arrangement the positional feedback is not taken directly from the position of the table but from the angular position of the leadscrew, and it will not register any discrepancies between the two. For instance, all pitch errors of the leadscrew will affect the position of the table, but they will not be discovered by the encoder. Any thermal expansion of the leadscrew will also affect the position of the table but not be sensed by the encoder. More important, there will still be a certain "dead zone" on the table; this is perceived as a difference in the actual positions of the

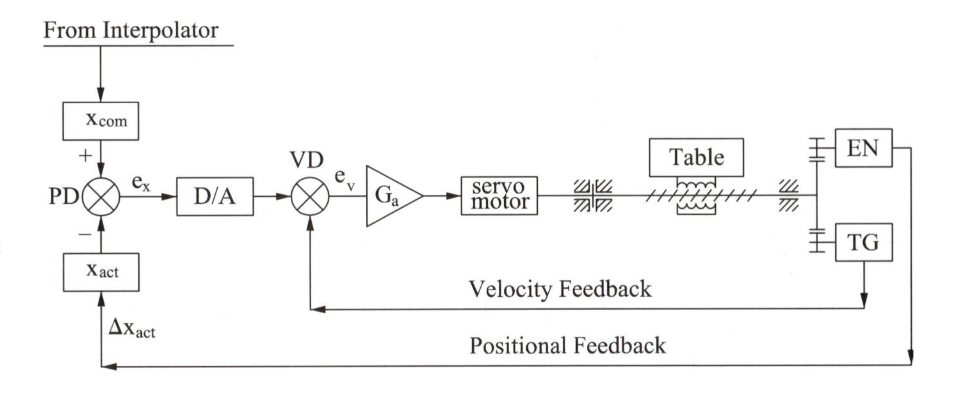

FIGURE 10.18
The positional servomechanism. The digital command is compared with actual position in the positional discriminator PD, and the difference, the error of position e_x after digital to analog conversion, is the command for velocity that gets compared with the feedback from the tachogenerator TG to give the velocity error e_v. This signal is amplified and drives the servomotor that drives the table via a leadscrew and nut transmission. The leadscrew angular position is measured by the encoder EN. Other arrangements are possible.

table when it is commanded into the same nominal position, once in the $+X$ direction and next in the $-X$ direction. Even if there is no backlash between the leadscrew and nut, the reversing direction of the friction forces in the guideways of the table will produce deflections in opposite directions of the bracket carrying the nut and of the thrust bearing of the leadscrew. When properly designed, these elements are stiff enough to reduce the dead zone to a few micrometers. In order to entirely eliminate all these effects, positional feedback must be derived directly from the table.

The positional encoder is shown in a little more detail in Fig. 10.19. Diagram *(a)* represents the rotary encoder, which was considered as a part of the servo just described. It consists of a disc with gratings, that is, with alternately transparent and opaque radial stripes. Two stationary sections with the same kind of gratings are located behind the disc. Light from a lamp is passed through the rotor and the stator onto a photocell. As the disc rotates, the gratings of the rotor and those of the stator open and close the passage of light, once per pitch of the gratings. This produces an alternating voltage in the photocell, as shown in *A*. This voltage is then amplified and truncated to become a square wave *B*. This is led to differentiating circuits that produce pulses *C*, one pair of a positive and negative pulse per pitch of the rotor gratings. For example, counting the positive pulses only and assuming a leadscrew with a pitch of 2 mm and an encoder disc with 10,000 gratings attached to the leadscrew, one pulse is obtained per BLU = 0.002 mm. The two sectors of the stator are shifted one against the other by one-quarter of the pitch of the grating; thus, they produce two signals mutually phase-shifted by a quarter of a cycle. This allows for a distinction of the direction of rotation to count the pulses as positive or negative increments, respectively.

Diagram *(b)* indicates the form of a linear (not rotary) positional transducer to be attached directly to the table. A very popular make of this transducer is the Inductosyn. It consists of a linear "scale" with a flat "winding" and a two-phase slider with identical flat windings. It acts as a transformer, like an unrolled synchro resolver, and it gives a signal of a carrier frequency phase modulated by the relative motion between the scale and the slider. This phase-modulated signal is subsequently converted into pulses. The scale is assembled of sections usually 250 mm long, and by slightly adjusting the individual sections, very high accuracy over long distances may be obtained. The Inductosyn is used, on one hand, on machines requiring very high accuracy indepen-

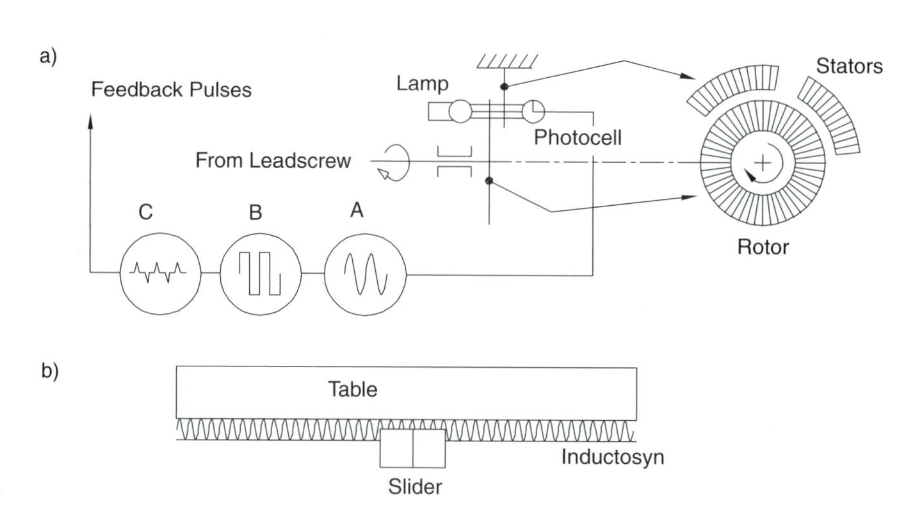

FIGURE 10.19
Positional feedback transducers. a) The rotary encoder with gratings on both the rotor and two parts of stator is the most popular device. It produces pulses per units of travel. A linear (straight) version is also used. b) Both rotary and linear analog phase-modulating devices are used: the synchro resolver or (shown) the Inductosyn. Their signal is then digitized.

dent of the friction in the guideways and, on the other hand, for very long travels. In these latter cases the leadscrew drive is not practical, and a rack-and-pinion drive may be used instead, which is not very suitable for driving a rotary encoder. Instead of the Inductosyn a linear scale with gratings is now commonly used, too.

10.3.2 Adaptive Control

It is appropriate to include in this introductory description of the basic characteristics of NC, at least very briefly, the concept of adaptive control (A/C). This concept deals with the problems arising because the machining process may vary due to the variations in the form and dimensions of the blank of the workpiece, which may be a casting or a forging. An NC program that was prepared with certain assumptions of the depth of material to be removed may have to adapt to the variations of the cut. Also, NC programs are prepared under certain assumptions of the tool-wear rate, and corresponding cutting speeds are selected. However, the actual wear rate may vary considerably and may differ greatly from the one assumed. Consequently, the cutting speed may have to be adapted to the actual tool-wear rate. Other phenomena that are difficult to predict, such as chatter vibrations, may occur, and the cutting conditions need to be changed to stabilize the cut.

In all these instances it is conceivable to use sensors to measure the significant parameters of the actual cutting process and to change the NC program (the feed rate f, the spindle speed n, the depth of cut, etc.) automatically during its execution so as to optimize the cutting operation. This is diagrammatically expressed in Fig. 10.20, which depicts as an example a face-milling operation. The process parameters to be sensed may be the torque on the spindle, the cutting force, vibration, and tool wear. They may be fed back into the CNC controller, which contains in its software the corresponding adaptive control strategies in the form of algorithms to optimize the conditions of the cut for maximum metal removal rate and for economically optimum tool life. A simple A/C system can be conceived according to Fig. 10.21. In this system the actual cutting force F_{act} is measured and compared with a nominal force F_{nom}, which the cutter can safely take. Their relative difference is established as the force error e_f. For $F_{act} < F_{nom}$, e_f is positive, and for $F_{act} > F_{nom}$, it is negative. For $F_{act} = F_{nom}$, $e_f = 0$. The next action to take is to change the velocity v_x of the motion so as to eliminate the force error. So, if $F_{act} < F_{nom}$, it is necessary to accelerate, and for $F_{act} > F_{nom}$, it is necessary

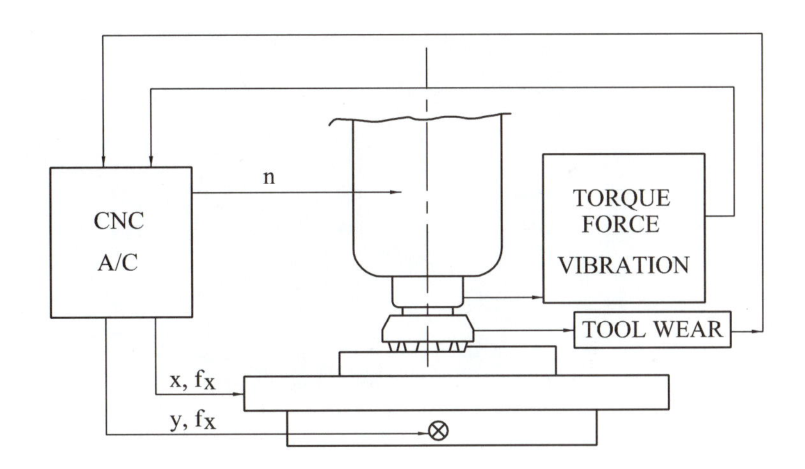

FIGURE 10.20 Adaptive control in milling, in general. Process characteristics are sensed and processed in the CNC (computer numerical control) A/C (adaptive control) system to adapt cutting conditions for a desired performance.

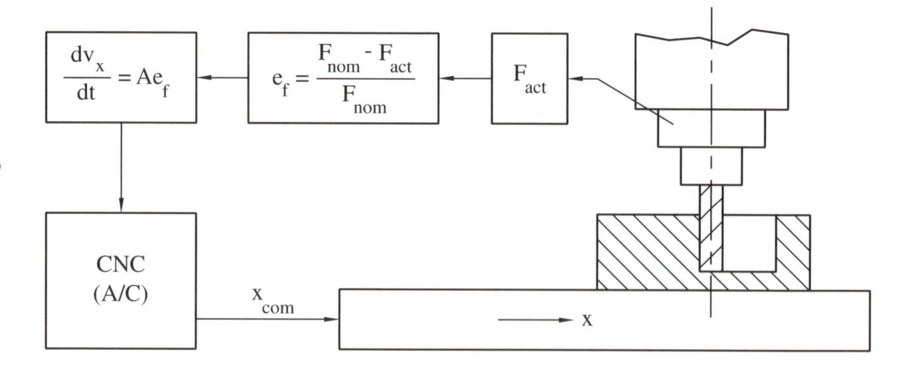

FIGURE 10.21
Adaptive control for constant milling force. The cutting force signal is processed in the CNC(A/C) that controls the feed rate so as to maintain the force at a desired level.

to decelerate. This is expressed by establishing that acceleration dv_x/dt be proportional to the force error. This action is directed into the CNC controller, which will correspondingly start changing the velocity of the commanded travel x_{com} and do so until $F_{act} = F_{nom}$. In this way the cutter will move rapidly where there is little to cut and slowly where there is a lot to remove. In the design of this system it is necessary to solve a number of problems that are mainly due to the requirements of a speedy reaction to the changes of the depth of cut. These problems will be analyzed later.

Before any of the more detailed analyses of NC and A/C are undertaken, it will be useful to discuss and describe the basic types of NC machine tools as they exist today. The two most common types are the turning center and the machining center.

10.3.3 Turning Centers

Turning centers are used for turning, boring, and drilling of a rotating workpiece with tools that carry out the feed motions. The workpieces are essentially of the axisymmetrical (rotational) shape. An example of a typical NC turning center will be described in some detail. The term *turning center* instead of simply *lathe* is used mainly because of the machine's rather versatile tooling and several other accessories.

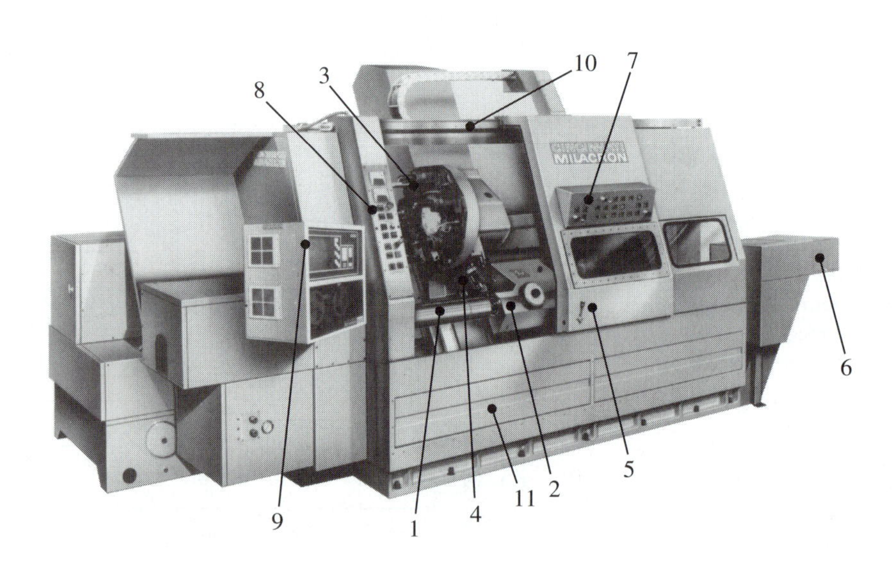

FIGURE 10.22
An NC turning center with one turret head for internal and another one for external turning. It can be used for both shaft- and disk-type workpieces. [MILACRON]

The machine in Fig. 10.22 is equipped to process both shaft-type workpieces, which are clamped in the chuck on their left end and supported by a tailstock center on their right end and short workpieces, which are clamped in overhang only (they are called "chucker parts"). The shafts are machined on outer diameters only (OD work), while the chucker parts are mostly machined both on the outside and on internal diameters (ID). A shaft-type workpiece (1) is supported by the center, which is mounted in the quill of the tailstock (2). The quill is pressed against the workpiece by a predetermined, hydraulically produced force. The tailstock can be moved and clamped on the bed so as to adjust for the various workpiece lengths. The left hand end of the workpiece and the chuck are not visible in this view. Tools are clamped in two turret heads that are mounted in one housing on the slide, which executes the two-coordinate (radial and axial, respectively) motions X, Z. The turret head (3) is equipped with drills, boring bars, and other types of tools for the ID work. This turret indexes about a horizontal axis that is parallel to the spindle axis. The turret head (4) serves for the OD work; it indexes about an axis that is inclined by about 15° to the radial direction. Each of the heads carries seven tools. The two heads are located so as not to interfere with the work.

The detail of the design of the bed of the machine is presented in Fig. 10.23. Unlike the bed of a universal, manually operated lathe (Fig. 7.12), where the cross slide moves in a horizontal plane to make it easy for the operator to watch the tool while moving towards the workpiece, the design of the automatic lathe is not so constrained, and the cross slide moves on guideways on the surface of the bed that is slanted towards the vertical. Not only is this bed more rigid, but it also makes all the chips fall down into the chip pan at the bottom of the machine. The headstock and the tailstock are attached to the bed as well as the X and Z guideways. The detail of the work from the outer turning turret head in machining a shaft is shown at the top of Fig. 10.24. The work of this head on a chucked workpiece is shown at the bottom left of Fig. 10.24, and the work of the turret head with the horizontal axis equipped with internal turning tools is shown at the bottom right of Fig. 10.24.

10.3.4 Machining Centers

Machining centers are used for drilling, milling, and boring operations carried out with rotating tools. The feed motions are executed either all by the tool carrier, or all by the

FIGURE 10.23
The bed of the turning center. It is slanted towards vertical to facilitate the fall of chips away from work into the chip pan at the bottom. Its triangular profile is stiff in bending and torsion. [MILACRON]

FIGURE 10.24
a) Outer turning of a shaft; b) outer turning of a short workpiece; c) inner turning operations such as boring and drilling, on the machine of Fig. 10.22.

workpiece carrier, or by a combination of the two. The workpieces are box-type or plate-type, and the surfaces to machine are planes, holes, and three-dimensional surfaces (the latter are often called sculptured surfaces). Thus, essentially, machining centers are milling and boring machine tools that are equipped with a tool magazine and an automatic tool change between the magazine and the spindle. They exist in two basic configurations: with horizontal spindle and with vertical spindle. The former type is usually equipped with a rotary table that either rotates in an NC coordinate or just indexes by 90°. It is most suitable for box-type workpieces (gear boxes, housings), and the spindle can access them from four sides. The vertical-spindle machines access the workpiece from one side only; thus, they are mostly used on plate-type workpieces or on housings to be machined from one side in a lengthy operation: contouring, die sinking. Cavity milling (die sinking), however, is often done with a horizontal spindle because the chips fall out of the cavity and do not clog the milling cutter.

Horizontal-spindle machining centers exist in a great range of sizes and in various configurations depending on how many axes are controlled, how the motions are distributed among the basic structural parts, how the table loading with the workpiece is executed, what the type of the tool magazine is and where it is located, and so on. The machine in Fig. 10.25 is one of the most basic types. It could be classified as medium to heavy; its table diameter is 750 mm, the spindle power is 18.75 kW, and the thrust force on the spindle is 54,000 N. In its basic form it has 3 NC coordinates: the *X*-axis is the horizontal, longitudinal motion of the table; the *Y*-axis is the vertical motion of the headstock on the front of the column, and the *Z*-axis is the horizontal motion of the column in the direction of the spindle axis. Thus, the *X, Y* axes represent a plane perpendicular to the spindle, and the *Z* motion is the drilling and boring motion. The table can be indexed into four positions accurately square to each other. The control system is located in a cabinet standing next to the machine. The two tool magazines are of a chain type and they carry 36 tools.

A different make of machining center (MC) is presented in Fig. 10.26. This is a five-axis NC machine. The spindle head carrying a small-diameter tool is tilted about 45° down from the horizontal and is capable of rotating around a horizontal *A* axis. The spindle head travels vertically on the column of the machine in the *Y*-axis. In front of the spindle is a workpiece in the form of a vertical plate with a pattern of ribs, and it is attached to a pallet clamped on the top of the table. The table rotates around a vertical axis *B* and this makes it possible to machine the ribs that may stand at another angle than perpendicular to the plane of the plate. Farther away from the spindle head

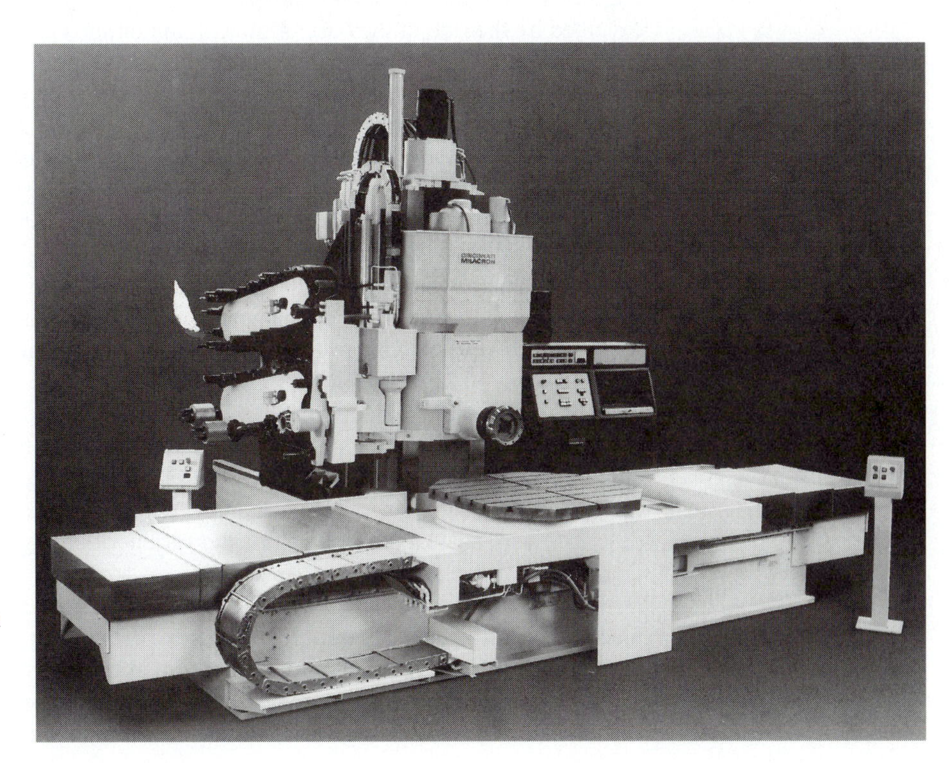

FIGURE 10.25
CNC horizontal-spindle machining center with a double chain-type tool magazine at the side of the column. It is intended for drilling, boring, end milling and face milling of plate- and box-type workpieces, with automatic change of many tools to be used in sequence. The table may be of indexing type to provide access to the workpiece from four sides, and it also may have continuous control to the additional rotary coordinate. [MILACRON]

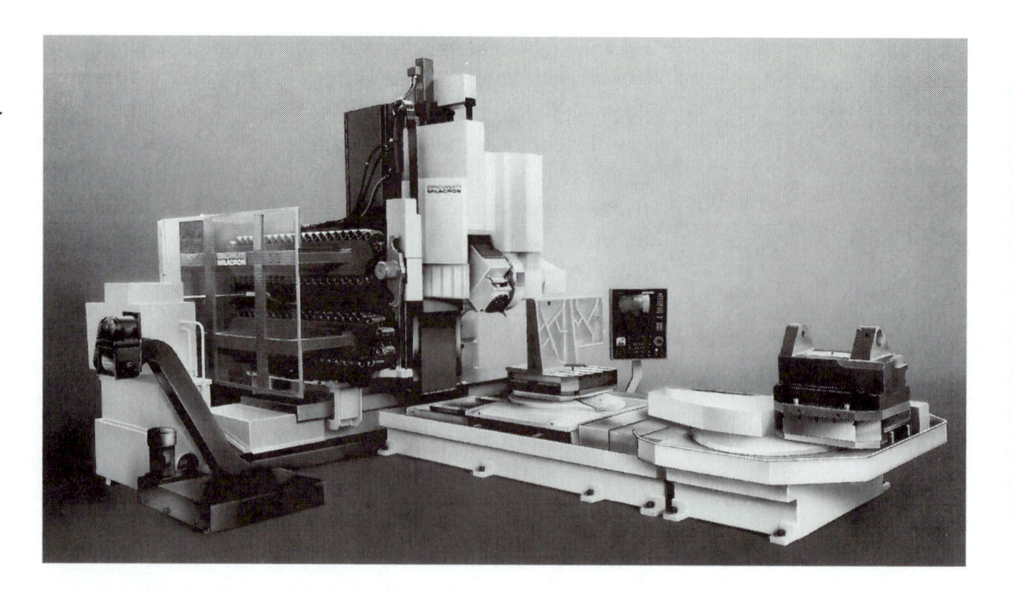

FIGURE 10.26
Machining center with five NC axes and a pallet shuttle. There are three straight motions mutually perpendicular, plus rotation of the table around a vertical axis and rotation of the spindle housing around a horizontal axis. In the front is a rotary pallet shuttle table. [MILACRON]

another, different workpiece is clamped on a pallet, to be shuttled to replace the previous workpiece when it is completed. The rotary table moves to and from the spindle with its base on a saddle that rides on the guideways located on the top of the front bed, and this is the Z motion. The waiting workpiece is on a rotary table that will accept the first workpiece and index by 180° to let the waiting workpiece shuttle onto the table, and this is also executed by the Z motion. Looking in the Z direction towards the spindle, the X coordinate motion is horizontal and perpendicular to the Z motion and it is executed by the ride of the base of the column on the bed of the machine that is square to the front bed. On the side of the machine is a double chain tool magazine and the tool-changing arm, and at the left end of the bed is the chip transporter that delivers the chips into a bin (not shown).

One design of the tool-changing mechanism is shown in Fig. 10.27. It is just about to move axially to pull the reamer out of the spindle, swing around the center of the double arm, position the face mill against the spindle and push it into the spindle. Then the rectangular column that carries the swingarm moves left, turns around its vertical axis and inserts the reamer into the tool magazine. While the face mill is working and the swingarm is out of the way, the magazine moves to prepare the next programmed tool for the reach of the swingarm. The designs of the machining centers shown in Figs. 10.25 through 10.27 date from about ten years back. Some of the latest designs are more compact and use recirculating ball or roller guideways; in larger sizes they use hydrostatic guideways (see Slocum), and they are equipped for higher feed rates, higher spindle speeds and powers, and have very fast tool changers with tool-to-tool time on the order of 5–7 sec; an example is the machine in Fig. 1.4. However, the machines described here at least partly illustrate the variety of configurations, each serving a particular area of application.

One particular concept has been used extensively in the aircraft industries for machining wing panels and similar complex parts of the fuselage of airplanes. These parts, made either from aluminum or titanium alloys, have to be light and strong at the same time. Therefore, they are made as "integral structures" with thin walls and thin ribs obtained by end milling out up to 95% of the material of a solid block from which

FIGURE 10.27
Details of tool magazine and of the tool-changing mechanism. The tool-changing double arm is shown holding at its left end a face mill just taken out of the tool magazine and at its right hand ready to pull an end mill from the spindle and swap the two. [MILACRON]

they are made. Their form is complex and often requires five-coordinate geometry. Because of the amount of material to be removed, the machines are made with three or four spindles working simultaneously on three or four identical parts. Generally, these machines are called 5-axis profilers. A three-spindle machine of this type is shown in Fig. 10.28. The individual, continuously controlled coordinate axes are *X*—the motion of the gantry on the longitudinal guideways arranged on the rails attached to the foundation; *Y*—the motion of the milling headstocks horizontally on the cross-rail of the gantry; *Z*—the up and down motion of the gantry; *A*—the swiveling motion of the headstocks around an axis parallel with *X*; *B*—the swiveling motion of the headstocks around an axis parallel with *Y*. The latter swivel is clearly visible in the picture.

For many years NC was applied to metal-cutting machine tools only. However, presently it is used also in forming machine tools, EDM wire cutting, plasma arc cutting and other thermal machining processes and, in a modified form, in the controls of robots. These applications are discussed in subsequent sections.

10.4 COMPUTERIZED, FLEXIBLE MANUFACTURING SYSTEMS

The designation of flexible manufacturing systems (FMS) has been commonly adopted for computerized combinations of NC machine tools with part (workpiece) handling

FIGURE 10.28
A three-spindle, five-axis NC profiler intended for milling of large aluminum aircraft parts. Three end mills are seen sticking out of the three tilted spindle heads. The whole large gantry travels on guideway rails along the stationary table of the machine. [MILACRON]

systems, and it essentially indicates computer-controlled equipment that operates on a variety of parts and includes, in an automated system, more activities than the individual NC machine tools.

The name and concept is separated from the NC machine tool mainly for historical reasons. The NC machine tool has existed since about 1950 and developed from the performance of a computer-controlled machining operation, including automatic tool change, with the NC programs inserted one by one on a punched tape. During the period 1965–1975 part programs were being stored in CNC (computer numerical control), where the mini- or micro-computer at the machine executed them through the controller. Around 1970, the transportation and allocation of parts to a group of NC machine tools was also included in the computer control, and this was given the designation of FMS. During further development various types of these systems were created with names like the flexible manufacturing center (FMC), computer automated factory (CAF), unmanned machining center (UMC), and so on. Other designations are also used, such as computerized manufacturing systems.

Several stages of these systems may be recognized, depending on the number of machine tools and tools involved and, generally, on the size of the overall computerized automated activity. Five such stages are outlined in the diagrams of Fig. 10.29. Diagram A represents the FMC (flexible manufacturing cell). It is a machining center (MC) with a tool magazine (TM) and automatic tool change and a part magazine (PM) that contains several parts on pallets. The part magazine includes an indexing motion that, after a part has been completed, brings the next pallet and part to the machining center. It also includes a pallet-changing mechanism (PC) that pulls the pallet with the completed part onto the waiting empty station of the part magazine and, after indexing, pushes the next part and pallet onto the table of the machining center. There is also a pallet-loading station (PL) where parts are taken from part storage (PS) and located and

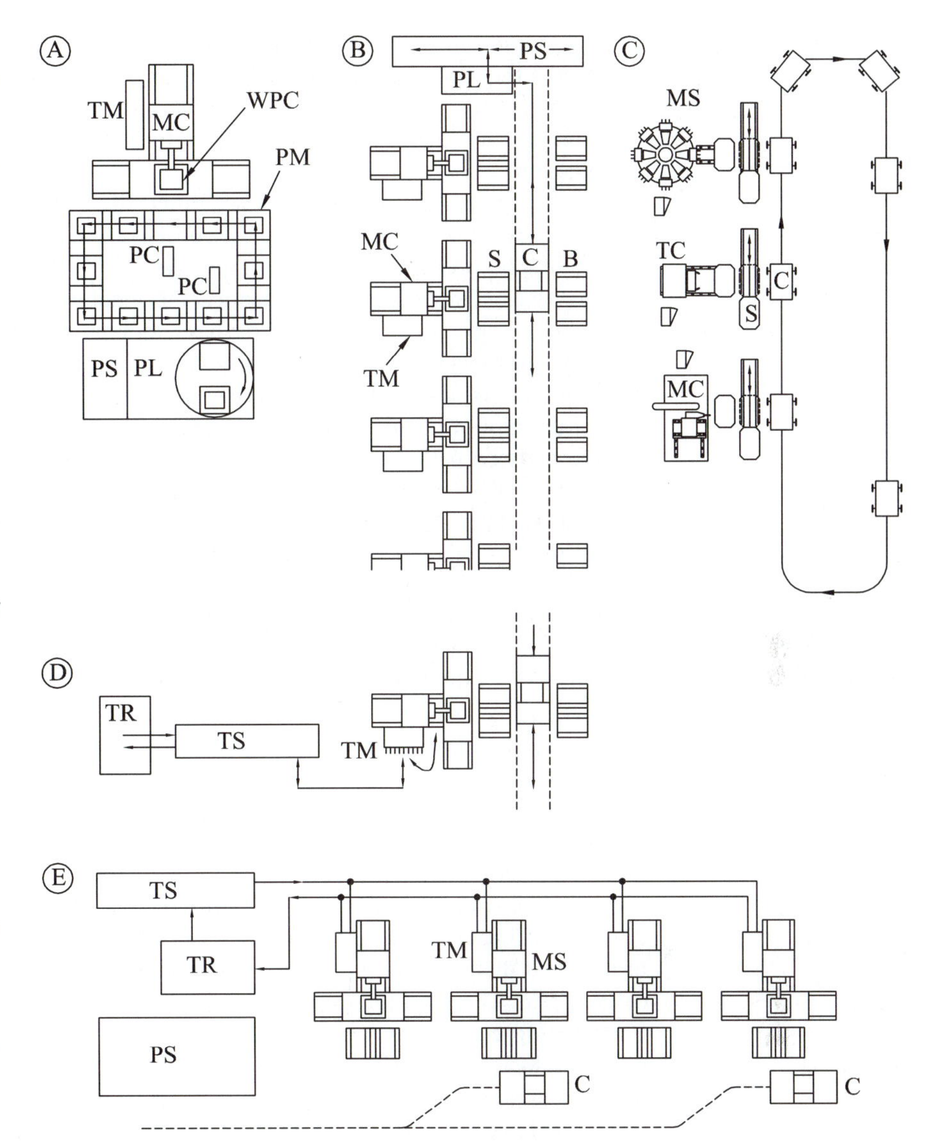

FIGURE 10.29
Basic types of computerized flexible manufacturing systems. A) A cell consists of a machining center *MC,* a part magazine *PM* and part storage *PS.* B) This system consists of several identical machining centers, a rail transportation system with carriages *C* for parts from and to part storage *PS.* For parts change from and onto the machine table, local shuttle table *S* and buffer *B* are used. C) A system in which different NC machine tools are interconnected by a transportation system. Shown is a machining center *MC,* a turning center *TC* and a multispindle head equipped machine *MS.* D) In this system there is an additional central tool storage *TS* that supplies the machine local tool magazines *TM.* The tool reconditioning shop maintains the whole tool inventory. E) This is a larger scale system of the D type. For part transport, carriages *C* travel independently on the floor, mostly in the form of automatically guided vehicles (AGV).

clamped on pallets. Instead of a rectangular part magazine, it may contain a rotary indexing table. The fixturing activity (pallet loading) is manual. The part magazine may be filled during only a part of the working time (e.g., a part of one working shift) and it may hold a sufficient supply of work for two or three shifts. Under certain conditions, which we will discuss later, this may then be an unmanned operation. Although a random operation is possible, the parts are mostly processed in this cell in a predetermined, fixed sequence. Usually, all the parts in the magazine are different, and they may represent a group of parts for one assembly. The same mix of parts may then be processed the next day or any other time. One of the reasons why the magazine is not filled with a batch of identical parts is that it would be necessary to manufacture identical fixtures (pallets) for all the parts in that batch.

The FMC is a very universal piece of equipment. It can operate on a great variety of parts. At any given time the number of different parts is limited by the number of tools in the tool magazine. Therefore, from time to time it is necessary to change the contents of the tool magazine to accommodate the needs of new parts. Apart from that, it is of course necessary to replace any tool as soon as it has been worn. Instead of a machining center the FMC can be applied to a turning center or any type of an NC machine tool.

The next stage is what is properly called the FMS (flexible manufacturing system), as shown in diagrams B and C. It consists of a group of NC machine tools connected by the workpiece transportation system. The FMS in B usually contains between two and eight essentially identical NC machine tools of a very universal nature (e.g., machining centers). Each of these has a tool magazine and automatic tool change. The types of tools in the individual magazines are mostly the same. There is a central part storage (PS) and pallet-loading (PL) station. A cart (C), which may be riding on rails or on the floor while guided by an underground wire and a radio transmitter, carries the parts on pallets to the individual machine tools. In front of each of these is a pallet shuttle system that carries out the exchange of workpieces between the machine tool and the cart. The shuttle (S) in this case consists of a pair of fixed slide stations. One of them accepts the new part. After the previous part is finished, the table of the machine moves to the other station and unloads the part on it, which is moved further on the cart. The table then returns to the first station and accepts the new part. The pushing and pulling of the parts through the shuttle is driven from the cart. On the opposite side of the cart track are buffer storage stations (B).

The main distinction between the FMS and the FMC of diagram A is that parts may be supplied to the individual machine tools in a random sequence that can be flexibly changed according to the actual state of production and according to varying production needs and strategies. The number of different parts being processed is larger and the allocation of parts to the individual stations is flexible and can adapt to variations in performance of the stations.

The system in diagram C differs from the one in B in that it consists of a mix of different NC machine tools. The diagram is a simplified representation of several FMSs of this kind that have been designed and installed; they contain several machine tools of each kind. The diagram shows only one of each of three kinds of NC machine tools: a machining center (MC), a turning center (TC), and a multispindle-head indexing machine (MS). Obviously, the universality of the system, as expressed by the number of different parts that are economically acceptable, is narrowed down. Because the various parts have differing needs for the three kinds of operations, it might happen that not all the machines can always be in use, which leads to a lower degree of utilization of the system. The inclusion of a multispindle drilling station indicates the intent of high-performance machining of larger batches of a limited number of different parts. The design of the shuttle (S) in this example is different from that in system B. It has a freedom of a linear motion. It can move opposite the table of the machine, accept the finished part, and move out of the way so that a new part can be loaded. The number of carts and pallets in circulation in the transportation system is larger. In this way, the transportation system also plays the role of intermediate buffer storage. The routing may include branching points to increase the flexibility of the system.

Diagram D shows a station of an FMS with an increased degree of automation. It involves a large tool storage (TS) with a computer-controlled link to the local tool magazine (TM) of the station. Thus changes in the tool mix in the TM are also included in the overall computer control. These changes are necessary to accommodate a larger variety of parts.

Diagram E presents what might be called the computer automated factory (CAF). It consists of a larger number of NC machining stations (MS), for example, machining or turning centers. These stations have their local tool magazines (TM). There is both a central part storage (PS) and a central tool storage (TS). Corresponding transportation systems supply parts and tools to the stations. After use, the tools return via the tool reconditioning shop (TR) where they are either reground or their inserts are indexed, changed, and reset. The part transportation system is assumed here to consist of carts guided by underground wires and radio transmission. Their routing is very flexible and they may move independently from part storage to any of the machine tools. They carry parts on pallets and may contain the mechanism that drives the pallets over the shuttle rails to and from the table of the machine tool.

Some examples follow of computerized flexible manufacturing systems executed and implemented for practical use. Historically, the first systems were of type C shown in Fig. 10.29, that is, systems consisting of groups of several NC machine tools with computerized part allocation and transportation. Perhaps the first of these systems was installed in 1967. It was concerned with machining cast magnesium alloy housings and mating covers for accessory units used in commercial and military aircraft. About 70 different parts in batch sizes between 25 and 300 were processed.

In 1978 two different extremes of computerized manufacturing were successfully installed. One was the FMS indicated as type A in Fig. 10.29. Although it might be considered as a modest option, it became evident that it was a very significant unit for any future manufacturing plant. By incorporating a single, universal machine tool like a machining center, the important parameter of utilization is upgraded to virtually 100%. Universality is also very high because there is no other less universal machine in the system. By providing a part magazine that could be loaded in one shift with enough work for three shifts, the unmanned operation in the afternoon and night shifts led to a significant increase in the labor-saving parameter. This is a computerized unit with excellent evaluation parameters. One of the first executions is shown in Fig. 10.30. In

FIGURE 10.30
The unmanned machining center. This is an FMS of type A of the classification scheme in Fig. 10.29. The rotary part magazine holds work for an entire unmanned night shift. [KEARNEY & TRECKER]

the foreground is the pallet loading station. The center is occupied by an annular part magazine with nine pallet-fixtures and a pallet shuttling mechanism in its center. In the background is the machining center with its work area enclosed by a complete box of glass walls to contain chips and coolant. The chips are transported away from the work zone by a conveyor discharging into a container seen on the left from the glass enclosure. The tool magazine located at the side of the machining center is not well visible in this picture.

Simultaneously with the introduction of the unmanned machining center in 1978, another extreme of the FMS principle was put into operation in the form of a very large system of type E in Fig. 10.29, which could be called a computerized automated factory. The system is described in [14]. A part of it is shown in the photograph in Fig. 10.31. It is a plant for manufacturing the central portion of a swing-wing fighter plane fuselage that consists of rather complex parts made of titanium alloys. The various parts are transported on carts, two of them seen in the right hand part of the picture, which are guided by underground wires and radio transmission and controlled from the central computer. The plant contains 28 large NC milling machines of two different types. The first is a gantry-type, three-spindle, five-coordinate, milling machine. Three of these machines are seen in a row in the central part of the picture. Between the first and second machines it is possible to see the end of a long table on which an operator is clamping parts while the gantry moves over the front part of the table and machines the parts located there. After completing this operation, the gantry will move to the rear end of the table and operate there while finished parts are removed from the front end. The other type of milling machine used is a column-type carrying pallets with parts and moving opposite two to four horizontal spindles. These machines move in three coordinate axes and are equipped with an automatic pallet-changing system. The gallery behind and above the machines contains the tool transportation system. There are several thousand tools included in the system. They circulate from a large central part stor-

FIGURE 10.31
A part of a computerized automated factory. This is an FMS of the E type. It produces parts of the fuselage of a fighter plane. It contains 28 large portal-type, three-spindle NC milling machines. The upper gallery houses the tool transport system to and from the local tool magazines. Parts are moved on AGVs with the striped bumpers seen in the foreground. [MESSERSCHMITT-BOLKOW-BLOHM]

age through the general transportation system to the machine tools and return to a tool reconditioning shop and back to the central storage. From the main transportation line on the gallery, elevators lead down to the local tool magazines of the individual machine tools. From these magazines the three tools are simultaneously exchanged in the three spindles. All the parts and tools and their movements are monitored and controlled by the central computer, which also takes care of the NC programs for all the machine tools. Actually the computing hardware consists of five interconnected computers. The computer-controlled manufacturing function is only a part of the overall CAD/CAM system. This factory is impressive because in addition to the computerized, automated shop, it contains only a few manually operated machine tools. Thus, the whole production depends almost entirely on computer automation. It runs smoothly and reliably.

In 1980 and 1981, many FMSs were introduced in the Japanese machine-tool factories. It is significant that they have made it a policy to install their newest products first in their own factories. In this way they gain direct experience to be used in further improvements of the product. Moreover, the manufacture of machine tools in medium lot sizes and with strong fluctuations in the demand for the individual products can make good use of the advantage of flexibility obtained by the computerization of production. The employer can better cope with drastic drops or increases in sales (due to the well-known "bust and boom cycle" characteristic of capital goods industries), because production can be increased without increasing the number of employees. Having an unmanned third shift, the employer can cancel it without laying anyone off. This is important in Japan, where there is a strong bond between a company and its workers.

The machine-tool manufacturer Yamazaki has transformed almost all of its manufacturing plants into computerized automated factories. One such shop is shown in Fig. 10.32. It consists of 10 machining centers that are serviced by one transportation pallet-loading cart that can hold two 8-ton workpieces. It includes a loading station that

FIGURE 10.32
The Yamazaki FMS. This is an FMS of the B type. It is used for machining the castings of machine-tool structure parts such as beds, columns, and tables. The rotary tool magazines (white circles in the picture) are all identical, each containing enough tools for all the operations. Tools are not replenished in the magazines individually; rather, whole magazines are exchanged. Workpieces are designed to require a minimum variety of tools. [YAMAZAKI]

communicates between part storage and the cart. There is also a toolroom where the circular tool magazines are loaded with tools. A crane carries the tool magazine drums to the drum-loading stations that execute the exchange of drums on the machine. The shop is operated on a three-shift basis. Two operators are present on the first and second shifts but none on the third shift. As seen in the photo, the parts in this shop are cast iron parts of machine-tool structures. The company claims, apart from the large labor savings, a reduction of in-process time from 90 to 3 days on average and a 50% reduction of the floor area. The Yamazaki company has built a machine-tool factory in the United States in Florence, KY, under the name Mazak. A substantial part of its production occurs on their own FMS installations.

Two other Japanese machine-tool factories have been using FMS methods in their plants. The installation at the Okuma factory, shown in Fig. 10.33, is of the B type of Fig. 10.29. The installation at the Toshiba factory shown in Fig. 10.34 is of type A, except that it contains two cells serviced from a common part magazine instead of only one.

The FMS concept is one of a high degree of automation, leading in some instances to completely "unmanned" work. In order to assure high quality in these operations, it is necessary to employ supervision systems that use a variety of sensors. A commonly used sensor is the "touch-trigger" probe. It is stored in the tool magazine; when called into action, it is inserted into the spindle and moved to check the proper position of the workpiece, the presence or absence of features such as bores, and also to ascertain the amount of thermally induced shift between spindle and workpiece. Other sensors used in these systems measure forces or power on the spindle, sense wear and breakage of tools, and warn of an incipient failure of bearings.

FIGURE 10.33
The Okuma FMS. Another FMS of the B type in a Japanese machine-tool factory. Similar systems have been installed in their U.S. factory. [OKUMA]

FIGURE 10.34
A cell of type A with two machining centers. [Toshiba]

From the vantage point of the end of the twentieth century, the FMSs developed twenty years ago and described here still retain their technical significance; they remain as examples of excellent planning and manufacturing control. It is true the FMSs installed in that period in the United States have mostly not satisfied economic expectations. The reasons have not been properly analyzed and explained. It may be that the companies lacked the capital base, vision, and courage, but many have expressed the opinion that most of the installations were not well conceived for each particular application. The manufacturing management science has progressed in the meantime to the concept of overall computer integrated manufacturing mentioned in Chapter 1. We can expect the FM developments of the past to be successfully adapted to fit into the CIM system.

10.5 POSITIONAL SERVOMECHANISM: REVIEW

This part presents a review of material that may have been presented in an introductory controls course. We will concentrate on some problems specific to machine tools and robots. These deal with the errors of two-dimensional motions and with the effects of flexibilities within the servoloop.

10.5.1 Characteristics of the Servomotor

A typical positional servomechanism was presented in Fig. 10.18. Here we will derive its transfer function and its response to a step input and to a ramp input. The central part of the positional servomechanism is the servomotor, and we will first derive its dynamic characteristics. The most common type of servomotor used in NC machine tools is a DC servomotor with permanent magnets (Fig. 10.35). Its speed is controlled by the input voltage e to the armature. This voltage is brought in usually through brushes and the commutator to the armature winding, which has a resistance R and inductance L. The armature current is denoted i. The mass moment of inertia of the rotor is J, and its angular speed is $\omega = 2\pi n$, where n is the number of revolutions/second. Opposed to the rotation is the load torque T_L. The friction in the bearings combined with the torque to drive the fan and its resistance to the rotations is approximately expressed as being proportional to the speed ω with a coefficient B. The rotation of the armature in the permanent magnetic field produces an induced back-electromotoric voltage proportional to ω, $e_b = K_E\omega$.

The equations describing the dynamics of the servomotor will now be derived. The imposed input voltage e is spent on producing the armature current i against the resistance R, and against the inductance L of the armature as well as on overcoming the back-electromotoric voltage.

$$e = Ri + L\frac{di}{dt} + K_E\omega$$

The second term is usually small, and for simplicity, it will be neglected. Correspondingly, the input voltage is:

$$e = Ri + K_E\omega \tag{10.1}$$

Usually a limit is imposed on the current i to prevent the demagnetization and overheating of the stator. At this stage we will neglect it in order to keep the model of the motor linear. Later on, in the simulation exercise, the nonlinearity of the current saturation will be included.

The current in the armature produces the useful torque of the motor:

$$T = K_T i \tag{10.2}$$

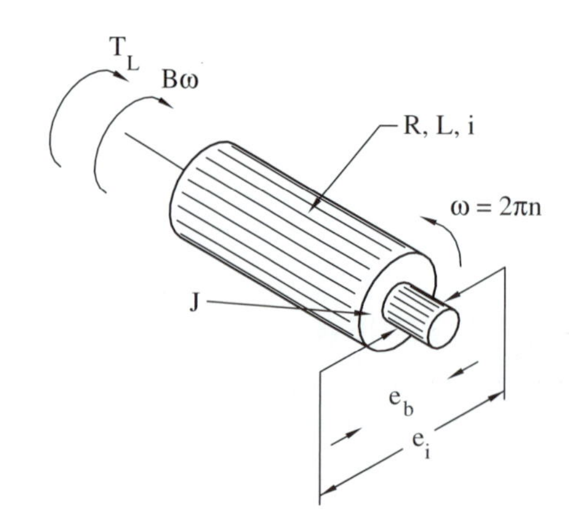

FIGURE 10.35
Diagram of a permanent-magnet, DC servomotor. The rotor and collector are shown with the electrical and mechanical parameters and the variables i (current in rotor winding), the input voltage e_i, back electromotoric voltage e_b, and rotational speed ω.

The torque is used to accelerate the rotor and to overcome frictional losses and external load T_L:

$$K_T i = J\dot{\omega} + B\omega + T_L \tag{10.3}$$

The external inertia consists of that of the leadscrew and of the table. The latter is driven through a large down-transmission ratio and can be neglected. The inertia of the leadscrew may be included in that of the motor. We will neglect the ventilation losses $B\omega$ without substantially affecting the outcome. Equation (10.3) can also be written with torque as input:

$$\dot{\omega} = \frac{T - T_L}{J} = \frac{K_T i - T_L}{J} \tag{10.4}$$

This is the form customary in papers dealing with robot control. In order to obtain a relationship between the input voltage e and the output speed ω, we will express i from (10.1) as

$$i = \frac{e - K_e\omega}{R} \tag{10.5}$$

and use it in (10.3):

$$\frac{K_T}{R}(e - K_E\omega) = J\dot{\omega} + T_L$$

The relationships of Eqs. (10.1) through (10.5) can be followed in the block diagram of Fig. 10.36, which depicts the transfer function of (Ω/E) with the electromotoric constant K_E as feedback. The load torque T_L is depicted as another external input. In another form, shown at the bottom of the figure, the load torque is multiplied by (R/K_T) and subtracted from the input voltage.

Now, proceeding further from Eq. (10.5) and rearranging, we have

$$\frac{K_T}{R}e = J\dot{\omega} + \frac{K_T K_E}{R}\omega + T_L$$

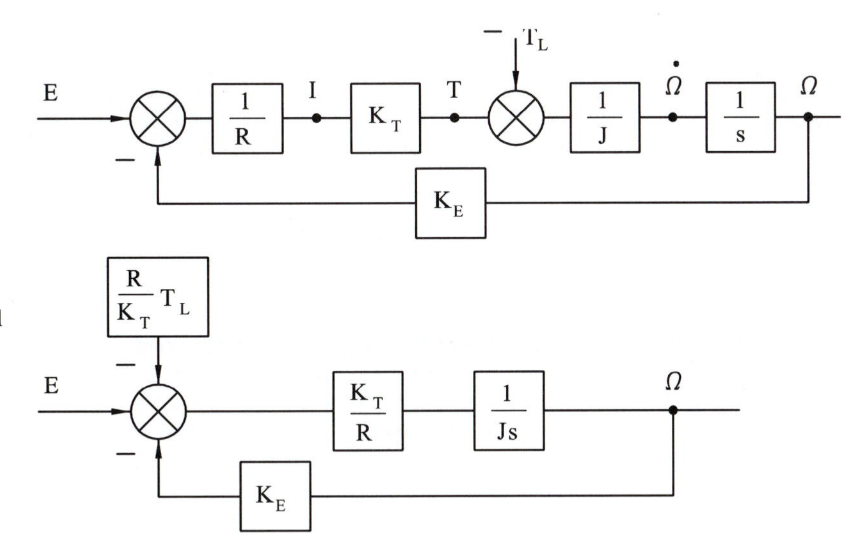

FIGURE 10.36
Block diagram of the servomotor. It relates output speed ω to input voltage e. There is internal feedback due to the back emf. (Capital letters are used in the block diagram, which is represented in the Laplace s-domain. In the text we understand them as time functions and use lowercase notations.)

and

$$\frac{JR}{K_T K_E}\dot{\omega} + \omega = \frac{1}{K_e}e - \frac{R}{K_T K_E}T_L \tag{10.6}$$

Denoting

$$\frac{JR}{K_T K_E} = \tau_{\mathrm{m}}$$

and

$$\frac{1}{K_E} = K_m \tag{10.7}$$

thus we have

$$\tau_{\mathrm{m}}\dot{\omega} + \omega = K_m\left(e - \frac{R}{K_T}T_L\right) \tag{10.8}$$

The parameter τ_{m} (sec) is the motor time constant and the parameter K_m is the motor gain. Both are determined from the design parameters of the motor. The transfer function (Ω/E) is obtained by Laplace transforming Eq. (10.8) or directly from the block diagram of Fig. 10.36:

$$\tau_{\mathrm{m}}\, s\Omega + \Omega = K_m\left(E - \frac{R}{K_T}T_L\right)$$

$$\Omega(s) = \frac{K_m}{1 + \tau_{\mathrm{m}}s}\left[E(s) - \frac{R}{K_T}T_L(s)\right] \tag{10.9}$$

This shows that in our formulation the servomotor is a first-order system, described by the first-order differential equation (10.8) with constant coefficients.

We will look separately at the *steady-state* characteristic and at the *transient behavior* of the motor. The *steady state* is one of constant input and output; angular acceleration is $\dot{\omega} = 0$. The effect of load T_{L} on speed ω is of interest. From Eq. (10.8), we have:

$$\omega = K_m\left(e - \frac{R}{K_T}T_L\right) \tag{10.10}$$

Obviously, gain K_m is defined as speed per input of one volt with no load. The (ω,T_L) characteristic is linear, as shown in Fig. 10.37. The load at which the speed drops to zero is the *stall torque* T_{st}:

$$T_{st} = \frac{K_T}{R}e \tag{10.11}$$

10.5.2 Step Input Response of the Servomotor

In the transient the external load torque may be neglected while the torque is spent on accelerating the inertias of the motor and of the driven parts. Let us consider a *step input transient,* as a response to $e = 0$ for $t < 0$ and $e = e_1$ for $t \geq 0$. From Eq. (10.8),

$$\tau_{\mathrm{m}}\dot{\omega} + \omega = K_m e_1 \tag{10.12}$$

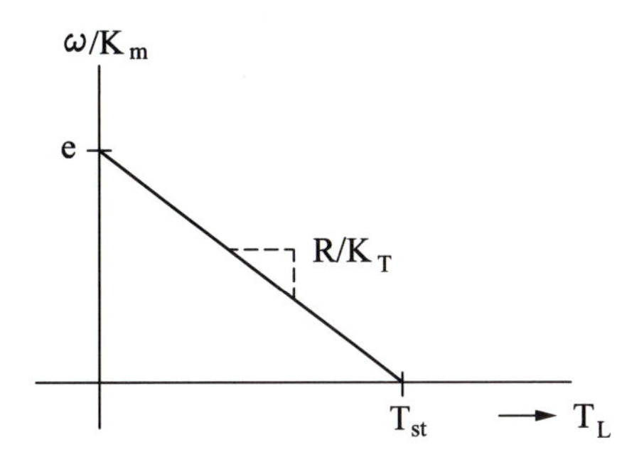

FIGURE 10.37 Speed/load characteristic of the servomotor. Speed is indirectly proportional to load, down to the stall torque T_{st}.

The general integral is obtained from the homogeneous equation

$$\tau_m \dot{\omega} + \omega = 0 \tag{10.13}$$

Suggesting $\omega = Ce^{st}$, $\dot{\omega} = s\, Ce^{st}$, the characteristic equation is obtained:

$$(\tau_m s + 1)Ce^{st} = 0$$

$$s = -\frac{1}{\tau_m}$$

and the general solution is

$$\omega = Ce^{-t/\tau_m}$$

Going back to Eq. (10.12) and suggesting the particular solution of $\omega = K_m e_1$, the complete solution is

$$\omega = Ce^{-t/\tau_m} + K_m e_1 \tag{10.14}$$

For the initial condition of $t = 0$, $\omega = 0$, the value of C is obtained from Eq. (10.14):

$$C = -K_m e_1 \tag{10.15}$$

and finally,

$$\omega = K_m e_1 (1 - e^{-t/\tau_m}) \tag{10.16}$$

$$\dot{\omega} = \frac{K_m e_1}{\tau_m} e^{-t/\tau_m} \tag{10.17}$$

The value of the initial acceleration is obtained from Eq. (10.17) for $t = 0$:

$$\dot{\omega}_0 = \frac{K_m}{\tau_m} e_1 \tag{10.18}$$

The response of Eq. (10.16) is graphically expressed in Fig. 10.38. This response could also have been obtained from the transfer function of Eq. (10.9) by setting $T_L(s) = 0$, $E(s) = e_1/s$, and inverse Laplace transforming.

The servomotor is commonly enclosed in a tachogenerator feedback loop. Since James Watt's time and his "ball governor," it has been known that a velocity feedback makes the static characteristic rather insensitive to load and it also improves the transient behavior. The tacho-loop is shown in the block diagram in Fig. 10.39. The speed ω is measured by the tachogenerator with gain H (V/(rad/sec)) and this signal is subtracted from the input voltage v. The difference is the velocity error e_v (V), which must

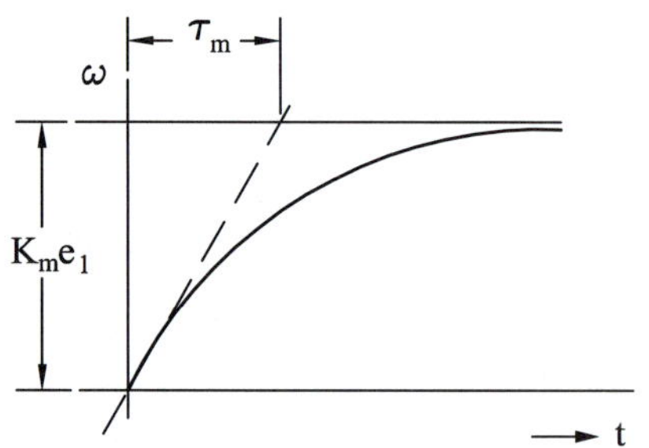

FIGURE 10.38
Response of the servomotor to a step input. The slope (acceleration) decreases exponentially with time. The gain K_m (rad/sec/V) and time constant τ_m (sec) are defined.

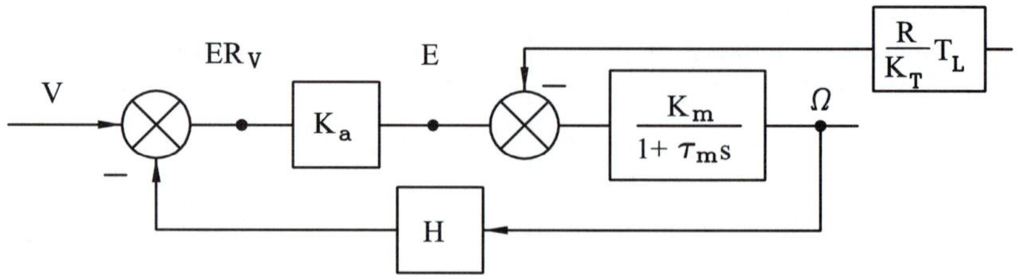

FIGURE 10.39
Block diagram of the servomotor with tachogenerator feedback. Command voltage v is compared with tachogenerator feedback voltage $H\omega$, and the error of velocity er_v drives the motor represented by the first-order block. The signal er_v is small as a result of subtraction of the feedback signal from the command signal, and it has to be amplified with factor K_a.

be amplified to become the input voltage e to the servomotor. The gain K_a is usually rather large so as to keep the error small. The servomotor is represented by its transfer function, Eq. (10.9). The "velocity closed-loop" transfer function (Ω/V) is derived by expressing

$$e = K_a(v - H\omega) \tag{10.19}$$

and using this in Eq. (10.8):

$$\tau_m\dot{\omega} + \omega = K_m\left[K_a(v - H\omega) - \frac{R}{K_T}T_L\right],$$

$$\tau_m\dot{\omega} + (1 + K_mK_aH)\omega = K_m\left(K_av - \frac{R}{K_T}T_L\right),$$

$$\frac{\tau_m}{1 + K_mK_aH}\dot{\omega} + \omega = \frac{K_mK_a}{1 + K_mK_aH}v - \frac{K_m}{1 + K_mK_aH}\frac{R}{K_T}T_L$$

Here

$$\tau_{CL}\dot{\omega} + \omega = K_{CL}\left(v - \frac{R}{K_aK_T}T_L\right) \tag{10.20}$$

we used

$$\tau_{CL} = \frac{\tau_m}{1 + K_mK_aH} \qquad K_{CL} = \frac{K_mK_a}{1 + K_m K_a H} \tag{10.21}$$

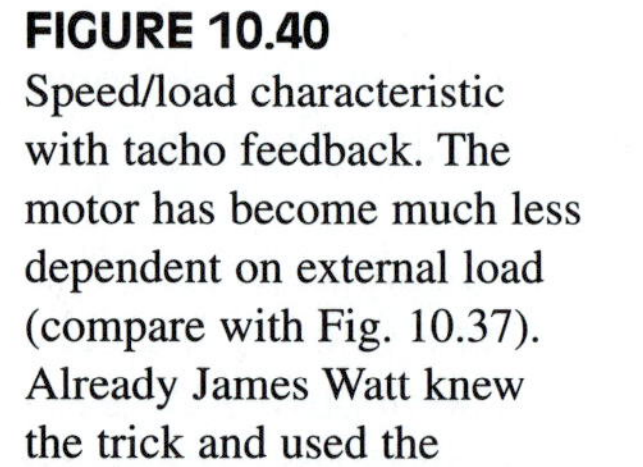

The number K_mK_aH may be rather large; the "closed-loop time constant τ_{CL}" may be much shorter than τ_m. The "closed-loop gain K_{CL}" is approximately (1/H) if the 1 in the denominator is neglected, and it may be kept about the same as that of the motor alone, K_m.

Note that the steady-state characteristic, Fig. 10.40, is much flatter; setting $\dot{\omega} = 0$ in Eq. (10.20), we have

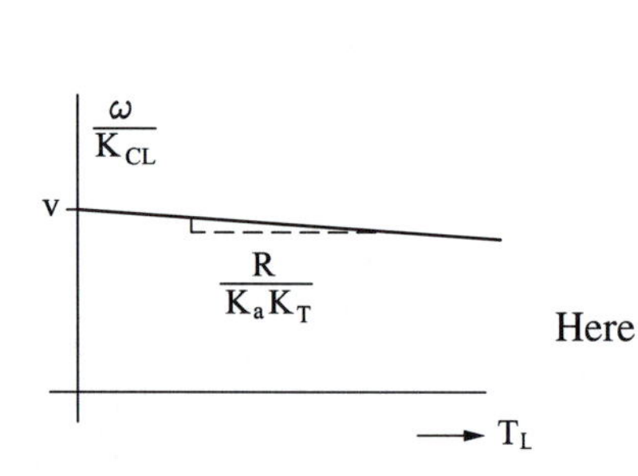

FIGURE 10.40
Speed/load characteristic with tacho feedback. The motor has become much less dependent on external load (compare with Fig. 10.37). Already James Watt knew the trick and used the famous "ball governor." It is a sin to use a servomotor without velocity feedback.

$$\omega = K_{CL}\left(v - \frac{R}{K_a K_T} T_L\right)$$

In comparison with Eq. (10.10) the negative slope of the effect of the load is now K_a times smaller; as mentioned before, K_a is chosen large (e.g., $K_a = 200$). This is the most important effect of the tachogenerator feedback: it makes the motor insensitive to (static) load. That was exactly what James Watt achieved by inventing the "ball governor." The tacho feedback also improves the transient, as is evident from the decrease of the time constant. However, for fast response the motor would need a large torque obtainable by a large current. In reality this effect is limited by the necessary limitation imposed on the current.

The transfer function between the input $V(s)$ and the output $\Omega(s)$ is formally again one of a first-order system:

$$\frac{\Omega(s)}{V(s)} = \frac{K_{CL}}{1 + \tau_{CL}s} \tag{10.23}$$

The response to the load torque T_L is

$$\frac{\Omega(s)}{T_L(s)} = \frac{K_{CL}}{K_a}\frac{R}{K_T}\frac{1}{1 + \tau_{CL}s} \tag{10.24}$$

However, we will mostly use just Eq. (10.23) because of the rather small effect of the load.

The step input response is written in a way similar to Eq. (10.16). For a step v_1:

$$\omega = v_1 K_{CL}(1 - e^{-t/\tau_{CL}}) \tag{10.25}$$

10.5.3 Time-Domain Simulation of the Servomotor

The response of the servomotor to a velocity step input was derived above under the assumption of linearity of the system. It was also mentioned that the current limiter makes the system nonlinear. This nonlinearity is easily considered, and the response of the servomotor to any form of input $v(t)$ is obtained by following the variables in small steps dt and running a program loop. Follow the block diagram of Fig. 10.39:

```
Input τm, Km, Ka, H, KT, R, KE, J, dt
Define initial state, for example, for t1 = 0, ω1 = 0.
For     n = 1:N
        vn = f1(tn)
        erv = vn − Hωn
        e = Ka erv
C Use Eq. (10.5)

               i = (e − KE ωn) / R

        If i > ilim, i = ilim
C Use Eq (10.4)

               dom = KT i / J

        omn+1 = omn + dom * dt
        tn+1 = tn + dt                                         (10.26)
```

The function f_1 can be a step, a ramp, or any other type, variation of the command voltage v.

10.5.4 The Positional Servomechanism

The overall block diagram of the positional servomechanism is presented in Fig. 10.41, which corresponds to Fig. 10.18. The input is the commanded position x_{com}. In the positional discriminator *PD* (top diagram), the actual position x_{act} is subtracted as it is generated by the positional feedback transducer. The two signals are numbers of BLUs. However, in our analyses, we will express them simply in millimeters. The positional error er_p,

$$er_p = x_{\text{com}} - x_{\text{act}}$$

is then converted to an analog voltage *v*. This conversion is represented as an amplification with the "positional" gain K_p (V/mm). The voltage *v* is the command for the servomotor to move. Obviously, a positional error er_p is necessary if the servomotor should move.

The response to *v* is the speed ω of the servomotor with its internal tachogenerator feedback. The internal function of the velocity loop with the gain factors K_a and K_m and tacho-gain *H* has been described in Section 10.5.1. The whole loop has been summarized using the gain K_{CL} and time constant τ_{CL}, and it has been expressed by the transfer function Eq. (10.23) and by Eq. (10.25). This compact form is used in the bottom diagram of Fig. 10.41. The servomotor drives the table through the leadscrew and nut transmission *r*. Thus the velocity of the table is

$$\dot{x}_{\text{act}} = r\,\omega \tag{10.27}$$

and it is integrated, see the integration symbol (1/*s*), into the actual position x_{act}. In practice, this integration is obtained by counting the pulses from the output of the encoder. The result is in BLU but we will read it in millimeters. The feedback is then represented by the unity 1 in the diagram.

The positional loop transfer function is obtained from the forward loop transfer function:

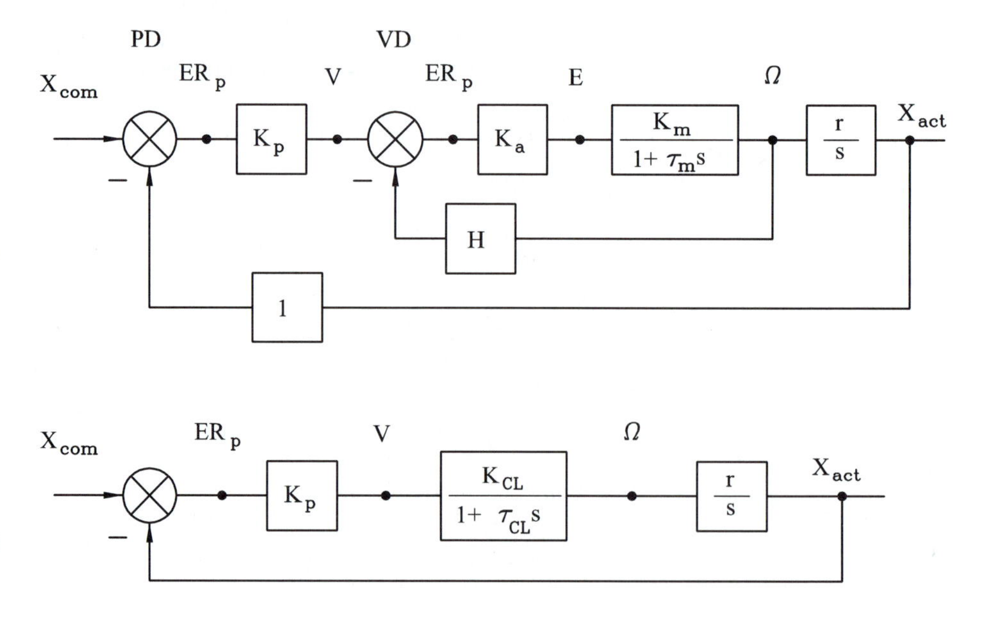

FIGURE 10.41 Block diagram of the positional servomechanism. Top: with details of the internal tacho-loop. Bottom: the servomotor with tacho-feedback as a single block. Rotational speed ω is passed through a transmission ratio *r* (e.g., a leadscrew), and the feedback transducer measures its integral, that is, the travel x_{act} that goes back to be subtracted from the command to produce the error of position er_p that tends to cancel out by being amplified by K_p to command the motion of the motor.

$$G(s) = X_{\text{act}}(s) \,/\, ER_p(s) \tag{10.28}$$

$$G = \frac{X_{\text{act}}}{X_{\text{com}} - X_{\text{act}}}$$

$$GX_{\text{com}} - GX_{\text{act}} = X_{\text{act}}$$

$$GX_{\text{com}} = X_{\text{act}}(1 + G)$$

$$\frac{X_{\text{act}}}{X_{\text{com}}} = \frac{G}{1 + G} \tag{10.29}$$

It is

$$G = \frac{K_p\, K_{CL}\, r}{(1 + \tau_{\text{CL}} s)s} = \frac{K}{(1 + \tau_{\text{CL}} s)s} \tag{10.30}$$

where $K = K_p\, K_{CL}\, r$ is the overall open-loop gain of the system.

Using Eq. (10.30) in Eq. (10.29):

$$\frac{X_{\text{act}}}{X_{\text{com}}} = \frac{K/(1 + \tau_{\text{CL}} s)s}{1 + K/(1 + \tau_{\text{CL}} s)s}$$

$$\frac{X_{\text{act}}}{X_{\text{com}}} = \frac{K}{\tau_{\text{CL}} s^2 + s + K},$$

$$\frac{X_{\text{act}}}{X_{\text{com}}} = \frac{K/\tau_{\text{CL}}}{s^2 + s/\tau_{\text{CL}} + K/\tau_{\text{CL}}} \tag{10.31}$$

Eq. (10.31) is the standard form of a second-order system. It expresses the amplitude of the harmonic response to a harmonic input if we set $s = j\omega$:

$$\frac{X_{\text{act}}(j\omega)}{X_{\text{com}}(j\omega)} = \frac{\omega_n^2}{\omega_n^2 - \omega^2 + 2j\zeta\omega\,\omega_n} \tag{10.32}$$

or else

$$\frac{X_{\text{act}}}{X_{\text{com}}} = \frac{1}{1 - \omega^2/\omega_n^2 + 2j\zeta\omega/\omega_n} \tag{10.33}$$

The natural frequency of the system is

$$\omega_n = \sqrt{\frac{K}{\tau_{\text{CL}}}} \tag{10.34}$$

and the damping ratio is

$$\zeta = \frac{1}{2\tau_{\text{CL}}\omega_n} = \frac{1}{2\sqrt{K\tau_{\text{CL}}}} \tag{10.35}$$

We could have derived the transfer function of Eq. (10.32) from the differential equation of the positional servomechanism with the block diagram of Fig. 10.41. The central part of it is the differential Eq. (10.20) expressing the servomotor action:

$$\tau\dot{\omega} + \omega = K_{CL}\, v$$

where

$$v = (x_{\text{com}} - x_{\text{act}})\, K_p \tag{10.36}$$

and also, according to Eq. (10.27),

$$\omega = \frac{\dot{x}_{\text{act}}}{r} \qquad \dot{\omega} = \frac{\ddot{x}_{\text{act}}}{r}$$

Combining these expressions, we get

$$\begin{gathered} \frac{\tau_{\text{CL}}\ddot{x}_{\text{act}}}{r} + \frac{\dot{x}_{\text{act}}}{r} = K_{CL}K_p\,(x_{\text{com}} - x_{\text{act}}) \\ \tau_{\text{CL}}\ddot{x}_{\text{act}} + \dot{x}_{\text{act}} + Kx_{\text{act}} = Kx_{com} \end{gathered} \tag{10.37}$$

Using Eqs. (10.34) and (10.35),

$$\ddot{x}_{\text{act}} + 2\zeta\omega_n\,\dot{x}_{\text{act}} + \omega_n^2\,x_{\text{act}} = \omega_n^2\,x_{\text{com}} \tag{10.38}$$

and setting

$$x_{\text{com}} = X_{\text{com}}e^{j\omega t} \text{ and } x_{\text{act}} = X_{\text{act}}\,e^{j\omega t} \tag{10.39}$$

we obtain Eq. (10.32).

10.5.5 Step Input Response of the Positional Servo

From basic linear control systems theory, we know that the response of a second-order system of Eq. (10.32) to a step input A, for the "underdamped" case ($\zeta < 1$) is

$$x_{\text{act}} = A\left[1 - \frac{e^{-\zeta\omega_n t}}{\sqrt{1-\zeta^2}}\sin\left(\omega_d t + \tan^{-1}\left(\frac{\sqrt{1-\zeta^2}}{\zeta}\right)\right)\right] \tag{10.40}$$

where the damped natural frequency is

$$\omega_d = \omega_n\sqrt{1-\zeta^2} \tag{10.41}$$

Equation (10.40) is expressed graphically later on in Fig. 10.42 for four different values of the damping ratio. The displacement x_{act} overshoots the commanded step, and the magnitude of the "overshoot" is

$$o = Ae^{-\pi\zeta/\sqrt{1-\zeta^2}} \tag{10.42}$$

and it occurs at the time

$$t_o = x/\omega_d \tag{10.43}$$

The natural frequency ω_n, Eq. (10.34), is decisive for the speed of the response, and the damping ratio ζ, Eq. (10.35), is decisive for both the speed of response and the overshoot. For $\zeta \geq 1.0$, there is no overshoot, but the response is sluggish. Often the value $\zeta = 0.7$ is preferably chosen because the overshoot is small, $o = 0.05$ A, and the response is reasonably steep. Going back to the design parameters of the motor and of the servo, and using Eq. (10.21), we can understand better how to influence the two basic parameters ω_n and ζ:

$$\omega_n = \sqrt{\frac{K}{\tau_{\text{CL}}}} = \sqrt{\frac{K_pK_{CL}r}{\tau_{\text{CL}}}} = \sqrt{\frac{K_prK_mK_a}{1+K_mK_aH}\,\frac{1+K_mK_aH}{\tau_{\text{m}}}} \tag{10.44}$$

$$\omega_n = \sqrt{\frac{K_p K_a K_m r}{\tau_{\mathrm{m}}}}$$

$$\zeta = \frac{1}{2\,\tau_{\mathrm{CL}}\,\omega_n} = \frac{1 + K_m K_a H}{2\tau_{\mathrm{m}}}\sqrt{\frac{\tau_{\mathrm{m}}}{K_p K_a K_m r}}$$

Neglecting the 1 in the numerator of the first term,

$$\zeta = \frac{H}{2}\sqrt{\frac{K_a K_m}{\tau_{\mathrm{m}} K_p r}} \tag{10.45}$$

Interpreting, we see that the natural frequency ω_n increases with the square root of all the static gain factors K_p, K_a, K_{m}, r, and with the square root of the decrease of the motor time constant. The damping factor ζ is proportional to the tachogenerator gain H and increases with the square root of the motor gains K_a, K_m; inversely, it decreases with the square root of the time constant τ_{m} and of the positional gains K_p and r.

10.5.6 Time-Domain Simulation of the Positional Servo

The response of the servo to any form of input $x_{\mathrm{com}} = f(t)$ may be obtained by time-domain simulation executed in small time steps dt. The computer program expands on the one written in Eq. (10.26) for the servomotor and follows the relationships as modeled in the block diagram of Fig. 10.41, including the differential equation (10.27):

$$\omega = \frac{\dot{x}_{\mathrm{act}}}{r} \qquad \dot{x}_{\mathrm{act}} = r\omega$$

As an alternative to Eq. (10.26), we may, for the sake of some of the following exercises, leave out the current limitation and use the servomotor performance parameters τ_{m}, K_m instead of the motor design parameters. In the program we may then make use of the motor differential equation (10.8):

$$\tau_{\mathrm{m}}\,\dot{\omega} + \omega = K_m e$$

We will no longer subscript the motor-related variables. For simplicity, we use x instead of x_{act}.

The program structure corresponds to the diagram of Fig. 10.41:

```
Input data: τm, Ka, Km, Kp, r.
Initial state: for example, for t1 = 0, x = 0, om = 0.
For n = 1:N
xcom_n = f (t_n)
erp = xcom_n − x_n
v = Kp erp
erv = v − H om
e = Ka erv
dom = (Km e − om)/τm
om = om + dom dt
ẋ = r om
x_n+1 = x_n + ẋ dt
t_n+1 = t_n + dt
end.
```

(10.46)

In this program, for simplicity and to correspond to the block diagram in Fig. 10.41, the current limit has not been considered. Such a consideration could be embedded by inserting the lines written between the two comments C in the listing of Eq. (10.26).

EXAMPLE 10.1 Performance of Positional Servos ▼

Determine the natural frequency ω_n, the damping ratio ζ, and the overshoot o of a servomechanism corresponding to Fig. 10.41 with the following parameters: $K_p = 200$ (V/mm), $K_a = 25$, $K_m = 2$ (rad/sec/V), $\tau_m = 0.500$ sec, $H = 0.48$ (V/rad/sec), $r = 0.5$.

According to Eq. (10.21),

$$\tau_{CL} = \frac{\tau_m}{1 + K_m K_a H} = \frac{0.5}{1 + 24} = 0.020 \text{ sec}$$

$$K_{CL} = \frac{K_m K_a}{1 + K_m K_a H} = \frac{50}{25} = 2 \text{ rev/sec/}V$$

According to Eqs. (10.34) and (10.35),

$$\omega_n = \sqrt{\frac{K}{\tau_{CL}}} = \sqrt{\frac{200}{0.02}} = 100 \text{ rad/sec} = 15.92 \text{ Hz}$$

$$\zeta = \frac{1}{2\tau_{CL}\omega_n} = 0.25, \quad \omega_d = 96.82 \text{ rad/sec}$$

The overshoot is, according to Eq. (10.42),

$$o = Ae^{-\pi\zeta/\sqrt{1-\zeta}} = e^{-0.811} = 0.444 \text{ mm}$$

The response to a step input of $A = 1$ mm, as in Eq. (10.40), is

$$x_{act} = 1 - 1.0328\, e^{-25t} \sin(96.82\, t + 1.318) \tag{10.47}$$

Checking: For $t = 0$, $x_{act} = 1 - 1.0328 \sin 1.318 = 1 - 1 = 0$. ▲

Exercise

How will the natural frequency ω_n and the damping ratio ζ change if the following occur?

1. As in Ex. 10.1.
2. The positional gain is increased to $K_p = 800$ (V/mm).
3. The velocity feedback H is increased to $H = 1.4$ (V/rad/sec).
4. The velocity feedback is eliminated, $H = 0$.

The parameters are summarized in Table 10.2. The transmission ratio is in all cases $r = 0.5$ mm/rad.

The values of the derived parameters, the natural frequency, and the damping ratio ζ are entered in the table. The results of the step $x_{com} = 1$ mm input response are plotted in Fig. 10.42. It is seen that the most important parameter is the damping ratio ζ. In case c), with the rather large damping ratio $\zeta = 0.71$, the overshoot is very small, and the motion is not vibratory. The lack of velocity feedback in case d), with $H = 0$, leads

TABLE 10.2 Parameters of the Four Test Cases

Case	H(V/rad/sec)	K_p(V/mm)	ω_n (rad/sec)	ζ
a.	0.48	200	100	0.25
b.	0.48	800	200	0.125
c.	1.4	200	100	0.71
d.	0	200	100	0.01

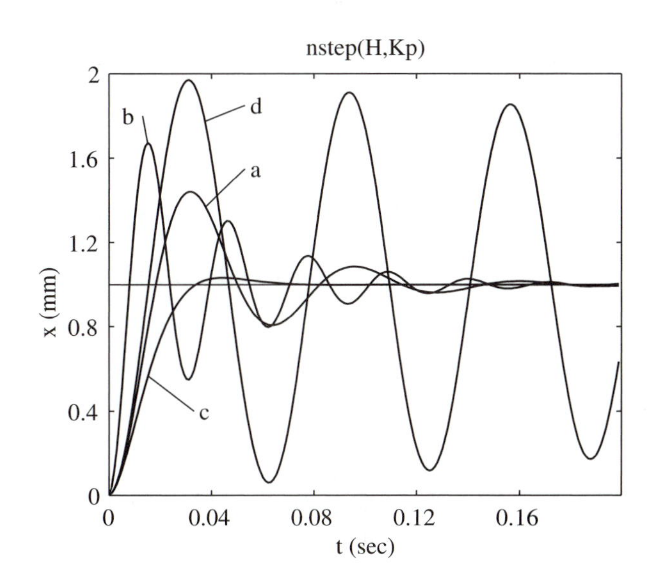

FIGURE 10.42 Step input response of servos with various gains and damping ratios. The values of ζ are a) 0.25, b) 0.125, c) 0.71, d) 0.01. The text explains how the frequency of the transient and its damping ratio are obtained. Of the four, *(c)* is commonly considered best; it is still reasonably fast and overshoots only a little.

to a very small damping ratio of $\zeta = 0.01$ and to a vibratory motion that decays very slowly. Eventually, all the responses settle at $x = 1.0$, as commanded.

10.5.7 Response to a Ramp Input of the Positional Servo

A positional ramp input is equivalent to a velocity step input, that is, to a command to start and move with a constant velocity (feed rate, mm/sec):

$$x_{\text{com}} = f_{\text{com}} \qquad t = Bt$$

This is, of course, the most common case rather than commanding a sudden positional step. It is then usual to request a finite distance x_1 to be reached, that is, to stop the motion after a time $t_1 = x_1/f$; see Fig. 10.43.

This is equivalent to superimposing, at time t_1, a negative ramp command $f_{\text{com}} = -Bt$ onto the original command. The positional command is then

$$\begin{aligned} x_{\text{com}} &= x_{\text{com},1} - x_{\text{com},2} \\ x_{\text{com},1} &= 0 \text{ for } t < 0 \\ &= Bt \text{ for } t > 0 \\ x_{\text{com},2} &= 0 \text{ for } t \le t_1 \\ &= -B(t - t_1) \text{ for } t > t_1 \end{aligned} \tag{10.49}$$

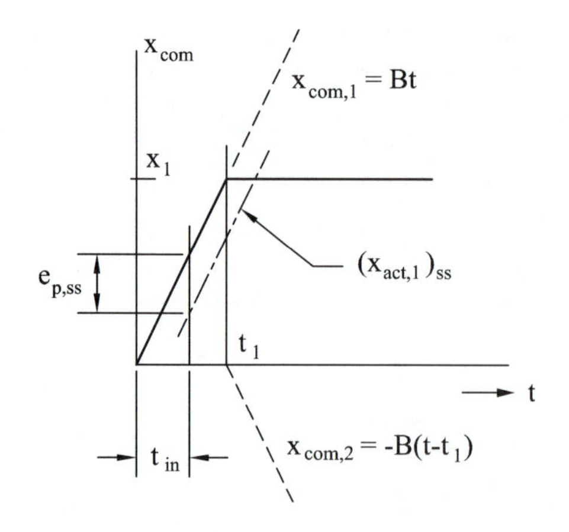

FIGURE 10.43
Response of the servo to a ramp input, steady state. After the initial transient, the output velocity becomes equal to the constant velocity of the input. At any instant the output position lags behind the input position by the "following error" $e_{p,ss}$ ("velocity lag").

The response to the signal $x_{\text{com},1}$ is, according to linear control systems theory,

$$x = B\left[t + \frac{e^{-\zeta\omega_n t}}{\omega_d}\sin(\omega_d t + 2\phi) - \frac{2\zeta}{\omega_n}\right] \tag{10.50}$$

where

$$\phi = \tan^{-1}\frac{\sqrt{1-\zeta^2}}{\zeta} \tag{10.51}$$

The graphical presentation of Eq. (10.50) is given later on. However, it is useful to note here in this general discussion that the second term decreases with time. After a sufficiently long time, the "steady-state" response is

$$(x_{\text{act},1})_{ss} = B\left(\frac{t - 2\zeta}{\omega_n}\right) \tag{10.52}$$

This is illustrated in Fig. 10.43, where time t_{in} represents the initial period during which the transient term sufficiently decays. After this period the actual motion is carried out with the same velocity $f = B$ as the commanded motion (it has the same slope in the graph), but it is lagging behind the commanded motion by the steady-state positional error $e_{p,ss}$, which is also called the "velocity lag," that is, the positional lag proportional to velocity:

$$e_{p,ss} = B\frac{2\zeta}{\omega_n} \tag{10.53}$$

Using the formulas of Eqs. (10.34) and (10.35), we have

$$e_{p,ss} = \frac{B}{\tau_{\text{CL}}\,\omega_n^2} = \frac{B}{K} \tag{10.54}$$

where $K = K_p K_{CL}\, r$ is the feedforward gain of the positional loop. This result is easy to understand by referring to the block diagram of Fig. 10.41. For steady state, set all $s = 0$. Then, if the velocity $x_{\text{act}} = B$, going back in the upper branch of the bottom diagram, it follows that $\omega = B/r$, and that $v = B/(rK_{CL})$; consequently, $er_p = v/K_p = B/(rK_{CL}K_p) = B/K$.

In most instances, the velocity lag does little harm. It only means that a commanded motion is executed with a slight delay. However, if several coordinate motions are executed simultaneously, it is important that the gains K be equal in all axes. In such a case of the lags being proportional to the component velocities, the direction of the resulting lag is in the tangent to the motion, and it does not affect the path of the motion. This will be discussed later. Actually, otherwise the velocity lag is useful; for example, if a motion is coming to an end, in a corner, it does not need to overshoot because the stop command comes while the table still has some distance to go, and it has time to decelerate.

10.6 ERRORS OF TWO-DIMENSIONAL TOOL PATH

In this section the positional servomechanism, defined in the preceding text as a second-order system, is investigated so as to determine the steady-state and transient errors of a path produced by simultaneous X and Y motions. One should still be aware of the fact that real servos are higher than second-order, and the results here are idealizations. The most significant deviation from reality of this treatment is due to flexibilities in feed drives, which are discussed in Section 10.8. The discussions in this part are carried out by means of examples using typical parameters of servomechanisms that are used in machine tools.

EXAMPLE 10.2 The Velocity Lag ▼

It has been explained, and it is obvious from the block diagram in Fig. 10.41, that in order to move the table with a steady velocity f (mm/sec), a positional error $e_{p,ss}$ must be maintained, which, after multiplication by K_p, becomes the control signal for the velocity servo. This error is called the velocity lag. It depends on the parameters of the servo according to Eq. (10.54), and it is proportional to the velocity $f_{\text{com}} = (f_{\text{act}})_{ss}$.

Most of our examples will use a feed rate of 25 mm/sec. This feed rate would typically correspond to end milling with a four-fluted cutter of diameter 20 mm using feed per tooth of $f_t = 0.1$ mm, and cutting speed $v = 235$ m/min; this results in a spindle speed of $n = 3750$ rpm. This may commonly be used for milling aluminum with an HSS cutter or milling steel with a carbide cutter. For milling aluminum, even much higher cutting speeds and feed rates could be used.

Evaluate the velocity lag $e_{p,ss}$ for a velocity of $f = 25$ mm/sec and the four combinations of servomechanism parameters of Example 10.1, as given in Table 10.2. Using Eq. (10.54), for $B = 25$ mm/sec, we have

$$K_{CL} = \frac{K_m K_a}{1 + K_m K_a H} = 2 \quad K = K_p K_{CL} r = 200, \qquad e_{p,ss} = 0.12 \text{ mm}$$

$$K_{CL} = 2, \quad K = 800, \quad e_{p,ss} = 0.03125 \text{ mm}$$

$$K_{CL} = 0.704, \quad K = 70.4, \quad e_{p,ss} = 0.355 \text{ mm}$$

$$K_{CL} = 50, \quad K = 5000, \quad e_{p,ss} = 0.005 \text{ mm}$$

The smallest "velocity lag" is obtained for servo d), which (see Fig. 10.42) had the worst transient behavior. The largest lag is found for servo c), with the least overshoot.

Assume now the servomechanisms in X and Y are being commanded to produce the same feed rate, 25 mm/sec. The corresponding commanded path is a straight line L under 45° in the X, Y coordinate system (see Fig. 10.44).

First, assume both servomechanisms have equal parameters, say those of c) above, and the motion is in the direction of positive X, positive Y. When the commanded position is Point 1, the x_{act} lags by 0.355 mm, and the y_{act} also lags by 0.355 mm, and the actual position is at Point 2, still exactly on the line L. Thus, the velocity lag does not in this case affect the contour of the part. Now assume that while the servo X has the parameters of case c), servo Y has the parameters b). The lag in Y is now only 0.03125 mm. When moving in the positive direction, the actual position is now at Point 3, and when moving in the negative direction, it is at Point 4. Due to the inequality of gains, or also of ζ and ω_n, in the two axes, the actual paths, now L' and L'', differ from the proper path L, and they are distant by an error Δ to one and the other side of the line L, respectively. As shown, the error is:

$$\Delta = (0.355 - 0.03125)\cos 45° = 0.229 \text{ mm}$$

What would be the magnitude of the error if servo X had the parameters of case a) and servo Y had parameters of case d)? The answer is

$$\Delta = (0.125 - 0.005)\cos 45° = 0.049 \text{ mm}$$

This example shows the importance of balancing the parameters of the servos in all axes. ▲

EXAMPLE 10.3 Ramp Input Response: Overshoot on Stopping a Motion ▼

Calculate and plot the motion of the table for servomechanisms *(b)* and *(c)* of Fig. 10.42 as a response to a command of motion with feed rate $f = 25$ mm/sec starting at $t = 0$. The command will stop after $t_1 = 0.070$ sec; this means that a total motion of 1.75 mm is commanded with the above velocity. Calculate the actual motion for a total of 0.140 sec.

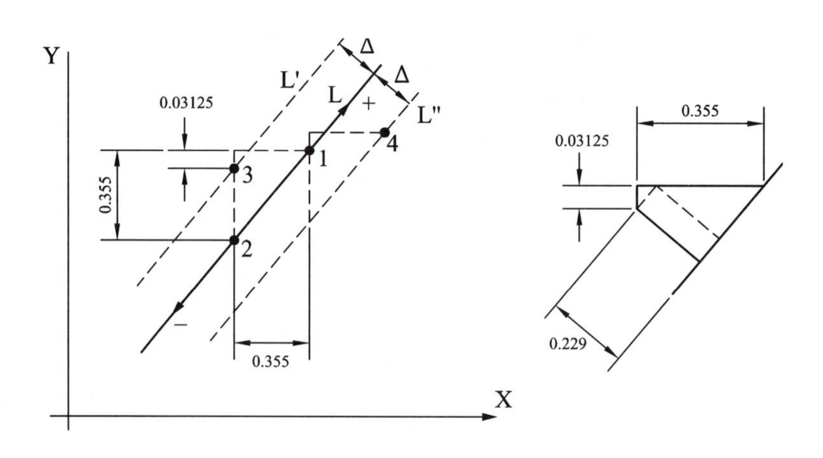

FIGURE 10.44
Form error due to unequal gains in two axes. If the gains in the two axes are not equal, the following errors in the two axes are not in the same proportion to their velocities, and the output path differs from the commanded track.

For the first period of $0 < t < t_1$ the motion will be the response to an input with a slope 25 mm/sec given by Eq. (10.50): for $0 < t < t_1$,

$$x_{act} = f\left[t + \frac{e^{-\zeta\omega_n t}}{\omega_d}\sin(\omega_d t + 2\phi) - \frac{2\zeta}{\omega_n}\right]$$

where

$$\phi = \tan^{-1}\frac{\sqrt{1-\zeta^2}}{\zeta}$$

The parameters of the two servomechanisms are given in the following table:

	ω_n (rad/sec)	ζ	$e_{p,ss}$ (mm)	ϕ (rad)	ω_d (rad/sec)	ω_d (Hz)
(b)	200	0.125	0.03125	1.318	193.65	30.82
(c)	100	0.71	0.355	0.781	70.42	11.2

After $t = t_1$ when the command stops, the motion can be represented as the sum of the response Eq. (10.50) and a response to a negative ramp $x = -f(t - t_1)$: for $t > t_1$,

$$x_{act} = f\left[t - (t - t_1) + \frac{e^{-\zeta\omega_n t}}{\omega_d}\sin(\omega_d t + 2\phi) - e^{-\zeta\omega_n(t - t_1)}\sin(\omega_d(t - t_1) + 2\phi)\right]$$

which is, for $t > t_1$,

$$x_{act} = f\left[t_1 + \frac{1}{\omega_d}\left(e^{-\zeta\omega_n t}\sin(\omega_d t + 2\phi) - e^{-\zeta\omega_n(t - t_1)}\sin(\omega_d(t - t_1) + 2\phi)\right]\tag{10.55}$$

After the commanded stop at $t = t_1$ the stopping transient sets in, which is essentially given by the third term in Eq. (10.55). After it dies out, we have, as commanded:

$$x_{act} = ft_1$$

The results are plotted in Fig. 10.45. All the calculations and plots were done for a commanded velocity $f_{com} = 25$ (mm/sec). If another feed rate $f_{com} = B$ (mm/sec) is commanded, all the responses, actually the values of x_{act} in Eqs. (10.50) and (10.55), should be multiplied by $B/25$. The stopping transient involves an overshoot of the commanded position. Its maximum occurs approximately at the time $t_m = t_1 + \Delta t$. The third term of Eq. (10.55) is a damped negative sine function. Its positive maximum is reached approximately when the argument reaches $(3\pi/2)$:

$$\omega_d\,\Delta t + 2\phi = \frac{3\pi}{2} \qquad \Delta t = \frac{3\pi/2 - 2\phi}{\omega_d}\tag{10.56}$$

The value of the overshoot is then obtained by introducing $t - t_1 = \Delta t$ into the third term of Eq. (10.55):

$$o = f\frac{e^{-\zeta\omega_n\Delta t}}{\omega_d}\tag{10.57}$$

It can be seen that a servo with smaller damping and higher natural frequency exhibits a smaller velocity lag but a larger overshoot than a servo with higher damping and lower natural frequency. ▲

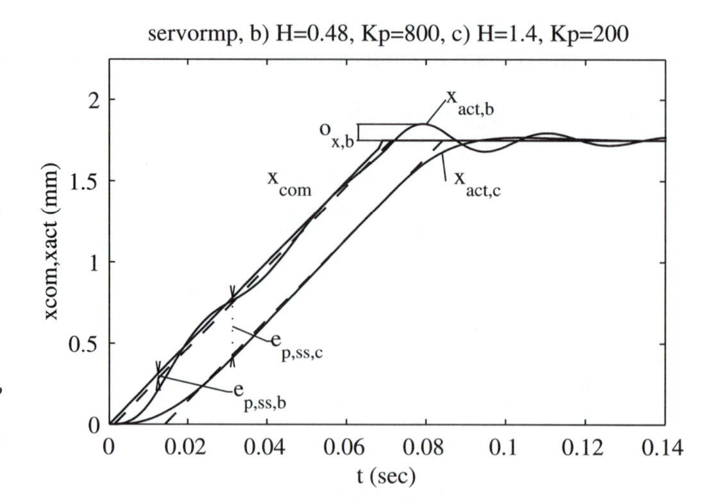

FIGURE 10.45
Response of the servo to a distance-limited ramp command. The plot has been obtained by time-domain simulation. It shows all the characteristics of the starting and stopping transients as well as of the steady-state motion in between. Cases *(b)* and *(c)* of Table 10.2 have been used. *(b)* has four times higher forward gain K_p than *(c)*, but only about half the velocity feedback gain H and consequently much less damping. Case *(b)* has less following error and more overshoot. As explained in the text, *(c)* is preferable for most applications.

EXAMPLE 10.4 Two-Coordinate Motions: Corner Motion ▼

Investigate the actual paths in (X, Y) coordinates, which are obtained using servomechanisms with various characteristics if a path is commanded in the form of a sharp corner. The results are plotted as indicated in Fig. 10.46. Starting from point S (0, 0) the Y motion is commanded with constant velocity $f = 25$ mm/sec (positive ramp). The y_{com} stops increasing when it reaches the value $y_{\text{com}} = 2.5$ mm (negative ramp). The actual displacement y_{act} will, of course reach $y_{\text{act}} = 2.5$ mm with a delay and will overshoot as shown in Fig. 10.45. At the same instant ($t =$ 0.1 sec) the X-motion command is issued, with the same feed rate $f = 25$ mm/sec, and this command stops when $x_{\text{com}} = 2.5$ mm. The actual motion will overshoot and continue for some time after that.

Actually, this is an example of linear interpolation in the command generation. Choose $\mathrm{d}t = 0.0005$ sec. The command will be generated in loops numbered n, and in the same loops the simulation of the X and Y servos will be carried out, as described in Eq. (10.46). The part of the command generation is

- Initial values and data: $n = 1$, $t_1 = 0$, $x_{\text{com},1} = 0$, $y_{\text{com},1} = 0$, $\mathrm{d}x = 0.0125$, $\mathrm{d}y = 0.0125$
- Command generation: Y motion: $n = 1$ to 200, $x_{\text{com,n}} = 0$, $y_{\text{com,n}} = n\,\mathrm{d}y$
  X motion: $n = 201$ to 400, $x_{\text{com,n}} = n - 200)\mathrm{d}x$,
  $y_{\text{com,n}} = 2.5$

After the whole motion is commanded, it is necessary to add some time for sufficient decay of the overshoot in X. Let us add 0.100 sec: $n = 401$ to 600, $x_{\text{com,n}} = 2.5$, $y_{\text{com,n}} = 2.5$. Generate the $x_{\text{act,n}}$ and $y_{\text{act,n}}$ values using the above command formulations and the servo simulations as derived in Eq. (10.46). Plot y_{act} versus x_{act}. Carry out for all four servo specifications of Table 10.2. The same specifications are used for the X and Y motions.

The results of the simulations for the same parameters as used in generating Fig. 10.42 are presented in Fig. 10.46. Note that none of the servomechanisms can execute a sharp corner truly enough. Servos *(a)*, *(b)* and *(d)* overshoot by 0.1–0.23 mm. The actual values of the overshoots correspond to those derived in Eq.

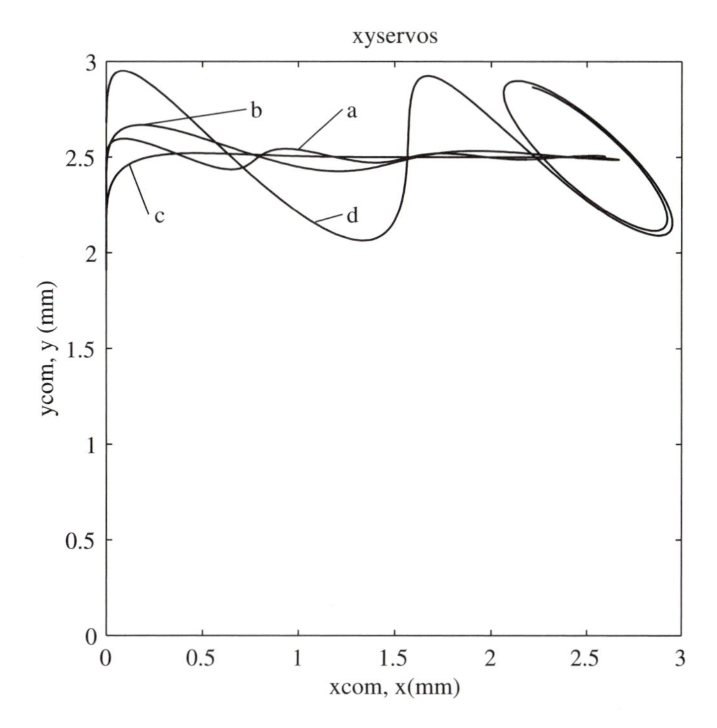

FIGURE 10.46
Results of Example 10.4. Motion through a corner of *X, Y* servos with identical parameters. Four different specifications according to Table 10.2 were used, the same as those that produced the motions in Fig. 10.42. Case *(a)* has moderate overshoot; *(b)* has more overshoot and slow decay of vibration; *(c)* has no overshoot and reasonably small round-off of the corner; *(d)* is absolutely unacceptable, with high overshoot and low vibration-decay rate. Obviously, *(c)* is best.

(10.56). With servo *(d),* which has the lowest damping $\zeta = 0.11$, the surface is rather strongly undulated for the whole length of the path of 2.5 mm after the corner. Servos *(c),* which have the highest damping of $\zeta = 0.67$ and the largest velocity lag of $e_{p,ss} = 0.38$ mm (see Eq. [10.54]), produce a large radius of about 0.12 mm inside of the corner; it is said that they "undershoot."

These errors would exceed the tolerances of most workpieces. In milling aluminum the feed rate may often be much higher than the 25 mm/sec used in our example. In high-speed milling it might be as high as 500 mm/sec. Correspondingly, the errors would be of the order of 2–5 mm. Servos with much higher gains must then be used.

On the other hand, there is hardly ever any need to execute a sharp corner, as explained in Fig. 10.47. Diagram *(a)* is the case of milling an outer sharp corner. The center of the milling cutter moves around this corner on a circular path with a radius equal to the radius of the cutter, path A. Alternatively, it may be programmed to move on path B. In this case the cutter stops generating the workpiece surface at Point 1 and does not start doing it again until Point 2. Any transients around Point 3 do not affect the machined surface as long as they die out before Point 2. For an inner corner there always has to be a radius provided. It is then recommended to use a cutter with a radius smaller than that of the workpiece, $R < r$. The resulting cutter center path is then again circular.

Most important, commands of the kind applied in the example where starts and stops change from zero to full velocity and back are never used in real NC programming. They would represent infinite accelerations and decelerations. Finite values of these are inserted in the commands, as is shown later in Fig. 10.74. In this way a rather accurate inner corner can be achieved even with a cutter

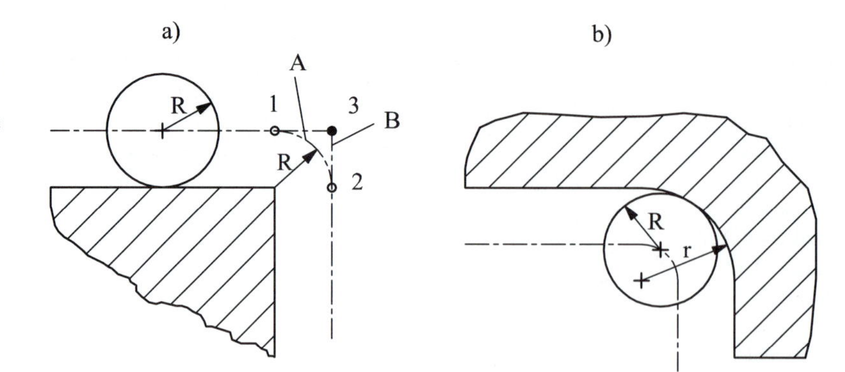

FIGURE 10.47
Strategies for programming corner motions. Effects of over- or undershoots can be moderated by a quarter-circular instead of sharp-corner command.

that has the same radius as that of the corner. However, in pocketing operations, a large number of inner corners must be passed through and, with severe slowdowns, considerable time loss may result. Therefore, it is necessary to improve the servos as much as possible and to introduce various feedback or feedforward compensation techniques so as to achieve high accuracies in corners, in addition to high feed rates. Cross-control techniques are currently being developed. Some of these techniques are explained in Section 10.8. ▲

EXAMPLE 10.5 Effect of the Dead Zone ▼

Let us now discuss the "dead zone." Upon a commanded reversal of motion, the actual motion remains "dead" over a certain zone *D*. This may be caused by friction in guideways combined with the flexibility of the thrust bearing of the leadscrew, or simply by play or backlash. In our representation it is considered as a play *D* in the drive between the positional feedback and the table. Thus it is located outside of the feedback loop. It might be located within the feedback loop if the positional transducer is directly on the table. In this case it causes problems with the stability of the servo. Here only the former case with the dead zone outside the loop is considered.

In order to describe the behavior of the link between x_{act} and x_{table} through the dead zone *D,* an auxiliary variable z is introduced. Its meaning is shown in Fig. 10.48. The initial position in the dead zone will be taken such that $z = 0$. This is the position indicated. Let us assume, for the start, that x_{act} is increasing, that is, moving to the right. The table will not follow the movement, and we will have

$$z = z + f\mathrm{d}t$$

There are three possible situations leading to three corresponding expressions for x_{table}:

1. $0 < z < D$: no change in x_{table}
2. $z \geq D$: hold $z = D$, and $x_{table} = x_{act} - D$
3. After reversal, for $f < 0$, we may have $z < 0$: hold $z = 0$, and $x_{table} = x_{act}$

Write the corresponding program and obtain the response to a step input $x_{com} = 1.0$ mm for the following servomechanism in two alternatives. (1) without a dead zone: $K_p = 1000$ (V/mm), $K_{CL} = 10$ (rev/sec/V), $\tau_{CL} = 0.050$ (sec), $r = 0.1$ (mm/rev); and (2) same as (1) but with a dead zone of $D = 0.15$ mm.

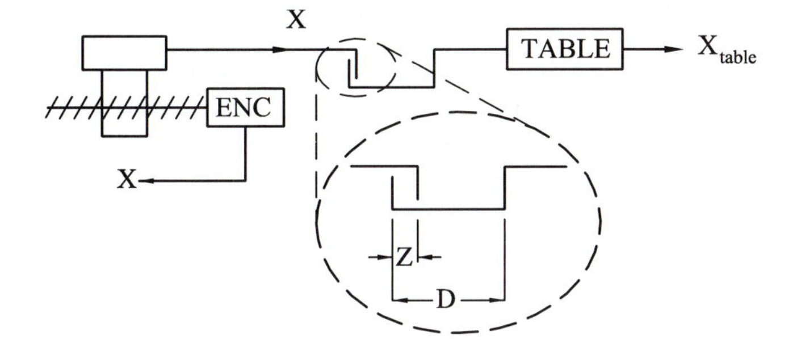

FIGURE 10.48 Definition of the dead zone as a difference in position reached from opposite directions. It may be the result of either backlash in the transmission or deformation due to friction in guideways.

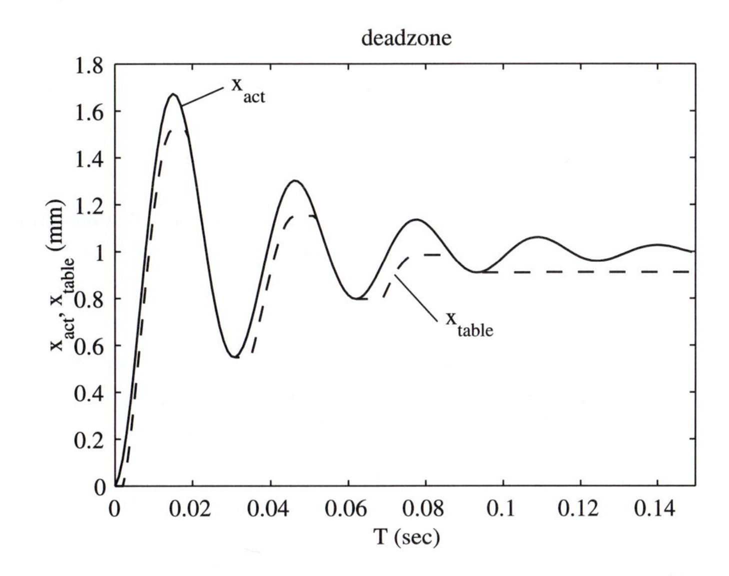

FIGURE 10.49 Result of Example 10.5. Step input response of a servo with a dead zone.

Plot the variables x_{act} and x_{table} versus time t for $0 < t \leq 0.150$ sec. Choose $dt = 0.001$ sec. The plots are presented in Fig. 10.49.

The curve x_{act} is identical with curve *(b)* in Fig. 10.42. The curve x_{table} corresponds to a dead zone D in the drive of the table. The table motion does not start until x_{act} reaches the value $D = 0.15$ mm. From there on it lags by D behind x_{act} until the reversal of the x_{act} motion when it stops. After x_{act} has moved back by D, the two motions go together until the next reversal, when x_{table} stops until x_{act} has moved forward by D. On the next reversal of x_{act}, the motion x_{table} stops for good because x_{act} has overshot by less than D and will now continue moving within the dead zone D. Later on, we will see how such a dead zone distorts the shape of a circle. ▲

EXAMPLE 10.6 Distortions of a Continuous Path ▼

In this exercise it is necessary to combine the simulation of the command generator with the simulation of the X and Y servomechanisms to plot out the actual path consisting of straight and circular sections and see the errors due to transients in corners and errors due to unequal gains and their effect on the distortion of a circular path as well as errors due to dead zones.

Simulation of Linear and Circular Interpolation

The purpose of the command generator (interpolator) is to repeatedly determine the increments in the individual coordinates over constant increments of time Δt. Our exercise will be limited to two coordinates X and Y. The tool path consists of linear and circular sections.

The *linear interpolation* is very simple. The input consists of the coordinates (x_s, y_s) of the starting point and (x_e, y_e) of the end point of the segment and of the desired feed rate f (mm/sec). First, the length of motion L and the component velocities of feeds f_x and f_y are determined:

$$L = \sqrt{(x_e - x_s)^2 + (y_e - y_s)^2} \tag{10.58}$$

$$f_x = \frac{f(x_e - x_s)}{L} \tag{10.59}$$

$$f_y = \frac{f(y_e - y_s)}{L} \tag{10.60}$$

Next the constant increments Δx and Δy are determined:

$$\Delta x = f_x \, \Delta t \tag{10.61}$$

$$\Delta y = f_y \, \Delta t \tag{10.62}$$

These increments are expressed to a finite number of digits; thus for every step an error is committed, and these accumulate over the total motion over the section. The maximum error per step is equal to the unit value U of the last digit, and there are a total of

$$m = \frac{L}{f\Delta t}$$

increments in the segment. In the worst case the total error may be up to

$$e_x = e_y = mU$$

For instance, if the length of the segment is 50 mm, the feed rate is 2.5 mm/sec, the time interval is $\Delta t = 4$ msec, and the computation is carried out with $U = 1 \times 10^{-8}$ mm, the maximum error will be

$$e_x = e_y = \frac{LU}{f\Delta t} = 50 \times \frac{10^{-8}}{2.5 \times 0.004} = 5 \times 10^{-5} \text{ mm.}$$

The circular interpolation is based on the diagram in Fig. 10.50. The task is to determine the increments $\Delta x_n, \Delta y_n$ when proceeding from point (x_n, y_n) to point (x_{n+1}, y_{n+1}) in one step of the interpolation. This step is executed in the time increment Δt. The coordinates of the center of the circle are (x_c, y_c), its radius is R, and the peripheral velocity of the motion (feed rate) is f. In the case shown the motion is counterclockwise (CCW). The polar coordinates are (R, ϕ), and the increment of the angle ϕ over the interval Δt is $\Delta\phi$.

The procedure can best be carried out in the following coordinates

$$x^* = (x - x_c), y^* = (y - y_c) \tag{10.63}$$

which are relative to the center C.

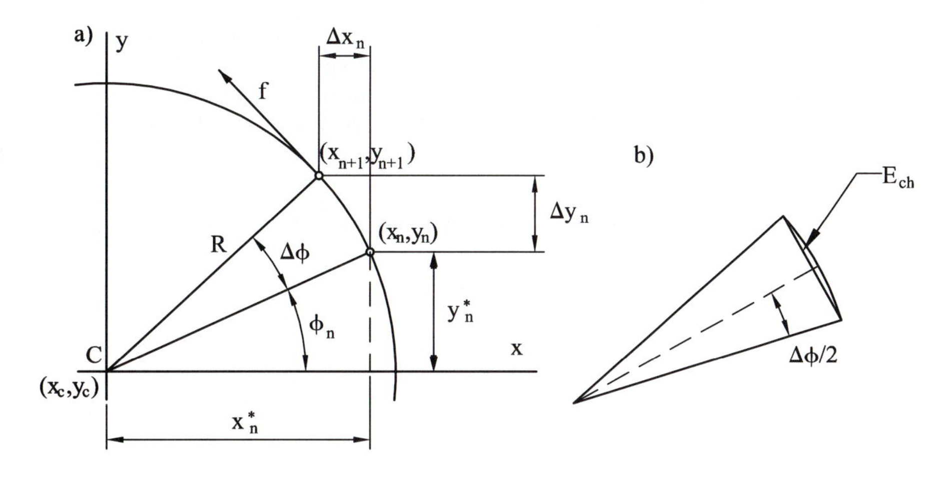

FIGURE 10.50 Generation of the circular interpolation command. Geometry for deriving the formulas for increments in x and y and also for defining the chordal error.

We have

$$x_n^* = R\cos\phi_n \qquad y_n^* = R\sin\phi_n$$

$$x_{n+1}^* = R\cos(\phi_n + \Delta\phi) = R(\cos\phi_n \cos\Delta\phi$$

$$-\sin\phi_n \sin\Delta\phi) = x_n^*\cos\Delta\phi - y_n^*\sin\Delta\phi \tag{10.64}$$

$$y_{n+1}^* = R\sin(\phi_n + \Delta\phi) = R(\sin\phi_n\cos\Delta\phi$$

$$+\cos\phi_n\sin\Delta\phi) = x_n^*\sin\Delta\phi + y_n^*\cos\Delta\phi$$

For small values of $\Delta\phi$, the first terms and first and second terms only, respectively, of the expansions of the sine and cosine functions can be taken:

$$\sin\Delta\phi = \Delta\phi \qquad (\text{error} < (\Delta\phi)^3/6)$$

$$\cos\Delta\phi = 1 - (\Delta\phi)^2/2 \qquad (\text{error} < (\Delta\phi)^4/24) \tag{10.65}$$

The increments Δx_n and Δy_n can be expressed as follows:

$$\Delta x_n = x_{n+1} - x_n = x_{n+1}^* - x_n^* = x_n^*(\cos\Delta\phi - 1) -$$

$$- y_n^*\sin\Delta\phi = -x_n^*(\Delta\phi)^2/2 - y_n^*\Delta\phi \tag{10.66}$$

$$\Delta y = y_{n+1} - y_n = y_{n+1}^* - y_n^* = x_n^*\sin\Delta\phi$$

$$+ y_n^*(\cos\Delta\phi - 1) = x_n^*\Delta\phi - y_n^*\frac{(\Delta\phi)^2}{2} \tag{10.67}$$

We have

$$x_{n+1} = x_n + \Delta x_n$$

$$y_{n+1} = y_n + \Delta y_n \tag{10.68}$$

Thus the command generation successively proceeds from the starting point S of the circular section with coordinates (x_s, y_s) to its end point with coordinates (x_e, y_e).

Two types of errors are associated with this procedure. One is due to the substitution of the truncated expansions for sine and cosine, and the other is due to the command actually connecting the two subsequent points by a chordal straight line instead of a circular arc. Let us evaluate their significance. ▲

---

1. Error Due to Approximate Expressions for Sine and Cosine

$$\begin{aligned} R_{n+1}^2 &= x_{n+1}^{*2} + y_{n+1}^{*2} \\ &= (x_n^* + \Delta x_n)^2 + (y_n^* + \Delta y_n)^2 \\ &= \left(x_n^* - x_n^* \frac{(\Delta\phi)^2}{2} - y_n^* \Delta\phi\right)^2 + \left(y_n^* + x_n^* \Delta\phi - y_n^* \frac{(\Delta\phi)^2}{2}\right)^2 \\ &= R_n^2 \left(1 + \frac{(\Delta\phi)^4}{4}\right) \end{aligned} \tag{10.69}$$

$$R_{n+1} = R_n \left(1 + \frac{(\Delta\phi)^4}{4}\right)^{1/2} \simeq R_n \left(1 + \frac{(\Delta\phi)^4}{8}\right) \tag{10.70}$$

In each step the radius increases by $R_n\ (\Delta\phi^4/8)$. The number of steps per circle is $2\pi/\Delta\phi$. Thus, the total error of the radius per full circle is

$$E_r = \left(\frac{2\pi}{\Delta\phi}\right)\left(\frac{R(\Delta\phi)^4}{8}\right) = \frac{\pi R(\Delta\phi)^3}{4} \tag{10.71}$$

Example

Let us choose the number of steps per full circle $n = 1000$. Correspondingly, $\Delta\phi = 2\ \pi/1000$.

$$E_r = \pi R \left(\frac{2\pi}{1000}\right)^3 /4 \simeq 2 \times 10^{-7} R$$

If $R = 100$ mm, $E_r = 2 \times 10^{-5}$ mm.

This is extremely small. Usually, the number of steps per circle is much larger than 1000 and consequently $\Delta\phi$ and E_r are still much smaller.

2. Chord-Height Error

Refer to Fig. 10.50b. The chord-height error E_{ch} is the difference between the radius and the distance of the chord from the circle center in the middle of the step, at $\Delta\phi/2$:

$$E_{ch} = R - R \cos \frac{\Delta\phi}{2}$$

Using Eq. (10.65),

$$E_{ch} = R\left(1 - 1 + \frac{(\Delta\phi)^2}{8}\right) = R\left(\frac{(\Delta\phi)^2}{8}\right) \tag{10.72}$$

Example

For the same parameters as in the above case,

$$E_{ch} = R\frac{4\pi^2}{8} \times 10^{-6} \simeq 5 \times 10^{-6} R$$

for $R = 100$ mm, $E_{ch} = 0.0005$ mm. The chordal error is larger than that due to the approximations of sine and cosine and is decisive for the choice of $\Delta\phi$.

In order to simulate the interpolator on a computer, Eqs. (10.61), (10.62) are used for linear interpolation, and Eqs. (10.66), (10.67), (10.68) are used for circular interpolation. The choice of the magnitude of the interpolation steps is dictated by, on one hand, Eqs. (10.71), (10.72) for the magnitudes of errors and, on the other hand, by the response time of the servo. The time increments must be small enough so as not to cause non-uniform motions.

3. Simulation of the Interpolation for the Tool Path as Given in Fig. 10.51

The path consists of three segments. The input data are $\Delta t = 0.001$ sec, $f = 200$ mm/sec, number of segments $m = 3$. Section data: the number of steps in each segment is N_i:

$$i = 1, x_{s1} = 0, y_{s1} = 0, x_{e1} = 25, y_{e1} = 60, N_1 = 650$$
$$i = 2, x_{s2} = 25, y_{s2} = 60, x_{e2} = 50, y_{e2} = 60, N_2 = 250$$
$$i = 3, x_{s3} = 50, y_{s3} = 60, x_{e3} = 50, y_{e3} = 60.$$
$$x_c = 100, y_c = 60, R = 50; N_3 = 3142$$

Initial values are $x_1 = 0$, $y_1 = 0$. The total length of the path is $L_{tot} = (25^2 + 60^2)^{1/2} + 25 + 100\pi = 404.16$ mm, $\Delta s = f\Delta t = 0.2$ mm, $n_{tot} = 2021$.

Command the motion along the path as given in Fig. 10.51. The command generation routines are immediately followed, in every loop, by the routines simulating the

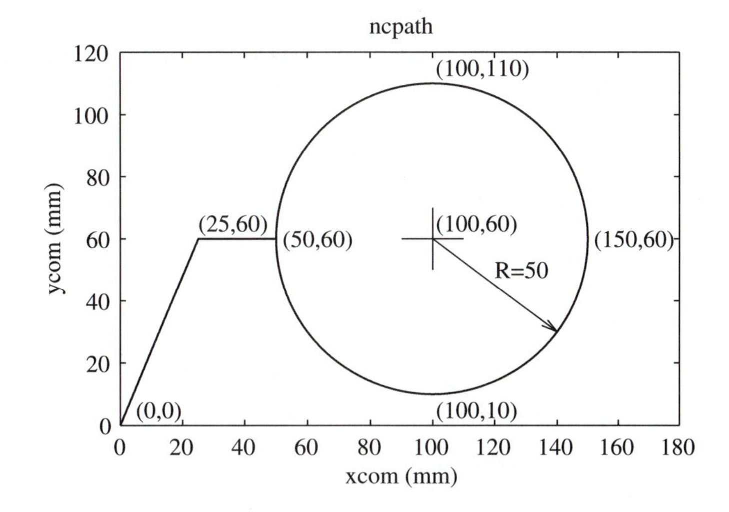

FIGURE 10.51
Geometry of the path in task Example 10.6.

servomechanisms. When the whole path has been commanded, the servos will not yet have finished the execution of the whole motion, and even after they have reached the final positions, it is necessary to continue the simulation to allow for the stopping transient to decay. Correspondingly, we hold the commands at the final position and continue for an additional period of about 0.1 sec. This is equivalent to choosing the total number of simulation loops $n_{\text{tot}} = 4242$. Carry out the above task for four different specifications of the servomechanisms:

(a) Both the X and Y servomechanisms identical, with these parameters: $K_p = 400$ (V/mm), $K_m = 10$ (rev/sec/V), $H = 0.5$ (V/rev/sec), $K_a = 1$, $r = 0.1$ (mm/rev), $\tau_{\text{m}} = 0.05$ (sec), no dead zones. These servos are reasonably well damped, $\zeta = 0.67$.

(b) Both the X and Y servos identical, with very low damping: $\zeta = 0.11$, with all parameters the same as in *(a)*, but $H = 0$.

(c) The two servos with different positional gains and correspondingly different velocity lags: the X servo with parameters as in *(a)* (the velocity lag $e_{p,ss} = 0.015f$), and the Y servo with all other parameters equal except for $K_p = 1000$, and $H = 0.25$ ($e_{p,ss} = 0.0035f$).

(d) Both servos equal to *(a)*, except that both of them have dead zones, $D_x = D_y = 8$ mm. These dead zones are unrealistically large in order to clearly show the corresponding distortion of the circle.

The listing of the corresponding Matlab program "ncpath" is printed in Fig. 10.52. The results are presented in Fig. 10.53a–d.

```
function[xcom,ycom,x,y,t,xa,ya,xb,yb,xc,yc,xtab,ytab]=ncpath
%The XY servos are commanded along the path defined in Fig.10.51
%four different sets of servo parameters are alternatively used
%f is feed rate(mm/sec),xstar and ystar are relative coordinates
%in the circular interpolation according to Eq.10,delfi is the angular step
%simulation of the path is formulated according to Eq.10.46
dt=0.001;f=200;t(1)=0;x(1)=0;dx=0;y(1)=0;dy=0;
delfi=2*pi/1571;xstar=50;ystar=60;cx=100;cy=60;
xcom(1)=0;ycom(1)=0;omx=0;omy=0;zx=0;zy=0;xtab(1)=0;ytab(1)=0;
for k=1:4
%first, regular servo
  if k==1
    Kpx=400;Kpy=400;Km=10;Ka=1;Hx=0.5;Hy=0.5;
    r=0.1;tam=0.10;D=0;
%second,no velocity feedback, Hx=Hy=0
  elseif k==2
    Kpx=400;Kpy=400;Km=10;Ka=1;Hx=0;Hy=0;
    r=0.1;tam=0.10;D=0;
%third, unequal gains Kpx and Kpy and different tacho
%calibrations Hx,Hy
  elseif k==3
    Kpx=400;Kpy=1000;Km=10;Ka=1;Hx=0.5;Hy=0.20;
    r=0.1;tam=0.10;D=0;
%fourth, introduce dead zones D=6 mm
  else
    Kpx=400;Kpy=400;Km=10;Ka=1;Hx=0.5;Hy=0.5;
    r=0.1;tam=0.10;D=6;
end
```

FIGURE 10.52
Listing of the Matlab program "ncpath" for simulating the motions according to the task of Example 10.6.

FIGURE 10.52
(continued)

```
    for n=1:2121
%formulate command
      t(n+1)=t(n)+dt;
%first the initial sloped straight section
      if n<=325
        dxcom=0.2*25/65;dycom=0.2*60/65;
%next the straight horizontal section
      elseif n>325 & n<=450
        dxcom=0.2;dycom=0;
%then the circular path
      elseif n>450 & n<=2021
          dxcom=-xstar*delfi^2/2-ystar*delfi;
          dycom=xstar*delfi-ystar*delfi^2/2;
%and final dwell between n=2021 and n=2121
        else
          dxcom=0;dycom=0;
        end
        xcom(n+1)=xcom(n)+dxcom;
        ycom(n+1)=ycom(n)+dycom;
        xstar=xcom(n+1)-cx;
        ystar=ycom(n+1)-cy;
%simulate the XY servos acc. Eq. 10.46 for the four different specifications,
%denote coordinates xa,ya; xb,yb; xc,yc;
errx=xcom(n)-x(n);
erry=ycom(n)-y(n);
vx=errx*Kpx;
vy=erry*Kpy;
ex=(vx-Hx*omx)*Ka;
ey=(vy-Hy*omy)*Ka;
domx=(ex*Km-omx)/tam;
domy=(ey*Km-omy)/tam;
omx=omx+domx*dt;
omy=omy+domy*dt;
x(n+1)=x(n)+r*omx*dt;
y(n+1)=y(n)+r*omy*dt;
if k==1
xa(n)=x(n);ya(n)=y(n);
elseif k==2
xb(n)=x(n);yb(n)=y(n);
elseif k==3
xc(n)=x(n);yc(n)=y(n);
else
%now comes the case with the dead zone resulting in coordinates xtab,ytab
  zx=zx+r*omx*dt;
if zx>=D
  zx=D;xtab(n+1)=x(n+1)-D;
elseif zx<0;
  zx=0;xtab(n+1)=x(n+1);
else
  xtab(n+1)=xtab(n);
  end
zy=zy+r*omy*dt;
if zy>=D
  zy=D;ytab(n+1)=y(n+1)-D;
  elseif zy<0
    zy=0;ytab(n+1)=y(n+1);
  else
    ytab(n+1)=ytab(n);
    end
end
end
end
%plot(xa,ya,xb,yb,xc,yc,xtab,ytab)
```

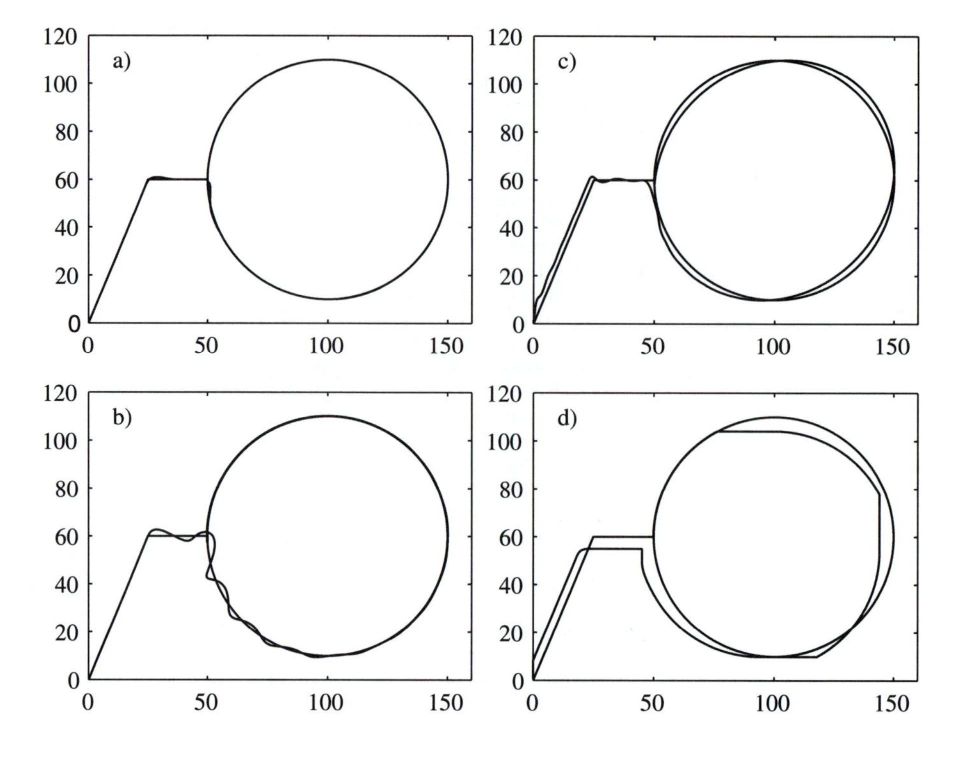

FIGURE 10.53
Results of Example 10.6. Servos with a) appropriate damping; b) little damping; c) different gains in the two axes causing nonmutually proportional velocity lags; instead of being circular the path is elliptical; d) dead zones produce flats at extremes in each axis.

The path as obtained in Fig. 10.53a corresponds to well-damped ($\zeta = 0.67$) servos with equal gains. The distortions are very small, and they are limited to sharp corners. In this example they extend rather far beyond the corners because of the very high feed rate. In most cases they would die out over a short distance.

The path obtained in *(b)* results from low damping of both servos ($\zeta = 0.11$). Correspondingly the overshoots and the transients in the sharp corners are excessive. Servomechanisms with low damping should normally not be used.

Path *(c)* corresponds to the two servos having different gains and producing different velocity lags. This leads to a sideways error on the straight path and to an elliptical distortion on the circular path. It is one of the basic rules to adjust the lag coefficients (the gains) on all axes to equal values.

Graph *(d)* shows the path for the case of equal dead zones in both axes, outside the positional loops. A typical distortion in this case is the occurrence of flats at the extreme coordinate values of the circle. The design of the drives and guideways of the machine tool should minimize the dead zones.

10.7 ADAPTIVE CONTROL FOR CONSTANT FORCE IN MILLING

10.7.1 Analysis of Stability

In this exercise, classical control system theory is used with gain-margin stability analysis based on the phase and magnitude of the open-loop transfer function.

The system to be analyzed is the one briefly described in relation to Fig. 10.21, which is now redrawn in a more detailed block diagram in Fig. 10.54a, where the milling process is also represented by a corresponding block. The part from the command displacement x_{com} to the displacement of the table x_{act} is the standard positional servomechanism, as discussed in the preceding exercises. The block of the milling process that converts the table position x_{act} into the milling force F_{act} is based on the following formula:

$$F_{act}(t) = CBf_t(t) \tag{10.73}$$

In this formula we do not take into account the periodic character of the milling force; we may be considering full-immersion slotting with a four-fluted cutter. Parameter B is the axial depth of cut, and the coefficient C expresses the workpiece material as well as the diameter of the cutter. The variable $f_t(t)$ is the feed per tooth, which can be expressed as the difference in the table positions over a time period τ_t during which two subsequent teeth pass (the tooth period). Then we have

$$F_{act}(t) = CB\left[x_{act}(t) - x_{act}(t - \tau_t)\right] \tag{10.74}$$

Laplace transforming Eq. (10.74),

$$F_{act}(s) = CBX_{act}(s)[1 - \exp(-\tau_t s)] \tag{10.75}$$

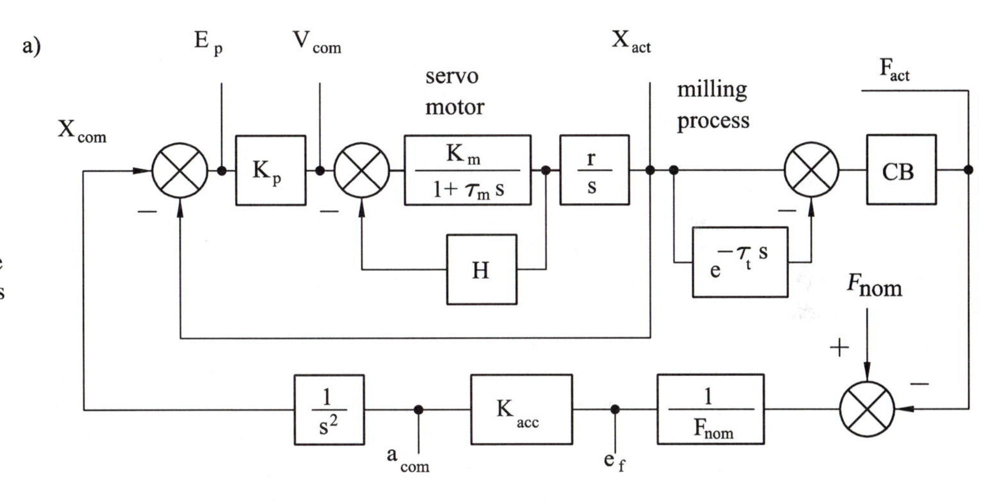

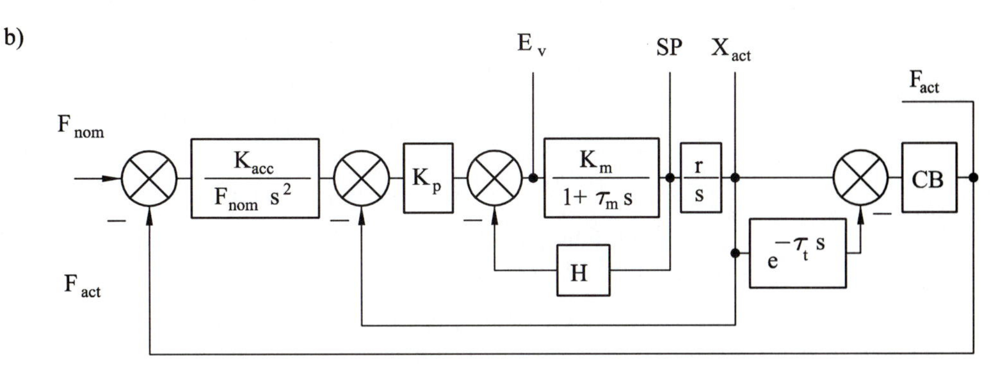

FIGURE 10.54 Block diagram of adaptive control for constant milling force. a) Position as input and force as output. The force F_{act} in the milling process is proportional to "chip load"' as the difference between the table positions x_{act} over a cutter tooth period τ_t. F_{act} is compared with the desired force F_{nom}, and the difference related to F_{nom} expresses force error e_f that requires a "change" of feed rate, that is, an acceleration a_{com}. After double integration this becomes the position commanded and closes the loop. b) The same block diagram is turned around for F_{nom} as input.

Once the milling force is obtained, the force error is established:

$$e_f = \frac{F_{\text{nom}} - F_{\text{act}}}{F_{\text{nom}}} \tag{10.76}$$

where F_{nom} is the desired cutting force, which should be kept constant by adapting the feed rate and, correspondingly, the feed per tooth f_t. This means changing the velocity of the motion.

This request is expressed by setting an acceleration proportional to the force error:

$$a_{\text{com}} = K_{\text{acc}} e_f \tag{10.77}$$

and obtaining the command feed rate f_{com} and, eventually, the commanded displacement x_{com} by double integration $1/s^2$.

It is more convenient to consider the nominal force F_{nom} as the input to the system and turn the block diagram correspondingly around, as done in Fig. 10.54b. The positional and velocity closed loops as well as the milling-process loop are internal loops to the overall closed-loop, force-controlled, adaptive system.

The task now is to investigate the stability of the system. For this purpose the open-loop transfer function (OLTF) has to be formulated and its phase and magnitude relationship investigated. The OLTF is

$$\frac{F_{\text{act}}(s)}{F_{\text{nom}}(s)} = \left(\frac{K_{\text{acc}}}{F_{\text{nom}} s^2}\right)\left(\frac{\omega_n^2}{s^2 + 2\zeta\omega_n s + \omega_n^2}\right)\{CB\,[1 - \exp(-\tau_t s)]\}$$

$$[a] \quad \times \quad [b] \quad \times \quad \{c\} \tag{10.78}$$

where the factor b is the closed-loop transfer function of the positional servo, and the factor c is the transfer function of the milling process.

Setting $s = j\omega$, the magnitudes and phase angles of the individual factors are

$$|a| = \frac{-K_{\text{acc}}}{F_{\text{nom}}\omega^2} \qquad \phi_a = -180°$$

$$|b| = \left|\frac{\omega_n^2}{\omega_n^2 - \omega^2 + 2j\zeta\omega\,\omega_n}\right|$$

$$= \frac{\omega_n^2}{[(\omega_n^2 - \omega^2)^2 + (2\zeta\omega\,\omega_n)^2]^{1/2}}, \qquad \phi_b = \tan^{-1}\frac{-2\zeta\omega\,\omega_n}{\omega_n^2 - \omega^2}$$

$$|c| = CB|1 - \exp(-j\tau_t\omega)\,| = CB|1 - \cos\tau_t\omega + j\sin\tau_t\omega|$$

$$= CB\,(2 - 2\cos\tau_t\omega)^{1/2}$$

$$\phi_c = \tan^{-1}\frac{\sin\tau\omega}{1 - \cos\tau\omega} \tag{10.79}$$

Now, first, the frequency ω_{lim} has to be determined, at which the total phase shift becomes $-180°$:

$$\phi = \phi_a + \phi_b + \phi_c = -180° \tag{10.80}$$

and, subsequently, the gain K_{acclim} for which the OLTF is

$$\left|\frac{F_{\text{act}}(j\omega)}{F_{\text{nom}}(j\omega)}\right| = 1 \tag{10.81}$$

EXAMPLE 10.7 Limit of Stability of the Adaptive Control System

Let us choose the parameters of the servomechanism as in case *(a)* of Example 10.2 and of Table 10.2, case *(a)*:

$$\omega_n = 100 \text{ rad/sec} \qquad \zeta = 0.25$$

The parameters of the milling process will correspond to milling low-carbon steel with an end-milling cutter diameter 20 mm at an axial depth of cut $B = 15$ mm. The corresponding value of the coefficient C in the force formula (10.73) may, in this case, be $C = 2000$ N/mm^2, and a reasonable nominal force will be $F_{\text{nom}} =$ 6000 N.

The tooth period is obtained by assuming four teeth and a cutting speed $v =$ 45 m/min $=$ 750 mm/sec:

$$\tau_t = \frac{\pi d}{4v} = 0.021 \text{ sec}$$

Using Eqs. (10.80) and (10.79),

$$-180° = -180° + \underbrace{\tan^{-1}\frac{-2\zeta\omega\,\omega_n}{\omega_n^2 - \omega^2}}_{\phi_b} + \underbrace{\tan^{-1}\frac{\sin \tau_t\omega}{1 - \cos \tau_t\omega}}_{\phi_c}$$

$$\phi_b + \phi_c = 0 \tag{10.82}$$

where, for the given parameters,

$$\phi_b = \tan^{-1}\frac{-50\,\omega}{100^2 - \omega^2} \tag{10.83}$$

$$\phi_c = \tan^{-1}\frac{\sin 0.021\,\omega}{1 - \cos 0.021\,\omega} \tag{10.84}$$

We must keep in mind that the argument of sine and cosine in Eq. (10.84) is in (rad).

In order to find the ω_{lim} for which Eq. (10.82) is satisfied the values of ϕ_b and ϕ_c are plotted in Fig. 10.55 as functions of ω. The value of ω_{lim} is found as

$$\omega_{\text{lim}} = 77.13 \text{ rad/sec} \tag{10.85}$$

For stability, the gain at this frequency should be less than 1. The limit value of K_{acc} is obtained from Eq. (10.81):

$$|\,G(j\omega)\,| = |\,F_{\text{act}}\,(j\omega)/F_{\text{nom}}\,(jw)| = K_{\text{acc}}\,|a|\,|b|\,|c|$$

where

$$|a| = \frac{CB/F_{\text{nom}}}{\omega^2} \tag{10.86a}$$

$$|b| = \frac{\omega_n^2}{[(\omega_n^2 - \omega^2)^2 + (2\zeta\,\omega\,\omega_n)^2]^{1/2}} \tag{10.86b}$$

$$|c| = (2 - 2\cos \tau_t\,\omega)^{1/2} \tag{10.86c}$$

and, for limit of stability

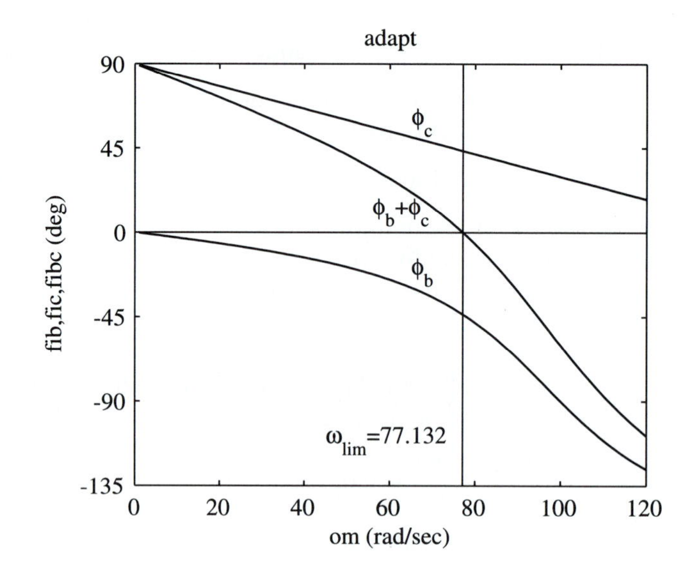

FIGURE 10.55
Bode plot of phase shifts for analyzing stability of the adaptive control system. Critical frequency for stability limit $\omega_{\lim} = 77$ rad/sec is found as the basis for determining the limit gain.

$$| G(\omega_{\lim}) | = K_{acc} \times |a\,(\omega_{\lim})| \times |b\,(\omega_{\lim})| \times |c\,(\omega_{\lim})| = 1$$

$$\text{or} \qquad |a_{\lim}| \;\; |b_{\lim}| \;\; |c_{\lim}| = \frac{1}{K_{acc}} \tag{10.86}$$

At $\omega = \omega_{\lim} = 77.13$ rad/sec, the value of $|abc| = 0.00218$ is found. This determines the value of $K_{acc\ \lim} = 459$ mm/sec^2. ▲

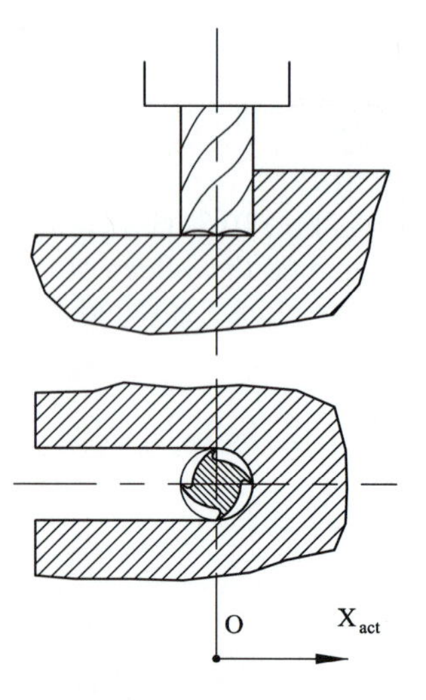

FIGURE 10.56
Starting position for the process in Example 10.8. This is a difficult choice because there is no cutting-in period; as soon as the tool (or table) starts to move it is in full cut.

EXAMPLE 10.8 Simulation of a Stable and an Unstable A/C System ▼

Now, as the limit gain for the A/C system of Example 10.7 has been established, it will be interesting to simulate the action of the system under stable and unstable conditions. This will be done for a rather simple case of a straight slotting cut that has come to rest and will be restarted under the A/C control, as indicated in Fig. 10.56. While the cut was stopping, the cutter was rotating and then stopped, so there was no force left. This state of standstill can be characterized by the following initial conditions: $F_{actl} = 0$, $x_{actl} = 0$, $om_1 = 0$, $t_1 = 0$, $f_{com} = 0$, $x_{com} = 0$. The input command of $F_{nom} = 6000$ N is given and held. All the parameters of the system are the same as in Example 10.7. Under these conditions of the start of the cut, the cutting force is given by the formula of Eq. (10.74) where, however, the term $x_{act}\,(t - \tau_t) = 0$, for $t < \tau_t$.

The simulation is done in a loop repeated while incrementing time t by dt. Choose dt = 0.001 sec. The simulation follows the diagram in Fig. 10.54b. The variables f (table feed rate) and x_{act} as well as F_{act} and t are subscripted by n, which also represents the count of the loops. It is necessary to choose the total number m of the loops so as to permit sufficient total time $t_{tot} = m\mathrm{d}t$ for the development of the action, whether stable or unstable. Let us choose $m = 1000$. The sequence of the program is:

Loop for $n = 1$ to m:

1. Determine the positional command:
   $e_f = (F_{nom} - F_{act,\,n})/F_{nom}$
   $a_{com} = K_{acc}\, e_f$
   Integrate twice:
   $f_{com} = f_{com} + a_{com}\, dt$
   $x_{com} = x_{com} + f_{com}\, dt$
2. Simulate the positional servo.
3. Determine the actual cutting force $F_{act,\,n}$.

   The time interval τ_t between the engagement of subsequent teeth is expressed by the corresponding number i_t of time steps dt: i_t = integer (τ_t/dt). For the first i_t loops, $n = 1$ to i_t, see Eq. (10.74):

$$F_{act,\,n} = CB\, x_{act,\,n} \tag{10.87}$$

   for $n > i_t$,

$$F_{act,\,n} = CB\left[x_{act,\,n} - x_{act,\,(n-i_t)}\right] \tag{10.88}$$

   When the system becomes unstable, the feedrate oscillates vehemently and may become negative for part of the cycle; the table is returning instead of moving forward and cutting is interrupted, no force is generated if

$$F_{act,n} < 0, \qquad F_{act,\,n} = 0 \tag{10.89}$$

4. Increment time: $t_{n+1} = t_n + dt$.

The Task

Write the program and execute it for two values of the gain factor K_{acc}. One of them is chosen below the limit gain, as obtained in Eq. (10.86), and the other above it:

(a) $K_{acc} = 200$ (mm/sec^2)
(b) $K_{acc} = 700$ (mm/sec^2)

Use these servo parameters: $K_p = 200$ (V/mm), $K_m = 50$ (rev/sec/V), $K_a = 1$, $\tau_m = 0.5$ (sec), $r = 0.5$ (mm/rev), $H = 0.48$ (V/rev/sec).

Use these force parameters: $F_{nom} = F_{lim} = 6000$ N, $C = 2000$ (N/mm^2), $B = 15$ (mm), $\tau_t = 0.020$ (sec).

Plot f and F_{act} versus t for both *(a)* and *(b)*, for $0 < t < 0.4$ sec. The listing of the Matlab program "adaptsim" is printed in Fig. 10.57, and the results of the simulation are given in Figs. 10.58 and 10.59.

In the stable case, Fig. 10.58, both the velocity of the table (feed rate f) and the milling force F_{act} increase from the start and reach the steady-state values with very small overshoots in about 80 msec. The steady-state force equals the nominal force $F_{nom} = 6000$ N. In the unstable case, Fig. 10.59, both the feed rate and the force F_{act} oscillate with the frequency, as obtained in Eq. (10.85), $f_{lim} = \omega_{lim}/2\pi = 12.3$ Hz, that is, with a period of about 81 msec, and they increase rather rapidly. However, this growth settles soon because feed f starts to become negative for a part of the cycle. For this interval, the force F_{act} remains zero. The unstable motion saturates and continues vibrating with a constant amplitude of

FIGURE 10.57
Listing of the Matlab program "adapsim" for simulation of the adaptive control system.

```
function[f,xact,Fact,t]=adaptsim(Kacc)
%The program executes EX10.8 where the parameters of the case
%are as follows;
Fnom=6000;C=2000;B=15;taut=0.020;Kp=200;Km=50;taum=0.5;r=0.5;
H=0.48;dt=0.001;om=0;t(1)=0;xact(1)=0;Fact(1)=0;fcom=0;xcom=0;
f(1)=0;
%f is feedrate(mm/sec),ef is relative force error,taut(sec) is the
%tooth period,taum is(sec) the motor time constant,acom(mm/sec^2)is commanded
%acceleration,om is motor speed(rad/sec),dom is angular acceleration,
%r is transmission ratio(mm/rad),C and B are cutting force parameters
%with taut=0.020 and dt=0.001 the number of steps between subsequent cuts
%(teeth of the milling cutter) is 20;this enters into the expression for Fact
for n=1:999
  ef=(Fnom-Fact(n))/Fnom;
  acom=Kacc*ef;
  fcom=fcom+acom*dt;
  xcom=xcom+fcom*dt;
  erp=xcom-xact(n);
  v=Kp*erp;
  erv=v-H*om;
  dom=(Km*erv-om)/taum;
  om=om+dom*dt;
  f(n+1)=r*om;
  xact(n+1)=xact(n)+f(n+1)*dt;
  t(n+1)=t(n)+dt;
  if n<21
    Fact(n+1)=C*B*xact(n+1);
  else
    Fact(n+1)=C*B*(xact(n+1)-xact(n-20));
  end
  if Fact(n+1)<0,Fact(n+1)=0;
  end
end
%plot(t,Fact),(t,f)
```

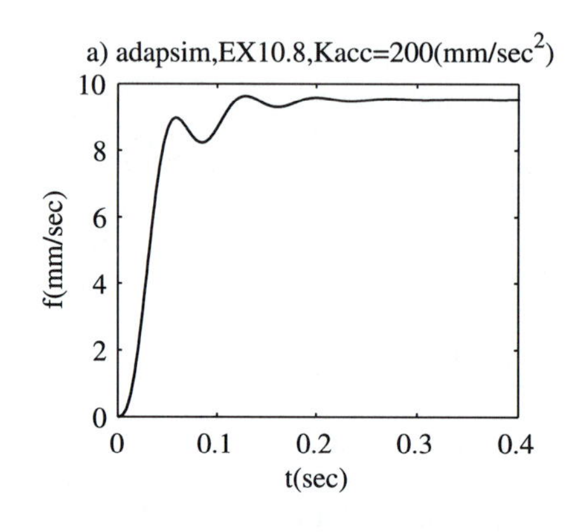

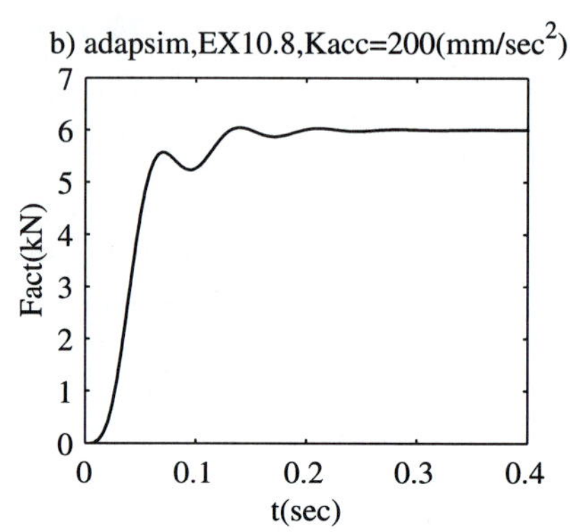

FIGURE 10.58
Result of Example 10.8. Stable A/C system. a) feed rate; b) cutting force—rather fast; after about 0.15 sec the required $F_{nom} = 6$ kN is reached.

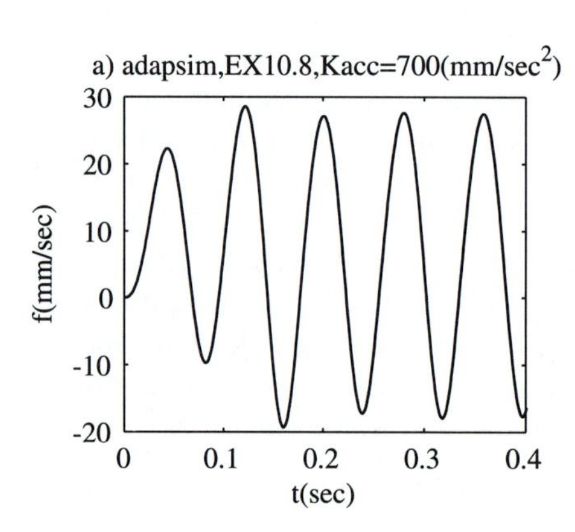

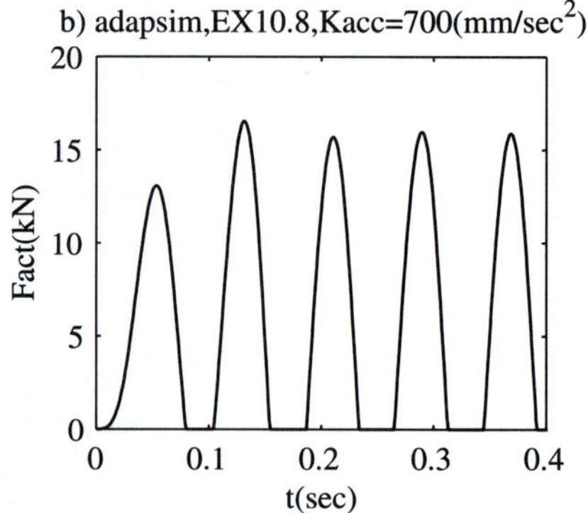

FIGURE 10.59
Result of Example 10.8. The result given in Fig. 10.58 may not be fast enough, and the tool may break. Attempting a higher gain, the system becomes unstable: a) feed rate—the motion starts to vibrate, b) cutting force reaches peaks of 15 kN, three times the desired F_{nom}.

velocity and of force. The peak force reaches periodically 15,000 N. These time-domain simulations confirm and illustrate very well the analysis of the limit of stability presented in Example 10.7. ▲

10.7.2 Summary of the Analyses of Numerical and Adaptive Control

Adaptive control is one of the modes of control used to improve the efficiency and accuracy of machining. The example of A/C discussed here was an elementary one. However, it was included to indicate the problems of stability and of transient behavior associated with the higher forms of control systems. It gives students an introductory insight into the capabilities as well as the limitations of these systems and prepares them for further studies of the extremely important field of advanced computer controls in manufacturing.

10.8 POSITIONAL SERVO DRIVING A SPRING-MASS SYSTEM

10.8.1 Two Basic Specifications: MT and ROB

In the preceding exercises it was shown that the errors of the tool path are especially serious in the cornering-type transients. We have mentioned that various kinds of dynamic corrections or compensations may be applied to minimize these errors. These correction techniques are necessary because in reality the servomechanism is not as simple as the second-order model used in the analytical exercises so far. If it were, then simple improvements could be achieved by increasing the gains (K_aK_m) and also H and K_p to increase the natural frequency ω_n and damping ζ; see Eqs. (10.44) and (10.45). There would of course be limits due to the current limit on what can be achieved. However, the main problem is that there are flexibilities in the transmissions of the table drive (those of the thrust bearing of the leadscrew, of the bracket for the nut, and of the leadscrew itself). These, together with the mass of the driven table or column, represent a spring-mass system that increases the order of the overall drive system. A higher-order system becomes unstable at higher gains. The problems with flexibility driving an inertia are much more serious in robots than in machine tools. Therefore, we will discuss some of the basic compensation techniques while keeping in mind *both machine tools and robots* and analyze them first in application *to a single coordinate motion.* The servomotors for both machine tools and most robots are about the same. The difference between the two groups of applications is in the spring-mass system. Flexibilities and inertias are larger in robots, and commanded motions are faster; transient errors are therefore larger. On the other hand, accuracy requirements are lower. The differences may best be briefly expressed by quoting the natural frequencies f_n of the structures. In machine tools the driven mechanical system will have f_n in the range of 20–100 Hz; in common industrial robots, the range is 4–20 Hz. The concerns about a servo driving a flexible structure extend to space structures—such as the articulate arm on the space shuttle—with natural frequencies between 0.1 and 2 Hz.

Subsequent to this section, the functioning of these servos is illustrated in application to robots as the analysis is extended to two-coordinate motions. In reality, of course, the same principles apply to motions in three or more coordinates. For illustration, we will use two different sets of parameters:

1. ***MT:*** The "machine tool" case with a slightly slower servo, with $\tau_{CL} = 0.04$ sec. and faster mechanical system: $m = 20$ kg, $k = 3.2 \times 10^5$ N/m, $c = 100$

N/(m/sec); $\omega_n = 126.5$ rad/sec, $f_n = 20$ Hz, and $\zeta = 0.02$. The natural frequency $\omega_n = 126.5$ rad/sec of the SMD system is *5 times higher* than the "corner frequency" of the drive, $(1/\tau_{CL}) = 25$ rad/sec.

2. ***ROB:*** The "robot case" with a slightly faster servo, $\tau_{CL} = 0.020$ sec, and a slow mechanical system: $m = 100$ kg, $k = 4 \times 10^4$ N/m, $c = 60$ N/(m/sec); $\omega_n = 20$ rad/sec, $f_n = 3.2$ Hz, and $\zeta = 0.015$. The natural frequency $\omega_n = 20$ rad/sec of the SMD system is *2.5 times lower* than the "corner frequency" $(1/\tau_{CL}) = 50$ rad/sec of the drive.

10.8.2 The Two Basic Alternatives, A and B

The combination of the positional servo with the spring-mass system may be designed in two fundamentally different modes depending on whether the spring-mass system is or is not included within the positional feedback (see Fig. 10.60). This figure is drawn as a mixture of block and structural diagrams. In alternative *A* the servomotor *SM* drives the transmission *r*; its position is denoted x_i (the "intermediate" *x*). This could be the angular position of the leadscrew as it might be ascertained by the encoder sitting on the shaft of the motor or on the end of the leadscrew. *This signal is used for positional feedback in alternative B* but not in *A*. Instead, in *A the feedback is taken from the table via a linear measuring device such as the Inductosyn or a linear scale with gratings.* The flexibility of the leadscrew and nut transmission is represented by the spring with stiffness *k;* the table mass is *m,* and the dashpot represents the damping coefficient *c* of the drive. Similarly, a servomotor *SM* may be driving a rotary coordinate of an arm of a robot via the transmission *TR* (see Fig. 10.61). *In alternative B the encoder sits on the shaft of the motor, and there is flexibility between it and the shaft of the arm. In alternative A the encoder is attached to the shaft of the arm.*

Thus in alternative *B* the system does not know the actual motion *x*. In alternative *A* it does, but the flexibility within the loop may lead to instability of the servo. To keep the system stable, the gains must be kept low, which leads to poor performance. We will analyze and discuss the pros and cons of the two solutions and introduce possible compensation measures both in the feedback and feedforward manner.

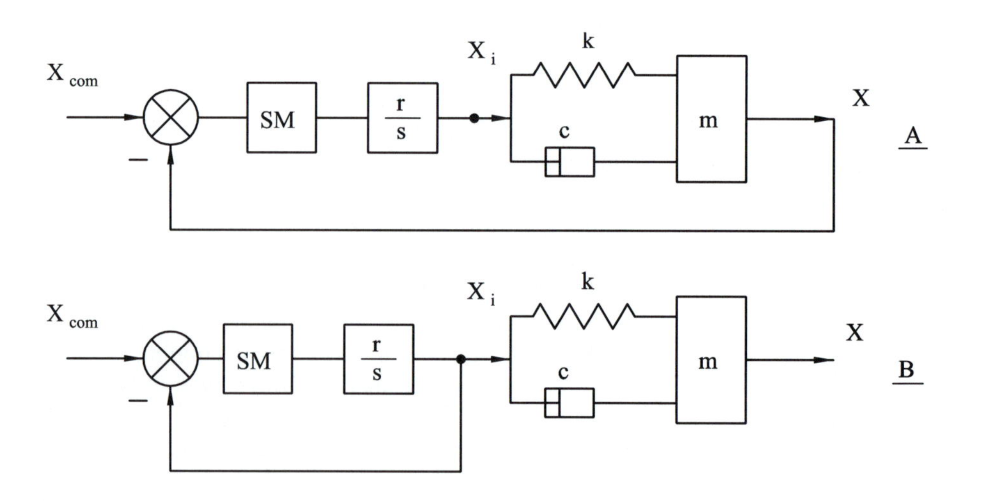

FIGURE 10.60
Two basic alternatives of the servo driving a spring-mass system (SMD). A: Flexible system SDM is inside of the positional loop. The actual motion *x* of the mass *m* is measured and fed back. This, however creates a fourth-order system, liable to instability. B: The SMD system is outside of the positional loop. Feedback is taken from the output x_i of the servomotor. The control system does not know the position of the mass and has no control over it. However, it remains of second order; it may behave badly, but it cannot become unstable. Later on we will see that a feed-forward strategy may be very beneficial.

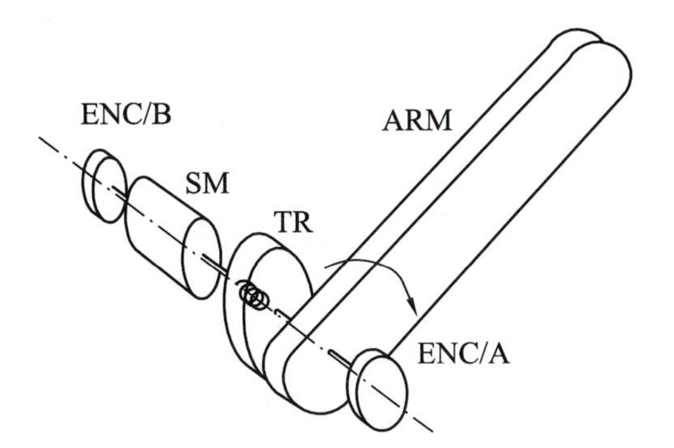

FIGURE 10.61
The two alternatives of the positional loop for a robot arm. The encoder ENC is either attached to the arm (A), or it is mounted in the shaft of the servomotor SM (B).The flexible transmission TR is between SM and ENC/A but not between SM and ENC/B.

Recapitulating the two alternatives, we focus on the following features:

- ***A:** spring-mass-damper (SMD) system is included in the feedback loop.*
- ***B:** SMD system is outside of the feedback loop.*

Formally, case A is represented by a fourth-order system, and it may become unstable. B is represented by a second-order system and thus cannot become unstable. Therefore, the treatments of the two cases will differ; the more complicated case of A is discussed first.

10.8.3 The "Machine Tool" Case with SMD System in the Feedback Loop: MT/A

This is alternative *A* of Fig. 10.60. The basic problem is the stability of the system. Let us first derive the differential equation and the transfer function across the structural part. The balance of forces on mass *m* is obtained as follows:

$$m\ddot{x} + c(\dot{x} - \dot{x}_i) + k(x - x_i) = 0$$

$$m\ddot{x} + c\dot{x} + kx = c\dot{x}_i + kx_i \tag{10.90}$$

By Laplace transforming, we obtain

$$\frac{X(s)}{X_i(s)} = \frac{cs + k}{ms^2 + cs + k} \tag{10.91}$$

$$\frac{X(s)}{X_i(s)} = \frac{k(1 + \tau_a s)}{ms^2 + cs + k} \tag{10.92}$$

where $(c/k) = \tau_a$ is the time constant of the "arm" (or of the structure).

Setting $s \rightarrow j\omega$,

$$\frac{X(\omega)}{X_i(\omega)} = \frac{k[1 + j(c/k)\omega]}{m[-\omega^2 + (k/m) + j(c/m)\omega]},$$

$$\frac{X(\omega)}{X_i(\omega)} = \frac{\omega_n^2}{\omega_n^2 - \omega^2 + 2j\zeta\omega\omega_n}\left(1 + 2j\zeta\frac{\omega}{\omega_n}\right)$$

$$\frac{X(\omega)}{X_i(\omega)} = \frac{1}{1 - \dfrac{\omega^2}{\omega_n^2} + 2j\zeta\dfrac{\omega}{\omega_n}}\left(1 + 2j\zeta\frac{\omega}{\omega_n}\right) \tag{10.93}$$

where

$$\omega_n^2 = \frac{k}{m}, \qquad \zeta = \frac{c}{2\sqrt{km}}$$

The block diagram of the system is then as shown in Fig. 10.62. It is a fourth-order system. Our first task must be to determine the limit of stability of this system. We will use the Nyquist criterion and draw the accompanying Bode plot. The Nyquist criterion is applied to the open loop transfer function (OLTF), and it states that if at a frequency ω_{lim} at which the phase ϕ becomes $-180°$ the magnitude $|G|$ of the OLTF is equal to 1, then the system is at the limit of stability. For $|G| > 1$ the system is unstable. Once again, the limit of stability is attained at

$$|G_{OL}(\omega_{\text{lim}})| = 1 \tag{10.94}$$

The Bode plots of the individual terms *(a)* through *(d)* in Fig. 10.62 have forms that are well known from control systems theory. Their forms are assembled in Fig. 10.63.

The individual blocks in the OLTF have the following phase shifts and magnitudes:

$$\phi_a = \tan^{-1}(-\tau_{CL}\omega) \qquad |G_a| = \frac{1}{\sqrt{1 + \tau_{CL}^2\omega^2}}$$

$$\phi_b = \frac{-\pi}{2} \qquad |G_b| = \frac{1}{\omega}$$

$$\phi_c = \tan^{-1}\left(\frac{-2\zeta(\omega/\omega_n)}{1 - (\omega^2/\omega_n^2)}\right)$$

$$|G_c| = \frac{1}{\sqrt{[1 - (\omega^2/\omega_n^2)]^2 + 4\zeta^2(\omega/\omega_n)^2}}$$

$$\phi_d = \tan^{-1}(\tau_a\omega) \qquad |G_d| = \sqrt{1 + \tau_a^2\omega^2}$$

and we have

$$\phi_{OL} = \phi_a + \phi_b + \phi_c + \phi_d \tag{10.95}$$

$$|G_{OL}| = K_p K_{CL} r \cdot |G_a| \cdot |G_b| \cdot |G_c| \cdot |G_d| \tag{10.96}$$

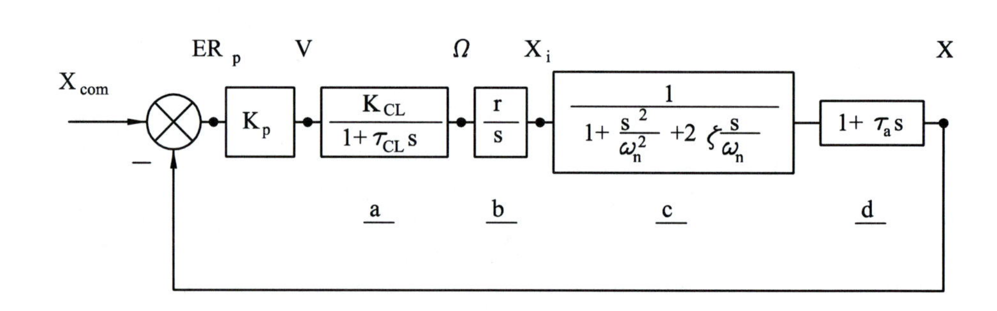

FIGURE 10.62 Block diagram of system A of Fig. 10.60. The SMD system is represented by the blocks *c* and *d*. Position feedback is taken from the output of these blocks.

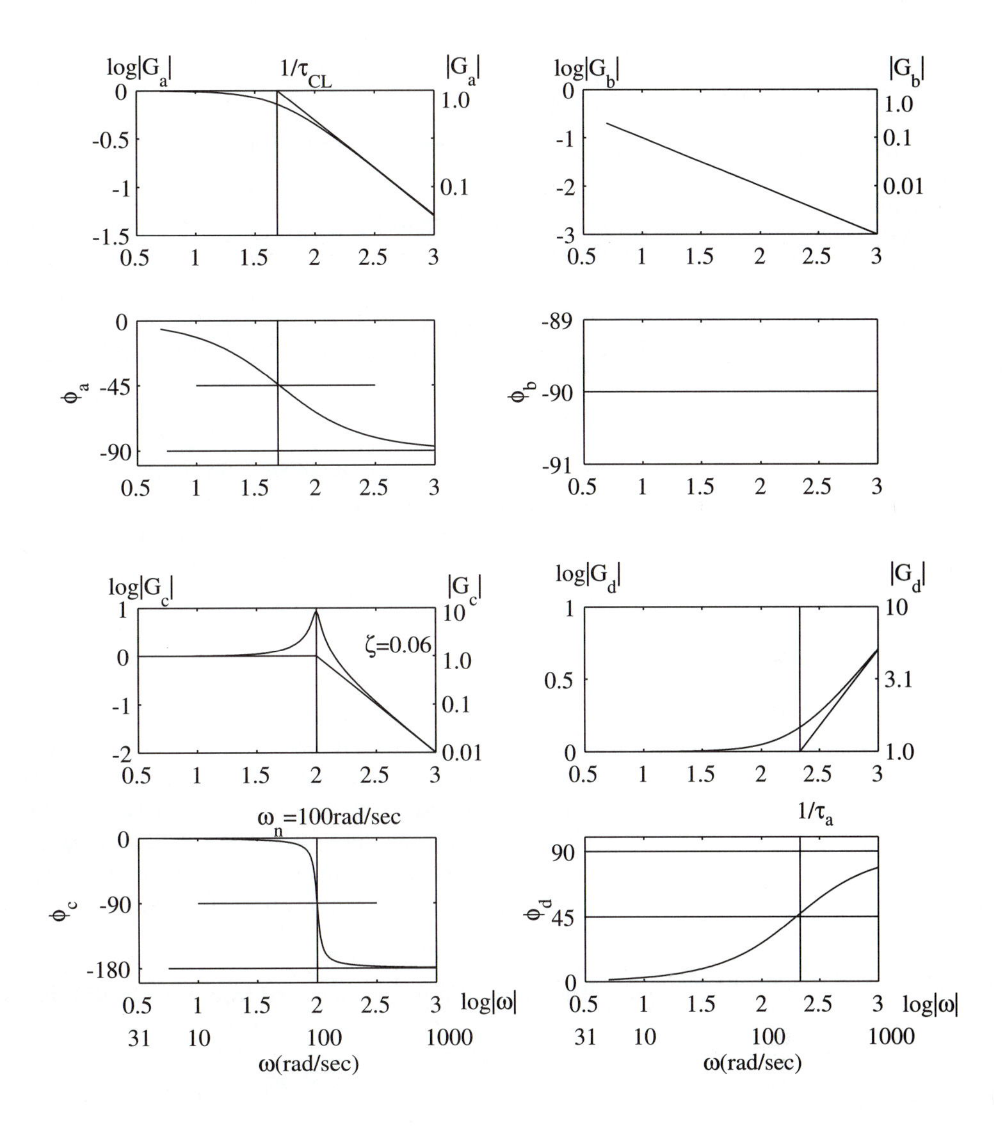

FIGURE 10.63 Schematics of the Bode plots of magnitude and phase of the individual blocks of the system in Fig. 10.62. The parameters used : $K_{CL} = 1.0$, $\tau_{CL} = 0.02$ sec, $r = 1.0$, $\omega_n = 100$ rad/sec, $\zeta = 0.06$, $\tau_a = 0.005$ sec.

EXAMPLE 10.9 Limit of Stability: The "Machine Tool" Case: MT/A ▼

Determine the limit of stability for case MT/A. Assume $K_{CL} = 40$ rad/sec/V, $r = 0.005$ m/rad, determine $K_{p,\,\mathrm{lim}}$. First, it is necessary to determine the frequency ω_{lim} at which the phase ϕ_{CL} becomes $-180°$. Referring to Eq. (10.95) and the above input data, $\tau_{CL} = 0.04$, $\omega_n = 126.5$, $\zeta = 0.02$, $\tau_a = c/k = 3.125\mathrm{e}{-4}$, we have

$$-\tan^{-1}(0.04\,\omega) - \frac{\pi}{2} - \tan^{-1}\frac{3.162\mathrm{e} - 4 \times \omega}{1 - 6.25\mathrm{e} - 5 \times \omega^2}$$

$$+ \tan^{-1}(3.125\mathrm{e} - 4 \times \omega) = -\pi$$

and ω_{lim} is obtained by plotting ϕ_{a}, ϕ_{c}, ϕ_{d}, and their sum ϕ_{acd} versus ω, and finding at which ω the sum of the three terms becomes $-90°$. It can be estimated by consulting Fig. 10.63 that ω_{lim} will be below ω_n. Therefore, the range of the plot for ω between 0 and 200 rad/sec is more than sufficient; see Fig. 10.64, where the

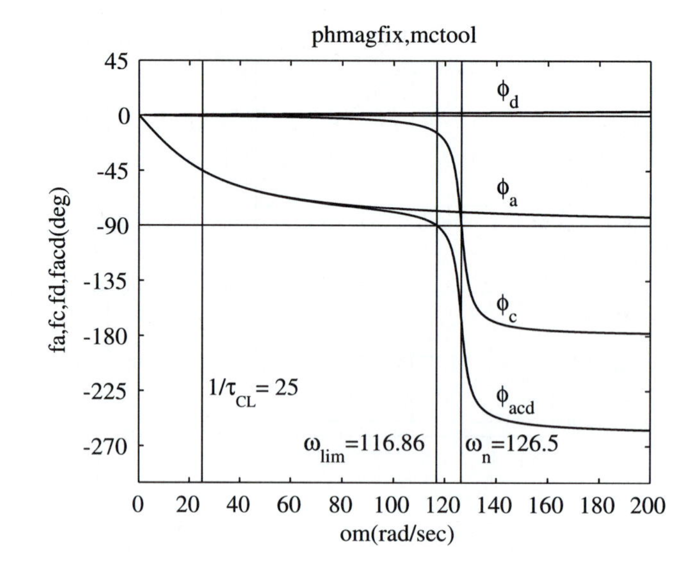

FIGURE 10.64 Phase plot for the system MT/A of Example 10.9. The critical frequency is $\omega_{\text{lim}} = 116.8$ rad/sec. The limit gain is then evaluated as $K_{p,\text{lim}} = 460$ V/m.

phase angles are given in degrees. Alternatively, Eq. (10.95) could be solved using any numerical method for solving nonlinear equations, such as the Newton-Raphson method. From Fig. 10.64 we can see that, indeed, at $\omega = 1/\tau_{\text{CL}} = 25$, the phase $\phi_{\text{a}} = 45°$, and at $\omega = \omega_n = 126.5$, the phase $\phi_{\text{c}} = -90°$. The phase ϕ_{d} is positive and increases rather slowly with ω. It will reach 45° at $\omega = 1/\tau_{\text{a}} = k/c = 3200$ rad/sec. Obviously, its effect on stability is negligible. It is found as follows:

$$\omega_{\text{lim}} = 116.8 \text{ rad/sec}$$

The value of $K_{p,\,\text{lim}}$ is obtained from Eq. (10.96); for $\omega = \omega_{\text{lim}}$,

$$|G_a| = \frac{1}{\sqrt{1 + 0.04^2\omega_{\text{lim}}^2}}$$

$$|G_b| = \frac{1}{\omega_{\text{lim}}}$$

$$|G_c| = \frac{1}{\sqrt{(1 - 6.25\text{e} - 5 \times \omega_{\text{lim}}^2)^2 + 1\text{e}{-7} \times \omega_{\text{lim}}^2}}$$

$$|G_d| = \sqrt{1 + 9.766\text{e} - 8 \times \omega_{\text{lim}}^2}$$

and

$$|G_{OL}| = 0.2 \times K_{p,\,\text{lim}} \times 0.209 \times 0.0086 \times 6.067 \times 1.0007 = 1.0$$

$$|G_{OL}| = 0.002176\, K_{p,\,\text{lim}} = 1.0$$

$$K_{p,\,\text{lim}} = 460 \text{ V/m}$$

EXAMPLE 10.10 Response to a Ramp Command of the Case MT/A (SMD System in the Loop) ▼

Now plot the response $x(t)$ to a ramp input command of $x_{\text{com}} = 50$ (mm/sec)t. Plot also the intermediate variable x_i. Carry this out for two cases: $K_p = 0.4\,K_{p,\,\text{lim}}$, and $K_p = 1.6\,K_{p,\,\text{lim}}$.

The simulation program will consist of the part from x_{com} to x_i, which will be the same as Eq. (10.46) except that the internal velocity servo will, for this exercise, be considered as one block described by Eq. (10.20), (with $T_L = 0$), and this section ends with x_i instead of x. We follow the block diagram of Fig. 10.62.

Input data: $\tau_{\text{CL}} = 0.04$ sec, $K_{CL} = 40$ rad/sec/V, $r = 0.005$ rad/m, $k = 3.2e^5$ N/m, $m = 20$ kg, $c = 100$ N/(m/sec).
Initial state: $t_1 = 0$, $x_{i,1} = 0$, $x_1 = 0$, $om = 0$, $\dot{x} = 0$.
for $n = 1{:}N$
 $xcom_n = 50t_n$
if $xcom_n > 5$
 $xcom_n = 5$
 $erp = xcom_n - x_n$
 $v = K_p erp$
 $omd = (K_{\text{CL}} v - om)/\tau_{\text{CL}}$
 $om = om + omd\,\text{d}t$
 $\dot{x}_i = r\,om$
 $x_{i,n+1} = x_{i,n} + \dot{x}_i\text{d}t$

This will be followed by a section representing the SMD system as expressed by Eq. (10.90):

$$\ddot{x} = \frac{c\,\dot{x}_i + kx_{i,n+1} - c\dot{x} - kx_n}{m}$$

$$\dot{x} = \dot{x} + \ddot{x}\,\text{d}t \tag{10.97}$$

$$x_{n+1} = x_n + \dot{x}\text{d}t$$

Use dt = 0.001 sec and run the simulation for a total time of 0.5 sec.
The two values of the gain are:

1. $K_p = 0.4\,K_{\text{lim}} = 184$ V/mm
2. $K_p = 1.6\,K_{\text{lim}} = 736$ V/mm

The listing of the corresponding Matlab program "machtool1" is printed in Fig. 10.65. The corresponding plots are in Fig. 10.66a and b. In case *(a)* the overshoot of 0.76 mm settles down rather rapidly into a final commanded position. In case *(b)* the vibratory motion increases rapidly. The case is strongly unstable. ▲

10.8.4 Flexibility Outside of the Loop: Case MT/B

The parameters of the machine tool case are still considered as used in Examples 10.9 and 10.10, but configuration *B* of Fig. 10.60 is used with positional feedback from an encoder on the motor shaft. The program "machtool1" of Fig. 10.65 must now be

FIGURE 10.65 Listing of the Matlab program "mchtool1" for simulation of a servo with an SMD system within the feedback loop (type A). It can also simulate a type B system with the SMD outside of the loop by substituting the output of the drive x_i for the output x from the SMD system in the expression for the position error: ($erp = x_{\text{com}} - x_i$).

```
function[xcom,xi,x,t]=mchtool1(Kp)
%This program executes EX10.10. It is a case with the (k,m,c)
%system included in the positional loop; the positional feedback
%is taken from the motion x of the mass; see the expression for erp.
t(1)=0;x(1)=0;om=0;xi(1)=0;dx=0;
Kcl=40;r=0.005;m=20;dt=0.001;
k=3.2e5;c=100;tacl=0.04;
xcom(1500)=5;
for n=1:1499
  xcom(n)=50*t(n);
  if xcom(n)>5
    xcom(n)=5;
  end
  erp=xcom(n)-x(n);
  v=Kp*erp;
  dom=(v*Kcl-om)/tacl;
  om=om+dom*dt;
  dxi=om*r;
  xi(n+1)=xi(n)+dxi*dt;
  ddx=(c*dxi+k*xi(n+1)-c*dx-k*x(n))/m;
  dx=dx+ddx*dt;
  x(n+1)=x(n)+dx*dt;
  t(n+1)=t(n)+dt;
end
%plot(t,xcom,t,xi,t,x)
```

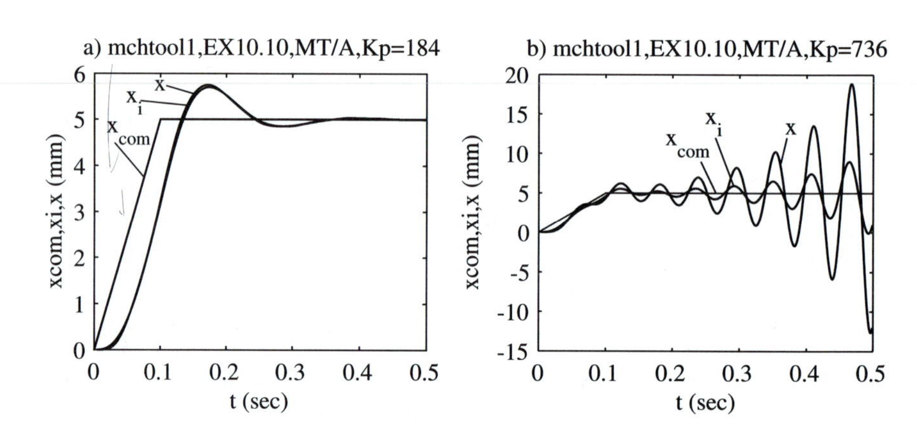

FIGURE 10.66 Response to a ramp command of servo MT/A: Example 10.10. a) Gain $K_p = 0.8K_{p,\text{lim}} = 184$ V/mm; system is stable but with a considerable overshoot. Gain should be further reduced. b) Gain $K_p = 1.6K_{p,\text{lim}} = 736$ V/mm, system is strongly unstable, absolutely unusable.

modified on the line defining the positional error. Instead of ($erp = x_{\text{com}} - x$), we now have ($erp = x_{\text{com}} - x_i$).

The characteristics of this second-order servo are, according to Eqs. (10.34) and (10.35),

$$\omega_n = \sqrt{\frac{K_p K_{CL} r}{\tau_{\text{CL}}}} = \sqrt{5K_p}, \qquad \zeta = \frac{1}{2\sqrt{K_p K_{CL} r \tau_{\text{CL}}}} = \frac{5.59}{\sqrt{K_p}}$$

Using the K_p values *(a)* 184 V/mm and *(b)* 736 V/mm as given in Example 10.10 and $K_{CL} = 40$ rad/sec/V, $\tau_{\text{CL}} = 0.04$ sec, $r = 0.005$ m/rad, we have

(a) $\omega_n = 30$ rad/sec, $\zeta = 0.41$

(b) $\omega_n = 60.6$ rad/sec, $\zeta = 0.206$

These parameters, when compared with the characteristics of the spring-mass system, $\omega_n = 126.5$ rad/sec and $\zeta = 0.02$, show that a rather "lazy" servo drives a fast though lowly damped system. In spite of the very low damping of this system, it follows the motion provided by the servo very well; see Fig. 10.67. Because the servo frequency is much lower in both cases, it filters the sudden commands and reproduces them in relatively slow transients in x_i, which drive the much faster spring-mass system. Consequently the displacement x of the mass of the system follows the drive x_i of the spring end with only small deviations. The difference that may be observed between the transients of case *(a)* and case *(b)* does not reflect the difference in damping between the two cases, and the die-out times are almost equal because the higher frequency of *(b)* makes up for it. The whole commanded motion is rather short and lasts only 0.1 sec. During this time the starting transient has not yet died down, and the stopping transient is affected by the phase with which it combines with the starting transient. The overshoots are of the same order as in case MT/A *(a)* with the lower gain.

One can conclude that in both cases the motion x is not much different from the stable case of alternative A in Fig. 10.66a. This shows that in the case in which the structural natural frequency is substantially higher than the natural frequency of the drive, very little improvement of the transients can be obtained by using the overall positional feedback. On the contrary, one runs the danger of producing an unstable system. It is then as well to use configuration B of the system.

10.8.5 The "Robot" Case with SMD System within the Loop: ROB/A

Let us now deal with the "robot case." The flexible structure is "slower" than the drive.

EXAMPLE 10.11 Limit of Stability of the Robot Case: ROB/A

Determine the stability conditions for configuration A of Fig. 10.60 of the case defined above in Section 10.8.1 as ROB; the spring-mass system's natural frequency is now 126.5/20 = 6.3 times lower than in the MT case of Examples 10.9 and 10.10, while the servomotor time constant is two times shorter than in case *(a)*.

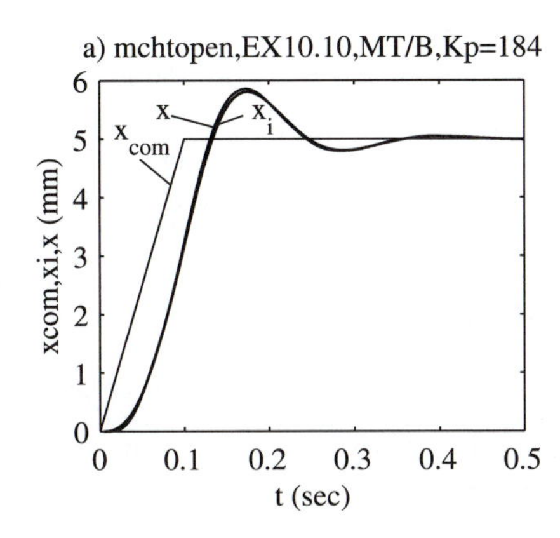

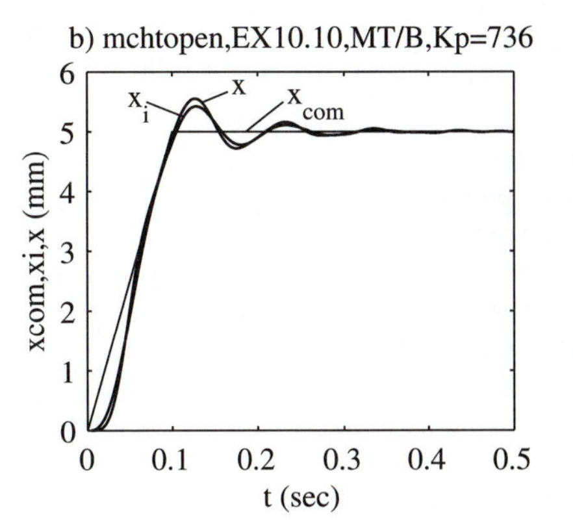

FIGURE 10.67
The servo of MT/B with flexibility outside the loop. a) Gain K_p the same as in Fig. 10.66a; system is stable. b) Gain K_p the same as in Fig.10.66b; system is still stable. In both *(a)* and *(b)* the overshoot is about in the range 0.6–0.9 mm and it decays rapidly. It is not worse than in case MT/A with the feedback in the loop and the smaller gain, but even with the higher gain it avoids the violent instability shown in Fig. 10.66b. As it is now, alternative B of Fig. 10.60 (SMD outside of the loop) looks preferable to A.

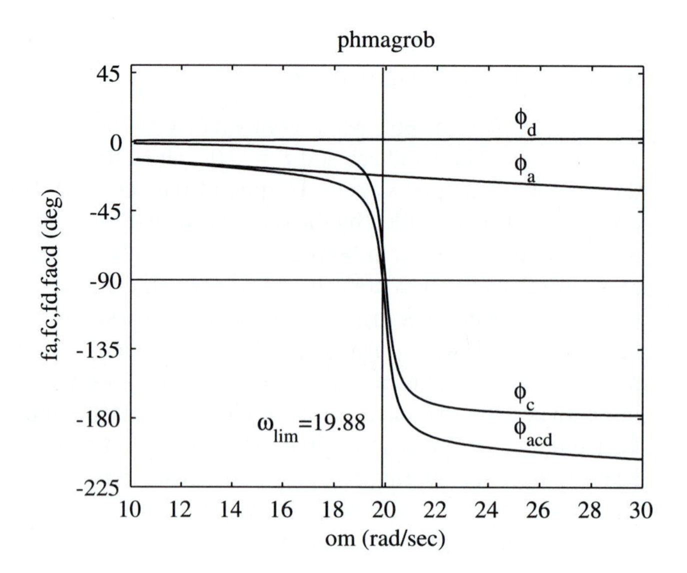

FIGURE 10.68
Open-loop phase plot for the ROB/A alternative of Example 10.11 ("robot" case, flexibility in the loop). Critical frequency is low: 19.88 rad/sec.

The phase plot is presented in Fig. 10.68. The critical frequency is obtained as $\omega_{\text{lim}} = 19.88$ rad/sec. The limit gain is obtained by using this frequency in Eq. (10.96). Choose $K_{CL} = 40$ and $r = 0.005$. The result is $G_a = 0.93$, $G_b = 0.0507$, $G_c = 20.486$, $G_d = 1.0004$, $G_{OL} = 0.2\ K_p \times 0.9659 = 1$, $K_{p,\text{lim}} = 5.18$ V/m.
Simulate a ramp motion for

(a) $K_p = 0.8\ K_{p,\text{lim}} = 4.14$
(b) $K_p = 3.86\ K_{p,\text{lim}} = 20$ V/m

The x_{com} is defined by a velocity $f = 100$ mm/sec over a period of 1.0 sec and then stopped for another 2.0 sec. Simulation is carried out with dt = 0.002 sec for a total time of 3.0 sec.

The results are presented in Fig.10.69a and b. They were obtained with the program of Fig. 10.65, inputting the different parameters of the current task. Case *(a)* with the very low gain K_p needed for stability is one of a "lazy system." It is stable, but due to the low gain, it has a very large following error. By the time the command stopped at $x_{\text{com}} = 100$ mm, the actual position only reached $x = 22$

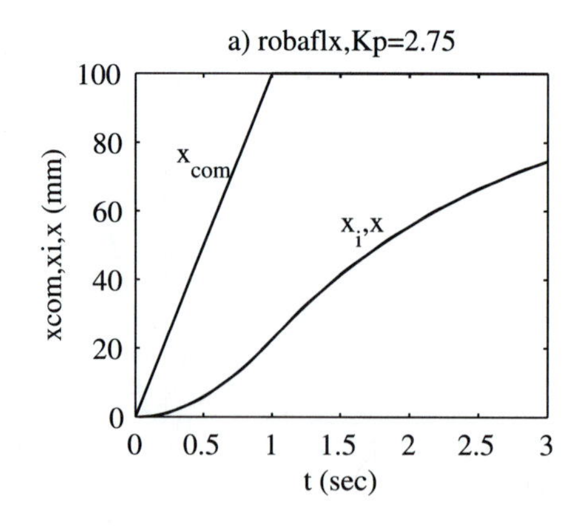

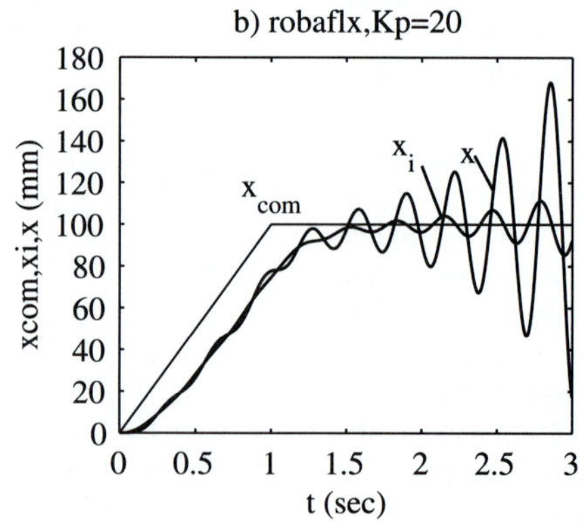

FIGURE 10.69
The case of ROB/A, Example 10.11. a) Ramp input response with gain $K_p = 0.53\ K_{p,\text{lim}} = 2.75$ V/mm; the system is stable but very slow ("lazy"): 2 sec after the command of a 1 sec motion the move is only 74% complete. b) Gain $K_p = 3.86\ K_{p,\text{lim}} = 20$ V/mm. System is very unstable. Obviously, alternative *A* does not work well with an SMD system with such high flexibility and inertia.

mm, and 2.0 sec after the command stopped it is still only $x = 75$ mm. Case *(b)* with the large gain is agile but strongly unstable.

Neither *(a)* or *(b)* give acceptable performance. The system, with its high flexibility and heavy mass, is very difficult to design. One of the possible measures to apply is the use of an accelerometric feedback, as discussed in the following section. ▲

10.8.6 Accelerometric Feedback Applied to the ROB/A System

In order to improve the stability of the system, it is necessary to apply some kind of lead term that uplifts the phase shift. Inspecting the phase plot in Fig. 10.68, we see that the decisive phase shift is due to the *(c)* term of Fig. 10.62, the second-order term related to the spring-mass system. The first-order term *(a)* of the servomotor plays only a minor role because its corner frequency of $1/\tau_{CL} = 50$ rad/sec is much higher than the natural frequency $\omega_n = 20$ rad/sec of the term *(c)*. The term *(d)* is again of no significance at all. As seen in Fig. 10.68, just below ω_n the combined phase ϕ_{acd} reaches $-90°$; the total phase shift when the $-90°$ of the term *(b)* is added is $-180°$ at this point. Right above ω_n the total phase reaches $-270°$ and then $-290°$ and from there on it decreases further due to the increasing effect of term *(a)*. At the corner frequency of $\omega = 50$ rad/sec, the ϕ_{acd} will be close to $-305°$. A simple lead term of the form $(1 + bs)$ would lift the phase up by 45° at $\omega = 1/b$. Thus, if we used it with $1/b = 20$ and $b = 0.05$, we would get about $-90° -90° + 45° = -135°$ at $\omega = 20$ rad/sec; soon thereafter, however, the *(c)* term drops by another 90°, and after crossing the $-180°$ mark, it would reach $-90° -180° + 45° = -225°$. We need something more powerful. This can be obtained by attaching an accelerometer to the robot arm. A piezoelectric one is small, inexpensive; and easy to attach.

The Nyquist criterion deals with the open loop of the system. Obviously, for phase shift it is irrelevant if the corrective block is included in the otherwise feedforward part or the feedback part of the loop. Its inclusion in the feedback is shown in Fig. 10.70. The *(e)* term is added:

$$e = 1 + as^2, \qquad s \rightarrow j\omega, \qquad e(j\omega) = 1 - a\omega^2 \tag{10.98}$$

and the signal arriving at the point OL of interrupting the loop is the feedback signal x_f:

$$x_f = xe = x(1 + as^2)$$

and in the time domain, we have

$$x_f = x + a\ddot{x} \tag{10.99}$$

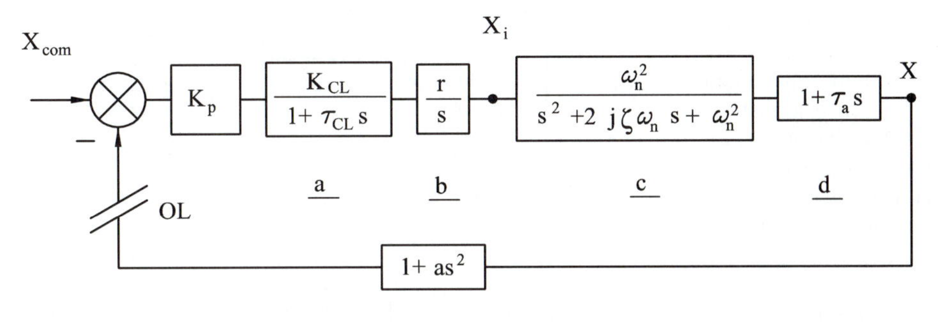

FIGURE 10.70
Stabilizing the system type ROB/A by using an accelerometer signal as^2 in the feedback.

This is a combination of the encoder signal to which the accelerometer signal is added after (A/D) conversion and scaling by the positive factor a.

The phase shift and gain factor of this term are obtained as follows. The term is purely real. For $\omega^2 < (1/a)$ it is positive; its phase shift is zero. For $\omega^2 > (1/a)$ it is negative; this amounts to shifting by $-180°$. Of course, once a phase shifts by 360°, it is back to 0° and may continue decreasing from there on. Thus we have

$$\phi_e = 0°, \qquad \text{for } \omega^2 < 1/a \qquad |G_e| = \sqrt{(1 - a\omega^2)^2}$$

$$\phi_e = -180°, \qquad \text{for } \omega^2 > 1/a$$

Therefore, if we applied the factor

$$a_{cr} = 1/\omega_{\text{lim}}^2 \tag{10.100}$$

where ω_{lim} is the frequency at which $\phi_{\text{abcd}} = -180°$, that is, $\phi_{\text{acd}} = -90°$ it would become $\phi_{\text{OL}} = -180° - 180° = -360° = 0$. From there on the phase would continue growing negative; but in the system at hand it would never reach $-180°$. This amounts to an absolute stabilization of the system. Accounting for inaccuracies, it is best to choose to turn the phase by $-180°$ a little below the ω_{lim} of the original system. In our case it was determined that $\omega_{\text{lim}} = 19.88$ rad/sec. This gives $a_{cr} = 0.00253$; a slightly larger a is chosen:

$$a = 0.0028 = 1.088a_{cr} \tag{10.101}$$

In order to test the effect of this corrective action, the simulation program of Fig. 10.65 may be taken for the case ROB/A, as described in Section 10.8.5, and modified on the positional error line:

$$erp = x_{\text{com},n} - (x_n + a\ddot{x}) \tag{10.102}$$

EXAMPLE 10.12 Using Accelerometer Feedback ▼

Let us take Example 10.11 and use the acceleration corrective term of Eq. (10.102) with the value of a as determined in Eq. (10.101) and apply it first to the strongly unstable case of Fig. 10.66b, with $K_p = 20$. The result is shown in Fig. 10.71a. The case is completely stable, without any overshoot. Even if the gain is

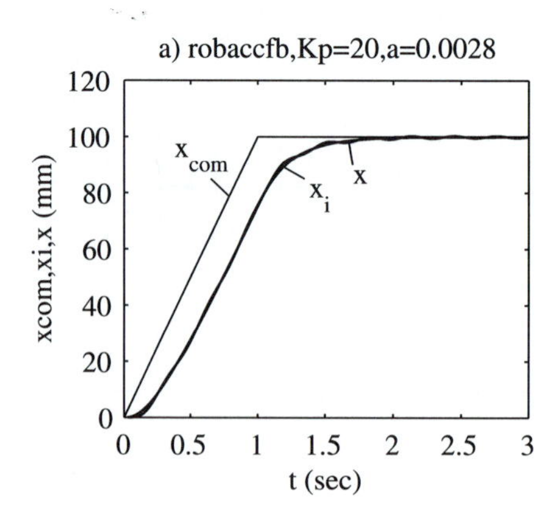

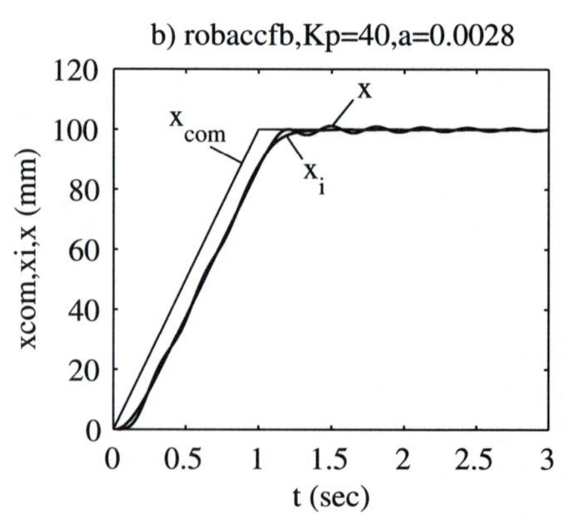

FIGURE 10.71
a)The case ROB/A of Example 10.11 and of Fig. 10.69b, stabilized in Example 10.12 by the accelerometric feedback. b) The same case as in *(a)*, with two times higher gain. There is a small overshoot, but the case is stable. The stabilizing action of the feedback is very powerful, and it can cure a system that in the preceding exercise was judged hopeless; see the comment on Fig. 10.69.

further increased to $K_p = 40$ (see Fig.10.71b), the system is still stable, although with a slight overshoot. However, it is now rather fast, with a rather small tracking error. ▲

EXAMPLE 10.13 The "Robot Arm": Configuration *B*

Finally, let us check the performance of the "robot arm" in configuration *B*, leaving the spring-mass system outside of the loop and, of course, not applying any corrective feedback term. The drive, with the feedback from x_i, is a second-order system. Recapitulating from Section 10.8.1, the structural SMD system parameters are $\omega_n = 20$ rad/sec, $\zeta = 0.015$.

First, let us check the characteristics of the second-order, positional feedback drive system. The parameters of the servo are $K_{CL} = 40$ rad/sec/V, $\tau_{\text{CL}} = 0.02$ sec, and $r = 0.005$ m/rad. As shown, Eqs. (10.34) and (10.35) can be applied to determine

$$\omega_{n,d} = \sqrt{\frac{K_p K_{CL} r}{\tau_{\text{CL}}}} = \sqrt{10K_p}, \quad \zeta_{\text{d}} = \frac{1}{\sqrt{K_p K_{CL} r \tau_{\text{CL}}}} = \frac{7.91}{\sqrt{K_p}}$$

where the subscript *d* means drive.

Let us try the higher of the two gains used in Example 10.12, $K_p = 40$ V/mm. In this case $\omega_{n,d} = 20$ rad/sec, and $\zeta = 1.25$. The drive has a natural frequency equal to the $\omega_n = 20$ of the spring-mass system. Its damping is rather high. The simulation program differs from the one in Example 10.11 in the *erp* term:

$$erp = x_{\text{com},n} - x_{i,n} \tag{10.103}$$

The result of the simulation is in Fig. 10.72. The motion x_i is without an overshoot, but x overshoots by 1.5 mm and takes a long time to decay. The velocity lag error is 12.5 mm. This is only slightly worse than the case of Fig. 10.71b of configuration *A* with the accelerometric feedback.

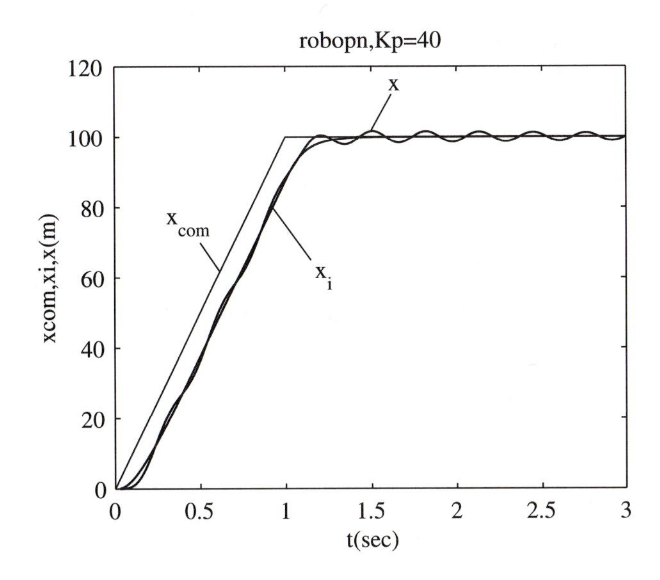

FIGURE 10.72 The case of Example 10.13. Ramp input response of the "robot arm," system ROB/B with the SMD system outside of the loop (recall Fig. 10.61, ENC/B), with gain two times higher than in the highly unstable ROB/A configuration of Fig. 10.69b. The case is stable, and response is acceptable.

For systems with even more flexibility and lower natural frequency of the spring-mass system as well as for systems with parameters similar to those just used for the "robot arm" but with a higher demand on accuracy, it is best to use configuration B and a "feedforward" compensatory term added to the command. This will be discussed next. However, a warning must be added right now, and it will be repeated: This marvelous approach only works if the parameters of the spring-mass system are accurately known. In robotic systems, both flexibility and inertia keep changing during a path, which complicates that task considerably. ▲

10.9 FEEDFORWARD COMPENSATION

In the preceding section, we analyzed a servomotor driving the spring-mass system in configuration A, in which everything was enclosed by feedback, and tests have shown how a feedback corrective term can be used to improve the stability and overall performance of the system. The corrective term was of the acceleration-term type. It has further been investigated how the system behaves in configuration B, in which the spring-mass structure is left outside the positional loop; in the case where the natural frequency of the structure is substantially higher than that of the positional servomechanism driving it, the system behaves reasonably well. However, when the structure is more flexible and slow, larger errors arise in the transient.

We now describe how to improve the behavior of these very flexible structures by using configuration B and including a compensatory term in the command. This technique, called "feedforward compensation," presumes a good knowledge of the characteristics of the driven structure.

Initially, we deal with the extreme in which the driving servomechanism is so much faster than the driven structure ($\omega_{n,\mathrm{d}} >> \omega_{n,\ \mathrm{structure}}$) that its errors can be neglected when compared with those generated in the structure. An "ideal servodrive" is assumed that follows the command immediately and exactly, $x_i = x_{\mathrm{com}}$.

10.9.1 Ideal Servodrive

So as not to have to change notations in the second step when the errors of the servodrive are acknowledged, the command will now be called $x_{i,\mathrm{com}}$. The case is illustrated in Fig. 10.73. The differential equation (10.90) expressing the balance of forces on mass m applies:

$$m\ddot{x} + c\dot{x} + kx = c\dot{x}_i + kx_i$$

Our question is this: If the desired form of x is expressed as a function of time, can we determine a form of $x_{i,\mathrm{com}}$ that, through the distortions due to the flexibility of the

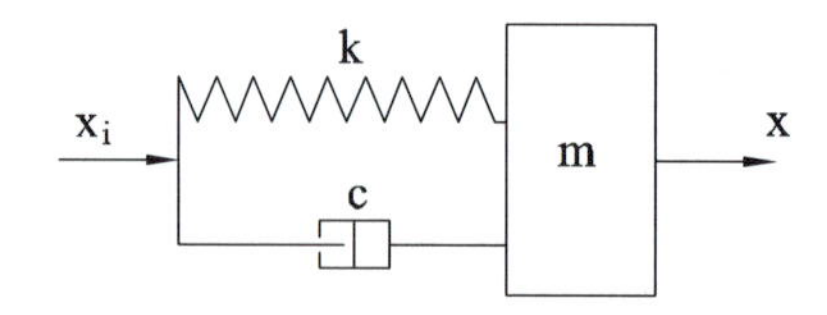

FIGURE 10.73 Physical diagram of the spring-mass-damper (SMD) system.

driven structure, will produce the desired x_{des}? We will find that this can be done if our desire is reasonable; this essentially means that we cannot ask for infinite acceleration or deceleration. As an example, let us define x_{des} as shown in Fig. 10.74. The motion should consist of period I with constant acceleration *a*, period II with constant velocity *v*, and period III with constant deceleration $(-a)$. The preceding equation is now rewritten with new notation:

$$c\dot{x}_{i,\text{com}} + kx_{i,\text{com}} = m\ddot{x}_{\text{des}} + c\dot{x}_{\text{des}} + kx_{\text{des}}. \tag{10.104}$$

This is a differential equation where the form of the input $x_{i,\text{com}}$ must be solved for the known output x_{des}.

First, consider phase I:

$$x_{\text{des}} = \frac{a}{2}t^2, \dot{x}_{\text{des}} = at, \qquad \ddot{x}_{\text{des}} = a \tag{10.105}$$

$$c\dot{x}_{i,\text{com}} + kx_{i,\text{com}} = ma + cat + \frac{ka}{2}t^2 \tag{10.106}$$

The solution will consist of the general term corresponding to the homogenous equation (with zero on the right side) and a particular term. For the homogenous equation

$$c\dot{x}_{i,\text{com}} + kx_{i,\text{com}} = 0$$

we propose the following:

$$\begin{aligned} x_{i,\text{com}} &= Xe^{st}, \\ cs + k &= 0 \\ s &= -\frac{k}{c} \\ x_{i,\text{com}} &= Xe^{-(k/c)t} \end{aligned} \tag{10.107}$$

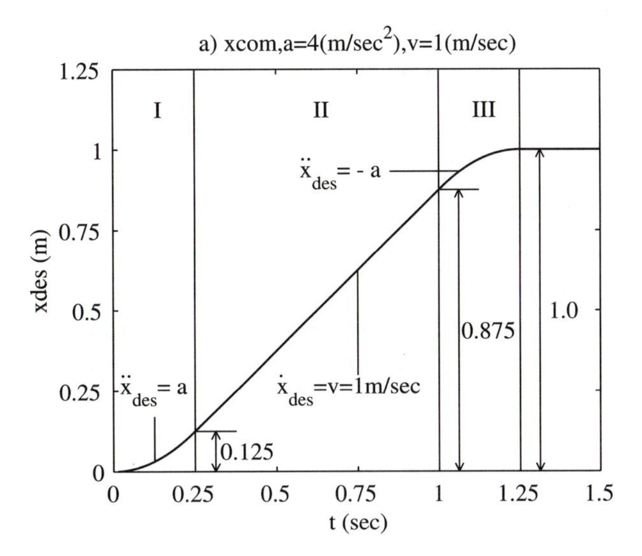

FIGURE 10.74
A typical motion command with constant acceleration, constant velocity, and constant deceleration.

For the particular solution, we propose

$$x_{i,\text{com}} = b_0 + b_1 t + b_2 t^2$$

The complete solution is

$$x_{i,\text{com}} = Xe^{-(k/c)t} + b_0 + b_1 t + b_2 t^2 \tag{10.108}$$

$$\dot{x}_{i,\text{com}} = -\frac{k}{c} Xe^{-(k/c)t} + b_1 + 2b_2 t \tag{10.109}$$

Using Eqs. (10.108) and (10.109) in (10.106):

$$-k\,Xe^{-(k/c)t} + b_1 c + 2b_2 ct + kXe^{-(k/c)t} + b_0 k + b_1 kt + b_2\,kt^2 = ma + cat + \frac{ka}{2}t^2 \tag{10.110}$$

The two exponential terms cancel out. By comparing terms with equal degree of t, we determine the coefficients b_0, b_1, b_2; starting with the highest degree:

$$b_2 = \frac{a}{2} \tag{10.111}$$

$$2b_2 ct + kb_1 t = cat$$

$$ac + kb_1 = ac \tag{10.112}$$

$$b_1 = 0$$

$$b_1 c + b_0 k = ma \tag{10.113}$$

$$b_0 = \frac{ma}{k}$$

Returning to Eq. (10.108):

$$x_{i,\text{com}} = Xe^{-(k/c)t} + \frac{ma}{k} + \frac{a}{2}t^2$$

for $t = 0$ and $x_{i,\text{com}} = 0$:

$$0 = X + \frac{ma}{k}, \qquad X = -\frac{ma}{k}$$

$$x_{i,\text{com}} = \frac{ma}{k}(1 - e^{-(k/c)t}) + \frac{a}{2}t^2$$

and recalling Eq. (10.105):

$$x_{i,\text{com}} = \frac{ma}{k}(1 - e^{-(k/c)t}) + x_{\text{des}} \tag{10.114}$$

The first term on the right side is the feedforward compensation term. Its physical meaning is obvious: the mass m resists acceleration by a force (ma). This force compresses the spring by

$$\delta_s = \frac{ma}{k} \tag{10.115}$$

but not immediately; the dashpot prevents this. However, the second term in the parentheses of Eq. (10.114) dies out rather quickly, with a time constant $\tau_i = (c/k)$. So, in a simplified way, neglecting the effect of the dashpot, it is necessary to add δ_s to the command to compensate for the compression of the spring.

The form of the corrected $x_{i,com}$ is shown in Fig. 10.75. It shows that at the beginning it is necessary to command a fast increase to a step δ_s and then keep commanding constant acceleration a plus δ_s. At the end of phase I, at time t_1, the command pulls back quickly to the constant v motion equal to the desired x. At the beginning of the third period, at time t_2, the drive pulls back quickly by δ_s and continues moving ahead with constant deceleration. At the end of this period, at t_3, the pull back is returned so that both the command (driving motion $x_{i,com}$) and the end motion x come to rest at the same time. Now we will look at the form in which a realistic servodrive could provide the desired driving motion x_i.

10.9.2 Real Servodrive

The ideal drive of the preceding section is now replaced by a servomechanism with a feedback from x_i. This corresponds to alternative B of Fig. 10.60. In its simplest form the driving servo corresponds to the block diagram of Fig. 10.76a. In this diagram the overall gain is $K = K_p K_{CL}\ r$, and τ is a brief form of notation for what we have called τ_{CL}. We will recall Eqs. (10.34) and (10.35) and express the two basic characteristics of the drive as follows:

$$\omega_d = \sqrt{\frac{K}{\tau}} \qquad \zeta_d = \frac{1}{\sqrt{K\tau}} \tag{10.116}$$

The whole block of the servomechanism is denoted as D for "drive." The whole system is then depicted in Fig. 10.76b as two blocks D and S, where the latter is the spring-mass-damper "structure" shown in Fig. 10.73 and described by the differential

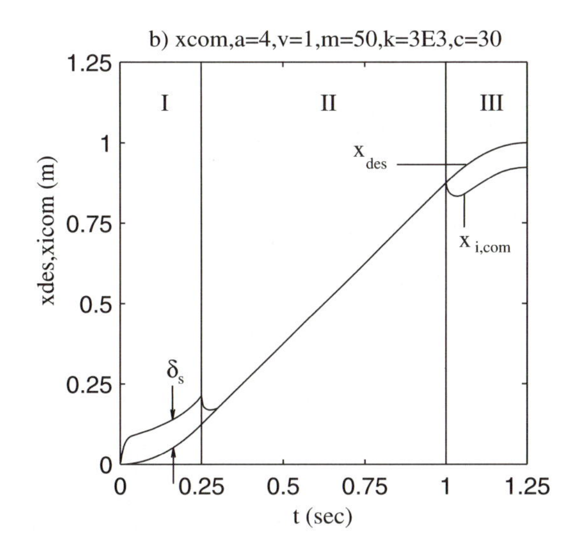

FIGURE 10.75 The feedforward corrected command using an ideal servo and an SMD system. In the acceleration phase I the drive must suddenly move ahead by δ_s and suddenly return at the end of the phase. This lead motion compresses the spring and dashpot to provide the acceleration force for the mass. In phase II no driving force is necessary; the mass is coasting with constant velocity. Phase III is the reverse of phase I.

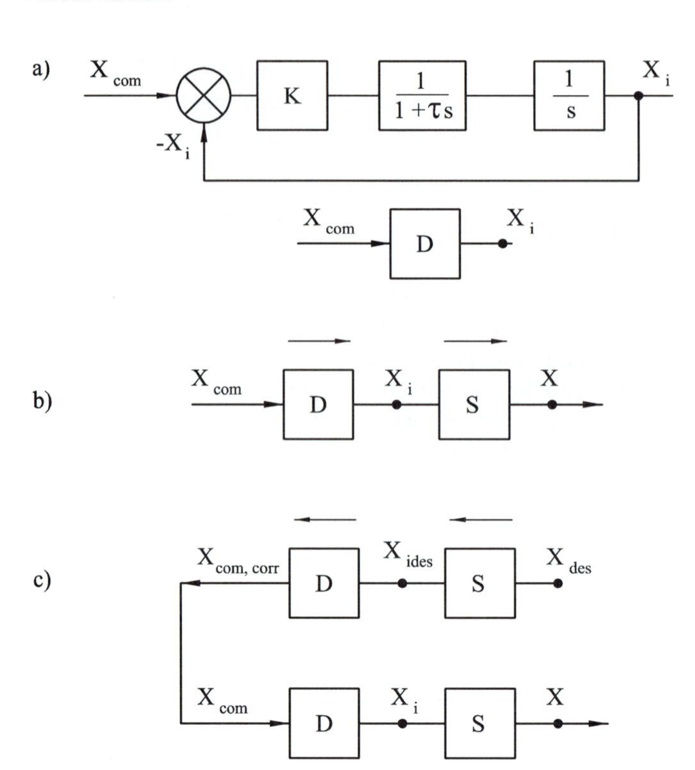

FIGURE 10.76
a) Block D (drive) represents the positional servomechanism.
b) Drive D provides the input to the spring-mass system S. c) Deriving the corrected command from the desired output and then driving the DS blocks. d) Changed notations z for x_{des} and y for $x_{i,des}$.

equation (10.90). In the direction of signal flow as indicated in Fig. 10.76b, the two blocks are described by recapitulating the differential equations in the common arrangement where the right side is the input and the left side is the dependent variable:

$$D\colon \ddot{x}_i + \frac{1}{\tau}\dot{x}_i + \frac{K}{\tau}x_i = \frac{K}{\tau}x_{com} \qquad (10.117)$$

$$S\colon m\ddot{x} + c\dot{x} + kx = c\dot{x}_i + kx_i \qquad (10.118)$$

In the "feedforward" treatment it is necessary to first go in the opposite direction (see Fig 10.76c) by defining the desired output x_{des}, deriving a corresponding desired intermediate motion $x_{i,des}$ and, furthermore, the corresponding corrected command $x_{com,corr}$. Subsequently (see the bottom path), the actual system is driven from the command as derived above. The top branch is the computer-based command generation; the bottom part is just a model of a real physical system. We will follow the command-generation path analytically for a single-coordinate system, assuming the desired motion in the form shown in Fig. 10.74. In order to simplify the notation, we use z for x_{des}, y for $x_{i,des}$, and x_{cc} for $x_{com,corr}$, (see Fig. 10.76d).

Equations (10.117) and (10.118) will be reversed:

$$S\colon c\dot{x}_{i,\text{des}} + kx_{i,\text{des}} = m\ddot{x}_{\text{des}} + c\dot{x}_{\text{des}} + kx_{\text{des}}$$

and rewritten with the new notations:

$$c\dot{y} + ky = m\ddot{z} + c\dot{z} + kz \tag{10.119}$$

where z is now a well-defined function of time,

$$D\colon x_{\text{com,corr}} = \frac{\tau}{K}\ddot{x}_{i,\text{des}} + \frac{1}{K}\dot{x}_{i,\text{des}} + x_{i,\text{des}}$$

Rewriting with the new notations, we have

$$x_{\text{cc}} = \frac{\tau}{K}\ddot{y} + \frac{1}{K}\dot{y} + y \tag{10.120}$$

It is seen from Eq. (10.120) that in order to determine x_{cc}, we not only need $\dot{y}$ and y obtainable from (10.119) but also $\ddot{y}$. Let us differentiate Eq. (10.119):

$$c\ddot{y} + k\dot{y} = m\dddot{z} + c\ddot{z} + k\dot{z} \tag{10.121}$$

Now, it is possible to proceed on the basis of the graph in Fig. 10.74, redrawn as Fig. 10.77 with the new notations. For $0 < t < t_1$,

$$z = \frac{a}{2}t^2 \qquad \dot{z} = at \qquad \ddot{z} = a \qquad \dddot{z} = 0$$

For $t_1 < t < t_2$,

$$z = z_1 + v(t - t_1) \qquad \dot{z} = v \qquad \ddot{z} = 0 \qquad \dddot{z} = 0$$

For $t_2 < t < t_3$,

$$z = z_2 + v(t - t_2) - \frac{a}{2}(t - t_2)^2 \tag{10.122}$$

$$\dot{z} = v - a(t - t_2) \qquad \ddot{z} = -a \qquad \dddot{z} = 0$$

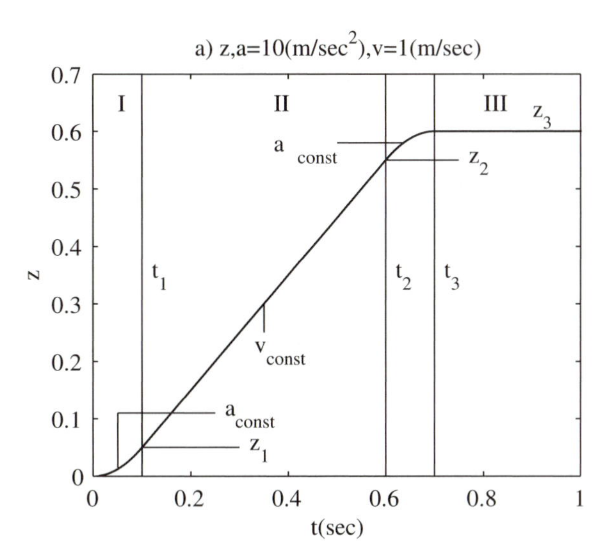

FIGURE 10.77
The desired motion z. The desired accelerations for Example 10.14 and Example 10.15 are set to $a = 10\ \text{m/sec}^2 \simeq 1\ \text{g}$.

Now continue as in Section 10.9.2, in which the solution of Eq. (10.119) was presented for phase I. For $0 < t < t_1$,

$$y = \frac{ma}{k}(1 - e^{-(k/c)t}) + \frac{a}{2}t^2,$$

$$\dot{y} = \frac{ma}{c}e^{-(k/c)t} + at \qquad (10.123)$$

$$\ddot{y} = -\frac{mak}{c^2}e^{-(k/c)t} + a$$

EXAMPLE 10.14 Feedforward Compensation in One Coordinate ▼

Express Eq. (10.123) graphically for x_{cc} versus time, together with z and y, for the particular case of the servomechanism specification for the robot (ROB/B) case formulated in Sections 10.8.1 and 10.8.2. Assume the following:

$k = 40{,}000$ N/m, $m = 100$ kg, $c = 60$ N/(m/sec)
$\omega_s = 20$ rad/sec, $f_s = 3.2$ Hz, $\zeta_s = 0.015$

where subscript s means "structure," that is, the SMD system. We choose a servomechanism "faster" than the structure; given $\tau_{CL} = 0.02$ sec, apply $K = 40$/sec, resulting in [see Eq. (10.116)] $\omega_d = 44.72$ rad/sec and $\zeta_d = 1.11$, where the subscript d means "drive." For the desired motion, choose $a = 10$ m/sec^2, $v = 1$ m/s, $t_1 = v/a = 0.1$ sec, and $x_1 = 0.05$ m.

Plots of the intermediate coordinate y for these parameters are shown in Fig. 10.78. They have the form determined and shown in Fig. 10.75. According to the preceding text, we can now use Eq. (10.120) to determine the proper x_{com}. However, there is a fundamental difficulty here. Once an x_{com} has been determined and the signal flow is actually sent in physical reality, the system will initially be at standstill; this means that the initial conditions at the intermediate stages will be $x_i = 0$, and $\dot{x}_i = 0$. However, here we have at $t = 0$ an initial desired velocity formulated in Eq. (10.123):

$$\dot{y}_0 = ma/c = 16.66\ m/\text{sec}$$

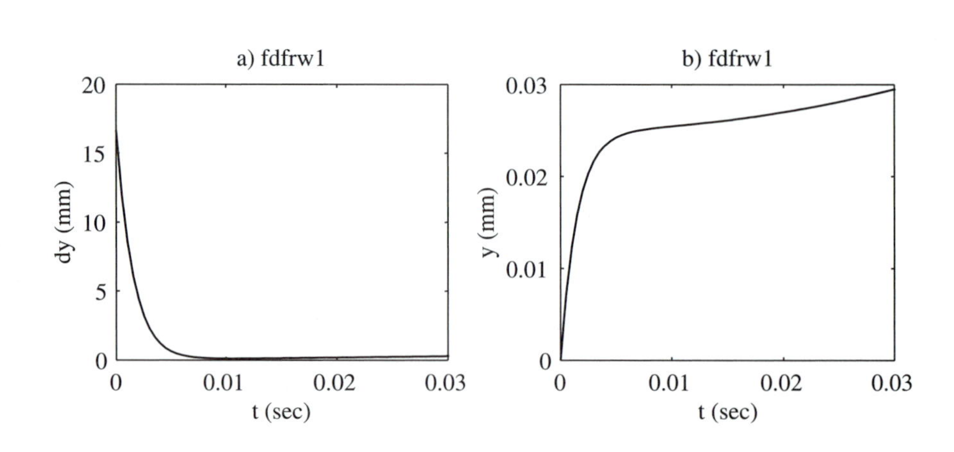

FIGURE 10.78 The "robot case," ROB/B, with feedforward, as in Example 10.14. Plot of the desired intermediate motion y and of its velocity dy, for the acceleration phase. As a time function, it is similar to the drive x_i in the case with the ideal drive in Fig. 10.75.

This actually implies an infinite positive initial acceleration, not a negative value:

$$\ddot{y}_0 = -mak/c^2 + a = -11{,}121\ m/\text{sec}^2$$

as postulated above.

In order to achieve at least approximately the rather high starting velocity, servo D would need an initial impulse. These problems can best be overcome by assuming a slight delay in $y(t)$, which would imply $\dot{y} = 0$. Analytically, this would lead to manipulations of a complexity beyond the scope of this text and the current section on servos. It would also not serve any practical purpose, since in practice the determination of x_{com} is carried out in the command-generating numerical controller. Therefore, let us end this discussion here and proceed to the next section in which we use a numerical procedure. ▲

10.9.3 Numerical Derivation of the Feedforward Compensation

Here, use Eqs. (10.119) and (10.120) by differentiating numerically over small time steps dt. In order to be able to do so, we need some preview of the input, at least over one time step. This does not pose any difficulty, since the desired motion has been determined beforehand, as the whole NC part program, or in the off-line programming function for robots. It is therefore possible to prepare the $x_{com,corr}$ also beforehand, off-line, by reading the input one step ahead. Even if the feedforward correction was required in "real-time," it would simply delay the command by one time step. We will now discuss the structure of the program to be used to derive x_{cc} from the known formulation of $z(t)$. The program will be run in loops by incrementing t_n by dt. Assume the following initial state of the system at standstill:

$$y_1 = z_1, \qquad \dot{y}_0 = 0, \qquad \dot{z}_0 = 0 \tag{10.124}$$

Often, we will simply also have $y_1 = z_1 = 0$. The initial zero velocities eliminate requests for infinite accelerations. The program structure is as follows:

Determining x_{cc}. Verification of the Effect

Inputs:

System parameters: K, τ, k, m, c

Motion parameters. For the motion as defined in Fig. 10.77, they are: a, v, t_1, t_2, t_3.

Procedure parameters: dt, total number of steps N.

Section 1: Defining the desired motion.

For $n = 1{:}N$

Define z_n.

Alt 1:

Define also $\dot{z}_n$ and $\ddot{z}_n$ from analytical expressions such as in Eq. (10.122). However, in most instances, analytical expressions may not be available, and the z derivatives will be derived numerically in Alt 2, in Section 2 and Fig. 10.79.

Section 2:

Derive derivatives of z. Determine y and its derivatives. Determine x_{cc}.

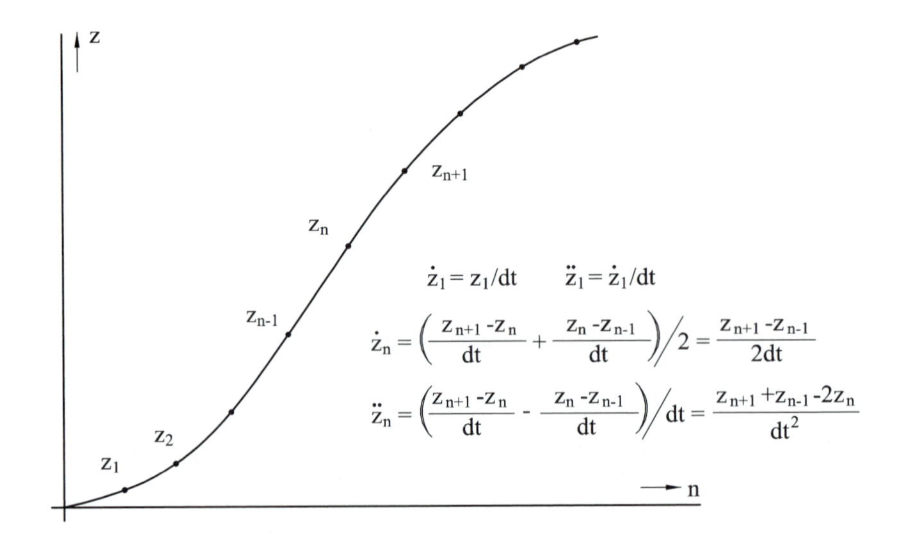

FIGURE 10.79 Scheme for obtaining the derivatives of the desired motion numerically as part of command generation that includes the feedforward strategy.

Initial conditions:

Assume $\dot{z} = 0$, for the derivation of the first $\ddot{z}$.

Assume $\dot{y} = 0$ by setting "$\dot{y}old$" $= 0$; this will be used for $\ddot{y}$. Set $y_1 = z_1$.

Looking ahead by one step limits this section to:

$n = 1{:}\,(N - 1)$,

Alt 2:

For $n < 2$, $t_1 = \mathrm{d}t$

$\dot{z} = z_1/\mathrm{d}t$

$\ddot{z} = \dot{z}/dt$ (meaning: $\ddot{z} = (\dot{z} - 0)/\mathrm{d}t$).

else (see Fig. 10.79)

$\dot{z} = (z_{n+1} - z_{n-1})/2/\mathrm{d}t$

$\ddot{z} = (z_{n+1} + z_{n-1} - 2z_n)/(\mathrm{d}t)^2$

Using Eq. (10.119), determine $\dot{y}$, y; determine $\ddot{y}$

$\dot{y} = (m\ddot{z} + c\dot{z} + kz_n - ky_n)/c$

$\ddot{y} = (\dot{y} - \dot{y}old)/\mathrm{d}t$

$\dot{y}old = \dot{y}$; this provides for an update of $\dot{y}old$

$y_{n+1} = y_n + \dot{y}\mathrm{d}t$

Using Eq. (10.120), determine x_{cc}:

$x_{cc} = (\tau\ddot{y} + \dot{y} + Ky_n)/K$

Section 3

Verification of the effect of the feedforward correction. Simulating the physical action, accept x_{cc} (still in the same loop of the program) as the x_{com}. Using Eqs. (10.117) and (10.118), derive x_i and x.

Initial conditions: $x_{i,1} = z_1, \dot{x}_i = 0, x_1 = z_1, \dot{x} = 0$.

$$\ddot{x}_i = [K(x_{cc} - x_{i,n}) - \dot{x}_i]/\tau$$

$$\dot{x}_i = \dot{x}_i + \ddot{x}_i dt, \qquad x_{i,n+1} = x_{i,n} + \dot{x}_i dt$$

$$\ddot{x} = (c\dot{x}_i + kx_{i,n} - c\dot{x} - kx_n)/m$$

$$\dot{x} = \dot{x} + \ddot{x}dt, \qquad x_{n+1} = x_n + \dot{x}dt.$$

Plot: z, y, x_{cc}, x_i, x, versus t.

The results of this exercise may be compared with simulating separately the action without the feedforward correction. This means running Section 3 above, with an input of $x_{com}(t) = z(t)$ of Section 1 above.

EXAMPLE 10.15 Using Numerical Derivation of the Feedforward Command: Motion in One Coordinate ▼

No Feedforward Compensation

First, take a system similar to the one used in Example 10.14 as regards the drive, the driven structure, and the desired path, as outlined in Fig. 10.77. The case is first presented without feedforward correction, and it follows just Section 3 of the algorithm presented in the preceding paragraph, except that the compensated command x_{cc} is not used; instead, use the one corresponding to the desired form of the x motion, now designated as z. The parameters of the case follow: $a = 10$ m/sec^2, $v = 1.0$ m/sec, $t_1 = 0.1$ sec, $t_2 = 0.6$ sec, $t_3 = 0.7$ sec, $z_1 = 0.05$, $z_2 = 0.55$, $z_3 = 0.6$, $dt = 0.001$ sec, and $N = 1200$. The servo drive is specified by $\tau = 0.02$ sec, $K = K_p K_{CL} r = 40$/sec, $\zeta = 0.56$, $\omega_d = 44.72$ rad/sec, and the spring-mass-damper (SMD) structure has $k = 40{,}000$ N/m, $m = 100$ kg, $c = 60$ N sec/m, and $\zeta_s = 0.015$.

The result of the simulation of this case without feedforward compensation is shown in Fig. 10.80a, where the noncompensated command z (used as x_{com}), the intermediate displacement x_i, and the motion x of the mass of the structural system are plotted. Note that, although the system is not unstable, because the SMD structure is outside of the loop and positional feedback is taken from the

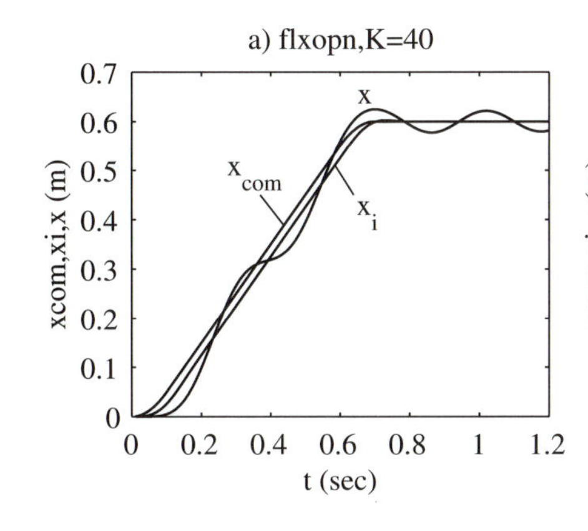

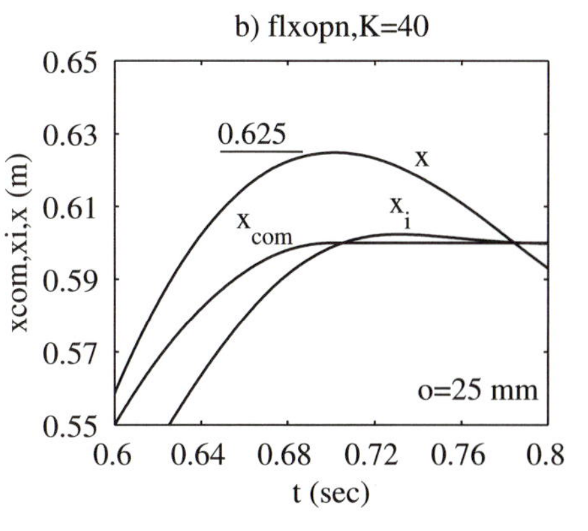

FIGURE 10.80 The case ROB/B without feedforward correction, as in Example 10.15 (similar to Fig. 10.72b). Detail of the end phase of the motions. The commanded position was 0.6 m. Overshoot is 25 mm, and vibratory motion with PTP magnitude 48 mm persists with very slow decay. Very bad performance.

encoder signal x_i, the mass is strongly vibrating with slow decay corresponding to the low damping, with an initial peak-to-peak (PTP) displacement of 0.048 m (48 mm). The initial overshoot in x of 25 mm is seen in detail in graph *(b)*, where a moderate overshoot of the intermediate point x_i is also clearly seen.

Feedforward Compensation Applied

Next, the feedforward compensation is derived and applied, and the result is illustrated in the following graphs. The listing of the Matlab program "fdfwd1" is printed in Fig. 10.81. The intermediate motion x_i is shown in a detail of its end phase in Fig. 10.82, together with the desired motion z and the actual motion x of the mass of the system. The characteristic pull-back at t_2 and return at t_3 is executed by x_i in the amounts of about 0.003 m (30 mm). Its effect is seen in the closeness of x to the desired motion z. The former is undershooting at $t_3 = 0.7$ sec by about 0.2 mm and settles down to an error of only 0.03 mm after 60 msec and then down to the desired value of 0.6 m.

The shape of the compensated command x_{cc} is presented in Fig. 10.83, together with x_i and x. The necessary compensating motions at $t = 0, t_1, t_2, t_3$ are large, about $\pm$ 4 m. It is obvious that these commands would impose extreme demands on the servomotor that would exceed the transient current limits. The simulation results presented here are not realistic because the servomotor current limitation was not included in the corresponding computer program, just to indicate how efficient feedforward compensation could be if it worked. In order to check, the servomotor accelerations called in this case have been plotted out in the scale of accelerations of the x_i coordinate (ddx_i) in Fig. 10.84. They reach $\pm$ 8000 m/sec^2. Considering a 25-mm-pitch leadscrew between the servomotor and the table displacement x_i, this translates into servomotor rotational accelerations on the order of $\omega = 2 \times 10^6$ rad/sec^2. For a typical rotor inertia of $J = 3e - 4$ kgm^2 this would require a torque $T = 600$ Nm and a corresponding current. However, a typical high-performance servomotor with the above inertia could deliver only about 40 Nm peak torque. Consequently, the excellent feedforward compensation performance could obviously not be achieved for the desired motion as specified above.

One way to solve the problem is to reduce the desired acceleration, say down to $a = 2$ m/sec^2. The case has again been run through simulation with this as the only difference. The plot of x_{cc}, x_i, and x is now as shown in Fig. 10.85. The compensatory changes in x_{cc} are reduced from 4 m to about 0.75 m, and the demands on the servomotor acceleration as reflected in $\ddot{x}_i$ *(ddxi)* are also reduced five times. This may still not be within the capability of a high-performance servomotor, and further reduction of the desired acceleration may be necessary.

Another way in which feedforward compensation can be successfully used for rather high-performance motions is discussed in the next exercise. ▲

EXAMPLE 10.16 Feedforward Compensation for Constant Jerk Motion ▼

Jerk is the word used for the third time derivative of motion, that is, the first derivative of acceleration. A motion in one coordinate may then have a different profile than the one that has been used according to Fig. 10.77. Instead of using a time block of constant acceleration at the beginning and end of the move, with constant

FIGURE 10.81
Listing of the Matlab program "fdfwd1" for simulation of feed-forward compensation of one coordinate motion. To run, choose acceleration (a).

```
function[z,dz,ddz,y,dy,xcc,ddxi,xi,x,t,t1]=fdfwd1(a)
%showing effect of feedforward compensation in one coordinate
%refer to Sec.10.9.3 and EX.10.15. The desired motion includes
% 1)constant acceleration a, 2)constant velocity v and 3)constant deceleration.
%Toatal motion is 0.6m, end of Sec.1 is at t1,x1, of Sec.2 is t2,x2 and end of Sec.3 is
%at t3,x3,followed by a settling section. Servo parameters are velocity, closed-
%loop time constant tau(sec) and open-loop gain K(1/sec). Desired motion is z,
%desired intermediate coordinate is y, compensated command is xcc, actual inter-
%mediate motion is xi and actual end motion is x. First derivatives are such
%as dx,second is ddx.
tau=0.02;k=4e4;m=100;c=60;dt=0.001;K=40;ddxi(1200)=0;
t(1)=0;xi(1)=0;x(1)=0;dxi=0;dx=0;v=1;
t1=v/a;x1=(a/2)*t1^2;t2=(0.6-2*x1)/v+t1;t3=t2+t1;
%defining desired motion as in Fig.10.77:
for n=1:1200
t(n)=(n-1)*dt;
if t(n)<=t1
z(n)=(a/2)*t(n)^2;
elseif t(n)>t1 & t(n)<=t2
z(n)=x1+1.0*(t(n)-t1);
elseif t(n)>t2 & t(n)<t3
z(n)=x1+1.0*(t(n)-t1)-(a/2)*(t(n)-t2)^2;
else
z(n)=0.6;
end
end
%initial conditions of y and z
y(1)=0;dyold=0;dy(1200)=0;dz(1200)=0;ddz(1200)=0;
%deriving dz,ddz
for n=1:1199
if n<2
dz(n)=(z(n+1)-z(n))/dt;
ddz(n)=dz(n)/dt;
else
dz(n)=(z(n+1)-z(n-1))/(2*dt);
ddz(n)=(z(n+1)+z(n-1)-2*z(n))/dt^2;
end
%deriving y
dy(n)=(m*ddz(n)+c*dz(n)+k*z(n)-k*y(n))/c;
ddy=(dy(n)-dyold)/dt;
dyold=dy(n);
y(n+1)=y(n)+dy(n)*dt;
%formulating xcc
xcc(n)=(tau*ddy+dy(n)+K*y(n))/K;
end
xcc(1200)=y(1200);
%driving the whole system by the xcc command
for n=1:1199
erp=xcc(n)-xi(n);
v=erp*K;
ddxi(n)=(v-dxi)/tau;
dxi=dxi+ddxi(n)*dt;
xi(n+1)=xi(n)+dxi*dt;
ddx=(c*dxi+k*xi(n+1)-c*dx-k*x(n))/m;
dx=dx+ddx*dt;
x(n+1)=x(n)+dx*dt;
end
%plot(t,z,t,y,t,xcc),(t,z,t,txi,t,x)
```

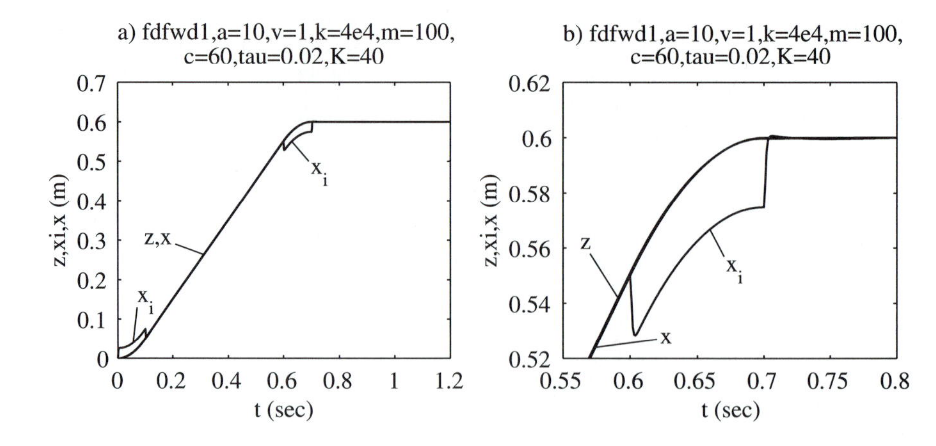

FIGURE 10.82
The desired motion x_{des} (z) and the motion x resulting from the feedforward compensated command in Example 10.15 are very close to one another; the error is negligible. The compensated x_i shows the typical down and up phase needed to cause compensation for the start and end of deceleration. However, this excellent performance is not realistically achievable because of hardware limitations.

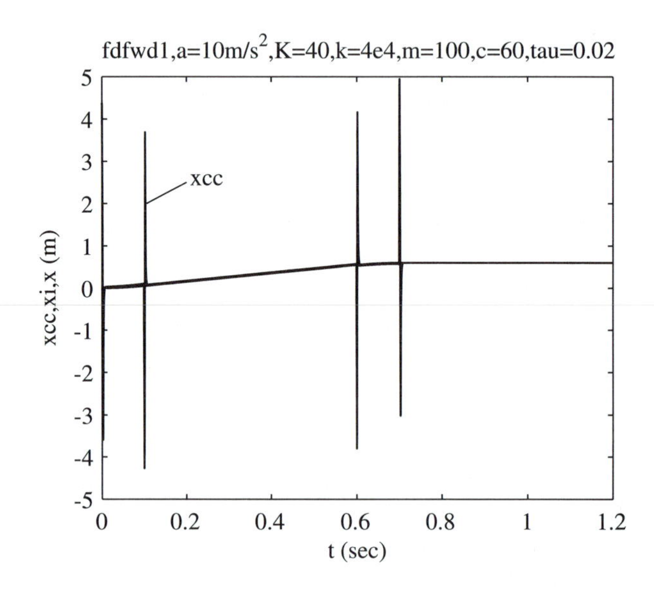

FIGURE 10.83
The feedforward compensated command x_{cc} of Example 10.15 contains large swings that tax the servomotor current.

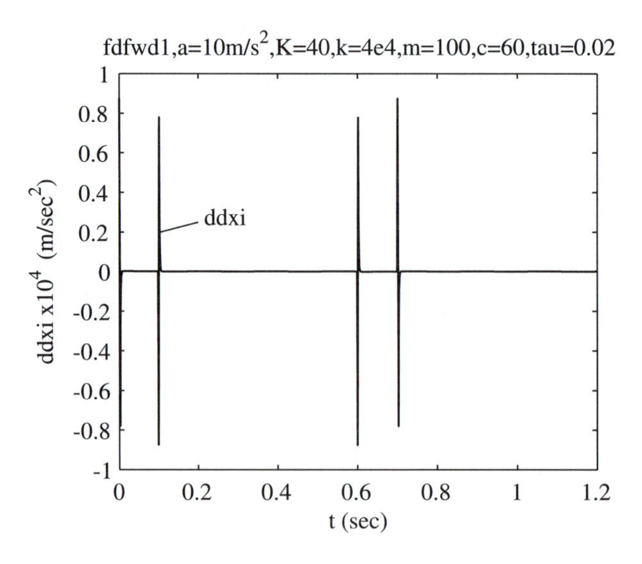

FIGURE 10.84
As a result of the swings in x_{cc}, the demanded accelerations ddx_i to be executed by the servomotor exceed the motor's capability.

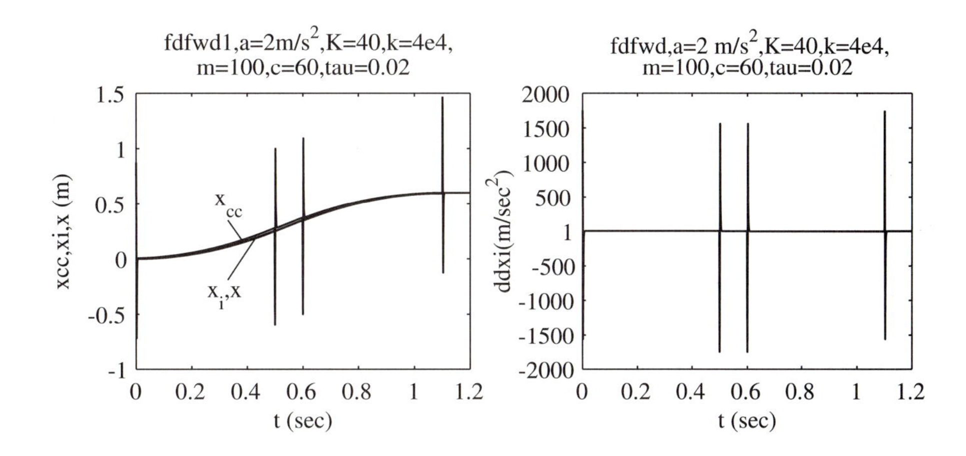

FIGURE 10.85 The compensated command x_{cc} and the corresponding motions x_i and x when desired accelerations are reduced five times to $a = 2$ m/sec^2. The swings in x_{cc} are also reduced five times. The plot on the right shows that the demanded acceleration ddx_i has also dropped about five times. The demanded acceleration calls for a motor torque that may still be rather high. Further reduction of the available acceleration may be needed unless the peaks in the demand are filtered down.

velocity in between, the initial (and final) blocks are executed with the acceleration rising (falling) uniformly through half of the block and then falling (rising) back to zero in the other half of the block. This is illustrated for the desired profile of motion z in Fig. 10.86. At the left is acceleration, and at the right is velocity. For the first 0.05 sec, acceleration (ddz) rises under constant jerk $q = 10/0.05 = 200$ m/sec^3, up to $a = 10$ m/sec^2. Then, from $t = 0.05$ sec to $t = 0.1$ sec, acceleration decreases under constant negative jerk, down to $a = 0$. Between $t = 0.1$ and $t = 0.6$ the driven slide moves with a constant velocity (dz) of $v = 0.5$ m/sec. At the end, in the last time block, the acceleration profile is reversed around the $a = 0$ axis, and velocity is mirrored around the $t = 0.1$ and 0.6 sec axes.

A computer program has been written in which the first part defines the acceleration profile $\ddot{z}(t)$, and the next steps determine $\dot{z}(t)$ and $z(t)$ by digital integration. From there on, the program represents the same servomechanism and the spring-mass system as in Example 10.15, driven by the x_{cc} command derived from the above-defined desired z motion.

First, the *uncompensated motions* are plotted in Fig. 10.87. If compared with those in Fig. 10.80 of Example 10.15, where constant acceleration blocks were used, and PTP vibration x value of 48 mm was recorded, the *PTPx* magnitude is now 26 mm. This just about corresponds to the 0.5 m/sec velocity, lower than the 1.0 m/sec, velocity of the former case. Also the overshoot of x_i is smaller in about

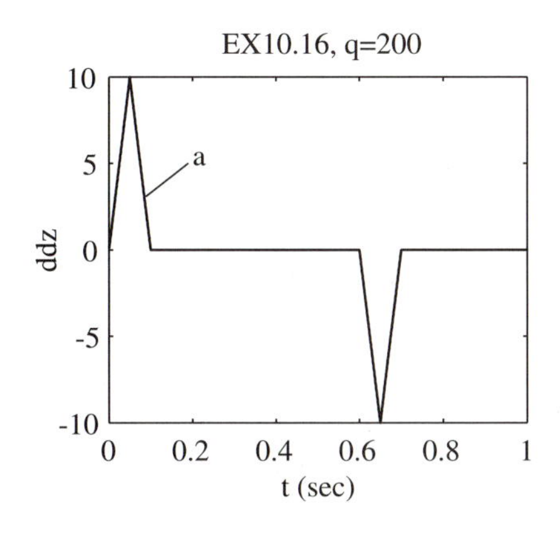

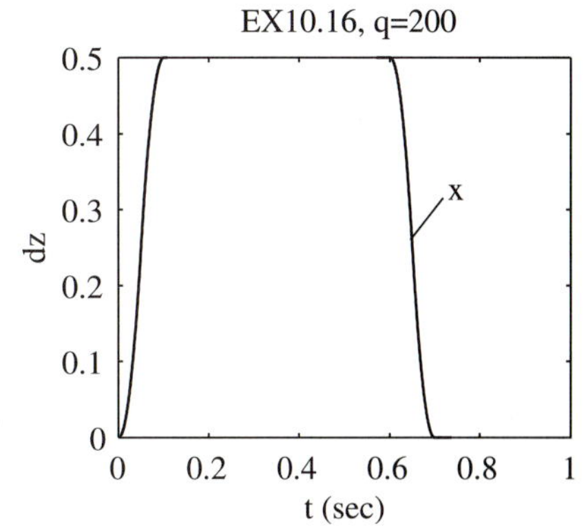

FIGURE 10.86 As an alternative solution, constant acceleration blocks in the desired motion are relaxed to variable accelerations with $a = 10$ m/sec^2 peak, under constant jerk: Example 10.16.

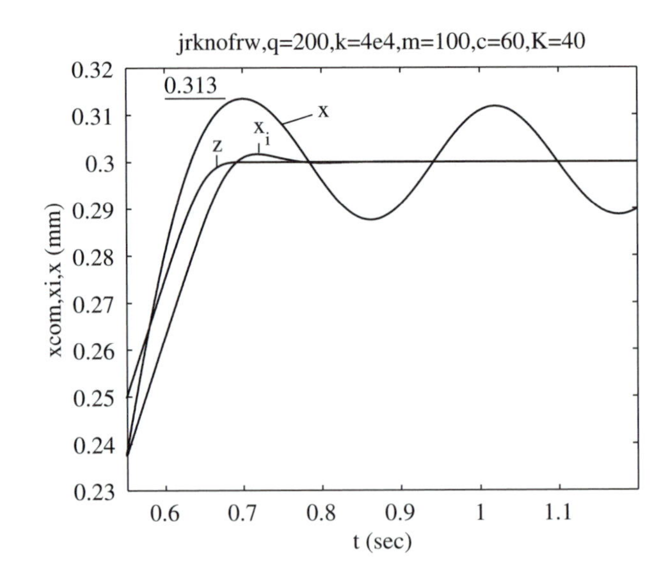

FIGURE 10.87
With the constant jerk command but without feedforward compensation, little improvement in the final motion x is achieved. The overshoot is reduced from 26 mm down to 13 mm, mainly due to the reduced average acceleration from 10 m/sec^2 to one half of that.

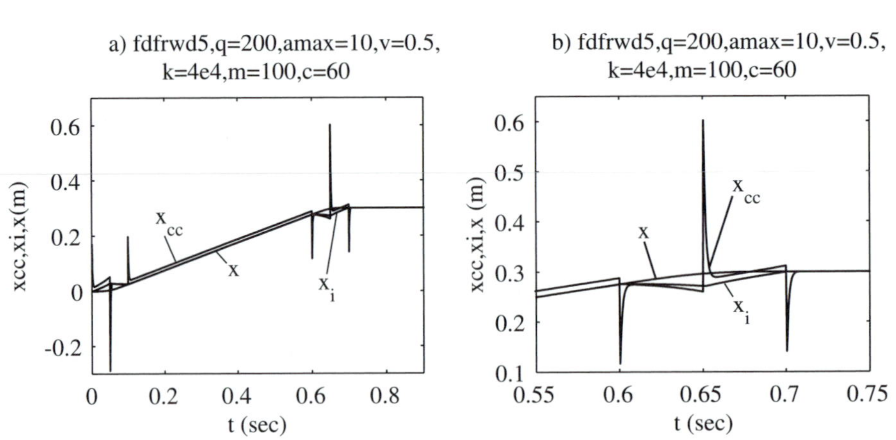

FIGURE 10.88
With constant jerk and feedforward compensation, the swings in x_{cc} are strongly reduced: Example 10.16. The resulting motion x of the mass of the SMD system is very accurate. At $t =$ 0.7 sec when the command stops, x is short of the target of 0.3 m by 0.08 mm, and at $t = 0.8$ sec it comes close to within 6 μm, without any overshoot. The swings in x_{cc} are reduced to 0.3 m.

the same ratio. The *compensated motions* are shown in Fig. 10.88. The resulting motion is rather accurate, without overshoot. The compensating transients in the x_{cc} commands peak at only about 0.3 m (300 mm), which is 13 times less than those in Fig. 10.80. This is a substantial reduction and results in substantially reduced demands on the servomotor accelerations as shown in Fig. 10.89. Their peaks are 600 m/sec^2 rather than the 8000 m/sec^2 values of Fig. 10.80. The corresponding peak torque of 45 Nm is within practical possibility. Also the compensated form of x_i is much more moderate.

It has now been demonstrated in Examples 10.15 and 10.16 that feedforward compensation may be very efficient in reducing the error motions of a driven SMD system and that this can be done in a realistic way, either by keeping the desired acceleration at an appropriate limit or by including constant jerk instead of constant acceleration blocks in the formulation of the desired motions, which achieves an even better performance. ▲

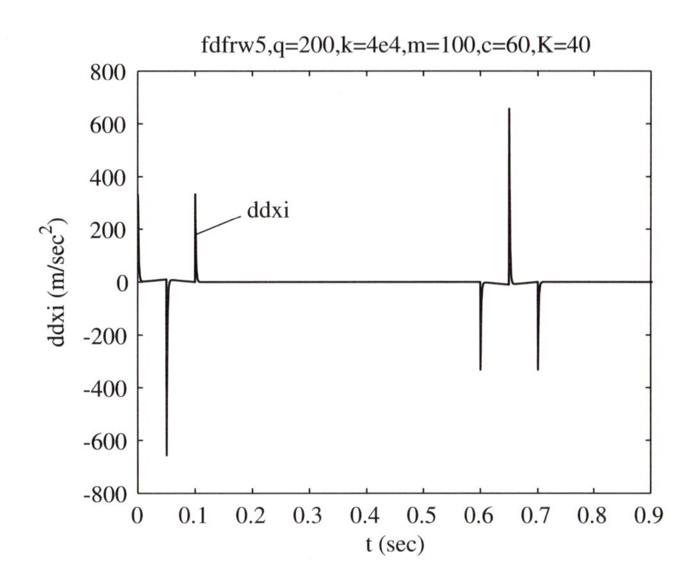

FIGURE 10.89 With constant jerk commands and feed-forward compensation as in Example 10.16, the peak instantaneous demand on acceleration ddx_i is down about 15 times from the original case of Example 10.15 to a value that is realistically achievable.

10.10 SIMPLIFIED ROBOT KINEMATICS AND DYNAMICS

10.10.1 Introduction: Types of Robots and Their Uses

Industrial robots are stationary, computer-controlled manipulators. They are distinguished from mobile robots used for earthly, submarine, or lunar explorations, nuclear power maintenance, and so on. The various applications of industrial robots in factory automation are listed and briefly characterized in Fig. 10.90. They are broadly classified into those tasks in which the robot moves a tool: paint spraying, spot welding, arc welding, light machining (deburring, drilling), and those where it moves a workpiece: loading of presses and of machine tools, die-casting machines, investment casting, palletizing, and so on. In the table of Fig. 10.90 they are also classified as regards the required accuracy, speed, path form, and useful load.

Most often these robots have six degrees of freedom, or coordinates (axes, joints): three in the base structure and three in the wrist attached at the end of the base structure. The three base joints are made in various combinations of translative (linear) and rotary motions. This is shown diagrammatically in Fig. 10.91. Examples of various makes and executions are shown in the following illustrations.

The joints of the robot in Fig. 10.92 are all of the rotary type. This kinematic arrangement has become known as "articulate." Its advantage is the high flexibility of its reach, which may wind around an obstacle; it is very apt for loading machine tools (Fig. 10.93), as well as for spot welding of car bodies (Fig. 10.94). In Fig. 10.93 three machined workpieces are seen in the front, one workpiece still to be machined is seen being loaded into the turning center at the right, and another unmachined workpiece is waiting at the end of the belt transporter at right bottom. Furthermore, a Zygo laser interferometer is available to measure the machined parts as the robot holds them in the instrument before putting them down in the pallet at left. Figure 10.94 illustrates the ability of the robot wrist to move the welding jaws along the door frame while it is dragging both the electric cables and the hydraulic hoses that deliver the clamping power. Still another articulate robot, shown in Fig. 10.95, is specialized for spray

	Positioning Error			Speed			Path Form			Useful Load		
	Large	Medium	Small	Large	Medium	Small		Contin. Path		Large	Medium	Small
Application Field	1mm (40 mil)	0.5mm (20 mil)	0.2mm (10 mil)	>5 ft/sec	>1 ft/sec	<1 ft/sec	PTP	Known	Un-known	>50 lb	>10 lb	<10 lb
Moving tool												
Spraying	X				X			X			X	
Spot welding		X		X			X			X		
Arc welding		X			X	X		X			X	
Machining		X				X		X	X	X*	X	
Assembly			X	X			X					X
Moving workpiece												
Press loading	X			X			X				X	
Large forging hammer, press	X			X			X			X		
Unload die casting mach.	X			X			X				X	
Large machine tools		X	X	X			X			X		
Palletizing	X			X			X			X	X	

*Machining Forces

FIGURE 10.90
Applications of industrial robots.

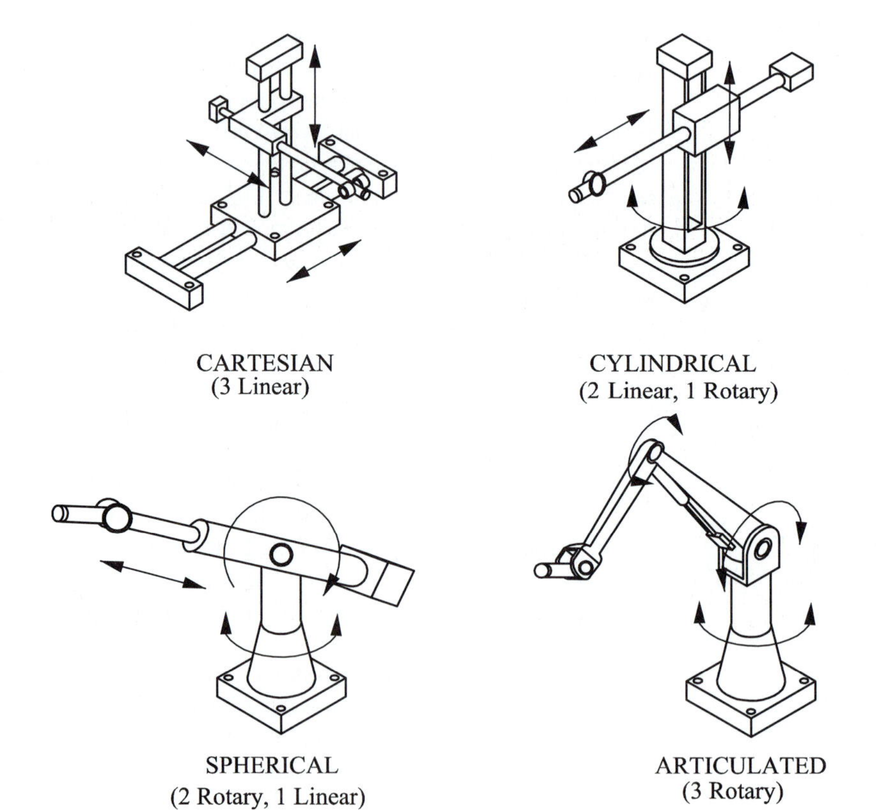

FIGURE 10.91
Four major coordinate systems for industrial robots resulting from combinations of translative and rotary motions.

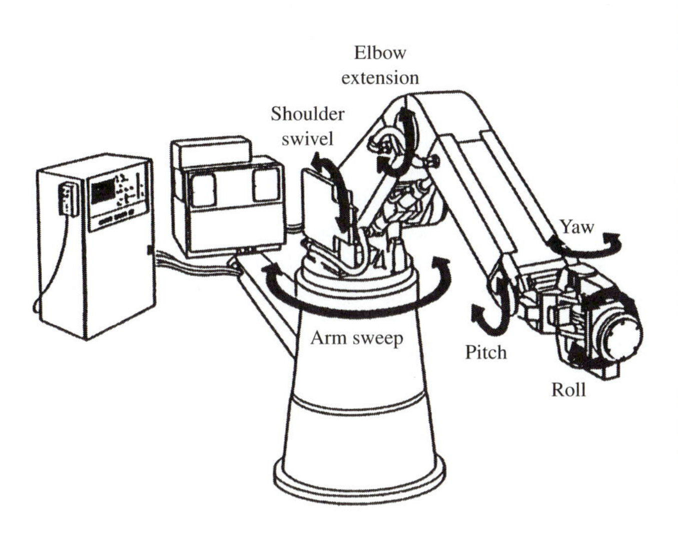

FIGURE 10.92
Robot with six rotational degrees of freedom, three in the main frame and three in the wrist; so-called articulate configuration. [MILACRON]

FIGURE 10.93
The articulate robot of Fig. 10.92 used for handling parts for two turning centers. [MILACRON]

FIGURE 10.94
The articulate robot of Fig. 10.92 used for spot welding automobile door frames. [MILACRON]

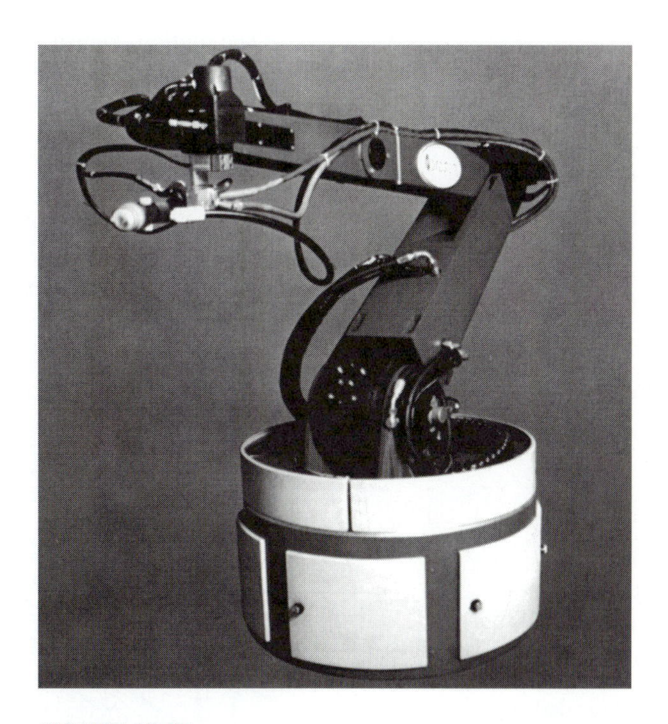

FIGURE 10.95
Robot for spray painting. This is one of the most popular tasks for robots. It works in a teach-playback mode. [NORDSON]

painting. Its work is based on the use of a teaching simulator, which has the same structure and kinematics, including position-reading encoders or synchros, but lacks the drives. The spray gun attached to the simulator is first manually moved by a skilled painter-operator, and all the moves are recorded. The record is used to control the actual robot; the drives repeat the same, often complex, motions in the automatic painting cycle.

Many types of robot kinematics and structural designs are used in a great variety of applications. The few pictures presented here showed rotary-joint machines, but this does not mean that they prevail. One supplier uses modular elements to produce different kinematics that may work in spherical, cylindrical, and Cartesian coordinates. Combinations of translative and rotary motions are used in many assembly tasks of the pick-and-place kind (see Chapter 11). Gantry-type Cartesian coordinate robots of a smaller but fast type have been installed by IBM to assemble printed circuit boards; they pick the various chips from the magazines and insert them in the proper locations. On a larger scale, a gantry system is used in the automobile underbody assembly, as shown in Fig. 10.96.

Except for the strictly Cartesian system, all the other kinematic configurations involve coordinate transformations. The programming tasks may by defined in the Cartesian "world coordinates," but the individual robot coordinates and their drives must be controlled in their own axes. Each axis or joint is driven by a positional servomechanism, in either the *A* or *B* arrangements of Fig. 10.60. Each coordinate is moved by means of some kind of a transmission that employs a positional feedback device (an encoder).

The motion-control alternatives are expressed in Fig. 10.97. Sensory feedback is used only in selected applications such as arc welding (a vision sensor to guide the arc along the seam) or insertion tasks in assembly (a touch sensor). Four different methods of programming are listed in the figure. The most common one is A: Teach by manual

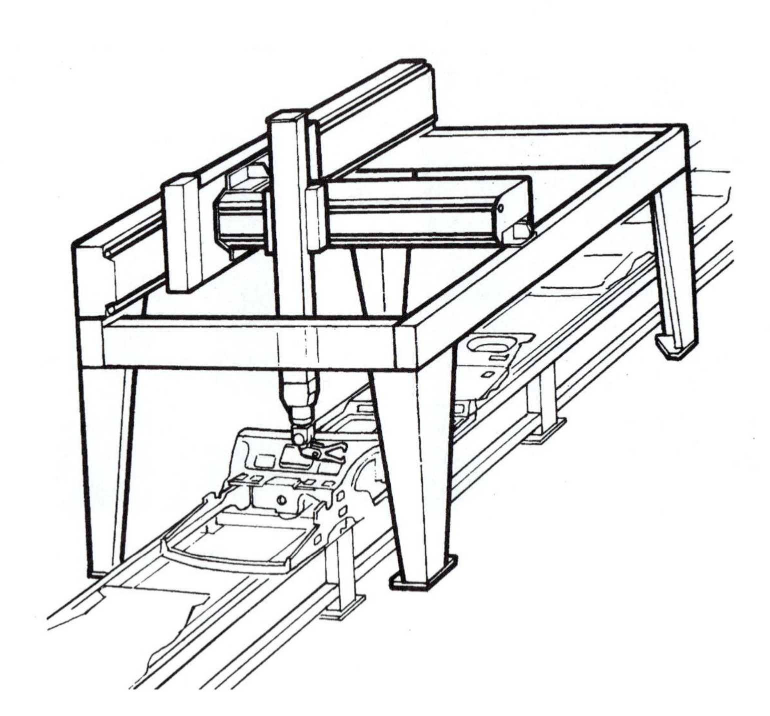

FIGURE 10.96
A gantry type Cartesian-coordinate robot used in car underbody assembly. (Reprinted from *Handbook of Industrial Robotics,* ed. S. Y. Nof [New York: J. Wiley & Sons, 1985].

FIGURE 10.97
Four different ways of programming robot motions.

Control System of Robots

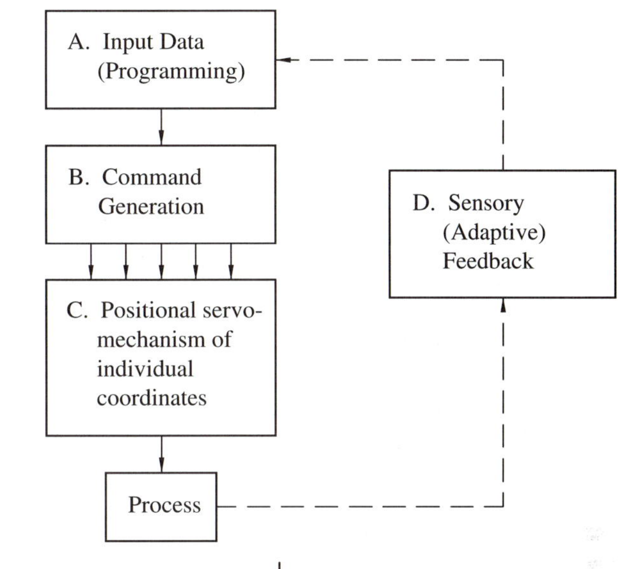

Programming

A. Teach by manual control and play-back.

B. Teach by guide-by-hand and play-back.

C. Off-line control (derive programming coordinates from drawings).

D. Tracing.

Command Generation

Check acceleration.

a) Point-to-point: Interpolation (quasi-linear, cubic) between points in joint coordinates.

b) Continuous path: Interpolation between a succession of points in X, Y, Z (special case: straight line) or joint coordinates.

control from a joystick panel (or toggle-switch panel, etc.) and play back. The most advanced method, still mostly under development, is C, the off-line programming mode. This is analogous to the task of NC programming of machine tools.

An example of method B is programming a robot for spray painting. The robot hand carries the spray gun. In the teaching mode an experienced spray painter holds the gun and does the spraying, executing the motions necessary to cover evenly all the surfaces of the workpiece. These surfaces may be complex and need combinations of rotary and linear motions. All the motions are recorded and then repeatedly played back when the robot alone does the work. The structure used in the recording phase is kinematically identical to that of the robot and includes identical feedback transducers but not the drives, so that it is easier to drive this simulator from the end of the kinematic chain. The initial designs used synchro resolvers for feedback, and these emit an analog, phase-modulated signal, which used to be recorded on magnetic tape. Lately, however, feedback is digital, as is the recording device. In method D a tracing sensor is used instead of the tool; it is driven along a template or a model while being driven in the basic three coordinates. It may be recording a signal for an additional coordinate on the wrist of the robot.

The command-generation task used in method C is similar to that used in NC machine tools; it employs linear, circular, and spline interpolation methods. Compared to machine tools, most of which work in three coordinates, the path generation for robots is more complex, generally involving six coordinates.

Due to the high inertia of robot arms and the rather high flexibility of the transmissions, the actual path of the end effector differs from the programmed path. The methods of suppressing static and dynamic errors such as those discussed in the preceding text, accelerometric feedback and feedforward compensation, are very important. A schematic illustration of the need for coordinate transformation is given in Fig. 10.98. The drawing at the top shows an articulate structure (three rotations) with additional degrees of freedom in the wrist, which carries a tool. The motion task in the world coordinates sounds simple: move the tool end on a straight line a while maintaining the tool shank in the (a, b) plane under a constant inclination α relative to the line a. Simple imagination reveals that this world-coordinate, 3D, straight motion

EXAMPLE: Joint coordinates A, B, C, D, E.

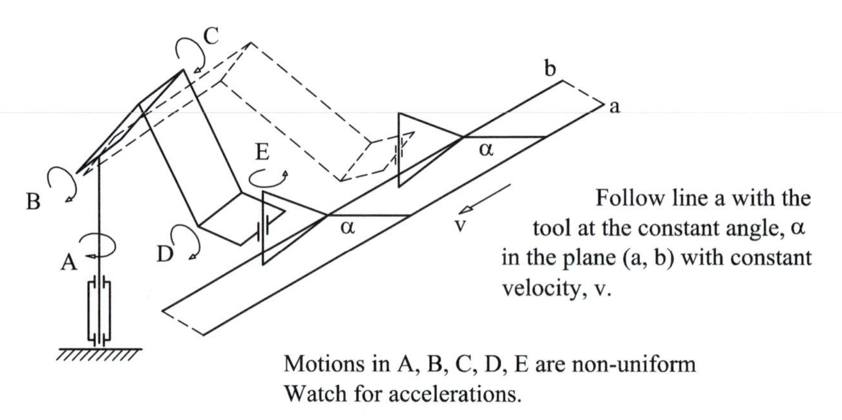

Simple example:

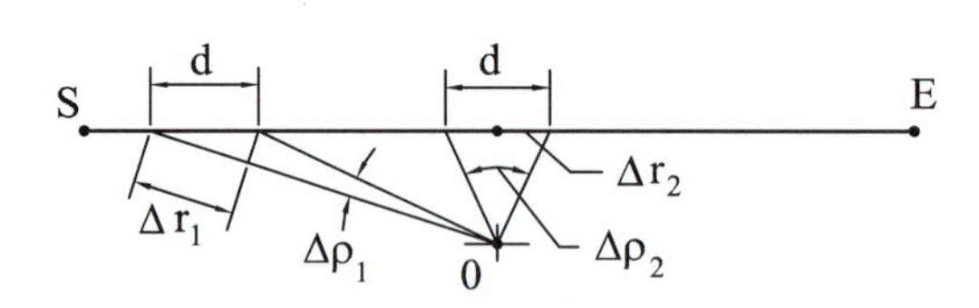

Extreme:

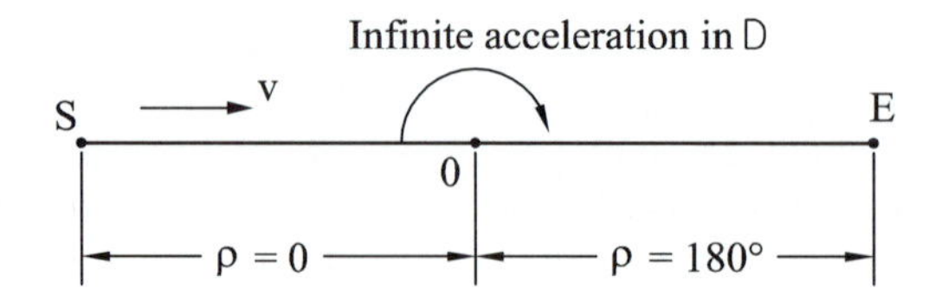

FIGURE 10.98
Joint versus world-coordinate system. Constant-velocity motion along a straight line in the world coordinates results in non-uniform motions in the robot coordinates. A 2D polar system example shows how increments in the radial and angular motions vary during constant-velocity motion in a straight line.

involves the action of all six coordinates. The bottom drawing shows a simplified, 2D, polar system with a constant-velocity movement on a straight line from S to E. In neither the rotary coordinate ρ nor the radial coordinate r does this represent a uniform-velocity motion.

10.10.2 Simplified Kinematics

This discussion is limited to a two-coordinate system with one rotary and one translative joint in the "polar" configuration (see Fig. 10.99). First, the so-called forward transformations, from the joint to the Cartesian world coordinates, are introduced:

$$\begin{aligned} x &= r \sin \phi \\ y &= r \cos \phi \end{aligned} \tag{10.125}$$

Inverse Transformations

These transformations are needed because our thinking about the desired path is all done in the (x, y) system from which we have to translate into the system of the robot joint coordinates. For path generation, we will actually be expressing coordinate increments, and we will need the transformation of the derivatives (the Jacobian). However, first we need the coordinate transformation:

$$\begin{aligned} r &= \sqrt{x^2 + y^2} \\ \phi &= \tan^{-1}\left(\frac{x}{y}\right) \end{aligned} \tag{10.126}$$

The Jacobian represents the "transmission ratio" for moments and forces or, kinematically, the ratio of velocities, or of increment per unit time increment.

$$\delta r = \frac{\partial r}{\partial x}\delta x + \frac{\partial r}{\partial y}\delta y$$

$$\delta \phi = \frac{\partial \phi}{\partial x}\delta x + \frac{\partial \phi}{\partial y}\delta y$$

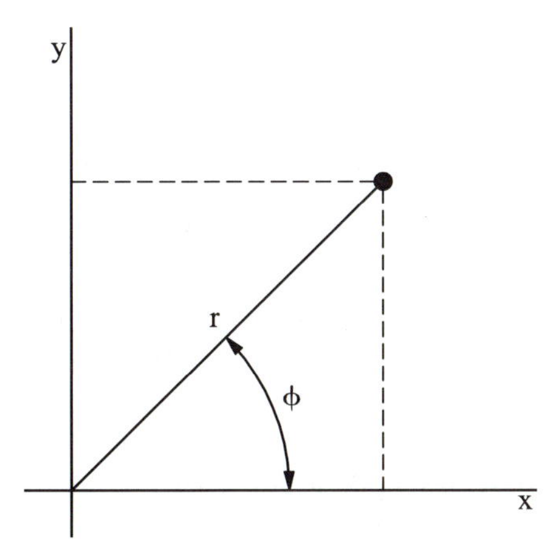

FIGURE 10.99
The 2D polar coordinate system.

This is the inverse Jacobian

$$\left\{\frac{\delta r}{\delta \phi}\right\} = J^{-1}\left\{\frac{\delta x}{\delta y}\right\}, \text{ where } J^{-1} = \begin{bmatrix} \frac{\partial r}{\partial x} \frac{\partial r}{\partial y} \\ \frac{\partial \phi}{\partial x} \frac{\partial \phi}{\partial y} \end{bmatrix}$$

We can obtain it either directly

$$\frac{\partial r}{\partial x} = \frac{x}{\sqrt{x^2 + y^2}}, \qquad \frac{\partial r}{\partial y} = \frac{y}{\sqrt{x^2 + y^2}}$$

$$\frac{\partial r}{\partial x} = \frac{x}{r} = \sin \phi, \qquad \frac{\partial r}{\partial y} = \frac{y}{r} = \cos \phi$$

$$\frac{\partial \phi}{\partial x} = \frac{\cos \phi}{r}, \qquad \frac{\partial \phi}{\partial y} = \frac{-\sin \phi}{r}$$

$$\left\{\begin{matrix} \delta r \\ \delta \phi \end{matrix}\right\} = \begin{bmatrix} \sin \phi & \cos \phi \\ \frac{\cos \phi}{r} & \frac{-\sin \phi}{r} \end{bmatrix} \left\{\begin{matrix} \delta x \\ \delta y \end{matrix}\right\} \tag{10.127}$$

or else, we could obtain J as the derivative of Eq. (10.125)

$$\frac{\partial x}{\partial r} = \sin \phi, \qquad \frac{\partial y}{\partial r} = \cos \phi$$

$$\frac{\partial x}{\partial \phi} = r \cos \phi, \qquad \frac{\partial y}{\partial \phi} = -r \sin \phi$$

$$\left\{\begin{matrix} \delta x \\ \delta y \end{matrix}\right\} = \begin{bmatrix} \sin \phi & r \ \ \cos \phi \\ \cos \phi & -r \ \ \sin \phi \end{bmatrix} \left\{\begin{matrix} \delta \mathrm{r} \\ \delta \phi \end{matrix}\right\} \tag{10.128}$$

and inverting:

$$\begin{bmatrix} \sin \phi & r \cos \phi \\ \cos \phi & -r \sin \phi \end{bmatrix}^{-1} = \frac{1}{-r \sin^2 \phi - r \cos^2 \phi} \begin{bmatrix} -r \sin \phi & -r \cos \phi \\ -\cos \phi & \sin \phi \end{bmatrix}$$

$$= \frac{-1}{r} \begin{bmatrix} -r \sin \phi & -r \cos \phi \\ -\cos \phi & \sin \phi \end{bmatrix} = \begin{bmatrix} \sin \phi & \cos \phi \\ \frac{\cos \phi}{r} & -\frac{\sin \phi}{r} \end{bmatrix}$$

which is the same as expression (10.127). We could now continue and express the relationships between second derivatives and express accelerations in joint coordinates as functions of accelerations in x and y.

Rather than using the closed-form expressions of continuous functions, we can start from some initial conditions of x and y as functions of time $x(t)$, $y(t)$ and start incrementing them. Then we can either use the relationships of Eq. (10.126) and then increment Δr and $\Delta \phi$, or use the relationship of Eq. (10.127) and from there obtain the increments of increments, $\Delta^2 r$, $\Delta^2 \phi$ as the accelerations.

Let us, however, formulate a rather simple case of $x(t)$, $y(t)$ as shown in Fig. 10.100. In this case the task is to move the end of the arm on a straight line from point

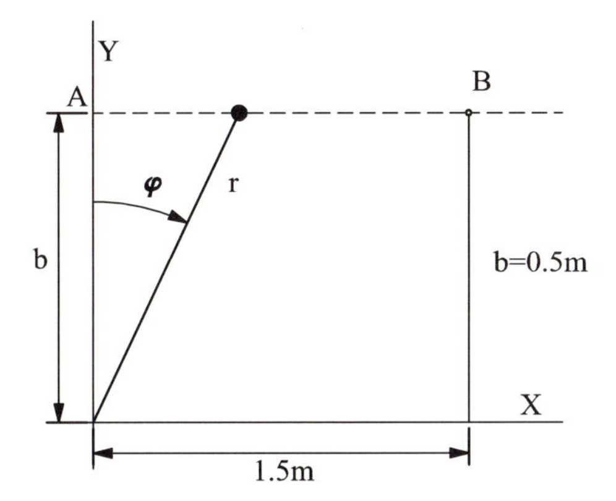

FIGURE 10.100 A simple path formulation. The mass at the end of radius r should move on a straight line parallel with the X axis, from A to B.

A to point B. This straight line is parallel with the X-axis. The form of the commanded motion and the flexibilities of the structure are discussed in the next section dealing with dynamics. Here we take advantage of the fact that in this case $y = \text{const}$. This simplifies the derivatives. Let us denote $y = b$. Then,

$$
\begin{aligned}
r &= \sqrt{x^2 + y^2} = \sqrt{x^2 + b^2} \\
\dot{r} &= \frac{1}{2}\frac{2x\dot{x}}{\sqrt{x^2 + b^2}} = \frac{\dot{x}x}{r} = \sin\phi\,\dot{x} \\
\ddot{r} &= \frac{(\dot{x}^2 + x\ddot{x})r - x\dot{x}\dot{r}}{r^2} = \frac{\dot{x}^2}{r} + \frac{x\ddot{x}}{r} - \frac{(x\dot{x})^2}{r^3} \\
&= \frac{x\ddot{x}}{r} + \frac{\dot{x}^2(r^2 - x^2)}{r^3} = \frac{x\ddot{x}}{r} + \frac{b^2\dot{x}^2}{r^3} = \sin\phi\,\ddot{x} + \cos^2\phi\,\frac{\dot{x}^2}{r}
\end{aligned}
\tag{10.129}
$$

$$
\begin{aligned}
\phi &= \operatorname{atan}\left(\frac{x}{y}\right) = \operatorname{atan}\left(\frac{x}{b}\right) \\
\dot{\phi} &= \frac{1}{1 + (x/y)^2}\frac{\dot{x}}{y} = \frac{b^2}{x^2 + b^2}\frac{\dot{x}}{b} = \frac{b\dot{x}}{r^2} = \cos\phi\,\frac{\dot{x}}{r} \\
\ddot{\phi} &= \frac{b\ddot{x}r^2 - b\dot{x}2r\dot{r}}{r^4} = \frac{b\ddot{x}}{r^2} - \frac{2bx\dot{x}^2}{r^4} = \frac{\cos\phi\,\ddot{x}}{r} - \frac{2\cos\phi\sin\phi\dot{x}^2}{r^2}
\end{aligned}
\tag{10.130}
$$

10.10.3 Dynamics of the 2D Polar Case

The simple path defined in Fig. 10.100 will be accepted for the following discussion and illustration. The system corresponds to the diagram in Fig. 10.101. Both coordinates are driven by positional servos in configuration B of Fig. 10.60. This means that the flexibilities and inertias of the two axes are outside of the positional loop.

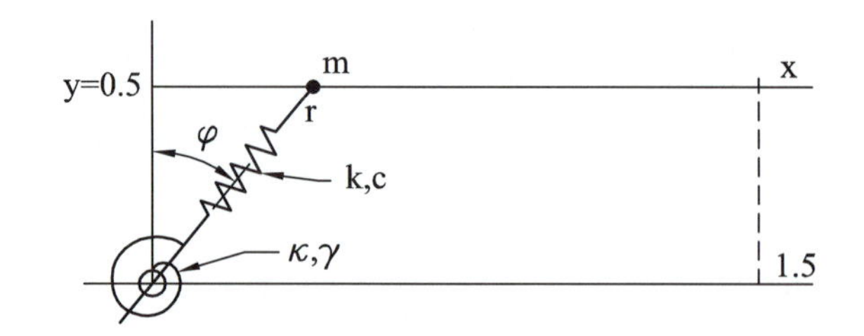

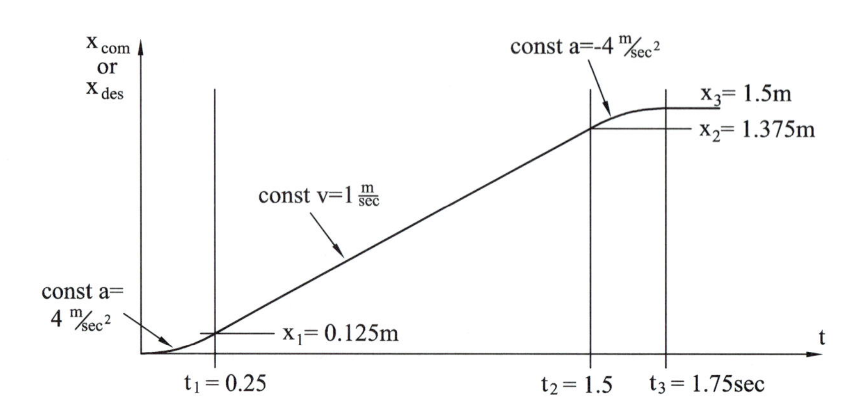

FIGURE 10.101 Top: springs and masses in the two joint coordinates. Bottom: desired motion in the x coordinate. A sequence of constant acceleration, constant velocity, and constant deceleration is prescribed.

Moreover, "perfect" servodrives will be assumed, as in the discussion in Section 10.9.1. We are then left with just the problems of driving the flexible and inertial systems that have been previously symbolized in Fig. 10.73, except that one of them is rotary. Its stiffness κ has the dimension of (Nm/rad), and its inertia $J = mr^2$ is in (kgm^2).

The two coordinates are driven according to the following equations:

$$\begin{aligned} m\ddot{r} + c\dot{r} + kr &= c\dot{r}_{\text{com}} + kr_{\text{com}} \\ mr^2\ddot{\phi} + \gamma\dot{\phi} + \kappa\phi &= \gamma\dot{\phi}_{\text{com}}\ \kappa\phi_{\text{com}} \end{aligned} \tag{10.131}$$

However, the inputs r_{com} and ϕ_{com} are not directly defined. The motion is defined in the coordinates $x_{\text{com}}, y_{\text{com}}$. The function $x_{\text{com}}(t)$ is formulated in Fig. 10.101 as consisting of a constant acceleration, constant velocity, and constant deceleration sequence in the x coordinate; also, y_{com} remains constant and equal to 0.5 m. The simulation programs to derive the motions of the system for the uncorrected and the feedforward corrected commands are described for a particular set of specifications in the following exercise.

EXAMPLE 10.17 Feedforward Compensation in Two Dimensions ▼

Uncorrected Command

The input x_{com} is identical with the time function specified in Fig. 10.101.

Inputs: $k = 4000$ N/m, $m = 100$ kg, $c = 55$ Nsec/m, $\kappa = 5625$ Nm/rad, $\gamma = 15$ Nm sec, $b = 0.5$ m, $a = 4$ m/sec^2, $v = 1$ m/sec.

Initial values: $x_{\text{com},1} = 0$, $y_{\text{com}} = 0.5$, $\phi_1 = 0$, $\dot{\phi} = 0$, $r_1 = 0.5$, $\dot{r} = 0$, $r_{\text{com},N} = 1.5811$, $\phi_{\text{com},N} = \text{atan}(3)$.
Simulation parameters: $dt = 0.001$ sec, $N = 2000$.

Section 1
For $n = 1{:}\ (N - 1)$
Define $x_{\text{com},n}$ and $y_{\text{com},n}$ as specified in Fig. 10.101.
Define $\dot{x}_{\text{com}}$.

Section 2
Use Eqs. (10.129), (10.130):
Define $r_{\text{com},n}$, $\phi_{\text{com},n}$, $\dot{r}_{\text{com}}$, $\dot{\phi}_{\text{com}}$.

Section 3
Use Eq. (10.131):
Express $\ddot{r}$, integrating, obtain $\dot{r}$ and r_{n+1}.
Express $\ddot{\phi}$ integrating, obtain $\dot{\phi}$ and ϕ_{n+1}.

Section 4
Use Eq. (10.125):
Obtain x, y.

Section 5
Plot r_{com}, r versus time, ϕ_{com}, ϕ versus time, y versus x.

The Matlab program "rob2nfwd" that follows this algorithm is given in Fig. 10.102.

The plots of the commanded and actual radial (r) and angular (ϕ) coordinates are presented in Fig. 10.103. Note that both coordinates vibrate strongly. The resulting motion, which is plotted in the reconstituted Cartesian coordinates x and y in Fig.10.104, comes out very badly. *The deviation from the commanded straight line when measured as the error in y is about −0.091 m, + 0.135 m* (−91 mm, + 135 mm).

Applying the Feedforward Corrections

Next the case is processed with feedforward corrections in r and ϕ. The program consists of setting the x_{des} and y_{des} identical to the functions specified in Fig. 10.101 and by coordinate transformation deriving r_{des} and ϕ_{des}. These are then used to derive the desired commands; this is accomplished by reversing Eq. (10.131):

$$\begin{aligned} c\dot{r}_{\text{com,des}} + kr_{\text{com,des}} &= m\ddot{r}_{\text{des}} + c\dot{r}_{\text{des}} + kr_{\text{des}} \\ \gamma\dot{\phi}_{\text{com,des}} + \kappa\phi_{\text{com,des}} &= mr^2\ddot{\phi}_{\text{des}} + \gamma\dot{\phi}_{\text{des}} + \kappa\phi_{\text{des}} \end{aligned} \tag{10.132}$$

These commands can be used to drive the coordinates and to finally obtain x and y. The inputs to the program are the same as in the incorrected case, and its structure is as follows:

Initial values: $r_{\text{com},1} = 0.5$, $\phi_{\text{com},1} = 0$, $\dot{r}_{\text{des}} = 0$, $\dot{\phi}_{\text{des}} = 0$.

Section 1
For $n = 1{:}\ N$

FIGURE 10.102
Listing of the Matlab program "rob2nfwd" for simulation of the motion of the polar system defined in Fig. 10.101, without feedforward compensation. Ideal drive is assumed for the scheme shown in Fig. 10.73. There is no feedback involved. The program involves transformation of the command from Cartesian to polar coordinates.

```
function[xcom,ficom,fi,rcom,r,x,y,t]=rob2nfwd
%Driving a 2D robot of Fig.10.101. Ideal servos are assumed,the spring
%damper, mass systems are outside of the feedback loops. No feedforward
%compensation is applied. Parameters are specified in EX.10.17.
k=4000;m=100;c=55;kap=5625;gam=15;
t(1)=0;dt=0.001;a=4;v=1;r(1)=0.5;dr=0;fi(1)=0;dfi=0;
x1=(a/2)*(.25)^2;x2=x1+v*1.25;x(1)=0;y(1)=0.5;
rcom(2000)=1.5811;ficom(2000)=atan(3.0);
%Defining the cartesian xcom and its first derivative, the ycom is constant,0.5m.
for n=1:1999;
if n<=250
xcom(n)=(a/2)*(t(n))^2;
dxcom=a*t(n);
elseif (n>250)&(n<=1500)
xcom(n)=x1+v*(t(n)-0.25);
dxcom=v;
elseif n>1500 & n<=1750
xcom(n)=x2+v*(t(n)-1.5)-2*(t(n)-1.5)^2;
dxcom=v-a*(t(n)-1.5);
else xcom(n)=1.5;dxcom=0;
end
%Deriving the commands in the polar system,rcom and ficom,Eqs.(10.129) and (10.130).
rcom(n)=sqrt(xcom(n)^2+0.25);
ficom(n)=atan(xcom(n)/.5);
drcom=dxcom*sin(ficom(n));
dficom=dxcom*cos(ficom(n))/rcom(n);
%using Eq. (10.131), derive resulting motion derivatives ddr,ddfi and then
%obtain r and fi.
ddr=(c*drcom+k*rcom(n)-c*dr-k*r(n))/m;
dr=dr+ddr*dt;
r(n+1)=r(n)+dr*dt;
ddfi=(gam*dficom+kap*ficom(n)-gam*dfi-kap*fi(n))/(r(n)^2*m);
dfi=dfi+ddfi*dt;
fi(n+1)=fi(n)+dfi*dt;
%combine r and fi into the resulting cartesian coordinate motions x and y.
x(n+1)=r(n+1)*sin(fi(n+1));
y(n+1)=r(n+1)*cos(fi(n+1));
t(n+1)=t(n)+dt;
end
xcom(2000)=1.5;
%plot(x,y)
```

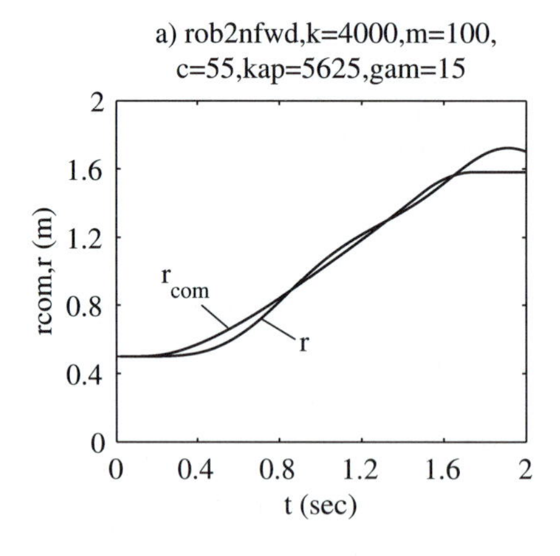

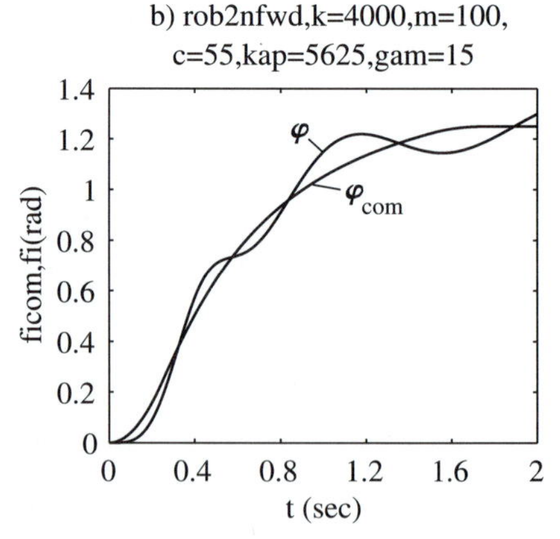

FIGURE 10.103
The commands and actual motions in the two polar system coordinates. Considerable vibrations arise.

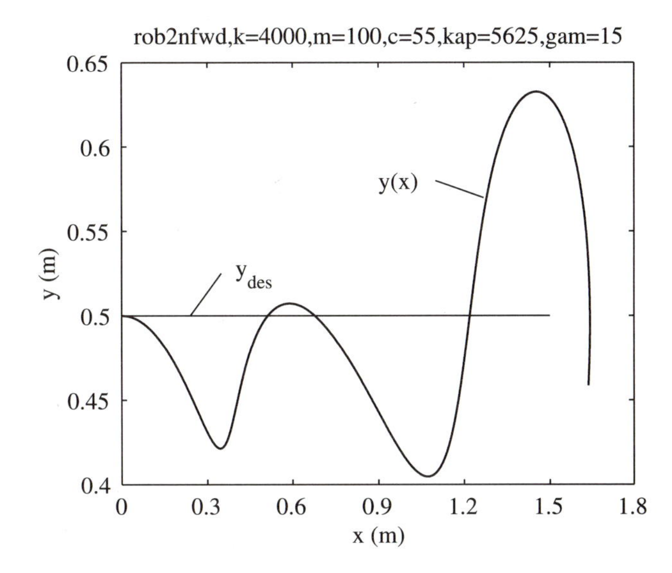

FIGURE 10.104 The resulting motion in the *x, y* coordinates, Example 10.17, no feedforward compensation. Due to the high flexibility and mass of the SMD system, the error is very large: 0.226 m, PTP.

Define $x_{\text{des},n}$, $y_{\text{des},n}$.
Use Eq. (10.126):
Determine $r_{\text{des},n}$, $\phi_{\text{des},n}$.

Section 2.
For $n = 1{:}N - 1$
Derive numerically the first and second derivatives of r_{des} and ϕ_{des}, as they will be needed for the next application of Eq. (10.132).

$$\dot{r}_{\text{des,new}} = (r_{\text{des},n+1} - r_{\text{des},n})/\mathrm{d}t$$

$$\ddot{r}_{\text{des}} = (\dot{r}_{\text{des,new}} - \dot{r}_{\text{des}})/\mathrm{d}t$$

$$\dot{\phi}_{\text{des,new}} = (\phi_{\text{des},n+1} - \phi_{\text{des},n})/\mathrm{d}t \qquad \ddot{\phi}_{\text{des}} = (\dot{\phi}_{\text{des,new}} - \dot{\phi}_{\text{des}})\mathrm{d}t$$

$$\dot{r}_{\text{des}} = \dot{r}_{\text{des,new}} \qquad \dot{\phi}_{\text{des}} = \dot{\phi}_{\text{des,new}}$$

Section 3 (continued in the loop of Sec. 2)
Using Eq. (10.132), determine the corrected input commands:

$\dot{r}_{\text{com,des}}$ and integrating, $\mathrm{r}_{\text{com,des}}$,

$\dot{\phi}_{\text{com,des}}$ and integrating, $\phi_{\text{com,des}}$.

Section 4
Using the corrected commands $r_{\text{com,des}}$ and $\phi_{\text{com,des}}$, proceed through Sections 3 and 4 of the preceding program used for the simulation of the uncorrected motions in Example 10.17.

Section 5
Plot: $r_{\text{com,des}}$, r versus t, $\phi_{\text{com,des}}$, ϕ versus t, y versus x.

The Matlab program "rob2frwd" that follows this algorithm is given in Fig. 10.105. The results obtained for the radial and angular coordinates are plotted in Fig. 10.106. The reconstituted resulting motion in coordinates, x and y is presented in Fig.10.107. Note that with the feedforward corrections the resulting

FIGURE 10.105
Listing of the Matlab program "rob2frwd" for simulation of the motions of the polar system of Example 10.17 under feedforward strategy.

```
function[xdes,fides,fi,ficom,rdes,r,rcom,x,y,t]=rob2frwd
%driving a 2D polar robot defined in Fig.10.101 and in EX.10.17.
%Ideal servos are assumed; feedforward compensation is applied.
k=4000;m=100;c=25;kap=5625;gam=15;
dt=0.001;a=4;v=1;r(1)=0.5;dr=0;fi(1)=0;dfi=0;
x1=(a/2)*(.25)^2;x2=x1+v*1.25;x(1)=0;y(1)=0.5;
rcom(1)=0.5;ficom(1)=0;
for n=1:2000;
t(n)=(n-1)*dt;
%Sec.1.Define the desired motion xdes, identical with the xcom in
%program "rob2nfwd.m"
if n<=250
xdes(n)=(a/2)*(t(n))^2;
elseif (n>250)&(n<=1500)
xdes(n)=x1+v*(t(n)-0.25);
elseif n>1500 & n<=1750
xdes(n)=x2+v*(t(n)-1.5)-2*(t(n)-1.5)^2;
else xdes(n)=1.5;
end
%Use Eq.(10.126),transform xdes into polar rdes and fides
rdes(n)=sqrt(xdes(n)^2+0.25);
fides(n)=atan(xdes(n)/.5);
end
%Sec.2.Using the scheme of Fig.10.79 derive derivatives of rdes and fides
drdes=0;
dfides=0;
for n=1:1999;
drdesnew=(rdes(n+1)-rdes(n))/dt;
ddrdes=(drdesnew-drdes)/dt;
dfidesnew=(fides(n+1)-fides(n))/dt;
ddfides=(dfidesnew-dfides)/dt;
%Sec.3.Determine the compensated commands rcom and ficom
drcom=(m*ddrdes+c*(drdesnew+drdes)/2+k*rdes(n)-k*rcom(n))/c;
rcom(n+1)=rcom(n)+drcom*dt;
subs=rdes(n)^2*m*ddfides+gam*(dfidesnew+dfides)/2+kap*fides(n);
dficom=(subs-kap*ficom(n))/gam;
ficom(n+1)=ficom(n)+dficom*dt;
%Sec.4.Use rcom and ficom to drive the SMD systems
%to obtain actual r and fi.
ddr=(c*drcom+k*rcom(n)-c*dr-k*r(n))/m;
dr=dr+ddr*dt;
r(n+1)=r(n)+dr*dt;
ddfi=(gam*dficom+kap*ficom(n)-gam*dfi-kap*fi(n))/(r(n)^2*m);
dfi=dfi+ddfi*dt;
fi(n+1)=fi(n)+dfi*dt;
drdes=drdesnew;
dfides=dfidesnew;
%combine the polar motions into the actual cartesian x and y
x(n+1)=r(n+1)*sin(fi(n+1));
y(n+1)=r(n+1)*cos(fi(n+1));
end
%plot(x,y)
```

motion is very accurate. *The y error is less than ± 0.0001 m (±0.1 mm).* This is more than a thousand times better than the case with no feedforward correction. Remember, however, that a perfect model of the system was used in determining the corrective term. In reality this is not possible. Also, with more degrees of freedom, the mathematics is much more complicated. Even so, substantial improvements are achievable.

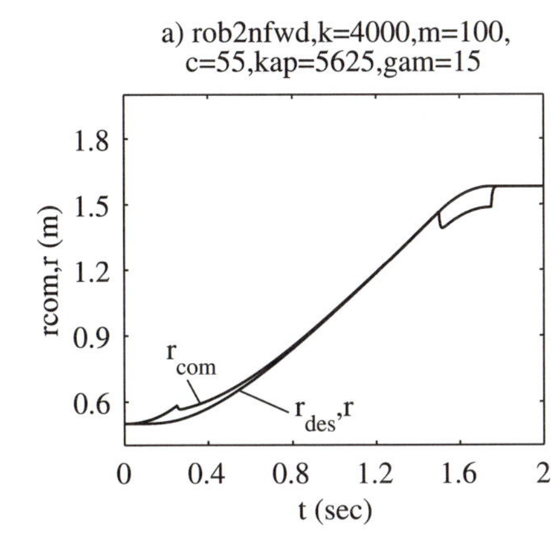

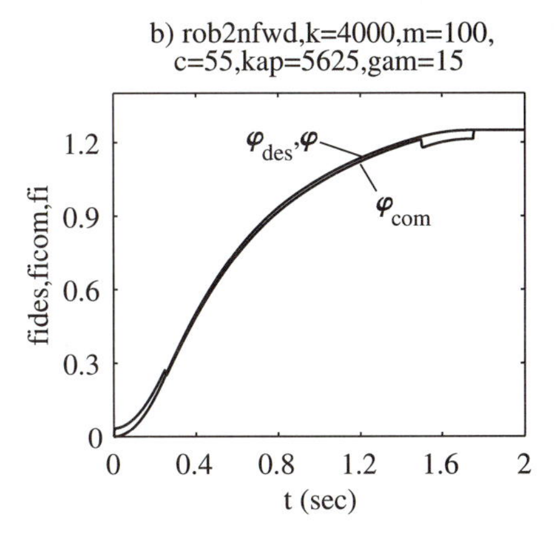

FIGURE 10.106 Feedforward compensation applied, as in Example 10.17a. The corrected command and the actual motion in *r*. b) The corrected command and the resulting motion in *N*.

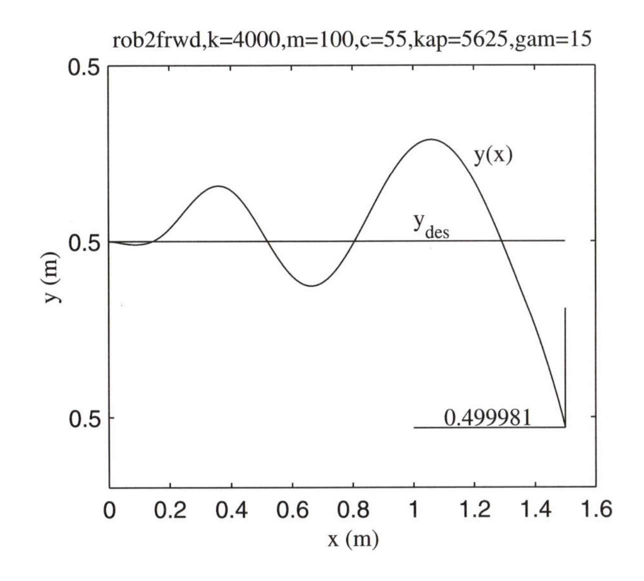

FIGURE 10.107 The resulting motion in the *x, y* coordinates when applying the feedforward compensations. Compare with Fig. 10.104. Notice the difference in the *Y* scale. Error is reduced to a very small value of < 0.0002 m.

10.11 CONCLUSION

The discussion of automation has been, to a large extent, involved with numerical control in general and with the design of the feed drive servomechanisms in particular. After the structural dynamics and the analysis of forced and self-excited vibrations in Chapter 9 and the resulting rules for the design of machine-tool structures, the topic of efficient and accurate drives of coordinate motions is the second half of a mechanical engineer's concern in the development of automated production machinery. It is true that the problems of CAD/CAM, especially the task of programming the NC tool path, are also very important for the success of NC machines, but that is a different topic, requiring a different expertise. Our discussion of that topic has been minimal, and students will have to consult additional specialized texts.

REFERENCES

1 Bollinger, J. G., and N. A. Duffie *Computer Control of Machines and Processes.* Addison-Wesley, 1988.

2 Kuo, B. C. *Automatic Control Systems.* 7th ed. Prentice-Hall, 1995.

3 Dorf, R. C. *Modern Control Systems.* 5th ed. Addison-Wesley, 1989.

4 Weck, M. *Handbook of Machine Tools.* Vol. 3. Automation and Controls. J. Wiley & Sons, 1984.

5 Koren, Y. *Computer Control of Manufacturing Systems.* McGraw-Hill, 1983.

6 Astrom, K. J., and B. Wittenmark. *Adaptive Control.* Addison-Wesley, 1989.

7 Gibbs, D., and T. Crandell. *CNC: An Introduction to Machining and Part Programming.* Industrial Press, 1991.

8 Groover, M. P. *Automation, Production Systems, and Computer-Aided Manufacturing.* Prentice-Hall, 1987.

9 Lynch, M. *Computer Numerical Control for Machining.* McGraw-Hill, 1992.

10 Miller, R. K. *Industrial Robot Handbook.* Van Nostrand Reinhold, 1989.

11 Luggen, W. W. *Flexible Manufacturing Cells and Systems.* Prentice-Hall, 1991.

12 Zeid, I. *CAD/CAM Theory and Practice.* McGraw-Hill, 1991.

13 Chen, Y. C., and J. Tlusty, "Effect of Low Friction Guideways and Leadscrew Flexibility on Dynamics of High Speed Machines" *CIRP Annals,* 44/1/1995.

14 Dronsek, M. "Technische und wirtschaftliche Probleme der Fertigung im Flugzeugbau." Messerschmitt-Bolkow-Blohm GmbH, Ottobrun, April, 1980.

Industrial Contributors

[BROWN & SHARPE] Brown & Sharpe Mfg. Co., Precision Park, P.O. Box 456, North Kingston, RI 02852.

[BULLARD] The Bullard Co., 286 Canfield Avenue, Bridgeport, CT 06609.

[F. J. LAMB] F. J. Lamb Co., 5663 East Nine Mile Road, Warren, MI 48091.

[FANUC] Fanuc Ltd., 5-1 Asahigaoka, 3-chome, Hino-shi, Chome, Tokyo 191, Japan.

[FUJITSU] FANUC Co., 1183-1 Shibokusa Aza-Marubigishi, Oshino-mura, Minamitsuru-gun, Yamanashi-Pref., 401-05, Japan.

[KEARNEY & TRECKER] Kearney & Trecker Corp., 11000 Theodore Trecker Way, Milwaukee, WI 53124.

[MESSERSCHMITT-BOLKOW-BLOHM] Messerschmitt-Bolkow-Blohm GmbH, Postfach 102104, 8900 Augsburg 21, Germany.

[MILACRON] Cincinnati Milacron, Inc., 4701 Marburg Avenue, Cincinnati, OH 46209.

[NATIONAL ACME] National Acme, Div. of Acme Cleveland Corp., 170 East 131 Street, Cleveland, OH 44108.

[NORDSON] Nordson Corp, 28061 Clemens Road, Westlake, OH 44145.

[OKUMA] Okuma Machinery Works Ltd., Oguchi-cho, Niwa-Gun, Aichi 480-01, Japan.

[OMNITRADE] Omnitrade Machinery, 78 Torlake Crescent, Toronto, Ont., Canada, M8Z 1B8.

[TOSHIBA] Toshiba Machine Co., Ltd., 2-11, 4-chome, Ginza Chuo-ku, Tokyo, Japan.

[WARNER & SWASEY] Warner & Swasey Co., University Circle Research Center, 11000 Cedar Avenue, Cleveland, OH 44106.

[WHITE SUNDSTRAND] White Sundstrand Machine Tool, Inc., 3615 Newburg Road, Belvidere, IL 61008.

[YAMAZAKI] Yamazaki Machinery Works, Ltd., 1 Norifune, Oguchi-cho, Niwa-Gun, Aichi-Pref., 480-01, Japan; Mazak Corp., 8025 Production Drive, Florence, KY 41042.

QUESTIONS

Q10.1 *Actions to automate: devices used.* Name the various actions to be automated. Describe ways of automating **(a)** tool change, **(b)** spindle speed change, **(c)** tool-path control.

Q10.2 *Savings achieved by automation.* Name and briefly explain the various categories of savings.

Q10.3 *Costs of automation.* Name the three basic components of automating an operation. Which of them is most significant for the degree of "flexibility of automation"?

Q10.4 *Single-spindle automatic lathes.* Describe the tool arrangements and directions of their feed motions on a Swiss-type and on a screw-type automatic lathe. Make simple sketches. How are the lengths and velocities of these motions controlled and driven?

Q10.5 *Multispindle automatic lathes.* Describe the arrangement and directions of motions of tool carriers on a six-spindle automatic lathe. How are the spindles mounted, and how is the transfer of the parts from station to station accomplished? Make simple sketches.

Q10.6 *Transfer line for machining automotive cylinder blocks.* Describe the structure of the transfer line. Name the various machining operations performed in the individual stations. What is the basic principle for achieving the high rate of production?

Q10.7 *Numerical control.* Explain the information flow in numerical control and the individual basic blocks of the system.

Q10.8 *Positional servomechanism.* Name the basic hardware components of the servo.

Q10.9 *Adaptive control for constant force.* Draw the diagram of the software and hardware components of the system.

Q10.10 *NC turning center.* Describe the basic subgroups of a typical NC turning center.

Q10.11 *NC machining center.* What are the types of work-pieces and machining operations? Make a sketch indicating the bodies performing the *X, Y, Z* motions. Describe the basic subgroups of the center, including tool and workpiece changing.

Q10.12 *NC machining cell.* What is the additional subgroup in an automatic cell with respect to a center as described in Question 10.11? Make a sketch.

Q10.13 *Flexible manufacturing systems.* Describe the various types of FMS. What is the basic computer-controlled function that makes the FMS more than just a group of NC machine tools?

Q10.14 *Servomotor: effect of tacho feedback.* Express the effects on steady-state and on transient performance.

Q10.15 *Positional servo: transient error.* Sketch out the step input response. Write out the formulas for natural frequency, for damping ratio, and for the magnitude of the overshoot.

Q10.16 *Steady-state error in a ramp motion.* Write out the formula for the "following error (the velocity lag)." How does it affect the contour error in a 2D, straight-line motion? in a circular motion?

Q10.17 *Positional servo driving a spring-mass system that is included in the position feedback.* Which part of the block diagram is decisive for the value of the critical frequency at which the system becomes unstable? What is the definite upper limit for the critical frequency? What kind of term may be added to the feedback to stabilize the system? Write out the formula of this term and the value of the coefficient involved.

Q10.18 *Positional servo driving a spring-mass system that is not included in the feedback.* What kind of compensation may be used to improve the motion of the mass of the added mechanical system? Where is the compensation term added? Write out the value of this term for a desired motion with constant acceleration (neglect the effect of damping).

Q10.19 *Industrial robots.* Name the various applications of industrial robots. Sketch the four kinematic arrangements of the three coordinates of the base structure carrying the wrist and hand.

Q10.20 *Robots: the two arrangements of the feedback.* Sketch out how the two arrangements, one including the transmission in the loop and the other one excluding it, are executed for a rotary joint and for a translative one.

Q10.21 *Robots: determining feedforward compensation.* Indicate the sequence of transformations in determining the compensated commands for a 2D, polar coordinate robot as well as of the simulations verifying the efficiency of the compensations.

PROBLEMS

10.1 *Servomotor performance.* See Fig. P10.1.

(a) Without tacho feedback. $R = 2\Omega$, $J = 0.0125$ kgm^2, $K_T = 0.5$ Nm/A, $K_E = 0.5$ V/(rad/sec). Determine: The time constant τ_m (sec) and the gain K_m (rad/sec/V), the initial acceleration $\dot{\omega}_0$ (rad/sec^2), at $t = 0$, $\omega = 0$, for a step input $e = 20$ V, and the initial current i_0 (A), and the stall torque T_{st} (Nm) for $e = 20$ V.

(b) With tacho feedback: $K_a = 20$. First determine the value of H for which the overall gain Ω/V will be the same as in *(a)*, $K_{CL} = K_m$. Disregarding any current limitation, determine τ_{CL} (sec), initial acceleration $\dot{\omega}_0$ (rad/sec^2) for $t = 0$, $\omega = 0$, and step input $v = 20$ V, and the corresponding initial current i_0 (A), and the stall torque T_{st} (N/m) for $v = 20$ V. Now assume current limitation to $i_{\lim} = 100$ A. What will be the initial acceleration $\dot{\omega}_0$ (rad/sec^2) ? As the speed increases the current remains $i_{\lim}$ until the speed is reached at which normally, without the limiting circuit, current would drop to 100A. What is the value of this speed ω_{100}? At which time t_1 will the current start dropping below 100 A? Plot the response to a step input $v = 30$ V for **(b1)** no current limitation; **(b2)** $i_{\lim} = 100$ A. Use $dt = 0.0001$ sec; total simulation time 0.05 sec.

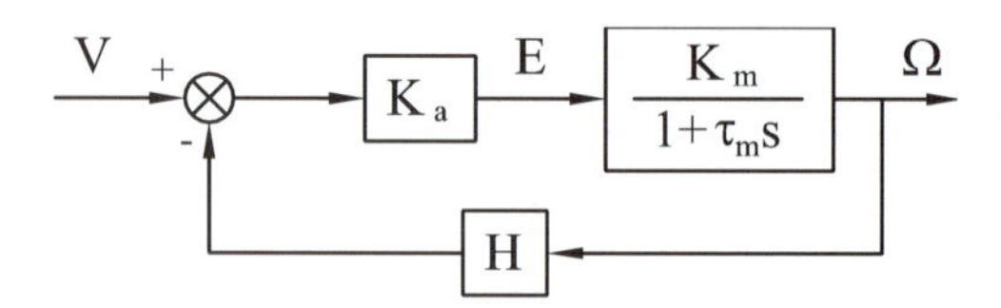

FIGURE P10.01

P10.2 *Servomotor: Effect of feedback.* See Fig. P10.1.

The following parameters apply: $R = 1.5\Omega$, $K_T = 0.4$ Nm/A, $K_E = 0.4$ V/(rad/sec), $J = 0.08$ kgm^2, $K_a = 50$, $H = 0.4$ V/(rad/sec).

(a) Determine τ_m (sec) and K_m (rad/sec/V) for the servomotor without tacho feedback and the corresponding stall torque T_{st} (Nm) for an input voltage of $e = 20$ V.

(b) Include the tacho feedback, neglect any potential current limitation. Determine τ_{CL} (sec) and K_{CL} (rad/sec/V) and the stall torque T_{st} (Nm) for an input voltage of $v = 20$ V.

(c) For step input $v = 20$ V, what is the value of the initial acceleration $\dot{\omega}_0$ (rad/sec^2), at $t = 0$?

(d) Assume current limit to $i_{\lim} = 150$ A. What will be the initial acceleration $\dot{\omega}_0$ (rad/sec^2)?

As the speed increases the current remains $i_{\lim}$ until such speed that would normally, without the limiting circuit, correspond to 150 A. What is the value of this speed ω_1? At which time t_1 will the current start dropping below 150 A?

10.3 *Servomotor and positional servomechanism.* Refer to Fig. P10.3.

Given: $\tau_m = 0.2$ sec, $K_m = 2$ (rad/sec), $H = 0.4$ V/(rad/sec), $K_p = 125$ V/mm, $K_a = 20$, $r = 0.1$ (mm/rad), determine the following.

(a) The parameters of the servomotor with the tacho feedback, K_{CL} (rad/sec/V) and τ_{CL} (sec).

(b) The natural frequency ω_n (rad/sec) and damping ratio ζ of the positional servo.

(c) The initial acceleration $\ddot{x}$ (mm/sec^2) at the start of a step command $x_{com} = 4$ mm. How many g's is it?

(d) The magnitude of the position error er_p (mm) in a steady-state motion with a velocity $\dot{x} = 5$ mm/sec.

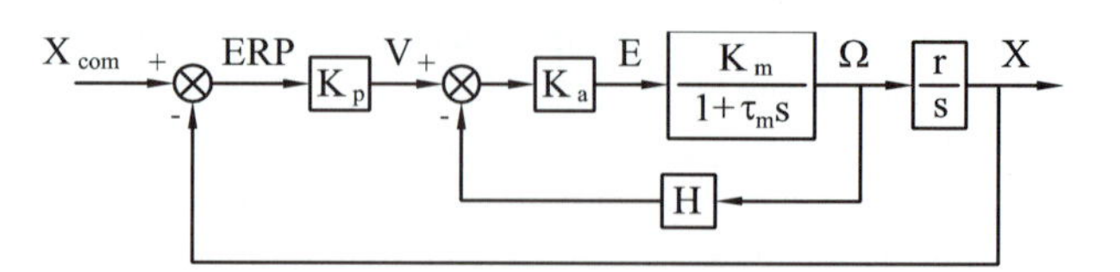

FIGURE P10.03

10.4 *Positional servomechanism.* Refer to Fig. P10.4.

$K_p = 100$ (V/mm), K_{CL} (rad/sec/V) $= 5$, $r = 0.2$ (mm/rad), $\tau_{CL} = 0.01$ (sec).

(a) Determine the natural frequency ω_n (rad/sec) and the damping ratio ζ.

(b) Determine the steady-state error $er_{p,ss}$ (mm) at velocity $\dot{x} = 20$ mm/sec.

(c) Determine the initial acceleration $\ddot{x}$ mm/sec^2 at a step input $x_{com} = 2$ mm and the overshoot o, and time t_1 at which it is reached.

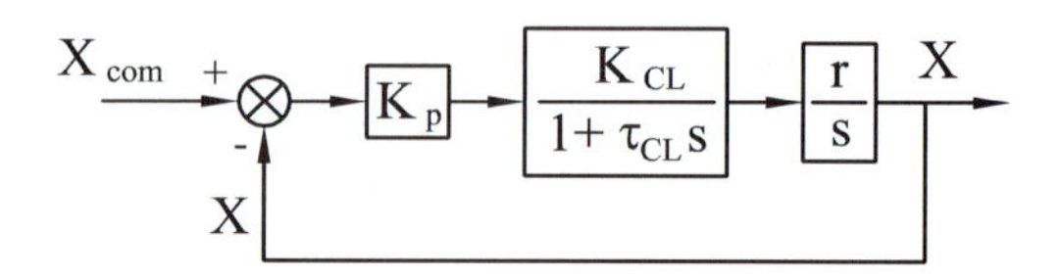

FIGURE P10.04

P10.5 *Understanding the positional servomechanism.* Refer to Fig. P10.5.

The solid line in this displacement graph is x_{com}, the dashed line is x. The lower part is the graph of velocities. Draw basic forms of $\dot{x}_{com}$ (solid) and of $\dot{x}$ (dashed line). Align characteristic points of this graph with those above. The motion proceeded with constant velocity $\dot{x} = 25$ mm/sec. Start the solutions at $t = 0$ in points O (origin of coordinates) and A. From this moment $x_{com} = 0$. The parameters of the servo follow: gain $K = K_p K_{CL} r = 62.5$ (1/sec), $\omega_n = 50$ (rad/sec), $\zeta = 0.4$.

To help with the velocity graph answer this question: As the displacement transfer function relates X and X_{com}:

$$X(s) = \frac{\omega_n^2}{s^2 + 2\zeta\,\omega_n s + \omega_n^2} X_{\text{com}}(s)$$

multiply both sides by s; the same transfer function will then represent a relationship between which variables? Therefore, for example, a response of velocity to a velocity step will have exactly the same form as the response of displacement to a displacement step. At which points will $\dot{x}$ be zero? At which time t_1 will velocity $\dot{x}$ reach minimum, and what will be the magnitude of this minimum?

(a) What is the value of x in point A?

(b) Assume that in point B, $x = 0.19$ mm. Shift the start of time to point B and forget about the past. How would the servo perceive its state at this point? Obviously, as a positional step input command. How big? If $t_B = 0$, what is t_c? What is the value of x_c?

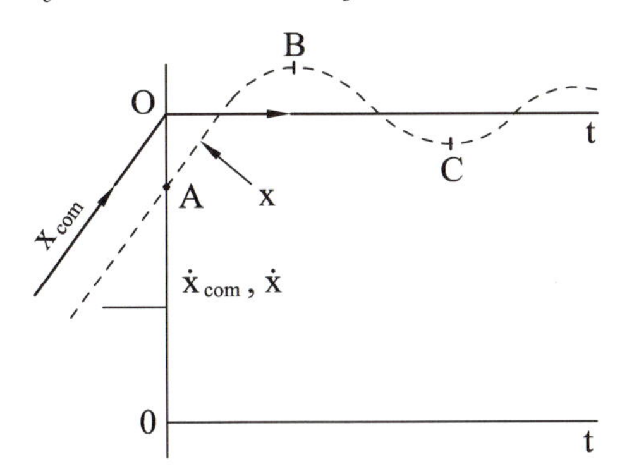

FIGURE P10.05

P10.6 *Positional servomechanism: ramp response.* Refer to Figs. P10.4 and P10.6.

$K = K_p K_{CL} r = 10$ (mm/sec/mm), $\tau_{\text{CL}} = 0.100$ sec. Determine ω_n (rad/sec), and ζ. Refer to Eq. (10.38).

(a) Concentrate on the motion after the command has stopped. At point O set $t = 0$, $x_{\text{com}} = 0$. Until this point the servo was moving with velocity $v = 50$ mm/sec. Assume that the position in point C is $x_c = 1.51$ mm, and in point D it is $x_D = 0.85$ mm. Determine: time $t_D =$ (sec), velocity $\dot{x}_{\text{D}}$ (mm/sec), displacement x_F (mm), and determine accelerations $\ddot{x}$ at points A, C, D.

(b) Determine the steady-state contour error. The gains are $K_x = 10$/sec, $K_y = 25$/sec. Moving with velocity $f = 50$ mm/sec on the contour, determine the error of the location of the actual path (measured normal to the commanded path) for the following two slopes and direction of motion; see Fig. P10.6b.

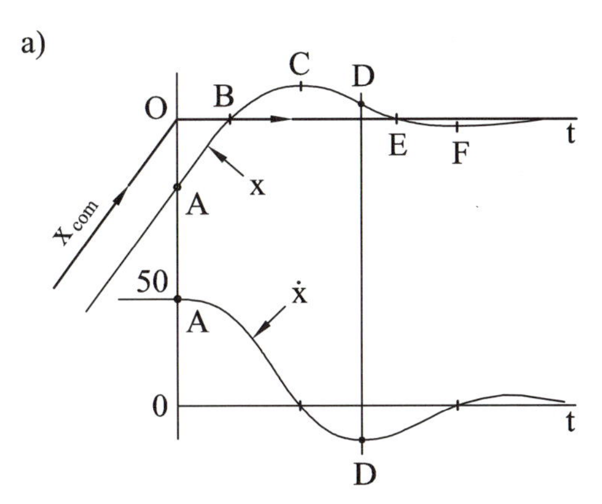

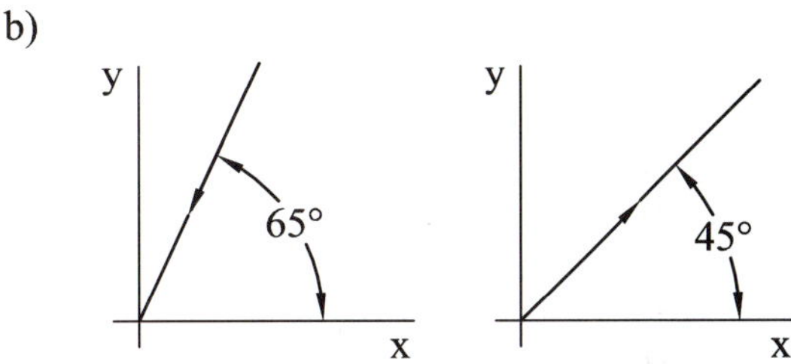

FIGURE P10.06

P10.7 *Steady-state error in an NC servo.* Refer to Figs. P10.4 and P10.7.

Consider the following positional servos: $K_{px} = 400$ (V/mm), $K_{py} = 200$ (V/mm), $K_{CL} = 5$ (rad/sec/V), $r = 0.05$ (mm/rad). Determine the error of location of the machined surface for the two cases shown in Fig. P10.7.

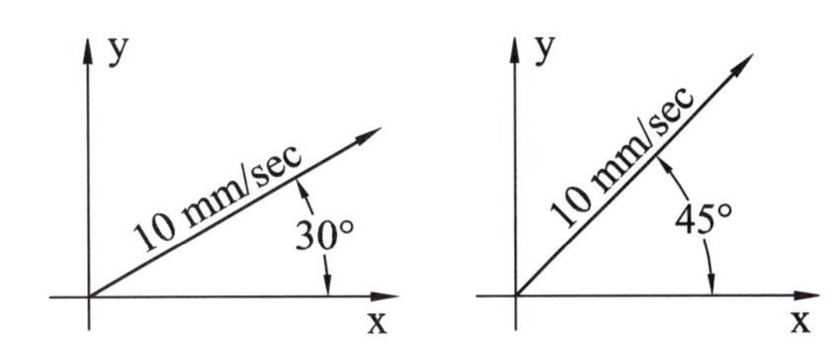

FIGURE P10.07

P10.8 *Steady-state error in a positional servo.* Refer to Figs. P10.4 and P10.8.

Assume positional servos in axes X and Y, for both $K_{CL} = 5$ (rad/sec/V), and $r = 0.2$ (mm/rad). The values for K_p differ: $K_{px} = 100$ V/mm, $K_{py} = 200$ V/mm. Determine the radius error e at points A, B, and C when circumferential velocity is

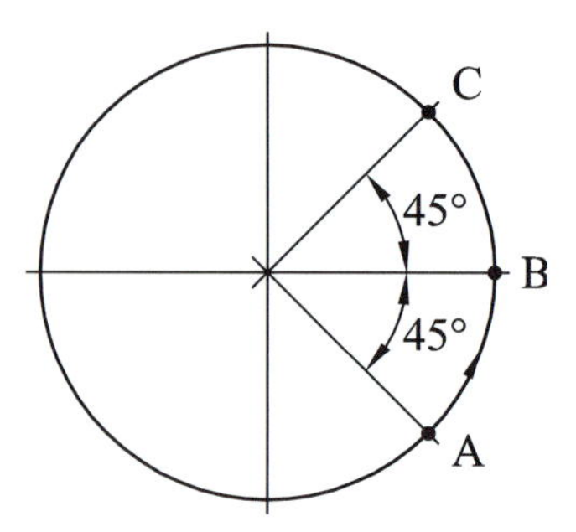

FIGURE P10.08

40 mm/sec. A larger radius means a positive error e; a smaller radius is marked as negative e.

P10.9 *Steady-state follow-up error.* Refer to Figs. P10.4 and P10.9.

The X servo: $K_p = 100$ (V/mm), $K_{CL} = 10$ (rad/sec/V), $r = 0.1$ (mm/rad); the Y servo: $K_p = 40$ (V/mm), $K_{CL} = 10$ (rad/sec/V), $r = 0.1$ (mm/rad). Contouring velocity is 50 mm/sec. Calculate the errors on sides a, b, and c and indicate the actual path in a dashed line. Leave out the corner transients.

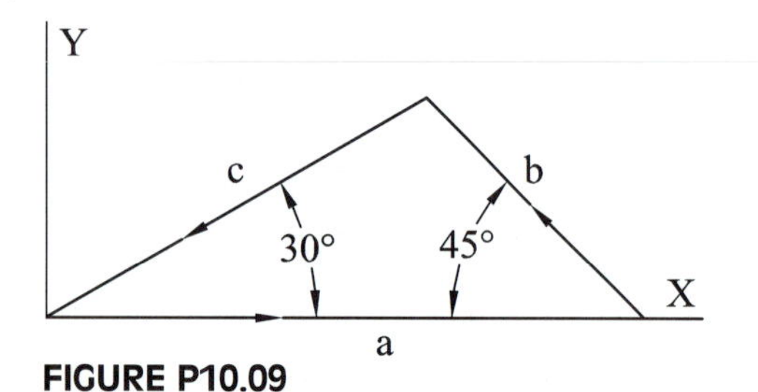

FIGURE P10.09

P10.10 *Positional servo.* Refer to Fig. P10.4.

The parameters of the servo follow: $K_{CL} = 7$ (rad/sec/V), $K_p = 100$ (V/mm), $r = 0.1$ (mm/rad) and we have two cases: **(a)** $\tau_{CL} = 0.045$ (sec); **(b)** $\tau_{CL} = 0.008$ (sec).

(i) Determine ω_n and ζ for both cases, as well as the following errors $e_{p,ss}$.

(ii) Use the same parameters for servos in X and in Y. Drive the Y servo for 0.1 sec with $\dot{y}_{com} = 50$ mm/sec and stop command Y and immediately start commanding the motion in X with $\dot{x}_{com} = 50$ mm/sec for the next 0.1 sec and stop. Follow the actual motions x and y for 0.2 sec longer. Write the corresponding computer program. Use $dt = 0.001$ sec. Do not consider motor current limitation. Plot the commanded and actual motions x and y versus time (as in Fig. 10.45) in the same graph. Plot out the corner motion as in Fig. 10.46 for both *(a)* and *(b)*, in the same graph.

P10.11 *Positional servos commanded in X, Y coordinates.* Refer to Fig. 10.46.

This task is analogous to Examples 10.1 and 10.4. Consider two different specifications, each valid equally to both X and Y servos. For both cases we have the following: $K_p = 125$ (V/mm), $K_a = 20$, $K_m = 2$ (rad/sec/V), $r = 0.1$ (mm/rad), $\tau_m = 0.2$ (sec). Furthermore, we have the following: **(a)** $H = 0.5$ (V/rad/sec); **(b)** $H = 0.1$ (V/rad/sec).

(i) Determine τ_{CL} (sec), K_{CL} (rad/sec/V), K (1/sec), ω_n (rad/sec), ζ.

(ii) Write the simulation program for a motion through a sharp corner with feed rate $f = 25$ mm/sec. The commands follow:

For $0 < t < 0.16$ sec, $x_{com} = 0$, $y_{com} = f\,t$.

For $0.16 < t < 0.32$ sec, $x_{com} = f\,(t - 0.16)$, $y_{com} = 4.0$ mm. Hold x_{com} and y_{com} for an additional 0.28 sec.

Use $dt = 0.001$ sec. Plot y versus x for both cases.

P10.12 *Adaptive control: limit of stability.* Refer to Fig. P10.12.

The positional servo parameters are $K_p = 200$ (V/mm), $K_{CL} = 5$ (rad/sec/V), $r = 0.1$ (mm/sec), $\tau_{CL} = 0.01$ (sec), $CB = 10{,}000$ N/mm, and tooth period $\tau_t = 0.040$ sec, $F_{nom} = 2000$ N. Determine the critical frequency ω_{lim} and the value of $K_{acc,lim}$. The cutter starts from a distance to approach the workpiece. With the given parameters and $K_{acc} = 0.7\ K_{acc,lim}$ determine the following.

(i) The initial commanded acceleration $\ddot{x}_{com}$.

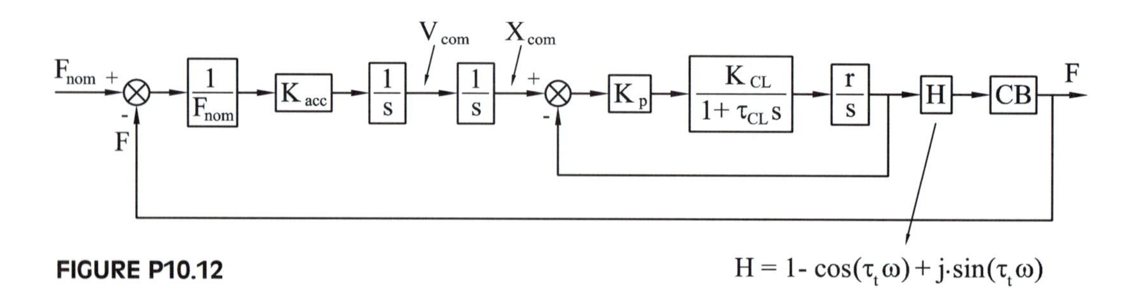

FIGURE P10.12

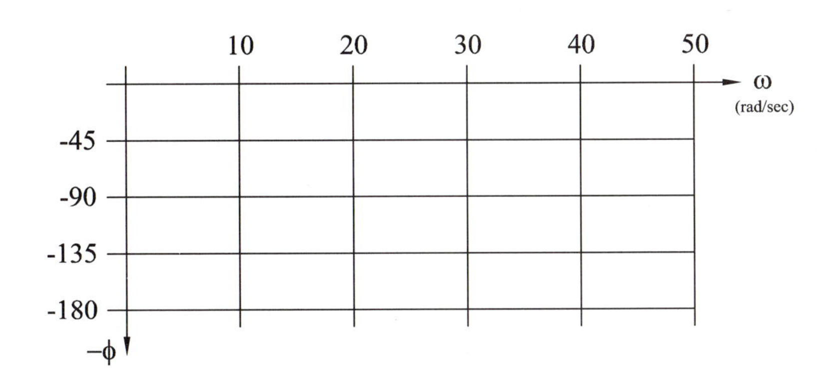

FIGURE P10.13

(ii) The steady-state feed rate f_{ss}, established long after the cutter penetrated the workpiece.

(iii) If the rapid approach velocity is limited to 15 mm/s, what will be the initial force F_{max} at the cutter entry into the workpiece?

P10.13 *A servomotor driving a spring-mass system: feedback from the motion of the mass m.* Refer to Fig. P10.13.

$\tau_{CL} = 0.05$ (sec), $\omega_n = 30$ rad/sec, $\zeta = 0.1$. Neglect the term *(d)* in the figure.

(a) We must first deal with the problem of stability, starting with the phase plot. Do not do that in the test in any detail. Just plot by hand the approximate shape of ϕ_a and ϕ_c. Determine between which two frequencies ω_1 and ω_2, you will find ω_{cr} (or ω_{lim}):

As a rather grossly approximate guess, assume that ω_{cr} is located at 65% of the interval (ω_1, ω_2).

(b) If the above is true, and we would like to try to apply an accelerometric feedback term, for the feedback line containing a block $(1 + as^2)$ as in Fig. 10.70, what will be the appropriate minimum value of a? How will it affect the phase at ω_{cr}? It will move it to $\phi_{\omega cr} = ?$.

P10.14 *Simple questions about stability in driving a flexible structure with feedback from the mass.* Refer to Fig. P10.13.

The gain K summarizes all the gains in the servo, $K = K_p K_{CL} r$. Its value is not needed here. Assume a coincidence such that the corner frequency of the servomotor $\omega_a = 1/\tau_{CL}$ is exactly equal to the natural frequency ω_n of the spring-mass system. Neglect completely the effect of term *(d)* in the figure.

(a) What is the value of the phase shift ϕ of the open-loop transfer function at the frequency $\omega_a = \omega_n$? Will the "critical frequency" be *lower* or *higher* than $\omega_a = \omega_n$? **(b)** What is the magnitude M of the Open Loop transfer function at $\omega_a = \omega_n$, neglecting *(d)*, assuming $K = 1.0$. Assume $\omega_a = \omega_n = 50$ and the damping ratio of the SMD system $\zeta = 0.4$.

P10.15 *Positional servo including a spring-mass system. stability.* Refer to Fig. P10.13.

Given: $\tau_{CL} = 0.1$ sec, $\omega_n = 40$ rad/sec, $\zeta = 0.1$. Neglect the *(d)* term. In order to save you tedious calculations of the critical frequency, check on two guesses: $\omega_{lim} = 28$ rad/sec and $\omega_{lim} = 30$ rad/sec. Evaluate the phase shift for both. Use linear interpolation to determine the actual ω_{lim}. Now that you have determined the approximately correct value of ω_{lim}, accept it as the true value and determine the value of K at the limit of stability.

P10.16 *A servo driving a spring-damper-mass system.* Refer to Fig. P10.13.

Consider two arrangements, A and B, as in Fig. 10.60. Carry out tasks similar to those in Examples 10.9, 10.10, 10.11, and 10.12. The servo parameter is $\tau_{CL} = 0.003$ sec. The SDM system parameters are $k = 2.5e7$ N/M, $m = 400$ kg, $c = 6000$ N/(m/sec); $\omega_n = 250$ rad/sec, $\zeta = 0.03$. The servo is fast, with corner frequency 333 rad/sec higher than ω_n of the SDM system.

Task 1. First, consider arrangement A. Determine the critical frequency of ω_{cr} and limit gain K_{lim}. Drive the system from the following command: with velocity 0.2 m/sec for 0.5 sec

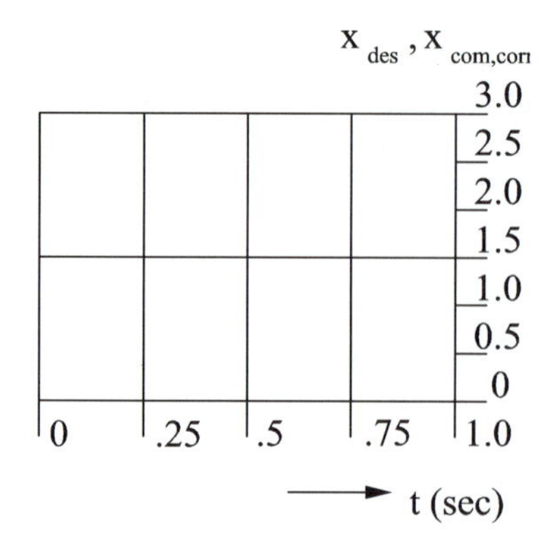

FIGURE P10.17

and stop at $x_{com} = 0.1$ m. Continue the simulation for a total time of 1 sec. Use $dt = 0.0005$ sec. Plot x_{com}, x_i, x versus t, for $K = 0.8$, 1.5, and 2 K_{lim}.

Task 2. Now use arrangement B with $K = 4\ K_{lim}$ of Task 1 and plot x_{com}, x_i, x versus t.

Task 3. Return to arrangement A and apply accelerometer feedback according to Eqs. (10.98) through (10.102) and to Fig. 10.70. Plot x_{com}, x_i, x versus t for $K = 2$ and 4 K_{lim} of Task 1.

P10.17 *Driving a flexible arm: feedforward compensation.*

Assume a perfect servomechanism; the task is analogous to Section 10.9.1, Fig. 10.73. The parameters of the case are $m = 200$ kg, $k = 4000$ N/m; damping will be neglected. The desired motion is $x_{des} = -4t^3 + 6t^2 + t$, for $0 < t < 1$ sec. Express $\dot{x}_{des}$ and $\ddot{x}_{des}$; neglect damping terms, and plot by hand: x_{des}, $\dot{x}_{des}$, and $\ddot{x}_{des}$. Express $x_{com,corr}$ and evaluate it for $t = 0$, 0.25, 0.5, 0.75, 1.0 sec. Plot x_{des} and $x_{com,corr}$. Use the plot grid as shown in Fig. P10.17.

P10.18. *Feedforward correction for a positional servomechanism.* Refer to Fig. P10.4 and to Fig. 10.74.

The task is limited to just the motion section with constant acceleration, $x_{des} = (a/2)\ t^2$, for $a = 4$ m/sec^2 and $t = 0$, 0.25 sec. We have $K = K_p K_{CL}\ r = 50$/sec, $\tau_{CL} = 0.010$ sec for the block D Fig. 10.76 and Eqs. (10.120). There is, however, no spring-mass system S to drive, we are compensating for the dynamics of the servo D, such as in Fig. P10.4 alone. By evaluating x_{des}, $\dot{x}_{des}$ and $\ddot{x}_{des}$ solve the expression $x_{com,corr}$. Write it out. Determine the value of $x_{com,corr}$ at $t = 0.2$ sec.

P10.19. *Feedforward compensation in a single-coordinate system of drive and structure, DS.* See Fig. 10.76, and Fig. P10.19.

This task is analogous to the exercise in Examples 10.14 and 10.15. It consists of the following:

(a) The case without compensation. Use x_{com} as defined in the graph P10.19. By simulation, obtain x_i and x. Use $dt =$ 0.001 sec, total number of steps $n = 1500$. Plot x_{com}, x_i, x versus time in one graph.

(b) Apply feedforward compensation. Use the notation as introduced in Fig. 10.76. Define z as the function in the graph of Fig. P10.19. Derive y and x_{cc}. Then use x_{cc} to drive the system and obtain x_i and x by simulation. Use $dt = 0.0005$ sec, $N = 600$. System specifications:

Drive: $K = 200$/sec, $\tau_{CL} = 0.03$

Structure: $k = 7e4$ N/m, $m = 50$ kg, $c = 90$ N/(m/sec)

Plot z, x_i, x versus time in one graph, x_{cc} versus time, and ddx_i versus time. This is the case of one coordinate that includes a realistic servo drive plus a spring-mass system.

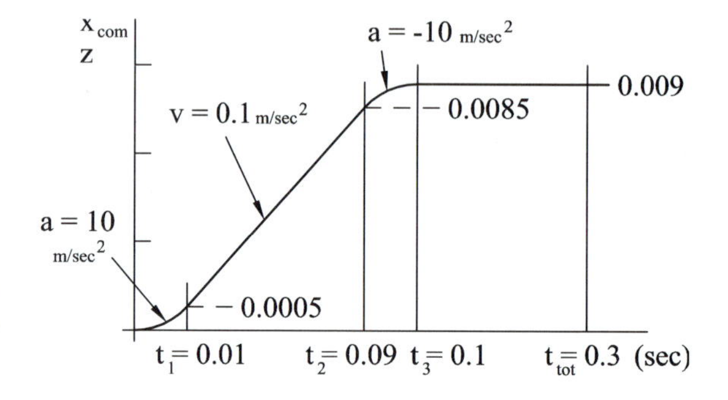

FIGURE P10.19

P10.20 *Robot with flexible arms, in polar coordinates: feedforward corrections.* Neglect the damping term; see Fig. P10.20. $k = 5000$ N/m, $m = 100$ kg, $\kappa = 5000$ Nm/rad. Assume a uniform motion on the line A, B with a velocity $\dot{x}_{com} = 2$ m/sec. Determine the feedforward corrective terms $\phi_{com,corr}$ and $r_{com,corr}$ at points $x = 0$ and $x = 1.0$ m.

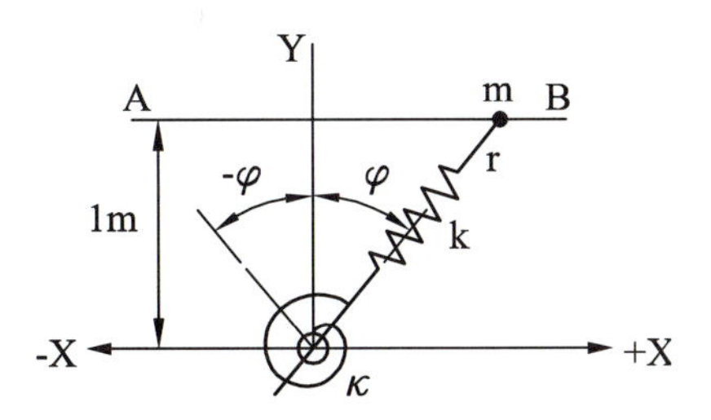

FIGURE P10.20

P10.21 *Robot kinematics and dynamics: Feedforward correction.* See Fig. P10.21.

A flexible arm and drive are commanded to move on a straight line starting at $x = 0$. Determine the corrective terms Δr_{corr} and $\Delta\phi_{corr}$ in the two positions x_1 and x_2, considering the "simple" feedforward correction only (neglecting the damping terms). Start with reproducing the corresponding acceleration formulas from Eqs. (10.129) and (10.130). For ϕ, they are in radians.

P10.22 *Feedforward compensation for a 2D robot.* See Fig. P10.22.

This is the case of a polar coordinate robot with flexible transmissions. Assume a perfect servomotor, and evaluate path errors due to the flexibilities given in Fig. P10.22a. Drive the mass on a straight path in the *x*, *y* coordinates from point *A* to point *B;* Fig P10.22b. At point *A* we have $t = 0$, $x = 0$. Path $x(t)$ is defined as shown in Fig. P10.22c. Carry out the following tasks. They are analogous to Example 10.17.

(a) First, no compensation will be used. Apply x_{com} and y_{com} as in Fig. P10.22b and c and through coordinate transformation, define ϕ_{com} and r_{com}. Through simulations of the two spring-mass-damper systems, one translative and the other one rotary, obtain ϕ and *r.* Finally, recombine ϕ and *r* into *x* and *y.* Plot ϕ_{com}, ϕ versus *t*, r_{com}, *r* versus *t*, and *y* versus *x*. Use $dt = 0.005$ sec for a total time of 4 sec.

(b) Apply feedforward compensation. Start with the functions as in Fig. P10.22b and c as x_{des}, y_{des}, and derive ϕ_{des}, r_{des}, (they will be the same as ϕ_{com}, r_{com} in *(a)*. Derive ϕ_{com}, r_{com} including feedforward transformations. Use these as inputs to obtain ϕ and *r* by simulation, and combine them into *x* and *y.* Plot the same kind of plots as in *(a)* and use the same d*t* and t_{tot}.

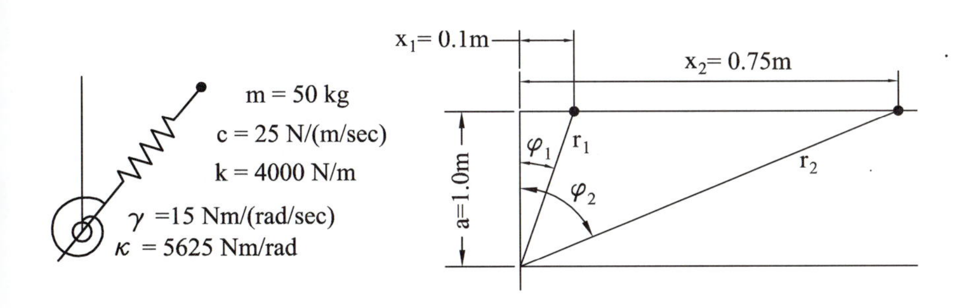

FIGURE P10.21

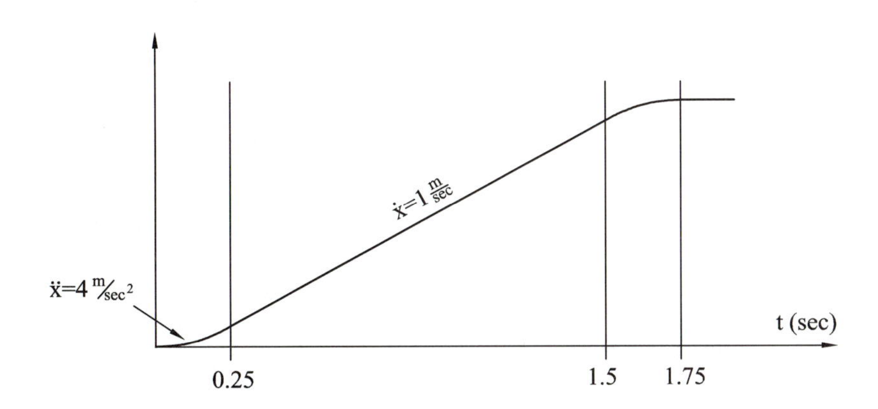

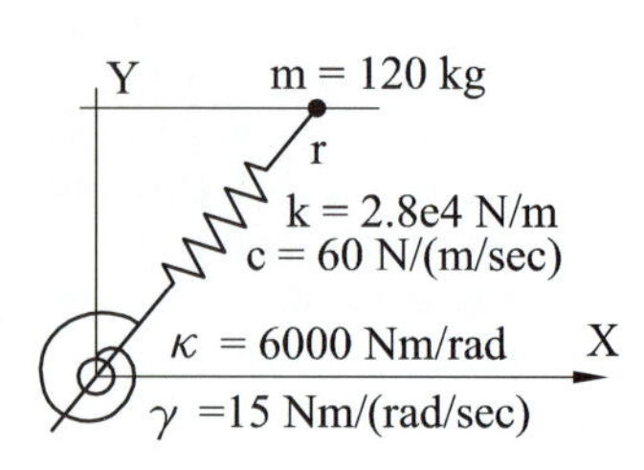

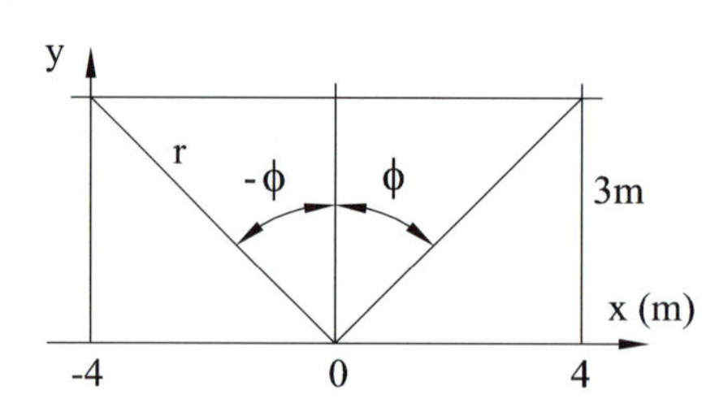

FIGURE P10.22

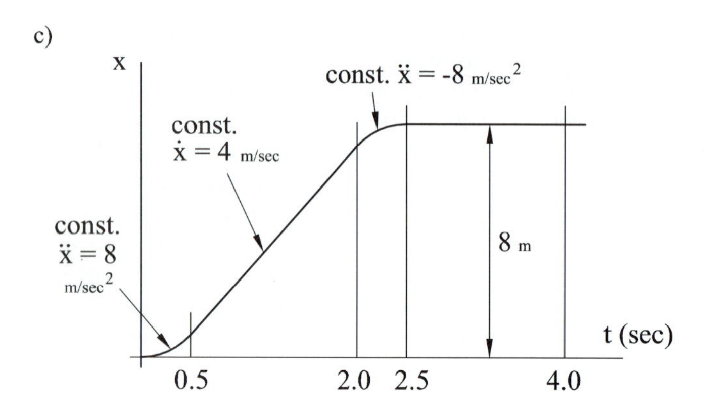

PART 4

Assembly and Nontraditional Processes

11 Assembly: Material Handling and Welding

11.1 INTRODUCTION

All the products of mechanical manufacturing are made of parts that are brought together in *subassemblies* and then in final *assemblies.* Most of the activities in manufacturing consist of receiving materials and converting them into component parts (e.g., housings, shafts, gears, bearings, bolts, etc.) by passing them through a succession of manufacturing operations. This involves moving the parts between the various pieces of production machinery and delivering them at the proper time, simultaneously with the other parts, into the assembly station. All of this movement is called *material handling.* It consists of both the transportation and the final handling during assembly. We will learn later about intermediate storage, and actually, we must include storage, retrieval, identification, counting, and reporting in material handing. In the final handling action, the part must be properly oriented. The core process in assembly is *joining* parts together. This may be done in a reversible way, for example by using bolts and nuts so that the assembly can, if required, be taken apart for repair or for exchange of worn parts. Assembly may also be permanent, using one of the many processes to produce permanent joints, such as riveting, brazing, or welding.

In this chapter all these activities are discussed, and substantial attention is devoted to welding. However, welding is not dealt with just as one of the joining processes in assembly; it is more substantial, more complex, and more costly, and it has a more significant influence on the strength of the assembly than, for instance, mechanical screwing or riveting actions. Thus it stands on a par with the other basic processes of mechanical manufacture, such as cutting or forming. There are pieces of machinery in which welding (and prior to it thermal cutting of plates into the pieces to be welded together) represents the bulk of the manufacturing effort: for example, excavators, cranes, and ships. Welding is also used in other areas beyond machinery production: in construction, or in building a pipeline. For these reasons, the discussion of welding and the analysis of one of its most significant modes, arc welding, occupy a major part of the chapter.

All the joining processes may be considered as components of the overall activities of assembling. Joining is the final step in a sequence of actions involving storing, moving, feeding, orienting, and manipulating material.

Mechanical joining is used in all products, although some of the subassemblies may have been produced by other processes, like welding. It is, therefore, a pervasive activity in mechanical manufactures. It may be based on a variety of designs of threaded bolts and nuts, or pins, retaining rings, and—for permanent joints—of seams. In preparing for these operations, machining processes of drilling, reaming, and tapping are used. For seaming and for pressing parts together, mechanical or hydraulic presses are used.

Welding is the most prevalent method of joining metals. It is carried out in many forms representing the various welding methods described in the following text. All of these have the common basic characteristic of binding together two pieces of essentially equal material, either directly, or with the addition of a filler of, again, essentially the same material. For instance, two pieces of steel are joined by using steel as the filler metal, although it may have more of the alloying elements than the base steel. These welding processes accomplish the joining action either by pure fusion, in which the base metals on each side of the joint are melted during the process, and the filler metal, if used, is also melted, or by a combination of heat and pressure. At the extreme opposite to fusion, pressure alone, without any heat, is used in solid-state welding. Between the two extremes, both heat and pressure are used, as in spot welding or friction welding.

The processes of brazing and soldering are similar to welding in that they use a molten filler metal to generate a joint. However, the filler metal is different from the metals being joined and has a lower melting point. The two processes are alike, but they are distinguished by the range of melting temperatures of the filler metals. In soldering, a lead-tin alloy known as solder is used, which, depending on its composition, melts between 183° and 275°C. The range of melting temperatures of the filler metals used in brazing is 450° to 800°C. Common filler metals are copper and silver.

In adhesive bonding, metallic or nonmetallic parts are joined by the application of an adhesive, usually one of the plastics. The advantages of adhesive bonding are that many compositions of similar or dissimilar materials can be joined, electrochemical corrosion is eliminated between dissimilar metals, and the process is carried out either at room temperature or at temperatures mostly lower than those used in soldering. The strength of the joint is generally rather low. However, if it is designed so that the stress is limited to shear over a large area the overall strength of the joint may be quite high.

Hard surfacing is a very economical way of providing wear-resistant surfaces on machine parts. A variety of alloy steels or tungsten carbides combined with a steel wire can be fused onto the substrate surface using various types of arc welding processes. An alternative method is thermal spraying, in which an electric arc or a gas flame is used to blast atomized molten metal or ceramics onto the surface of parts to be coated with a wear-resistant or corrosion-resistant layer.

Thermal cutting processes, including laser beam, air carbon arc, plasma arc, and oxygen flame cutting, are used to efficiently cut parts out of sheet metal and of plates from 1 mm to 1 m thick. These methods are based on cutting by melting and by oxidation. The heaviest plates are cut using the oxygen flame cutting process, where a stream of oxygen is used to burn iron. It is helped by preheating flames that burn an auxiliary fuel gas. Otherwise, an electric arc may be used with the oxygen stream. In plasma arc cutting, other gases, such as an argon-hydrogen mixture, are used to produce a high-velocity plasma jet to melt the metal and blow it away. The thermal cutting processes are discussed in more detail in Chapter 12.

11.2 MATERIAL HANDLING

Materials must be moved from the entry points at receiving docks where they have been delivered by trucks, through the processing and inspection stations on the factory floor to the assembly stations and from there to the shipping docks. On the way they are frequently picked up, loaded, and unloaded onto and from transporting devices or into and from the processing machines. They may have to wait in initial, intermediate, or final storage complexes. There may be some unpacking and, finally, packing involved.

Small parts are moved in boxes or in bins; larger parts are placed on pallets, either individually or as a group in a well-organized pattern that makes it easier to use a programmable pickup device for unloading them from the pallet. The bottom of a pallet (see Fig. 11.1) has openings for the forks of a forklift truck; it is easily lifted and deposited on the floor. The top of a pallet may be as simple as shown in the figure, or it may contain pins, partitions, or racks to hold particular types of parts. Universal pallets with plain tops exist in several standard sizes and may be made of wood or of plastic. Such pallets are used for transporting many different parts and must be distinguished from the pallets used in a flexible manufacturing system (FMS), as described in Section 10.4, where they serve as fixtures to precisely locate and hold workpieces on machining centers. These pallets are made of steel. Apart from the simple, small bins used as containers for small parts, such as bolts and nuts, which are made of sheet metal, plastic, or even cardboard, larger containers are made from corrugated metal or welded wire. They have feet on the four corners and can be stacked or placed in pallet racks like any other unit load.

The rather complex system of storage and movement of parts in mass-production plants is usually designed as a part of the manufacturing facility. Material handling is then an integral part of the manufacturing system and may sometimes be the major part of it. As an illustration, consider personal computers. These require almost no processing stations; production consists of assembling the computers. The individual parts, such as the casings, monitor tubes, printed circuit boards, connectors, keyboards, and so on, are manufactured elsewhere and delivered to be received and moved through the plant. Material handling may represent 80% of all the activity.

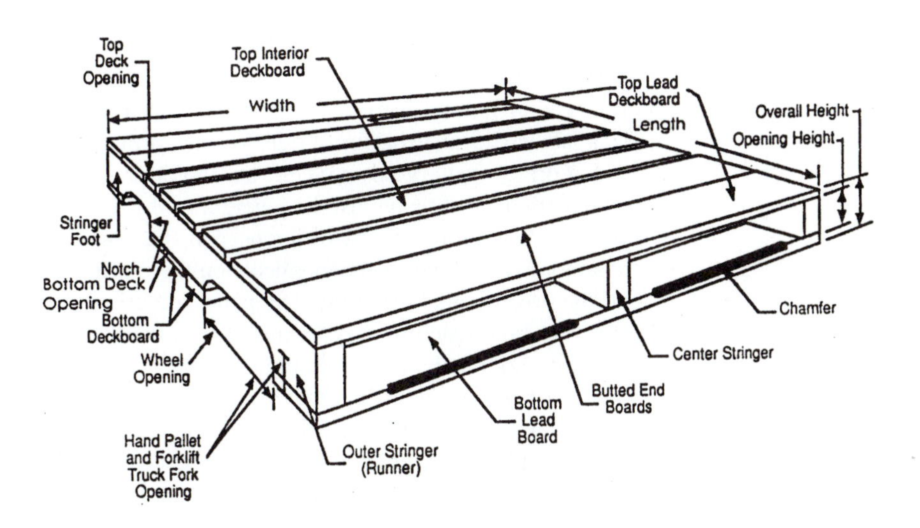

Figure 11.1
The common pallet for transportation of parts. The bottom is standardized to be carried on various types of trucks. [EMH]

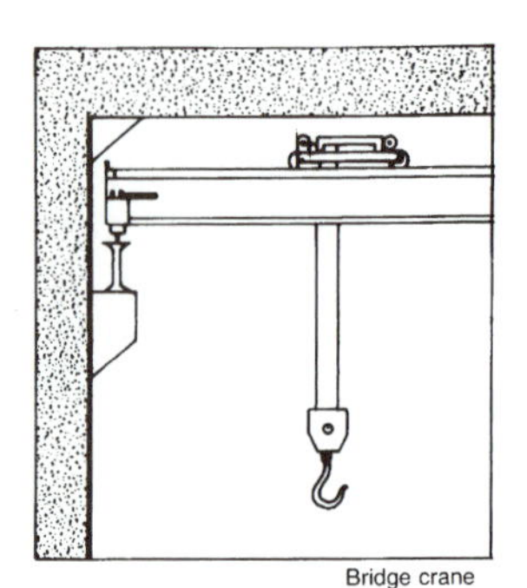

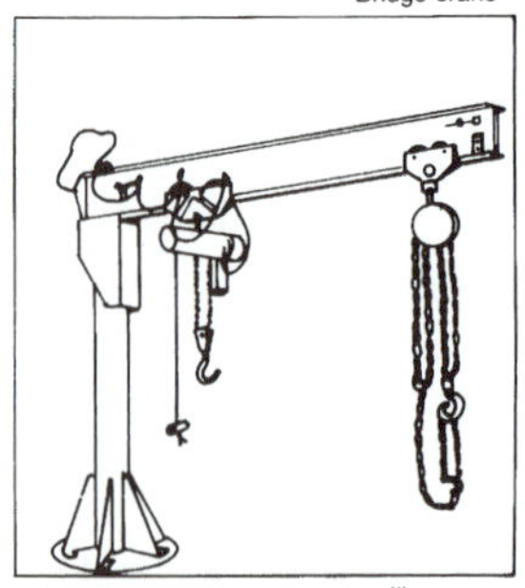

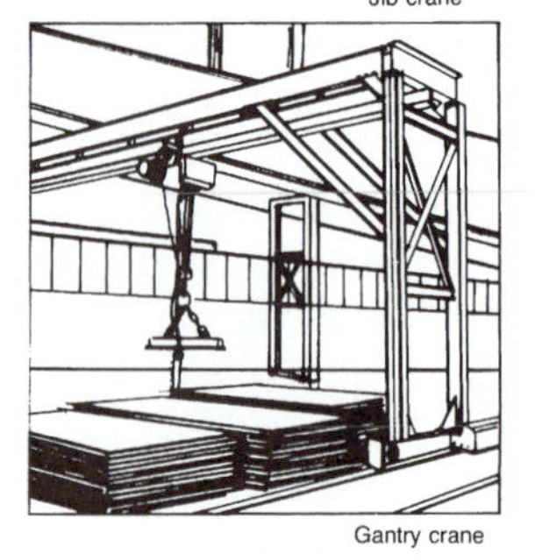

Figure 11.2
Three types of cranes. The bridge crane can service a whole factory bay. The jib crane is for local use. The gantry type can be moved to area of need. [BMH]

Most production, however, involves small and medium lots, and the movement of parts becomes a very complex and variable system that must be flexible. Such systems may be noncomputerized, as when a dispatcher directs forklift drivers between individual storage and processing stations, but computer control is becoming more common. The task of programming and controlling such a system is in the domain of industrial engineering. The various aspects of this task are discussed in Ward and White [1]. The rather vast field of material handling, which is supplied by a number of producers of many types of equipment and represented by a number of industrial associations, is also covered as regards information and education by the MHI institute.

The logistics of material handling is not dealt with in this book; further discussion is limited to the description of selected hardware. Figure 11.2 shows three basic types of cranes. The bridge crane may cover a whole large shop or a bay of a shop. It can lift and carry rather large loads and is typically installed in an assembly bay of products consisting of heavy parts (e.g., large machine tools). Local lifting and moving is provided by either a jib crane or a gantry crane. The gantry crane can move on rails; the lighter, double-leg gantry cranes run on wheels and can be moved to various locations for temporary use.

For transportation of parts in box-type containers or other types of unit loads over longer distances, wheel, roller, or belt conveyors are used (Fig. 11.3). The roller type used in most instances has rollers that are free to rotate. The load can then be moved manually or by gravity if the conveyor is sloped. The load can be stopped at any location, and the conveyor can divert by a right angle with a corner equipped with a field of free-rotating balls instead of rollers. It can also be branched and interfaced with a belt conveyor. The belt conveyor moves the load by friction in a synchronous way. Overhead and in-floor tow conveyors are also illustrated. The overhead conveyor has the great advantage that it does not occupy floor space and can be installed and take shortcuts above production machinery. An example is an installation in an automobile factory where crankshafts are turned and ground and then sent over to the engine-assembly transfer line. With a combination of belt and roller conveyors, push-type diverter devices are used for sorting parts and sending them on two different paths. Obviously, sensing of identification marks is involved; these are most commonly the well-known bar code labels. In general, identification of goods is used throughout a computerized material handling system.

Conveyors are essentially rigid systems producing movements along fixed paths. A rather limited flexibility is obtained by the use of branching. Therefore they are not used in midsize lot productions. Transportation in such shops is generally provided by various types of industrial trucks. A selection of them is shown in Fig. 11.4. The forklift truck labeled "counterbalanced rider" is one of many types that are distinguished by the kind of drive—battery-powered electric motor or combustion engine—whether the load is counterbalanced by an installed weight or the truck is equipped with trigger feet that reach under the fork, and by differences in size and load capacity. The "walkie" shown in the top right corner has two wheels on the outriggers and a third wheel at the back, and the driver walks behind it. The pallet truck is similar. The narrow-aisle truck may have very high reach; it may be installed in the aisle of a storage/retrieval system and guided on rails both at the bottom and at the top. The remaining illustrations represent various forms of tractors that may push or pull different trailers.

A vehicle that represents the utmost in flexibility is the automatic guided vehicle (AGV), one form of which is shown in Fig. 11.5. This is a pallet-handling vehicle, but it may be designed as a towing truck, unit load transporter, forklift truck, or an assem-

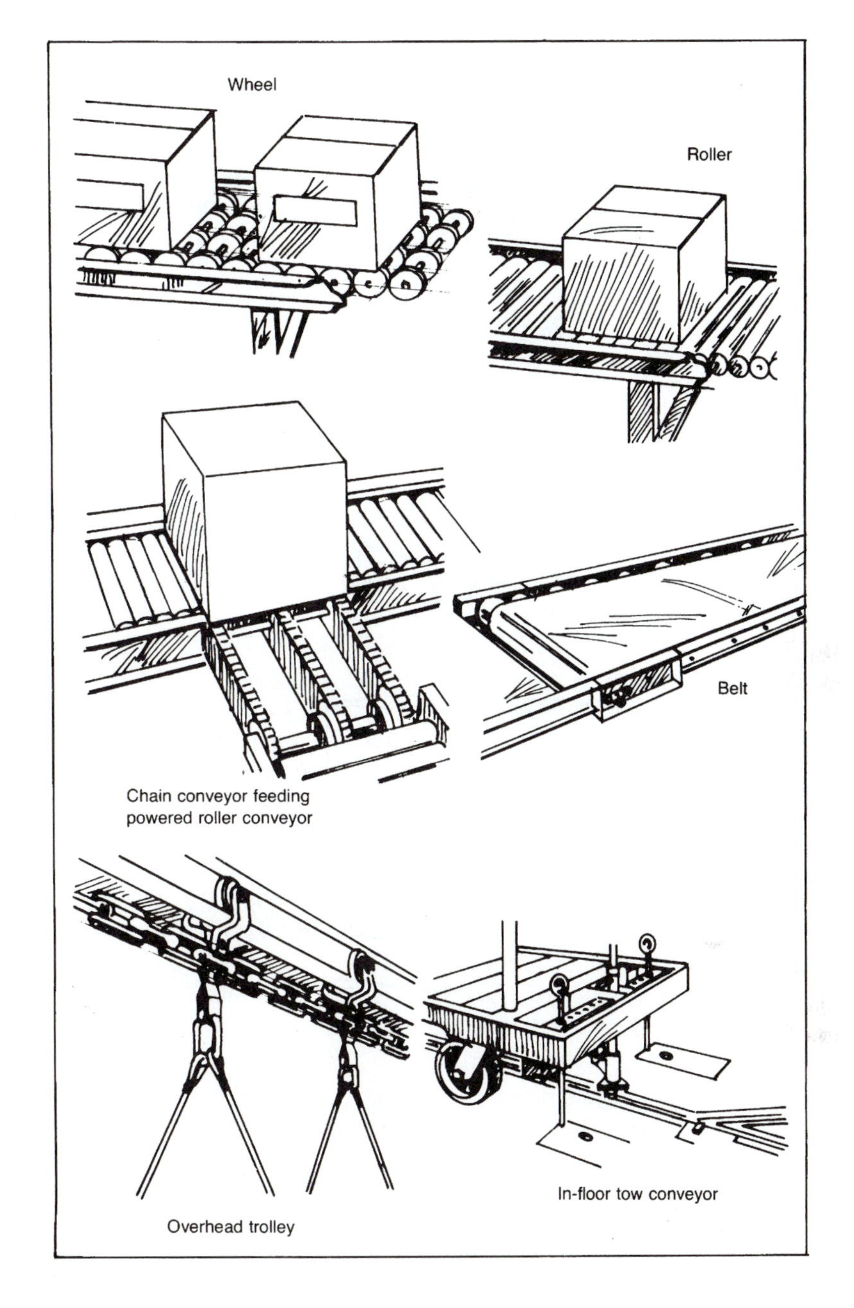

Figure 11.3
Various types of conveyors. The four types at top move parts in boxes or in pieces. The overhead trolley moves parts over and above machines located on the floor. The in-floor conveyor transports larger loads, perhaps in pallets, to individual destinations. [BMH]

bly line vehicle equipped with a robotic arm. Its motion is controlled by a computer, and it incorporates various types of guiding/sensing systems. Guidance may be provided by a wire embedded in the floor of the shop, by stripes of fluorescent particles painted on the floor, or by markings on the walls. In the first case the wire transmits a radio frequency, and the vehicle carries the receivers. In the second case photosensors are used. In the third case a laser beam scans the bar codes of the markings. Many other solutions are being developed that impart more and more flexibility to the form of the path. Some solutions rely on an inertial guidance system, interpolating between

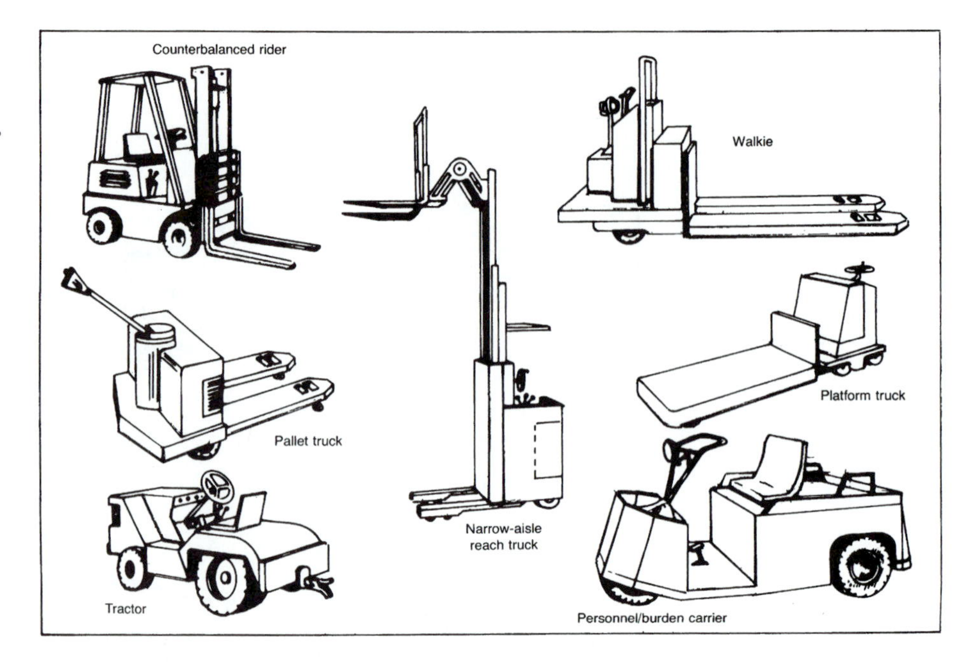

Figure 11.4
Industrial trucks are specialized for various types of loads. There are the forklifts, the pallet and platform trucks, and the tractor. All are motorized and manually steered. [BMH]

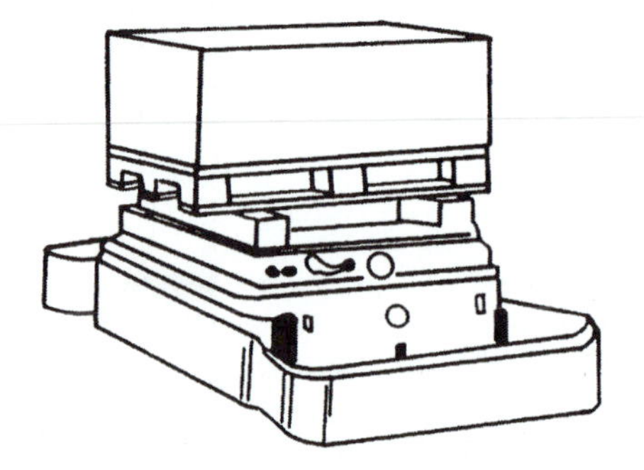

Figure 11.5
Automatic guided vehicle (AGV). These may have various tops. They are driverless, computer-remote-controlled vehicles with sensors and feedback from a net of wires in the floor or from other kinds of locator systems. [EMH]

beacons in the floor, or on a grid of wires. The AGV is protected against collision by sonar systems that reflect ultrasound from obstacles. An application of AGVs in airplane manufacture is shown in Fig. 10.31. AGVs are becoming more and more popular as the corresponding technology matures and the vehicles and systems get less expensive. Ultimately they may offer several advantages over other material handling systems: flexibility, reliability, operating savings, unobstructed movement, and easy interfacing with other systems.

No material handling system can exist without storage facilities. Let us briefly describe some of these, concentrating on the computerized automated storage and retrieval systems (AS/RS). First, however, we will look at the noncomputerized case (Fig. 11.6). There is a three-level, drive-through, rack-type system for storing unit loads on pallets. The forklift truck can drive into or back out of the rack structure between the vertical posts in order to pick up or deposit a pallet at any location in an aisle. Retrieval of loads is possible on a first-in first-out basis. There are no horizontal cross members except at the top of the rack structure. A load deposited in the deepest or for-

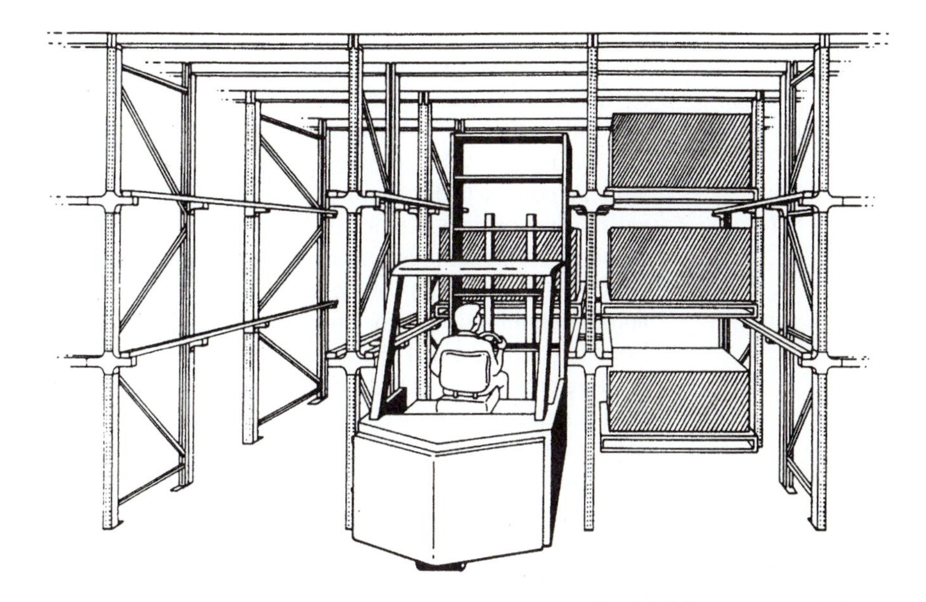

Figure 11.6
Drive-through pallet rack with forklift truck. [EMH]

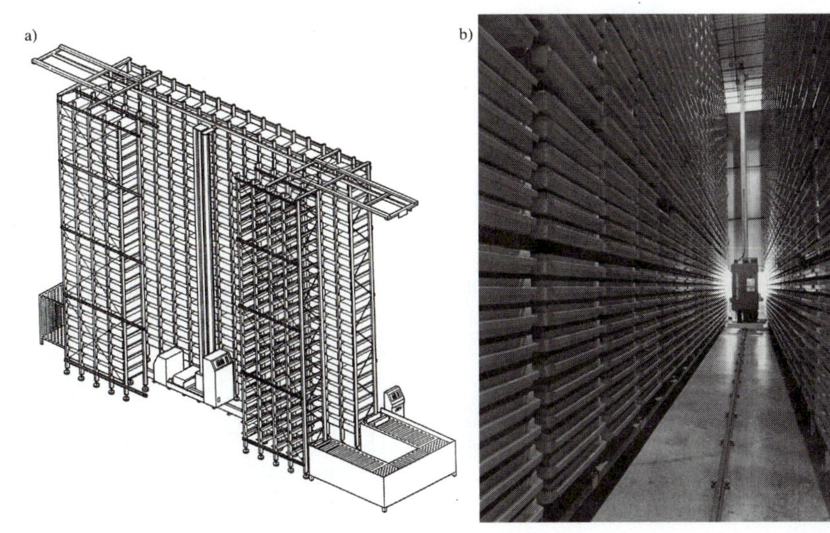

Figure 11.7
Single-aisle, mini-load automatic storage and retrieval systems (AS/RS) with conveyor pick-up/deposit. a) Drawing that includes the narrow-aisle truck. [EMH]; b) photograph of the system. [LITTON]

wardmost location can be retrieved as the first load from the other side of the aisle. AS/RS systems are often very large; Fig. 11.7 shows the so-called mini-load system, which stores parts in bins. At the left is the drawing of the system; at right is a photograph with a view down the middle track. The storage/retrieval machine moves in the aisle, as seen in the cut-out of the first row of racks. The bins are picked up and deposited on the roller conveyors in the front. The storage/retrieval machine is a structural frame or mast guided on rails at the bottom and at the top. Its servomotor drives have three axes of motion: along the aisle, up and down, and into and out of the racks. These motions can be executed fully automatically, moving pallets in and out and depositing them at the pickup/deposit station. Alternatively, the machine carries a cab from which an operator who controls the motions can retrieve a particular part of the contents of a bin. These illustrations show "single-deep" racks on each side of the aisle. Double and multiple deep lanes exist in high-density storage systems. Transverse motion in these deep lanes may be provided by gravity when the sides of the lanes contain rollers that slope down to one side of the structure; see Fig. 11.8. A different system has been developed by one of the leading manufacturers of AS/RS equipment, in which a compact, flat, autonomous vehicle is carried to the desired level by the vertical transfer vehicle, which can travel at the bottom of the lane to any location in the whole system. Examples of huge, high-density AS/RS installations are described in [1], which lists applications in the manufacture of self-developing film; in a central storage and distribution center for bolts, nuts, and other fasteners; in roller-bearing manufacture, in the storage and distribution of automotive subassemblies, in the manufacture of printing presses, in a warehouse for meat products, in storage of in-process components for aerospace electromechanical assemblies, in pharmaceutical production, and in the manufacture of heavy trucks. These applications illustrate the high level of development and acceptance of these impressive systems, which represent an important aspect of material handling activities.

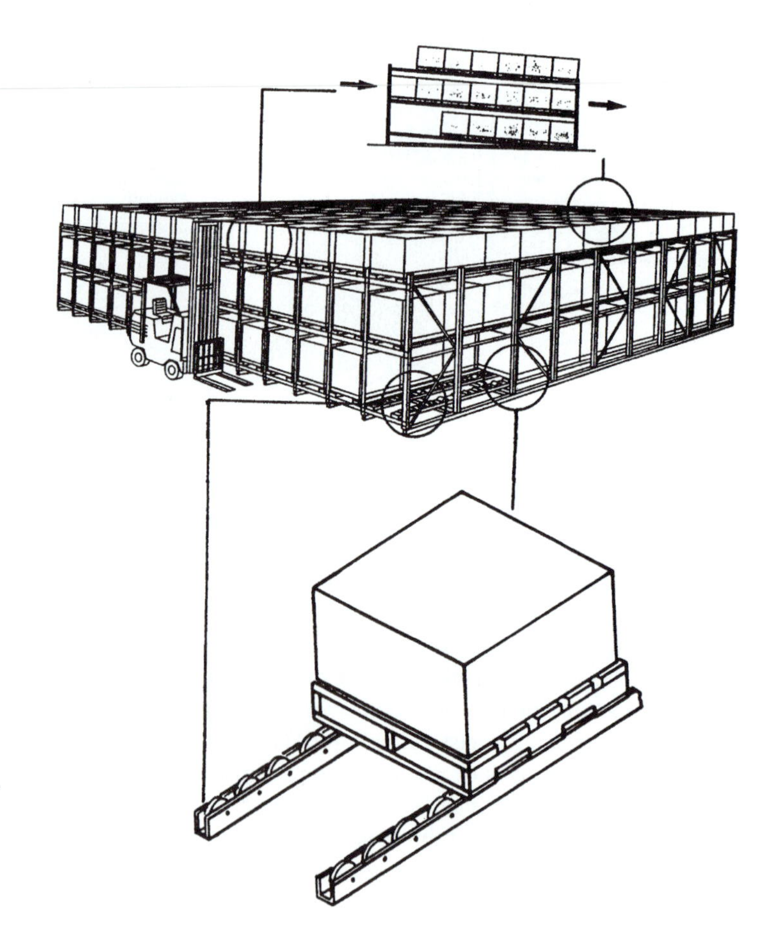

Figure 11.8
Pallet-load, flow-delivery system with multiple deep lanes. Motion along the lanes is provided by gravity. Alternatively, an autonomous vehicle can be carried to the desired level to travel in the lane. [EMH]

11.3 MECHANICAL JOINING

Let us distinguish between separable and permanent joints. In the first category are the threaded fasteners, snap fits, and retaining rings. They are intended to permit rather easy dismantling, but they should also provide a secure joint, safe against spontaneous release. In the second class are joints that involve a plastic change of shape or dimension, such as rivets, seams, beads, and crimps. In between the two classes are joints that can be taken apart with special effort such as interference and shrink-fit joints.

Threaded fasteners are screws that fit into a threaded hole and bolts that pass through a plain hole and are tightened by means of nuts. Various screws and bolt heads are shown in Fig. 11.9 and various nuts in Fig. 11.10. Instead of headed bolts, threaded studs may be used with nuts on both sides of the joint. The thin jam nut is used on top of the full nut. The latter carries the preload in the bolt, and the former is tightened opposite it to secure against release. Slotted nuts are used in combination with cotter pins or wires that lock the nuts in place. Two other types of screws deserve mentioning: set screws and self-tapping screws. Set screws are mostly made of hardened steel; they are used to hold bushings, gear hubs, and coupling hubs on shafts, and they are capable of transmitting a limited amount of torque. Self-tapping screws cut into sheet metal and eliminate the need of a thread-tapping operation. Apart from the basic types described here, there are many other types of threaded fasteners: thread bushing inserts, coil inserts, anchor and weld nuts, self-clinching and self-piercing fasteners, studs, quick-operating fasteners, expanding, self-sealing, and so on. Threaded fasteners are

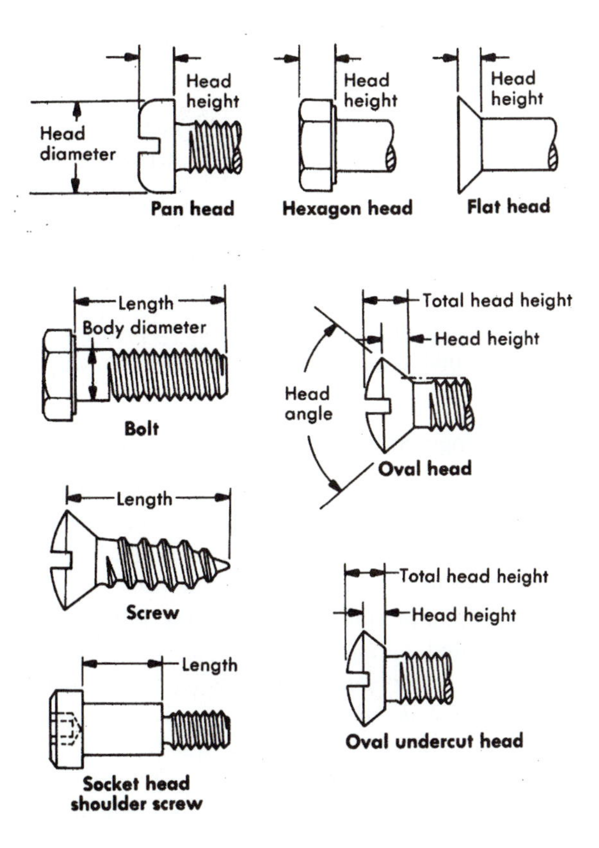

Figure 11.9
Various types of bolt and screw heads. [SME4]

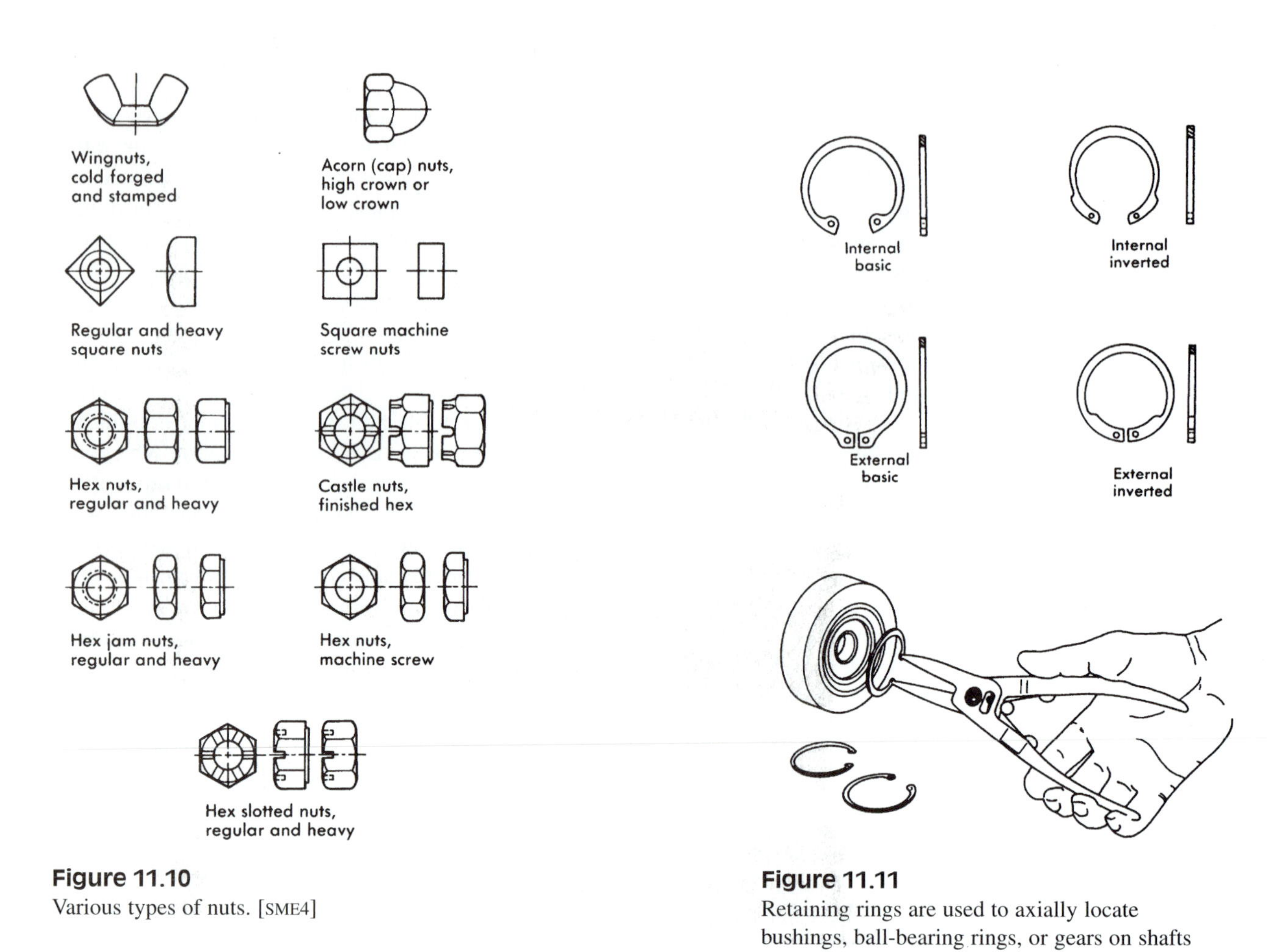

Figure 11.10
Various types of nuts. [SME4]

Figure 11.11
Retaining rings are used to axially locate bushings, ball-bearing rings, or gears on shafts or in housings. [SME4]

the most common elements in machinery. However the various subassemblies are held together, the final assembly mostly involves screws, bolts, and nuts.

Retaining Rings, Snap Fits, and Spring Clips

Retaining rings are used for axially locating bushings, gears, ball-bearing rings, and similar such parts on shafts (external rings) or in housings (internal rings); see Fig. 11.11. The rings fit in corresponding grooves and can be easily removed with simple tools.

Snap fits and spring clips use spring-type elements that are brought into place by a simple pushing motion; they hold a lid or cover firmly onto the base part; see Fig. 11.12. Compared with using screws this is a fast operation easily mechanized.

Semipermanent connections of a hub onto a shaft or of a pin in a hole are obtained by *press fitting* or *shrink fitting*. The two parts are manufactured with an interference fit and then joined either by applying a large force produced by a mechanical or hydraulic press or by heating the external part so that it expands sufficiently to allow insertion of the shaft with modest pressure. Shrink-fit clamping of an end-mill shank into a cylin-

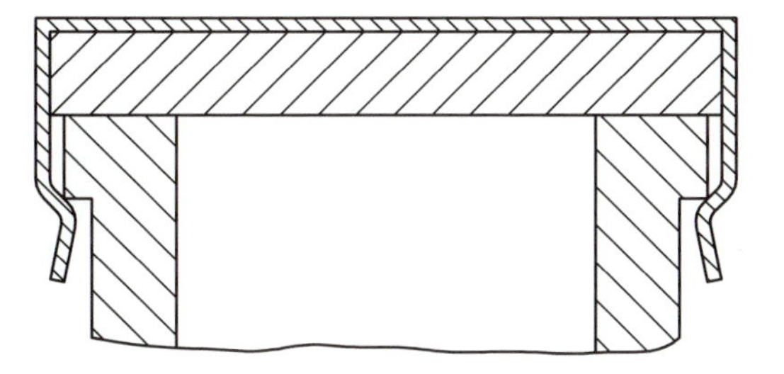

Figure 11.12
A spring clip used to attach a cover plate. Snap and spring clips are used for easy assembly of parts by a simple pushing motion.

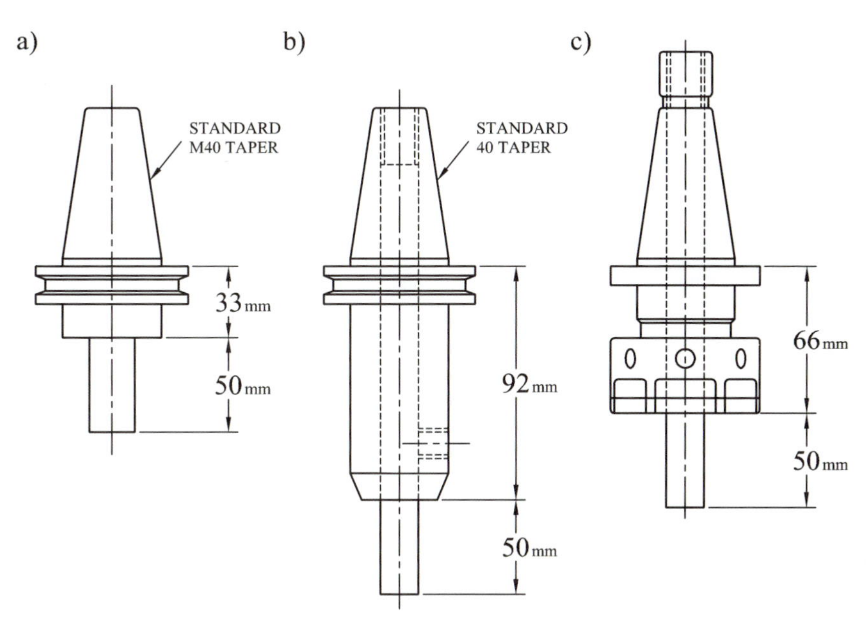

Figure 11.13
a) An end mill is shrink-fit into the holder. A special induction type furnace is used to clamp and to release the tool. The connection is accurate and rigid, and permits a shorter overhang of the tool than b) using set screws or c) collets.

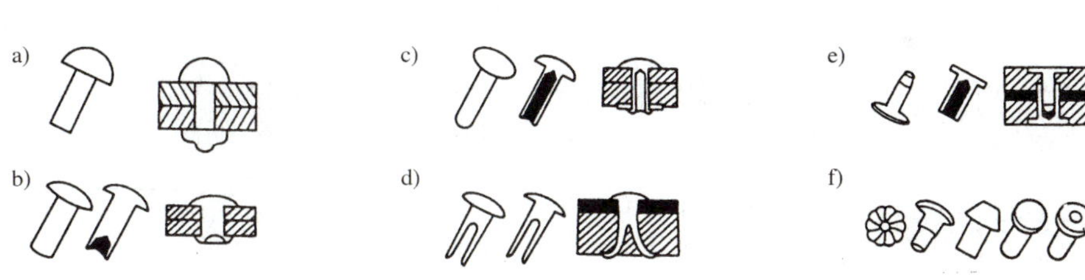

Figure 11.14
Various types of rivets. They are used for permanent joining of sheet and of plates. Riveting is a large operation in the assembly of aircraft, where it is highly automated. [SME4]

drical bore in the toolholder, as shown in Fig. 11.13a, is lately becoming a common practice. The main advantage gained is the much shorter overhang of the tool from the spindle compared to the use of set screw (Fig. 11.13b) or collet (Fig. 11.13c) connections, and the resulting higher stiffness on the tool. With press and shrink fits, the joint can be disassembled by using sufficient force.

Permanent mechanical connections are obtained by riveting and, for sheet metal, by the use of seams and beads. Various types of rivets are shown in Fig. 11.14. Shown are solid, tubular, semitubular, and compression rivets. Riveting is often done on special drilling and riveting machines and can be fully automated. Examples of seams are shown in Fig. 11.15a, and of crimping a bead to join two tubes and to join a hub onto a shaft are shown in Fig. 11.15b.

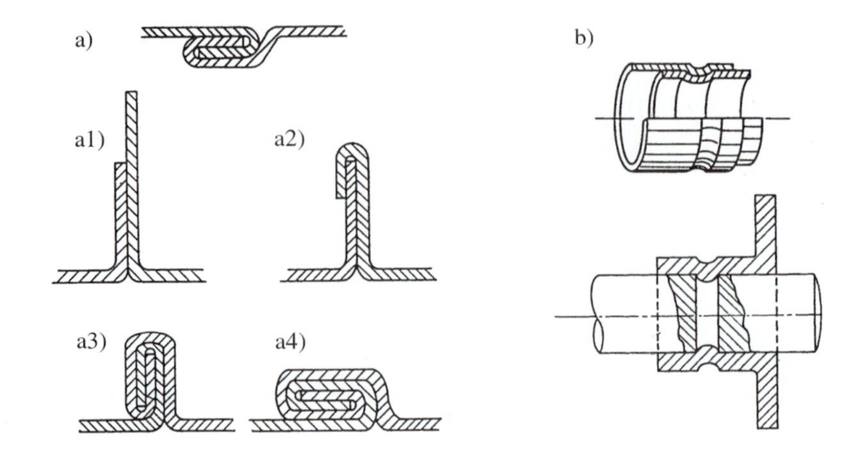

Figure 11.15
a) Single-lock seam and steps in forming a double-lock seam. [SME4]
b) Beads to connect two pipes and a hub onto a shaft. [SME4]

11.4 ASSEMBLY

Most of the assembly operations are carried out manually. This also applies to products made in large quantities, such as cars, although some of the operations are automated, such as spot welding of car bodies, and some of the subassemblies are also automated. On one hand this applies to small subassemblies such as those indicated in Fig. 11.16. This figure was prepared by one of the leading builders of machines for automatic assembly, which we will discuss later. Even some of the larger subassemblies are put together on automated systems (see Fig. 11.17). This is an example of a transfer line for assembly and testing of automobile differential gear boxes with a rate of 300 boxes per hour. The system can be set up for several sizes of the product. Examples of this kind are, however, rather rare.

Many products are made in much smaller quantities and manually assembled from subassemblies. An example of a partly assembled milling machine is presented in Fig. 11.18. The large structural parts, such as the base, the rear wall, the pedestals, and the horizontal saddle, have all been arc welded of platelike parts and attached to one

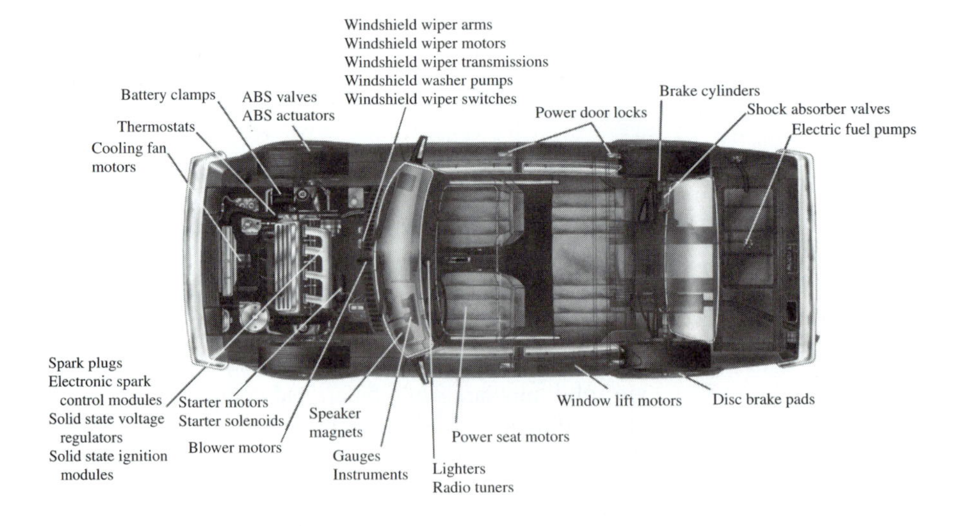

Figure 11.16
Automatic assembly of small groups of parts is used in mass production. As examples, small subassemblies used in an automobile that have been assembled automatically are shown. [BODINE]

Figure 11.17
Transfer line for automatic assembly and testing of differential gear transmission boxes. [INGERSOLL-RAND]

Figure 11.18
Welding and screw bolting are the most common assembly techniques. Shown is the example of a partially assembled milling machine using welded subassemblies joined by screws, bolts, and nuts. [MTRC]

another by means of screws and bolts to form the solid structure of the machine. During the assembly of this main structure it has been necessary to use well-established metrology to ensure parallel and perpendicular relations, as appropriate, of those surfaces to which the guideway rails and the leadscrew housings have been bolted. Their positions, in turn, have also been carefully aligned to be parallel and square, as applicable. Subassembled groups of parts, such as servomotors, leadscrew bearing housings, and so on, are also visible.

The cost of assembly represents a major part of manufacturing costs. Therefore, a lot of attention has been devoted to reducing this cost, and two fundamentally different approaches have been developed. The first one, called *design for assembly,* is based on achieving substantial savings by changing the design of the product to make it easier to assemble [4]. The other approach is based on *mechanization* and *automation.* It is difficult to design devices and mechanisms that could compete with the human sensory perception combined with the two-hand dexterity. If you watch an assembly operator grasp a graphite brush and spring and insert them into the body of a small DC electric motor to be safely secured in position and pressed against the collector, and try to design an automatic device to replace her, you will see what is involved. You would have to improve on not only the perception and dexterity of the operator, but her speed too. Nevertheless, many products are successfully assembled on automatic machines. The characteristics of such products follow: sufficient quantity, at least one million per year per shift (one product every ten seconds); product consists of a small number of parts, generally less than ten; high quality and uniformity of parts; stable design that permits at least four years of depreciation for the machine; exclusion of difficult assembly operations. Mostly it is a matter of a product consisting of small parts, less than 50 mm in their major dimension.

It has been mentioned above how difficult it is to economically replace the human operator by an automatic device. We should note also the aspect of the significance of quality and uniformity in the parts to be assembled. The parts may contain various kinds of defects such as screws without threads, chipped or discolored parts, parts out of tolerance, pieces of swarf, burrs, and so on. It is rather easy for an operator to identify these defects and reject such parts or, if they jam the operation, to fix it. The down time due to defective parts is usually negligible in manual assembly. On the other hand, defective parts can cause severe problems in automatic assembly machines by jamming in feeding devices, preventing workhead operation, or spoiling an otherwise acceptable assembly. This will cause substantial down time and spoil the whole cycle of operations on the machine.

Boothroyd and his colleagues at the University of Massachusetts have carried out extensive studies of the difficulty with which various parts can be assembled manually. They have coded parts, handling processes, and assembly processes, and devised a system that evaluates the difficulties of assembling a particular product. In doing so, the designer is encouraged to identify problems and eliminate them by redesigning the parts. They recognized the extreme importance of the latter aspect, and have since worked on the design for assembly (DFA) system. Let us summarize and illustrate their work.

Depending on annual production volume; number of parts in the assembly; difficulty of handling, inserting, and assembling the parts; and the number of variations (styles) of the product, various types of assembly systems should be selected (see Fig. 11.19). MA means manual assembly with the work carriers moved on a roller conveyor between the stations. The operator performs the handling and assembling operations and has the corresponding parts and tools at hand. MM means a mechanically aided manual assembly, which differs by the addition of automatic feeding and orienting devices.

AI means automatic assembly using an indexing (synchronous machine), and AF means an automatic assembly using a free-transfer machine (nonsynchronous). AP is a system using programmable workheads and part magazines, and AR uses two-arm robots. The latter two systems are in early development stages. Some of the assembly tasks are outlined in Fig. 11.20a. The various operations differ in the aspect of self-locking; in the ease of aligning and positioning; in the resistance to insertion; in the need for screwing, plastic deformation, or riveting; and in the number of directions of

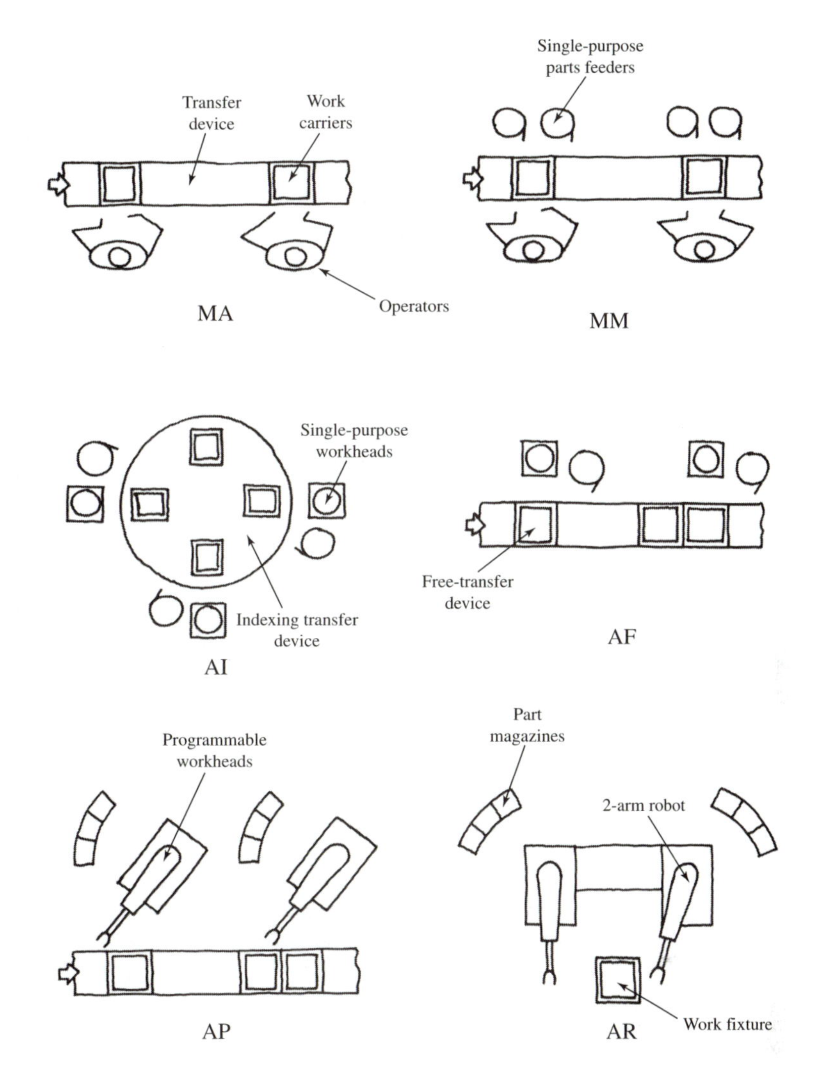

Figure 11.19
Various types and degrees of automated assembly are selected depending on production volume, number of parts, difficulty, and variations of the product. [BOOTHROYD1]

insertions, as well as in the use of metallurgical processes. A part of a coding system for automatic assembly processes is shown in Fig. 11.20b.

Automatic assembly machines consist of a number of stations and a transport system carrying the work carriers from station to station. This motion can be either continuous or intermittent. In the former case the workhead that carries the tools and performs the assembling actions travels along with the work carrier over the distance necessary to perform the operation and then returns and catches onto the next work carrier. Machines with intermittent transport are more common; they use stationary workheads for the rotary and the in-line types, as shown in Figs. 11.21a and b. In these figures only one magazine, feeder, and workhead are shown, but they are actually located in each station. The work carriers are moved from station to station, and the motion stops while the different operations take place at each station. The assembled product is completed during one revolution of the indexing table or else during one passage through the line. Thus, one product is completed after each index. Both machines illustrated work in the synchronous mode; that is, all the work carriers move at the same

a)

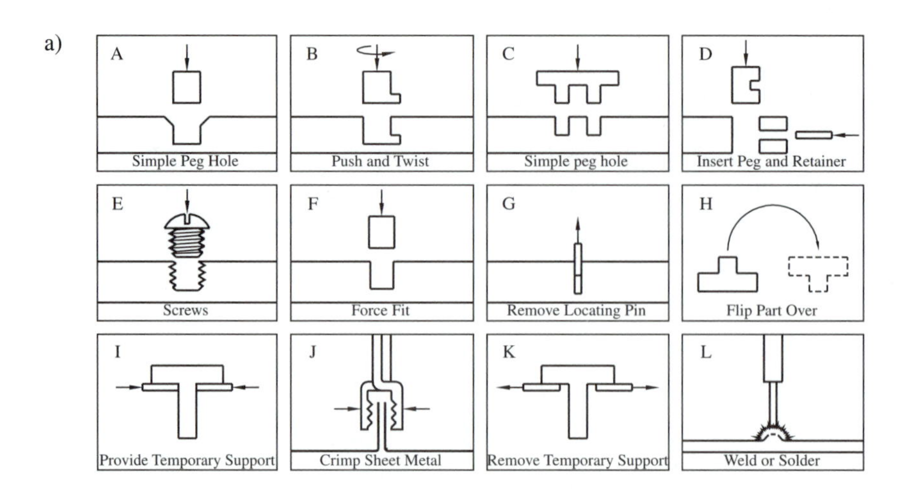

b)

				No screwing operation or plastic deformation immediately after insertion		Plastic deformation immediately after insertion		
						Plastic bonding		
				Part is easy to align and position	Part is not easy to align and position	Part is easy to align and position	Part is not easy to align or position	
							No resistance to insertion	Resistance to insertion
				0	1	2	3	4
Part secures itself only	Straight line initial insertion	From vertically above	3		Press fit			
Part secures itself only	Straight line initial insertion	Not from vertically above	4		Press fit			
Part secures itself only	Initial insertion not straight line motion		5		Press fit			

Figure 11.20
a) Various elementary assembly tasks; b) Part of a coding system for automatic assembly processes. [BOOTHROYD1]

time. This has the disadvantage that a breakdown of an individual workhead stops the whole machine. Production ceases until the fault has been cleared. An in-line, free-transfer machine, also called nonsynchronous, is shown in Fig. 11.22. In it the work carrier can be disconnected from a continuously moving conveyor and clamped in position as it arrives at a station. There is buffer space between stations. Once the operation at a station is successfully completed, the work carrier is released to the buffer, and the next carrier is admitted into the station. In this way fault times at individual stations average out over the whole machine, while the contents of the individual buffers vary to compensate for variations of the station times.

It is necessary to mention briefly some of the large variety of mechanisms for feeding and orienting the parts delivered into the workheads. The most common such device is the vibratory bowl feeder (Fig. 11.23). The bowl is attached to three suspension springs inclined under the angle ψ. On the inside wall of the bowl is a spiral of the ledge-shaped track that is sloped upward under the angle θ. Typically, $\psi = 40°$, and $\theta = 3°$. The coefficient of friction between the part and the track ranges from 0.15 for steel to 0.8 for a track lined with rubber. The bowl vibrates on the springs, driven by an

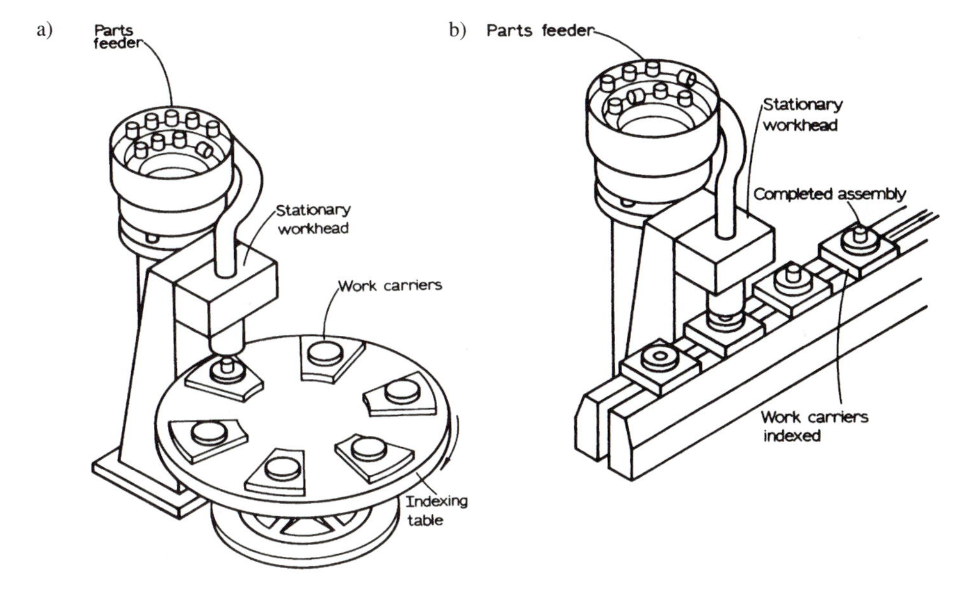

Figure 11.21
a) Rotary indexing assembly machine; b) In-line indexing assembly machine. [BOOTHROYD2]

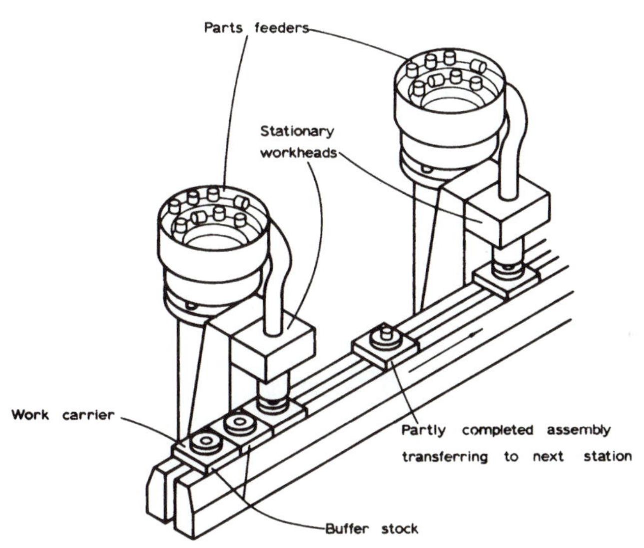

Figure 11.22
Free-transfer, in-line assembly machine. There is a buffer zone between stations. [BOOTHROYD2]

electromagnet acting between the bowl and the base. Commonly, the magnet is simply fed directly from the line power supply. The current is passed through a rectifier so that it produces a 60 Hz vibration. Without rectification it would generate 120 Hz, because both the positive and negative half-wave cause magnetic attraction. During one vibratory cycle the part is accelerated and slips ahead or even jumps (hops) ahead and the part travels up the track. Along the track various gates and escapements are located that allow only parts that are properly oriented to pass towards the outlet and down a chute to the workhead. These orienting devices exist in many forms. Four examples that deal with parts in the form of U, screw bolts, cups, and truncated cones are shown in Fig. 11.24. Apart from the vibratory bowl, other types of feeders exist: for example, the centerboard hopper feeder (Fig. 11.25a) and the elevating hopper feeder (Fig. 11.25b).

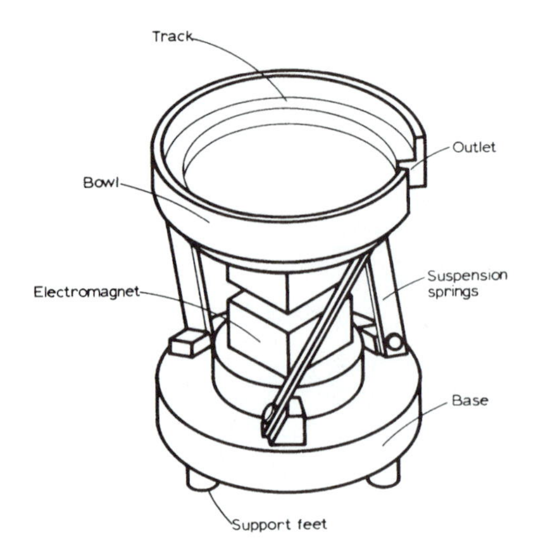

Figure 11.23
The vibratory bowl feeder is the most common device for storing, orienting and delivering small parts to an assembly station. [BOOTHROYD2]

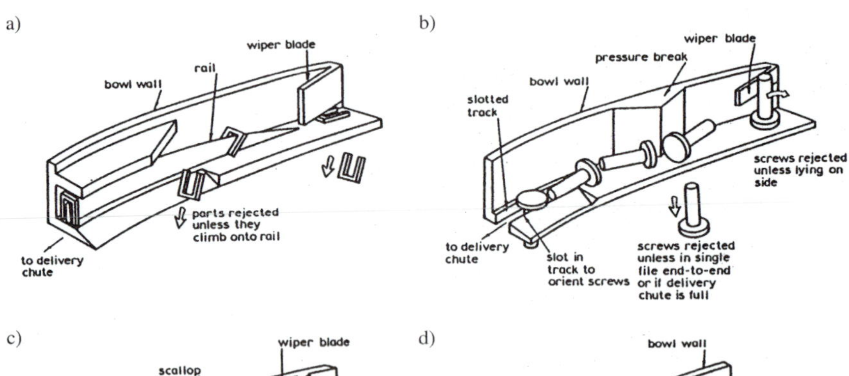

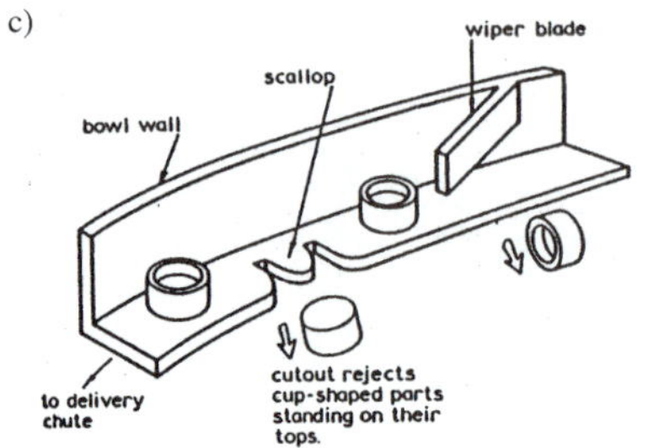

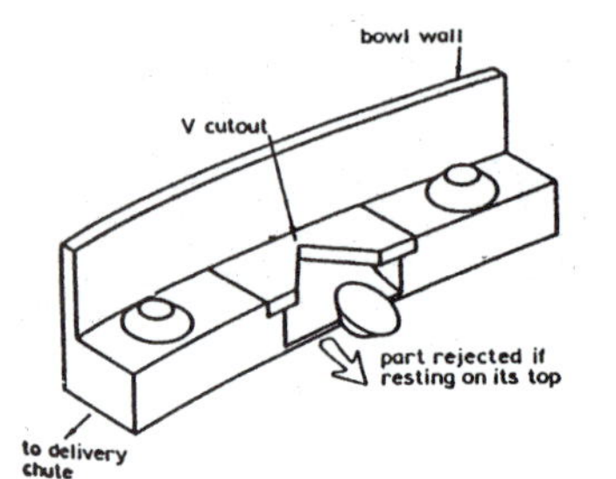

Figure 11.24
Examples of orienting devices in a vibratory feeder designed for: a) U-shaped parts, b) screws, c) cups, d) truncated cones. [BOOTHROYD2]

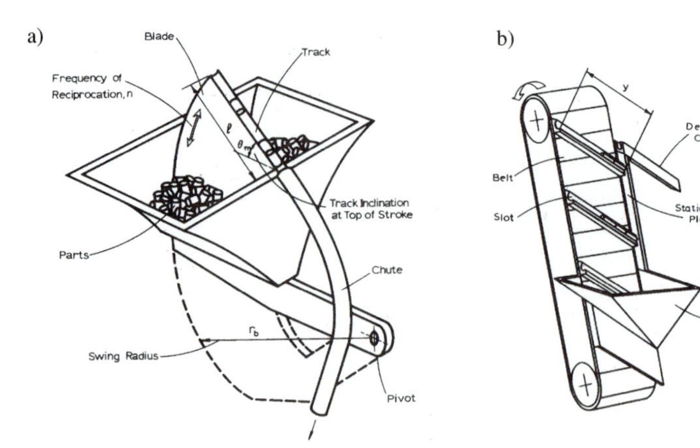

Figure 11.25
a) Centerboard hopper feeder, b) elevating hopper feeder. [BOOTHROYD2]

A variety of mechanisms are located at the end of the delivery chute or delivery track; their role is to feed the part from the chute, place it onto the work carrier, and actually insert it in the assembly. One of these used to insert a screw and set the screwdriver into action is shown in Fig. 11.26. In some instances a pick-and-place unit (Fig. 11.27) is used to lift the part from the feeder track and place it on the assembly. The device uses a mechanical, magnetic, or vacuum hand. Two variants of the unit are shown: in the first, the arm moves up, across, and down in straight-line motions; in the second, the transfer motion is rotational, and it is combined with a move that turns the part over.

In order to illustrate the design principles of an automatic assembly machine, a modular chassis used by one of the leading manufacturers of synchronous machines for in-line machines is shown in Fig. 11.28. It consists of 2–18 bays (8 shown) with each bay providing six potential workstations acting on three work carrier pallets. These are

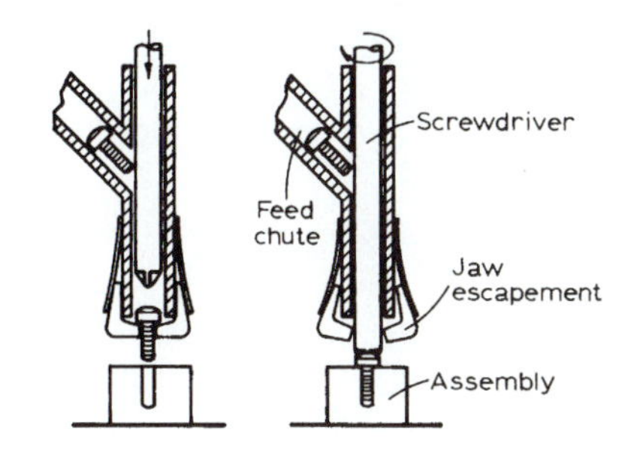

Figure 11.26
Parts-placing mechanism for automatic screwdriver. [BOOTHROYD2]

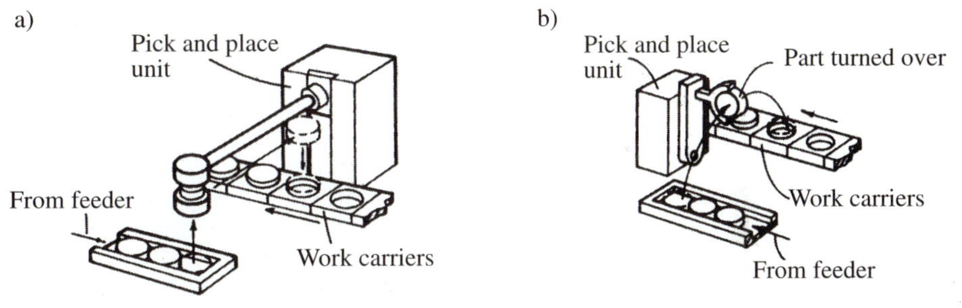

Figure 11.27
Pick-and-place units. Parts are lifted by a mechanical, magnetic, or vacuum hand and moved. A flip-over may be included. [BOOTHROYD2]

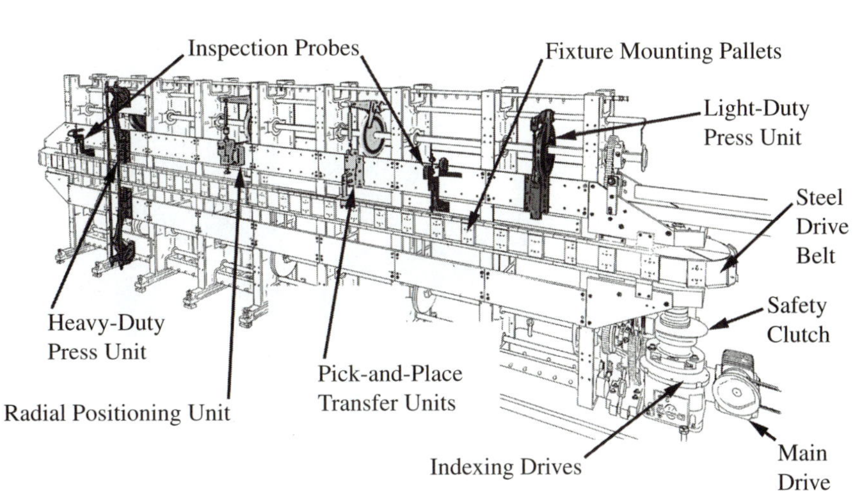

Figure 11.28
Modular chassis of an automatic assembly synchronous machine. Parts are held on pallets transferred from station to station by intermittent motion of the steel belt carrier. The statons include various types of assembly devices such as pick-and-place, orienting, inspection, and presses, and they are powered from an upper and a lower camshaft. [BODINE]

attached to a pretensioned steel belt. The indexing motion of the belt is driven from a crossover cam-type box via a safety clutch. Chain transmissions link the same main drive with the upper and lower camshafts that drive the tooling of the workstations. Examples of workstations are presented. The stations illustrated here are a heavy duty press unit, radial positioning unit, pick-and-place transfer unit, inspection probes, and a light-duty press unit. As shown in Fig. 11.29, the part magazines and feeders are located in front of the machine. Examples of products made on these machines are shown in Fig. 11.30. Taking the example of the doorknob, 13 different styles can be

Figure 11.29
Feeding units in front of the modular chassis assembly machine. [BODINE]

Figure 11.30
Examples of products assembled on the machine shown in Fig. 11.28. [BODINE]

accommodated using the variety of parts shown in Fig. 11.31. Changeovers are achieved rapidly due to drop-in fixture inserts and simple end-tooling changes. The assemblies are produced at a rate of 2400/hr.

The machine just described permits rather limited flexibility. Development work is being done to use robots and achieve a high level of flexibility that will permit economic automation of the assembly of families of products, each made in medium lots of several hundred. This process would extend the principles of group technology discussed in Section 1.4 of Chapter 1 from families of parts to families of part assemblies. Boothroyd has made studies of this problem and elaborated on three types of such systems: the one-arm, two-arm, and multistation robot assembly systems. The economy of these systems is based on the easy programmability of the robotic motions as well as on the fact that the robot itself, although a rather complex machine, is not one special type of machine but is very universal. As such, it can be manufactured in larger quantities and become much less expensive than a simple but special kind of pick-and-place unit. The basic idea of such a machine is illustrated in Fig. 11.32.

Figure 11.31
Different parts used to assemble various styles of doorknob assemblies. [BODINE]

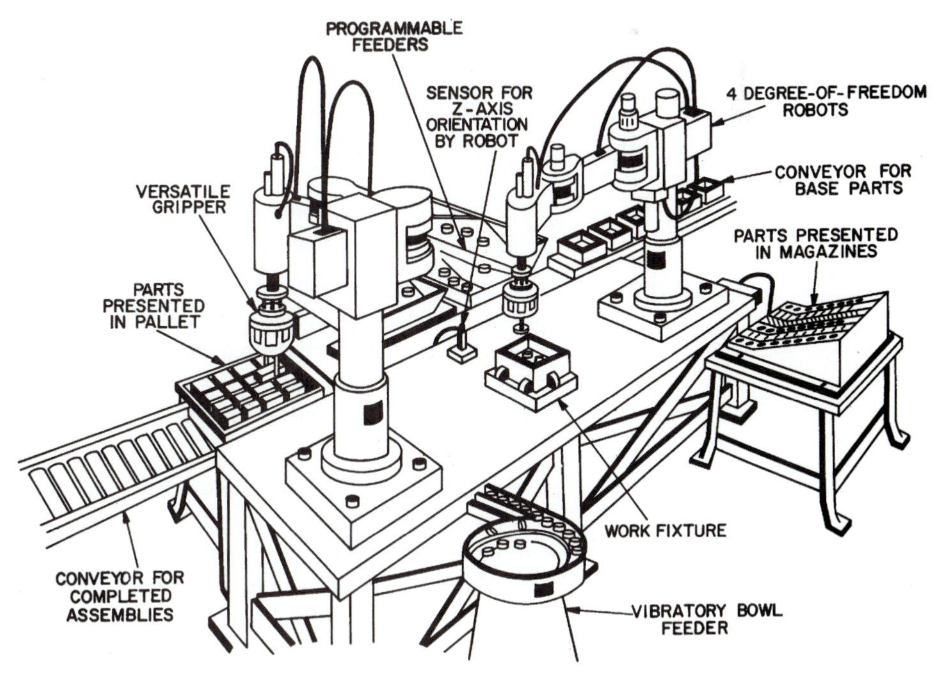

Figure 11.32
Design of a two-arm robot assembly station. This is an outline of a flexible assembly system, including programmable feeders and pick-and-place robots with programmable motions and versatile grippers. [REPRINTED BY PERMISSION FROM BOOTHROYD, G., DEWHURST, P., *DESIGN FOR ROBOT ASSEMBLY,* BOOTHROYD AND DEWHURST INC., 1985.]

Note that the robots have two rotary motions around vertical axes and a vertical up-and-down motion, plus a fourth axis that permits rotation of the hand, again around a vertical axis to permit orientation of the part. Three types of feeders are considered: the rigid-type, vibratory bowl feeder; the more flexible arrangement of parts in pallet-type magazines, where the robot can pick up in any position; and a third type known as programmable feeders. The latter are of a belt conveyor type, where parts pass by an optical sensor and can then be oriented by means of an actuator. Also the gripper is of the versatile type with interchangeable jaws and adjustable gripping motion. The two arms are so synchronized that while one is collecting a part, the other one does the inserting function, or else one arm can hold the part down while the other does the inserting. These systems are not yet fully operational, although simpler versions using portal-type robots for insertion of chips into printed circuit boards are rather common. This topic is discussed further in Chapter 12.

11.5 DESIGN FOR ASSEMBLY

We have seen that the largest savings in assembly are obtained by reviewing the design of the product and of its parts so as to simplify and facilitate the assembly. Rules and coding systems have been established to provide guidance in the review process.

1. The first and most basic rule is to minimize the number of parts in the product and to minimize the number of different parts (such as screw sizes). This rule is important not only in assembly but for the whole manufacture of the parts and products. Its effect is strongest in the material handling activity: there are fewer parts to store and move. This effort may yield more complex parts by combining several of them in one. Thanks to the spread of NC machining, there is no longer such a great difference in machining a complex housing rather than a simple one. Quite a few other rules are used to simplify assembly; some apply to both manual and automatic assembly, and some apply specifically to the latter. Furthermore, it is convenient to divide these considerations in two parts: ease of the assembling operations, and ease of feeding and orienting.
2. The next rule is to have a base part on which the assembly is built. This base part must have features that allow quick and accurate location on the work carrier.
3. Next, it is important to minimize the number of directions in which parts are inserted. Ideally, all parts are inserted directly from above, so that gravity assists the placing of the part. If insertion occurs from more than one direction, then once all insertions from one direction are completed, this subassembly is reoriented for the following insertions to be made again from the same direction.
4. Provide chamfers to help in insertions. The individual parts should be designed to prevent nesting or tangling; see Fig. 11.33. Parts should be symmetrical to provide a multiplicity of positions that satisfy assembly. On the other hand, provision of asymmetric features on the nonfunctional perimeter may assist in orientation of functional features such as holes. Provide conical screw ends to facilitate insertion.

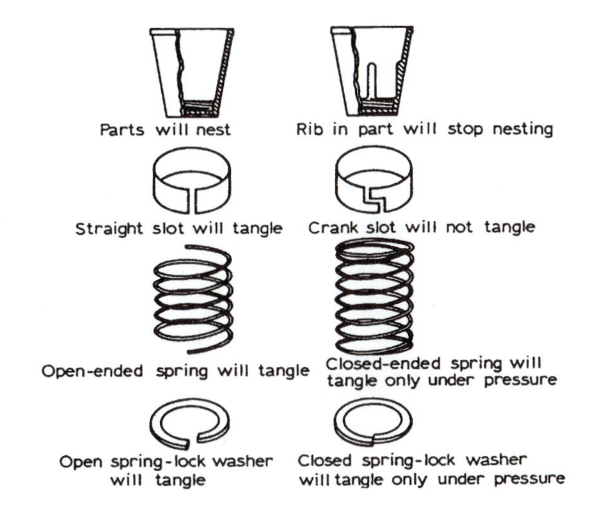

Figure 11.33
Examples of redesign to prevent nesting or tangling of parts. [FROM R. IREDALE, *AUTOMATIC ASSEMBLY—COMPONENTS AND PRODUCTS,* METALWORKING PRODUCTION, APRIL 8, 1964.]

11.6 WELDING PROCESSES

11.6.1 Introduction

Welding encompasses a great variety of processes. The description of these processes and, further, an analysis of arc and spot welding represent the substance of this chapter. The great number of different welding processes is justified by the variety of applications. The method chosen depends on the structural mass of components to be welded, or mostly on the thickness of the plates representing the most common components, and further on the material of the components, their quantity, and the need for automation, as well as on the availability and cost of the particular welding equipment. Obviously, for welding continuous seams in butt, lap, T and corner configurations of plates or pipes, one of the arc welding processes will be chosen, while joining the sheet metal parts of a car body is best achieved by spot resistance welding. For tiny welding spots on microcomponents, laser beam welding may be selected because of its accurate control; at the other end of the range of applications, railway tracks are welded using the thermit welding process. Single-pass welding of thick plates along a vertical joint is efficiently achieved by electroslag or electrogas welding. Cladding of flat plate is a very characteristic application of explosion welding. Rotational parts may best be joined axially by friction welding. We might continue illustrating the multitude of welding processes and their applications; however, we need to proceed in a systematic way to establish some kind of classification of welding processes. The criteria to use may be the kind of combination of heat and pressure, ranging from the fusion welding processes through those where both heat and pressure are applied, as in resistance welding and other processes, and on to solid-state welding, where it is almost solely pressure that is used, with little heat.

The *fusion* processes include the following welding methods; their names imply the energy source:

Oxyacetylene flame welding (OAW)

Arc welding

Shielded metal arc welding (SMAW)

Gas metal arc welding (GMAW)

Flux cored arc welding (FCAW)

Gas tungsten arc welding (GTAW)

Plasma arc welding (PAW)

Submerged arc welding (SAW)

Electroslag welding (ESW) and electrogas welding (EGW)

Thermit welding (TW)

Laser beam welding (LBW)

Electron beam welding (EBW)

The abbreviations in brackets are standard ones established by the American Welding Society.

Among the processes that involve both *heat and pressure,* we mention the following:

Flash and upset welding

Resistance welding: spot welding, seam welding, friction welding

The process in which it is essentially the *pressure* that produces the bond is

Explosion welding

Other welding processes to mention are

Ultrasonic welding

Diffusion welding

Let us now describe the individual processes and mention the distinguishing characteristics of each, including the nature of the welding action, the energy source and the necessary equipment, type of workpieces, material of workpieces, possible degree of mechanization and automation.

The *fusion processes* work mostly in a localized way by moving the weld along a path where the two parts must be joined. These joints are made either by simply melting the two parts locally without any added metal or, mostly, by also melting a filler rod or wire to provide a seam in the various situations shown in Fig. 11.34. The examples shown include the butt joints, corner, lap edge, and T joints and they can be made either without or with filler material. In the latter case the edges of the two parts are prepared beforehand in various ways, such as the single V, double V, single U, and so on. The parts to be joined are often sheet or plate metal, pipes, or other bulky bodies and housings. The position of the weld has a strong effect on the ease of welding and on the obtainable performance of the process. The illustrations in Fig. 11.35 express this very well, considering the effect of gravity acting on the filler metal that is being deposited in its molten stage and must be transferred from the electrode to the joint. Obviously the flat position is the easiest and allows the highest deposition rates.

Figure 11.34
Types of welded joints. [HOBART1]

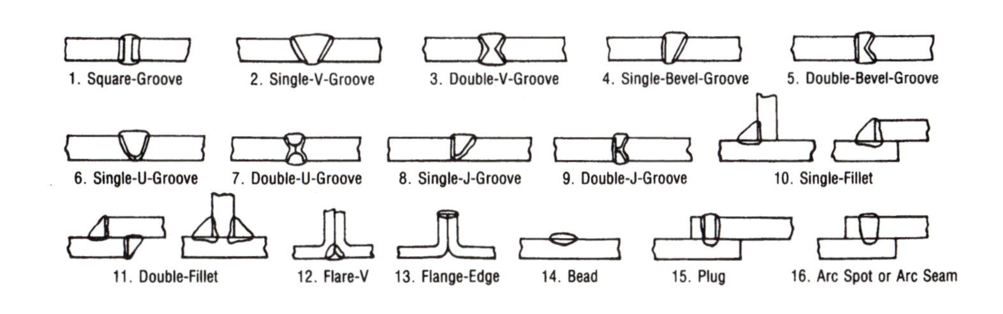

Figure 11.35
Welding positions. Difficulty of welding increases from left to right. Different types of arc welding processes and their operating conditions must be selected accordingly. [HOBART1]

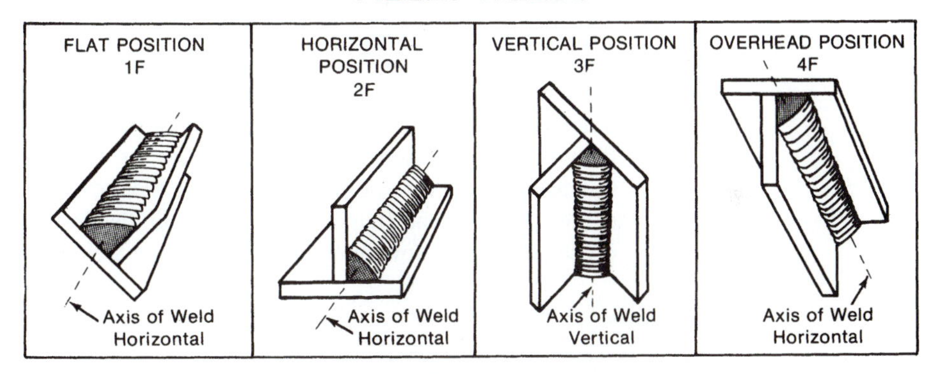

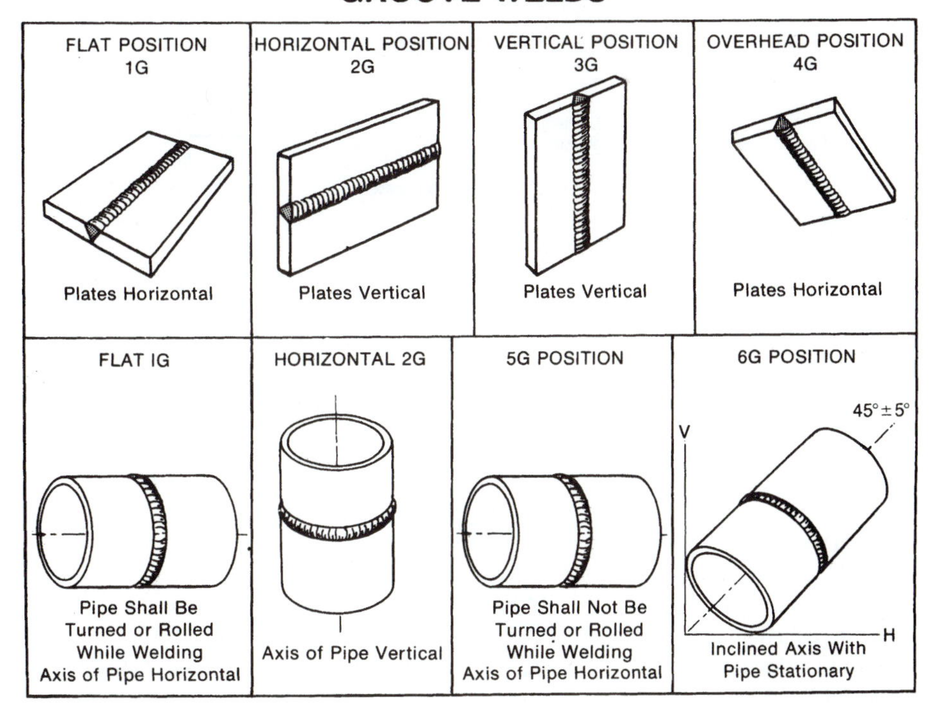

Increasingly difficult are the horizontal, vertical, and overhead positions. Pipe welding is easiest if the pipe can rotate around a horizontal axis and the weld is made at the top, in a flat position. The work gets much trickier on a nonrotating pipe in the positions 2G, 5G, or 6G. The different welding processes described further on are differently suited to the different tasks. Often the weld seam is produced in several passes (see Fig. 11.36), and the motion of the welding gun along the seam may be straight or carried out in a weaving pattern. Examples of lap and T joints of flat plates as well as of profile beams and of pipes are shown in Fig. 11.37 as part of our introductory description of the most common welding tasks.

The heat that is applied to the weld is spent on melting the volume of metal in the joint; it is then conducted away through the surrounding parts. A certain volume adjacent to the weld is heated to temperatures high enough to cause metallurgical changes. This is called the heat-affected zone (HAZ); see Fig. 11.38. Because some of the metallurgical changes may be detrimental, one aspect of the art of welding is to either avoid such changes or minimize the size of the HAZ. Metallurgical changes are tied to temperature-time profiles at the inner and outer boundaries of the zone. In any case, the more concentrated the heat applied to the weld, the smaller will be the HAZ and the more efficient the process, because the heat spent in the HAZ is unproductive. The *intensity* of the heat source is an important property, and most of the development of the welding processes has aimed to increase it. It is expressed in watts per square millimeter. The intensity of heat delivered by the arc welding processes is higher than that of the oxyacetylene flame, and it is still higher in the laser beam and electron beam processes. With a low-intensity source, some welding operations might be impossible

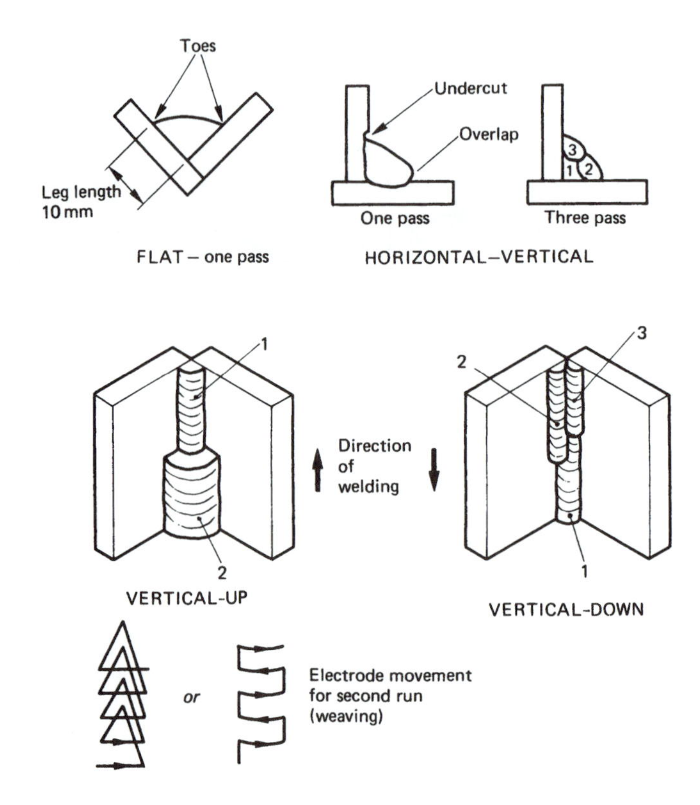

Figure 11.36
Deposition of a 10-mm leg-length fillet weld. The bead is often deposited in multiple passes or by using a weaving motion. [GOURD]

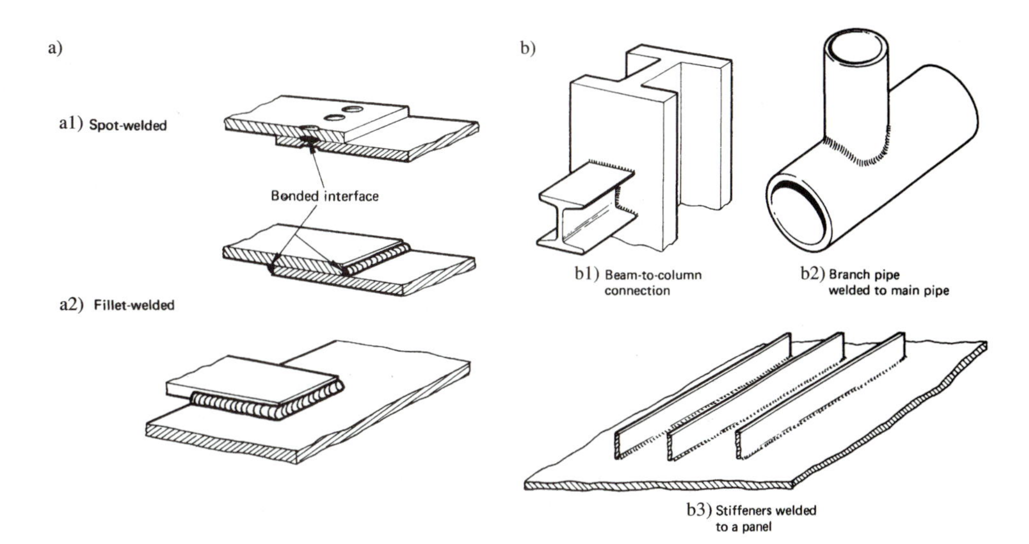

Figure 11.37
a) Examples of lap joints.
b) Examples of T joints that have been fillet welded.
[GOURD]

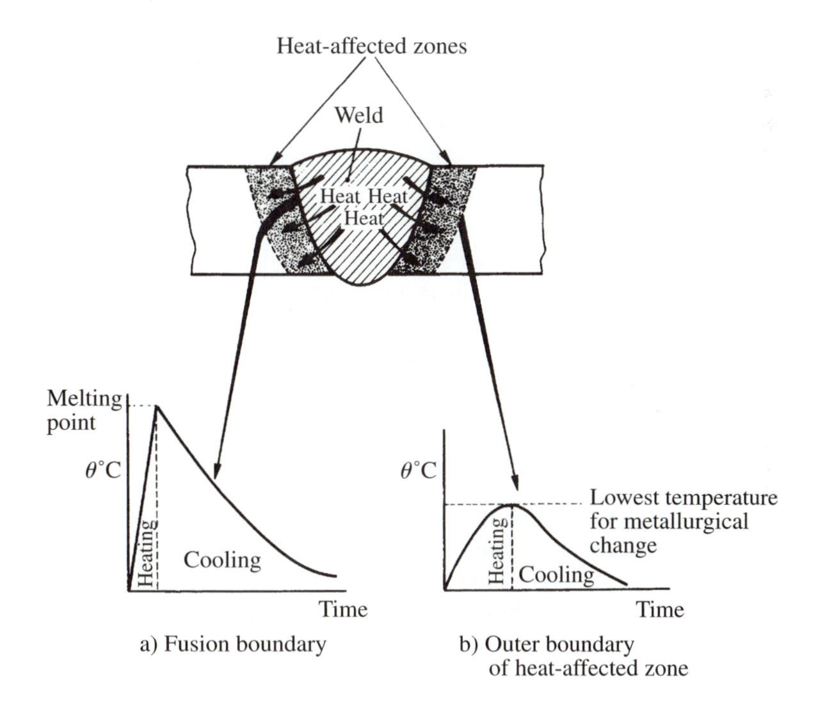

Figure 11.38
Heat-affected zone boundaries and the heating and cooling profiles decisive for metallurgical changes.
[GOURD]

because the heat is conducted away as fast as it is delivered without producing any melting. On the contrary, a high-intensity source like an electron beam can produce very narrow and deep welds with a very small HAZ.

These concepts are formalized in [9] by defining the *energy input H* as the energy per unit length of weld of a traveling heat source such as an arc:

$$H = P/V \tag{11.1}$$

where P is in watts and V in mm/sec. The actual dimension of H is then in J/mm and it represents the energy per unit length of weld. For an electric arc, $P = EI$. Whatever the energy source, some part of it is not used for welding and is lost to the surroundings. The heat transfer efficiency f_1 is used to arrive at the net input:

$$H_{\text{net}} = f_1 H \tag{11.2}$$

However, not all of this energy is used for the primary task of melting metal; some of it is conducted away and produces the above-mentioned heat-affected zone. In Fig. 11.38 the section through a completed weld is divided in two areas:

A_m: cross section of melted base metal and of melted filler metal

A_z: heat-affected zone

The heat-affected zone has so far not been clearly defined in our text; it will be discussed later. Here it is shown symbolically to indicate the nonmelted part of the base, which is, however, involved in the heat transfer. The heat Q_m necessary to melt a certain volume of metal is a property of the metal, and it consists of the heat needed to raise the temperature of the metal to its melting point and of the heat of fusion. It is obtained approximately as follows:

$$Q_m = \frac{(T_m + 273)^2}{300{,}000} (\text{J/mm}^3) \tag{11.3}$$

where T_m (°C) is the melting temperature. The melting efficiency f_2 is obtained as the ratio of the heat spent on melting over the total heat input:

$$f_2 = \frac{Q_m A_m}{H_{\text{net}}} \tag{11.4}$$

The melting efficiency depends on the intensity of the source and on the thermal conductivity of the metal. Thus, if oxyacetylene flame, which has low intensity, is used on aluminum, only about 2% of the heat is used to melt the metal, and the rest is conducted away. Very high-intensity sources such as electron and laser beam accomplish melting with almost 100 percent efficiency.

11.6.2 Oxyacetylene Welding

Several types of fuel gases can be combined with oxygen in a welding torch. However, acetylene is the most suitable and it is almost exclusively used in a process which is accordingly called oxyacetylene welding. Of all the gases that could be considered, it has the highest flame temperature, rather high combustion velocity, and the highest combustion intensity both in the primary and secondary cones of the flame. A welding torch is used, the principle of which is shown in Fig. 11.39. The flow rate of acetylene and of oxygen can be regulated by separate needle valves. They are mixed in the mixing chamber, and the mixture flows out through the torch tip. A variety of styles and sizes of torches are available, from smaller ones delivering about 0.01 to 1 m^3/h up to heavy-duty ones delivering as high a flow as 11 m^3/h of acetylene.

The chemical reaction between acetylene and oxygen is carried out in two stages. In the first stage, which provides the inner cone of the flame, acetylene reacts with oxygen according to the following equation:

$$C_2H_2 + O_2 \rightarrow 2CO + H_2 \tag{11.5}$$

One volume of acetylene and one volume of oxygen produce two volumes of carbon monoxide and one volume of hydrogen. The flame produced in the reaction with

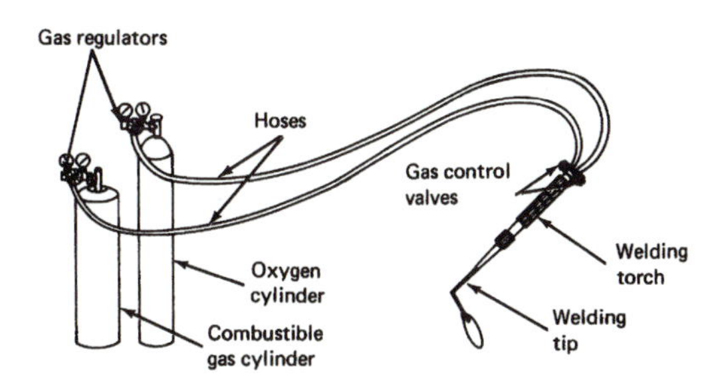

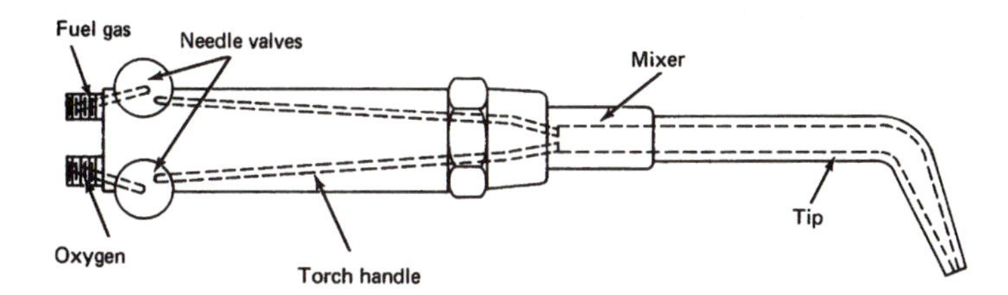

Figure 11.39 Oxyfuel gas welding equipment. The details of the torch are shown. [WH2]

exactly the one-to-one ratio of the supply gases is called neutral because there is no excess of either carbon or oxygen to either carburize or oxidize the metal. With a slight change in the proportions of the supply gases, a carburizing or an oxidizing flame can be obtained according to the needs of a particular application. The second stage of the reaction follows this equation:

$$2CO + H_2 + 1.5O_2 \rightarrow 2CO_2 + H_2O \tag{11.6}$$

The oxygen for this reaction is taken from the surrounding air; thus the molten metal is protected against oxidation. The heat generated in this reaction is greater than in the first stage, but its combustion intensity and temperature are low because of the larger area. The outer cone contributes substantially to the overall heat input and preheats the material in front of the puddle. The temperature of the flame in the primary cone is 3087°C, and the heat generated in the primary cone is 19 MJ/m^3; it is 36 MJ/m^3 in the secondary zone, providing for a total of 55 MJ/m^3 of acetylene. The combustion intensity in the primary cone is 114 W/mm^2, and it is 91 W/mm^2 in the secondary cone, for the stoichiometric mixture of acetylene with oxygen. The equipment for oxyacetylene welding is simple and inexpensive. As shown schematically in Fig. 11.39 it consists of the gas cylinders, each with pressure-regulating valves and pressure gages, hoses, and the welding torch. The acetylene cylinder contains porous material filled with acetone in which the acetylene is dissolved. The pressure must be held below about 100 MPa and the temperature below 99°C to prevent explosion. Instead of using acetylene cylinders the gas may be produced in a simple generator, where it is evolved in a reaction between calcium carbide and water. The whole equipment package is easily installed on a cart with wheels, so that it is light and mobile. Because of its low cost and easy transportability, oxyacetylene welding is still used in many repair and maintenance shops. It is an old method that originated around 1900, and in most applications it has been replaced by arc welding. It can be used for welding low-carbon and low-alloy steels, and with care also on higher-carbon steels, cast iron, aluminum, bronzes, and copper. A filler rod is held in the flame to supply material to fill the gap between the parts. Except for welding steel, fluxes are used to remove the oxide layer from the surface of the base metal and to protect the molten puddle from the atmosphere and from the gases in the flame. Flux may be used in the form of powder or paste, or as a coating on the welding rod. The oxyacetylene process is also used for surfacing operations and for thermal cutting.

11.6.3 Arc Welding Processes

Arc welding is the most important of all the welding processes. It is used universally in all kinds of mechanical manufacturing, from ship-building to car production, in the structures of machine tools, in agricultural and textile machinery, and so on. It is also important outside of machine shops, in building construction and in pipeline construction, for example. We will discuss the six different arc welding processes mentioned in the introduction to this chapter. Each of them is best in particular applications, depending on the base material, the size and type of weld, the accessibility of the weld, and the requirements for mechanization and automation.

In all of them the source of the welding energy is the electric arc, which is produced between an electrode and the work (base). The arc is an electric discharge through a column of ionized gas, the plasma. The temperature in the arc is very high, between 5,000 and 30,000 K. The arc is maintained at voltages between the terminals of 15–40 V, and it carries currents commonly in the range of 50–500 A, although currents as low as 0.3 to 10 A are encountered in plasma arc welding of thin-gage metals, and currents as high as 1600 A are used in submerged arc welding. The arc travels along the weld seam, melting the base metal and usually also the filler metal. The filler metal is provided either by melting the electrode, which has the form of a metallic rod or wire called a consumable electrode, or by being fed separately into the arc in processes using a nonconsumable electrode made of tungsten. In all the processes the arc and the puddle of the molten metal must be shielded against oxidation and against other kinds of contamination from the atmosphere.

The various processes differ in the way the shielding is provided. They further differ in the characteristics of the electric power supply. This may deliver DC, AC, or "pulsed" currents; it may have "constant-voltage" or "constant-current" characteristics and may be connected with the electrode negative (EN, also called "straight" polarity) or with the electrode positive (EP, also called "reversed" polarity). The type of material, type of shielding, and type of current influence the ways in which metal is transferred from the electrode to the base; therefore, a particular process may be more or less suitable for welding in the more difficult positions. The flat position is the easiest one, and any of the processes can be used. The overhead position is most difficult because the metal is transferred against gravity. Only some of the processes are suitable in this case, and even with these processes, much lower metal deposition rates are obtainable than in the flat, horizontal, and vertical positions, respectively. The mechanisms that drive the metal from the electrode to the base are discussed later. Let us now very briefly describe and summarize the six alternatives of arc welding and mention their distinctions. Each of them is discussed in more detail again.

The first process, illustrated in Fig. 11.40, is shielded metal arc welding (SMAW). Metallic electrodes in the form of rods are used that are covered by a layer of flux. The arc is generated between the lower end of the electrode and the work. The consumable electrode melts away and is transferred to the weld. The heat of the arc converts the cov-

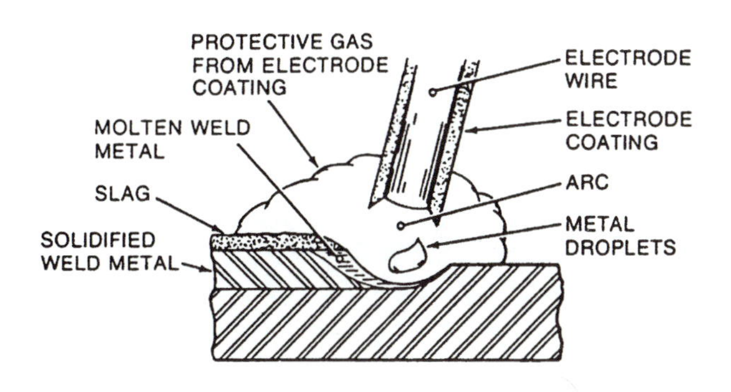

Figure 11.40
Shielded metal arc welding. The coating of the metal stick electrode provides gas for shielding of the arc as well as protective slag over the molten metal pool. [HOBART1]

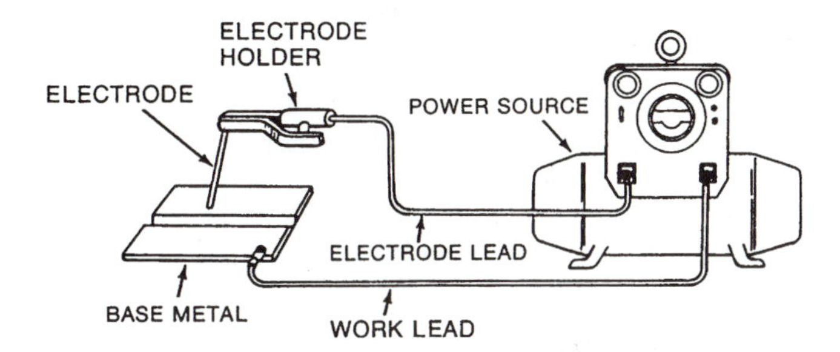

Figure 11.41
Equipment for SMAW is simple and portable. It can be moved to the work site; cables may be rather long. This makes it the most commonly used process. [HOBART1]

ering to gas and molten slag. The gas shields the arc, and the slag shields the molten metal as it solidifies. The upper end of the electrode is bare and it is clamped in a holder through which runs the end of the cable from the power supply. The material of the core of the electrode is similar to the material of the base metal. The coverings exist in a variety of compositions which affect the chemistry of the weld and the stability of the arc.

Depending on applications, both the EP and EN polarities are used with the DC current, and AC is used as well. The process is suitable for most of the commonly used metals and alloys and for all welding positions. The equipment is very simple, as shown in Fig. 11.41; no gas lines or cooling water lines are needed as in the other processes, just the electric power cables, which can be of considerable length to reach from the power source to the point of welding. The maximum current is limited by the heating up of the core of the electrode along whose whole length the current is passing. Overheating breaks down the covering. Therefore the deposition rates obtainable are lower than in GMAW or in SAW. Also, the need to change electrodes and to remove slag after each pass slows down the whole operation. It is a manual operation not suitable for mechanization or automation. However, because the equipment is simple and portable and the process is very flexible and adaptable to many situations and locations, both indoors and outdoors, it is used as well in production lines as in ship-building, bridge and building construction, in oil refineries, and on pipelines, and it is the most common arc welding process.

In the gas metal arc welding process, GMAW (Fig. 11.42), the electrode is in the form of wire that is continuously fed into the arc. Current is brought in through the contact tube in the wire guide in the torch. Shielding of both the arc and the molten metal is provided by an externally supplied gas. Depending on application, either the inert gases argon or helium are used, or carbon dioxide. The equipment (Fig. 11.43) is more complex than for SMAW because of the gas supply line and wire-supply components, but the process is amenable to mechanization and automation. The equipment then also includes

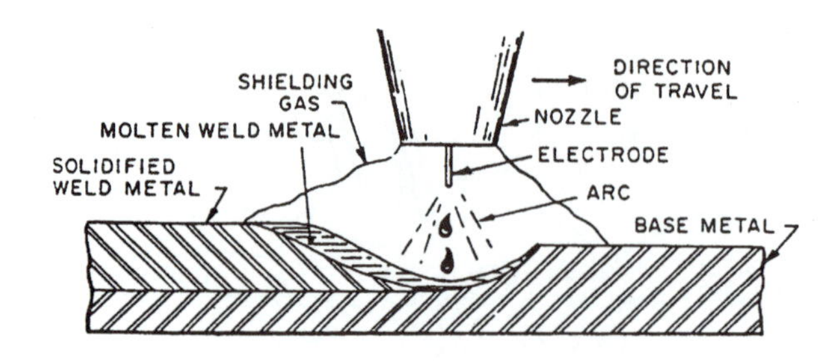

Figure 11.42
Gas metal arc welding. Shielding of arc and weld pool is provided by the gas. The electrode is thin wire. [HOBART2]

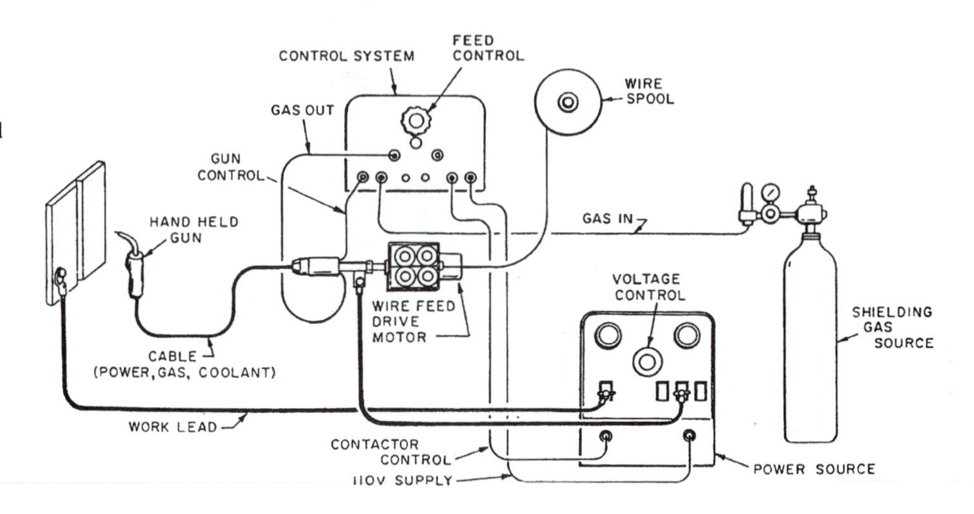

Figure 11.43
Equipment for GMAW is a little more complex than for SMAW because of the added gas line, but it is possible to mechanize or automate the process. [HOBART2]

machinery to guide the torch along the desired path. This may be a rail and carriage for simpler straight-line motions. For complex motions, it is common to use robots.

The GMAW process uses mostly a DC power source with EP polarity and the "constant-voltage" characteristic. As explained later, this is the basis for self-regulation of the process in which the burn-off rate adjusts itself automatically to any change in the wire feed rate. This aspect is very important for successful automation of the operation. The GMAW process is utilized in high-production welding operations. All commercially significant metals can be welded in all positions by choosing the appropriate parameters and conditions for the process.

The next process is flux cored arc welding (FCAW). It differs from the GMAW process in that the consumable, continuously fed electrode is tubular and contains a flux core that provides the shielding medium. This may or not be assisted by an additional gas shield. The FCAW process is used for welding carbon and low-alloy steels, stainless steels, and cast irons. Essentially, the FCAW process is a combination of the SMAW process, with the electrode wire and coating turned inside out, and the GMAW process.

The next two processes, gas tungsten arc welding (GTAW) and plasma arc welding (PAW), use nonconsumable electrodes, which may thus be used in situations where no filler metal is required. However, in most instances filler metal is added from a rod or wire fed into the arc. This permits better control of the ratio of added metal to melted base metal than with the processes using consumable electrodes. The weld quality is high because there is no slag, as in SMAW, that might be entrapped in the weld, and no filler metal is transferred across the arc as in GMAW, which could cause spatter. In the GTAW process (Fig. 11.44) the arc is generated between a tungsten

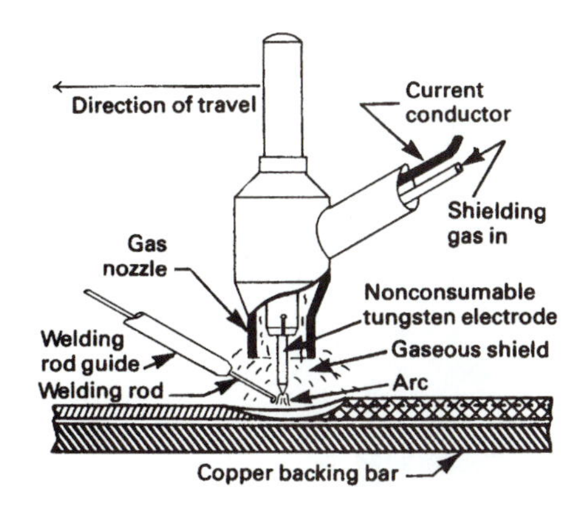

Figure 11.44
Gas tungsten arc welding. The nonconsumable electrode generates the arc. Metal is added separately. Separation of melting and filling actions makes for better control and quality of weld. [WH2]

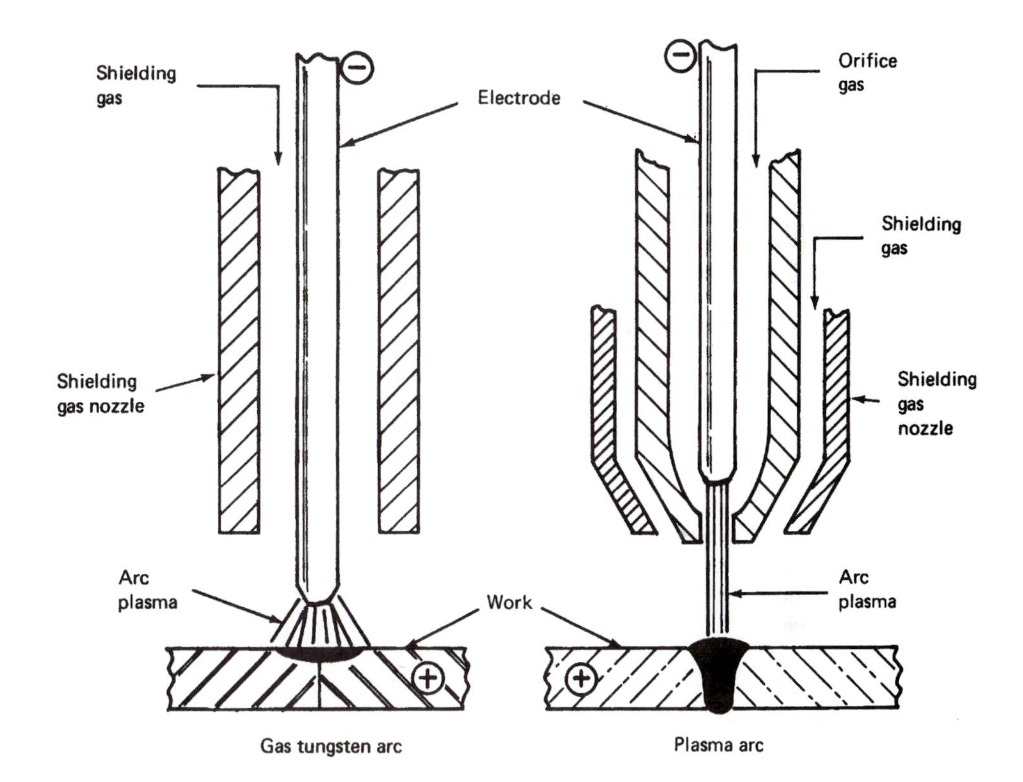

Figure 11.45
Comparison of the gas tungsten arc and plasma arc welding processes. Plasma in PAW is constricted, leading to high heat intensity, resulting in narrow welds with a minimized HAZ. [WH2]

electrode and the base, and inert gas is supplied through the torch to shield the arc and the molten metal. This process can be used to weld almost any metal, but it is the preferred process in the aerospace industry for welding stainless steel, titanium alloys, nickel alloys, and aluminum alloys. It can be used in all positions. However, it is rather slow and not good for thick sections, because deposition rates are low. In the PAW process (Figs. 11.45 and 11.46) the intensity of the heat input is very much increased by constricting the arc in a water-cooled copper orifice. Shielding is obtained primarily from the hot ionized gas issuing from the orifice. There is also an outer shielding gas supply to protect the weld. This arrangement, in which the nonconsumable tungsten electrode is recessed inside the orifice, leads to a very high directional stability of the arc and its insensitivity to variations in the tool standoff distance. The applica-

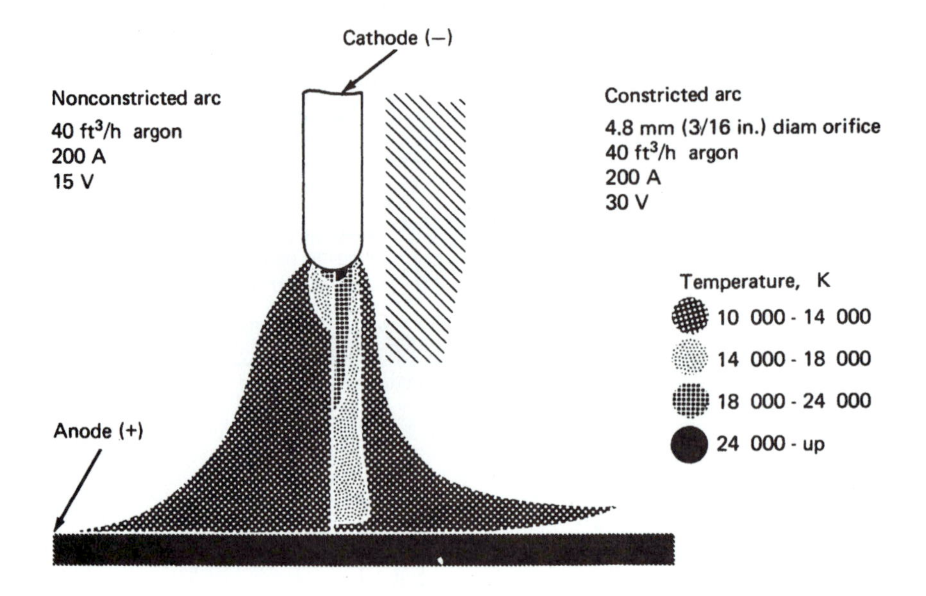

Figure 11.46
Effect of arc constriction on temperature and voltage. A narrow high-temperature zone is created in PAW. [WH2]

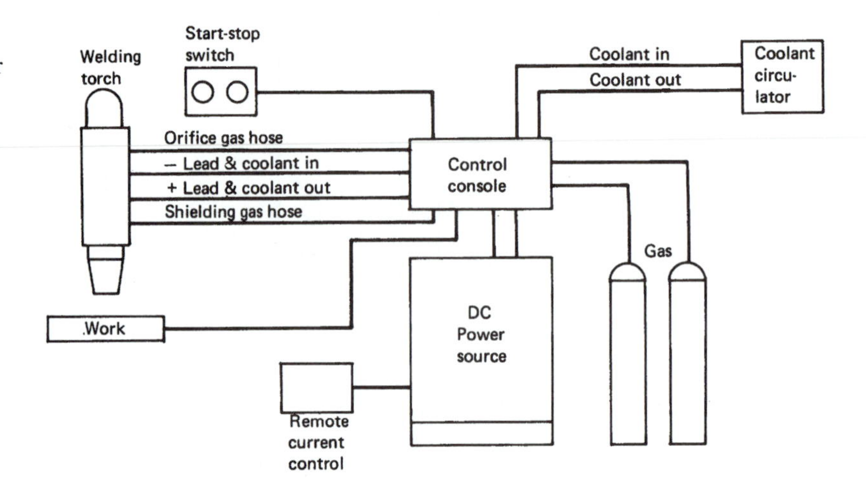

Figure 11.47
Equipment for PAW is rather complex as it includes two gas lines and cooling-water hoses. It also requires a special starting electric circuit. [WH2]

tions are of two kinds. The first one is low-current PAW, which is suitable for welding very thin sheet metal or welding small components. The second one is high-current PAW, which offers better penetration due to the higher heat intensity, and permits higher welding speeds than GTAW. However, the equipment shown in Fig. 11.47 is more expensive because of the need of a cooling water supply for the orifice and also because of the double inert gas supply lines. Because of the high heat intensity, this process is also used for thermal cutting.

The last process considered is submerged arc welding (SAW) (Fig. 11.48). A bare wire electrode is mechanically fed into the arc, as in GMAW, but shielding is provided from granular flux that covers the welding arc. This flux is supplied through a feeding tube located in front of the arc. The unmelted flux acts as an air shield, a heat insulator, and a radiation shield. The molten flux provides slag covering the molten metal, and the vaporized flux provides the plasma for the arc. Very high currents are possible,

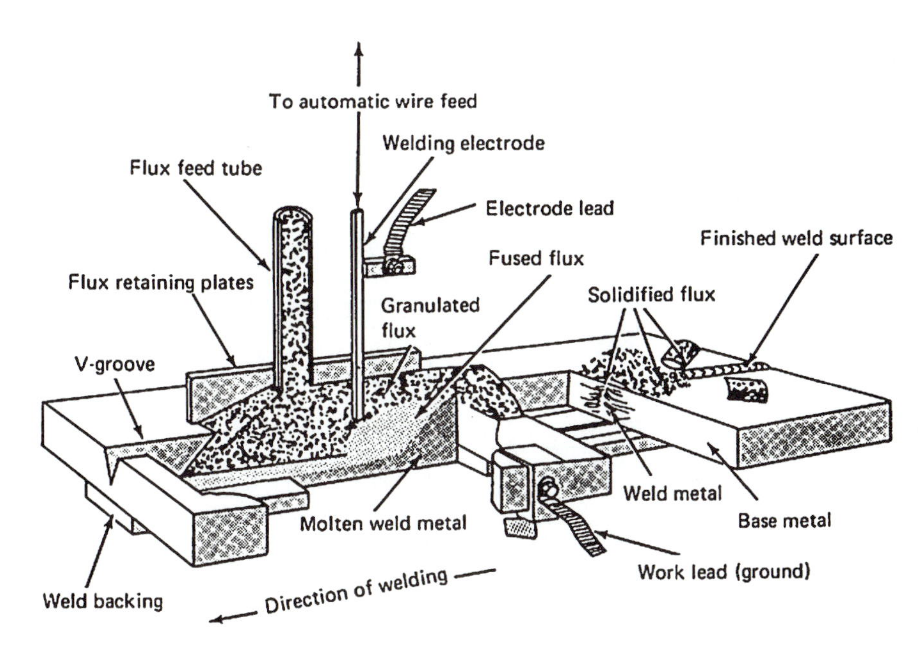

Figure 11.48
Submerged arc welding uses granular flux to efficiently cover the arc and pool. Very high currents can be used, but it is limited to horizontal flat positions. It must be mechanized using a servocontrol. [WH2]

and they result in very high deposition rates. The process has to be mechanized because the arc is not visible; it has to include automatic control of the wire feed rate, and it is limited to flat and horizontal welding. It is used for carbon and alloy steels and for nickel alloys.

Before we proceed with more detailed analysis of the various arc welding processes, let us discuss some of their common aspects, such as the characteristics of the arc and of the metal transfer.

The Arc

The welding arc represents an electrical discharge through a gaseous conductive medium. With nonconsumable electrodes (GTAW), this medium is the plasma, which is the ionized state of a gas composed of nearly equal numbers of electrons and ions. In GTAW the tungsten electrode is a good emitter of electrons. Tungsten has a very high melting point (3370°C), and it easily supplies electrons if it is the cathode in the circuit. This type of cathode is called thermionic. Pure electrode is seldom used in GTAW. An addition of 1–2% thoria or of 0.15–0.4% zirconia improves electron emissivity and extends the electrode life. Most structural metals used in the consumable electrode processes (SMAW, GMAW, SAW) form what is called a cold spot cathode, and their emissivity has to be helped by coatings containing oxides or alkali metals. Most GTAW welding is done with DC current and is electrode negative (EN). However, in welding aluminum, the metal is covered with a layer of aluminum oxide, which has a much higher melting point and is an insulator. Polarity must be reversed so that electrons emitted from the work disrupt the oxide layer. Therefore, GTAW welding of aluminum is done with EP or else with alternating current (AC) in which the oxide layer is disturbed during the half cycle in which the work is negative. Most of the heat is produced by the striking electrons; less is generated by the striking positive ions. The differences of the three situations are illustrated in Fig. 11.49. Note that

Current type	DC	DC	AC (balanced)
Electrode polarity	Negative	Positive	
Electron and ion flow Penetration characteristics	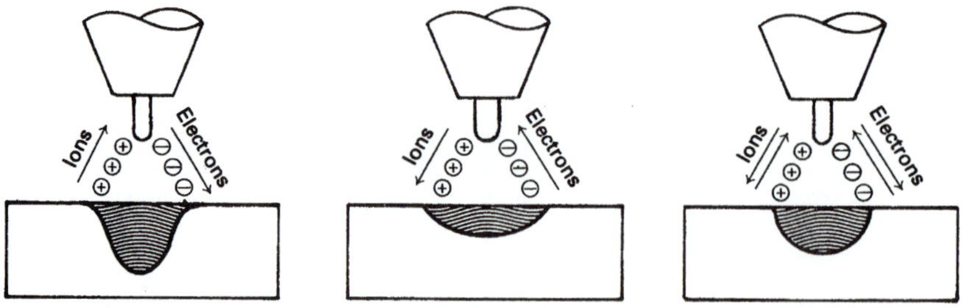		
Oxide cleaning action	No	Yes	Yes—once every half cycle
Heat balance in the arc (approx.)	70% at work end 30% at electrode end	30% at work end 70% at electrode end	50% at work end 50% at electrode end
Penetration	Deep; narrow	Shallow; wide	Medium
Electrode capacity	Excellent e.g., 3.18 mm (1/8 in.)—400 A	Poor e.g., 6.35 mm (1/4 in.)—120 A	Good e.g., 3.18 mm (1/8 in.)—225 A

Figure 11.49
Characteristics of current types and polarities for gas tungsten welding, showing their effects on penetration. [WH2]

with DCEN, high currents can be used with a rather small-diameter electrode without damaging it. With DCEP it is necessary to use a larger-diameter electrode and less current. The penetration is shallow. With AC, the performance is in between the two previous cases. Other aspects to consider in AW are considered in the discussion of the GTAW process in the following text.

The arc has a conical shape in most of the processes (less so in PAW), increasing in diameter toward the workpiece, regardless of polarity. Consequently, in the consumable electrode processes, magnetic force components drive the metal droplets toward the workpiece. Other mechanisms that contribute to metal transfer are discussed later.

The arc temperatures reach between 5,000 and 30,000 K depending on the nature of the plasma and the current. Thus, they depend on the type of shielding gas or on the composition of the electrode coating in SMAW. In the latter case, with coatings containing sodium and potassium, which ionize easily, the temperatures are about 6,000 K. In GTAW temperatures in an argon shielded arc reach 18,000 to 24,000 K, and in PAW they exceed 24,000 K. Isotherms of temperatures in a 200-A arc in argon between a tungsten cathode and a water-cooled copper anode were presented in Fig. 11.46, which illustrates how the temperatures decrease along and across the arc. Heat is lost from the arc by conduction, diffusion, convection, and radiation.

In processes with consumable electrodes the arc contains, apart from the plasma, droplets of molten metal and, in SMAW, droplets of molten slag and vapors from the coatings. Metal is transferred through the arc in several forms, which are discussed in the following text.

Metal Transfer

In the processes with consumable electrodes the metal is transferred in various ways, depending on the process and on current, voltage, and shielding gas. Some of the transfer modes are forceful and stable, and they can be used for "out-of-position" welding (positions other than flat); others are less directional and may be associated with considerable splatter of metal. The forces that propel the metal from the electrode to the anode are of several kinds: gravity, the "pinch effect" due to the neck of a liquid drop at the electrode, explosive evaporation of the neck, electromagnetic forces due to the conical shape of the arc and the divergence of the streams of electrons, and friction effects in the plasma jet.

We will now review the various modes of transfer, starting with the GMAW process, in which four different modes are distinguished: short-circuiting, globular, spray, and pulsed spray transfers. Almost all GMAW applications use DCEP, and this will be assumed unless otherwise stated. The cycle of the *short-circuiting transfer* is shown in Fig. 11.50. This mode is associated with the lowest range of currents and voltages and small wire diameters. As the wire is fed towards the work it touches the molten pool of weld causing a short circuit. The voltage drops and current surges, due to which a globule is pinched off, and an arc ignites across the gap. Due to the heat of the arc, the end of the wire melts and approaches the work again, and so on. This cycle repeats 20–200 times per second. A small, fast-freezing weld pool is created with very fine, small spatter. The type of shielding gas has little effect on this mode, which can also work with DCEN. The current increase at short circuit must be high enough to melt the drop off but not so high as to cause much spatter. The rate of current increase is controlled by the magnitude of inductance in the power supply. The short-circuiting transfer mode is suitable for joining thin sections of metal because of its low heat input, and it can be used in all positions due to the directionality of wire feed and to the fast freezing of the weld.

Globular transfer occurs at relatively low currents and voltages but higher than those in short-circuiting transfer. With CO_2 shielding it occurs at most current and voltage levels. As shown in Fig. 11.51a, metal is transferred in the form of larger droplets. This mode is suitable essentially for flat and horizontal positions because it depends on gravity, and there is a larger amount of spatter.

Spray transfer occurs when using argon as the shielding gas; the globular transfer then changes distinctly above a certain critical current level to the axial spray mode. In this mode tiny droplets of metal detach from the tip of the electrode at a high rate of several hundred per second. The driving force here is the pinch effect (Fig. 11.51b).

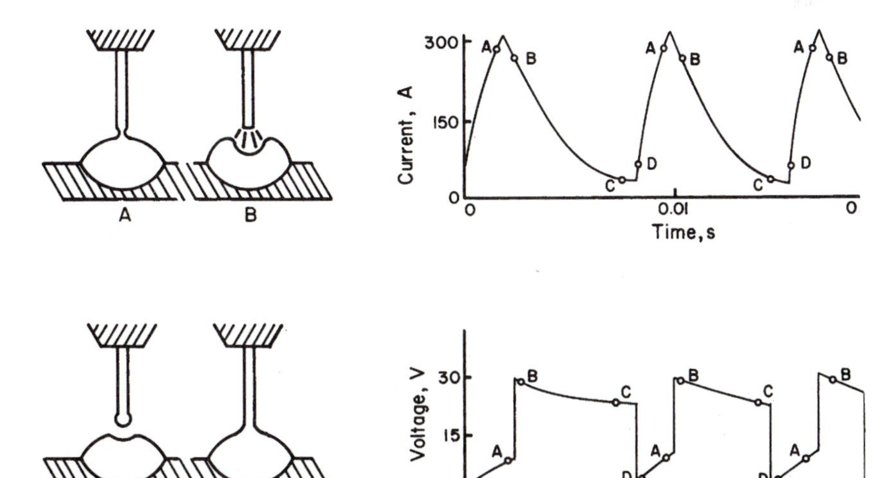

Figure 11.50
Short-circuiting transfer. It is obtained with low currents and voltages and small wire diameters. It cycles 20 to 200 times per second and gives rather small spatter joining thin sections in all positions. [WH3]

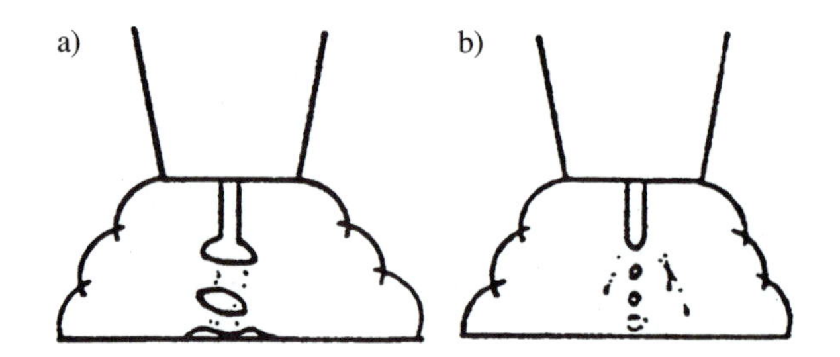

Figure 11.51
a) Globular transfer is used with moderate currents with CO_2 shielding, in flat and horizontal positions. Spatter is large. b) Spray transfer consists of tiny metal droplets at a rate of several hundred per second. It is obtained with argon shielding and currents above a threshold that varies with metal welded and electrode diameter. Used for all positions, but best for the flat one. [HOBART2]

The critical transition current shown in Fig. 11.52 depends on the metal involved, on the diameter of the electrode, and on the stick-out of the electrode, that is, the length between the point of current pickup and the arc. For example, for a steel electrode of diameter 1.6 mm, it is about 270 A. The spray transfer mode produces a very stable arc suitable both for flat and out-of-position welding of thick sections because of the high currents. However, for nonflat positions, the large molten puddle may be difficult to control. Spray transfer can only be obtained with argon or argon-rich gases, it is not obtainable with helium or CO_2 shielding.

The spray mode is achievable at currents below the transition values by using *pulsed current* (Fig. 11.53). The current is changed between a background level and a level above the transition current. Globules of metal are generated with the frequency of the current variation. One globule is ejected per cycle. There is no metal transfer during the low-voltage part of the cycle.

In SMAW the metal transfer encompasses spray transfer of fine globules up to a transfer of a sequence of larger globules that may periodically short circuit the arc. In order to achieve satisfactory transfer, it is necessary to properly combine the type of metal with the type of coating and polarity, depending on the position of welding. A large variety of coating compositions is available, and a suitable one is obtainable for almost any application, including out-of-position welding.

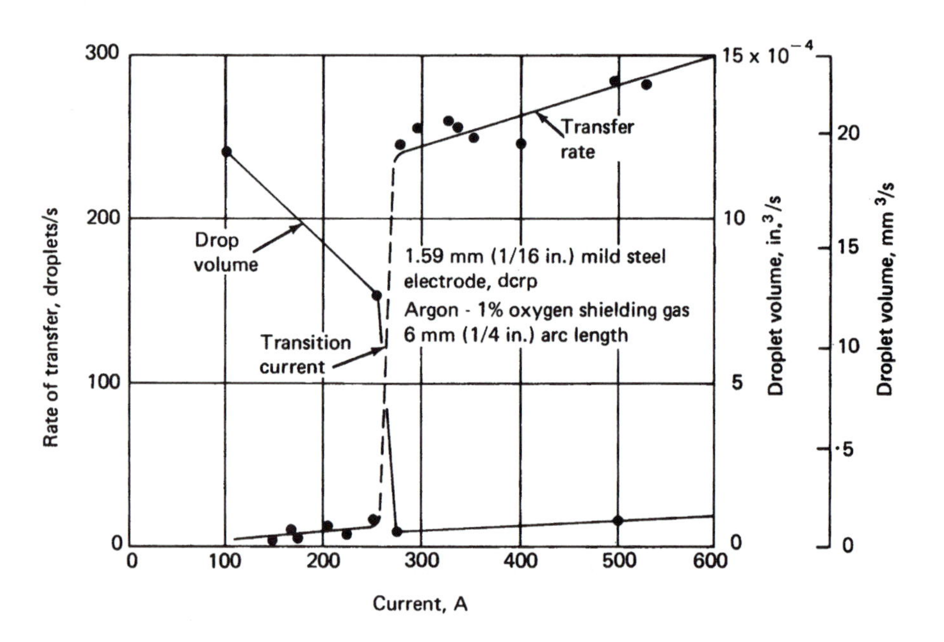

Figure 11.52
Variation in volume and transfer rate of drops in spray mode with welding current (steel electrode). [WH2]

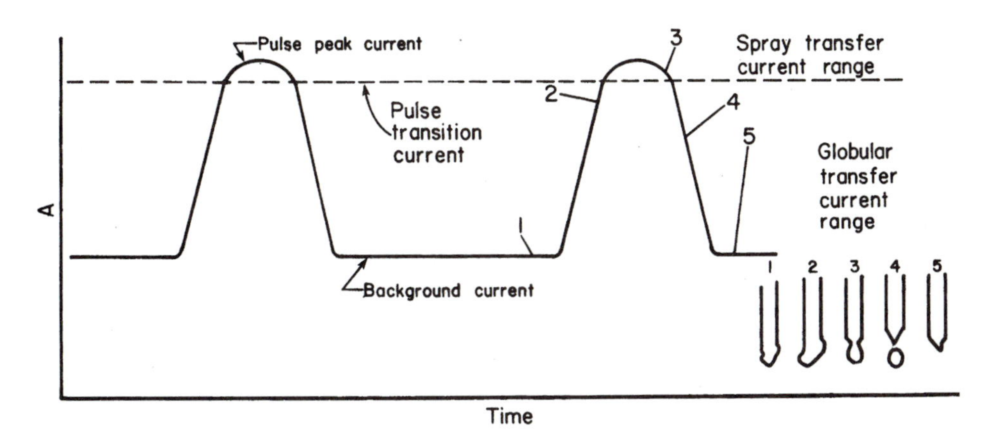

Figure 11.53
The output current wave form of the pulsed current power supply; the metal transfer sequence is also shown. One globule of metal is ejected per cycle. Pulsed mode is suitable for all positions. [WH1]

In SAW the arc is completely surrounded by a thick flux blanket, and direct observations of the arc and of metal transfer in the arc are not possible. Research was conducted using X-ray techniques to study the modes of metal transfer that are encountered. It was found that the transfer is in the form of globules and fine droplets depending on the current level. There are no special problems because SAW is always used in the flat position, and gravity helps in the transfer.

The various modes of metal transfer are individually more or less suitable to the different welding positions. In GMAW most of the spray-type welding is done in the flat position. Pulsed and short-circuiting modes can be used in all positions.

In conjunction with the question of metal transfer, it is useful to look also at the stability of the molten weld pool, at the shape of the weld cross-section as given by the ratio of weld width to depth of penetration, and at the profile of the weld—whether convex or concave. In GMAW the penetration depends strongly on the kind of shielding gas. Both in GMAW and SMAW even small quantities of source elements added to the electrode or to the coating may have strong effects on the shape of the surface of the weld. Very strong influence is exerted by oxygen contents in the shielding gas or in the flux respectively. Oxygen affects the surface tension in the weld pool. A low-oxygen weld has a high surface tension and the weld becomes convex. Conversely, a high oxygen level causes low surface tension and a concave bead. Neither of the two extremes is desirable. With high surface tension the metal flow in the seam groove may be restricted, and with low surface tension (i.e., high "wetting ability" of the weld), the metal may flow out of the joint. The medium case is mostly required.

In SMAW the flux plays an important role in determining the profile of the bead. The slag may act as a mold that keeps the molten metal in place. The slag must be sufficiently fluid; it should have high surface tension and should solidify rapidly. If the weld pool is not correctly controlled its shape may get distorted and create undercuts or overlaps. If nothing else can be done, it may be useful to carry out the weld in multiple passes.

Power Sources

The power sources are either DC or AC. For the former, either an engine-driven electric generator is used or a transformer and silicon rectifier. The latter requires only a transformer; however, additional circuits are provided to establish the desired voltage/current characteristics. These are distinguished as either "constant i" or "constant e," but in both cases the word *approximately* should precede the word *constant*. The characteristics in Fig. 11.54 are of the *const i* type and are typically used for the SMAW process. At $i = 0$ is the "open-circuit voltage," which is held to about 70–80 V for safety. At $e = 0$ the "short-circuit" currents are found. The four curves can be individually switched on and used in the indicated operating range for a choice of four current levels, resulting in four different deposition rates. Figure 11.55 shows the power source characteristic together with the arc (e, i) characteristic for four different arc lengths. The working point is found on the intersection of a power characteristic with an arc characteristic. For the arc, it may be observed that in a range of low currents the voltage drop across the arc decreases with current and reaches a minimum at about 50 A. From there on the voltage drop increases with the current. Here we have again the *const i* power source with a steep negative slope. Note that if the welder, in manually guiding the electrode over the work, moves up and down by a few millimeters, and the voltage changes by about 20%, the resulting current change is only about 8%. This means that the deposition rate remains fairly constant in spite of the unsteady hand of the operator.

The *const e* characteristic is shown in Fig. 11.56. It is used in GMAW to permit a wide choice of wire-feed rates that require a wide variation of burn-off rates and need a wide variation of the current i, while keeping the voltage and the arc length fairly constant.

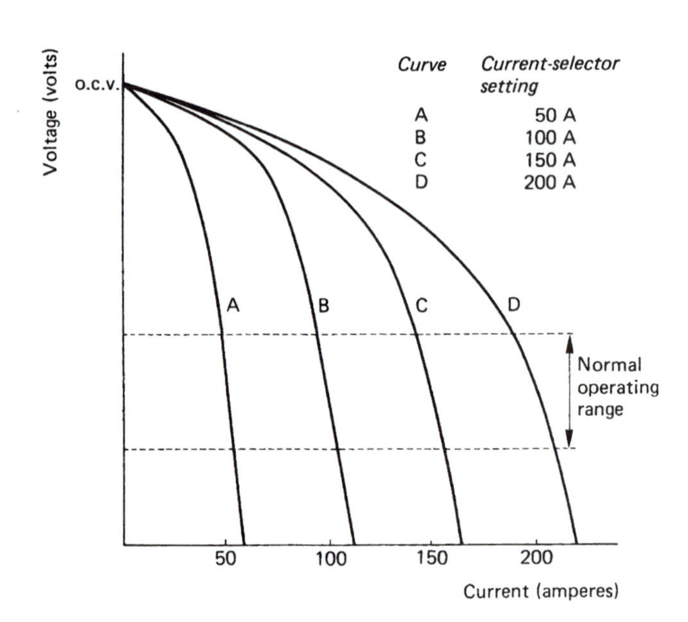

Figure 11.54
Output characteristics for various settings of the "constant current" power supply. Open-circuit voltage is held at about 70–80 V for safety. In the operating range current varies little with voltage change. Suitable for SMAW. [GOURD]

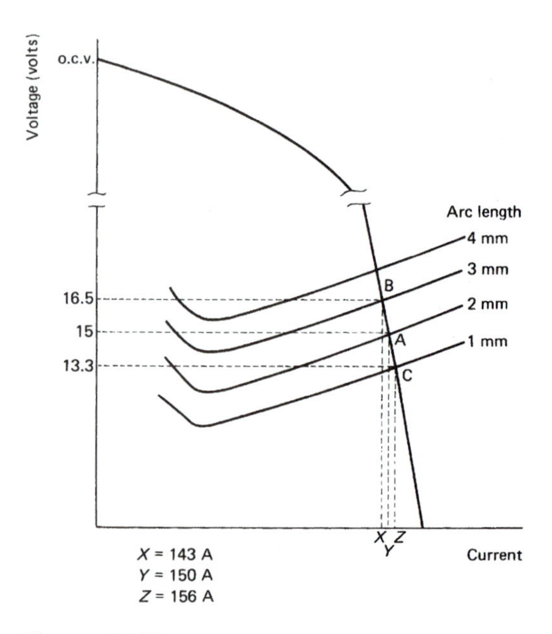

Figure 11.55
Variations in voltage and current with change in arc length. Operating points are obtained at intersections of arc and power-supply characteristics. [GOURD]

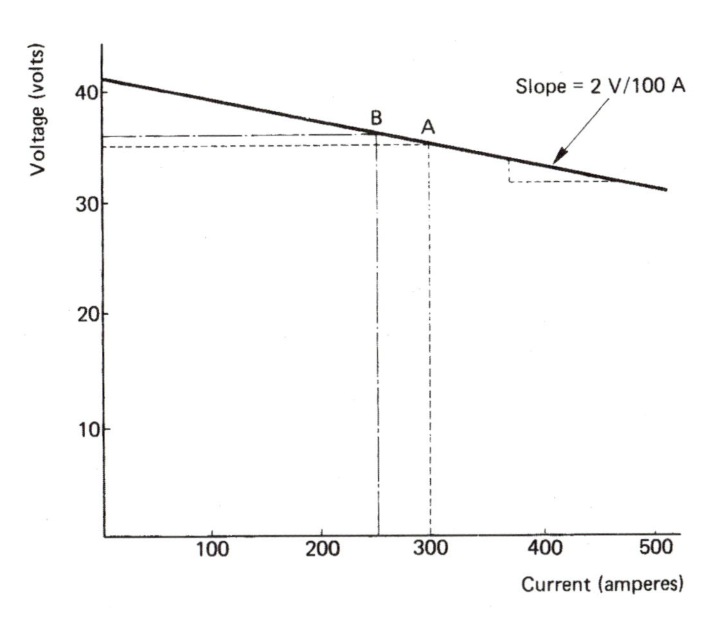

Figure 11.56
Output characteristics for a "constant-voltage" power supply. Suitable for GMAW. Deposition rates obtained over large range with various wire-feed rates require large current changes at small changes in arc voltage (and length). [GOURD]

11.6.4 Other Welding Processes

Later we will concentrate even more on the arc welding process and choose it as the analytical exercise of this chapter, including the aspects of controls, of heat transfer, and (in a simple way) of residual stresses. First, however, we complete the brief overview of the various welding processes.

Electroslag Welding (ESW)

Thick vertical welds on storage tanks, on ships, or on large press frames are produced by electroslag welding. It is a very powerful process. Deposition rates as high as 20 kg/hr per electrode can be achieved. Pressure vessel walls as thick as 400 mm may be welded. With voltages in the range of 30–40 V, currents of 600 A up to 1000 A are used. The equipment is rather simple, and it is mounted on the part being welded.

The filler metal is essentially provided by wire fed into the weld. Wires with diameter 2.4 mm and 3.2 mm are used with feed rates ranging between 20 and 150 mm/sec. Several parallel wire electrodes may be used simultaneously.

There are two variations of the process: the "conventional" method and the consumable guide method. In the conventional method the wire electrode is fed through nonconsumable guide tubes made of copper and insulated on their surfaces. These guides must be mechanically moved up as the flux bath rises. In the consumable guide method the guide tubes are made of the same material as the wire electrode, and they are held in a stationary head above the welded joint; thus they melt off just above the surface of the flux bath.

A diagram of the latter process is shown in Fig. 11.57. Although the process may look similar to the GMAW or the SAW process, which also use wire electrodes, there is no arc except at the beginning when there is not yet any molten metal. The heat is generated by the passing of the electric current through molten electrically conductive flux. The weld is created in the cavity formed by the two sides of the base metal and

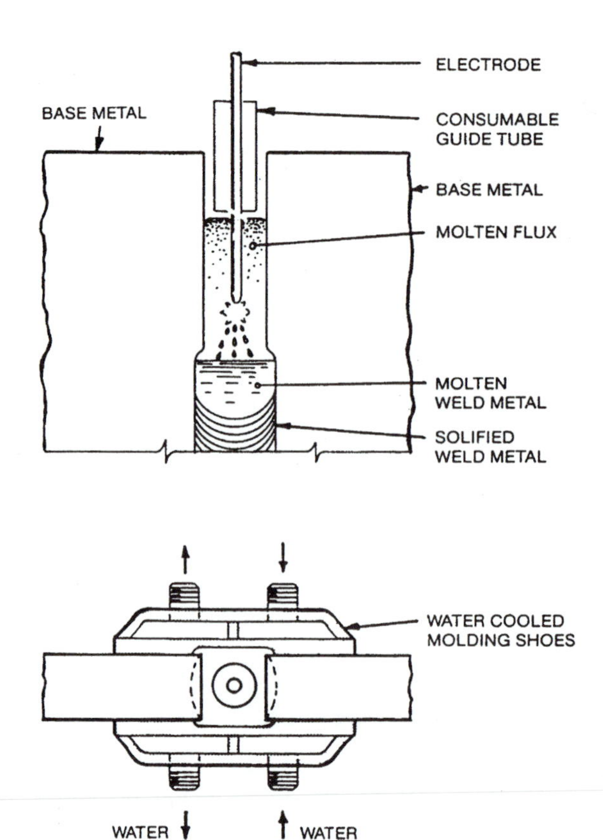

Figure 11.57
Consumable electrode method of electroslag welding. Wire electrode is submerged in flux. Used for very high metal deposition rates in joining thick sections along a vertical weld. [HOBART3]

by water-cooled molding shoes attached to the sides of the two plates. At the beginning the electrode wire reaches to the bottom of this cavity, where some iron powder or steel wool is placed so that an arc is struck. Once the arc stabilizes, more flux is added until a molten pool of flux forms, which floats on top of the molten weld pool. The two layers rise continuously while below them the weld metal solidifies. A constant-voltage power source is used, and the current adjusts itself to the wire feed speed. Figure 11.58 shows a diagram of the equipment required. Depending on the thickness of the weld, single or multiple wires are used, and they can be stationary for up to 60 mm or oscillating laterally for up to 100 mm thickness per wire. This is a mechanized or even automated single-pass process. Because of the high heat input, it is limited to welding low-carbon steels, low-alloy steels, some stainless steels, and some nickel alloys.

Electrogas Welding (EGW)

Electrogas welding (EGW) is a process similar to ESW, and it is used in similar applications—for vertical, single-pass, thick welds that are created essentially as castings into a mold represented by the two sides of the base metal and water-cooled retaining shoes, as shown in Fig. 11.59. However, the heat and the weld metal are supplied through an arc and wire feed. In this sense it is actually the GMAW process already discussed, except now the electrode wire is supplied vertically into the weld. A solid or flux-cored electrode may be used; the latter is shown in the drawing. For wider welds

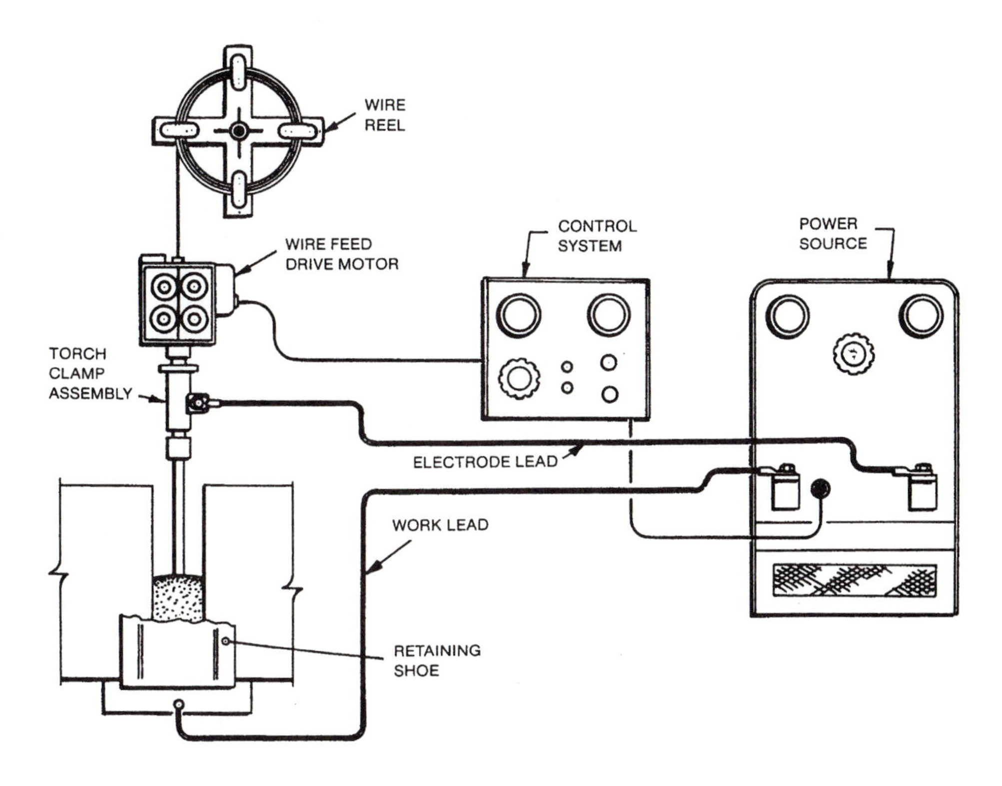

Figure 11.58
Equipment for consumable guide electroslag welding. Single or multiple wires are used for up to 60 mm width per wire or up to 100 mm with wire vibrating laterally. [HOBART3]

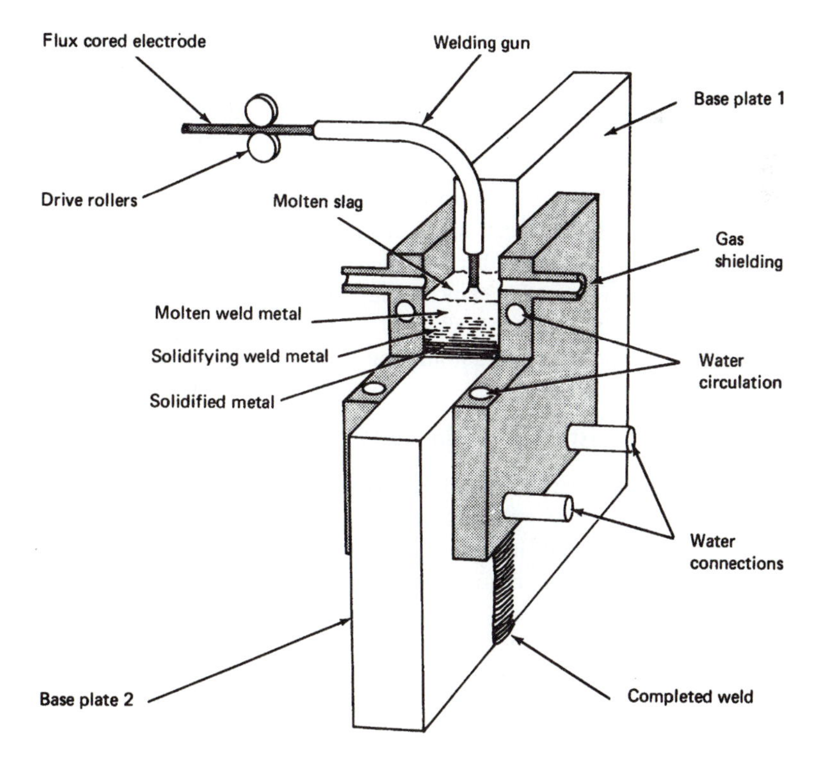

Figure 11.59
Electrogas welding with a flux cored electrode. Similar applications as for electroslag welding. [WH2]

several electrodes in parallel are used or a single one is oscillated across the weld width. Shielding of the arc is provided by a gas blown around the wire electrode. A constant-voltage power source is used.

Two other fusion welding processes are used for rather special applications that require their high power intensity: *electron beam welding* (EBW) and *laser beam welding* (LBW). The high power densities are derived from the capabilities of accurately focusing the beams. The equipment is essentially the same as that used in electron beam and laser beam machining respectively, as described in Chapter 12. Both processes are used in two types of operations: (1) welding of thin foils, where rather low, accurately metered power must be maintained, and (2) more powerful welding of narrow and deep welds.

The highest power densities are obtainable in EBW. Three variations of the process exist: high vacuum (HV), medium vacuum (MV), and nonvacuum (NV). In all of them the electron beam is generated in a high vacuum. From there the beam passes into the work chambers in HV and MV processes and out into the regular environment in the NV process. The HV process is unique, producing very high depth-to-width ratios of the weld, up to 50:1. Maximum weld-metal purity is possible due to the absence of contaminating gases such as oxygen and nitrogen. Power density up to 1.6e9 W/m^2 is attainable. The disadvantages are the limited size of the workpiece and the rather long pumping time for reaching the necessary vacuum (about one hour). Examples of extreme work parameters are (1) a weld in a 12.5-mm-thick 2219 aluminum plate made by a single pass, with a welding speed of 40 mm/sec and with 6 kW power and (2) a weld in a 100-mm carbon-steel plate made with 2 mm/sec speed. Depth-to-width ratios are lower in the MV and NV processes. However, the pumping time is reduced to seconds and zero, respectively. An example of an NV weld is given for a weld in 19-mm stainless steel plate using 12 kW. The HV applications are in the high-precision operations in the nuclear, aerospace, and electronics industries, whereas the NV applications are used where speed is important, primarily in the automotive industry. The LBW process does not require a vacuum. Power densities of 1.6 MW/m^2 are achievable, but the beam is partially reflected, depending on the smoothness of the surface. For example, absorption of a low-intensity CO_2 laser beam may be 40 percent for stainless steel and as low as 1 percent for polished aluminum or copper. The advantages of LBW are that (1) the HAZ is minimal, (2) the high power density permits welds between metals with different properties and different thicknesses, (3) electrical contact with the workpiece is not necessary, (4) it is possible to make spot welds "on the fly," with parts moving up to 3 m/min, and seam welds in thin sheet at speeds up to 80 m/min, (5) spot welds can be made with accurate positioning, and (6) the process is well suited to automation.

Resistance Welding (RW)

In resistance welding a combination of pressure and heat produced by an electric current passing through the contact of parts to be joined is used to produce a local weld. The parts are usually sheet metal. In contrast to arc welding, there is no shielding gas, no flux, and no filler metal used. The three basic kinds of operations are shown in Fig. 11.60: *spot, seam,* and *projection* welding. The operation most commonly used is spot welding (RSW). In significance it is comparable to arc welding. The popular case is joining the parts of car bodies; one can find several thousand spot welds on a single car. The process of RSW is explained in Fig. 11.61. The two sheet metal parts are combined

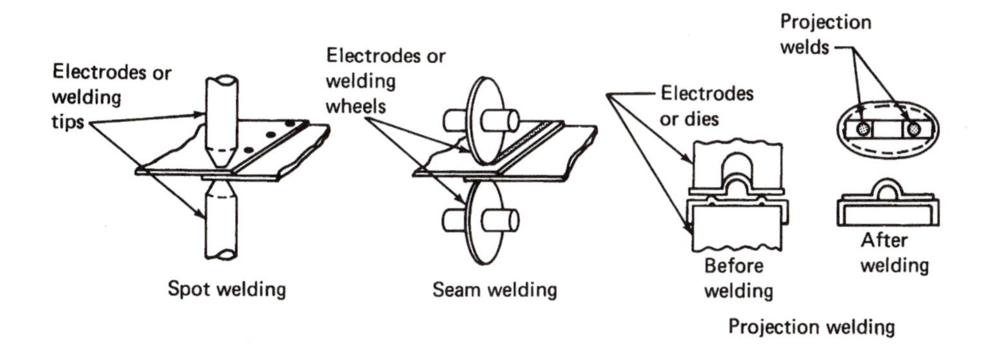

Figure 11.60
Three variations of resistance welding: spot, seam, and projection welding. [WH3]

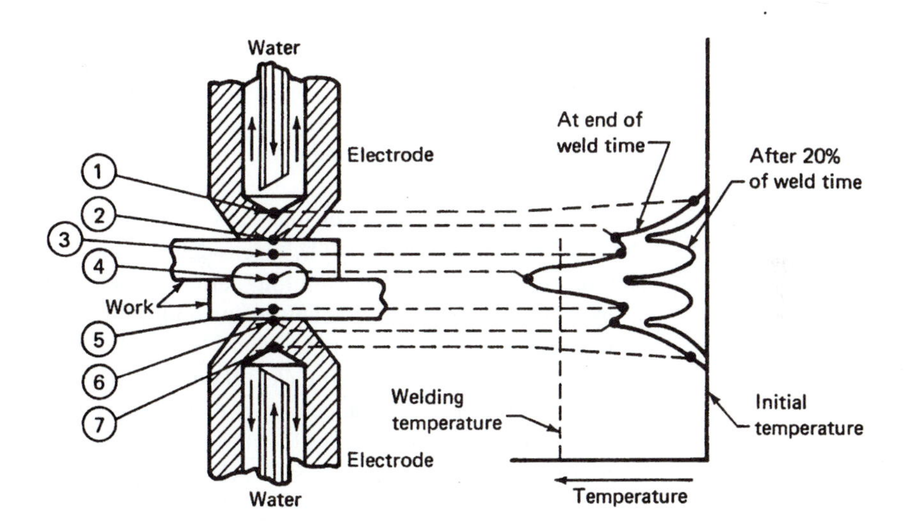

Figure 11.61
Temperature distribution at various locations in spot welding. Combination of pressure and heat produces a melted and frozen nugget (4) in the point of peak temperature. The process is controlled by time or current parameters. [WH3]

in a lap joint. They are squeezed locally by the action of two water-cooled electrodes, and an AC current is passed. The voltage is rather low, but currents are high, 5,000 to 15, 000 A for sheet metal thickness between 0.5 and 2 mm, in mild steel. The heat generated along the path of the current is

$$Q = I^2Rt$$

A typical force in the contact is about 1000 N. The resistant points are marked 1 through 7; obviously, the useful heat is generated at point 4, although points 3 and 5 contribute to the heating of the weld spot, too. All other heating points produce only losses. The highest temperature arises at point 4, and it exceeds the melting temperature for a fraction of a second. The welding cycle starts with the application of the pressure, followed by the start of the current, the switching off of the current, and after an additional interval, release of the pressure. The melted and frozen part is called the "nugget." It is important to maintain uniformity of the size and strength of the nugget by achieving uniform heat spent and temperature achieved. The cycle may be controlled by counting the cycles of the AC current. However, because of the variation of surface finish and contamination in the contact, with the resulting variation of the resistance in the contact, control is more often based on measurement of the temperature at a point close to the welding spot, or on measurement of the resistance in the circuit, or in some instances of the approach of the electrodes during the weld cycle.

The resistance of the weld point itself is small, on the order of 0.0001 Ω, and the resistance of all the other contact points must be even smaller. This applies mainly to the contact between the electrodes and the work. Electrodes must have good electrical conductivity as well as good strength and hardness to resist plastic deformation caused by repeated application of the clamping force. With a worn (deformed) electrode, the contact area increases and the nugget area increases as well. Current density thus decreases, and weld generation requires more time, or else the weld quality deteriorates. Electrodes are made of various copper alloys. Generally, the harder the alloy, the lower its electrical conductivity. The choice depends mainly on the work material. Heat is dissipated through the work material towards the electrodes and through the electrodes. Materials with good thermal conductivity, such as aluminum, need higher welding currents because a large part of the heat is conducted away from the weld. This happens much less in materials with low thermal conductivity, such as stainless steel: its thermal conductivity is nine times lower than that of aluminum (see Table 8.3), but it is four times as strong. Obviously, the clamping force will be higher. A strong copper alloy is suitable for the electrode in spite of its lower conductivity because of the lower current density used. A highly electrically conductive copper alloy is selected for the electrode in welding aluminum.

The spot weld cycle takes into account the cooling rates that may adversely affect the HAZ around the nugget. The electrodes remain under pressure after the current is shut off, to participate in the cooling period. However, it may be necessary to include a lower current preheat period before and a lower current period after to slow down the cooling after the actual welding period.

In *resistance seam welding (RSW)* one or both of the electrodes have the form of wheels (see Fig. 11.60). The current is then applied either intermittently at large intervals to produce spaced spot welds during continuous motion of the work sheets, or intermittently at short intervals or even continuously to produce continuous seam welds. The latter process produces gas- or liquid-tight joints in sheet metal tanks such as automobile gasoline tanks or in the manufacture of heat exchangers and radiators. In RSW cooling is accomplished by flooding the surfaces of the welded parts with a coolant.

Resistance projection welding (RPW) is used to join various kinds of parts to sheet metal parts. Less overlap and closer weld spacing are possible because the current concentrates on the projection and is not shunted through adjacent welds. Projection welds can be located more accurately than spot welds. Large, flat-face electrodes are used, which reduces electrode wear. On the other hand, an additional, projection-forming operation is necessary.

The machinery used in spot, seam, and projection welding must provide the motions and forces necessary for these operations. A common type used in spot welding is the air-operated rocker arm machine (Fig. 11.62). In automotive production, robots carrying multiple welding heads have been introduced on a large scale in the 1980s for spot welding of car bodies. Several such robots are located around the car in a number of welding stations arranged in a transfer line.

High-frequency welding exists in two forms: as a resistance welding process (HFRW) and as induction (resistance) welding (HFIW). In both instances it is resistance welding; the distinction is in the way in which the current is brought into the workpiece whether by contact in the former case or by a transformer coil in the latter case. Using high frequency, typically 10 kHz but also higher, up to 500 kHz, instead of the 60 Hz used in regular resistance welding, confines the current to the surface of the

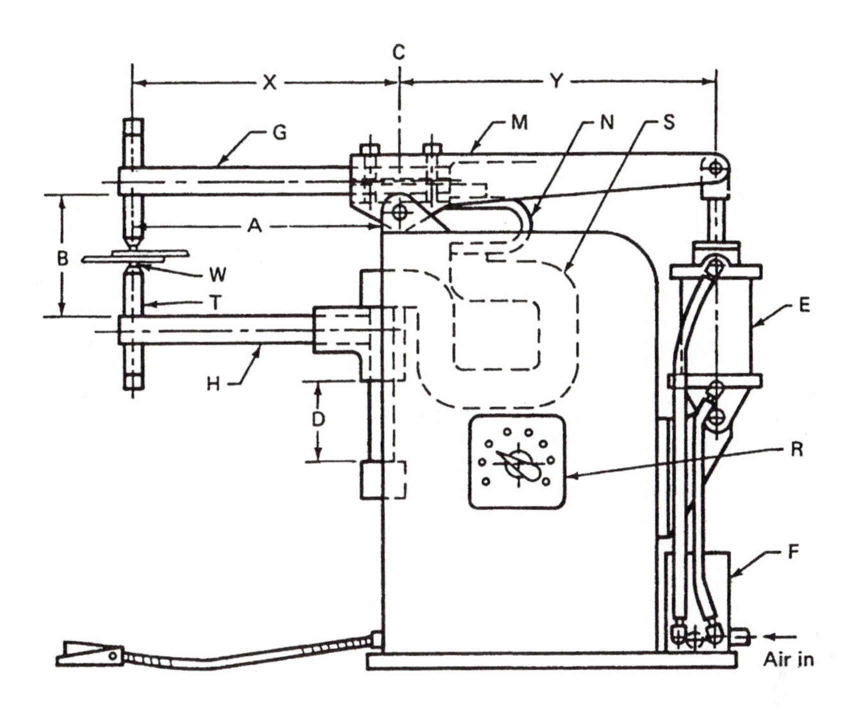

Figure 11.62
Air-operated, rocker-arm spot welding machine is a popular manually controlled type. The piston of air cylinder E acts on the upper arm G to squeeze the overlapping sheet between electrodes T. [WH3]

part, where it can further be confined to a narrow location by holding a part of the induction coil as a proximity conductor over the surface at the weld site. The various applications are indicated in Fig. 11.63.

The process has several advantages. The heat is concentrated in a narrow zone, and most of the molten metal is squeezed out of the joint during the upsetting portion of the cycle. In this way also the cast structure is eliminated. Very thin tubes, with 0.25–0.5 mm wall thickness, can be butt welded. The contact problems are minimized because introducing currents up to 2,000 A through rubbing contacts at 400 kHz overcomes the problems of penetrating the surface oxides. An example of a rather impressive application is shown in Fig. 11.64.

A variety of *additional welding processes* exist, which are briefly described below. In *flash welding* two solid parts, most simply two cylindrical rods, are joined at their ends while being pushed together and current produced at their contact. Heat is generated by resistance at minute contact points to the flow of the current as well as by minute arcs across these points. Upsetting follows the heating period, and the molten metal is extruded out. *Upset welding* is similar, but the force is applied simultaneously with the resistance heating phase. Temperature is held in the recrystallization range, below melting. This produces a clean weld. In *percussion welding* an arc is produced by a short pulse of electric energy produced by the discharge of a capacitor, and the parts are brought together in a percussive manner. Alternatively, the impact is produced by an electromagnet. The whole process is very short and produces very thin layers at the contact of the two parts. It is mostly used for joining electrical contacts to contactor arms.

Friction welding in its common form is used to join two round bars at their faces. The heat is generated by rubbing the two faces under pressure, must commonly by rotating one of them around their common axis. In one of two basic modes the part is attached to a flywheel that is brought up to speed and then the rotating bar is pressed against the stationary one. All the kinetic energy is converted into heat. In the other

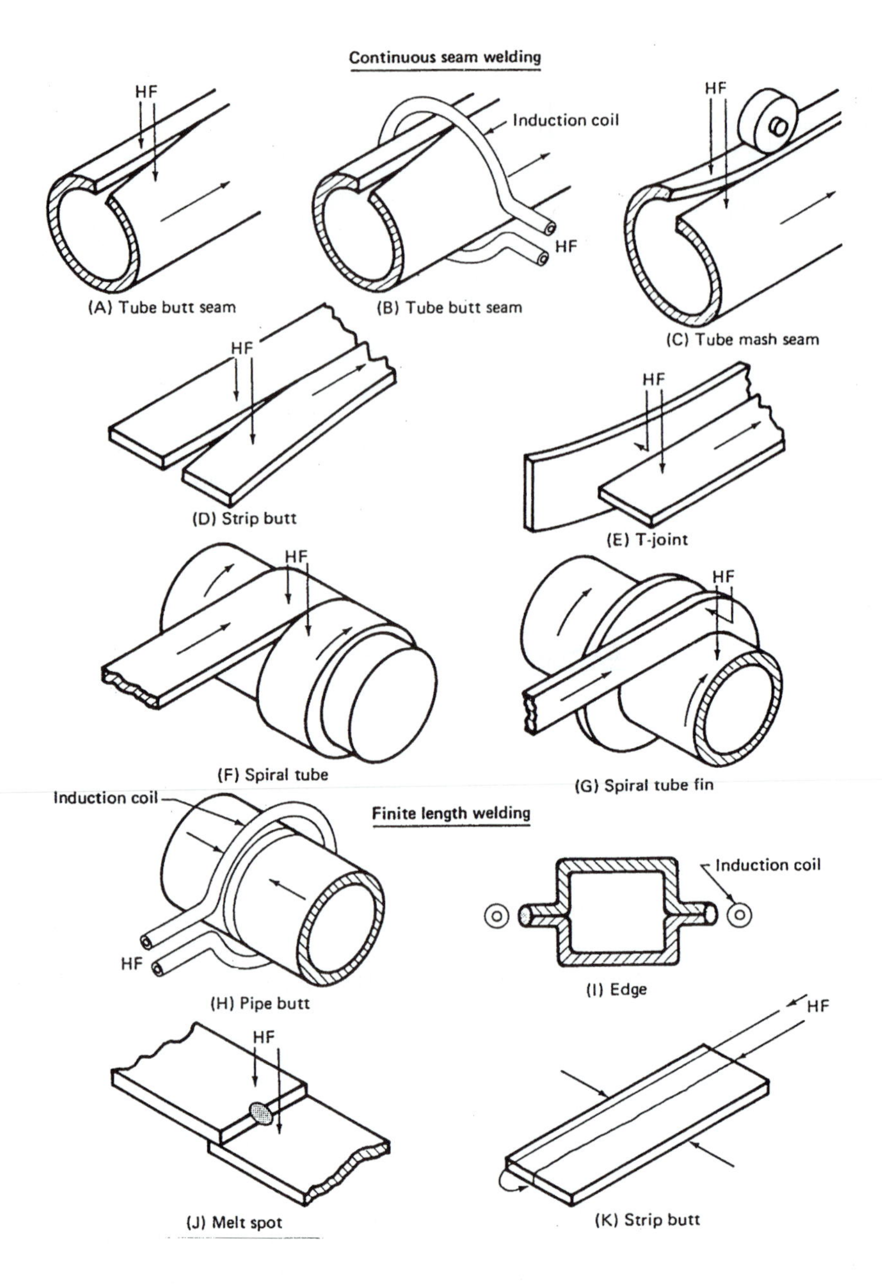

Figure 11.63
Basic high-frequency welding applications exist as either resistance- or induction-heated processes. [WH3]

mode the part is driven via a coupling that slips at a predetermined torque. No filler metal or flux is added. The heat-affected zone is rather thin, and the joint is as strong as the base metal. The process is easily automated. This process is used to join a wide range of similar and dissimilar metals.

In *explosion welding* a thinner plate or tube is attached over a large surface to a massive part by generating a high contact pressure between them, causing plastic flow and welding. The pressure is produced by the detonation of an explosive, and welding does not happen over the whole surface at once but progresses from one end to the other. It is used to join materials that could not be welded by other processes, such as carbon steel to stainless steel or aluminum or titanium to steel. The major commercial

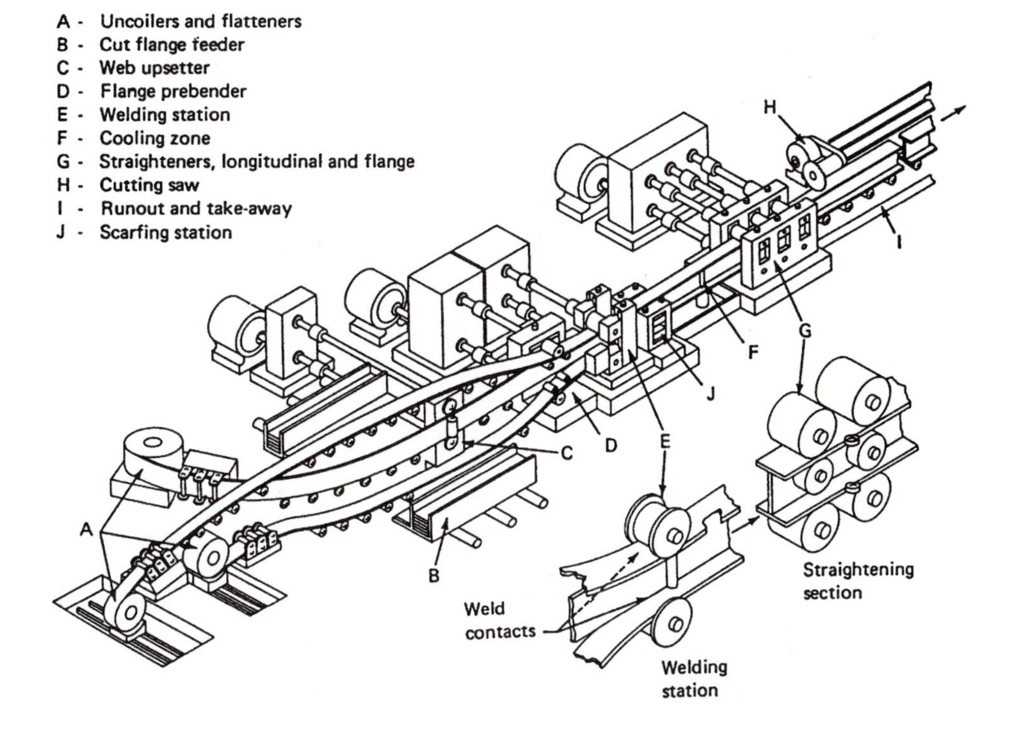

Figure 11.64
Mill arrangement for fabricating I beams by high-frequency resistance welding in an automated, continuous process. [WH3]

application of the process is cladding of flat plate. The bimetallic plate can then further be hot- or cold-formed by forging or rolling.

Ultrasonic welding is a process in which a solid-state bond, without melting, is produced by vibrating the two surfaces along one another under low normal pressure. It is used for joining similar and dissimilar metals in the form of foil, wire, and ribbons to flat surfaces in the electronic industries. It is useful for the encapsulation of materials such as explosives and reactive chemicals that require hermetic sealing and cannot be processed by heat or electric current. The vibratory shearing action disturbs surface films and permits virgin metal contact at many points. Atomic diffusion occurs across the contact area, and the metal recrystallizes to a very fine grain structure. The frequency of vibration is in the range of 10 to 75 kHz, and the driver is a piezoelectric or a magnetostrictive exciter.

Diffusion welding is based on solid-state diffusion under pressure at elevated temperature. No melting occurs, and a solid filler metal (diffusion aid) may or may not be used between the welded surfaces. The surfaces must be rather flat and smooth (machined, ground, and polished), and oxides and contaminants must be removed. All this is necessary to permit an intimate contact under pressure. Joining is produced by diffusion. The diffusion rate is exponentially affected by the temperature used. The highest metallurgically possible temperature is used, which is at 0.5 T_m, and in some cases up to 0.8 T_m. The process needs protective atmosphere, and the workpiece is located in a chamber in vacuum or inert atmosphere. Examples of application are titanium aircraft parts. Processing time of several hours is needed. Another use is in high-strength, heat-resistant nickel alloys. In this case an intermediate layer of pure nickel is used as the diffusion aid.

Thermit welding (TW) was invented at the end of the nineteenth century for welding of railway rails and crane rails. Superheated molten metal from a thermochemical reaction between aluminum and a metal oxide provides the heat and the filler. In the commonly used application, the reaction occurs between aluminum powder and iron oxide, and it produces molten iron, aluminum oxide, and heat. The aluminum oxide floats at the top as slag and protects the iron from air. The temperature of the molten iron is about 4,500°F. The reaction takes less than a minute to complete. For rail joining, a combination mold-crucible made of sand bonded with phenolic risers is built around the joint. After the operation the mold is broken and the excessive parts of the casting such as the gates and the resin are removed by an oxyacetylene torch. Thermit welding is also used to repair cracks in large steel castings and forgings.

11.7 CONTROL OF THE ARC

Arc control is discussed here in two parts: (1) the self-regulation of the GMAW process using thin wire, and (2) the servomechanism control of the SAW process using thicker wire. First the important relationships between the voltage e, current i, arc length l, and burn-off rate r will be discussed.

11.7.1 Melting Rates

The electrode in GMAW, SMAW, and SAW is melted away partly by the heat generated at its tip as a result of the annihilation of the kinetic energy of the electrons or ions, respectively, whether the electrode is positive or negative (the energy being higher in the EP case), and to a larger extent by the heat generated in the electrode wire by resistance heating due to the current passing through it. The heat generated in the arc does not reach the electrode except by radiation, which is a minor effect. Therefore, those parameters that affect the plasma, such as the shielding gas or flux, do not directly affect the melting rate r. The share of the heating contributed by the resistance and current combination is naturally greater for small wire diameters, long electrode extensions, and metals with high specific resistance. The decisive parameter for the melting rate is the current i. A general formula applies:

$$r = bi + cx\,i^2 \tag{11.7}$$

where the first term expresses the effect of electrode tip heating, and the second term corresponds to resistance heating; x is the electrode extension, that is, the length between the input of the current into the electrode and the tip of the electrode. The coefficients b and c both depend on the material of the electrode and on its diameter. The melting rate r (mm/sec) is the velocity with which the wire is burned off. Let us consider some experimental data and look at Fig. 11.65, taken from the handbook published by one of the largest makers of welding equipment. It applies to steel wire and covers a wide range of electrode diameters, from 0.9 to 3.2 mm. The individual curves in the diagram can be approximated by straight lines around the average working points.

We will be dealing with *changes* of the process parameters as *deviations* from some initial state. Obviously, the relationship between the change Δr of the melting rate and the change Δi of the current can be linearized and expressed as follows:

$$\Delta r = b\,\Delta i \tag{11.8}$$

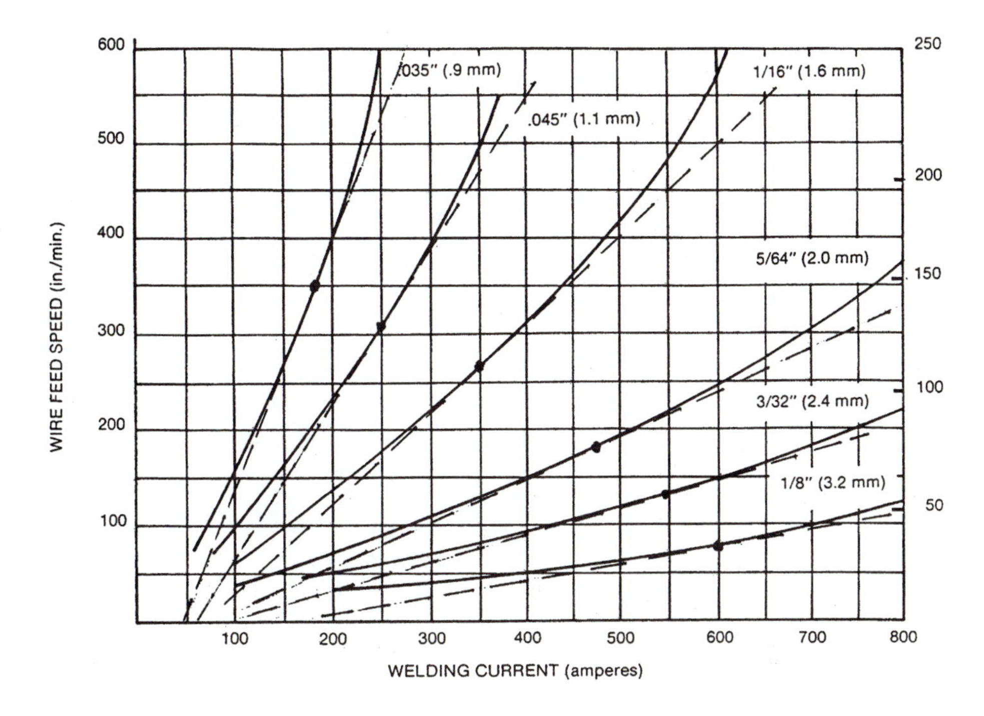

Figure 11.65 Wire feed speed versus welding current for steel electrode wire in GMAW. The individual curves may be approximated as linear relationships by choosing an average working point and drawing a tangent to it. [HOBART2]

(The vertical coordinate of the graph is marked as wire feed rate which, of course, in the steady state is equal to the melting rate.) The working points are expressed in Table 11.1 by the average currents used, and the values of the coefficient b are given as taken from the graph in Fig. 11.65. The selected working points correspond to current densities of 280–75 A/mm^2 for wire diameters 0.9–3.2 mm, respectively.

Actually, the graphs may also be interpreted in terms of volumes of metal instead of length of wire, and a relationship can be written:

$$\Delta r = g\ \Delta i \tag{11.9}$$

where Δr (mm^3/sec) and g (mm^3/sec/A). The values of g corresponding to Table 11.1 are, respectively; 0.7, 0.65, 0.72, 0.63, 0.54, and 0.6. Thus the specific volumetric rate is almost constant, regardless of wire diameter, and its average value is $g = 0.65$ (mm^3/sec/A) for carbon-steel wire.

A comparison of the various arc welding processes is presented in Fig. 11.66, where volumetric melting rates are plotted versus welding current for welding of steel with the SMAW and SAW processes. Note that, on average, the melting rates are proportional to current. If the weight rate is transformed to volumetric rate the graph yields again an average value $g = 0.65$ (mm^3/sec/A).

Table 11.1 Melting Rate Coefficients

Wire diameter d (mm)	0.9	1.1	1.6	2.0	2.4	3.2
Current i (A) at work point	180	250	350	475	550	600
Coefficient b (mm/sec/A)	1.1	0.68	0.36	0.2	0.12	0.075

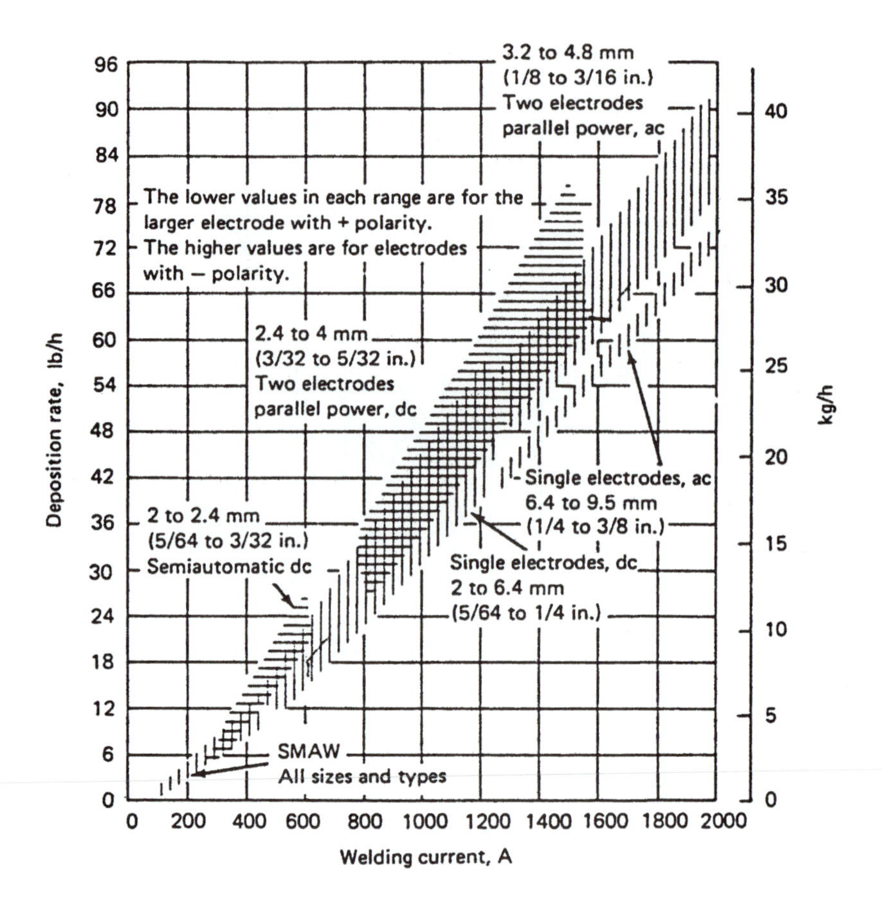

Figure 11.66
Approximate deposition rates of some SAW process variations compared to SMAW. Note that the linear characteristics of wire speed versus current drawn for various wire diameters may be approximately consolidated as volumetric or mass deposition rates versus current, within a range. The average rate for steel is about 0.65 mm^3/sec/A. [WH2]

The coated-electrode SMAW process gives the lowest deposition rates because of current limitations. Because of the large distance between the power supply at the upper end of the stick and the arc, the resistance heating would produce a large power loss, high temperature of the electrode, and peeling off of the coating. Since this is a manual process, it would be difficult anyway to handle higher deposition rates. The mechanized or automated GMAW and SAW processes produce much higher deposition rates.

Arc Control

We will discuss the arc from the point of view of a control system regulating the deposition rate. All the other influences, such as the type of coating or of shielding gas, have already been discussed. It has been explained that the melting rate of a given electrode depends primarily on the level of the current.

In SMAW, where the electrode is guided manually, the operator advances it as it is melted away and tries to maintain a constant arc length and, correspondingly, a constant bead profile. Inevitably, the arc length varies to an extent, and in order to keep the melting rate constant, a power source of the "constant-current" type is used. The operator changes the current setting in order to change the melting rate and adjusts the rate at which he or she advances the electrode to the work. The latter change is small because rather large electrode diameters are used as well as rather low melting rates.

In GMAW and SAW the processes are automated. Some kind of control is needed to achieve the following:

- Maintain the melting rate constant in spite of inevitable variations of the distance between the welding gun and the work
- Change the deposition rate to another desired value; both the current and the wire feed rate must be changed. The mechanisms used in this control are different for the two processes, mainly because of the difference in the wire diameter, an important point.

Self-Regulation of GMAW

When GMAW was first introduced in the late 1940s, it was soon recognized that, especially when using thin wire, there was a self-regulating capability inherent in the process. When the wire feed rate is increased, the arc length shortens, current increases, and melting rate increases to adjust to the new wire feed rate. This occurs with the constant-voltage power supply and, as regards steady states, it is easily understood. There is, however, also a transient state in which some disturbance of the process occurs. We will now investigate the inherent control scheme and derive the relationships for the transient. In order to describe the process, it is necessary to establish the relationships between its variables (Fig. 11.67). The welding gun is guided above the welding seam, and the machine, or the operator, attempts to keep the gap h between the nozzle and the weld constant. This distance is also called the tip-to-work distance. The electrode wire sticks out of the contact tube with an extension x, and the arc length is l.

11.7.2 Self-Regulation of the Arc in SMAW and GMAW

As mentioned above there is a "natural" control built into these processes. The demands on this control are different for the two processes. In SMAW the main requirement is to keep the deposition rate constant and insensitive to changes in the gap h. In GMAW the same requirement applies, and it is also required that the process adapt quickly to a command for a different deposition rate. These demands are best fulfilled if the power sources are different.

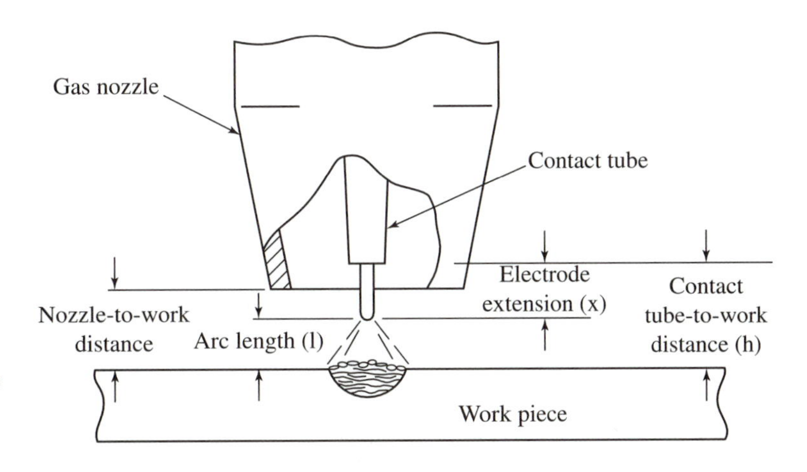

Figure 11.67
Gas metal arc welding terminology. The parameters x, h, l are involved in the control of the arc. [WH2]

In SMAW the "constant *i*" characteristic has to be used, while the "constant *e*" is used in GMAW (Fig. 11.68). In both instances the word *constant* is not meant in an absolute sense but only as an approximation. In the graph, A is the mean operating point, and the points P_i are the power-source characteristics, which are usually set manually, in steps. The L_i lines are the arc characteristics, each corresponding to a particular arc length l, and there is a continuous transition between them. As long as we have chosen a P line, we must enforce its (e, i) relationship. This has to be satisfied simultaneously with the L line relationship. Thus we have, for the L lines (*a*rc characteristics),

$$e = e_a + m_a i \tag{11.10}$$

and

$$e_a = al, \qquad e = al + m_a i \tag{11.11}$$

For increments, we have simply

$$\Delta e = a\Delta l + m_a \, \Delta i \tag{11.12}$$

Typically, the following values apply, and they will be used in our examples:

$$m_a = 0.03 \text{ V/A}, \qquad a = 3 \text{ V/mm} \tag{11.13}$$

For the P lines (*p*ower characteristics), we have, for "const *i*,"

$$i = i_o + \frac{e}{m_p} \tag{11.14}$$

or

$$e = m_p (i - i_o) \tag{11.15}$$

where i_o is associated with the choice of the power supply characteristic P and, typically,

$$m_p = -0.27 \text{ V/A} \tag{11.16}$$

For "const *e*," we have

$$e = e_o + m_p i \tag{11.17}$$

where e_o is associated with the choice of the P line and, typically,

$$m_p = -0.03 \text{ V/A} \tag{11.18}$$

In both cases, for increments, we have

$$\Delta e = m_p \, \Delta i \tag{11.19}$$

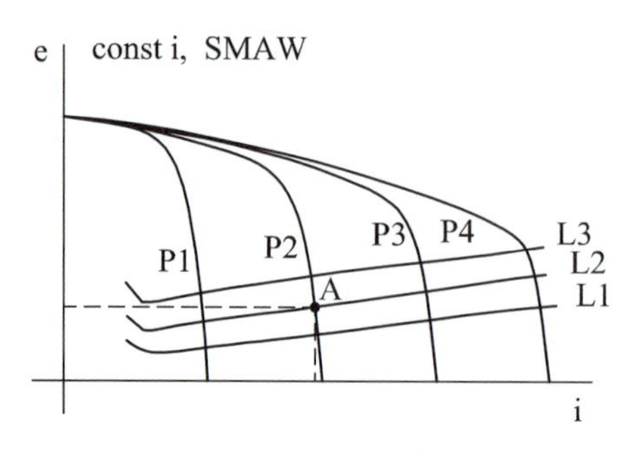

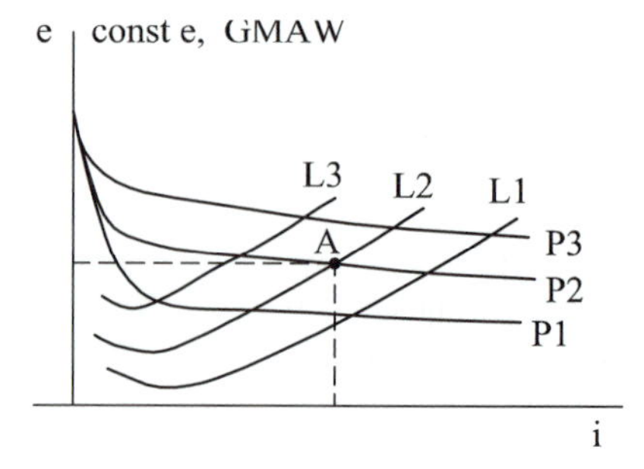

Figure 11.68
The power-supply characteristics versus arc characteristics. A working point A, when selected on a particular power line P_i, will also satisfy the particular arc length L_j.

Burn-off rates are considered simply in a direct proportionality relationship:

$$r = bi \tag{11.20}$$

where, typically the b values of Table 11.1 are used, as they depend on the wire diameter d. There is also an effect of the electrode extension x which, for our examples, we will neglect. As mentioned before, SMAW regulates automatically the effects of Δh, and GMAW regulates automatically the effects of Δf, where f is the wire-feed velocity in (mm/sec).

Let us now investigate the transient and steady-state changes and incorporate all the relationships in the block diagram of Fig. 11.69a. All the variables are actually deviations Δ from an average operating state. For simplicity, the symbol Δ is omitted. The diagram is drawn essentially for the input of Δh by assuming that $\Delta f = 0$. However, it can be turned around to explain the effect of Δf, with $\Delta h = 0$. From Fig. 11.67, we have

$$l = h - x \tag{11.21}$$

Further on, we use Eqs. (11.12) and (11.19) to go from l to e and i. The internal feedback loop gives

$$i = \frac{1/m_p}{1 - (m_a/m_p)} al = \frac{a}{m_p - m_a} l \tag{11.22}$$

as a relationship between (increments of) l and i. For further simplification of notation, we introduce

$$k_{il} = \frac{a}{m_p - m_a} \tag{11.23}$$

and Eq. (11.22) is simplified as

$$i = k_{il}\, l \tag{11.24}$$

where i and l are again, actually increments Δi, Δl. Henceforth we will not repeat this caution; it will apply until revoked later.

a)

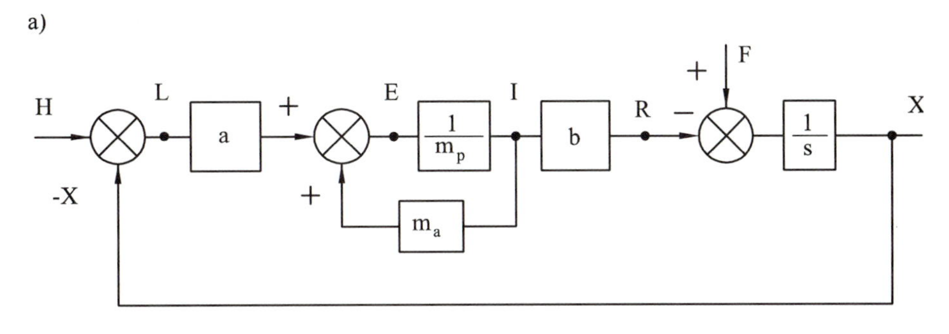

b)

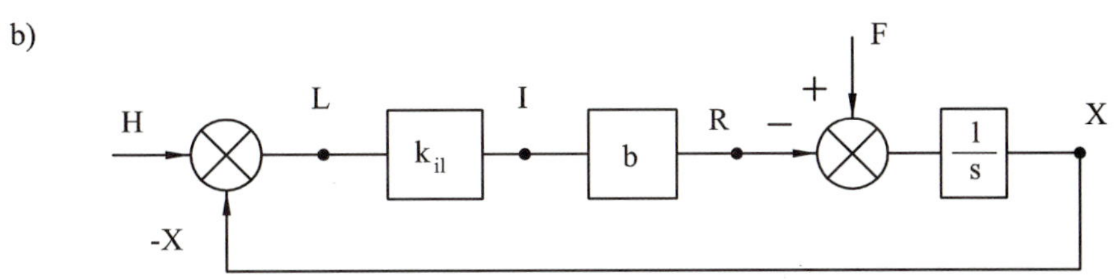

Figure 11.69
a) Block diagram of self-regulation of the GMAW process. Gap size h is input, wire extension x is output. Their difference defines arc length l that, in combination with the power characteristic (see Fig. 11.68), determines the current i. That gives the burn-off rate r that works against the wire feed rate f. The integral of the difference of the two rates is the extension length. The control system is of first order. b) Modification of the diagram: the inner loop from l to i is replaced by a gain factor k_{il}.

In the block diagram we finally arrive at the statement that the *rate of change* of the extension x is the result of the difference between feed rate f and burn-off rate r:

$$\dot{x} = f - r \tag{11.25}$$

The above scheme assumes that the transfers $1/m_p$ and m_a are instantaneous. This is close to true for the arc, m_a, but as for $1/m_p$ there may deliberately be an inductance in the line, for the short-circuiting and AC applications. Then we would have $E = m_p (1 + k_L s) I$, where k_L is the inductance. However, we will proceed here by assuming $k_L = 0$.

Thus also, following the block diagram, we have

$$\begin{aligned} e &= la + m_a i = \left(a + \frac{m_a a}{m_p - m_a}\right) l \\ &= \left(1 + \frac{m_a}{m_p - m_a}\right) la = \frac{am_p}{m_p - m_a} l \end{aligned} \tag{11.26}$$

Applying the above-chosen numerical values of the parameters, we have for "const i,"

$$i = \frac{3}{-0.25 - 0.029} l = \frac{-3}{0.279} l = -10.75\, l \tag{11.27}$$

$$e = 2.688\, l \tag{11.28}$$

For "const e,"

$$i = \frac{3}{-0.03 - 0.029} l = \frac{-3}{0.059} l = -50.85\, l \tag{11.29}$$

$$e = 1.526\, l \tag{11.30}$$

The block diagram of Fig. 11.69a is then redrawn as shown in Fig. 11.69b using the k_{il} notation. The transfer function X/H is evaluated as follows:

$$\frac{X}{H} = \frac{-b\, k_{il}/s}{1 - (bk_{il}/s)} = \frac{-bk_{il}}{s - bk_{il}} = \frac{1}{1 - (1/bk_{il})s} \tag{11.31}$$

$$\frac{X}{H} = \frac{1}{1 + \tau s} \text{ where } \tau = -1/bk_{il} \tag{11.32}$$

For a step input $H = H_1/s$

$$X = \frac{H_1}{(1 + \tau s)s} \tag{11.33}$$

and (see Fig. 11.70)

$$x(t) = h_1 (1 - e^{-t/\tau}) \tag{11.34}$$

For the arc length, we have

$$\begin{aligned} l &= h_1 - x \\ l &= h_1 - h_1 (1 - e^{-t/\tau}) = h_1 e^{-t/\tau} \end{aligned} \tag{11.35}$$

and for the burn-off rate,

$$r = b\, k_{il}\, l = -\frac{h}{\tau} e^{-t/\tau} \tag{11.36}$$

So, as a result of a stepwise change h_1 in the gap, the extension x, after the transient, will change by $x = h_1$, the arc length that instantaneously increases by $l = h_1$ will regulate into its initial length (no permanent change), and likewise the burn-off rate r will, after the transient, return to the original value.

EXAMPLE 11.1 SMAW, GMAW, SAW: Step Change Δh ▼

Determine the time constant, the instantaneous changes, and the steady-state changes of x, l, and r.

(a) SMAW, const i

Specifications: wire $d = 4$ mm, $b = 0.06$, $k_{il} = -10.75$, bead area $A = 15\ \text{mm}^2$, bead width $w = 5.48$ mm, $i_{mean} = 150$ A, $f_{mean} = r_{mean} = 9$ mm/sec, traversing speed $v = 7.54$ mm/sec.

A stepwise disturbance $h = 2$ mm is introduced.

$$\tau = -\frac{1}{bk_{il}} = \frac{1}{0.06 \times 10.75} = 1.55 \text{ sec} \tag{11.37}$$

The instantaneous effects of the disturbance are $\Delta l_{inst} = 2$ mm, $\Delta i_{inst} = k_{il} \times \Delta l = -21.5$ A, that is, 14.3% of i_{mean} and a corresponding change of $\Delta r_{inst} = b \times \Delta i = 1.29$ mm/sec. These instantaneous disturbances will be restored to 63% within 1.55 sec; during this interval the arc travels a distance $y = v \times \tau = 11.7$ mm.

The bead area reduces instantaneously by 14.3% to $A = 12.85\ \text{mm}^2$; correspondingly, the width reduces to $w_{min} = 5.07$ mm, that is, by 7.4%.

(b) GMAW, const e

Specifications: wire $d = 1.2$ mm, $b = 0.59$, $k_{il} = -50.85$, $f_{mean} = r_{mean} = 100$ mm/sec, $i_{mean} = 167$ A, $A_{mean} = 15\ \text{mm}^2$, $w_{mean} = 5.48$ mm, $v = 7.54$ mm/sec.

$$\tau = \frac{1}{0.59 \times 50.85} = 0.033 \text{ sec}$$

This time constant is 46 times shorter than in *(a);* the effect will be restored to 63% over a length of only $y = 0.25$ mm; *however, the instantaneous transient* change of the current is now $\Delta i_{inst} = 102$ A, that is, 60.9%; because of the very short duration of the disturbance, this should be fully acceptable.

(c) SAW, const e

Specifications: wire $d = 4.8$ mm, $b = 0.04$, $k_{il} = -50.85$, $f_{mean} = 18$ mm/sec, $i_{mean} = 450$ A, $A_{mean} = 30\ \text{mm}^2$, $w_{mean} = 6.9$ mm, $v = 11$ mm/sec.

$$\tau = \frac{1}{0.04 \times 50.85} = 0.490 \text{ sec} \tag{11.38}$$

The initial instantaneous disturbance of the current is 102 A, and it is restored to 63% after a travel of $11 \times 0.49 = 5.4$ mm. This is already a more *serious disturbance* than in GMAW (thin-wire process).

In SMAW any feed changes are accomplished by the welder. In GMAW such changes are set on the wire-feed drive. The process will "naturally" adapt to make a change. ▲

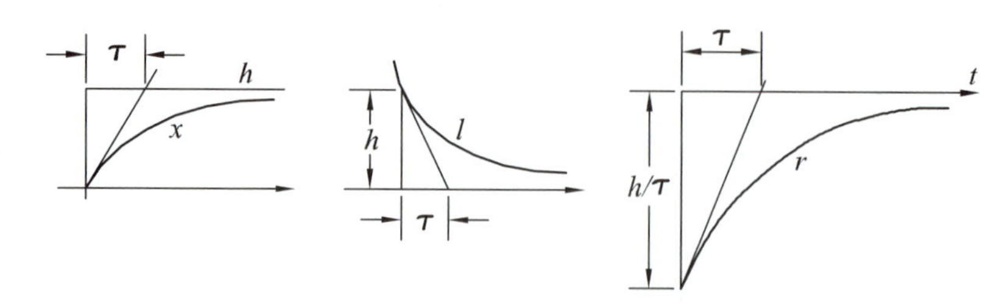

Figure 11.70
Response of process variables h, l, r to a step change in the gap h between the gun and the workpiece. With "const i" characteristic and SMAW, the time constant is rather long; see Example 11.1. It is much shorter with the "const e" characteristic and GMAW.

EXAMPLE 11.2 Step Change Δf: GMAW, Const e

Determine changes of f, l, e, and i.

Specifications: $d = 1.2$, $b = 0.59$, $f_{\text{mean}} = 100$ mm/sec, $A = 15$ mm^2, $v_{\text{mean}} = 7.54$ mm/sec, $\Delta f = 50$ mm/sec, $\tau = 0.33$ sec (see Fig. 11.71).

$$\frac{R}{F} = \frac{-k_{il}b/s}{1 - k_{il}b/s} = \frac{1}{1 + \tau s} \tag{11.39}$$

Eq. (11.39) is equivalent to Eq. (11.32); therefore, for $F = f_1 s$

$$r = f_1\,(1 - \mathrm{e}^{-t/\tau})$$

The burn-off rate adjusts, after a short transient, to the input of Δf.

$$(f_1 - r) = f_1 - f_1\,(1 - \mathrm{e}^{-t/\tau}) = f_1 \mathrm{e}^{-t/\tau}$$

$$x = f_1 \int_0^t \mathrm{e}^{-t/\tau} = -\tau f_1 \,|\, e^{-t/\tau}\,|_0^t = \tau f_1\,(1 - \mathrm{e}^{-t/\tau}) \tag{11.40}$$

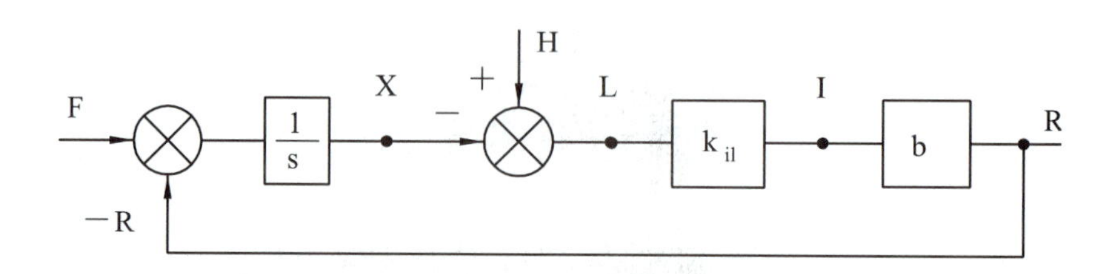

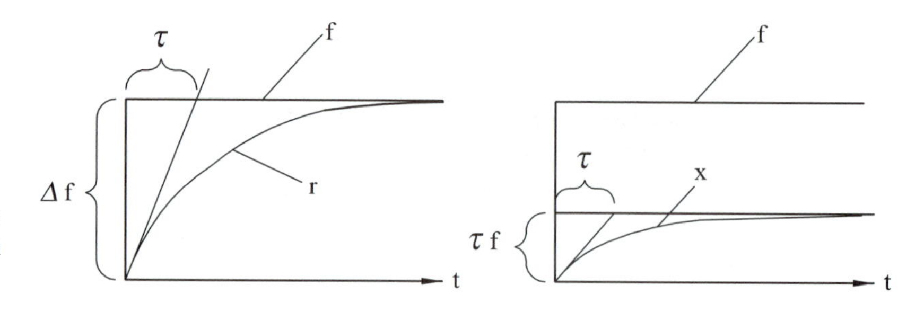

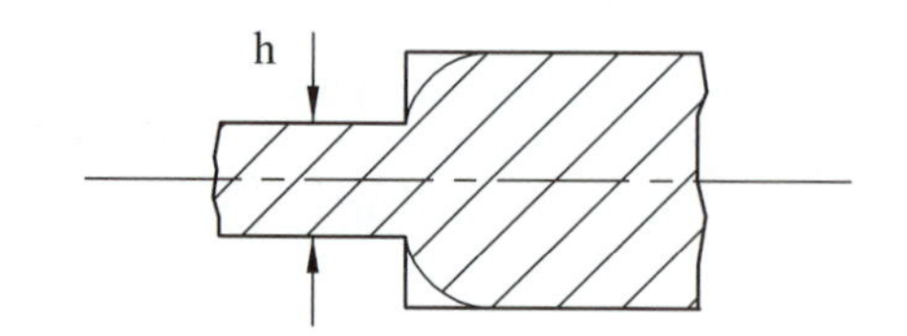

Figure 11.71
Block diagram and response of burn-off rate r (deposition rate) and of electrode extension x to a step change of the feed rate f. The deposition rate increases after a short transient to a steady-state value. The bead width h increases accordingly.

and $\Delta l = -\Delta x = -\tau f_1\ (1 - e^{-t/\tau})$. Final $\Delta l = -0.033 \times 50 = -1.65$, and $\Delta e = -0.03 \times b \times \Delta f = -0.885$ V. Final $\Delta i = 50/6 = 84.75$ A.

For "const i," it would not be possible to increase i by this much along the original P line (this would mean, for $m_p = -0.27$ V/A, a change of $\Delta e = 22.9$ V).

It has now been shown that the thin-wire GMAW process with the "const e" power source is naturally suitable for a mechanized feed rate. Any required change in the deposition rate is simply obtained by commanding the corresponding feed rate, and the process adapts automatically to it. This process behaves well also in reaction to changes in the gap; see Example 11.1b. Although a large instantaneous change of the current occurs it is of a rather short duration, and thus it is mostly considered acceptable.

The mechanized SAW process does not behave as well as GMAW. Thicker wires are used, resulting in longer time constants and larger disturbances. It was shown above that $\tau = 0.490$ sec, and the response to a Δh is not very favorable. Also, the change in wire-feed rate causes larger changes in the arc; for example, for a $\Delta f = 10$ mm/sec, $\Delta l = 4.9$ mm. For these reasons it is preferable to use servo control for this process. It is then also possible to avoid large current changes in response to Δh changes by choosing a "const i" power source. ▲

11.7.3 Servo Control in SAW

We have just arrived at a recommendation of using the "const i" source characteristic for SAW, where the wire-feed drive is mechanized as in GMAW in which, however, the use of the "const e" power source is common. In the latter the large swings of current i due to changes Δh are tolerable because of the short time constant of the "self-regulation" of the process and, on the other hand, the large changes of i needed to accommodate a command of a large change in wire-feed rate Δf do not require large changes in arc length l and arc voltage e because of the "const e" characteristic. These favorable circumstances do not apply to SAW, where much thicker wires and larger currents are used and longer time constants of the natural self-regulations are encountered.

The solution then is to provide fast response by using a suitable servo control to prevent large current swings due to Δh by using the "const i" characteristic, and to achieve the large current changes needed for large changes in wire-feed rate by switching over the P characteristic (see Fig. 11.72). If a Δi is needed, the process is transferred from point A to point B by selecting the P_3 characteristic on the power source.

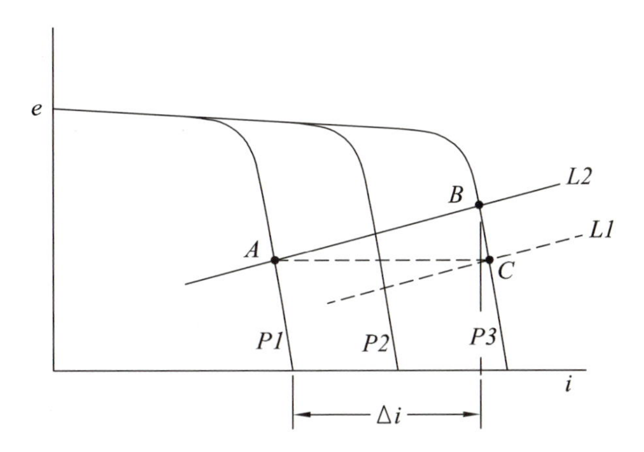

Figure 11.72
Change of deposition rate of SAW by switching over supply-current characteristics to move the process to a new operating point at a correspondingly higher current.

All the relationships between small changes of *e, i,* and *l* around the new working point *B* remain essentially the same as they were at point *A*. It is possible to use the servo to control for almost the same voltage *e* and arc length *l* as in point *A* and move to point *C*, that is, to automatically move from the arc characteristic L_2 to L_1.

Thus, whereas the change of feed rate and consequently of current in the self-regulated GMAW process has been initiated by changing the setting of *f* on the feed-drive control, now it is initiated by changing the setting of the power-source characteristic *P*. The immediate reaction of the process to this change is, obviously, a much larger current with the old sets of *l* and *e;* the burn-off rate *r* immediately increases, and the length *l* starts to increase in an attempt to regulate to a smaller current. However, the voltage starts to increase greatly, and the process may get out of hand. It is the purpose of the servo to take care of maintaining the process as required. This is especially important: unlike SMAW, where the operator who watches the process immediately starts to feed in faster to maintain the proper arc length under visual control, the SAW process is buried and not visible; the servo must therefore quickly start increasing the feed rate.

As before, we will check the reactions to both Δh change and to a request for a change of Δf. Let's start with the latter because it is the main task for the servo. As explained in the preceding paragraph, we are mainly interested in learning what happens as the *P* characteristic is changed, resulting in an introduction of a change Δi. Keeping to our usual notation, we omit the Δ sign.

The most practical variable to sense is the voltage across the arc, or, approximately, the voltage between the electrodes. The servo will try to maintain it unchanged. The variable to be controlled is, obviously, the feed rate *f*. Let us try to find out how well the feed rate will adapt to the Δi input and check how all the other variables change at the same time. First, however, let's draw the block diagram so as to express the feedback loop.

As before, all the variables shown in the block diagram of Fig. 11.73 are deviations from mean values; this applies also to E_{ref} and to the error *ERR* itself. An error is obviously needed for the feed drive, with its gain K_f and time constant τ_f, to move at all. The gain K_f expresses the feed rate (mm/s) per one *V* input; it simply includes the transmission between the motor and wire drive. Depending on the gain, the error may be large. Here we are showing what change in *f* would be caused by a change in E_{ref} and a corresponding change in *ERR*. The block *ERR*→*F* represents a DC servomotor, including an internal tacho feedback. The time constant τ_f may be made rather short. The internal loop may be expressed as follows:

$$\frac{E}{F} = \frac{-m_p k_{il}/s}{1 - (bk_{il}/s)} = \frac{-m_p k_{il}}{-bk_{il} + s} \tag{11.41}$$

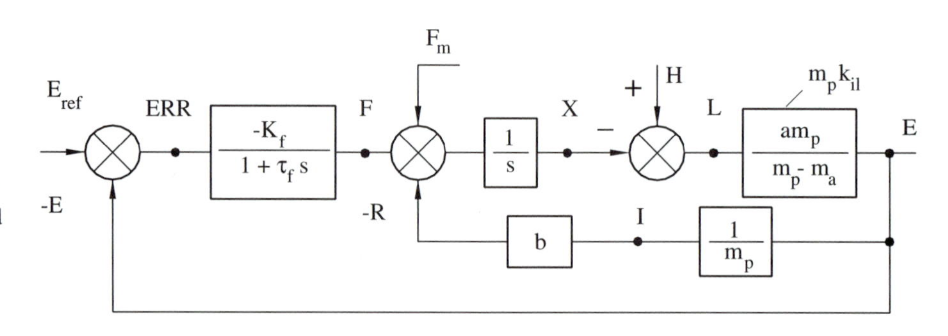

Figure 11.73
Block diagram of the servo control of the SAW process. It includes a feed-drive servomotor with gain K_f and time constant τ_f. The desired (reference) arc voltage is used as input. A second-order system is obtained.

Referring to Eq. (11.32), $bk_{il} = -1/\tau_a$, where now we add the subscript a denoting the internal time constant of the arc in order to distinguish it from the feed-drive time constant τ_f. Then we have

$$\frac{E}{F} = \frac{m_p}{b}\frac{1}{1 - (s/bk_{il})} = \frac{m_p}{b}\frac{1}{1 + \tau_a s} \tag{11.42}$$

and

$$\frac{E}{E_{\text{ref}}} = \frac{-\dfrac{K_f}{1 + \tau_f s}\dfrac{m_p}{b}\dfrac{1}{1 + \tau_a s}}{1 - \dfrac{K_f m_p}{b(1 + \tau_f s)(1 + \tau_a s)}} = \frac{-K_f m_p}{b(1 + \tau_f s)(1 + \tau_a s) - K_f m_p}$$

$$= \frac{-K_f m_p/b}{\tau_a \tau_f s^2 + (\tau_a + \tau_f)s + 1 - (K_f m_p/b)} \tag{11.43}$$

$$= \frac{K(1 - K_f m_p/b)/\tau_a \tau_f}{s^2 + (1/\tau_f + 1/\tau_a)s + (1 - K_f m_p/b)/\tau_a \tau_f}$$

This is a second-order response with a steady state gain K

$$K = \left(\frac{E}{E_{\text{ref}}}\right)_{ss} = -\frac{K_f m_p}{b - K_f m_p} \tag{11.44}$$

EXAMPLE 11.3 Servo Control of SAW ▼

Choose the time constant of the feed drive motor. Determine the natural frequency, damping ratio, and corresponding gain of the system, as well as the actual steady-state values of f, i, and e. Determine the effects of an input of Δf. So, for example, for the "const i" power source and a d = 4.8 mm wire, $m_p = -0.25$, $b = 0.04$, $k_{il} = -10.75$, $\tau_a = -1/bk_{il} = 2.33$ sec, and we choose $\tau_f = 0.020$. Then, the natural frequency of the system is

$$\omega_n^2 = \frac{1 - (K_f m_p/b)}{\tau_a \tau_f} = \frac{1 + 6.25\,K_f}{0.0466} \tag{11.45}$$

and

$$\zeta = \frac{1/\tau_f + 1/\tau_a}{2\omega_n} = \frac{50.043}{2 \times 4.632\sqrt{1 + 6.25\,K_f}} \tag{11.46}$$

Let us try to maintain a reasonable damping ratio and choose $\zeta = 0.7$. This leads to a choice of $K_f = 9.37$; let's choose $K_f = 10$ and get $\zeta = 0.678$. This also leads to $\omega_n = 36.91$, and $f_n = 5.875$ Hz. The steady-state gain becomes $K = 0.984$. The steady-state voltage adjusts to 98.4% of the reference voltage. All the values can easily be adjusted to suit some selected mean working state. Remembering that all our variables are deviations from such a mean state, we may decide, for example, that our mean working voltage, irrespective of the current used, will be $e_{\text{mean}} = 30$ V and choose the initial state such that $f = 24$ mm/sec and, correspondingly, $i = 600$ A. To satisfy the steady-state relationships expressed by the diagram of Fig. 11.72, and accepting $K_f = 10$ mm/sec/V, we will write, for the full values of variables (not the deviations): $e_{ref} = 30$ V; $err_{ss} = e_{ref} - e = (1 - 0.984)\ e_{ref} = 0.48$ V; $e_{ss} = 29.52$. Then we must accept: $f = f_o - [K_f/(1 + \tau_f s)] \times err$, and determine the bias f_o from the steady-state values:

$$f_{ss} = f_o - K_f\, err$$

$$24 = f_o - 10 \times 0.48$$

$$f_o = 28.8 \text{ mm/sec}$$

and

$$i = i_o + \frac{e}{m_p} = i_o - \frac{0.984\, e_{ref}}{0.25}$$

$$i_o = 600 + 3.936 \times 30 = 718 \text{ A}$$

For another choice, $K_o = 20$, we would have $f_o = 33.6$ mm/sec. Furthermore, we have

$$l_{ss} = \frac{m_p - m_a}{m_p a} e_{ss} = 0.372 \times 29.52 = 10.98 \text{ mm}$$

and choosing $h_o = 20$ mm, we have $x_{ss} = 9.02$ mm. ▲

The Effect of Input Δf

Now, we can investigate the effects of the input of a large step Δi leading to a new commanded current; see Fig. 11.74a. Here no change in reference voltage is considered, as the reference is kept unchanged during the change of the power-supply characteristic. First the internal loop:

$$\frac{I}{R} = \frac{-k_{il}/s}{m_p K_f k_{il}/(1 + \tau_f s)s} = \frac{-k_{il}(1 + \tau_f s)}{\tau_f s^2 + s + m_p K_f k_{il}}$$

and

$$\frac{I}{\Delta I} = \frac{\dfrac{bk_{il}(1 + \tau_f s)}{\tau_f s^2 + s + m_p K_f k_{il}}}{1 - \dfrac{bk_{il}(1 + \tau_f s)}{\tau_f s^2 + s + m_p K_f k_{il}}}$$

$$= \frac{bk_{il}(1 + \tau_f s)}{\tau_f s^2 + s + m_p K_f k_{il} - bk_{il} - bk_{il}\tau_f s} \tag{11.47}$$

$$= \frac{-(1 + \tau_f s)}{\tau_a \tau_f s^2 + (\tau_a + \tau_f)s + 1 - (m_p K_f/b)}$$

This result must be understood so that i is initially the original mean current. In our notation it is pursued as an increment, or deviation. Thus, originally $i = 0$, and we expect it to change very little. On the contrary, Δi is a large current input change. All the other parameters, such as e, l, and x, are also meant as deviations from the original mean values.

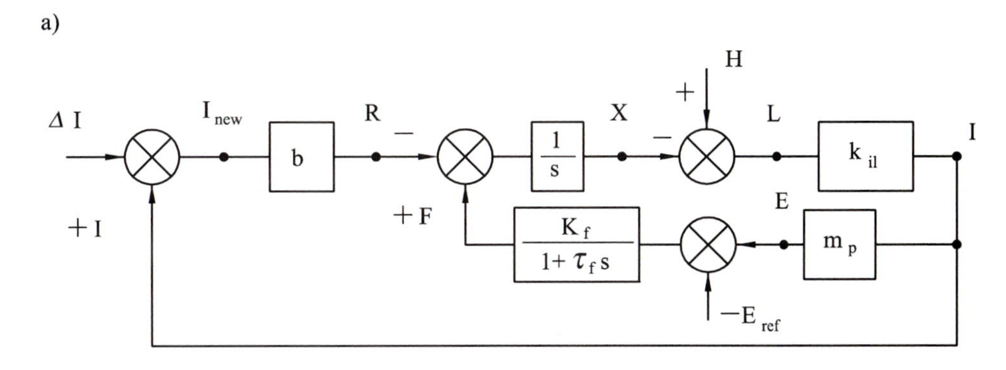

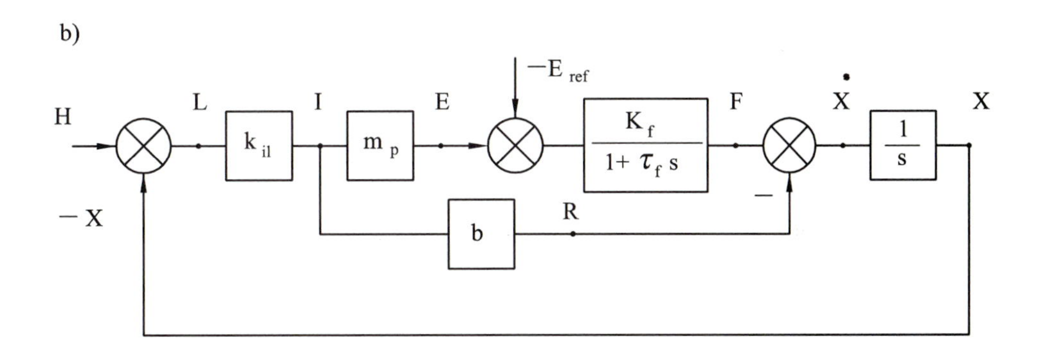

Figure 11.74
a) Block diagram of SAW servo control with current change as input; b) block diagram of SAW servo control with gap change as input. All variables are taken as deviations from mean initial values.

Interpreting and evaluating Eq. (11.47): Using $m_p = -0.25$, $b = 0.04$, $K_f = 10$, $t_f = 0.02$, $\tau_a = 2.33$, and further modifying, we have

$$\frac{I}{\Delta I} = \frac{-(1 + \tau_f s)}{1 - \dfrac{m_p K_f}{b}} \frac{\left(1 - \dfrac{m_p K_f}{b}\right) / \tau_a \tau_f}{s^2 + \left(\dfrac{1}{\tau_a} + \dfrac{1}{\tau_f}\right) s + \left(1 - \dfrac{m_p K_f}{b}\right) / \tau_a \tau_f} \tag{11.48}$$

The effects of the "disturbance" Δi can be determined according to the block diagram in Fig. 11.74a. In this diagram all variables except Δi are considered as deviations from a mean steady state as it existed before this significant current change. In this sense we may simply consider the original (before the introduction of Δi) variables to be as follows: $i_m = 0$, $r_m = 0$, $h_m = 0$, $l_m = 0$, $x_m = 0$, $e_m = 0$; or else we may choose some characteristic values like $i_m = 600$ A, $r_m = b \times i_m = 24$ mm/sec, $e_m = 30$ V, $l_m = 10$ mm, $x_m = 20$ mm, $h_m = 30$ mm, and the variables described by the relationships of Fig. 11.74a are then added to these mean values.

Eq. (11.48) indicates that current i (deviation of the current from i_m) and voltage e (deviation from e_m, etc.) respond to Δi in a second-order transient, for which the natural frequency and damping have already been determined above by Eqs. (11.45) and (11.46), and then they settle in the following steady states:

$$i_{ss} = \frac{-1}{1 - (m_p K_f / b)} \Delta i = \frac{-1}{1 + (0.25 \times 10/0.04)} \Delta i = -0.0157\, \Delta i \tag{11.49}$$

$$e_{ss} = m_p i_{ss} = -0.25 i_{ss} = 0.0039\ \Delta i \tag{11.50}$$

$$r_{ss} = b(\Delta i + i_{ss}) = 0.9843b\ \Delta i = 0.0039\ \Delta i \tag{11.51}$$

For example, if a change of the deposition rate, that is, a change of the burn-off rate r equivalent to a change of the feed rate of 6 mm/sec is required (25% of $f_m = 24$ mm/sec), one has to rely on the knowledge of the coefficient b for the given process and electrode, for example, $b = 0.04$, and command a current change of $\Delta i = 6/0.04 = 150$ A. Assuming then that the original mean current was $i_m = 600$ A, the new current is

$$i_{new} = i_m + \Delta i - 0.0157\ \Delta i = 600 + 150 - 2.35$$

$$= 600 + 147.65 = 747.65\ \text{A}$$

With reference to the previous i_m, the change is 147.65 A. The feed rate change will be $f_{ss} = r_{ss} = bi_{\text{new}} = 5.91$ mm/sec. The voltage will change to $e_{ss} = 0.0039\ \Delta i = 0.585$ V, and the arc length will change by $e/a = 0.585/3 = 0.195$ mm.

It is now obvious that the servocontrol based on the comparison of the voltage e on the arc with a reference voltage (here $e_{\text{ref}} = 0$, as a deviation) keeps the arc on almost the same arc length and voltage, while the feed rate and deposition rate change as commanded.

The servocontrol behaves very well also in response to a change in the gap, which might be caused by an unevenness of the welded surface. This will be shown next. Using the block diagram in Fig. 11.74b, the transfer function *X/H* is obtained:

$$\frac{X}{H} = \frac{\dfrac{k_{il}}{s}\left(\dfrac{K_f m_p}{1 + \tau_f s} - b\right)}{1 + \dfrac{k_{il}}{s}\left(\dfrac{K_f m_p}{1 + \tau_f s} - b\right)}$$

$$= \frac{\dfrac{k_{il} K_f m_p - k_{il} b(1 + \tau_f s)}{s(1 + \tau_f s)}}{1 + \dfrac{k_{il} K_f m_p - k_{il} b(1 + \tau_f s)}{s(1 + \tau_f s)}}$$

$$= \frac{k_{il} K_f m_p - k_{il} b(1 + \tau_f s)}{s(1 + \tau_f s) + k_{il} K_f m_p - k_{il} b(1 + \tau_f s)}$$

$$= \frac{\left(1 - \dfrac{K_f m_p}{b}\right) + \tau_f s}{\tau_a \tau_f s^2 + (\tau_a + \tau_f) s + 1 - \dfrac{K_f m_p}{b}}$$

$$\frac{1 - \dfrac{K_f m_p}{b}}{\tau_a \tau_f s^2 + (\tau_a + \tau_f) s + 1 - \dfrac{K_f m_p}{b}} \left(1 + \frac{\tau_f}{\left(1 - \dfrac{K_f m_p}{b}\right)} s\right) \tag{11.52}$$

Eq. (11.52) shows that x and l go through a second-order transient with f_n and ζ as derived earlier, with a small lead term, and the steady-state $x_{ss} = \Delta h$ and $l_{ss} = 0$.

The interesting question is now whether the current swing in this transient actually remains low as we expected once we have chosen a process with a "const i" power source.

EXAMPLE 11.4 Servo control of SAW: Effect of a change of the gap Δh ▼

Determine transient and steady-state changes of x, l, and i. Obviously, $i = k_{il} \times l$. It will first shoot down by $\Delta i = k_{il} \times \Delta h$ and then, soon settle down to $i_{ss} = 0$. For the parameters as chosen in Example 11.3, $k_{il} = -10.75$, and we choose $\Delta h = 2$ mm, and $\Delta i = -21.5$ A. Eq. (11.52) then becomes

$$\frac{X}{H} = \frac{1363}{s^2 + 2\zeta\omega_n + \omega_n^2}(1 + 0.000315s) \tag{11.53}$$

$$\omega_n = 36.9, \qquad \zeta = 0.678$$

The initial current decrease is $\Delta i = -21.5$A, which is negligible; assuming $i_m = 600$A, this is only 3.6%; see Fig. 11.75. In this illustration a system with a rather low damping of $\zeta = 0.22$ is assumed, leading to large initial overshoots. ▲

11.7.4 Time Domain Simulation

Now we will work through an example case using the time-domain simulation, which derives the variations of the individual parameters by proceeding in small time steps. In all the preceding discussions we have dealt with the relationships between the deviations of the variables. Thus we had $\Delta i = \Delta e / m_p$, $\Delta e = m_a \times \Delta i$. We will not deal with deviations of the variables anymore, but with their absolute values. Let us recapitulate the relationships to use.

From the power-source characteristic [see Fig. 11.76 and Eq. (11.15)]:

$$e = m_p (i - i_o) \tag{11.54a}$$

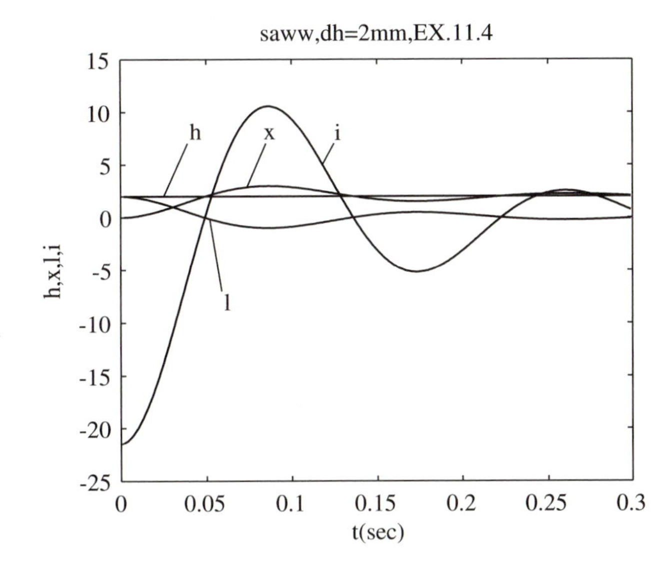

Figure 11.75
Servo control of SAW. Response of deviations of variables of the process to a step change of gap $h = 2$ mm. Arc length l increases stepwise by 2 mm, but it then regulates to the original length, and this is compensated by the increase of the extension x by 2 mm. The current initially drops by 21.5 A and, after an oscillatory transient it settles at the initial level.

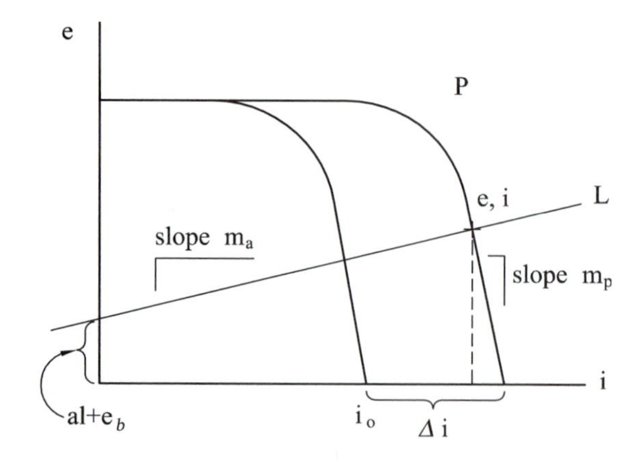

Figure 11.76 Servo control of SAW. Power and arc characteristics.

From the arc characteristic

$$e = a \times l + e_b + m_a \times i \tag{11.54b}$$

where e_b (the base voltage) determines the vertical location of the L characteristic in the graph, which then moves up or down with the changes in arc length. For the burn-off rate, we will use direct proportionality:

$$r = b \times i \tag{11.55}$$

and the feed rate will result from the servodrive

$$\tau_f \dot{f} + f = K_f(e - e_{\text{ref}}) + f_m \tag{11.56}$$

where f_m is the bias velocity for $(e - e_{\text{ref}}) = 0$.

EXAMPLE 11.5 Servo Control of SAW

Use simulation to determine the effect of $\Delta i = 200$ A and also of $\Delta h = 2$ mm on i, e, f, and l.

For our example we choose the following parameters: $a = 3$ V/mm, $m_a = 0.03$ V/A, $m_p = -0.3$ V/A, $b = 0.04$ mm/sec/A; for the servodrive we will have $K_f = 10$ mm/sec/V, and $\tau_f = 0.02$ sec.

Further, let us choose the initial values of the variables so as to start with a fully balanced state in which $e = e_{\text{ref}} = 30$ V. The basic P characteristic is anchored at $i_o = 800$ A; correspondingly [see Eq. (11.54a)], $i = 800 - 30/0.3 = 700$ A, and the burn-off rate is $r = b \times i = 0.04 \times 700 = 28$ mm/sec. For balance, the feed rate (the bias) is also $f_m = 28$ mm/sec.

Obviously, in this situation [see Eq (11.56)], $\dot{f} = 0$. We then introduce a step $\Delta i = 200$ A. The relationships are expressed in the block diagram of Fig. 11.77. For the arc, choose $h_o = 25$ mm, and the value of $e_b = -27$V; l_o will result from Eq. (11.54b) as $l_o = 12$ mm and $x_o = 13$ mm. Choose d$t = 0.001$ sec, total simulation time 0.4 sec and plot i, e, f, and l versus time. The listing of the simulation program written in Matlab is in Fig. 11.78.

The results of an input of $\Delta i = 200$ A are presented in the graphs of Fig. 11.79. First, let us look at graphs *(a)* and *(b)* of i and e versus time t. The current i starts with the value of 700 A and immediately, upon the switch to the *P3* char-

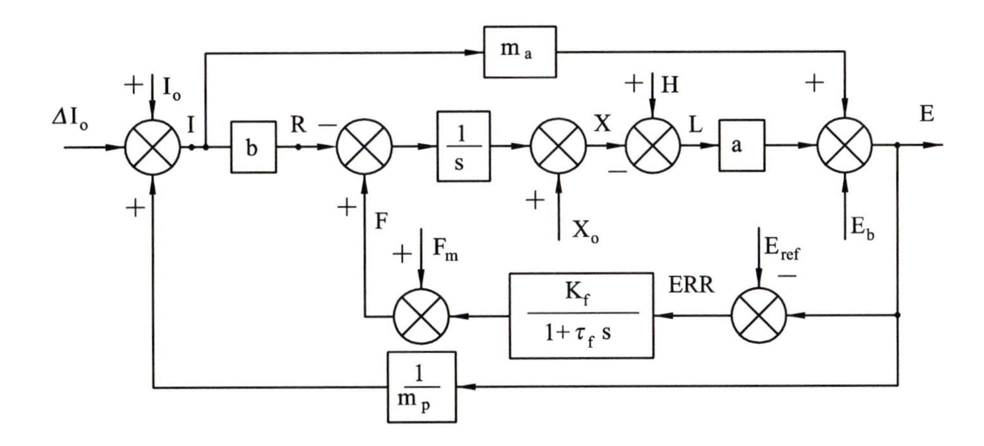

Figure 11.77
Servo control of SAW. The block diagram with full values of variables (not just deviations).

Figure 11.78
Matlab listing of the SAW servocontrol simulation program "weld11."

```
function[ee,ie,fe,le,te]=weld11(dli,dlh)
a=3;ma=0.03;fm=28;e=30;mp=-0.3;b=0.04;dt=0.00025;eb=-27;
xm=13;hm=25;x=xm;f=fm;taf=0.02;i0=800;eref=30;Kf=10;
%the initial (mean) values of the variables are xm,hm,fm,
%and the bias voltage is eb. taf is time constant and Kf the gain
%of the servomotor,i0 is the base current for the power
%supply characteristic.
%The following equations follow the block diagram Fig.11.77
%and equations (11.54-11.56)
for n=1:1600
i=i0+e/mp+dli;
r=b*i;
dx=f-r;
x=x+dx*dt;
l=hm+dlh-x;
e=a*l+ma*i+eb;
err=e-eref;
df=(err*Kf+fm-f)/taf;
f=f+df*dt;
t=(n-1)*dt;
%for plotting, variables are renamed and subscripted
ee(n)=e;
ie(n)=i;
fe(n)=f;
le(n)=l;te(n)=t;
end
%plot(te,ee),(te,ie),(te,fe),(te,le)
```

acteristic, which adds 200 A at $t = 0$, rises to $i = 900$ A. However, for the initial length of arc $l = 12$ mm, this gives rise to the voltage e to 35.4 V, upon which the current drops to 882 A (by about 2%).

Then the system starts to regulate, and it reaches the new steady state in about 0.15 sec. In the final state the current is 897 A (−3 A, that is, 0.3% below target), and the voltage is 30.8 V. Looking at graph *(c)*, we see that the feed rate temporarily increased from the initial $f_o = 28$ mm/sec to about 66 mm/sec in order to quickly increase the extension x and reduce the arc length, graph *(d)*. Within about 0.1 sec the feed rate dropped to almost the final value, which amounts to 35.9 mm/sec. The arc length regulated quickly down and reached the final value of 10.3 mm.

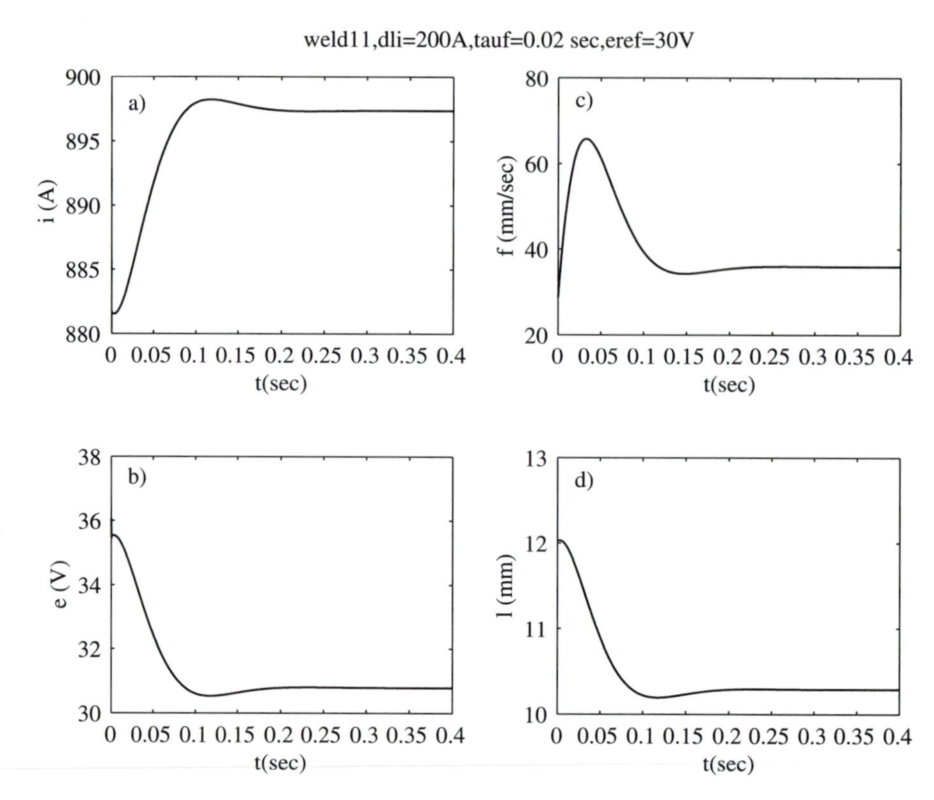

Figure 11.79
Servo control of SAW. Example 11.5. Response to a current step change by 200 A, by switching over of the *P* characteristic. Corresponding changes are shown for a) current, b) voltage, c) feed rate, d) arc length. All the parameters respond to the initial sudden rise of the current from 700 A to 900 A; the voltage changes, and current drops to 882 A. After the ensuing transient the current settles at 897 A, and the feed rate correspondingly increases from the initial 28 mm/sec to the final 35.9 mm/sec.

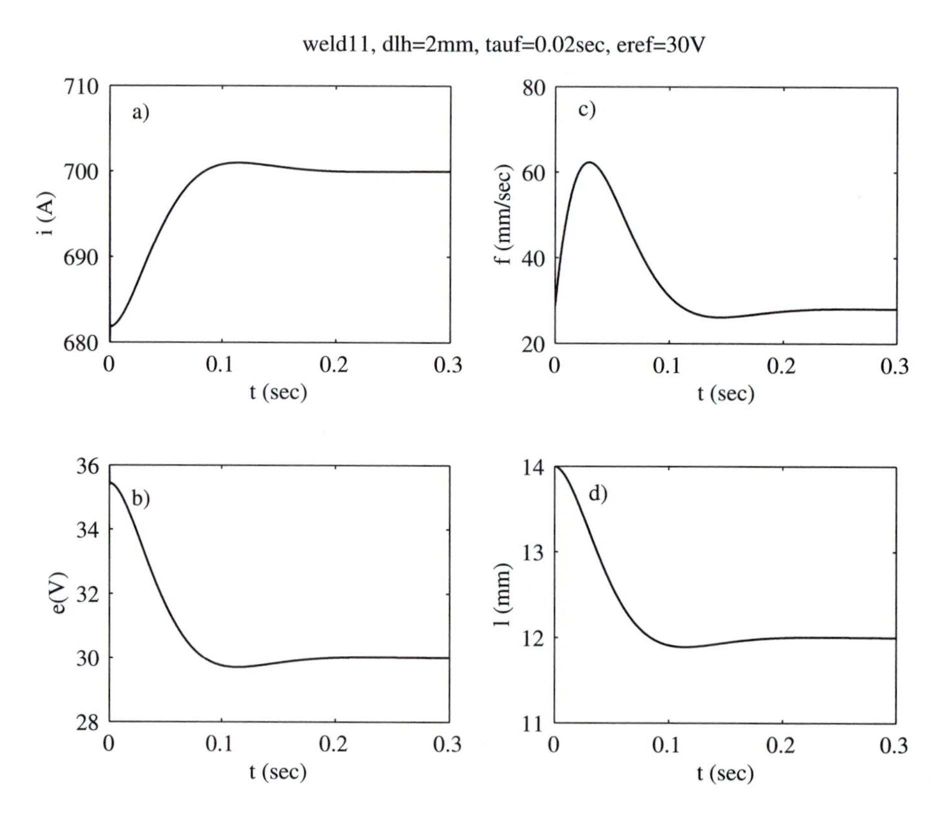

Figure 11.80
Servo control of SAW. Example 11.5. Response to a step change of gap *h* by 2 mm; a) change of current: it first drops suddenly to from 700 A to 682 A and then regulates back; b) change of voltage: it rises first to 35.5 V and then regulates back; c) change of feed rate; over 0.2 sec there is a transient; d) change of arc length is similar to that of the voltage. After the transient, the arc characteristics return to the initial state.

The effect of Δh is illustrated in Fig. 11.80. The effects as plotted in these graphs and as described in the following correspond to those derived in Fig. 11.75 with one main difference due to the difference in damping ratios: $\zeta = 0.22$ for Fig. 11.75, and it is 0.68 now. The variables *l*, *e*, and *i* are initially immediately disturbed: current *i* drops from 700 A to 682 A, voltage rises from 30 V to 35.5 V, and the arc length increases by the value of the step $\Delta h = 2$ mm. The feed rate increases temporarily to increase the extension *x*. All these initial disturbances are rather small. After about 0.2 sec the whole process (*i, e, l*) returns to the initial values. The system regulated the gap change out, quickly and completely.

These illustrations show that the servo control of SAW works fast without introducing large disturbances. ▲

11.8 HEAT TRANSFER IN ARC WELDING

11.8.1 Continuous Field Solution: Thick Plate Formulation

The general statement for a 3D heat transfer is based on Fig. 11.81. For one coordinate, *x*, the heat input into an element with sides $(\delta_x, \delta_y, \delta_z)$ is

$$dQ_x - dQ_{x+\delta x} = k\frac{\partial^2 T}{\partial x^2}\delta_x\,\delta_y\,\delta_z \tag{11.57}$$

The heat balance for the 3D element with unit sides is obtained as follows:

$$k\left(\frac{\partial^2 T}{\partial x^2}+\frac{\partial^2 T}{\partial y^2}+\frac{\partial^2 T}{\partial x^2}\right)+Q' = c\rho\frac{\partial T}{\partial t} \tag{11.58}$$

where Q' is the heat generated within the element, if any, k (N/(sec °C) is thermal conductivity, and ρc (N/mm^2 °C) is specific heat per volume. Introducing thermal diffusivity

$$\alpha = \frac{k}{\rho c} \tag{11.59}$$

Eq. (11.54) is rewritten as

$$\frac{1}{\alpha}\frac{\partial T}{\partial t} = \left(\frac{\partial^2 T}{\partial x^2}+\frac{\partial^2 T}{\partial y^2}+\frac{\partial^2 T}{\partial z^2}\right)+\frac{Q'}{k} \tag{11.60}$$

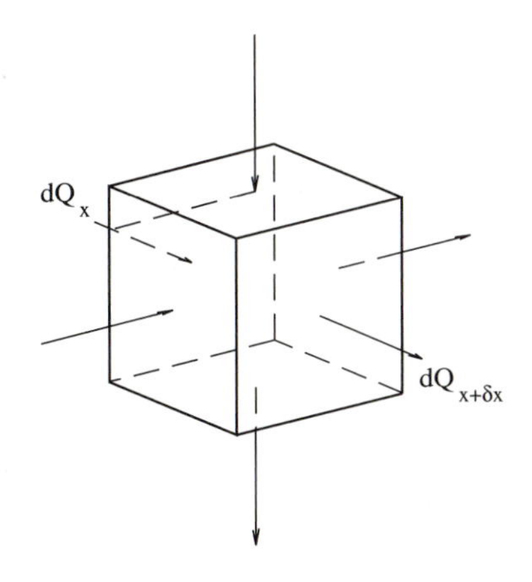

Figure 11.81
The 3D heat-transfer representation.

This is a *transient 3D* heat transfer. Rosenthal [14] has proposed a coordinate transformation that converts it into a *3D steady-state* case; see Fig.11.82a. He assumes a quarter-infinite block (infinite in $+Z$, $+X$, $+Y$, $-Y$) with a *point* heat source A (an idealized concentrated arc with infinite intensity) moving along the X-axis with velocity v. A point P (x, y, z) undergoes a variation of temperature as the arc is approaching, passing, and moving away. However, if the initial part of the arc motion that is close to the $x = 0$ origin is neglected, and the origin of coordinates is located in the arc, the field measured in the new coordinates ξ, y, z becomes stationary. This could be compared to waves on the surface of a lake produced by a moving boat. For an observer on the shore, the field in front of him undergoes a change as the boat is passing. But an observer sitting in the boat sees a field of waves in front, in back, and at the sides of the boat that travels with the boat and does not change. To move the origin of coordinates from point O to point A (the arc), the x coordinate transforms from x to ξ:

$$\xi = x - vt \tag{11.61}$$

and the derivatives transform as follows:

$$\frac{\partial \xi}{\partial x} = 1, \quad \frac{\partial \xi}{\partial t} = -v \tag{11.62}$$

With these transformations, Eq. (11.60) becomes

$$-\frac{v}{\alpha}\frac{\partial T}{\partial \xi} = \left(\frac{\partial^2 T}{\partial \xi^2} + \frac{\partial^2 T}{\partial y^2} + \frac{\partial^2 T}{\partial z^2}\right) + \frac{Q'}{k} \tag{11.63}$$

It may be noticed that now the variable t has been eliminated and Eq. (11.63) describes a *3D steady state.* Rosenthal provided the solution of this equation, both as is, neglecting any heat convection from the surface of the block, as well as including it by adding additional terms in Eq. (11.63). Only the former case will be used.

The solution, which in practice applies reasonably well to *welding a thick plate* (see Fig. 11.82b), is

$$T - T_o = \frac{q'}{2\pi kR}\, e^{-\lambda v(\xi + R)} \tag{11.64}$$

a)

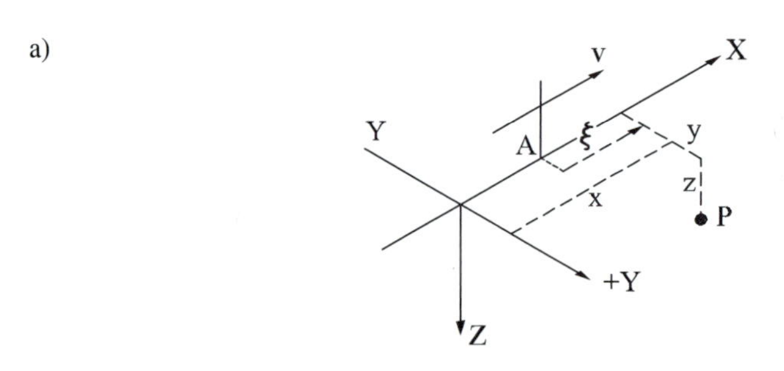

b)

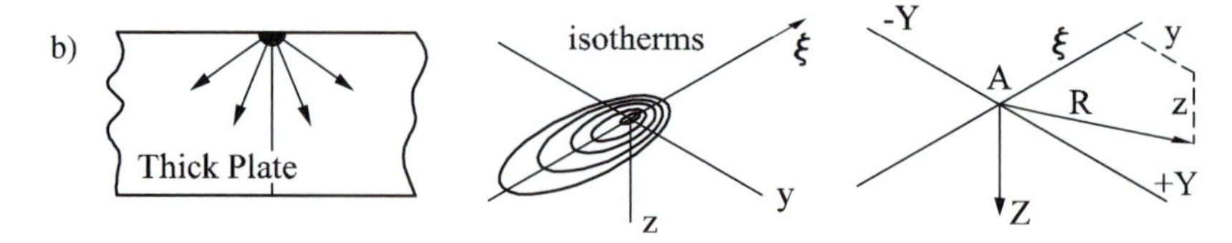

Figure 11.82
a) Coordinate transformation obtained by locating the origin of coordinates in the moving arc changes a 3D transient into a 3D steady-state heat transfer. Far enough from the initial edge of the plate, the field around the moving arc does not change. b) Coordinates of the 3D "thick plate" formulation. The X-axis is the line of the arc travel, the Y-axis is on the surface of the plate and perpendicular to the arc travel, and the Z-axis points into the depth under the arc. Point A is the moving arc.

where T_o is the initial temperature of the plate, q' is the arc power input, and

$$\lambda = \frac{1}{2\alpha}, \qquad R = \sqrt{\xi^2 + y^2 + z^2}, \qquad R > 0 \tag{11.65}$$

Let us investigate the temperature distributions in several sections through the plate. Consider the (Y, Z) plane, passing through the arc, that is, $\xi = 0$. Eq (11.64) simplifies to

$$T - T_o = \frac{q'}{2\pi kR} \, e^{-\lambda v R} \tag{11.66}$$

Next, let us concentrate on the surface of the plate, $z = 0$. Then we have

$$T_a - T_o = \frac{q'}{2\pi ky} \, e^{-\lambda v y}, \qquad y > 0 \tag{11.67}$$

and for $y < 0$, the temperature profile is symmetrical around $y = 0$. From Eq. (11.66) it is obvious that the temperatures in this plane depend on R only; hence, the isotherms are semicircular around point $y = z = 0$; see Fig. 11.83. Their values are found on the surface. The profiles of the surface temperature are plotted versus y, with the T-axis extending upwards from the Z-axis.

EXAMPLE 11.6 Temperature Profile on the *Y*-axis (see Fig. 11.83) ▼

The actual curves in this figure correspond to welding steel with an input of $q' =$ 1 kW = 1e^6 mW and welding velocity $v = 2$ mm/sec. Due to the idealization of the arc as point source with zero area of heat input into the plate at point A, an infinite temperature results at this point. However, we may delineate an area w wide where at $y = w/2$, $T_a = T_m$, the melting temperature of steel, $T_m = 1480$°C. In practice, due to fast streams in the weld pool, the temperature over this surface may be taken as constant. The cross-hatched area enclosed by the 1480°C isotherm indicates the cross-section of the weld pool.

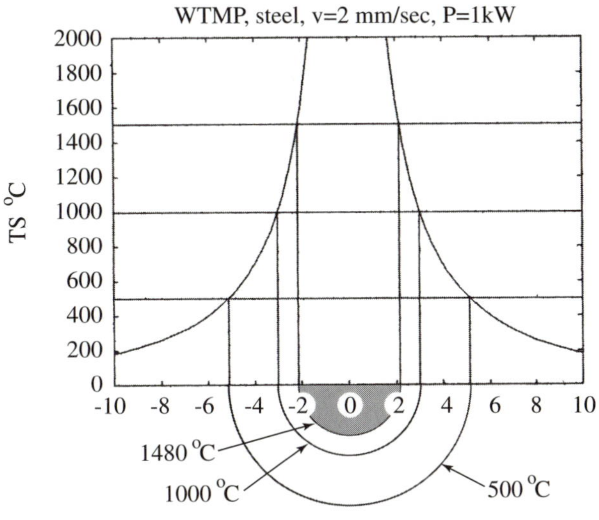

Figure 11.83
Results of Example 11.6. Temperatures T_a along the Y-axis of the work and isotherms in the YZ plane for carbon steel, 1000 W heat input and welding velocity 2 mm/sec. At melting temperature of 1480°C the molten pool extends ± 2.15 mm sideways; the weld bead is 4.3 mm wide. The vertical scale of the graph may be changed in proportion to heat input.

Assuming room temperature $T_o = 20°C$ and melting temperature $T_m = 1480°C$, the material inside of this isotherm would be liquid (neglecting the phase transformation heat) and, assuming the liquid flowing inside the pool, leading to very good heat conduction, we may arrive at a flat profile within $|y| < 2.15$ mm.

Assuming no heat loss by convection, the heat flow across any surface around the moving source (in steady state) is constant and equal to q'. So, whether the profile within $y = 2.15$ mm is flat or not, with the given gradients at this boundary, the heat flow is 1000 W. Thus we actually need more than $P = 1000$ W to generate this situation. The 1000 W is escaping into the surrounding material and maintains its temperature profile as shown and the power of $P_m = A_m\, v\, Q_m$ [$Q_m = 10.24$ J/mm^3 · sec; Eq. (11.3)] is needed to keep on melting the metal in the pool. In the given case the efficiency η of the heat expenditure is

$$\eta = \frac{P_m}{P_m + P} = \frac{141\text{ W}}{1141\text{ W}} = 0.0124$$

This is an approximate analysis, and it will be done in more detail later when the 2D case is discussed.

The plane (ξ, z) is the plane of symmetry. Therefore the positive axis y goes left and right, identically. In the direction z there is only half space, below the (ξ, y) plane. ▲

EXAMPLE 11.7 Temperature Profiles for Three Different Materials ▼

Here we illustrate the temperature distributions and the various parameters of the thermal field around the arc by considering three different materials as listed in Table 11.2.

The graphs of temperature profiles on the surface along the Y-axis for the three materials are shown in Fig. 11.84 for heat input $P = q' = 1\ kW = 1e^6$ mW and for traversing speed $v = 2$ mm/s. The melting temperatures are entered in the graph as well as the corresponding widths w of the weld. Note that, overall, for any given sideways distance from the arc, the temperatures are highest for stainless steel *SS* and lowest for aluminum *Al,* with those for carbon steel *CS* in between. Obviously, the most decisive parameter here is the thermal conductivity k, which is lowest for *SS* and highest for *Al,* or else the thermal diffusivity α. It should also be realized that the vertical scale is proportional; that is, the temperatures at any y are proportional to the power input, as is obvious from Eq. (11.67).

TABLE 11.2 Thermal Parameters of Three Materials

Material	α (mm^2/sec)	k (N/sec · °C)	ρc (N/mm^2°C)	λ (sec/mm^2)	T_m (°C)
Carbon steel	12	43	3.7	0.042	1480
Stainless steel	4.4	15	3.6	0.11	1425
Aluminum	60	140	2.3	0.008	546

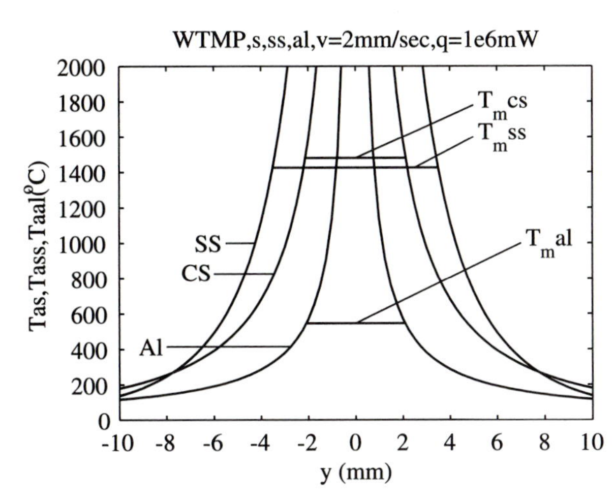

Figure 11.84
Results of Example 11.7. Temperatures T_a along the Y-axis of the "thick" work for carbon steel, stainless steel, and aluminum, for 1000 W heat input and $v = 2$ mm/sec. For any width, temperatures are highest for stainless steel and lowest for aluminum, due to increasing thermal diffusivity of the materials in that order.

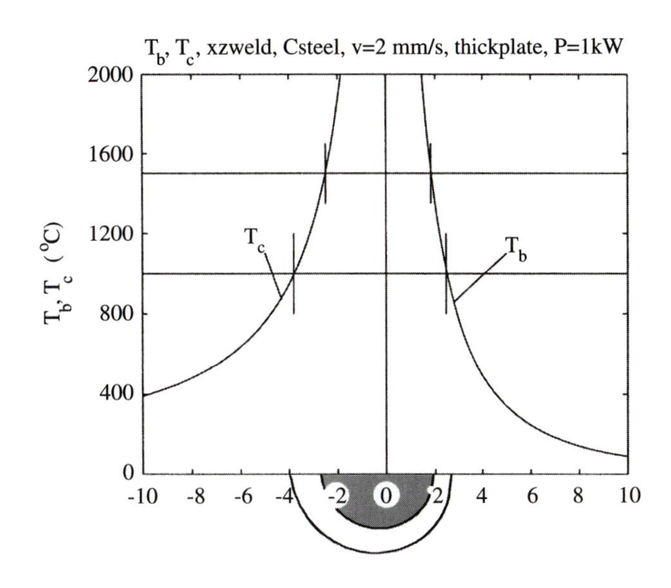

Figure 11.85
Temperatures T_b and T_c along the X-axis of the "thick work" for carbon steel, $P = 1000$ W, and $v = 2$ mm/sec. In front of the arc temperatures drop steeply and much less so behind the arc.

The temperature profiles in the (ξ, z) plane are governed by the following equation:

$$T - T_o = \frac{q'}{2\pi k R} e^{-\lambda v(\xi + R)} \tag{11.68}$$

which, on the surface of the plate, for $z = 0$ reduces to

$$T_b - T_o = \frac{q'}{2\pi k \xi} e^{-2\lambda v \xi} \tag{11.69}$$

for $\xi > 0$ (in front of the source on the welding line). For $\xi < 0$ (behind the source on the welding line) we have

$$T_c - T_o = \frac{-q'}{2\pi k \xi} \tag{11.70}$$

These profiles and also the isotherms corresponding to Eq. (11.68) are shown in Fig. 11.85. Note that the gradients are steeper in front of the source than behind the source. The isotherms into the plate are not circular in this section, and they are elongated in the $-\xi$ direction. All the gradients become steeper with an increase in the traversing speed v. This is illustrated in Fig. 11.86 for welding stainless steel with $v = 1$, 2 and 3 mm/sec. With $v = 3$ mm, the heat can hardly propagate in the direction of the arc motion, as it is caught up by the moving arc. In all the directions there is less spread of heat with increasing speed v, except for $-\xi$, behind the arc, where it is independent of v. The efficiency of the use of the

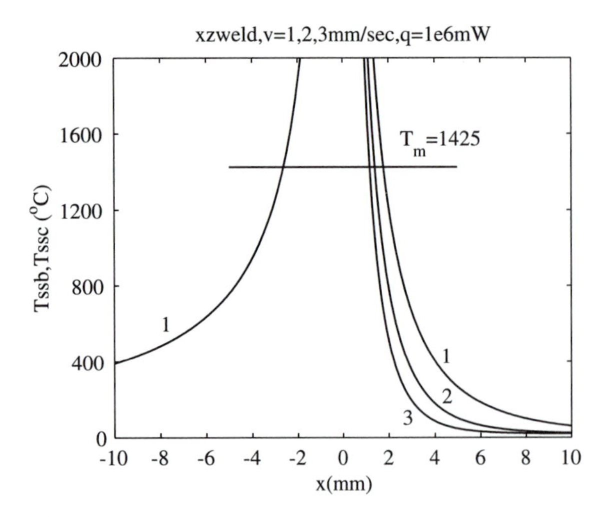

Figure 11.86
Effect of welding speed v on the T_b and T_c temperatures in welding stainless steel. The gradient in front of the arc gets steeper with welding speed. It is not affected behind the arc.

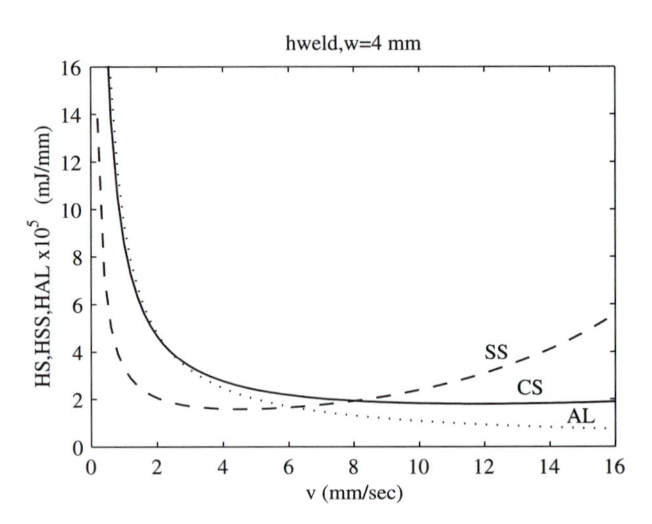

Figure 11.87
Heat H per unit length of weld as function of welding speed v for the "thick plate" case for three materials. For carbon steel it drops fast for speeds between 1 and 3 mm/sec and then the drop flattens out. For aluminum and lower speeds it is almost identical with that for carbon steel, but it keeps decreasing into the higher speed ranges. For stainless steel it reaches a minimum at about 3 mm/sec and then it increases with increasing speed. This is explained by the heat not being able to spread ahead of the arc faster than the motion of the arc.

heat input improves with the speed v. This is best expressed by the value of the specific heat input per unit length of weld,

$$H = P/v \tag{11.71}$$

P (mW), v (mm/sec), H (mJ/mm = J/m). Considering Eq (11.67), we have

$$H = \frac{q'}{v} = (T_a - T_o)\,(2\pi ky/v)e^{\lambda vy} \tag{11.72}$$ ▲

EXAMPLE 11.8 Heat Input Versus Welding Speed for Three Materials ▼

Choosing a particular weld width $w = 2y = 4$ mm for which $T_a = T_m$, Eq. (11.72) is graphically expressed in Fig. 11.87. The specific heat H is plotted versus v for the three materials of Table 11.2. Note that, within the range of speeds used in the graph, for carbon steel CS the value of H decreases strongly for speeds v up to about 5 mm/sec and then it flattens out. For aluminum Al the decrease is strong enough at least up to v = 10 mm/sec. Otherwise these two dissimilar materials have almost the same H. Although the melting temperature of Al is only about a third of that for steel, the heat spreads out much faster in Al because of the thermal diffusivity, which is about five times higher than for carbon steel. Therefore, as the heat conducts away rapidly, more of it must be put in to provide the melting power. The stainless steel needs much less heat input per unit length of weld than both the other materials, for speeds below 2 mm/sec. This is explained by its low thermal diffusivity, four times lower than for carbon steel and fourteen times lower than for aluminum. The value of H for SS bottoms out at $v = 4$ mm/sec, and for higher speeds it actually increases. This is obviously due to the "steep front wall" ahead of the arc, as shown in Fig. 11.86. ▲

11.8.2 Gradients: Cooling Rates

We have seen that the gradient dT/dy is indicative of the power losses. Taking

$$T_a - T_o = \frac{q'}{2\pi kR} e^{-\lambda vR}$$

the gradient is obtained as

$$\frac{dT_a}{dR} = \frac{q'}{2\pi k} \frac{-\lambda vRe^{-\lambda vR} - e^{-\lambda vR}}{R^2}$$

$$= \frac{-q'}{2\pi k} \frac{e^{-\lambda vR}}{R} \frac{1 + \lambda vR}{R} = (T_a - T_o) \frac{1 + \lambda vR}{R} \tag{11.73}$$

Of interest is the gradient at $T_a = T_m$ and at $R = y_m$, if y_m denotes half the width of the melted zone. So, for $T_a = 1480°C$, and $R = 2$ mm, and $\lambda = 0.042$ for carbon steel,

$$\frac{dT_a}{dR} = \frac{1460}{2}(1 + 0.084v)$$

So for $v = 2$ mm/sec,

$$\frac{dT_a}{dR} = 853°C/mm$$

and for $v = 4$ mm/sec,

$$\frac{dT_a}{dR} = 975°C/mm$$

The gradient is 1.14 times higher for the twofold increase in speed, but the energy input per unit length is actually only 1.14/2 = 0.57 of that for the lower speed.

The cooling rate is important for the martensite transformation as well as for tempering or otherwise affecting a heat-treated material. The steepest cooling gradient is behind the source, where

$$T_c - T_o = \frac{-q'}{2\pi k\xi}$$

and recalling Eq. (11.62):

$$\frac{dT_c}{dt} = -v\frac{dT_c}{d\xi} = -\frac{vq'}{2\pi k\xi^2} = \frac{v}{\xi}(T_c - T_o) \tag{11.74}$$

The decisive cooling rate is the one at $T_c = 550°C$, that is, at the temperature near the "pearlite nose" on the T-T-T diagram. This temperature is found at a particular distance behind the source, which can be obtained from Eq. (11.74):

$$\xi = \frac{-q'}{2\pi k (T_c - T_o)}$$

So, finally, Eq. (11.74) is transformed into

$$\frac{dT_c}{dt} = -\frac{2v\pi k (T_c - T_o)^2}{q'} \tag{11.75}$$

where the critical case is for $T_c = 550°C$.

In order to slow down (decrease) the cooling rate it is necessary to slow down the traverse speed v, or increase the power input q', or preheat the plate, that is, increase T_o.

11.8.3 The 2D Case: The "Thin Plate" Line Heat Source q'' (see Fig.11.88)

Here q''/g is the power per unit thickness. All our examples will assume $g = 5$ mm. There is no temperature gradient in z. This is a 2D case.

The Rosenthal solution [14] is

$$T - T_o = \frac{q''}{2\pi kg} e^{-\lambda v \xi} K_o(\lambda\, v\, R) \tag{11.76}$$

where K_o is a Bessel function, and $R = \sqrt{\xi^2 + y^2}$.

$$K_o(x) = -\left\{ \ln\left(\frac{x}{2}\right) + 0.57722 \right\} I_o(x) + \frac{x^2}{2^2} + \frac{x^4}{2^2 \times 2^4}\left(1 + \frac{1}{2}\right) + \frac{x^6}{2^2 \times 2^4 \times 2^6}\left(1 + \frac{1}{2} + \frac{1}{3}\right) + \cdots \tag{11.77}$$

and

$$I_o(x) = 1 + \frac{x^2}{2^2} + \frac{x^4}{2^2 \times 2^4} + \frac{x^6}{2^2 \times 2^4 \times 2^6} + \frac{x^8}{2^2 \times 2^4 \times 2^6 \times 2^8} + \cdots \tag{11.78}$$

In general, $K_o(x)$ decreases with x in a way similar to what we have seen in the 3D case.

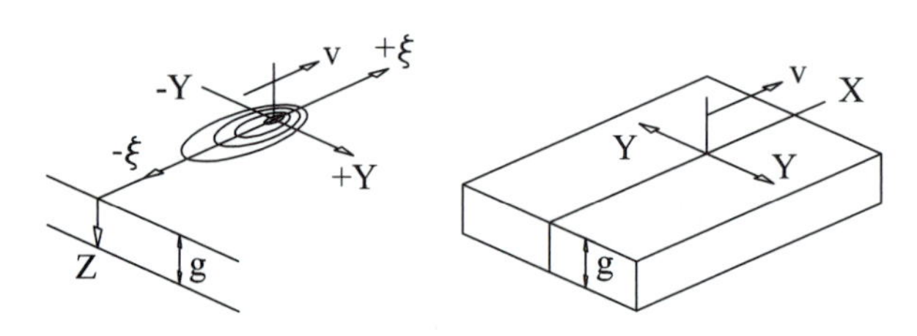

Figure 11.88
Coordinate formulations for the 2D "thin plate." There is no temperature variation in the Z direction.

We will examine now the effect of thermal parameters of the weld material and of the welding speed on the efficiency of welding. In a way analogous to the preceding analysis for thick plate welding, Eq. (11.76) may be interpreted for temperatures along the *Y*- and *X*-axes:

T_a expresses temperatures along the *Y*-axis.

T_b are the temperatures on the weld line in front of the source ($\xi > 0$).

T_c are the temperatures on the weld line behind the source ($\xi < 0$).

EXAMPLE 11.9 Temperature Profiles for the Thin Plate

First, some comparisons are made for an input of $P = 1000$ W $= q''$, and $g = 5$ mm. The temperature profiles of T_a are shown in Fig.11.89 for the three materials of Table 11.2. This graph may be compared with the one for the thick plate in Fig. 11.84. The general trends are the same but the temperature lines cross now at narrower welds. The temperatures T_a, T_b, T_c for carbon steel are plotted in Fig. 11.90. Again, the gradients are steepest along T_b and less steep along T_a and least steep along T_c. ▲

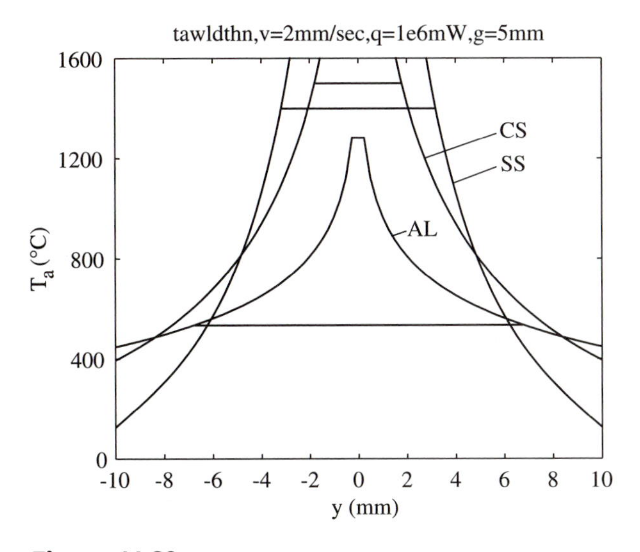

Figure 11.89
Results of Example 11.9. The T_a temperatures along the *Y*-axis in the "thin plate" for steel, stainless steel, and aluminum. Melting temperatures are marked by horizontal lines. For narrower welds stainless steel is hottest, carbon steel is next, and aluminum is coolest. However, at $y = \pm$ 5 to 6 mm the temperatures are almost the same for all three materials.

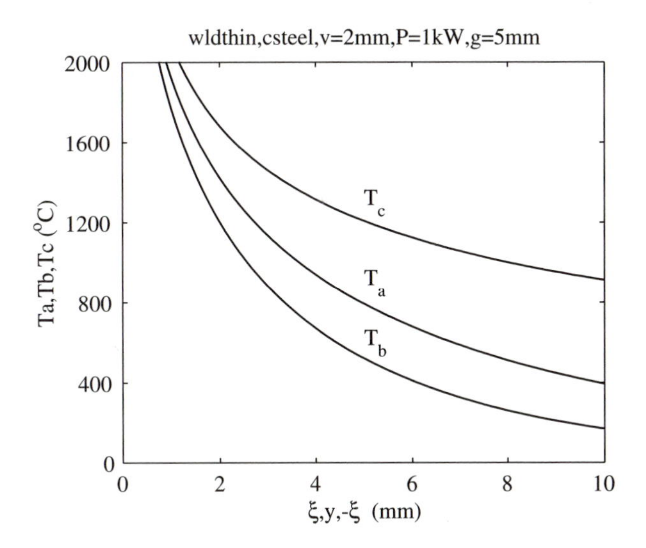

Figure 11.90
Example 11.9. The T_a (along *Y*), T_b (along + *X*), and T_c (along − *X*) temperatures for carbon steel in the "thin plate." The steepest gradient is in front of the arc, next is the one to the sides, and slowest is behind the arc.

EXAMPLE 11.10 Heat Energy per Unit Length Versus Welding Speed for the Thin Plate ▼

The values of specific heat H for the three materials, as functions of the welding speed v, are plotted in Fig. 11.91. This graph is drawn to the same scale as the graph of H versus v for the thick plate in Fig. 11.87, and both of them apply to the same weld width of $w = 4$ mm. However, in the former case the isotherms were circular into the material and then the heat was spreading further into the depth. Now, the surface temperatures apply fully through the thickness $g = 5$ mm of the material and the spread is two-dimensional. Nevertheless, the trends are similar, and the absolute values H do not differ much. For instance, at $v = 4$ mm/sec the ratios of H (thin)/H (thick) are 1.4 for carbon steel, 2.1 for stainless steel, and 0.8 for aluminum. The minimum for stainless steel is found at $v = 2.5$ mm/sec for the thin plate and between 3 and 4 mm/sec for the thick plate. For carbon steel, the minimum is found at $v = 7$ mm/sec for the thin plate; for high speeds, H increases gradually. For the thick plate, for $v > 8$ mm/sec, H is almost constant. For aluminum, in both cases H decreases with increasing v for the whole range of the graphs. In both cases H increases for stainless steel for $v > 5$ mm/sec. For the thin plate it rises steeply after $v = 8$ mm/sec, and at $v = 11.4$ mm/sec it drops to negative values from a rather high peak. This indicates that the heat in the front of the arc cannot spread (due to the low thermal diffusivity fo the material), and it accumulates like a wall; it is impossible to weld at these speeds.

The graph in Fig. 11.91 is significant, as it shows that at speeds of 2 to 4 mm/sec all three materials need approximately the same energy per unit length of the weld. They all need much more at speeds below $v = 1$ mm/sec, and they start to differ strongly for higher welding speeds. ▲

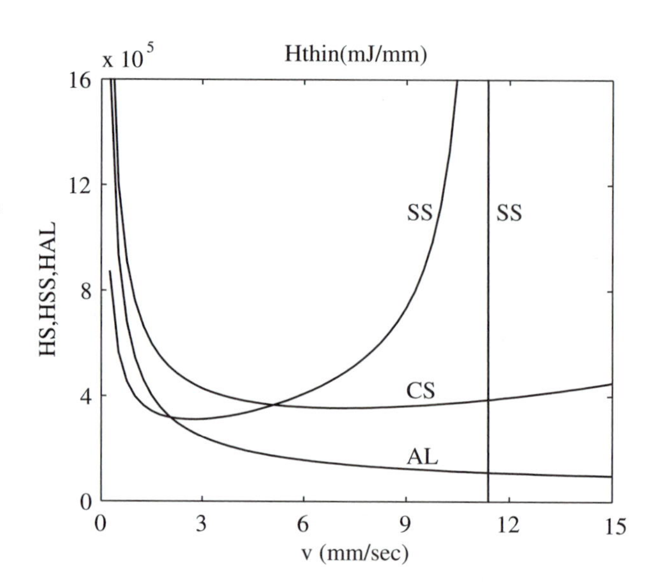

Figure 11.91
Example 11.10. Specific heat H per unit length of weld as a function of welding speed v for the "thin plate" case and the three materials. The trends are similar to those for the thick plate shown in Fig. 11.87. For stainless steel, minimum is at about $v = 2.5$ mm/sec and for about $v > 10$ mm/sec welding is not possible. At between 2 and 3 mm/sec, heat per unit length of weld is about the same for all three materials.

11.8.4 The Finite-Difference Approach: Thin Plate (2D)

The simplifications of the above solutions are due to assuming thermal properties constant, no heat losses by convection (good enough), and point or line sources. We can remove the last aspect and also avoid trying to obtain the above solutions, as well as accept final dimensions of the plates by using the finite-difference approach. The transient formulation would be (see Fig. 11.92)

$$\left(\frac{T_{i-1,j} - T_{i,j}}{\Delta x} - \frac{T_{i,j} - T_{i+1,j}}{\Delta x}\right.$$

$$\left. + \frac{T_{i,j-1} - T_{i,j}}{\Delta y} - \frac{T_{i,j} - T_{i,j+1}}{\Delta y}\right) (\Delta x g)k + Q \tag{11.79}$$

$$= \frac{T_{i,j}^{n+1} - T_{i,j}^{n}}{\Delta t} (\Delta x \Delta y g)\rho c$$

where n is the superscript increasing with time in increments Δt. Also $\Delta x = \Delta y$. The term Q applies for the single element where the heat is input.

Transforming coordinates of a moving source to obtain a quasi-steady state centered around the source amounts to accepting that, as we progress in the direction of motion, step by step, the temperature field remains constant as measured from the source.

Considering Eq. (11.79), this means that if in the xz plane the temperature distribution at instant n is as shown in Fig. 11.92, it will move in our geometric space, forward to $(n + 1)$. The temperature $T_{i'j}^{n}$ will thus become $T_{i'j}^{n+1} = T_{i-1'j}^{n}$ and $\Delta t = \Delta x/v$.

Thus, Eq. (11.79) will become one of steady state:

$$(T_{i-1,j} + T_{i+1,j} + T_{i,j-1} + T_{i,j+1} - 4T_{i,j})gk + P \tag{11.80}$$

$$= (T_{i-1,j} - T_{i,j})\ \Delta y g \rho c v$$

Modifying and acknowledging that $\Delta x = \Delta y$,

$$(T_{i-1,j} + T_{i-1,j} + T_{i,j-1} + T_{i,j+1} + 4T_{i,j})$$

$$+ \frac{P}{gk} = (T_{i-1,j} - T_{i,j}) \frac{\Delta x v}{\alpha}$$

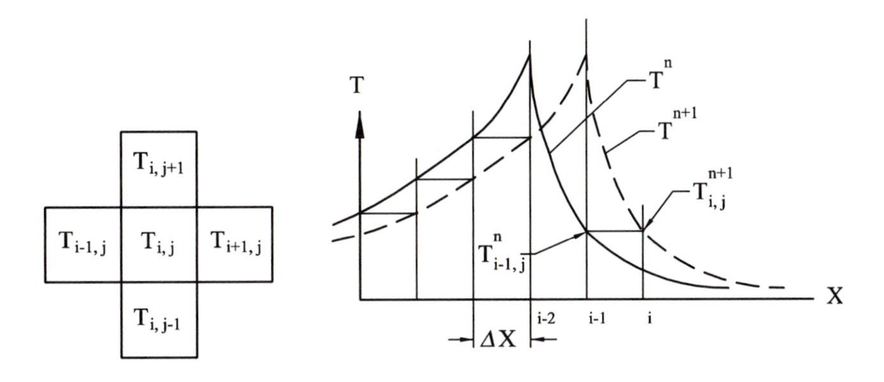

Figure 11.92
Finite-difference elements for a 2D field. Transformation of coordinates from the transient to the steady-state case is shown that is equivalent to the Rosenthal transformation for the continuous field indicated in Fig. 11.82a.

where $\alpha = k/\rho c$; denoting $b = \Delta xv/\alpha$, we have, finally,

$$T_{i-1,j}(1 - B) + T_{i+1,j} + T_{i,j-1} + T_{i,j+1} - T_{i,j}(4 - B) = -\frac{P}{gk} \tag{11.81}$$

where $P = 0$ for all elements except the one with the heat input. This equation applies to the whole field except for the boundaries, at which it has to be modified. Also, it has to be transformed as regards the subscripts to use a single subscript for each temperature by an appropriate choice of node numbering.

The field has $25 \times 100 = 2500$ nodes; see Fig. 11.93. The centerline c represents the line of welding and also the line of symmetry, below which there is a field of temperatures identical with those above it. The size of each element is $\Delta x \times \Delta y = 2.5 \times 2.5$ mm, and the thickness of the plate is $g = 5$ mm. Thus the actual plate is $2 \times 24.5 \times 2.5$ mm $= 122.5$ mm wide and $100 \times 2.5 = 250$ mm long.

It is assumed that there is no heat convection from the surfaces of the plate. The boundaries are so specified that, with the arc presently in the node 1726 where the power P (mW) is injected, on the left of boundary a is the final temperature T_f, and boundary b is still at room temperature, which is set at $T = 0$°C.

The numbering of the elements suggested in Fig. 11.93 helps to rewrite Eq. (11.81) with new subscripts. So, for example if $(i, j) \rightarrow n$, $(i + 1, j) \rightarrow n + 25$, $(i, j + 1) \rightarrow n + 1$.

Equation (11.81) becomes

$$T_{(n-25)}(1 - B) + T_{(n+25)} + T_{n-1} + T_{n+1} - T_n(4 - B) = 0 \tag{11.82}$$

or $-P/gk$ for the element $n = 1726$. At the boundaries shown in Fig. 11.93, Eq. (11.82) is modified.

Boundary ***a:***

We assume that left of this boundary is the final uniform temperature T_f such that the heat input had already spread over the plate:

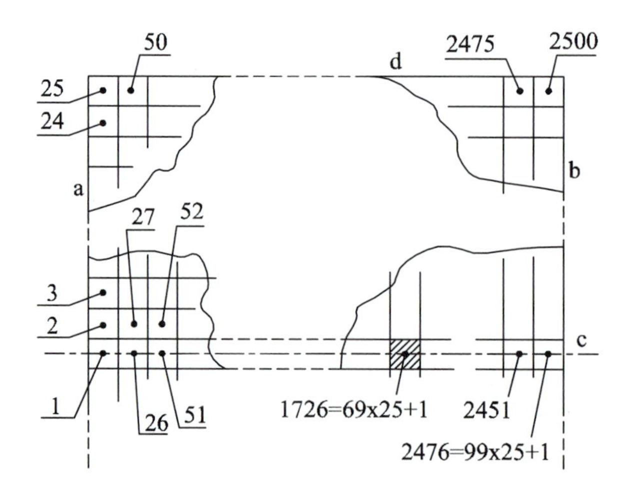

Figure 11.93
The coordinates of the thermal field for the finite-difference computations. Heat input from the arc is in element 1726.

$$T_f = \frac{P(\mathrm{W}) \times 1000}{vwg\rho c} \tag{11.83}$$

where v is traversing speed (mm/sec), w is plate width, $w = 49\ \Delta x$ (mm), and g (mm) is plate thickness.

Thus $T_{n-1} = T_f$, for $n = 1, 25$

Boundary ***b:***

$$T_n = 0,\ n = 2476,\ 2500$$

Boundary ***c:***

The field below this line is symmetrical with the one above it: $T_{n-1} = T_{n+1}$

Boundary ***d:***

No heat flux across the boundary: $T_{n+1} = T_n$. Eq. (11.82) becomes $T_{(n-25)}\,(1 - B) + T_{n+25} + T_{n-1} - T_n\,(3 - B) = 0$. Also, the heat input P is at $n = 1726$.

Solution Procedure

1. Start with an initial estimate: Assume all temperatures at the level of the power source ($n = 1726{,}1750$) and behind it to equal the final temperature; see Eq. (11.83). $T_n = T_f$ for $n = 1$ to 1750, and assume that all temperatures in front of the power source are zero: $T_n = 0$, for $n = 1751$ to 2500. Obtain the starting value of the parameter $TOLD = T_f$.

2. The following procedure will be repeated and new values of T_n will be repeatedly obtained until satisfactory convergence is achieved. This criterion will be based on the convergence of T_{1726}.

2.1 Solve the new temperatures on boundary *a*.

$$T_1 = T_{26} + 2T_2 + T_f(1 - B)/4 - B$$

$$\text{Do, } n = 2, 24\text{: } T_n = T_{n+25} + T_{n-1} + T_{n+1} + T_f(1 - B)/4 - B$$

$$T_{25} = T_{50} + T_{24} + T_f(1 - B)/3 - B$$

2.2 Proceed to the right, line by line, as far as the column with the heat source:

Do 3, $i = 1, 68$
$n = 25i + 1$
$T_n = (T_{n-25}\,(1 - B) + T_{n+25} + 2T_{n+1})/(4 - B)$
Do 2, $j = 2, 24$
$n = 25i + j$

2 $\quad T_n = (T_{n-25}\,(1 - B) + T_{n+25} + T_{n-1} + T_{n+1})/(4 - B)$
$n = 25i + 25$

3 $\quad T_n = (T_{n-25}\,(1 - B) + T_{n+25} + T_{n-1})/(3 - B)$

2.3 The point of welding, the element $n = 1726$. (Alternatively, the heat source could be spread over several elements.)

$$T_{1726} = [T_{1701}\,(1 - B) + T_{1751} + 2T_{1727} + (P(\mathrm{mW})/gk)]/(4 - B)$$

2.4 Upwards in the $i = 69$ column:
Do 4, $j = 2, 24$
$n = 1725 + j$

4 $T_n = (T_{n-25}(1 - B) + T_{n+25} + T_{n-1} + T_{n+1})/(4 - B)$
$T_{1750} = (T_{1725}(1 - B) + T_{1775} + T_{1749})/(3 - B)$

2.5 Right (in front) of the welding point:
Do 6, $i = 70, 98$
$n = 25i + 1$
$T_n = (T_{n-25}(1 - B) + T_{n+25} + 2T_{n+1})/(4 - B)$
Do 5, $j = 2, 24$
$n = 25i + j$

5 $T_n = (T_{n-25}(1 - B) + T_{n+25} + T_{n-1} + T_{n+1})/(4 - B)$
$n = 25i + 25$

6 $T_n = (T_{n-25}(1 - B) + T_{n+25} + T_{n-1})/(3 - B)$

2.6 Boundary *b:*
$T_{2476} = (T_{2451}(1 - B) + 2T_{2477})/(4 - B)$
Do 7, $j = 2, 24$

7 $T_{2475+j} = (T_{2450+j}(1 - B) + T_{2475+j-1} + T_{2475+j+1})/(4 - B)$
$T_{2500} = (T_{2475}(1 - B) + T_{2499})/(3 - B)$

2.7 Test convergence
Error = T_{1726} − TOLD
If Error > Error max
TOLD = T_{1726}
Loop back to 2.1

3. Print all T_n's in a matrix *j,i*.

EXAMPLE 11.11 Using the Finite-Difference Method to Compute Temperature Fields ▼

Compute the temperature field for

1. Steel, $v = 2$ mm/sec, $P = 586{,}000$ mW
2. Stainless steel, $v = 2$ mm/sec, $P = 261{,}000$ mW
3. Aluminum, $v = 5$ mm/sec, $P = 690{,}000$ mW
   $\Delta x = 2.5$ mm, $g = 5$ mm.

The half welding field has $j = 1{:}25$, $i = 1{:}100$.
The welding arc is in element 1726.

	k	ρc	Errormax
Steel	43	3.7	1
Stainless	15	3.6	1
Al	140	2.3	1
	(N/sec °C)	N/(mm^2 °C)	°C

(a) Print the temperatures in the field truncated to tens of degrees C.

(b) Plot the temperatures. ▲

Analysis of the Results of the Computations

The computed fields are more realistic than the analyses presented before. We have done computations at various speeds for the three materials. We can look at the isotherms in the (ξ, y) plane, at the effects of welding speed on P and H, and we can evaluate cooling rates.

The program "tmpfld1" used in the computations is in Fig. 11.94. The temperature profiles over the Y-axis, marked T_a and the T_{bc} over the X-axis are presented in

```
function[t,tas,tass,taal,tbcs,tbcss,tbcal,x,y]=tmpfld1(errmax)
%This program computes the temperature field of an arc
%weld for three materials. The procedure follows Sec.11.8.4,Eq.(11.81)
%(parameter B is marked as b) and (11.82).The results are presented as
%plots of temperature profiles along the X axis (bc) and
%along the Y axis (a).
%the field stretches for 25x100, dx=dy=2.5 mm, g=5mm
%roc is spec. heat per volume,alp is diffusivity
%cond is conductivity, tf is final temperature
%ii=1 is steel, ii=2 is stainless, ii=3 is aluminum
%arc is located at i=1, j=70,n=1726,
%index k counts the number of loops until err<errmax
dx=2.5;
for ii=1:3
if ii==1
P=586;v=2;roc=3.7;alp=12;cond=43;
elseif ii==2
P=261;v=2;roc=3.6;alp=4.4;cond=15;
else
P=690;v=5;roc=2.3;alp=60;cond=140;
end
b=dx*v/alp;
k=0;
err=errmax+1;
tf=1000*P/(612.5*roc*v);
%choosing initial temperature field
for n=1:1750;
t(n)=tf;
end
for n=1751:2500;
t(n)=0;
end
told=tf;
%setting the condition of convergence
while err>errmax
k=k+1;
%temp in weld point:
t(1726)=(t(1701)*(1-b)+t(1751)+2*t(1727)+P*1000/(5*cond))/(4-b);
%column above weld point
for i=2:24;
n=1725+i;
t(n)=(t(n-25)*(1-b)+t(n+25)+t(n-1)+t(n +1))/(4-b);
end
%top point in the weld column
t(1750)=(t(1725)*(1-b)+t(1775)+t(1749))/(3-b);
%all points on the weld line left of weld point
for i=1:68;
n=(69-i)*25+1;
t(n)=(t(n-25)*(1-b)+t(n+25)+2*t(n+1))/(4-b);
%all points left of the weld column except weld pt and top point
for j=2:24;
n=(69-i)*25+j;
t(n)=(t(n-25)*(1-b)+t(n+25)+t(n-1)+t(n+1))/(4-b);
```

Figure 11.94
Matlab program for the FE computation, "tmpfld1."

Figure 11.94
(continued)

```
end
%top points (boundary d)
n=(69-i)*25+25;
t(n)=(t(n-25)*(1-b)+t(n+25)+t(n-1))/(3-b);
end
%points right of weld point, on the weld line
for i=70:98;
n=25*i+1;
t(n)=(t(n-25)*(1-b)+t(n+25)+2*t(n+1))/(4-b);
%field right of weld point
for j=2:24;
n=25*i+j;
t(n)=(t(n-25)*(1-b)+t(n+25)+t(n-1)+t(n+1))/(4-b);
end
%the top line, boundary d, right of weld point
n=25*i+25;
t(n)=(t(n-25)*(1-b)+t(n+25)+t(n-1))/(3-b);
end
%right lower corner
t(2476)=(t(2451)*(1-b)+2*t(2477))/(4-b);
%boundary b
for i= 2,24;
t(2475+i)=(t(2450+i)*(1-b)+t(2475+i-1)+t(2475+i+1))/(4-b);
end
%right upper corner
t(2500)=(t(2475)*(1-b)+t(2499))/(3-b);
%left lower corner
t(1)=(t(25)+2*t(2)+tf*(1-b))/(4-b);
%first column, boundary a
for n=2:24;
t(n)=(t(n+25)+t(n-1)+t(n+1)+tf*(1-b))/(4-b);
end
%top of boundary a
t(25)=(t(50)+t(24)+tf*(1-b))/(3-b);
err=t(1726)-told
told=t(1726);
end
%[MARK]:FROM HERE ON THE PROGRAM "TMPFLD2" DEVIATES
%tbcs,tbcss,tbcal are temp profiles along the weld path X
%for steel, stainless steel and aluminum respectively
for n=1:100
x(n)=n*2.5;
if ii==1
tbcs(n)=t(25*n-24);
elseif ii==2
tbcss(n)=t(25*n-24);
else
tbcal(n)=t(25*n-24);
end
end
%tas,tass,taal are temp profiles on the axis Y passing through
%weld point
for n=1:25
y(n)=n*2.5;
if ii==1
tas(n)=t(1725+n);
elseif ii==2
tass(n)=t(1725+n);
else
taal(n)=t(1725+n);
% Plot (y,tas,y,tass,y,taal),(x,tbcs,x,tbcss,x,tbcal)
end
end
k
end
```

Figs. 11.95a, b, and c. The results in graphs *(a)* and *(b)* are based on different convergence requirements, of errmax = 1°C and errmax = 5°C, respectively. The heat input *P*(W) is chosen so as to obtain the temperature in the weld point close to the melting temperature. If it is found to be higher or lower, the value of *P* has to be proportionally changed, and the computation is repeated. The distribution of temperatures is similar to what has been obtained in the explicit formulas of the simplified computations. The finite-difference computation is still obtained under simplifying assumptions of constant thermal parameters of conductivity and diffusivity, not varying with temperature. However, the assumption of a point heat source is not used anymore, and the heat input area is assumed here equal to the area of one element. It can of course be formulated as covering several elements, and it is possible to use non-uniform distribution. Although no convection of heat into the air has been assumed, it would be rather easy to consider it. Also, no melting heat has been included, and it could be incorporated as well. The discrete character of the computation causes some small deviations, such as a local temperature drop in front of the arc for the T_b profile for the stainless steel case in graph *(c)*. This could be eliminated by choosing a smaller element size. Nevertheless, besides examining a computerized digital technique as an exercise in heat-transfer computations, the student has experienced the use of the Gauss-Seidel iterative method for solving large systems of linear equations and was able to appreciate the rather fast convergence of the computation. The condition for stopping the iterations is formulated by specifying an errmax as a difference in the temperature under the arc (n = 1726) between two subsequent iterations. Figure 11.95b is presented for the case where errmax = 5°C was chosen, which required 23 iterations for steel, 4 for stainless steel, and 15 for aluminum. In Fig. 11.95a with errmax = 1°C, 47 iterations were required for steel, 4 for stainless steel, and 50 for aluminum. Yet the differences

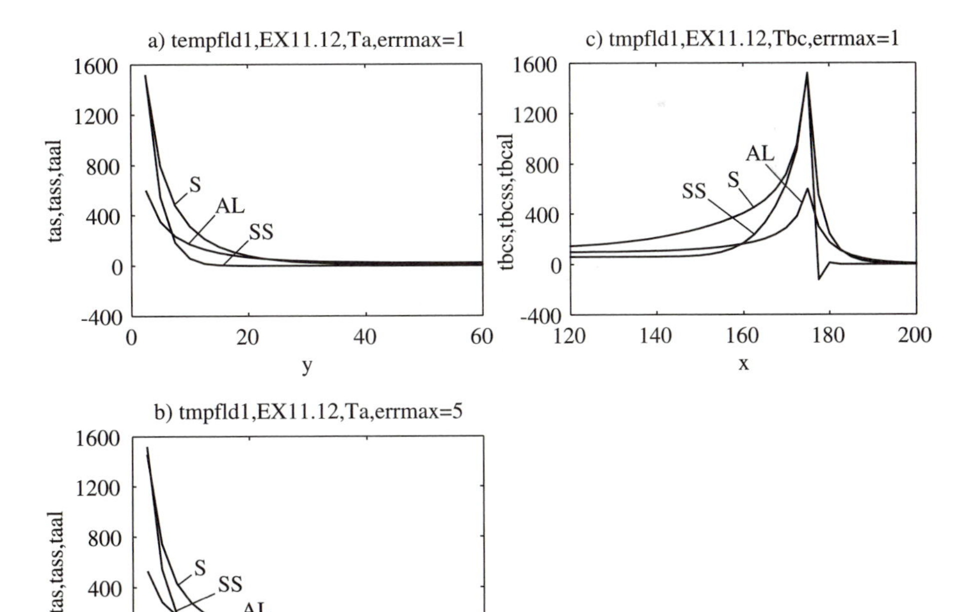

Figure 11.95
Results of Example 11.11. a) Temperature profiles T_a, for the three materials computed with convergence to errmax = 1; b) same as a) but with errmax = 3; c) temperature profiles T_{bc} for the three materials, errmax = 1.

between *(a)* and *(b)* are very small. Nevertheless, we have run the T_b, T_c computation (Fig. 11.95c) for the smaller error of 1°C. The program "tmpfld1" has been written for the T_a, T_{bc} profiles as outputs. The program "tmpfld2," Fig. 11.96, is identical except it provides an output for values of the temperatures that makes it possible to obtain easy printouts (see Fig. 11.97) and also other types of graphic output. These are presented as examples for steel as the only material. In Fig. 11.98 the contours of isotherms in

```
function[t,tt,tc,tds,tdss,tdal,tcs,tcss,tcal]=tmpfld2(errmax)
%This program computes temperature fields of an arc weld
%for three different materials. The procedure follows Sec.11.8.4, Eq.(11.81)
%parameter B is marked as b).The results can be printed out
%as well as plotted in the form of contours or of a mesh.
%the field stretches for 25x100, dx=dy=2.5 mm, g=5mm
%roc is spec. heat per volume,alp is diffusivity
%cond is conductivity, tf is final temperature
%ii=1 is steel, ii=2 is stainless, ii=3 is aluminum
%arc is located at 1=1, j=70,n=1726, index k counts number of loops
%for convergence until err<errmax.
%FROM HERE ON AS FAR AS [MARK] THIS PROGRAM IS IDENTICAL WITH "TMPFLD1":
dx=2.5; etc, CONTINUE WITH LISTING "TMPFLD1" AS FAR AS ITS [MARK]
%[MARK] FROM HERE ON THIS PROGRAM DEVIATES FROM "TMPFLD1"
for n=1:2500
tr(n)=round(t(n));
end
k
%express temperatures in the i,j coordinates;tr are
%rounded off to integers, tc are full numbers
for i=1:25
for j=1:100
n=i+(j-1)*25;
tt(i,j)=tr(n);
tc(i,j)=t(n);
end
end
%for printing select every 3rd point in i and every 4th one in j
for m=1:9
for k=1:25
a(m,k)=tt(3*m-2,4*k-2);
td=a';
end
end
%distinguish individual materials
for i=1:25
for j=1:100
if ii==1
tts(i,j)=tt(i,j);
tcs(i,j)=tc(i,j);
tds=td;
elseif ii==2
ttss(i,j)=tt(i,j);
tcss(i,j)=tc(i,j);
tdss=td;
else
ttal(i,j)=tt(i,j);
tcal(i,j)=tc(i,j);
tdal=td;
%print:tds,tdss,tdal;plot:contour(tcs,15),(tcss,15),(tcal,15)
%plot:mesh(tcs),mesh(tcss),mesh(tcal)
end
end
end
end
```

Figure 11.96
Matlab program for the computation, with different output form, "tmpfld2."

Figure 11.97
Printout of the temperature field for Example 11.11, *tds* for steel. Temperature under the arc reaches 1467 °C; three steps ahead of the arc the work is still without temperature increase; far behind the arc it reached the final temperature 129 °C.

129	129	129	129	129	129	129	129	129
129	129	129	129	129	129	129	129	129
129	129	129	129	129	129	129	129	129
129	129	129	129	129	129	129	129	129
129	129	129	129	129	129	129	129	129
129	129	129	129	129	129	129	129	129
129	129	129	129	129	129	129	129	129
129	129	129	129	129	129	129	129	129
129	129	129	129	129	129	129	129	129
129	129	129	129	129	129	129	129	129
129	129	129	129	129	129	129	129	129
130	130	130	129	129	129	129	129	128
132	132	131	130	129	128	128	127	126
140	140	136	130	126	124	122	121	118
168	164	146	128	117	111	108	105	100
245	219	161	116	94	85	81	77	71
432	309	158	84	56	47	44	41	37
1467	279	88	36	21	16	15	14	13
47	33	16	7	4	3	3	2	2
2	2	1	1	0	0	0	0	0
0	0	0	0	0	0	0	0	0
0	0	0	0	0	0	0	0	0
0	0	0	0	0	0	0	0	0
0	0	0	0	0	0	0	0	0
0	0	0	0	0	0	0	0	0

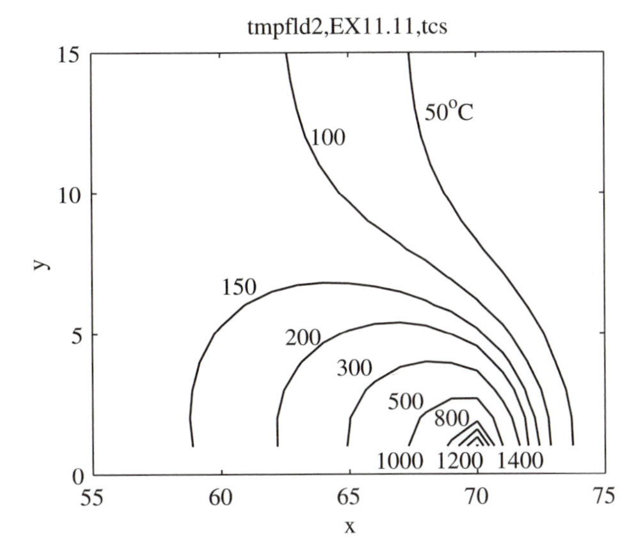

Figure 11.98
Temperature field of Example 11.11, for steel, produced as "contour."

the field are shown; also the temperature field surface produced by the Matlab graphics program "mesh" is shown in Fig. 11.99. The variation of the specific heat with welding speed has been evaluated from a number of computations repeated for different welding speeds, and it is presented in Fig 11.100. It is quite remarkable to see how little Fig. 11.100 differs from Fig. 11.91, obtained from the explicit formulas. They differ in absolute values because the latter one is for weld width 4 mm while in the former it is $w = 2.5$ mm. However, the variations of H with welding speed are the same, except that the graph of Fig. 11.100 does not include higher speeds for steel and stainless steel because of the limited computational effort spent.

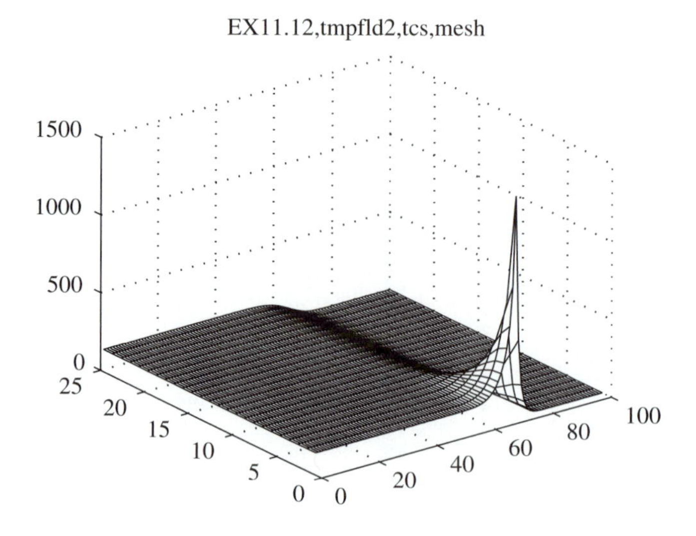

Figure 11.99
Temperature field of Example 11.11, for steel, produced as "mesh."

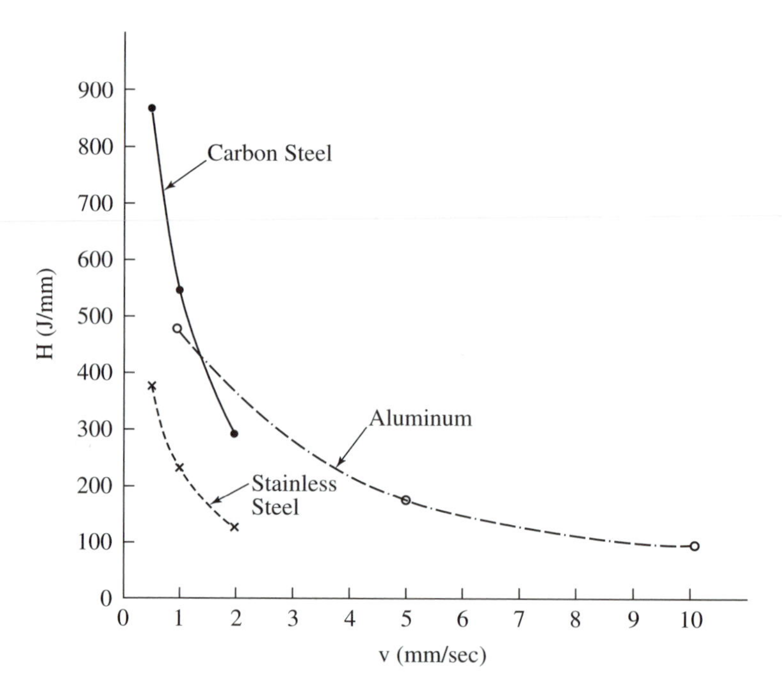

Figure 11.100
Heat input per unit length of weld, Example 11.11. The results compare approximately with those for a continuous field in Fig. 11.91.

11.9 RESIDUAL STRESSES AND DISTORTIONS

In welding heating is produced locally, which causes local stresses to exceed the yield strength of the material. Consequently, local residual stresses remain in the weldment after the process is completed and temperatures equalize to room temperature. Simultaneously, shrinkages develop locally and lead to distortions of the weldment. In the areas around a weld bead, such as a butt weld, the stresses will be recognized as longitudinal (L) and transversal (T), and distortions occur as indicated in Fig. 11.101. The top of the figure depicts the simple case of a free set of two plates. This is rarely the case; such a set is usually a part of a more complex structure, such as a box, where

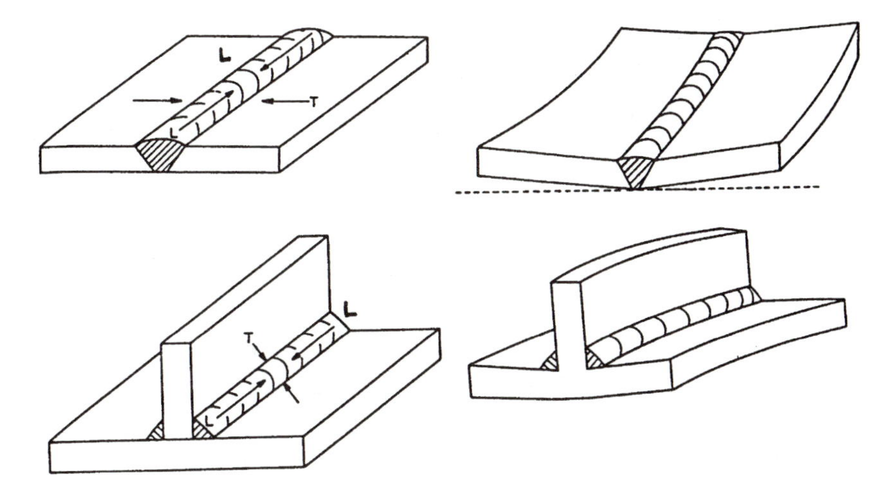

Figure 11.101 Longitudinal and transversal shrinkage stresses in butt and fillet welds. [WH3]

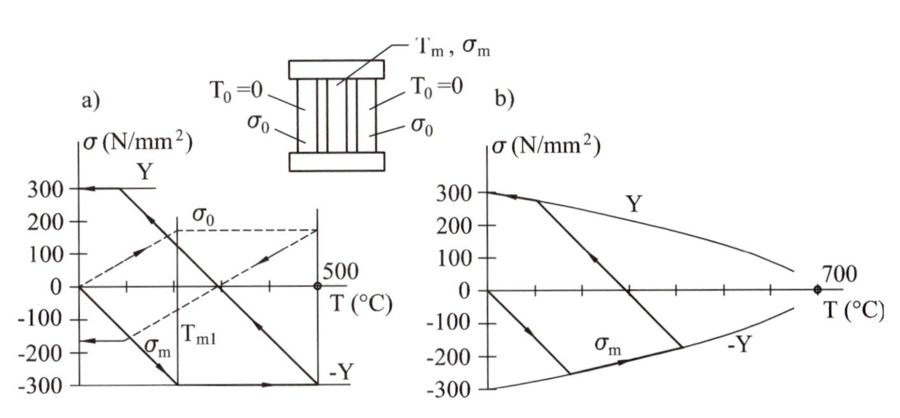

Figure 11.102 Model of residual longitudinal stresses in a butt weld. The middle column represents a strip in the middle of the weld bead and the outer columns model strips along the sides. Simultaneous heating of the whole length is assumed and strains are constrained as equal in the three strips. a) Stresses in the middle and the outer columns on heating up to 500 °C and cooling down assuming constant yield stress. b) Same as (a) except yield stress decreases with temperature.

both fillet welds and butt welds are used. The stresses, mainly the transversal ones, and distortions will be affected by the constraints of the whole structure. For reasons to be discussed later, the residual stresses may have detrimental effects, and obviously the distortions will also have to be minimized. We can derive a basic understanding of these two phenomena from simplified situations.

In order to get a general assessment of the level and distribution of the longitudinal stresses, a rather simple model presented in [9] is used; see Fig. 11.102. The model consists of three bars of equal cross-sections and of equal length with their ends connected by cross-bar heads that enforce a condition of maintaining equal lengths. In this model the middle bar represents the weld bead and the outer bars represent the surrounding metal of the two plates welded by means of the bead. Although in the actual process the heating of the bead proceeds in time, we will assume that the middle bar is heated uniformly to temperature T while the outer bars remain at the initial temperature, which is taken as reference, $T_o = 0$. It is further assumed, at first, that the yield stress Y remains constant, independent of temperature. As we start to raise T, the middle bar is elongating and drags the two outer bars along. Tensile stress is produced in the outer bars, and compressive stress in the middle bar. The balance of forces acting on the cross heads is expressed as

$$\sigma_m + 2\sigma_o = 0 \tag{11.84}$$

Thermal expansion and elastic strain combine to produce resulting strain:

$$\varepsilon = \alpha T + \frac{\sigma}{E} \tag{11.85}$$

where a is thermal expansion coefficient and E is modulus of elasticity. Recalling that $T_o = 0$, the condition of equal length is expressed as

$$\alpha T + \frac{\sigma_m}{E} = \frac{\sigma_o}{E} \tag{11.86}$$

Expressing σ_o by σ_m, from Eq. (11.84):

$$\sigma_o = -\frac{\sigma_m}{2}$$

and inserting this in Eq. (11.85), we have

$$\alpha T + \frac{3}{2}\frac{\sigma_m}{E} = 0$$

and

$$\sigma_m = -\frac{2}{3}\alpha T E \tag{11.87}$$

This confirms that the stress in the middle bar is compressive, and it increases in proportion with the temperature. Let us choose an example for steel, using thermal expansion coefficient a = 1e−5/°C, E = 2 e 5 N/mm^2, Y = 300 N/mm^2, and plot σ versus T. Eq (11.87) becomes

$$\sigma_m = -1.33\, T \tag{11.88}$$

which is a straight line under the slope of −1.33 N/(mm^2/°C). The stress σ_m reaches yield strength $-Y$ at

$$-300 = -1.33T_1$$
$$T_1 = 225\ °C \tag{11.89}$$

When increasing the temperature T beyond T_1, stress σ_m does not increase anymore; $-\sigma_o$ remains at $-\sigma_m/2$, and there is no more expansion because all the bars are held at the extension of the outer bars caused by their still elastic stress. Let us stop heating up at T = 500°C and start cooling off. The stress $|\sigma_m|$ drops below Y; the deformation becomes elastic, and the absolute value of σ_m decreases under the slope of 1.33 N/mm^2/°C. At 275°C the stress vanishes, and on further cooling it reverses to tensile and increases under the same slope. At 50°C it reaches the yield stress in tension Y, and it remains at this value through $T = 0$. At the end there is a remaining tensile residual stress $\sigma_m = Y$ and a compressive residual stress $\sigma_o = -Y/2$. The system of the three bars has shrunk by $\varepsilon_o = \sigma_o/E$, that is, by $\Delta L = \sigma_o L/E$. For bars 100 mm long this would amount to 75 μm. The model shows two important features resulting from the heating and cooling cycle on the middle bar, representing the welding bead: there remains a residual tensile stress in the bead which is as high as the yield stress of the material, and a longitudinal shrinkage has occurred. This all happens also in the real weld. In the areas surrounding the bead a compressive residual stress has been produced. It is easily seen from the graph that the results are the same for whichever maximum temperature has been reached in the cycle as long as it exceeded $2T_1$. This is of course always the case within the width of the welded seam where the maximum temperature equals the melting temperature, which is 1480°C for steel.

Graph *(b)* releases the idealized assumption of yield strength independent of temperature. In this graph a rather arbitrary function $Y(T)$ is assumed, where Y decreases with T and reaches zero at close to the recrystallization temperature, say at 700°C. The final result is the same. In the middle beam the residual stress is $\sigma_m = Y$, the yield stress at room temperature, and it is $\sigma_o = -Y/2$ in the outer beams.

Similar results are obtained even if we assume some other profile of the temperature distribution across the weldment (see Fig. 11.103). Here the model consists of five columns. It is symmetrical about the middle column, which has temperature T, the intermediate columns have temperature $T/2$ and the outer columns remain at $T = 0$. The stresses are σ_m, σ_i, and σ_o respectively. In order to make it easier to produce and understand the graph, constant Y is assumed.

$$\sigma_m + 2\,\sigma_i + 2\,\sigma_o = 0 \tag{11.90}$$

and

$$\alpha T + \frac{\sigma_m}{E} = \frac{\alpha T}{2} + \frac{\sigma_i}{E} = \frac{\sigma_o}{E} \tag{11.91}$$

Combining Eqs. (11.90) and (11.91), we have

$$\begin{aligned} \sigma_m &= -0.6\,\alpha TE \\ \sigma_i &= -0.1\,\alpha TE \\ \sigma_o &= 0.4\,\alpha TE \end{aligned} \tag{11.92}$$

The stress in the middle column reaches the yield stress at T_1:

$$-Y = -0.6\,\alpha\,T_1\,E$$

$$T_1 = \frac{1.667\,Y}{\alpha E}$$

For the parameters chosen before for our example, we have

$$T_1 = 250°C$$

At this stage, we have

$$\sigma_m = -Y = -300\ \text{N/mm}^2$$

$$\sigma_i = -Y/6 = -50\ \text{N/mm}^2$$

$$\sigma_o = 2Y/3 = 200\ \text{N/mm}^2$$

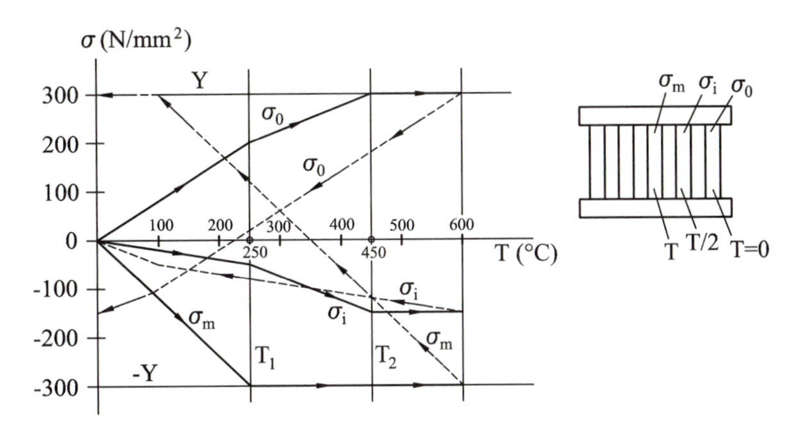

Figure 11.103
Model of residual stresses with three temperature zones (five strips).

As the temperature T of the middle column is increased, the stress $\sigma_{\rm m}$ remains at the $-Y$ value. Equations (11.90) through (11.92) are then replaced by

$$-Y + 2\sigma_i + 2\sigma_o = 0 \tag{11.93}$$

and

$$\frac{\sigma T}{2} + \frac{\sigma_i}{E} = \frac{\sigma_o}{E} \tag{11.94}$$

Combining Eqs. (11.93) and (11.94), we obtain

$$\begin{aligned} \sigma_i &= -0.25\ \alpha TE + 75 \\ \sigma_o &= 0.25\ \alpha TE + 75 \end{aligned} \tag{11.95}$$

For our example parameters, we have

$$\sigma_i = -\frac{T}{2} + 75$$

$$\sigma_o = \frac{T}{2} + 75$$

The outer column stress will then reach the yield stress at T_2:

$$Y = \frac{T_2}{2} + 75$$

$$T_2 = 450°\text{C}$$

At this stage, we have $\sigma_m = -Y$, and $\sigma_i = -Y/2$; for the parameters of the case, it is $\sigma_m = -300$, $\sigma_o = 300$, $\sigma_i = -150$. For any further increase of T, these values do not change. On cooling, all the stresses immediately become elastic and return along the original elastic slopes. So, for instance, as shown in the graph, if heating stopped at $T = 600°$C, stresses return with decreasing temperature along the dashed lines. The values of residual stresses at $T = 0$ can be read off the graph:

$$\sigma_m = +Y = 300\ \text{N/mm}^2$$

$$\sigma_i = 0$$

$$\sigma_o = -Y/2 = -150\ \text{N/mm}^2.$$

Again, the residual stress in the bead itself is as high as the yield stress of the material at room temperature. The residual stresses discussed above occur in unrestrained members. Additionally, different residual stresses occur in the transversal direction that are caused by external restraint.

The rather high residual stresses and the corresponding distortions may very often be unacceptable. The tensile residual stresses may cause brittle fractures at rather low stresses caused by external loading, at values far below the yield stress of the material. These occur in low- and medium-carbon steels at lower temperatures. In the past a number of ships broke in half in the cold waters of northern Atlantic. Since then measures have been instituted to prevent this. Also the fatigue strength may be adversely affected by the residual stresses.

The basic means used to reduce or eliminate residual stresses are twofold. Either the structure is subjected to high loads or "stress-relieving" temperatures. The reason for relief is fundamentally the same in both cases, and it consists of loading the whole structure over the yield point either by high loads or low or zero loads at temperatures at which the yield stress becomes very low or vanishes completely.

To understand the principle, see Fig. 11.104. Imagine a butt weld that will be loaded in longitudinal tension. The longitudinal residual stress after the welding operation is plotted against the distance from center of weld as the curve marked 0. The distribution corresponds to the derivations conducted with reference to Figs. 11.102 and 11.103. In the center there is high tensile stress, close to the yield stress. It drops down to zero with the move away from the center and becomes compressive, then reaches a compressive maximum and decays away. The curves 1, 2, and 3 indicate stress distribution when, over the stress of curve 0, higher and higher uniform tension is applied. At level 3 general yielding is reached. When subsequently the stresses σ_1 or σ_2 are released, the residual stress distributions 1′ and 2′ are obtained; the peak stress is substantially decreased. When the general yielding stress is removed, there remain no residual stresses because of the uniformity of the yield. A similar effect is obtained if the whole weldment is put in a furnace and heated uniformly just below the recrystallization temperature, at which the yield stress drops to zero and general yielding occurs. After slow, uniform cool-down, residual stresses have been removed.

As regards distortions, rather complex situations may arise depending on the complexity of the structure. Various techniques have been developed to minimize distortions. Several simple cases of distortions are shown in Fig. 11.105, and several simple measures to prevent distortions are shown in Fig. 11.106. Once distortions have developed, various methods are applied to correct them. One of the most popular is flame straightening, which consists of locally heating the structure to about 600°C and spray-cooling it. Another technique is to straighten the distorted structures under a press, causing a local yielding and permanent deformation back to the correct shape.

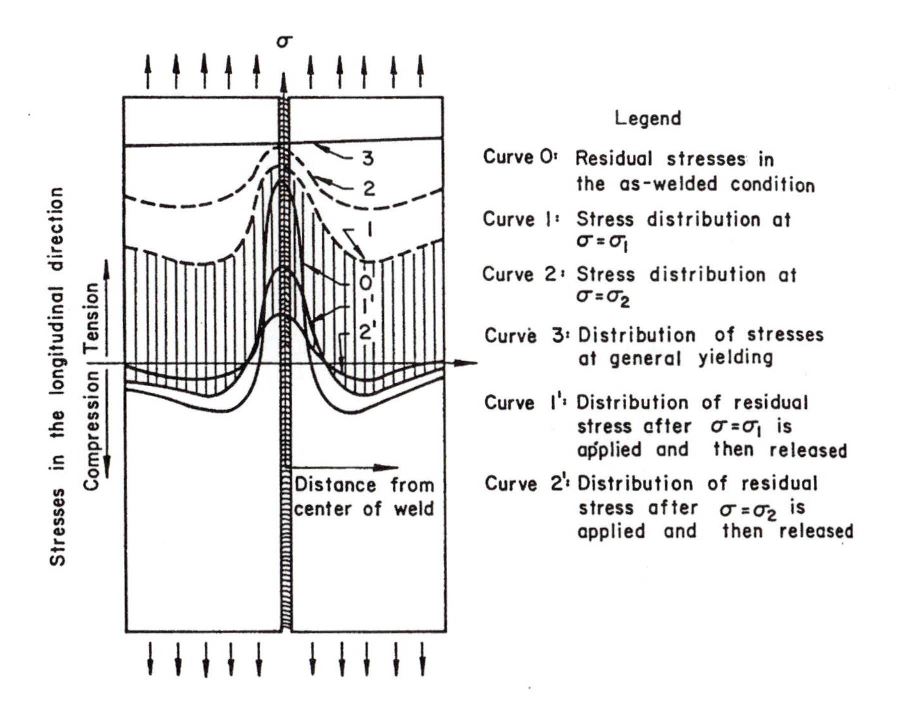

Figure 11.104
Releasing residual longitudinal stresses by uniform loading. [WH3]

Figure 11.105
Basic types of distortions in a butt and a fillet weld. [WH3]

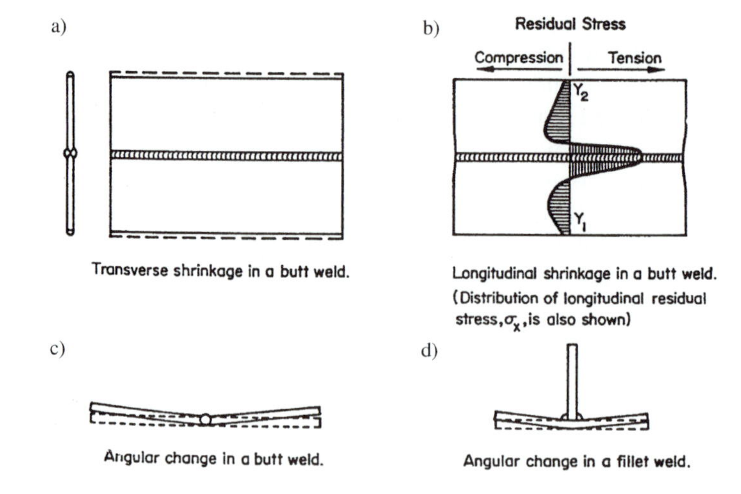

Figure 11.106
Examples of techniques for preventing distortions. [WH3]

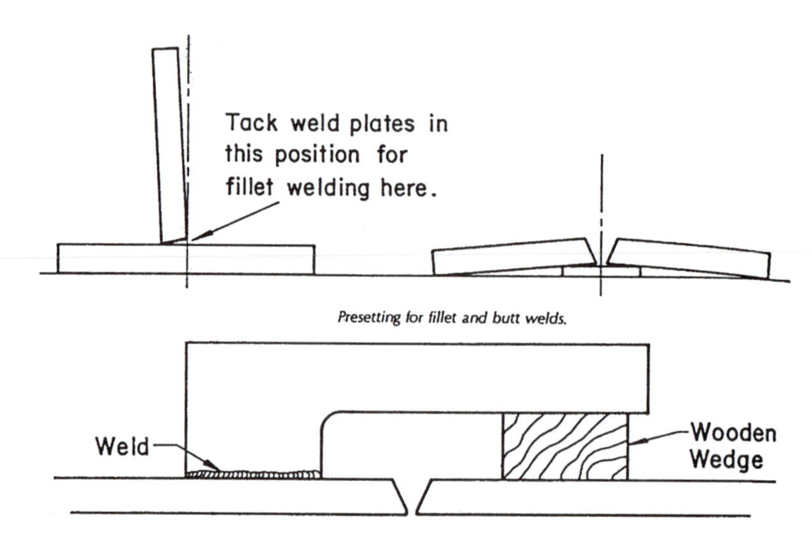

REFERENCES

1 Ward, R. E., and J. A. White. *The Essentials of Material Handling*. Charlotte, NC: Material Handling Institute, 1996.

2 Parmley, R. O., ed. *Standard Handbook of Fastening and Joining*. 2nd ed. New York: McGraw-Hill, 1989.

3 Society of Manufacturing Engineers. *Tool and Manufacturing Engineers Handbook, Quality Control and Assembly.* 4th ed. Vol. 4. Dearborn, MI: Society of Manufacturing Engineers, 1987.

4 Boothroyd, G. *Design for Assembly.* Department of Mechanical Engineering, Univ. of Massachussetts, 1979.

5 Boothroyd, G., C. Poli, and L. E. Murch. *Automatic Assembly.* New York: Marcel Dekker, 1982.

6 Boothroyd, G., C. Poli, and L. E. Murch, *Handbook for Feeding and Orienting Techniques for Small Parts.* Department of Mechanical Engineering, Univ. of Massachussetts, 1978.

7 Boothroyd, G., *Assembly Automation and Product Design.* New York: Marcel Dekker, 1991.

8 Boothroyd, G., P. Dewhurst, and W. Knight. *Product Design for Manufacture and Assembly.* New York: Marcel Dekker, 1994.

9 ASM International. *ASM Handbook.* Vol. 6. *Welding, Brazing, and Soldering.* Materials Park, OH: ASM International, 1993.

10 American Welding Society. *Welding Handbook.* Vol. 1. *Fundamentals of Welding.* 7th ed. Miami, FL: American Welding Society, 1976.

11 American Welding Society. *Welding Handbook.* Vol. 2. *Welding Processes.* 7th ed. Miami, FL: American Welding Society, 1978.

12 American Welding Society. *Welding Handbook.* Vol. 3, *Welding Processes.* 7th ed. Miami, FL: American Welding Society, 1980.

13 Gourd, L. M. *Principles of Welding Technology.* London: Edward Arnold, 1995.

14 Rosenthal, D., "The Theory of Moving Sources of Heat and Its Application to Metal Treatments," Trans. ASME, 68:849–866, 1946.

Illustration Sources

[BMH] Reprinted with permission from R. A. Kulwiec, *Basics of Material Handling,* Material Handling Institute, Charlotte, NC, 1981.

[BOOTHROYD1] Reprinted with permission from G. Boothroyd, *Design for Assembly,* Department of Mechanical Engineering, Univ. of Massachussetts, 1979.

[BOOTHROYD2] Reprinted with permission from G. Boothroyd, *Assembly Automation and Product Design,* Marcel Dekker, New York, 1992.

[EMH] Reprinted with permission from R. E. Ward and J. A. White, *The Essentials of Material Handling,* Material Handling Institute, Charlotte, NC, 1996.

[GOURD] Reprinted with permission from L. M. Gourd, *Principles of Welding Technology,* Edward Arnold, London, 1995.

[HOBART1] Reprinted with permission from *Technical Guide for Shielded Metal Arc Welding,* 1980, Hobart Brothers Co., Troy, OH 45373.

[HOBART2] Reprinted with permission from *Technical Guide for Gas Metal Arc Welding,* 1980, Hobart Brothers Co., Troy, OH 45373.

[HOBART3] Reprinted with permission from *Technical Guide for Electroslag Welding,* 1980, Hobart Brothers Co., Troy, OH 45373.

[SME4] Reprinted with permission from *Tool and Manufacturing Engineers Handbook, Quality Control and Assembly,* 4th ed., Vol. 4, Society of Manufacturing Engineers, Dearborn, MI 1987.

[WH1] Reprinted with permission from *Welding Handbook.* Vol. 1, *Fundamentals of Welding,* 7th ed., 1976, American Welding Society, 550 NW Le Jeune Rd, Miami, FL 33126.

[WH2] Reprinted with permission from *Welding Handbook.* Vol. 2, *Welding Processes,* 7th ed., 1978, American Welding Society, 550 NW Le Jeune Rd, Miami, FL 33126.

[WH3] Reprinted with permission from *Welding Handbook.* Vol. 3, *Welding Processes,* 7th ed., 1980, American Welding Society, 550 NW Le Jeune Rd, Miami, FL 33126.

Industrial Contributors

[BODINE] The Bodine Corp., 317 Mountain Grove Street, P.O. Box 3245, Bridgeport, CT 06605-0903.

[INGERSOLL-RAND] Ingersoll-Rand, 8 Bartles Corner Park, Suite 101, Flemington, NJ 08822.

[LITTON] Litton Unit Handling Systems, 7100 Industrial Road, Florence, KY 41042.

[MTRC] Machine Tool Research Center, University of Florida, Mechanical Engineering Bldg, Gainesville, FL 32611.

QUESTIONS

Material Handling

Q11.1 Describe briefly the stations on the general path of a part or a subassembly through a factory.

Q11.2 Which types of transportation equipment of parts are suitable in mass production? What is the advantage of overhead conveyors? Which types of transportation means are used in small-lot, flexible production? Which of them is an automated type?

Q11.3 What kind of sensor do AGVs use as position feedback in the control of their path?

Q11.4 Describe a pallet rack storage system. What is an AS/AR system? Give an example of how it is interfaced with the transportation system in the plant.

Mechanical Joining

Q11.5 Name elements used for separable joints and those used for permanent joints. What are set screws used for? What is the advantage of shrink-fitting of end mills in their holders? How is it done? Does it not take too much time to get the tool in and out? Name one large-scale application of rivets.

Q11.6 Sketch a seam used to join the lid to a soda can.

Assembly

Q11.7 For automated assembly of small products, what are the typical conditions for an economically successful application of assembly automation? Name five examples of products assembled automatically.

Q11.8 List various assembly tasks.

Q11.9 Explain the difference between a synchronous and an asynchronous (free transfer) assembly machine.

Q11.10 Sketch an example of an orienting device on the ledge of a vibratory bowl feeder.

Q11.11 What is the role of a pick-and-place unit?

Q11.12 What would be the elements of a flexible, programmable, automated assembly system?

Q11.13 Summarize briefly the basic rules of design for assembly.

Welding

Q11.14 Name two welding processes for each of these conditions:

Only heat applied, no force

Only force applied, no heat

Force and heat equally significant

Q11.15 Name five welding processes applied along a line, in the order of increasing heat intensity. How is heat intensity defined? What are the effects of high intensity on heat efficiency, on HAZ, on penetration?

Q11.16 Name the types of arc welding carried out manually. Name those carried out mostly by mechanized or automated means. What type of arc welding is best suited for field work and why?

Q11.17 How much heat is needed to melt metals? Give Q_m (J/mm^3) for steel 1035, stainless steel 302, Ti alloy, and aluminum. Which thermal parameter of the work material is most decisive for thermal efficiency f_2 of welding (or for heat losses)?

Q11.18 Name at least four beneficial roles of electrode coating in SMAW.

Q11.19 Which arc welding processes use consumable electrodes? nonconsumable electrodes?

Q11.20 How is the heat H per unit length affected by welding speed? Explain why. Give the figure numbers in which this effect is shown.

Q11.21 Name three processes that do not use either shielding or filler metal.

Q11.22 Which modes of metal transfer in the arc can be used in the vertical and overhead welding positions? Which mode is limited to the flat position? Under which conditions does the latter mode occur? Which of the two polarities, DCEN or DCEP, gives deeper penetration, and why?

Q11.23 Name the kinds of supply lines, such as electric cable, shielding gas hose, and others, needed in the following processes: SMAW, GMAW, GTAW, PAW, and SAW.

Q11.24 Why is the natural arc control that is so useful in GMAW not suitable for SAW?

Q11.25 How high is the residual stress left in the weld bead? Explain methods used to relieve this stress.

PROBLEMS

P11.1 *Self-regulation in GMAW;* see Fig. P11.1.

$k_{il} = -50$ A/mm, $b = 0.5$ mm/sec/A, initial $h = 16$ mm, $l = 8$ mm.

Introducing a sudden, stepwise change Δf in feed rate, $\Delta f = 50$ mm/sec.

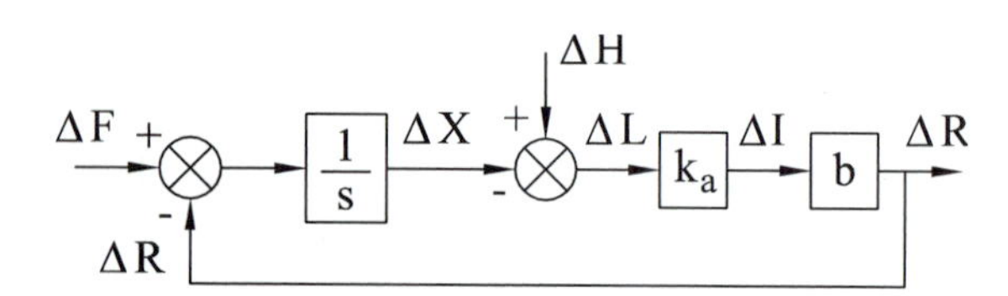

Figure P11.1

Determine the time constant τ of the process. Write out the equation for the Δr and for Δl as a function of time. Determine the final, steady-state values of Δr and Δl.

P11.2 *Self-regulation in GMAW;* see Figs. P11.1 and P11.2.

$m_p = -0.025$ V/A, $m_a = 0.026$ V/A, $b = 1.0$ mm/sec/A, $a = 3$ V/mm.

Determine the time constant of the process τ. Introduce a step change in the wire-feed rate $\Delta f = 35$ mm/sec. Write out the expressions for $\Delta r(t)$ and $\Delta x(t)$. Draw the plots Δr and Δx, Δl (by hand approximately), as in Fig. P11.2. What will be the final steady-state change of the current Δi?

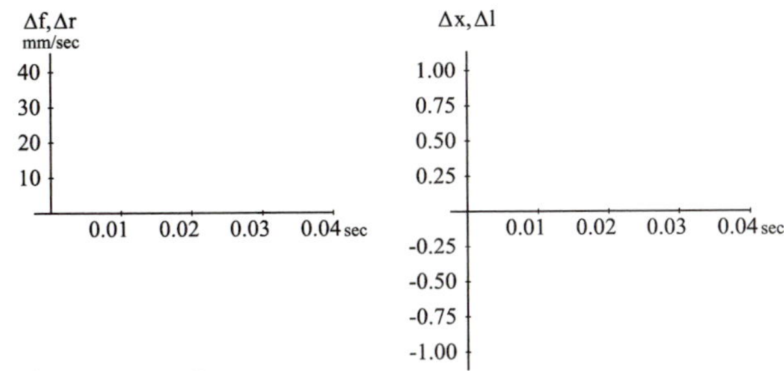

Figure P11.2

P11.3 *Self-regulation in GMAW;* see Fig. P11.1 and P11.3.

$a = 3$ V/mm, $m_a = 0.022$ V/A, $m_p = -0.03$ V/A, $b = 0.85$ mm/sec/A.

Determine the time constant τ_a. Introduce a step change of the gap h, $\Delta h = 2.5$ mm. Write out the formula for current i as a function of time, and plot i versus t in Fig. P11.3. Determine the initial values at $t = 0$ and the final values at $t = \infty$ for x_0, l_0, r_0, e_0, x_f, l_f, r_f, e_f.

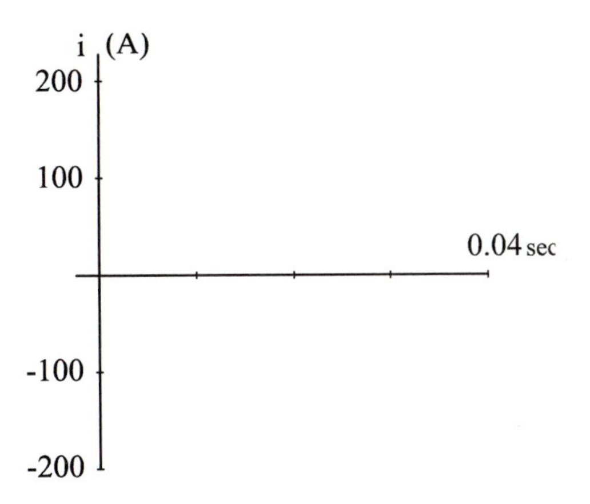

Figure P11.3

P11.4 *"Natural," inherent, self-regulating control: comparing three processes, SMAW, GMAW, and SAW*

Common for all three: $a = 3$ V/mm, $m_a = 0.03$ V/A

The burn-off coefficients and wire diameters are:

	SMAW	GMAW	SAW
d(mm)	3.0	1.0	6.0
b(mm/sec/A)	0.1	0.85	0.0625

For the "const i" power source, $m_p = -0.3$ V/A.

For the "const e" power source, $m_p = -0.03$ V/A.

(a) For the following processes and power sources, determine the arc time constant τ_a (for the self-regulating mode).

1. SMAW, const i
2. GMAW, const e
3. SAW, const i
4. SAW, const e

(b) Introduce a step change $\Delta h = 2$ mm.

Draw the block diagram $\Delta h \rightarrow x$ and either redraw it as $\Delta h \rightarrow i$ and write out the transfer function and derive the solution i(t), or use the solution for $x(t)$ as derived in the text and, also for l(t), and then write i(t). Determine, for all four cases, the value of the initial change of the current.

(c) Take the cases of GMAW, const e, and SAW, const e, and plot out the transient of the current i in response to a commanded change of feed rate: $\Delta f = 50$ mm/sec for GMAW, and $\Delta f = 5$ mm/sec for SAW. Consider self-control only; write out the corresponding expression, and plot by hand.

P11.5 *Servo control in SAW.* Refer to Figure 11.74 and Example 11.3.

$d = 5$ mm, $b = 0.035$, $m_p = -0.35$, $m_a = 0.03$, $a = 3$, $K_f = 12$, $\tau_f = 0.020$.

(a) Input a step of $\Delta i = 50$ A.

Determine the final, steady-state values of the deviations e and i.

(b) Determine the natural frequency ω_n (rad/sec), f_n (Hz), and the damping ratio ζ of the system.

(c) Now, instead of a current change, input a step $\Delta h = 2$ mm. Plot out the approximate variation of the deviation $x(t)$. Neglect the lead term $\tau_f s/(1 - K_f m_p/b)$ in the corresponding equation (11.52). Indicate the value of the peak of the overshoot and the time at which it occurs.

P11.6 *Servomechanism control in submerged arc welding;* see Fig. 11.77 and Fig. P11.6

Input: change in current. Use the block diagram of Fig. 11.77 as guidance and write a simulation program. The initial steady-state in point O on the 809 A line is characterized by $x_0 = 9$ mm, $h_0 = 20$ mm, $i_0 = 809$ A, $f_0 = 17.725$ mm/sec. We have also $m_p = -0.3$, $m_a = 0.03$, $a = 3.0$, $b = 0.025$, $\tau_f = 0.015$ sec, use bias voltage $e_b = -24.27$ V. Use dt = 0.001 and run the simulation for 0.3 sec. Determine K_f so as

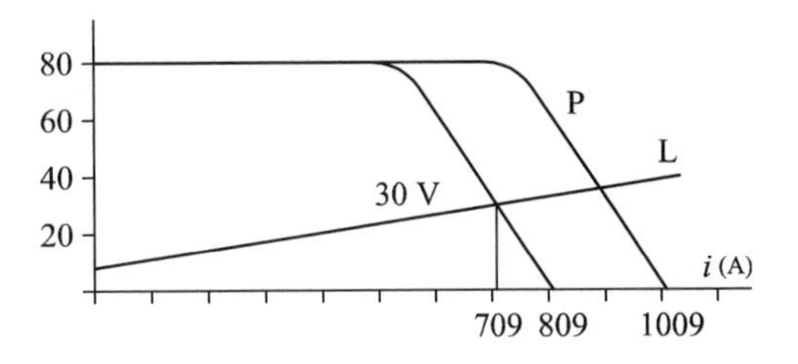

Figure P11.6

to obtain damping ratio $\zeta = 0.7$. Obviously, the equation for current i is

$$i = i_o + \frac{e}{m_p} + \Delta i$$

Input $\Delta i = 200$ A and use reference voltage $e_{ref} = 30$ V. Plot i, l, e, and f versus time.

P11.7 *Thermal conditions in welding*

Consider thick plate welding using the corresponding explicit formulas for temperature profiles. Welding alloy steel, $\lambda = 0.049$ (sec/mm^2), $k = 38$ (N/(sec · °C)). Traversing velocity $v = 5$ mm/sec, melting temperature $T_m = 1510$ °C, initial temperature $T_0 = 20$°C. The weld width is 10 mm.

(a) Determine the power input q' (mW).

(b) Determine the distance ξ behind the arc, at which the temperature cools down to the critical value of 550°C (of T-T-T diagram).

(c) Determine the cooling rate at this point.

(d) If the maximum permissible cooling rate is 27 °C/sec, what will be the necessary preheat temperature $T_{0,pr}$?

P11.8 *Temperatures in Welding*

(a) Using the closed-form expressions for "thick plate" welding, determine the power input q' (mW) for welding steel with $v = 3$ mm/sec, stainless steel with $v = 2$ mm/sec, and aluminum with $v = 5$ mm/sec; the width of the bead $w = 6$ mm in all three cases. Thermal properties for these materials are given in Table 11.2.

(b) Calculate the heat input H (kJ/mm) for the three cases as well as the thermal efficiency η.

(c) Plot the heat input H (kJ/mm) versus speed v (mm/sec) in the range 0.1–10 mm/sec for a weld of $w = 6$ mm, for the three materials. Plot all three in one plot as in Fig. 11.84.

P11.9 *Temperature profiles, thin plate welding*

(a) Using closed-form expressions for "thin plate" welding, plot graphs such as in Fig. 11.89 using different speeds for the three metals; welding steel with $v = 3$ mm/sec, stainless steel with $v = 2$ mm/sec, and aluminum with $v = 5$ mm/sec; the width of the bead is $w = 6$ mm, and power input is $q' = 1\text{e}6$ mW in all three cases.

(b) Check the temperatures at $y = 3$ mm and change the input powers so as to obtain melting temperatures at these points. Plot again to confirm. List the values of q' that result in the required width $w = 6$ mm.

P11.10 *Temperature field in welding: finite-difference computation;* see Fig. P11.10

Welding steel: $\rho c = 3.7$(N/°C · mm^2), $k = 43$(N/sec · °C), $\alpha = 12$ mm^2/sec.

Plate thickness $g = 4.5$ mm, $\Delta x = \Delta y = 3$ mm, $v = 2$ mm/sec, $P = 750$ W.

Write the equation expressing the relationship between the field temperatures, for $T_{i,j} = T_{1601}$. Assume that at a certain stage of the computation, $T_{1626} = 800$°C, $T_{1602} = 569$°C, $T_{1576} = 604$°C; at this stage what is T_{1601}?

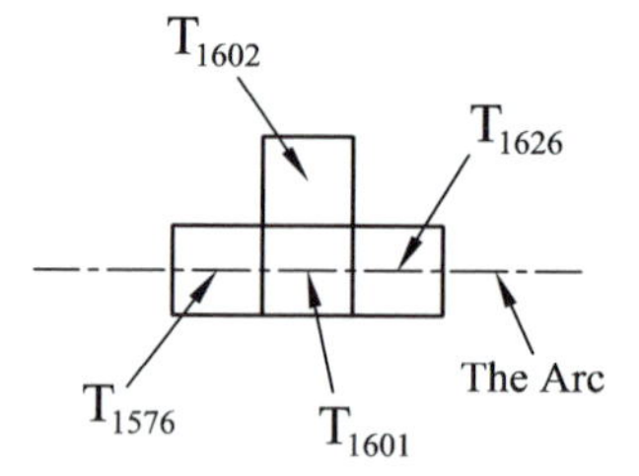

Figure P11.10

P11.11 *Finite-difference computation of a thermal field of a weld*

This task is analogous to Example 11.11. The size of the field is different, as is the way of presenting the results. The size of the field is given by $i = 1{:}20$, $j = 1{:}150$. The transformed independent variable T has the dimension $n = 1{:}3000$. We have $\Delta x = 3$ mm, $\Delta y = 3$ mm, $g = 4.5$ mm. The arc is in location $n = 2001$. The welded material is steel with parameters $k = 43$ (N/sec · °C), $\alpha = 12$ (mm^2/sec), $\rho c = 3.7$(N/°C · mm^2).

(a) The starting parameters are power input $P = 800$ W, welding speed $v = 2$ mm/sec. Calculate the final temperature T_f. Write the computer program to compute the 3000 values of T and plot a graph of temperature profiles along lines parallel with the weld line: those in rows $i = 1, 2, 3, 10, 20$. Use errmax = 2°C. Determine the number of iterations k. Read the maximum temperature T_{2001}. Plot all five lines in one graph, versus distance x.

(b) Change input P so as to obtain $T_{2001} = 1490$°C, with the same welding speed and obtain new plots. What is the value of the new P?

(c) Increase the speed to $v = 4$ mm/sec and run the computation with the original value of P. Obtain new plots, and read maximum temperature T_{2001}.

Attach the three graphs.

P11.12. *Residual stresses;* see Fig. 11.103.

Carry out an exercise like those in Section 11.9, but assume a different gradient of temperature in the direction away from the weld line, $T_i = T_m/4$, $T_0 = 0$. Use yield strength $Y = 300$ N/mm^2 and, for simplicity, keep it constant, independent of temperature; thermal expansion coefficients $\alpha = 1\text{e}{-5}$/C, and modulus of elasticity $E = 2\text{e}5$ N/mm^2.

Plot graphs like those in Fig. 11.103 and determine temperatures T_1 and T_2 at which first the middle bar and then the outer bar become plastic. Determine the residual stresses σ_m, σ_i, σ_o.

12 Nontraditional Processes

12.1 INTRODUCTION

Until about the 1940s production of machine parts relied solely on the two traditional processes of forming and machining. In both of them, the first one changing the shape of the workpiece and the second removing selected portions of the workpiece material, the actions were mechanical, involving substantial forces and corresponding mechanical energies. These processes have been developed to a high degree of sophistication to carry out the many different operations needed to produce the many variations of machinery. These two traditional processes still remain today the backbone of machinery production. The marvelous new processes discussed in this chapter, which have called upon the use of chemical, electrical, and electrochemical methods have not displaced them to any substantial degree. We will see that all these nontraditional processes are much less efficient than mechanical cutting with a well-defined tool edge or even grinding using a conglomeration of less well-defined edges mounted in the grinding wheel. They are less capable of high material removal rates (MRR), and they use more (in most instances much more) energy per unit of MRR. In general, they do not surpass the accuracy achievable with the traditional processes, where a diamond turning or grinding machine can be built to produce surfaces with a tolerance of 10 nanometers. Neither can they deliver better surface finish than that achievable by turning, milling or grinding, honing, or polishing. However, in quite a number of particular operations, they excel. It is possible to produce *smaller-diameter holes* and with a *higher length-to-diameter ratio* by electrochemical machining (ECM), electro-discharge machining (EDM), laser beam machining (LBM), and electron beam machining (EBM) than by conventional drilling, and it can be done *in hard steel,* in a *hard ceramic,* or in a *superalloy.* These processes can produce *non-round holes* in all those materials, and the holes do not have to be through holes; they can be *blind* if so desired. They can *cut plates* from paper, cloth, plastic, composites, ceramics and all metals better than a saw, with a *narrower kerf,* along a *curved path,* with minimal corner radii or with sharp corners and some of them, plasma arc cutting (PAC) and oxyfuel gas cutting (OFC) can *cut very fast* in very thick material. OC can cut steel as thick as 2 m. Two electrical processes, ECM and EDM, are capable of producing *3D sculptured cavities in hard steel.*

Here we briefly introduce these processes individually; each one is discussed in more detail in the following sections of the chapter.

There are two purely mechanical processes, *ultrasonic machining (USM)* and *water jet machining (WJM),* the latter with a variation of Abrasive Jet Machining (AJM). The first seems to have been derived from the way watchmakers made bearing seats in a corundum insert for the tiny axles of the tiny gears, using a drilling motion with an abrasive. In USM the high-frequency (20–100 kHz) vibration of a tool separated from the workpiece by a slurry with an abrasive makes the tool sink into the workpiece, which may be one of many brittle materials such as glass, quartz, carbide, tool steel and others. WJM uses a high-velocity water jet. Water driven by very high pressure through a nozzle acquires supersonic speed and cuts paper, cloth, plastics and other relatively soft materials. The motion of the table with the work material can be NC controlled. Addition of abrasive grains improves the action in cutting plastics reinforced with strong fibers and metals.

Electrochemical machining (ECM) uses electrolysis to remove material from a workpiece made into an anode in an electrolyte bath. The process exists in a number of modes: sinking, drilling, deburring, polishing, and grinding. In each mode it finds special applications. For all of them the common limitation is that the workpiece must be electrically conductive. This means that it is limited to metals, but the advantage is that the hardness of the metal does not matter; soft steel is machined as well and fast as hard steel. In the sinking mode ECM is used to make 3D molds and dies in hardened steel. A water solution of sodium chloride is commonly used as the electrolyte, and the tool may be made of copper. High currents of several thousand amps (A) are used, and relatively high MRR can be achieved, on the order of 1–10 cm^3/min. This approaches the MRR obtained in finish milling of nonhardened steel. The tool does not wear at all and very good surface finish is obtained, but the tool shape is not easy to determine because it is not a replica of the desired shape. The high electric power source is expensive, and the high flow rates of the electrolyte need rather high pressures. The machine must be made strong and rigid to stand up to the forces between the tool and work, and all the surfaces exposed to the fluid must be made of stainless steel or plastics. It is difficult to dispose of the used erosive electrolyte with various heavy metal elements contained in it. For these reasons, the process has not found as wide a use as its many positive characteristics would indicate. The manufacture of mold and die cavities is instead mostly done using the EDM process.

The other modes of ECM are used with advantage for special operations described in the following text. The drilling of small diameter round (d = 0.5–6 mm) and unround holes with up to 200:1 length-to-diameter ratios in stainless steel, Ti, and Mo alloys and superalloys, or else EC grinding of sintered carbide tools are good examples.

Chemical machining (CHM) is used to etch patterns into surfaces of various materials such as aluminum, titanium, copper, and silicone into prevailingly shallow depths or else to completely etch through thin sheets in a mode called chemical blanking. The patterns are determined by a mask made of material resistant to the etching fluid. The mask is formed by cutting or scribing, or photochemically. No forces are involved, and therefore very fine details in thin materials can be blanked. In the form of photolithography the process is also used in the manufacture of electronic circuits (see Sec. 12.11).

In *electrodischarge machining (EDM)* the metal is removed by sparks generated with a high frequency (10–500 kHz) between tool and workpiece through a dielectric fluid. The tools are made of graphite or brass or other metals. Their shape is a replica of the desired shape of the cavity being produced but they wear rather strongly. The ratio of volume of metal removed from the workpiece to the tool wear volume is only between 2 and 10; thus several tools are used before a part is finished. In the sinking mode the process is the most popular one for the production of hardened steel dies and molds. Before EDM and ECM, such cavities were machined in the steel block that first had to be normalized into the soft state and then, after machining, it was hardened and, finally hand-finished by honing. It is considered preferable to use EDM, although NC milling has to be used anyway for making the electrode, and a layer of the EDMed surface is thermally damaged to a depth of several tenths of a millimeter. Recently it was shown that hardened steel can be high-speed milled using carbide and CBN tools, and this operation may seriously challenge EDM's almost complete current monopoly of this business. The supremacy of EDM over ECM in the sinking process, in spite of more then ten times smaller MRR, is explained not only by the nonagressive oil used as the working fluid but mainly by the ingenious combination of NC motion routines with simple tool shapes to generate complex shapes.

The dominance of the mode wire EDM (WEDM) in the production of punching and blanking dies in hardened tool steel remains unchallenged. It uses thin brass wire fed from reel to reel as the EDM cutting tool. The kerf width is 20–50 microns. The control resolution is usually 1 micron. The productivity is enhanced by features such as automatic threading, robot palletization, collision protection, and others.

Laser beam machining (LBM) and welding (LBW) is a very versatile and precise process because the beam can be narrowly focused, and the power can be accurately controlled. Its effectiveness is affected by the fact that for many metals, a large part of the light energy is reflected and wasted. LBM is also used for heat treatment of selected areas of the workpiece, and for welding, drilling, and cutting. The penetration of the energy is rather small because the speed of the heat conduction does not match the pulse duration. However, for high-energy laser welding (3 to 15 kW), the beam produces a hole in the material, and the laser energy is focused at the bottom; much deeper penetration is thus possible. Two kinds are distinguished, microwelding and macrowelding. The former is very precise. Drilling of holes with diameters down to 75 microns is possible to a depth of up to 15 mm in stainless steel, W, Ta, beryllium, uranium, and in superalloys. Cutting of sheet steel is enhanced by a stream of oxygen directed to a small spot of the laser beam.

In *electron beam machining (EBM) and welding,* an electron beam is used to produce excellent welding and cutting operations to large depths if done in a high vacuum. Holes down to 0.05 mm in diameter in a 200:1 L/d ratio can be produced in steel, tungsten, quartz, and other materials. Very narrow and deep welds, such as a butt weld 3 mm wide and 150 mm deep with a welding speed of 2 mm/sec can be accomplished. The process works also in a medium vacuum and at atmospheric pressure, but with strongly diminished performance. EBM operates in two distinct ranges of power. At low power, high-precision and delicate welding, drilling, and cutting of thin foil can be precisely controlled. At the high end, powerful welding and cutting is accomplished.

Oxygen cutting (OC, OFC) is the most powerful cutting operation for low-carbon, low-alloy steels and titanium. It severs the metal by a high-temperature exothermic reaction of iron or titanium with oxygen. The most effective version of the process is oxyfuel gas cutting. The fuel gas flame preheats iron to the ignition temperature of 870°C when the exothermic reaction of iron starts. Thick plates of steel can be cut: at

0.25 m thickness, the cutting feed rate is about 2 mm/sec. Plates 2 m thick have been cut. The process is much less efficient in cast iron, stainless, and other alloy steels.

Plasma arc cutting (PAC) is used for fast cutting of medium to thick plates of many metals, including carbon steel, stainless steel, and aluminum. Cutting steel plate less than 75 mm thick is faster than with OFC, and below 25 mm it is five times faster. Various plasma gases are used. The currents range from 70 to 1000 A.

The nontraditional processes are well presented in references [25], [26], [30]. The processes discussed above plus other chemical and physical processes as applied to electronic circuit manufacture and to NC moldless material consolidation (rapid prototyping) are reviewed in Sections 12.11 and 12.12.

12.2 ULTRASONIC MACHINING (USM)

USM uses a tool vibrating at an ultrasonic frequency, 20–100 kHz, to impact an abrasive slurry in a gap of 25–100 μm and gradually sink into the surface of the workpiece or even produce holes in thin sheets of the workpiece material. It can be applied to a wide variety of workpiece materials, whether electrically conducting or not: glass, quartz, diamond, carbides, silicon, tool steel, and graphite. Although ductile materials can also be machined, the bulk of the operations involve brittle materials. The overall setup is illustrated in Fig. 12.1. The driving force is generated in a piezoelectric transducer fed from an electric generator of alternating voltage with the corresponding chosen frequency. The vibration produced in the transducer is amplified in a horn with such shape and dimension as to resonate with the driving force. As shown in Fig. 12.2, the mode of these resonant vibrations is such that it may have two nodal points between the driver and the tool end, while the amplitude at this end is several times larger than that at the root of the horn. The vibration is longitudinal, in the direction of the axis of the horn; as depicted, it is in the vertical direction. The mode shape in the picture is drawn in the direction perpendicular to that of the vibration, to be visualized.

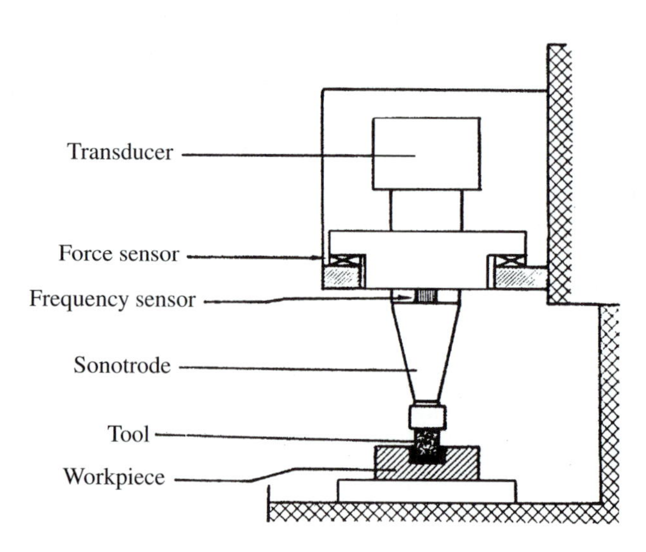

Figure 12.1
Main components of the ultrasonic machining (USM) system.

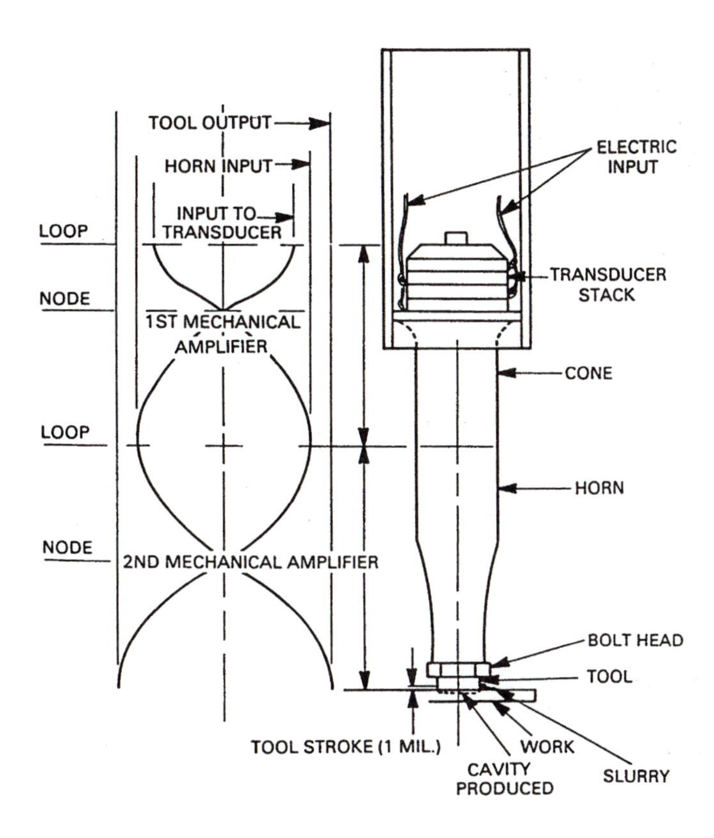

Figure 12.2
Transmission and amplification of vibration from the driver to the tool. The horn (sonotrode) is excited at resonance with the second mode of axial vibration so as to obtain maximum amplification at the working end. [WELLER]

The powers used range between 0.1 and 40 kW, and vibrations range to just about the size of the gap, peak-to-peak. Various kinds of abrasives used are boron carbide, aluminum oxide, silicon carbide, in the form of sharp, angular grains of diameters in the range of 25–60 μm, and water is used to transport them below the tool and away from it. Tools are made of soft carbon or stainless steel.

The USM process is limited to rather small MRR over areas up to 50–100 cm^2 with penetration rates between 0.2 and 6 mm/min. The tool gets worn, too, and depending on the work material, the ratio of stock removed to tool wear may range between 100:1 for glass, silicon, and graphite down to 2:1 for tungsten carbide, boron carbide, and synthetic ruby. Applications include coining operations, in which the tool form is imprinted in the workpiece and production of round and unround holes. Multiple holes may be cut simultaneously. The smallest holes are on the order of 0.1 mm, and the largest hole reported was 10 cm in diameter. The large holes, however, are preferably made by trepanning, in which the tool is limited to a tube with wall thickness of 0.5 to 1.0 mm. One example of such an operation produced a hole 12.7 mm diameter in a 4.75 mm thick tungsten carbide wire-drawing die. Such dies are made for drawing wire with round or nonround cross sections of various kinds of shapes.

The process is purely mechanical; unlike the thermal processes, it does not leave behind a damaged or heat-transformed surface layer. The equipment is relatively simple, and the vibratory heads may be attached to conventional machine tools. The USM process is also used to enhance traditional cutting processes such as turning, boring, and twist drilling of hard and brittle materials. It improves metal removal rates and increases tool life.

12.3 WATER JET CUTTING (WJC)

With water jet cutting, the material is removed by the impact of a high-velocity water jet. Very high pressure, on the order of up to 400 MPa (50,000 psi) produces a jet velocity greater than the speed of sound. The small nozzle opening, 0.1 to 0.4 mm diameter produces a very narrow kerf (width of cut). The otherwise much more productive and competitive process of wire electro-discharge machining is limited to electrically conductive workpiece materials. Other competing processes, such as laser beam and electron beam machining require much more complex and expensive equipment. Thus WJC is the process of choice in cutting sheets of paper, cloth, kevlar, glass epoxy, and fiber-reinforced plastics; even materials such as titanium and boron can be cut. Brittle materials are not suitable, because they crack or break under the action of the water jet. Soft and friable materials are easiest to cut. It is a cool process, unlike others, it does not burn or otherwise thermally affect the cut surfaces.

It is not a powerful process and cuts rather thin sheets. Moreover, because of the limited transverse stability of the jet, grooves may result on the surface, and jet deflection may cause the fibers in the composite material to be jumped over. To avoid this, rather slow feed rates must be used. The water jet may be attached to the wrist of a robot or it may be used over an NC coordinate table. An example of the equipment is shown in Fig. 12.3. A pressure intensifier in combination with an accumulator is used to achieve the very high water pressures used. Alternatively, multiple piston pumps are used. Examples of performance data are given in Table 12.1.

Adding abrasive grains to the water jet greatly increases the cutting efficiency. The above-mentioned behavior of "jumping" over hard material zones in composites is eliminated, and cut surfaces are smoother. Impressive performance is obtained in cutting titanium plates up to 200 mm thick.

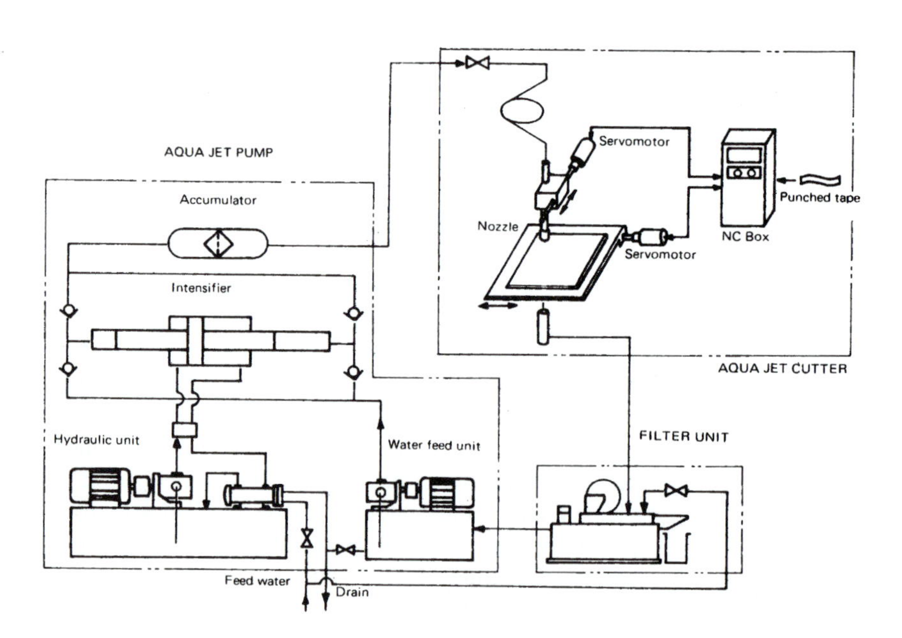

Figure 12.3
Water jet cutting (WJC) system. The intensifier of pressure is used to obtain the very high pressure at the nozzle resulting in ultrasonic speed of the water. [COURTESY SUGINO MACHINE LTD.]

TABLE 12.1 Performance of WJC Cutting for Various Materials

Material	Thickness (mm)	Feed Rate (m/min)
Leather	2.2	20
Vinyl chloride	3	0.5
Polyester	2	150
Kevlar	3	3
Graphite	2.3	5
Gypsum board	10	6
Corrugated board	7	200
Pulp sheet	2	120
Plywood	6	1

12.4 ELECTROCHEMICAL MACHINING (ECM)

For a detailed treatment of this process see Ref. [25]. Several modes of ECM are distinguished: die sinking, drilling, deburring and polishing, and grinding. The process is based on the anodic dissolution during electrolysis. Since the work of Michael Faraday (1791–1867), it has been known that if two metal electrodes are submerged in a conductive electrolyte bath and connected to a DC current, metal is deplated from the anode and plated on the cathode. Not until the 1960s was this process utilized for machining, in which the metal removed from the anode (workpiece) is prevented from deposition on the cathode (tool) and flushed away by the flow of the electrolyte.

All the metal removal is achieved by electrolytic action. The shape of the workpiece is determined by the shape of the tool and by its motion towards the workpiece, see Fig. 12.4. As shown in (*a*), initially the highest current densities and consequently the highest metal removal rates occur at the peaks of the tool where the gap is narrow. Later, as shown in (*b*), removal rates become fairly uniform over the tool surface. The process is powerful, and metal removal rates on the order of 10–20 cm^3/min per minute are possible, regardless of the hardness or any other physical property of the work material, provided that it is an electrical conductor. Good surface finish is obtainable,

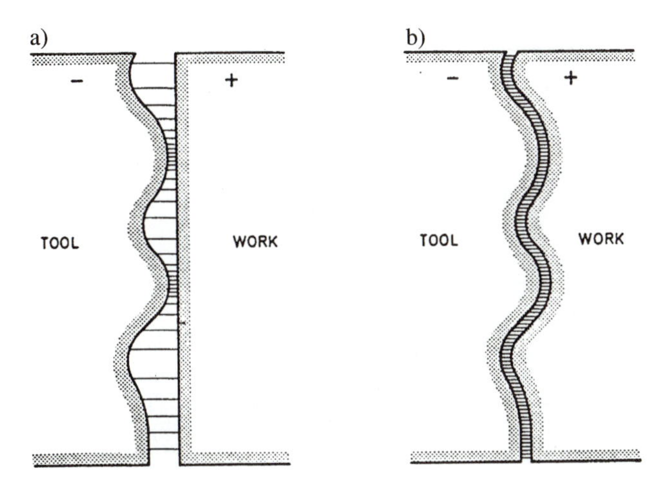

Figure 12.4
Principle of electrochemical machining (ECM). a) Most of the electrolytic action occurs at the peaks of the tool until b) eventually it evens out. [DEBARR]

and no residual stresses are left in the machined surface, nor is any thermal damage done to it. No tool wear develops. Thus, the process is very attractive. On the negative side, the rather large forces between the tool and workpiece require a strong and rigid machine structure, and anticorrosive materials such as stainless steels and plastics must be used for all the machine parts in contact with the electrolyte. A very powerful source with high currents is needed. The electrolyte pumping and filtering system is also rather complex and must include electrolyte regeneration as well as processes dealing with the disposal of spent electrolytes.

The die-sinking ECM is illustrated in Fig. 12.5. The example shown involves the generation of a generally vertical cylindrical cavity. The electrolyte is pumped through the hole in the tool that is fed with a constant velocity into the workpiece. This tool is insulated on its periphery and acts essentially by its bottom surface only, where almost all of the current passes between tool and workpiece. Some of the action occurs also at the sides of the bottom plate, generating a hole with a diameter larger than the tool by an overcut. The workpiece is clamped in a fixture attached to the base of the machine, from which it is insulated. The metal particles removed from the workpiece are flushed away by the flow of the electrolyte; they are first separated as sludge in the centrifuge and then also in the filter. Another important function of the electrolyte is cooling of the tool and workpiece. Because the conductivity of the fluid depends on its temperature, it is useful to keep the temperature constant, and a heat exchanger may be included in the hydraulic system.

In a typical application, the current may be as high as 10,000 A, producing a metal removal rate of 20 cm^3/min using electric power of 200 kW. The gap between tool and workpiece may be 0.1 mm and feed rate of the tool towards the workpiece 3 mm/min. A common electrolyte is a 10% per weight solution of sodium chloride (common salt) that is pumped into the gap between tool and workpiece at a flow velocity of 5 m/sec with an average hydrostatic pressure of 2 MPa (280 psi). In this operation a hardened steel workpiece may be machined, and the tool electrode is made of copper.

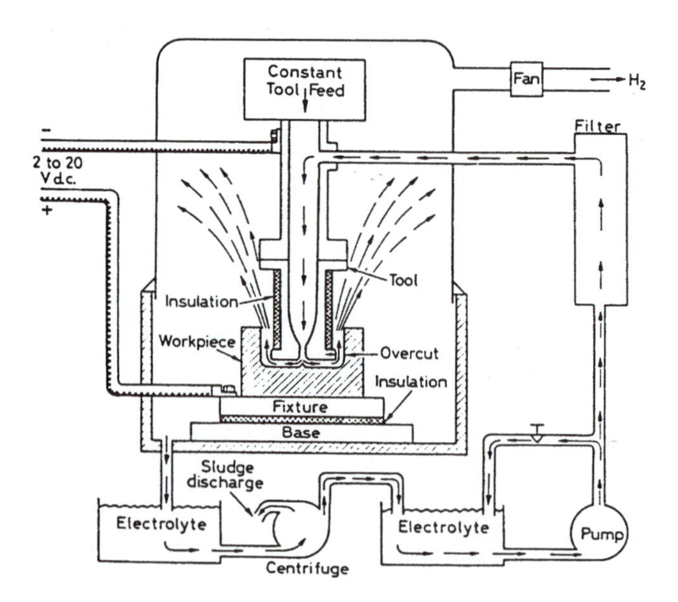

Figure 12.5
The ECM die-sinking machine. [DEBARR]

12.4.1 Metal Removal Rate: Working Gap

The laws of electrolysis were stated by Faraday as follows:

1. The amount of chemical change produced by an electric current, that is the amount of any substance deposited or dissolved, is proportional to the quantity of electricity passed (current × time).
2. The amounts of different substances deposited or dissolved by the same quantity of electricity are proportional to their chemical equivalent weights.

Because of the common shortage of notation letters and traditional usage of notations from which we do not want to deviate, the reader is asked to carefully distinguish in the following derivations between *A* for area and A for ampere and *V* for volume and V for volts.

Theoretically, the amount removed by 1 Faraday = 96,500 Coulombs = 96,500 Asec is 1 gram equivalent weight of the metal. Then the removed volume is

$$V = \frac{ItN}{96{,}500n \times \rho} \tag{12.1}$$

where V (cm^3) is the volume removed, I (A) is current, t (sec) is time, N (g) is atomic weight, n is valence, and ρ (g/cm^3) is specific weight.

For illustration, consider the electrochemical dissolution of bivalent iron, $N = 55.85$, $n = 2$, $\rho = 7.8$ g/cm^3. For a current of $I = 1000$ A,

$$V = \frac{1{,}000 \times 55.85}{96{,}500 \times 2 \times 7.8} = 0.037 \text{ cm}^3/\text{sec} = 2.2 \text{ cm}^3/\text{min}$$

Assuming 20 V as the voltage used, this metal removal would require 20 kW power. Let us compare it with the power needed in the classical chip-forming, metal-cutting operation. In Table 8.1 we find the typical specific force for mild steel $K_s = 2400$ N/mm^2, and if we cut with a typical cutting speed $v = 120$ m/min = 2m/sec, we will remove the 0.037 cm^3/sec with a chip cross section of 0.037/2 = 0.0185 mm^2 and a cutting force of 0.0185 × 2400 = 44.4 *N*, resulting in 89 W of power, which is 225 times less than for the ECM process. This illustrates how efficient classical metal cutting is compared even with one of the most effective electrical methods. Yet, ECM has a role to play in machining hardened steel and other hard metals and in producing complex shapes such as nonround holes and sculptured surfaces found in forging dies and injection molds, operations that cannot be commonly performed by traditional cutting.

Formula (12.1) can, correspondingly, be written as

$$V = V_s I t \tag{12.2}$$

$$V_s = \frac{60\,N}{96{,}500\,n\,\rho}$$

is the specific volume in cm^3 per one ampere and one minute for the particular material. Values of V_s for selected materials are given in Table 12.2.

The table shows that the values of V_s differ very little for most of the metals using a valence of 2 and again when using a valence of 3. For Cr, Co, Cu, Fe, Ni, and valence 2, V_s ranges between 2.05e−3 and 2.24e−3. Larger differences are found for the very light Mg (4.34e−3) and the very heavy W (valence 6, V_s = 0.99e−3).

In the electrolytic cell a number of chemical reactions occur at the cathode (the tool) and the anode (the workpiece). An example of a typical chemistry is the machining

TABLE 12.2 Specific Volume Vs for Selected Materials

Metal	N(g)	n	ρ (g/cm^3)	1000 V_s cm^3/(Amin)
Al	27	3	2.67	2.21
Cr	52	2/3	7.19	2.24/1.5
Co	59	2/1	8.85	2.07
Cu	63.5	2	8.96	4.41/2.20
Fe	55.8	2/3	7.86	2.21/1.47
Mg	24.3	2	1.74	4.34
Mo	95.9	3/4	10.22	1.94/1.46
Ni	58.7	2/3	8.90	2.05/1.37
Si	28.1	4	2.33	1.87
Ti	47.9	3/4	4.51	2.20/1.65
W	183.9	6/8	19.3	0.99/0.74
Zn	65.4	2	7.13	2.85

of iron in an NaCl (sodium chloride) electrolyte. The iron ions Fe^{++} leave the surface of the anode and are attracted to the negative ions that exist in the electrolyte:

$$Fe^{++} + 2(OH)^- \rightarrow Fe(OH)_2$$

The ferrous oxide mixes with air and oxidizes to Fe $(OH)_3$, a red-brown sludge. The complete reaction is

$$2Fe + 4\,H_2O + O_2 = Fe(OH)_3 + H_2$$

The hydrogen gas evolves on the cathode. The material removed from the workpiece is flushed by the flow of the electrolyte.

The processes at the electrodes require an electric potential both as the decomposition voltage and polarization voltage that cannot be used for driving the current through the electrolyte. This additional voltage ΔE must be subtracted from the voltage E imposed on the electrodes. This is commonly expressed by introducing a current efficiency η, which may range between 0.75 and 0.9.

The electrolyte is essential for the chemical reactions, and beyond that, it carries heat and reaction products from the machining zone. An effective electrolyte should have good electric conductivity, be inexpensive, readily available, nontoxic, and the less corrosive, the better. The most common electrolyte is sodium chloride, and next is sodium nitrate, which is less corrosive but also less conductive. The resistivity of an electrolyte depends on its temperature and concentration. For a 12% NaCl solution in water at 60°C, a typical value of resistivity is $r = 4\ \Omega$cm. Temperature control is necessary for high-precision work because the variation in conductivity causes variation of the gap.

The resistance R to current is proportional to the thickness g of the gap and to the resistivity r of the electrolyte, and inversely proportional to the area A:

$$R = \frac{gr}{A} \tag{12.3}$$

Then the current I obtained with voltage E between electrodes is

$$I = \frac{E}{R} = \frac{\eta EA}{gr} = JA \tag{12.4}$$

where $J = I/A$ is current density. The dimensions involved are I (A), g (cm), A (cm^2), r (Ω cm), $R(\Omega)$, and J(A/cm^2). The efficiency η expresses the fact that there are voltage losses for the discharge potential at the electrode and also losses of current to side reactions such as gas evolution. A typical current density in English units is 1000 A/in^2; thus, typically, $J = 155$ A/cm^2.

Equation (12.2) expresses the metal removal rate $M = V/t = V_s I$ as it is produced by the electrolytical action. In steady state the feed rate f times the area A equals M:

$$f_{ss}A = V_s I \tag{12.5}$$

and

$$f_{ss} = V_s J \tag{12.6}$$

For $J = 155$ A/cm^2 and iron with $V_s = 2.2\text{e}{-3}$ cm^3/(Amin), $f = 0.342$ cm/min. The size of the steady-state gap is obtained from (12.4):

$$g = \frac{\eta E}{Jr} \tag{12.7}$$

or else it is obtained as a function of the feed f:

$$g = \frac{\eta E V_s}{fr} \tag{12.8}$$

The gap, for a given standard voltage, say 20 V, a given resistivity $r = 4$ Ωcm, and efficiency $\eta = 85\%$ is smaller the higher the current density and also smaller as feed rate increases:

$$g(\text{cm}) = \frac{4.25}{J} = 4.25\frac{V_s}{f}$$

For example, for $J = 155$A/cm^2 and $V_s = 2.21\text{e}{-3}$ cm^3/(Amin) for iron, and $f =$ 0.342 cm/min, $g = 0.026$ cm.

Equations (12.7) and (12.8) can also be interpreted so as to state that current density for a given voltage E and electrolyte resistivity r increases as the gap decreases.

The gap self-regulates with changes of feed rate. If f is increased, the gap decreases and the current increases, producing a higher metal removal rate; see Eq. (12.2) as it corresponds to the higher feed rate. This can be shown by describing the transient between parallel plates of the tool and the workpiece.

The rate of the gap change is equal to the actual linear removal rate d minus the feed rate f, as provided by the drive to the tool motion:

$$\frac{\text{d}g}{\text{d}t} = d - f \tag{12.9}$$

where $d = M/A = V_s\, I/A$, and, according to Eq. (12.4)

$$I = \frac{\eta EA}{gr}$$

Then

$$\frac{\text{d}g}{\text{d}t} = \frac{\eta E V_s}{gr} - f \tag{12.10}$$

Denoting

$$C = \frac{\eta E V_s}{r} \tag{12.11}$$

Eq. (12.8) becomes

$$\frac{dg}{dt} = \frac{C}{g} - f \tag{12.12}$$

Inverting Eq. (12.10) and separating the variables,

$$dt = \frac{g}{C - fg} dg$$

and substituting $u = C - fg$, $du = -fdg$, $g = (C - u)/f$, $dg = -du/f$,

$$dt = \frac{u - C}{f^2 u} du$$

$$dt = \frac{1}{f^2}\left(1 - \frac{C}{u}\right) du$$

and integrating

$$t = \left| \frac{u}{f^2} - \frac{C}{f^2} \ln u \right|_{u_0}^{u}$$

where $u_0 = C - fg_0$, $u = C - fg$ and, finally,

$$t = \frac{1}{f}(g_o - g) + \frac{C}{f^2} \ln\left(\frac{C - fg_0}{C - fg}\right) \tag{12.13}$$

The solution (12.13) of the differential equation (12.12) can also easily be obtained by digital time-domain simulation. This is done in the following example.

EXAMPLE 12.1 Transient Gap in ECM ▼

The parameters of the case are $V_s = 0.0021$ cm^3/(min A); specific resistivity of the electrolyte $r = 4\ \Omega$ cm, feed $f = 0.342$ cm/min, voltage $E = 20$ V, efficiency $\eta = 0.85$. With these parameters, it has been established in Eqs. (12.6) and (12.7) that the steady-state gap is $g_{ss} = 0.026$ cm.

Let us compute the transient for two different initial gaps: (*a*) $g_{in} = 0.04$ cm, (*b*) $g_{in} = 0.01$ cm. The Matlab program is called "ecm2." It starts from Eq. (12.12).

```
function [t,g1,g2]=ecm2(gin1,gin2,f)
dt=0.001;g1(1)=gin1; g2(1)=gin2;
t(1)=0;eta=0.85;E=20;Vs=0.0021;r=4;
C=eta*E*Vs/r;
for n=1:599
dg1=(C/g1(n)-f)*dt;
dg2=(C/g2(n)-f)*dt;
g1(n+1)=g1(n)+dg1;
g2(n+1)=g2(n)+dg2;
t(n+1)=t(n)+dt;
end
end
```

The results of the computations are shown in Fig. 12.6. Note that in both cases the gap self-regulates to the steady state value of 0.026; it takes about 0.3–0.4 min (18–24 sec). In that time the mechanical feed traversed a distance of about 0.12 cm.

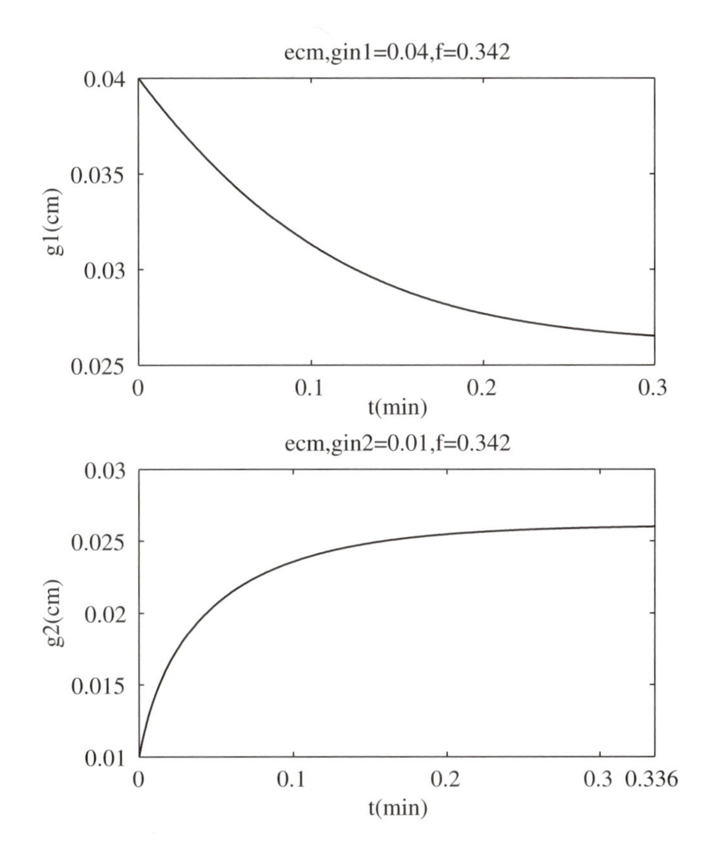

Figure 12.6
Transient of gap self-regulation in ECM. The equilibrium gap corresponding to feed rate of 0.342 cm/min is 0.026 cm. Starting from a larger gap (top) or from a smaller gap (bottom) the steady state is reached in about 0.4 min.

This example illustrates the excellent capability of the process to adapt to a chosen feed rate. For higher feed rates, the gap is smaller, precision is higher, and experience shows that the surface finish is also better. The feed rate is limited by the current available, and by the fact that with a smaller gap the pressure in the electrolyte increases which may slow down its flow. This leads to a corresponding increase in temperature, which may produce boiling of the electrolyte and disruption of the process. In addition, with smaller gaps there is a danger of sparking and damage to the machined surface. ▲

EXAMPLE 12.2 Electrolyte Flow ▼

Assuming that all the power of the operation converts to heat that remains in the electrolyte, the temperature rise ΔT in passing through a length Δx of the gap g between parallel plates is obtained as

$$\Delta T = \frac{RI^2}{gwv\,\rho\,c}$$

where w is the width of the gap, v (cm/sec) is the mean flow velocity of the electrolyte, and ρ and c are the density and specific heat of the electrolyte. Further,

$$\Delta T = \frac{grJ^2 w\Delta x}{gwv\,\rho\,c} = \frac{rJ^2\Delta x}{v\rho c} \tag{12.14}$$

The resistivity r is temperature-dependent:

$$r = \frac{r_o}{1 + \alpha \Delta T} \tag{12.15}$$

where r_o is resistivity at 25°C, typically $r = 7\ \Omega$ cm, and $\alpha = 0.02/°C$, and ΔT is the temperature change. Also $\rho = 1$ g/cm^3 and $c = 4.18$ J/(g°C) = 418 Nm/(g°C). Combining (12.14) and (12.15) and expressing v, we have

$$v = \frac{r_o J^2 \Delta x}{\Delta T(1 + \alpha \Delta T)\rho c} \tag{12.16}$$

Using the above-given parameters, current density $J = 155$A/cm^2, and length of the gap $\Delta x = 5$ cm, and $\Delta T = 75$°C, we have

$$v = \frac{7 \times 155^2 \times 5}{75\,(1 + 0.02 \times 75) \times 1 \times 4.18}$$

$$= 1072 \text{ cm/sec} = 10.72 \text{ m/sec}$$

At this velocity the electrolyte would boil. Therefore, it is better to use a velocity of about 30 m/sec. Obviously, it is primarily the need to avoid the overheating of the electrolyte that determines the electrolyte flow velocity required.

Now, let us determine the pressure needed at the entrance to the gap to produce this flow. It is preferable to use turbulent rather than laminar flow because it is necessary to vigorously flush the hydroxide away from the anode and also maintain the flow at the cathode to release the bubbles of the hydrogen. With the rather high flow velocity, rather high Reynolds numbers are obtained both for the flow between parallel plates, typical for die sinking, and the flow through a hollow drilling electrode.

The Reynolds number is obtained as follows:

$$Re = \frac{\rho v D}{\eta}$$

where D is the "hydraulic mean diameter," which is $D = d$, the inner diameter of a cylindrical tube and $D = 2g$ for flat channels. Typically, viscosity $\eta = 1.2$ cP at 20°C and 0.6 cP at 60°C, where 1 cP = 0.01 g/(cm sec). So, for $v = 3000$ cm/sec and a gap $g = 0.025$ cm, and $D = 0.05$ cm,

$$Re = \frac{3000 \times 0.05}{0.01} = 15{,}000$$

For $Re > 3000$, flow is usually turbulent, and for the above value it certainly is so. The pressure required to overcome the viscous forces at these values of R is given as

$$p = \frac{0.3164\ \rho v^2 x}{2DRe^{0.25}} \tag{12.17}$$

In this formula both the coefficient 0.3164 and Re are dimensionless, $\rho = 1000$ kg/m^3 = 1000 N sec^2/m^4, $v = 30$ m/sec, $x = 0.05$ m, and $D = 0.0005$ m

$$p = 2.573e^6 \text{ N/m}^2 = 2.573 \text{ MPa} \tag{12.18}$$

The average pressure over the length of $x = 5$cm is $p_{av} = 1.286$ MPa. For an area $A = 5 \times 3$ cm, the force between the tool and the workpiece will be $F = 1929$ N

(428 lbf). For larger workpiece areas, the force may be rather high, requiring substantial stiffness of the frame of the machine and of the feed drive. ▲

Tool Shape

In ECM the work surface does not reproduce exactly the surface of the tool. Actually, they may differ substantially. Let us take a tool in the form of a cylinder being sunk axially into the workpiece (see Fig. 12.7a). Most of the work is done by the face of the tool where the gap is smallest. However, the peripheral surface of the cylinder also participates in the process, and although the current at the top of the hole is smallest because of the largest gap, and the metal removal rate is much lower than at the bottom, the process operates there for the entire time of the feeding of the tool, and the removed amount accumulates there. For some specific electrolytes, the overcut is reduced by an oxide layer on the anode that reduces metal removal rates for smaller current density; see Fig. 12.7b. The best accuracy of the hole is achieved by insulating the outer surface of the tool by means of a thin silicon carbide and silicon nitride coating; see Fig. 12.7c. Figure 12.8a shows a tool with three plane regions inclined by 0°, θ, and 90° to the feed direction. The equilibrium gap is g_1, $g_2 = g_1/\cos\theta$, and for the surface parallel with the feed, the gap g_3 has a parabolic shape. For a curve B in the XY plane on the tool, the curve A develops on the workpiece; see Fig. 12.8b. Both surfaces are cylindrical in the Z direction. A derivation of the shape of the tool $y_1 = f(x_1)$ is presented in [25] while assuming the shape of the workpiece as

$$y = a + bx + cx^2$$

The solution is obtained in the form

$$y_1 = a + b\,x_1 + cx_1 - g\left[1 + \frac{(b + 2cx_1)^2}{1 + 2cg}\right]$$

where g is the equilibrium gap for a pair of surfaces parallel with the X-axis. Figure 12.9 presents three examples of tool and work shapes. In (*a*) the work surface form shows a transition between the equilibrium gaps at the sides. In (*b*) and (*c*) examples of shapes are given for tools with a plane of symmetry along the feed direction. It is obvious that a computer program may be used to derive the shape of the tool from the desired shape of the work surface. The best approach is to use a simplified computation and try a tool and then correct it for errors on the work surface.

The materials commonly used for the tool are copper, brass, bronze, and stainless steel, and most are easy to machine on an NC milling machine. However, it is also possible, if a prototype workpiece is available, to use it for producing the tool electrochemically. Then the need for cut-and-try corrections is minimized. Subsequently, holes for the flow of the electrolyte are drilled into the tool. Once a satisfactory tool

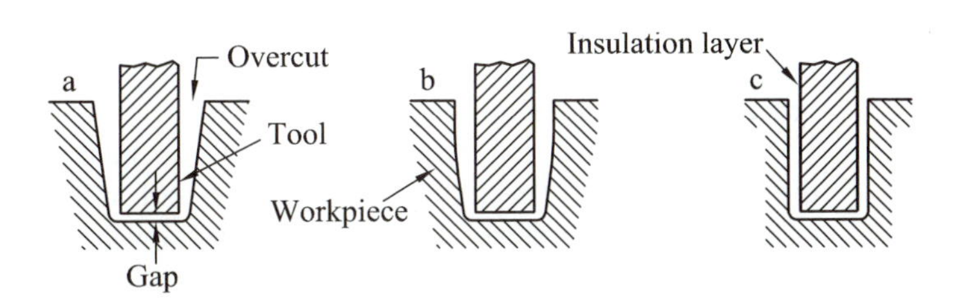

Figure 12.7
a) A large overcut in ECM of a hole is generated by the action on the outer cylindrical surface of the tool. It can be reduced by b) passivation of the tool surface or c) insulation coating on the cylindrical part to leave active just the face of the tool. [SNOEYS]

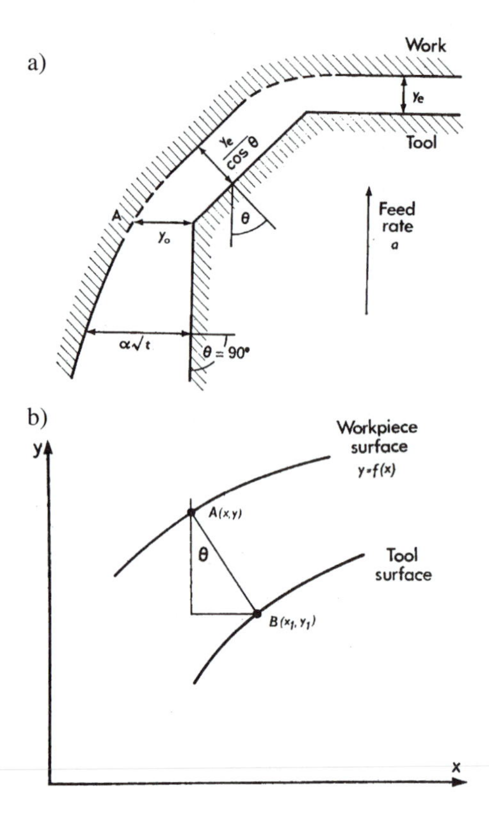

Figure 12.8
a) Equilibrium gap for three regions with different slopes in ECM. b) Point *A* on workpiece surface transforms to point *B* on tool surface. [DEBARR]

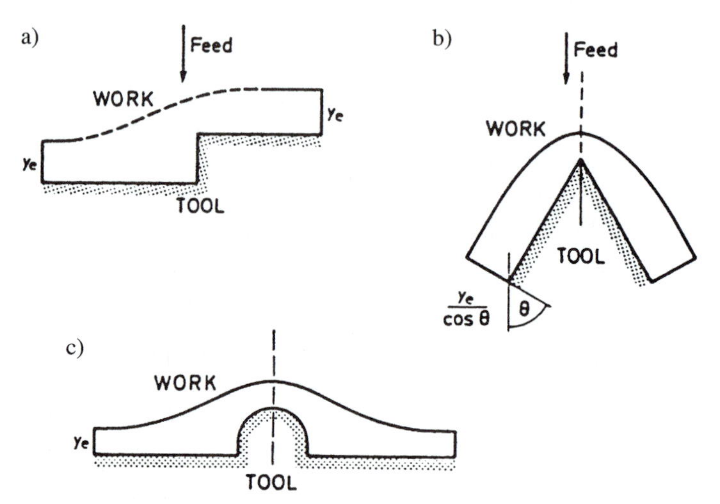

Figure 12.9
Comparison of equilibrium shapes of tool and workpiece in ECM. a) Effect of a step between two flats; b), c) shapes for tools symmetrical along an axis parallel with feed. [DEBARR]

has been produced, it can be used many times, because it will not wear at all if the ECM process is properly controlled.

Let us make a few comments regarding the materials used in the construction of the machine tool. The fixtures holding the tool and workpiece, the work tank, the enclosure, the reservoirs, and piping are made of plastic such as PVC, polyethylene, solventless epoxy, chlorinated rubber, or composites such as fiberglass. All these materials have good chemical resistance to most of the electrolytes. The structure of the machine has to be strong and rigid so as not to distort under the rather high forces between tool and workpiece, and it is made of steel weldments painted with a strong layer of corrosion-resistant paint. The electrolyte pump and some other parts of the hydraulic circuit are made of stainless steel.

The ECM process is used in several well-established forms. We have been describing *die sinking,* and Figs. 12.10–Fig. 12.12 show examples of this kind of work. Machining time for the connecting rod forging die in Fig. 12.10 is given as 18 min, and tolerances of form as well as from part to part have been maintained within 0.05 mm. The advantage of the fact that the electrode does not wear means that it can be used repeatedly. It can also be used for repair work as the forging die gets worn. If a new electrode is eventually needed it can be cast from the prototype cavity.

The photograph in Fig. 12.11 illustrates the high surface quality machined on a stainless steel part. The identical reflection patterns indicate the excellent repeatability of form. In Fig. 12.12 the two halves of an electrode are shown for a turbine blade produced by ECM die sinking. A selection of jet engine compressor blades with platforms and leading and trailing edges finished in a single operation is presented in Fig. 12.13. Blades with a surface area up to 200 cm^2 are machined with a current density of 100 A/cm^2. Lower current density would extend the machinable area at the cost of surface finish and of metal removal rate. A number of other applications of ECM are described below.

Electrolytic deburring is based on the ability of ECM to attack most forcefully the surfaces closest to the electrode, and this applies especially to burrs. The workpiece can simply be suspended in the electrolyte near to a plane cathode, but more effective action is achieved by shaping the cathode according to the form of the workpiece and to the location of the burrs to be removed. An example is shown in Fig. 12.14. Unlike the sinking operation, in deburring the cathode is stationary. The electrolytic action removes the burrs without affecting the other surfaces of the workpiece.

Figure 12.10
Impression for connecting rod forging die, ECM-produced in a solid blank in 18 min. [ANOCUT]

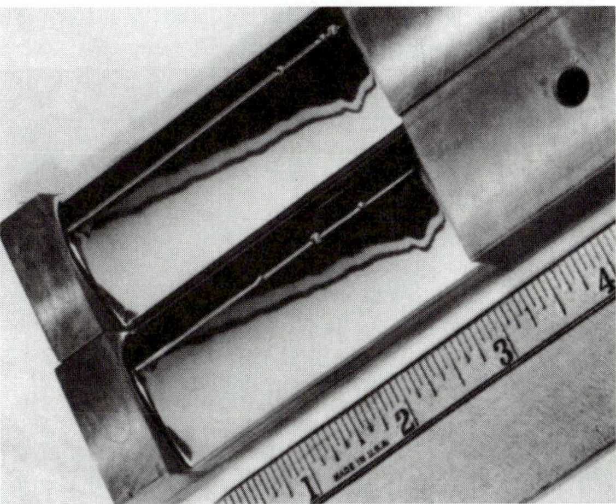

Figure 12.11
Stainless steel parts ECM-produced with the same electrode to show high surface quality and repeatability. [ANOCUT]

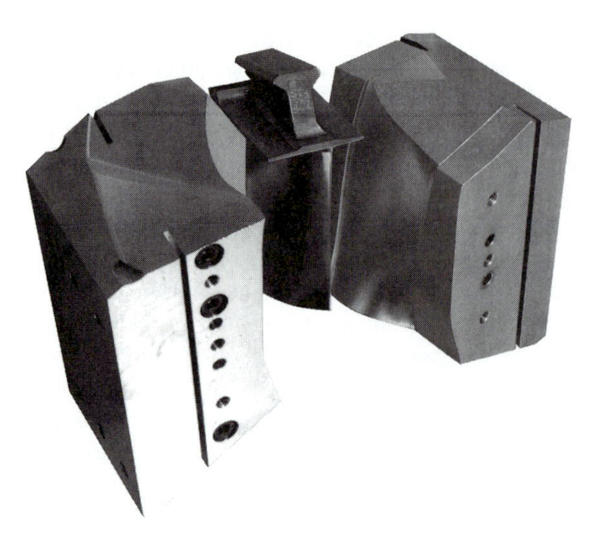

Figure 12.12
A turbine blade made by ECM and the two half-electrodes used. [AMCHEM]

Figure 12.13
Selection of blades with platforms and leading and trailing edges finished in a single operation. [AMCHEM]

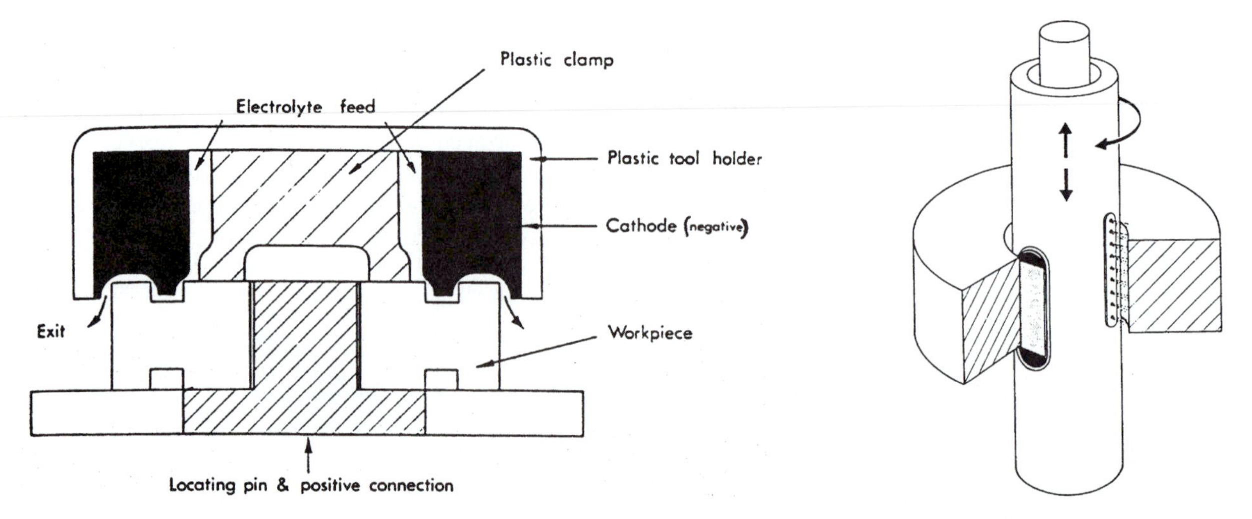

Figure 12.14
Electrolytic deburring. The cathode is stationary. Burrs are removed at the edges of the workpiece. [DEBARR]

Figure 12.15
Electrolytic honing is a combination of abrasive and electrolytic actions resulting in higher accuracy and shorter machining time. [DEBARR]

In *electrolytic honing,* the ECM action (see Fig 12.15) is used to increase several times the metal removal rate, especially on hard materials. The action is cooler, and higher accuracy is obtained on thin walls because the honing pressure is lower. Conventional honing stones are used, and current is passed to the workpiece via electrodes mounted between the stones with a small gap versus the surface being machined. The electrolyte is supplied through holes drilled in the electrodes. The result is a combination of abrasive and ECM processes.

With *electrolytic grinding,* the ECM process enhances the abrasive process. The original application, which is still an excellent example, is grinding of sintered carbide inserts for cutting tools, where a diamond grinding wheel is used in a metallic bond. The ECM process speeds up the metal removal rate and increases the life of the diamonds. Electrolyte is supplied into the gap that is created by the diamond grains separating the body of the wheel and the ground surface, as shown in Fig. 12.16. The grains also remove an electrically resistant film from the surface of the workpiece. In the carbide grinding operation savings are quoted of 90% of the wheel cost and 50% of labor cost. In general, the process is applied to hard materials, including hardened steel. Grinding of fragile parts such as honeycomb structures is possible. Titanium and alloy steels can be ground with wheels with aluminum oxide grains, resulting in higher removal rates and longer life of the grinding wheel. No thermal damage is imparted to the ground surface. Surface finish of 0.2 μm is obtainable. Metal removal rates with a 25-mm-wide wheel are on the order of 2.5 mm/min/1000 A. Current densities of 125 A/cm^2 are used.

Mirror grinding with ECM dressing is one of the latest developments in grinding mirror-quality surfaces. It is called Elid (Electrolytic In-Process Dressing) [2]. It is stated that no efficient grinding technique can exist without an efficient dressing technique. This successful technique differs from electrolytic grinding in that it uses a conventional superabrasive, metal-bonded grinding wheel, and instead of any of the corrosive electrolytes, it uses a conventional coolant. The electrolytic action is provided between a negative copper electrode and the metallic bond of the wheel as the anode. This is just the reverse of the electrolytic grinding process, in which the grinding wheel is the cathode and the workpiece is the anode, and the electrolytic action is between wheel and workpiece.

The grinding wheel grains used are either diamond or CBN (cubic boron nitride). For mirror-surface grinding, very fine grain sizes of 2 μm down to 0.3 μm diameter are

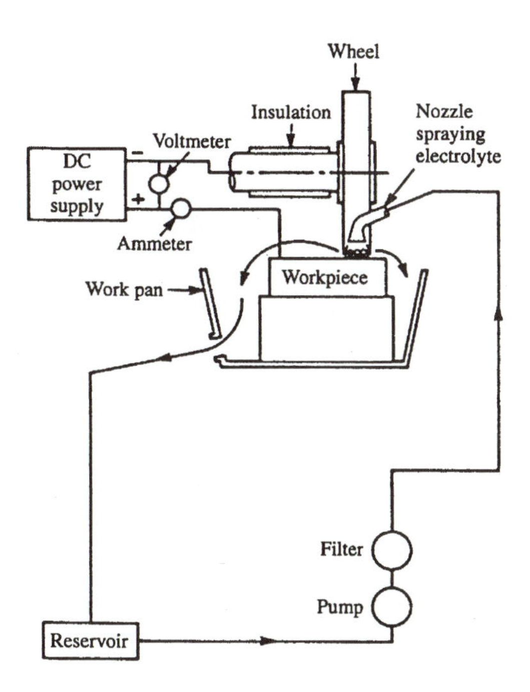

Figure 12.16
Electrochemical grinding (ECG). Grinding wheel with diamond grains and metallic bond is used. Much improved life of the grains results, and higher metal removal rate is obtained. Grinding of sintered carbide tools is a typical application. [WELLER]

applied. The bond materials used are cast iron or cobalt. Workpiece materials to which this process has been applied include silicon, silicon nitride, zirconia, silicon carbide, and tungsten carbide. This illustrates another difference from electrolytic grinding, which can only be used on metallic materials. In all those applications, mirror-quality surfaces are obtained without the necessity of subsequent lapping. Surface roughness values are in the range of $R_{max} = 15$ to 100 μm and $R_a = 2$ to 10 μm. The process is used to produce lenses and mirrors. Another significant application is slicing and surface grinding of silicon wafers from the monocrystal ingot for the production of VLSI electronic chips.

Two specialized processes of *electrochemical drilling* have been developed for drilling small-diameter, long holes in superalloys used for jet engine compressor and turbine blades and in refractory metals. The first is called shaped tube electrolytic machining (STEM). It is used for drilling round or shaped holes using acid electrolytes. Commonly, multiple holes are drilled simultaneously; see Fig. 12.17. The electrolyte is chosen relative to the work material and is generally either 10–20% sulphuric acid or 15–30% nitric acid. The electrolyte is fed through an acid-resistant (usually titanium) cathode tube with the desired shape of the cross section or at least with a shaped end. The tube is insulated along its whole length except for the tip. The electrolyte returns via a narrow gap between the tube insulation and the wall of the hole. The electrode stems are guided through two- or three-tier plastic guide blocks, which may slightly modify the direction of the holes. Hole diameters ranging from 0.5 mm to 6 mm with up to 200:1 depth-to-diameter ratio can be produced with tens or several hundred holes made in parallel. Drilling feed rates of 2–4 mm/min are achieved. Acid is used as the electrolyte because it keeps the removed metal in solution or suspension and thus does not clog the passages.

The process is primarily used for machining air-cooling and weight-reduction holes in jet engine blades. Nickel, cobalt, molybdenum, titanium, and stainless steel superalloys are the successfully drilled materials.

Capillary drilling also uses acid electrolytes. The tubes in this case are fine glass capillaries with a noble metal wire, such as platinum, passing through the center of the tube. The tube carries the acid to the work area, and the wire carries the electric power.

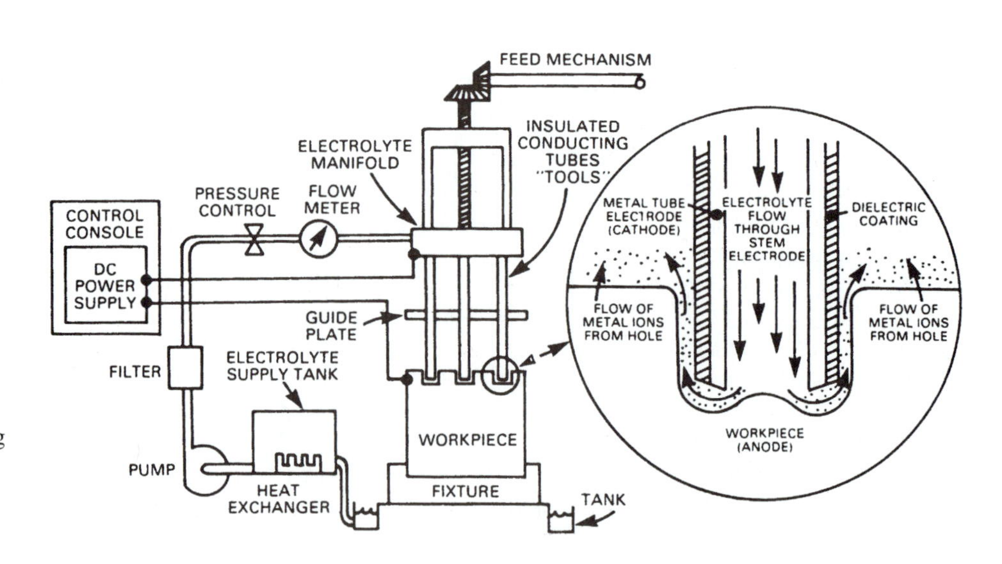

Figure 12.17
Shaped tube electrolytic machining (STEM). Acid electrolyte is used in drilling multiple holes with length-to-diameter ratio up to 200:1. [WELLER]

The process is used to make holes of less than 0.4 mm diameter with a length-to-diameter ratio of more than 10:1. Special emphasis is given to acid filtration to prevent debris from blocking the small bore of the glass tubes. As in STEM drilling, the metal finish is of high integrity without a recast layer. Feed rates are similar to those of the STEM process. Figure 12.18 shows examples of holes drilled in turbine blades by both processes described here.

12.5 CHEMICAL MACHINING (CHM), PHOTOCHEMICAL MACHINING (PCM)

Chemical machining is the controlled dissolution of workpiece material (etching) by means of a strong chemical reagent (etchant). For more information see Ref. [31]. The principle of the operation is shown in Fig. 12.19. The material is covered by a protective coating resistant to the etchant, which is usually a liquid although, exceptionally, in micromachining a gas etchant may be used. The covered workpiece is fully submerged in the etchant liquid, which is moderately stirred and often also heated to 40°–80°C to speed up the process. The protective mask has openings of the form that is to be machined, differing by the amount of the "overcut." The process produces surfaces sunken below the initial surface to a very uniform depth. Rather shallow depths are produced up to a maximum of about 12.5 mm, but mostly much less, on the order of 0.05–1.0 mm. The size of the workpieces ranges from very small to several meters square, and the equipment is simple, consisting of a tank of corresponding dimensions, and pumping and ventilation equipment. No machining forces are involved. Good surface finish is obtained without residual stresses and without any burrs. Thin sheets of materials may be machined through in a process called chemical blanking.

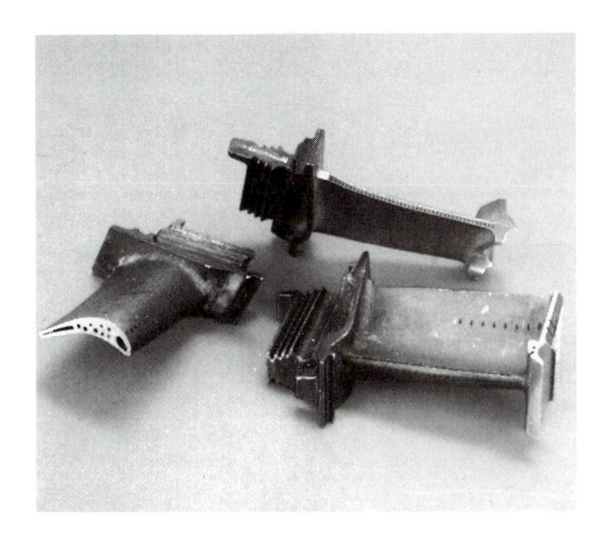

Figure 12.18
Turbine blades with edge holes made by capillary drilling and radial holes by STEM drilling. [AMCHEM]

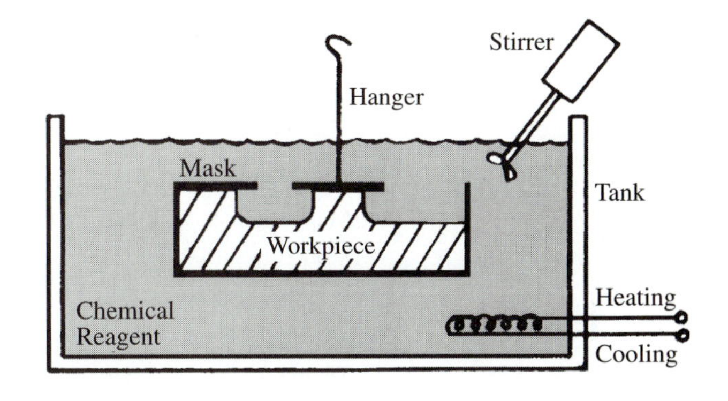

Figure 12.19
Chemical machining (CHM) is a process of controlled etching of workpiece material by means of a strong chemical agent over areas not protected by the mask. [BELLOWS]

Multilevel depth can be achieved by successively stripping masks and resubmerging the workpiece. Controlled rate of submersion or withdrawal from the bath will produce sloped surfaces.

Many different materials can be processed by CHM. Large, shallow areas are especially suitable for this process. Short-run, quick-change, low-cost tooling offers process flexibility. Because of the strong chemicals used, operators must be protected against splashes of the etchant and fumes. Some of the combinations of work material, maskants, etchants, and etch rates are listed in Table 12.3.

It is obvious from the table that the process is relatively slow, especially if larger depth is required. The etch factor is the ratio of the undercut u over the depth of cut d; see Fig. 12.20.

The CHM processes are classified according to the type of application as *chemical milling,* where areas of surface depressions are produced, or else the whole surface of a part is machined; and *chemical blanking,* where intricate shapes are produced out of thin sheets of the material. They are further distinguished according to the method by which the maskant openings are produced, either by *scribe-cut-peel* or by a photographic technique—*photochemical machining*—or else by *silk-screen printing.*

Chemical Milling

The chemical milling process originated in the aircraft industry, where it is applied to reduce the thickness of large aluminum panels over selected areas so as to reduce the weight of the part while not substantially affecting the strength and stiffness of the panel. In these operations, in spite of the slow etching rates, the overall volumetric metal removal rates may be relatively high. For instance, with an etching rate of 0.025 mm/min over an area of $5 m^2$, the *MRR* is 125 cm^3/min, which would just about be equivalent to the mechanical machining operation of end milling of aluminum using 2.6 kW of power. An example of chemical milling of an aluminum aircraft wing skin with a total surface of 11.9 m^2 with two-thirds of it being etched is given in [26]. The maskant used was an elastomeric material that was flow-coated and then scribed with the required pattern. Due to the depth tolerances required, a standard, proprietary alkaline etchant was used, and the part submerged in a deep tank. The process went on in

TABLE 12.3 Maskants and Etchants

Workpiece Material	Etchant	Maskant Operating	Etch rate mm/min	Etch factor
Aluminum	$Fe\,Cl_3$	Polymers	0.013–0.025	1.5–2.0
	NaOH	Neoprene	0.020–0.030	
Magnesium	HNO_3	Polymers	1.0–2.0	1.0
Copper	$Fe\,Cl_3$	Polymers	2.0	2.5–3.0
	$Cu\,Cl_2$		1.2	
Steel	$HCl: HNO_3$	Polymers	0.025	2.0
	$Fe\,Cl_3$		0.025	
Titanium	HF	Polymers	0.025	1.0
	$HF: HNO_3$			
Nickel	$Fe\,Cl_3$	Polyethylene	0.013–0.038	1.0–3.0
			0.013–0.038	
Silicon	$HNO_3:HF:H_2O$	Polymers	very slow	—

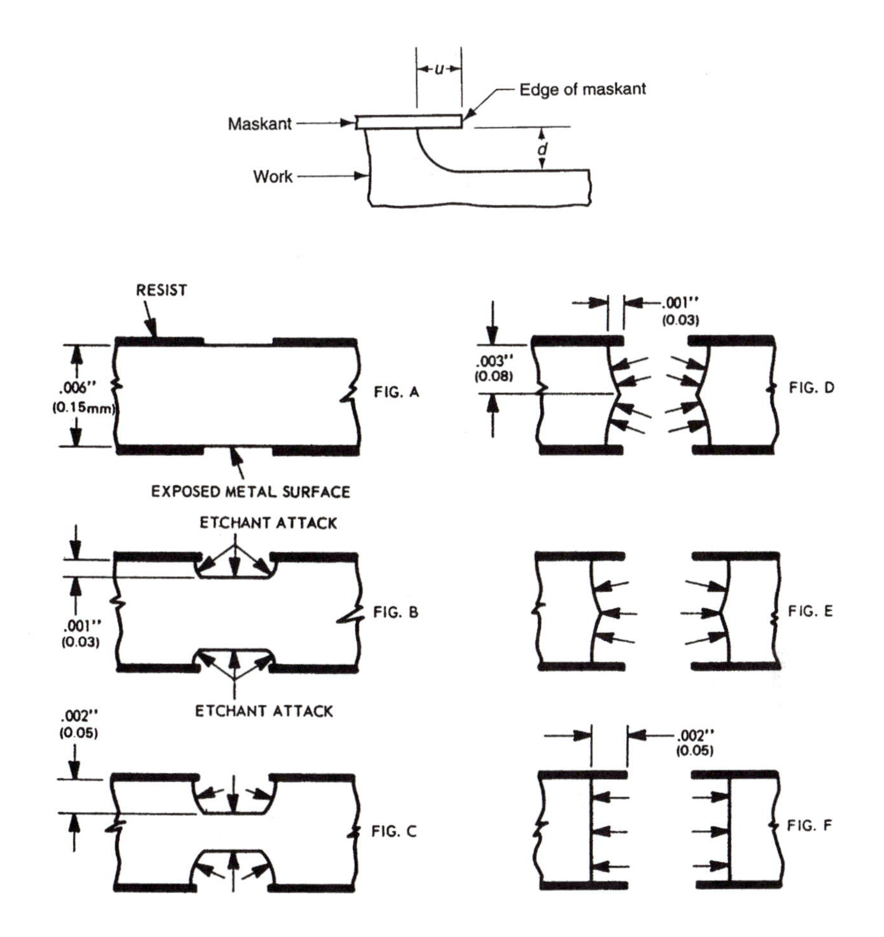

Figure 12.20
Undercut in CHM. An example of the progress of a two-sided operation (chemical blanking) is shown in steps A to F. [WELLER]

several steps because various parts of the surface had to be machined to different depths. Another example is given of an aluminum vent screen used in a helicopter. Low-volume production of 15–30 parts was needed. The material was 1.6 mm thick and was etched through (blanked) simultaneously from both sides. A lacquer-based maskant was applied through a silk screen on a semiautomatic printing machine, and conveyorized spray etching was used with hydrochloric acid as the etchant. A rather high etch rate of 0.13 mm/min was achieved.

In these operations the typical sequence of the CHM process is used. First the surface of the workpiece is cleaned to ensure the adhesion of the maskant: vapor degreasing, alkaline cleaning, and deoxidizing. The second step is the application of the maskant, by dipping, flow coating, or spraying. Several coats are applied. Then the maskant is cured for several hours in air or, in the case of steel, titanium, and refractory alloys, oven-baked for half an hour at 107°C. The next step is scribing, which is often done with the help of templates machined of fiberglass or aluminum. The mask is cut out manually using a knife or else a "hot knife" similar to a soldering iron but with a sharp tip; it melts the mask. More recently, the need for templates has been eliminated, and the mask is cut by an NC laser beam; see Fig. 12.21. Subsequently, the etching process is carried out. Finally, the mask is removed by hand stripping or by immersing the masked part in a suitable demasking solution, followed by rinsing.

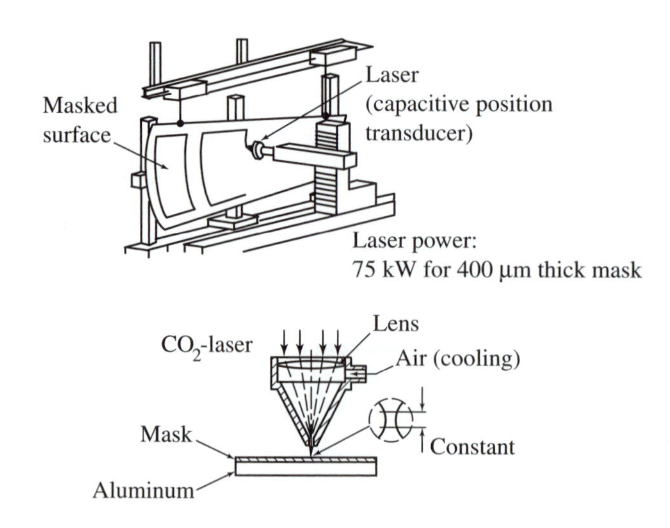

Figure 12.21
NC laser cutting of masks on large surfaces to be chemically milled. [FROM MASKOW, REF. [11]]

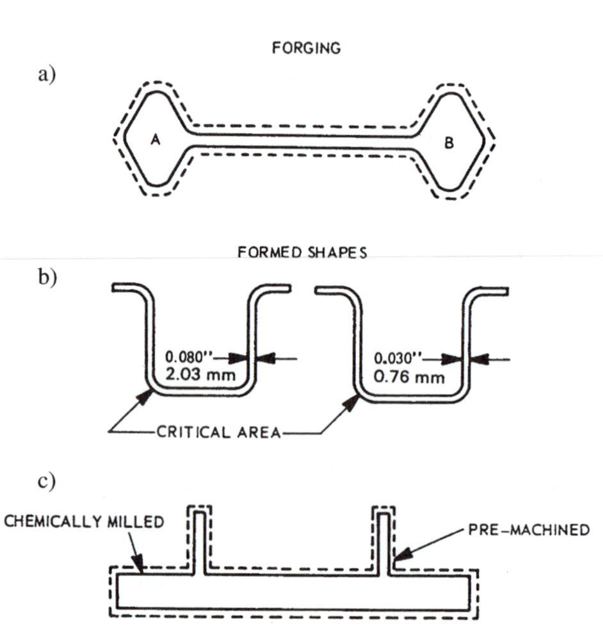

Figure 12.22
Thinning of parts by CHM: a) A forged connecting rod is made lighter; b) a molding that could not be made on a press brake to the desired thickness of 0.76 mm was easily made 2.03 mm thick and subsequently thinned; c) thin walls of aluminum casting were premachined and then thinned by CHM. [WELLER]

In a different kind of application CHM is used to thin out walls, webs, and ribs of parts that have been produced in an operation such as machining, forging, casting, or sheet metal forming in which the walls, webs, or ribs could not be made as thin as required. Three examples of such cases are presented in [26] and shown in Fig. 12.22. Part (*a*) is a forging in which the central portion must be thinner than practically obtainable in the die-forging operation. It is then best to make the heavier ends also thicker by the amount we wish to remove from the central part and then submerge the whole forging into the etching bath. This eliminates the need of masking and scribing. The case illustrated involved a connecting rod that had to be light for use in high-speed engines. The case depicted as (*b*) is one of a molding with a hat-shaped cross section produced on a press brake from a sheet 2 mm thick. However, the molding is requested with material thickness of only 0.75 mm, which cannot easily be formed because of the difficulty of handling such a thin strip and the consequent probability of warping. Therefore, the CHM process is applied by submerging the whole part in the etching bath. The third case, shown in (*c*), is the case of an aluminum casting in the form of a panel with deep pockets separated by stiffening ribs. These can be rough pre-milled and the final wall thickness achieved by subsequent CHM.

Photochemical Blanking (PCB)

The photochemical blanking process replaces the shearing operations of punching and blanking carried out on mechanical presses, as described in Chapters 4 and 5. The PCB process is often preferable because it is so much simpler to produce such a tool as the photo-resist mask than producing the punching and blanking dies, which are economically justifiable only for large production quantities. The PCB process is preferable for very thin parts and parts made of hard and brittle materials that would break in the mechanical operation, where high forces and stresses are generated that do not exist in PCB. Another advantage is that no burr is produced in PCB.

The maskant used in the process, called the photo resist, is such that it polymerizes if exposed to ultraviolet light. It is subsequently developed with the result that the unexposed part of the mask is dissolved and removed and becomes the opening through which the etchant acts on the workpiece.

The steps of the process are illustrated in Fig. 12.23. The workpiece material is first degreased and cleaned using an appropriate chemical. Then it is coated with photo resist on both sides. The coating is applied by dipping, spraying, flow coating, roller coating, or laminating. The choice depends on the shape and size of the part and on the type of resist being used. After coating, the part is baked to drive off the solvents. Then it is exposed to UV light through negatives of the mask pattern. These negatives have been photographically produced from a drawing of the pattern of the mask. This drawing may be prepared to a larger scale and reduced photographically many times so as to diminish inaccuracies. After exposure (printing), the resist is developed to produce the mask. After development, all the remnants of the unexposed resist are washed away. Any residual solvents may be removed by baking under infrared lamps or in air ovens, and the mask is left to cool down. Next follows the etching operation. Then the mask is removed either mechanically or by chemical action, and the workpiece is washed and dried.

Examples of parts produced by photochemical blanking are shown in Fig. 12.24. The accuracy of the mask depends mainly on the accuracy of the original drawing from which the negatives of the mask are photographically produced. The tolerances of the drawing depend on the accuracy of the plotter, which may be on the order of 0.05 mm. For small parts the drawing may be reduced 20 times using a high-class camera, and tolerances of 0.01 mm may be achieved on the print.

An alternative to the photochemical way of making the mask is the use of screen-printed resist. The clean blank of the workpiece is placed in a screen printer, and acid-resistant ink is screened onto it to produce the mask. The screen printing process is less accurate than the one using the photo resist, but it is simpler and substantially cheaper, and it is preferable whenever the required tolerances permit.

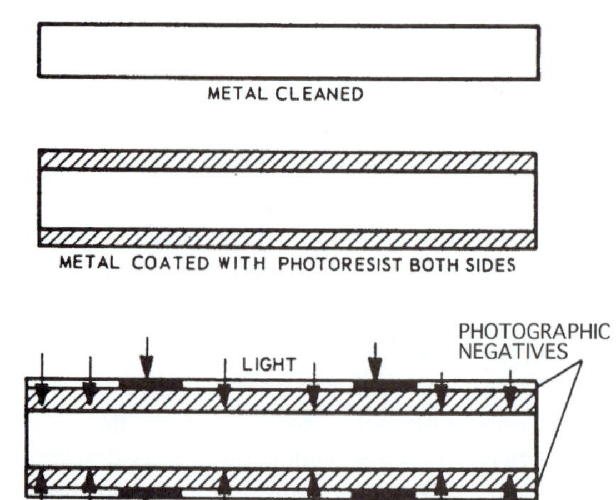

RESIST EXPOSED THROUGH NEGATIVES (DOUBLE SIDED)

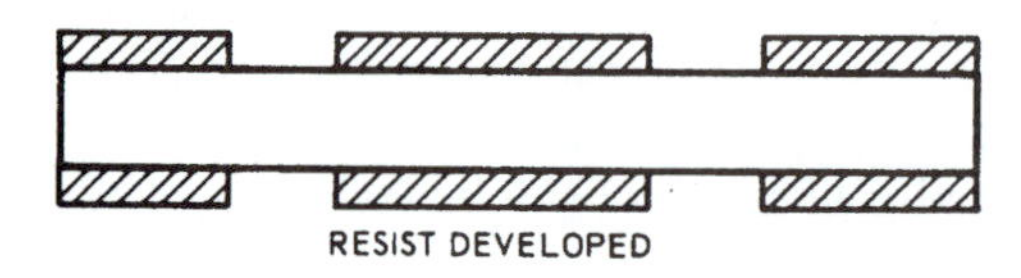

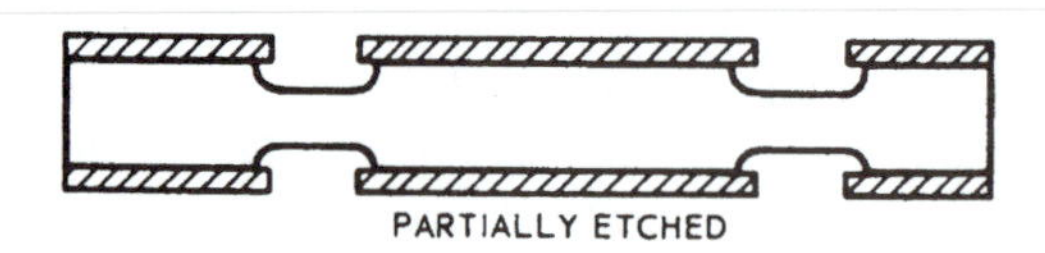

Figure 12.23
Steps in photochemical blanking (PCB). [WELLER]

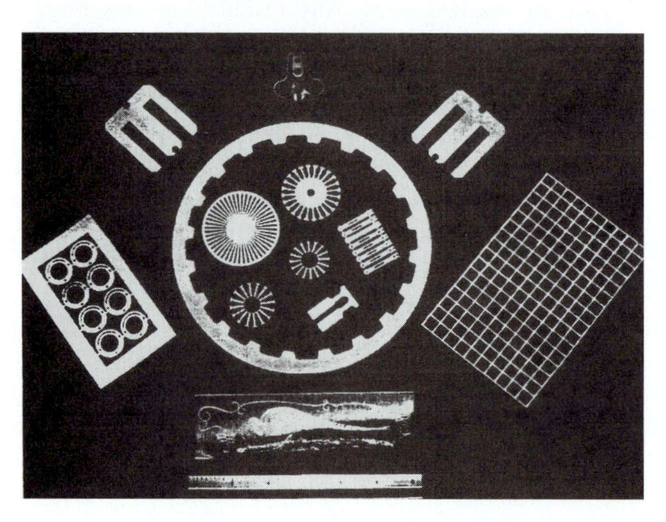

Figure 12.24
Parts produced by photochemical blanking. [CHEMCUT]

12.6 ELECTRO-DISCHARGE MACHINING (EDM)

EDM is one of the most commonly used nontraditional machining processes. It is based on fast, repetitive spark erosion of the workpiece and, like ECM, can only be used on electrically conductive work materials. It is popular in spite of much lower metal removal rates, rather high tool wear (compared to zero tool wear in ECM), and the production of a heat-affected zone on the machined surface. One of its main modes of operation, die sinking is in its application analogous to ECM and we will explain why, in spite of the two mentioned disadvantages, it is more widely used. The second mode of operation, wire EDM is unique in its application to the production of punching dies and of extrusion dies. The two basic modes will be discussed separately.

EDM Die Sinking

The principle of the metal removal in EDM die sinking is explained in Fig. 12.25. The tool is being fed into the workpiece under a servomechanism-type drive that maintains a constant gap within the range of 0.01 mm for finishing and 0.04 mm for roughing. The gap between tool and work is filled with an insulating dielectric fluid. DC pulsed voltage is applied across the gap. Electrical spark discharges occur in rapid succession between tool and workpiece, at points of the currently smallest gap. The discharge creates a small cavity in the surface of the workpiece and another one in the tool surface. Depending on the conditions of the process, the ratio *W/R* between the wear of the workpiece (volume removed) and that on the tool varies between 1 and 100. This means that the tool wear ranges from equal to volume removed down to only one hundredth of it. The discharges repeat with the frequency of the power supply voltage, which is controlled in a range between 10 and 500 kHz. The voltage level is between 50 and 100 V (of open circuit), and average currents range between 30 and 1500 A, the latter value corresponding to a machine with a 2m^2 table area. The details of the process as presented by one of the pioneering producers of EDM equipment are shown in Fig. 12.26. Instantaneous values of voltage U and current I are plotted under the pictures of the individual phases of one discharge. The frequency of the pulses has to adapt to the rise and fall time of the current, and these in turn depend on the power-source characteristics, on the type of the dielectric, on the size of the gap, and on the magnitude of the current. All these parameters affect metal removal rate and the wear ratio *W/R*.

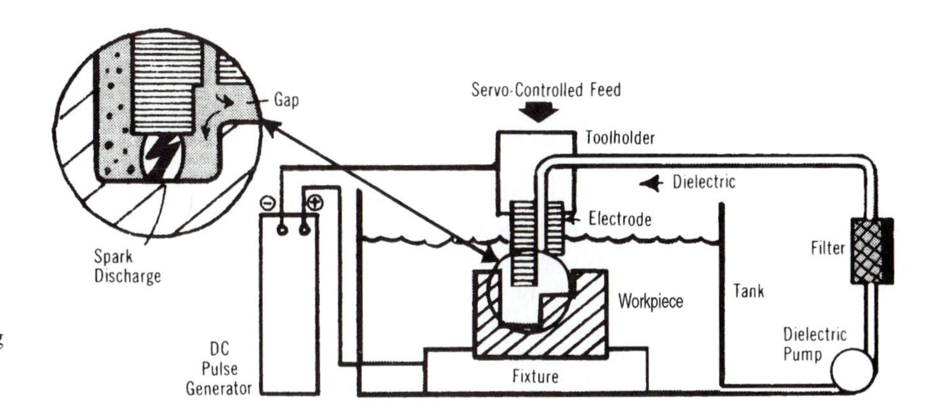

Figure 12.25
Electro-discharge machining (EDM) as used for die sinking. [BELLOWS]

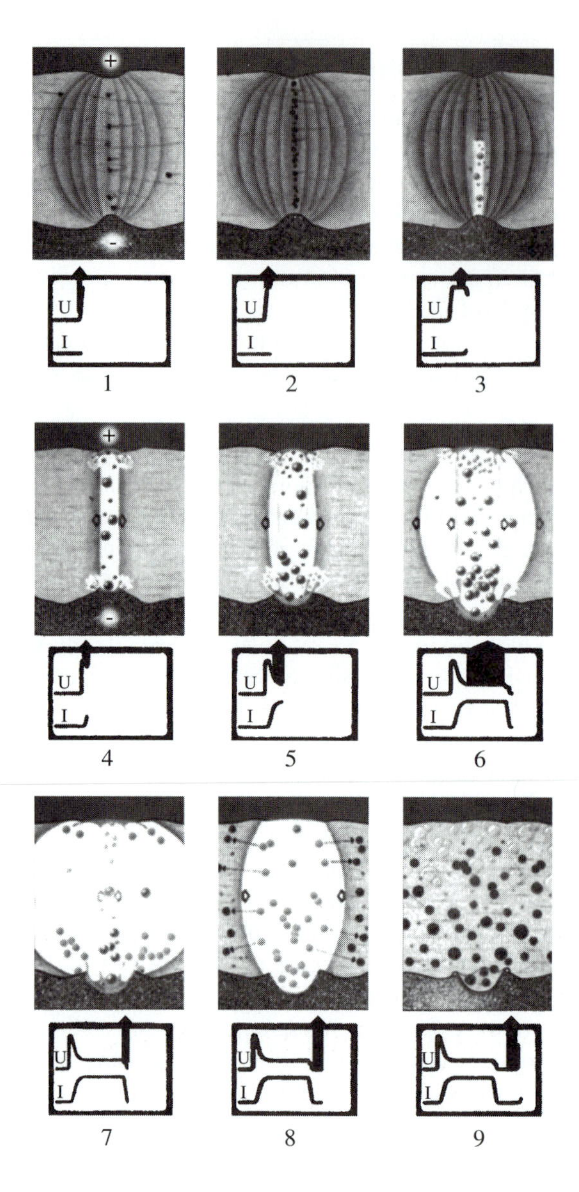

Figure 12.26
Spark discharge in EDM. 1. Building-up of an electric field. 2. Formation of a bridge by conductive particles. 3. Beginning of discharge due to an emission of negative particles. 4. Flow of current by means of negatively and positively charged particles. 5. Development of a discharge channel due to a rise in temperature and pressure. 6. Formation of a vapor bubble. 7. Reduction of the heat input after a drop in the current, associated with an explosion-like removal of material by vaporization and fusion. 8. Collapse of the vapor bubble. 9. Residues: metal particles, carbon, and gas. [AGIE]

The dielectric fluid acts as an insulator, generator of the plasma for the discharge, coolant, and flushing medium for chip removal. Other desirable characteristics are low viscosity, high dielectric strength, high flash point, freedom from acid or alkaline components, and well-understood and controlled levels of toxicity. Comparison of dielectric fluids for brass electrodes and tool steel workpieces appears in Table 12.4. The tool materials must have good machinability because the tools are mostly produced by machining, and good wear ratio, *W/R*. Some of the tool materials are compared in Table 12.5.

Because the tools wear rapidly, it is necessary to use several, mostly identical tools in succession during one EDM operation or else, in repetitive work, to use a finishing tool from one operation as a roughing tool in the next one. The tools are produced by cutting; sculptured tools are produced by NC milling. For the manufacture of a forging die or of an injection-molding die the tool may be cast by using the model of the part to be forged as part of the casting mold.

TABLE 12.4 Comparison of Dielectric Fluids for Brass Electrodes and Tool-Steel Workpieces

EDM Fluid	Machining Rate	W/R Ratio
50 viscosity S.S.U hydrocarbon oil	4.1	2.8
Distilled water	5.7	2.7
Tap water (typical)	6.1	4.1
Triethylene glycol H_2O (40*)	10.8	6.8
Tetraethylene glycol H_2O (30*)	7.0	11.3

Notes: Machining Rate = mm^3 work removed/amp min; *W/R* Ratio = Volume work removed/Volume tool removed
*Volume of H_2O percent

TABLE 12.5 Comparison of Tool Materials

Tool Material	Wear Ratio *W/R*	Machinability	Work Material
Graphite	5–100	A	steel
Brass	1–7	B	all metals
Steel	1–1.5	B	steels
Copper	1–2	B	all metals
Zinc	2–7	cast	steels
Aluminum	5–7	A	steels

A typical specific metal removal rate obtainable in EDM die sinking using a graphite tool and steel workpiece, is $MRR_s = 10$ mm³/min per 1 A. So, for a 50 A nominal current, $MRR = 500$ mm³/min $= 0.5$ cm³/min. To compare with ECM, we may read the specific removal rate from Table 12.2, for steel, $V_s = 2.24$ mm³/min per 1 A. However, in ECM commonly 1000 A current is used, resulting in $MRR = 2.24$ cm³/min. Thus, typically ECM may remove metal about 5–20 times faster for very high currents. Comparing further, let us consider machining a large die by milling; remember, however, that the steel workpiece would have to be annealed for machining and subsequently hardened, whereas in ECM and EDM the workpiece can be machined in the hardened state. Let us use 5 kW of power for roughing and 0.5 kW for finishing. We would then be able to achieve (see Table 8.1) $MRR = 26$ cm³/min in roughing and 2.5 cm³/min in finishing. Both the EDM and ECM values used above applied to roughing; machining is 50 times and 10 times faster, respectively. In addition, in both the electrical processes it is also necessary to produce the tool first, which is mostly done by machining. The advantage of the electrical processes is their ability to machine hardened steel as well as the difficult-to-machine nickel-based alloys.

Let us return to comparing die ECM with EDM for die sinking. In ECM the tool does not wear, but it has to be developed in several steps of trial and error; this is not needed in EDM. However, the final tool in ECM can be used again and again; in EDM the tool has to be made again and again.

As for the machined surface integrity, no residual stresses are generated in ECM, surface finish of 0.1 to 2.5 μm rms is obtained, and it improves with higher cutting rates. In EDM there is a heat-affected zone 0.125–0.5 mm thick with high residual stress. This may affect the fatigue strength of the workpiece. Surface roughness ranges from 2.5 μm for finishing to 40 μm rms in roughing.

Ultimately, what gives EDM the advantage is first the rather easy to use dielectric fluid, as compared with the aggressive electrolytes in ECM, and the lower flow rates, resulting in lower pressures in the dielectric fluid and consequently lower forces between tool and workpiece. Second, and perhaps most important, is the combination of NC motions with simpler tool shapes to producing more complex shapes in the workpiece, which is the reason behind the widespread acceptance of EDM machines for die sinking. This is illustrated in Fig. 12.27, reproduced from a catalog of an EDM machine manufacturer. A typical machine is configured with a horizontal table moving in *X, Y,* with a vertical ram carrying the tool and moving in *Z*. The ram may contain a spindle rotating as the C axis under NC.

Examples of die sinking work are presented in the following illustrations. Fig. 12.28 shows how a threaded tool can be simultaneously fed in *Z* and rotated around the *Z*-axis to produce a nut. An example of the manufacture of a large steel die for stamping of an automobile fender is shown in Fig. 12.29.

Wire EDM

Wire EDM (WEDM) is a unique operation in which the tool electrode is represented by a thin wire that is reeled off one spool onto another, so that new wire is always entering the cut, and the worn wire is disposed of. The operation resembles band sawing, except that the kerf is very narrow and the path of the cut under NC may have a very complex shape, including rather sharp corners. Wire diameters of 0.1–0.3 mm are used; the wire is mostly made of strong brass with $UTS = 900/\text{mm}^2$, which makes it possible to keep it strongly tensioned and straight. There are actually no cutting forces acting on it. Copper, tungsten, or molybdenum wires are also used. The diagram of the process is in Fig. 12.30. The basic NC motion axes are *X* and *Y* associated with the table

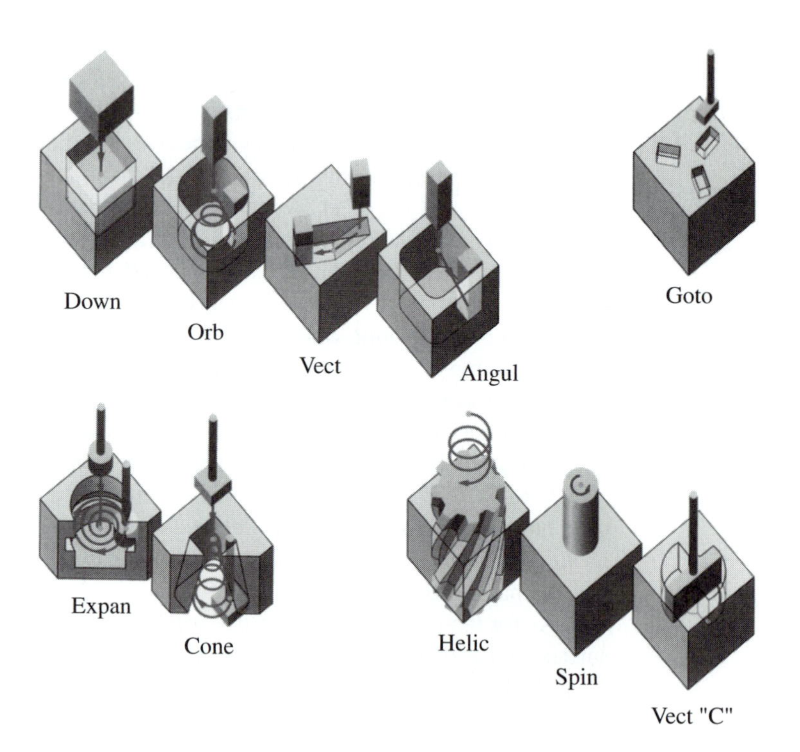

Figure 12.27
Combination of tool shape and NC controlled motion in EDM. Simple tool shapes are used for a variety of complex workpiece shapes. [CHARMILLE]

Figure 12.28
EDM of an internal thread. The electrode in the form of external thread is simultaneously rotated and fed down axially. [CHARMILLE]

Figure 12.29
EDM of an injection mold for the manufacture of automobile fenders. [CHARMILLE]

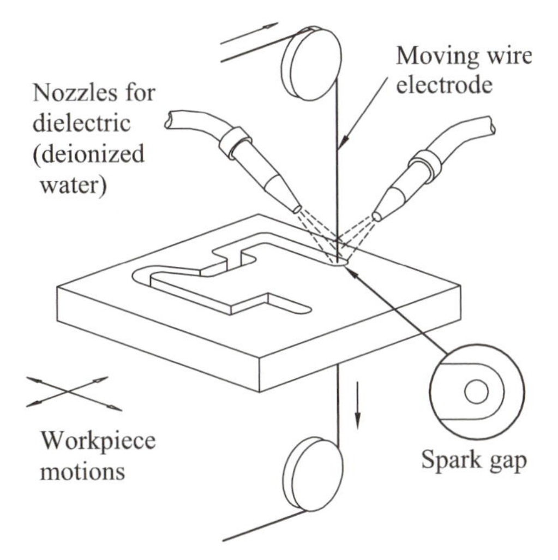

Figure 12.30
Wire electric discharge machining (WEDM). [WELLER]

of the machine. The wire is guided through two diamond wire guides, Fig. 12.31, located on two different vertical levels. The upper guide is moved in two additional NC axes, u and v, so as to provide tapering of the cut surface needed for the relief angle in punching dies and in extrusion dies. Often it is possible to produce the punch and die in one cut.

A typical power supply has open-circuit voltage of 100 V, pulse time range 1 to 100 μsec with up to 300 kHz pulse frequency and current of 50–500 A. Deionized

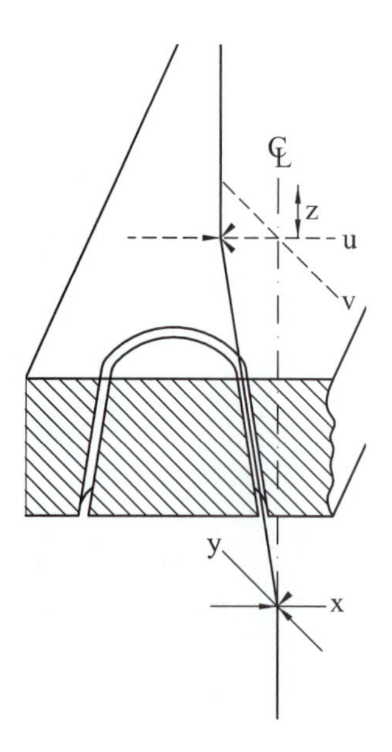

Figure 12.31
Generating the tapered relief on a punching die. [AGIE]

water is the commonly used dielectric fluid. Kerf width is 20–50 μm. The control resolution is typically 1 μm. Cutting performance is measured as the area of the cut (length of cut $l \times$ workpiece thickness h) where the thickness h may typically be up to 150 mm and the performance is $A_s = 300$ mm^2/min for $h = 75$ mm.

The productivity of the machine is enhanced by a number of features, such as automatic threading and rethreading, automatic feed-rate control with varying width of cut, manual and robotic palletization, and collision protection. High accuracy is obtained by means of using automatic dielectric temperature control. The NC system may also include a Z-axis moving the ram vertically to adapt to nonflat workpieces and an A or B rotary axis on the workpiece head.

Examples of work are in Figs. 12.32, 12.33, 12.34. The example of punching and blanking dies in Fig. 12.32 is an impressive demonstration of multiple cutting and rethreading the wire. The operations of edge finding, wire cutting and rethreading are programmed in the NC cycle. The automatic wire threader works with a water jet and a round diamond guide system. Wire cutting, threading and restart operations are performed automatically if an incidental wire breakage occurs. The retry function permits wire threading to be designated so that sections may be remachined if unsatisfactory results are observed due to burring of starter hole locations or other wire difficulties. The jump function may be used to move the machining wire to the next starter hole. In Fig. 12.33 is an example of a die for manufacturing electronic components and in Fig. 12.34 is the punch and die for key blanks. The pair of tools is made in one EDM operation.

12.7 LASER BEAM MACHINING (LBM)

The word *machining* in the title of this section is only partially correct although widely used. Lasers are used for a variety of processes, such as welding, heat treating, and material removal and they can be applied to many kinds of materials such as metals,

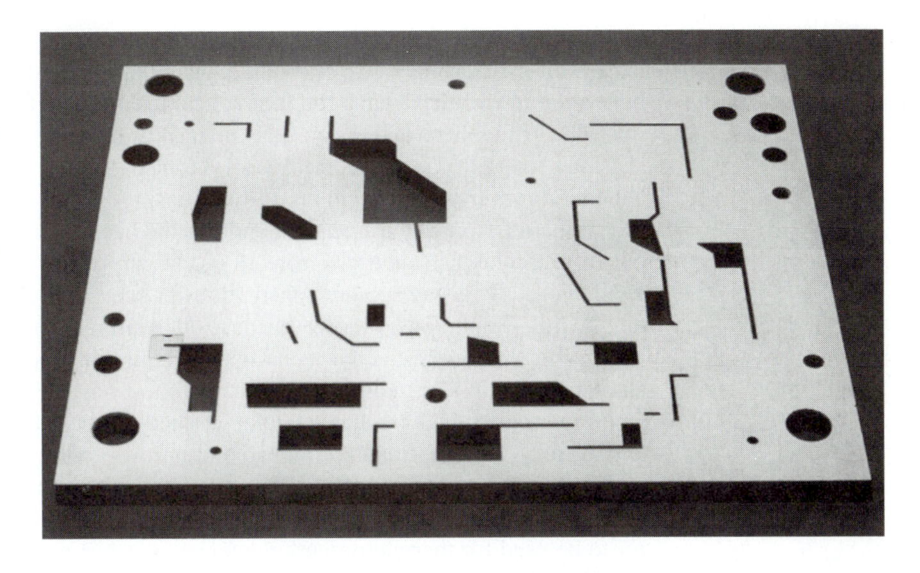

Figure 12.32
View of a multiple blanking die made by EDM. The cutting of the many separate openings in this workpiece involves repeated starting hole drilling, wire re-threading, edge finding, and wire cutting in an automatic NC cycle. [CHARMILLE]

Figure 12.33
WEDM-made blanking die for an electronic component. [CHARMILLE]

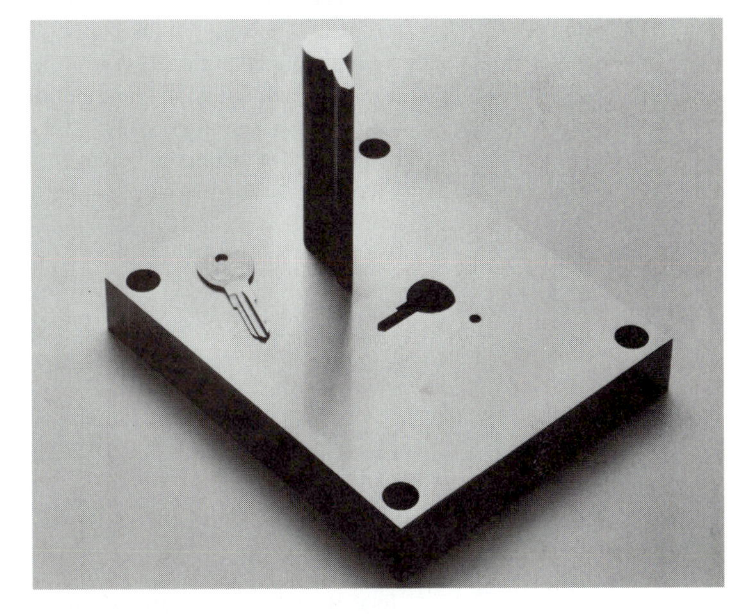

Figure 12.34
Punch and die for key blanks, produced by WEDM. [CHARMILLE]

ceramics, plastics, and composites. It is a rather universal and flexible technique. All these processes involve the use of narrowly concentrated heat in the form of electromagnetic radiation, or light. The heat can be controlled to produce just heating of the solid, or melting, or even vaporization of the material. The word *laser* stands for light amplification by stimulated emission of radiation, and it denotes a source of monochromatic, coherent, parallel beams that may be focused on a very small spot, providing very high power density on the order of 10^6 to 10^7 W/mm^2. The word *monochromatic* means that the radiation has a single frequency, and *coherent* means that there is a constant phase relationship between any two points of the radiation: the energy travels in well-synchronized waves. Many kinds of lasers exist based on gas or solid media, but only a few of these have found practical use in manufacturing.

The principle of lasing can be explained as shown in Fig. 12.35. The device consists of a tube filled with gas or of a solid transparent crystal as the active medium, the excitation power source, and two mirrors. Lasing is produced between the mirrors, but one of them is partially transparent and lets the beam escape, and this is the actual output from the laser. Bound electrons in atoms can be excited to upper levels, and they would normally decay spontaneously to lower energy levels and give up energy in randomly directed photons. If however the excited atom is immersed in an electromagnetic wave with a frequency equal to its natural emission frequency, the atomic system is stimulated to emit its radiation in phase with the wave. The energy of the wave is thus amplified. An optical resonator is then created by the feedback of reflection from the two mirrors. The amplitude of this radiation grows coherently on each pass through the medium until it saturates when the rate of stimulated emission is equal to the rate of atomic excitation. Some of the radiation escapes through the partially transmissive mirror to be used for thermal processing as the laser beam.

The arrangement of the CO_2 gas laser source is shown in Fig. 12.36. The gas that circulates through the laser tube is actually a mixture of carbon dioxide (CO_2), which supplies the lasing energy, nitrogen (N_2), which keeps the upper energy levels populated through collisions, and helium (He), which provides intracavity cooling. A typical composition is 5% CO_2, 15% N_2, and 80% He. The excitation energy is supplied by an electric discharge between electrodes located at the ends of the tube. The tube is water-cooled on the outside. The laser beam is transported by a system of mirrors and

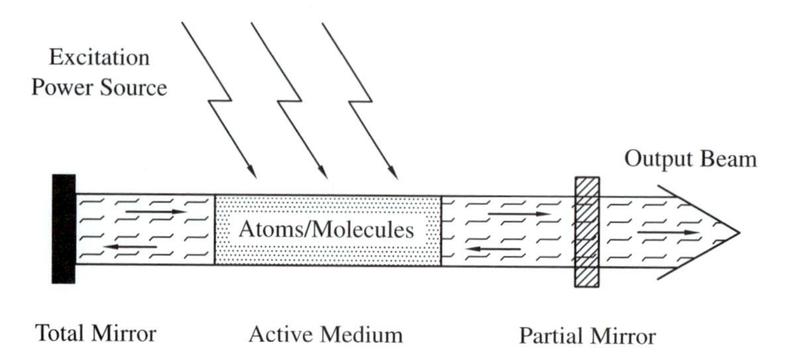

Figure 12.35 Basic components and action of a laser (light amplification by stimulated emission of radiation). [REPRINTED WITH PERMISSION FROM D.R. WHITEHOUSE, REF. [1]]

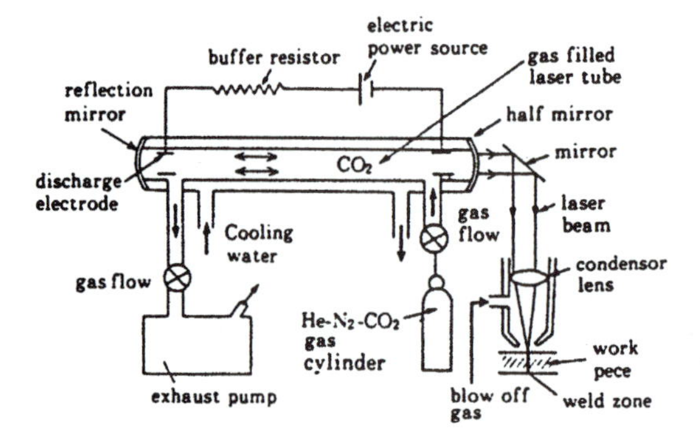

Figure 12.36 Design of a coaxial CO_2 gas laser. [FROM TANIGUCHI, REF. [9]]

focused by a lens onto a small spot. A gas stream is supplied around the beam and directed toward the workpiece. Either an inert gas is used to protect the operation, or in the case of cutting, oxygen is used to enhance the melting and evaporation of the material. It is important to realize that in a case of a metallic workpiece, a small part of the radiation is absorbed and used for the process, while a major part is reflected and lost. The reflectivity is associated with electrical conductivity, and it is very high for silver, copper, and aluminum. These metals absorb only a small fraction of the radiation energy. Reflectivity depends further on the surface roughness of the workpiece and on the laser wavelength. The CO_2 laser wavelength is 10.6 microns, which is in the far infrared spectrum and is much more reflected than the radiation of lasers with shorter wavelengths as is characteristic of the solid-state lasers. The most common of these is the Nd:YAG laser, which uses a neodymium-doped yttrium aluminum garnet crystal as the active medium. Its wavelength is 1.06 microns.

The schematic of the laser with a crystal rod as the active medium is shown in Fig. 12.37. In this case the excitation is provided by krypton or xenon lamps located in an elliptical reflector that focuses the light onto the laser rod. In solid-state lasers removal of waste heat is a fundamental problem, and the radius of the rod is limited by the need to conduct the surplus heat to its cooled periphery. This also sets a practical limit to the power that can be extracted from the system. The laser systems may operate either in a continuous mode CW (continuous wave) or in repeated short pulses.

The characteristics of three laser systems that are most significant for industrial application are presented in Table 12.6. There we find two entries for power: the "average power" is meant as either CW or the average of the pulsed power; "peak power" means the power of an individual pulse. The CO_2 laser can deliver 10 or in some cases up to 15 kW, and the Nd:YAG up to 1 kW. The power is adjustable and adaptable to each particular operation. While the average power of the excimer lasers is rather low, very high peak power is obtainable at very short wavelengths. The TDL (times diffraction limited) number refers to the amount of divergence of the beam, which may be caused by the laser being excited by other modes than the principal coaxial mode; the best quality is expressed by TDL = 1. The Nd:YAG and excimer beams may be delivered by means of fiber optics.

The effect of the focused beam incident on the work surface depends on the time scale and on power density. For very short pulses, the spread of heat into the work

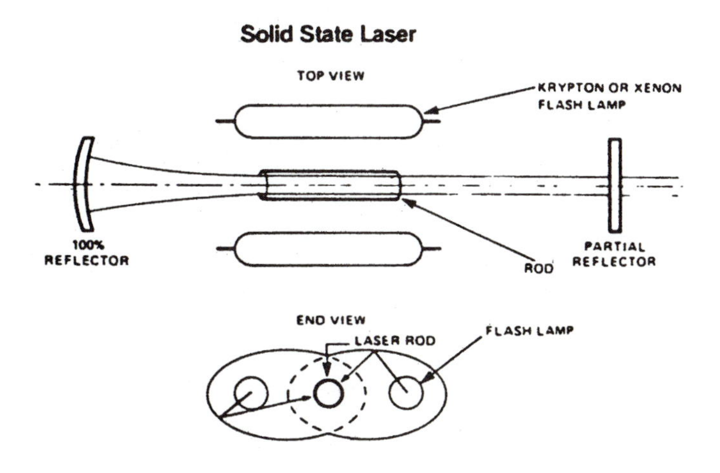

Figure 12.37
Top and end view of a solid state Nd:YAG laser. [WELLER]

TABLE 12.6 Operational Comparison of CO_2, Nd:YAG, and Excimer Lasers

	CO_2	Nd:YAG	Excimer
Active Medium	CO_2, N2, He gases	Nd:YAG crystal	Noble and halogen gases
Excitation	Electrical discharge	Arc lamp	Electrical discharge
Wavelength (Microns)	10.6	1.06	0.19/0.25/0.31
Average power (kW)	1–10	0.1–1	0.1–0.2 (pulse)
Peak power (kW)	10	50	50,000
Rep rate (kHz)	10	20	0.5
Efficiency (%)	5–15	1–3	1–3
Beam size (mm)	10–30	2–10	6 × 20–10 × 30
Beam quality (TDL)	1–3	1–3 limited 10–100 full	1–100 limited 100 full
Consumables	CO_2, N2, He gases	Arc lamps	Noble, halogen and buffer gases
Transmissive optics	ZnSe	Quartz	UV quartz
Reflective optics	Metal	Metal or dielectrics	Dielectrics
Fiber delivery	None	Quartz	UV quartz

material is small. The heat of the beam is absorbed in a thin layer, on the order of 0.1 μm. Whitehouse [1] calculates that with an absorbed intensity of 10^5 W/mm^2 at the surface of a typical steel, the surface temperature will increase at the phenomenal rate of 10^{10}°C/sec with a spatial temperature gradient of 10^5°C/cm. These numbers show why it is so easy to vaporize a local area of the material and leave behind a small heat-affected zone (HAZ). Various processes occur as a result of the beam striking the surface. The absorption of the radiation depends exponentially on depth. Heat is conducted from the surface. If the heat intensity is high enough, the surface will begin to melt rapidly, and a fusion front penetrates into the workpiece. For welding purposes it is desirable to achieve deep penetration before the surface vaporization begins. With higher powers, metal removal is obtained for drilling and cutting. Thus, by regulating the radiation intensity and the lengths of pulses up to continuous wave, and depending on the workpiece material, lasers are used for surface heat treatment, for welding, drilling, and cutting. The workpiece material characteristics that are significant are the absorption coefficient, which is inversely dependent on electrical conductivity (aluminum has poor absorption; ceramics and plastics are excellent absorbers); thermal conductivity and diffusivity, which determine the depth of penetration; and specific heat and latent heats, which affect fusion and vaporization.

EXAMPLE 12.3 Power in Laser Drilling ▼

As an example, let us calculate the laser energy needed to drill a hole of diameter 1.0 mm in a steel sheet 1 mm thick. This represents a volume V and mass m: $V =$ 0.785 mm^3, $m = 0.006$ g. The heat needed to melt 1 g of steel is $H_m = 1381$ J/g [see Eq. (11.3)]. The heat to bring the temperature of the liquid steel to the boiling point of 3000°C is obtained by using specific heat $c = 0.474$ $J/(g°C)$, $H_b =$

(3000 − 1530) × 0.474 = 1694 J/g. The vaporization heat is H_v = 6342 J/g. The total energy for vaporization of 1 g of steel is H = 8417 J.

Assuming the use of an Nd laser with a short wavelength and an absorption coefficient of 50%, we will need for this drilling operation the energy of E = 2 × 8417 × 0.006 = 101 J. Assuming a pulse length of 10^{-5} sec the required instantaneous power will be P = 1.01e7 W, and the power intensity $I = P/A$ = 1.28e7 W/mm^2.

This simple calculation shows that the major part is the evaporation, which is the mode needed in the material removal operation. It also shows that the pulsing mode of laser control is very suitable for drilling.

The parameters used for such energy calculations are assembled in Table 12.7 for several metals. ▲

Laser Beam Welding (LBW)

Let us now briefly discuss the various applications of the LBW processes. There are several advantages to the process: no material is in contact with the workpiece, so that there is no contamination. Laser welding is done in atmosphere, in contrast to electron beam welding, which must be done in vacuum. The heat-affected zone is very small. This is especially important in cases where a weld must be made near a heat-sensitive element such as a glass-to-metal seal.

The depth of penetration in LBW is rather small because it depends on the speed of conduction of the heat from the energy imparted on the surface of the workpiece. However, for high-energy laser welding, the beam produces a hole in the material, and the laser energy is focused at its bottom; much deeper penetration is thus possible.

There are two ranges of LBW applications. One, which might be characterized as microwelding, takes advantage of the achievable precision and control. The Nd laser is usually used, either in the pulsed or CW mode. In [3] an example is given that shows dramatically how the localized heating capability of the laser was utilized. It involved hermetic sealing of an automotive air bag miniature detonator with the weld made within 1 mm of a temperature-sensitive primary explosive. The detonator was a stainless steel cylindrical can 0.25 mm in diameter and 6 mm long (see Fig. 12.38a). The laser was used to attach a header to the can containing the high explosive by producing overlapping spots 0.125 mm in diameter and 0.075 mm deep [4]. A CO_2 laser was operated in a repetitively pulsed mode with energies as low as 0.2 J per 40-microsecond pulse. This can be evaluated as instantaneous power of 5000 W and as 4e5 W/mm^2. A continuous seam 0.25 mm wide was produced. The welds were hermetic and

TABLE 12.7 Physical and Thermal Properties of Selected Metals

	Aluminum	Titanium	Molybdenum	Tungsten	Iron
Melting temperature T_m (°C)	660	1668	2610	3410	1536
Boiling temperature T_b (°C)	2450	3260	5560	5930	3000
Specific heat c (J/g/°C)	0.90	0.527	0.255	0.134	0.474
Heat to melt H_m (J/g)	1075	2784	2716	2343	1381
Heat of vaporization H_v (J/g)	10,536	9305	5609	4211	6342
Heat to evaporate H_e (J/g)	13,222	12,928	9077	6891	8417

satisfied a 28-day temperature and humidity cycling test between −18°C and +70°C and 13000-g shocks. Another microwelding example from the same company is a razor blade (Fig. 12.38b), on which 30 microwelds were made in less than 1 second.

For macro, high-output welding, the graph in Fig. 12.39 offers good informative data. It presents traverse rates for single-pass, narrow, deep welds in 304 stainless steel as a function of CO_2 laser power in the CW mode. These represent very impressive performance.

Drilling

Lasers are used for drilling small and relatively deep holes in metals, ceramics, plastics, and composites. The metals involved are stainless steel, tungsten, tantalum, beryllium,

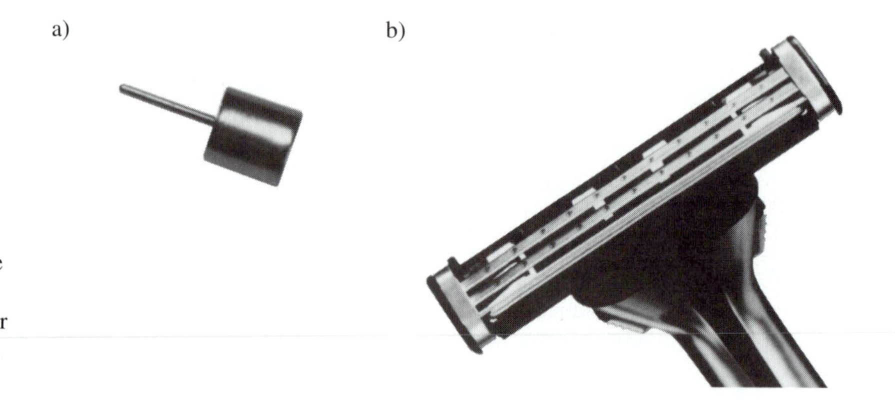

Figure 12.38
a) Laser welded automobile air bag detonator; b) 30 microwelds made on a razor blade in less than 1 second. [LUMONICS]

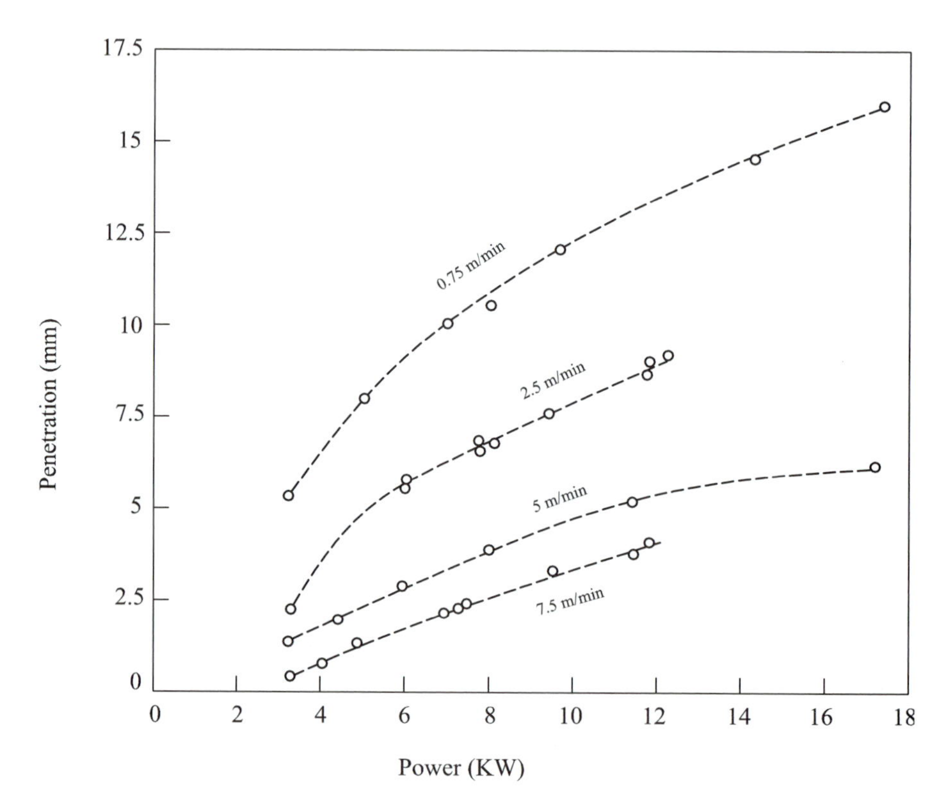

Figure 12.39
Penetration and traverse rates for CO_2 laser welding of 304 stainless steel related to laser power. [FROM READY, REF. [4]]

and uranium. The range of these operations is indicated in Fig. 12.40, where the process is compared with EDM and ECM capabilities. Maximum depth up to about 15 mm is shown for holes with diameters from 75 μm up. The small-diameter holes are drilled in "percussion mode," that is, by repeated pulses, each of which advances the depth of cut as it evaporates and ejects the corresponding amount of material. Larger holes may be made by contour cutting a cylindrical slug out of a sheet or plate of material, in the "trepanning" mode. This is achieved by using a special head with an eccentric end sleeve. An example of a drilling operation is shown in Fig. 12.41—a case of drilling 194 holes with 1.5-mm diameter, with a tolerance of ±0.075 mm in a cooling frame made of a high-temperature-resistant nickel alloy Nimonic, used in land-based turbines. A "Laserdyne" 550 Beam Director and a Lumonics JK704 Nd:YAG laser were used.

It has been mentioned that at room temperature most metals reflect a high percentage of the infrared radiation of the CO_2 laser. The absorption of this radiation is enhanced by introducing oxygen at the point of laser beam impingement. The oxygen

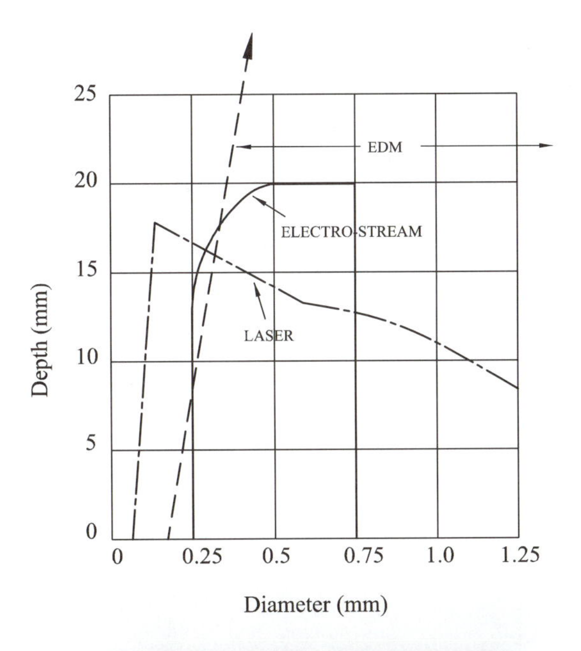

Figure 12.40
Comparison of drilling holes using laser beam and using ECM and EDM. [From Bolin, ref. [5]]

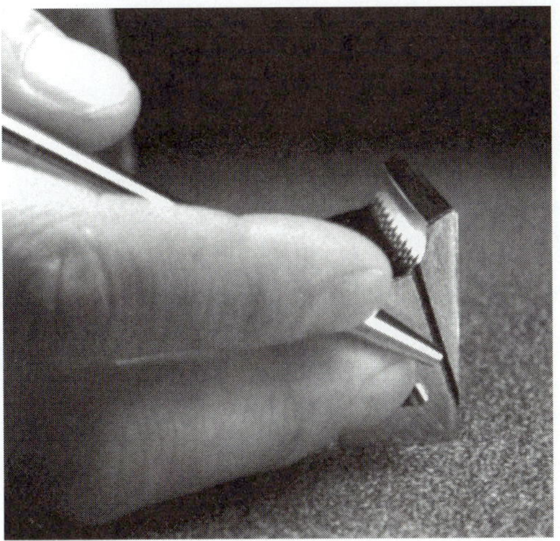

Figure 12.41
Laser drilling of 1.5-mm-diameter holes 12° angle to the surface in a cooling frame of a Nimonic superalloy turbine blade. 194 holes have been made in 4 hours using a YAG laser. [Lumonics]

jet serves several purposes. First, it reduces the reflectivity of the metal surfaces. Further, once the surface reaches high temperature, the oxygen produces an exothermic reaction, causing a molten puddle and enhancing the cutting process. Furthermore, the gas blows the metal vapors or even a plasma away from the cutting area, which otherwise would prevent the laser beam from fully reaching the surface. Finally, at the bottom of the cut the gas sweeps away the molten slag, producing a clean cut. For optimum operation, the gas must be delivered as a stream over an area only slightly larger than the laser beam spot, converging at sonic velocity. A low pressure flow of oxygen produces a self-burning effect which very badly affects the quality of the cut. With optimum conditions, the balance of heat produced by the laser and the exothermic reaction is such that only the metal traced by the laser beam is cut. Approximately 30% of the heat is supplied by the laser and 70% by the reaction with oxygen. Comparing this cut with oxyacetylene cutting, the width of the cut is typically 0.2 mm instead of about 3 mm for the latter process. Under these circumstances the gap between the nozzle and the cut material (typically 1mm) must be maintained rather accurately. Most systems include a sensor and automatic regulation of the gap.

In some cases, dependent on the workpiece material, an inert gas is used instead of oxygen. Such a gas does not contribute to heat generation but serves all the other purposes mentioned above. Laser cutting has a number of advantages, such as the narrow kerf, no wear of tools, as in shearing or in routing (end milling), rather smooth edges are produced, and no cutting forces are involved. It is an operation in which the relative motion between the laser beam and the sheet metal is mostly numerically controlled in two axes or else in three or more axes when a 3D sheet metal part is involved. A typical machine for large, flat, sheet metal parts is shown in Fig. 12.42. In this case the X coordinate motion is produced by the gantry that houses the laser system, and the Y (transversal) motion is produced by the laser head moving on guideways inside of the gantry. The working width is 4 m, and it can be shared by two cutting heads that may be moved

Figure 12.42
A machine for NC controlled laser beam cutting. The laser path is shown using mirrors. [ESAB]

symmetrically in $+Y$ and $-Y$ to simultaneously cut mirror images of parts. The system of mirrors used to direct the path of the beam is shown in the inset of the figure.

Maximum cutting speed of 20m/min is achievable. CO_2 lasers with rated power of 1–3 kW are available. Operation of the laser can be controlled from the CW (continuous mode) to pulsed mode up to 5 kHz frequency with a 20-microsecond rise time. A typical thermal efficiency of the laser is about 7%; the 3 kW laser machine needs 46 kW input power, and the laser is water cooled by a chiller delivering 57 liters/min. The lasing gases are supplied by a 12,000 rpm turbo compressor delivering 25 liters per hour of the gas mixture. An additional system is needed for the assist gases, such as oxygen or nitrogen.

Other types of machines may use different configurations of the motion axes. For instance, both X and Y motions may be associated with the table carrying the workpiece. Examples of cutting performance are given in Fig. 12.43 for cutting of mild steel with oxygen assist, for cutting stainless steel with nitrogen assist, and for cutting aluminum with air assist. These data are obtained using a CO_2 laser with 3500 W of continuous power. It is shown that low-carbon steel with oxygen assist can be cut up to a thickness of 25 mm. For the various plate thicknesses, the focal length of the laser must be adjusted for best performance. Dominance of the exothermic reaction of burning steel must be avoided. A wide kerf and very rough cut surface would result. In cutting stainless steel, nitrogen or oxygen can be used, with the oxygen enabling a higher traversing speed. The use of nitrogen results in a cleaner cut. There is not much difference in the performance of the two cases because stainless steel is oxidation-resistant. Due to the much lower thermal conductivity than that of carbon steel, less heat escapes by conduction, and this results in the higher feed rate in thin materials. However, it also prevents deep penetration into the material and hence limits the maximum thickness to 15 mm. Aluminum is difficult to cut because of its high thermal conductivity and also because of its high reflectivity of the laser light. Therefore, both the feed rate in thin material and maximum thickness that can be cut are lower than for steel.

Cutting performance data for thinner sheets have been published by another company [6]. In Fig. 12.44 we reproduce data for the application of a 1000-W CO_2 laser. Often, at the lower thickness range, 500 W or 250 W of power are used successfully. The graph shown indicates again that aluminum must be cut at lesser thickness values than low-carbon steel. Titanium is very oxidation-prone and must be cut under a pro-

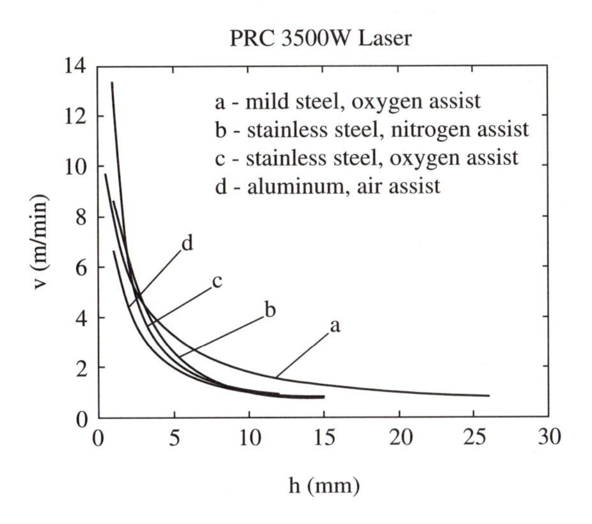

Figure 12.43
Cutting performance of a 3500W laser. [COMPILED FROM DATA BY PRC CORP.]

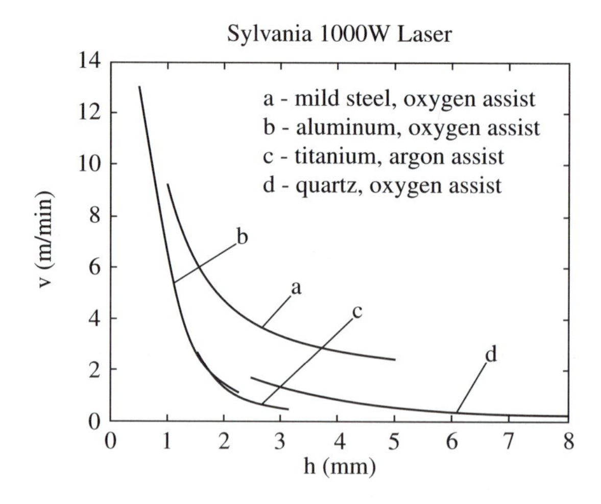

Fig 12.44
Performance data for laser beam cutting of thin sheet. [COMPILED FROM ENGEL, REF. [6]]

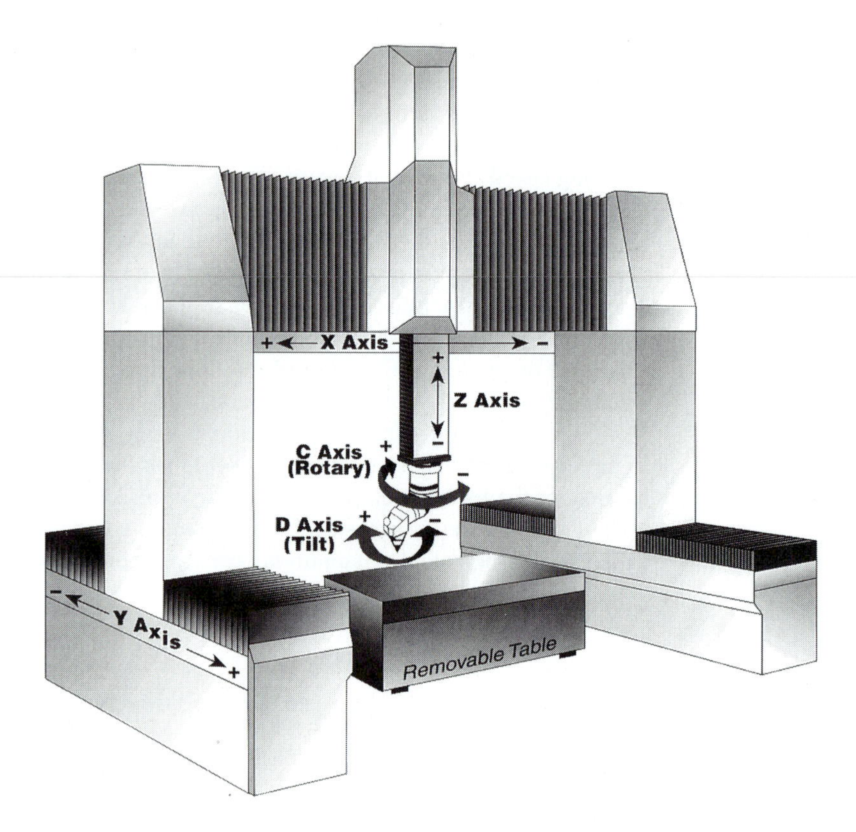

Figure 12.45
Five-axis NC laser processing machine. Fiber optics is used to guide the laser beam. The machine is used for welding and cutting operations on large sheet metal parts, such as trimming of stamped automotive body parts in an assembly line. [LUMONICS]

tective gas atmosphere, in this case argon, although helium and nitrogen are used as well. The graph shows that quartz can be very efficiently cut by a laser beam, and it is a common method applied in the optical industry. An example of a five-axis NC laser processing machine is shown in Fig. 12.45. It can be used for welding and cutting operations on large parts, especially sheet metal parts. Many examples have been quoted for trimming stamped automotive body parts.

Some informative data for the other extreme of the laser power range have been prepared as described in [7]. Table 12.8 presents selected conditions for cutting alloy

TABLE 12.8 High-Power Laser Cutting

Material	Thickness (mm)	Power (kW)	Gas	Feed Rate (m/min)
4340	20	10	Air	0.235
	20	10	O_2	1.125
	20	15	He	0.375
Ti-6Al-4V	6	10	He	2.0
	20	10	He	0.5
	25	15	He	0.375
	50	15	He	0.125

steel 4340 and titanium Ti-6-Al-4V with high-power CO_2 lasers. At these high powers, good-quality cuts are only obtained in an inert atmosphere, although obviously oxygen assist gives the best cutting rate on the alloy steel. Titanium is cut preferably with helium assist.

12.8 ELECTRON BEAM MACHINING (EBM)

EBM is similar to laser beam machining in several characteristics: it delivers highly and accurately focused energy onto a small spot on the surface of the workpiece; it can be used for heat treatment, welding, and cutting; it is applied to a great variety of workpiece materials, from metals to ceramics and plastics. However, EBM is most effective when carried out in high vacuum of the order 10^{-3} to 10^{-6} mm of mercury, whereas LBM works in air. The limitations of laser power due mainly to problems with cooling the active medium and the loss of energy by reflection of the radiation do not apply to the electron beam. Continuous power levels up to 10 kW have been accomplished. The accelerating voltage of the electron beam gun between 30 and 170 kV is used with currents between 50 and 1000 μA. With the beam focused to a spot diameter between 0.025 and 0.25 mm, power density can reach values up to 10^8 W/mm^2.

The diagram of the EBM process is in Fig. 12.46. The source of a directional stream of electrons is a cup-type, negatively charged electrode with a hot tungsten filament. Downstream is located the anode, at ground potential. The electrons emitted from the electron gun are accelerated towards the anode and pass through its opening. They are further focused by electromagnetic lenses. This only works in high vacuum. In the basic mode the workpiece is also located in the vacuum chamber. The electrons travel at more than half the speed of light (300,000 km/sec) and are not impeded by collisions with any gas ions until they strike the workpiece. This heat source exceeds any other process as regards power density and precision.

Holes of very small size, down to 0.05 mm diameter with very high depth-to-diameter ratios, up to 200:1, can be produced. Very narrow and deep welds can be made in a single pass, such as a butt weld 3 mm wide and 150 mm deep in carbon steel, with a welding speed of 120 mm/min. Aluminum plates up to 450 mm thick have been welded. Because of the concentration of the energy to narrow zones, shrinkage and distortion are minimized. Welding in vacuum produces maximum weld purity because of the absence of contaminating gases.

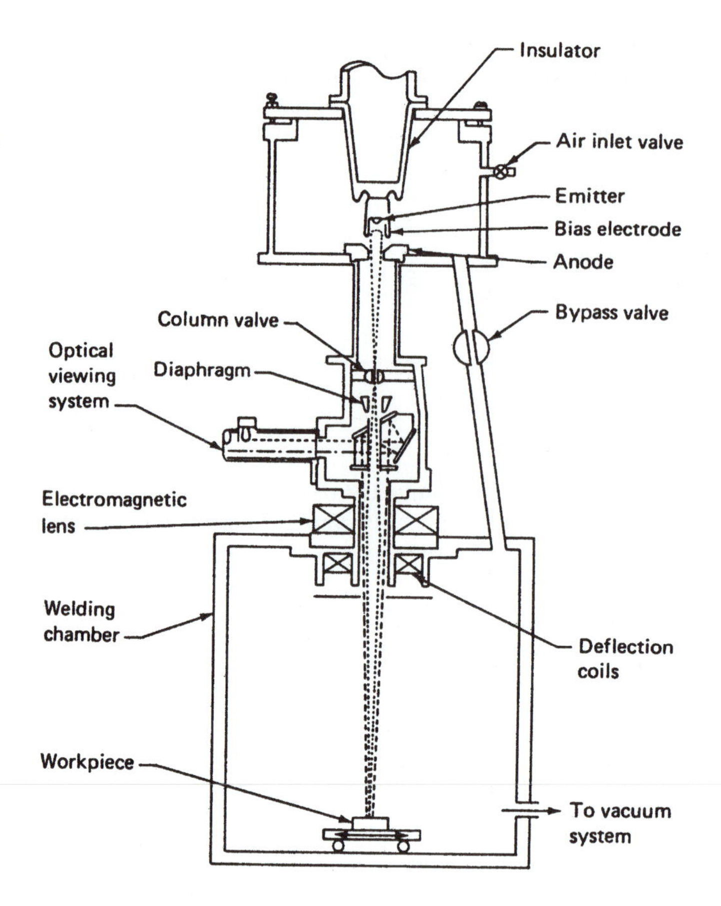

Figure 12.46
Electron beam machining (EBM) and processing. A typical high-voltage electron beam welding machine arrangement. [WH3]

Pumping down to high vacuum may take up to an hour, depending on the size of the workpiece and consequently on the size of the chamber, which leads to a considerable slowdown of production; therefore, this mode of operation is only used in special applications where it can be justified. In practice two other modes of EBM have been introduced, the *medium vacuum* process, and *nonvacuum* EBM. The beam generation is done in high vacuum in all three modes, but in the latter two the beam passes through orifices large enough to pass the beam but too small for any significant back diffusion of gases. The pressure in the work chamber of the medium vacuum process is maintained at the level of 10^{-3} to 25 mm Hg (torr), the level below 1 torr is called "soft" vacuum, and levels between 1 and 25 torr are called "quick" vacuum. The pump-down time for the latter is a matter of seconds.

EBM operations are essentially practiced in two distinct ranges of power. At the low end, high-precision and delicate welding, drilling, and cutting of thin foils is carried out, and it is achieved thanks to the capability of precise control of the energies and motions involved. At the high end, powerful welding and cutting are accomplished.

Drilling Small Holes and Cutting Thin Slots

Drilling small holes is performed in a variety of materials, mostly rather thin. Typical examples are listed in Table 12.9.

TABLE 12.9 Holes Drilled by EBM in Various Materials

Work Material	Work Thickness (mm)	Hole Diameter (mm)	Drilling Time (sec)	Voltage (kV)	Average Current (μA)	Pulse Width (μsec)	Frequency (Hz)
Stainless steel	0.25	0.013	<1	130	60	4	3,000
Tungsten	2.00	0.13	10	140	100	80	50
Aluminum	0.25	0.025	<1	150	30	20	50
Alumina	0.41	0.076	<1	130	100	80	50
Al_2O_3	2.5	0.13	10	140	100	80	50
Quartz	0.76	0.30	30	125	60	80	50
Si_2O_3	3.18	0.025	<1	140	10	12	50

TABLE 12.10 Slot Cutting by EBM in Various Materials

Work Material	Work Thickness (mm)	Slot Width (mm)	Cutting Rate (mm/min)	Voltage (kV)	Average Current (μA)	Pulse Width (μsec)	Frequency (Hz)
Stainless steel	0.18	0.1	50	130	50	80	50
Tungsten	0.05	0.05	100	150	30	20	50
Alumina	0.05	0.025	175	150	30	80	50
Al_2O_3	0.75	0.1	610	150	200	80	200

Holes with diameters larger than 0.13 mm can be drilled in a trepanning mode by deflecting and rotating the beam. Examples of cutting slots are given in Table 12.10.

Another example of slot cutting is dicing of a silicon wafer into individual chips.

EXAMPLE 12.4 Power in EBM Cutting ▼

Let us check the case of cutting a thin sheet of tungsten as listed on line 2 of Table 12.9. The average power delivered by the beam is

$$P_{av} = 1.5e5\text{V} \times 3e - 5\text{A} = 6.5\ \text{W}$$

The metal removal rate is

$$\begin{aligned} MRR &= 0.05 \times 0.05 \times 100 = 0.25\ \text{mm}^3/\text{min} \\ &= 0.0042\ \text{mm}^3/\text{sec}\ 5\ 4.2e{-}6\ \text{cm}^3/\text{sec} \\ &= 8.33e{-}5\ \text{g/sec} \end{aligned}$$

From Table 12.7, the total heat to evaporate 1 g of tungsten is 6342 J, which, for the above *MRR*, would require 0.52 W average power. The two power values, the one delivered and the one required, show that the efficiency of the process is 8.1%. That means that in spite of the short duration of the pulse and its high intensity, 91.9% of the heat is dissipated into the material.

Considering the 20-μsec pulse per 0.02 sec (50 Hz) cycle, the cutting takes only 0.001% of the time. The instantaneous delivered power is 6.5/0.001 = 6500 W, and assuming the diameter of the beam impingement is equal to the cut width, $d =$ 0.05 mm, the power density is 3.31e6 W/mm^2.

The high accuracy of the EBM process is utilized in making custom large-scale integrated circuits (LSIs), reticles, and masks. High-speed writing systems and highly sensitive resist processes have also been developed [8]. The diagram of such a system is shown in Fig. 12.47. It is a variable-shape E-beam system for 0.5-μm pattern writing, using a rectangular beam shape. The motion of the highly focused beam is produced by a system of electrostatic deflectors. Using 2 μA current and 30 kV accelerating voltage, a beam edge resolution of 0.2 μm is achieved. A field size of 2.6 by 2.6 μmm is used, with motion resolution of 0.02 μm. With 100-mm diameter of the wafer, a throughput of 20 wafers per hour is accomplished.

It has been mentioned above that, apart from high-precision, low-power EBM, there is also the high-power range of applications in welding. All three vacuum modes, high, medium and nonvacuum, are used. In the high-vacuum mode, maximum weld penetration and minimum width are achieved, in ratios up to 50:1 with the butt weld in plates up to 450 mm thick in aluminum and up to 150 mm in steel. In medium vacuum, the beam size is larger, resulting in lower power density, lower penetrations, and wider widths of welds are obtained; however, it is successfully used in batch productions in operations such as welding of a gear to a shaft.

In the nonvacuum mode the beam is dispersed much more; consequently the gun-to-work distance must be kept short, less than 40 mm. This limits the shape of the workpiece near the joint. An example of the welding performance of nonvacuum welding of steel is presented in Fig. 12.48. Penetrations up to 25 mm are possible. ▲

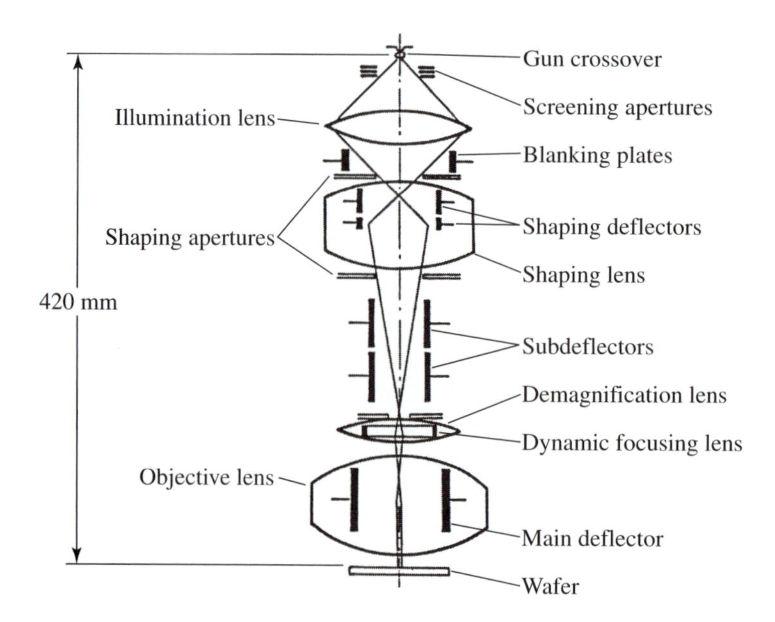

Figure 12.47
Electron optical column of a variable-shape E-beam system used in making large-scale integrated (LSI) electronic circuits. [FROM HARADA, REF. [8]]

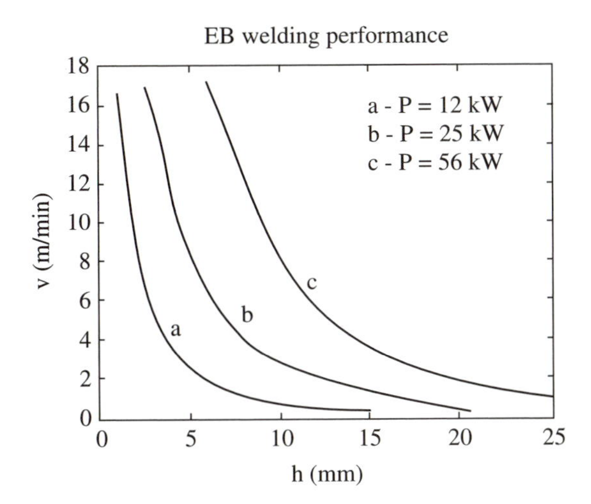

Figure 12.48 Welding performance in steel of a nonvacuum EBM system.

12.9 OXYGEN CUTTING (OC)

Oxygen cutting (OC) is not exactly a nontraditional manufacturing process. It has been practiced for a long time and has been the fundamental process for cutting thick steel plates and preparing them for welding. We have included it in this chapter because it is a chemical process, and also because it is the precursor of the plasma arc cutting process, which is the most significant alternative and which by itself is commonly considered as one of the nontraditional ones.

OC severs or removes metal by a high-temperature exothermic reaction with oxygen. This is the case in massive cutting of low-carbon and low-alloy steels, and in cutting titanium. Steel as thick as 2 m has been cut. Cast iron, alloy steels, and especially stainless steels and other oxidation-resistant metals are more difficult to cut, and the process must be aided by the use of a chemical flux or metal powder. The OC process allows for several alternatives, with oxyfuel gas cutting (OFC) as the most significant and most commonly used one. The fuel gas–oxygen flame is needed to bring the temperature of iron up to and maintain it at the ignition temperature of 870°C, when the basic exothermic reaction of iron takes place. The OFC process uses a torch with a nozzle in which there are two parts, one for the delivery of the preheating gas mixed with oxygen, and the other for the supply of high-purity oxygen (see Fig. 12.49a). The pure oxygen reacts rapidly with the hot metal in a narrow section and blows the metal out of the cut (see Fig. 12.49b).

Three chemical reactions are involved:

1. $Fe + O \rightarrow FeO + 267$ kJ
2. $3Fe + 2O_2 \rightarrow Fe_3O_4 + 1120$ kJ
3. $2Fe + 1.5O_2 \rightarrow Fe_2O_3 + 825$ kJ

The most important is the reaction 1) in which, theoretically, 0.29 m^3 of oxygen reacts with 1 kg of iron to produce Fe_2O_3. Practically, more or less oxygen is consumed depending mainly on the thickness of the cut. The latter case occurs if not all of the molten metal fully oxidizes but is removed by the kinetic energy of the oxygen stream.

The alloying elements in steel are generally oxidized or dissolved in the slag without affecting the process. If some oxidation-resistant elements such as nickel and

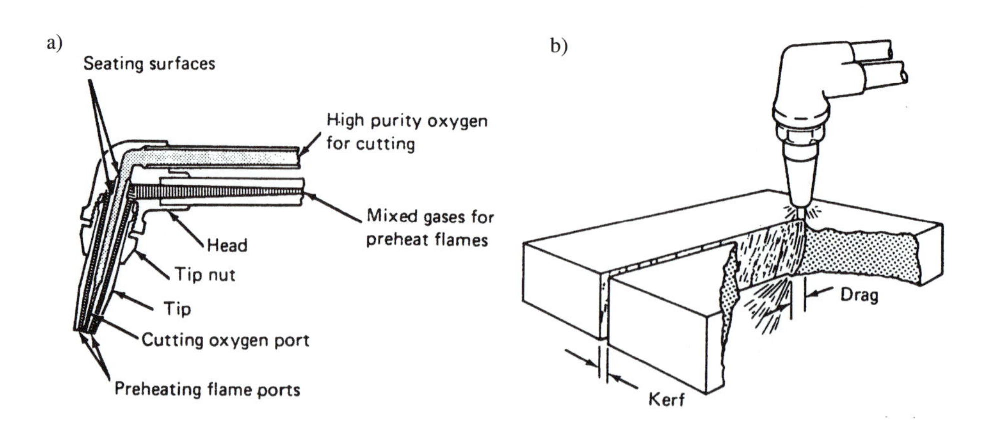

Figure 12.49
a) Oxygen fuel cutting torch. Two components are delivered: the gas and oxygen mixture for preheating the work, and high-purity oxygen to feed the exothermic reaction of burning iron; b) the oxygen fuel gas cutting process (OFG). [WH2]

chromium are present in higher percentages, modifications of the cutting technique are required. This is certainly so in the case of stainless steel. Among special techniques we find torch oscillation, use of a waste plate of carbon steel on top of the plate to cut, delivery of a stream of iron-rich powder into the cut.

Various gases are used with the oxygen to provide the preheating flames: acetylene, natural gas, propane, and propylene. Gasoline is also used with a special torch in which it vaporizes before burning.

In many applications OFC is done manually with the operator manipulating the torch. The equipment is rather simple; it consists of the cylinders of oxygen and of the preheating gas, hoses, and pressure regulators. Different tips are interchangeably attached to the torch for cutting different thicknesses of material, and the design of the torch varies depending on the type of the preheating gas. Gaseous oxygen cylinders are available in various sizes up to 8.5 m^3, and they are usually pressurized to 16MPa. Larger supplies are obtained from liquid oxygen cylinders equipped with liquid-to-gas converters. Larger supplies may also be obtained by connecting a number of cylinders by manifolds. Alternatively, gaseous oxygen may be transported from the producer plant to the user in long, high-pressure tubes mounted on truck trailers. Acetylene is supplied in multiple cylinders or produced in the consumer's plant in large generators. In such a generator, small lumps of calcium carbide are discharged from a hopper into a body of water.

In many industries OFC is carried out on large gantry-type machines that are most often numerically controlled and use multiple torches for simultaneous cutting of several identical parts. Several torches may be combined to produce edge preparation of plates for welding, as shown in Fig. 12.50.

Manual cutting is used on site to cut pieces for production, or to cut up scrap, or for salvage operations and disassembly. In steel foundries, gates and risers are cut off the castings. Machine cutting is used in plants involved in constructing machinery frames and housings assembled by welding and in the production of ships built of welded steel plates.

Typical performance data for cutting low-carbon steel using methacetylene-propadiene stabilized preheating gas are presented in Fig. 12.51. The consumption of gases is typically, for a 50-mm-thick plate, 85 l/min of cutting oxygen at 350 kPa, 15 l/min of preheating oxygen at 100 kPa, and 5 l/min of MPS gas at 50 kPa.

Heavy cutting of steel over 300 mm and up to 1500 mm thick is done by using heavy-duty torches, such as in ingot cropping and salvaging operations. Oxygen pres-

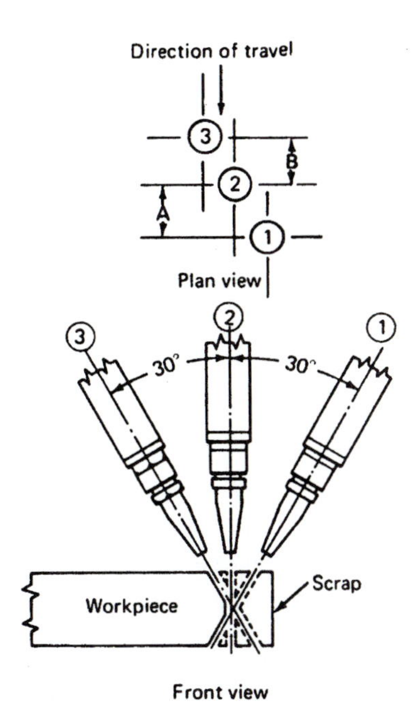

Figure 12.50
Cutting a double bevel edge preparation with a root face for subsequent butt welding of the plate on this edge. [WH2]

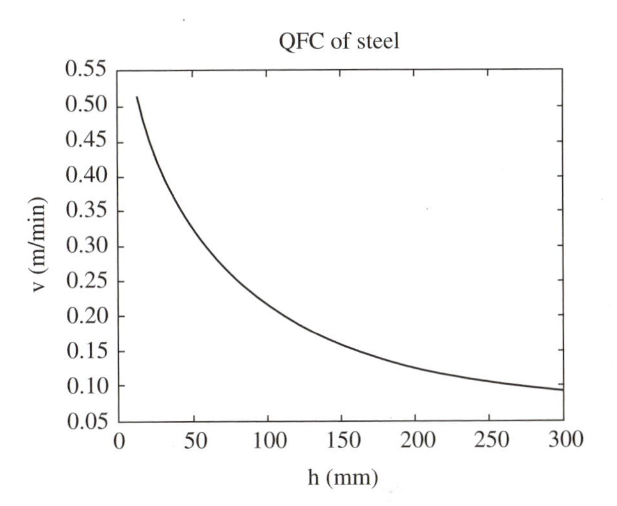

Figure 12.51
Performance of oxygen fuel gas cutting in low carbon steel using methacetylene-propadiene stabilized (MPS) gas.

sures of 150 to 380 kPa at the torch are used, with a flow in the range of 500–2000 l/min, achieving speeds between 30 and 100 mm/min.

12.10 PLASMA ARC CUTTING (PAC)

PAC is used for fast cutting of medium-to-thick plates of many metals, including carbon steel, aluminum, and stainless steel. Cutting of carbon plate less than 75 mm thick is faster than with OFC, and below 25 mm thickness, it is up to five times faster. The smoothness of the cut surface is comparable to OFC, but there is almost no surface oxidation if water injection or water shielding is used.

The plasma arc cutting torch is similar to the one used in plasma arc welding, described in Chapter 11. The difference consists in a higher velocity of the plasma gas jet than in welding, which melts the metal and blows it away. The process uses a DC electrode-negative transferred arc between a tungsten electrode and the workpiece. The plasma is concentrated into a narrow beam of high velocity and high temperature in a constricting orifice. Various plasma gases are used for the various workpiece materials: nitrogen, nitrogen-hydrogen, or argon-hydroxen mixtures. Titanium is cut with pure argon. Carbon steels are cut by using a mixture of 80% N_2 and 20% O_2, or else just Ni_2 when water shielding is used. There is a separate shielding gas nozzle that may use CO_2 or Ni_2. Instead of the shielding gas nozzle, a water-injection nozzle may be used. In

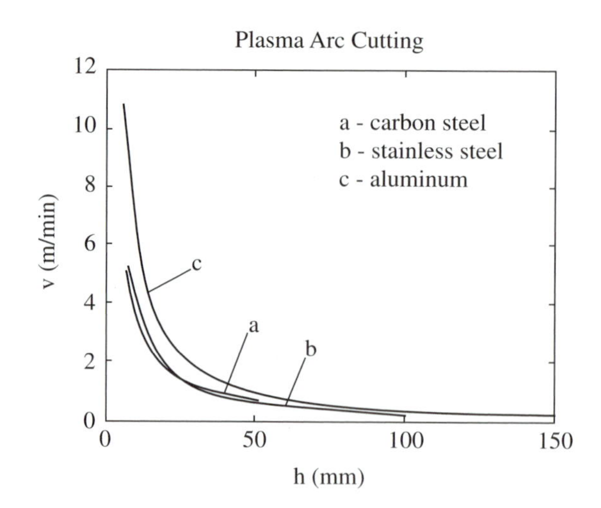

Figure 12.52
Performance of plasma arc cutting (PAC). For steel thickness below 75 mm it is faster than OFG; below 25 mm it is five times faster. It can cut stainless steel and is very efficient for cutting aluminum.

another arrangement water is injected near the constricting orifice to further constrict the flame. Voltages of 120–200 V are used for thinner plates and 400 V for plates up to 50 mm thick. The currents range between 70 and 1000 A depending on material and cutting speed.

Typical performance data are presented in Fig. 12.52. Note that aluminum is easy to cut. Thicknesses up to 150 mm (6 in) can be cut, and at $h = 5$ mm, a traversing velocity of 10 m/min is achieved. Thin carbon steel is cut faster than with OFC, but the maximum practical thickness is only 50 mm. Performance in cutting stainless steel is almost the same as on carbon steel, but plates can be cut up to 100 mm in thickness.

In mechanized cutting it is common to fill the table of the machine with water reaching up to the bottom surface of the plate being cut. The gases escaping from the plasma produce turbulence in the water, which traps the fume particles. In addition, a nozzle is attached to the torch that produces a water curtain around the arc. This reduces the noise generated in the cutting operation.

12.11 ELECTRONICS MANUFACTURING

A vast field of electronics manufacturing activities has developed over the past two decades as a notable example of human creativity and technological capability. It is an excellent illustration of the basic tenet of modern engineering that postulates the interrelationship between design and manufacturing, or concurrent engineering. The performance of the core components of a computer, the integrated circuits (ICs), has been increasing rapidly, while at the same time their cost was being reduced due to continuous miniaturization. Progress in the design of ICs went hand in hand with the progress of manufacturing techniques. The number of components, diodes, transistors, and capacitors per chip has increased a hundred-thousandfold between 1965 and 1995, and hundreds of millions of components can now fit on a single chip. The industry went from small-scale integration to medium, large, very large, and ultra-large-scale integration (VLSI, ULSI). The trend is towards one billion transistors per chip (gigascale integration, GSI). The thickness of connecting lines has decreased to the submicron level. Many unique processes and pieces of production equipment have been developed along the way. Some of the processes are based on methods described earlier in this chapter, and others are extensions or outgrowths of the various physical and chemical procedures. Altogether, these techniques are summarized as planar technology because they consist of adding, subtracting, or modifying parts of thin layers of a plane silicon chip. Repeated use of photolithography is at the center of this technology.

Several hundred identical chips are manufactured from a thin silicon wafer that has been sliced from a rather large (up to 200 mm in diameter, and 1–3 meters long) cylindrical monocrystal of very pure silicon (see Fig. 12.53).

Electronic grade silicon (EGS) is a polycrystalline silicon of such high purity that impurities are in the order of parts per billion. It starts as metallurgical grade silicon (MGS) produced from quartzite (SiO_2) in an electric arc furnace. The MGS is ground to powder, which reacts with HCl gas in a fluidized bed reactor at 300°C and is reduced in a subsequent reaction with hydrogen at 1000°C to produce the EGS. This is followed by the production of the single-crystal ingot called boule, which is pulled out from molten EGS in a vacuum induction or resistance-type electric furnace. In this process the pure silicon is doped to produce either the *p*- or *n*-type silicon.

Pure silicon is a poor conductor. However its conductivity can be dramatically altered by introducing very small amounts, on the order of 1 part per million, of foreign elements called dopants. There are two kinds of these. Pentavalent elements such as N, P, As, or Sb will form a substitutional solid solution with Si, adding surplus electrons to some of the Si atoms, which become negative charge carriers. This is the *n*-type semiconductor. The *p*-type semiconductor is created by doping silicon with trivalent elements such as B, Al, and Ga that produce atoms with one electron missing. Such holes can be considered positive charge carriers.

The boule of the silicon monocrystal is ground on the outer diameter using a cup-type diamond grinding wheel and then sliced into wafers using a thin cut-off, ring-type diamond abrasive saw. The wafers are cut to a thickness of 0.3 to 0.4 mm and subsequently ground on the periphery to round off the rim so as to prevent chipping. Then they are polished to perfect flatness and smoothness. Finally, the wafer is chemically cleaned to remove residues and organic films. The wafer then goes through a long series of processes in which hundreds of identical chips or dies are produced, tested,

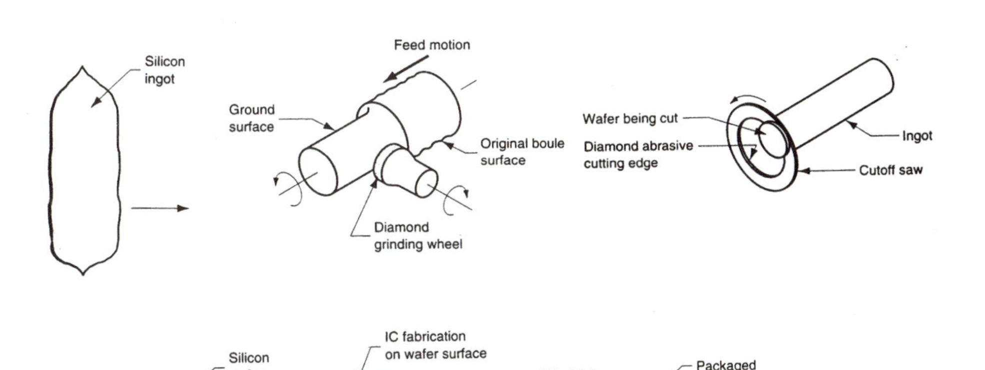

Figure 12.53
Processing the pure silicon ingot into a chip. [From M. P. Groover, ref. [33]]

and finally cut off and separated. The common size of a chip is a square with 5-mm sides that carries the integrated circuits. The chips are packaged in subsequent operations and inserted into printed circuit boards (PCBs).

Photolithography (see Section 12.5) is used between the individual processing steps to determine the geometry of the part of the plane that should be processed, either etched away or covered by the chosen material. A typical procedure is illustrated in Fig. 12.54. It shows the individual steps used to dope a selected area so as to convert a part of an *n* field to *p* as part of the process of creating a bipolar *p-n-p* junction transistor. The procedure starts from the initial state 1 of an *n* zone that has been provided on a *p* substrate. In step 2 a layer of silicone dioxide is created either by thermal oxidation or by chemical vapor deposition. SiO_2 is commonly used for various purposes, most often as an insulator. In this case it acts as barrier to the doping process. An opening in the SiO_2 layer will have to be provided over the area to be doped. Therefore, in step 3 a layer of photoresist is applied, and then a mask is located over it. The mask is a thin, nontransparent film deposited on a glass plate. The photoresist is exposed to UV light through the glass plate with the mask and is hardened over the unprotected area. In step 5 the resist is developed in such a way that the developing solution dissolves the unexposed areas, and then these parts of the resist are washed away. The hardened part of the resist is baked onto the substrate, which is then exposed to hydrofluoric acid (HF) that has been chosen as an etchant that does not affect the resist and has a selective

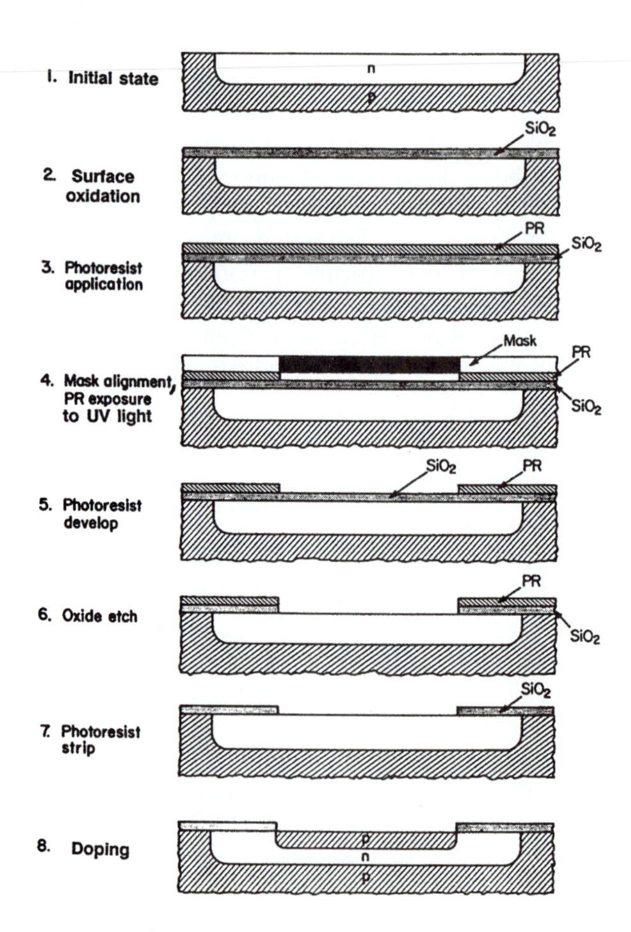

Figure 12.54
Application of photolithography. Sequence to obtain a *p-n-p* transistor. [HUMMEL]

effect by dissolving the SiO_2 but not the silicon base (step 6). The resist coating is stripped away by a corresponding chemical, and the just-created surface is washed clean. Finally, the exposed substrate is doped by ion implantation. A succession of similar procedures is used to generate the multitude of electronic components and their interconnections on the chips created on the wafer.

Some of the steps in the generation of a MOSFET transistor are presented in Fig. 12.55. The processes used in the individual steps are discussed later in more detail. The design of the transistor is at the top of the figure. It is built on a block of *p*-type silicon that is covered by an insulating layer of SiO_2. This layer is removed at S (source) and D (drain), where it is replaced by aluminum connectors AL. Under these contacts there are heavily *n*-doped zones of the source and the drain. The magnitude of the current that will flow between them is controlled by the gate G, which consists of a bar of *n*-doped polysilicon PS (polycrystalline silicon) that is conductive and extends from the aluminum input strip to reach over a thinned layer (less than 100 nm) of the insulator. By applying positive voltage to the gate, the electric field repels holes from the channel C between the source and the drain and, when the control voltage exceeds a threshold value, an electron cloud is produced in the channel C. An electron current will flow to the positive charge of the drain. It is now obvious that the designation of this component stands for metal oxide silicon field effect transistor.

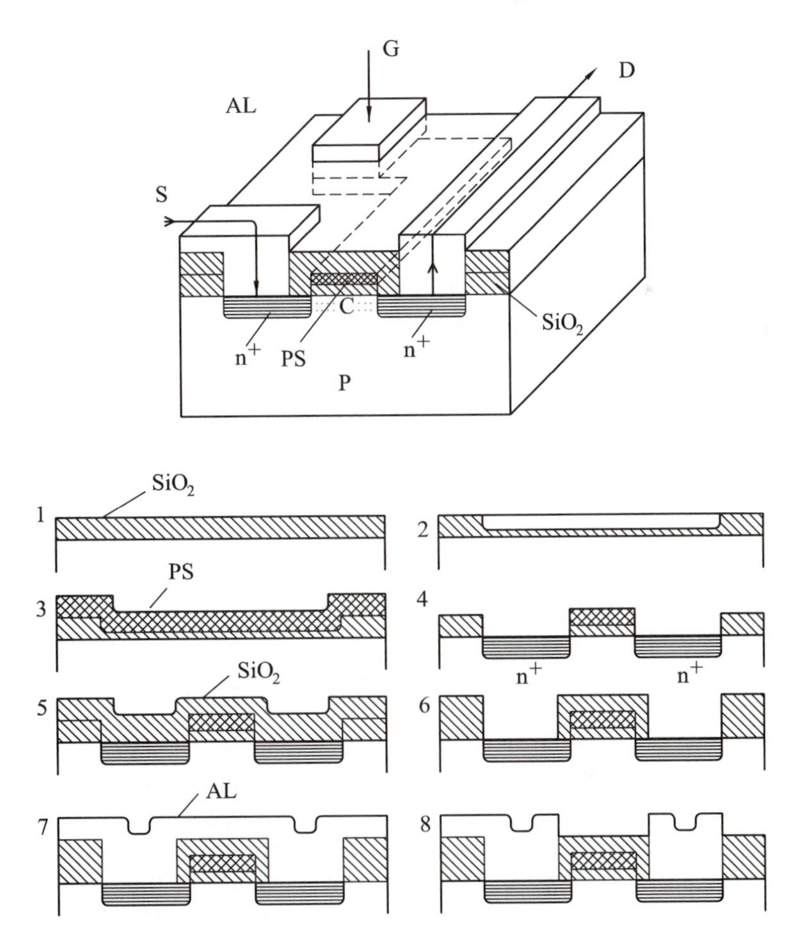

Figure 12.55
Steps in the generation of a MOSFET transistor. It will have three aluminum strips for external connections to source S and drain D over the n^+-doped areas and to gate G over the channel C. The latter is connected via the inverted L-shaped polysilicon (PS) layer.

A brief expression of the manufacturing steps is presented in the figure. In each of the steps photolithography is used by applying a photoresist layer (PHR), locating a mask (MSK) over it, exposing the photoresist (EX), removing the unexposed photoresist (REM), then carrying out the process of either adding, removing, or modifying a layer of the naked area, and washing away the remaining resist (WASH): PHR-MSK-EX-REM-process-WASH. In the eight parts of the process in Fig. 12.55 these photolithography procedures are repeated but they are not included in the description; they should be implied. For simplicity of the graphics, too, only the frontal plane views of the 3D progression of creating the MOSFET element are given. It should not be difficult to imagine their extrapolation into the 3D picture at the top of the figure. The individual processes are described below.

(1) The thin silicon dioxide layer that will become the insulating barrier between the gate and the channel C is generated. Except for a border at the periphery of the element, the SiO_2 layer is etched down by either applying HF or by some other technique.The etching action is stopped when the desired depth is reached. The result is shown in (2).

In (3) the polysilicone gate is produced, and the *n*-doped areas of the source and the drain are created. First, a layer of polysilicone is applied over the whole surface. It will serve as the gate. It is then etched away except over the inverted L-shaped area shown in the top figure. In this etching operation the thin SiO_2 layer created in (1)–(2) is also removed except under the polysilicone. This exposes the substrate between the thick border layers of SiO_2 and the polysilicone. These areas are then *n*-doped. The result is seen in (4).

Next, in (5) another layer of SiO_2 is applied and then etched away over the areas above the source and the drain and above the rear end of the inverted L of the polysilicon gate. This will provide access to the three aluminum connectors. The result is seen in (6).

Finally, in (7) the whole area is first covered with a layer of aluminum that is then etched away except for the three strips that act as connections to the other parts of the circuit; see (8).

At the end, a layer of P-glass (phosphorus-doped silica) is deposited over the whole surface for protection.

The simplified diagram in Fig. 12.56 shows how different electronic circuit elements such as transistors, diodes, resistors and capacitors are located side by side in the planar design of the IC.

In the example of Fig. 12.55 the following processes were described:

Coating the silicon substrate with SiO_2

Etching SiO_2

Coating SiO_2 with polysilicon

Etching polysilicon

n-doping the silicon substrate

Coating silicon, polysilicon, and SiO_2 with aluminum

Etching aluminum

The techniques that may be used in these operations are thermal oxidation, chemical vapor deposition (CVD), physical vapor deposition (PVD), thermal diffusion, ion implantation, sputtering, wet chemical etching, and dry plasma etching.

Thermal oxidation may commonly be used for the generation of a silicon dioxide layer on the silicon substrate in an atmosphere of oxygen or of steam at tempera-

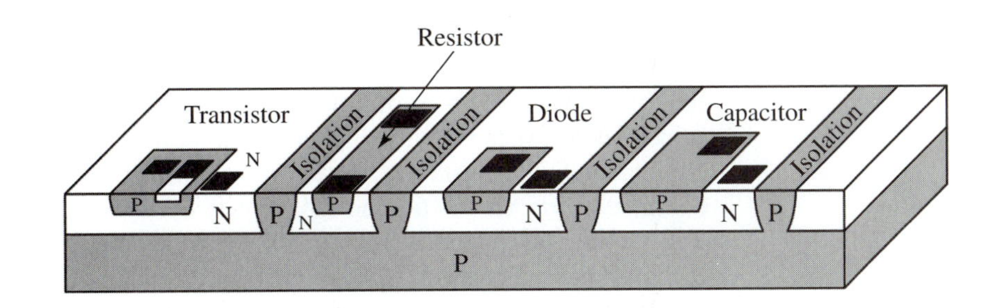

Figure 12.56 Basic components of integrated circuits. [HUMMEL]

tures between 900° and 1300°C. *CVD* is used in a variety of deposition processes and is based on chemical reactions of gases at elevated temperatures. So, for instance, polysilicon is deposited as a result of reduction of silane (SiH_4) at 600°C. The *PVD processes* are carried out in vacuum at slightly elevated temperatures. In the basic process, the metal or ceramic to be deposited is heated to a high temperature at which it vaporizes. Resistant heating is provided by a tungsten filament, or induction heating of a metal in a crucible is used, or else an electron beam bombards and evaporates the material. The emitted atoms travel in the vacuum chamber in straight lines from the source and when they strike the cold surface of the substrate they condense on it. In a modified process a DC electric field is applied with the metal to deposit as cathode and the substrate as anode. In yet another modification the chamber is filled with argon at a moderate pressure (3–6 Pa) and an electric field of several kV is produced that ionizes the argon to create plasma. Heavy positive ions are accelerated against the surface of the cathode and dislodge atoms from its surface that travel across the electric field and deposit on the anode. This method is called *sputtering.* The process of *ion implantation* is used for controlled doping of silicon. Atoms of the selected material are ionized, accelerated, and filtered in a deflecting magnetic field; they are further accelerated in an electric field of several hundreds of kV and finally directed by scanner plates towards the target, where they penetrate to a certain depth under the surface. *Thermal diffusion* is another method used for doping the silicon substrate. Boron may be applied for *p*-doping, and phosphorus, arsenic, and antimony for *n*-doping. This process is carried out at high temperatures between 800 and 1200°C. In *wet etching* a solution of an acid is used in a process similar to chemical machining (Section 12.5). It is important to choose the acids for selective action. So, for instance, HF dissolves SiO_2, but not aluminum, and phosphoric acid (H_3PO_4) dissolves Al without attacking SiO_2 or Si. Neither of them attacks photoresist, which is an organic polymer and can be removed (stripped) with a solvent such as acetone. Wet etching has several disadvantages, such as disposal problems and the problem of undercuts (see Fig. 12.20). Therefore, it is commonly replaced by *plasma etching,* in which only small quantities of affluents are produced and the process is anisotropic so that walls perpendicular to the wafer surface can be generated without overcuts. An electric field is generated in the processing chamber. The reactive gas is pumped into the chamber, where it is ionized. So, for instance, carbon tetrafluoride gas (CF_4) produces plasma that contains active fluor ions. This is used to etch Si, SiO_2, and Si_3N_4. Etching rates of silicon can be increased by adding 10% of O_2 to CF_4. Very fine details can be produced, with 1-μm width. After the plasma etching is completed, the etching gas is replaced by oxygen to strip the resist.

The technology of IC manufacturing has become a highly specialized field. It is beyond the scope of our text to try and discuss the theoretical details of the processes requiring advanced levels of physics and chemistry. However, mechanical engineering principles are very usefully applied in the design and construction of the specialized equipment. So, for instance, NC and servos similar to those discussed in Chapter 10 are being used for *XY* positioning tables, called stages, requiring submicron accuracies. The motions may be driven by piezoelectric or electrodynamic actuators. Positional feedback is derived from the laser interferometer (see Chapter 9) or, for short motions, from capacitive gages, and the slides float on aerostatic or magnetic-suspension guideways. Such stages are applied to the production of masks for photolithography (stepper stages) for multiplying the photographic pattern designed for one chip. Similar systems, called probers, are used to move the wafer under an array of probes to check the quality of the individual circuits.

Once the individual chips have been cut off the wafer, they must be suitably packaged in a plastic or ceramic enclosure and connected to the package leads. The dual-in-line package (DIP) is a very commonly used type; see Fig. 12.57. These components are subsequently mounted on printed circuit boards (PCB). The PCBs are made of polymer composites reinforced with glass fiber. The insulating board is covered by a thin layer of copper, which is then selectively etched away to provide the interconnections. Packaged chips and other electronic components are mounted on one side of the board by inserting their leads in predrilled holes (PIH—pin-in-hole technology) and soldered on the bottom side or else soldered on the top surface (SMT—surface-mount technology); see Fig. 12.58. Automatic insertion machines are used in these operations. The connections between the elements are provided on one side of the board, or on both sides, or even in multiple layers when one layer could not avoid intersections of the connection lines. These connections are produced by photolithography, which

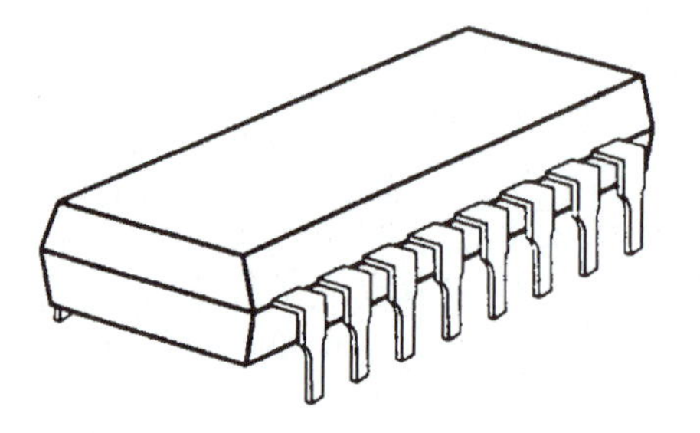

Figure 12.57
Dual-in-line package of a chip.

Figure 12.58
A printed circuit board (PCB). Packaged chips and other electronic components are mounted on the board that is made of an insulating material that has been covered by a thin layer of copper etched so as to provide connecting paths between the components.

again includes producing a mask defining the circuit pattern and exposing a layer of a photoresist, followed by etching away the copper cladding, leaving behind just the rather thin copper lines. The accuracy required in the production of these connections is lower than in the production of patterns on the chips, and often a less-expensive method of screen printing is used instead of photolithography.

12.12 ADDITIVE CNC MANUFACTURING (RAPID PROTOTYPING)

In 1988 the stereolithography (SLA) system was introduced. It was soon followed by a variety of other systems developed during the early 1990s that have all been called rapid prototyping, and the acronym RP has become generally accepted. By 1996 about 950 RP systems were installed worldwide. Research activities sprang up that were devoted to this field, and a tremendous amount of journal papers and conference proceedings have been published to date [10]. The fascination with the field has two main causes.

For one, these techniques differ fundamentally from the three basic traditional processes used to produce 3D parts, including those that have well-defined, fully enclosed cavities. The only one of the classical techniques able to do this, including the cavities, is casting (Chapter 3). However, a casting is made in a *mold* that, for the cavity, also contains a *core,* and the new RP processes do not need either of these. They are *moldless* and *coreless.* The other two traditional processes, forming and cutting, cannot produce the enclosed cavities at all. The first just *changes the shape* of a bulk part, for which it mostly also needs a *die,* and the latter (machining, EDM, ECM) *removes* unwanted portions of a bulk. The RP processes *add* material in layers to produce the part. Therefore, they could be called *additive processes.* In essence, 3D parts are obtained by consolidation from powders or from liquid, or by ink jet printing, without the use of molds or dies. Other fitting names for these processes could be *moldless NC consolidation* or *layered NC manufacturing.*

Secondly, the RP processes use the computer input in a very immediate way, much more so than, for instance NC machining. They represent a *direct manufacturing extension of CAD.* Many of the activities necessary to link CAD with CAM in NC machining, those that constitute process planning (see Chapter 1), such as choice of tools, speeds, feeds, depth of cut, choice of the sequence of tool passes, use of " features" in CAD-CAPP, are thus not needed. Once the outer and inner boundary surfaces are determined in the CAD stage, they just have to be sliced into 2D layers, and the tool, such as a laser beam, is moved to fill out the surface of a sequence of layers. This is a relatively simple programming task. Right at the beginning of the SLA system development, a programming language was included called STL that has become a standard subsequently used by most of the other RP systems.

However, the name "rapid prototyping" is misleading. The parts produced by processes such as SLA are *not real prototypes* of machine parts because they are made of materials whose properties differ from and are mostly inferior to the materials of the actual machine parts. In the case of SLA and of several other processes, acrylic polymers are used that can be rapidly cured by ultraviolet light, and a UV laser is used for this purpose. Their strength and stiffness are rather low, and their glass-transition temperature T_g (see Chapter 6) is in the range of only 40–75°C. Above these temperatures the material becomes viscoelastic and quickly loses its strength. It is therefore not possible to test these parts for functional behavior in their intended use. Nevertheless, these parts are made rather accurately, within 0.1 up to 0.5 mm, and they offer the designer an early opportunity to check the geometric form of rather complex 3D shapes—a visual and dimensional verification. The product is a rather accurate geometric model that can be used by all the participants of the concurrent engineering process to assess the suitability of the design. The claim is made that validating and communicating complex solid design concepts with co-workers,vendors, and customers on the basis of drawings only is difficult and prone to mistakes. It is a significant advantage to use a solid physical model. In some cases this model offers also partial functional verification. An example has been given of a car engine intake manifold made of a clear transparent material to check the flow of gas through it. This, however is limited to cold gases, and no mechanical loads can be applied. It may then be useful to understand the RP process as *rapid 3D modeling* rather than prototyping.

The shortcomings of the applicability of most of the RP processes has led, in the second half of the 1990s, to developments of methods additional to the original processes that make it possible to obtain parts in realistic materials. To illustrate, the original SLA process offers a modification in which the product made in a polymer is used as a pattern for the investment casting process, instead of the traditional wax pattern. Correspondingly, prototypes of parts made of steel can be obtained. This and other additional processes are now called *rapid tooling (RT)*. The following text briefly describes various RP and RT processes that currently exist; they are discussed and compared, and examples of their applications are given. They are all relatively new, and certainly other variants will be developed.

12.12.1 Rapid Modeling

The first *stereolithography (SL)* machine was presented on the market in 1988, and it is still the most popular one [12]. The process is based on photopolymerization, that is, a property of some special plastics to cure and solidify from a liquid when exposed to UV light. Thus it is not a thermal process, and therefore a relatively low laser power can be used, and a rather fine resolution of the pattern is possible. The process follows the scheme shown in Fig. 12.59. A vat is filled with the liquid polymer; its level is adjusted to keep its surface in the plane of focus of the laser beam. An He-Cd laser with 325-nm wavelength or a solid-state Nd:YVO_4 laser with 354.7-nm wavelength is used. The workpiece is generated in layers 0.05 to 0.2 mm thick. A platform on which the workpiece will be built is submerged in the liquid, first just by one layer thickness below the surface; then it is stepwise lowered, layer after layer. A special technique is used to spread a uniformly and accurately thin layer. The laser beam motion is controlled to stop and generate a "voxel" (3D pixel) and move to the next one. The size of the voxel is determined by the power of the laser, its focus and the stopping time, and it should guarantee a proper overlap with the underlying layer (see Fig. 12.60). To save time, the workpiece cross section is only partially scanned and solidified; it is filled mainly along the outer and inner contours of the section together with some cross-hatching over the area in between and fully filled on the top and bottom faces. The liquid contained within the cross-hatching lines will be solidified after all the stereolithography work, which ends with the "green part" and is then moved to a post-curing, broadband UV oven. Before post-curing, the part is cleaned and removed from the platform using a sharp blade. SL machines are available in three sizes, for parts up to 250 × 250 × 250 mm, 350 × 350 × 400 mm, and 508 × 508 × 580 mm.

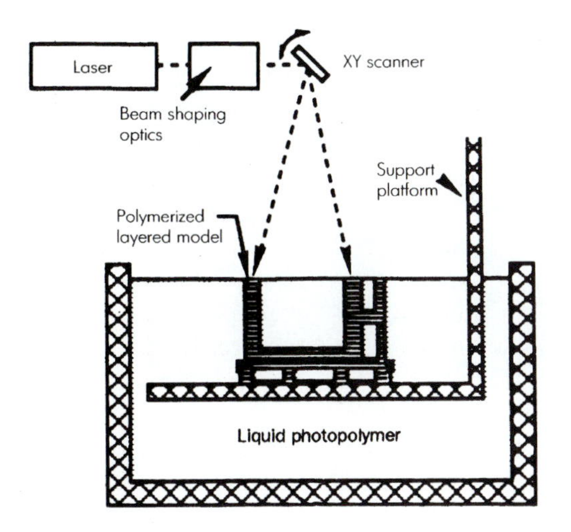

Figure 12.59
The stereolithography (SL) process. A UV laser beam produces fast photopolymerization of special liquid resin in NC travel over layer after layer, with the support platform stepping down between layers. The resins used have low mechanical properties. [FROM JACOBS, REF. [12]]

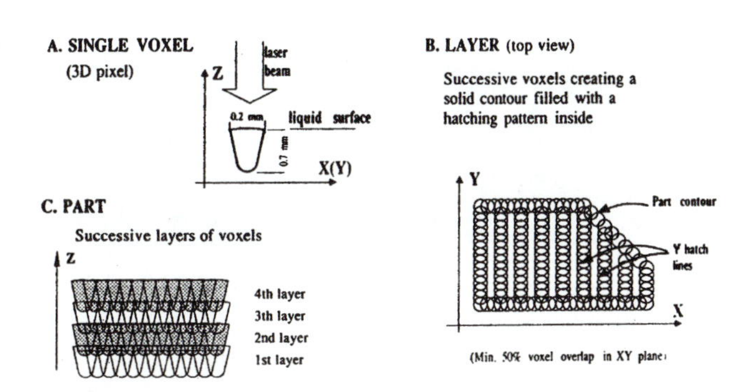

Figure 12.60
Point-by-point stereolithography procedure. Solidification is produced in individual voxels during stops of the UV beam movement. A solid structure is produced filled with liquid polymer to be subsequently cured in an oven. [FROM HULL, REF. [32]]

Scan speed is 500 mm/sec for a 15-mW laser and 5000 mm/sec for a 216-mW laser. The performance figure of 65 cm^3/hour is given for the former type. Post-curing may take several hours.

The NC program is derived from the CAD model of the part using special software that generates the layering. The STL language is becoming common for this task. It defines the external and internal boundaries of the solid body as consisting of flat triangles and then slices them, layer by layer, to determine 2D areas with their inner and outer borders. The program then formulates the scanning path of the beam. It also takes care of generating supports for overhanging parts of the workpiece. The manufactured workpiece is finished by polishing, painting, or spray-metal coating. The supports that played a role similar to that of a fixture in a machining operation are cut off. Additional operations of drilling, boring, and tapping may be performed.

Sample parts are shown in Fig. 12.61. The first one is the model of an intake manifold to be used for flow studies to help finalize the design. The second is a turbine blade model that will, after inspection and potential modification, be further processed into an aluminum part for further testing.

An alternative system, called *solid ground curing (SGC)* [13], uses a layer-by-layer instead of point-to-point technique (see Fig. 12.62). The process consists of two phases for each layer, the production of a mask, steps A to E, and the solidification of the layer pattern in the photopolymer, steps 1 to 6. The mask may be produced by photolithography applied to photosensitive plastic foils, one for each layer, or by charging a glass plate electrostatically with a toner in a process that is commonly used in photocopying machines. The latter case is the one shown in our figure. The manufacture of the part is done differently from the SLA process. Instead of using a liquid bath, the

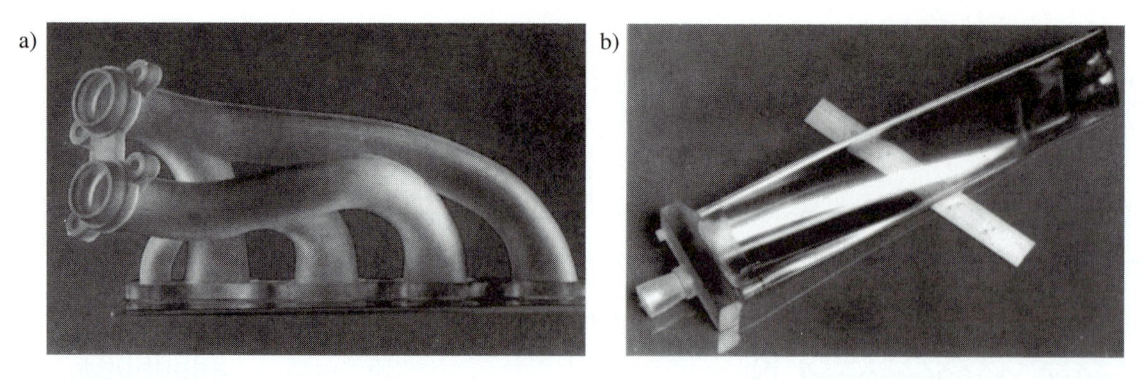

Figure 12.61
Examples of models produced by the SL process: a) car engine intake manifold; b) turbine blade. [FROM JACOBS, REF. [12]]

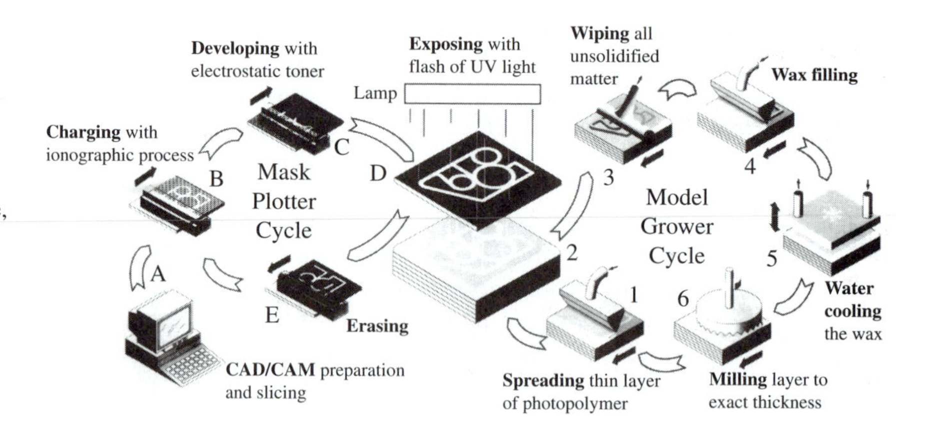

Figure 12.62
The layer-by-layer stereolithography, solid ground curing (SGC) process: Creating the mask, A to E, and solidifying the pattern over the whole layer at once, 1 to 6. Photopolymers are used, as in SL. [CUBITAL].

part is embodied in wax. Any new layer of the photopolymer is spread over the surface and exposed to UV light through the mask as a whole. Mercury lamps emitting a broad UV spectrum are used, with 2500 W power. Next, the unsolidified polymer is wiped off, and the void is filled with molten wax that is then water-cooled. Finally the surface of the just-produced layer is milled flat, the piston with the platform and workpiece is lowered by the amount equal to layer thickness, and it is then ready for the next step 1. Layer thickness is chosen between 0.05 and 0.15 mm, and production time of 65 to 75 sec/layer is claimed. A performance figure of 24 cm^3/min is claimed. Obviously, this process is much faster than the point-to-point one described before, and no post-curing is needed. There is also no need for generation of supports, because the polymer part is supported by the wax during the process. Examples of toy car models are shown in Fig. 12.63. Figure 12.64 shows a model of a new design of ice hockey skate blade. The base of the blade is made of a high-strength, cold-resistant, fiber-reinforced plastic that is bonded to the shoe sole. The metal blade is encased in a profiled plastic beam that is bolted to the base. The geometric models of the two plastic parts are claimed to have been prepared and made using the SGC method within 14 hours from obtaining the CAD files. The parts were then sanded and painted to provide realistic appearance

Figure 12.63
Models of a toy car produced by the SGC process. [CUBITAL]

Figure 12.64
Model of ice hockey skating blade produced by the SGC process. [CUBITAL]

models for inspection and modification by the designers. A second set of models was then made and found satisfactory for the manufacture of hard tooling for injection molding of the actual product.

The processes described in the preceding paragraphs are limited to the processing of photopolymers. A much broader range of polymers, specifically thermoplastics such as PVC, polycarbonate, ABS, nylon, and also investment-casting wax can be 3D-consolidated by *selective laser sintering (SLS).* The process [14] is diagrammatically presented in Fig. 12.65. The powder of a polymer is preheated to a temperature below its melting point. Selective solidification is produced by further heating, layer by layer, by means of a scanning laser beam. Sintering occurs when the viscosity of the grains drops causing the surface tension to be overcome, and the grains fuse without fully melting.

In conclusion, for the three RP processes producing parts made of polymers, it is useful to review the matter of the *materials.*The photopolymers used in stereolithography (SLA) and in solid ground curing (SGC) exist as acrylated resins and also as epoxy resins. The materials used in selective laser sintering (SLS) include a variety of polymers such as ABS, nylon, polycarbonates, and also investment-casting wax. The acrylated resins have *UTS,* tensile strength, and modulus of elasticity, *E,* only about half of

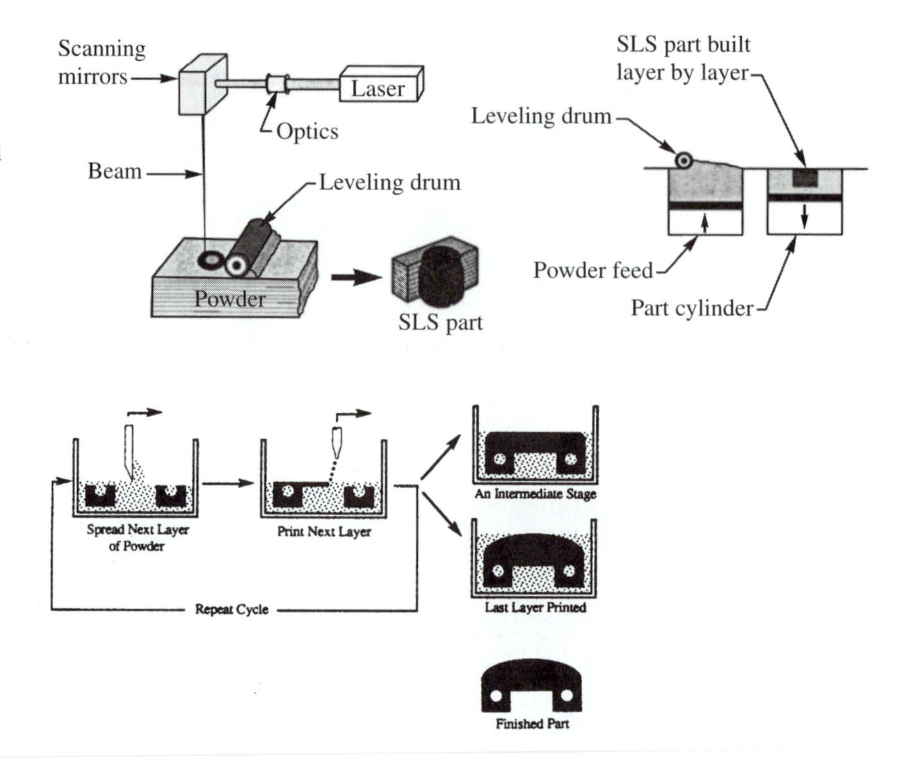

Figure 12.65
The selective laser sintering (SLS) process. A rather wide variety of polymeric materials are used, preheated below melting. The laser beam adds local heat, causing solid fusion of grains. [FROM DTM CORP., REF. [14]]

Figure 12.66
The 3D printing (3DP) process. Different powders, including ceramics, metals, and polymers, are used and selectively mixed with droplets of binder material delivered by the ink-jet technique. Multiple nozzles are used. Polymeric binder may subsequently be burned out in a furnace, while the grains of the product are sintered. The porous body is then infiltrated by a metallic binder in another oven. [FROM SACHS, REF. [15]]

those of nylon, and also a much lower elongation at break, 4–14 as compared to 6–32 for nylon. Their glass-transition temperature T_g is 53°C. They provide fine geometric detail, excellent surface finish, accurate form, and exceptional dimensional stability. However, because of the low strength and low useful temperature range, they cannot be used for snap fitting and other functional tests involving loads. The epoxy resins have better strength and E modulus, but again a rather low T_g of 65–90°C. They cannot be exposed to moisture, nor to almost any chemicals, and exposure to temperatures above 40°C causes the epoxy to soften, warp, and distort. Thus their use is limited to models to be used as visual aids. The SLS process produces a coarser surface finish due to the size of the sintered particles: 50 μm for fine nylon and 90 μm for regular nylon and polycarbonate. For both the SLA and SLS products, surface finishes can be improved by light sanding for the former and sealing and sanding the surface for the latter. The SLS parts made of nylon or of polycarbonate can be exposed to moisture, many chemicals, and temperatures up to 180°C. Therefore they can be subjected to mechanical or thermal loading, and they are useful for functional review.

The *3D printing (3DP)* system is explained in Fig. 12.66 [15]. Like the SLS, it uses the sintering process but not by involving a laser beam. Instead, it works with layers of powder materials and injects a polymeric binder into the powder over those areas that are the sections through the part being made. Many different powder materials are being considered including ceramics, metals, and polymers. The process can be applied to metallic or ceramic grains mixed with grains of a binder material, such as iron and copper or tungsten carbide and cobalt. The injecting operation is done in the ink-jet printing manner. To speed up the process, heads with multiple nozzles are used, arranged in linear or rectangular arrays of various sizes, up to the limit of a machine that would print the whole layer in one pass. The driving force for the formation of droplets is the reduction of surface energy resulting from an instability of a cylindrical stream of the molten binder material. The droplet formation is supported by piezo-induced vibration of the tip. The binder has some conductivity and gets charged by passing through the control electrode. The stream of droplets can be shut off by the deflection electrode, which deviates it against the knife edge. The process as applied to a stainless steel or tool-steel powder consists of the following steps: spread powder, print a colloidal latex binder into the powder, remove loose powder to reveal the green part that is held together by the bonding agent. Typically, it consists of 58% steel, 10% polymer, and 32% open porosity. Further, it is necessary to burn out the polymeric binder in a furnace and lightly sinter the part at 120°C for several hours to about 63% density; then infiltrate the part with a copper alloy in a second furnace operation at 1100°C. At this stage full density is achieved.

The authors discuss the speed potential of the method by assessing the time needed for creating the layers of the workpiece material, for injecting the binder and its solidification. If the basic powder material is dispersed in a liquid vehicle, drying times of 0.1–10.0 sec per layer are expected. Continuous jet printing with 20-μm droplets at 1 MHZ would result in writing speed of 20 m/sec. A 0.5-m by 0.5-m layer could be printed in 0.13 sec. Taking also the binder drying time, a total time per layer of 5 sec is estimated. An example of the ability of the process to manufacture complex ceramic parts is illustrated in Fig. 12.67. This example part is made in alumina powder using collidal silica as a binder. It consists of 284 layers, each 178 μm thick. This gives a total height of 50 mm. The use of various materials leads to the manufacture of molds and dies. This kind of application is discussed among others in the following section.

Several more processes will now be mentioned. The first one is called *fused deposition modeling* [16]. It is assumed that the process may be applicable to a variety of materials from wax to thermoplastics and metals. The principle is shown in Fig. 12.68. A filament of the material is delivered into a nozzle where it is heated and melted. By NC motion of the nozzle, it is deposited in the desired pattern while it solidifies in contact with the cold preceding layer of material. An example was given of a polymer filament with 1.25 mm diameter heated just 1 degree above melting temperature, which ensured cooling within 0.1 sec. The flow rate was controlled by precision volumetric pumps. The nozzle traveling speed of 380 mm/sec was used in producing layer thickness in a range between 0.025 to 1.25 mm. Another process called *ballistic particle*

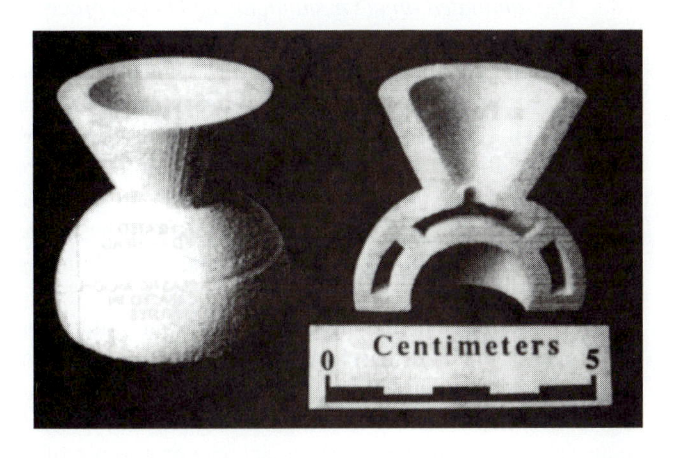

Figure 12.67
Ceramic part made by the 3DP process from aluminum powder with colloidal silica as binder. The part was created in 284 layers 178 μm thick each. [FROM SACHS, REF. [15]]

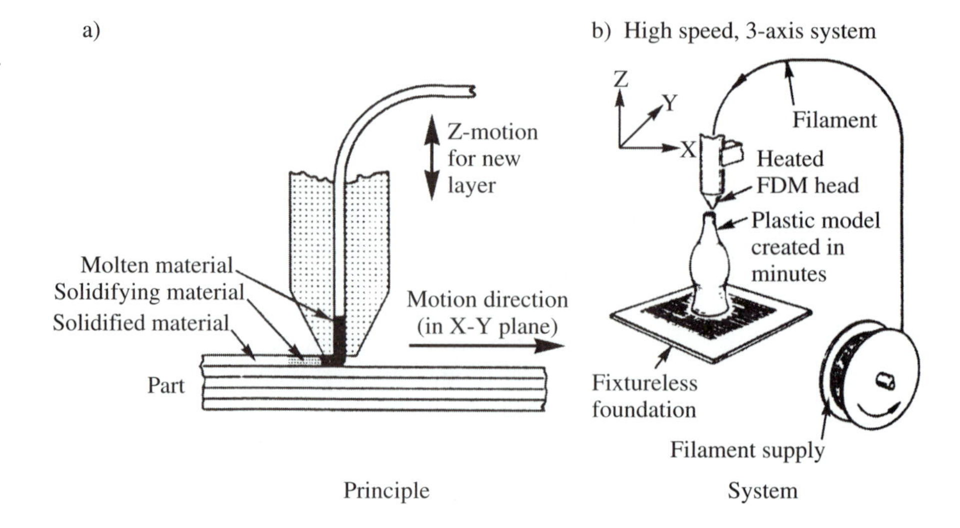

Figure 12.68 Fused deposition modeling. A filament of a polymer is melted in the nozzle and deposited under NC in the *XY* axes. [FROM CRUMP, REF. [16]]

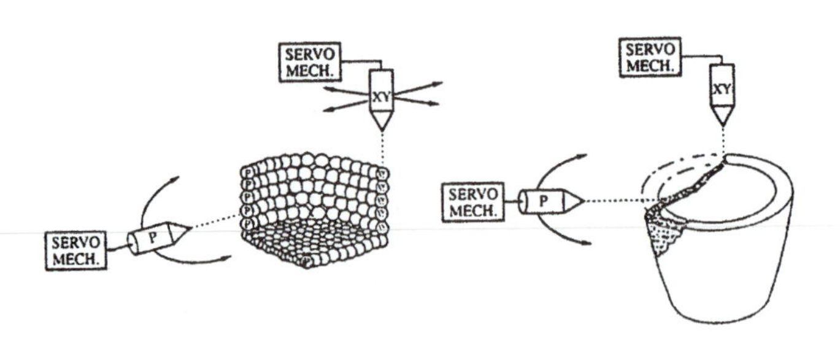

Figure 12.69 Ballistic particle manufacturing. Parts are made by shooting droplets of molten material on top of each other. Ink-jet printing multiple nozzles are used operating at 10 kHz. Material is mostly wax for investment-casting patterns. [FROM RICHARDSON, REF. [17]]

manufacturing produces parts by shooting droplets of molten material on top of each other (see Fig. 12.69). The use of ink-jet printing heads with an array of 32 parallel nozzles operating at 10 kHz allows for high deposition rates. The technique has been applied to the manufacture of wax models for investment casting, but it is intended to be used for thermoplastics and metals. A prototype of a machine has been built [17] that deposits up to 1 kg of aluminum droplets in one hour. A six-axis robot machine may be developed to create 3D complex parts.

The *laminated object manufacturing (LOM)* process [18] works from foil material in the form of rolls or stacks of cut sheets; see Fig. 12.70. Many different materials that can be cut by a laser beam (see Section 12.7) include paper (cellulose), plastics, metals, ceramics, composites. Paper is most commonly used because it is inexpensive and also because with many layers glued together it resembles wood, and the model naturally offers itself for use as a pattern for sand casting. The process consists of bringing the sheet material onto the platform, bonding it to the layer underneath, cutting the layer out, and discarding the rest. As shown at the left in the figure, for the bonding operation a mask is prepared and laid on top of the sheet. The bonding polymeric material is spread on its underside and, using the UV lamp, the selected area of the layer is heated to activate the bonding material. The area that should not be bonded is shielded from the heat by the mask. As shown at the right, the laser beam is scanned to produce cuts along the outer and inner contours. The operation is fast, because only the outline is traced: a 50 W CO_2 laser is used. In the following section, it will be indicated how the different materials are handled by LOM to achieve prototypes of real parts.

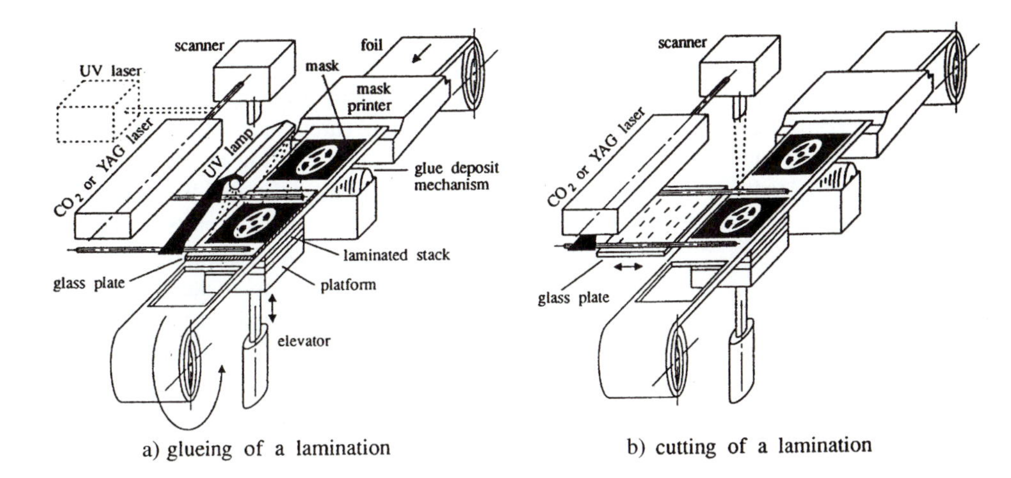

Figure 12.70
The laminated object manufacturing (LOM) process works from foils of paper, plastics, ceramics, metals, or composites by laser beam cutting the pattern of each layer. [FROM FEYGIN, REF. [18]]

12.12.2 Rapid Tooling

The extended goal of rapid prototyping is to produce prototypes in the actual materials of the production part and, possibly, by the same production process. So, for instance, models that have been obtained by using the LSA or SGC process in a photopolymer should be replaced by a part of the same geometry but in a polymer with good mechanical and thermal properties or simply in the same polymer as the production part. For processes such as SLS or FDM that make models in common thermoplastics, it may be desired to obtain them by injection molding that delivers superior quality. Furthermore, it is of interest to rapidly obtain prototypes in aluminum, cast iron, or a steel alloy, and also to rapidly prepare production tooling for these parts.

Let us first discuss methods that have been developed as extensions of the oldest and most widely used RP process, stereolithography (SLA) [19]. Let us recall that its primary product is a rather accurate model of a part, and it is made in a photopolymer by the technique illustrated in Fig. 12.59. In the laser curing process only the outer and inner surfaces of the model are made dense, while the inside solid structure is more or less porous, filled by liquid polymer to be solidified in the post-process curing operation. Two other filling modes have been developed. One is called ACES for accurate clear epoxy solid and it uses numerous, closely spaced laser scans that create glasslike surface finishes and virtually transparent parts. Obviously, it takes a longer time to generate these structures; however, they can be used in several additional processes. In the first instance this model can be reproduced as a vacuum casting made of a polyurethane so as to improve the mechanical and thermal properties. Secondly, it can lead to either soft or hard tooling for injection molding (IM).

In *vacuum casting,* the ACES technique is used to produce a rather accurate model of the final part. It is then sanded and polished to fine finish. This is important because even very small surface defects will be reproduced in the following operations. The model is used as a pattern to make a mold in silicone RTV (room temperature vulcanizing) rubber. This is done so that a sprue may be glued to the pattern that is then suspended in a corrugated paper box. The silicone RTV material is poured into the box to fully surround the pattern. The whole is placed in a vacuum chamber and degassed at room temperature. Subsequently the RTV mold is cured at 50°C for about four hours. Then the mold is cut in two halves using a scalpel and reassembled. Finally, any of a wide range of two-part polyurethanes can be vacuum cast in the mold. These resins provide a broad choice of properties such as hardness, strength, tensile modulus, elongation at break equal to and mostly surpassing those of nylon 6 and ABS. An example is shown in Fig. 12.71, where at lower left is the SL pattern, the two RTV mold halves are at the top, and three polyurethane castings of a boom box are at lower right, in three different colors. RTV vacuum casting is relatively slow. Only about 4–8 parts can be made per day, and the mold has a limited life and can produce only about 10–20 parts. Although it is a technique that produces parts in materials with good properties that can be subjected to functional tests, and these parts are obtainable in a span of a few weeks, other techniques were developed to make it possible to produce parts by injection molding. These are obtainable over a time period shorter than that needed when the steel molds are made in the established ways by machining or EDM.

Direct AIM (ACES injection molding) [20] means direct injection of thermoplastics into molds built by stereolithography. This is not an intuitive step, since the SL resin has a glass transition temperature $T_g = 75°C$, and the thermoplastics are injected at temperatures as high as 240°C. However, with proper cooling and slowdown of the IM action, small numbers of moldings of PE, ABS, PP, and PS can be successfully produced. The injection pressures range from 12 MPa for LDPE to 24 MPa for ABS. The process starts with stereolithographic design and manufacture of the cavity and core for the part to be made. These are assembled in a standard MUD (master unit die) frame to be fitted in the IM machine. However, because of the low thermal conductivity of the SL resin, these mold halves are made as 5–10-mm-thick shells used as inserts in a frame to be back-filled by aluminum-powder-filled, granulate epoxy resin mixed with aluminum shot. Copper tubing is wound through the back fill for "conformal" cooling.

Injection is carried out slowly, and in order to minimize damage to the mold at ejection, the hold time of the molding after injection is rather long to permit cool-down of the molding to about 43°C. Cycle times of 4–5 minutes are used in contrast to conventional IM cycle times of 5–15 seconds for steel tools. The direct AIM molds are claimed to be good for producing up to 100 parts. For larger productions, it is necessary to use hard molds, and the next process discussed has been designed to do that.

3D Keltool™ is a process that replicates the ACES master in fused powdered steel. It may be carried out in two alternative ways indicated in Fig. 12.72. One starts with the ACES model produced as a "positive master" and creates the cavity and the core of

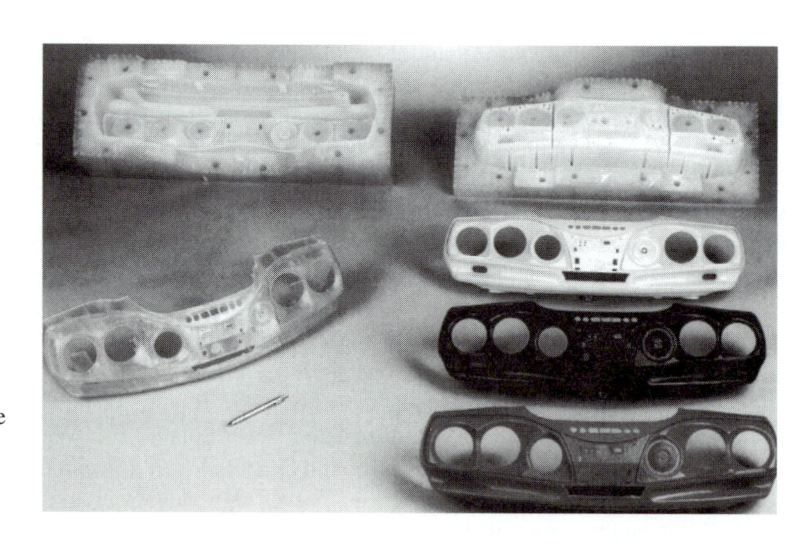

Figure 12.71
Example of use of rapid tooling. The SL pattern (lower left), the vacuum casting mold halves made of room temperature vulcanizing (RTV) rubber (top), and three polyurethane castings of a boom box housing. [FROM JACOBS, REF. [19]]

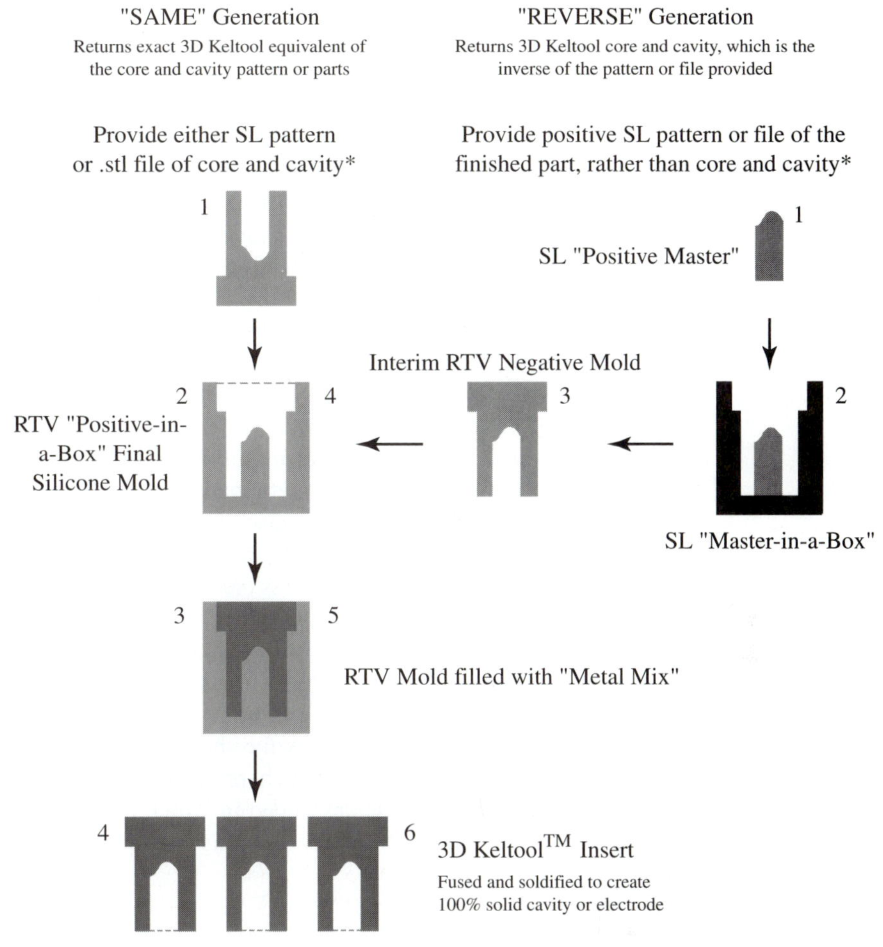

Figure 12.72
The 3D Keltool process. Two different ways of producing powder-metal injection mold inserts. The sequence at left follows steps 1, 2, 3, 4 and the one at right passes through 1, 2, 3, 4, 5, 6. [From Jacobs, ref. [19]]

the mold by "reverse generation," while the other starts by CAD of the cavity and core and their production as the ACES model and through "same generation" replicates them in sintered steel (e.g., the A6 type of tool steel, with about 30% of its volume infiltrated by copper). Let us follow the reverse generation method. In step 1 the ACES model of the final part (the master) is produced by stereolithography in the SL 5170 epoxy resin (a photopolymer); its shape is shown in Fig. 12.72 as the light gray picture. The ACES stereolithography is relatively slow, and this step may take 30–40 hours for larger parts. In step 2 the master is placed in a box and vacuum-degassed silicone RTV is poured around it and fills the box. After curing the RTV rubber, the master and the mold are taken out of the box (see the shape of the mold in 3). It is then used in step 4 to produce an RTV positive in a box mold the shape of which is drawn light gray in 4. This mold is then filled with a mixture of steel powder and thermoplastic binder powder (see the darker gray shape in 5). This mixture is cured to obtain the "green mold insert" that is taken out of the box and fired in a furnace to fuse the metal particles and burn out the binder. Finally, the 30% porous part is infiltrated with copper. A hard

(35–55 Rc) injection mold insert results, as shown in 6. It may be even better for IM than solid steel molds because it has superior thermal conductivity due to the Cu contents. It is claimed to be good for regular injection molding of 1 to 10 million thermoplastic products.

Currently the 3D Keltool inserts are limited to the parts within 100 mm cube. Accuracy of 25 μm is claimed. An example of the "rapidity" of this way of preparing IM hard tooling a water flow valve is shown in front of Fig. 12.73. Both the positive and negative masters are also shown as well as the hard die inserts. It is claimed that the ACES positive prototype build time was 5 hours, and it was ready in 2 weeks from purchase order. Tool design was complete and approved in another week, and in another 2 weeks time the mold insert masters were ready and sent out for Keltool generation of hard tooling that was returned after 3 weeks. The mold was tested and approved 2 months after PO. It is claimed that this saved 10 weeks against the common way of producing the IM mold. So far, our discussion has concentrated on production of polymeric products, either in polyurethanes or thermoplastics. Next, processes are discussed for production of metal parts.

The *Quickcast* process is also an extension of stereolithography. The SLA process is used to prepare a model of the part. This model is then used instead of a wax pattern in the investment casting process (see Section 3.7 and Fig. 3.32) to produce the desired part in aluminum or in various types of steel. Instead of melting and pouring the wax pattern out of the mold, the pattern made of the SL resin is burned out of the ceramic mold. However, because the thermal expansion coefficient of the photopolymer is much higher than that of the ceramic shell of the mold, pattern expansion during the burnout would normally crack the shell. An additional problem is the ash regenerated by burning the polymer, which produces inclusions in the casting. Therefore, the Quickcast process was developed in which only the external surfaces and an internal lattice structure are cured, representing only about 15% of the mass of the model. Holes

Figure 12.73
Positive and negative polyurethane masters, powder-metal inserts in the injection mold, and, in the front, the final molding of a water flow valve resulting from the Keltool process. [FROM JACOBS, REF. [19]]

in the bottom of this model allow the uncured resin to drain out. The resulting model to be used as a pattern is largely hollow but still strong enough for the investing process. During burnout the pattern collapses before the shell can crack, and it leaves very little ash, which can be blown out.

An excellent illustration of such a pattern is shown in Fig. 12.74, taken from an article describing the manufacture of prototypes of an access panel on the tail of the Harrier QR7 aircraft [21]. This complex shape part, approximately 300 mm square, should be light and, therefore, preferably have a wall thickness of only 2 mm, to be cast in aluminum. This was achieved by the Quickcast process. It is claimed that cost reduction of 69% was achieved, and a time reduction from 24 down to 11 weeks, when compared with the conventional way of producing metal molds for wax patterns. Obviously, it is a good example of true rapid prototyping.

Other examples are presented in Fig. 12.75, which shows the resin pattern, the ceramic shell, and the metal casting of an impeller, and in Fig. 12.76, showing the resin pattern and the casting of a 2.4-liter car engine block.

The Quickcast process, replacing the wax patterns by those made by the SLA process, is clearly economical only for the production of a small number of prototypes. If a large number of investment castings is needed, the Quickcast process or anyone of the other RP processes capable of the task can be used to produce molds in aluminum or in steel for the manufacture of wax patterns.

Other RP processes can be used to produce wax patterns for investment casting, the selective laser sintering (SLS) process, and the fused deposition modeling (FDM) process. These are straightforward applications, and they can be applied if only a small number of prototype metallic parts is needed because for every one of them, a new pattern must be made, requiring a number of hours for the RP process.

The laminated object manufacturing (LOM) process is versatile in that it can handle sheets of a wide variety of materials and offers several possibilities for rapid tooling [22]. One of these is to use paper sheets with polymer bond to produce patterns for

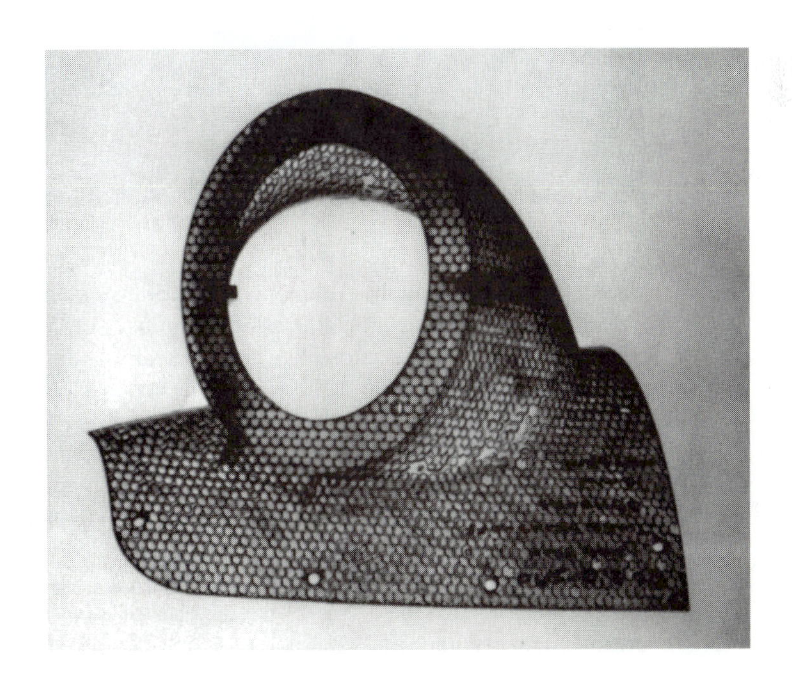

Figure 12.74
A Quickcast model of an access panel for the Harrier QR7 aircraft to be used as pattern for investment casting of the aluminum part. The pattern will be burned out of the ceramic mold. [From Throup, ref. [21]]

Figure 12.75
Example of the resin pattern of an impeller, the ceramic shell, and the metal casting produced by a combination of Quickcast and investment casting. [From "Quickcast" by 3D Systems]

Figure 12.76
Example of the Quickcast pattern and the shell mold casting of a 2.4 liter Mercedes Benz engine block. [From "Quickcast" by 3D Systems]

sand casting, replacing the commonly used patterns made of wood. Typically, it can be used for up to 50 sand molds. If a higher number of castings are required, the LOM pattern is used to cast a match plate pattern in aluminum or cast iron, which is used for all the subsequent casting. There is, however, a size limitation: the largest LOM machine currently on the market has a working envelope of 560 × 813 × 500 mm.

Another application is to produce tools for hydroforming of sheet metal (see Section 4.4). Aluminum, steel, copper, nickel, and titanium sheets are hydroformed at up to 83-MPa pressures. For thin aluminum or copper sheets, the paper LOM is satisfactory, but for the more demanding applications, the LOM builds up the die from sheets of fiberglass-epoxy composite using the CO_2 laser to cut them.

LOM can also be used to build parts of high-density, monolithic ceramics, ceramic matrix composites, and polymer matrix composites. The process can use sheet preforms such as monolithic ceramic tapes, whisker-reinforced ceramic tapes, and ther-

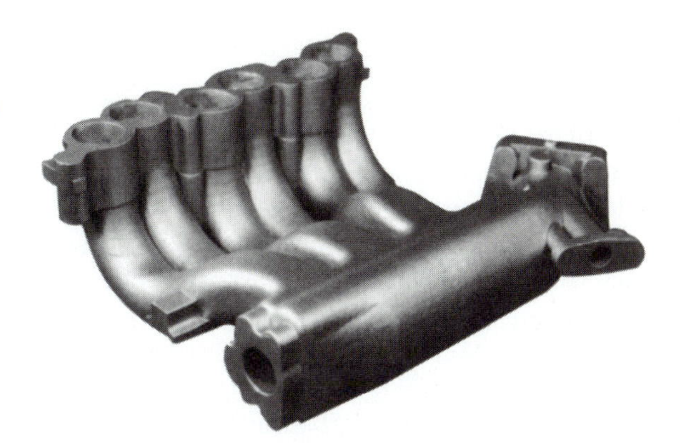

Figure 12.77
A 3-liter automobile engine intake manifold cast in metal by the direct shell production casting (DSPC) process. [FROM UZIEL, REF. [23]]

moset polymer prepregs. For example, tapes of SiC powder bonded by a polymer form flexible thin sheets that can be used to laminate and produce monolithic parts. After LOM these parts are post-processed by burning the polymer binder out and infiltrating by liquid or vapor silicon at temperatures above 1400°C. The silicon fills the pores and reacts with residual carbon, forming additional SiC. In another instance, an SiC matrix with SiC fibers has been processed. Composites with continuous fiber are not successfully cut using the CO_2 laser. Good results have been achieved with a 20-W pulsed, copper-vapor laser. Glass-fiber-epoxy composites, glass-fiber polypropylene and graphite-fiber polypropylene prepregs have also been successfully laminated.

The *3DP* process uses a multinozzle printing head to inject binder in ceramic powder; it has been described in [23] in its use as *direct shell production casting (DSPC).* It is used to directly produce a ceramic mold (patternless molding and casting) for casting prototype parts in aluminum, cast iron, or steel. It uses CAD to design the mold from the part CAD geometry by adding the gating system. It may include integral ceramic cores. An example is given in Fig. 12.77 of an intake manifold for a 3-liter automobile engine. It illustrates the complexity of both the outer form and of the cavities that are obtainable by the process. From the same CAD file that was used to produce the prototype casting, it is possible to cast all the production tooling, such as patterns and tool boxes for sand casting.

12.13 CONCLUSION

The nontraditional processes have been firmly established over the past two to four decades as essential elements of the manufacturing technology. Among them, solid and wire cutting EDM and laser cutting, welding, and heat treatment have found widespread use. Their recent extension into the special fields of electronics manufacturing and into rapid prototyping and rapid tooling represent new and continuously developing computer-based technologies. The development of methods for producing computer chips opened the door to a whole new world of the information age. The RP/RT processes are still young, and it is not yet possible to assess their full potential. However, they will undoubtedly strongly influence the manufacturing systems of concurrent engineering and of flexible manufacturing.

REFERENCES

1 Whitehouse, D. R. "Lasers in Fabricating: Achieving Maximum Productivity." Landing, NJ: PRC Corp., 1992.

2 Ohmori, H., and T. Nakagawa. "Mirror Surface Grinding of Silicon Wafers with Electrolytic In-Process Dressing." *CIRP Annals* 39 (January 1990).

3 "Advanced Manufacturing." Lumonics, Inc., August 1996.

4 Ready, J. F. "Laser Applications in Metalworking." Paper no. MRR75. Soc. of Manufacturing Engineers, Dearborn, MI, 1975.

5 Bolin, S. "Lasers Light the Way to Low-Cost Drilling and Cutting." *Manufacturing Engineering,* December 1981.

6 Engel, S. L. "Laser Cutting of Thin Materials." In *Thermal Machining Processes.* Dearborn, MI: SME, 1979.

7 Thompson, A. "High-Power Laser Operating Costs Plummet." Landing, NJ: PRC Corp., 1995.

8 Harada, K. "Current Status of E-Beam Lithography." *Bull. Japan Society of Precision Engineering* 22, no. 4 (December 1988).

9 Taniguchi, N. "Research and Development of Energy Beam Processing of Materials in Japan." *Bull. Japan Society of Precision Engineering* 18, no. 2 (1984).

10 Kruth, J. P. "Material Increase Manufacturing by Rapid Prototyping Techniques." *CIRP Annals* 40 (February 1991): 603–14.

11 Maskow, J. "Neue Fertigungstechnologien beim Automatisierten Chemischen Abtragen." *ZwF* 80 (1985): 275–77.

12 Jacobs, P. *Stereolithography: From Art to Part.* Valencia, CA: 3D Systems, 1993.

13 "Think 3, Think Fast, Think Cubital." Cubital, Inc. 1997.

14 "The Selective Laser Sintering Process, Third-Generation Desktop Manufacturing." Nat. Conf. on Rapid Prototyping, DTM Corp., 1990.

15 Sachs, E., M. Cima, and J. Cornie. "Three-Dimensional Printing: Rapid Tooling and Prototypes Directly from a CAD Model." *CIRP Annals* 39 (January 1990).

16 Crump, S. "Fused Deposition Modeling: Putting Rapid Back in Prototyping." 2nd International Conference on Rapid Prototyping, 1991.

17 Richardson, K. E. "The Production of Wax Models by the Ballistic Particle Manufacturing Process." 2nd International Conference on Rapid Prototyping, 1991.

18 Feygin, M. "Laminated Object Manufacturing." Nat. Conf. on Rapid Prototyping, 1990.

19 Jacobs, P. *Recent Advances in Rapid Tooling from Stereolithography.* Valencia, CA: 3D Systems, 1997.

20 Decelles, P. *Direct AIM Prototype Tooling.* Valencia, CA: 3D Systems, 1997.

21 Throup, S. C., and T. Hiatt. "Applying Rapid Manufacturing Techniques to the Production of a Thin-Wall Aluminum Cavity." *Rapid News* 2, no. 7 (December 1997).

22 Chartoff, R. P., and D. R. Tolin. "Tooling and Low-Volume Manufacture through LOM." Prototyping Technology International, UK & International Press, 1997.

23 Uziel, Yehoram. "Seamless CAD to Metal Parts." Prototyping Technology International, UK & International Press, 1997.

24 Hull, C. "Rapid Prototyping in the 1990s." 2nd International Conference on Rapid Prototyping, 1991.

25 DeBarr, A. E., and D. A. Oliver. *Electrochemical Machining.* London: McDonald & Co., 1968.

26 Weller, E. J., and M. Haavisto. "Nontraditional Machining Processes." 2nd ed., Dearborn, MI: Soc. of Manufacturing Engineers, 1984.

27 Hummel, R. E. "Electronic Properties of Materials." 2nd ed. New York: Springer Verlag, 1985, 1993.

28 "Welding Processes, Arc and Gas Welding and Cutting." In *Welding Handbook,* vol. 2, 7th ed. Miami, FL: American Welding Society, 1978.

29 "Welding Processes, Resistance and Solid State Welding." In *Welding Handbook,* vol. 3, 7th Ed. Miami, FL: American Welding Society, 1980.

30 Bellows, G. "Nontraditional Machining Guide." Publication No. MDC 76-101, Machinability Data Center, a DoD Information Analysis Center.

31 Bellows, G. "Chemical Machining." Publication No. MDC 77-102, Machinability Data Center, a DoD Information Analysis Center.

32 Snoeys, R., F. Staelens, and W. Dekeyser. "Current Trends in Non-Conventional Material Removal Processes." *CIRP Annals* 35 (February 1986): 467–80.

33 Groover, M. P. *Fundamentals of Modern Manufacturing.* Prentice Hall, 1996.

Illustration Sources

[BELLOWS], Reprinted from G. Bellows, *Nontraditional Machining Guide.* Publication no. MDC 76-101, Machinability Data Center, a DoD Information Analysis Center.

[DEBARR], Reprinted from A. E. DeBarr and D. A. Oliver, *Electrochemical Machining.* McDonald & Co. Publishers, London, 1968.

[HUMMEL], R. E. Hummel, *Electronic Properties of Materials,* 2nd ed., Springer Verlag, New York, 1985, 1993, with permission by R. E. Hummel.

[SNOEYS], Reprinted from R. Snoeys, F. Staelens, and W. Dekeyser, "Current Trends in Non-Conventional Material Removal Processes," *CIRP Annals* 35 (February 1986): 467–80.

[WELLER], Reprinted with permission from E. J. Weller and M. Haavisto, *Nontraditional Machining Processes,* 2nd ed., Soc. of Manufacturing Engineers, Dearborn, MI, 1984.

[WH2], Reprinted with permission from *Welding Handbook, Vol. 2, 7th ed., "Welding Processes",* 1978, American Welding Society, 550 NW Le Jeune Rd, Miami, FL 33126.

[WH3], Reprinted with permission from *Welding Handbook, Vol. 3, 7th ed., Welding Processes,* 1980, American Welding Society, 550 NW Le Jeune Rd, Miami, FL 33126.

Industrial Contributors

[3D] 3D Systems, Inc., 26081 Avenue Hall, Valencia, CA 91355.

[AGIE] AGIE Industrial Electronics Ltd., Losone-Locarno, Switzerland.

[AMCHEM] AMCHEM Co. Ltd, 28 Roman Way, Coleshill, Birmingham B46 1HQ, England; in the USA: 2850 S. Industrial Highway, Ann Arbor, MI 48104.

[ANOCUT] Anocut Inc., Elk Grove Village, IL.

[BPM] BPM Technology, Inc., 1200 Woodruff Road, Suite A-19, Greenville, SC 29607.

[CAMATTINI] Camattini-North America, Div. of NEST Technologies, 3849 Ridgemore Drive, Studio City, CA 91604.

[CHARMILLES] Charmilles Technologies S. A., P.O. Box 373, 8–10, rue du Pres-de-la-Fontaine, CH-1217 Meyrin 1-Geneva, Switzerland; in the USA: Charmille Technologies Corp., 560 Bond Street, Lincolnshire, IL 60069-4224.

[CHEMCUT] Chemcut Corp., State College, Pennsylvania.

[CUBITAL] Cubital 13 Ha'Sadna St., Industrial Zone North Raanana 43650 Israel; in the USA: Cubital America, Inc., 1307F Allen Drive, Troy, MI 48083.

[DTM] DTM Corp., 1611 Headway Circle, Building #2, Austin, TX 78754.

[ESAB] ESAB L-TEC Cutting Systems, P.O. Box 100545, Ebenezer Road, Florence, SC 29501-0545.

[HELISYS] Helisys, Inc., 24015 Garnier Street, Torrance, CA 90505.

[INGERSOLL] Ingersoll Hansen GmbH, Postfach 1140, D-57291 Burbach, Germany; in the USA: Ingersoll GmbH, 1301 Eddy Ave., Rockford, IL 61103.

[LUMONICS] Lumonics, Inc., 105 Schneider Road, Kanata, Ontario, Canada K2K 1Y3; in the USA: 19776 Haggerty Road, Livonia, MI.

[PRC] PRC Corp., North Frontage Road, Landing, NJ 07850.

QUESTIONS

Nontraditional Processes

Q12.1 In general, in which applications are the N-T processes used with substantial advantage compared to traditional cutting and forming?

Q12.2 Which processes are limited to machining electrically conductive materials? How is their performance affected by the hardness of the metal?

Q12.3 Which of the processes affect adversely the processes of the workpiece on the surface (how deep), or in depth? What kinds of damage occur?

Q12.4 Name typical applications for USM , WJM, AJM.

Q12.5 Compare ECM and EDM in the basic operation of producing (sinking) a cavity, such as one for an injection mold or casting die as regards tool material, tool wear, metal removal rate, power source, current level, and cutting fluid.

Q12.6 Compare the difficulty of determining the 3D shape of the tool for ECM and for EDM.

Q12.7 Name and illustrate the various specialized applications of ECM.

Q12.8 Compare and give application examples for drilling holes by ECM, EDM, LBM, and EBM.

Q12.9 How is chemical milling applied on a large scale? What three different methods are used to produce the mask for such an application? Which material of the mask and which etchant can be used in CHM of aluminum skins?

Q12.10 Explain the steps in photochemical blanking and give examples.

Q12.11 Sketch out examples of combining simple tool shape with NC generating motion to produce, while sinking in the vertical direction, (a) constant square section with corner radii; (b) rectangular cross section with one side sloped; (c) several circular cylinders with different diameters, one above the other.

Q12.12 How are blanking dies machined by wire EDM? How is the transfer of the wire accomplished from hole to hole in a multiple hole pattern? How is a matching set of a punch and die made?

Q12.13 Name the various processes based on the application of the laser. How do they differ in intensity? Name typical tasks. What main types of lasers are used in these operations?

Q12.14 Which material properties affect the performance of laser cutting and laser drilling? How is the necessary power determined?

Q12.15 Compare the applicability and performance of LBM and EBM in cutting sheets and plates of various materials. Which of the three modes of EBM are used?

Q12.16 Compare the applicability and performance of OC and PAC in cutting various materials.

Electronics Manufacture

Q12.17 Name the processing stages starting from a boule of EG silicon to a packaged VLSI chip. What three different actions may be carried out in a photolithography step?

Q12.18 Explain the procedure of using photolithography in an etching action. Make a 3D sketch of a block of silicon before and after a step in which a top layer of silicon dioxide is removed over an area in the shape of the letter H. Include the layers of the mask, the resist, and the SiO_2 layer.

Q12.19 Explain the alternative processes used to dope silicon and also the alternative processes used to remove a layer of silicon.

Q12.20 Which kind of radiation is used in the common micron-level photolithography, and which kind of the photoresist? Which kind of radiation may be used to produce finer resolution of the ULSI patterns? What kind of action will then be used instead of photolithography?

Q12.21 How are the lines connecting the individual chips on a PCB made?

Rapid Modeling

Q12.22 Name five different RP processes and the materials of the models they produce.

Q12.23 Explain the steps used in stereolithography and those in solid ground curing.

Q12.24 Explain the difference between selective laser sintering and 3D printing.

Q12.25 Which sheet materials can be processed in laminated object manufacturing?

Q12.26 Which is the only process mentioned in Section 12.12.1 that does not work in 2D layers? How does it work?

Rapid Tooling

Q12.27 What are the steps to use to reproduce the model made in a photopolymer by SLA or SCG in a polymer with much better mechanical and thermal properties? What kind of polymers are obtainable, and with which polymers commonly used in injection molding is their strength, hardness, and elongation comparable?

Q12.28 How does the direct ACES injection molding process work? What is generated in the CAD part of it? What is the material of the injection mold, and what is its T_g temperature? What is the injection temperature of an ABS or PP polymer? How is the mold built to overcome this disparity in properties? How many parts can be made in one such mold?

Q12.29 How does the 3D Keltool process work, and how is it related to the SLA process? What is the material of the injection mold, and how many parts can be produced in one mold?

Q12.30 How does the Quickcast process work? What advantage does it offer over the conventional investment casting process? What are its limitations?

Q12.31 Which RP and RT processes can be used to prepare a metallic injection molding die? Which ones can be used to make a durable mold for wax patterns? By which processes can the wax patterns be directly made?

Q12.32 What are the various uses of the 3DP process? Of the LOM process?

PROBLEMS

P12.1 ***Steady state and transient in ECM.*** You are machining steel with current $I = 2000$ A, a voltage of 22 V, machined area $A = 10$ cm^2, electrolyte resistivity $r = 5$ Ωcm, and process efficiency $\eta = 0.75$. Determine the steady-state gap g and steady-state feed rate f. Start the process from a gap $g_1 = 0.05$ cm. Plot out the current and the gap transients; choose dt = 0.001 and time span 0.5 sec.

P12.2 ***Electrolyte flow in ECM.*** For the steady-state conditions of Problem 12.1 and flow across the longer side of the 2.5 cm × 4 cm area (length of channel 2.5 cm), determine the Reynolds number for velocity 35 cm/sec, viscosity of the fluid $\eta = 0.8$ cP, $\rho = 1$ g/cm^3, and determine the pressure p driving the flow and the temperature increase ΔT (°C).

P12.3 ***Power in laser drilling.*** You are drilling molybdenum, 1.5 mm thick, hole diameter $d = 0.5$ mm, absorbtion 50%, pulse length 2e−5 sec. Determine the instantaneous power P (W).

Index

B

C

N

S

T